International Conference on Advances in Sustainable Construction Materials & Civil Engineering Systems (ASCMCES-17)

MATEC Web of Conferences Volume 120 (2017)

Sharjah, United Arab Emirates
18-20 April 2017

Part 1 of 2

Editors:

A. Shanableh
S. Barakat
S. Al-Toubat
K. Hamad
M. Maalej
M. Omar
R. Al-Ruzouq

ISBN: 978-1-5108-4704-0

Printed from e-media with permission by:

Curran Associates, Inc.
57 Morehouse Lane
Red Hook, NY 12571

Some format issues inherent in the e-media version may also appear in this print version.

Printed by Curran Associates, Inc. (2017)

For additional information, please contact EDP Sciences – Web of Conferences at the address below.

EDP Sciences – Web of Conferences
17, Avenue du Hoggar
Parc d'Activité de Courtabœuf
BP 112
F-91944 Les Ulis Cedex A
France

Phone: +33 (0) 1 69 18 75 75
Fax: +33 (0) 1 69 28 84 91

contact-edps@webofconferences.org

Additional copies of this publication are available from:

Curran Associates, Inc.
57 Morehouse Lane
Red Hook, NY 12571 USA
Phone: 845-758-0400
Fax: 845-758-2633
Email: curran@proceedings.com
Web: www.proceedings.com

TABLE OF CONTENTS

PART 1

PART 2

MATEC Web of Conferences **120**, 00001 (2017) DOI: 10.1051/matecconf/201712000001
ASCMCES-17

INTERNATIONAL CONFERENCE ON ADVANCES IN SUSTAINABLE CONSTRUCTION MATERIALS & CIVIL ENGINEERING SYSTEMS (ASCMCES-17)

University of Sharjah
Sharjah, United Arab Emirates

April 18-20 2017

Preface

The first international conference on Advances in Sustainable Construction Materials & Civil Engineering Systems (ASCMCES-17) was organized by two research groups in the Research Institute of Sciences and Engineering (RISE); the Sustainable Civil Infrastructure Systems (SCIS) Group and the Sustainable Construction Materials & Structural Systems (SCMSS) Group, both of which are hosted by the Department of Civil and Environmental Engineering at the University of Sharjah. The theme of the conference "Advances in Sustainable Construction Materials & Civil Engineering Systems" was chosen to reflect the visionary and sustained development of the United Arab Emirates and to underpin the importance of discussing the latest developments and innovation in the field. The conference aimed to bring researchers from around the world to share their research findings and address issues related to research in areas of sustainable construction materials, and structural and infrastructural systems.

International Conference on Advances in Sustainable Construction Materials & Civil Engineering Systems (ASCMCES-17)

CONFERENCE TOPICS

Sustainable Structural Systems

- Structural Control and Health Monitoring
- Sustainable Systems Design
- Performance-Based Design
- Seismic Protection and Vibration Control
- Numerical Computation, Simulation, and Modelling

Sustainable Concrete Technology

- Advanced Cementitious Materials
- Engineered Cementitious Composites
- Nano Materials Applications
- Advances in Laboratory Testing

Recycling for Sustainable Concrete

- Materials Recycling
- Green Construction Materials

Fiber Reinforced Construction Materials

- Low Carbon Binders
- High Performance Materials

Geotechnical & Geo-Environmental

- Soil & Groundwater Contamination and Remediation
- Ground improvement and Reinforced Soil Structures
- Soil Dynamics and Seismic Hazards
- Soil-Structure Interaction
- Centrifuge Modelling of Geotechnical Construction
- Saturated and Unsaturated Soil Mechanics

Construction, Performance & Management

- Condition Rating and Deterioration Modelling
- Sustainable Construction
- Maintenance and Rehabilitation Techniques
- Budget Allocation and Optimization
- Maintenance Scheduling
- Contractual Management
- Asset Management

Sustainable Transportation & Pavement

- Green Roads and Transportation
- Smart Vehicles and Intelligent Transportation Systems
- Highway Safety and Geometric Design
- Transportation Operations & Planning
- Airports and Advances in Pavement Techniques
- Pavement Design, Modelling, Performance, & Management
- Asphalt Binder and Mixture Characterization
- Accelerated Testing of Pavement Structures and Materials

Water & Environment

- Water and Wastewater Treatment and Reuse
- Drainage, Flooding, and Drought Management
- Modeling & Simulation of Water and Environmental Systems
- Integrated Coastal Zone Management
- Air Pollution Control
- Environmental Impact Analysis and Risk Reduction
- Waste Management and Resource Recovery
- Global Warming Impacts and Mitigation

Geographic Information Systems & Remote Sensing Applications

- Remote Sensing of Environment
- GIS and Water Resources
- GIS for Transportation (GIS-T)
- Natural Hazard Monitoring and Mapping
- Photogrammetry and 3D Modeling
- Building Information Modeling (BIM)

Sustainable Solutions for Buildings, Construction and Infrastructures

- Green Building Materials
- Energy, Water Efficiency, and Renewable Resources
- Durability and Resiliency
- Risk Analysis and Reliability
- Life-Cycle Assessments

International Conference on Advances in Sustainable Construction Materials & Civil Engineering Systems (ASCMCES-17)

Edited By:

Prof. Abdallah Shanableh
Prof. Mohamed Maalej
Prof. Samer Barakat
Dr. Maher Omar
Dr. Salah Al-Toubat
Dr. Rami Al-Ruzouq
Dr. Khaled Hamad

RESEARCH INSTITUTE OF SCIENCES & ENGINEERING
معهد البحوث للعلوم والهندسة

Research Institute of Sciences and Engineering (RISE)
University of Sharjah, P.O. Box 27272, Sharjah, United Arab Emirates

Conference Committees

Chairs

Prof. Abdallah Shanableh
Prof. Mohamed Maalej

Organizing Committee

Dr. Salah Al-Toubat, Co-Chair
Dr. Rami Al-Ruzouq, Co-Chair
Dr. Mohsin Siddique, Member
Dr. Waleed Zeiada, Member
Dr. Moussa Leblouba, Member
Dr. Saleh Abu Dabous, Member
Dr. Muhammad Talha Junaid, Member

Scientific Committee

Prof. Samer Barakat, University of Sharjah, UAE (Co-Chair)
Dr. Maher Omar, University of Sharjah, UAE (Co-Chair)
Prof. Abdallah I. Husein Malkawi (Chancellor, Fahad Bin Sultan University, KSA)
Prof. Nemy Banthia (University of British Columbia, Canada)
Prof. Surendra P. Shah (Northwestern University, USA)
Prof. Ahmed Al-Kaisy (Western Transportation Insti., Montana State University, USA)
Prof. Ardeshir Faghri (Delaware Center for Transportation, USA)
Prof. Muhammed Basheer (University of Leeds, UK)
Prof. Omar S. Baghabra Al-Amoudi (King Fahad University, KSA)
Prof. Surendra Manjrekar (India Chapter of ACI, Sunandaglobal)
Prof. David Lange (University of Illinois at Urbana Champaign, USA)
Prof. Christian Heipke (Leibniz University Hannover, Germany)
Prof. Ayman Habib (Purdue University, USA)
Prof. Obada Kaylai (University of New South Wales, Australia)
Prof. Tarek Zayed (Concordia University, Canada)
Prof. Mohammed Dulaimi (The British University in Dubai, UAE)
Prof. Udechukwu Ojiako (The British University in Dubai, UAE)
Prof. Ghazi Abu-Farsakh (Jordan University of Science and Technology, Jordan)
Prof. Bashar Al-Omari (Jordan University of Science and Technology, Jordan)
Prof. Ali Alharthy (Sultan Qaboos University, Oman)
Prof. Mohamed Maalej (University of Sharjah, UAE)
Prof. Abdallah Shanableh (University of Sharjah, UAE)
Prof. Radhi Muhsin AlZubaidi (University of Sharjah , UAE)
Dr. Kamil Kaloush (Arizona State University, USA)
Dr. Shane Underwood (Arizona State University, USA)
Dr. Sherif Elbadawy (Mansoura University, Egypt)
Dr. Mohamed Elbasyouny (Parsons, UAE)
Dr. Mena Souliman (The University of Texas at Tyler, USA)
Dr. Liberato Ferrara (Politecnico di Milano, Italy)
Dr. Alex Rieder (GCP Applied Technologies Inc., USA)
Dr. Muntasir Abbas (Infrastructure and Systems Engineering, Virginia Tech, USA)
Dr. Ghassan Chehab (American University of Beirut, Lebanon)

International Conference on Advances in Sustainable Construction Materials & Civil Engineering Systems (ASCMCES-17)

Sponsors

American Concrete Institute	United Arab Emirates Ministry of Infrastructure Development	Sharjah Electricity and Water Authority
Sharjah Research Academy	International Society for Photogrammetry and Remote Sensing	SAHARA Communications
Fahad Bin Sultan University	Middle East Drymix Mortar Association	Holiday International Hotel Sharjah

Statement of Peer review

In submitting conference proceedings to MATEC Web of Conferences, I certify to the Publisher that I adhere to the **Policy on Publishing Integrity** of the journal in order to safeguard good scientific practice in publishing.

1. All articles have been subjected to peer review administered by the proceedings editors.
2. Reviews have been conducted by expert referees, who have been requested to provide unbiased and constructive comments aimed, whenever possible, at improving the work.
3. Proceedings editors have taken all reasonable steps to ensure the quality of the materials they publish and their decision to accept or reject a paper for publication has been based only on the merits of the work and the relevance to the journal.

Title, date and place of the conference

Title: International Confrence on Advances in Sustianable Construction Materials & Civil Engineering Systems, ASCMCES-17

Date: 18-20 April, 2017

Place of the Confrence: Al-Razi Hall, Medical and Health Science Colleges, University of Sharjah

Proceedings editor(s):

Prof. Abdallah Shanableh
Prof. Mohamed Maalej
Prof. Samer Barakat
Dr. Maher Omar
Dr. Salah Al-Toubat
Dr. Rami Al-Ruzouq
Dr. Khaled Hamad

Date and editors' singature

MATEC Web of Conferences **120**, 01001 (2017) DOI: 10.1051/matecconf/201712001001
ASCMCES-17

Effect of nonlinear material behavior of laminated composite plates with central rectangular hole subjected to out -of- plane loading

Ghazi Abu-Farsakh[1],* *Yasser Hunaiti*[2], *and Asaad Al Bustami*[2]

[1]Jordan University of Science and Technology, Civil Engineering Department, Irbid, Jordan
[2]Jordan University, Civil Engineering Department, Amman, Jordan

Abstract. The purpose of this paper is to investigate the effect of nonlinear material behavior on four layered, symmetric; angle-ply laminated composite plate with various fiber-orientation angles; $(\theta = 30^\circ, 45^\circ \text{and } 60^\circ)$. The plate has a central square-hole and subjected to out-of-plane uniformly distributed load. The effect of Stress Concentration Factor (SCF) resulting from redistribution of in-plane stresses (σ_x, σ_y, τ_{xy}) around the hole was taken into consideration. Square plates with simply supported boundary conditions were considered in the present study. The analysis was carried out utilizing the ANSYS-computer program. The presence of a central hole was found to concentrate the maximum stresses at the corners of the hole. The nonlinear material behavior was found to redistribute the in-plane stresses more reasonably and smoothly around the hole-perimeter and hence resulting in smaller SCF-values.

1 Introduction

In engineering science, composite materials can be generally defined as any material that has been physically assembled from two or more materials to form one single bulk material with new properties. The resulting composite material will have the advantages of good characteristics of each constituent material.

A common type of composite materials is known as the fibrous composites; which mainly consist of two separate components; the matrix and the fibers. The matrix component will act as the adhesive material that holds the fibers together to form the bulk of the composite material. It usually consists of various types such as: epoxy, polymer, metal and thermoplastic. The fibers are of different types such as: Carbon, Boron, Graphite and Kevlar. The matrix and fibers together form what is known a lamina. Combination of several laminas form what is known as a laminated composite. Lamination is used to

* Corresponding author: ghazi@just.edu.jo

combine the best aspects of the constituent layers in order to achieve a more useful material. The properties that can be emphasized by lamination are strength, stiffness, low weight, corrosion resistance, etc., as indicated in [1]. Jones, 1975.

It is well known that a rectangular plate with a rectangular hole has been widely used in industrial designs such as: Civil, Mechanical, Marine and Aeronautical Engineering. Composite materials as: Graphite/Epoxy, Boron/Epoxy, and Glass/Epoxy are widely used in structural components in military (Stealth fighter F-117A Nighthawk and B-2 bomber) and commercially (for example: the forthcoming Boeing 787 Dreamliner airplanes), because of the advantage of weight saving, hence, achieving high strength-to-weight ratios in various components. Furthermore, providing enhanced corrosion resistance and improved stealth characteristics. Best composite design was achieved through the variation of stacking pattern and fiber-orientation.

Linear and nonlinear structural behavior of composite plates and shells was investigated by several researchers using the finite element method with a variety of approaches. Most of the researches were concerned with the problem of geometric nonlinearity of large or moderately large deformations. Han et al. [2] used the hierarchical finite element method to carry out the geometrically nonlinear analysis of laminated composite rectangular plates, utilizing the first order shear deformation theory and Timoshenko's laminated composite beam functions. Zinno et al. [3] used the Lagrange approach to develop a three-dimensional element with two-dimensional kinematic constraints for the geometric nonlinear analysis of laminated composite plates.

Ganapathi et al. [4] presented an eight-node quadrilateral finite element based on the Reissner–Mindlin plate theory to analyze thick laminated composite plates accounting for moderately large deformations. Polit et al. [5] studied the geometrically nonlinear behavior using a high order triangular finite element plate model having six nodes. The element ensured continuity of transverse interfacial shear stresses of multi-layered composite plates.

Zhang et al. [6-8] developed a four-node quadrilateral elements for linear and geometrically nonlinear analysis of thin to moderately thick laminated composite plates. Khoa et al. [9] developed a rectangular non-conforming element based on Reddy's higher order shear-deformation plate theory to analyze the laminated composite plates. Further studies were carried out on the subject of geometric nonlinearities by other investigators; Salehi et al. [10], Choudhary et al. [11] and many others.

Several researches were carried out on the effect of fiber-orientation, number of plies and laminate-sequence in reducing weight of symmetric and anti-symmetric laminated composite plates subjected to uniform pressure loads; Naghipour et al. [12] developed an analytical procedure to investigate the bending-stretching significance of anti-symmetric and cross-ply laminated composite plates based on a higher order shear displacement model using zigzag-function.

A limited research was carried out on the problem of material nonlinearity and its effect on the behavior of composite laminated plates, especially, in the case of a plate with holes. Abu-Farsakh et al. [13] studied the problem of stress-concentration around a central circular hole in fibrous composite laminated plates subject to in-plane tensile loading. The main aim of the research was to investigate the effect of material nonlinearity on stress-concentration and stress-redistribution around the hole.

For studying the behavior of composite materials with holes, the ANSYS-computer program [14] will be used to account for the stress-concentration and resulting stress-redistribution caused by the presence of holes. Stress-concentration is an important parameter to be taken into consideration in structural design, because the point near maximum stress-concentration zone is often the location of damage initiation in the structure

2 Scope and objectives

The main objective of the present paper is to investigate the effect of linear and nonlinear material behavior on the induced in-plane stresses (σ_x, σ_y and τ_{xy}) in simply supported square plates and then to obtain the corresponding Stress-Concentration Factors (SCF). Plates have symmetrical angle-ply laminates composed of four layers $[\Theta/\text{-}\Theta]_s$. The effect of fiber-orientation angle (Θ) is taken into consideration using three values; 30, 45 and 60-degrees. All plates are subject to out-of-plane uniformly distributed loads of intensities 2.5 MPa and 5 MPa.

3 Application of ANSYS finite element model

The ANSYS-computer program is used for both linear and nonlinear analysis. In linear analysis; element-type SHELL99 is used while for nonlinear analysis element-type is SHELL91. Each of the elements have eight nodes with six degrees-of-freedom at each node; three translations and three rotations. A summary description of element-type, element-geometry, composite-material, number of plies, fiber-orientation angles, plate dimensions and boundary conditions are as indicated in Table (1).

Table 1. Summary description of ANSYS applications.

Title	Description
Number of Layers	Four layers, with ply thickness 2.5 mm each. The fiber-orientation angles are 30^0, 45^0 and 60^0 having staking-sequences in the laminate as $[+\theta°/\text{-}\theta°]_s$. Angle ($\theta$) is taken with respect to the positive x-axis, where x and y are body global axes. 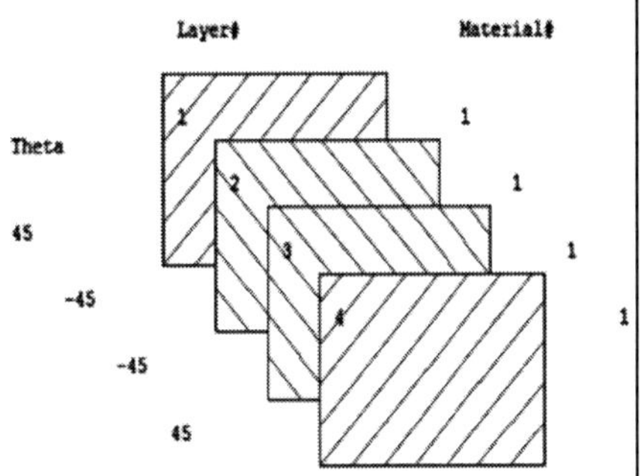
Plate Geometry	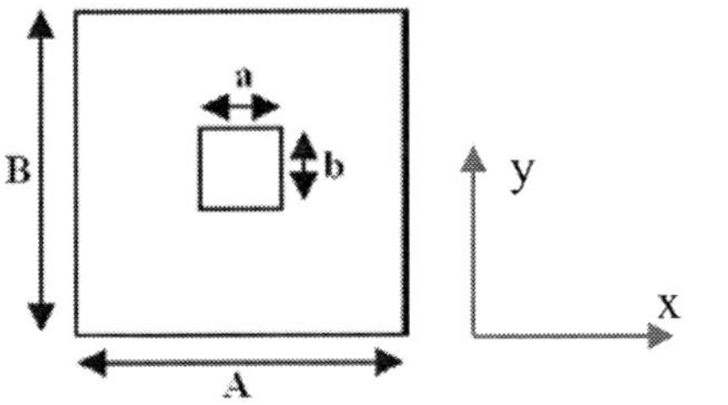
	A=B= 100 mm a=b= 10 mm

Mesh Generation	Due to using shell element, the mapped mesh on the area is created as indicated in the figure, and then the number of divisions required for mapped mesh is specified.	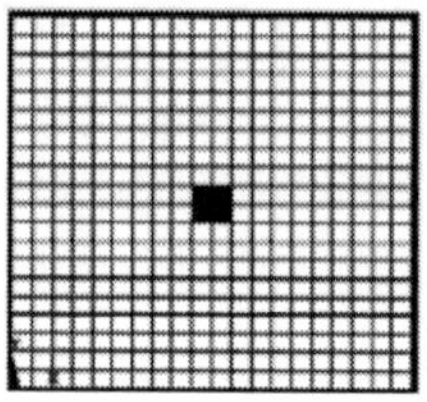
Boundary Conditions	All edges simply supported: $U_x = U_y = U_z = 0$, $\Theta_x \neq 0$, $\Theta_y \neq 0$, $\Theta_z \neq 0$	The simply supported conditions for all edges are used.
Loads	An out-of-plane uniformly distributed load of intensity 5 MPa.	An out-of-plane uniformly distributed load of intensities 2.5 and 5MPa.

4 Nonlinear material model

The nonlinear mechanical properties of the composite material are expressed as [15-17]:

$$M_i = M_o\left[1 - B_i\left(\overline{U}_p\right)^{C_i} + D_i\left(\overline{U}_p\right)\right] \quad (1)$$

Where,
M_i, M_o: Secant and tangential mechanical properties
B_i, C_i, D_i: Material-property constants for mechanical property (i).
Three in-plane mechanical properties; E_1, E_2 and G_{12}, are expressed using Eq. 2. The initial values of mechanical properties and the corresponding material constants are as indicated in Tables 1 to 2. The plastic strain energy term $\left(\overline{U}_p\right)$ is normalized as:

$$\left(\overline{U}_p\right) = \frac{U_p}{U_o} = \frac{U_s - U_e}{U_o} \quad (2)$$

For in-plane stress case, the total secant strain energy density $\left(U_s\right)$ may be expressed in terms of principal material directions (1 & 2) as:

$$U_s = \frac{1}{2}\left(\frac{\sigma_1^{\,2}}{E_1} + \frac{\sigma_2^{\,2}}{E_2} + \frac{\tau_{12}^{\,2}}{G_{12}}\right) \quad (3)$$

Where, E_1, E_2 and G_{12} are the secant moduli at each stress level.

Similarly, the initial elastic moduli E_{o1}, E_{o2} and G_{o12} can be used to obtain the elastic strain energy density, such that:

$$U_e = \frac{1}{2}\left(\frac{\sigma_1^{\,2}}{E_{01}} + \frac{\sigma_2^{\,2}}{E_{o2}} + \frac{\tau_{12}^{\,2}}{G_{o12}}\right) \quad (4)$$

The plastic strain energy density (U_p) is obtained (as shown in Fig. 1) as:

$$U_p = U_s - U_e \tag{5}$$

The term (U_o) is used to normalize the plastic strain energy density, and it is usually taken equal to unity (i.e. 1 MPa).

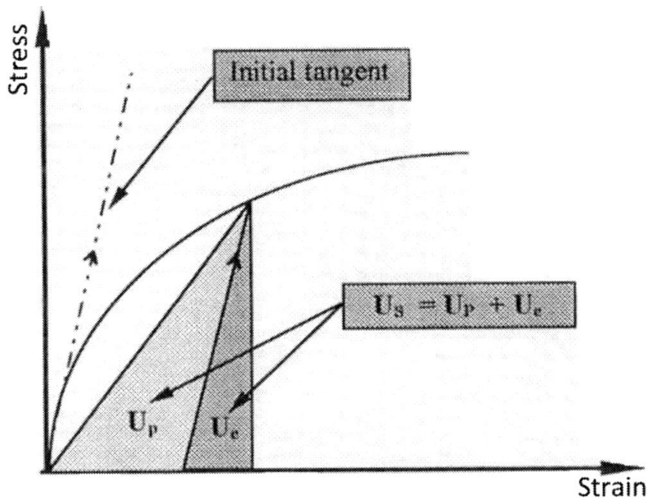

Fig. 1. Typical non-linear stress-strain curve for the *ith*-mechanical property showing the strain energy density terms.

The normal-stress versus normal-strain relation $(\sigma_x - \varepsilon_x)$, see Table-1 for x-axis, was predicted using the nonlinear material model which was originally developed by the first author [15]. In order to account for the nonlinear material behavior, a computer program MCOMP was developed for this purpose. Every layer in the composite laminate was defined as a new material-type having a specific orientation-angle (Θ). Accordingly, the material model was properly incorporated in the ANSYS-computer program in the material-property section. The stress-strain relationship was dealt with as a multi-linear fitting technique for the provided stress-strain graph. The computer program utilized the **Newton-Raphson** technique to achieve step-convergence at each increment.

5 Analysis of results and discussion

The stress-distribution profile was greatly affected due to the existence of a central square hole. Thus, allowing for the redistribution of stresses around the hole to take place. Hence, allowing the stresses to change sign from tension to compression in the vicinity of hole-corners with increased maximum stresses in the nearby region. In this paper the simply supported boundary conditions will be considered only, as indicated earlier in Table 1.

The stress-distribution profiles (σ_x, σ_y, and τ_{xy}) due to uniformly distributed load intensity of 5 MPa for the case of simply supported square plate (A/B=1) and a square hole (a/b=1), versus x-distance are illustrated in Figures 2-a, 2-b and 2-c, respectively. For the case of Θ=30^0. In the linear analysis case, the stress-distribution (σ_x) showed a maximum value for Θ=30^0 (about 205 MPa at x=35 mm). On the other hand, the maximum compressive stresses reached 135.38 MPa, on the hole-perimeter (x=45 mm) due to the redistribution of stresses process. With respect to the nonlinear analysis, the maximum tensile stress (σ_x) reached 174.22 MPa at x=25 mm, while the maximum compressive stress was 9.887 MPa at x=45 mm.

In the present paper, the Stress Concentration Factor (SCF) is defined as:

$$\mathrm{SCF} = \frac{\text{Maximum stress with hole}}{\text{Maximum stress without hole}} \tag{6}$$

This factor is affected by hole-size (a/b), plate-size (A/B), fiber-orientation angle (θ), and boundary conditions of the plate.

The Stress-Concentration-Factor (SCF) as affected by the fiber-orientation angle due to in-plane stresses (σ_x, σ_y and τ_{xy}) for the case A/B=1 with a central square hole (a/b= 1.0), is illustrated in Figure 3. Three fiber-orientation angles are taken in to consideration; 30, 45 and 60-degrees. A well-known composite material is considered herein in order to compare linear with nonlinear behaviors using Narmco-5605 Graphite/Epoxy composite material. The nonlinear in-plane shear properties according to Eq. (1) are given as: B_{12}=0.562625, C_{12}=0.395999 and D_{12}=0.112317. Material properties in 1 and 2 principal material directions are considered linear with initial moduli E_{10}=1.062x10^4 MPa, E_{20}=1.276x10^5 MPa and G_{120}=5.861x10^3 MPa.

With respect to SCF, obtained linear values ranged approximately between (2.2 to 2.4), while the corresponding nonlinear values ranged approximately between (1.1 to 1.5) at all Θ-values for all stresses, as shown in Figures 3-a, 3-b and 3-c. As indicated, the maximum stress-values obtained from nonlinear analysis at the hole-sides are decreased and stresses are subjected to redistributions around the hole-perimeter and become smoother (i.e. no sharp peaks). Hence, the resulting SCF obtained from nonlinear analysis are smaller than those obtained from linear analysis.

Other comparisons using the nonlinear analyses under out-of-plane uniformly distributed loads of 5 and 2.5 MPa intensities are shown in Figures 4-a, 4-b and 4-c. The stress distributions (σ_x, σ_y and τ_{xy}) are exhibited for the plate case (A/B=1), hole-ratio (a/b=1) and Θ = 30-degrees. As the load is increased from 2.5MPa to 5 MPa, maximum σ_x-stresses are increased accordingly in a variable manner. That is, at the load 5 MPa the rate of increase was higher at Θ = 30-degrees and decreased gradually at bigger angles (Θ = 45 and 60-degrees); see Figure 4-a. On the other hand, the maximum σ_y-stresses are increased gradually with increased fiber-orientation angle (Θ), as illustrated in Figure 4-b. In-plane shear stresses (τ_{xy}) showed almost a linear decreasing-rate with increased fiber-orientation angles but with steeper slope for the 5 MPa load-intensity compared to the 2.5 MPa intensity, as indicated in Figure 4-c.

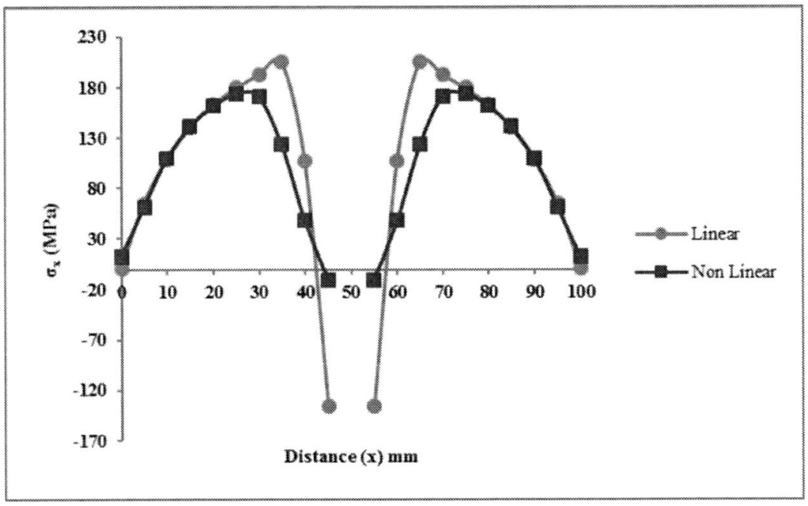

Fig. 2-a. In-plane stress-distribution (σ_x) versus distance (x); A/B=1, a/b=1 and $\Theta = 30^0$.

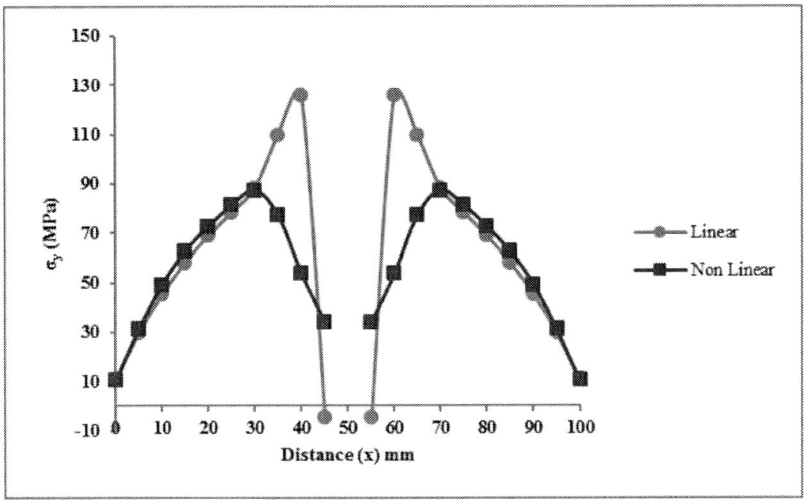

Fig. 2-b. In-plane stress-distribution (σ_y) versus distance (x); A/B=1, a/b=1 and $\Theta = 30^0$.

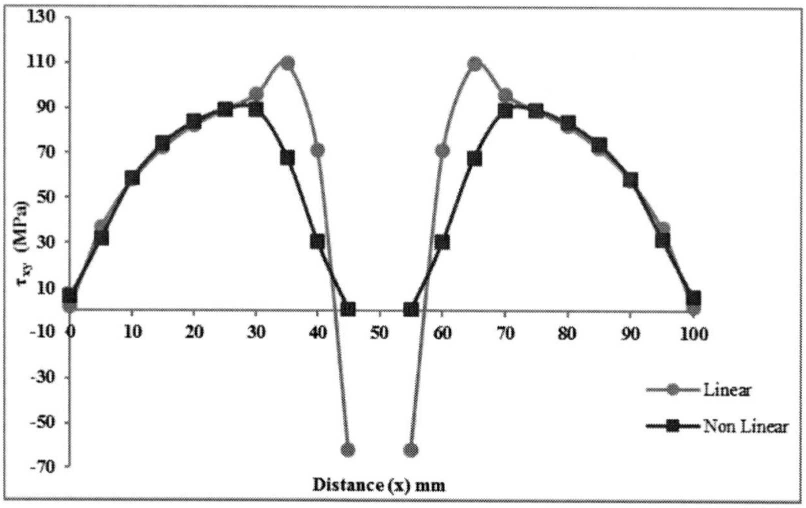

Fig. 2-c. In-plane stress-distribution (τ_{xy}) versus distance (x); A/B=1, a/b=1 and $\Theta = 30^0$.

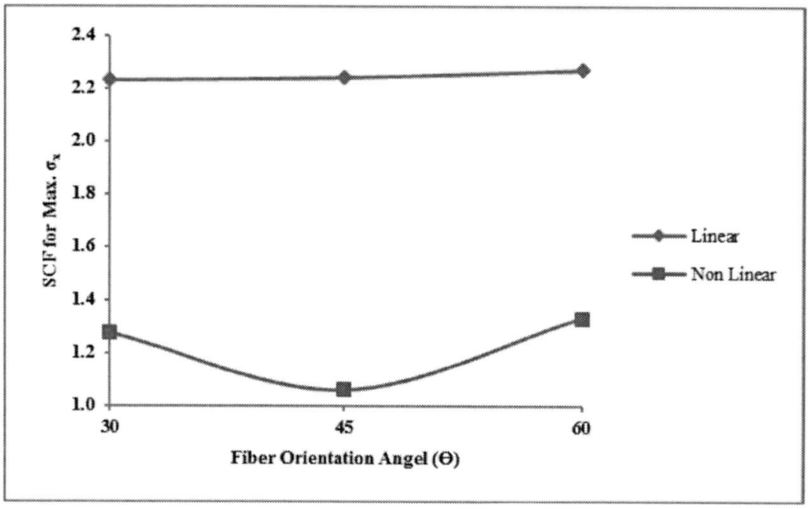

Fig. 3-a. SCF for maximum (σ_x) versus Θ; A/B=1, a/b=1 for the uniformly distributed load of 5 MPa.

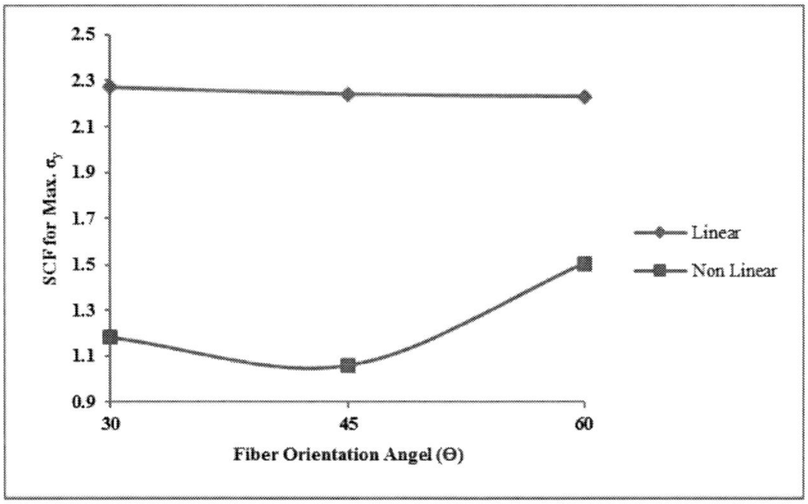

Figure 3-b. SCF for maximum (σ_y) versus Θ; A/B=1, a/b=1 for the uniformly distributed load of 5 MPa.

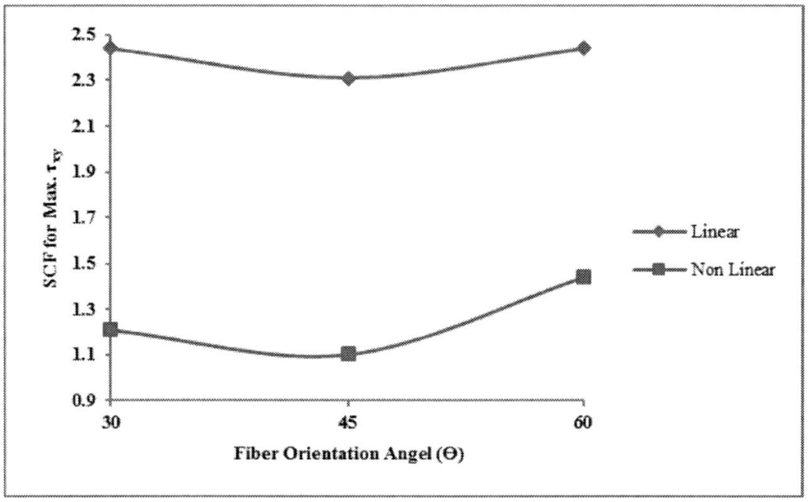

Figure 3-c. SCF for maximum (τ_{xy}) versus Θ; A/B=1, a/b=1 for the uniformly distributed load of 5 MPa.

DOI: 10.1051/matecconf/201712001001

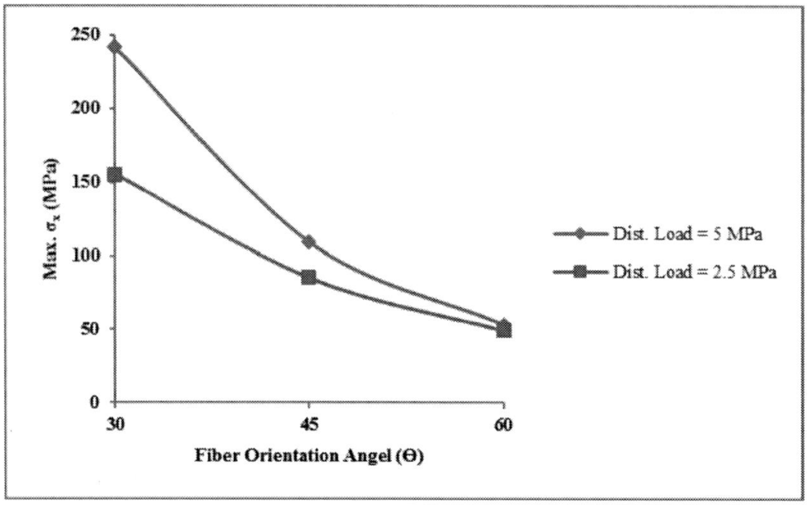

Figure 4-a. Maximum (σ_x) versus Θ; A/B=1, a/b=1 for the uniformly distributed loads of 5 and 2.5 MPa.

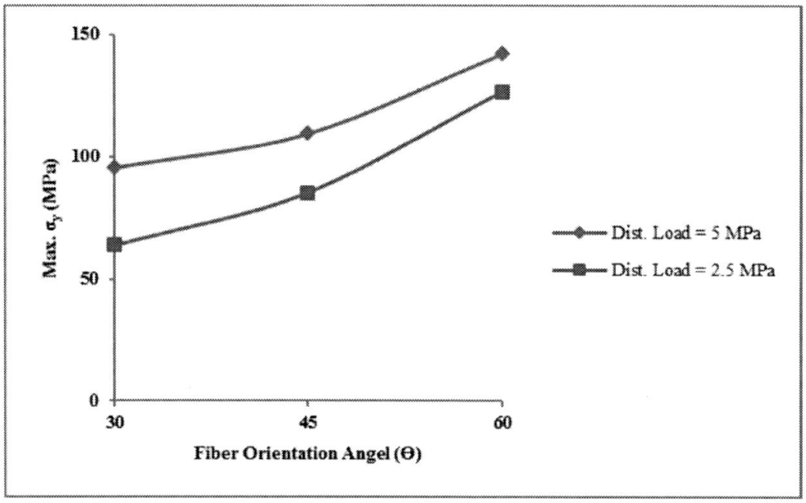

Fig. 4-b. Maximum (σ_y) versus Θ; A/B=1, a/b=1 for the uniformly distributed loads of 5 and 2.5 MPa.

DOI: 10.1051/matecconf/201712001001

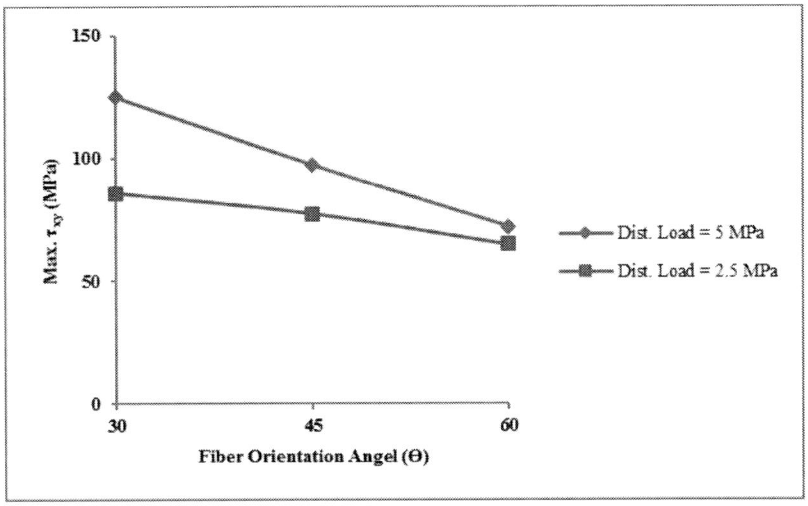

Fig. 4-c. Maximum (τ_{xy}) versus Θ; A/B=1, a/b=1 for the uniformly distributed loads of 5 and 2.5 MPa.

As illustrated in Figure 5, stress values are decreased as we move away from the hole, and the nodal results in the nonlinear analysis showed discontinuity at corner nodes of the hole. Thus, the allowable stresses will be exceeded in the hole-vicinity.

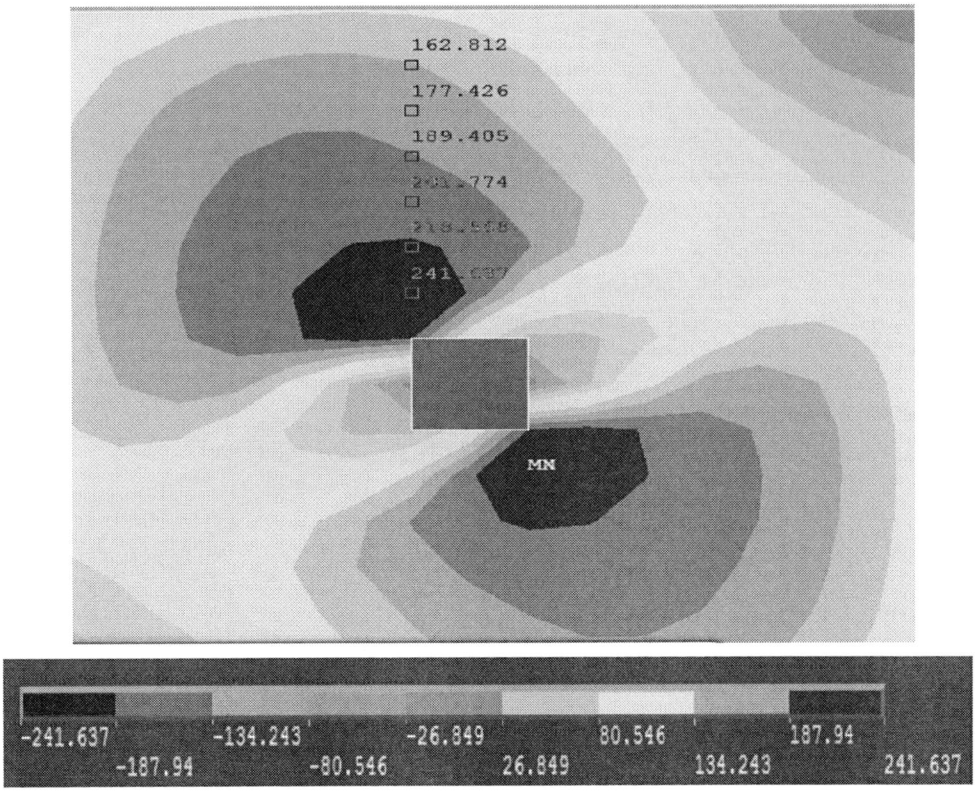

Fig. 5. A representative figure showing stress-σ_x contours for the square plate with a square hole for the case Θ = 30°.

6 Conclusion

Several parameters showed an increasing effect on the stress-concentration phenomenon in rectangular plates with central square hole and subjected to uniformly distributed loading, such as: plate aspect-ratio, hole aspect-ratio, fiber-orientation angle, plate boundary conditions and method of analysis; linear or nonlinear. In the present paper two factors were discussed; fiber-orientation angle and method of analysis due to space limitations.

Maximum stresses around the hole-perimeter resulted from linear analysis showed larger values than those corresponding to nonlinear analysis. This was attributed to the stress-redistribution process which occurred in the vicinity of the hole due to the nonlinear material behavior.

For the considered fiber-orientation angles (30, 45 and 60-degrees), the range of Stress Concentration Factors (SCF) using linear analysis was approximately in the range of (2.2 to 2.4), while the corresponding range for nonlinear analysis was approximately (1.1 to 1.5).

References

1. R. M. Jones. *Mechanics of Composite Materials*. New York: McGraw-Hill (2011)
2. W. Han., K. M. Petyt, F. E. Anal. Des. **18**(1–3), pp. 273–288 (1994)
3. R. Zinno, E. J. Barbero, Comp. Struc., **57**(3), pp. 455–466 (1995)
4. M. Ganapathi, O. Polit, M. Touratier. *Int. J. Num. Meth. Eng.*, **39**(20), pp. 3453–3474 (1996)
5. O. Polit, M. Touratier, Compos. Struc., **58** (1), pp. 121–128 (2002)
6. Y. X. Zhang, K. S. Kim, Int. J. Num. Meth. Eng., **61**, pp. 1771–1796 (2004)
7. Y. X. Zhang, K. S. Kim, Comp. Meth. Appl. Mech. Eng., **194**, pp. 4607–4632 (2005)
8. Y. X. Zhang, K. S. Kim, Compos. Struc. **3**, pp. 301–310 (2006)
9. N. N. Khoa, T. I. Thinh, Viet. J. Mech., VAST, **29** (1), pp. 47–57 (2007)
10. M. Salehi, S. R. Falahatgar, Trans. B: Mech. Eng. **17** (3), pp. 205-216 (2010)
11. S. S. Choudhary, V. B. Tungikar, Int. J. Eng. Sci. Tech., **3** (6), pp. 4897-4907 (2011)
12. M. Naghipour, H. M. Daniali, K. S. H. A. Hashemi, W. App. Sci. J., **4** (5), pp. 681-690 (2008)
13. G. Abu-Farsakh, A. Almasri, D. Qa'dan, Sci. Eng. Compos. Mat. (SECM), *22* (1), pp. 31-36 (2015)
14. *ANSYS Tutorial*, Release 9.0.
15. G. A. Abu-Farsakh, Compos. Sci. Tech., **20** (4), pp. 349–360 (1989)
16. G. A. Abu-Farsakh, Y. A. Abdel-Jawad, J. Compos. Tech. Res., (JCTRER), **16** (2), pp. 138-145 (1994)
17. G. A. Abu-Farsakh, Y. A. Abdel-Jawad, J. Compos. Tech. Res. (JCTRER), **17** (2), pp. 90-98 (1995).

An approximate procedure for estimating seismic demands for buildings

Benazouz Chikh[1,*], *Moussa* Leblouba[2], *Youcef* Mehani[1] and *Mohamed* Hadid[3]

[1]National Earthquake Engineering Research Center, Earthquake Engineering Division, 16005 Hussein Dey Algiers, Algeria
[2]Department of Civil & Environmental Engineering, College of Engineering, University of Sharjah, P.O. Box 27272 Sharjah, United Arab Emirates
[3]National School of Built and Ground Works Engineering, Kouba, Algiers, Algeria

Abstract. Methods of the estimation of the seismic demands of nonlinear structures require a sort of iterative procedures, no matter the elastic or inelastic response spectra are used in the procedure. Many studies dealt with the development of different inelastic spectra with the aim to simplify the evaluation of inelastic deformations and performance of structures. Recently, the concept of inelastic spectra has been adopted in the global scheme of the performance-based seismic design through capacity-spectrum methods (CSM).For instance, the modal pushover analysis (MPA) has been proved to provide accurate results for inelastic buildings to a similar degree of accuracy to the response spectrum analysis (RSA) in estimating peak response for elastic buildings. In this paper, a simplified nonlinear spectral analysis is made to the original MPA is proposed. The proposed procedure, which is applicable to multi-degree-of-freedom MDOF systems uses an inelastic spectrum and gives peak responses consistent with those obtained when using the nonlinear time history analysis (NL-RHA).The accuracy of the proposed procedure is verified against the nonlinear time history analysis results of two buildings. The comparison showed that the new method is capable to provide accurate peak response.

1 Introduction

As an alternative to the complex nonlinear dynamic analysis to estimate the seismic demands of structures, several simple evaluation methods have been proposed (Gulkan and Sozen 1974; Freeman 1975; Newmark and Hall 1982; Fajfar and Fischinger 1988; Kowalsky 1994; Sasaki et al. 1998; Gupta and Kunnath 2000; Albanesi et al. 2000; Priestley and Kowalsky 2000; Miranda 2001; Chopra and Goel 2001; Lin and Chang 2003; Maja and Fajfar 2012; Benazouz et al. 2012). Such methods are based on the characterization of the structural capacity and the good quantification of seismic intensity,

*Corresponding author: cheikhpbsd@gmail.com

such as the Capacity Spectrum Method CSM (Freeman 1975) and its different implementations.

Since the development of the CSM Method, many response spectra have been proposed to replace the conventional elastic spectra in order to achieve accurate evaluation of the inelastic response of structures, (Sheng and Biggs 1980; Iwan 1980; Newmark and Hall 1982; Kowalsky 1994). Fajfar (1999) have proposed a N2 method for nonlinear analysis of seismic damage of reinforced concrete buildings. This method employs an inelastic response spectrum and the capacity curve is obtained from a single-degree-of-freedom-system (SDOF) equivalent to the multi-degree-of-freedom-system (MDOF). It is implemented in Eurocode 8 (CEN 2005) and it was used in a comparison between traditional non-linear static methods in the evaluation of asymmetric structures.

The Modal Pushover Analysis (MPA), developed by Chopra and Goel (2001), which includes the effects of higher modes of vibration, provides a good estimation of the seismic demand while maintaining a simplistic concept and computing attraction due to the invariability of the distribution of forces. After, the modal pushover analysis undergoes changes by Chopra et al (2004), for which an extension of this analysis was developed under the name (Modified Modal Pushover Analysis, MMPA) that combines the elastic influence of higher modes with inelastic response of the first mode using several combinations such as the Square Roof of the Sum of Squares (SRSS).

In this paper, we first present the development of a new method based on ductility demand response spectrum (Benazouz *et al.*, 2012) for the evaluation of seismic demands of MDOF systems. This procedure, is developed to consider the contributions of higher modes of vibration, is detailed below in this paper.

2 Modal pushover analysis procedure (MPA)

The modal Pushover analysis (Chopra and Goel, 2001), is based on structural dynamics theory, its accuracy and reliability in estimating the peak response of inelastic MDOF systems has been evaluated extensively by its authors. For convenience we summarize the MPA as a series of steps used to estimate the peak inelastic response of MDOF (Chopra 2007):

1. Compute the natural frequencies, ω_n and modes, ϕ_n, for linearly elastic vibration of the building.
2. For the nth-mode, develop the base shear-roof displacement, $V_{bn} - u_{rn}$, pushover curve for force distribution

$$s_n^* = m\phi_n \tag{1}$$

Where m is the mass matrix of the structure.

3. Idealize the pushover curve as a bilinear curve (Fig. 1a). If the pushover curve exhibits negative post-yielding stiffness, idealize the pushover curve as elastic-perfectly-plastic.
4. Convert the idealized pushover curve to the force-displacement, (Fig. 1b) for the nth–mode inelastic SDOF system by utilizing

$$\frac{F_{sny}}{L_n} = \frac{V_{bny}}{M_n^*} \quad D_{ny} = \frac{u_{rny}}{\Gamma_n \phi_{rn}} \tag{2}$$

In which M_n^* is the effective modal mass, $F_{sn}$is the lateral forces, ϕ_{rn} is the value of ϕ_n at the roof, $\Gamma_n = \phi_n^T m1/\phi_n^T m\phi_n$ and $L_n = \phi_n^T m1$

5. Compute peak deformation D_n of the nth-mode inelastic SDOF system defined by the force-deformation relation of Fig. 1b and damping ratio ξ_n. The elastic vibration period of the system is

$$T_n = 2\pi \left(\frac{L_n D_{ny}}{F_{sny}}\right)^{1/2} \tag{3}$$

6. For an SDOF system with known T_n and ξ_n, D_n canbe computed by nonlinear response history analysis (NL-RHA) or from the inelastic design spectrum (Chopra 2007, Section 7.11).
7. Calculate peak roof displacement u_{rn} associated with the nth-mode inelastic SDOF system from

$$u_{rn} = \Gamma_n \phi_{rn} D_n \tag{4}$$

8. From the pushover database (Step 2), extract values of desired responses r_n: floor displacements, storydrifts, plastic hinge rotations, etc.
9. Repeat Steps 3-7 for as many modes as required for sufficient accuracy. Typically, the first two or three modes will suffice.
10. Determine the total response (demand) by combining the peak "modal" responses using the SRSS rule:

$$r = \left(\sum r_n^2\right)^{1/2} \tag{5}$$

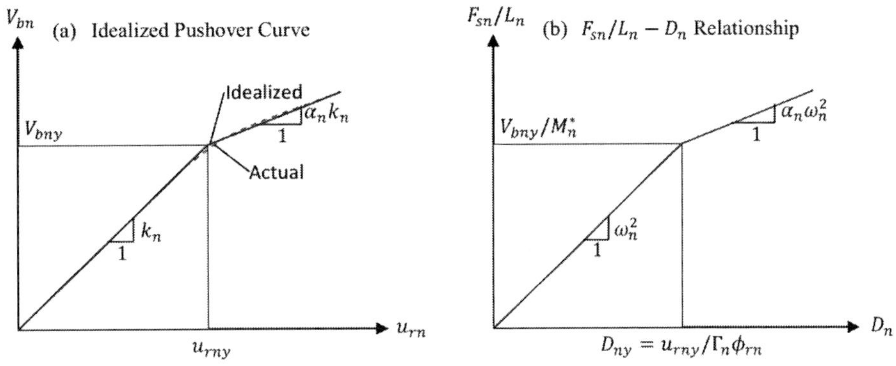

Fig.1. Properties of the nth-mode inelastic SDOF system from the pushover curve (Chopra and Goel 2001).

3 Proposed procedure

A simplified method is proposed for estimating the seismic response of MDOF systems. However, instead of using inelastic design spectra we use the ductility demand response spectrum (Benazouz et al., 2012). Summary of the implementation of this procedure is presented next.

The construction of the capacity curve starts with obtaining the multimodal pushover curves of the MDOF system subjected to lateral forces distributed over the building height according to S_n^*, where the movements will be decomposed in the form of a series of normal modes. The application of the force vector F over the height of the building for each time step t gives forces and stresses on the elements in a static analysis. This force distribution, F, can be obtained with a decomposition of the system inertia (Chopra 2007):

$$M1 = \sum_n \Gamma_n M\phi_n = \sum_n S_n \tag{6}$$

Mainly the mode shapes control the capacity curve obtained using above force distribution. The capacity curve undergoes a bilinear representation as illustrated in Figure 2(a). Then, using Eq. 2 we transform the curve into AD format (Figure 2b). The obtained curve is known as the capacity diagram of the structure.

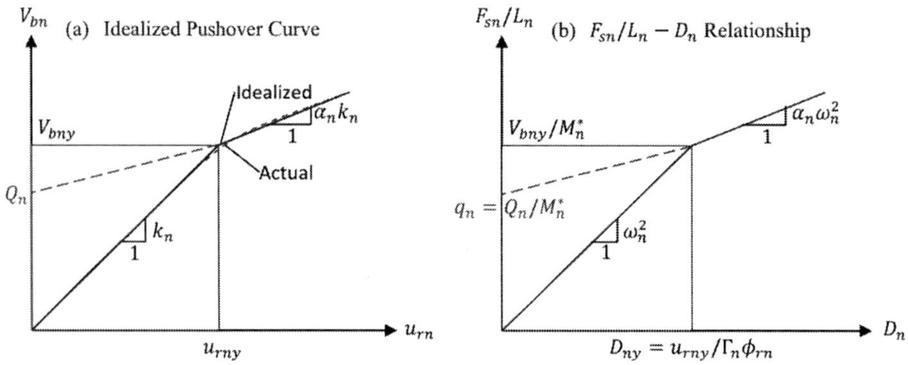

Fig.2. Properties of the nth-mode inelastic SDOF system from the pushover curve with yield strength coefficient.

For the nth- mode, this idealization is used to calculate the yield strength Q_n and the yield strength coefficient q_n (Fig 2b) as follows:

$$q_n = \frac{Q_n}{M_n^*} \tag{7}$$

In which M_n^* is the effective modal mass.

This coefficient will be used for evaluating the seismic demands that are represented by an inelastic response spectrum that is summarized in the next part.

The seismic demand in this paper is determined by using the Ductility-Demand Response Spectrum (DDRS) (Benazouz *et al.*, 2012), which will be introduced in detail hereafter. In constructing the response spectra we consider an inelastic SDOF system as shown in Fig.3, its motion when subjected to an earthquake ground motion is governed by the following equation:

$$m\ddot{x} + c\dot{x} + f(x,\dot{x}) = -m\ddot{x}_g(t) \tag{8}$$

Where m, c, and f represent the mass, damping, and the resisting force of the system, respectively. $\ddot{x}_g(t)$ denotes the earthquake acceleration input to the SDOF.

The resisting force f is defined as the sum of the linear part and the hysteretic part:

$$f = k_p x + Qz \tag{9}$$

In the above equation, k_p is the postyield stiffness, Q is the yield strength (it has been determined through the capacity curve), and z represents the dimensionless variable that characterizes the Bouc-Wen model of hysteresis (Wen 1976), it is given by:

$$\dot{z} = \frac{\dot{x}}{x_y}\left[A - |z|^{\lambda}(B\text{sign}(\dot{x}z) + \beta)\right] \tag{10}$$

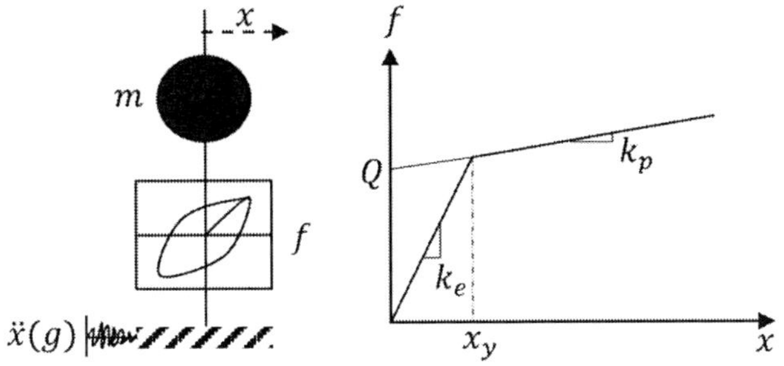

Fig. 3. Capacity curve of a (SDOF) bilinear system.

In Eq. (10), x_y is the yield displacement; A, B, λ, and β are the parameters that control the shape of the hysteresis loop.
Substituting Eq. (9) into Eq. (8) and dividing by m yields:

$$\ddot{x} + 2\xi\omega\dot{x} + \alpha\omega^2 x + qgz = -\ddot{x}_g(t) \tag{11}$$

In which: ξ, ω, α, and q represent the damping ratio, circular frequency, post-to-preyield stiffness ratio, and the yield strength coefficient, respectively. From Fig 2b, q is defined as the yield strength divided by the system weight w, $w = mg$ where g stands for the gravity, respectively.
Next, Eq. (11) is rewritten in terms of ductility factor,μ. Substituting: $x = x_y\mu$, $\dot{x} = x_y\dot{\mu}$, and $\ddot{x} = x_y\ddot{\mu}$ in Eq. (11) and dividing by x_y gives:

$$\ddot{\mu} + 2\xi\omega\dot{\mu} + \alpha\omega^2\mu + \omega^2(1-\alpha)z = -\frac{\omega^2(1-\alpha)}{qg}\ddot{x}_g(t) \tag{12}$$

It can be observed from Eq. (12) that for a given ground acceleration, $\mu(t)$ depends on ξ, ω, α and q. To obtain meaningful system response to an ensemble of ground motions, the system yield strength coefficient has to be defined relative to the intensity of individual ground motions. Using the parameter η introduced by Mahin and Lin (1983) as:

$$\eta = \frac{qg}{PGA} \tag{13}$$

where, PGA stands for the Peak Ground Acceleration.
Incorporating η into Eq. (12) results:

$$\ddot{\mu} + 2\xi\omega\dot{\mu} + \alpha\omega^2\mu + \omega^2(1-\alpha)z = -\frac{\omega^2(1-\alpha)}{\eta}\bar{\ddot{x}}_g(t) \tag{14}$$

In which, $\bar{\ddot{x}}_g(t)$ represents the ground acceleration normalized with respect to the PGA.
The procedure to construct the ductility demand response spectrum for inelastic systems corresponding to specified levels of normalized yield strength η, is summarized in the following steps (Benazouz *et al.*, 2012) :

1. Define the ground motion $\ddot{x}_g(t)$;
2. Select and fix the damping ratio ξ and the post-to-preyield stiffness ratio α.
3. Specify a value for η;
4. Select a value for elastic period T;
5. Determine the ductility response $\mu(t)$ of the system with, T, ξ and α equal to the values selected by solving Eq. (14). From $\mu(t)$ determine the peak ductility factor μ;

6. Repeat steps 4-5 for a range of T, resulting in the spectrum values for the η value specified in step 3.

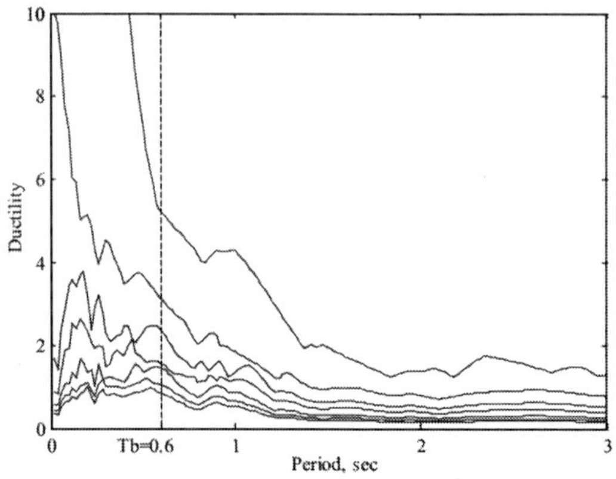

Fig. 4. DDRS for inelastic system computed for El Centro 1940 (N/S) component ($\alpha = 10\%, \xi = 5\%$ $\eta = 0.25, 0.5, 0.75, 1, 1.5, 2, 2.5$ from top line to bottom line).

The value of the ductility factor is read from the spectrum developed by the above procedure and multiplied by x_y to obtain the peak deformation x_m.
The spectrum was divided logically into three period regions according to the last work described in (Benazouz *et al.*, 2012). In Fig. 4, T_b marks the transition from the acceleration-sensitive region to the velocity-sensitive region which ends around T = 3 sec.
For the nth-mode, the maximum displacement of the original structure based on ductility is given by:

$$x_n = \mu x_y \tag{15}$$

This is determined from the ductility demand response spectrum (DDRS). At this level, the pushover analysis provides a good estimation of the maxima r_{no} (peak modal response) of any response (displacement profile, story drifts, joint rotations, etc.)
Combining those modal responses using any of the proved modal combination methods (Eq. 5) we obtain a good estimation of the peak value r_o of the total structural response.This technique lacks theoretical basis, however it is proved to provide meaningful results.

4 Application and discussion

Two examples of eight and three stories reinforced concrete plane frame structures (regular structures) are considered in order to evaluate the efficiency of the proposed method. They have been chosen for representing the behavior of low and medium-rise buildings. These examples are an eight (8St) and three-story (3St) reinforced concrete buildings as illustrated in Fig. 5. These structures are designed according to the Algerian code RPA-2003 (CGS 2003). Each structure consists of three bays frame, spaced at 4m and a uniform story height of 3m with no significant height irregularities. The purpose of this study is to confirm the application of the proposed method for each frame structure under a design earthquake. The geometric properties of the components are presented in Fig. 5. Vibration modes and periods of these structures for linearly elastic case are summarized in Table 1. The capacity curves were obtained by applying the Adaptive Pushover Analysis (Reinhorn et al. 2006).

These curves were transformed to the capacity diagram ($S_{an} - D_n$) format, and gave the curves shown in Fig. 6. The numerical models were simulated using IDARC program (Reinhorn 1997). The supports were modeled as infinitely rigid to avoid the soil-structure interaction. The periods of the selected structures were: T = 0.81 sec for the 8St building, in the velocity-sensitive spectral region, and T = 0.32 sec for the 3St building, in the acceleration-sensitive spectral region. Results obtained show that the proposed procedure give peak displacements close to those that obtained by NL-THA. Concrete compressive strength was 25MPa. Concrete density is $2.5t/m^3$, concrete Young modulus is $3.21\ 10^7$ kN/m^2 and reinforcements yield strength is 400 MPa.

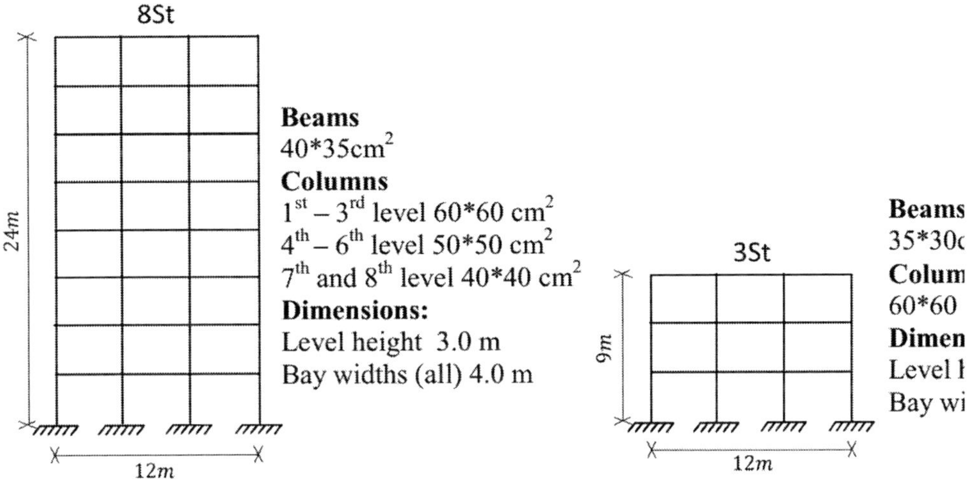

Fig. 5. Elevation view of the 8St and 3St structures.

Table 1. Dynamic characteristics

	Mode	f_n (Hz)	T_n (sec)	Γ_n	L_n(%)
8St	1	1,22	0,81	1,558	77,10
	2	3,60	0,27	-0,611	11,87
	3	6,40	0,15	-0,386	4,74
3St	1	3.13	0.32	0.461	85.63
	2	10.54	0.094	0.168	11.38
	3	20.07	0.05	0.086	2.3

4.1 Ground motion records

Four ensembles of ground motions, each with 20 records were considered in this application. The ensembles, denoted by LMSR, LMLR, SMSR, and SMLR represent four combinations of large (M = 6.6-6.9) or small (M = 5.8-6.5) magnitude and short (R = 13-30 km) or long (R=30-60 km) epicentral distance. These motions were selected from the PEER (Pacific Earthquake Engineering Research Center) Strong Motion Database. More details about this database can be found elsewhere (Chopra *et al.*, 2003). These values were used for both horizontal directions of ground motion in the proposed procedure in order to be able to compare its results with those obtained by NL-THA.

4.2 Capacity curve

First, the Pushover curves of the two structures were transformed to the equivalent SDOF system (see Fig. 6). For example the Pushover curve of the eight-story structure is transformed to the equivalent SDOF capacity curve. In this case, the transformation factor amounts of $\Gamma_1 = 1.558$, and the effective mass is equal to $L_1 = 40912454.4$ kg. The capacity diagram shown in Fig. 9 was obtained by dividing the forces corresponding to the idealized SDOF system by the equivalent mass. For 8St structure, the yield strength ratio is equal to $q_1 = Q_1/L_1 = 0.795$.

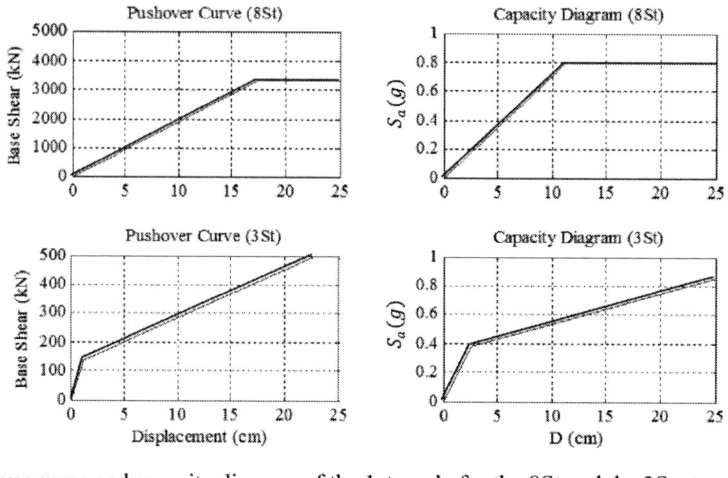

Fig. 6. Pushover curve and capacity diagram of the 1st mode for the 8St and the 3St structures.

4.3 Seismic demand

The simplified procedure was evaluated by comparing the computed displacements (displacement divided by the building height), and the inter-story drift ratio (relative drift between two consecutive stories normalized by story height). Since the time-history results were based on a set of 80 simulations, both the mean and the dispersion (standard deviation σ) about the mean value are shown in the plots.

Peak displacement profiles and inter-story drift ratio profiles estimated by the NL-THA analyses and those obtained by the proposed procedure for the buildings are shown in Figs. 7 - Fig.8. These figures show the means and the standard deviations of the peak displacement profile estimated by NL-THA analyses and the predictions using the proposed procedure for each building sorted by type of record.

The displacement demands along the heights of the buildings showed that the proposed method gives good results and generally yield to better estimates of the peak displacement profiles particularly for the 8St building. Comparing the time-history responses for the different accelerations indicates that the difference between the ground motion of far field records and near field records generally produce more variability in the demands. It is interesting to mention that the story displacement demands from the proposed method are always conservative.

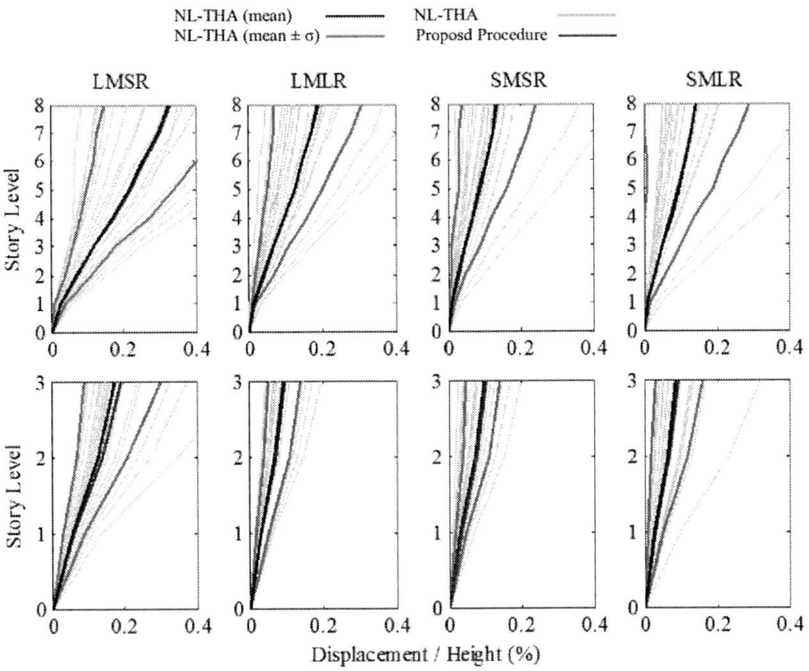

Fig. 7. Displacements obtained by the proposed procedure and NL-THA for the four ensembles.

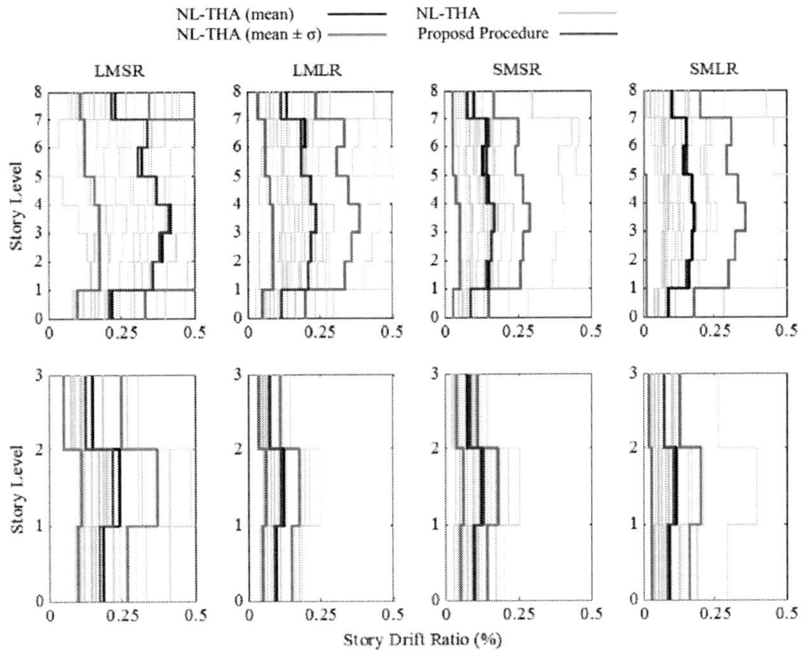

Fig. 8. Inter-story drift ratio obtained by the proposed procedure and NL-THA for the four ensembles.

5 Conclusions

An approximate procedure for seismic demands assessment of MDOF system has been developed and its accuracy was verified by examples. An inelastic modal decomposition in terms of ductility has been developed to construct the simplified procedure. That was verified using the seismic response of an example steel frame structure for which capacity curve data is available. The results indicated that more reliable displacement predictions are obtained from the proposed method.

The base shear–roof displacement ($V_{bn} - x_{rn}$) curve is developed from a Pushover analysis. This Pushover curve is idealized as a bilinear force–deformation relation for the nth mode of inelastic SDOF system. This idealization is used to determine the normalized yield strength coefficient η_n and the post-to-preyield stiffness ratio α_n to estimate the ductility demand. The peak deformation of this SDOF system, determined by the design spectra DDRS, is used to determine the target value of roof displacement at which the seismic response is determined by the Pushover analysis. The total demand is determined by combining the responses of the first three modes by the SRSS combination rule.

The efficiency of the simplified procedure is evident; the designer needs only to have the Pushover curve of the structure and the design earthquake(s) to determine peak response of any structure, namely, base displacement and base shear. This method is applicable to a variety of uses such as a rapid evaluation technique for a large inventory of buildings, a design verification procedure for new construction, an evaluation procedure for an existing structure to identify damage states. The ductility demand is given by the direct estimation where the ductility calculated from the design spectra diagram matches the value associated with the period of the system.

References

1. Albanesi, T., Nuti, C. and Vanzi, I. (2000), "A simplified procedure to assess the seismic response of nonlinear structures", Earthq. Spectra, 16(4), 715-734.
2. Benazouz, C., Moussa, L. and Ali, Z, (2012), "Ductility and inelastic deformation demands of structures", J. Struct. Eng. Mech, 42(5), 631-644.
3. CEN (2005) Eurocode 8 — design of structures for earthquake resistance. Part 3: assessment and retrofitting of buildings. European standard EN 1998-3. June 2005. European Committee for Standardization, Brussels
4. CGS (2003), Algerian seismic building code, National Center of Applied Research in Earthquake Engineering CGS, Algiers.
5. Chikh B, Mehani Y and Leblouba M. (2016). Simplified procedure for seismic demands assessment of structures. Structural Engineering and Mechanics. 59(3):455-473.
6. Chopra, A.K. and Goel, R.K. (2001), A Modal Pushover Analysis Procedure to Estimate Seismic Demands for Buildings: Theory and Preliminary Evaluation, PEER, Berkely, California, 2001/03.
7. Chopra, A.K. and Chatpan, C. (2003), Inelastic deformation ratios for design and evaluation of structures: single-degree-of-freedom bilinear systems. EERC, Berkely, California, 2003/09.
8. Chopra, A.K. and Chatpan, C. (2004), "Inelastic deformation ratios for design and evaluation of structures: single-degree-of-freedom bilinear systems", J. Struct. Eng, 130(9), 1309-1319.
9. Chopra, A.K. (2007), Dynamics of Structures - Theory and Applications to Earthquake Engineering, (3rd Edition), Prentice Hall, New Jersey.

10. Fajfar, P. and Fischinger, M. (1988), "N2 a method for nonlinear seismic analysis of regular structures" Proceedings of 9th World Conf on Earth Eng, Tokyo-Kyoto, Japan, 5:111-116.
11. Freeman, S.A., Nicoletti, J.P. and Tyrell, J.V. (1975), "Evaluation of existing buildings for seismic risk-a case study of Puget Sound Naval Shipyard", Proceedings of 1st U S National Conference on Earthquake Engineering, Berkeley, USA, pp 113-122.
12. Iwan W D (1980) Estimating inelastic spectra from elastic spectra. Earthq Eng Struct Dyn. 8(4):375-388.
13. Gülkan, P. and Sözen, M. (1974), "Inelastic response of reinforced concrete structures to earthquake motions". ACI J Proceedings. 71(12), 604-610.
14. Gupta, B. and Kunnath, S.K. (2000), "Adaptive spectra-based pushover procedure for seismic evaluation of structures", Earthq. Spectra, 16(2), 367-392.
15. Kowalsky, M.J. (1994), "Displacement-based design-a methodology for seismic design applied to RC bridge columns", Master's thesis, University of California at San Diego, La Jolla, California.
16. Lin, Y.Y. and Chang, K.C. (2003), "An Improved Capacity Spectrum Method for ATC-40, Earthq. Eng. Struct. D, 32(13), 2013-2025.
17. Maja, K. and Fajfar, P. (2012), "The extended N2 method considering higher mode effects in both plan and elevation", Bull. Earth. Eng, 10(2), 695-715.
18. Miranda, E. (2001) "Estimation of inelastic deformation demands of SDOF systems", J. Struct. Eng., 127(9), 1005-1012.
19. Newmark, N.M. and Hall, W.J. (1982), "Earthquake spectra and design", Earthquake Engineering Research Institute, Berkeley, California.
20. PEER Strong Motion Database [Online]. http://peer.berkeley.edu/smcat.
21. Priestley, M.J. and Kowalsky, M.J. (2000), "Direct Displacement-Based Seismic Design of Concrete Buildings", Bull of the New Zealand Society for Earth Eng, 33(4), 421-444.
22. Reinhorn, A.M. (1997), "Inelastic analysis techniques in seismic evaluations", Fajfar, P. and Krawinkler, H. (eds.), "Seismic design methodologies for the next generation of codes", Balkema, Rotterdam, pp: 277-287.
23. Reinhorn, A.M., Valles, R.E. and Kunnath, S.K. (2006), "IDARC 2D version 6.1-User's Guide", State University of New York, Buffalo, NY.
24. Sasaki, K.K., Freeman, S.A. and Paret, T.F. (1998), "Multimode pushover procedure (MMP)—A method to identify the effects of higher modes in a pushover analysis", Proceeding of the 6th U.S. Nat. Conf. on Earthq. Eng, Seattle, Washington.
25. Shih-Sheng P L, John M B (1980) Inelastic response spectra for a seismic building design. ASCE J Struct Div. 106(6):1295-1310.
26. Wen, Y.K. (1976), "Method for random vibration of hysteretic systems", J. Eng. Mech., 102(2), 249-263.

Coordinated robotic system for civil structural health monitoring

Uvais Qidwai[1], * and *Muhammad* Ali Akbar[1]

[1]College of Engineering, Qatar University, Doha, Qatar

Abstract. With the recent advances in sensors, robotics, unmanned aerial vehicles, communication, and information technologies, it is now feasible to move towards the vision of ubiquitous cities, where virtually everything throughout the city is linked to an information system through technologies such as wireless networking and radio-frequency identification (RFID) tags, to provide systematic and more efficient management of urban systems, including civil and mechanical infrastructure monitoring, to achieve the goal of resilient and sustainable societies. In this proposed system, unmanned aerial vehicle (UAVs) is used to ascertain the coarse defect signature using panoramic imaging. This involves image stitching and registration so that a complete view of the surface is seen with reference to a common reference or origin point. Thereafter, crack verification and localization has been done using the magnetic flux leakage (MFL) approach which has been performed with the help of a coordinated robotic system. In which the first robot is placed at the top of the structure whereas the second robot is equipped with the designed MFL sensory system. With the initial findings, the proposed system identifies and localize the crack in the given structure.

1 Introduction

The advancement in modern civil architecture along with the considerable number of natural disasters in the last decades creates an alarming situation for the reliability of the infrastructures. If only weather related natural disasters are considered then total 6,457 events has been recorded from 1995 to 2015, according to the United Nations International Strategy for Disaster Reduction (UNISDR) [1]. It has also been claimed in [1] that every year 205 million people on average were affected by such natural disasters. These disasters may not only caused large number of causalities but also damages the stability of the infrastructure which creates alarming situation for those who survives. Therefore, a regular and highly precise monitoring system is required to provide the quality health assurance of the civil structures [2]. In current tradition a team of experts is required to physically examine the buildings and then prepare the report accordingly. However, this approach is not very feasible because of the recent structural complexities along with their heights which are much more than the old traditional buildings. The recent advances in sensors,

* Corresponding author: uqidwai@qu.edu.qa

robotics, unmanned aerial vehicles, communication, and information technologies, can be adopted to fulfil the dream of ubiquitous cities where virtually everything throughout the city is linked to an information system by means of wireless networking and radio-frequency identification (RFID) tags. Thus, a more efficient and systematic management of urban systems monitoring including civil and mechanical infrastructure is become possible to achieve the goal of resilient and sustainable societies.

In the past, many different approaches have been adapted to provide frequent/long-term monitoring of the structures. Most common of these approaches is the vibrational analyses, in which an accelerometer will be connected to different joints of the structure and the vibrations recorded by the instrument will be used to determine the damage in the structure [3-6]. However, vibration-monitoring approach cannot be used to assess the slow degradation of the structure [7]. Moreover, increasing the uncertainties caused by different factors like the environmental condition will also decrease the reliability of damage detection using vibration approach [8].

A magnetic flux leakage (MFL) based structural health monitoring (SHM) approach is one of the predominant approach for metallic structures due to their excellent reproducibility and reliability [15]. The magnetic sensor reads constant values of magnetic flux produced by permanent magnet but when it comes in contact with a crack the leakage in magnetic field will be used to identify the crack. This approach has widely been used for cable inspection [15], steel pipes [16] etc.

Digital imaging is considered to be among the promising approach for automated vision based inspection (VBI) which can help to remotely perform the crack analysis. It is currently being the predominant approach to detect the crack on structure [9-10]. The performance of digital imaging is broadly dependent on the way of capturing images as well as the algorithm used for image processing. In [11] Zhu et al. used a personal camera to take the sight images of the bridge. These images will then be processed using edge detection followed by Hough transform to provide the crack identification. The research work has been conducted using the assumption that the concrete columns of each building have a pair of vertical lines with uniform texture and colour. However, the assumption is not applicable for the real world structure.

The similar approach using marker instead of vertical lines has been used for crack identification using unmanned aerial vehicle (UAV) by Ellenberg et al. in [12]. The need of markers can be overcame by using invariant features based image stitching approach. In this approach the set of images obtained before and after the crack will be stitched separately by using the scale invariant feature like SURF, SIFT etc. Thereafter, the final stitched image of the structure before and after crack will be compared to perform the SHM. The approach has been investigated for bridge inspection [11] and tunnel inspection [13]. The image stitching based SHM seems to be promising for the identification of physical and mechanical defects on the surface of the structure. However, the key aspect of such approach lies in the selection of feature detection algorithm which best suits for image stitching and crack identification.

Although the VBI is seems to be the most effective approach for crack identification. However, it can only provide limited information for maintenance planning [14]. The MFL sensor on the other hand can be used to localize the crack but the response time is slower due to the settling time required to from stable magnetic field during movement [15]. Therefore in this research an integration of VI and MFL has been proposed to detect, verify and localize the crack or structural deformation. The VBI is used to provide early identification of the crack presence whereas the MFL will help to quantitatively analyze and localize the crack. The whole system is automated to get more accurate and robust SHM.

The task of VBI has been achieved with the help of UAV followed by an automated image stitching and anomaly detection algorithm. Whereas the MFL inspection has been achieved with a modular robotic design. In which the first robot is placed at the top of the structure whereas the second robot is equipped with the designed MFL sensory system. The proposed approach can easily be adopted to any metallic structure.

The remaining sections of this paper are organized as follows. Section 2 covers the experimental setup. The proposed approach is described in Section 3. Section 4 is concerned with the simulation results and their discussion. Section 5 concludes the paper.

2 Experimental setup

Since the goal of this research work is to develop a coordinated SHM which uses the advantages of visual inspection for early detection along with the Magnetometer based verification approach. Therefore the experimental setup has also been setup to adopt these two approaches and are described in this section.

2.1 UAV specification

Visual inspection is considered to be the predominant approach for crack detection and identification [10]. This approach is mainly performing the similar operation as that of a sight inspector and that is the reason it has also been used to assist him in SHM. The images for the vision based approach can either be achieved by a fixed camera as in [11] or by means of a flying UAV as adopted in [12].

In our research the images has been obtained using a commercially available UAV (Phantom Professional-3), as shown in Fig. 1(a). The reason of selecting this UAV is because of its high stability and control mechanism which is ensured through different control and sensing modules such as gyroscope, 3-axes accelerometer and an ultra-sonic sensor [17]. The UAV during inspection process is shown in Fig 1(b).

Phantom drone, with its reinforced structure offered a highly safe and robust design in spite of having a net weight of only 1280g. The powerful air braking mechanisms of phantom enables it to be stop instantly such that it can hover in one place as soon as the control sticks are released. The most important feature of Phantom Pro. III lies in its 12 megapixels "4K" camera, which can be used to take high-quality images and videos, with a 3-axis gimbal control based advance image stabilization technology which insure highly stable and clear images even during the flight [17].

(a) (b)

Fig. 1. Phantom drone in (a) stationary and (b) during inspection.

2.2 Hall effect based verification

The MFL is considered to be the most economical approach for inspecting the metallic structures [16]. Traditionally a hand-held device is used for performing the MFL inspection. These devices are normally huge in sizes and can mostly be used for inspecting the pipelines or the sheets. However, it is very difficult to inspect a building using these devices because of the size and weight. Therefore, in this research an autonomous MFL inspection robot has been presented which is small, flexible and remotely accessible.

The robotic design for this purpose used the concept of abseiling, in which the cliff-climber tighten the rope at the top and then move up and down with the support of the rope. Similarly, the designed robot have two modules connected with each other. The first module of robot (FMR) is placed at the top of the structure and is responsible to assist the other module in inspection by performing all computation tasks related to positioning control, data acquisition etc. The second module of the robot (SMR) is having its own control circuitry, permanent magnets and an array of MFL sensor. The permanent magnets were not only used to provide the magnetic field but are also used to keep a uniform contact between the MFL sensor and the structure specially when the second module moving vertically.

The FMR should be strong enough to hold the SMR and other environmental effects like strong wind, gravitation pull etc. Moreover, it should have a precise motion along with other features like obstacle detection to avoid any collision. However, in this research, a Lab grad scaled version of FMR has been used which is designed using a VEX robotic kit [18], as shown in Fig. 2(a).

The FMR is also equipped with ultra-sonic sensors which can help the FMR to change its path in case of obstacle. The control mechanism of FMR is done with the help of Arduino Mega2560 Microcontroller. It is also responsible to coordinate between the FMR and SMR. A compass and GPS module is also placed in FMR which helps to estimate the target location with respect to the current location.

The SMR is designed in such a way that the size of the robot is small enough to move on all parts of the target structure. It has its own motors and controller as well but the motors are small enough to be easily driven by a 9V battery. The permanent magnet along with the MFL sensor are placed at the lower base of the robot, as shown in Fig. 2(b). The SMR is connect with a wire controlled by the wrench motor placed on the FMR.

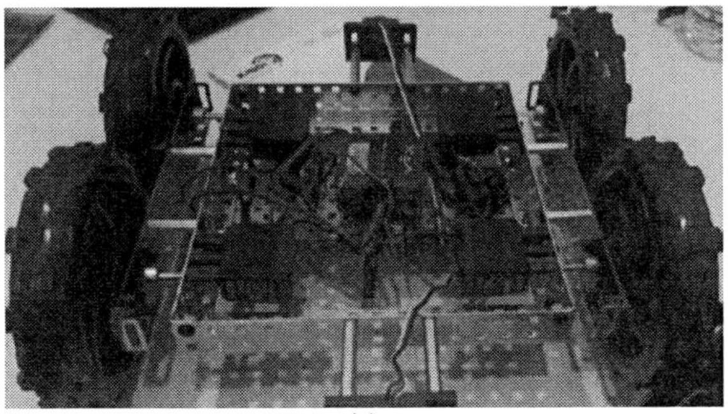

(a)

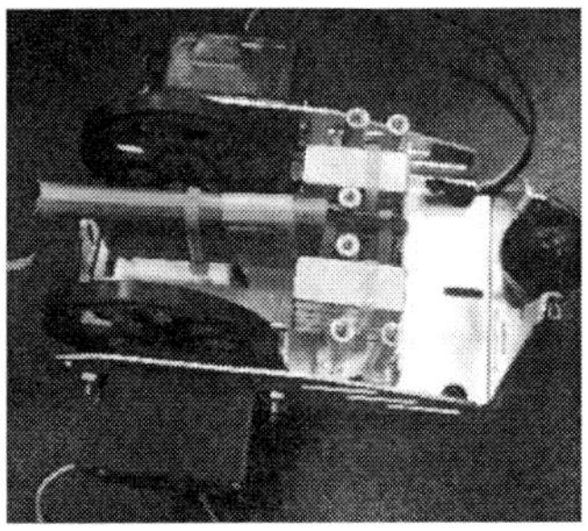

(b)

Fig. 2. The coordinated robot for MFL (a) FMR and (b) SMR.

3 Proposed approach

The proposed approach used the advantages of the visual and MFL inspection approach to improve the efficiency of the SHM. Therefore, the usage of both approaches should be done in a way that the whole inspection is carried out in an optimal time period. Thus, due to the fast processing of visual inspection, it is done first followed by an MFL based verification approach. The overall inspection process is described in Fig.

3.1 Visual inspection process

The visual inspection has been carried out such that the drone will take-off from a fixed point and take images at different heights without changing the GPS coordinate values of start point during flight. After completing the first scan, the coordinates of the GPS will be shifted and same procedure of taking images at different heights will be conducted. The process remain continue until the drone reaches to the starting GPS coordinates. The images which were taken at different heights for particular coordinates are considered as a single set.

Thereafter the image stitching (IS) is applied on individual sets. The process of IS involves a series of steps which were applied on the consecutive images of a particular set, such that one of the image is taken as a reference image (RI) whereas the other one is termed as the current image (CI). The resultant stitched image will be RI for the next consecutive image and then the whole stitching process is applied. The process remain continue for each set until a final stitched image has been obtained from them. The stitched result will be saved in the database with its corresponding GPS values.

The initial step in the IS algorithm is to determine the point of interest (POI) in both RI and CI. There are different algorithm to perform the detection of POI, among them some are already present as MATLAB built-in library function. In this paper we adopted Speeded Up Robust Features (SURF) the most commonly used approaches for POI. It is a scale and rotation invariant point of interest detector which was introduced by Herbert et al. in 2006 [14]. Thereafter, the feature has been extracted around each POI. The whole process of POI and feature extraction is done separately on both images which needs to be registered.

After determining the POI and their corresponding feature vector. The common features between the images have been computed which will help to determine the changes between the CI with respect to the RI. Since the matched features are in actual corresponding to some SURF points in both images. Therefore, after feature matching the corresponding point of SURF which are common in both images is determined with the help of matched features. These points are termed as matched points (MP). It should be noted that the MPs may have some outliers as well, in order to reduce them the RANdom SAmple Consensus (RANSAC) algorithm is performed to exclude the outlier and compute the homogeneousness between two images. Finally, the affine transformation has been applied on CI using the reduced MP so as to align CI with respect to RI. The transformed CI is fused in to RI in a way that only the portion of CI which is not present in RI will be added in to it.

The same procedure of taking and stitching the images of the same structure will be repeated again after few months, depending upon the structural sensitivity as well as the severity of the weather condition around it. The current results will be compared with the stitched images present in the data base and if some anomaly is detected then the HP coordinates (i.e. the GPS coordinates) along with the estimated height for that particular location will be sent to the FMR to proceed the crack verification using MFL.

3.2 MFL inspection process

The GPS module present in the FMR will guide the robot about its own location. As soon as Arduino Mega2560 Microcontroller receives the GPS coordinates from the system. It will translate them and compare them with its current location. The need of translation is because the FMR is present at the top of the building whereas the drone is flying at particular distance from the building. In order to obtain a correct translation the drone should remain at particular distance form in structure during the whole scanning process. The robot will take its direction based on the comparison result between its current GPS coordinates and the translated received GPS coordinates. As the robot moves it will keep checking the current GPS values and take decision accordingly.

Since there might be some temporary or permanent obstacle present on the roof for decoration purpose. Therefore an ultrasonic range sensor has been used such that when the robot come close to an obstacle at defined distance the sensor will guide the robot to change its path and as soon as the obstacle is disappeared from the sensor range the robot will again start checking the GPS value to reach to its target destination.

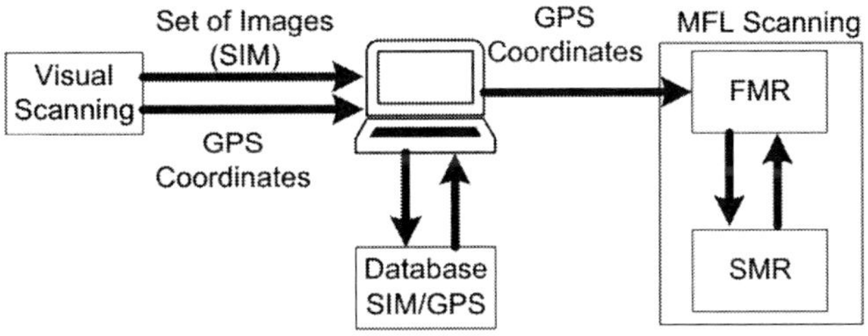

Fig. 3. Overall Inspection approach.

As it reaches to the target destination it will instruct the wrench motor to allow the SMR to reach to the location and obtain the current MFL reading of that place. These reading will be sent to the System. If an anomaly is detected then it is verified that the structure is having deformation at that particular location. If in vision based approach multiple anomalies have been detected then the robot will perform same procedure to determine the faults.

4 Simulation results

4.1 Results for UAV

The VBI from UAV has been performed by taking multiple images before and after the induced anomaly. The first set of images which was taken without anomaly are termed as images without crack (IoC), whereas the images taken after anomaly are termed as images with crack (IwC). These images are first stitched together and then the final stitched result of IoC and IwC are subtracted from each other to observe the presence of anomaly.

In this research the anomaly has been created by removing a screw from the structure. The stitched result for IoC and IwC along with the difference between them are shown in Fig. 4. It should be noted that the displaced screw has been clearly identified.

DOI: 10.1051/matecconf/201712001003

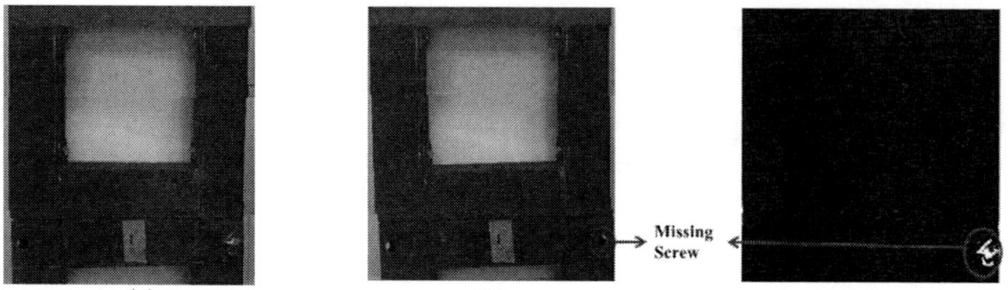

(a) (b) (c)

Fig. 4. Stitching results using stationary UAV for (a) IoC (b) IwC, and the crack identification (c) after filtering the subtraction result of (a) and (b).

4.2 Results for MFL testing

After detecting the presence of anomaly the respective GPS location along with the expected height has been sent to the FMR which directed the wrench motor and the SMR to move to that location and verify the presence of anomaly using MFL. The obtained result has been shown in Fig. 5, it should be noted that the graph has two peaks one is at the negative direction while the other is at the positive direction. Also, the negative peak comes before the positive one.

The reason of having a negative peak is because of the junction between two structural blocks as indicated clearly in Fig. 5(a). The structure which is used in this research consist of three cubes placed on each other. The place where the cubes boundaries touching each other have an inherent cavity at that point. At every junction the metal thickness will increased which decreases the magnetic field. Whereas, the positive peak is actually representing the absence of screw because the leakage in magnetic flux increase if there is an opening.

Although the negative field is false indication but this false indication will be marked for future so that the field change in these points will be ignored in future inspection. However if VBI indicates problem at the points of false indication then an inspector will be required to investigate that area.

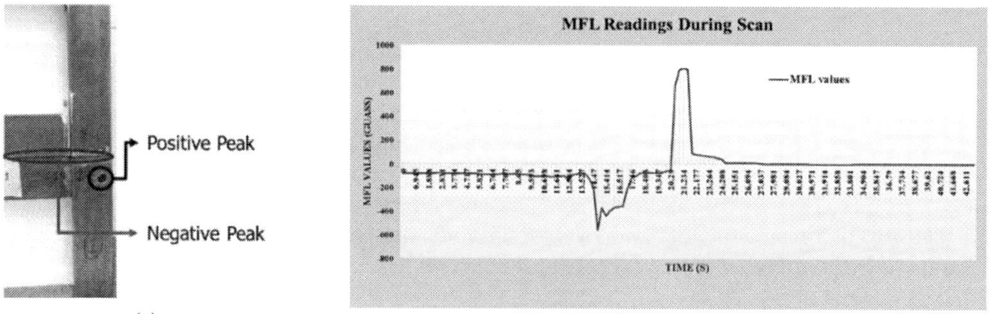

(a) (b)

Fig. 5. Inspection results from MFL sensor.

5 Conclusion

The aim of this research is to develop coordinated robotic system for civil SHM. In order to achieve this objective, a UAV based VBI approach followed by a MFL based verification and localization has been adopted on a lab-grade sample structure.

In the first round of scan, the UAVs ascertain the coarse defect signature using panoramic imaging. The structure has been scanned before and after the induced defect and the stitched image has been obtained for both scans. These two images after alignment has been compared and the defect which was in this research a missing screw has been identified. The GPS coordinates along with the height information of the defect has been sent to the FMR which after positioning itself with respect to the sent GPS values, will direct the SMR to reach to the location and perform the MFL based scanning of that region. The results obtained using MFL verified the defect which was located by the VBI.

With the initial findings, the proposed system appears to be a robust and inexpensive alternative to current approaches for automated inspection of civil/mechanical systems. The combination of VI and MFL approach provided the opportunity to detect, verify and localize the deformation in the structure.

This paper was made possible by National Priorities Research Program (NPRP) grant No. 7-234-2-109 from the Qatar National Research Fund (a member of Qatar Foundation). The statements made herein are solely the responsibility of the authors.

References

1. The Human cost of weather related disasters, http://www.unisdr.org/files/46796_cop21weatherdisastersreport2015.pdf
2. F. N. Catbas, D. L. Brown, & A. E. Aktan, "Parameter estimation for multiple-input multiple-output modal analysis of large structures", *Journal of engineering mechanics*, **130**, 8, pp. 921-930 (2004)
3. S. W. Doebling, C. R. Farrar, & M. B. Prime, "A summary review of vibration-based damage identification methods", *Shock and vibration digest*, **30**, 2, pp. 91-105 (1998)
4. S. S. Ivanovic, M. D. Trifunac, & M. I. Todorovska, "Ambient vibration tests of structures-a review", *ISET Journal of Earthquake Technology*, **37**, 4, pp. 165-197 (2000)
5. J. M. Caicedo, E. Clayton, S. J. Dyke, M. Abe, & J. Tokyo, "Structural health monitoring for large structures using ambient vibrations", *Proceedings of ICANCEER Conference*, Hong Kong, pp. 379-384 (2002)
6. D. F. Mazurek, & J. T. DeWolf, "Experimental study of bridge monitoring technique", *Journal of Structural Engineering*, **116**, 9, pp. 2532-2549 (1990)
7. B. Glisic, D. Inaudi, J. M. Lau, & C. C. Fong, "Ten-year monitoring of high-rise building columns using long-gauge fiber optic sensors", *Smart Materials and Structures*, **22**, 5, pp. 055030 (2013)
8. H. R. Kess, & D. E. Adams, "Investigation of operational and environmental variability effects on damage detection algorithms in a woven composite plate", *Mechanical systems and signal processing*, **21**, 6, pp. 2394-2405 (2007)
9. M. R. Jahanshahi, & S. F. Masri, "A new methodology for non-contact accurate crack width measurement through photogrammetry for automated structural safety evaluation", *Smart materials and structures*, **22**, 3, pp. 035019 (2013)
10. Chen, ZhiQiang, and Jianfei Chen. "Mobile Imaging and Computing for Intelligent Structural Damage Inspection." *Advances in Civil Engineering*, **2014** (2014)
11. Z. Zhu, S. German, and I. Brilakis. "Detection of large-scale concrete columns for automated bridge inspection." *Automation in construction*, **19**, 8, pp. 1047-1055 (2010)
12. A. Ellenberg, L. Branco, A. Krick, I. Bartoli, & A. Kontsos. "Use of Unmanned Aerial Vehicle for Quantitative Infrastructure Evaluation". *Journal of Infrastructure Systems*, **21**, 3, (2014)
13. Chaiyasarn, K., Kim, T. K., Viola, F., Cipolla, R., & Soga, K. "Distortion-Free Image

Mosaicing for Tunnel Inspection Based on Robust Cylindrical Surface Estimation through Structure from Motion". *Journal of Computing in Civil Engineering*, 04015045 (2015)

14. Bay, H., Tuytelaars, T., & Van Gool, L., "Surf: Speeded up robust features", In *Computer vision–ECCV 2006*, Springer Berlin Heidelberg, pp. 404-417 (2006)
15. Park, Seunghee, et al. "Magnetic flux leakage sensing-based steel cable NDE technique." *Shock and Vibration*, **2014** (2014)
16. Mandal, K., D. Dufour, and D. L. Atherton. "Use of magnetic Barkhausen noise and magnetic flux leakage signals for analysis of defects in pipeline steel." *IEEE transactions on magnetics*, **35**, 3, pp. 2007-2017 (1999)
17. Phantom Pro-3: http://www.dji.com/product/phantom-3-pro. Accessed 4 Feb 2016
18. Vex Robotic Kit: http://www.vexrobotics.com/

Suggesting alternatives for reinforced concrete deep beams by reinforcing struts and ties

Khattab Saleem Abdul-Razzaq [1*], and *Sarah* Farhan Jebur [2]

[1] Department of Civil Engineering, Diyala University, Iraq.
[2] Department of Civil Engineering, Diyala University, Iraq.

Abstract. This paper studied reinforcing struts and ties in deep beams based on the Strut-and-Tie Model (STM) of ACI 318M-14. The study contained testing 9 simply supported specimens, divided into 3 groups. The difference between the groups was the loading type which was 2-concentrated forces, 1-concentrated force and uniformly distributed load. Each group contained three specimens; the first specimens in each group were conventional deep beams as references which had a length of 1400 mm, a height of 400 mm and a width of 150 mm. The second specimens were the same as references in dimensions, but with removing shoulders. In addition, only the paths of struts & ties of STM were reinforced in the second specimens as compression and tension members, respectively. The third specimens were the frames that took their dimensions from STM of ACI 318M-14. From the experimental work, it is found that the proposed frames were good alternatives for the references despite the small loss in ultimate capacity. However, these proposed frames already carried loads greater than the factored design loads of STM. In comparison with the references, these proposed frames provided 41-51% reduction in weight, 4-27% reduction in cost besides providing front side area about 46-55%.

1 Introduction

The reinforced concrete deep beam is a very important structural element in various types of concrete structures. It is generally distinguished as short and deep, having a small thickness in comparison with their span or depth. Typical applications of deep beam comprise foundation pile caps, transfer girders, tanks, folded plates of roof structures, shear walls, offshore structures and foundation walls. It is commonly received many small loads in their own plane and transferring them to a small number of reaction points through compressive struts that connected by ties [1]. The strut and tie models are particularly appropriate for designing the discontinuity in the stress distribution regions where the distribution of strain is nonlinear within the deep beam cross section [2]. Modeling the deep beam as strut-and-tie has been confirmed a safe method, unified and rational for designing

* Corresponding author: dr.khattabsaleem@yahoo.com

the reinforced concrete members under variable loadings [3] despite the fact that it is conservative in some cases [4].

The problem of filling areas between crowded reinforcement is critical because the deep beam heavy reinforcement. In addition, the deep beam has a heavy weight, a big volume and its wide front side intercepts the cables, pipes or ducts from passing through. The present study was then an attempt to reinforce struts and ties paths in order to find alternatives for the conventional RC deep beams. The aim was to reduce cost and weight in addition to gain a front side area that can be used for service utilities.

2 Experimental program

The experimental program consisted of constructing and testing nine simply supported SCC specimens, divided into three groups (A, B and C) as shown in Figure 1. Each group consisted of three specimens. The first specimens in each group were conventional deep beams (DB.2P, DB.1P and DB.U specimens). The second specimens in each group were deep beams in which only the paths defined by STM of ACI 318M-14[5] were reinforced (STM.2P, STM.1P and STM.U specimens) besides removing the shoulders of the specimens STM.1P and STM.U. Finally, the third specimens in each group were the reinforced concrete frames (FR.2P, FR.1P and FR.U specimens) that took their dimensions from the load paths defined by STM, then reinforced according to ACI 318M-14.

It is worth to mention that the struts in the specimen STM.2P, STM.1P, STM.U, FR.2P, FR.1P and FR.U were reinforced as compression members using the (Interaction Diagram of Rectangular Columns) while the ties were reinforced as tension members (22.4.3.1) of ACI 318M-14. It is also worth to mention that the section of every frame was unified; the width was b=150mm (the width of the reference beams) and height was w_{sb} (STM section height at the bottom of the inclined struts). In order to recognize specimens designation easily, Table 1 shows the way followed in this designation.

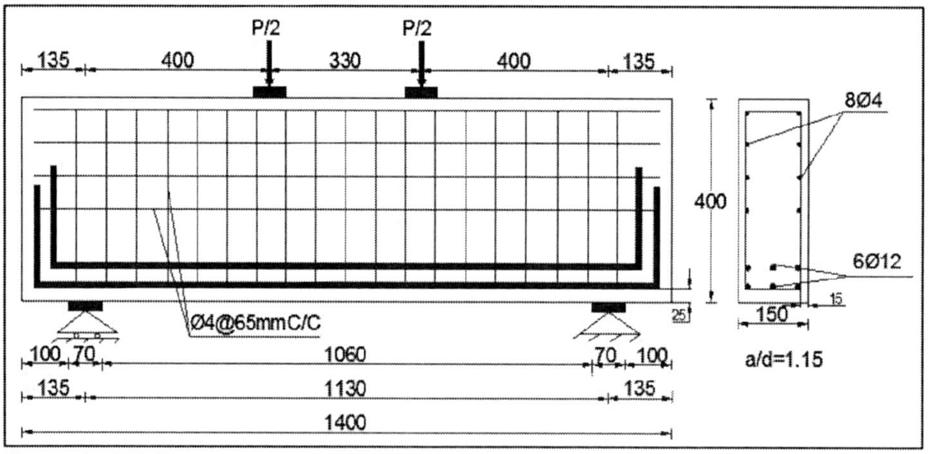

Group A : DB.2P

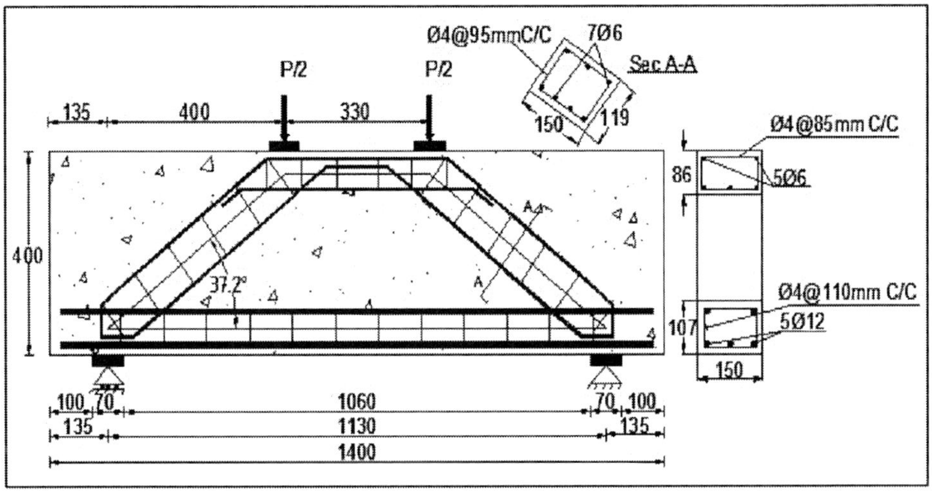

Group A : STM.2P

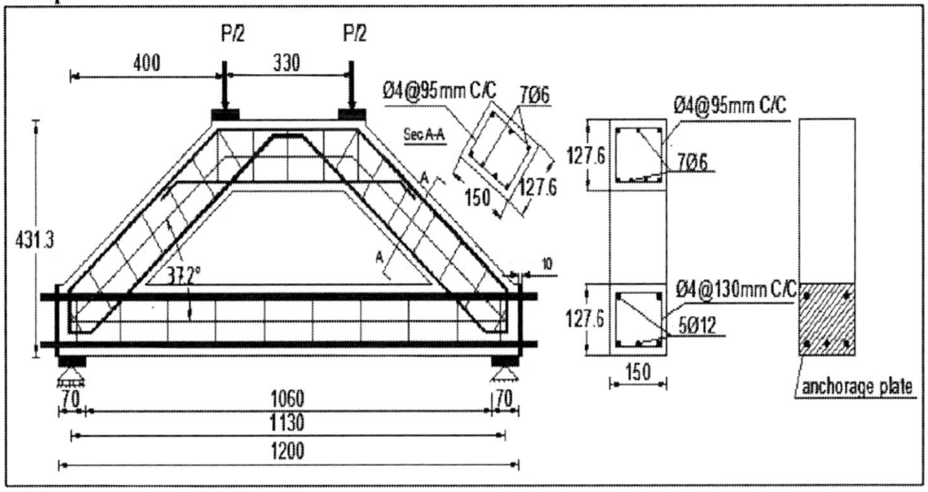

Group A : FR.2P

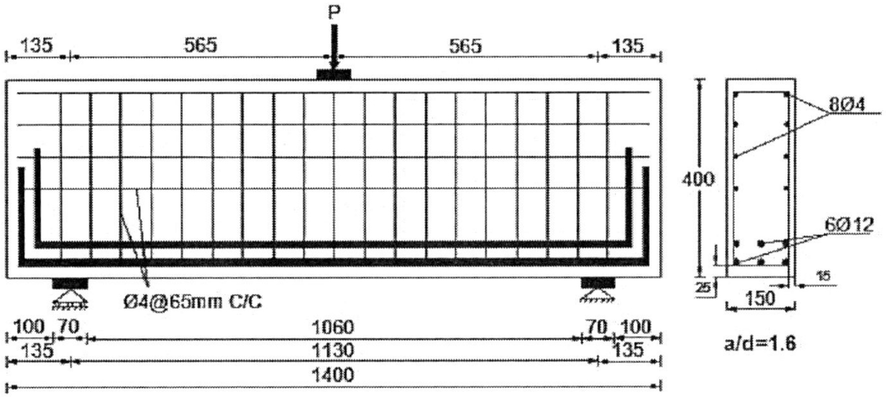

Group B : DB.1P

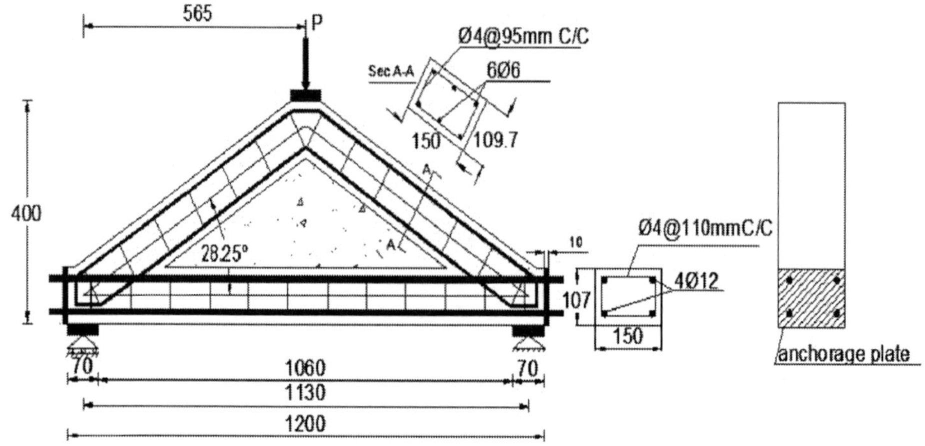

Group B :STM.1P

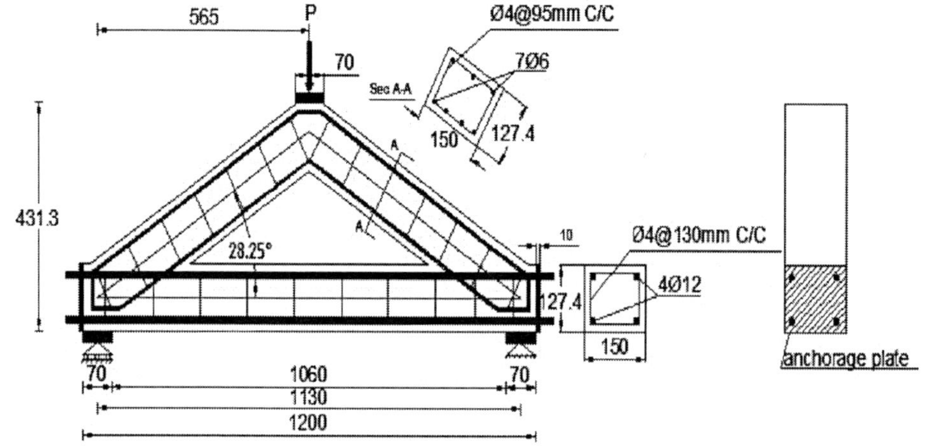

Group B : FR.1P

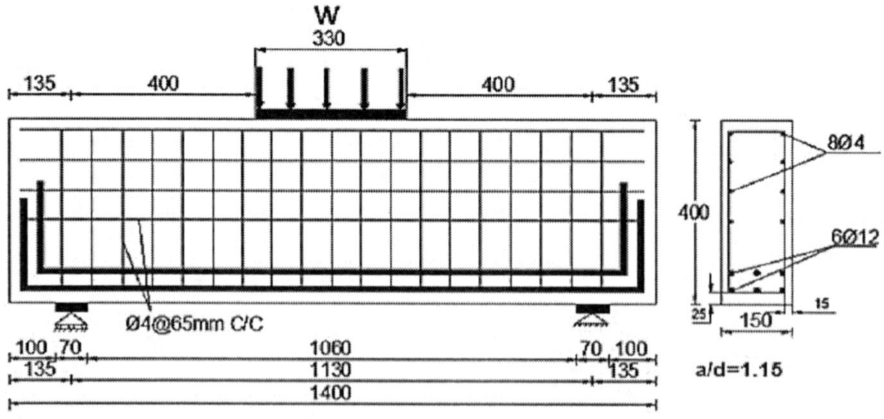

Group C : DB.U

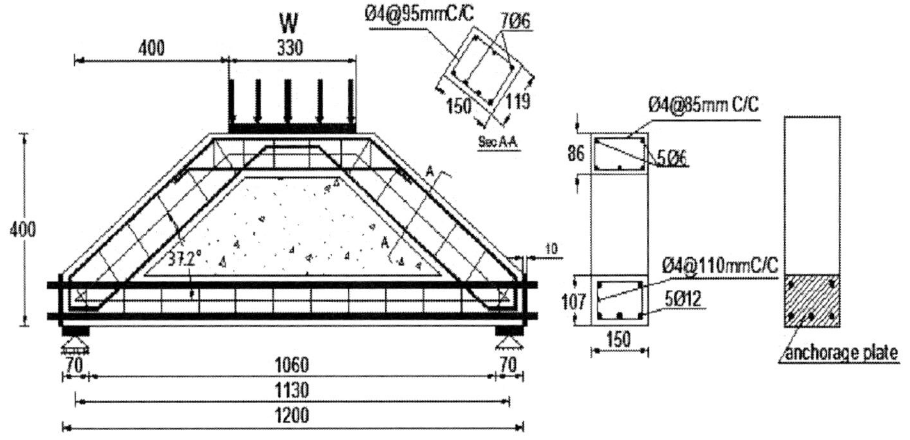

Group C : STM.U

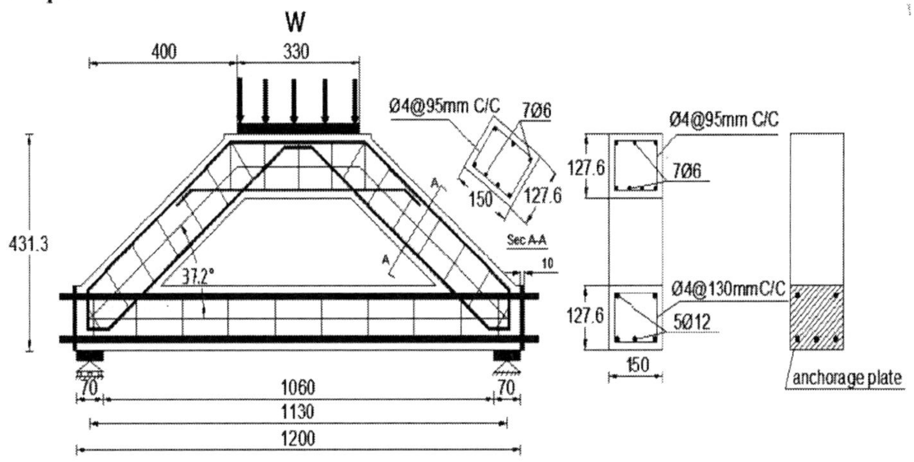

Group C : FR.U

Fig. 1. The Nine Specimens, all dimensions are in mm.

Table 1. Specimens designation way.

Letter	Meaning
DB	Conventional **D**eep **B**eam
STM	Only paths drawn by **STM** were reinforced in deep beam
FR	Reinforced Concrete **FR**ame that took its geometry from STM stress paths
1P	Subjected to **1**-concentrated force (**P**)
2P	Subjected to **2**-concentrated forces (**2P**), which means actually (2*0.5P)
U	Subjected to **U**niformly Distributed load

2.1 Material properties

The properties of the materials such as cement, aggregates, additives and reinforcing steel used for preparing the reinforced self-compacted concrete (SCC) specimens tested in this study were as follows:

Ordinary Portland cement type I of Tasluja factory was used for producing SCC. Local Al-Ukhaider graded natural sand with 2.63 fineness modulus. The coarse aggregate (10 mm maximum particles size of crushed gravel) was used in SCC mixtures. According to EFNARC [6], the limestone powder particle size was less than 0.125 mm. Turbid liquid Sika ViscoCrete 5930 was used in SCC mixtures which submitted with ASTM C 494/C 494M-99a [7] types G and F (free from chlorides). Deformed steel bars of diameter 12 & 6 mm ($f_y = 550\ MPa\ and\ f_y = 572\ MPa,$ respectively) were used as longitudinal reinforcement for the specimens. Deformed steel bars of diameter 4 mm ($f_y = 604\ MPa$) were used for the vertical and horizontal shear reinforcement of the specimens.

During constructing every group, six standard 150mm x 300mm cylinders were cast. Three cylinders were used to measure the compressive strength (f'_c) and the remaining three for split tensile strength (f_{ct}). In addition to that, three 500mm x 100mm x 100mm prisms were cast and used for testing the concrete modulus of rapture (f_r). All specimens were demolded about 24 hours after casting, and then cured by using dump blanket (cover) and sprinkled continuously with water for 28 days. Then, the specimens were white painted to help in the observation of crack propagations during testing. The mixed concrete properties are summarized in Table 2.

Table 2. Mix proportions of SCC.

Group	f'_c (MPa)	f_{ct} (MPa)	f_r (MPa)	E_c^* (MPa)
A	35	3.838	6	27805
B	34.4	3.824	5.75	27566
C	34.1	3.812	5.7	27445

*The values of $E_c$were calculated by ($4700\sqrt{f'_c}$) provided by (ACI 318M-2014)

2.2 Test set-up and instrumentation

The specimens were tested using a hydraulically universal testing machine of 2000kN capacity (Jet materials Ltd. Company). The tests were conducted under monotonic static loading up to failure at the Structural Laboratory / College of Engineering / University of Diyala. The tested specimens were simply supported at the ends and loaded with three different types of loadings; 2-concentrated forces, 1-concentrated force and uniformly distributed load. One dial gauge of 0.01 mm accuracy was fixed under the specimen to calculate the deflection at mid-span. A micro-crack meter device with an accuracy of 0.02 mm was used to detect and measure first crack width development at all loading stages for all specimens.

3 Experimental results and discussions

3.1 General behavior of the specimens

Nine SCC specimens were subjected to increasing static monotonic loads until failure as shown in Plates 1 to 9. Table 3 summarizes the test results for all specimens.

Table 3. Summary of test results for tested specimens.

Group	Specimen	$P_{cr-flex.}$ (kN)	% Change in $P_{cr-flex.}$	$P_{cr-diag.}$(kN)	% Change in $P_{cr-diag.}$	$P_{exp.}$(kN)	P_{STM}(kN)	$P_{exp.}/P_{STM}$	$P_{cr-flex.}/P_{exp.}$	$P_{cr-diag.}/P_{exp.}$	$\Delta_{cr-flex.}$(mm)	$\Delta_{cr-diag.}$ (mm)	$\Delta_{fail.}$ (mm)	$\Delta_{cr-flex.}/\Delta_{fail.}$	$\Delta_{cr-diag.}/\Delta_{fail.}$
A	DB.2P	150	-	240	-	562	447.2	1.25	0.27	0.43	2.25	3.8	8.62	0.26	0.44
	STM.2P	100	-33.3	225	-6.25	470		1.05	0.21	0.47	1	2.16	5.5	0.18	0.39
	FR.2P	110	-26.6	230	-4.16	522		1.16	0.21	0.44	1.22	2.3	5.2	0.23	0.44
B	DB.1P	115	-	225	-	355	286.5	1.23	0.32	0.63	1.17	2.5	5.73	0.2	0.43
	STM.1P	75	-34.7	175	-22.2	300		1.04	0.25	0.58	1.8	3		0.34	0.57
	FR.1P	100	-13	225	0	325		1.13	0.3	0.69	1.92	3.54	4.73	0.4	0.74
C	DB.U	125	-	225	-	547.8	435.7	1.25	0.23	0.41	1.53	2.64	7.57	0.2	0.34
	STM.U	100	-20	175	-22.2	420.2		0.96	0.23	0.41	1.9	2.82	6.58	0.29	0.42
	FR.U	150	+20	225	0	505		1.15	0.29	0.44	1.76	2.26	6.3	0.27	0.35

The general crack behaviour is presented in detail as follows:

a. The reference specimens DB.2P, DB.1P and DB.U: the first visible flexural cracks appeared in the region of maximum bending moment and extended nearly vertically upward. These cracks were observed at 23-32% of the ultimate load. Then, first shear cracks appeared at a loading level 41-63% of the ultimate load.

b. STM.2P, STM.1P and STM.U specimens: the first visible flexural cracks took place in the specimen mid-span at 21-25% of the ultimate load. These flexural cracks extended vertically upward. After that, the first shear cracks appeared at 41-58% of the ultimate load.

c. FR.2P, FR.1P and FR.U: the first visible straight cracks appeared at 21-30% of the ultimate load along the bottom, then the back and front sides of the specimen tie. These flexural cracks were vertical and approximately parallel to each other. While at 44-69% of the ultimate load, the first shear cracks were developed at the nodal support zones.

DOI: 10.1051/matecconf/201712001004

Group A

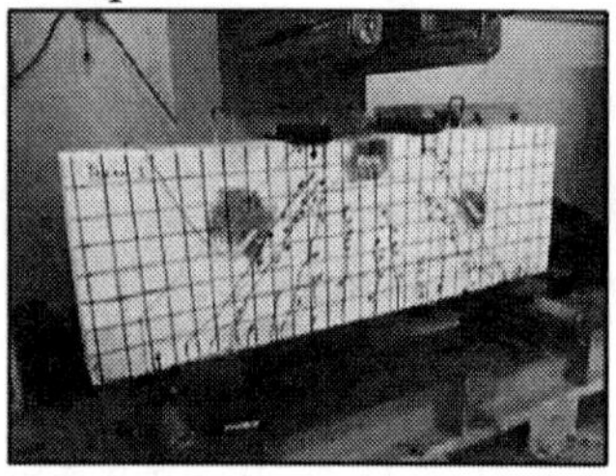

Plate 1. DB.2P
(flexural-shear failure)

Plate 2. STM.2P
(diagonal splitting failure)

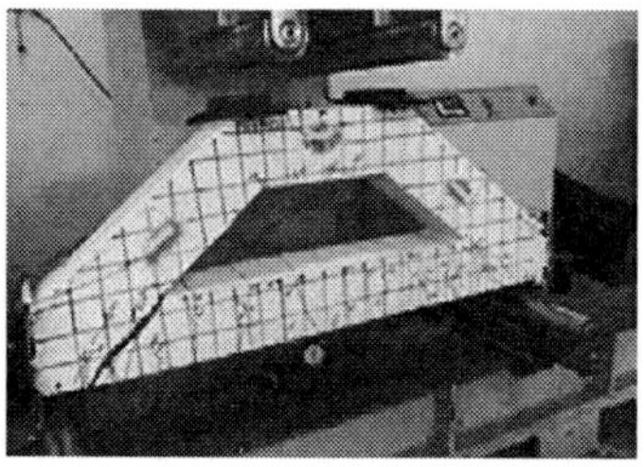

Plate 3. FR.2P
(diagonal splitting failure)

Group B

Plate 4. DB.1P
(nodal failure)

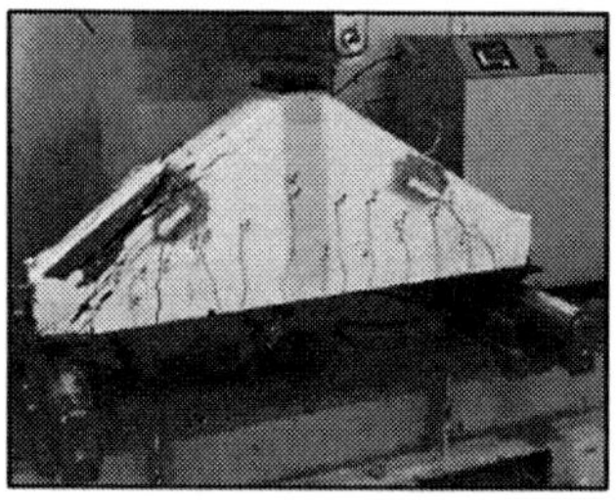

Plate 5. STM.1P
(diagonal compression failure)

Plate 6. FR.1P
(diagonal splitting failure)

Group C

Plate 7. DB.U
(flexural-shear failure)

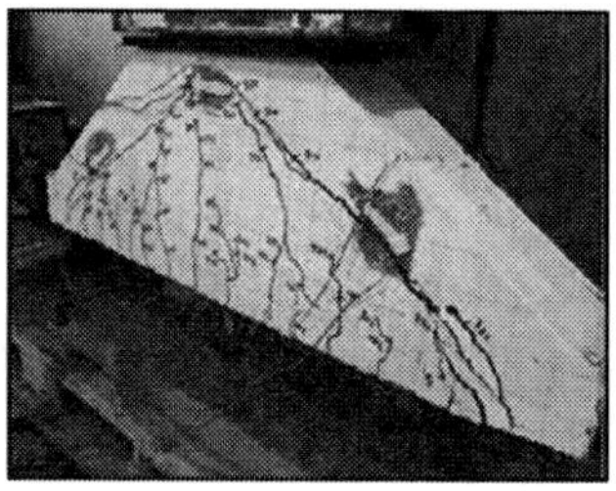

Plate 8. STM.U
(diagonal splitting failure)

Plate 9. FR.U
(diagonal splitting failure)

Fig. 2. The general crack behaviour.

It has been observed that the width of $P_{cr-flex}$ did not exceed the maximum crack width of 0.5 mm which is specified by [8-10]. While the width of $P_{cr-diag}$ had exceeded the maximum width of 0.5 mm specified in [8-10]. Table 4 shows more details about crack characteristics of the experimental SCC specimens at failure loads in which it had also been observed that the widths of the diagonal shear cracks were much wider than those of the flexural cracks. This fact sounded reasonable because the shear failure was dominant.

Table 4 shows that the number of the flexural cracks in the proposed specimens STM.2P, STM.1P, STM.U, FR.2P, FR.1P and FR.U was more than those in the reference specimens DB.2P, DB.1P and DB.U. That took place because there was less concrete around the ties of the proposed specimens in addition to the less main reinforcement that were put in these ties. While the number of the diagonal cracks in STM.2P, STM.1P, STM.U, FR.2P, FR.1P and FR.U specimens was less than those in DB.2P, DB.1P and

DB.U specimens. That took place because the compressive forces in the reinforced strut splitted the tie at its end (near the node) which prevented the dispersion of shear forces.

Table 4. Crack characteristics of experimental SCC specimens at failure.

Group	Specimen designation	$P_{exp.}$ (kN)	1st flexural crack width (mm)	1st shear crack width (mm)	Number of flexural cracks at failure	Number of shear cracks at failure
A	DB.2P	562	0.3	0.83	6	10
	STM.2P	470	0.22	0.7	7	9
	FR.2P	522	0.26	0.75	14	8
B	DB.1P	355	0.2	0.64	5	11
	STM.1P	300	0.14	0.5	12	4
	FR.1P	325	0.16	0.55	10	7
C	DB.U	547.8	0.29	0.8	12	16
	STM.U	420.2	0.26	0.7	10	5
	FR.U	505	0.22	0.74	14	5

3.2 Failure modes

The conventional reinforced reference deep beams exhibited generally two modes of failure, flexural-shear failure (DB.2P and DB.U) and nodal failure (DB.1P). That is why, during loading, it was seen that the inclined shear cracks developed which preceded by vertical flexural cracks at mid-span. In case of DB.2P and DB.U, the main cracks that caused the final failure propagated upwards from the supports towards the load application points. While in case of DB.1P, high stresses developed under the loaded area which led to a more sudden nodal failure than the flexural-shear failure of DB.2P and DB.U.

While all the proposed specimens in which the paths of struts and ties were reinforced as compression and tension members, respectively failed by diagonal splitting failure (except STM.1P). This was attributed to the horizontal and the vertical components of the reinforced strut compressive force that caused tension and shear, respectively in the tie. These two components led to split the tie at its end. The specimen STM.1P failed by a more sudden destruction of the right strut (which led to diagonal compression failure) than the diagonal splitting failure of the other proposed specimens.

3.3 Load-deflection behaviour

The load–deflection curves for the specimens of groups A, B, and C are shown in Figures 3 to 5, respectively. Generally, in both the conventional deep beams and the proposed specimens, the load-deflection curves were roughly linear in the greater portion of the loading and then the curves started to bend slightly. This general behaviour happened because the shear deformation was the predominant behaviour which led to the brittle failure. This brittle failure decreased the strength of the specimens below the flexural capacity and noticeably decreased the ductility of them.

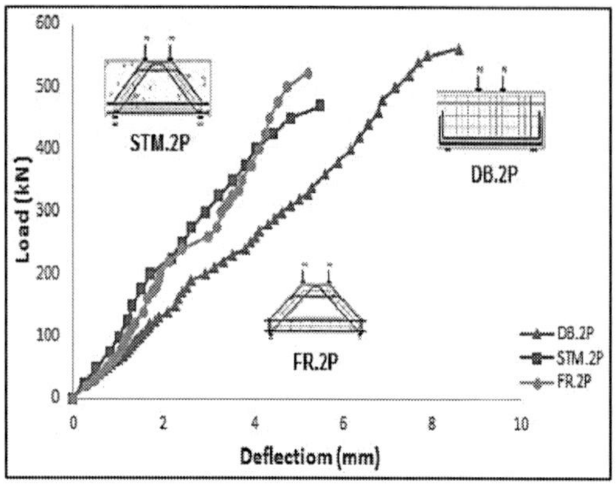

Fig. 3. Load-mid span deflection curves for group A.

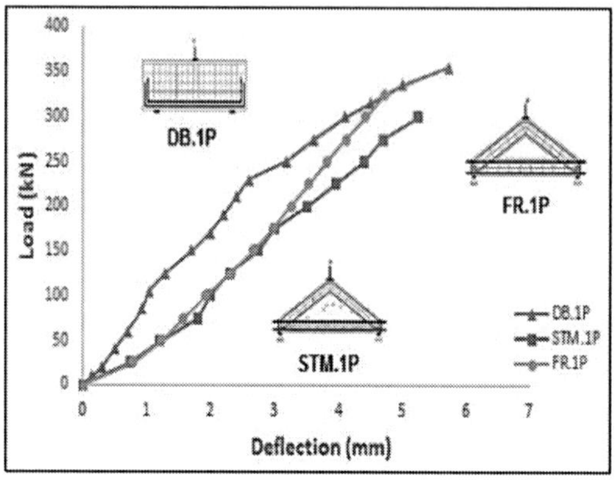

Fig. 4. Load-mid span deflection curves for group B.

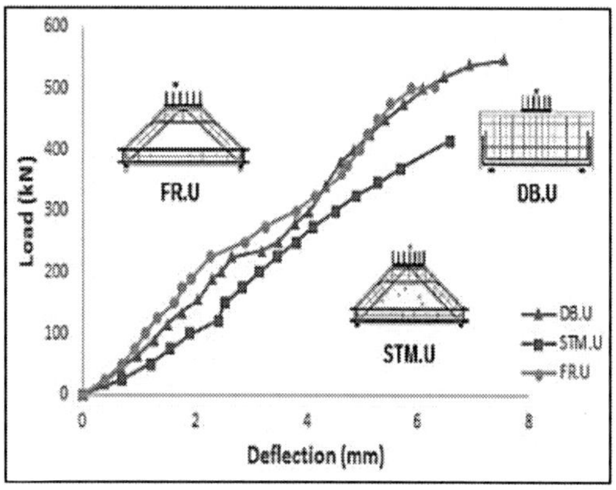

Fig. 5. Load-mid span deflection curves for group C.

3.3.1 Shear strength

The design of reference specimens were developed by using the Strut and Tie Method of Appendix A, ACI 318M-14. Test results indicated that each reference specimen DB.2P, DB.1P and DB.U resisted loads greater than the theoretical design load of Strut and Tie Method (P_{STM}). In more details, experimental results P_{exp} showed that the Strut and Tie Method is conservative. In this study, in case of central single concentrated force, P_{exp} was greater than P_{STM} by 19%. While P_{exp} was greater than P_{STM} by about 20% in the cases of two central concentrated forces and uniformly distributed load.

At the same time, P_{exp} was greater than P_{STM} in the cases of the specimens STM.2P, STM.1P, FR.2P, FR.1P and FR.U. The following comparisons between P_{STM} and P_{exp} for the proposed specimens are summarized as follows:

1-Group A: P_{STM} was less than P_{exp} by 4.8% and 14.3% for the cases of STM.2P and FR.2P, respectively.

2-Group B: P_{STM} was less than P_{exp} by 4.5% and 11.8% for the cases of STM.1P and FR.1P, respectively.

3-Group C: P_{STM} was greater than P_{exp} by 3.6% for the case of STM.U while it was smaller than P_{exp} by 13.7% for the case of FR.U.

It should be noted here that after test the specimen STM.2P, it was noticed that its shoulders had no cracks at all. Therefore, the shoulders of the specimens STM.1P and STM.U were removed later in order to save cost, weight and front side area. Table 5 shows the comparisons between the specimens concerning cost, weight and front side area.

Table 5. Comparison between all specimens.

Group	Specimen	P_{STM} (kN)	$P_{exp.}$ (kN)	$P_{STM}/P_{exp.}$	% Decrease in						
					Weight		Total Weight	Cost		Total Cost	Front side area
					Conc.	Reinf.		Conc.	Reinf.		
A	DB.2P	447.2	562	0.79	-	-	-	-	-	-	-
	STM.2P		470	0.95	-2.4	38.6	0	-2	-9	-6.17	0
	FR.2P		522	0.85	43.2	17.3	41.5	43.1	-20	4.45	46.4
B	DB.1P	286.5	355	0.8	-	-	-	-	-	-	-
	STM.1P		300	0.95	47.6	48.3	47.6	47	20.8	30.65	51.78
	FR.1P		325	0.88	51	45.7	50.6	49.6	13.5	27	55.35
C	DB.U	435.7	547.8	0.79	-	-	-	-	-	-	-
	STM.U		420.2	1.03	34.3	38.6	34.2	34.4	-8.5	8.3	35.7
	FR.U		505	0.86	43.2	17.3	41.5	43.1	-20	4.45	46.4

4 Conclusions

Removing the concrete where the stress paths of struts and ties do not pass through and then reinforcing these struts and ties did not make significant effects on the behaviour of these proposed specimens in comparison with the conventional reference deep beams. This took place because the stresses from the loading to the supporting points actually pass through these paths that defined by the STM of Appendix A, ACI 318M-14. Based on the experimental test results, the following conclusions can be summarized:

1-Mode of failure:

a. The conventional reinforced reference deep beams failed generally in the well-known flexural-shear failure due to the role of the main, vertical and horizontal reinforcement. That is why, during loading, diagonal cracks appeared at the end of flexural cracks which clarified that combined flexural and shear failure happened in the deep beams.

b. While the proposed specimens in which the paths of struts and ties were reinforced as compression and tension members, respectively failed generally by diagonal splitting failure. This was attributed to the compression force in the reinforced strut that caused tension and shear at the ends of the tie (which led to split the tie at one of these ends).

2-In predicting shear strength of conventional reference deep beams, STM of Appendix A, ACI 318M-14 was conservative and showed lower-bound design when compared with experimental work. STM predicted strengths for the conventional deep beams subjected to 1-concentrated force, 2-concentrated forces and uniformly distributed load were lower than experimental strengths by 19.2%, 20.4% and 20.4%, respectively.

It is wealthy noting that the ultimate capacities of the proposed specimens did not much differ from the theoretical design load of the STM. These negligible differences in ultimate capacities made the proposed specimens good alternatives for the conventional deep beams.

3-The proposed frames were good alternatives for the conventional deep beams because they saved cost, weight and provided a front side area for service utilities:

a. In case of 2-concentrated forces, the gain in cost, weight and front side area was about 4.45%, 41.5% and 46.4%, respectively.

b. In case of 1-concentrated force, the gain in cost, weight and front side area was about 27%, 50.6% and 55.35%, respectively.
c. In case of uniformly distributed load, the gain in cost, weight and front side area was about 4.45%, 41.5% and 46.4%, respectively.
4-Generally, in both the conventional deep beams and in the proposed specimens, the load-deflection curves were approximately linear in the larger part of the load application and then the curves somewhat bent. Therefore, it was obvious that the shear deformation was the predominant behavior which led to brittle failure.
5-There was a general crack behavior for both the reference conventional deep beams and the proposed specimens. In almost specimens, the first visible flexural cracks appeared in the midspan region and extended nearly vertically upward. Then, first shear cracks appeared near the supports. Generally, it had been observed that the first flexural cracks in all specimens were not critical and they did not exceed the maximum width limits. While the first diagonal cracks were more critical and they exceeded the maximum width limits.

Notations:

$P_{exp.}$	Experimental load, kN
P_{STM}	Theoretical load according to Appendix A, ACI 318M-14 Strut and Tie method, kN
W	Uniformly distributed load, kN/m
$P_{cr-flex.}$	First flexural cracking load, kN
$P_{cr-diag.}$	First diagonal cracking load, kN
$\Delta_{cr-flex.}$	Deflection corresponding to the 1st flexural crack load, mm
$\Delta_{cr-diag.}$	Deflection corresponding to the 1st diagonal crack load, mm
$\Delta_{fail.}$	Deflection corresponding to the ultimate load of the tested specimen, mm
w_{sb}	Width of inclined strut at support node, mm
f'_c	150mm*300mm Cylinder compressive strength of concrete, MPa
f_r	500mm x 100mm x100mm prisms Modulus of rapture, MPa
f_{ct}	150mm*300mm Cylinder Indirect tensile strength (splitting tensile strength), MPa
f_y	Yield Stress of Steel reinforcement, MPa
E_c	Modulus of elasticity of concrete, MPa

References

1. Ashour, A. F., & Yang, K. H. Application of Plasticity Theory to Reinforced Concrete Deep Beams (2008)
2. Schlaich, J., Schäfer, K., & Jennewein, M. Toward a consistent design of structural concrete. *PCI journal*, **32**(3), 74-150 (1987)
3. Liang, Q. Q., Xie, Y. M., & Steven, G. P. Topology optimization of strut-and-tie models in reinforced concrete structures using an evolutionary procedure. *ACI Structural Journal*, **97** (2), 322-330 (2000)
4. Hwang, S. J., Lu, W. Y., & Lee, H. J. Shear strength prediction for deep beams. *Structural Journal*, **97**(3), 367-376 (2000)
5. ACI Committee, American Concrete Institute, and International Organization for Standardization. Building code requirements for structural concrete (ACI 318-14) and commentary. American Concrete Institute (2014)
6. EFNARC, S. Guidelines for self-compacting concrete. London, UK: Association House, 32-34 (2002)
7. ACI Committee. American Concrete Institute, and International Organization for

Standardization. Self-consolidating concrete (ACI 237-07). American Concrete Institute (2007)
8. Machida, A. Recommendation for design and construction of concrete structures using continuous fiber reinforcing materials. Tokyo: Research Committee on Continuous Fiber Reinforcing Materials, Japan Society of Civil Engineers (1997)
9. Canadian Standards Association. Design and construction of building components with fibre-reinforced polymers. Toronto: Canadian Standards Association (2012)
10. Canadian Standards Association, CSA Group & Canadian Standards Association. Canadian highway bridge design code (2014)

Dynamic characteristics of old railway steel bridges in Egypt

Maha M. Hassan [1,*] and *Hussein* H. Abbas [2]

[1]Structural Engineering Department, Cairo University, Giza, Egypt
[2] Structural Engineering Department, Al-Azhar University, Cairo, Egypt

Abstract. Dynamic response of existing railway bridges is influenced by many factors including structural properties, railway traffic speed, and train-to-bridge frequency. Excessive vibrations cause passengers' discomfort, rail elements deterioration, higher derailment risks, fatigue deformation problems, stress amplification, and eventually high maintenance costs. In such cases, applied remedies include limiting traffic speed or changing bridge mass or stiffness. Traffic induced vibrations can also be mitigated by enhancing structural damping through application of Tuned Mass Dampers. Egypt has one of the oldest railway transportation network with several steel bridges crossing the Nile River, in addition to several canals. In a running project, dynamic properties of selective bridges on the Egyptian railway networks infrastructures are investigated through extensive field testing. The considered bridges cover a wide range of structural properties including simply-supported and continuous bridges, spans ranging between 40 to 90 m, single and double track lines, and open timber bridges. Field tests results are studied; hence, key factors affecting bridge dynamic characteristics are discussed and highlighted. Moreover, results are compared with those extracted from finite element analyses. The project aims at evaluating the possibility of excessive vibration problems while considering different values of train velocities. In the current study, focus is given to two example bridges. Field and numerical results are exhibited and discussed. Eventually, possibility of excessive vibration problems are determined considering different values of train velocities. It was observed that available formulas in current design codes do not capture warping and torsional modes that present fundamental modes for truss bridges. In addition, existing railway bridges do not satisfy serviceability limit states when considering high speed trains.

1 Introduction

Dynamic properties of steel bridges affect the railway transportation network performance. Induced vibrations due to train passages over open timber steel bridges can result in discomfort to passengers. In addition, excessive vibrations can result in increasing the impact factor values compared to the ones considered in design. Historically, the serviceability deflection limits were used as a sign for any expected undesirable motion.

* Corresponding author: mahamoddather@eng.cu.edu.eg

However, undesirable vibrations may be linked to several parameters including bridge geometry, speed of trains, and specifications of bearings. Building simulation models that can accurately capture the effect of these parameters is complicated. Hence, field tests are considered the most reliable method for determining the dynamic characteristics of bridges. A steel bridge is a massive and elastic structure and, as such, is able to vibrate in a number of different ways, with different frequencies. In a structure as complex as a bridge, the possible frequencies are so numerous and varied that the noise emitted may not indicate a dominant mode; and, the important vibrations, which strain the main members and have natural frequencies approaching those of any probable periodic forces acting on the structure, may lie below the audible range.

In addition to human comfort limits, dynamic properties influence the response of bridges under seismic loads. Dynamic properties of steel bridges have been extensively investigated due to its effects on the structural response of bridges. Different researchers focused on building reliable models that can provide accurate predictions for the dynamic response of bridges [1-3]. Different parameters were considered including speed of train, train-to-bridge frequency, mass and span ratios, as well as bridge damping. Field testing was also employed to determine mode shapes and associated natural frequencies [4-6]. These properties were used to pinpoint defects and damage in the investigated bridges through damage indices [7-10].

The railway transportation network in Egypt incorporates steel bridges crossing the Nile River in addition to several canals. Some of these bridges are constructed since over 100 years. During such period, the design of steel bridges was governed by the dead and live loads in addition to wind loads. Impact factors were commonly applied to the live loads in order to account for the uncertainties related to the dynamic moving loads. Besides the check on stresses within members, deflection limitations were imposed to limit bridge vibration under the different load effects. In this paper, results of field tests on two steel open-timber truss bridges with different systems are presented. The study focuses on the dynamic properties of the considered bridges and relates them to the probability of induced vibrations. The work in this paper represents a research effort exerted to reach a formula relating the bridge dimensioning and system for open timber through truss bridges to their dynamic properties based on the gathered data and the performed analyses. The aim of the current paper is to focus on the dynamic properties of two bridges through both experimental and analytical studies.

2 Description of bridges

In Egypt, an extensive assessment campaign has been initiated for a large group of steel railway bridges [11-13]. Part of the assessment includes performing dynamic tests which are used to determine the fundamental mode shapes and frequencies of the studied bridges. These were mainly used to validate the built finite element models. In the following sections, focus will be given to dynamic test and finite element results for two example bridges: Marazek and Mansoura Bridges. These bridges were chosen as they represent two different structural systems and usage of bridge. Figures 1 and 2 show a general view of the two studied bridges. Marazek bridge was constructed between 1955-1959. The bridge is a combined railway and roadway bridge over the Nile River near Helwan City. It consists of 9 equal spans with a total length of 810 m. Each three successive spans are continuous. The bridge cross section includes one railway track in addition to a traffic lane and two sidewalks. Mansoura bridge was constructed in 1913. The bridge consists of 4 spans representing simple steel open floor truss. Each span is 68.8 m length with a total length of 280 m. At the time of its construction, the bridge had double track lines. The bridge has three fixed simple spans and one movable (swing type) bay. The main structural system for

DOI: 10.1051/matecconf/201712001005

each bay is a simply supported riveted truss with variable depth: 5.0 m at support and 9.0 m at middle of the bay. The main trusses are connected with transverse secondary beams, horizontal upper and lower bracing system, and lower beams. Figures 3 and 4 show the general layout of the bridge.

The two bridges are chosen to present different structural systems, different span length, and different degree of indeterminacy. Different tests were performed on both bridges. The current paper will focus on dynamic test procedures and their results.

Fig.1. Marazek Bridge, 1959.

Fig.2. Mansoura Bridge, 1913.

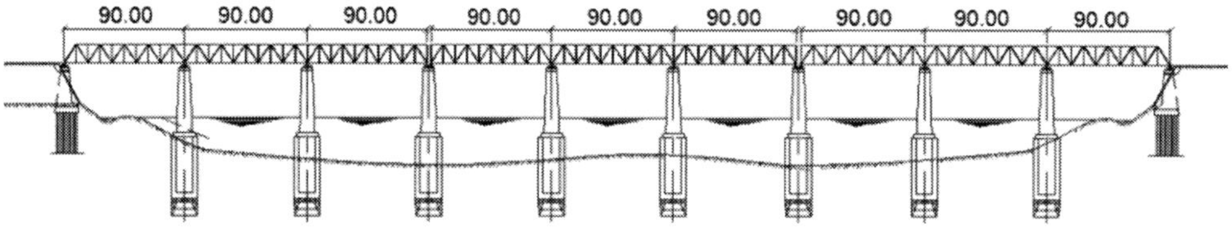

Fig.3. General Layout of Marazek Bridge.

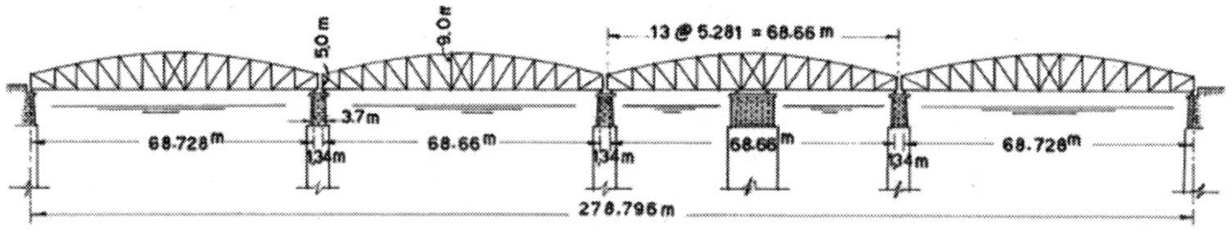

Fig. 4. General Layout of Mansoura Bridge.

3 Dynamic test procedure and results

The two bridges were subjected to passage of train as an excitation method in order to facilitate observing the investigated modes (Figure 5). Accelerometers are attached to the bridge at suitable arrangement to capture the shape of the most relevant modes. Measurements were recorded for 120 seconds using sampling rate of 200 reading/second. Readings were also taken three times in order to consider average of results. Figure 6 shows the instrumentation and data acquisition system. Acceleration for Mansoura bridge are exhibited in Figure 7 after removing the part recorded during passage of train in order to preclude the influence of the large mass of the train.

Table 1 summerizes the main dimensional properties of the considered two bridges. Meanwhile, Table 2 lists the concluded frequencies from the performed tests. For Mansoura bridge, it was observed that the observed acceleration at bottom chord level are higher than the vertical movement. It was also generally observed that the main movement of the bridge are combined lateral movement at upper and lower bracing levels combined with torsional movement. For Marazek bridge, the observed modes included lateral mode followed by flexural mode then several combined flexural and torsional modes.

Fig. 5. Mansoura Bridge Dynamic Test: Excitation Trains and Used Accelerometer.

Fig. 6. Data Acquisition System and Registration of Data.

Table 1. Geometric Properties of Studied Bridges.

Bridge	Construction Year	Span (m)	Height (m)	Width (m)	Structural System
Mansoura Bridge	1913	68.66	5 - 10	8.9	Simply Supported
Marazek Bridge	1959	90.00	11	12.5	Continuous – 3 spans

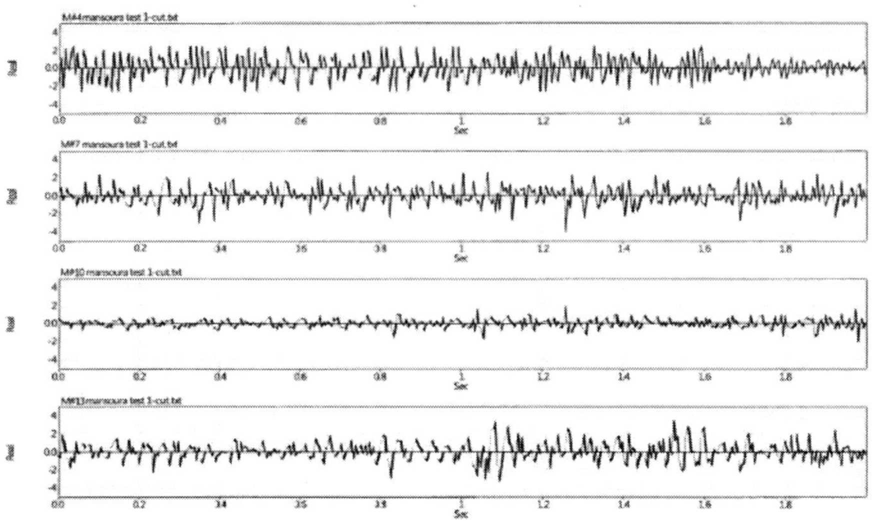

Fig. 7. Samples of Registered Acceleration.

Table 2. Frequencies (Hz) of Measured Modes.

Mode	Mansoura Bridge	Description of Modes	Marazek Bridge	Description of Modes
1	2.030	Lateral Movement at Upper Level + Torsion	0.634	Lateral Movement
2	3.980	Lateral Movement at Lower Level + Torsion	1.950	Flexural Movement
3	----	----	2.260	Combined Flexural + Torsional Movement
4	----	----	2.760	Torsional Movement
5	----	----	3.600	Combined Flexural + Torsional Movement
6	----	----	7.690	Second Flexural Movement

4 FE models

Three dimensional models were built using SAP 2000 [14] program for the two studied bridges. Figure 8 shows the general layout of the models. Sectional properties as deduced from the available as-built drawings are defined for the different frame elements. For Mansoura bridge, several members were added to the bridge during repair procedures applied through the bridge service life. These modifications were accounted for in the FE model [15]. Connections between truss members and different floor beams are simulated using rigid and pinned connections. Models are built considering the sectional properties extracted from the as-built drawings or measured during the visual inspection phase. Frame elements are used to simulate the different bridge components. Rigid and pinned joints are used to simulate the connection at the truss and floor beam levels as per the observed details. Model considered in the current study accounts for all the applied repair procedures along the service life of the bridge.

For both bridges, the resulting modes coincide with the experimentally observed values; however, for Marazek bridge, large difference in frequency value is observed. Source of deviation are under investigation. Possible sources include the defects observed at one of the bearings and loss of one of bridge verticals due to a derailment accident of a train in 2003. It is also observed that frequency for Mansoura bridge is higher than that of Marazek bridge. This may be attributed for the larger depth of Marazek bridge which leads to higher stiffness. In addition, the continuity of Marazek bridge would increase its stiffness.

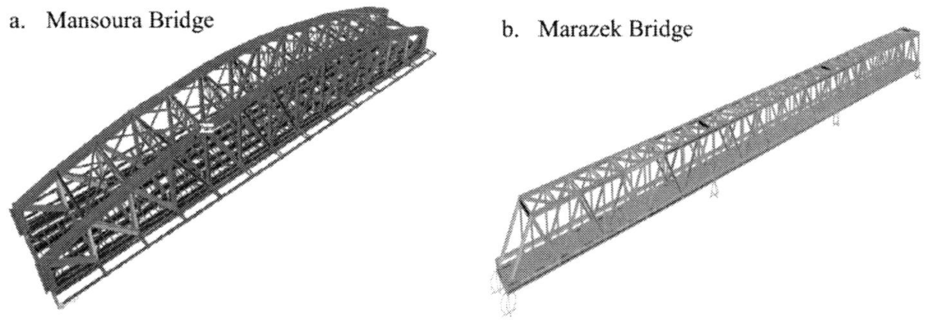

Fig. 8. FE Models for Mansoura and Marazek Bridges.

The Egyptian code for bridges [12] specifies a range for acceptable bridge frequencies according to the following:

Upper Bound: $n_o = 94.76\ L^{-0.748}$ [Hz] (2)

Lower Bound:

$4\ m \leq L \leq 20\ m$ $\quad n_o = 80 / L$ [Hz] (3)

$20\ m < L \leq 100\ m$ $\quad n_o = 23.58\ L^{-0.592}$ [Hz] (4)

As can be seen, the formula only relates the frequency only to bridge length regardless of other influencing factors. In addition, these formulas consider that the governing mode is vertical flexural mode. Meanwhile, the Egyptian code [16] does not provide formulas for torsional modes. Table 4 summerizes the resulting values considering effective spans for the considered bridges. As can be seen, the measured values do not fall within the calculated range. Moreover, the percentage of error is calculated considering the average of the two bound values and the flexural mode frequency. It can be noted that code formula does not capture bridge performance. Hence, for truss bridges, Egyptian code requires performance of dynamic analysis to evaluate dynamic bridge performance.

Table 3. Frequencies (Hz) of Measured Modes.

Mode	Mansoura Bridge	Description of Modes	Marazek Bridge	Description of Modes
1	2.433	Lateral Movement at Upper Level + Torsion	1.780	Lateral Movement
2	4.049	Vertical Flexural + Local Deformation of Upper Portal Bracing Mode Shape	2.517	Flexural Movement

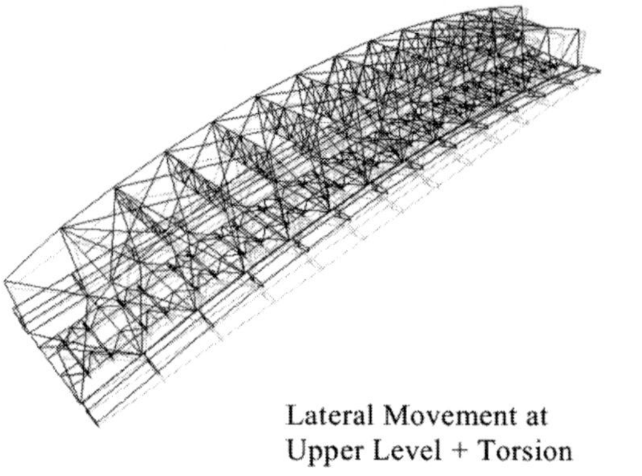

a. First Mode Shape – Mansoura Bridge.

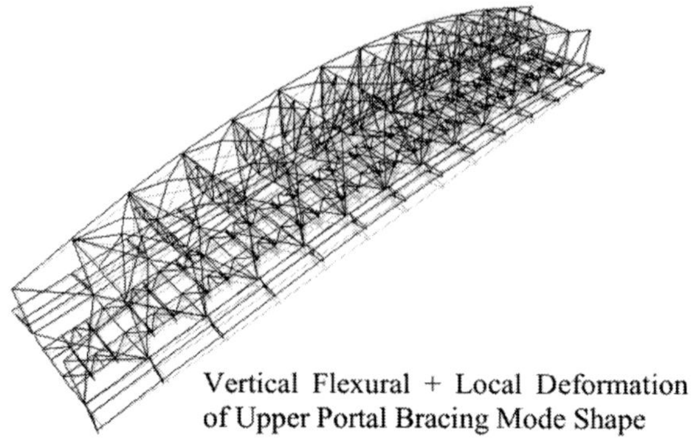

b. Second Mode Shape – Mansoura Bridge.

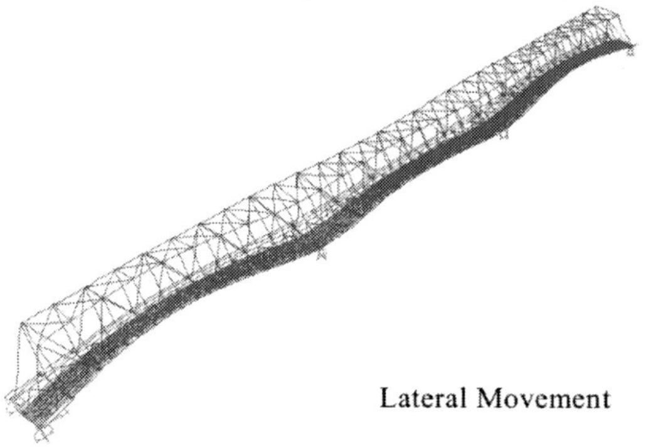

c. First Mode Shape – Marazek Bridge.

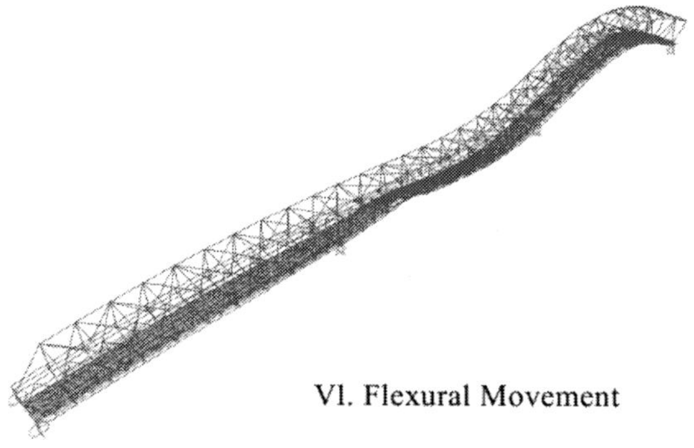

d. Second Mode Shape – Marazek Bridge.

Fig. 9. Calculated Mode Shapes

Table 4. Calculated Upper and Lower Frequencies using Egyptian Code Formulas [16].

Frequency [Hz]	Mansoura Bridge	Marazek Bridge	Percentage of Error (%)
Lower Bound	1.907	3.949	27%
Upper Bound	1.875	3.867	14%

Train velocity in Egypt usually ranges between 60 – 90 km/hr. The input frequency (ff) can be calculated according to the following:

$$f_f = v/3.6/D \text{ [Hz]} \quad (1)$$

Where:
v: Train speed [km/hr]
D: Characteristic wagon length (m)
For the used wagons in Egypt, the range of input frequencies is listed in Table 5.

Table 5. Input Frequency Range for Different Train Velocities.

Train Velocity (km/hr)	90	80	70	60
Input Frequency Range (Hz)	1.1 – 2.1	0.9 – 1.9	0.8 – 1.7	0.7 – 1.4

In case bridge dynamic properties coincide or fall in ranges within these values, resonance phenomena may occur. Truss bridges are generally susceptible to undergoing transverse vibrations under resosnant conditions, leading to excessive vertical acceleration. This may lead to exceeding serviceability limit states and discomfort of users. Figure 10 shows the transmissibility ratio for Mansoura bridge while considering the flexural mode frequency. The frequency ratio for the different considered train velocities are plotted in order to evaluate the possibility of resnonace. As can be seen, the current structural systems does not allow for high speed velocities.

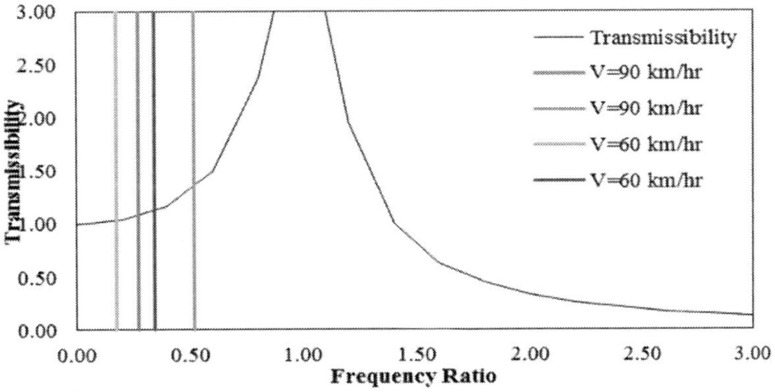

Fig. 10. Transmissibility versus Frequency Ratio for Mansoura Bridge.

5 Conclusions

Dynamic properties of two example bridges have been studied in the current work, both experimentally and analytically, in order to better understand the bridge dynamic properties and the influence of different parameters. This work is part of a comprehensive assessment campaign of a large group of steel bridges located along the railway network in Egypt. Focus is given to truss bridges with different spans and structural statistical systems. Truss bridges are generally susceptible to transversal lateral modes at the upper and lower bracing levels followed by vertical flexural mode shapes. Results were examined against formulas provided by local code provisions. These provisions link bridge fundamental frequency to its length regardless of its width or statical system. Meanwhile, formulas for torsional modes are not presented. In addition, input frequencies by current wagons are evaluated against bridge properties in order to assess probability of resonance phenomenon. Results are studied in order to deduce a suitable formula for calculating steel truss bridges, however, the discussion is beyond the scope of this paper.

The authors wish to express their sincere appreciation to the structural team at EHAF Consulting Office for the technical support during the inspection campaign including over 75 railway bridges. The authors also would like to acknowledge the support provided by the Egyptian National Railways (ENR).

References

1. M. Turek, C.E. Ventura, J. Centeno, B. Pandey, S. Huffman, *In IOMAC: 3rd International Operational Modal Analysis* (2009)
2. P. Keller, C. Higgins, S.C. Lovejoy, J. Brdg. Eng. **20**, 9 (2014)

3. S.H. Ju, H.T. Lin, J. Eng. Struct. **25**, 2, 241-250 (2003)
4. C. Gentile, Saisi, J. Constr. Build. Mat. **25**, 3709-3720 (2011)
5. C.E. Ventura, J.C. Carvajal, E. Dascotte, *In Dynamics of Bridges Conference Series: IMAC-XXVIII* **5**, 46-56 (2010)
6. M. Majka, M. Hartnett, J. Comp. Struct. **86**, 6, 556-572 (2008)
7. K.C. Chang, C.W. Kim, J. Eng. Struct. **122**, 156-173 (2016)
8. G. Yan, S.J. Dyke, A. Irfanoglu, J. Mech. Sys. Sign. Process. **28**, 195-211 (2012)
9. S.C. Siriwardane, J. Case Stud. Eng. Fail. Analy. **4**, 50-58 (2015)
10. M. Montazer, S.M. Seyedpoor, J. Shock Vib. (2014)
11. H.H. Abbas, M.S. Saleh, M.M. Hassan, *In 2nd HBRC-ISHMII: Bridge Testing, Monitoring and Assessment, SHM19* (2016)
12. H.H. Abbas, M.S. Saleh, M.M. Hassan, *In 2nd HBRC-ISHMII: Bridge Testing, Monitoring and Assessment, SHM20* (2016)
13. H.H. Abbas, M.S. Saleh, M.M. Hassan, *In 2nd HBRC-ISHMII: Bridge Testing, Monitoring and Assessment, SHM21* (2016)
14. CSI (2003). SAP2000: Integrated Software for Structural Analysis and Design. Computers and Structures, Inc., Berkeley, CA, U.S.A.
15. H.H. Abbas, M.M. Hassan, *In 19th IABSE Congress Stockholm: Challenges in Design and Construction of an Innovative and Sustainable Built Environment*, 2502-2509 (2016)
16. ECP 207 (2015). The Egyptian Code of Practice for Loads and forces in bridges and intersections. Housing and building research center. Giza, Egypt: Building and Physical Planning.

MATEC Web of Conferences **120**, 01006 (2017) DOI: 10.1051/matecconf/201712001006
ASCMCES-17

Analysis of the punching behavior of RC flat slabs with horizontal and vertical shear reinforcement

Rasha T. S. Mabrouk[1,*], and *Ali* A. Hegab[1]

[1]Faculty of Engineering, Cairo University, Giza, Egypt

Abstract. Punching failure is one of the major drawbacks of using flat slabs in the design of reinforced concrete structures. Punching shear analysis is very complicated which appears clearly in the large variation of the equations used by the different codes in calculating the punching shear capacity. There are many factors that affect the punching shear capacity of an element. This current research aims to investigate the contribution of two of these factors which are the horizontal flexural reinforcement, and the vertical shear reinforcement in the form of stirrups. Following the experimental program conducted at the reinforced concrete laboratory of the Faculty of Engineering at Cairo University, Egypt and verification using the finite non-linear analysis software (ANSYS 15), good correlation was observed between the analytical models and their corresponding experimental data. This proves that ANSYS 15 allows for studying the punching behavior of the flat slab in more details which could not be monitored in the laboratory. As a result, a detailed parametric study as well as comparison with some design codes was conducted using thirteen full scale flat slab specimens in order to simulate real life problems. The specimens had a loaded span of 2000 mm and a total thickness of 250 mm. The variables under study were the flexural reinforcement ratio and stirrups spacing and diameter in order to further enhance the understanding of the behavior of flat slabs in punching.

1 Introduction

Reinforced concrete flat slab is a popular structural system due to its construction advantages. For architectural engineers it has the advantage of freeing them from any constraints in partitioning areas, also, it gives them larger clear floor height. It is preferred by mechanical engineers because it helps them avoid making sleeves in beams to pass mechanical ducts. For site structural engineers, it gives them the benefit of making easier formworks and the ease in placing slab reinforcements. On the other hand, the flat slab has the major drawback of brittle punching shear failure at the slab column connections. The analysis of the flat slab is very complicated due to the large number of variables involved, and the fact that the transfer of loads and stresses from columns to slab goes through

[*] Corresponding author: yrasha@yahoo.com

different elements. Therefore, several researchers have given special attention to the punching behavior of slab column connections.

Analytical, numerical, as well as, extensive experimental studies have been conducted in the past decades to investigate this problem. Moe [1] concluded that the punching shear strength is related to the square root of the concrete compressive strength and then he proposed equations for the calculation of the ultimate nominal punching shear strength. The model presented by Kinnunen and Nylander [2] is one of the first analytical models that depend on the mechanical behavior and properties of slab column connection. This was followed by Kinnunen [3] then Andresson [4] who updated the model to consider the effect of vertical punching reinforcement.

This complexity in design appears clearly in the large variation of the different equations adopted by the design codes in calculating the punching shear capacity. Some design codes take very large factor of safety and other codes neglect some of the elements that can contribute effectively to the punching shear resistance. For example, in the Egyptian Code of practice [5], the effect of the vertical reinforcement and the horizontal reinforcement is not considered in the punching capacity calculation. While, the American code ACI 318R-14 [6] neglects the effect of the horizontal reinforcement. However, the Eurocode [7] considers all of the three parameters.

Hegab and Mabrouk [8] and Hegab [9] studied six half-scale column-slab specimens cast in the concrete laboratory at the Faculty of Engineering, Cairo University. The six specimens had the same concrete dimensions; 1100×1100 mm slab with thickness 100 mm, and 150×150 mm column with height 250 mm. The specimens were supported at the four edges and the loaded span was taken as 1000 mm. The specimens were divided into two groups. Group I consisted of three specimens with top and bottom mesh reinforcement with flexural reinforcement ratios of 0.40%, 0.55% and 0.785%. While group II was reinforced with vertical reinforcement in the form of stirrups in addition to the top and bottom mesh used in group I. The details of the specimens are shown in Table 1.

Table 1. Reinforcement details of the six specimens tested [8, 9].

Group	Specimen	Tension RFT	Compression RFT	Stirrups	Plan
I	S1-1	5Ø10/m	5Ø10/m	---	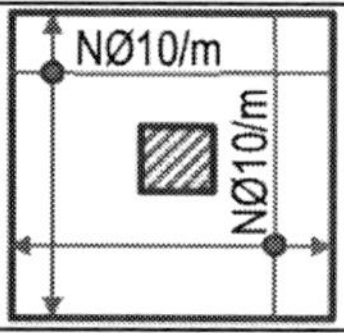
	S1-2	7Ø10/m	7Ø10/m	---	
	S1-3	10Ø10/m	10Ø10/m	---	
II	S2-1	5Ø10/m	5Ø10/m	Ø8@100 mm	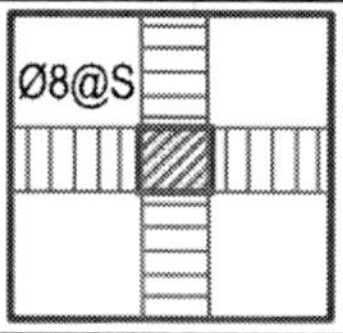
	S2-2	7Ø10/m	7Ø10/m	Ø8@70 mm	
	S2-3	10Ø10/m	10Ø10/m	Ø8@50 mm	

They concluded that a significant increase in the punching capacity due to the increase in the flexural reinforcement mesh is observed. The punching ultimate capacity increased by 13.2% due to the increase in the flexural reinforcement by 40.3%, and it increased by 20% by doubling the flexural reinforcement ratio. Moreover, there is a noticeable increase in the punching load capacity in group II specimens compared to their corresponding specimens of group I due to the addition of stirrups. This increase is more pronounced in specimen S2-3 (19.5%) which asserts that the spacing between stirrups is preferred to be limited to a small value [8, 9]. However, more detailed investigation is needed with more variables as well as an in depth study of each of the parameters affecting the punching behavior of flat slabs in order to better understand this kind of behavior.

The main aim of this research is to further study the role of the flexural horizontal reinforcement and the vertical shear reinforcement in the form of stirrups in resisting the punching shear acting on reinforced concrete flat slabs.

2 Finite element analysis

The finite element analysis program ANSYS 15 was used in this research. Concrete was modelled using the SOLID65 element which has eight nodes with three degrees of freedom at every node. These are translations in the x, y, and z directions. The element is capable of plastic deformation, cracking, and crushing in three orthogonal directions [10]. Figure 1 shows the typical elements used to model both concrete and steel. Concrete material properties are defined in ANSYS as poisons ratio 0.2 and modulus of elasticity 25,200 MPa. The value of the tensile strength is input and an idealized stress strain curve was used to model concrete as shown in Figure 2.

Subsequently, the reinforcement rebar was defined as the discrete element LINK180 which is a three dimensional truss element with two nodes; each node has three degrees of freedom in the form of translation in the three global directions [10] as shown in Figure 1. The material definitions for this steel element are assumed linearly as poisons ratio of 0.3 and modulus of elasticity 200,000 MPa. The idealized stress strain curve for steel is shown in Figure 3. While, Figures 4 and 5 show the mesh used for concrete as well as for horizontal steel and vertical stirrups.

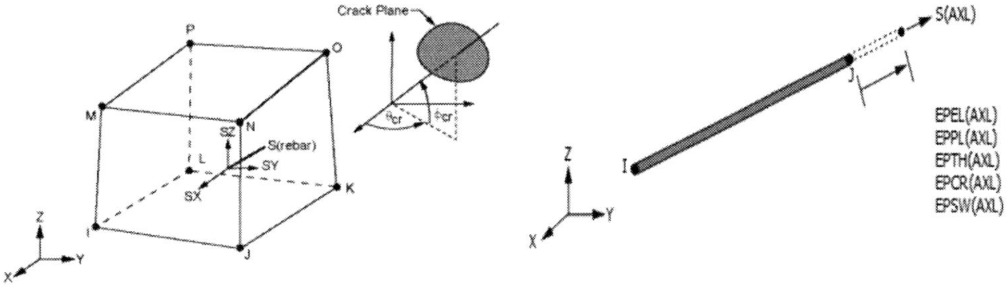

(a) Concrete (element Solid 65) (b) Steel (element Link180)

Fig. 1. Elements used for modelling concrete and steel.

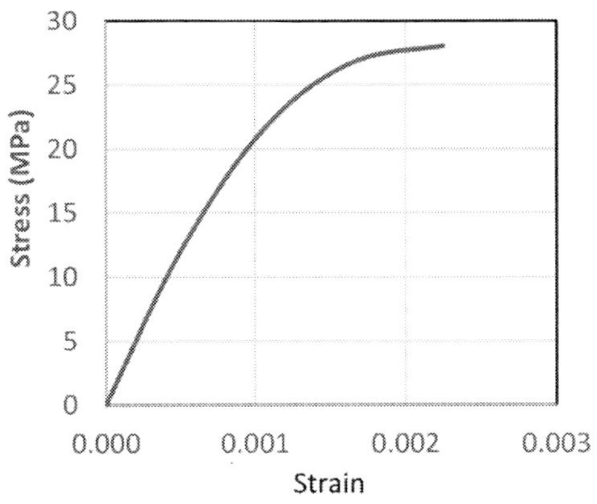

Fig. 2. Idealized compression stress strain curve for concrete.

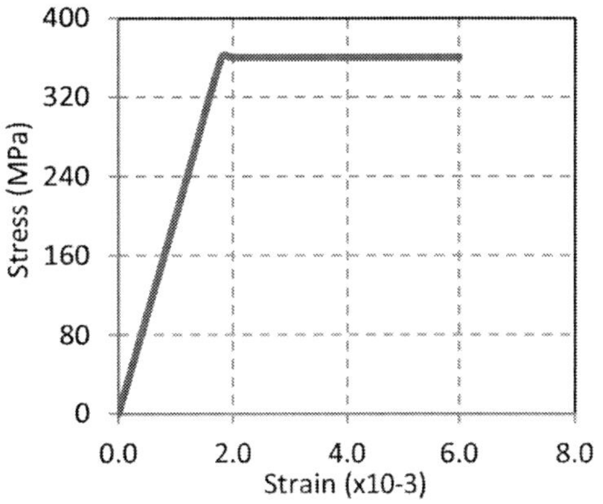

Fig. 3. Idealized stress strain curve for steel.

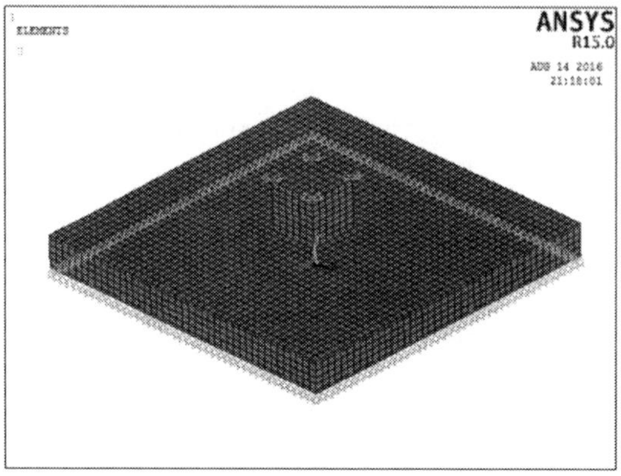

Fig. 4. Mesh used for concrete and boundary conditions.

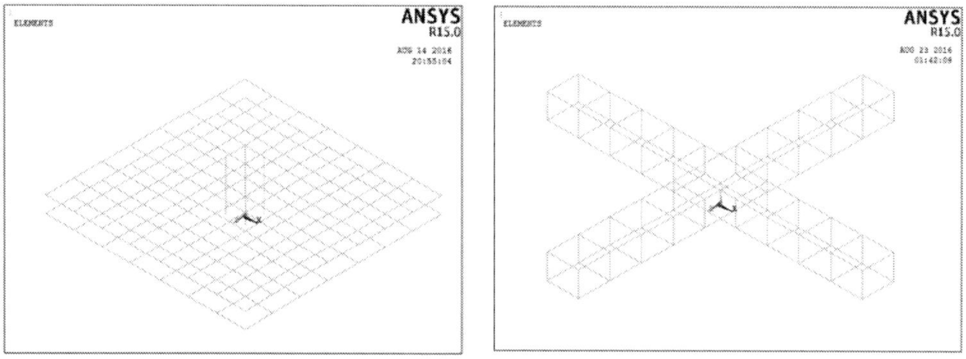

(a) Horizontal reinforcement (b) Stirrups

Fig. 5. Mesh used for steel reinforcement.

This study was conducted in two steps. First, the results estimated using the ANSYS program had to be verified against experimental data. Second, a more detailed parametric study was performed to further assess the punching behaviour of the slab-column connections in flat slabs

2.1 Experimental verification

As a first step the six slab specimens mentioned in Table 1 were analysed using the non-linear finite element software (ANSYS 15) [8, 9]. Figure 6 and 7 show an example of the load deflection relations for a specimen from Group I and another specimen from Group II, respectively where S refers to the experimental specimens and M refers to the corresponding models using ANSYS 15. Hegab and Mabrouk [8] and Hegab [9] concluded that good correlation can be seen between both the analytical and the experimental data. Comparing the error percentage of the ultimate load and the corresponding deflection between the experimental data and the analytical model, the percentage of error did not exceed 10% for the ultimate load and 16% for the deflection which is quite acceptable.

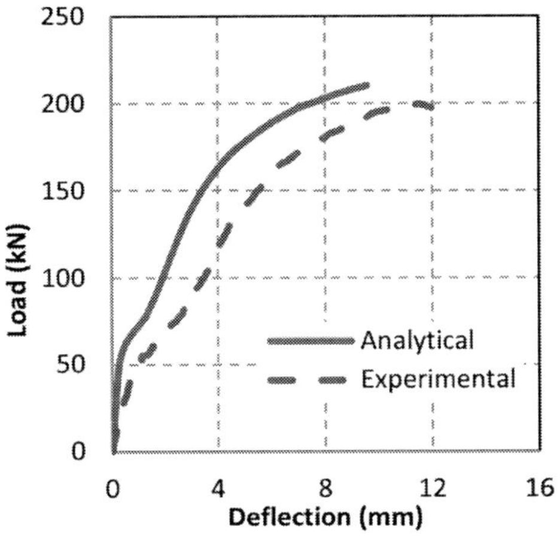

Fig. 6. Load-Deflection curves for S1-3 and M1-3

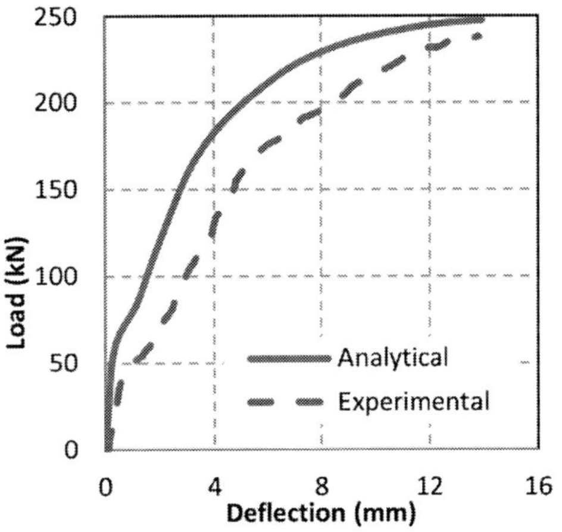

Fig. 7. Load-Deflection curves for S2-3 and M2-3

2.2 Parametric study using full scale models

Based on the results in section 2.1, ANSYS 15 can be utilized to further investigate the effect of the horizontal and vertical reinforcement on the punching behavior of flat slabs. For a better simulation of real life problems, thirteen full scale specimens were studied. All the thirteen specimens were slabs of total dimensions 2100×2100 mm with a thickness of 250 mm, and 300×300 mm column with height 250 mm supported at the four edges as shown in Figure 8. The loaded span was taken as 2000 mm and the concrete compressive strength as 35MPa. High grade steel (360/520) was used. Table 2 and Figure 9 show the details of the thirteen specimens analyzed in this study. The specimens had flexural reinforcement ratios varying from 0.67% to 2.45% as well as different arrangements of vertical reinforcement. Specimens S01 to S05 were reinforced with a horizontal mesh with spacing 100 mm. However, the bar diameter was changed from 16 to 25 mm. S06 to S10 had bar spacing of 150 mm and also varying bar diameters from 16 to 25 mm. All of these ten specimens had no vertical stirrups. In the last three specimens (S11, S12 and S13), stirrups were added. The spacing of the stirrups was changed between S11 and S12 while the diameter of stirrups was taken 12 mm for S12 and 10 mm for S13 in order to study the effect of these two parameters on the behavior of the slabs.

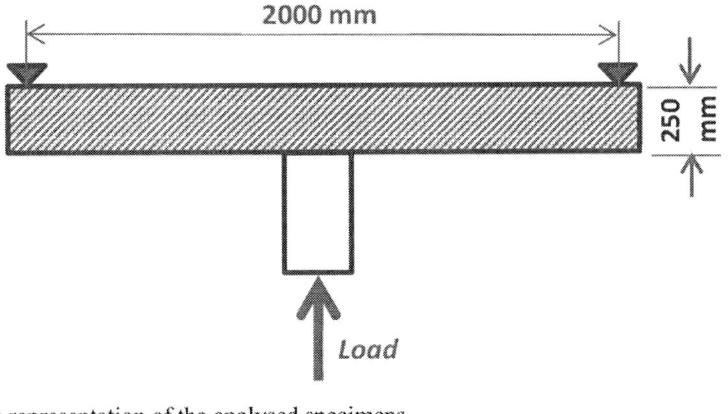

Fig. 8. Schematic representation of the analysed specimens.

Table 2. Details of all the specimens.

Specimen	Reinforcement			Horizontal RFT ratio (ρ) %
	Top mesh	Bottom mesh	Stirrups	
S01	Ø16@100 mm	Ø16@100 mm	---	1.00
S02	Ø18@100 mm	Ø16@100 mm	---	1.27
S03	Ø20@100 mm	Ø16@100 mm	---	1.57
S04	Ø22@100 mm	Ø16@100 mm	---	1.90
S05	Ø25@100 mm	Ø16@100 mm	---	2.45
S06	Ø16@150 mm	Ø16@150 mm	---	0.67
S07	Ø18@150 mm	Ø16@150 mm	---	0.85
S08	Ø20@150 mm	Ø16@150 mm	---	1.05
S09	Ø22@150 mm	Ø16@150 mm	---	1.27
S10	Ø25@150 mm	Ø16@150 mm	---	1.64
S11	Ø16@150 mm	Ø16@150 mm	Ø12@100 mm	0.67
S12	Ø16@150 mm	Ø16@150 mm	Ø12@50 mm	0.67
S13	Ø16@150 mm	Ø16@150 mm	Ø10@50 mm	0.67

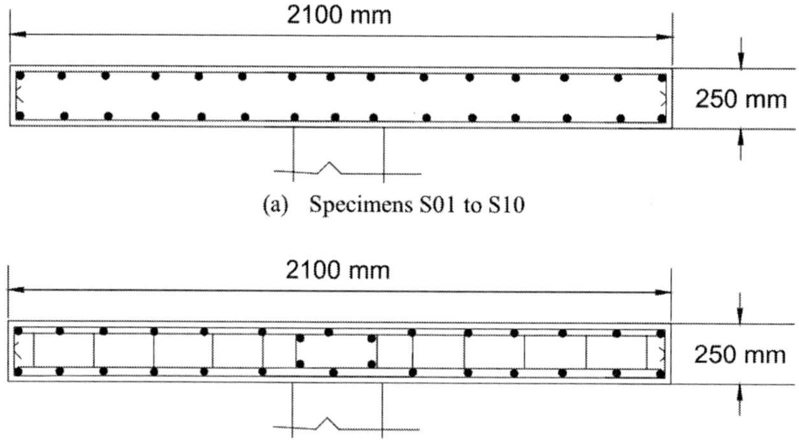

(a) Specimens S01 to S10

(b) Specimens S11 to S13

Fig. 9. Reinforcement details of the specimens.

3 Results and discussion

The specimens were divided into four groups A, B, C and D; each focusing on one parameter under study. This section discusses the results of the specimens computed using ANSYS 15 and in addition the results obtained were compared with the values calculated using E.C.P (203-2007) [5], ACI 318R-14 [6], and Eurocode2 [7]. Typical punching failure occurred in all the specimens as seen in Figure 10.

3.1 Effect of horizontal flexural reinforcement

Figures 11 and 12 show the load deflection relationships for groups (A) and (B). It can be seen from the figures that the horizontal flexural reinforcement ratio had a significant effect on the punching capacity. As the flexural reinforcement ratio was increased, the punching shear capacity increased which is in accordance with earlier experimental findings. Comparing specimens S01 and S05, the ultimate punching capacity increased by 18% when the reinforcement ratio was increased from 1% to 2.45%. For group B, the punching capacity increased by 35% when the reinforcement ratio was increased from 0.67% to 1.64%. In addition, the stress in the flexural reinforcement reached the first yield stress (P_y) at larger load when the reinforcement ratio was increased. However, the value of the ultimate deflection decreased as the flexural reinforcement ratio increased indicating less ductility. Figure 13 shows the effect of flexure reinforcement ratio where the ultimate capacities of the ten slabs are plotted. It can be seen that the rate of increase in the punching capacity decreased as the reinforcement ratio increased. After reaching a certain reinforcement ratio, which in this case is 2%, no significant increase in the ultimate punching capacity can be seen.

Table 3 shows the output results from the ANSYS as well as the comparison with the three design codes; E.C.P (203-2007) [5], ACI 318R-14 [6], and Eurocode2 [7]. The three design codes generally underestimated the punching capacity compared to that calculated using finite element analysis. The punching capacities estimated according to the E.C.P (203-2007) were the most conservative .The E.C.P (203-2007) [5] as well as the ACI 318R-14 [6] do not take the effect of horizontal reinforcement into consideration which resulted in a fixed ultimate capacity value among the ten specimens despite the change in the reinforcement ratios. However, the output values using the Eurocode followed a similar pattern to the output obtained from ANSYS.

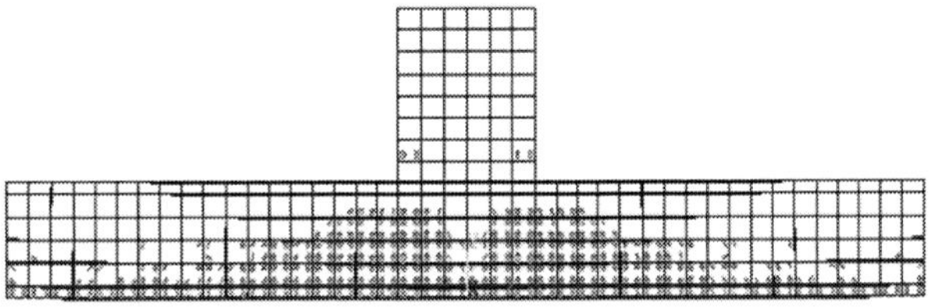

Fig. 10. Typical punching failure for specimen S06.

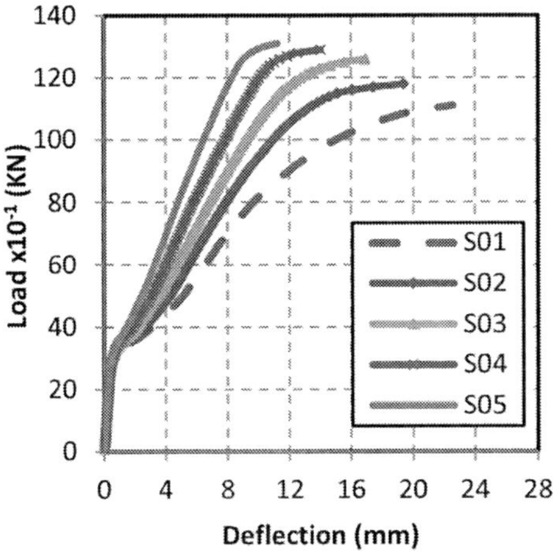

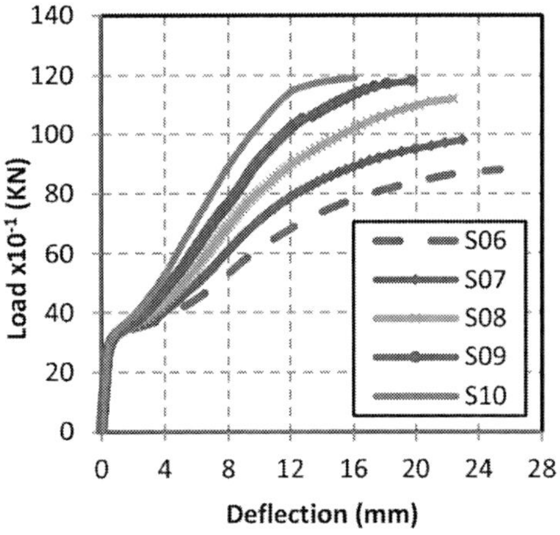

Fig. 11. Load – deflection curve for Group (A) **Fig. 12.** Load – deflection curve for Group (B)

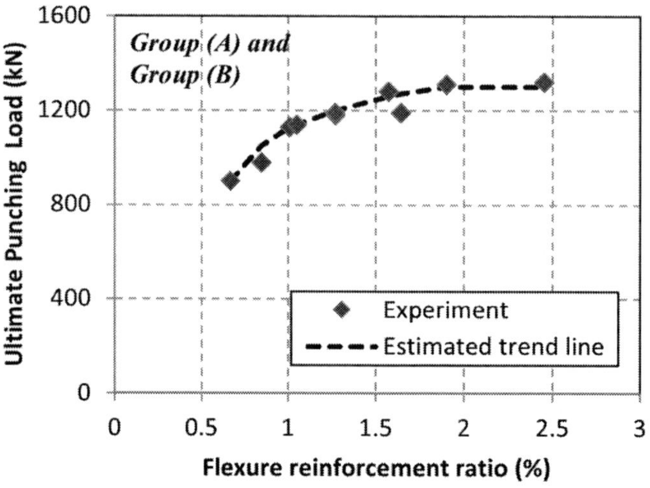

Fig. 13. Effect of horizontal reinforcement ratio on punching capacity.

Table 3. Analytical results – effect of horizontal reinforcement.

Group	Specimen	Horizontal RFT ratio (ρ) %	ANSYS			ACI	Euro	ECP
			P_y* (KN)	Δ_u** (mm)	P_u (KN)	P_u (KN)	P_u (KN)	P_u (KN)
(A)	S01	1.01	690	23	1110	708.4	649.5	619.2
	S02	1.27	830	19	1180	708.4	701.1	619.2
	S03	1.57	980	17	1260	708.4	752.4	619.2
	S04	1.90	1180	14	1290	708.4	801.8	619.2
	S05	2.45	no	11	1310	708.4	872.7	619.2
(B)	S06	0.67	480	25	880	708.4	566.5	619.2
	S07	0.85	590	23	980	708.4	613.2	619.2
	S08	1.05	700	22	1120	708.4	658.0	619.2
	S09	1.27	840	20	1180	708.4	701.1	619.2
	S10	1.64	1040	16	1190	708.4	763.4	619.2

* P_y Load corresponding to first yield stress. **Δ_u Deflection at ultimate load.

3.2 Effect of vertical shear reinforcement (Stirrups)

Using vertical shear reinforcement increased the punching capacity of the tested slabs. It should also be noted here that the ACI computation in case of using stirrups gave ultimate punching load values closest to that obtained from the ANSYS.

3.2.1 Effect of stirrups spacing

The results for group (C) can be seen in Table 4 and Figure 14. Adding stirrups with spacing 100 mm which corresponds to 0.5d resulted in an increase in the punching shear capacity of 4.5 % compared to specimens with no shear reinforcement. While in case of spacing 50 mm (0.25d) the increase was 10.2 % compared to S06. This can be attributed to the fact that smaller stirrups spacing give a higher area of steel resisting the punching stresses at the critical section

Table 4. Analytical results – effect of stirrups spacing

Group	Specimen	Stirrups		ANSYS			ACI	Euro	ECP
		S (mm)	Ø (mm)	P_y (KN)	Δ_u (mm)	P_u (KN)	P_u (KN)	P_u (KN)	P_u (KN)
(C)	S06	--	--	480	25	880	708.4	566.5	619.2
	S11	100	12	520	31	920	1015.8	1224.0	N/A
	S12	50	12	520	44	970	1073.3	1224.0	N/A

3.2.2 Effect of stirrups diameter

Similarly, increasing the bar diameter of the stirrups led to an increase in the load capacity as can be seen in Table 5 and Figure 15. Using diameter of 10 mm and 12 mm led to an increase in the punching capacity of 3.4 % and 10.2 %, respectively compared to S06.

Table 5. Analytical results – effect of stirrups diameter.

Group	Specimen	Stirrups		ANSYS			ACI	Euro	ECP
		S (mm)	Ø (mm)	P_y (KN)	Δ_u (mm)	P_u (KN)	P_u (KN)	P_u (KN)	P_u (KN)
(D)	S06	--	--	480	25	880	708.4	566.5	619.2
	S13	50	10	520	37	910	1073.3	1224.0	N/A
	S12	50	12	520	44	970	1073.3	1224.0	N/A

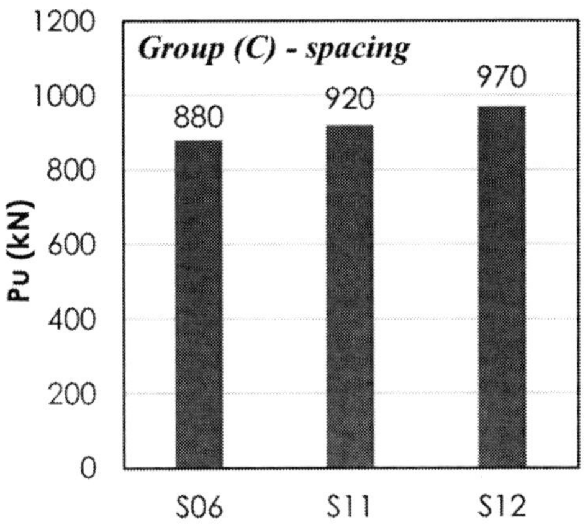

Fig. 14. Effect of stirrups spacing.

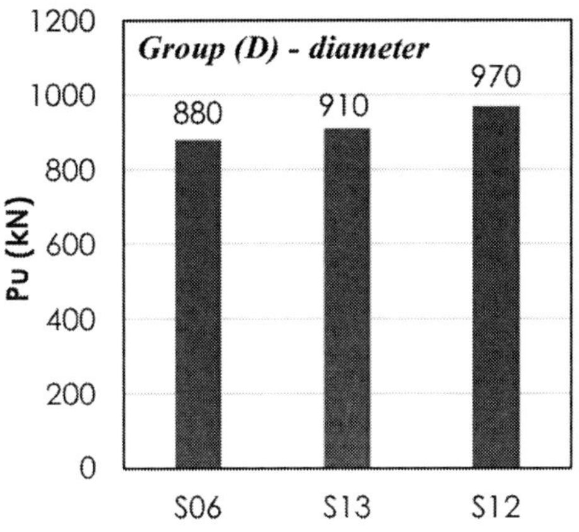

Fig. 15. Effect of stirrups diameter.

5 Conclusions

Based on the results of the study conducted on thirteen full scale flat slab specimens, the following conclusions can be drawn:

1. The ANSYS program is an effective tool that can be used to stimulate the punching behaviour of flat slabs.

2. Studying full scale flat slab specimens with horizontal mesh spaced at 100 mm (0.5d), the ultimate punching capacity increased by 18% when the reinforcement ratio was increased from 1% to 2.45%. While for the mesh spaced at 150 mm (0.7d), the ultimate punching capacity increased by 35.2% when the reinforcement ratio was increased from 0.67% to 1.64%.
3. As the flexural reinforcement ratio increased, the value of the load at first yield increased but the ultimate deflection decreased.
4. The flexural reinforcement ratio has a significant effect on the ultimate punching capacity of flat slabs. However, the rate of increase in the punching capacity decreases as the reinforcement ratio increases.
5. For the vertical shear reinforcement, using bigger bar diameter and smaller spacing between stirrups increase the punching capacity. The increase in ultimate load capacity reached about 10% in the specimens studied.
6. When using vertical shear reinforcement, the spacing between stirrups is preferred to be limited to small values, 0.5d or smaller, which is in accordance with the international design codes.
7. Results obtained using the E.C.P (203-2007), ACI318-14, and Eurocode2 differ according to the parameters considered in computation.
8. E.C.P (203-2007) has proven to be the most conservative among the three design codes while ACI318-14 gave reasonable results in case of using stirrups.

Deep gratitude and special thanks are given to Prof. Dr. Magdy E. Kassem for his continuous support and help throughout this research.

References

1. J. Moe, "Shearing Strength of Reinforced Concrete Slabs and Footings under Concentrated Loads," Development Department Bulletin D47, Portland cement Association, Skokie, Apr. (1961).
2. S. Kinnuen, H. Nylander, "Punching of Concrete Slabs without Shear Reinforcement," Transactions, **No.158**, Royal Institute of Technology, Stockholm, Sweden, (1960)
3. S. Kinnuen, "Punching of Concrete Slabs with Two-Way Reinforcement with Special Reference to Dowel Effect and Deviation of Reinforcement from Polar Symmetry", Transactions, **No.198**, Royal Institute of Technology, Stockholm, (1960).
4. J.L. Anderson, "Punching of concrete slabs with shear reinforcement", Transactions of the Royal institute of technology, **No. 212**, Stockholm, Sweden, (1963).
5. Egyptian Code for design and Construction of reinforced concrete structures, ECP 203-2007 ed., Egypt, (2007)
6. ACI Committee 318, 'Building Code Requirements for Structural Concrete (ACI 318-14) and Commentary (318R-14)," American Concrete Institute, Farmington Hills, MI, (2005).
7. Euro-Code 2, "Design of Concrete Structures, Part 1-1: General Rules and Rules for Buildings," CEN, EN 1992-1-1, Brussels, Belgium, (2004).
8. A. Hegab, R. Mabrouk, *Ninth Alexandria International Conference on Structural and Geotechnical Engineering, AICSG9*, RC-01, (2016).
9. A. Hegab, "Behavior of RC Flat Slabs with Horizontal and Vertical Shear Reinforcement in Punching". *Master thesis, Structural Engineering Department, Faculty of Engineering, Cairo University, Egypt,* (2016)
10. SAS IP, "ANSYS Theory Reference Release 5.6", ANSYS, Inc. Canonsburg, Pennsylvania, USA, (1999).

MATEC Web of Conferences **120**, 01007 (2017) DOI: 10.1051/matecconf/201712001007
ASCMCES-17

Effect of web openings on the structural behavior of RC beams subjected to pure torsion

Taha Abdo[1], *Rasha Mabrouk*[1,*]

[1]Faculty of Engineering, Cairo University, Giza, Egypt

Abstract. In the construction of modern buildings, the presence of openings became a necessity nowadays in order to accommodate the many pipes and ducts required for the different services. On the other hand, the presence of these openings in RC beams affects their structural behavior. One important behavior that needs further study would be torsion. Currently, there are no guide lines available to analyze or design RC beams with web openings under pure torsion. Thus, the main objective of this research is to investigate the behavior of simply supported RC beams with openings subjected to pure torsion. In the first part of this paper, an experimental study is conducted on seven beams subjected to pure torsion loading in order to investigate the effect of the number of openings, the spacing between stirrups, and beam depth. All beams have a constant clear span length of 1800 mm and a constant beam width of 150 mm. The beams have varying opening number and stirrups arrangement as well as varying beam depth in order to investigate the effect of these parameters on the behavior of the beams. Systematic measurements such as the cracking torque, the ultimate torque, the angle of rotation of the beam at cracking, and the angle of rotation of the beam at the ultimate load are conducted for further understanding of the beam behavior under torsion. In the second part of the paper, the experimental results are compared with the numerical results obtained using the non-linear finite element analysis program, ANSYS. Good agreement between the experimental and numerical results is found.

1 Introduction

In most buildings, utility pipes and ducts are accommodated in the space within the floor ceiling sandwich. Passing these pipes and ducts through transverse openings in the floor beams saves a significant amount of headroom and results in a more compact design. In multistory buildings, the total height can be reduced due to presence of the transfer openings that allow the passage of pipes and ducts through it [1]. The presence of transverse openings will transform the simple beam behavior into a more complex behavior. Currently, there are no recommendations for analyzing or designing reinforced concrete beams with web openings subjected to pure torsion in the design codes. Some standard methods for design are available but much work is still needed.

* Corresponding author: yrasha@yahoo.com

There are previous studies conducted such as those done by Mansur, et al. [2] which studies reinforced concrete beams with large rectangular openings under pure torsion where the opening sizes and their locations were varied. Also, there is the research carried out by Elbadawy [3] to study the effect of the opening dimensions, beam width, and the presence of the vertical stirrups on the structural behavior of beam with opening under torsion. Akhtaruzzaman [5] and Mansur and Tan [6] developed equations to calculate the torsional moment resistance of reinforced concrete beams with rectangular and circular opening. Fawzy [7] studied the use of externally bonded fiber reinforced polymers to strengthen concrete beams with a large web opening subjected to pure torsion. Venugopal M. [8] studied the behavior of GFRP retrofitted rectangular RC beams with small web openings under torsion.

The design codes such as ACI 318-08 [9] and ECP-203-2007 [10] do not give any explicit guidance for designing these elements with openings. The main objectives of this research is to study the effect of openings on the structural behavior of beams subjected to pure torsion in an attempt to obtains some specific guidelines to help structural engineers when dealing with this problem. The research is divided into two parts; an experimental part and as analytical part as shown below.

2 Experimental program

Table 1 shows the details of the seven beam specimens tested in this research. All the test samples had an overall span of 2000 mm and width of 150 mm. As shown in Figure 1, the overall length of 2000 mm was divided into three parts; the middle part was 1600 mm which was the test zone. The two end parts were 200 mm each, and heavily reinforced to be safe against failure during testing. Three parameters were studied namely; the effect of the number of openings (N), the effect of spacing between stirrups (S) and the effect of the beam height (H). Figure 2 and Figure 3 show the details of all the beams under study.

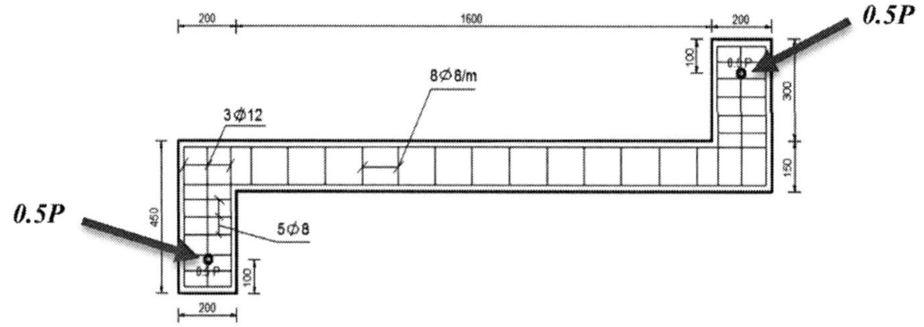

Fig. 1. General plan of the beams under study.

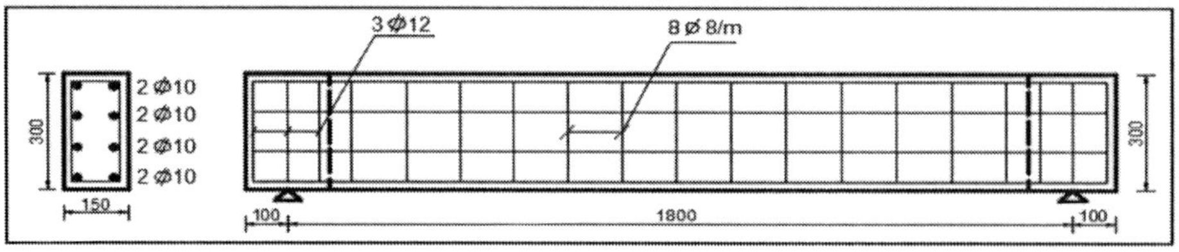

Fig. 2. Dimensions and reinforcement details of the control beam B1 (N0-S125-H300).

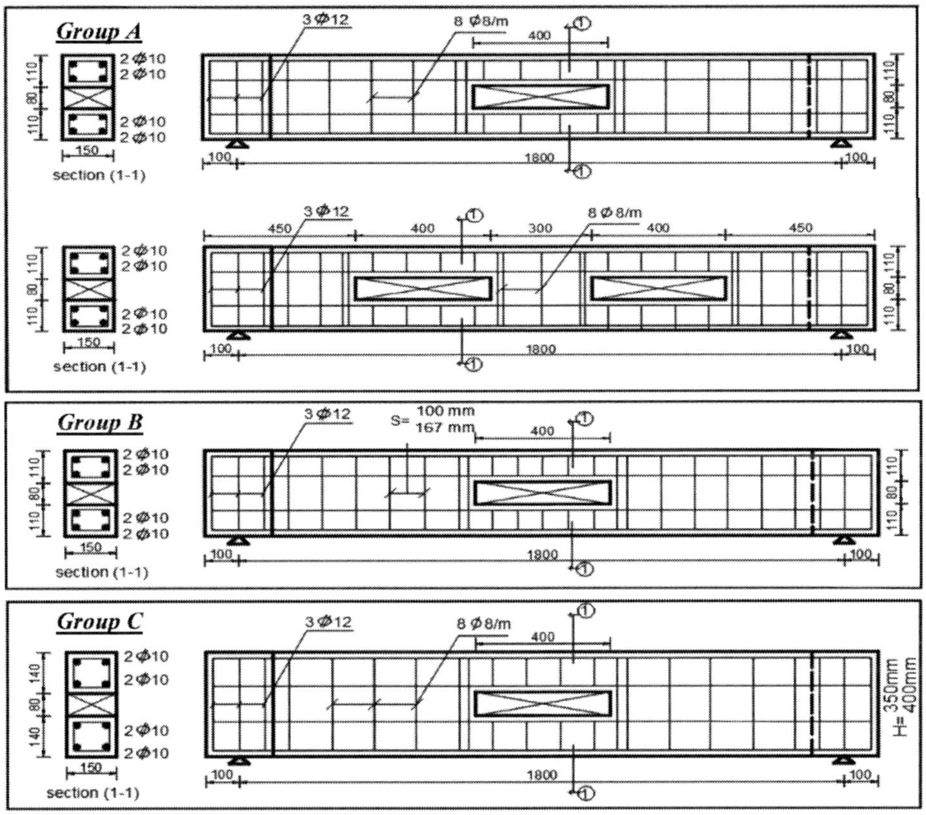

Fig. 3. Dimensions and reinforcement details of the tested specimens with openings.

Table 1. Outline of the experimental program

Group	Beam	Notation	Studied Parameter	Opening Variable			Longitudinal Reinforcement	Stirrups
				N	S (mm)	H (mm)		
	B1	N0-S125-H300	Control Beam	0	125	300	8Φ10	8 Ø 8/m
A	**B2**	N1-S125-H300	Number of openings (N)	1	125	300	8Φ10	8 Ø 8/m
	B3	N2-S125-H300		2	125	300	8Φ10	8 Ø 8/m
B	**B4**	N1-S100-H300	Spacing between stirrups (S)	1	100	300	8Φ10	10 Ø 8/m
	B5	N1-S165-H300		1	167	300	8Φ10	6 Ø 8/m
C	**B6**	N1-S125-H350	Beam Height (H)	1	125	350	8Φ10	8 Ø 8/m
	B7	N1-S125-H400		1	125	400	8Φ10	8 Ø 8/m

2.1 Materials used

Normal strength concrete using local materials was used in casting all the specimens. The concrete mixture was designed for a characteristic compressive strength of 30 MPa after 28 days. The concrete mix consisted of 350 kg cement, 175 liter of water, 1250 kg of coarse aggregate, 650 kg fine aggregate, and 3.5 liter of B.V.F admixture. The form work was

 DOI: 10.1051/matecconf/201712001007

removed after 72 hours from casting. All beams were cured by water sprinkling twice a day for 14 days, and then left in the laboratory atmosphere until the day of the test. The results of the compression test for concrete cubes on the day of the test gave an average compressive concrete strength of 35 MPa which is higher than the required value. High grade steel bars (36/52) of 10 mm and 12 mm were used for the longitudinal reinforcement while mild steel bars (24/36) were used for stirrups with 8 mm diameter. The tension normal test was performed to evaluate the yield and ultimate strength for the steel used. The results of the yield strength for bars of diameter 8, 10 and 12 were (f_y = 343, 497, 613 N/mm^2, respectively) and the ultimate strength were (f_u = 500, 596, 710 N/mm^2, respectively).

2.2 Test setup

The test set-up was designed so that pure torsional moment is applied to the tested specimens. The test specimens were simply supported on movable supports specifically made for the test. Loading was applied using a 1000 kN capacity hydraulic jack. The applied vertical load was equally distributed on the two cantilevers at both ends of the beam using a rigid steel I-beam in order to produce torsional load. Two LVDTs were attached to the two ends of the cantilevers to measure the vertical displacement at various loading stages. Figure 4 shows the general arrangement of the test setup for all the specimens.

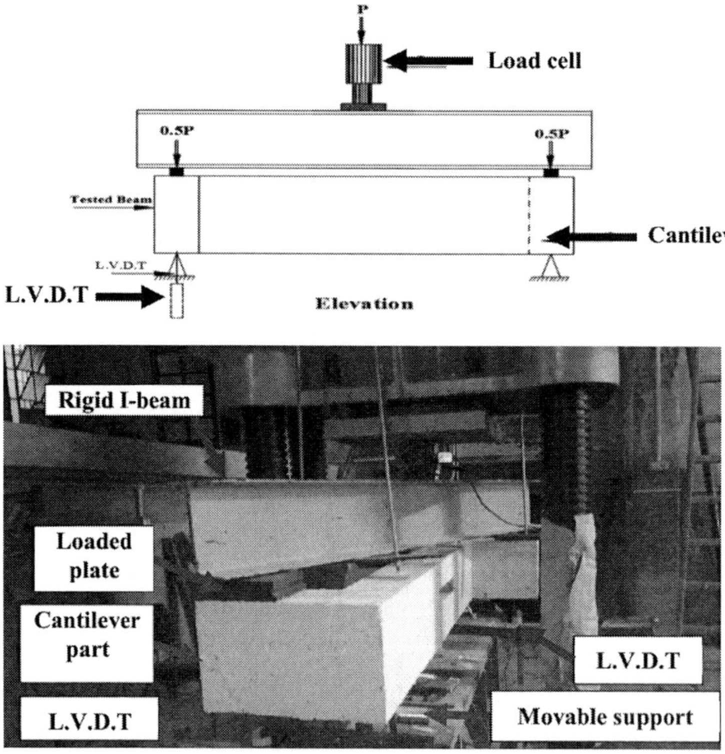

Fig. 4. Arrangement of the test setup for all the tested specimens.

The load on each cantilever was applied in increments. After each increment, all LVDTs readings were recorded. The development and propagation of cracks were marked up to failure. The rotation angle was calculated by dividing the deflections under the two loaded points by the distance between the centerline of the tested beams and loaded point.

The torque is obtained by multiplying the applied load on the cantilever by the distance between the centerline of tested beams and loaded point.

3 Analytical study

In order to assure this study and give the opportunities for future parametric studies, the seven specimens were verified by the non-linear finite element software; ANSYS16 [11]. Eight-node solid brick element SOLID65 was used to model the concrete. SOLID65 is used for 3-D modeling of solids with or without reinforcing bars. The solid is capable of cracking in tension and crushing in compression. The element is defined by eight nodes having three degrees of freedom at each node: translations in the nodal x, y, and z directions. The most important aspect of this element is the treatment of nonlinear material properties. The concrete is capable of cracking (in three orthogonal directions), crushing, plastic deformation, and creep. Figure 5 shows the elements used in modeling reinforced concrete using ANSYS16 as well as the compression stress-strain used for concrete. The reinforcement steel was defined using the three dimensional truss element; LINK180 with two nodes; each node also has three degrees of freedom. The mesh used for the specimens as well as the boundary conditions are defined as shown in Figure 6.

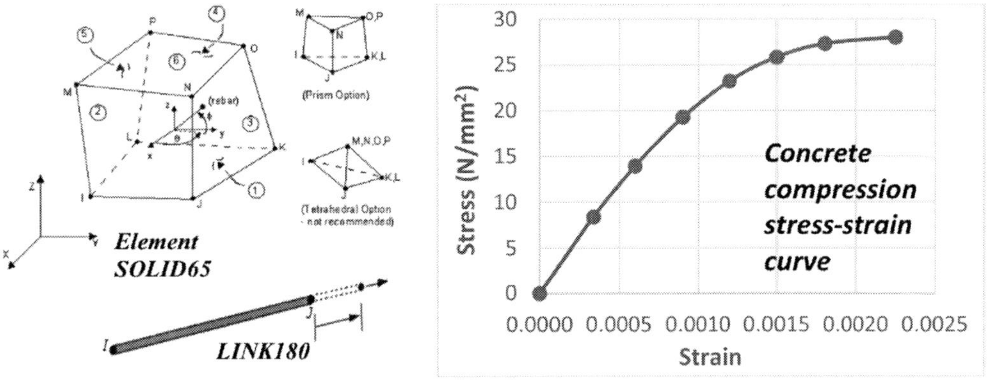

Fig. 5. Modeling of the beams using ANSYS16

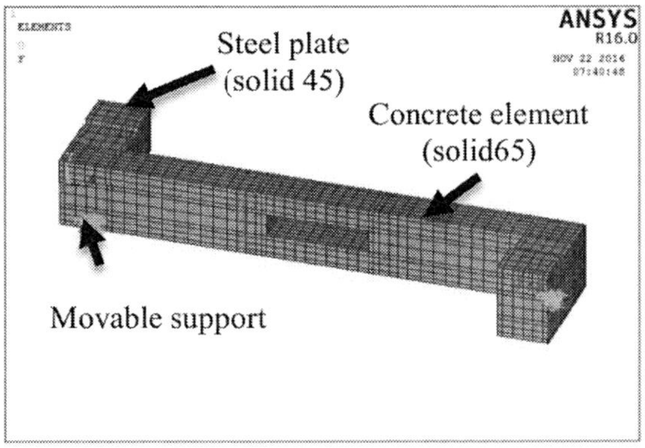

Fig. 6. Mesh used in the analysis

4 Results and discussions

4.1 Crack patterns

Figure 7 shows the crack patterns for three of the beams under study as obtained from experiment as well as from the analysis using the ANSYS program. It is noticeable that all the tested beams with openings have the same crack patterns. The first crack appeared at the corner of the opening then at the top chord member. These cracks were inclined with an angle around 45 degrees to the longitudinal axis of the beam. The angle of inclination of the compression diagonals with respect to the beam axis, θ, depends on the ratio of the force carried by the longitudinal reinforcement to that carried by the stirrups [12]. In addition, good agreement can be seen between the FEM analysis using ANSYS16 and the experimental results.

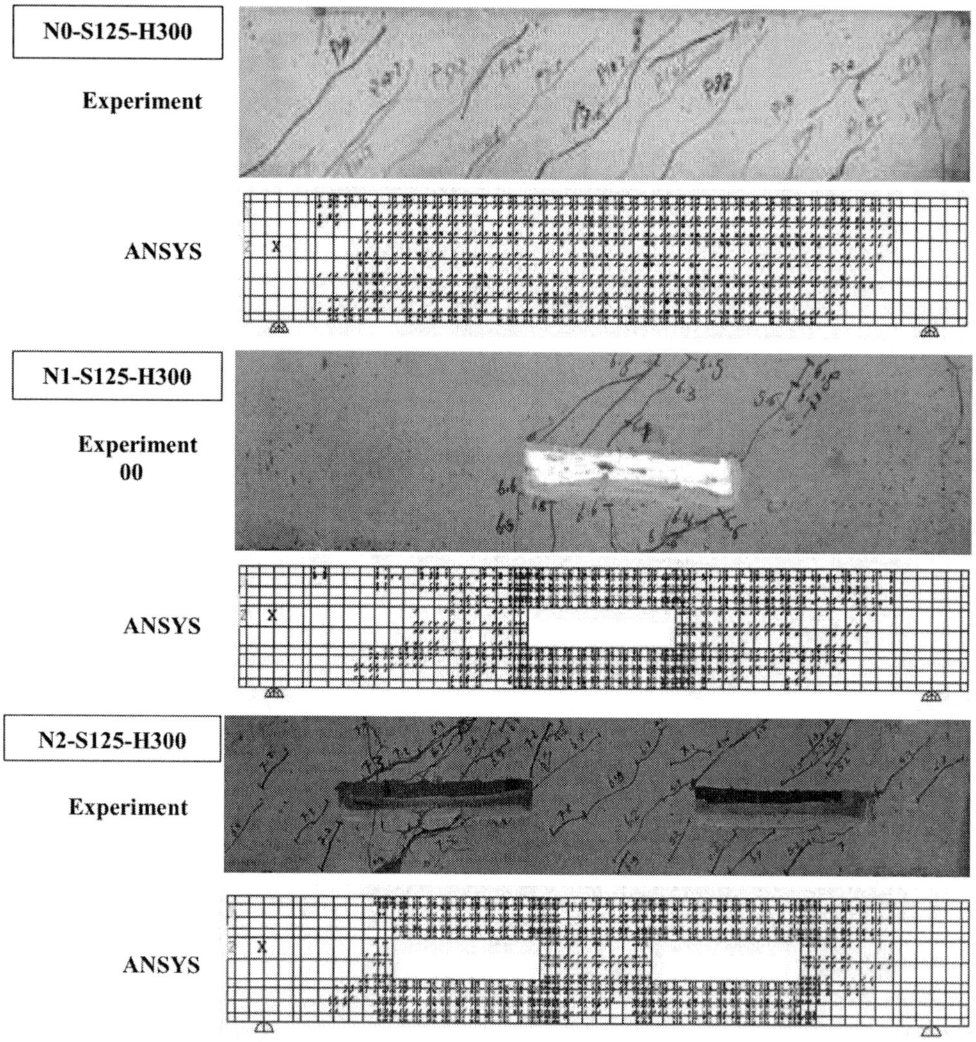

Fig. 7. Analytical and experimental crack patterns for specimens N0-S125-H300, N1-S125-H300 and N2-S125-H300

4.2 Torque - Rotation curves

Comparison between the experimental and analytical outputs of the torque - rotation curves for all specimens can be seen in Figures 8 to 10. The three figures also show the individual effects of the studied parameters. The shapes of the curves are divided into two parts: the first part is linear up to the cracking torque and the second part after cracking which is nonlinear. The slope of the curve decreases with the increase in the torque indicating the post cracking behavior.

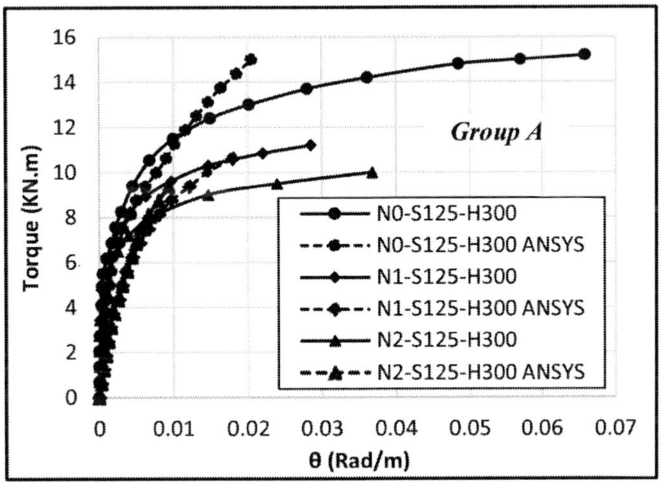

Fig. 8. Torque - rotation curve: Number of opening (Group A).

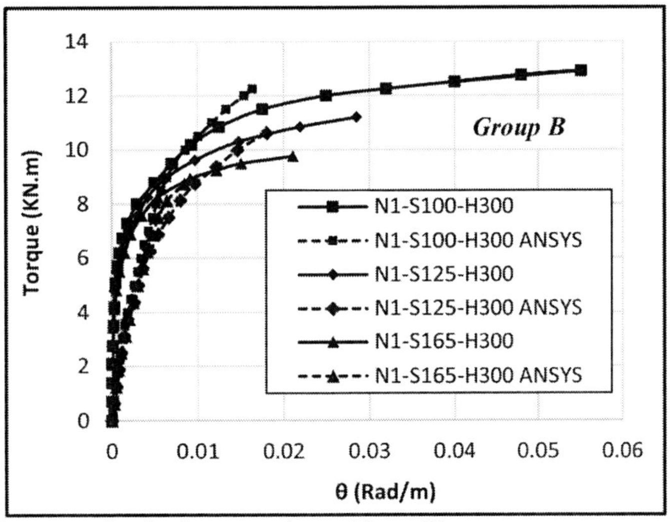

Fig. 9. Torque - rotation curve: Spacing between stirrups (Group B).

In Figure 8, it can be seen that both the ultimate torsional capacity as well as the ductility of the beams are significantly reduced by the presence of openings. Using one opening caused a reduction in the ultimate capacity of about 26 % while the use of two openings caused a reduction of about 34%. Figure 9 shows that decreasing the spacing between the stirrups to 100 mm caused a significant in improvement in ductility over that with S=165 mm as well as an increase in the ultimate capacity with 32%. Comparing specimens, N1-S125-H300 and N1-S100-H300, both beams had one opening and the same beam height but it can be seen that reducing the spacing between the stirrups improved the behavior of the beam. The ultimate capacity was slightly higher but the ultimate angle of rotation was significantly improved almost close to that of the beam with no opening (N0-S125-H300).

Increasing the height of the beam from 300 mm to 400 mm caused an increase in the ultimate capacity of the beam by 27.7 % but not a very noticeable change in the angle of rotation is recorded as shown in Figure 10. On the whole, a reasonable agreement can be seen between the experimental and the analytical results.

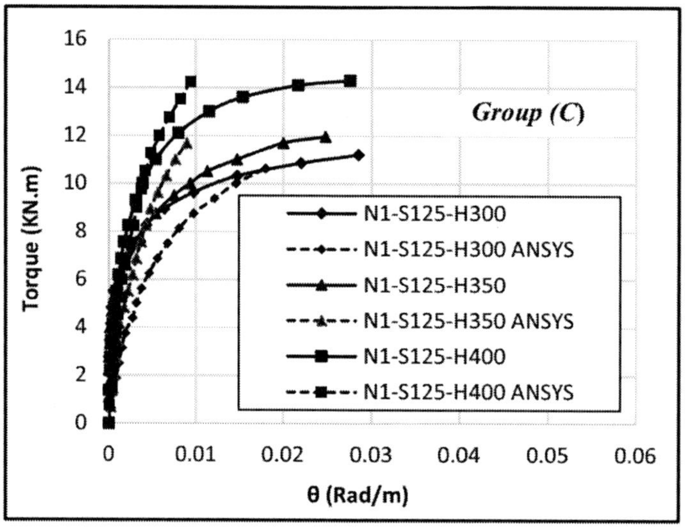

Fig. 10. Torque - rotation curve: Beam height (Group C).

4.3 Cracking and ultimate torque

The cracking and ultimate torques (T_{cr}, T_u) and the corresponding angles of rotation (Θ_{cr}, Θ_u) for each specimen as well as a comparison with the ultimate torque obtained from ANSYS are summarized in Table 2. It can be seen that the results of the FEM analysis and the output from the experiments agree well together. The difference between them does not exceed 11% which is below the acceptable value.

Table 2. Experimental and analytical results of the tested specimens

Group	Specimens	Experiment				ANSYS	$\frac{T_u \text{ Ansys}}{T_u \text{ Exp}}$ (%)
		T_{cr} (kN.m)	Θ_{cr} (rad/m)	T_u (kN.m)	Θ_u (rad/m)	T_u (kN.m)	
(A)	N0-S125-H300	12.375	0.0122	15.20	0.0658	15.00	98.68
	N1-S125-H300	7.975	0.0038	11.20	0.0285	10.63	94.86
	N2-S125-H300	7.150	0.0040	10.00	0.0367	9.38	93.75
(B)	N1-S100-H300	8.525	0.0046	12.90	0.0550	12.25	94.96
	N1-S125-H300	7.975	0.0038	11.20	0.0285	10.63	94.86
	N1-S165-H300	7.700	0.0035	9.76	0.0211	8.75	89.93
(C)	N1-S125-H300	7.975	0.0038	11.20	0.0285	10.63	94.86
	N1-S125-H350	9.215	0.0074	11.96	0.0248	11.69	97.72
	N1-S125-H400	9.763	0.0093	14.30	0.0276	14.25	99.65

Figure 11 shows the results of both cracking and ultimate torque for Group A which studies the effect of the number of openings. While, Figure 12 shows the results for Group B which studies the effect of the spacing between stirrups. Finally, Figure 13 shows the results for Group C which studies the effect of the height of the beam. As discussed previously in section 4.2, the three figures show that the three parameters have a significant effect on the behavior of the beams under pure torsion load.

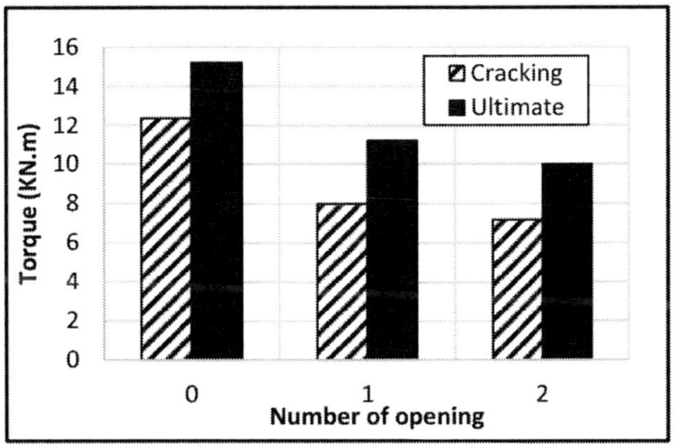

Fig. 11. Cracking and ultimate torque for Group (A)

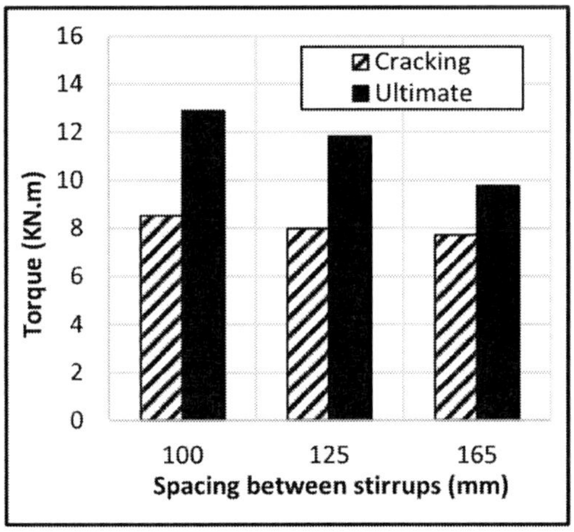

Fig. 12. Cracking and ultimate torque for Group (B)

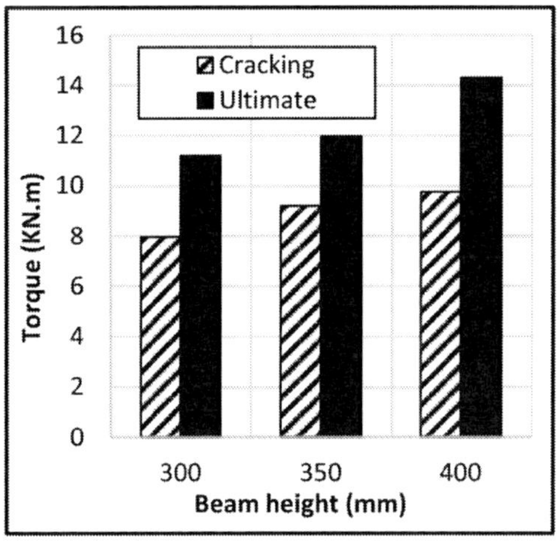

Fig. 13. Cracking and ultimate torque for Group (C)

5 Conclusions

Seven reinforced concrete beams were studied in this research. The beams were tested under pure torsion load and then verified using FEM analysis program ANSYS16. The outputs of the experimental program were compared and contrasted with each other as well as with the analytical results and the following conclusions were drawn:

1. The torque-rotation relationship for all the beams under study was linear up to the cracking torque and after that it became nonlinear. The slope of the curves after cracking was very small indicating reduction of beam stiffness after cracking.
2. When the number of openings was increased, the torsional capacity decreased and the ultimate angle of rotation increased. Compared to the control beam, the ultimate torque

for the beam having one opening and that with two openings was reduced by 26 % and 34 %, respectively. While, the cracking torque decreased by 35% and 42 %, respectively.

3. When the spacing between stirrups changed from 165 mm to 100 mm, the torsional capacity was increased by about 11% for the cracking torque and 32% for the ultimate torque.
4. When the beam height was increased by 25% (from 300 to 400 mm), the torsional capacity increased by about 22.4% for the cracking torque and by about 27.7% for the ultimate torque.
5. The ultimate angle of rotation increased by 40 % when the spacing between stirrups was decreased by 25%. However, almost no change was seen when the height was increased by 25%.
6. Based on the above, to increase the torsional capacity of a beam that has openings, structural engineers may increase the depth of the beam, and / or decrease the spacing between stirrups around opening. However, the increase in the beam height increases the ultimate torsion capacity with almost no change in the angle of rotation. While, decreasing the spacing between the stirrups cause a noticeable increase in the ultimate capacity but a significant improvement in the ultimate angle of rotation.
7. The analytical results obtained from (ANSYS 16) agree well with the experimental results in terms of crack pattern as well as the ultimate load values which showed differences of no more than about 11%.

Deep gratitude and special thanks are given to Prof. Dr. Magdy E. Kassem for his continuous support and help throughout this research. Without his help, this paper possible would not have been possible.

References

1. M. A. Mansur, S. K. Ting, S. L. Lee, J. Struct. Eng., **109(8)**, 1887-1902 (1983)
2. M. A. Mansur, S. K. Ting, S. L. Lee, J. Struct. Eng., **109(8)**, 1780-1791 (1983)
3. M. El-Badawy, "Torsional Behavior of RC Beams with Web Opening ", *MSc. Thesis, Faculty of Engineering, Ain-shames University, Cairo, Egypt* (2010)
4. A. Salama, "Behavior of Flange Concrete Beams with Web Opening Under Pure Torsion", MSc. Thesis, Faculty of Engineering, Cairo University, Giza, Egypt (2014)
5. A. A. AKHTARUZZAMAN, ACI Code Torsion Equations Modified for Rectangular Concrete Beams with an Opening, Engineering Sciences, **2(1)** (1990)
6. M. A. Mansur, K. H. Tan, "Concrete Beam with Opening Analysis and Design," U.S.: CRC Press LLC, (1999)
7. K. Fawzy, M. M Hashem, A. M. Elnady, Int. J. Eng. and Innovative Technology (IJEIT), **Vol. 4, Issue 1**, pp.197-204 (2014)
8. M. Venugopal, "Behavior of GFRP Retrofitted Rectangular RC Beams with Small Web Openings under Torsion", *Thesis, Department of Civil Engineering, National Institute of Technology, Rourkela, Odisha, India* (1990)
9. A. C. Institute, Building Code Requirements for Structural Concrete .manual of concrete practice, USA, Michigan: ACI Committee 318-08, (2008)
10. Egyptian Code for design and Construction of reinforced concrete structural, ECP 203-2007 ed., Egypt, (2007)
11. SAS IP, "ANSYS Theory Reference Release 5.6", ANSYS, Inc. Canonsburg, Pennsylvania, USA, (1999)
12. M. Mansur, A. Hasnat, ASCE Journal, **Vol. 105, No. STII**, pp. 2433-2447 (November 1979)

Repair of reinforced concrete beams using carbon fiber reinforced polymer

Abdul Saboor Karzad[1,*], *Salah* Al Toubat[1], *Mohammed* Maalej[1], and *Pierre* Estephane[2]

[1]Department of Civil & Environmental Engineering, Sustainable Construction Materials and Structural Systems Research Group, University of Sharjah,P.O.Box 27272, Sharjah, UAE
[2]GCP Applied Technologies, P.O.Box 5006, Dubai, UAE

Abstract. This research paper is part of an ongoing research on the behaviour of Reinforced Concrete (RC) beams retrofitted with Externally Bonded Carbon Fiber Reinforced Polymer (EB-CFRP). A total of 5 large-scale rectangular beams, previously damaged due to shear loading, were repaired and strengthened with EB-CFRP and tested in this study. The major cracks of the damaged beams were injected with epoxy and the beams were wrapped with 2 layers of EB-CFRP discrete strips with 100mm width and 150mm center to center spacing. The beams were instrumented and tested to failure under three points loading in simply supported configuration. The measured test parameters were the beams deflection, maximum load, and the strain in the FRP strips. The failure mode was also observed. The results showed that applying EB-FRP strips increased the shear strength significantly relative to the original shear capacity of the beam. The results demonstrate that the application of EB-FRP strips used in this study is an effective repair method that can be used to repair and strengthen damaged beams.

1 Introduction

Most of the Existing structures such as bridges and buildings are made of reinforced concrete (RC). RC is an economical and versatile construction material and in most cases, it is strong and durable. However, certain circumstances; for instance, poor design and construction, low-quality construction materials, design and construction according to old earthquake resistant design guidelines, environmental effects (e.g. corrosion, freeze-thaw, etc.), and other deteriorating factors can decrease the service life of the structures [1]–[3]. Therefore, repair and retrofitting of these deteriorated structures is essential to maintain and extend their designed service life [4], [5].

The traditional rehabilitation methods (i.e. adding new structural members, enlarging the existing section, bonding steel plates and etc.) are in line with several disadvantage such as difficult and time consuming application, and lack of durability [5]. To overcome these disadvantages, the application of EB-FRP sheets has emerged as an innovative rehabilitation technology in response to the increasing demand for retrofitting of existing structures[6]–[8]. The EB-FRP technology has several advantages over traditional rehabilitation techniques. The advantages are high tensile strength, noncorrosive,

* Corresponding Author akarzad@sharjah.ac.ae

nonmagnetic, lightweight, quick and easy installation and etc.[1]–[7], [9]–[13]. It is worth nothing that several experimental and analytical researches have been conducted since 1990s, to investigate the technique and behaviour of this new technology and establish its effectiveness [12]–[18]. Based on these studies some guidelines are developed [21]–[23].

EB-FRP repair and strengthening can be executed in three types of applications: confinement, flexural and shear strengthening[3], [19]–[21].The behaviour of the first two types of applications (i.e. confinement and flexural strengthening) is well understood however the shear strengthening using EB-FRP (i.e. full wrapping, U-jacketing, side bonding and near surface mounted (NSM) in vertical or inclined orientation) is still a matter of further research [12], [18], [22]–[24].

Reinforced concrete structures are designed such that flexural failure is the governed mode of failure and shear failure is prevented. To prevent the sudden failure (shear) and assure the ductile failure (flexural), the RC beams should acquire adequate shear strength[25]. The later requirement is the key impetus for conducting this research which is investigating the repair and strengthening of shear deficient beams using EB-CFRP.

2 Experimental program

2.1 Materials properties

The EB-CFRP discrete strips used in this study, are unidirectional woven carbon fiber flexible sheet. The CFRP sheet has a thickness (t_{frp}) of 0.167mm, a tensile modulus of 230GPa, tensile strength of 4.9GPa, and ultimate elongation of 2.1%. The CFRP sheets were encapsulated and bonded to the specimens using a compatible polymeric encapsulation resin through the wet application process of bonding.

Prior to wrapping the pre-damaged beams with EB-FRP, an appropriate epoxy (Denepox^{TM-40}) was used to inject the existing major cracks of the beams. Denepox is an ultra-low viscosity, 2-component epoxy resin for structural injections in concrete that after mixing both components cure into a rigid compound.

The concrete mix used for casting the beams had a 28th day compressive strength (fc') of 38.4 MPa according to ASTM C39. The beams longitudinal reinforcement had a yield strength of fy = 550 MPa, and the transverse reinforcement yield strength was fy = 290 MPa.

2.2 Test specimens and matrix

A total of 5 full-scale damaged RC beams (DB) in two types: with minimum shear reinforcement and without shear reinforcement, were repaired (RB) and retested. Each type of specimen consisted of pairs of identical slender pre-damaged beams tested in an earlier research work (Ardavan Yazdanbakhsh 2008). The beams are rectangular cross-section (390 x 230 mm) with effective depth of 330 mm, length of 2700 mm, span of 2300 mm, and shear span to depth ratio (a/d) equal to 3.5. Photos of typical specimens previously tested by Ardavan are shown in Fig.1 [26].

All tested beams were heavily reinforced in flexure by placing three longitudinal steel bars with a nominal diameter of 32mm. Heavy flexural reinforcement was used to ensure that the shear failure is the dominant mode of failure. In addition, two longitudinal steel bars with a nominal diameter of 10mm were provided in compression zone of the beams. Moreover, 2 out of 5 beams were reinforced in shear with minimum shear reinforcement ratio (according to ACI 318) using 6mm steel stirrups spaced at 200mm with a clear cover of 20mm.

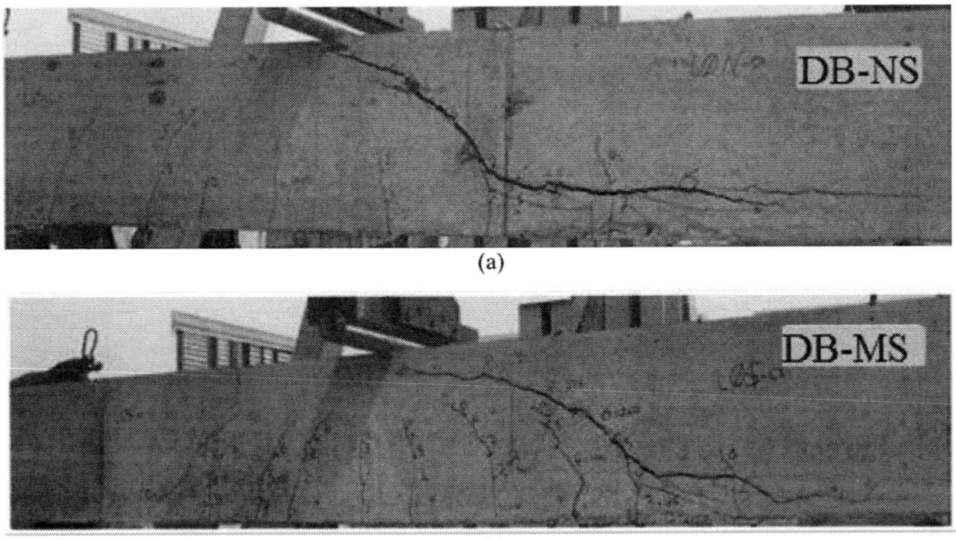

(a)

(b)

Fig. 1. Pre-damaged beams (source [26]), a) without stirrups, b) with minimum ratio of stirrups

The test matrix of the experimental program is summarized in Table 1. Each specimen is identified by a label that shows the condition of beams, ratio of conventional shear reinforcement, number of EB-CFRP layers, and ends with a small letter showing identity of duplicate beams. The label of damaged beams starts with two letters DB while the repaired beams starts with RB. For instance, the beam DB-NS-a means damaged beam (DB) without steel stirrups (NS), while DB-MS-a means damaged beam with minimum ratio of steel stirrups (MS) according to ACI-318. Similarly, specimen RB-NS-2L-a means Repaired Beam (RB) without stirrups (NS) and wrapped with 2 layers of EB-FRP (2L), and specimen RB-MS-2L-a means Repaired Beam with minimum ratio of steel stirrups and wrapped with 2 layers of EB-FRP (1L). While RB-NS-0L-c means Repaired beam with epoxy injection only and without stirrups.

Table 1. Test matrix.

Label	Condition	Test Objective
RB-NS-2L-a	Repaired / Wrapped	Extent of shear strength recovery using EB-FRP retrofitting.
RB-NS-2L-b	Repaired / Wrapped	Extent of shear strength recovery using EB-FRP retrofitting.
RB-MS-2L-a	Repaired / Wrapped	Extent of shear strength recovery using EB-FRP retrofitting.
RB-MS-2L-b	Repaired / Wrapped	Extent of shear strength recovery using EB-FRP retrofitting.
RB-NS-0L-c	Repaired / Unwrapped	Injected epoxy contribution to shear strength recovery.

 DOI: 10.1051/matecconf/201712001008

2.3 Repair methodology

Repair process of the beams shown in Fig.2, included steps which are: cleaning the major crack with air blower, sealing the cracks and attaching the packers with an appropriate mortar (Nitomortar FC^{TM}), injecting and filling the major cracks using epoxy (Denepox^{TM}), surface grinding with diamond grinder, and surface preparation with Nitomortar^{TM} and primer-coat (Nitowrap^{TM}), and finally hand lay-up application of U-shaped CFRP discrete strips (Nitowrap FRC 300^{TM}). It should be mentioned that before bonding CFRP strips to the specimens, the corners of the beams were chamfered to prevent stress concentration in the sharp bottom edges of the beams.

Four out of five beams were wrapped with 2 layers (2L) of EB-CFRP in U-jacketing scheme and one beam was repaired with epoxy injection only (0L). It is worth nothing that prior to bonding the CFRP strips, the pre-existing major cracks in the beams were injected with epoxy. The epoxy was injected using the specific pressure pump through the attached vessels (packers) on the surface of the specimens. The EB-CFRP strips bonded to the beams, had a width (wfrp) of 100 mm and a center to center spacing (sfrp) of 150 mm and a net spacing of 50mm as shown in Fig. 3.

(a) (b)

(c) (d)

Fig. 2. . a) Pre-damaged beam with attached packers on the main crack, b) Injecting epoxy using pressure pump, c) Beam ready for wrapping after surface grinding and primer coat application, d) Wrapping the beam with CFRP strips using compatible encapsulation resin.

2.4 Test setup and instrumentation

The test setup consisted of a simply supported loading configuration with roller supports and proper instrumentation. The specimens were tested under monotonic static loading in a displacement control mode with 0.01mm/sec loading rate. Meanwhile, strain development, cracking sequence and deflection of beams were monitored during testing. Deflection of the beams were measured using linear voltage displacement transducers (LVDTs) at three different points along the span (i.e. under the loading point and in the middle of both shear spans). Electrical resistance strain gauges were attached to longitudinal reinforcement and CFRP strips to measure the strain development. Fig. 3 shows the test setup, point of measuring the deflection and strain gauges locations. As shown in the Fig. 3, total of 4 external FRP strain gauges (FS1-FS4) were attached vertically to EB-CFRP strips, and one embedded flexural strain gauge (FLS5) was bonded horizontally to the surface of the middle longitudinal reinforcement bar.

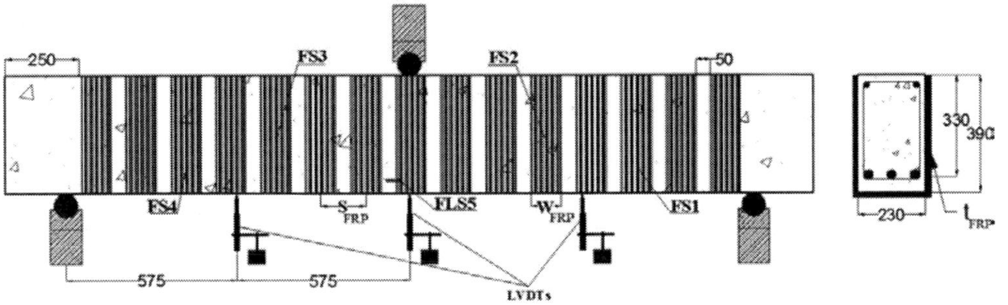

Fig. 3. Test setup and layout of the beam with EB-CFRP and instrumentations (all units are mm).

3 Experimental results and discussions

3.1 Failure modes

Fig. 4 shows photos of typical tested repaired specimens after failure. As can be seen in Fig.4, all specimens exhibited either full debonding or local debonding of CFRP sheets accompanied by separation of the concrete cover from the beams lateral faces. The beams without minimum stirrups failed in shear by delamination of EB-FRP followed by collapse. However, beams with minimum ratio of steel stirrups experienced a flexural failure followed by local debonding of EB-FRP. The delamination of CFRP strips in beams without stirrups, was quite sudden while in beams with minimum ratio of stirrups it was happening gradually and slower.

Specimen RB-MS-2L-a exhibited yielding of its longitudinal reinforcement recorded by embedded strain gauge attached to it, indicating flexural failure. As for the Specimen RB-MS-2L-b, the flexural failure was indicated by exhibiting longitudinal splitting cracks parallel to the compression zone steel reinforcement. It is worth noting that some of the repaired beams exhibited a change in the location of major shear crack after repairing. In other words, the major shear crack was switched from one shear span to the other shear span after repair of the pre-damaged beams. Furthermore, it can be seen in the photos that debonding occurred at the shear span that has experienced a wide diagonal shear crack. Whereas, the other shear span remained intact.

3.2 Load-deflection curves

The load-deflection curves of the reference beams and repaired beams are plotted in Fig. 5. As can be inferred from Fig. 5, repair and strengthening of damaged beams increased the initial stiffness of the beams and reduced the relative deformation at peak load compared to the reference beams which is in agreement with the findings in the literature [27]. The exception was the RB-MS-2L-a that exhibited reduction in stiffness when retested the reason for which was the level of damage existed in this particular beam (a chunk of concrete was missing at the bottom of existing shear crack before repair). The load-deflection diagrams also show that the repaired beams without steel stirrups exhibited very brittle failure by a sudden drop in load resisting. Which is also evident in striking the LVDTs after sudden release of energy as can be inferred from Figure 5a. Whereas, the repaired and strengthened beams that were also reinforced with steel stirrups exhibited relatively ductile behaviour shown in Figure 5b compared to the beams without stirrups.

(a)

(b)

Fig. 4. Repaired beams after failure (a) without Stirrups, (b) with minimum ratio of stirrups.

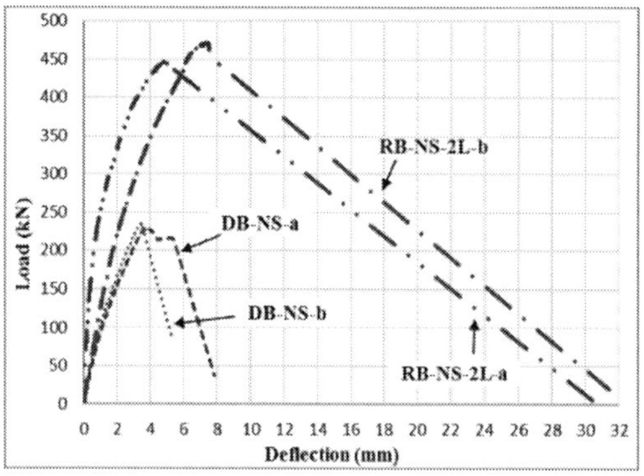

(a)

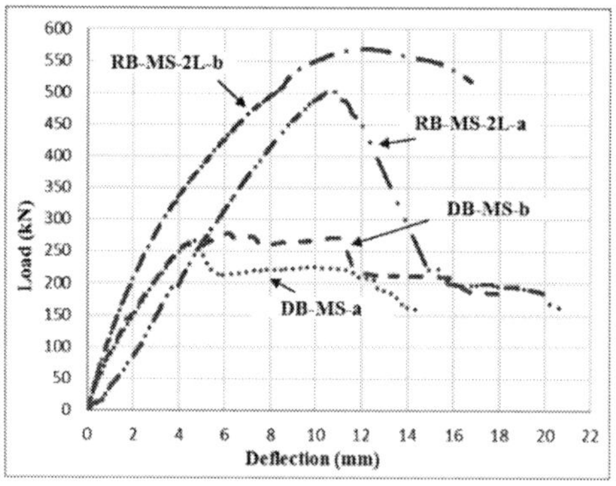

(b)

Fig. 5. Load Vs. Deflection of tested beams, a) Reference and repaired beams without steel stirrups, b) Reference and repaired beams with minimum ratio of steel stirrups.

3.3 Strain response

The maximum measured strain responses in CFRP strips are plotted against the average shear stress (Vu / bd, Vu=P/2) of the repaired beams in Fig. 6. The maximum measured CFRP strains plotted in Fig. 6, are not necessarily the absolute maximum CFRP strain as the strain in CFRP is affected by location of strain gauge relative to the location of crack. The EB-CFRP strips (in specimens namely RB-NS-2L-a, RB-NS-2L-b, and RB-MS-2L-a) did not experience significant strain until flexural shear crack initiated at around 2 MPa shear stress. The crack initiation is inferred by the change in the pattern of stress-strain curve as can be noticed in Fig. 6. Which indicates that the crack started to incline and propagate upward but intercepted by the EB-CFRP strips that started engaging in carrying the load. This process continued until the major shear crack was created and opened significantly resulting in debonding of the EB-CFRP and followed by collapse of the beam. However, the exception was the gauge response in specimen RB-MS-2L-b which shows that FRP started engaging in load resisting at around 2.5 MPa shear stress. This particular beam exhibited local FRP debonding and ended to highest contribution by EB-CFRP compared to the other beams. Moreover, it can be inferred from the curves in the Figure that the debonding was sudden in specimens without steel stirrups (sudden drop) while it was gradual in the specimens with minimum ratio of steel stirrups (dropping and resisting).

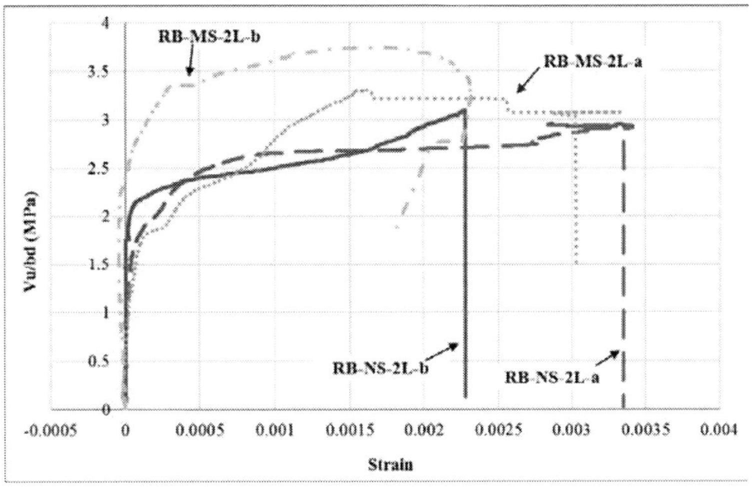

Fig. 6. Average shear stress Vs. maximum CFRP strain responses of repaired beams.

3.4 Shear Capacity

The original shear failure load (unstrengthen), as well as the failure load after repair (strengthened), the gain in shear capacity added to the corresponding unstrengthen beam capacity, and the failure mode of the tested beams are summarized in Table 2. It is worth mentioning that one of the damaged beams (namely DB-NS-0L-c) was repaired with epoxy injection only (i.e. RB-NS-0L-c). The reason for testing this particular specimen was to investigate the effect of injected epoxy on shear strength recovery of the repaired beams. From the comparison of test results, it was revealed that the original capacity of the damaged beams was reinstated by epoxy injection only. Indicating that the additional gain in shear capacity of the repaired beams with respect to their original shear strength, is the net contribution of EB-CFRP as tabulated in Table 2.

The contribution of EB-CFRP to the shear strength (VF), is calculated based on subtraction of the shear load resisted by concrete (Vc), and stirrups (Vs) from the total resisted load after repair as follows:

$$V_f = V_u - (V_C + V_S) \qquad (1)$$

From the results presented in Table 2, it can be found that repair and strengthening of damaged beams using 2 layers of CFRP, resulted in 95 % increase of the shear strength. It is worth noting here that the additional shear strength provided by EB-CFRP allowed two of the beams (namely RB-MS-2L-a and RB-MS-2L-b) to develop its full flexural capacity. The flexural failure was revealed either from yielding of flexural rebar (i.e. strain response) or the development of the longitudinal splitting cracks at the top surface of the beams parallel to the steel reinforcement.

Table 2. Test results.

Beam Type	**Ultimate Shear Capacity**		**Strength Gain Using FRP V_F,exp (kN)**	**FailureMode**
	Original Strength Vu,exp (Vc+Vs) (kN)	After Repair Vu,exp (Vc+Vs+V_f) (kN)		
RB-NS-2L-a	115	224.5	109.5	Shear + Debonding
RB-NS-2L-b	118	235.5	117.5	Shear + Debonding
RB-MS-2L-a	138.5	251.5	113	Flexural + Debonding
RB-MS-2L-b	139	284	145	Flexural
RB-NS-0L-c	100	102	NA	Shear

4. Conclusions

The aim of this study was to investigate the performance of EB-CFRP repair technique in shear strength recovery of shear deficient beams. Two types of repaired specimens were tested: beams without stirrups and beams with minimum steel stirrups. 4 beams were repaired and strengthened using EB-CFRP and epoxy injection while one beam was repaired with epoxy injection only. From the experimental results and analysis, it was revealed that repair of severely shear-damaged RC beams using simultaneous application of EB-CFRP and epoxy injection, not only reinstated but also significantly enhanced the original shear capacity of the beams. Based on the results and discussions the following conclusions are drawn;

- The main type of failure was debonding of EB-CFRP where the shear stresses were high and the major shear crack developed, that triggered the delamination of strips and ended by collapse.
- The additional shear strength provided by EB-CFRP system to the shear-deficient beams with minimum ratio of shear reinforcement, resulted in change of mode of failure from pure debonding to flexure failure followed by debonding.
- The use of 2 layers of EB-CFRP discrete strips with 100mm width and 150mm spacing resulted in approximately 95% increase in shear strength compared to the original capacity of the reference beams.

In summary, the aforementioned consequences indicate that this technology is an effective approach to improve the shear capacity of shear critical beams. This outcome suggests that EB-CFRP repair technique is a viable alternative for repair and strengthening of existing concrete structures in need of repair.

References

1. D. Baggio, K. Soudki, and M. Noël, "Strengthening of shear critical RC beams with various FRP systems," *Constr. Build. Mater.*, vol. **66**, pp. 634–644, (2014).
2. G. C. Manosr, K. Katakalos, M. Theofanous, E. Kozikopoulos, and S. Upgrade, "Experimental And Numerical Investigation Of Rectangular R / C Beams Retrofitted Against Shear Utilizing Frp Strips," no. pp. 12–14,June,(2013).
3. C. Pellegrino and M. Vasic, "Assessment of design procedures for the use of externally bonded FRP composites in shear strengthening of reinforced concrete beams," *Compos. Part B Eng.*, vol. **45**, no. 1, pp. 727–741, Feb. (2013).
4. A. A. Shash, "Repair of concrete beams – a case study," vol. **19**, pp. 75–79, (2005).
5. T. C. Triantafillou, "Shear Strengthening of Reinforced Concrete Beams Using Epoxy-Bonded FRP Composites," no. **95**, pp. 107–115, (1999).
6. O. Benjeddou, M. Ben Ouezdou, and A. Bedday, "Damaged RC beams repaired by bonding of CFRP laminates," vol. **21**, pp. 1301–1310, (2007).
7. J. F. Chen and J. G. Teng, "Shear capacity of FRP-strengthened RC beams: FRP debonding," *Constr. Build. Mater.*, vol. **17**, no. 1, pp. 27–41, (2003).
8. A. Bousselham and O. Chaallal, "Shear Strengthening Reinforced Concrete Beams with Fiber-Reinforced Polymer : Assessment of Influencing Parameters and Required Research," no. **101**, pp. 219–227, (2005).
9. C. Deniaud and J. J. R. Cheng, "Shear Behavior of Reinforced Concrete T-Beams with Externally Bonded Fiber-Reinforced Polymer Sheets," no. **98**, (2001).
10. H. Nikopour and M. Nehdi, "Shear repair of RC beams using epoxy injection and hybrid external FRP," pp. 1865–1877, (2011).
11. M. R. Islam, "Shear strengthening of RC deep beams using externally bonded FRP systems," vol. **27**, pp. 413–420, (2005).
12. I. a. Bukhari, R. L. Vollum, S. Ahmad, and J. Sagaseta, "Shear strengthening of reinforced concrete beams with CFRP," *Mag. Concr. Res.*, vol. **62**, no. 1, pp. 65–77, Jan. (2010).
13. A. Mofidi and O. Chaallal, "Tests and Design Provisions for Reinforced-Concrete Beams Strengthened in Shear Using FRP Sheets and Strips," *Int. J. Concr. Struct. Mater.*, vol. **8**, no. 2, pp. 117–128, (2014).
14. O. Chaallal, "Reinforced-concrete beams retrofi tted in shear," vol. **15**, no. 4, (2014).
15. S. Dirar, J. M. Lees, and C. T. Morley, "Precracked reinforced concrete t-beams repaired in shear with prestressed carbon fiber-reinforced polymer straps," *ACI Struct. J.*, vol. **110**, no. 5, pp. 855–865, (2013).
16. X. Z. Lu, J. F. Chen, L. P. Ye, J. G. Teng, and J. M. Rotter, "RC beams shear-strengthened with FRP: Stress distributions in the FRP reinforcement," *Constr. Build. Mater.*, vol. **23**, no. 4, pp. 1544–1554, Apr. (2009).
17. A. Mofidi, O. Chaallal, and M. Asce, "Shear Strengthening of RC Beams with Externally Bonded FRP Composites : Effect of Strip-Width-to-Strip-Spacing Ratio," no. pp. 732–743, October,(2011).
18. C. Pellegrino and C. Modena, "Fiber-reinforced polymer shear strengthening of reinforced concrete beams: Experimental study and analytical modeling," *ACI Struct. J.*, vol. **103**, no. 5, pp. 720–728, (2006).
19. N. Banthia, L. Bisby, R. Britton, R. Cheng, G. Fallis, R. Hutchinson, A. Mufti, K. W. Neale, J. Newhook, K. Soudki, and L. Wegner, "An Introduction to FRP Strengthening of Concrete Structures Composites for Construction," (2004).
20. A. C. I. Committee, Guide for the Design and Construction of Externally Bonded FRP Systems, (2008).
21. Fib bulliton 14 and T. Group 9.3, Externally bonded FRP reinforcement for RC

structures, (2001).
22. A. Bousselham and O. Chaallal, "Behavior of Reinforced Concrete T-Beams Strengthened in Shear with Carbon Fiber-Reinforced Polymer — An Experimental Study," no. **103**, pp. 339–347, (2006).
23. C. Pellegrino and C. Modena, "Fiber Reinforced Polymer Shear Strengthening of Reinforced Concrete Beams with Transverse Steel Reinforcement," *J. Compos. Constr.*, vol. **6**, no. 2, pp. 104–111, May (2002).
24. A. Mofidi and O. Chaallal, "Shear Strengthening of RC Beams with EB FRP: Influencing Factors and Conceptual Debonding Model," *J. Compos. Constr.*, vol. **15**, no. 1, pp. 62–74, (2011).
25. J. Sim, G. Kim, C. Park, and M. Ju, "Shear Strengthening Effects with Varying Types of FRP Materials and Strengthening Methods," pp. 1665–1680.
26. Yazdanbakhsh A, "Shear Behavior of Synthetic Fiber Reinforced Concrete Beams," MSc Thesis, Univrsity of Sharjah, UAE,(2008)
27. M. Ekenel and J. J. Myers, "Durability performance of RC beams strengthened with epoxy injection and CFRP fabrics," *Constr. Build. Mater.*, vol. **21**, no. 6, pp. 1182–1190, (2007).

Performance of wire rope isolators in the seismic protection of equipment

Moussa Leblouba[1,*] and *Samer* Barakat[1]

[1]University of Sharjah, Department of Civil & Environmental Engineering, P.O.BOX 27272, University City, Sharjah, United Arab Emirates

Abstract. Different isolation systems are currently employed to protect sensitive equipment and their hosting structures from undesirable vibrations originating from earthquake ground motions. Wire rope isolator (WRI), a type of passive isolator is known to be effective in isolating the vibrations and shock, can be used for vibration isolation of structures and equipment. The aim of the present work is to study the performance of WRIs in mitigating the vibrational response of equipment mounted on the upper floor of a multi-story structure using numerical simulations. In multistoried buildings, the upper floors experience significantly amplified accelerations than the lower floors and hence the sensitive equipment requires an effective vibration isolation system. The present study shows that the WRI can provide an effective isolation by reducing the equipment acceleration through its lateral flexibility while satisfying the displacement demand. The Performance of the WRI was assessed considering six different earthquake excitations in a series of numerical simulations. The influence of flexibility of the structure on the performance of the WRI is also reported.

1 Introduction

The vibrations from the sources such as an earthquake or due to operations of heavy machineries can affect the functionality and cause structural damage to the structures and surrounding equipment. This signifies the application of vibration isolators to isolate the system from the vibration environment [1-4]. In general, the mitigation of vibrational response of the equipment and structures can be performed through three approaches [5, 6]: (1) Isolation of the entire housing structure, (2) isolation of single equipment, (3) isolation of raised-floor system. The first approach is suitable for new construction where isolators are installed between the structure and foundation to reduce the damage to both primary structure and equipment [7]. The second approach concentrates only on the equipment; the isolation system is inserted at the base of the equipment to reduce the transmission of vibration. However, both the approaches are confronted with a difficulty to achieve the desired long period within an affordable displacement [8].

* Corresponding author: mleblouba@sharjah.ac.ae

In the third approach, the equipment is fixed to a secondary floor, which decouples from the structural floor by an isolation system consisting of several isolators. This approach provides the potential benefit to both new and existing constructions. Raised floor approach also possesses advantages such as low cost when compared with first approach and it can also provide long period while maintaining the affordable displacements [3, 9]. Therefore, raised floor approach is well-established as an effective technique for the vibration mitigation of equipment [6], hence, the present work focuses on the third approach. Different types of passive isolation devices are used and they can be broadly categorized as linear and non-linear isolators. Linear vibration isolators are useful only if the excitation frequencies are higher than their natural frequency. Hence, their application is limited to moderate environmental disturbance [2]. This limitation can be overcome by using non-linear vibration isolators [2]. Several researchers have developed different types of nonlinear vibration isolators and have investigated their unique dynamic behaviors. A detailed survey on the developments of nonlinear vibration isolators is provided by Ibrahim [2], in which many cited studies [10-11] reveals that the introduction of nonlinear damping and nonlinear stiffness is of great benefit in vibration isolation. Recently, Wire Rope Isolators (WRI), a type of passive vibration isolation which exhibits non-linear behavior, has become the subject of intensive studies [12-17].

2 Modeling of WRI in shear

The geometrical characteristics and specification of the WRI used in this study are shown in Figure 1 and Table. 1 respectively. The test results showed that the WRI exhibits non-linear hysteresis behavior when subjected cyclic shear loading (Figure 2). This type of behavior is simulated using the Bouc-Wen model of hysteresis [8, 12, 18-19]. The Bouc-Wen model is the first-order nonlinear differential algebraic equation, which hysterically relates the input displacement and the output restoring force, its general form is given by:

$$F = \alpha \frac{F_y}{x_y} x + (1-\alpha) F_y z \tag{1}$$

where F is the restoring force, x is the displacement, F_y is the yield force, x_y is the yield displacement and z is a dimensionless hysteresis quantity in the range of [-1, +1] and is given by:

$$x_y \dot{z} = A\dot{x} - \{\beta |\dot{x}| z |z|^{n-1} + \gamma \dot{x} |z|^{n}\} \tag{2}$$

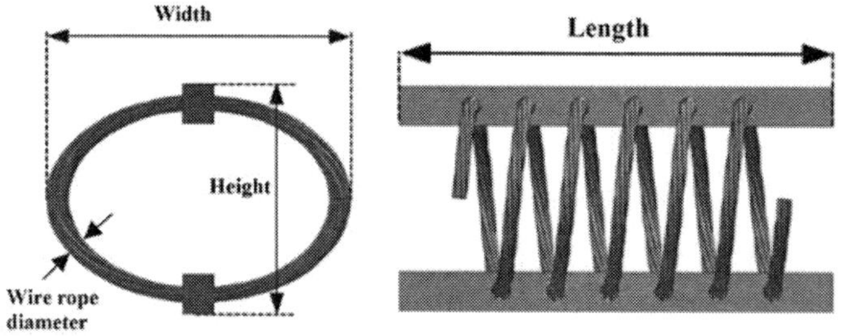

Fig. 1. Geometry of the wire rope isolator.

The dimensionless parameters A, β, γ, and n control the shape of the hysteresis loop. The above hysteresis model is implemented in SAP2000® program as the Plastic Wen NL-Link element, which is recommended for 2D static and dynamic analysis. The parameters of the Bouc-Wen model were used to mathematically define the behavior of the WRI in the simulation model to determine the response. The parameters of the Bouc-Wen model were determined using the Least-Squares method (Table. 1). Figure 2 shows that the Bouc-Wen model with the identified parameters can predict, with enough accuracy, the experimental force-displacement relationship.

Table 1. Geometrical characteristics and Bouc-wen parameters of the WRI used in the present study.

Geometrical characteristics	
Wire rope Diameter	9.5 mm
Length of WRI	215 mm
Width of WRI	105 mm
Height of WRI	76 mm
Number of Turns	8
Bouc-Wen Parameters	
Loading direction	Lateral
x_y	0.433mm
F_y	151.42 N
α	0.127

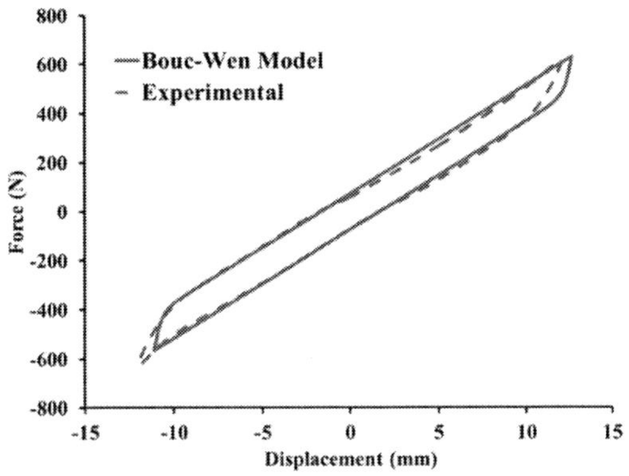

Fig. 2. Force-displacement curve of the WRI and its simulation using Bouc-Wen model.

3 Case study buildings

To assess the performance of wire rope isolators in protecting equipment against earthquake induced vibrations, we consider the eight-story structure shown in Figure 3, which is taken from Ismail *et al.* [6]. The equipment is mounted on a raised floor, which is connected to the 7th floor through a wire rope isolator system. At this level, it is expected to have a strong amplification of seismic acceleration with strong component at long periods [6]. In this study, 36 WRIs were used to isolate the raised floor and equipment. The study has been conducted for the two cases: case (1) equipment fixed to the hosting floor (Figure 4(a)) and case (2) Equipment is mounted on an isolated raised floor (Figure 4(b)). The hosting structure is a reinforced concrete Moment Resisting Frame (MRF) modeled as a linear shear building. The nonlinearity is concentrated only at the isolation system level. For all structural vibration modes, a 2% of critical damping is considered.

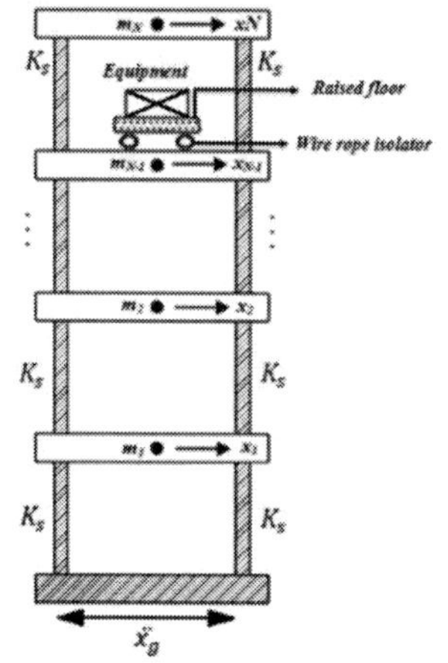

Fig. 3. Idealized structure with equipment mounted on the upper floor through raised floor approach.

The raised floor is modeled as rigid in the horizontal direction while the equipment is modeled as a linear single-degree of freedom system (SDOF) system with mass m_e, stiffness k_e and damping c_e. The equipment's damping is equal to the structural damping. The effective isolation can be achieved by increasing the period of the isolated equipment to large values [8]. Accordingly, the equipment is designed to achieve a fundamental period of 0.03775 s when fixed and 1.35 s when isolated. Consistent with the work done by Ismail *et al.* [6], three example structures having different lateral stiffness matrices were also considered to study the influence of flexibility of the structure.

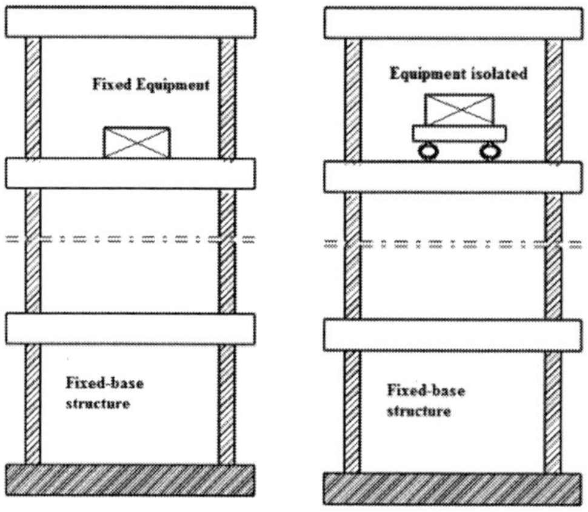

Fig. 4. (a) The case of fixed-base equipment; (b) the case of equipment isolated from the structure through WRI using raised floor approach.

4 Numerical simulation models

Nonlinear earthquake response history analyses (NLRHA) were performed for the structure and equipment using the structural analysis program SAP2000®. The super structure was modeled as a linear shear building with different lateral stiffness matrices (0.5 K_s, K_s, and K_s2) to study the influence of the superstructure flexibility on the performance of the seismic isolation system. A 2% damping ratio is assumed for all structural modes. The WRI is modeled in SAP2000® as a nonlinear link (NL-Link) element having a behavior governed by the Bouc-Wen model of hysteresis (Plastic Wen) (see Section 2). The properties of the selected WRI along with its Bouc-Wen parameters were given in Table. 1. The simulation models were subjected to six different earthquake records. These records have been categorized based on the peak ground acceleration (PGA) to Peak ground velocity (PGV) ratio [20] as shown in Table 2. The NLRHA was performed for the two cases (fixed-base equipment and base-isolated equipment). The reinforced concrete MRF superstructure is an eight-story building having each story height and bay width, 3000 mm x 5000 mm respectively. The cross sections of the columns and beams were modeled as 1000 mm x 250 mm and 750 mm x 250 mm, respectively.

Table 2. Earthquake records used in the present study.

Record#	Earthquakes	PGA (g)	PGA/PGV ratio	Class
1	San Francisco California , 1957	0.085	1.67	High
2	Banja Luka Yugoslavia,1981	0.074	2.31	High
3	Kern County California,1952	0.179	1.01	Intermediate
4	Mexico City,1985	0.123	1.17	Intermediate
5	San Fernando California,1971	0.132	0.61	Low
6	Near E. Coast of Honshu, Japan,1968	0.226	0.68	Low

5 Results and discussion

Considering first the case of the superstructure with a lateral stiffness K_S and an equipment having a mass of 1% of the total structural mass. Figure 5 compares the equipment's peak acceleration and peak relative displacement for the two cases considered in this study. Figure 5(a) demonstrates the effectiveness of the WRI in reducing the equipment acceleration; the WRI isolation system can reduce up to 55% of the equipment acceleration on an average for all earthquake excitations. This reduction is mainly due to the reduced base shear force transmitted from WRI to the raised floor. The reduction in shear force is due to the decoupling offered by the nonlinear behavior of the WRI. A maximum reduction of 60 % was observed for the Coast of Honshu 1968 earthquake record, which is categorized as high level.

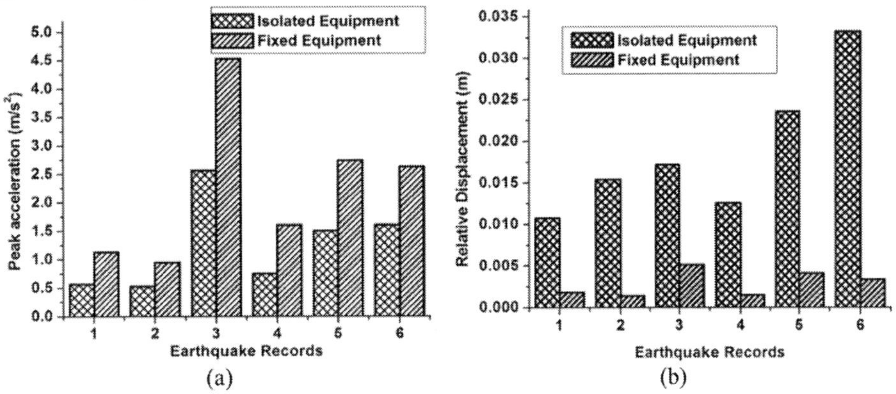

Fig. 5. (a) Peak acceleration and (b) and peak relative displacement response of the fixed-base and base-isolated structures.

The decrease in acceleration is achieved at the expense of increased lateral displacement. Figure 5(b) shows the equipment's peak lateral displacement relative to the hosting floor under each earthquake excitation. This is a property of most passive isolation systems; the isolation system decouples the equipment from the structure resulting in a much larger lateral displacement. A maximum displacement of 35 mm was observed for the Honshu, Japan, 1968 earthquake, which can be due to the strong motion at lower frequencies close to the isolated equipment's frequency.

Figure 6 compares the relative displacement time history (between the equipment and the hosting floor) for the case of Mexico City 1985 earthquake. The figure shows that the WRI provides 46% reduction in the acceleration while maintaining an acceptable lateral displacement.

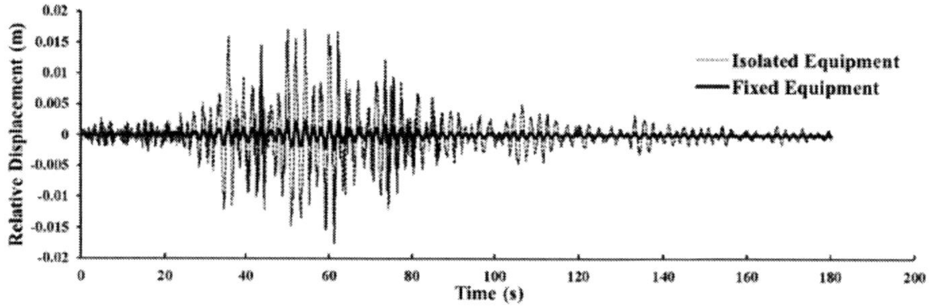

Fig. 6. Relative displacement time history of the equipment of the fixed and isolated equipment under Mexico City earthquake.

The influence of flexibility of the structure on the WRI performance is studied by simulating the response of the three structures having a lateral stiffness of $0.5K_s$, K_s and 2 K_s subjected to harmonic base excitations having a magnitude of $0.5g$ and different frequencies ranging from 1 Hz to 20 Hz.

The peak accelerations ratio between the fixed and isolated equipment under different harmonic frequencies are shown in Figure 7. The WRI was effective in reducing the acceleration even under harmonics having frequencies close to the structural dominant frequencies, which are 2.796 Hz for the $0.5K_s$, 3.664 Hz for the K_s, and 5.203 Hz for the $2K_s$. For different harmonic base excitations, larger displacements of the equipment are observed in the most flexible structure (i.e., the $0.5K_s$) (Figure 8). However, the WRI system could maintain the displacement of the equipment within an acceptable range for all excitation frequencies.

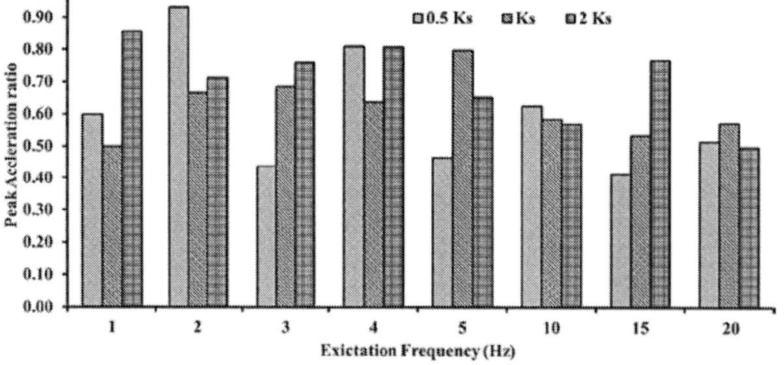

Fig. 7. Peak acceleration response under different harmonic base excitations.

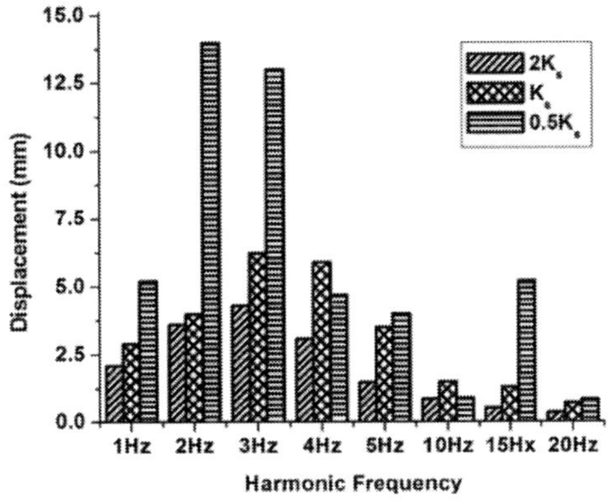

Fig. 8. Displacement response under different harmonic base excitations.

5 Conclusions

The present work investigated the performance of wire rope isolators (WRIs) in isolating equipment mounted on the upper floor of a moment resisting frame structure. It was shown that the WRI is effective in reducing the equipment acceleration, hence, minimizing structural damages. However, the lateral displacement of the equipment is larger in this case. In addition, the following conclusions are drawn:

1. WRI reduced the acceleration of the equipment by 55% on an average and a maximum of 60 % reduction was observed for Honshu, Japan 1968.
2. The maximum displacement demand among the applied earthquake excitations is 35 mm. Furthermore, the present simulation model also provides a way to determine the displacement demand required for the application and hence suitable WRI can be selected to provide the required displacement demand.
3. The WRI isolation system is effective and practically useful for different structures having different properties.

This study was partly funded by the SEED fund by the University of Sharjah. This support is highly appreciated.

References

1. J. Alanoly, S. Sankar, J. Mec Des-T ASME **109**, 2 (1987)
2. R.A. Ibrahim, J. Sound Vib. **314**, 3 (2008)
3. M. Ismail, J. Rodellar, F. Ikhouane, Struct Control Htlth **19**, 1 (2012)
4. Rivin, Eugene I., *Passive vibration isolation*, New York, ASME press (2003)
5. M. Hamidi, M.H. El Naggar, Earthq. Eng. Struct Dyn. **36**, 11 (2007)
6. M. Ismail, J. Rodellar, F. Ikhouane, J. Sound Vib. **326**, 3 (2009)
7. F. Naeim, J.M. Kelly, *Design of Seismic Isolated Structures: From Theory to Pra*ctice, Wiley, (1999)
8. G.F. Demetriades, M.C. Constantinou, A.M. Reinhorn, Eng. Struct **15**, 5 (1993)
9. M. Ismail, J. Rodellar, G. Carusone, M. Domaneschi, L. Martinelli, Acta Mech. **224**, 3 (2012)
10. D. Xu, Q. Yu, J. Zhou, S.R. Bishop, J. Sound Vib. **332**, 14 (2013)
11. B. Ravindra, A.K. Mallik, J. Sound Vib. **170**, 3 (1994)

12. P. Paolacci, R. Giannini, "*Study of The Effectiveness of Steel Cable Dampers for The Seismic Protection of Electrical Equipment*," in Proc. 14th World Conference on Earthquake Engineering, Beijing, China (2008)
13. G. Massa, S. Pagano, E. Rocca, S. Strano, Meccanica **48**, 7 (2013)
14. P. S. Balaji, M. Leblouba, Muhammad E. R., L. H. Ho, Mech Based Des. Struc. **44**, 4 (2016)
15. P. S. Balaji, M. Leblouba, Muhammad E. R., L. H. Ho, J Mech Sci Technol. **30**, 1 (2016)
16. P. S. Balaji, M. Leblouba, Muhammad E. R., L. T. Vuia, J Mech Sci Technol. **29**, 4 (2015)
17. P. S. Balaji, M. Leblouba, Muhammad E. R., L. H. Ho, IOP Conf. Ser.: Mater. Sci. Eng.. **78**, 1 (2015)
18. M. Leblouba, S. Altoubat, Muhammad E. R., P. S. Balaji, Shock Vib. **2015**, ID 482063 (2015)
19. Y.K. Wen, J. App. Mech. **47**, 1 (1980)
20. N. Naumoski, A.C. Heidebrecht, W.K. Tso, M.U.E.E.R. Group, "A Selection of Representative Strong Motion Earthquake Records Having Different A/V Ratios," McMaster University, Earthquake Engineering Research Group (1988)

Artificial earthquake record generation using cascade neural network

Khaldoon A. Bani-Hani[1,*], *and Mu'ath* I. Abu Qamar[2]

[1]Civil Engineering, Jordan University of Science & Technology, Irbid, Jordan
[2]Civil Engineering, Yarmouk University, Irbid, Jordan.

Abstract. This paper presents the results of using artificial neural networks (ANN) in an inverse mapping problem for earthquake accelerograms generation. This study comprises of two parts: 1-D site response analysis; performed for Dubai Emirate at UAE, where eight earthquakes records are selected and spectral matching are performed to match Dubai response spectrum using SeismoMatch software. Site classification of Dubai soil is being considered for two classes C and D based on shear wave velocity of soil profiles. Amplifications factors are estimated to quantify Dubai soil effect. Dubai's design response spectra are developed for site classes C & D according to International Buildings Code (IBC -2012). In the second part, ANN is employed to solve inverse mapping problem to generate time history earthquake record. Thirty earthquakes records and their design response spectrum with 5% damping are used to train two cascade forward backward neural networks (ANN1, ANN2). ANN1 is trained to map the design response spectrum to time history and ANN2 is trained to map time history records to the design response spectrum. Generalized time history earthquake records are generated using ANN1 for Dubai's site classes C and D, and ANN2 is used to evaluate the performance of ANN1.

1 Introduction

Design of structures to withstand seismic activity is mostly depends the results of seismic hazard analysis and site response analysis [1]. This analysis is conducted on sites of interest by calculating and measuring the factors that affect the study site conditions by the seismic waves that initiated at the bedrock level [2]. In general, the seismic activity originated on bedrock is amplified due to the soil geology and soil profile. This might produce divesting seismic waves with significant impacts on structures. The site investigation and seismic hazard analysis are the most critical phases in earthquake engineering design and analysis [3, 4].

UAE has minor seismic activity and few of them were recorded, however, the neighbouring areas are described as active seismic areas, such as Iran, Afghanistan and Oman. The importance of considering the seismic design in UAE is increasing due to the

* Corresponding author: khaldoon@just.edu.jo

remarkable expansion and investment in huge infrastructure and mega constructions like the Palm Island, Jumeira, and Burj Khalifa. For this reason UAE municipality and structural engineers in UAE starts to include the seismic design part in their projects and monitoring and recording of seismic activities.

Municipality of Dubai, for example, provides the construction sites by reports of soil investigation that were performed at selected sites to be utilized for seismic hazards analysis. In this study, the soil column profile and boreholes reports, for Dubai, are adopted from recent work by Irfan [1]. Around 36 boreholes from Dubai area in this study were most of the boreholes depths stretched between 15 m to 30 m.

This paper presents an inverse mapping problems solution. Most engineering problems are fundamentally inverse problems. The inverse problems is to determine the input from the output of a system or to determine the system characteristics from the measured input and output of the system (system identification). Artificial Neural Networks are employed in this study to perform inverse mapping problem to obtain generalized time history for design response spectra using ANN.

2 Methodology

As mentioned above, soil conditions have major effect on seismic waves propagated from bedrock. The soil characteristics are responsible of earth vibration differences for different sites even though the source is the same at the bedrock. The process of analysing the waves at this level is called the site response analysis [3]. The response characteristics of specific site undergone site response analysis mainly depends on the soil column geology. Therefore, results' variation are expected for site response analysis from one site to another. This result suggests conducting a site-specific response analysis for each structures to be constructed. To do so, time, money and efforts would be compromise and it is really would be unpractical to be carried out for each site under study. To simplify the work with reliable and affordable method, building codes introduce 'site amplification factors' that can replace and account for the amplification or de-amplification of the seismic waves due to site geotechnical properties. The National Earthquake Hazard Reduction Program [5] presents maps for North American region that give an estimate to the site amplification factors

Strong ground motions records are essential to execute a comparable linear and nonlinear dynamic analysis of critical structures. The response of the structures under several time history records for different earthquakes are estimated using DeepSoil [6]. The ground time histories motion are tuned to match the target response spectrum for specific site with particular site condition. Probabilistic Seismic Hazard Analysis (PSHA) is a stochastic method that can produce the target response spectrum for this goal.

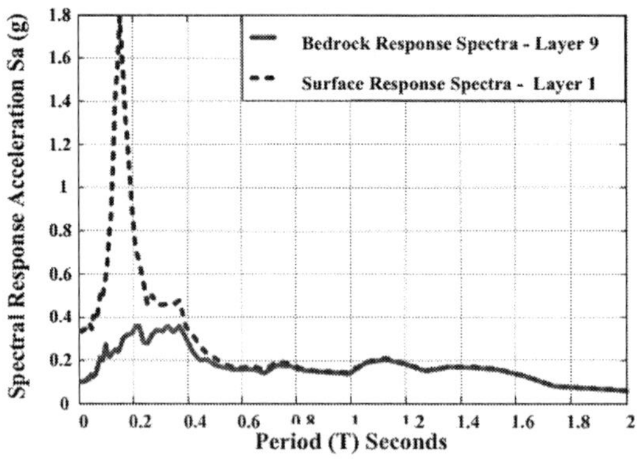

Fig. 1. A comparison between response spectra at surface and bedrock to quantify local site effect.

The site response analysis is usually conducted for particular sites using the estimated ground motions at the bedrock level. The amplification or de-amplification factors can be measured as the ratio between the bedrock spectrum and the estimated surface response spectrum as shown in Figure 1 and Figure 3.

2.1 Spectral matching

Spectral matching is to tune the selected records to match target response spectrum. This was conducted using SeismoMatch [7]; a software available freely for research The spectral matching conducted in the time domain where the source record is calibrated to be compatible with the site-specific response spectrum [8]. Spectral matching has been under research for several years; several approaches are proposed and discussed [4]. Some of these methods perform the matching in time domain and others do the work in frequency domain. Some current software that are commonly used to perform this such as EZFRISK, RSP Match EDT and SeismoMatch use the time domain method to adjust the accelerograms.

To achieve spectral matching two inputs are required: (a) Target response spectrum computed by seismic hazard analysis. This spectrum is referred to as target because the generated accelerograms are tuned and adjusted to have a response spectrum compatible with this response spectrum. (b) A number of accelerograms chosen carefully to be tuned and customized to the target response spectrum. Eight historical records were selected for the Dubai study area. The response spectrum of these accelerograms were generated and matched to the target spectrum to find the best record that provide the most compatible response spectrum concerning spectrum shape and confidence in de-aggregation results [1].

The eight time history records selected here are available in PEER (Pacific Earthquake Engineering Research). Dubai's target response spectrum was defined and SeismoMatch v.2.1 was utilized to generate the required matched records. The historical records names and corresponding information used for matching are given in Table 1.

Table 1. Time histories selected for spectral matching Magnitude (Source PEER).

Earthquake	PGA (g)	Distance (km)	Mag.
Chi-Chi, Taiwan-02, 17:57, 09-19-79	0.1173	27	5.90
Morgan Hill, 1984-04-24, 21:15	0.1017	27	6.19
Irpinia, Italy-02 1980-11-23 19:35	0.1062	23	6.20
Coalinga-01 1983-05-02 23:42	0.1127	29	6.36
N. Palm Springs 1986-07-08 09:20	0.1056	39	6.06
Whittier Narrows-01 1987-10-01 14:42	0.1126	30	5.99
Griva, Greece 1990-12-21 06:58	0.1038	33	6.10
Chi-Chi, Taiwan-06 1999-09-25 23:52	0.1190	26	6.30

Acceptable spectral matching results was achieved with the eight ground motion records for Dubai city. The target response spectrum and the generated matched spectrum are in perfect agreement and have relatively closer trends as depicted in Figure 2.

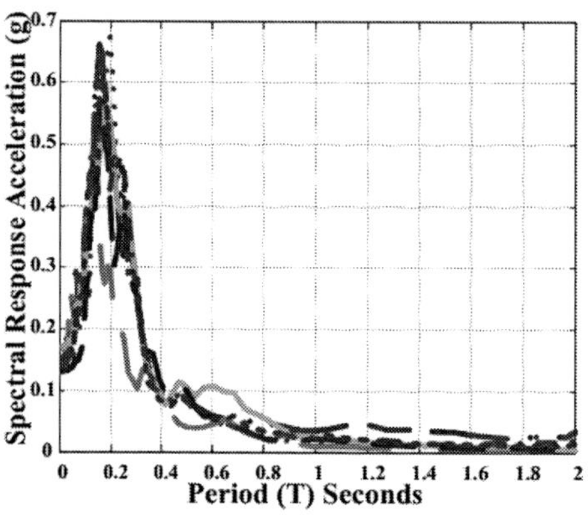

Fig. 2. Matching response of selected records on Dubai response spectrum.

2.2 Site response analysis

One dimensional linear site response analysis was conducted for 36 boreholes from Dubai city using DeepSoil software. Considering the spatial distribution sites for Dubai, the borehole logs were chosen for site analysis [1]. The propagated ground motion through soil profiles defined using the boreholes data in Dubai, are used to find the response of soils and their effect such as amplification factors. Site amplification factors were estimated following the recommendations of IBC2012, where the spectral response at 0.2 and 1s periods are required as shown in Figure 3. Finding the ratio between two response spectra, Fa and Fv can be calculated for all sites used in this study. To show the effect of soil profile on seismic waves propagating from the bedrock, such as response spectra comparison between bedrock and surface spectra, PGA of input record through soil column, and shear wave velocity through soil column. These effects are presented in Figure 3.

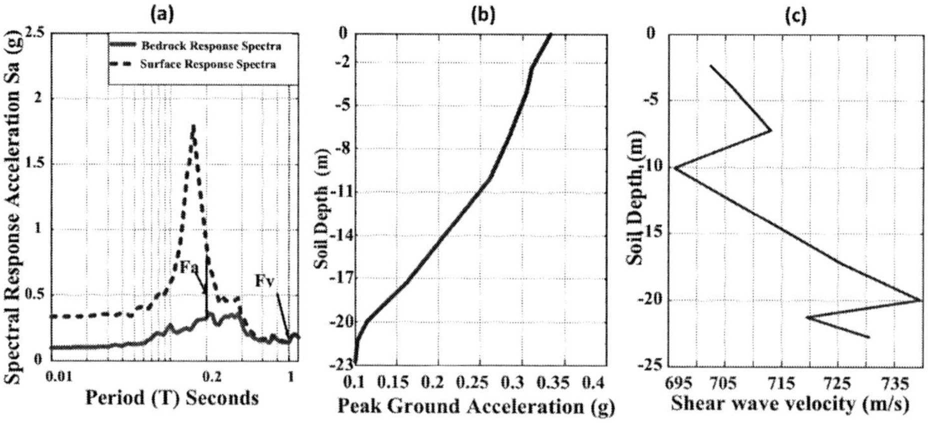

Fig. 3. (a) Response spectra on surface and half space, (b) effect of soil on PGA of input earthquake record, (c) Shear wave velocity (m/sec) Versus Depth (m).

The amplification factors are calculated for each time history and the average value was determined as shown in Table 2. Additionally, the response spectrum of each matched records are computed and the average spectrum is estimated to compute the corresponding short (0.2s) and 1-second periods (S_s and S_1). Figure 4 and 5 show how the S_s and S_1 for the records are estimated for the two soil classes.

Table 2. Amplification factors based on selected time histories for Dubai.

Amplification Factor	F_a	Fv
DeepSoil – Class D	1.390	4.350
DeepSoil– Class C	0.961	1.812

Based on the response spectrum generated and the estimated S_s and S_1, the short periods (S_{MS}) and at 1 second (S_{M1}) are computed for site class C and D using IBC 2012 recommendations. The final design response spectrum for sites that classified as "C & D" at Dubai city is depicted on Figure 4 and 5.

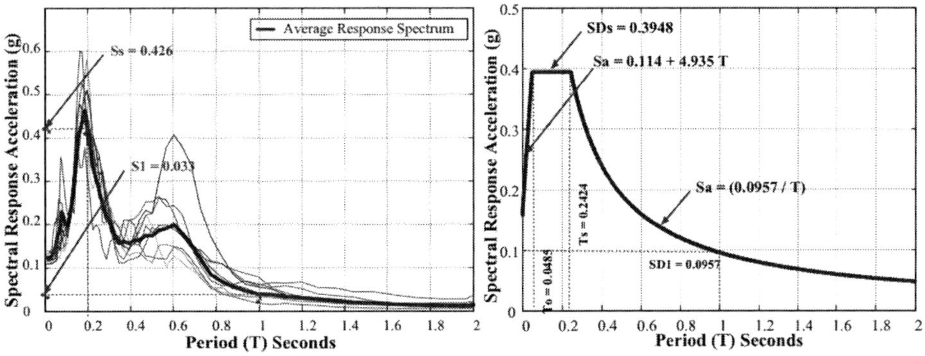

Fig. 4. Average response spectrum and the corresponding design response spectrum for Dubai – Site Class C.

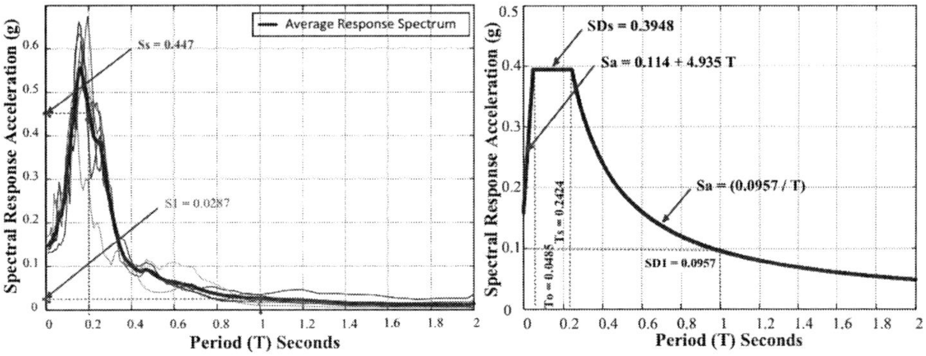

Fig. 5. Average response spectrum and the corresponding design response spectrum for Dubai – Site Class D.

2.3 Generation of artificial earthquake records using (ANN)

Construction of an artificial neural network capable of producing a reasonable artificial accelerogram is the theme of this research. Artificial neural network is an intelligent system that realizes the general relation between set of inputs and outputs points through training process. The idea of using artificial neural network is inspired by the nerve cells in the human brain. The evolution of ANN modelling go through three stages: training, validation, and testing. Neural network training mechanism is shown in the Figure 6.

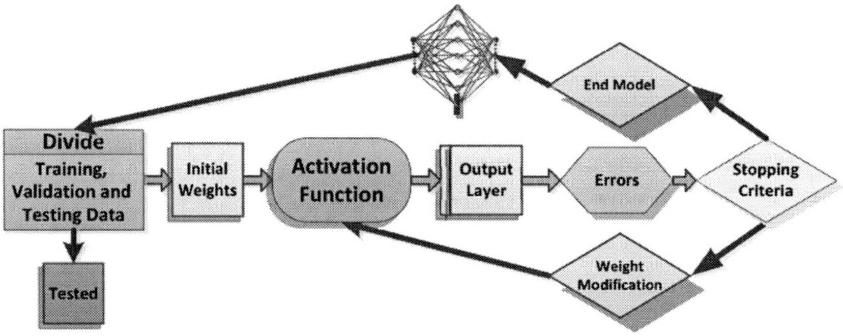

Fig. 6. Schematic chart showing the neural network training mechanism.

To train ANN that can learn the inverse problem between the design response spectrum and the corresponding accelerograms, thirty historically novel records are utilized and verified. The time histories records employed in the training both neural networks are described in Table 3. The training data were prepared by discretizing the accelerograms at 0.02 seconds with strong motion durations of 10 seconds. To control the duration for all accelerograms, points with zero amplitude were embedded to all accelerograms to achieve similar number of discrete points in training data. All the acceleration response spectra were computed using DeepSoil for 5% damping. The design response spectra also discretized with same number of points used in accelerograms.

Table 3. Earthquake accelerograms used in training neural networks.

No.	Earthquake	PGA (g)	No	Earthquake	PGA (g)
1	Chi-Chi, Taiwan-02	0.1173	16	San Fernando 1971	0.2994
2	Morgan Hill 1984	0.1017	17	Mammoth Lakes-01	0.2818
3	Irpinia, Italy-02	0.1062	18	Norcia, Italy 1979	0.1856
4	Coalinga-01	0.1127	19	Taiwan SMART1(5) 1981	0.1124
5	N. Palm Springs 1986	0.1056	20	Nahanni, Canada 1985	0.1512
6	Whittier Narrows-01	0.1126	21	Kalamata, Greece-01 1986	0.2649
7	Griva, Greece 1990	0.1038	22	Tabas, Iran 1978	0.1089
8	Chi-Chi, Taiwan-06 1999	0.1190	23	New Zealand-02 1987	0.2926
9	Helena, Montana-01	0.1674	24	Superstition Hills-02 1987	0.1349
10	Imperial Valley-02	0.2584	25	Spitak, Armenia 1988	0.2071
11	Trinidad 1980	0.1474	26	Cape Mendocino 1992	0.1668
12	Upland 1990	0.2006	27	Landers 1992	0.1298
13	Loma Prieta 1989	0.2488	28	Northridge-01 1994	0.1061
14	Kobe, Japan 1995	0.1426	29	Duzce, Turkey 1999	0.1305
15	Northern Calif-03 1954	0.1861	30	Manjil, Iran 1990	0.1703

Two Cascade forward backward neural networks (ANN1 and ANN2) are constructed. ANN1 trained to learn the inverse mapping for the design response spectra obtained using DeepSoil for 5 % damping ratio for the corresponding 30 prescribed accelerograms. ANN2 trained to perform forward mapping that relate earthquake accelerograms to the design response. The methodology in training the inverse problem and the feedforward mapping and working stages are schematically depicted in Figure 7. The proposed method implemented using thirty accelerograms with PGA (0.1 – 0.3) g and design response spectra developed using Dubai amplification factors for 5% damping ratio to train the neural networks. The neural networks robustness is demonstrated when new design response spectrum is used in the ANN model.

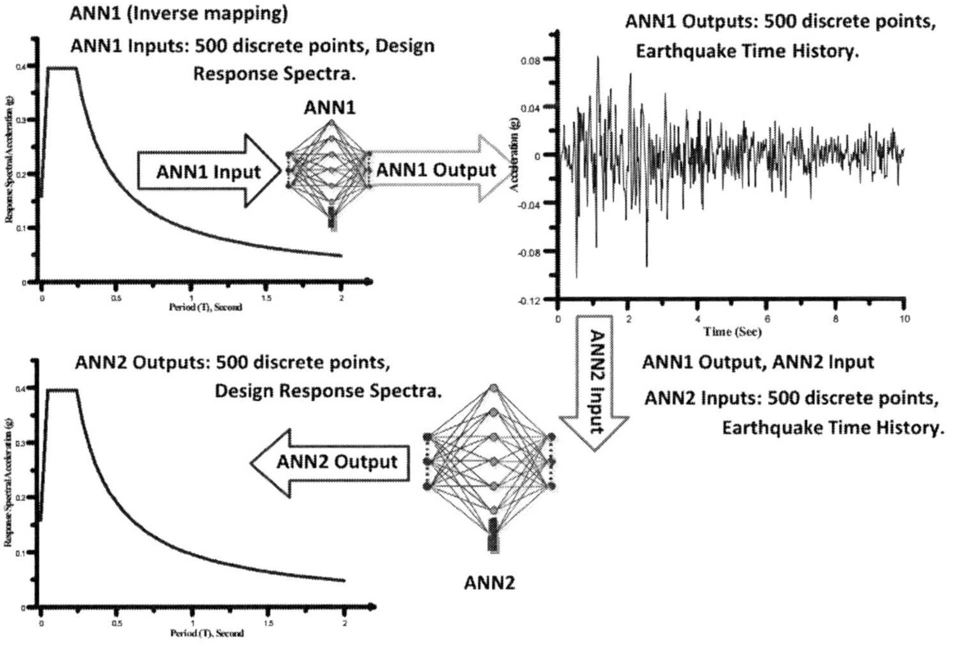

Fig. 7. ANN1 and ANN2 used in generating the accelerograms.

In general, the accelerogram predicted using ANN1 is expected to have same characteristics of the accelerograms group used in training the neural networks, or to map the response spectrum to one of the accelerograms used in the training set. In either case, the NN predicted accelerograms are expected to provide a response spectra similar to that used as input to ANN1. If none of the previous cases true, the ANN1 will generate artificial earthquake accelerograms matched to one of the records used in the training group. Therefore, it is expected that this ANN1 to be a good replacement to many software for generating an artificially spectral-matched accelerogram. The artificially generated accelerogram retains features of accelerograms exist in the training set and it can be called generalized time history earthquake record because its response spectra is close to the response spectra that developed from a group of time histories for specific site conditions and PGA range, magnitude, and distance range.

3 Results

As mentioned above two cascade neural networks, ANN1 and ANN2 are trained and built to map the relationship between the accelerograms and the response spectra in both direction. Evaluation of the developed ANN capability in replicating reasonable accelerograms matched design response spectra is investigated for different earthquake records. In Figure 8, a comparison between the generated accelerogram and the original accelerogram is shown for Palm Springs time history, as can be seen the ANN1 is greatly successful in generating the accelerograms from the response spectrum of the original record. Similar results were found for other historical records.

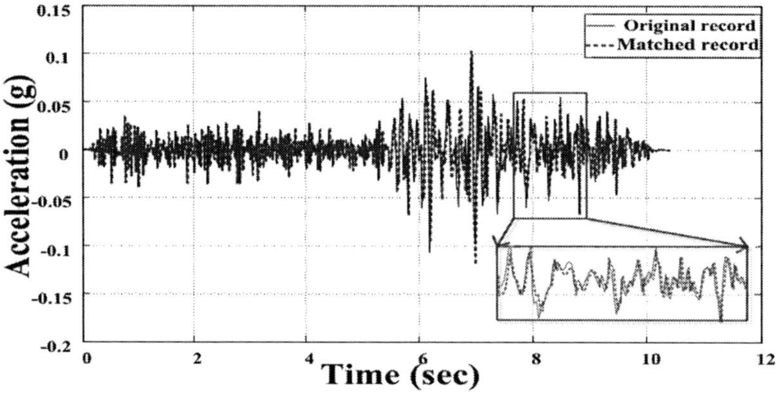

Fig. 8. Earthquake Palm Springs, time history record before and after spectral matching.

Additionally, both neural networks are tested for generating the corresponding accelerograms from design response spectra generated using the method discussed earlier by the help of the software DeepSoil. The generalized time history record and the corresponding design response spectrum for Kobe, Japan 1995 earthquake using ANN1 and ANN2, is depicted in Figure 9. It is very clear that both cascade neural networks model are capable of producing soundly correct accelerograms, using ANN1, and the corresponding design response spectrum using ANN2

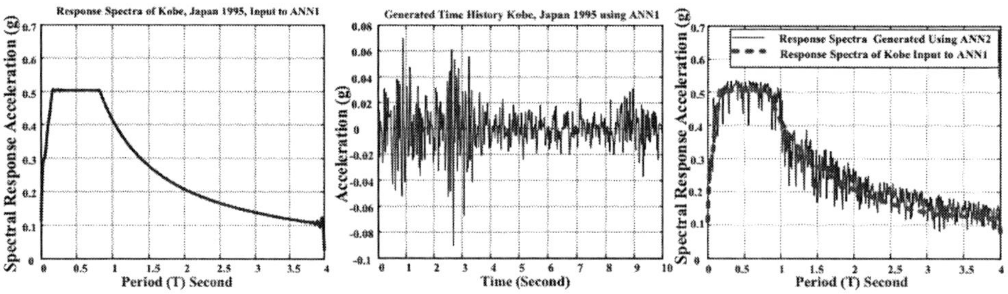

Fig. 9. The generalized time history record and its response spectrum, Kobe, Japan 1995 using ANN1 & ANN2.

In Figure 10, the trained neural networks are fed with a Dubai design response spectrum classes C and D, as input to trained neural network, and the generated accelerogram used as input to ANN2 and its response spectrum was obtained and shown. This results show that the generated accelerograms by ANN have the same characteristics of the training group set and the corresponding response spectrum is closely matched to the required design response spectrum.

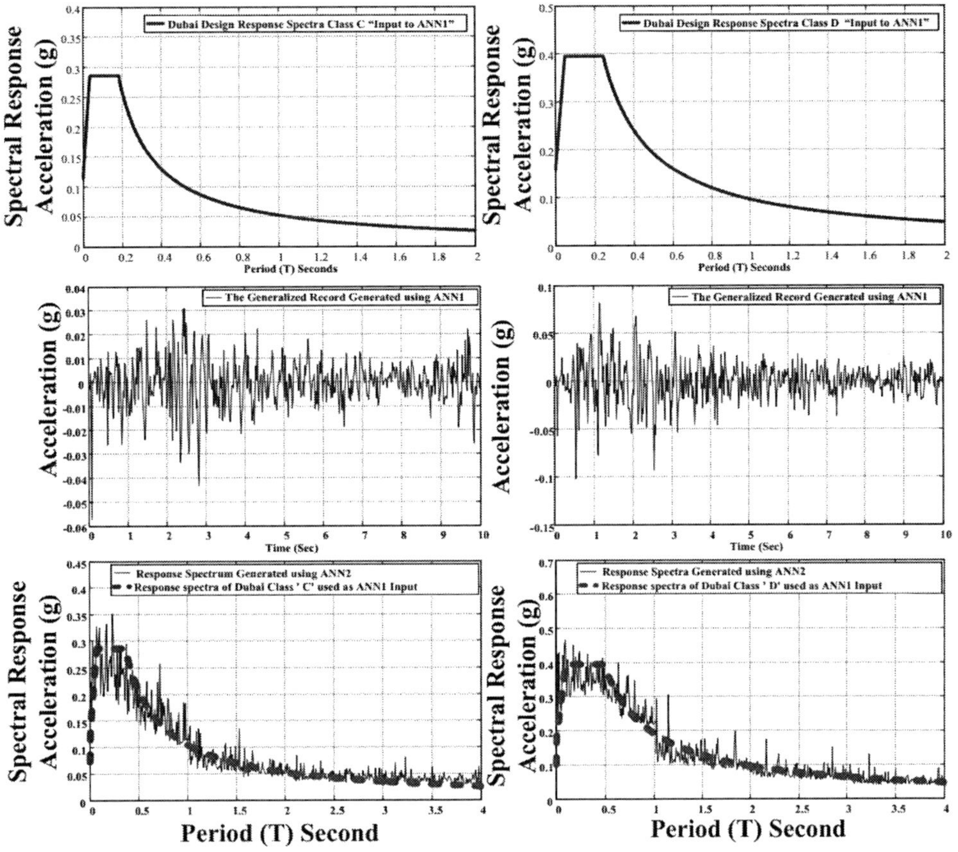

Fig. 10. The generalized time history record and its response spectra, Dubai class C and D. using ANN1 & ANN2.

ANN1 and ANN2 were verified with earthquake response spectra not used in the training set group to demonstrate the generality and soundness of the method. In other words, when new input spectrum that was not enclosed in the training data used as input, the NN will act in two different strategies. The neural network selects an accelerogram from the training pool if its response spectrum was close to the input spectrum, or when there is no similar response spectrum matched accelerogram in the training set, the neural network model will generate artificially matched spectrum accelerogram based on the characteristics it learned from training set. The above discussed examples embrace both categories mentioned above. Finally it is safe to say that practically the trained neural network is successful of producing artificial earthquake time history records corresponds to any novel design response spectrum maintain the desired response spectrum characteristics and close to the input response spectrum.

4 Conclusion

Spectral matching is performed for eight selected earthquake records to match Dubai response spectra using SeismoMatch software. The soil columns are obtained by performing nonlinear and equivalent linear site response analysis using DeepSoil software. Acceleration parameters S_s and S_1 are obtained by average response spectra at 0.2 and 1s respectively. Next, the design acceleration parameters, the design response spectrum are developed for two site soil classes C and D for Dubai. Furthermore, generation of a generalized time history earthquake record is obtained by solving the inverse mapping problem using cascade artificial neural network for any design response spectrum. Artificial neural network is trained using 30 design response spectra generated by DeepSoil software. The response spectra of the generalized time history compared to input spectra were in good agreement. In general, using ANN1 and ANN2, it is possible to generate time histories accurately with hassle free and no software with no loss of generality. The generated accelerograms have preserved the actual profile of original accelerograms with sound and representative characteristics. The generated accelerograms are general in the sense of comprising the knowledge of the accelerograms obtained by the neural network during training.

References

1. M. Irfan. *Seismic site response analysis and characterization of major cities of U.A.E.* M.S. Thesis, University of Sharjah, Sharjah, UAE, (2011)
2. N. A. Abrahamson, W.J. Silva. *Empirical response spectra attenuation relations for shallow crustal earthquakes.* Seismological Research Letters, **68**, 1, 94-127 (1997)
3. J. P. Stewart, A. H. Liu, Y. Choi. *Amplification factors for Spectral Acceleration in Tectonically active Regions.* Bulleting of Seismological Society of America, **93,** 1, 332-352 (2003)
4. K.A. Bani-Hani, A.I. Malkawi. *A Multi-step Approach to Generate Response-Spectrum-Compatible Artificial Earthquake Accelerograms.* Soil Dynamics and Earthquake Engineering, accepted for publication, (2017)
5. PEER (*Pacific Earthquake Engineering Research Center, NGA Database*), http://peer.berkeley.edu/nga/search.html
6. Y.M.A Hashash, D.R. Groholski, C. A. Phillips, D. Park, M. Musgrove. *DEEPSOIL5.1.* User Manual and Tutorial. **107** (2012)
7. Seismosoft Ltd. *SeismoMatch Version 2.1.0.* Available from: http://www.seismosoft.com/en/HomePage.aspx (2009)
8. F. Cotton, F. Scherbaum, J.J. Bommer, H. Bungum .*Criteria for selecting and adjusting ground-motion models for specific target applications: applications to Central Europe and rock sites*. Journal of Seismology, **10**, 2, 137-156, (2006)

Derivation of minimum steel ratio: based on service applied stresses

Humam AL Sebai[1,*], *Salah* Al Toubat[1]

[1]University of Sharjah, Sharjah city, UAE

Abstract. Reinforced concrete beams undergo several design stages starting with the immediate application of loading and till the ultimate failure of the beam. In many practical applications, the beam is designed to function in the service stage during which steel plays a major role in the determination of the location of the Neutral axis. The ACI code implements several provisions on the minimum steel ratio in order to encourage ductile behaviour of the beam and to prevent against the unfavourable scenario of a sudden failure. The ACI 318-08 provisions encompass a derived expression to calculate the minimum amount of flexural reinforcement that is independent of concrete strength. This paper suggests an expansion to the formula by deriving the minimum steel ratio based on the modulus of rupture and the applied service stresses. The special case for a cracking bending moment is extrapolated from the suggested formula and the result is compared to different minimum steel area formulas.

1 Introduction

Tensile steel reinforcement is indispensable to the concrete beam should the applied loads produce stresses that exceed the modulus of rupture. Steel becomes active then providing the necessary tensile strength and forming the cracked transformed section of the beam. The transformed section consists of uncracked concrete that is limited to the area above the neutral axis and an "equivalent" area of concrete to the steel located below the neutral axis. Upon the determination of the depth of the neutral axis, the cracked moment of inertia of the beam is calculated, and consequently the stresses in both concrete and steel are calculated and compared to the applied service stress limits. The process is usually conducted with the area of steel given from the start or based on the ACI code provisions [1].

However, the equation provided according to the ACI code is intended of the case of a prevailing concrete strength. Salmon [2] have updated and expanded the formula to account for the modulus of rupture and introduced his equation into ACI 318-95 [3]. In addition to the ACI several codes, specifications and researches have come up with their criteria for the minimum steel area. Some of these are listed in Table 1.

* Corresponding author: halsebai@sharjah.ac.ae

Table 1. Minimum Steel area for reinforced concrete.

Source	Year	Minimum Steel Area	Comments
Proposed	2017	$A_{s,min} = \frac{k_{\min}{}^2}{2n(1-k_{\min})} b_w d_s$ $k_{\min} = \max\begin{cases} k_1 \\ k_4 \end{cases}$	$k_1 = \frac{3-\sqrt{9-4n_c n_h^2}}{2}$ $k_4 = 1 + C_n(\sqrt{3}sin\alpha - cos\alpha)$
ACI 318-08[3]	1963	$A_{s,min} = \frac{3\sqrt{f_c'}}{f_y} b_w d_s \geq \frac{200}{f_y} b_w d_s$	
Freyermuth and Aalami [4]	1997	$A_{s,min} = \frac{3\sqrt{f_c'}}{f_{su}} b_w d_s$	
ASBI [5]	1964	$A_{s,min} = \frac{0.3\, f_{ct}}{f_y} b_w h$	$f_{ct} = 7.3\sqrt{f_c'}$
Salmon[2]	1995	$A_{s,min} = \frac{f_r}{f_y}\left(\frac{h}{d_s}\right)^2\left(\frac{C}{5.1}\right) b_w d_s$	C is a multiplier hat adjusts the section modulus for different beam shapes

Note: $A_{s,min}$ = minimum area of nonprestressed flexural tension reinforcement; h = height of the section; b_w = web width; d_s = distance from extreme compression fiber to centroid of nonprestressed flexural tension reinforcement; f_c' = specified compressive strength of concrete; f_{ct} = direct tensile strength of concrete; f_{su} = specified tensile strength of nonprestressed flexural tension reinforcement; f_r = modulus of rupture of concrete; f_y = specified minimum yield strength of nonprestressed flexural tension reinforcement; n_c = is the ratio of modulus of rupture to allowable compressive strength; n_s = is the ratio of modulus of rupture to allowable tensile strength; n_h = is the ratio of section height to depth of reinforcement; C_n and α are variables defined in the context of this paper.

2 Methodology of derivation of minimum steel ratio

The derivation of minimum steel ratio in this paper proceeds in the following order:

a. Derivation of the Neutral axis depth ratio ($\boldsymbol{k}$) and solving for the corresponding steel ratio
b. Expressing the cracked moment of inertia in terms of $\boldsymbol{k}$
c. Solving for $\boldsymbol{k}$ in the flexural stress formula of concrete and steel and dismissing the unrealistic/unfeasible solutions
d. Extrapolating the special cracking moment case and obtaining the minimum feasible limit of $\boldsymbol{k}$
e. Substituting $\boldsymbol{k}$ in the steel ratio formula and solving to obtain expression for minimum steel ratio

The derived formula assumes a linear stress-strain relation and a perfect bond between the concrete and steel. However, the procedure is limited to the case of a rectangular beams or T beams with neutral axis in the flange.

3 Location of neutral axis

Once the beam has beam cracked, the neutral axis of the beam immediately deviates from the centroid of the beam, and the transformed section is formulated as shown in Figure 1. The neutral axis location is defined as the point of which the first moment of the section about the neutral axis is zero. This leads to the formulation of the following Equation 1:

$$b\frac{x^2}{2} = n\rho bd(d - x) \qquad (1)$$

Solving for x leads to: -

$$x = \left(\sqrt{(n\rho)^2 + 2n\rho} - n\rho\right)d = kd \qquad (2)$$

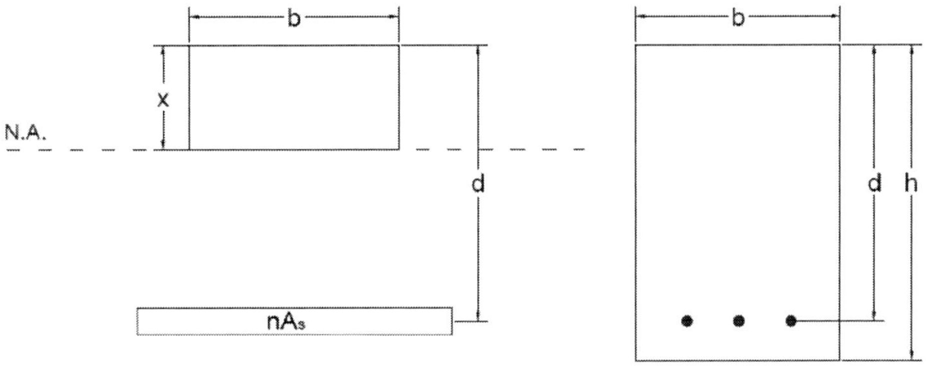

Fig. 1. Transformed area of the cross-section.

Equation 2 determines the depth of the neutral axis as a some ratio k (called the N.A ratio) multiplied by the depth of the reinforcement steel d. Solving Equation 2 for steel ratio result in Equation 3: -

$$\rho = \frac{k^2}{2n(1-k)} \qquad (3)$$

Despite the fact that k is limited to a maximum value of 1, the steel ratio in unbounded, this is evident in Figure 2 as shown.

DOI: 10.1051/matecconf/201712001011

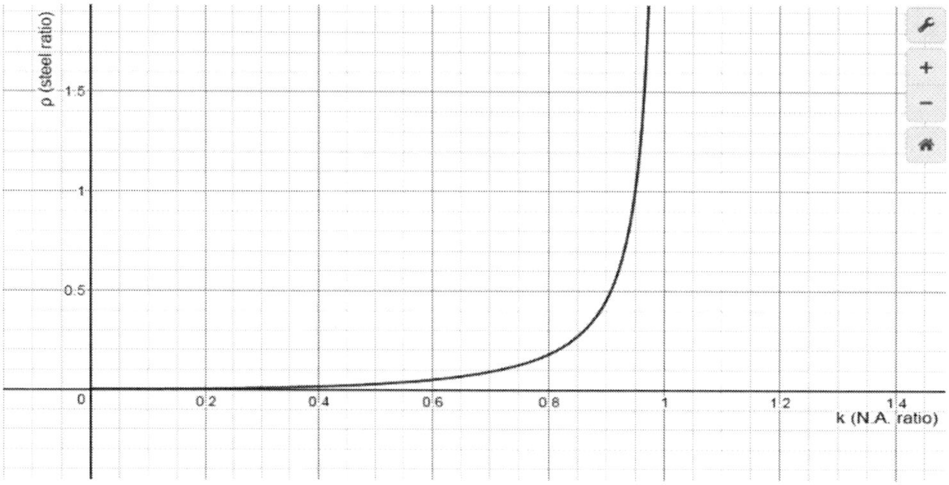

Fig. 2. Relationship between steel ratio and N.A. ratio.

4 Evaluating cracked moment of inertia

The cracked moment of inertia is then evaluated according to Equation 5 which is obtained by substituting Equation 2 and Equation 3 into Equation 4: -

$$I_{cr} = \frac{1}{3}bx^3 + n\rho bd(d-x)^2 \quad (4)$$

$$I_{cr} = \left(\frac{k^2}{2} - \frac{k^3}{6}\right) bd^3 \quad (5)$$

5 Applying the concrete service stress limit

The stresses applied in the extreme compression fibers of concrete must be limited to the service stress limit as displayed according to Equation 6: -

$$f_c \geq \frac{M_c x}{I_{cr}} \quad (6)$$

Substituting Equation 2 and Equation 5 into Equation 6 and rearranging the terms, a quadratic form in terms of k results as Equation 6 dictates: -

$$k^2 - 3k + \frac{6M_c}{f_c bd^2} \leq 0 \quad (7)$$

Solution of Equation 7 produces two solutions:

$$k_1 = \frac{3 - \sqrt{9 - \frac{24M_c}{f_c bd^2}}}{2} \quad (8)$$

$$k_2 = \frac{3 + \sqrt{9 - \frac{24M_c}{f_c bd^2}}}{2} \quad (9)$$

The second solution is dropped because it yields k that is always greater than one. Therefore, the feasible solutions of the inequality presented in Equation 7 lie within the range $[k_1, 1]$. It must be noticed that while the interval $[k_1, 1]$ is a solution to the inequality presented in Equation 7, the interest of this paper is the value of k_1 itself because according to Figure 2 the minimum feasible value of k, results in the corresponding minimum feasible value of ρ. Furthermore, by imposing the limitation that $f_c \geq f_r$, it can be shown that k_1 is limited to not exceed an approximate value of 0.53.

This obtained feasible range however, is based solely on the compression limit of the concrete. Thus, the steel tensile limit must be treated similarly.

6 Applying the steel service stress limit

The stresses applied in the tensile reinforcement steel must be limited to the service stress limit as displayed according to Equation 10: -

$$f_s \geq \frac{M_s(d-x)n}{I_{cr}} \qquad (10)$$

Substituting Equation 2 and Equation 5 into Equation 10 and rearranging the terms, a cubic form in terms of k results as Equation 11 dictates: -

$$k^3 - 3k^2 - \frac{6M_s n}{f_s b d^2}k + \frac{6M_s n}{f_s b d^2} \leq 0 \qquad (11)$$

Solution of **Equation 11** produces three solutions: -

$$k_3 = 1 + 2C_n cos\alpha \qquad (12)$$

$$k_4 = 1 + C_n\left(\sqrt{3}sin\alpha - cos\alpha\right) \qquad (13)$$

$$k_5 = 1 - C_n\left(\sqrt{3}sin\alpha + cos\alpha\right) \qquad (14)$$

Where

$$C_n = \sqrt{1 + \frac{2M_s n}{f_s b d^2}} \qquad (15)$$

$$\alpha = \frac{1}{3}arcsin\sqrt{\frac{\left(1+\frac{2M_s n}{f_s b d^2}\right)^3 - 1}{\left(1+\frac{2M_s n}{f_s b d^2}\right)^3}} \qquad (16)$$

The first solution is dropped because it yields k that is always greater than one. The last solution is also dropped because it turns out to be negative for all practical cases. Therefore, the feasible solutions of the inequality presented in Equation 11 lie within the range $[k_4, 1]$.

Likewise, the interest of this paper is the value of k_4 itself because it is less than 1. The value of k_4 is then compared to the value of k_1 and the maximum of the two is considered to control as it satisfies both of inequalities Equation 7 and Equation 11.

7 Extrapolating the special case of the cracking moment

Modifying both Equations 8 and 13 for the cracking moment defined according to ACI in Equation 17 results in the following: -

$$M_{cr} = \frac{f_r b h^3}{6} \quad (17)$$

$$k_1 = \frac{3-\sqrt{9-4n_c n_h^2}}{2} \quad (18)$$

$$k_4 = 1 + C_n(\sqrt{3}sin\alpha - cos\alpha) \quad (19)$$

Where

$$C_n = \sqrt{1 + \frac{n n_s n_h^2}{3}} \quad (20)$$

$$\alpha = \frac{1}{3}arcsin\sqrt{\frac{\left(1+\frac{n n_s n_h^2}{3}\right)^3 - 1}{\left(1+\frac{n n_s n_h^2}{3}\right)^3}} \quad (21)$$

$$n_c = \frac{f_r}{f_c} \quad (22)$$

$$n_s = \frac{f_r}{f_s} \quad (23)$$

$$n_h = \frac{h}{d} \quad (24)$$

This formulation presented in Equations 18-24 renders the definition of the minimum k dependent upon the three dimensionless ratios n_c, n_s, and n_h exclusively. k_{min} is then defined as: -

$$k_{min} = max\begin{cases}k_1\\k_4\end{cases} \quad (25)$$

Finally, the minimum steel ratio is computed as: -

$$\rho_{min} = \frac{{k_{min}}^2}{2n(1-k_{min})} \quad (26)$$

Accordingly, the minimum area of steel is determined by multiplying ρ_{min} with the section width and effective depth.

8 Comparison of proposal with other formulas

In order to validate the derivation of the minimum steel ratio formula that has been provided in this paper, a comparison to other models is made. The comparison assumes fixed parameters as indicated in Table 2 and evaluates the minimum area of steel according to each formula that has been presented in Table 1. Accordingly, the results are tabulated as shown in Tables 3 and 4.

Table 2. Section geometric and material properties of the beam.

Geometry			Material		
Section property	**Dimension**	**Unit**	**Material property**	**Dimension**	**Unit**
h	20	in	$f_y = f_s$	60000 / 40000	psi
b	12	in	f_r	$7.5\sqrt{f'_c}$	psi
d	17	in	f_{ct}	$7.3\sqrt{f'_c}$	psi
n	9	-	f_c	$\frac{1}{2}f'_c$	psi

Table 3. Minimum steel area results in in^2 for $f_y = 60000\ psi$.

$f'_c\ (psi)$	**1500**	**2000**	**2500**	**3000**	**3500**	**4000**	**4500**	**5000**	**5500**	**6000**
ACI 318-08	0.6800	0.6800	0.6800	0.6800	0.6800	0.6800	0.6842	0.7212	0.7565	0.7901
ASBI	0.3393	0.3918	0.4380	0.4798	0.5182	0.5540	0.5876	0.6194	0.6497	0.6785
Salmon	0.2680	0.3095	0.3460	0.3790	0.4094	0.4377	0.4642	0.4893	0.5132	0.5361
Proposed	0.5100	0.3630	0.3098	0.3402	0.3682	0.3944	0.4190	0.4423	0.4646	0.4859
Freyermuth and Aalami	0.3792	0.4379	0.4896	0.5363	0.5793	0.6193	0.6569	0.6924	0.7262	0.7585

Table 4. Minimum steel area results in in^2 for $f_y = 40000\ psi$.

$\acute{f}_c\ (psi)$	**1500**	**2000**	**2500**	**3000**	**3500**	**4000**	**4500**	**5000**	**5500**	**6000**
ACI 318-08	1.0200	1.0200	1.0200	1.0200	1.0200	1.0200	1.0264	1.0819	1.1347	1.1851
ASBI	0.5089	0.5876	0.6570	0.7197	0.7774	0.8310	0.8815	0.9291	0.9745	1.0178
Salmon	0.4020	0.4642	0.5190	0.5686	0.6141	0.6565	0.6964	0.7340	0.7698	0.8041
Proposed	0.5100	0.4190	0.4700	0.5163	0.5591	0.5990	0.6366	0.6723	0.7063	0.7388
Freyermuth and Aalami	0.5689	0.6569	0.7344	0.8045	0.8690	0.9290	0.9853	1.0386	1.0893	1.1377

The results are plotted against each other in **Figures 3 and 4**.

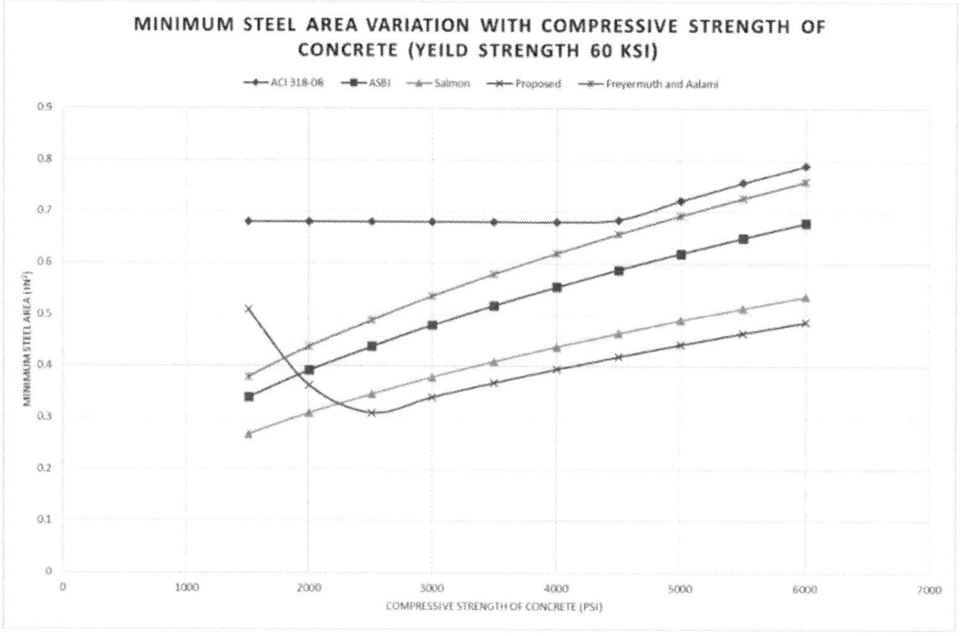

Fig. 3. Results for a yield strength of steel $f_y = 60000\ psi$.

 DOI: 10.1051/matecconf/201712001011

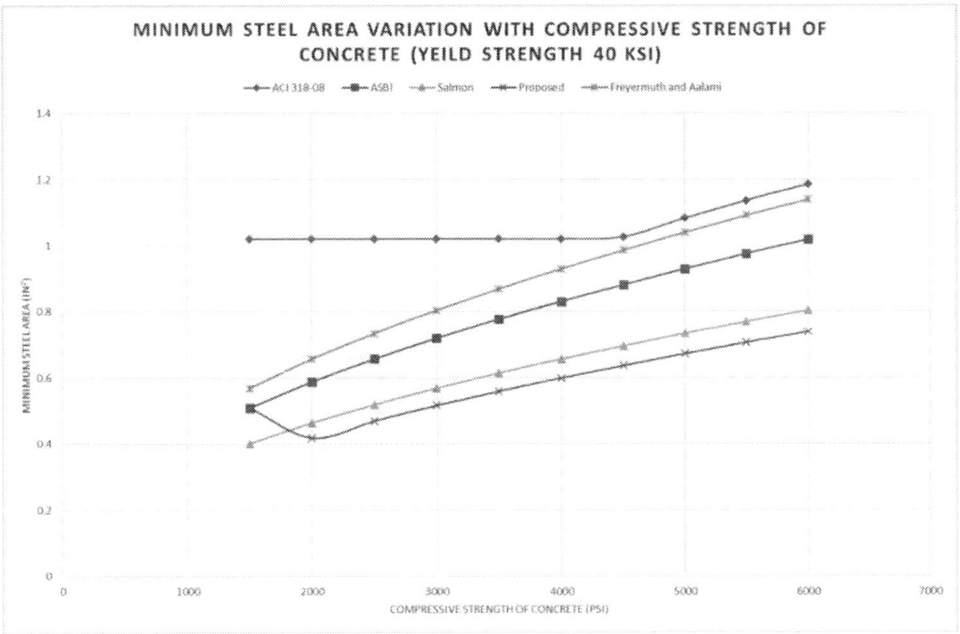

Fig. 4: Results for a yield strength of steel $f_y = 40000\ psi$

9 Discussion

The pattern of the results as displayed in Figure 3 indicates complete equivalence in the general trend of the minimum steal area resulting from all the formula, with the ACI 318-08 being the most conservative of all. Among the investigated formulas, only two of them are defined as a compound function of two criteria: steel and concrete stress requirement. Namely, the ACI-318-08 and the proposed formula. From the comparison, it can be noted that the first criteria of the ACI-318-08 formula namely $\frac{200}{f_y}\ b_w d_s$ is far more conservative than the second criteria.

Furthermore, the proposed formula seems to deviate from the general trend at values of concrete compressive strength roughly around $2000\,psi$, throughout which the concrete stress limit controls. It should be noted that the proposed formula is the only one that clearly accounts for both the stress limits of concrete and steel resulting in the possibility to define the best match of concrete and steel stress limits. **Figures 3 and 4** indicate that $\mathrm{f_y} = 40000$ psi is matching more harmonically with the results as the trend of the proposed formula exhibits less disparity when compared to $\mathrm{f_y} = 60000$ psi.

10 Conclusion

The proposed formula is clearly in match with the other formulas but it results in a minimum area of steel that is less than the ones resulting from all the other formulas. The computations of the proposed formula is a little more complicated, however, this issue is of no concern to programs as the formula can be computed easily in a straight forward algorithm without any need for any sort of cyclic routines. The formula is however (like the rest), limited to the assumptions presented in the methodology section. Further work is expected to expand the range of applicability of the proposed formula.

References

1. Stephan J., Richard Brice, and Bijan Khaleghi. 2010. Making sense of minimum flexural reinforcement requirements for reinforced concrete members. Chicago. PCI journal. Summer. pp. 65–68 (2010).
2. Wang, C., and C. G. Salmon. Reinforced Concrete Design. **6th** ed. pp. 558–560. Hoboken, NJ: John Wiley and Sons Inc (2002).
3. ACI Committee 318. Building Code Requirements for Structural Concrete (ACI 318-95) and Commentary (ACI 318R-95). Detroit, MI: ACI. (1995)
4. Freyermuth, C. L., and B. O. Aalami. Unified Minimum Flexural Reinforcement Requirements for Reinforced and Prestressed Concrete Members. ACI Structural Journal, **V. 94**, No. 4 (July–August): pp. 409–420 (1997).
5. Leonhardt, F.. Prestressed Concrete Design and Construction. 2nd ed. [In German.] Translated by C. Armerongen. Berlin-Munich, Germany: Wilhelm, Ernst & Sohn (1964).

MATEC Web of Conferences **120**, 01012 (2017) DOI: 10.1051/matecconf/201712001012
ASCMCES-17

Effect of shear span-to-depth ratio on the shear behavior of BFRP-RC deep beams

Siyam Alhamad[1] , *Yasser* Al Banna [1] , *Ahmad* Al Osman[1] , *Jihad* Mouthassseeb[1] , *Suliman* Abdalla[2] , and *Farid* Abed[3,*]

[1]Bachelor in Engineering, Department of Civil and Environmental Engineering, AUS, Sharjah, UAE.
[2]PhD Candidate, Department of Building, Civil & Environmental Engineering, Concordia University, Quebec, Canada.
[3]Professor, Department of Civil and Environmental Engineering, AUS, Sharjah, UAE.

Abstract. This study investigates the shear behavior of deep concrete beams reinforced with basalt fiber reinforced polymer (BFRP) bars for flexure without web reinforcements. The experimental testing performed herein consisted of a total of 4 short beams, three of which were reinforced with BFRP and one beam was reinforced with steel bars. The primary test variable was the shear-span-to-effective-depth ratio (*a/d*) and its influence on the beams' mid-span deflections, shear capacity, load-deformation relationships and the failure modes.

1 Introduction

Infrastructure deterioration due to corrosion of steel reinforcement is one of the biggest challenges facing the construction industry. Fiber Reinforced Polymer (FRP) bars are good candidate to replace steel bars in construction. The number of studies on the response of RC structural elements reinforced with FRP bars is rising. Current research on BFRP bars has proven their efficiency in bond durability [1-2] and to reinforce slender beams [3-4]. However, the study of short (deep) beams reinforced with BFRP bars is lacking in literature.

Slender beams and flexural structural elements reinforced longitudinally with FRP bars suffers in practice from large deformations and wide cracks due to reinforcements low modulus of elasticity. On the contrary, load transfer in short beams differs greatly from that in slender ones. Tied arch action is the mechanism by which short beams transfer loads where the concrete extending from the loading point to the nearest support act as compression diagonal struts and the longitudinal reinforcements act as the tie. Given the high tensile strength of FRP bars compared to regular steel ones, it is believed that FRP bars would greatly enhance the tied arch mechanism in short beams.

* Corresponding author: fabed@aus.edu

Omeman et al. [5] reported the results of testing twelve short beams reinforced with steel and carbon-FRP (CFRP). The study examined variables such as *a/d*, reinforcement ratio (ρ), concrete compressive strength (f'_c) and beams effective depth (d). It was found that CFRP-beams exhibited higher shear capacities in comparison to their steel counterparts. Also, CFRP-beams had larger crack width than steel-beams near failure. Abed et al. [6] experimentally investigated the response of nine GFRP-reinforced short beams. The study considered the effect of variables such as *a/d* ratio, ρ, f'_c and d on the beams shear capacity and overall response. It was noted that GFRP-beams exhibited lower stiffness and larger mid-span deflections when compared to their steel counterparts. This was attributed to the low modulus of elasticity of the GFRP bars compared to regular steel. Moreover, twelve short beams reinforced with GFRP bars were tested and analyzed by Andermatt and Lubell [7]. The study examined the influence of *a/d* ratio, f'_c and d on the short beams response and on the accuracy of Strut-and-Tie Models (STM). The STM method adopted by ACI 318-08 [8] and CSA S806-02 [9] were utilized to predict the capacity of tested beams. It was pointed out that all tested beams developed tied arch mechanism. Such an observation was derived from the uniform strains along the reinforcements and the concrete crack width and orientation. Moreover, the CSA S806-02 STM provided good predictions for the tested specimens shear capacity in comparison to ACI 318-08 STM.

In the present work, the shear response of short beams reinforced with BFRP is experimentally investigated. The effect of shear span-to-depth ratio (*a/d*) on beams' mid-span deflections, shear capacity, load-deformation relationships and the failure modes are highlighted.

2 Experimental program

Three BFRP-reinforced beams were tested in addition to one steel-reinforced beam that served as a reference for the BFRP-beams. The test matrix of the experimental program is shown in Table 1.

Table 1. Test matrix and specimen details

Beam	*a/d*	Load (kN)	Deflection (mm)	Failure mode
B1-BFRP	1.15	213.0	16.28	Concrete Crushing at Horizontal Strut
B2-BFRP	1.48	159.2	14.65	Splitting at the Diagonal Strut
B3-BFRP	1.82	79.5	7.85	Splitting at the Diagonal Strut
B4-STL	1.15	206.9	9.69	Splitting in the Diagonal Strut

2.1 *Test specimen and setup*

The test specimen was 2000 mm long with a rectangular cross section of 140 mm wide. The effective depth of the beams was kept the same for all four beams at 260 mm. Different span-to-depth ratios, *a/d* = 1.15, 1.48, or 1.82, were attained by changing the distance between the loading points. A concrete cover of 45 mm measured from the soffit of the beam to the center of the reinforcing bars was maintained in all of the tested beams.

The beams were tested under four-point loading configuration with a clear span of 1000 mm and 500 mm overhang on each side to provide sufficient anchorage length for the longitudinal bars. A schematic of the test setup with cross-section dimensions and reinforcement details of the test specimen is shown in Fig. 1. All beams were instrumented at mid-span with 5 mm strain gauges bonded to the tensile bars. The beams were loaded at a constant rate of 0.6 kN/s in load control until failure occurred using an Instron actuator of 2000 kN capacity. Cracks initiation and propagation were manually recorded for each specimen at each stage of loading. Beams deflections were measured by means of linear variable differential transducers (LVDTs) mounted at beams mid-span. A data acquisition captured the readings of strain gauges and LVDTs at all stages of loading.

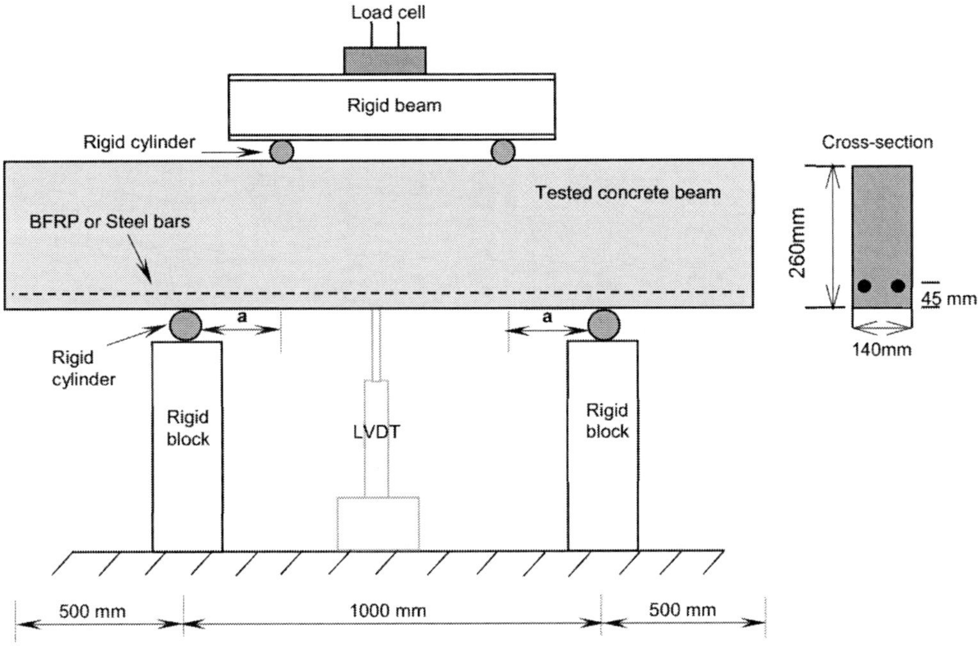

Fig. 1. Schematic of test setup with cross sections and reinforcement details

2.2 *Materials*

A ready-mix concrete with compressive strength of 40 MPa was considered. The concrete compressive strength was measured by standard concrete cubes of 150 mm tests. Steel bars of nominal yield strength of 460 MPa and elastic modulus of 200 GPa were used to reinforce the control specimens.

The actual diameters of the BFRP bars used in this study were measured by volume displacement as recommended by the ACI 440.3R-04 committee [8]. The 12 mm BFRP bars have a uniform sand coating with shallow spiral indentations spaced at 2.75 mm along their surface. The mechanical properties of these BFRP bars were obtained by conducting tensile tests for 5 specimens. The average ultimate stress and elastic modulus were measured as 1230 MPa and 46.2 GPa, respectively.

3 Results and discussion

3.1 Failure modes

Fig. 2 shows sample photos of the failed beams (B1-BFRP and B2-BFRP) where all specimens collapsed in shear. A small crack propagating from one of the supports to the nearest loading point diagonally resulted in the beam failure. This typical shear failure was expected due to the absence of shear reinforcements in the beams.

Shear failure of the BFRP-reinforced beams B1-BFRP was accompanied by the top concrete crushing between the two point loads. In beam B2-BFRP, concrete in compression crushed under the point load that connected the shear failure crack to the support. It is worth noting that beams that showed top concrete crushing exhibited a more ductile failure at ultimate than other beams that failed with only diagonal shear cracks.

Fig. 2. Failure modes for selected beams

3.2 *Load–deflection relationships*

The load-deflection relationships of the tested BFRP-beams and the control Steel-beam are shown in Fig. 3. The BFRP-reinforced beams exhibited almost linear load-deflection curves after cracking unlike the steel beams that showed elasto-plastic load-deflection curves until failure occurred. It can be noted from Fig. 3 that increasing the *a/d* ratio increases the mid-span deflection measured at any stage of loading. This finding was explained by the tendency of beams having large *a/d* ratios to behave as a flexure-critical member rather than a shear-critical one. Smaller deflection was exhibited by the steel-beam which is noted from the higher post-cracking stiffness compared to BFRP counterpart. The small post-cracking stiffness encountered in the BFRP-reinforced beams was attributed to the low modulus of elasticity of the BFRP bars compared to that of the steel bars, which resulted in wider cracks.

All of the BFRP-beams showed an abrupt failure when their ultimate shear capacities were attained as illustrated in Fig. 3. The brittle failure of the BFRP-specimens can be noticed from the load-deflection response of beam B3-BFRP. The sudden drop at ultimate indicated the collapse of the beams once their ultimate strength was achieved without showing any residual deformations. On the contrary, B1-BFRP showed a more ductile behavior that was attributed to the concrete crushing in compression between the two point loads prior to collapse. The load-deflection response was also consistent with the strain gauge readings along the bars. It is worth pointing out that the recorded strains showed no sign of anchorage problems in any of the beams. Prior to cracking, the strains in the tensile reinforcement were insignificant. However, after cracking the strains increased significantly as a result of the formed cracks. Strains recorded in the steel-reinforced bars were less than those recorded in their BFRP-reinforced counterparts at all stages of loading

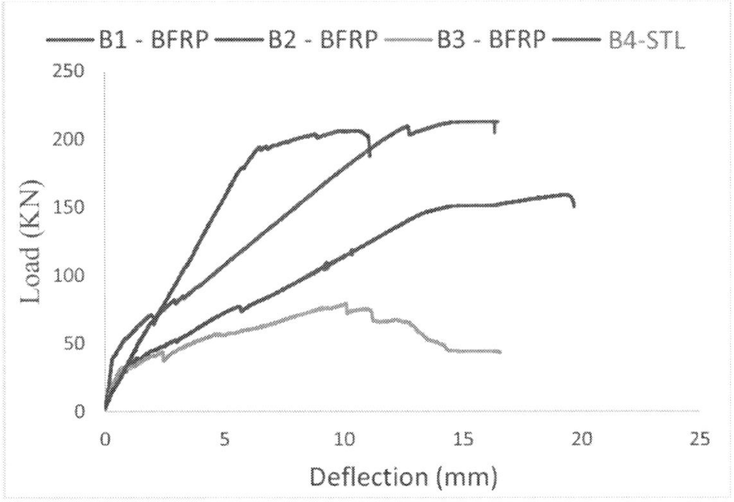

Fig. 3. Effect of *a/d* on the Load-Deflection Relationship

3.3 *Strength analysis*

The tested beams shear capacities and maximum deflections are summarized in Table 1. The results suggest that BFRP-beams recorded slightly higher ultimate loads than the steel counterpart. It should be emphasized that the deflection at ultimate load of the BFRP-beams was higher than that of their steel counterpart as a result of the low modulus of the BFRP bars compared to steel ones.

Fig. 4 correlates the tested beams shear strength to their *a/d* ratio. A linear inverse proportionality can be noted between shear strength to their *a/d* ratio. Also, decreasing *a/d* ratio by 19% from 1.82 for specimen B3-BFRP to 1.48 for specimen B2-BFRP lead to an increase in the shear capacity by 100% from 79.5 to 159 kN, respectively. Additional reduction of the *a/d* ratio by 22% to a ratio of *a/d* = 1.15 in B1-BFRP resulted in 34% capacity increase.

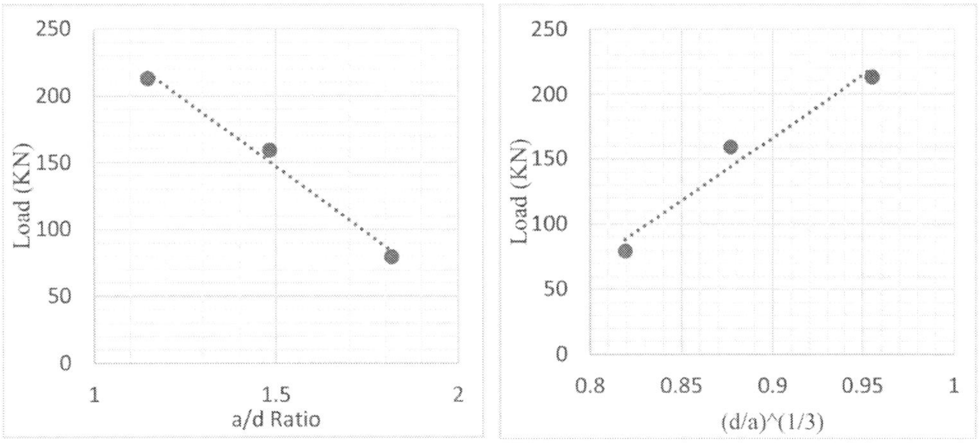

Fig. 4. Effect of *a/d* on the shear capacity of FRP-reinforced concrete short beams

It is also observed from Fig. 4 that the shear strength of BFRP-RC beams is proportional to the cubic root of the span to depth ratio [(d/a)1/3]. This linear proportionality is in agreement with the recommendations of the Canadian standard S806-02 (CSA, 2002) [9] as given by the following equation 1.

$$V_c = 0.05\lambda\phi_c k_m k_r k_a (f_c')^{\frac{1}{3}} b_w d_v \tag{1}$$

Where

$$k_m = \sqrt{\frac{V_f d}{M_f}} \leq 1.0 \tag{2}$$

$$k_r = 1 + (\rho_w E_F)^{\frac{1}{3}} \tag{3}$$

$$k_a = \frac{2.5}{M_f/V_f d} \geq 1.0 \tag{4}$$

Here λ is the concrete density factor, ϕ_c is the concrete resistance factor, and ρ_w is the longitudinal reinforcement ratio percentage, d_v is the effective shear depth taken as the greater of 0.9d or 0.72h (mm), E_F is the modulus of elasticity of FRP bars (MPa), V_f is the factored shear (N), M_f is the factored moment (N.mm), d is the effective depth (mm), and the quantity $V_f d/M_f$ is the inverse of a/d. The correlation between V_c and *a/d* in equation (1) can be derived from the multiplication of factors $k_m = \sqrt{V_f d/M_f} = 1/\sqrt{a/d}$ and $k_a = 2.5/(M_f/V_f d) = 2.5/(a/d)$.

4 Conclusions

This paper reports on the experimental testing of Basalt-FRP reinforced deep beams and a steel RC counterpart for reference. The main variable in consideration is the shear span to effective depth ratio (*a/d*). It was found that all beams failed by a diagonal crack that extended from the support to nearest loading point. Also, higher ductility was noted near failure for beams that showed top concrete crushing in addition to typical diagonal shear crack. The overall BFRP-RC beams exhibited higher mid-span deflections and less stiffness compared to their steel counterparts. This is attributed to the low modulus of elasticity of the BFRP bars in comparison to steel bars. Increasing the *a/d* ratio increased the mid-span deflection due to the tendency of beams with large *a/d* ratios to behave as a flexure-critical member rather than a shear-critical one. Moreover, a linear correlation between the shear capacities and the cubic root of *a/d* ratio was evident from experimental result which agrees with the CSA S806-12 code. This correlation is attributed to the inclination of the angle between the diagonal compression struts and the horizontal tie which resulted in improving the beam's arch-action mechanism.

References

1. Z. Omeman, M. Nehdi, H. El-Chabib, Can. J. Civil. Eng. **35**, 1 (2008)
2. F. Abed, H. El-Chabib, M. AlHamaydeh, J Reinf. Plast. Comp. **31**,1063 (2012)
3. A. Al-Tamimia, F. Abed, A. Al-Rahmani, Adv. Concrete. Const. **2**, 1 (2014)
4. American Concrete Institute. ACI 440.3R-04 (2004)
5. Canadian Standards Association. CSA S806 (2012)
6. A. Altalmas, A. El Refai, F. Abed, Constr. Build. Mater. **81**, 162 (2015)
7. A. El Refai, F. Abed, A. Altalmas, J. Compos. Const.-ASCE **19**, 04014078 (2014)
8. A. El Refai, F. Abed, J. Compos. Const.-ASCE, **20**, 040150824 (2015)
9. A. El Refai, F. Abed, A Al-Rahmani, Constr. Build. Mater. **96,** 518 (2015)

MATEC Web of Conferences **120**, 02001 (2017) DOI: 10.1051/matecconf/201712002001
ASCMCES-17

An overview on the research on self-healing concrete at Politecnico di Milano

Liberato Ferrara[1,*]

[1]Department of Civil and Environmental Engineering, Politecnico di Milano, Italy

Abstract. Self-healing cement based materials, by controlling and repairing cracks, could prevent "permeation of driving factors for deterioration", thus extending the structure service life, and even provide partial recovery of engineering properties relevant to the application. The author's group has undertaken a comprehensive investigation focusing on both experimental characterization and numerical modelling of the self-healing capacity of a broad category of cementitious composites, including high performance cementitious composites reinforced with different kinds of fibres. Both autogenous healing has been considered and self-healing engineered techniques, including the use of pre-saturated natural fibres and of crystalline admixtures. Tailored methodologies have been employed to characterize the healing capacity under different exposure conditions and for different time spans, ranging up to two years. The healing capacity has been quantified by means of suitably defined "healing indices", based on the recovery of mechanical properties correlated to the amount of crack closure, measured by means of optical microscopy. A predictive modelling approach, based on modified micro-plane model, has been formulated. The whole investigation represents a step towards the reliable and consistent incorporation of self-healing concepts and effects into a durability-based design framework for engineering applications made of or retrofitted with self- healing concrete and cementitious composites.

1 Introduction

The porous structure of concrete is one of the main causes of its being prone to degradation; even if it is generally accepted that a well-proportioned and properly cured concrete, produced using a low water-to-cement (w/c) ratio, leads to obtain a finished product with good durability performance, no concrete material and structure can be made absolutely waterproof. Because of the porous structure of concrete, water can penetrate through pores and micro-cracks, due to either capillary absorption and/or hydrostatic pressure. Concrete professionals typically adopt the term "permeability" to describe the resistance of concrete to water ingress and/or passage under actual service conditions. This includes the overall permeability issue of concrete including possible existing cracks, due, e.g., to restrained drying shrinkage, thermal deformations as well as sustained service loads.

* Corresponding author: liberato.ferrara@polimi.it

In view of the aforesaid statements, it would be highly desirable and advantageous to be able not only to "design" and cast a concrete as compact (and impervious) as possible, but also to be capable to provide the material some capacity to reduce its permeability in the cracked state. In this framework, (autogenous or engineering triggered) self-healing cement-based construction materials, able to control and repairing early stage cracks, would represent an exceptional asset to the 21st Century Civil Engineering. Self-sealing/healing of cracks could prevent permeation of driving factors for deterioration and even provide a partial recovery of the engineering properties of the material relevant to the intended applications, thus contributing to extend the service life of the engineering works [1,2].

Discovered as early as in 1836 by the French Academy of Science, and attributed to either delayed hydration of cement/binder, carbonation or the combination of both, many researchers demonstrated that concrete, under favourable conditions, can heal naturally without any particular additives [1,2]. As a matter of fact, because such a capacity turned out to be quite randomly scattered and thus neither reliable nor predictable in an engineering application perspective, a paramount effort is currently challenging the concrete research community of investigating "engineering" self-healing capacity of concrete and cement-based construction materials. Several different techniques, along different main directions of investigation, have been conceived and explored: i.e., self-healing engineered with fibre reinforcements, mineral-producing bacteria, super absorbent polymers, healing agents contained in shell and tubular capsules and other proprietary chemical admixtures or even mineral additions already traditionally employed as cement substitutes. The supply of water (moisture) is essential, especially in the case of addition of chemical agents able to promote the deposition of crystals inside the crack, but since most infrastructures are exposed to rain or underground water, usually this is an easily satisfied requirement. Several other variables can affect the phenomenon of self-healing, such as the mix proportions, the stress state along the cracks, thermo-hygrometric conditions etc. [1, 2].

The authors' research groups have undertaken for about a 5 years a collaborative comprehensive research project, focusing on both experimental characterization and numerical modelling of the self-healing capacity of a broad category of cementitious composites, ranging from normal strength concrete to high performance cementitious composites reinforced with both steel and natural fibres. Both autogenous healing capacity and self-healing engineered techniques have been investigated, including the use of pre-saturated natural fibres as well as of crystalline admixtures, including their synergistic effects. Tailored methodologies have been employed to characterize the healing capacity of different investigated cement-based materials, which are based on a comparative evaluation of the mechanical performance measured through 3- or 4- point bending tests or indirect tension tests. Tests have been performed on pre-cracked specimens (to a target value of the crack opening), after scheduled conditioning times to selected exposure conditions (e.g., water immersion, wet and dry cycles, humid exposing and dry climates). The healing capacity has been quantified by means of suitably defined "healing indices", based on the recovery of the permeability and of the mechanical properties, and correlated to the amount of crack closure, measured by means of optical microscopy and also "estimated" through suitable indirect methodology. Chemical characterization of the healing products has been performed to understand the different mechanisms governing the observed phenomena. As a further step, predictive modelling approaches, based on either a poro-plasticity and fracture approach or on a modified micro-plane model, have been formulated. On the one hand, a zero-thickness interface model has been proposed to include the effects of crack healing, through suitably defined porosity-based functions which affect the time evolution of mechanical properties of concrete. On the other hand, a micro-plane model incorporates the self-healing effects through a delayed cement hydration rule and the effects of cracking on the diffusivity and the opposite repairing effect of the self-healing.

2 Experimental programme: materials and methodology

Employed mix-designs are summarized in Table 1.
For Normal Strength Concrete (NSC) the self-healing evaluation methodology employed un-notched prismatic beam specimens (500 mm × 100 mm × 50 mm) pre-cracked in 3-point bending up to two different crack opening levels, respectively equal to 0.15 and 0.30 mm, and then either stored in water or left exposed to open air up to 12 months. 3-point bending tests were repeated up to failure after scheduled conditioning periods. Self-healing was thus evaluated by comparing the peak-load bearing capacity exhibited by the specimens after conditioning to the residual one the same specimens featured upon pre-cracking when unloaded at the prescribed crack opening [3].

For steel- and steel-sisal fibre reinforced High Performance Cementitious Composites [4-8] beam specimens of 500 mm x 100 mm x 30 mm were employed, cut from larger cast slabs (1m × 0.5 m). Slabs were cast taking advantage of the self-compacting features of the fresh mixtures, also in order to achieve a tailored alignment of the fibres, and beam specimens were cut with their axis either parallel or orthogonal to the casting flow direction (i.e. the preferential fibre alignment), in order to discriminate the effect of flow inducing fibre alignment on the performance of the HPFRCC. Specimens were pre-cracked in 4-point bending (4pb) and the Crack opening Displacement (COD), was measured by means of two LVDT transducers over a gauge length equal to 200 mm. Specimens featuring a deflection softening response (i.e. with fibres perpendicular to the beam axis) were pre-cracked up to a COD value equals to 0.5 mm. On the other hand specimens with fibres parallel to the axis, featuring a deflection hardening response, were pre-cracked in the post-peak regime equals to (COD_{peak} + 0.5 mm), where COD_{peak} denotes the measured value of the COD in correspondence of the peak stress, under the assumption of the same opening of the localized crack. After, pre-cracking specimens were "conditioned" in four different "environments": (i) immersion in water; (ii) dry (20°C and 50% RH); (iii) moist (20°C and 90% RH); (iv) wet and dry cycles, alternating one day in water, at T = 20 °C, and one day of exposure in dry environment. At the end of the scheduled conditioning period (1, 3, 6 and 24 months), specimens were tested up to failure according to the same 4pb set-up employed for pre-cracking. "Superposition" between pre-cracking and post-conditioning stress-crack opening curves allowed the self-healing capacity to be evaluated.

For ordinary Fibre Reinforced Concrete (FRC), double edge notched 150 mm side square specimens were employed and pre-cracked, according to the "Double Edge Wedge Splitting (DEWS)" test methodology [9], up to a crack opening value equal to 0.25mm; specimens were then subjected to different exposure conditions: immersion in tap water, open air exposure and wet/dry cycles (4 days in water and 3 days exposed to open air). After one-month exposure to the different conditioning environments, specimens were tested again up to or further a crack opening of 0.25mm, and, after that, subjected again to the same exposure conditions for two further months and finally, they were re-tested again and pre-cracked up to further 0.25mm. The following cases were also considered:

- specimens continuously healed for three months after pre-cracking at 0.25 mm, re-cracked up to further 0.25 mm, further healed for one month and then re-cracked again up to further 0.25 mm;
- specimens continuously healed for six months after pre-cracking at 0.25 mm and then re-cracked up to further 0.25 mm.

Reference specimens were monotonically tested as well up to a crack opening width equal to 2.5mm. The healing efficiency was evaluated by comparing the crack-cycling curves with the monotone curve for a similar reference specimen.

Correlation with crack closure, evaluated through image analysis of optical microscopy garnered crack images, was also always performed.

Table 1. Mix designs of investigated cementitious composites (in kg/m³).

Constituent	NSC		FRC		HPFRCC (PoliMI/UFRJ)		
Cement CEM II 42.5 A/L	300	300	360	360			
CEM I 52.5 R					600	600	600
Slag					500	500	500
Crystalline admixture		3		3		3	
Water	180	180	180	180	200	200	200
	w/c = 0.6		w/c = 0.50		w/(c+s) = 0.18		
Superplasticizer	3	3	3.5	3.5	33	33	33
fine aggr. 0-2 mm					982	982	982
fine aggr. 0-4 mm	1080	1077	814	811			
coarse aggr. 4-16 mm	880	880	1081	1077			
Steel fibers Dramix 5D 65/60BG			20	20			
Steel fibers Bekaert OL 13/0.16					100	100	50
Sisal fibers							7

3 Experimental results: analysis and discussion

3.1 Normal strength concrete

In Figure 1a the results of a typical test, in terms of load vs. COD curves are shown: the graphs are built up to compare the curves pertaining respectively to the pre-cracking test and to the post-conditioning up-to-failure test for the same specimens. From the values of nominal bending stresses an Index of Load Recovery (ILR) has been defined:

$$ILR = \frac{P_{max\ reloading,} - P_{unloading}}{P_{max,uncracked} - P_{unloading}} \tag{1}$$

An estimation of the crack closure due to the self-healing can be provided, by operating a "backward" shifting along the COD axis, of the stress-COD curve representative of the behaviour of each pre-cracked specimen after t conditioning, until the stress-COD curve of the same specimen, as measured during the pre-cracking test on the virgin undamaged sample is met. The new value of the "origin" COD can then estimated and an index of crack healing (ICH) was defined as:

$$ICH = \frac{COD_{pre-cracking} - COD_{post-conditioning}}{COD_{pre-cracking}} \tag{2}$$

Figures 2 a-c show the trend of the ISR and ICH vs. the exposure time for different exposure conditions and the related correlation:

- specimens immersed in water and made with concrete containing the crystalline additive exhibited an almost immediate and quite significant recovery, which even upon prolonged exposure, showed continuing improvement of the recovered performance; on the other hand, specimens made with plain concrete and immersed in water showed a more gradual recovery, which anyway, even after six months, barely attained half the level achieved by concrete with the crystalline additive;
- specimens exposed to open air and made with concrete containing the crystalline additive showed a gradual recovery capacity, as high as the one exhibited by plain concrete specimens in water; on the contrary a scant recovery capacity at all was exhibited by specimens without the additive.

The results highlighted in Figures 3 a-d and the SEM images with the related EDS analysis in Figures 4 a-b confirm the aforementioned statements and are coherent with the EDS spectrum of the admixture [3].

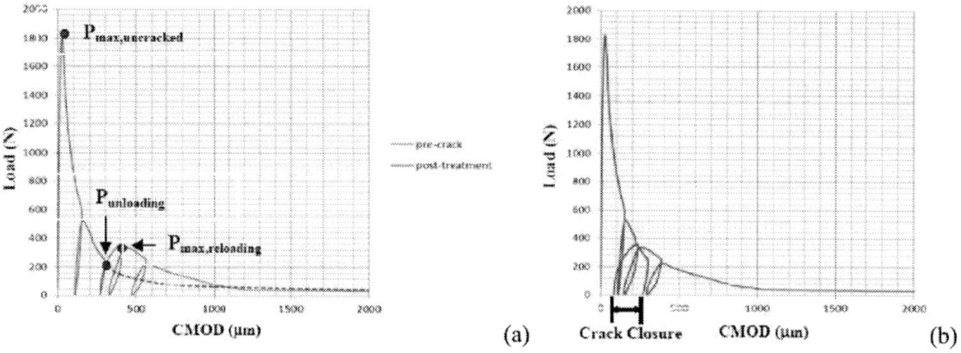

Fig. 1. example of load-COD curve obtained from 3pb tests on the same specimen before and after T-RH conditioning (a); proposal of a procedure to evaluate crack closure (b).

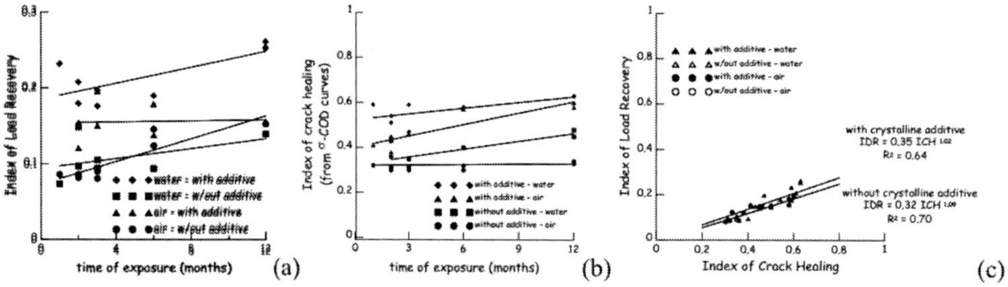

Fig. 2. Indices of Load Recovery (a) and of Crack Healing (b) and related correlation (c).

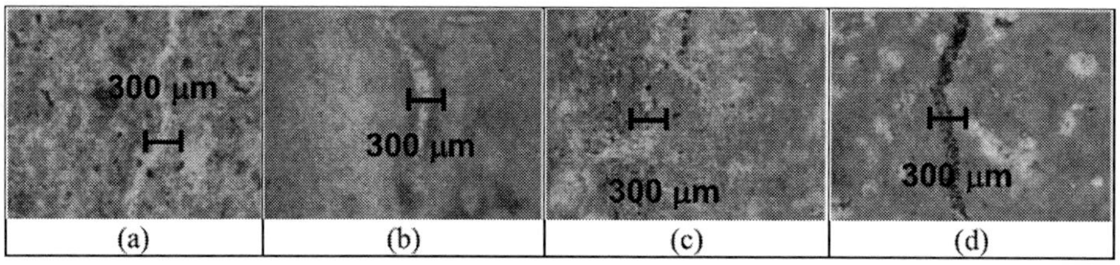

Fig. 3. Healed/healing cracks for specimens with (a,c) and without (b,d) crystalline additive after six months of immersion in water (a,b) and exposure to air (c,d).

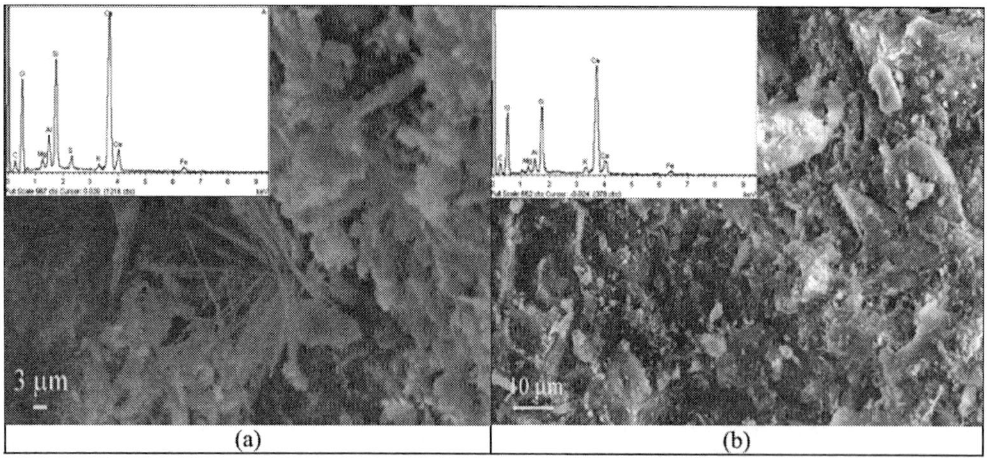

Fig. 4. SEM images and EDS analyses for specimens with (a,c) and without (b,c) crystalline additive after three months of immersion in water.

3.2 Fibre reinforced concrete and effect of cracking/healing cycles

As a first proof of the occurred healing, the maximum crack opening was measured before and after each healing period with a digital optical microscope at different locations along the crack for both concretes. An Index of Crack Healing was calculated as above (Eq. (2)).

The results are shown in Figures 5 a-b, first comparing the effect of a one cycle healing duration (1, 3 and 6 months) and then comparing the performance after the repetition of cracking-healing cycles (1+2 months vs. 3 months).

The following statements hold:

- prolonged exposure to healing conditions, whatever it is, results in an improved crack healing; the presence of crystalline admixture in the mix results in a faster crack sealing, mostly under less favourable exposure conditions (wet/dry cycles and even exposure to open air) (Figure 5a);
- the performance after the second cracking/healing cycle, is generally comparable with that after the corresponding first cracking-healing cycle, and somewhat better when a crystalline admixture is added to the mix; the longer the duration of the first healing period (3 vs. 1 month), the better the performance after the second cracking/healing cycle (Figure 5 b);
- for the same cumulative healing period, an intermediate cracking/healing cycles (1+2 months vs. 3 months continuing exposure) results in a slightly worse healing performance (Figure 5 b).

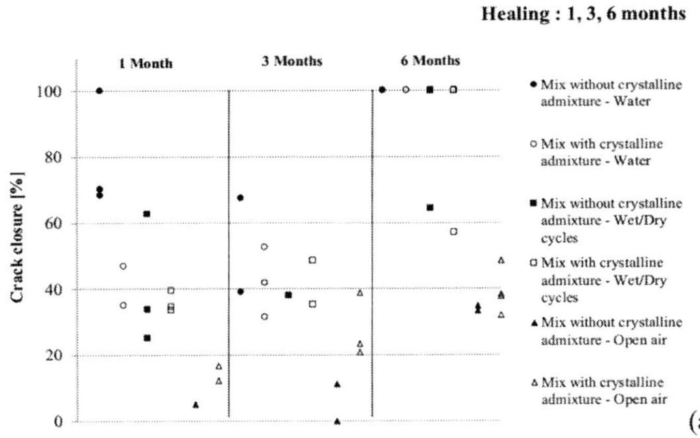

(a)

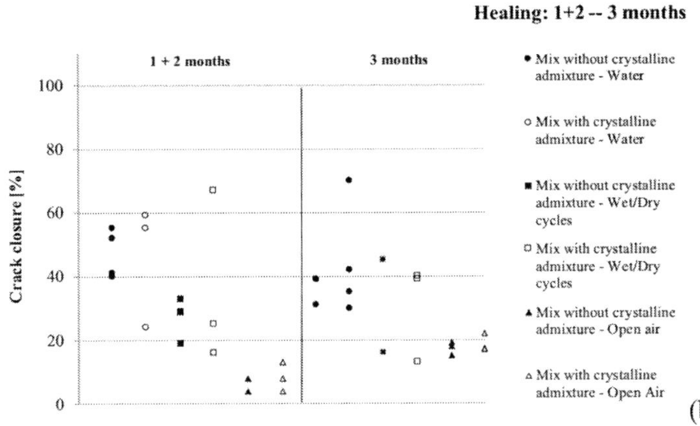

(b)

Fig.5. Index of Crack Healing (crack closure %) after different healing and cracking/healing exposures for ordinary Fibre Reinforced Concrete.

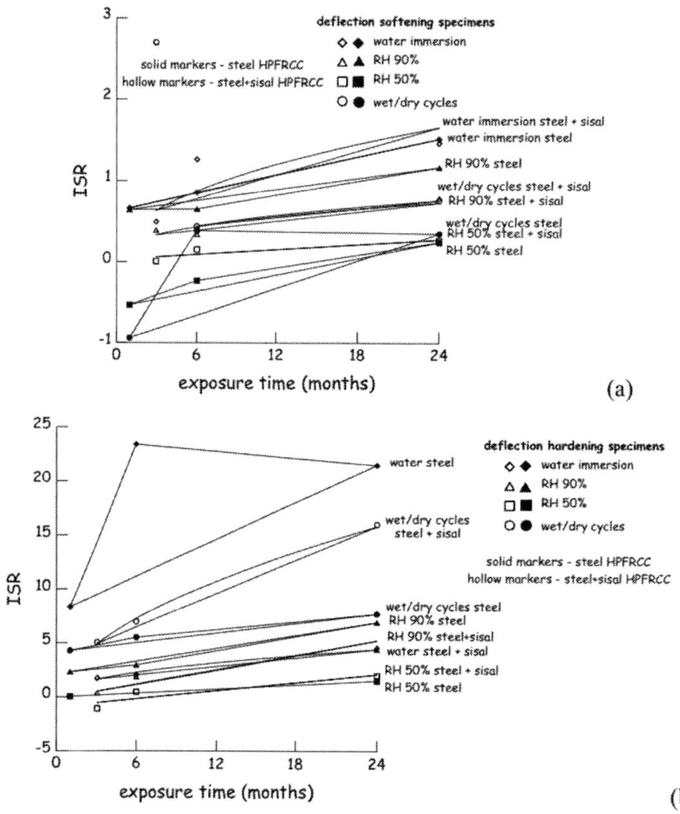

Fig.6. Index of Strength Recovery for deflection softening (a) and hardening (b) specimens made with steel only and hybrid steel+sisal HPFRCCs.

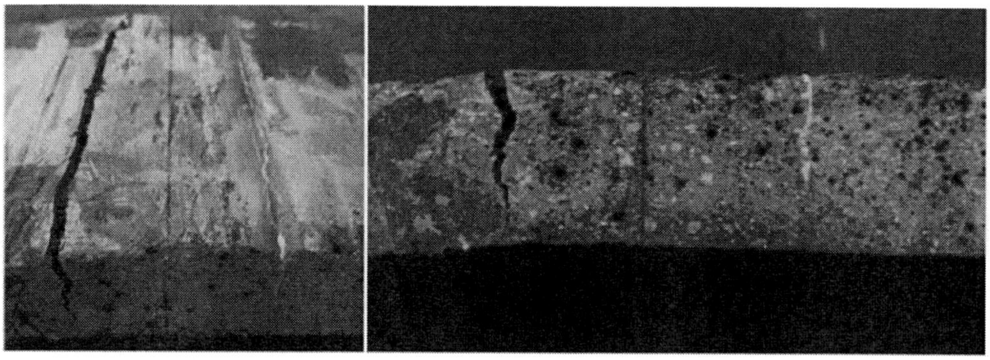

Fig. 7. Healed crack and new crack formed in post-healing tests in sisal + steel HPFRCC.

3.3 High performance fibre reinforced cementitious composites

Quantitative evaluation of the effects of self-healing has been performed once again calculating the Index of Stress Recovery through Equation (1) from the experimental results. Indices, plotted in Figure 6 for steel-only and steel+sisal HPFRCC specimens, highlight, on the one hand, the easily predicable effect of the exposure conditions and, on the other, the effects of conditioning duration.

The effectiveness of natural sisal fibres as facilitators and enhancers of the healing capacity also clearly appears, mainly with reference to the performance of specimens exposed to wet/dry cycles, which actually represents the most challenging condition in this respect. As a matter of fact, the natural fibres at a cracked site absorb the water in the wet stages of the cycles and transport it throughout the bulk material in the vicinity of the same crack. This contributes to an enhancement of the delayed hydration reactions, which are, together with $CaCO_3$ precipitation under wet and dry cycles, responsible of self-healing, thanks to the higher and longer availability of water, as retained by the fibres all along the drying stage of each cycle. This may also like to contribute to "shadow" the negative effect of calcium carbonate dissolution, precipitating during the same drying stages of each cycle, which is also responsible for self-healing and which may dissolve upon prolonged exposure, mainly in wider cracks, as observed.

For a deflection hardening specimen immersed in water for six months, in the post-conditioning test, the sealed crack did not reopen and a new crack was formed elsewhere (Figure 7). As a matter of fact, natural fibres also promote a better and more thorough healing at the fibre-matrix interface level, which acted synergistically with the reconstructed through-crack material continuity justifying the observed results.

Specimens immersed in water and containing also the crystalline admixture showed an even better performance (Figure 8). This may lead to hypothesize that, because of the expansion caused by the crystalline admixture reactions, some kind of "internal chemical pre-stressing" of fibres may have occurred [6]. SEM analyses of the fractured healed surfaces in specimens both without and with the crystalline admixture showed, in the latter case, a rather amorphous structure of the products covering the aggregate grains (Figure 9a) with typical composition of cement hydration products, thus confirming that healing is mainly due to delayed hydration of un-hydrated binder particles. In the former, the presence of crystals is observed (Figure 9b) with chemical composition compatible with the composition of the admixture. Deposition of healing products on the surface of the fibres is also evident (Figure 9c). The observed fibre rupture is attributable to an improved bond rightly resulting from the deposition of the aforementioned products, confirming the hypothesis about chemical pre-stressing resulting from the admixture action.

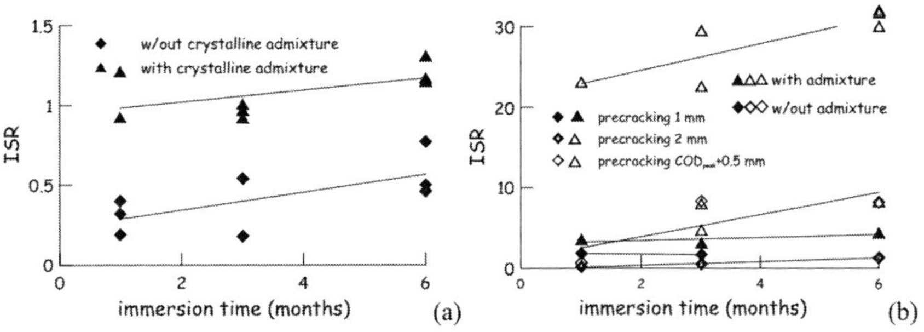

Fig. 8. Index Strength Recovery: deflection softening (a) and hardening (b) steel only HPFRCC specimens without and with the crystalline admixture immersed in water.

Fig. 9. Microscope image of healed fractured surfaces of specimens without (a) and with (b) the crystalline admixture; healed products on a ruptured fiber in HPFRCC with crystalline admixture (c).

4 Numerical modelling

Two modelling approaches have been proposed by the authors' group. A first one, based on micro-mechanics, incorporates the self-healing effects olf modeling the delayed cement hydration, as well as the effects of cracking on the diffusivity and the opposite repairing effect of the self-healing on the proposed micro-plane model constitutive laws [10].

The second one is a discontinuous-based porosity model and represents an extension of a fracture energy elastic-plastic interface formulation which now includes porosity evolution induced by self-healing mechanisms. The formulation accounts for the characterization of concrete failure behaviour in mode I and II fracture types. The post-cracking response is considered by means of specific work softening rules in terms of work spent and porosity evolution [11].

Both models have been calibrated on the NSC self-healing campaign detailed above and the comparison between the experimental and the numerical results (Figures 10a-b) highlights their reliability.

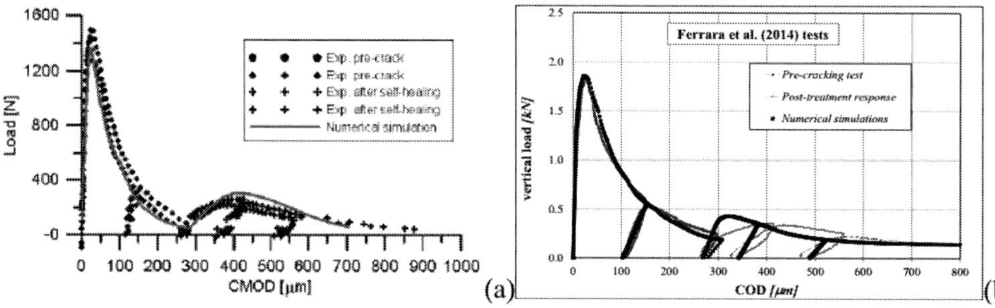

Fig. 10. Experimental vs. numerical results for self-healing capacity of NSC [10 and 11].

5 Conclusions

In this paper, the most significant results have been summarized of a comprehensive investigation on the self-healing capacity of different kinds of cement based materials, currently on-going for more than five years at the author's home institution, also in cooperation with other prominent institutions worldwide. The following statements hold:

- in presence of water NSC possesses a scattered autogenous self-healing capacity;
- tailored "self-healing activating" admixtures, such as crystalline ones, may enhance and make more reliable this capacity, even in case of other less favourable exposure conditions, and for cracks up to a few hundred microns (200-300 μm) wide;
- the same holds also for ordinary FRC, whose residual post-cracking strength can be significantly retained upon healing, also under repeated cracking/healing cycles, and mostly in the presence of crystalline admixtures;
- because of their peculiar composition HPFRCCs exhibit a significant autogenous healing capacity, always favoured by water and high relative humidity, which can lead to completely heal cracks up to several hundred microns (up to 0.5 mm) wide even occurring at quite later ages;
- the use of crystalline admixtures in fibre-reinforced concrete and cementitious composites can result in significant enhancement of the healing performance. This could be caused by some kind of internal "chemical pre-stressing" of fibres, triggered by the expansive reactions of the admixtures, which needs anyway to be further and more systematically investigated;

- the combined use of industrial and natural fibres enhance the aforementioned autogenous healing capacity of HPFRCCs thanks to the porous structure of fibres which can absorb water and promote its diffusion not only on the crack faces but through the bulk matrix of the composite;
- in all fibre reinforced cementitious composites the effects of healing on recovery or conservation of post-cracking mechanical performance upon exposure to different environments is attributable to both reconstruction of through-crack matrix continuity and healing of fibre/matrix bond, which also benefits from the aforementioned factors;

Further fundamental experimental investigations are needed to clarify the nature of self-healing products and their role in the sealing of the cracks and recovery of the material mechanical properties. Nonetheless, the results herein shown and the partially conclusive statements drawn from their analysis encourage to continue with the research on this topic to pave the way for the diffusion of self-healing cement-based materials into engineering practice with reference to both the construction of new buildings and the retrofitting of existing damaged or deteriorating structures. To this aim, it is fundamental in the authors' opinion to orient the investigation towards the characterization of healing capacity of cement based materials under sustained through-crack tensile stresses, as well as towards assessing the effectiveness of healing persistence and repeatability under alternating healing-loading/cracking cycles, as herein shown with reference to preliminary results.

A fundamental gap has to be filled which deals with the availability of reliable modelling tools able to reproduce the self-healing mechanisms and predict their effects on the performance of the materials. In this way, self-healing concepts could be consistently incorporated into durability design frameworks and life-cycle assessment design approaches. In this paper, the main results have been shown with reference to the calibration of two numerical models, which represent a first step in the aforesaid direction.

Self-healing of NSC and FRC (in cooperation with dr. V. Krelani) at PoliMi has been partially supported by Penetron Italia Ltd. and Penetron International Ltd. Self-healing of HPFRCCs has been funded by Politecnico di Milano-Young Researchers grant 2011 and partly performed in cooperation with Universidad Politecnica de Valencia, Spain (prof. P. Serna Ros). Investigation on NFRCCs and self-healing modeling has been developed in cooperation with Universidade Federal do Rio de Janeiro, Brasil (in cooperation with proff. R.D. Toledo Fliho, F. Silva and S.R. Ferreira) and Univesidad de Buenos Aires, Argentina (in cooperation with dr. A. Caggiano) and Universidade Nacional de Tucuman, Argentina (in cooperation with Prof. G. Etse), in the framework of EnCoRe" Project (FP7-PEOPLE-2011-IRSES n. 295283) funded by the European Union.

References

1. H. Mihashi, T. Nishiwaki, T., J. Adv. Concr. Technol., **10** (2010).
2. K.van Tittelboom, N. De Belie, Materials, **6** (2013).
3. L. Ferrara, V. Krelani, M. Carsana, M., Constr. Buildg. Mats., 68 (2014).
4. L. Ferrara, V. Krelani, F. Moretti, M. Roig Flores, P. Serna Ros, Cem. Concr. Comp. (submitted).
5. L. Ferrara, V. Krelani, F. Moretti, Cem. Concr. Comp, **73** (2016).
6. L. Ferrara, V. Krelani, F. Moretti, Smart Mats. Structs., **25** (2016).
7. L. Ferrara, S.R. Ferreira, T.N. da Cunha Moreira, F. de A. Silva, R.D. Toledo Filho, Cem. Concr. Comp. (part I, submitted).
8. L. Ferrara, S.R. Ferreira, T.N. da Cunha Moreira, F. de A. Silva, R.D. Toledo Filho, Cem. Concr. Comp. (part II, submitted).
9. M. di Prisco, L. Ferrara, M.G.L. Lamperti, Mats. & Structs., **46** (2013)
10. G. di Luzio, L. Ferrara, V. Krelani, Comp. & Structs. (submitted).
11. A. Caggiano, V. Krelani, L. Ferrara, G. Etse, Comp. & Structs. (submitted).

Self-sealing concrete mortar, concrete mortar mixed with super absorbent polymer

Moayyad Al-Nasra[*]

Civil and Infrastructure Engineering Department, American University of Ras Al Khaimah, AURAK, Ras Al Khaimah, UAE.

Abstract. Researchers tried almost every conceivable material to be mixed in concrete to alter some of its properties. In this research superabsorbent polymer is used as an admixture to improve its water tightness properties especially in concrete tanks, aeration basins, and retaining structures. The concrete tries to seal the leak by blocking the water path by semi-solid gels. This gel is the product of the interaction of the superabsorbent polymer with water. Several samples have been prepared to study the effect of the superabsorbent polymer on the concrete strength, and on the concrete capability to block the water flow. Concrete cubes, concrete beams, and concrete tension samples were prepared to study the concrete strength of concrete mixed with superabsorbent polymer. Also concrete short cylindrical samples were prepared to study the concrete ability to block the water flow. These cylindrical samples were intentionally broken into two approximately equal pieces to create an induced crack. The artificially cracked samples are then subjected to water pressure to study the concrete ability to stop the water flow though the induced concrete crack. Two kinds of water pressure are used in this research; the constant head, and the falling head.

1 Introduction

There are several applications of the superabsorbent polymer (SAP) are available commercially. One of the most known use of the SAP is in some agricultural applications. The SAP absorbs water at very high rate and releases it slowly with time. This property is very useful for plants that needs constant supply of water. Other application of this product is in the baby diaper products. It is used for its capacity to absorb water and to convert it into gel. This property is very useful for any application that requires to retain water in a semi-solid form. When the SAP is subjected to drying process, it loses the water slowly, and eventually go back to its crystal dry form.

It has been long struggle to improve the properties of concrete by adding all kinds of admixtures. Mixing super absorbent polymer with concrete not only improves its water tightness capability, but it also provides the concrete with internal curing moisture it needs to improve its strength. This property is very useful in cases where curing of concrete is

[*] Corresponding author: malnasra@gmail.com

difficult. The fresh concrete plasticity and consistency change as well. This property is yet to be further explored. This self-sealing property of concrete can also be used in any tank or container that holds and retains any kind of liquid. The super absorbent polymer absorbs water at a rate of 300 to 500 times its own weight, in a process transforming the polymer dry powered into gel. Samples of different amount of superabsorbent polymer were prepared to study the effect of the increase of the superabsorbent polymer on the concrete strength and on the concrete water tightness capability.

The concrete is mixed with SAP at different percentages to study its effect on the concrete properties. Several samples of different SAP quantities are prepared. Samples of 0% SAP are also prepared to be used as control samples. The amount of SAP used in this study is expressed as a percentage by weight of the Portland cement. This percentage ranges from 0% to 0.3% by weight of Portland cement. Previous study by the author showed that the concrete is very sensitive to the SAP amount.

Normal weight concrete is produced by mixing Portland cement, aggregate, and water at moderate temperature. The mixing proportion is the key to good quality durable concrete. The quality of aggregates is very important factor to be considered when targeting quality concrete. The aggregate forms approximately 70% of the total mass of the hardened concrete. Portland cement is the most expensive ingredient, and the most important ingredient. It is the binding agent that starts its binding function the moment the water is added to the mix. The chemical reaction is triggered by the water, and continues with time. There two major well distinguished stages in the fresh concrete phase; the initial setting and the final setting. The water plays significant role in the concrete quality, and it is the cheapest ingredient. In the normal strength concrete, where ingredients are mixed without the help of any admixtures, the water is added at a rate of almost twice as much as needed for the chemical reaction. The excess water beyond the amount needed for the chemical reaction plays a negative role in the concrete. It increases the drying shrinkage, and increases the potential to propagate cracks in the mass of the hardened concrete. The water is necessary for concrete workability, and for the chemical reaction, but its amount must be under control. Nowadays, several admixtures are used to reduce the negative effect of the excess water on the hardened concrete without jeopardizing the concrete workability. Water / cement ratio became a significant parameter to be considered in the concrete mix design.

In this study, SAP is the only admixture used. The SAP starts absorbing water within minutes, and converts it into gel. Several masses of gel are then mixed with the rest of the concrete ingredient. These masses will release water slowly giving the chemical reaction the water it needs with time. This process has a great positive effect on the concrete final product. Unfortunately, the gel occupies large volume in the concrete mass. When the gel loses the water, it shrinks leaving large voids in the concrete mass. These voids play role in reducing the concrete strength.

AlNasra studied the effect of adding SAP on the concrete insulation property [1]. He proved that the SAP has the potential to save energy by improving the concrete insulation property. He showed that the increase in the SAP quantity added to the concrete mix increases the insulation degree of the hardened concrete.

The effect of the SAP on the concrete shrinkage was studied by several researchers. Also the effect on the SAP on the concrete strength is also investigated [2]. It is proven that the SAP has a significant role in increasing the concrete shrinkage. The loss of water with time increases the concrete shrinkage. Autogenous type of shrinkage is also studied in concrete [3]. The studies showed the uncontrolled shrinkage leads to concrete cracking. This can be of great challenge when high strength concrete is considered. The SAP played a role reducing the built-up stresses responsible of generating cracks in concrete during the process of restrained hardening of high performance concrete [4].

Al-Nasra used special kind of SAP called Sodium Polyacrylates in concrete as an admixture [5]. His study focused on optimizing the amount of SAP to be mixed in concrete. The optimizing process focused on the concrete strength as the main parameter. Several samples of different sizes were used. AlNasra concluded that approximately 0.11 percent of SAP by weight of Portland cement is the most effective percentage. This small percentage has little effect on reducing the concrete strength due to generating extra voids.

Excess amount of SAP added to the concrete mix increases the potential of generating cracks and weakens the concrete. Several studies suggested that the potential use of SAP as crack healing agent [6]. When concrete cracks it exposes larger surface areas to unfriendly chemicals. These unfriendly chemical accelerates the concrete deterioration from the inside. The super absorbent polymer helps in protecting the concrete by expanding and swelling in the face of the unfriendly chemicals, reducing the exposed concrete areas and limiting the access of these unfriendly chemicals to the concrete.

Al-Nasra, and Daoud studied the plasticity of concrete mixed with SAP [7]. The gel forming SAP changes the texture of the fresh concrete. They showed that adding SAP changes the concrete workability, plasticity, and consistency. They also concluded that a relatively small amount of SAP has the capability to make substantial changes in the behaviour of the fresh concrete.

2 Super absorbent polymer

The super absorbent polymer is composed of charged long chains of polymer. It contains multiple and repeated cross linked chains. These chains of polymer have the capability of absorbing water up to 500 times its own weight. Fig. 1 shows an actual picture of the super absorbent polymer used in this study. The figure shows the polymer before absorbing water, and the polymer after absorbing water in the amount of approximately 500 times its own weight. It can be noticed from this figure that the polymers took well distinguished crystal look alike shape. The ruler shows the scale in inches, reflecting the actual size of the super absorbent polymer used in this study.

Fig. 1. The super absorbent polymer before and after absorbing water.

This type of super absorbent polymer has been used widely in many commercial products mainly for its great ability to absorb water or liquid several hundred times its own weight. This property found its way into many agricultural applications. It is usually mixed with soil forming small water reservoirs. These reservoirs release water slowly with time. The constant supply of water keeps the plants and flowers fresh.

3 Concrete strength

Two types of samples were prepared; samples for strength tests and samples for water flow tests. The strength test samples were used to test the concrete in compression, tension, and flexure. Only one mould is used to produce water flow test samples. Four different percentages of the SAP were used; 0%, 0.10%, 0.20%, and 0.30%. The 0% sample is used as a control sample in order to study the effect of adding SAP to the concrete mix. Table 1 shows the concrete ingredients used in this study. The water/cement ratio is taken in this study as 0.5. The dimensions of the compression cubes are 5 cm X 5 cm, while the dimensions for the beam samples are 5 cm X 5 cm X 21 cm. Direct tension samples were also prepared.

Table 1. Concrete ingredients used.

Sample ID	Sand (gm)	Cement (gm)	SAP (gm)
Plain	400	200	0.0
0.10%SAP	400	200	0.20
0.20%SAP	400	200	0.40
0.30-SAP	400	200	0.60

Fig. 2 shows the effect of the SAP on the compressive strength of the concrete. The strength of concrete increases with time at a rate decreases with time. This applies for all of the samples within the time of the test. The percentage of SAP used has a substantial effect on the compressive strength. The 0.10% SAP samples performed well compared to the other samples. The increase in the SAP quantity beyond the 0.10 % has negative effect on the compressive strength of the concrete. This is believed to be due to the excessive voids in the concrete solid mass generated by the SAP gel.

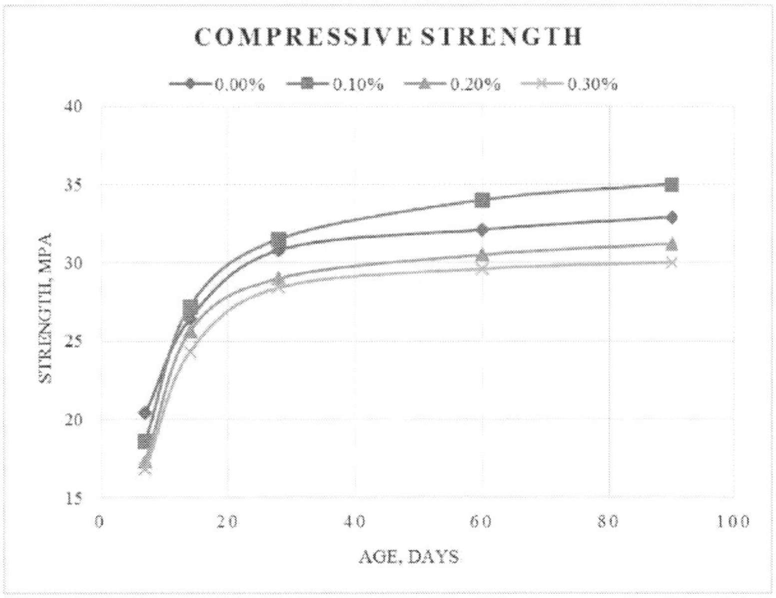

Fig. 2. Compressive strength of concrete cubes mixed with SAP.

DOI: 10.1051/matecconf/201712002002

The fresh concrete mixed with SAP showed little or no shiny surface. The shiny surface is the result of concrete bleeding. Also the fresh concrete of samples mixed with SAP showed a different texture and colour compared to the plain concrete samples. Gel like firm texture became more visible with the increase in the amount of the SAP added to the concrete mix.

Fig. 3 shows the test set up of the compression cubes. Fig. 4 shows the direct tension test. The direct tension concrete samples are called dog bone samples due to its shape. Fig. 5 shows the flexural beam test set up. The test results of the direct tension samples were very close to each other. The same can be said about the flexural beam samples. The inaccuracy of the testing is larger than the difference between samples results. The size of the samples also played significant role in making the test inconclusive. However, the general trend of the results seems to enforce the trend of the compression samples.

Fig. 3. Compression test set up.

Fig. 4. Direct tension test set up.

Fig. 5. Flexural test set up.

4 Water flow test

The samples of the water flow test are of cylindrical shape. Its height is 10 mm, and its diameter is 35 mm. The objective of the test is study the ability of the concrete to block the water flow through an induced artificial crack. Each water flow sample is broken into almost two equal parts, then these parts are put together again. The assembled sample represents a crack in the concrete of 10 mm depth, and width of 35 mm. Four groups of samples were prepared. Each of different SAP percentage. The amount of SAP used in this study ranges from 0% to 0.3% by weight of Portland cement. The effect of increasing the amount of SAP mixed in the concrete mix is also the focus of this study. The first sample tested was the one with 0% of SAP. This sample is used as a control sample.

Fig. 6 shows the drawing of a typical water flow sample. The induced crack is a results of applying flexural stress at the middle of the cylinder. This procedure is standardized to make sure all samples were subjected to similar conditions. Special mould was used to produce the water flow samples. The mould is basically a rubber ring of the inside dimensions of 10 mm in height, and 35 mm in diameter. The same mould is used again during testing to hold the broken pieces together.

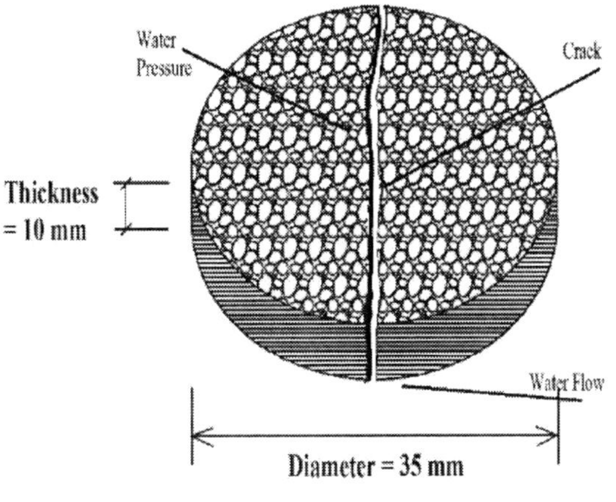

Fig. 6. Typical water flow concrete sample.

The fresh concrete is poured into a rubber mould of 10 mm height and 35 mm inside dimensions. The concrete top is levelled and smoothened. The fresh sample was then covered in plastic sheets to maintain moisture in the concrete sample. The sample is then cured by

 DOI: 10.1051/matecconf/201712002002

submerging it in water bath for additional 6 days. The sample is then removed from the water bath and left to dry for additional 7 days. The water flow test started on the day 14. On day 14 all samples were broken into approximately identical two pieces. Fig. 7 shows one of the samples which is broken into two pieces. Fig. 8 shows the same sample ready to be tested. The water flow sample is placed back in its mould, and then placed in a steel casing. The steel casing is then placed in the testing chamber in such way that the water has no path to flow but to flow though the concrete sample and in particular through the concrete crack.

Fig. 7. Typical water flow concrete sample with induced crack.

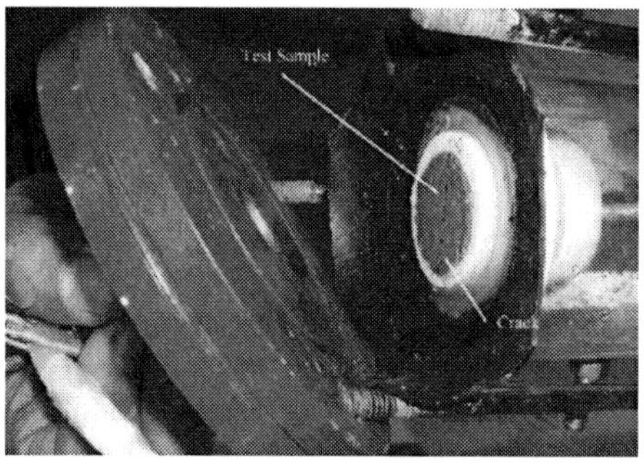

Fig. 8. Prepared water flow sample in the testing chamber.

Water pressure is applied at the top side of the water flow sample. The samples were subjected to both constant water pressure (constant head) and decreasing water pressure (falling head). These two different types of pressure are used in this experiment to study the effect of the water pressure on the sealing capability of concrete. The amount of water flowing through the concrete is then measured. This amount is indicative to the concrete sealing capability. The smaller the amount of water flowing through the concrete sample the better water tightness capability, and the better sealing capability of concrete mixed with SAP. The water discharge can be determined by equation 1.

$$Q_{SAP} = \frac{V_W}{t} \qquad (1)$$

Where;
Q_{SAP}= water flow rate through the concrete sample in ml/min
V_w= volume of water flowing through the sample in millilitre
t = time measured in minutes.

Fig. 9 shows the test set up for the water flow test. The same set up is used for both the falling head experiments, and the constant head experiments. The sample is cased inside the testing chamber, and subjected to water pressure at the top side only. The perimeter of the sample is sealed by epoxy to prevent any water flow through the sides.

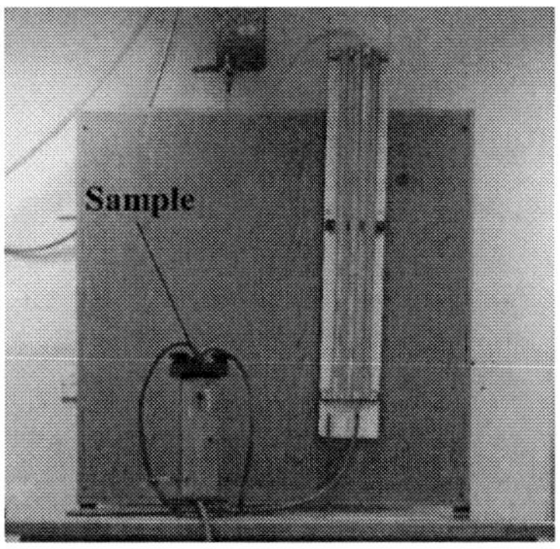

Fig. 9. Test set up for falling head and constant head tests.

5 Test results

This study focused on the effect of the SAP on the concrete in terms of strength and in terms of sealing capability. Several SAP percentages were used to study the effect of the increase of SAP on the targeted concrete property. The experiments took several months to reach to distinctive conclusions. The water flow though the concrete samples were subjected to both constant head water pressure and falling head water pressure.

The constant head test is conducted over several months around the clock. The broken concrete sample is put together in the same mould, then placed in water pressure chamber. The sample is subjected to 105 cm water pressure at all the time. Constant supply of water is provided to maintain the water at the same level in order to maintain constant water pressure. The flow of the water is measured in terms of water flow discharge in millilitre per minute, where the volume of water is measured per unit time. The first sample tested was the plain concrete sample. The water flow was almost at a constant rate ranges from 26.8 ml/min to 29.2 ml/min. The tendency is that the flow rate increases with time but at a very small rate. Within the time constraint of this test, the increase in the flow rate of the plain sample was not significant. Fig. 10 shows the flow rate of the three SAP concrete samples of 0.10 %, 0.20%, and 0.30 by weight of Portland concrete. The plain concrete sample has a very large water flow value and almost constant to be included in this figure. The other three samples showed significant changes in flow rates. The flow rate decrease with time. This applies for all of the SAP samples, but the one with high percentage showed faster decrease in water flow rate with time. Eventually the water flow stops. The water flow through the sample with high SAP content stopped sooner compared to the one with low SAP content. Fig. 11 shows the stoppage time as it is related to the amount of SAP used in the concrete mix. The curve depicts high nonlinearity relationship. The stopping time is very sensitive to the amount of SAP used. For cases where the concrete sealing property is needed, higher percentage of SAP could be used. Equation 2 shows the relationship between the water flow stopping time and the amount of the SAP used in the concrete mix. This relationship is derived for this particular case, and can't be extrapolated or interpolated to match any other case.

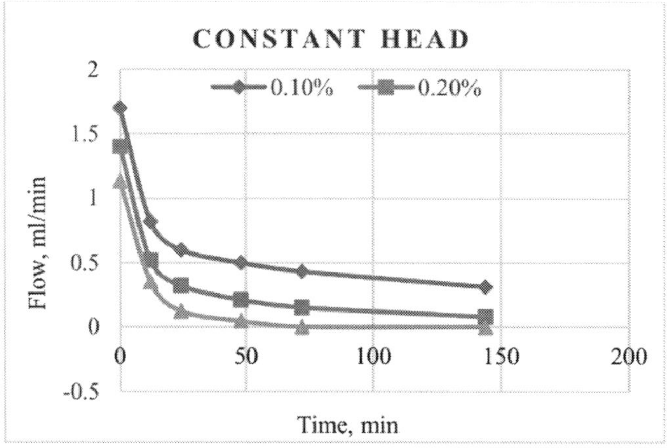

Fig. 10. Water flow rate at constant head water pressure.

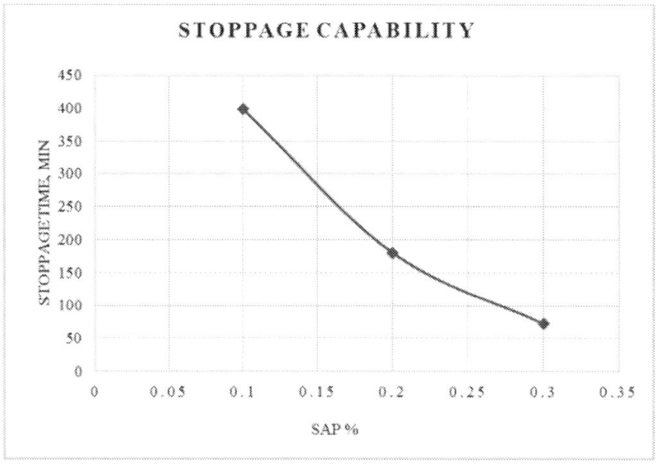

Fig. 11. Effect of SAP quantity on the water flow stoppage time.

$$t_{st} = 960\ e^{-8.6\,\eta} \tag{2}$$

Where,
t_{st} = the water flow stoppage time in minutes
η = super absorbent polymer amount expressed in terms of percentage by weight of the Portland cement

The falling head test is conducted in a very similar way. The water pressure started at 105 cm of water pressure and ended, in many cases, few centimetre above the sample level. The total volume of water flow is measured in this case. The plain concrete sample showed large water flow at the very beginning and trickled down to few drops at the end when the water pressure approaches the zero mark. Fig. 12 shows only the plain concrete sample and the 0.10% SAP sample. Showing the other two on this figure will make the figure clustered and difficult to read. Fig. 13 shows the effect of the amount of SAP on the water flow rate. The increase in the SAP amount in the concrete mix decreases the water flow rate and improves the chances of the concrete to seal cracks. The water pressure has also significant effect on the water flow. The increase in the water pressure increases the water flow rate, which is expected conclusion. This study used low water pressure range.

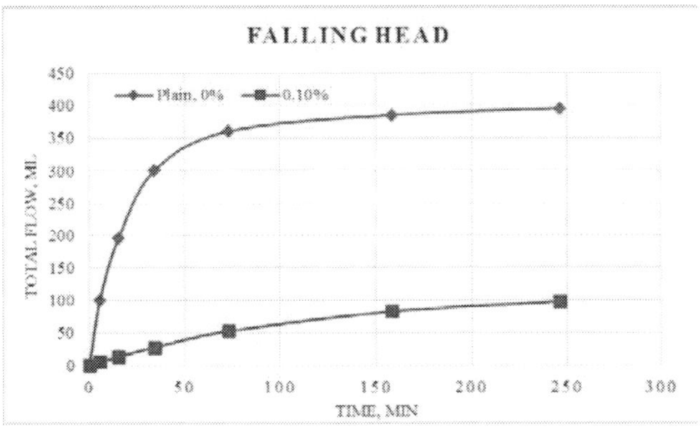

Fig. 12. Effect of SAP quantity on the water flow rate.

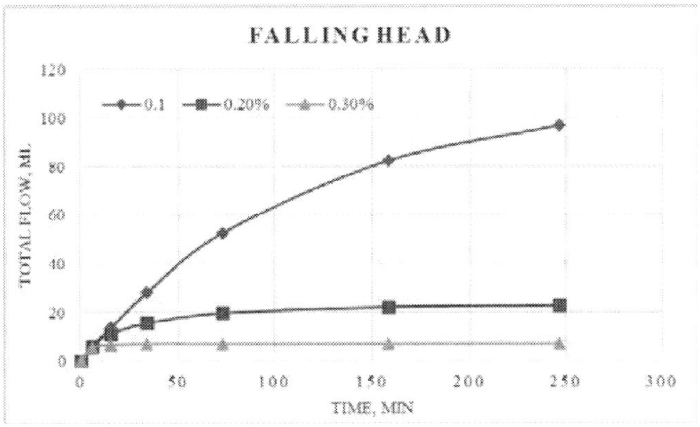

Fig. 13. Effect of the amount of SAP on the water flow rate.

6 Conclusions

The super absorbent polymer has a significant effect on the behaviour of concrete in its two stages, fresh concrete and hardened concrete. The texture of the concrete changes substantially, while the plasticity of the concrete improves. The bleeding water reduces with the increase in the amount of the super absorbent polymer added to the concrete mix. There is an optimum value of SAP to be added to the concrete as an admixture as far as the concrete strength is concern. The increase beyond this optimum amount will affect the concrete strength negatively. The SAP also provides the concrete with internal curing by releasing water slowly with time inside the concrete mass. This internal curing helps increase the concrete strength especially in areas where it is difficult to cure the concrete with water. The SAP absorbs water up to 500 times of its own weight and coverts it into gel. This gel expands with water, and occupies spaces and cavities in the concrete mass. This property can be utilized to seal the concrete cracks, and to stop the leaks in the cracked concrete structures subjected to water pressure. The increase in the amount of SAP added to the concrete mix improves the concrete sealing capability. This property is demonstrated by the ability of the cracked water flow concrete samples to stop the water flow. The crack is induced artificially in the water flow concrete samples. These water flow samples are then subjected to water pressure of constant value and of decreasing value. The samples of high SAP content was able to stop the water flow sooner when subjected to certain water pressure.

References

1. M. Al-Nasra, Concrete Made for Energy Conservation Mixed with Sodium Polyacrylates, Int. J. of Engineering Research and Application, **3**, Issue 5, ISSN: 2248-9622, pages 601 – 604, Sept-Oct 2013.
2. M. Jensen, Use of superabsorbent polymers in concrete, Concrete International, **35**, No. 1, January 1, 2013.
3. M. Jensen, O, Hensen, Autogenous deformation and RH-Change in prospective, Cement and Concrete Research, **31**, No. 12, Dec. 2001.
4. M. Jensen, O. Hensen, Water-Entrained cement-based materials: Experimental observations, Cement and Concrete Research, **32**, No. 6, June 2002.
5. M. Al-Nasra, Optimizing the Use of Sodium Polyacrylate in Plain Concrete, Int. J. of Engineering Research and Applications (IJERA) ISSN: 2248-9622,**3**, Issue 3, May-Jun 2013.
6. S. Didier, K. Tittelboom, N. De Belie, S. Steuperaert, P. Dubruel, The use of superabsorbent polymers as a crack sealing and crack healing mechanism in cementitious materials, *3rd International conference on Concrete Repair, Rehabilitation and Retrofitting*, Cape Town, South Africa, 2012.
7. M. Al-Nasra, and M. Daoud, Investigating the Use of Super Absorbent Polymer in Plain Concrete, Int. J. of Emerging Technology and Advanced Engineering, **3**, Issue 8, ISSN 2250-2459, 2013

3D printing cement based ink, and it's application within the construction industry

Zhu Jianchao*, *Tao* Zhang *Mansour* Faried, and *Chen* Wengang

China State Construction Engineering Corporation (Middle East)
Engineering Department 1 Lake Tower Plaza, Dubai, United Arab Emirates

Abstract. The 3D printing technology is the engine key of the third industrial revolution, after introduction of the automation in the eighteenth century and the concept of mass production in early of twentieth century. 3D printing technology now offers the magic solution to balance both the benefits, and overcome the major associated problem with the previous concept which was the need of repetition. The 3D printing technology has two main critical success factors: the printing machine and the printing material (ink). This paper focusses on cementitious-based materials and the ability to utilize the technology in the construction industry. The research took a qualitative approach based on previous literature reviews as well as in-house research results carried out by the authors' employer Research and Development Center. The paper summarizes the approach towards to an appropriate mix design which can achieve the requirement of the printing process, and overcome the current constraints which are hindering the wide application of 3D print in construction industry. The authors believe that the research topic and result will have great impact on pushing the construction industry forward towards achieving the UAE Government's strategy and target to achieve twenty-five percent (25%) of the buildings in Dubai by the year of 2030 relying on the 3D printing methodology. The research also concluded that even though the technology is adding a great value to the construction industry, it must be remembered that the technology is still in its infancy, and further research is required to achieve even higher strength printing materials that would be workable in multi-story buildings without the need of additional steel reinforcement.

1 Introduction

The three-dimensional (3D) printing technology concept has been known about since the 1980's. It is always promoted as a major form of innovation in many industries, but in fact, research on the topic to date extends for almost 4 decades–[1]. The technology application with the construction Industry is attracting field specialist interest day after day for the last few years, aiming to develop a giant scale 3D printer which could print the whole or a part of construction buildings, theoretically in any form or design [2].

3D printing technology is a form of Additive Manufacturing (AM), where a three dimensional object solid will be formed through the sequence-laying of horizontal layers

* Corresponding author: zhu_jianchao@chinaconstruction.ae

one on top of other [3]. There are hundreds of trials of deploying that concept in different forms, different materials and different machines.

According to Literatures, the two main parameters controlling the process of the 3D printing are the printing machine, and the printing ink, which accordingly could determine what kind of product the process could produce [4]. However, for construction purposes, there were 3 main categories that all the trials will fall under: Contour Crafting Method, D-Shape method, and Concrete Printing Method [5].

Each of the above processes will have a different objective, and then different forms of application.

The Contour Crafting Method is targeting the printing speed and productivity, so it is a kind of low resolution printer that is the most common method and what experts are aiming to develop further for mass production and mass construction [6], due to high speed of printing, a certain limitation over material strength and building shapes are associated with this method.

The D-Shape Method on the other hand, is targeting the complexity of the printed element; it works in different methodology as the machine is laying a full layer of fine sand and loose material, then the printer nozzle will go over and apply a form of glue as a binder on the locations which is required to be solid. That method is very slow, but it can produce both high resolution, and complicated objects, especially as it uses the loose material as a bed to support the suspended and horizontal surfaces, which the other 2 machines cannot do [7].

The third category of machines are what called as concrete printers or "Concjet". Those machines are using the same method of printing as the Contour Crafting Method as it pre-mixes the printing material before printing, but the difference here that it goes in smaller thickness of layers with lower speed of printing which allows to use higher strength of concrete. However, with this process, the low printing speed is putting a major limitation on the productivity and object size. [5]

This paper will focus on the Contour Crafting Method, and will represent some of laboratory test results carried out by the Author's in-house research and development center in Beijing - China, which aimed to develop a machine and printing material that could be used in rapid construction and 3D printing for building. The paper's scope will be more focused on the cement-based printing ink and how to control its variable parameters in order to fit the purpose in terms of fluidity, workability, setting time, initial strength and final strength.

2 Cement-based printing ink

For the last 100 years cement has been considered as the master raw material for construction, the cement used is mainly one of the 3 major types: ordinary Portland cement (OPC), calcium-aluminate Cement (CAC) and sulphoaluminate cement (SAC). The three types of cements have a different characteristics and accordingly different applications. OPC is the most commonly-used cement material and recognized widely for its high strength and stability as well as its low cost, while its disadvantage is the longer setting time comparing to other types. The other 2 types provide a shorter setting time with a high early strength capacity which make them more fit to several special engineering applications such as high temperature application and quick repair in emergency situations.

This paper mainly focusses on both of OPC and SAC as a principal raw material for the cementitious 3D printing Ink, while CAC has been excluded due to its lower final strength [9] which requires additional researches and tests and will be the scope of a future paper, currently under progress. The research intends to examine the above-mentioned materials (OPC & SAC) for their different properties such as setting time (initial and final), workability and durability, and mechanical properties such as compressive strength and flexural strength.

SAC has been extensively used in China for wide variety of building works, especially for those buildings located in very low temperature zones, ports projects, underground construction projects and precast concrete members it has a proven track record of successful application and it is fit for the purpose it has been planned for. The SAC has a short setting time, high strength at early stage, and steadily developing strength at later stage. These properties are well-suited and desirable for the 3D printing process, as well as the architectural requirement of the technology. Therefore research into the enhancement of SAC's setting time and workability, has resulted in the development of an SAC-based 3D printing material. Furthermore, these material properties and performance have been analyzed and tested.

OPC is the most widely-used cement worldwide due to its low cost, availability of manufacturers and suppliers, stable performance and many other advantages. OPC has also been used to produce the 3D printing materials, However OPC, characterized by its slow hydration, longer setting time and lower strength when compared to SAC hinders its application in the 3D printing technology.

In principle, the 3D printing process requirement in terms of materials can be summarized as follows; 1) controllable and flexible setting time, 2) have a certain level of workability and fluidity to allow material to flow through the printing head while still having the ability to bear the above layer's weight after printing, 3) to have a proper bonding between the Various layers. That bonding should be uniform in all directions. 4) Should have high early strength to ensure that material can support own weight and super-imposed layers as, with the progress of printing, the bottom layers will be required to take the load of the whole wall height in 2-3 hours.

The above 4 criteria have been set as the material development research objective in order to reach to the specific mix design which can fit the purpose.

3 Selection of raw materials and method

3.1 Cement

For the test purpose, samples for both materials have been obtained for the available commercial materials in Chinese Local Market, the chemical composition for the tested samples and their performance parameters could be seen-below in table 1 and table 2

Table 1. Chemical composition of tested samples.

T	Al2O3	CaO	SO3	SiO2	Fe2O3	MgO	Loss on ignition
SAC	35.17	42.54	10.79	6.13	1.53	1.24	2.10
OPC	6.65	58.93	2.54	24.12	3.78	0.79	3.19

Table 2. Physical and mechanical properties for the samples.

Varieties	Normal consistency	Setting time / min		The stability of	Compressive strength /MPa		Flexural strength /MPa	
	/%	Initial setting	Final setting		3d	28d	3d	28d
SAC	27.9	45	73	Qualified	47.6	55.7	7.7	8.6
OPC	28.1	183	237	Qualified	28.4	52.5	5.3	8.1

3.2 Aggregate

in the process of 3D printing, the 3D printer will be fed by material through a pump, high pressure pipe and a nozzle, the smallest diameter in this route is the nozzle diameter which is about 50mm. Accordingly, and in order to ensure proper flow of material the maximum size of aggregate cannot exceed 5mm. The experiment found that the most suitable material for that could be the very fine sand corresponding to 20-40, 40-70 quartz sand which is available widely in Middle East region.

3.3 Mixing ingredients (mineral additives)

For this experiment, activated blast furnace slag produced by a manufacturing firm in China has been used as mineral additive in order to improve the workability without effecting the strength. This material is either gray or white color and it has a specific surface area of 600 m2/Kg, it also compliant with Chinese standard GB/T18046-2000. This material is found to be of high quality to fit its purpose with pozzolanic properties. The chemical composition of that material is as per below Table 3.

Table 3. Chemical composition of the slag powder additive.

Al2O3	CaO	SO3	SiO2	Fe2O3	MgO	Loss
15.62	32.81	3.86	36.11	3.08	5.79	2.73

3.4 Setting time controllers

The key technology in the 3D printing process is to control the setting time, either by accelerating the setting time using accelerators additives, or by delaying the setting time using retarders. In the case of SAC, different admixtures has been used as accelerators: lithium carbonate (later will be referred to as J1), Lithium Hydroxide (J2), Sulfate (J3) and various type of retarders: boric acid (H1), sodium gluconate (H2), tartaric acid (H3), citric acid (H4), sodium tetra borate (H5) and tri-sodium phosphate (H6).

For OPC, the SAC itself can be used as additive for acceleration and certain amount of slump-component admixture as a retarder.

3.5 Water reducing agents (superplasticizers)

A locally manufactured product in China has been used as superplasticizers for both types of cement which mainly composed of poly-carboxylic acid powder.

3.6 Volume stabilizer

Several components have been developed in house to maintain the concrete volume and minimize the shrinkage has been used with both types of cement.

4 Measurements standards

Chinese standards have been used as a reference to measure and evaluate the material parameters and performance: Chinese standard GB/T 1346-2011 has been used to test the water requirement for normal consistency, setting time and soundness of OPC. Chinese standard GB/T 17671-1999 & JGJ70-2009 has been used to measure the cement strength. Chinse standard GB/T 2419-2005 has been used to assess the fluidity of the cement mortar. Lastly, Chinese standard GB/T 50082-2009 has been used to measure the long-term performance and durability of ordinary concrete as well as material chemical resistance and anti-carbonation durability. For those tests, commercial equipment has been deployed to measure the concrete viscosity and concrete behavior in terms of hydration and development of strength across the time.

The material curing has been made using a curing box at temperature of 20 °C, and humidity of 95%. Testing cubes have been made in the size of 25mm × 25mm × 280mm.

5 Setting time control

The material that has been developed and tested is still considered as a cement mortar, although it has been enhanced and improved to fit the technology requirement through chemicals additives. The research tested both SAC and OPC, and tried to adjust the setting time, workability, and early strength to meet the desired requirement of the 3D printing process. There is a lot of previous research which concluded that cement-sand ratio of 1:1 can lead to the best workability and mechanical properties of the mortar. The same ratio has been considered in this research as well.

A super plasticizer agent has been added to reduce the water content, and therefore higher final strength could be achieved accordingly a water/cement ratio of 0.35 – 0.38 has been used. Furthermore, by adding mineral admixture, composite volume stabilizer and other additives it is possible to improve material mechanical properties and setting time control for SAC.

5.1 Selection of retarder

Since SAC is characterized by its rapid hardening, the addition of retarder to the mix is strongly invited, especially when the surrounded environment has a high temperature, the retarder in concept will generate an insoluble thin film around the cement particles which will act as a barrier and reduce the hydration rate of cement and subsequently will delay the hardening process. For this study, 6 different types of retarders have been tested individually: boric acid (H1), sodium gluconate (H2), tartaric acid (H3), citric acid (H4), sodium tetra borate (H5), tri-sodium phosphate (H6), The effect of each of the those retarders on the SAC setting time can be seen as shown in Figure 1 & Figure 2 below.

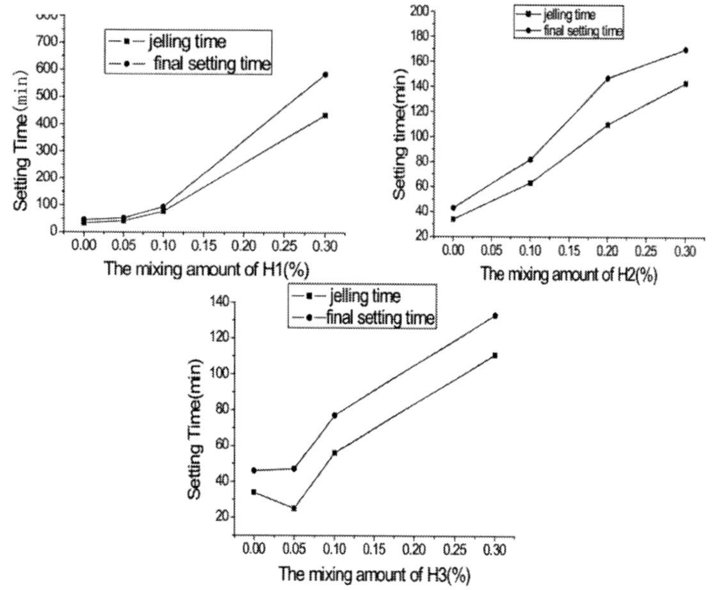

Fig. 1. Effect of retarders (H1, H2, and H3) on the setting time of SAC.

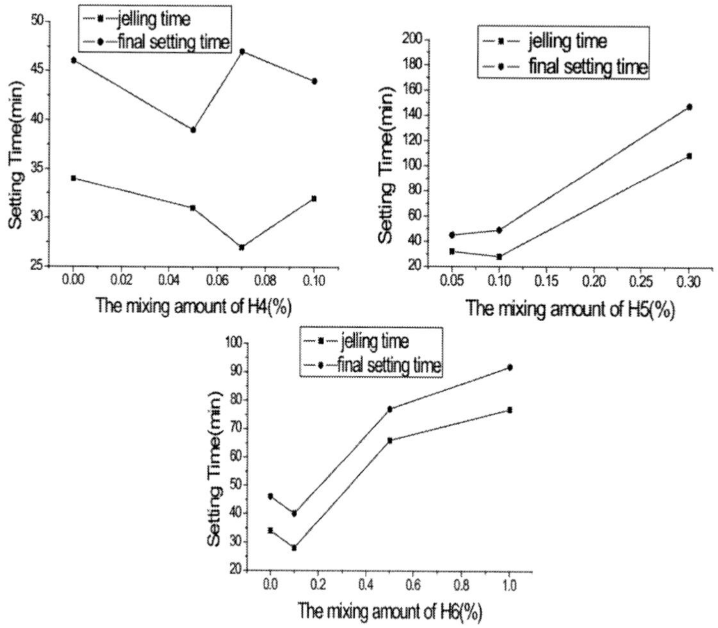

Fig. 2. Effect of retarders (H4, H5, and H6) on the setting time of SAC.

Figure 1 and Figure 2 above explain the behavior of mortar setting time with the change of the quantity of the retarding agent added for each material separately. From the analysis of the above, some of those agents have been excluded due to undesired and inconsistent development of setting time delay and accordingly they are not suitable for the 3D printing purpose.

5.2 Selection of accelerating agent

The accelerating agents is just an opposite of the retarder in its action, and it is essential that it is used specially in a very low temperature environment to order to gain higher early strength, and also in cases where certain engineering mixes are required to achieve high early strength in a short time. For this study, three different materials have been tested as an accelerating agents: lithium carbonate (J1), lithium hydroxide (J2) and calcium chloride (J3). The effect of those agents on SAC setting time are shown in Figure 3.

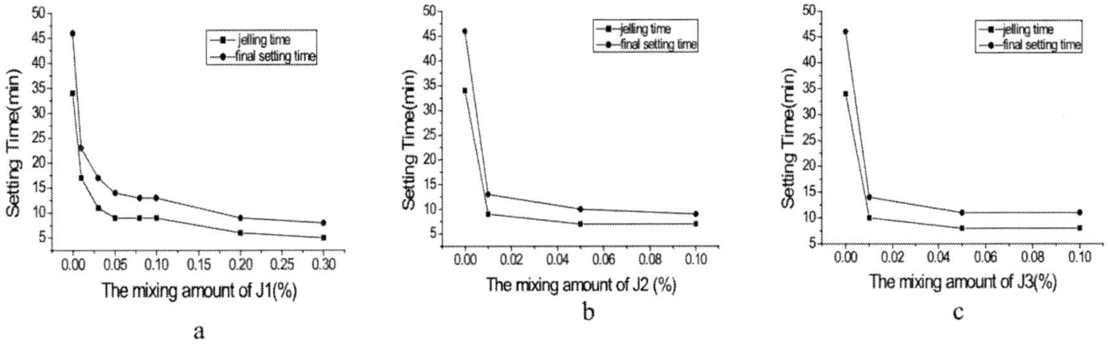

Fig. 3. Effect of accelerating agents (J1, J2, and J3) on the setting time of SAC.

By analysing the above graphs, it can be concluded that J1 can reduce the setting time to half when using it at the ratio of 0.01% (17 minutes), which can be reduced further to 9 minutes at ratio of 0.05% and to 5 minutes when the ratio is increased to 0.3%. For J2, as it has strong base, accordingly it will have a stronger effect on reducing the initial setting time comparing to J1 and J3, so the setting time can be shortened to 9 minutes with the ratio of 0.01%. For J3, it has a good performance with OPC, however its preference for SAC is not favourable.

In term of developing the strength, the below Figure 4 will show the difference between the three accelerators comparing with the case when the cement is used with any accelerators over different duration of time.

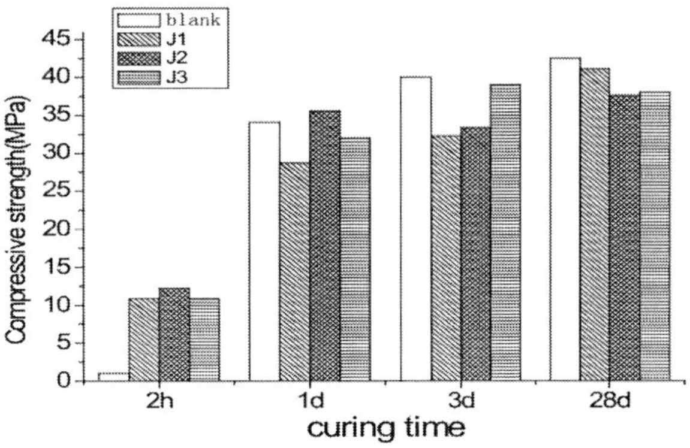

Fig. 4. Effect of early strength agents on the strength of SAC.

5.3 Testing results

Several trials have been made using each of those chemical agents and different performance measures have been taken to verify the impact of each agent separately. The experiments have shown the following:

A) Some of those tested retarders were found to have a counter-productive reaction at the early stage where, at certain dosage it acted as an accelerator. That is totally unsuitable for the 3D printing process and accordingly they were excluded.
B) With some other retarders, the increment of the dosage would dramatically increase the setting time disproportionally to the dosage increment value, and that also is not preferable to 3D printing process and accordingly they have been excluded.
C) Some of the tested retarders were found to be characterized with unstable and unpredictable performance with the dosage increment that also has been excluded.
D) Some retarders found to be fit for application with a certain amount of minimum dosage, since the reaction performance below that dosage is not stable. Although it is not highly preferred, that particular retarder was not excluded as it could act better when mixing with other types of retarders.
E) Some of the accelerator agents found to be with a minor impact on SAC comparing to OPC and since complete control over material setting time is targeted then that particular accelerator has been excluded.
F) Generating a composite retarding agent by certain mix of different retarders has delivered a more desired performance compared to the performance of each component acting individually.
G) To obtain the ultimate time-control agent, a further composite solution has been generated (referred to later as JH) by adding an accelerating agent to previously concluded composite retarding agent. Using several experiments on the dosage, the relationship between the JH dosage and setting time of concrete and early strength of concrete was calculated.
H) Different compositions for JH agent were developed (JH1- JH6) with different ingredients of retarders and accelerators, the below table 4 reflects the value of initial setting time and final setting time of each.

Table 4. Initial and final setting time for composite agent JH1-JH6

Use the adjustable solidification agent JH Composition	Setting time	
	Initial setting time	Final setting time
JH-1	15	26
JH-2	35	46
JH-3	75	90
JH-4	24	38
JH-5	45	59
JH-6	87	109

6 Evaluation parameters of tested materials

6.1 Mechanical properties

The samples for test is collected by cutting it off from a full size printed walls, the samples have been tested for the compressive and flexural strength, and Table 5 below represents the best results obtained when using JH-2 and a setting time control agent.

Table 5. Flexural and compressive strength of specimen.

Strength	Flexural strength /MPa		Compressive strength/MPa	
	2h	**28d**	**2h**	**28d**
Molding specimens	2.4	8.3	13.5	58.6

According to results in Table 5 above, the early compressive strength after 2 hours is more than 10MPa, which is sufficient to support the wall full weight. Figures 5 to 8 below represent different directions of flexural strength, compressive strength and flexural failure morphology.

Fig. 5. Transverse specimens Flexural strength.

Fig. 6. Vertical specimen flexural strength.

Fig. 7. Transverse specimens compressive strength.

Fig. 8. Vertical specimen flexural damage Morphology.

Figure 9 below show a comparison between the different directions of Flexural strength and compression strength. The results proposed a considerable difference between different directions of loading due to higher bonding strength within each layer compared to bonding between different layers, which need to be taken in account during the design. However, the vertical compressive strength can still reach up to 44.6 MPa, which is good enough to ensure safety of the structure.

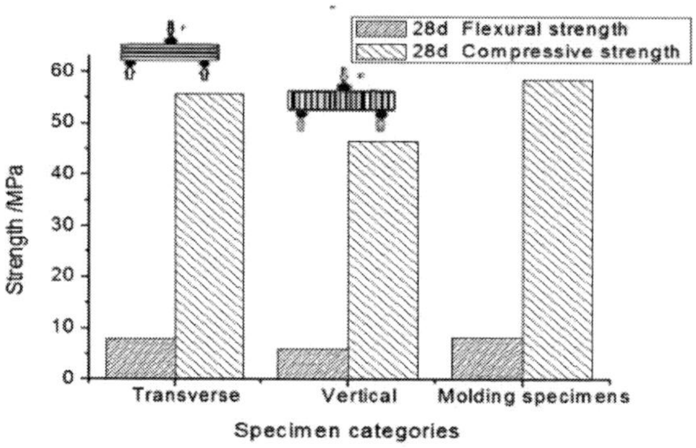

Fig. 9. Comparison of the compressive strength of cutting specimens and molding specimens.

6.2 Chemical properties

The durability of the material is directly related to its chemical properties and its resistance to various chemicals. The printed material has been tested for various chemical tests as following:

- Chloride permeability Rapid chloride permeability test method to measure the migration of the chloride ions within the material. The below Table 5 represent the relation between chloride ion permeability of the concrete and electric flux. Table 6 represents measurements taken for 3 different samples of 3D print materials, according to both Tables 5 and 6, the permeability of the 3 D printed materials can be classified as "medium".

Table 5: The corresponding relationship between electric flux and chloride ion permeability of concrete.

Electric flux/C	Chloride ion permeability	Concrete
＞4000	High	High water-cement ratio (0.6)
2000～4000	medium	Medium water cement ratio (0.4 ~ 0.5)
1000～2000	low	Low water cement ratio (0.3 ~ 0.4)
100～1000	very low	Emulsion modified concrete
＜100		silica fume concrete etc.

Table 6: Results of electric flux for the 3D printing materials.

NO.	3D-1	3D-2	3D-3
Electric flux/C	2441Q	2332Q	2249Q

- Resistance to freeze-thaw cycle performance. The rapid freeze-thaw method has been used to test the thermal performance of the 3D printing materials, as it measures the loss in weight and the relative dynamic elastic modulus after exposing the material to 200 cycles of Freeze-thaw. The results of the test are summarized in table 7 below.

Table 7: The freeze-thaw cycle experimental results of building 3D printing materials.

Times	0 time	100 times	200times	200 times

Item	Weight /g	Dynamic modulus/ GPa	Weight /g	Dynamic modulus/ GPa	Weight /g	Dynamic modulus/ GPa	Weight Loss /%	Relative dynamic modulus /%
	9171	33.39	9187	32.86	9201	32.94	0	97.3%

The result of table 7 above indicates that the materials didn't encounter weight loss after 200 cycles of Freeze-thaw. Accordingly the materials can be used safely in cold environments.

- Impermeability and carbonation resistance. The concrete has six grades to classify its Impermeability, which are P4, P6, P8, P10, P12, and more than P12 [8], which basically represent the materials ability to deal with water pressure in the impermeability test, and accordingly the waters ability to penetrate the material. Figure 10 below indicates that the tested material can be classified as "greater that P12 with average seepage height of 24.5 mm.

Fig. 10. Water penetration height under pressure P12.

Drying shrinkage of building 3D printing materials Figure 11 below represents a comparison between the 3D printing material and the normal cement mortar in terms of volume shrinkage:

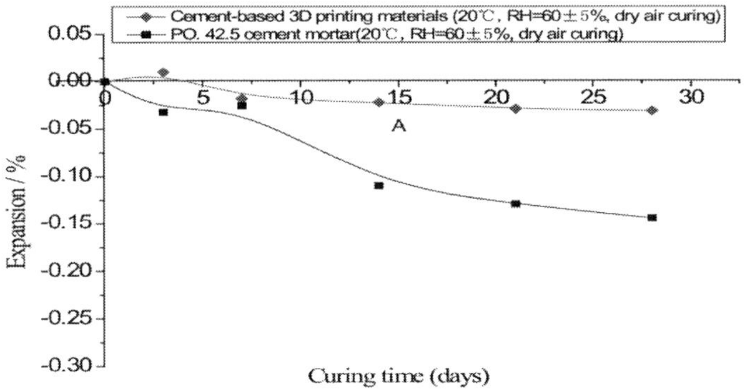

Fig. 11. Comparison of shrinkage of PO42.5 standard cement mortar and building 3D printing materials.

From Figure 11 above, it is obvious that the shrinkage of 3D printing mortar is much less compared to normal mortar which is a considerable added-value feature.

7 Application of cement-based building 3D printing materials

 DOI: 10.1051/matecconf/201712002003

Printing Ink development can't go alone with the development of printing machine and printing software, The Technical Center of CSCEC in Beijing has developed various versions of printing machines in different sizes as shown below in Figures 12 to Figure 16

Fig. 12. Small-sized building 3D printer.

Fig. 13. Large-sized building 3D printer.

Fig. 14. Process of printing wall member used cement-based 3D printing materials.

Fig. 15. Steel of 3D printing wall .

Figure 14 above shows how the early setting time plays an essential role in the 3D printing process, as the bottom layers has got enough strength to support the top layers within the time used by machine to print up to the top layers.

Figure 15 shows the possibility of adding steel members (manually) during the printing process to improve the structural performance of the printed members.

8 Conclusion

The paper represents several researches made by the In-house Technical Center of the Author's Company on the development of the 3D printing Mortar (Ink) in order to use the 3D technology in Construction Industry. The Paper examined two types of cement, OPC and SAC for the production of 3D printing mortar. It also examined different setting time controllers, plasticizers and volume stabilization agents. The paper proposes that using SAC will be more appropriate for the 3D printing mortar due to its shorter initial setting time as well as higher early strength properties. The paper also proposes that for the comprehensive time setting time control, a compound of retarder and accelerator are required to be used together, the type of each and quantity are variable upon the printing conditions, temperature, ratio between width and height of the printing surface, printing speed and the desired final strength.

The paper also highlights the importance of the development of the printing machine along with the development of printing materials, and the possibility of using Steel members to improve the structural properties of the printed members, which is still a subject for further research.

References

1. Khoo, Z., Teoh, J., Liu, Y., Chua, C., Yang, S., An, J., Leong, K. and Yeong, W. (2015). 3D printing of smart materials: A review on recent progresses in 4D printing. *Virtual and Physical Prototyping*, **10**,3 pp.103-122.
2. Canessa, Enrique, Carlo Fonda, and Marco Zennaro. *Low-Cost 3D Printing For Science, Education & Sustainable Development*. ICTP—The Abdus Salam International Centre for Theoretical Physics, (2013). Print.
3. Le, T., Austin, S., Lim, S., Buswell, R., Gibb, A. and Thorpe, T. (2012). Mix design and fresh properties for high-performance printing concrete. *Materials and Structures*, **45**,8 pp.1221-1232.
4. Wong, K.V. and Hernandez, A., (2012). A review of additive manufacturing. *ISRN Mechanical Engineering, 2012.*
5. Lim, S., Buswell, R., Le, T., Austin, S., Gibb, A. and Thorpe, T. (2012). Developments in construction-scale additive manufacturing processes. *Automation in Construction*, **21**, pp.262-268.
6. Bhatia, U., (2015). 3D printing technology. *Int. J. Eng. Tech. Res*, **3**, pp.327-330.
7. Le, T., Austin, S., Lim, S., Buswell, R., Law, R., Gibb, A. and Thorpe, T. (2012). Hardened properties of high-performance printing concrete. *Cement and Concrete Research*, **42**,3 pp.558-566.
8. Chumnanklang, R., Panyathanmaporn, T., Sitthiseripratip, K. and Suwanprateeb, J., (2007). 3D printing of hydroxyapatite: effect of binder concentration in pre-coated particle on part strength. *Materials Science and Engineering: C*, **27**,4 pp.914-921.
9. Heikal, M., Radwan, M.M. and Morsy, M.S., (2004). Influence of curing temperature on the Physico-mechanical, Characteristics of Calcium Aluminate Cement with air cooled Slag or water cooled Slag. *Ceramics-Silikáty*, **48**,4 pp.185-196

The effects of silica fume and hydrated lime on the strength development and durability characteristics of concrete under hot water curing condition

Ali Hamza[1,*], *Shahram* Derogar[2] , and *Ceren* Ince[1]

[1]Civil Engineering Department, Middle East Technical University, Northern Cyprus Campus, Kalkanlı, Güzelyurt, North Cyprus via Mersin 10, Turkey.
[2]Civil Engineering Department, European University of Lefke, Lefke, North Cyprus via Mersin 10, Turkey.

Abstract. Sustainability is considered to be highly important for preserving continued industrial growth and human development. Concrete, being the world's largest manufacturing material comprises cement as an essential binding component for strength development. However, excessive production of cement due to high degree of construction practices around the world frames cement as a leading pollutant of releasing significant amounts of CO2 in the atmosphere. To overcome this environmental degradation, silica fume and hydrated lime are used as partial replacements to cement. This paper begins with the examination of the partial replacement levels of hydrated lime and silica fume in concrete and their influence on the mechanical properties and durability characteristics of concrete. The effect of hot water curing on concrete incorporated with both silica fume and hydrated lime is also investigated in this paper. The results reported in this paper show that the use of silica fume as a partial replacement material improved both the mechanical properties and durability characteristics of concrete due to the formation of calcium silica hydrate crystals through the pozzolanic reaction. Although the hydrated lime did not significantly contribute in the development of strength, its presence enhanced the durability of concrete especially at long-term. The results also showed that hot water curing enhanced the strength development of concrete incorporated with silica fume due to the accelerated rate of both the hydration and pozzolanic reaction that takes place between silica fume and calcium hydroxide of the cement matrix particularly at early times. The results reported in this paper have significant contribution in the development of sustainable concrete. The paper does not only address the use of alternative binders as a partial replacement material in concrete but also suggest proper curing conditions for the proposed replacement materials. These practices play a vital role in the development of sustainable concrete.

[*] Corresponding author: ali.hamza@metu.edu.tr

1 Introduction

Recent developments in the construction industry has promoted concrete as 'highly efficient material with improved mechanical properties and durability' [1]. However, such advancements have also highlighted cement and concrete manufacturing as carbon and energy intensive processes. It has been claimed that up to 5 – 9% of the total global anthropogenic carbon emissions [2, 3] are due to manufacturing of cement and concrete. Furthermore, concrete is not an everlasting building material when analyzed against time. Eventually, the life cycle of concrete will end up as limestone, sand and clay within our earth's environment [1]. Therefore, the significant issue of improving durability and service life of concrete is always crucial to have its attention [4]. From the idea of sustainability in concrete manufacturing, there is a need to develop an environmentally friendly type building material which ensures a good balance between performance and sustainability concern [5, 6]. With the extensive use of Portland cement, it is inevitable that such idea of sustainable concrete can be accomplished. However, if the cement manufacturing is tend to shift with the mixture of 'second generation' or 'low carbon' binders [7], the idea of achieving sustainable concrete may emerge to existence.

In various recent studies, it has been addressed that hydraulic lime and supplementary cementitious materials (SCMs) do possess essential properties that leads towards strength development and enhancement of durability of concrete. [8-11]. Hydraulic lime, being one of the oldest binders contributes substantial history in construction practices and manufacturing [7]. Furthermore, SCMs broadly pozzolans, have also been considered as a viable approach to be used as an alternative binder [7]. Pozzolans normally results in poor performance and low strength development at early stages due to absence of adequate quantities of $Ca(OH)_2$ [12]. To overcome such issue of strength development and durability, possibilities for improving the pozzolanic reaction by adding hydrated lime have been mentioned in the literature [13-19]. Silica fume, being one of the most popular pozzolans is considered to play a vital role in developing strength, durability characteristics and consuming calcium hydroxides during the hydration reaction of Portland cement [18, 20, 21]. Therefore, such pozzolanic reactions may contribute to lower heat liberation, adequate early strength development, effective consumption of hydrated lime and enhancement in durability characteristics [6, 18].

Replacing Portland cement with hydrated lime and pozzolans such as silica fume does not solve this complex issue. Since, each binder has its own chemical reaction, it is important to highlight the significance of curing time and conditions. Cement hydration completes in 28 days as cement attains its compressive strength. However, lime carbonation is comparatively a slower reaction. Even though, pozzolanicity does have the ability to consume calcium hydroxide to form calcium silicate hydrate (C-S-H) which is essential for strength development, especially at early stages [17], the rate of reaction is also comparatively slower. In addition to that, it has also been criticized that various experimental studies conducted on concrete incorporated with hydrated lime and pozzolans have been cured for short interval of time. The results may vary if the curing period was much longer [22] and/or the curing conditions have been provided in a different way such as the introduction of hot water curing or steam curing conditions [23].

Therefore, the aim of this paper is to offer a holistic approach of using hydrated lime and silica fume as partial replacement of Portland cement as binders. The idea is to investigate the possible role of hydrated lime and silica fume used as a partial replacement to cement on the strength development and durability characteristics. Strength development includes compressive strength test as a key parameter being investigated. Tests such as water permeability and freezing/thawing resistance for determining the pore structure of concrete are included in the durability characteristics. The possible effect of hot water curing conditions is also highlighted addressing the significant role on strength development and durability characteristics of the concrete.

2 Materials and mix design

The experimental study was conducted on freshly-mixed and hardened state concrete using Portland cement, hydrated lime (CL80) and Silica Fume as binders. Portland cement was acquired from Adana Cimento Ltd. and complies with [24]. CL80 was obtained from Nur Kirec Ltd. and complies with [25]. The designation of CL80 indicates the calcium lime content is not greater than 80%. Silica Fume was acquired from Cyprus Environmental Enterprises Ltd. (CEE) and complied with [26]. Apart from binders, other elements of concrete prepared include fine aggregates, coarse aggregates and water. Particle size distribution of fine aggregates and coarse aggregates have been made accordance with [27]. Various replacement levels of CL80 and silica fume were used to investigate the influential role on the fresh and harden state properties of Portland cement concrete. For mix design calculations, the slump value has been kept constant to 100 mm. Therefore, the water/binder (w/b) ratio have been adjusted accordingly. The masses of binders required to produce the necessary mix proportions by volume were calculated from carefully determined values of relative density. For fine and coarse aggregates, specific gravity was determined according to [28, 29].

3 Experimental procedures

3.1 Measurement of slump

The slump of fresh concrete prepared containing various volume fraction replacement levels of CL80 and silica fume was measured using the standard slump test apparatus. The slump value has been determined as per [30]. The slump value has been kept constant to 100 mm. By keeping such consistency of slump value, the water/binder ratio has been kept between the ranges of 0.50 to 0.70.

3.2 Curing conditions

Two types of curing conditions have been investigated for curing processes of concrete. A normal portable water curing condition having water temperature of 20°C is maintained. Furthermore, hot water curing condition having water temperature to 40°C is additionally kept up. Curing conditions have been standardized as per [31].

3.3 Measurement of compressive strength

Compressive strength testing was conducted on various hardened replacement levels of concrete including CL80 and silica fume respectively. The testing has been conducted from 28 days to 180 days on 150 mm standard cubes. Samples of each replacement level were tested using a compressive testing machine. The loading rate has been kept constant to 0.6 MPa/sec. An average compressive strength has been taken among all the values from each replacement level of concrete produced in accordance with [32].

3.4 Measurement of water permeability

Standard cubes of 150 mm were used for determining the water permeability for each replacement level of concrete produced. Each specimen was fixed into the permeability testing machine and pressurized water has been applied from the bottom of the specimen. Adjustment of the sample is important on the permeability machine to avoid water leakage from the sides of the specimen. All specimens were subjected to 1 atmospheric pressure (0.1 MPa). It is important to mention that the mass measurements before and after the water permeability test should be taken to understand each specimen's water infiltration capacity. After 72 hours, each specimen was removed from the permeability machine and split into two equal halves for determining the water penetration depth. Average depth was taken from four equidistant spots along the bottom of the specimen from where the water pressure has been applied. The experiment is done based on [33].

3.5 Measurement of freeze and thaw resistance

For freeze and thaw cycles, cube samples of 150mm have been placed in freeze and thaw cabinet. Similar procedure has been followed as reported by [34]. It includes freezing and thawing phenomenon for the cured samples of various replacement levels at -30°C to 30°C. Two cycles have been maintained on a daily basis. Within 12 h each day, the samples have been experiencing freezing and thawing automatically when the temperature reaches -30°C or 30°C. The cycle keeps on going twice a day from the last 180 days. The relative humidity is maintained ranging from 85% to 95%. It is important to mention that the surface of the samples have not been covered during the freezing and thawing, so that the sample is fully exposed to moisture changes.

4 Experimental results

4.1 Mechanical properties

4.1.1 Influence of silica fume on the compressive strength of concrete

The influence of the use of silica fume as a partial replacement to cement on the compressive strength of concrete is investigated in this section. Replacement levels of silica fume were 10 and 20%. The samples under investigation were cured under water at the laboratory conditions for 180 days. The influence of the use of silica fume on the compressive strength of concrete is shown in Figure 1.

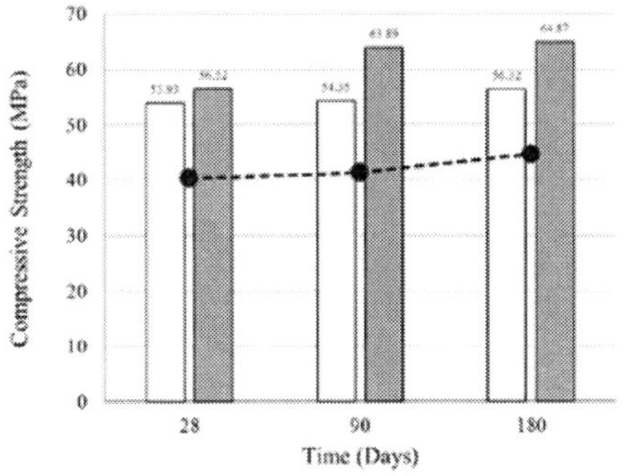

Fig. 1. Influence of silica fume on the compressive strength of concrete □; 20% silica fume, ▨; 10% silica fume and ●; concrete control specimen.

It is shown in Figure 1 that the use of silica fume resulted in an increase in the compressive strength of concrete over the 180 days when compared to the control concrete specimen. However, the increase in the replacement levels of silica fume resulted in a decrease in the compressive strength. A very hydraulic binder, cement, is being replaced with a pozzolanic material that reduces the rate of hydration reaction specifically at short terms. The development of the concrete strength after 28 days is solely depending on the pozzolanic reaction between the calcium hydroxide and the silica fume. It can be seen in Figure 1 that at each replacement level, the compressive strength of concrete is increasing with time. It must be underlined that both replacement levels of silica fume resulted in a greater compressive strength of concrete with time.

4.1.2 The effect of hot water curing on the strength development of concrete incorporated with silica fume

The effect of hot water curing on the strength development of concrete incorporated with silica fume is investigated in this section. The effect of hot water curing is used particularly for the early strength development of pozzolanic reaction which has much slower rate than cement hydration. The effect of hot water curing on the compressive strength of concrete with 10% replacement level of silica fume is shown in Figure 2.

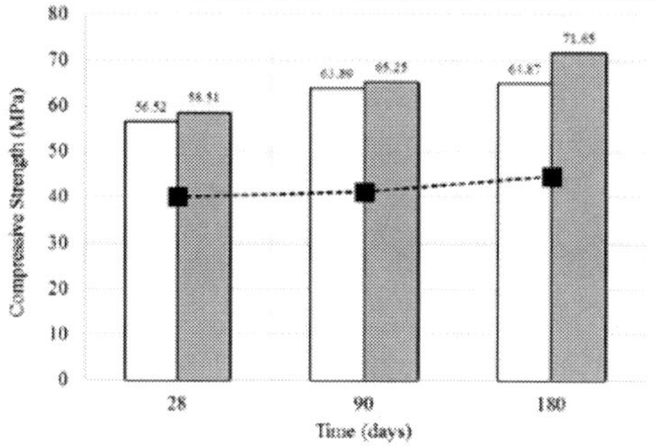

Fig. 2. The effect of hot water curing on the strength development of concrete incorporated with 10% silica fume □; cold water cured, □; hot water cured, •; concrete control specimen.

It can be seen in Figure 2 that hot water curing enhanced the strength development of pozzolanic concrete. The increase in the compressive strength of concrete incorporated with 10% silica fume is more markedly seen in 6 months. It is well known that pozzolanic reaction is depended on the hydration reaction and hence the formation of calcium hydroxide. The effect of hot water curing on the strength development of concrete therefore cannot yet be seen at 28 days. It must be underlined however that hot water curing enhanced the compressive strength of concrete incorporated with 10% silica fume at 3 and 6 months significantly.

4.1.3 The effect of using hydrated lime on the strength development of concrete incorporated with silica fume

In this section, cement is partially replaced with silica fume and hydrated lime. Replacement level of silica fume was kept constant to 20% while the replacement levels of hydrated lime were 10 and 20%. The effect of using hydrated lime on the compressive strength of concrete incorporated with silica fume cured under both cold and hot water conditions is shown in Figure 3.

It can be seen in Figure 3 that the use of hydrated lime as a replacement to cement resulted in a decrease in compressive strength of concrete incorporated with silica fume. Compressive strength of concrete replaced with hydrated lime and silica fume, however is increasing with time. The increase in the compressive strength is mainly due to the presence of calcium hydroxide in the cement matrix that accelerated the rate of the pozzolanic reaction. It can be seen in Figure 3 that at 6 months, the compressive strength of concrete at all combinations are higher that the compressive strength of the control specimen at 28th days. Hot water curing also resulted in an increase in the compressive strength of concrete at both replacement levels of hydrated lime.

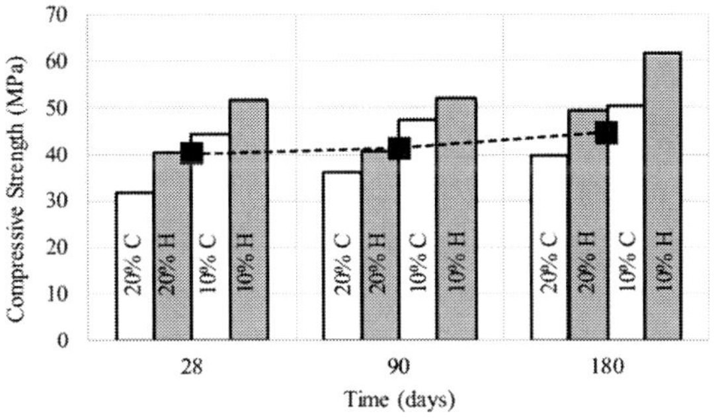

Fig. 3. The effect of using hydrated lime on the compressive strength of concrete incorporated with silica fume cured under both cold and hot water conditions •; concrete control specimen, C; cold water curing, H; hot water curing.

4.2 Water permeability

4.2.1 The effect of using hydrated lime on the water permeability of concrete incorporated with silica fume

The effect of using hydrated lime and silica fume on the water permeability of concrete is investigated in this section. Concrete specimen with 20% replacement level of silica fume is considered first in this study. Concrete specimen with 20% replacement level of silica fume and 20% replacement level of hydrated lime is then used to study the freeze and thaw resistance of concrete. Concrete control specimen is also subjected to the analysis for the comparison reason. Water penetration depth (mm) versus time (days) for concrete samples are shown in Figure 4.

It can be seen in Figure 4 that the use of 20% silica fume as a replacement to cement resulted in a dramatic decrease in water penetration depth of concrete at 3 and 6 months. These results are in good agreement with the strength development of concrete incorporated with 20% silica fume. The great reduction in water penetration depth of concrete and hence the increase resistance to water permeability can be attributed to the pozzolanic reaction and therefore the consumption of calcium hydroxide and thus the formation of more of calcium silica hydrate gels that enhances the resistance to water permeability of concrete at long-term.

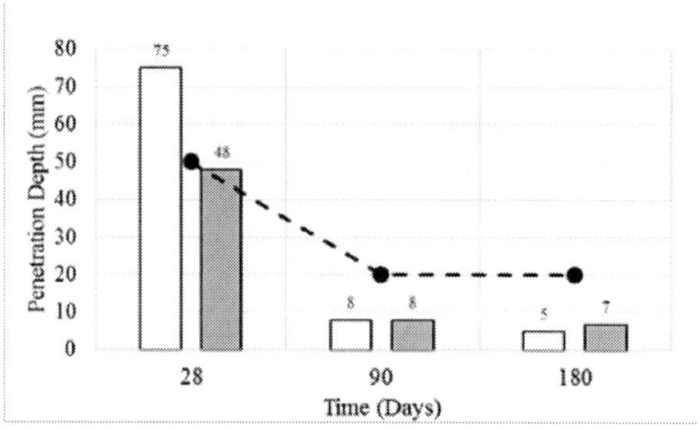

Fig. 4. Effects of using hydrated lime and silica fume on the water permeability of concrete □; 20% silica fume, □; 20% silica fume with 20% hydrated lime, •; concrete control specimen.

It can also be seen in Figure 4 that the use of 20% hydrated lime in concrete incorporated with 20% silica fume is also leading a great reduction in water penetration depth of concrete. Even at 28th days, there is a remarkable decrease in water penetration depth of concrete when hydrated lime is used in concrete incorporated with silica fume. Since the particle size of silica fume and hydrated lime are much finer than the cement binder, filler effect must played a significant role in the resistance to water penetration of concrete. It must be underlined that at 180 days, the concrete specimen prepared using 20% silica fume and 20% hydrated lime resulted in a great reduction in water penetration depth and hence resulted in an enhanced resistance of water penetration of concrete.

4.3 Freeze and thaw resistance

4.3.1 The effect of silica fume on the freeze and thaw resistance of concrete

The effect of silica fume on the freeze and thaw resistance of concrete is investigated in this section. Cement replacement level of 10 and 20% of silica fume is used to study the freeze and thaw resistance to concrete. Following cured at laboratory conditions for 28 days, the specimens were subjected to cycling freeze and thaw for 28, 90 and 180 days.

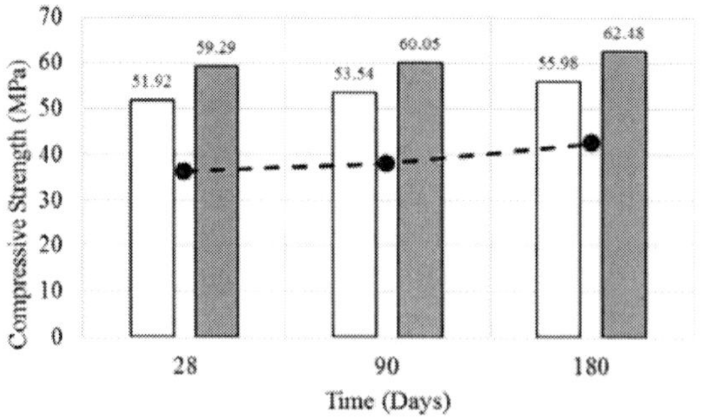

Fig. 5. Effects of silica fume on the freeze and thaw resistance of concrete □; 10% silica fume, □; 20% silica fume, •; concrete control specimen.

Figure 5 shows the compressive strength of concrete incorporated with silica fume that are subjected to freeze and thaw cycles. The concrete control specimen that is subject to freeze and thaw is also shown in Figure 5 for comparison reason. It can clearly be seen in Figure 5 that the increase in the replacement level of silica fume resulted in an increase resistance to freeze and thaw. In increase in the replacement level of silica fume resulted in an increased rate of pozzolanic reaction that takes place between the calcium hydroxide and the silica and hence resulted in great formation of calcium silicate hydrate gels. The increase in compressive strength lead to a higher resistance to freeze and thaw of concrete.

4.3.2 The effect of using hydrated lime on the freeze and thaw resistance of concrete incorporated with silica fume

The effect of using hydrated lime on the freeze and thaw resistance of concrete incorporated with silica fume is investigated in this section. Figure 6 show compressive strength of concrete specimens subjected to freeze and thaw for 28, 90 and 180 days. It can be seen in Figure 6 that the use of 10% hydrated lime with 20% silica fume resulted in a greater resistance to freeze and thaw resistance however, when the replacement level of hydrated lime is increased to 20%, the resistance to freeze and thaw was identical to the concrete control specimen. The main reason for this behavior is attributed to the decrease in the overall strength of concrete when 20% hydrated lime is used as a replacement level.

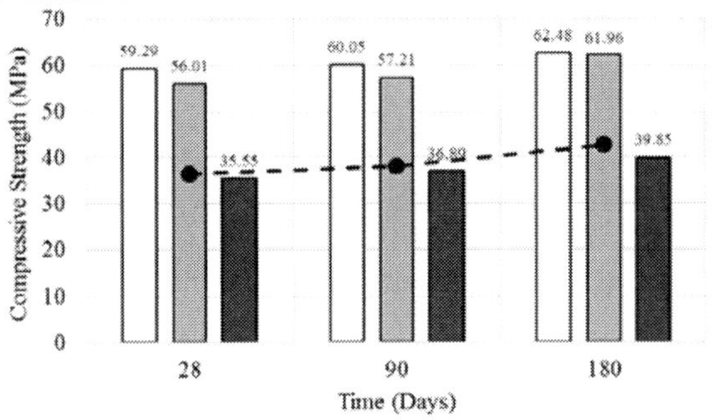

Fig. 6. Effects of using hydrated lime and silica fume on the freeze and thaw resistance of concrete □; 20% silica fume, □; 20% silica fume with 10% hydrated lime, □; 20% silica fume with 20% hydrated lime, •; concrete control specimen.

5 Conclusion

The paper investigates the use of silica fume and hydrated lime as a replacement material to cement on the strength development and durability of concrete. The effect of hot water cure condition on the strength development of concrete incorporated with silica fume and hydrated lime is also investigated in this paper. The results show that the use of silica fume enhance the compressive strength of concrete especially at long-term. It is also shown in this paper that the hot water cure condition played a significant role in the strength development of concrete incorporated with silica fume. Although the use of hydrated lime might result in a decrease in compressive strength of concrete at short-terms, the compressive strength of concrete with all combinations of silica fume and hydrated lime at 6 months are higher that the compressive strength of the control specimen at 28th days. It is also shown in the paper that the hot water curing also resulted in an increase in the compressive strength of concrete at both replacement levels of hydrated lime. The use of silica fume and hydrated lime decreased the water penetration depth of concrete and enhance the water permeability of concrete dramatically. The increase in the replacement level of silica fume resulted in an increased resistance to freeze and thaw resistance of concrete. Although the use of 10% hydrated lime enhanced the freeze and thaw resistance of concrete incorporated with silica fume, freeze and thaw resistance of concrete with 20% replacement level of lime was equivalent to the resistance of concrete control specimen. Thus, by these results, we can summarize that the role of hydrated lime, silica fume along with hot curing conditions not only contribute towards required mechanical properties and enhanced durability characteristics. It also creates a new way to use these neglected materials and conditions to promote the idea of sustainable construction.

References

1. Aıtcin, P.C., The durability characteristics of high performance concrete: a review. Cement and Concrete Composites, **25**(4), 409-420 (2003)
2. Shi, C., Jiménez, A.F. and Palomo, A., New cements for the 21st century: the pursuit of an alternative to Portland cement. Cement and concrete research, **41**(7), 750-763 (2011)
3. Harrison, A.J.W., Low carbon cements and concrete in modern construction. *In UKIERI Concrete Congress–Innovations in Concrete Construction*, Jalandhar, 723-746 (2013).
4. Sun, W., Zhang, Y.M., Yan, H.D. and Mu, R., Damage and damage resistance of high strength concrete under the action of load and freeze-thaw cycles. Cement and Concrete Research, **29**(9), 1519-1523 (1999)
5. Sharma, A. and Rani, A., Green concrete and its demand on the sustainable construction industry. International Journal of Engineering Science Invention Research and Development, **2**(2), 472-480 (2016)
6. Garg, C. and Jain, A., Green Concrete: Efficient & Eco-Friendly Construction Materials. International journal of research in engineering and technology, **2**(2), 259-264 (2014)
7. Grist, E.R., Paine, K.A., Heath, A., Norman, J. and Pinder, H., The environmental credentials of hydraulic lime-pozzolan concretes. Journal of Cleaner Production, **93**, 26-37 (2015)
8. Hooton, R.D. and Bickley, J.A., Design for durability: the key to improving concrete sustainability. Construction and Building Materials, **67**, 422-430 (2014)
9. Bentz, D.P., Hansen, A.S. and Guynn, J.M., Optimization of cement and fly ash particle sizes to produce sustainable concretes. Cement and Concrete Composites, **33**(8), 824-831 (2011)
10. Zhang, T., Yu, Q., Wei, J. and Zhang, P., Efficient utilization of cementitious materials to produce sustainable blended cement. Cement and Concrete Composites, **34**(5), 692-699 (2012)
11. Barbhuiya, S.A., Gbagbo, J.K., Russell, M.I. and Basheer, P.A.M., Properties of fly ash concrete modified with hydrated lime and silica fume. Construction and Building Materials, **23**(10), 3233-3239 (2009)
12. Hardtl, R. and Zement, H., The pozzolanic reaction of fly ash in connection with different types of cement. *In Proceedings of the 10th International Conference on Chemistry of Cement,* **101**, 8-15 (1997)

13. Loughborough, M.T., Mukherjee, P.K. and Malhotra, V.M., Development of high-strength concrete incorporating a large percentage of fly ash and superplasticizers. Cement, Concrete and Aggregates, **4**(2), 81-86 (1982)
14. Mehta, P.K., Pozzolanic and cementitious by-products in concrete--another look. Special Publication, **114**, 1-44 (1989)
15. Ghosh, S.N., Sarkar, S.L. and Harsh, S., *Mineral admixtures in cement and concrete.* Abi Books (1993)
16. Mira, P., Papadakis, V.G. and Tsimas, S., Effect of lime putty addition on structural and durability properties of concrete. Cement and concrete research, **32**(5), 683-689 (2002)
17. Juenger, M.C. and Siddique, R., Recent advances in understanding the role of supplementary cementitious materials in concrete. Cement and Concrete Research, **78**, 71-80 (2015)
18. Mazloom, M., Ramezanianpour, A.A. and Brooks, J.J., Effect of silica fume on mechanical properties of high-strength concrete. Cement and Concrete Composites, **26**(4), 347-357 (2004)
19. Ozturk, Y., 2011. *The use of supplementary cementing materials in the manipulation of mix water transport in mortar-masonry systems.* MPhil Thesis, The University of Manchester, UK
20. Fidjestøl, P. and Dåstøl, M., The history of silica fume in concrete–From novelty to key ingredient in high performance concrete. *In Proceedings of the Congresso Brasileiro do Concreto.* 4th–9th Sept 2008. Salvador: El Salvador (2008)
21. Elahi, A., Basheer, P.A.M., Nanukuttan, S.V. and Khan, Q.U.Z., Mechanical and durability properties of high performance concretes containing supplementary cementitious materials. Construction and Building Materials, **24**(3), 292-299 (2010)
22. Toutanji, H., Delatte, N., Aggoun, S., Duval, R. and Danson, A., Effect of supplementary cementitious materials on the compressive strength and durability of short-term cured concrete. Cement and Concrete Research, **34**(2), 311-319 (2004)
23. Chithra, S. and Dhinakaran, G., Effect of hot water curing and hot air oven curing on admixed concrete. International Journal of ChemTech Research CODEN (USA): IJCRGG, 1516-1523 (2014)
24. ASTM C150/C150M-16e1 Standard Specification for Portland Cement, ASTM International, West Conshohocken, PA (2016)
25. BS EN 459-1. Building Lime. Definitions, specifications and conformity criteria. London: BSI (2015)
26. ASTM C1240-15 Standard Specification for Silica Fume Used in Cementitious Mixtures, ASTM International, West Conshohocken, PA (2015)
27. ASTM C33/C33M-16e1 Standard Specification for Concrete Aggregates, ASTM International, West Conshohocken, PA (2016)
28. ASTM C127-15 Standard Test Method for Relative Density (Specific Gravity) and Absorption of Coarse Aggregate, ASTM International, West Conshohocken, PA (2015)
29. ASTM C128-15 Standard Test Method for Relative Density (Specific Gravity) and Absorption of Fine Aggregate, ASTM International, West Conshohocken, PA (2015)
30. ASTM C143/C143M-15a Standard Test Method for Slump of Hydraulic-Cement Concrete, ASTM International, West Conshohocken, PA (2015)
31. ASTM C192/C192M-16a Standard Practice for Making and Curing Concrete Test Specimens in the Laboratory, ASTM International, West Conshohocken, PA (2016)
32. BS EN 12390-3. Testing hardened concrete. Compressive strength of test specimens. London: BSI (2002)

33. BS EN 12390-8. Testing hardened concrete. Depth of penetration of water under pressure. London: BSI (2009)
34. Khoury, N.N. and Zaman, M.M., Environmental effects on durability of aggregates stabilized with cementitious materials. Journal of Materials in civil Engineering, **19**(1), 41-48 (2007)

Effect of elevated temperatures on mechanical performance of cement mortar with nanoclay

Mohammad R. Irshidat[1,*], *and Mohammed H.* Al-Saleh[2]

[1]Department of Civil Engineering, Jordan University of Science and Technology, Irbid – Jordan
[2]Department of Chemical Engineering, Jordan University of Science and Technology, Irbid – Jordan

Abstract. In this research, the effect of nanoclay addition on mechanical strengths of cement mortars subjected to elevated temperatures was investigated. Mortars with three different replacement dose of montmorillonite nanoclay (0, 1 and 2% of cement weight) were prepared and then subjected to 200 °C, 400 °C, and 600 °C for 2 hrs. Compressive and flexural strengths of the heated and unheated specimens were obtained. The results showed that nanoclay addition to cement mortar caused marginal enhancement in its compressive and flexural strengths at room temperature. But, the efficiency of nanoclay to enhance the flexural strength of cement mortar increased at higher temperatures. The maximum relative improvement in flexural strength due to 2% nanoclay addition was achieved at 400°C.

1 Introduction

Elevated temperatures usually cause severe deterioration to concrete elements and affect their durability due to the different thermal properties of concrete constituents. Elevated temperature may affect the concrete's strengths and modulus of elasticity [1-3]. To minimize the harmful effects of elevated temperatures on concrete, suitable supplementary cementing materials (SCM) can be used. Various literatures showed that adding fly ash, slag, or silica fume significantly affect the performance of concrete and enhance its thermal properties [4-8]. On the other hand, the small size and tremendous properties of nanomaterials make them a good choice to be used as an additive to enhance the behaviour of cement mortar and concrete. Previous work showed that adding nanoclay into cement mortar enhanced its mechanical strengths, chloride penetration resistance, and reduced its permeability [9-12]. Chang et al. [13] showed that adding small percentages of nanoclay (montmorillonite) into cement paste enhanced its compressive strength and decrease its permeability. The optimum compressive strength and permeability were achieved by adding 0.6 and 0.4% nanoclay, respectively. They also reported that adding nanoclay caused higher rate of C–S–H production in samples containing nanoclay. Morsy et al. [14] reported that adding 8% nanoclay (metakaolin) enhanced the compressive and tensile strengths of cement mortar by 7% and 49%, respectively. Morsy et al. [15] investigated the

* Corresponding author: mrirshidat@just.edu.jo

effect of using CNT and nanoclay on the mechanical properties of cement mortar. They reported that adding 6% nanoclay and 0.02% CNT improved the compressive strength of the cement mortar by 29%. Farzadnia et al. [16] studied the mechanical strengths and flowability of cement mortar with halloysite nanoclay. Their results showed that adding 3% and 2% nanoclay enhanced the compressive strength and gas permeability of cement mortar by 24% and 56%, respectively. A. Hakamy et al. [17] studied the influence of nanoclay and calcined nanoclay on the properties of cement paste. Their results showed that adding 1% nanoclay into cement paste decreased its porosity and water absorption by 31% and 34%, respectively, but increased its density by 10%. They also found that adding the addition of 1% nanoclay into cement paste enhanced its compressive strength, flexural strength, and impact strength by 40%, 43% and 34%, respectively. In this work, mechanical strength degradation of cement mortar with montmorillonite nanoclay due to high temperatures was investigated.

2 Experimental program

The cement mortar was prepared using Portland cement (Type II), silica sand, nanoclay and water. The properties of silica are shown in Table1. Montmorillonite nanoclay was used as received from Nanocor® Inc., USA. Three mixes of cement mortar were prepared, one control (NC0) and two with nanoclay (NC1 and NC2). The nanoclay was first dispersed in water then used to mix the mortar. Water to cement ratio of 0.55 and cement to sand ratio of 1:3 were used. 50 mm cubic molds were used to cast the compressive strength specimens, whereas prisms of 40mm*40mm*160mm were used to cast the flexural strength specimens as shown in Figure 1. After casting, the specimens were cured for 28 days. Then, the specimens were subjected to temperatures of 200°C, 400°C, and 600°C for two hours. Finally, three specimens of each mix were tested using the ASTM standards.

Table 1. Silica sand properties.

Property	Value
Maximum particle size	1.18 mm
Absorption	1.5 %
Fineness modulus	1.72
Specific gravity	2.6

(a) (b)

Fig. 1. Steel molds used to cast mortar samples for (a) compressive test (b) flexural test.

DOI: 10.1051/matecconf/201712002005

3 Results and discussions

Compressive strengths of all tested specimens are shown in Figure 2. The compressive strength of unheated specimen without nanoclay was equal to 37.5 MPa. Heated the cement mortar at 200°C increased its compressive strength by 45%. Elevated the temperature to 400°C reduced the strength compared to specimens heated at 200°C. At 600°C, huge reduction in the compressive strength was noticed. On the other hand, adding nanoclay increased the compressive strength of all test specimens (heated or unheated). The maximum enhancement was noticed in the case of specimen with 2% nanoclay heated at 200°C. The enhancement in the compressive strength in the presence of nanoclay may be attributed to the higher rate of production of C–S–H and more consumption of CH in samples containing nanoclay [14]. More over, the nanoclay with their small size may fill the voids and made the microstructure of the mortar denser.

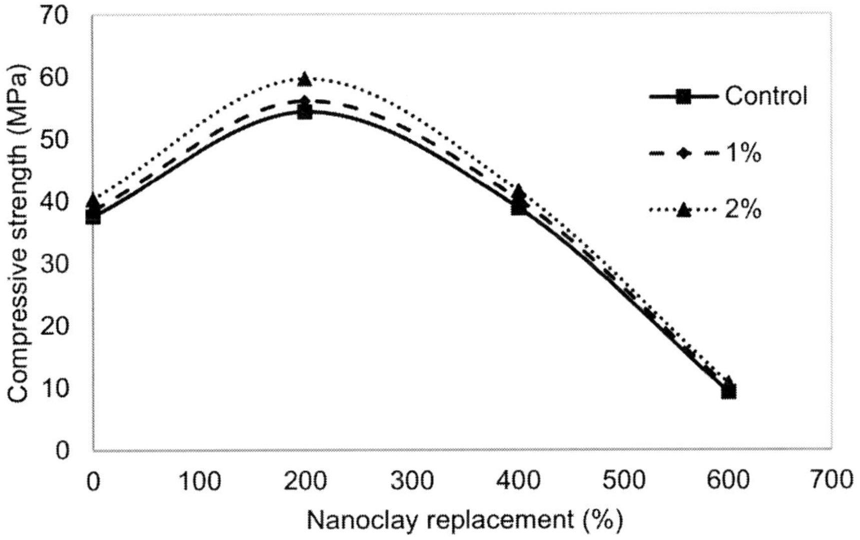

Fig. 2. Compressive strength of cement mortar exposed to elevated temperatures.

On the other hand, the flexural strengths of all specimens are shown in Figure 3. It is clear that adding nanoclay to cement mortars and test them at room temperature insignificantly enhance their flexural strengths. When exposing to elevated temperatures, control specimens and those with nanoclay own same performance. The flexural strength of all specimens increased when exposed to 200°C, then decrease when the temperatures elevated to 400 °C and 600°C. Moreover, maximum enhancement in flexural strength due to nanoclay addition was noticed when the specimens heated at 400°C as shown in Figure 3. The enhancement could be attributed to the fact that the consumption of CH and the extra formation of C–S–H in presence of nanoclay could fill up pores and led to denser microstructure of cement mortars [15].

In general, the enhancement in compressive and flexural strength of cement mortar when heated at 200°C could be attributed to the activation of the hydration process due to heating. When the temperature increased to 400 °C and 600 °C, increasing of internal vapor pressure and degradation of hydration products could be happened caused reduction in the strengths.

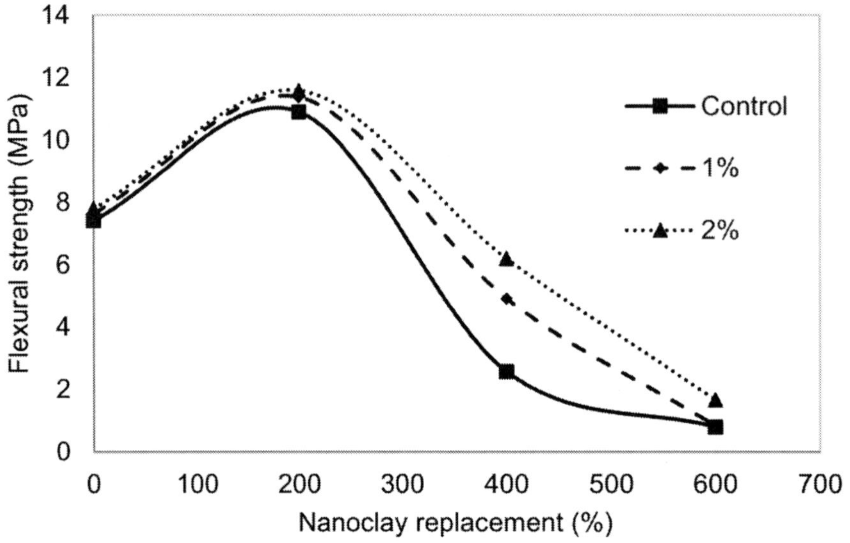

Fig. 3. Flexural strength of cement mortar exposed to elevated temperatures.

4 Conclusions

The following conclusions can be drawn:

1. Exposing cement mortar specimens to elevated temperature of 200°C increased their compressive and flexural strength. Elevated the temperature to 400°C and 600°C decreased the strengths.
2. Adding 1% and 2% of nanoclay to cement mortar enhanced their compressive and flexural strength.
3. Maximum enhancement in compressive strength was achieved in the case of heated cement mortar with 2% nanoclay at 200 °C.
4. Maximum enhancement in flexural strength was achieved in the case of heated cement mortar with 2% nanoclay at 400 °C..

The work presented in this article was funded by the Deanship of Research at Jordan University of Science and Technology.

References

1. S.K. Handoo, S. Agarwal, S.K. Agarwal, Cem. Concr. Res. **32** 1009–1018, (2002)
2. S. Aydın, Fire Saf. J. **43**, 610–617, (2008)
3. A. Al-Ostaz, M. Irshidat, B. Tenkhoff, P.S. Ponnapalli, J. Mater. Civ. Eng. **22,** 136–144, (2010)
4. A. Nadeem, S.A. Memon, T.Y. Lo, Constr. Build. Mater. **38**, 338–347, (2013)
5. M. Heikal, H. El-Didamony, T.M. Sokkary, I.A. Ahmed, Constr. Build. Mater. **38**, 1180–1190, (2013)
6. A.M. Rashad, M.H. Khalil, Constr. Build. Mater. **40**, 522–532, (2013)
7. A.M. Rashad, D.M. Sadek, H.A. Hassan, J. Clean. Prod. **112**, Part 1, 1086–1096, (2016)

8. Antoni, M., Rossen, J., Martirena, F., and Scrivener, K. Cement and Concrete Research, **42**(12), 1579–1589, (2012)
9. Chang, T.-P., Shih, J.-Y., Yang, K.-M., and Hsiao, T.-C. Journal of materials science, **42**(17), 7478–7487, (2007)
10. He, X., and Shi, X. Journal of the Transportation Research Board, **2070**(1), 13–21, (2008)
11. Kuo, W.-Y., Huang, J.-S., and Lin, C.-H. Cement and Concrete Research, **36**(5), 886–895, (2006)
12. Wei J, Meyer C. J Mater Sci,49 (21), 7604, (2014)
13. T.-P. Chang, Jeng-Ywan Shih, Kuo-Ming Yang, Tien-Chin Hsiao. J. Mater. Sci. **42**, 7478–7487, (2007)
14. M.S. Morsy, S.H. Alsayed, M. Aqel. Int. J. Civic. Environ. Eng **10**, 23–27, (2010)
15. M.S. Morsy, S.H. Alsayed, M. Aqel. Constr. Build. Mater. **25**,145–149, (2011)
16. Farzadnia N, Abang Ali AA, Demirboga R, Anwar MP. Cem Concr Res, **48**, 97–104, (2013)
17. Hakamy A, Shaikh FUA, Low IM. Compos Part B Eng,**78**, 174–84, (2015)

A comprehensive investigation into the effect of temperature variation on the mechanical properties of sustainable concrete

Abdulkader El Mir[1,*], and *Salem* Nehme[2]

[1]PhD student, Budapest University of Technology and Economics, Department of Construction Materials and Technologies, Műegyetem rkp. 3, Budapest, Hungary
[2]Assocaiate Professor, Budapest University of Technology and Economics, Department of Construction Materials and Technologies, Műegyetem rkp. 3, Budapest, Hungary

Abstract. Minimizing the production energy and resources consumption are the key principle for engineering sustainability. In the case of concrete structures, this concept can be achieved by the use of materials in the most efficient way considering in the mix design the optimal mechanical and durability properties. The substitution of ordinary Portland cement for other supplementary cementitious materials is assessing the possibility of enhancing the sustainability and decreasing the environmental impact of concrete. Mass concrete is rich in cementitious materials which results in high temperature within the concrete, hence several hazards such as cracking or temperature differences between the interior and the surface of concrete could be prevented. An experimental study evaluated on several one cubic meter sized concrete elements in which during the primary phase of hydration, the temperature variation is recorded in several location offsets with respect to time. Thermal variations results are analyzed in accordance with the cement type, CO_2 emission production of cement, compressive strength, water tightness, drying shrinkage and rapid chloride migration coefficient. The results indicate that slag cement CEM III/B 32.5, that incorporates highest amount of slag, ensured improved mechanical, thermal and durability properties in comparison with ordinary Portland cement CEM I 32.5.

1 Introduction

Over the few next decades, the production of concrete has been growing in the less developed countries (in Asia, Africa and parts of South America). Therefore, the problem of greenhouse gas emissions and the reduction of resources are likely to become a severe issue [1]. The concept of sustainability helps to improve the future of concrete production, providing encouraging potentials in reducing environmental impact. The main environmental problem in concrete production is the CO_2 emission during cement creation. According to the U.S. Geological Survey, 5% of the global CO_2 emissions are directly related to the production of

* Corresponding author: abdelkader.elmir@hotmail.com

3.3 billion tons of cement per year [2]. Thus, substituting clinker is probably the most effective strategy to decrease the energy consumption in cement plants. For the concrete industry, when it comes to sustainability, simply it means providing more strength with lower emissions of CO_2. In other words: more clinker made with less fuel, less implementation of cement in concrete, more strength and durability with less cement. Several earlier studies have already been carried out on this field [3-5]. This work was mainly based on clinker replacement by large amount of slag, fly ash or other supplementary cementitious materials, aiming for the achievement of a green concrete.

Mass concrete is widely applied in large projects. Hence, it becomes a necessity that concrete meets the requirements without cracking. Generally, massive concrete elements possess large section sizes. Therefore, after concrete placing, internal temperature rises drastically due to the heat of hydration.

Controlling the hydration heat from mass concrete during construction is a critical issue to obtain the targeted concrete quality. Blast furnace slag cement is mostly applied for construction projects where the strength undergoes a slow rate of development. By blending blast furnace slag with ordinary Portland cement, slag cement is produced with different classification depending on its mass replacement level of Portland cement. This paper aims to perform an objective comparison between three different types of local commercial cements in which the focus goes to the slag cement. The latter is analyzed in terms of the environmental impact, mechanical, thermal and durability properties of massive concrete.

2 Experimental program

2.1 Materials

In this study, three types of cement were used in order to produce three independent concrete mixtures. One ordinary Portland and two Blast furnace slag cement types were applied with different clinker and slag proportions (see Table 1).

Table 1. Cement properties.

Cement type	**Blaine, cm²/g**	**Cement composition (%)**			**Heat of hydration (J/g), 41 hours**
		Clinker	**Blast furnace slag**	**Lime-stone**	**MSZ EN 196-9**
CEM I 32,5 N-LH	2419	97	-	3	241.0
CEM III / A 32,5 N-MSR	3795	53	43	4	180.4
CEM III / B 32,5 N-LH/SR	4505	32	65	3	169.8

Materials were the same as the ordinary Portland cement which are produced in a local plant and modified depending on the percentage required for the slag into the cement composition. Danube quartz sand and gravel were used as aggregates for concrete preparation (maximum aggregate size was 16 mm). Nominal grading fractions were used: sand 0/4 mm, small gravel 4/8 mm and medium gravel 8/16 mm. Water to cement ratio was set to be w/c = 0.46 with constant cement amount of 330 kg/m2. A relative low water to cement ratio was selected in order to reach enhanced durability properties. The superplasticizer "BASF

Masterease 5007" was applied in order to obtain an F4 flow diameter classification with respect to European Standards [6].

Table 2. Mixtures proportion.

Mixture ID	CEM I-32,5	CEM III/A 32,5	CEM III / B-32,5
cement type	CEM I 32,5 N-LH	CEM III / A 32,5 N-MSR	CEM III / B 32,5 N-LH/SR
w/c ratio	0.46	0.46	0.46
Cement (kg/m^3)	330	330	330
Sand (kg/m^3)	769	769	769
Small gravel (kg/m^3)	480	480	480
Medium gravel (kg/m^3)	673	673	673
Superplasticizer (kg/m^3)	4.65	5.34	5.55

2.2 In-situ and laboratory tests

Three concrete cube blocks of 1 m^3 were constructed with various types of cement. The concrete was delivered from a local ready-mix plants following by the common practice for placement and curing. A total of seven temperature sensors were implemented inside the concrete cube in several locations as indicated in Figs. 1a-b. Outdoor ambient temperature is recorded also along with the concrete ones. Total duration of temperature measurement from concrete placement took 7 days.

Fig. 1a. Instrumentation of the 1 m^3 cube.

Several mechanical and durability tests were performed. Steel molds were placed on site and tested later in the laboratory. Cube specimens (150 × 150 × 150 mm) were tested for compressive strength using a universal closed-loop hydraulic testing machine performed based on European Standard [7] at a constant loading rate of 11.25 kN/s. The testing age was of 2, 7, 28 and 56 days. Prismatic specimens (75 × 75 × 250 mm) were used to determine the drying shrinkage. After their removal from steel molds, the studs were glued on the four sides of the longitudinal surface of the concrete prism using a special adhesive. Demountable mechanical strain gauge (DEMEC), that can read up to 0.001 mm, was the shrinkage measurement tool. Thereby the change in length of the specimens was measured between 1 and 77 days of curing time.

 DOI: 10.1051/matecconf/201712002006

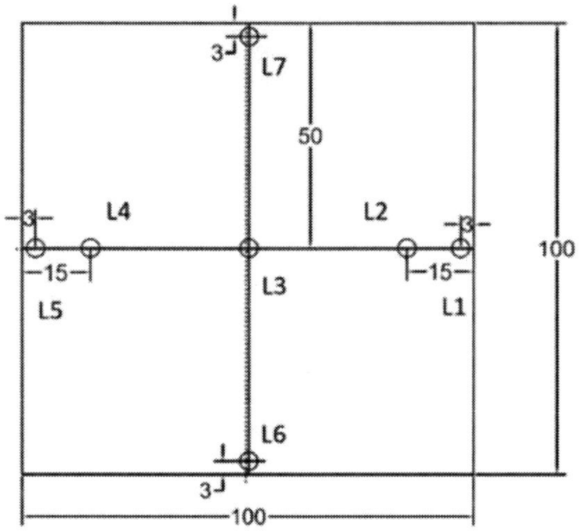

Fig. 1b. Schematic cross section of the sensors location in centimeters.

Referring to following European Standard [8], water tightness tests were performed on cubes specimens (150 × 150 × 150 mm) at the age of 72 days. Nevertheless, rapid chloride migration test was carried out on cylindrical specimens (Ø100 × 50 mm) according to the Nordtest Method NT Build 492 [9]. Tests were carried out at specimen age of 180 days.

3 Results and discussion

3.1 Compressive strength

The compressive strength values were determined using the cubes casted on-site. Fig. 2. shows the compressive strength values versus the testing time at different ages. It can be seen that at the same conditions of curing and water to cement ratio, CEM III/B 32.5 exhibited the maximum compressive strength reaching a value of 68 MPa at 56 days of age. Moreover, all cements having 32.5 grade showed almost similar compressive strength values at 56 days, ranging between 66 and 70 MPa. However, CEM I 32.5 provided a higher compressive strength at 7 and 28 days in comparison with CEM III/A 32.5 and CEM III/B 32.5. This behavior is explained by the effect of higher clinker content developing more the hydration reaction over the slag cement type (Higher strength rate development). The reduction in clinker content in case of CEM III/A 32.5 and CEM III/B 32.5 did not lead to a decrease in the compressive strength with respect to CEM I 32.5. Similar values were achieved at 28 days with clinker reduction up to 65% of slag. Hence, adequate strength values are obtained despite the low level of clinker.

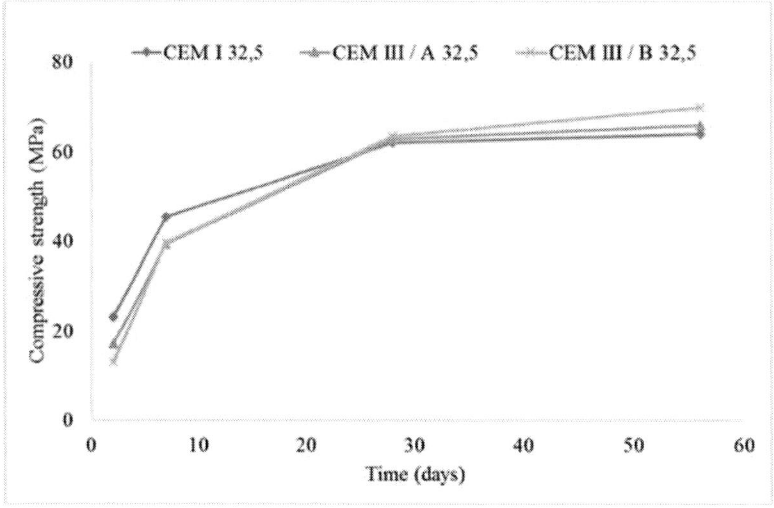

Fig. 2. Development of compressive strength with time in terms of applied cements.

3.2 Drying shrinkage

Shrinkage is a sensitive phenomenon resulting from the variation in the concrete volume and temperature, hence cracking problems could occur, in which are directly connected to corrosion and deterioration of the reinforcements [10]. Drying shrinkage arises when capillary water evaporates from the hardened concrete due to temperature variations with the ambient one. Fig. 3. shows the development of shrinkage of all concrete mixtures with different cement types over time. Test results shows that the drying shrinkage values of all mixtures, which is recorded up to 66 days, fell below 1000 microstrains at 56 days of age, which is considered adequate according to standards of Australia [11].

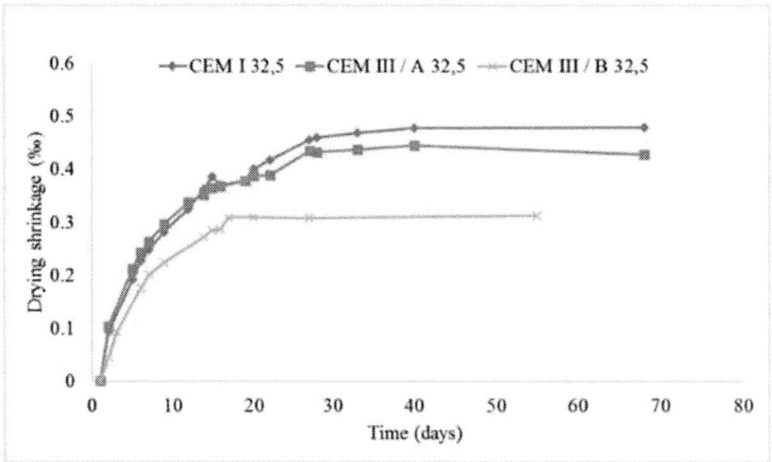

Fig. 3. Variation of drying shrinkage with time of the three applied cements.

At earlier ages, high shrinkage rates up to 16 days can be noticed for all mixtures. Afterwards, these values increase gradually till 30 days, and then slowly reduce after 30 days. The shrinkage of CEM I 32.5 reached the maximum value among all studied cements of 480 microstrains. When the slag was integrated into the concrete, drying shrinkage value decreased with the increase of slag content by 35% at 65% replacement level by slag referring to CEM I 32.5. Hence less drying shrinkage is obtained when the clinker is replaced by slag material.

3.3 Effect of temperature on cement type

The heat generated from cement hydration along with the variation of ambient temperature could cause cracks in the concrete [12-13]. It was also considered that the cement type directly affects the hydration heat [14].

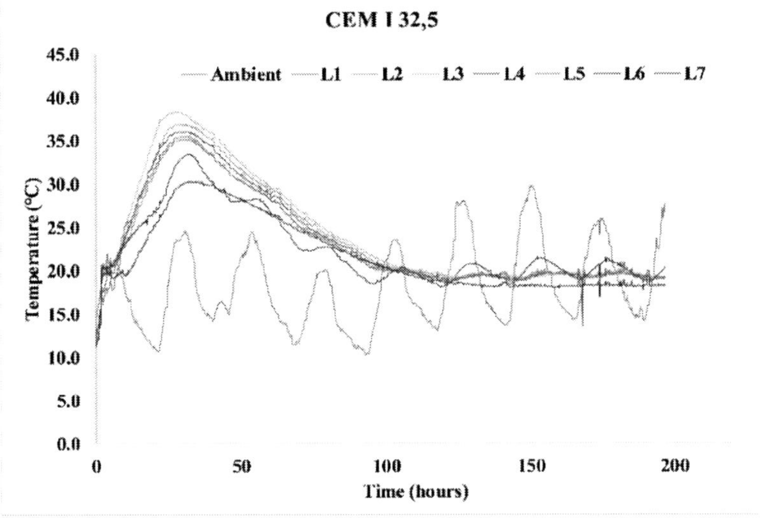

Fig. 4a. Temperature-time history for CEM I 32.5 at various locations in the cube.

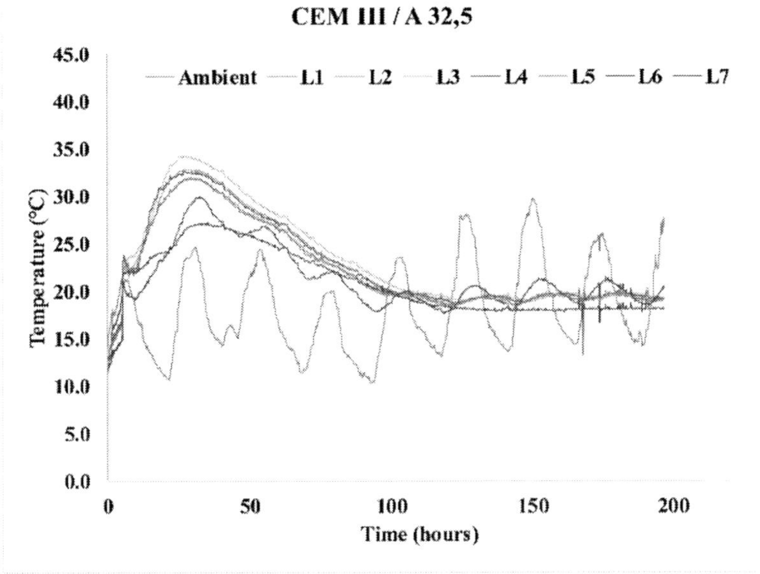

Fig. 4b. Temperature-time history for CEM III/A 32.5 at various locations in cubic cube.

 DOI: 10.1051/matecconf/201712002006

In order to demonstrate the influence of the size on the hydration heat, temperature sensors mentioned earlier in section 2.2 were located at different position in the cube element (Fig. 1a-b.). The variation of temperature with time at multiple locations were recorded as illustrated in Fig. 4a-c.

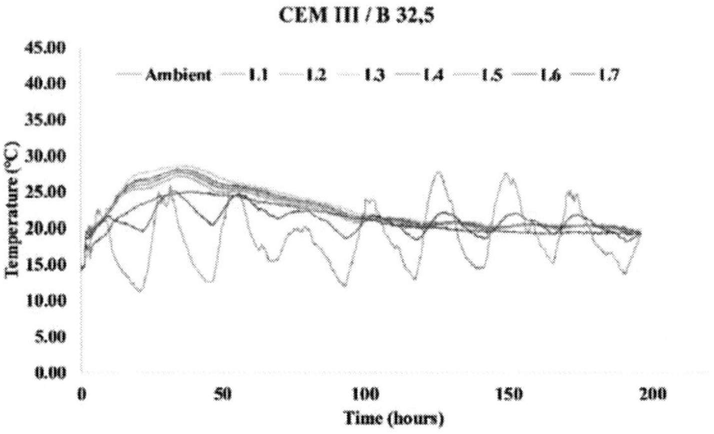

Fig. 4c. Temperature-time history for CEM III/B 32.5 at various locations the cube.

Highest temperature values were noted at the center of the 1 m^3 element. Portland cement with high amount of clinker showed the maximum peak of temperature reaching 38 °Cin case of CEM I 32.5. With the partial replacement of the Portland cement by slag, this value drops with the increase of slag content, reaching 27 °C in case of CEM III/B 32.5 (highest slag content). Regarding the locations of the sensors, temperature values recorded near to the surface, away from the center, showed how differences in temperature occur in mass concrete. However, in case of CEM III/ B 32.5 containing the highest amount of slag, temperature differences between inner and outer locations were not significant. Hence crack hazards and less tensile stresses are expected to appear inside the concrete element.

3.4 Water tightness and rapid chloride migration coefficient

Water penetration depth under pressure leads to some observations regarding the permeability of tested concrete. Fig. 5. shows that CEM III/ B 32.5 mixture has a low permeability, with a maximum penetration depth of 9 mm. However, in case of CEM I 32.5 and CEM III/A 32.5, at same level of water cement ratio w/c = 0.46, their relevant concrete mixtures have higher penetration depth reaching 18 mm. According to Neville, impermeable concrete falls into the 50 mm or less of water penetration depth. For values less than 30 mm, concrete is classified under impermeable concrete for aggressive environmental exposure conditions [15]. On the other hand, the rapid chloride migration coefficient was recorded for all tested concretes. It can be clearly observed that the difference between slag and ordinary Portland cement in resisting the chloride migration. The latter showed a better response, along with the increase of the slag content in the cement composition. The CEM III/B 32.5 shows relatively better performance in resisting the chloride (Fig. 5.).

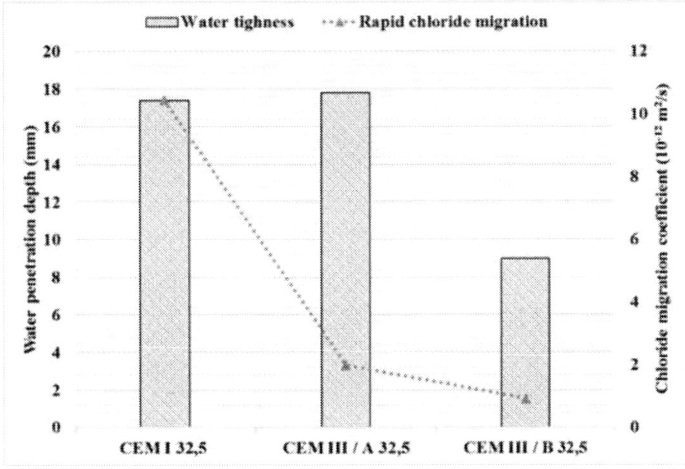

Fig. 5. Water penetration depths and chloride migration coefficients of mixtures.

4 Environmental performance evaluation

The reduction of clinker content in the mixture composition could lead to a major reduction of the impact of clinker production on the environment in comparison with mixtures incorporating a higher amount of clinker.

This enhancement is due to clinker replacement by supplementary cementitious materials such as fly ash, slag and limestone powder. However, using such materials with their current economic value and availability in the country may be challenging. Special Slag cement reaching 75% of slag content, gypsum and other additional materials is produced in Hungary. This composition has an enormous ecological advantage which is clearly distinguished from traditional Portland cement. Thereby, CO_2 emission and energy costs are reduced. Fig. 6. shows a comparison between CEM I, CEM III/A, CEM III/B and Slagstar in terms of CO_2 emission and energy consumption reduction [16]. The impact of slag cement in reducing the CO_2 can be compared with ordinary Portland cement. By slag replacement of the clinker, CEM/B 32.5 reaching 20% clinker provided 212 kg per ton of CO_2 emission. Thus critical savings of CO_2 emission are achieved.

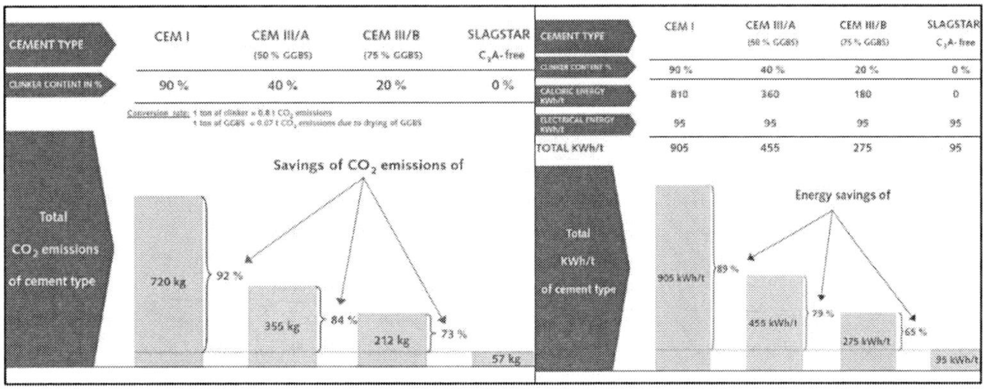

Fig. 6. Comparison of CO_2 emissions per ton of cement and the energy consumption in kWh/t.

5 Conclusions

The purpose of this study was to investigate a low-carbon commercially used cement type (Blast furnace slag). This paper reports the mechanical, thermal and durability properties of concretes containing different cement types. From the experimental observations on the temperature variation, mechanical and durability properties of tested concretes, the following conclusion can be drawn:

1. The introduction of cement with high amount of slag (65%) was enabled reaching similar compressive strength values in comparison with cement containing clinker (93%) at 28 days of age;
2. Concrete containing a high amount of slag achieved shrinkage values reductions up to 35% in comparison to the control concrete, with the similar grade and at the same water cement ratio levels;
3. The water tightness results were low in CEM III/B, which indicates an improved and dense concrete microstructure;
4. The chloride impermeability is highly enhanced with the use of slag cement. The chloride permeability was reduced by 90% with respect to the control mixture;
5. Slag cement are considered environmental friendly providing great reduction in CO_2 emission, in comparison with ordinary Portland cement.

References

1. P. Aitcin, S. Mindess, *Sustainability of Concrete* (Taylor & Francis e-Library, 2011)
2. U.S. Department of the Interior, U.S. Geological Survey, *Mineral Commodity Summaries 2011* (U.S. Geological Survey, Reston, Virginia, 2011)
3. M. Shariq, J. Prasad, H. Abbas, Cem. Concr. Compos. **68**, 35-45 (2016)
4. X. Wang, K. Wang, J. Li, N. Garg, S.P. Shah, J. Sustain. Cem. Mater. **3**, 245-255 (2014)
5. S.G. Nehme, Epa. - J. Silic. Based Compos. Mater. **67**, 28-33 (2015)
6. *BS EN 206:2013, Concrete-specification, performance, production and conformity, Br. Stand.* (2013)
7. *BS EN12390-3:2009, Testing hardened concrete. Part3: Compressive strength of test specimens, Br. Stand.* (2009)
8. *BS EN12390-8:2009, Testing hardened concrete. Part 8: Depth of penetration depth of water under pressure, Br. Stand.* (2009)
9. *NT BUILD 492, nordtest method* (1999)
10. S. Miyazawa, T. Yokomuro, E. Sakai, A. Yatagai, N. Nito, K. Koibuchi, Constr. Build. Mater. **61**, 90–96 (2014)
11. AS 1379—2007, Specification and supply of concrete (2007)
12. X. Zou, A. Chao, Y. Tian, N. Wu, H. Zhang, T.Y. Yu, X. Wang, Meas. J. Int. Meas. Confed. **45**, 1077–1082 (2012)
13. P. Kara, A. Borosnyói, O. Fenyvesi, Epa. - J. Silic. Based Compos. Mater. **66**, 18–22 (2014)
14. M.H. Lee, B.S. Khil, H.D. Yun, Indian J. Eng. Mater. Sci. **21**, 536–542 (2014)
15. A. Neville, *Properties of concrete* (fourth ed. Pearson, 1995)
16. R. Novak, W. Schneider, E. Lang, *Material Science*, **58**, 70-78 (2005)

Experimental study of the effect of limestone grading on some mechanical properties of concrete

Madiha Z. J. Ammari[1,*] and *Richard Fowler* [2]

[1]Civil & Infrastructure Engineering Department, American University of Ras Al Khaimah, Ras Al Khaimah, U.A.E
[2]Stevin Rock L.L.C, Ras Al Khaimah, U.A.E

Abstract. This study reports the experimental work undertaken to investigate the optimum grading of limestone to be used in concrete mixes. 36 concrete cubes were prepared for testing. Four different Fineness Moduli and grading were tested 2.4, 2.6, 2.92 and 3. For all tests, the cubes were left in curing until testing at the age of 3, 7 and 28 days respectively. Samples were loaded to failure and the average compressive strength was used for comparison purposes. Flow table test has been performed on the fresh concrete directly after mixing to measure workability and the average of the maximum concrete spread parallel to the two edges of the table was recorded. Results revealed that the optimum Fineness Modulus for the limestone to be used as fine aggregate in the concrete mix to get maximum compressive strength is 2.78. The flow table tests revealed an increment in the workability of fresh concrete with higher Fineness Modulus of limestone used in the concrete mix as fine aggregate. The workability of the optimum Fineness Modulus, 2.78, was found to be 412 mm which is a mix with considerable workability. An ideal grading has been recommended in this research study and checked to match the ASTM grading requirements for fine aggregate.

1 Introduction

Concrete is the most commonly used material in construction industry all over the world. The conventional concrete is produced using natural sand from marine, river or dune as fine aggregate. Digging sand from river bed or marine in excess quantity is hazardous to environment. The cheapest and the easiest way of getting substitute for natural sand is obtained from manufactured sand from aggregate quarries known as limestone. Concrete made with limestone filler as replacement of natural sand in concrete was investigated by Omar M. Omar et al. [1] measuring its compressive strength, tensile strength, permeability and modulus of rupture. Concrete using various combinations of manufactured sand and quarry dust as complete or partial replacement for conventional river sand was investigated

* Corresponding author: madiha.ammari@aurak.ac.ae

by a group of researchers such as Joseph O. Ukpata et al. [2], Rana Burhan Abdurrahman Alshahwany [3] and A. Jayaraman and V. Senthil kumar [4].

UAE is located in the middle of the gulf desert, and with the majority of its cities located close to the sea, there are two main possible sources of sand that can economically be used in the construction work, which are the desert or the sea. Desert sand is very fine which adversely affects the different properties of the produced concrete mixes in terms of compressive strength, workability and bond. On the other hand, the salt in the marine sand would affect the durability of reinforced concrete structures.

The sand in UAE is excavated from the bottom of the sea or ocean for different purposes including the excavation done for different offshore construction work. This sand is then transported to different local suppliers. The suppliers have the responsibility of washing the sand from salt and sieving to remove the large sized sea shells. Even though the marine sand, which is the source of sand used in UAE, is washed before distribution, it is not possible to get rid of all salts adhered to sand particles. Anyway, the marine sand is fine graded and cannot used lonely as fine aggregate in the concrete mixes.

Limestone, which is considered as aggregate quarry waste, could add the required properties to the concrete mix as fine aggregate in terms of compressive strength, workability, durability and sustainability. This study reports the experimental work undertaken to investigate the optimum grading of limestone to be used. The effects of limestone as fine aggregate on fresh and hardened properties of concrete had been investigated. The investigation included testing of compressive strength and workability of different concrete mixes.

While concrete made with limestone filler as replacement of natural sand in concrete was widely investigated for the best percentage of limestone to be used in the mix, no significant research work has been done to examine the effect of the Fineness Modulus and grading of limestone used as fine aggregate in concrete mixes and its effect on the different properties of concrete.

2 Research methodology and data analysis

2.1 Materials used in experiments

2.1.1 Cement

Ordinary Portland Cement (OPC) locally available in UAE was used in the experiment.

2.1.2 Water

Portable water available in the Civil and Infrastructure Engineering laboratory was used.

2.1.3 Coarse aggregate (gravel)

Crushed limestone from the quarry Stevin Rock L.L.C. located in Ras Al Khaimah was used in the experiment with specific gravity 2.59, Bulk density 1480 kg/m3 and grading shown in Fig (1).

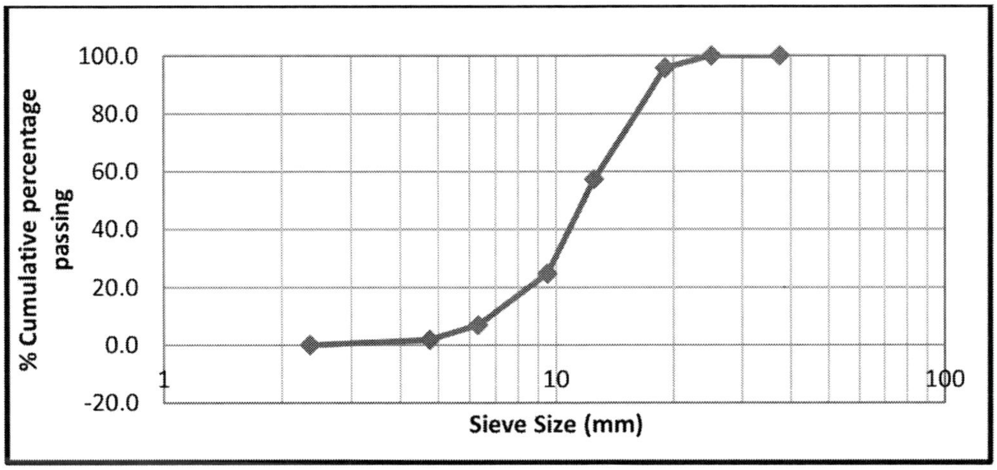

Fig. 1. Grading of coarse aggregate used in the experiments.

2.1.4 Fine aggregate

The fine aggregate used in the mixes was totally a limestone with different grading used in the different concrete mixes. The limestone used for testing all samples was provided by Stevin Rock L.L.C, which is the only aggregate quarry available in Ras Al Khaimah, one of the southern emirates of the United Arab Emirates. The limestone used for all mixes was dry and stored in the same place. The specific gravity of the limestone was found to be 2.8.

Limestone with different Fineness modulus was used for each concrete mix. The required fineness modulus was achieved by sieving the limestone using automatic sieve shaker for a specified enough time to maintain the required quantity of retained limestone size on each standard sieve. Fineness modulus of 2.4, 2.6, 2.92 and 3 were prepared for the different concrete mixes. Tables (2) through (5) show the different grading of limestone used to achieve the different Fineness modulus used in each concrete mix. All used limestone grading matches the ASTM grading requirements of fine aggregate as mentioned in Neville and Brooks [5]. Table (1) below shows the ASTM standard grading requirements.

Table 1. ASTM requirements for grading of fine aggregate.

Sieve size (mm)	% cumulative Passing
10	100
5	95 - 100
2.36	80 - 100
1.18	50 - 85
0.6	25 - 60
0.3	10 – 30
0.15	2 - 10

2.2 Comparing the limestone grading used in the tests with ASTM requirements

The grading of the limestone filler with different grading and fineness moduli 2.4, 2.6, 2.92 and 3 are all compared to the grading of fine aggregate required by ASTM. Tables (2) through (5) shows that all limestone grading used in the tests met the ASTM fine aggregate grading requirements.

Table 2. grading of limestone with fineness modulus 2.4.

Sieve size (mm)	Mass retained (g)	% retained	% cumulative Retained	% cumulative Passing	ASTM % cumulative Passing required
10	0	0.0	0.0	100.0	100
5	0	0.0	0.0	100.0	95 - 100
2.36	70	7.0	7.0	93.0	80 - 100
1.18	190	19.0	26.0	74.0	50 - 85
0.6	190	19.0	45.0	55.0	25 - 60
0.3	260	26.0	71.0	29.0	10 – 30
0.15	200	20.0	91.0	9.0	2 - 10
0.075	90	9.0		0.0	
		F.M	**2.4**		

Table 3. grading of limestone with fineness modulus 2.6.

Sieve size (mm)	Mass retained (g)	% retained	% cumulative Retained	% cumulative Passing	ASTM % cumulative Passing required
10	0	0.0	0.0	100.0	100
5	0	0.0	0.0	100.0	95 - 100
2.36	35	3.8	3.8	96.2	80 - 100
1.18	235	25.6	29.4	70.6	50 - 85
0.6	260	28.3	57.7	42.3	25 - 60
0.3	186	20.3	78.0	22.0	10 – 30
0.15	121	13.2	91.2	8.8	2 - 10
0.075	81	8.8		0.0	
		F.M	**2.6**		

Table 4. grading of limestone with fineness modulus 2.92.

Sieve size (mm)	Mass retained (g)	% retained	% cumulative Retained	% cumulative Passing	ASTM % cumulative Passing required
10	0	0.0	0.0	100.0	100
5	0	0.0	0.0	100.0	95 - 100
2.36	76	8.1	8.1	91.9	80 - 100
1.18	337	35.8	43.8	56.2	50 - 85
0.6	225	23.9	67.7	32.3	25 - 60
0.3	130	13.8	81.5	18.5	10 – 30
0.15	87	9.2	90.8	9.2	2 - 10
0.075	87	9.2		0.0	
		F.M	**2.92**		

Table 5. grading of limestone with fineness modulus 3.

Sieve size (mm)	Mass retained (g)	% retained	% cumulative Retained	% cumulative Passing	ASTM % cumulative Passing required
10	0	0.0	0.0	100.0	100
5	0	0.0	0.0	100.0	95 - 100
2.36	120	12.0	12.0	88.0	80 - 100
1.18	337	33.7	45.7	54.3	50 - 85
0.6	225	22.5	68.2	31.8	25 - 60
0.3	144	14.4	82.6	17.4	10 – 30
0.15	87	8.7	91.3	8.7	2 - 10
0.075	87	8.7		0.0	
		F.M	**3.0**		

2.3 Fineness modulus (F.M) calculations

The Fineness modulus (F.M) is calculated for each grading by finding the sum of the cumulative percentages retained on the sieves of the standard series, divided by 100.The standard series consists of sieves, each twice the size of the preceding one, 150µm, 300µm, 600µm, 1.18mm, 2.36mm, 5mm and 10mm. Higher value for Fineness Modulus indicates a coarser grading.

2.4 Mixes used in experiments

A fixed concrete mix proportion with effective water cement ratio 0.64 was used for all mixes with cement-fine aggregate-coarse aggregate mix proportions of 1: 2.4: 1.8 and the materials mixed thoroughly using pan mixer.

2.5 Tests used in experiments

2.5.1 Flow table test

Standard flow table, cone and tamping wooden rod were used in the whole tests in order to determine the workability of the fresh concrete mix. The inside of the cone and the table top were moistened at the beginning of each test. The cone is filled with concrete in two layers. Each layer is tamped 25 times with the standard tamping rod, and the top surface of the cone is surfaced by means of screeding. Immediately after filling, the cone is slowly lifted, and the unsupported concrete slumped. Before lifting the cone, excess concrete is removed, and surrounding table is cleaned. The table top then lifted and allowed to drop 15 times. The concrete spreads and the maximum spread parallel to the two edges of the table is measured. The average of these two values was recorded representing the flow of the fresh concrete mix in millimeters (mm).

2.5.2 Compressive strength of concrete

Standard metal cubes 150X150X150 mm^3 were used to cast concrete. For the purpose of testing the ideal grading and fineness modulus of limestone, 36 cubes were prepared for testing. Four different fineness moduli and grading were tested 2.4, 2.6, 2.92 and 3. For all tests, the cubes are left in curing until testing at the age of 3, 7 and 28 days respectively. For each curing time 3 samples were loaded to failure using MATEST compression testing machine and the average is recorded for comparison purposes. Table vibrator was used to compact all the samples for a specified time.

3 Results and discussions

The optimum Fineness Modulus and grading of limestone used as fine aggregate has been examined to improve the compressive strength and workability of concrete mixes. Table (6) and Figure (2) shows the average compressive strength for each Fineness Modulus of limestone used as fine aggregate to prepare standard specimens of the concrete mixes tested at the age of 3, 7 and 28 days.

Table 6. average compressive strength (in MPa) of standard concrete cubes prepared using limestone with different Fineness Modulus as fine aggregate.

F.M. of Limestone	3 days	7 days	28 days
2.4	17.39	21.76	22.8
2.6	22.89	24.37	31.09
2.92	20.73	27.22	30.64
3	18.15	20.87	26.24

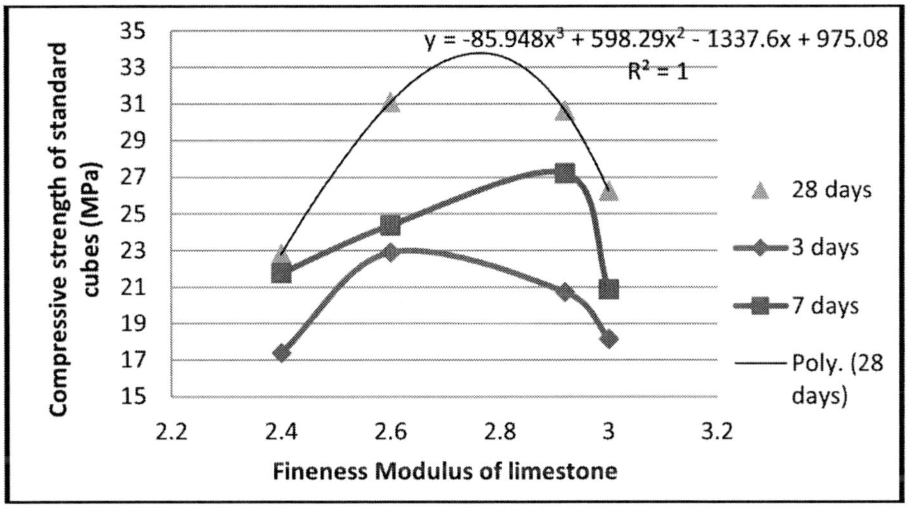

Fig. 2. Average compressive strength (in MPa) of standard concrete cubes prepared using limestone with different Fineness Modulus as fine aggregate.

As shown in Figure (2) a polynomial interpolation to the third degree is used for the 28 day testing results to determine the optimum Fineness Modulus for the limestone to be used as fine aggregate in the concrete mixes in order to improve the compressive strength of hardened concrete. To find the optimum Fineness Modulus of the lime stone, the derivative of the cubic equation is found and solved. See equation (1) below.

$$\frac{dy}{dx} = -257.24x^2 + 1196.58x - 1337.6 = 0 \quad (1)$$

Solving this derivation, the optimum fineness modulus for the limestone to be used as fine aggregate in the concrete mix to get maximum compressive strength is found to be 2.78.

Table (7) and Figure (3) shows the average flow table test results for different Fineness Modulus of limestone used as fine aggregate to prepare the concrete mixes tested directly after mixing. The results reveal that there is an increment in the workability of fresh concrete with higher Fineness Modulus of limestone used in the concrete mix as fine aggregate. In other words, the workability of the fresh concrete mix increases with the coarser grading of the limestone. It is good to indicate that for the lowest Fineness Modulus of limestone, the standard cubes were honeycombed and the mix was obviously not workable which also leaded to lowest compressive strength recorded value.

Table 7. workability of fresh concrete mixes measured by the average record of the flow table test (mm) for different Fineness Modulus of limestone used in the concrete mix.

F.M. of Limestone	flow table (mm)
2.4	300
2.6	388
2.92	420
3	425

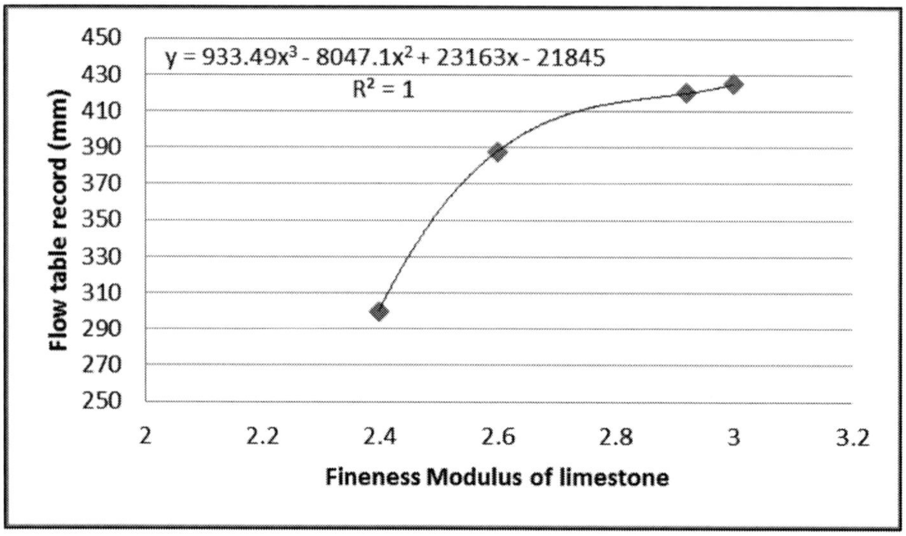

Fig. 3. workability of fresh concrete mixes measured by the average record of the flow table test (mm) for different Fineness Modulus of limestone used in the concrete mix.

As shown in Figure (3) above, a polynomial interpolation to the third degree is used in order to determine the expected workability of the optimum Fineness Modulus for the limestone to be used as fine aggregate in the concrete mixes which is found previously as 2.78. Applying the Fineness Modulus value of 2.78 in the cubic equation, the expected flow table record would be 412 mm and this is a mix with considerable workability.

The recommended grading then can be found for the optimum Fineness Modulus for the limestone as shown in Table (8) below. Also, the recommended grading was checked and compared with the ASTM requirements for the fine aggregate in Table (8) below.

Table 8. Recommended grading of limestone with fineness modulus 2.78.

Sieve size (mm)	Mass retained (g)	% retained	% cumulative Retained	% cumulative Passing	ASTM % cumulative Passing required
10	0	0.0	0.0	100.0	100
5	0	0.0	0.0	100.0	95 - 100
2.36	57.7	6.2	6.2	93.8	80 - 100
1.18	295.9	31.6	37.8	62.2	50 - 85
0.6	243.2	26.0	63.8	36.2	25 - 60
0.3	145	15.5	79.3	20.7	10 – 30
0.15	109	11.6	90.9	9.1	2 - 10
0.075	85	9.1		0.0	
		F.M	**2.78**		

4 Conclusions

The optimum Fineness Modulus of limestone to be used as fine aggregate in concrete mixes was found to be 2.78 and recommended grading was given in the research findings Table (8).

The calculated workability for the recommended grading of limestone is found to be 412 mm which is considered as a workable mix. Any grading of fine aggregate should be checked and compared with the ASTM grading requirements. The recommended grading of limestone with the optimum Fineness Modulus of 2.78 met the ASTM grading of fine aggregate requirements as shown in Table (8) in the research findings above.

The findings of this research work is valuable for limestone manufactures, aggregate quarries, and suppliers in order to improve the different properties of the concrete mixes in terms of compressive strength and workability without extra cost by providing the limestone to the construction market with optimum Fineness Modulus and grading that meets the ASTM grading requirements for fine aggregates.

The authors would like to thank the School of Graduate Studies & Research at the American University of Ras Al Khaimah for their financial support to this research work. The authors would also like to thank Stevin Rock L.L.C for providing this research with required coarse aggregate and lime, in addition to their support to the American University of Ras Al Khiamh with all equipment required to perform the tests.

References

1. Omar M. Omar, Ghada D. Abd Elhameed, Mohamed A. Sherif and Hassan A. Mohamadien, "*Influence of limestone waste as partial replacement material for sand and marble powder in concrete properties*". Housing and Building National Research Center (HBRC Journal), June 2012, Volume **8**, pp 193-203 (2012)
2. Joseph O. Ukpata, Maurice E. Ephraim and Godwin A. Akeke, "*Compressive strength of Concrete using Lateritic Sand and Quarry Dust as Fine Aggregate*". ARPN Journal of Engineering and Applied Sciences, January 2012, Volume **7**, No.1, pp 81-92. (2012)
3. Rana Burhan Abdurrahman Alshahway, "*Effect of Partial Replacement of Sand with Limestone Filler on some properties of Normal Concrete*".Al-Rafidain Engineering ,June 2011, Volume**19**, No.3 (2011)
4. A. Jayaraman and V. Senthil kumar, "*Optimization of fully replacement of natural sand by m-sand in high performance concrete with nanosilica*". International Journal of Emerging Technology and Advanced Engineering, November 2013, Volume **03**, Issue 11, pp 497-502 (2013)
5. A.M. Neville and J.J. Brooks, "*Concrete Technology*", Prentice Hall, (2010)

South African construction and demolition waste procedure and its sourced material effects on concrete

Johannes Bester*, *Deon* Kruger, and *Broadley* Miller

University of Johannesburg, PO Box 524, Auckland Park, 2006, Johannesburg, South Africa

Abstract. This paper presents the results of a study into the use of fine aggregate from construction and demolition waste (C&DW) in the South African construction industry, and the impact this has on the important fresh and hardened properties of concrete when used as partial replacement for virgin materials. The C&DW were tested at replacement levels of 0%, 25%, 50%, 75% and 100%. The recycled fine aggregate was pre-soaked to mitigate the expected effect of higher water absorption of recycled concrete aggregate (RCA). However, pre-soaking of the fines attributed to the poor performance of the fresh and hardened properties of the concrete. The pre-soaking procedure used was problematic when used with fine aggregates and caused a large amount of excess water to be added to the concrete mix as a result. This study highlights the importance of planning the demolition process of a structure to ensure that the best possible quality of RCA can be extracted. The study revealed that the use of recycled fine aggregate as a replacement of virgin materials in the manufacture of concrete has a much greater negative impact than the use of coarse recycled aggregates. In addition, the pre-soaking procedure of fine aggregates warrants special attention.

1 Introduction

For many years and across the globe, the use of natural resources and the abuse of the environment have been seen as necessary "costs of development", but recently the drive for a greener, more sustainable construction industry has been on the rise worldwide. The introduction of green building rating with systems such as those employed by Leadership in Energy and Environmental Design (LEED) in the United States of America and the Green Star rating system in South Africa (SA) have changed the global view of green technology. This change in public perception has led to the demand, and subsequently, the value of including of "green" properties in materials has being on the rise.

This international trend has resulted in a rethink of the construction industry as a whole worldwide, resulting in what is known as sustainable development, which the Bruntland Report defines as: "Development that meets the needs of the present, without compromising

* Corresponding author: jannesb@uj.ac.za

on the ability of the future generations to meet their needs." According to the triple bottom line theorem of Sustainable Development, Sustainable Development comprises of three main parts, namely: Economic-, Environment- and Social factors [1]. All three of these factors need to be considered and balanced in order to make a project or product as sustainable as possible and thus meet the "Triple Bottom Line" of sustainable development.

2 The construction and demolition waste industry

The idea of recycling of construction and demolition waste (C&DW) is not a new one, but one that many developed countries such as those in the European Union and Japan have implemented very successfully for a number of years now. Various studies have shown that the C&DW waste streams in Europe alone in the late 90's stood at around 200-330 million tons per year. This type of waste was generally made-up of around 67% of masonry and old concrete rubble and it is assumed that the waste make-up is the same in developed countries worldwide. Recycling of this C&DW generated across Europe however stood at only 28%, whereas some countries such as the Netherlands recycle up to 95% of their construction and demolition waste [2]. Various international regulatory bodies the world over has noticed this possibility for growth of C&DW recycling. The EU even has short term goals in place to enforce the recycling of between 50-90% of all the construction and demolition waste generated in their jurisdiction [3].

The C&DW stream in South Africa (SA) alone is estimated at around 5-8 million tons [4]. This shows that there is a massive opportunity for growth in the recycling industry worldwide as well as in SA with regards to the recycling of C&DW. SA still has vast open spaces and natural aggregate resources are still available, therefore it is not surprising that SA is lagging behind in terms of its development of measures to promote the recycling of C&DW. However, as the available area for landfill sites are reduced, natural resources that are depleting and the pressures from global markets increase, such as with the need for ISO 9000 and ISO 14000 requiring more companies to have quality environmental management systems, SA might soon see the need for effective recycling of C&DW [5].

3 Aspects of construction and demolition waste recycling

To truly partake in the sustainable development process one has to look at all the aspects of waste minimisation throughout the project life-cycle. The 4 R's of waste minimisation are [6]:

1. Recovery
2. Reuse
3. Recycling
4. Reduction

The recycling of construction and demolition waste could have a number of positive aspects, such as:

1. Reduced impact on the environment, through lowering the demand for virgin materials;
2. Reduced cost of materials, as recycled materials are usually less expensive than virgin materials;
3. Reduced cost of transportation of material, as demolition waste could be used on the same site or in a nearby area;
4. Reduced impact on the environment, through a decrease in the need to landfill the waste, the lifespan of landfills could be increased;
5. Reduced landfilling costs.

At present the major uses of recycled construction and demolition waste has been for roadfills and backfills [3]. This unfortunately is considered more of a "down-cycling" process resulting from the perception that recycled aggregates are of inferior quality and thus virgin materials are still used for the majority of new concrete production. This is because past studies by institutions such as RILEM (The International Union of Testing and Research Laboratories for Materials and Structures), ACI (American Concrete Society) and Building Contractors Society of Japan (B.C.S.J) have shown a large degree of variability in the quality of the recycled material and the concrete made therewith.

Tests on the effects of recycled aggregates when used in the production of new concrete have been conducted internationally for over 30 years and from this some properties and effects have been proven to be quite common. These general findings have shown that concrete mixes using recycled concrete aggregate (RCA) can be designed in much the same way as those using natural aggregates, provided that the higher water absorption in RCA is appropriately accounted for when determining the water content.

In an effort to try and minimise the great variation in the fresh properties and strength results obtained when using recycled aggregates in the production of new concretes, most countries that allow the use of recycled aggregate in structural concrete have adopted necessary criteria or classification standards that the recycled aggregate must adhere to in order to be deemed safe for the use in the production of new concrete. These classification measures have seemed to be quite effective in minimising the risk when recycled aggregates are used, as well as giving the users a guide to the effects that the aggregates would have on the concrete as well as ways in which to mitigate them.

4 Experimental design

The RCA that was used in this research was sourced from the demolition of an unknown structure. As the case generally would be in South Africa, the RCA that was produced had little control over the recycling process, thus it contained various contaminants (Figure 1). After crushing and sieving the material in the laboratory, it was found to contain mostly fine particles and hence the subsequent tests will be on the use of the waste material for the replacement of the fines part of the concrete mix as described by the Fulton's Concrete Technology handbook [7].

Fig. 1. Contaminants in the RCA.

A control mix (Table 1), simulating a ready-mix concrete, was made by the Afrisam Roodepoort Plant. Replacement levels of 25%, 50%, 75% and 100% by mass was used to determine how the level of replacement affected the fresh and hardened properties of the concrete.

Table 1. Control concrete mix design.

Description	Control mix
w:c	0.67
Water (L)	195
CEM II A-M (S-V) 42.5R (kg)	*290*
Eikenhof crusher sand (kg)	*990*
22mm Eikenhof stone (kg)	*1 075*
Slump (mm)	*60*

The 25%, 50%, 75% and 100% replacement mixes were made in accordance with the procedure set out in SANS 5861-1: Concrete tests – Mixing fresh concrete in the lab. After casting of the moulds, the test cubes was cured in water in accordance with the standards set out in SANS 5861-3: Concrete tests – Making and curing of test specimens. A total of 15 test cubes and 3 test beams of each of the different mixes were manufactured. It must be mentioned at this point in time that no economic comparisons are considered for this project.

4.1 Fresh concrete properties

The tests used to evaluate the effect of the recycled fine aggregate on the fresh properties of the concrete compared to that of the control mix were the slump test (SANS 5862-1: 2006), slump retention and total bleed test (ASTM C232/C232M) and the setting time test (SANS 501 963-3: 2006).

4.2 Hardened concrete properties

In order to generate a strength gain curve 15 cubes sized 100mm x 100mm x 100mm were crushed at 1, 3, 7 and 28 days from the day of casting, in accordance with SANS 5863: Concrete tests – Compressive strength of hardened concrete. The flexural strength was determined on 28 days, in accordance with SANS 5863: Concrete tests – Flexural strength of hardened concrete. These specimen sizes were 500mm x 100mm x 100mm, and there were 15 specimens tested. A single load was applied at mid-span.

4.3 Pre-soaking

The aggregates were pre-soaked in an attempt to mitigate most of the negative effects caused by recycled aggregates when used in concretes. The pre-soaking procedure consisted of submerging the material in water an hour before being used. After 40 minutes the material was removed from the water and allowed to air dry for 20 minutes. This was done to allow excess water to drain off and evaporate, in order to have the material as close as possible to a saturated surface dry condition.

Pre-soaking on the aggregates could also have a negative effect on the results. The very fine dust particles could be washed away during draining after the pre-soaking, which could have an effect the bleeding characteristics. In addition, if the material washed away consisted of un-hydrated cement particles as expected, it could also have an effect on the compressive and flexural strengths achieved.

5 Results

One of the material classification tests that was performed on the sand, was the relative density (RD) test. The RD of the sand used in the control mix was 2.89, whereas the RD of the RCA sand was 2.57. The lower RD could be attributed to the usually more porous nature of RCA.

Figure 2 shows the gradings of the sand used in the control mix and the demolition waste sand (DW) used to manufacture the RCA. It also indicates the envelope used in the concrete mix design. The results indicates that the RCA fines are very fine, and falls mostly outside the fines limit, as set out by The Concrete Institute [2]. A possible reason for this is the amount of masonry present in the RCA.

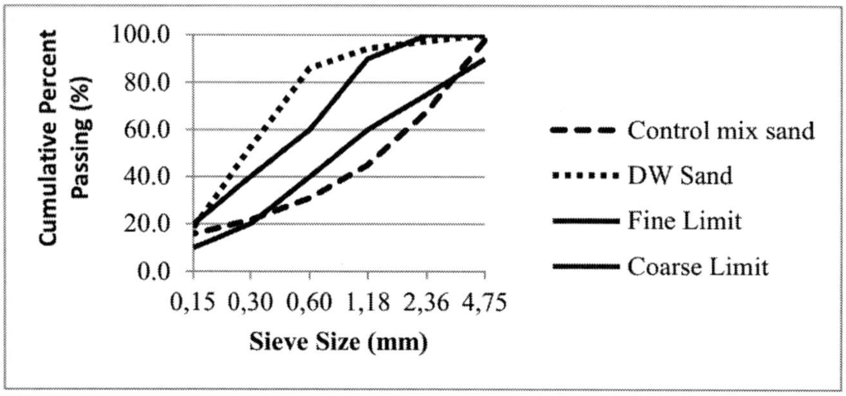

Fig 2. Grading envelope, control sand and DW sand gradings.

5.1 Fresh properties results

5.1.1 Slump

Figure 3 indicates that here is a massive increase in the slump value of the mixes utilizing the DW aggregate as compared to the control mix. The RCA mixes had a small positive increasing trend in the slump value as the percentage DW replacement was increased. Unfortunately, due to time constraints, the material was not left to air dry completely to the saturated surface dry condition desired. The effect of the increased capillary action of the finer material that would make it difficult for excess water to drain off after the pre-soaking, the material was not in a proper saturated surface dry condition and most likely had a large degree of surface water present. This surface water is exacerbated by the material's small particle size, which means that the percentage water retained by the recycled material is much higher for the fine material than it would have been for a larger particle size. All the mixes however still possessed good cohesion.

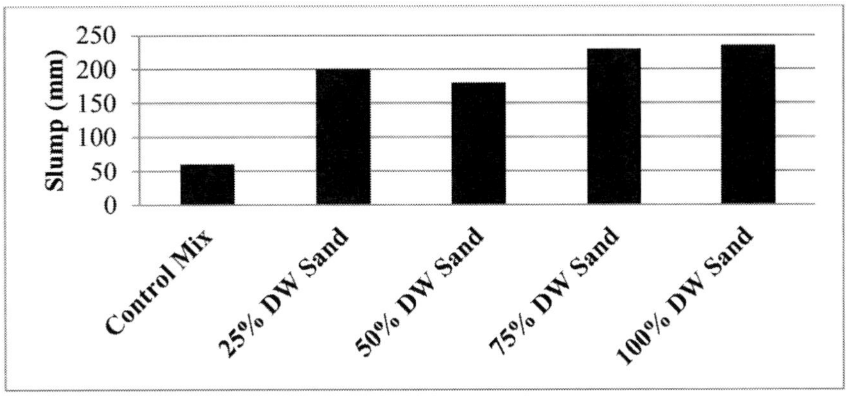

Fig. 3. Slump results.

5.1.2 Slump retention

When the trend lines produced by the slump retention results of the different mixes containing the demolition waste is compared with that of the control mix (Figure 4) there is a slight trend of reduced slump retention or faster setting rate than expected. However the trend does not increase as the level of replacement increases. The reason for this reduced slump retention could be as a result of the higher water absorption of the demolition waste material, thus effectively taking water from the mix. Thus, the pre-soaking procedure used for this material may have been inadequate and a longer pre-soak time could have been necessary.

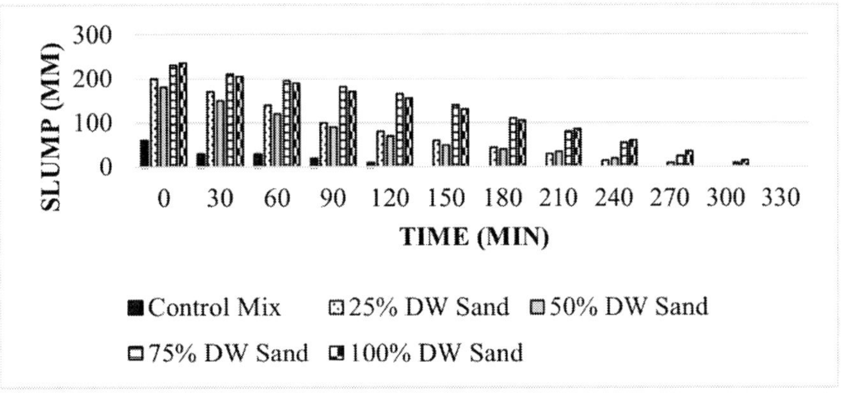

Fig. 4. Slump retention.

5.1.3 Setting time

The setting and hardening process of Portland cement can be divided into 3 distinct stages [8], these stages are:

1. Dormant Stage – paste remains plastic and workable
2. Setting stage – the paste is stiff and unworkable
3. Hardening stage – the paste is a rigid solid and gains strength with time

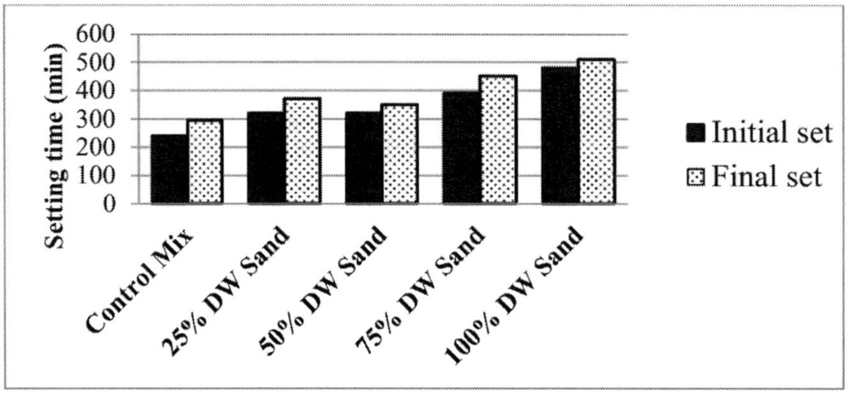

Fig. 5. Setting time.

5.1.4 Bleed volume

The transition between the dormant and setting stage is known as the initial set and the transition between the setting and hardening stages is known as the final set. As can be seen from the Figure 5, there is an almost linear increase in the setting times achieved for the mixes containing the demolition waste sand material compared with that of the control mix. It can therefore be deduced that an increase in the amount of recycled fines would result in an increase in the setting time of the mix. The high workability in conjunction with the increased setting times of the mix could be an indicator of a high degree of free water in the mix. If the setting stage of the 50% replacement is taken as an outlier, then the results in Figure 5 would suggest that the setting stages of the mixes containing the demolition waste material reduces as the level of replacement increases.

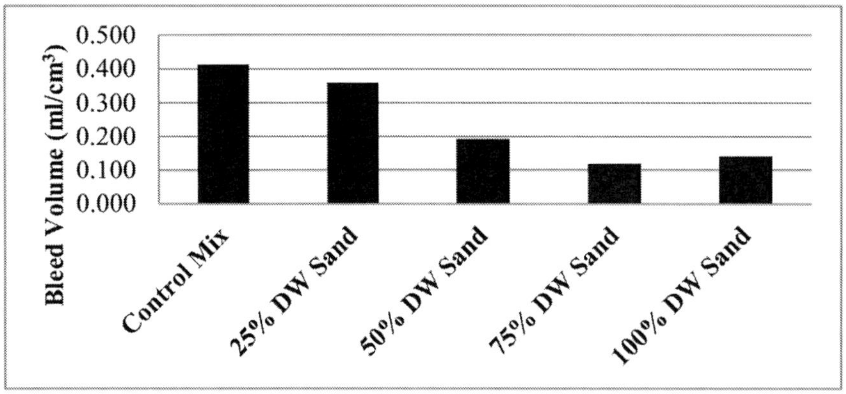

Fig. 6. Bleed volume.

It is important to note that the bleed volume results in this study were only the bleed volumes measured on the surface of the concrete after the 280 minute test period, and not the total bleed volume of the concrete. This is because weight measurements of the test samples after being oven-dried were not measured and therefore, the total bleed volumes could not be calculated. The bleeding of concrete is affected by 2 main factors, the relative density of the aggregates in the concrete mix as well as the amount extremely fine material present in the concrete mix. As can be seen in Figure 6, the bleed volumes of the concretes containing the demolition waste material showed lower bleeding than that of the reference mix, and the results also showed a trend of decreasing bleed volumes as the level of replacement increased.

This was unexpected as the slumps earlier pointed to excess water present in the concrete mixes and the grading of the material, in Figure 1, also showed a reduced amount of very fine or dust particles present, as well as the assumption that much of the finer particles would have been washed away during the pre-soaking procedure. All of these factors should have increased the bleed volumes experienced. However, the lower relative density of the recycled aggregate as well as the higher water absorption of the demolition waste aggregate seems to have been sufficient to surpass all of these factors and thus effectively reduced the amount of bleeding of the concrete mixes containing the demolition waste sand material.

5.2 Hardened properties results

5.2.1 Compressive strength development

It can be seen from Figure 7 that as the replacement of the virgin aggregate with that of recycled demolition fines increases, there is a drastic reduction in the compressive strength development of the concrete. This progressive loss in compressive strength could be attributed to the gradual increase in the lower quality demolition waste sand. This effect could have been exacerbated by the high volume of pores in the DW aggregate present in the concrete mixes leading to an increase in the voids of the concrete and thus a loss in the concrete's strength. The mixture, as it stands, is not really advisable for any use, due to the severely reduced strength for the amount of cement added, effectively making the concrete uneconomical to use.

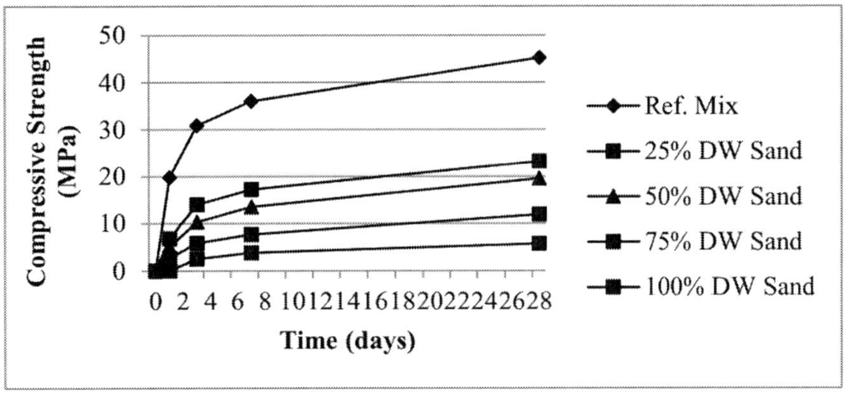

Fig. 7. Compressive strength development curves.

5.2.2 Flexural strength

It can be seen from Figure 8 that the flexural strength of the concrete decreased with the increase in sand replacement. It is said that concrete mixes made with virgin aggregates typically have a compressive strength to flexural strength ratio of around 1:10, and the control mix stays true to this assumption. The recycled demolition waste sand showed ratios ranging between 1.33:10 to 2.4:10. The ratio increase with an increase in the percentage replacement. This could be as a result of the better interfacial bond strength of the material to the new cement mortar because of its rougher texture caused by the attached cement mortar paste.

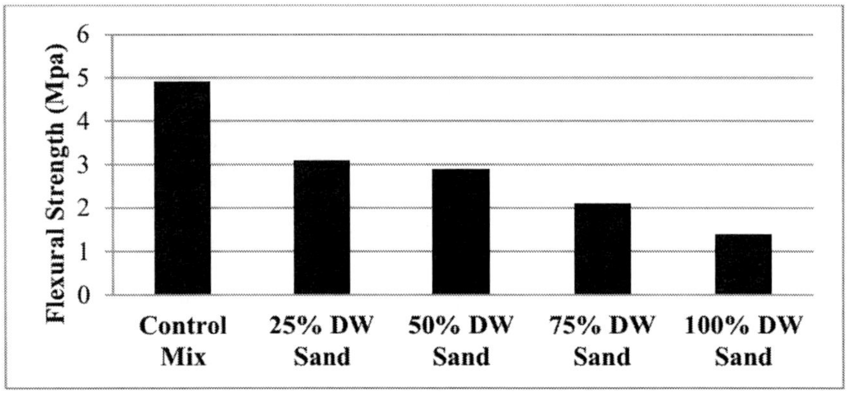

Fig, 8. Flexural strength.

6 Conclusion

The masonry contaminants found in the C&DW material caused the concrete to react in a different manner to that which is usually expected from a clean recycled aggregate concrete. The fresh properties of the concrete showed a marked increase in the slump, and a large drop in the slump retention. The initial set, and final set increased as the replacement of natural aggregate increased. The bleed volume also showed a marked decrease. There was a marked decrease in the compressive strength development of the concrete, and the flexural strength also decreased. There was a marked increase in the ratio of concrete flexural strength to compressive strength.

It can be concluded that to use of recycled aggregate from the demolition of an unknown structure, which has little control over the recycling process will negatively affect the properties, as well as the durability. To address these negative effects, adjustments needs to be made to the mix design, thus increasing the cost of the concrete.

References

1. G. Owens (Editor), *Fundamentals of Concrete,* The Concrete Institute, Midrand (2013)
2. P. K. Mehta, Reducing the environmental impact of concrete. Concrete International. (2007)
3. E. K. Lauritzen, Recycling Concrete – An Overview of Challenges and Opportunities. In Liu, T.C. & Meyer, C., eds. Recycling Concrete and Other Materials for Sustainable Development., ACI International – SP – 219. (2005)
4. K. Benjamin, Performance of concrete made with commercially produced recycled coarse and fine aggregate in Cape Peninsula. Master's Thesis. UCT. (2004)
5. G. Grieve, *Aggregates for Concrete*. Chapter 3 in Fulton's concrete technology. Midrand: The Cement & Concrete Institute, pp.55-56. (2009)
6. P.A. Vesilind, W. Worrell, D. Reinhart, *Solid waste engineering*. California. Thomson Learning. (2002)
7. B. Addis, J. Goodman, *Concrete mix design.* Chapter 11 in Fulton's concrete technology. Midrand: The Cement & Concrete Institute, pp. 219-224. (2009)
8. J. Kellerman, S. Crosswell, *Properties of fresh concrete*. Chapter 6 in Fulton's concrete technology. Midrand: The Cement & Concrete Institute, pp. 83-94. (2009)

Evaluation of pozzolime mixtures as a sustainable binder to replace portland cement in structural concrete

Nahida Kadum[1], *Tareq* al-Attar[2],* and *Zyiad* Al-Azzawi[2]

[1]MSc, Ministry of Youth and Sport, Baghdad, Iraq
[2]PhD, University of Technology, Baghdad, Iraq

Abstract. This study aimed to evaluate the properties of concrete mixes produced mainly of hydrated lime and different types of Pozzolan. This mixture, which is nominated as *PozzoLime*, is presented as an alternative sustainable binder to replace Portland cement in concrete. The study tends to optimize the Pozzolan to lime proportions with respect to workability and strength of concrete. The investigated parameters are; water to binder ratio, aggregate to binder ratio and type of Pozzolan. The effects of these parameters on properties of the mixture in fresh and hardened states are studied. The included tests were; temperature profile, slump, fresh and oven-dry density, compressive, splitting tensile and flexural strength and dynamic modulus of elasticity. Silicafume, Fly ash and Metakaolin were the used Pozzolan in this study .The temperature profile through hydration (coffee cup test) shows that all *PozzoLime* mixtures have significantly less environmental impacts (evolved heat) than Portland cement. The results also show that structural concrete with more than 28 MPa compressive strength at 28 days age can be produced by a binder consisted of 1:1 by weight ratio of Silicafume to hydrated lime, , and with a total binder content of 333 kg/m^3.

1 Introduction

The manufacture of Portland cement is an energy and carbon intensive process and it is responsible for 5 to 8 percent of the total global carbon emissions. Pozzolana-Lime cements are one of a number of 'second generation' low-carbon cements being developed to meet the demand for concrete in construction [1].

Lime was known to the Greeks and was widely used by the Romans. The Romans also knew how to make hydraulic binder, i.e. lime-Pozzolan cement by adding materials, such as volcanic ash or powdered bricks, tiles and pottery to lime [2].

Lime could be considered as a sustainable binder due to lower production energy needs, lower CO_2 emission during production and CO_2 absorption by carbonation on setting [3].

* Corresponding author: 40076@uotechnology.edu.iq

Pozzolans, such as: Silica fume, Fly ash, Slag which are also known as supplementary cementitious materials, are in most by-products for many industrial processes. Their applications in concrete industry are well-known and had been well investigated. The definition of Pozzolanic reaction could be summarized by that it is the chemical reaction between the finely grounded siliceous or siliceous aluminous phases in Pozzolans and $Ca(OH)_2$, which is a by-product of the formation of calcium silicate hydrates CSH of Portland cements, to form compounds possessing cementitious properties [4].

For sustainability requirements, Meyer [5] listed many potential tools to face the challenges resulted from using Portland cements in concrete. Some of these relevant tools are:

1. Since the main challenges are associated with the production of Portland cement, whereas concrete itself is basically environmentally friendly, a primary goal is to use as much concrete with as little Portland cement as possible.
2. Replace as much Portland cement as possible by supplementary cementitious materials, especially those that are by-products of industrial processes. This reduces the amount of CO_2 generated and the amount of embedded energy.

According to that the present study was carried out as a part of a research program to produce *PozzoLime* concrete as a sustainable binder. This binder will contain no Portland cements and could be used as structural concrete.

2 Experimental work

2.1 Materials

The used Pozzolanic materials throughout this study were: Silica fume (ASTM C1240-10), Fly ash and Metakaolin (ASTM C618-08). A hydrated lime conforming to the ASTM C821-09 was utilized to produce the *PozzoLime*. In addition to that, an ordinary Portland cement (ASTM C150-11, Type I) was used for comparison. Table 1 shows the chemical composition and fineness of these used materials.

Natural sand, with fineness modulus of 2.9 and specific gravity (SSD) of 2.62, was used in mixes of this work as fine aggregate. Crushed gravel, with maximum size of 14 mm, was used as coarse aggregate. This aggregate has a bulk density of 1565 kg/m^3 and specific gravity (SSD) of 2.65. Tap water was used for mixing and curing of the produced concrete mixes. Glenium 54 (BASF) was the used high-range water reducing admixture. It is free from chlorides and complies with the ASTM C494-10, Types A and F.

Table 1. Chemical composition and fineness of the used binders.

No.	Property		OPC*	L*	SF*	FA*	MK*
1	Oxide Content, %.	CaO	61.50	86.44	1.21	0.98	3.00
		SiO_2	21.80	---	90.64	65.65	51.35
		Al_2O_3	6.50	---	0.02	17.69	33.40
		Fe_2O_3	2.20	---	0.01	5.98	2.30
		MgO	1.40	2.12	0.01	0.72	0.17
		SO_3	2.50	---	0.23	0.19	0.15
		Na_2O	0.28	---	0.22	1.35	---
		K_2O	0.51	---	0.15	2.99	---
		LOI	2.40	---	2.86	3.10	7.80
2	Surface Area (Blaine), m^2/kg		240	1200	20000	773	1100

*: OPC=ordinary Portland cement; L= Lime; SF= Silica fume; FA= Fly ash; MK=Metakaolin

2.2 The investigated parameters and details of concrete mixes

A reference mix was made with ordinary Portland cement and proportioned according to the ACI 211 Committee method, ACI 211.1-92 [6]. The minimum specified compressive strength at 28 days for this mix was 30 MPa. Another nine mixes were produced in which the Portland cement was totally replaced by different Pozzolan-Lime mixtures. In all *PozzoLime* mixes the ratio of Pozzolan to Lime was 1: 1 by weight. Table 2 lists the details of the all produced mixes.
The experimental part of this study was designed to investigate the influence of the following parameters on the properties of fresh and hardened concrete mixes:

1. Water to binder ratio by weight, W/B.
2. Aggregate to binder ratio by weight, A/B.

As shown in Table 2, five mixes were prepared by using Silica fume and Lime as a binder. Therefore, mixes LS1, LS2 and LS3 are made with W/B ratio of 0.5, 0.55 and 0.6 respectively but with a constant A/B ratio of 4. Meanwhile, mixes LS4, LS1 and LS5 are mixes made with A/B ratio of 3, 4 and 5 respectively and with a constant W/B ratio of 0.5.
With respect to Fly ash, the mixture of Fly ash-Lime has failed to set and hardened for many trials, therefore, it was decided to use a ternary system of Silica fume, Fly ash and Lime to overcome this problem. According to that, three different FA to SF ratios were adopted in preparing the mixes LSF1, LSF2, LSF3. They were 10: 40, 20: 30 and 30: 20 percent by weight respectively.
For Metakaolin-Lime mix, LMK, the W/B ratio was 0.5 by weight and the A/B was 4 by weight.
All casted specimens were moist-cured in water till the age of testing.

Table 2. Details of investigated mixes.

Mix	Binder *, kg/m³					Sand kg/m³	Gravel kg/m³	Water kg/m³	HRWR % by wt. of binder	W/B ratio	A/B ratio
	OPC	L	SF	FA	MK						
Ref	400	0	0	0	0	704	1056	200	0	0.5	4
LS1	0	200	200	0	0	640	960	200	2.5	0.5	4
LS2	0	197.5	197.5	0	0	635	950	220	2	0.5	4
LS3	0	195	195	0	0	630	940	240	1	0.6	4
LS4	0	244	244	0	0	570	900	244	2.5	0.5	3
LS5	0	166.5	166.5	0	0	680	1020	166.5	2.5	0.5	5
LSF	0	200	160	40	0	640	960	198	2	0.4	4
LSF	0	200	120	80	0	640	960	160	1.5	0.4	4
LSF	0	200	80	120	0	640	960	160	1	0.4	4
LM	0	200	0	0	200	640	960	198	2.5	0.5	4

*: OPC=ordinary Portland cement; L= Lime; SF= Silica fume; FA= Fly ash; MK=Metakaolin

2.3 Testing program

The following tests were conducted on concrete mixes in their fresh state:

- Slump test (ASTM C143-10).
- Fresh density test (ASTM C138-10).
- Setting time of test (ASTM C191-08).
- Coffee cup test (temperature rise profile) [7].
- Compressive strength test (BS 1881:P.116- 1983).

- Splitting tensile strength test (ASTM C496-04).
- Flexural strength test (ASTM C78-10).
- Dynamic modulus of elasticity (ASTM C597-09).
- Oven-dry density and absorption tests (ASTM C642-06).

3 Results and discussion

3.1 Slump, density, and absorption

Slump and fresh density records of OPC and *PozzoLime* concrete mixes were taken immediately after mixing. The results of these two tests are listed in Table 3.
All the binders that were used in producing *PozzoLime* have higher specific surface area than the used Portland cement in producing the reference mix. Therefore, this has necessitated the use of Glenium 54 in *PozzoLime* mixes to keep them workable. The slump values were inversely proportional to the increase in aggregate content due to the increase in the water requirement. Also the slump improved with the increase in the Fly ash content due to the decrease in the surface area of the binder.
PozzoLime mixes had always lower fresh and oven-dry densities than the reference mix. The reduction ranged between 4 and 11 and between 6 and 9 percent for fresh and oven-dry densities respectively. The Silica fume and Metakaolin mixes show almost identical reduction. Meanwhile, the inclusion of Fly ash in the binder system causes this reduction to be lower. This could be due to that finer particles would pack better than coarser ones. However, this could be considered as an advantageous characteristic for these mixtures with respect to their impact on the formwork structure and dead load calculations.

Table 3. Result of slump, density and absorption tests.

Mix	W/B ratio	A/B ratio	Slump, mm	Fresh density, kg/m^3	Oven-dry density at 28-day, kg/m^3	Absorption at 28-day, %
Ref	0.5	4	98	2464	2224	7.5
LS1	0.5	4	135	2237	2048	4.1
LS2	0.55	4	122	2235	2036	4.8
LS3	0.6	4	133	2232	2030	5.4
LS4	0.5	3	138	2200	2020	4.9
LS5	0.5	5	126	2270	2100	3.9
LSF1	0.4	4	131	2342	2035	8.6
LSF2	0.4	4	138	2350	2061	7.9
LSF3	0.4	4	135	2357	2089	7.1
LMK	0.5	4	139	2272	2088	8.2

3.2 Setting time

Table 4 shows the setting time, initial and final, for the refrence, LS1, LSF3 and LMK mixes. The results demonstrate that in general, all *Pozzolime* mixes exhibit longer setting time than the reference mix, this may be due to the absence of C_3A and C_3S and the stiffening is taking place only due to the reaction of $Ca(OH)_2$ with Pozzolanic materials which is slow in nature [8]. Some of the delay could be caused by the absence of ettringite as indicated by XRD test [9].

Set retardation is an advantage during hot weather, allowing more time to place and finish the concrete especially in Iraq where there is a need for retarders in summer. However, during cold weather, pronounced retardation can occur with *PozzoLime* mixture, significantly delaying finishing operations. Longer setting time could cause delay in formwork dismantling. Accelerating admixtures can be used to overcome this issue.

Table 4. Setting time results.

Mix	Setting time, hours	
	Initial	Final
Ref	2.0	7.5
LS1	4.5	35.5
LSF3	5.0	37.0
LMK	2.5	29.0

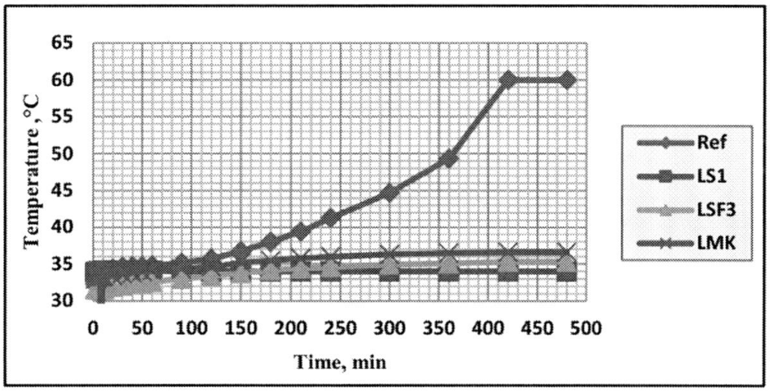

Fig. 1. Temperature rise due to hydration profile for reference, LS1, LSF3 and LMK mixes.

3.3 Temperature rise profile due to hydration (coffee cup test)

Cementitious paste mixtures generate heat as they hydrate. Although the coffee cup test procedur [7] is not capable of identifying specific chemical changes in the cementitious materials, it could be employed to monitor the heat generated by paste mixture materials.

This test is suitable to monitor the temperature rise due to hydration of *PozzoLime* paste and compare it with that generated from Portland cement hydration.

Figure 1 displays the temperature rise for pastes of four selected mixes (which could be used as an indication for the generated heat of hydration). The change in shape of the temperature profile and the difference between maximum temperatures observed for *PozzoLime* mixtures, as compared to Portland cement, have a lower heat of hydration than Portland cement pastes. The LS1 paste shows the lowest temperature values and seems to be constant all over the testing period. There is a continuous slight increase in temperature with time for LSF3 mixture and the LMK paste shows a higher temperature than other Pozzolans.

The observed reduction in temperature rise is beneficial in producing concrete for massive structures and hot weather [10].

3.4 Strength development

As shown in Table 5, the results demonstrate that, in general, all mixes exhibit continuous increase in strength with the increase in curing age. It is also observed that always the reference mix has higher strength than other mixes. This could be due to the slow reaction of Lime-Pozzolana with water.
With respect to compressive strength, the Silica fume-Lime and Silica fume-Fly ash-Lime mixes have proved themselves suitable for producing structural concrete. All of these mixes, which were cast in cubic specimens, have attained more the 22 MPa at 28 days age (17 MPa for cylinders). On the other hand, Metakaolin-Lime mixture could be recommended for mortar mixes and uses.

Table 5. Results of different types of strength tests for the investigated mixes.

Mix	W/B ratio by weight	Compressive strength, MPa, at age:			Splitting tensile strength, MPa, at age:			Flexural strength, MPa, at age:	
		7 d	28 d	90 d	7 d	28 d	90	28 d	90 d
Ref	0.5	23.1	35.8	40.1	1.90	2.50	3.25	3.1	4.5
LS1	0.5	15.6	27.1	36.6	1.60	2.20	2.80	2.3	3.1
LS2	0.55	12.8	23.0	31.1	1.42	1.90	2.50	2.0	2.8
LS3	0.6	10.2	18.2	24.4	1.23	1.60	2.15	1.8	2.5
LS4	0.5	14.9	26.5	34.9	1.50	2.10	2.70	2.2	2.7
LS5	0.5	19.4	28.4	38.6	1.80	2.35	3.10	3.4	3.7
LSF1	0.4	17.2	32.1	38.2	2.16	2.90	3.60	3.7	4.6
LSF2	0.4	15.3	30.4	36.7	2.10	2.80	3.30	----	----
LSF3	0.4	13.7	27.4	34.2	1.80	2.60	3.10	----	----
LMK	0.5	5.2	6.1	7.2	0.41	0.49	0.66	0.5	0.6

The evaluation of compressive strength gain with time is of great concern to structural engineers. ACI Committee 209 [11] recommended the following relationship for moist-cured concrete made with ordinary Portland cement (ASTM C150 - Type I):

$$\mathbf{f}_{\mathbf{cmt}} = f_{c28}\left[\!\left[\frac{t}{4.0+0.85\,t}\right]\!\right] \tag{1}$$

Where:
f_{cmt} = mean compressive strength, MPa, at age t in days.
f_{c28} = mean 28-day compressive strength, MPa.

The *PozzoLime* compressive strength results have not complied with this relationship. The compressive strength values were overestimated at early age, 7 days, and underestimated in later age, 90 days. Therefore, upon the present work results, the numerical coefficients Eq. (1) may be adjusted to make the relationship fits for modeling the rate of strength gain for *PozzoLime* concrete, as shown in Eq. (2):

$$\mathbf{f}_{\mathbf{cmt}} = f_{c28}\left[\!\left[\frac{t}{7.17+0.74\,t}\right]\!\right] \tag{2}$$

When examining the percentages of increase in the splitting tensile strength, the following remarks could be concluded:

1. Mix LS1has a similar rate of strength gain to the reference mix and this could be resulted from the high specific surface area of Silicafume, $20m^2/g$.

2. For the ternary system Silica fume-Fly ash- Lime, the percentages of strength increase in the interval 7 – 28 days were higher than those in the interval 28 – 90 days. The main variable for mixes LSF1, LSF2 and LSF3 was the Fly ash content as a replacement for the Silica fume. Fly ash is known for its slow hydration process [12].

3. The LSF mixes showed higher tensile strength than that of the others, even higher than the reference mix, and that may be resulted from adopting lower W/B ratio in producing them.

3.5 Dynamic modulus of elasticity, E_d

Based on the measurement of the ultrasonic pulse velocity, the dynamic modulus of elasticity was calculated [4]. Table 6 lists the calculated values.

The E_d for reference mix was always higher than those for other mixes. LMK mix showed the lowest value. It could be concluded that the same trend for E_d was observed in development as the compressive strength but with slower rates for different mixes.

Table 6. Calculated values of dynamic modulus of elasticity for the investigated mixes.

Mix	Dynamic modulus of elasticity, GPa, at age:		
	7 days	28 days	90 days
Ref	57.0	65.7	67.9
LS1	37.0	42.8	44.3
LS2	35.2	41.2	42.8
LS3	33.0	38.2	42.1
LS4	34.3	44.7	46.9
LS5	46.0	47.6	52.2
LSF1	33.9	34.0	44.4
LSF2	34.8	34.9	45.2
LSF3	35.9	36.1	46.0
LMK	33.5	33.6	33.7

3.6 Effect of W/B ratio on strength development of PozzoLime concrete

For mixes LS1, LS2 and LS3, Figures 2a and 2b illustrate the relationships between the water to binder, W/B, ratio and compressive and splitting tensile strength respectively at a given moist-curing age. It is clear that the strength decreases with the increase of W/B ratio. And it could be outlined here that, within the investigated range of W/B ratios, strength of *Pozzolime* concrete, just like the Portland cement concrete, is a function of the W/B ratio and the degree of hydration. At a given temperature of hydration, the degree of hydration is time dependent and so is the strength. Within the investigated range of W/B ratios, Abram's rule is applicable and the W/B ratio is inversely proportional to *Pozzolime* concrete strength. This behavior can easily be explained as the natural consequence of a progressive weakening of the matrix caused by increasing porosity with the increase in W/B ratio [13]. To generalize this conclusion, more research is needed to investigate values of W/B ratio outside the present range.

 DOI: 10.1051/matecconf/201712002009

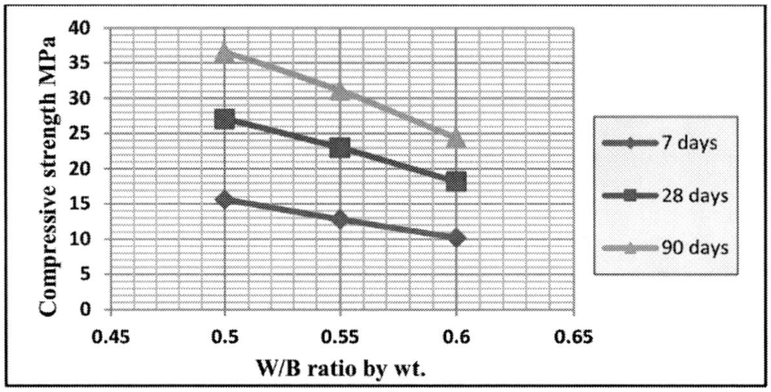

(a)

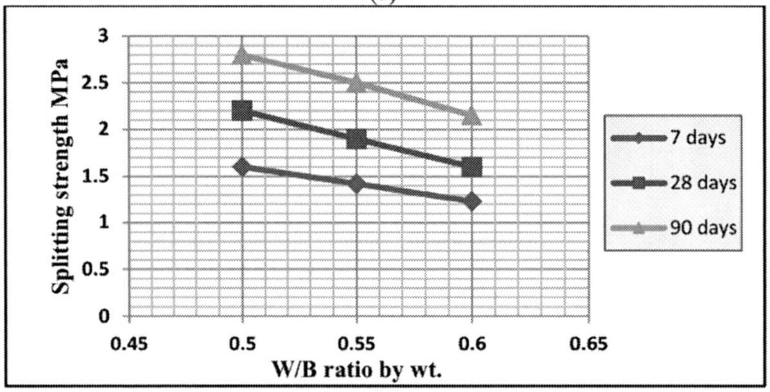

(b)

Fig. 2. Effect of W/B ratio on strength development of LS1, LS2 and LS3 mixes; **a**: compressive strength, **b**: splitting tensile strength.

3.7 Effect of A/B ratio on strength development of PozzoLime concrete

For mixes LS1, LS4 and LS5, Figures 3.a and 3.b illustrate the relationship between aggregate to binder, A/B, ratio and strength at a given moist-curing age. These Figures show that strength increases with the increase of A/B ratio at the same age, higher aggregate content gives higher strength. The main explanation of this influence lies in the total volume of voids in the concrete. Clearly, if the paste represents a smaller proportion of the volume of concrete, then the total porosity of the concrete is lower, and hence its strength is higher [14]. It could be observed also that curves in Figure 3 have low rate of increase (small slope) at different ages, and this could mean that aggregate content has a slight effect on strength development.

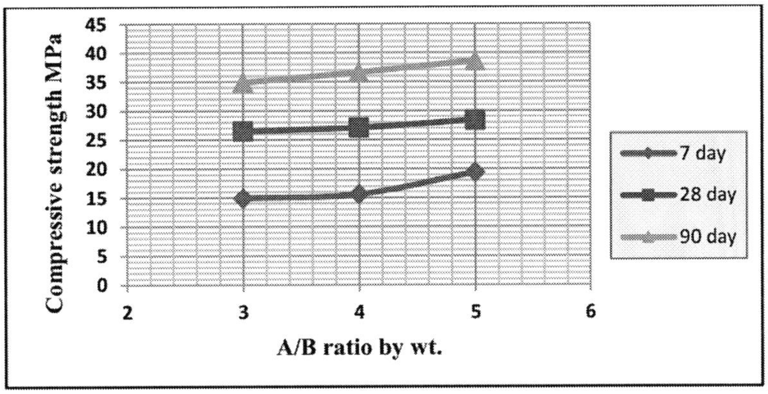

(a)

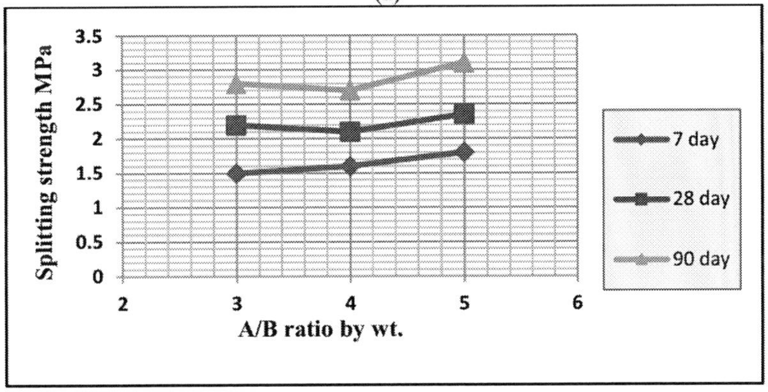

(b)

Fig. 3. Effect of A/B ratio on strength development of LS1, LS4 and LS5 mixes;
a: compressive strength, **b**: splitting tensile strength.

4 Conclusion

According to the present experimental results, the following conclusions could be extracted:

1. The Silica fume-Lime and Silica fume-Fly ash-Lime mixes have proved themselves suitable for producing structural concrete. All of these mixes have attained more the 22 MPa at 28 days age. On the other hand, Metakaolin-Lime mixture could be recommended for mortar mixes and uses.

2. *PozzoLime* mixes have always lower fresh and oven-dry densities than Portland cement mix. This could be considered as an advantageous characteristic for these mixtures with respect to their impact on the formwork structure and dead load calculations.

3. *PozzoLime* mixes exhibit longer setting time than the Portland cement mix. Set retardation is an advantage during hot weather but needs attention in cold regions and causes delay in formwork dismantling.

4. *PozzoLime* pastes, as compared to Portland cement, have a lower heat of hydration than Portland cement pastes. The Silica fume-Lime paste shows the lowest temperature rise and rate that seemed to be constant all over the testing period. There is a continuous slight increase in temperature with time for Silica fume-Fly ash-Lime paste and the Metakaolin-Lime paste shows a higher temperature rise than other Pozzolans.

5. Within the investigated range of W/B ratios, Abram's rule is applicable and the W/B ratio is inversely proportional to *PozzoLime* concrete strength. To generalize this conclusion, more research is needed to investigate values of W/B ratio outside the present range.

6. Different types of *PozzoLime* concrete strength have increased with the increase of A/B ratio at the same age but the increase in aggregate content has a slight effect on strength development.

References

1. E. R., Grist, K. A., Paine, A., Heath, J., Norman, H., Pinder, Clea. Prod. **93**, 26 (2015)
2. O., Kirca, *Structural Analysis of Historical Constructions* (2005)
3. A., Velosa, R., Veiga, Sust. Const., Mat. & Prac. 775 (2007)
4. A. M., Neville, *Properties of Concrete* (2011)
5. C., Meyer, *ASEM13*, Korea, 79 (2013)
6. ACI 211.1-92, *ACI Man. Conc. Prac.* (2010)
7. NCPT Center, Coffee Cup Test, Iowa State University (2008)
8. V. Indrawati, A., Manaf, J Phys. Sci., **19**, 51 (2008)
9. N. M. Kadum, *MSc Thesis,* University of Technology, Iraq (2016)
10. M. A., Caldarone, K. A., Gruber, R. G., Burg, J ACI, **91**, 37 (1994)
11. ACI 209R-92, *ACI Man. Conc. Prac.* (2010)
12. J., Newman, B. S., Choo, *Adv. Conc. Tech.: Mat.* (2003)
13. P. K., Mehta, P. J., Monteiro, *Conc.: Microst. Prop. & Mat.* (2005)
14. H., Lee, V., Vimonsatit, P., Chindaprasirt, K., Boonserm, J Adv. Agri. & Env. Eng. **1**, 6 (2014)

MATEC Web of Conferences **120**, 02010 (2017) DOI: 10.1051/matecconf/201712002010
ASCMCES-17

Mechanical properties of lightweight aerated concrete with different aluminium powder content

Rana Shabbar[1,2,*], *Paul* Nedwell[1], and *Zhangjian* Wu[1]

[1]School of Mechanical, Aerospace and Civil Engineering, University of Manchester, M13 9PL, UK
[2]School of Engineering, University of Kufa, Najaf, Iraq

Abstract. Aerated concrete is produced by introducing gas into a concrete, the amount dependent upon the requirements for strength. One method to achieve this is by using powdered aluminium which reacts with the calcium hydroxide produced upon hydration of the cement. The aim of the current study was to investigate the influence of the powder content on the mechanical properties of aerated concrete namely; compressive and flexural strengths, modulus of elasticity, density and porosity. The results indicated that an increase in aluminium content caused a decrease in the compressive and tensile strengths. It also produced a decrease in the modulus of elasticity. When the aluminium content increased, the density decreased and the porosity increased.

1 Introduction

Aerated concrete (AC) is produced by introducing gas bubbles into a conventional concrete which then produces a material of lower density. The introduction of powdered aluminium (a foaming agent) reacts with the calcium hydroxide formed on hydration of cement to produce hydrogen gas bubbles. It is not only used to insulate for sound and heat [1, 2] but is also fire resistance [3, 4]. AC can be used to decrease the dead load, earthquake effect and the size of building members. Thus, foundations become more economical and the building cost is decreased [5, 6]. However, this type of concrete has lower mechanical properties and additional cement is required for the same strength as a normal concrete [7]. Typical lightweight concrete has densities from 1000 to 2000 kg/m^3 and compressive strengths from 1 to 100 N/mm^2 [7].

Several investigations have been carried out regarding the influence of aluminium (Al) content on the compressive strength of aerated concrete [8-10]. Modulus of elasticity and flexural strength of lightweight concrete with different percentages of pumice aggregate replacement instead of sand [11] and AC with varying blocks and bricks (aerated concrete, solid and hollow block, moulded and wire cut brick) were also studied [12, 13]. Some studies into the density and porosity of AC have also been reported [10, 13, 14]. The current study,

* Rana Shabbar: rana.shubber@uokufa.edu.iq / rana.shabbar@postgrad.manchester.ac.uk

using laboratory testing, aims to provide a better understanding of the effect of Al concentration on the compressive and flexural strengths, modulus of elasticity, density and porosity of aerated concrete. A control of 0% will be compared with 0.25%, 0.5%, 0.75% and 1% content of Al.

2 Experimental details

2.1 Materials

AC composed of one part of CEM I/52.5N with two parts of Leighton Buzzard sand together with Al with the purity of 99.7%. The mix proportions of varying AC are shown in Table 1.

Table 1. Mix proportions of the AC with different Al contents.

Material	kg/m^3				
	Mix 1	Mix 2	Mix 3	Mix 4	Mix 5
Cement	350	350	350	350	350
Sand	700	700	700	700	700
Water	175	175	175	175	175
Aluminium powder	0	0.87	1.75	2.63	3.5
Superplasticizer (SP)	4.2	4.2	4.2	4.2	4.2

2.2 Procedure

First, cement and sand were mixed together. Then aluminum powder was added and mixed for 30 sec. Finally, warm water (55 ± 1 °C) with the superplasticizer were added and mixed for 2 min.

2.3 Casting and testing

100 mm cube was used for compression test. The compressive strength of the AC was determined according to the BS EN 1881 -116 [15]. Flexural strength was measured by using prisms with dimensions 40 mm cross section and 160 mm length according to the BS EN 1351 [16]. Cylinders with diameter 100 mm and length 200 mm were used for determined modulus of elasticity (E_s) according to BS EN 1352 [17]. The dry density of the AC was measured according to the BS EN 992 [18]. Porosity was measured by vacuum saturation method according to Hall [19].

3 Test results and discussion

3.1 Compressive strength

In this study, the compressive strength of AC mixtures was a function of different Al contents, which are plotted in Figure 1, and presented lower strengths when compared with the control. The lowest compressive strength was 18.6 N/mm^2 with 1% Al due to largest number of voids and lowest rate of calcium silicate hydrate (C-S-H) gel. Thus, it did not contribute sufficiently to the compressive strength. The highest strength was for control (0% Al) at 53 N/mm^2 which then decreased significantly on introduction of 0.25% Al to 32.3 N/mm^2 and then decreased slightly as the Al increased with 26 N/mm^2 for 0.5% and 23.4

N/mm² for 0.75% Al. Similar results were obtained by Guglielmi *et al.* [8] who reported that when the Al content increased from 0.2 to 0.4%, the compressive strength decreased because of the higher homogeneous distribution and quantity of the pores and the inter-pore struts are thinner, conferring a lower strength to the material.

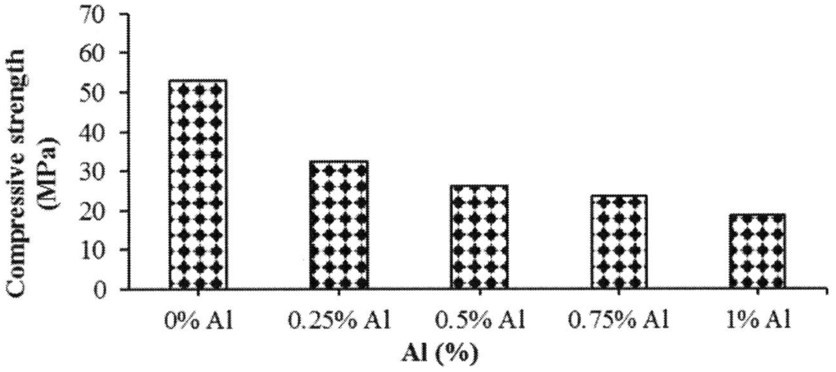

Fig. 1. Compressive strength of cube AC specimens as a function of Al content.

3.2 Flexural strength

The variations of the flexural strength of AC are presented in Figure 2 which shows that it decreases when the Al content increases. The lowest flexural strength was 3.3 MPa with 1% Al. The highest strength was for control (0% Al) at 5.5 MPa which then decreased slightly on introduction of 0.25% Al to 4.6 MPa, 4.2 MPa for 0.5% and 3.7 MPa for 0.75% Al. Similar results were obtained by [20, 21], when the flexural strength increased due to an increase in compressive strength.

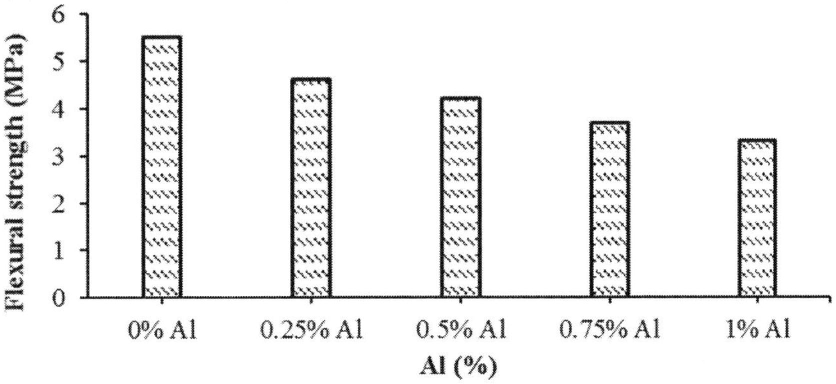

Fig. 2. Flexural strength of AC as a function of Al content.

3.3 Modulus of elasticity

The modulus of elasticity (Es) is mainly effected by the nature of the aggregates and cement paste. Additional effects are provided by the bond and arrangement between the particles [22]. Figure 3 presents the modulus of elasticity of AC with different Al content. In comparison to control, AC specimens have less stiffness due to the presence of the Al, which caused larger and more pores to be formed and reducing strength. It is clear that the AC which is obtained from high Al powder content caused a greater loss in the modulus of elasticity than that with lower content. The lowest Es was 7.8 MPa with 1% Al. The highest Es was for control (0% Al) at 23MPa which then decreased slightly on introduction of 0.25% Al to 18.9 MPa, 14.1 MPa for 0.5% and 9.7 MPa for 0.75% Al.

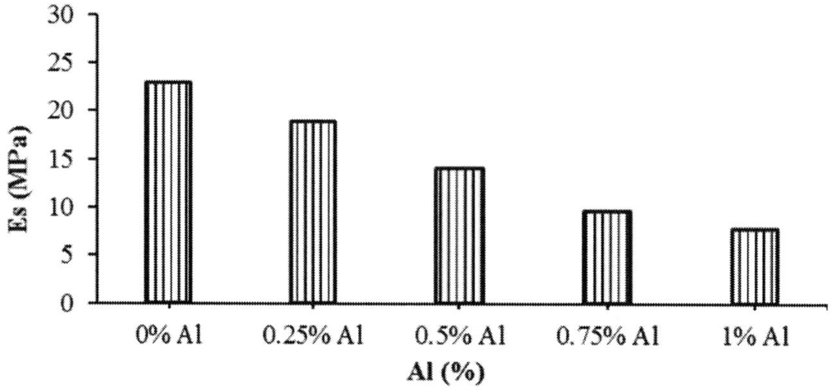

Fig. 3. Modulus of elasticity of AC as a function of Al content.

3.4 Density and porosity

The dry density and the porosity of AC samples are both a function of Al powder content as shown in Figure 4. The result revealed that the dry density varied as expected for the different Al contents. It decreased as the foaming agent percentage increased because of the higher number of pores created. The density decreased from 2102 Kg/m^3 for the control to 1841 Kg/m^3 with 0.25 % Al powder. Then it decreased gradually with the increasing Al content. The lowest density was obtained with 1% Al which was 1489 Kg/m^3. The same behaviour was observed for the porosity of the AC samples which increased from 10 % to 26 % when the Al content increased from 0% to 1%. AC samples with Al powder from 0.5 to 1% revealed the present of pores with a non-uniform shape. These were larger than those observed for AC with 0.25% Al, which indicates the coalescence of the pores in mixtures with higher Al content. Scheffler and Colombo explained the coalescence of the pore to the high reactivity of the Al powder [23]. The fracture surfaces of the AC samples with different Al contents in this study are shown in Figure 5.

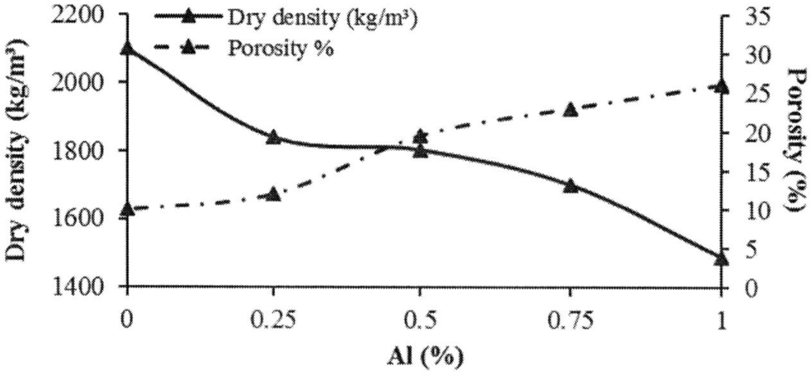

Fig. 4. Apparent density and porosity of AC as a function of Al content.

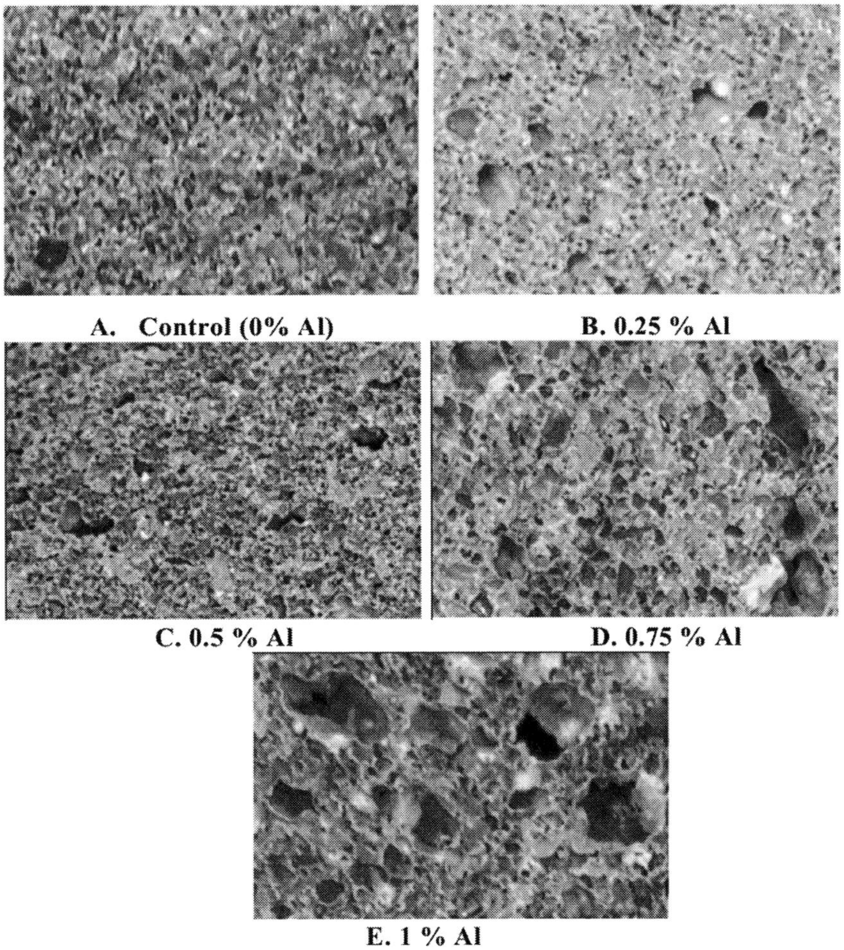

Fig. 5. Fracture surface of AC samples with different Al content.

4 Conclusion

This study has shown that by adding Al powder:

1. The compressive strength of the AC specimens decreases with increasing the Al powder content. Due to the high reactivity of Al, the rate of production calcium silicate hydrate (C-S-H) gel may decrease which corresponds with the reduction in the strength.

2. The flexural strength of the AC has direct relationship with compressive strength results and it decreases as the foaming agent increases.

3. The modulus of elasticity of the AC specimens declined according to the increase of the content of the Al.

4. When the Al powder content increased, the density decreased gradually. However, the porosity increased significantly due to higher number of pores formed (Fig.5).

5. The fractured surfaces of the AC samples with the Al powder from 0.5 to 1% revealed the present of the pores with a non-uniform shape, which were larger than those observed for AC with low content 0.25%.

The authors are grateful for financial support of the Iraqi ministry of high education and scientific research (MOHESR), my colleague Mr. Mohammed Al-Taee and technician support of Mr. John Mason in the school of MACE, University of Manchester.

References

1. A. Aidan, et al., Preparation and properties of porous aerated concrete. Scientific Works of the University of Ruse, **48**, 9 (2009)
2. A. J. Hamad, Materials, Production, Properties and Application of Aerated Lightweight Concrete: Review. International Journal of Materials Science and Engineering, **2**, 2 (2004)
3. N. Narayanan and K. Ramamurthy, Structure and properties of aerated concrete: a review, Cement and Concrete Composites, **22**, 5 (2000)
4. A. Keyvani1, Thermal performance & fire resistance of autoclaved aerated concrete exposed humidity conditions. International Journal of Research in Engineering and Technology, **3**, 3 (2014)
5. D. H. Lim and B.H. Oh, Experimental and theoretical investigation on the shear of steel fibre reinforced concrete beams. Engineering Structures, **21**, 10 (1999)
6. O. A. Düzgün, R. Gül, and A. C. Aydin, Effect of steel fibers on the mechanical properties of natural lightweight aggregate concrete. Materials Letters, **59**, 27 (2005)
7. A. M. Neville and J. J. Brooks, *Concrete technology*. 2nd ed. Harlow: Longman Scientific & Technical (2010)
8. P. O. Guglielmi, et al., Porosity and Mechanical Strength of an Autoclaved Clayey Cellular Concrete. Advances in Civil Engineering (2010)
9. I. S. Raj and E. John, A Study on the Properties of Air-Entrained Concrete for Masonry Blocks. International Journal of Scientific Engineering and Technology. **3**, 11 (2014)
10. A. A. Aliabdo, A.-E.M. Abd-Elmoaty, and H. H. Hassan, Utilization of crushed clay brick in cellular concrete production. Alexandria Engineering Journal. **53**, 1 (2014)
11. R. Şahin, et al., The effects of different cement dosages, slumps and pumice aggregate ratios on the compressive strength and densities of concrete. Cement and Concrete Research. **33**, 8 (2003)
12. A. Ahmed and A. Fried, Flexural strength of low density blockwork. Construction and Building Materials. **35**, p. 516-520 (2012)
13. T. M. Prakash, et al., Properties of Aerated (Foamed) Concrete Blocks International Journal of Scientific & Engineering Research. **4**, 1 (2013)
14. K. H. Yang and K. H. Lee, Tests on high-performance aerated concrete with a lower density. Construction and Building Materials. **74**, p. 109-117 (2015)
15. BS EN 1881-116, *Testing concrete. Method for determination of compressive strength of concrete cubes* (1983)
16. BS EN 1351, *Determination of flexural strength of autoclaved aerated concrete* (1997)
17. BS EN 1352, *Determination of static modulus of elasticity under compression of autoclaved aerated concrete or lightweight aggregate concrete with open structure* (1997)
18. BS EN 992, *Determination of the dry density of lightweight aggregate concrete with open structure* (1996)
19. C. Hall, *Water transport in brick, stone and concrete*, ed. W.D. Hoff. London: E. & F. N. Spon (2000)

20. N. B. Eden, et al., Autoclaved aerated concrete from slate waste Part 1: Some property/density relationships. International Journal of Cement Composites and Lightweight Concrete. **2**, 2 (1980)
21. T. M. Prakash, B. G. N. Kumar and Karisiddappa, Strength and elastic properties of aerated concrete block masonry. International Journal of Structural and Civil Engineering Research. **2**, 1 (2013)
22. R. V. Silva, J. de Brito and R. K. Dhir, Establishing a relationship between modulus of elasticity and compressive strength of recycled aggregate concrete. Journal of Cleaner Production. **112**, Part 4 (2016)
23. M. Scheffler and P. Colombo, *Cellular ceramics: structure, manufacturing, properties and applications*, ed. M. Scheffler and P. Colombo. Weinheim: Chichester: Weinheim: Wiley-VCH Chichester: John Wiley distributor (2005)

The role of microsilica in sustainable concrete

Robert Lewis*

Elkem Silicon Materials

Abstract. This paper reviews at the way we approach sustainability with concrete construction. It looks at the current thoughts on how to achieve this and asks if we are trying to be *too* green and actually potentially causing problems for the future. It reviews projects around the world that are using multiple cementitious blends to achieve long term durability and reduced construction costs. The point is raised that despite a drive to be green and be sustainable, the criteria for many projects is still the cost per cubic metre.

1 Introduction

Sustainability in construction is not just about the concrete – although concrete is the main target for criticism levelled at the building industry. It is about getting the best use of all our construction materials, with less damage to the environment as we build.

It is about building something as efficiently as possible, so that will do its job, for as long as possible, and with materials that have the least effect on our environment. It's all about doing that with cost efficiency. That does not mean *as cheaply as possible*, but at the price that gives the best lifetime value.

There are many people in this world who simply do not like concrete. The reasons range from the practical to the aesthetic. The major point always cited, is that concrete uses Portland Cement and production of cement is the worst source of 'manufactured' CO_2 on the planet. QED all concrete is bad for the environment. Concrete uses Natural resources: Aggregates and Drinking Water: QED all concrete is bad for the environment. Concrete is grey and ugly. Yet it is the worlds most used construction material and some of the world's most magnificent buildings are constructed in concrete.

There is also an overview that all concrete is the same. It is made out of cement, stone, sand and water, it's grey and sticky and goes hard, and when it has set it has no beauty – it's just like grey boulders or building blocks. This simply adds to the thought that it is all the same product and therefore should all cost about the same

* Corresponding author: robert.lewis@elkem.no

and therefore, the cheapest is probably going to do the job anyway, so why pay more? Sadly, that has often been reflected in the collapse of buildings and other constructions that should have had more resilience. A number of times, the author has worked on specifications for high durability and long lifetime concretes, only to have the designs turned around by the time construction starts, because: "Well, we decided we don't really need that performance" and "We can get almost the same from this mix, and it saves $5 a cubic metre". If that performance was needed in the first place, then it probably still is needed – and it is not price per cubic metre you need to consider – it is total construction and lifetime costs.

2 Concrete needs to be 'Engineered'

The concrete we use has to be designed to give the right performance, not just when it has set and is 'in the building'. It needs to be thought of from the very point of when it arrives on site:

In the plastic state: how long will it last at a certain workability; how is it going to be conveyed / placed / finished / cured? These are all criteria for the rheology of the mix.

As it sets: how fast / slow does the setting need to be; how much temperature rise can be tolerated?

When it has set – how soon does it need to reach a critical strength; what ultimate strength is required; flexural, tensile, creep values?

For the conditions it is placed in: what resistance to the ingress of chlorides, sulfates, water, other chemicals; abrasion or water erosion, physical impact, temperature changes?

How long is it expected to last before repair or replacement: 50 years, 100, 200, 800?

All of these factors need to be taken into consideration to design the concrete mix. It is no good designing something in the laboratory that is either impractical to produce or impractical to handle. The last line on any concrete site is the person who works to place, compact, finish and cure it. If that person is not able to do any of those things easily, then chances are they won't get done properly – and a 'perfect' concrete is destroyed. The better the design, the easier the concrete is to use, the less concrete consumed and the longer it lasts. The longer it lasts, the less resources we use because we don't have to re-build it in 10 or 20 years' time. This is where the engineering – to design for purpose – comes in, to enable us to get the optimum for what we want to do. All these things add up – to give less cost – while achieving the best performance:

Improved Plastic Properties: ease of placing, faster construction, less machinery
= Better Economics.

Increased Strength: faster construction, less volume
= Better Economics.

Increased Durability: increased lifetime, fewer repairs
= Better Economics.

Combining the key points: less machinery, less volume, fewer repairs = less impact on the environment
= Sustainability.

3 Sustainable resources for concrete

Cement: Improvements to production processes have reduced the carbon footprint of Portland cement. Higher qualities are now being designed and blended to give different qualities for varying applications. However, pure Portland cement, no matter the current process, will still yield around 800 to 900kg of CO_2 for every tonne of cement produced. Therefore, if we can use less pure Portland we will reduce the impact of the environment.

Supplementary Cementitious Materials (SCMs): These are by products from other industrial processes, such as fly ash from coal fired power stations; ground granulated blastfurnace slag (GGBS) from steel making; microsilica from ferrosilicon production and metakaolin from clay refining. These are the major Type II additions – latent hydraulic and pozzolanic binders. The use of these materials can replace large volumes of Portland cement and contribute differing levels of effect on rheology, and long term properties. Designed use can significantly improve such characteristics as strength and impermeability leading to much higher performance concrete. By needing less Portland cement to achieve the requirements of the concrete, the carbon footprint can be greatly reduced.

Use of Aggregates: These are an obvious natural resource and we need to slow down the rate of quarrying, bringing down mountains and dredging the sea, to get fresh new aggregates each time we need to make some concrete. The use of recycled aggregates (RCA) has been shown to be perfectly effective in new concrete. So much so that trials with 50% RCA have given equal performance to concretes with 100% new aggregates. As buildings and such have to come down to allow new ones to go up, we should be improving the processing and use of recycled aggregates to reduce our consumption of new material. Lightweight aggregates are often made from waste material and can be used very effectively in concrete. Improved designs mean that structural strength lightweight concretes – circa 60MPa + at around 1900kg/m3 can be produced. This means bridges and larger projects can use less normal weight concrete – giving less dead weight to the structure and enhancing insulation in buildings. This latter point can have a further impact by reducing the need for heating and cooling.

Water: Is precious, especially drinking water, and that is what we use for making concrete. Admixtures are now made that can drastically reduce the amount of water needed per cubic metre. Better design of the concrete, using particle packing can also mean less water. Greater use of recycling units on plants and major projects means that a lot of water can be recovered and re-used, saving more fresh water.

Optimising the design: Engineering the concrete for the particular qualities needed can mean reductions in volumes produced: Using higher strength to reduce the column and slab sizes; better rheology to enable less machinery; durability for the lifetime expected, meaning less repairs and no rebuilds, means less volume of concrete overall, which impacts on all the above points.

4 Cementitious steps towards sustainability

To reduce our carbon footprint, the first and easiest way is to remove cement, but this can only be taken so far. To make concrete we still need cement, or more correctly, the free lime from the cement hydration, to activate the SCMs that we are putting in to replace the cement taken out. We then have the decision of how much we replace. The more we replace with fly ash and ggbs, the slower the concrete gets, in terms of setting and strength gain – and durability performance. If we want high performance, or high level properties at an early age, we need a good balance of cement and SCM's.

It is nice to think of cement free materials – geopolymers and such – but these materials are not in mass production and, just because they don't have portland cement in them, does not mean they are 'greener':

(When all things are taken into consideration) "....it appears that geopolymer concrete has a similar impact on global warming to standard concrete."[1]

As such, we need to consider the designs of our concrete mixes, and engineer them to achieve ALL the characteristics we want – and we need to do this while also considering the total life costing of the project. Use of software programs such as Life365 has increased rapidly over the last few years – and the data produced has shown that most projects may seem to cost more to start with, but cost less in terms of total value. Some projects, though using more expensive concrete, have shown that even the initial construction cost is reduced, let alone the long term costs.

5 Supplementary cementitious materials

Most people are aware of the three major materials here. Metakaolin and Limestone Fines are sometimes included in this list. However, metakaolin does have a processing energy cost and limestone fines are only marginally pozzolanic, so the concentration here is on low environmental impact and normal to high pozzolanic reactivity. To run a basic recap:

Fly Ash is collected from coal-fired power stations. It has a silicon dioxide (SiO_2) content of around 40% ~ 50% and is a similar particle size to cement. The replacement rate is normally from 15% - 50%, to achieve improvements in rheology, heat reduction, and give long-term durability benefits.

GGBS is a by-product from the steel industry and is normally ground to a particle size roughly the same as cement. For some applications it can be ground finer than cement and this will increase the reactivity. The SiO_2 content is usually 35% ~ 40%.

As ggbs is weakly hydraulic as well as pozzolanic, the replacement rate can be from 25% - 90%. It is used for major heat reduction in large elements, improved long-term durability benefits.

Microsilica (also known as silica fume) is collected from the ferrosilicon industry. It has a high SiO_2 content, normally between 90 ~ 95%. Due to the nature of the production process for the initial product the particle size for microsilica is approximately 100 times finer than Portland cement. This gives a much higher reactivity than fly ash or ggbs and it is usually added to an original or recalculated cementitious content at between 5% - 15% (max 25%). The size, spherical shape and reactivity can have a range of influence on the concrete mix, improving the rheology, reducing permeability, improving strength and durability characteristics. It is often used in the cementitious blends for creating Ultra High Performance Concrete (UHPC).

Metakaolin is a similar product to microsilica. It is a fine, highly pozzolanic material produced from calcined waste clays. As such, it has an energy footprint and care should be taken in considering its sustainable aspects when used in a concrete mix.

5.1 Use of fly ash and GGBS in high volumes

The first thought then, if we want to reduce the CO_2 our concrete causes, is to replace as much of the cement as we can with either fly ash or ggbs. Well, that is a 'yes *and* no'. You can't just throw them in 'blanket fashion'; the performance of the concrete, plastic, setting and final characteristics have to be taken into the design process.

It is no good designing a 'perfect' green concrete in the laboratory, if it will not work in practical terms in the real world. The design of the concrete for the towers on the Tsing Ma bridge in Hong Kong, was originally for 30% opc and 70% ggbs. This was to give low heat and high chloride resistance. As a design, on paper, technically this was good – the concrete would have low heat when produced and would, in the long term, develop high chloride resistance. The drawback was the practicality of construction, as the mix was to be slipformed to create the towers. Due to the slow rate of setting, hardening and strength development of the mix with 70% ggbs, slipforming was found to be impossible. Replacing 5% of the ggbs with microsilica improved the hardening rate and the slipforming was able to run.

A number of places, including Hong Kong and Dubai, have stipulated that concrete shall contain ggbs, to reduce the carbon footprint. The additional point to that statement is that 'equivalence' [to a Portland mix] can be achieved at 56 to 90 days. This is a point of great concern with regards to the early age and ultimate performance of the concrete. The key point here is ***time***. Larger quantities of FA or GGBS can have great effects on setting time, hardening rate, strength gain and *real time durability*. While in some cases, waiting – and curing the concrete – for another two to three months – could be feasible, it is not on most projects.

6 Real time durability

Projects want to move forward, usually as fast as possible, but at least within reason. Waiting 56 or 90 days to see if a concrete has achieved the required strength would not be sensible on most projects – and what do you do if you've progressed 6 or 7 floors upwards and the concrete in the columns of floor one has not achieved the required strength?

It is not only strength, but it is also the durability characteristics. The 'equivalence' at 56 or 90 days is tested on lab samples. These are concretes that have been carefully cured for that length of time. These are the best results you can get for that concrete, kept under the best conditions. What is the actual, in-situ, concrete achieving? Has it been cured for the same length of time? Has it developed the impermeability you require? If it is facing the ground conditions where it needs a water resistance of less than 5mm, but is known only to reach that level of resistance at 56 days, how much has it been affected in the 56 days it has already been exposed? If you add chlorides and sulfates in to that equation as well, and the potential of 90-day equivalence, the possible failure of the concrete becomes very likely. A concrete in situ, once 'attacked' by the environment in the early ages, will never achieve the same results as that 'perfectly cured' concrete in the lab.

In some case, yes, there may be a time window where concrete can be stored and cured, or cured in-situ for a length of time, but the likelihood of that being 56 to 90 days is very small.

So, yes, while we can use good quantities of these products – but the mixes have to be engineered to give the required performance, from the start. This is where the addition of microsilica to such mixes can result in major improvements.

6.1 Being green for green's sake

While it is nice to think about how we can use 'green' environmentally friendly products, there is no point in going for such materials if they aren't sustainable! It defeats the object:

The First Platinum LEED project on the East Coast of the USA, had to replace the cladding on the building in less than 5 years, because it wasn't durable enough. It was replaced with concrete…

At a previous conference, a speaker from China said they wanted construction to use more bamboo cladding, *because it grows faster than current use*. When asked "Have you calculated the balance point – when accelerated new use, and subsequent repair, exceeds the regrowth rate?" the answer was "I don't know – I don't believe we've thought about that" …

Unless we use these products wisely, we actually stand more chance of doing harm rather than good. The same goes for our concrete. We want it to have higher performance, but we also want it to be a sustainable as possible. We have to find the balance.

6.2 Balance in the design

If we want to achieve practical results and keep our CO_2 and environmental impact low, then we need to achieve the balance between levels of additions that will give us these factors. That is, without compromising the setting, hardening and 'real time' early age durability we can use the following levels of replacement and addition:

Between 20% and 30% FA – in combination with 8 to 10% MS.[2]

Or 40% to 60% GGBS – in combination with 6 to 10% MS.[3]

We can also use 20% to 50% percentage of recycled aggregates and the recycled wash down water. The use of particle packing software to blend the aggregates and cementitious materials will give us the best rheology – and can often achieve a reduction in the total cementitious content.

There is nothing new to Pozzolanic Concrete – the Romans used volcanic ash and aggregates, mixed with lime over 2,000 years ago, and the Pantheon and similar buildings still stand today, so we know how to do this. We have been using binary blends – opc/fa and opc/ggbs – for almost a century. In the last 50 years that has expanded to opc/ms binaries and then to ternary and quaternary blends, using all the

 DOI: 10.1051/matecconf/201712002011

SCM's. We are using the synergistic effects of these materials to engineer even better concretes. It is no longer the case of asking "Should I use FA, or GGBS, or MS?" It is "What's the best combination to give me the performance I need?"

7 Research and real life

In 1991, Elkem undertook a study of comparative concretes in the UK – with and without the addition of microsilica. The trials were overseen and testing conducted by Messrs Sandberg.
Based on a 400kg/m^3 cement content at around 0.42 w/c ratio the mixes were opc; opc+ms (40kg addition); opc/ggbs at 50:50; opc/ggbs+ms; and an src mix to show comparative sulfate resistance. Two sets of variation on the opc mixes were done, one with a change of aggregate and one with steel fibres. The full results are compiled in the Sandberg Report, which was presented as a paper in 1993[4]. The basic strength and durability results are shown in the graphs, listed 1 to 5. The results showed that even when using a 50:50 blend of ggbs – which achieved good results – the performance could be significantly improved by the addition of the microsilica especially at early ages as shown by the graph on compressive strength. While only performed in this run with ggbs, subsequent 'real-life' projects have shown that the triple combination with fly ash is equally effective. Following on from the research, based on its findings, a number of projects were guided to use triple blend cementitious materials and have achieved very high levels of durability performance.

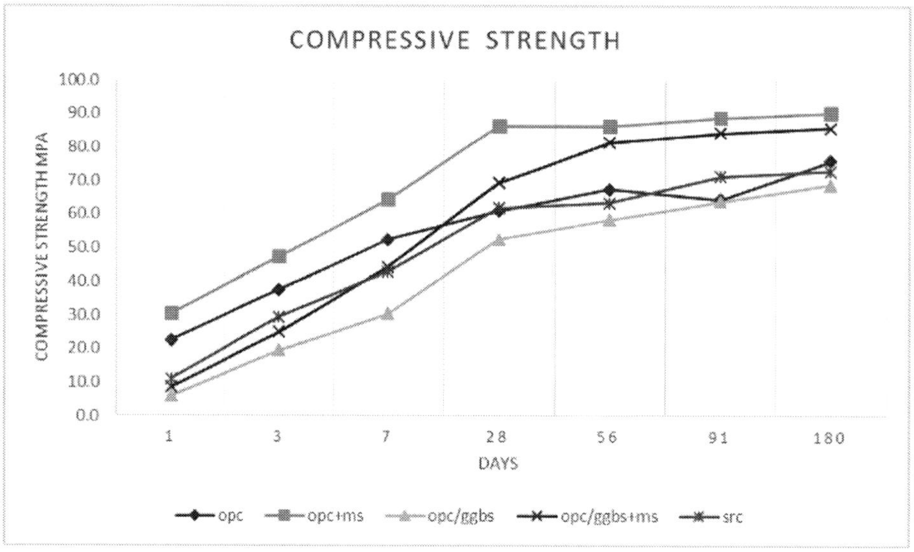

Fig. 1. Compressive strengths of test cubes from Sandberg; ref 4.

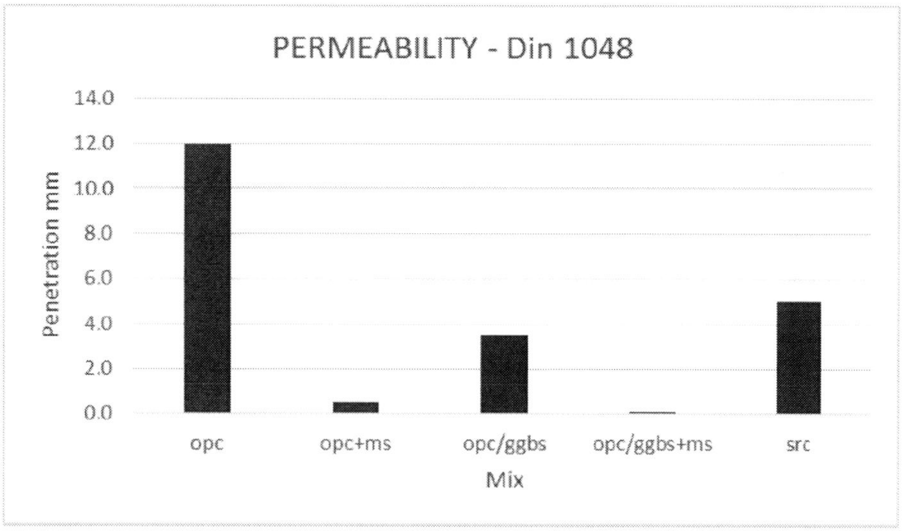

Fig. 2. Water permeability at 28 days, to DIN 1048, from Sandberg, ref 4.

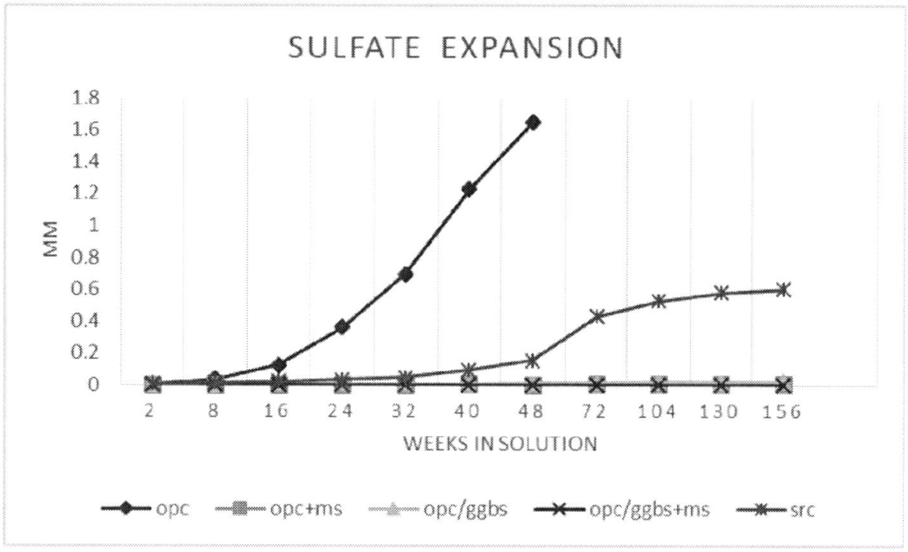

Fig. 3. Expansion of mortar prisms, from Sandberg, ref 4.

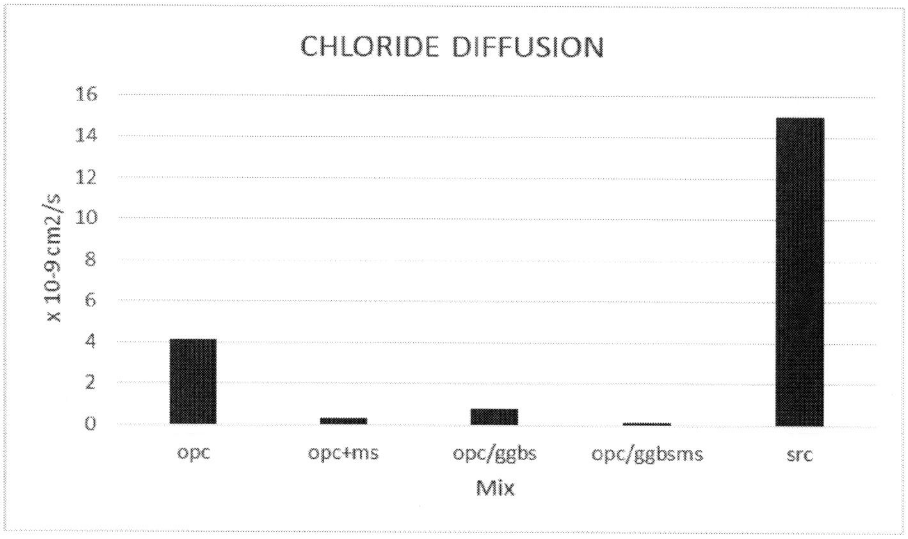

Fig. 4. Chloride Diffusion, 90 day ponding test, from Sandberg, ref 4.

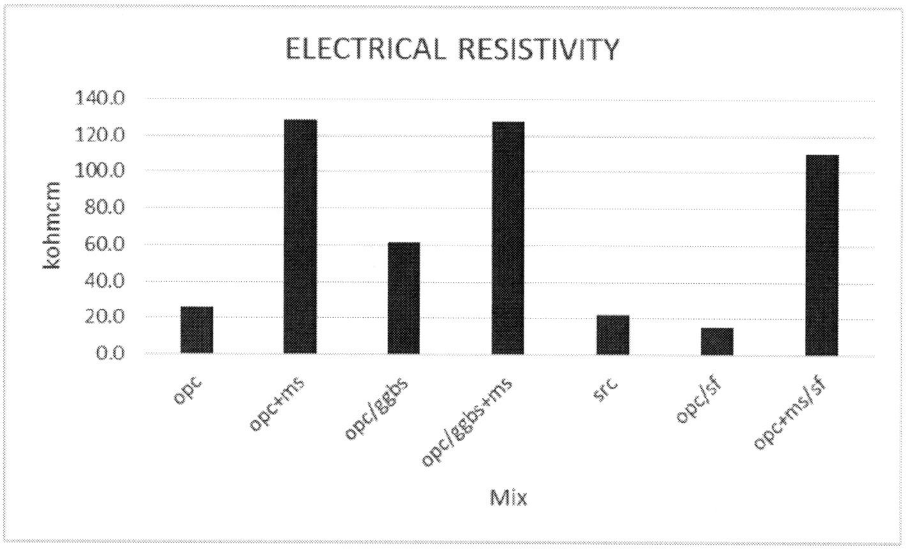

Fig. 5. Electrical Resistivity, kohm/cm, from Sandberg, ref 4. (it should be noted here that this test meter only read up to 128. Real life testing has shown values over 500).

8 Project references

8.1 Singapore metro and DTSS

The twin tunnels between Somerset and Newton Circus Stations on the MTR were cast in 1986. One tunnel was cast using a Sulfate Resistant Cement concrete with 400kg/m3. To protect this concrete from being attacked by the sulfate rich acidic groundwater, it was covered with a membrane. The other tunnel was cast with an opc mix using 10% microsilica at the same cementitious content. This tunnel was cast without a membrane. After 15 years it was clear that the sulfate resistant concrete was failing, while the microsilica concrete was showing no signs of attack. The success of this 'in-situ' trial lead to the use of microsilica concrete for the Deep Tunnel Sewage System that would cover the whole island.

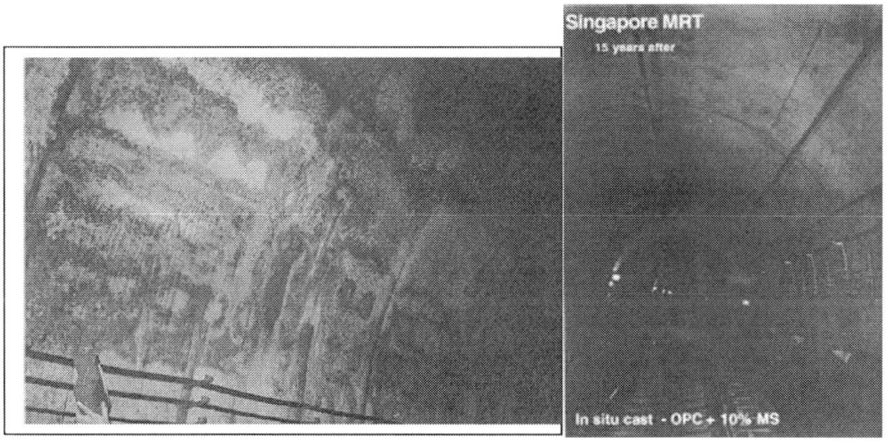

Fig. 6. Pictures of the two tunnels after 15 years in use.

8.2 Offshore and bridges

Marine environments were an early target for the use of microsilica concrete. A small addition, up to 3%, would improve the rheology of pumped concretes so that great distances – often vertically in the case of slipforming oilrigs – could be achieved. The high chloride resistance and synergy with fly ash and ggbs meant that mass pours could be done with low heat evolution and the high strength capabilities meant that even lightweight concrete could be produced with structural strengths.

Fig. 7. The Troll Platform being towed out into the North Sea

 DOI: 10.1051/matecconf/201712002011

Fig. 8. The Oresund Bridge, between Denmark and Sweden and Bandra-Worli Sea Link in India. Both these projects used binary blends of opc+ms and triple blends of opc+fa+ms.

Fig. 9. Nordhordland Bridge – lightweight concrete: 55MPa at 1900kg/m3 was used to construct this floating bridge. The East Sea Bridge in Shanghai (right) is 32 km long and used triple and quad cementitious blends in the two Marine Concretes used for the construction.

Fig. 10. Tsing Ma bridge, Hong Kong, 1997.

8.3 Tsing Ma bridge.

This suspension bridge was one of the first to use a low heat ggbs mix, while slipforming the two 206m tall suspension towers. The original design was for 70% ggbs as a binary mix with opc, but when it was found that this did not set and harden fast enough, 5% ms was added to give the concrete a 'kick' and allowed slipforming to proceed. Other concretes on this project also included fly ash at 25% with ms at 5%. The success of the concrete used for Tsing Ma lead the Civil Engineering Department of the Government to create a focussed Marine Concrete Specification, that included microsilica as a specified component, in 2002. In 2013, this specification was upgraded to stipulate triple blends, with a compulsory addition of microsilica:

8.4 Recommended specification for reinforced concrete in marine environment

The main features of the specification are summarised as follows:
The minimum characteristic strength of the concrete mix shall be *45 MPa.*
The maximum water/cementitious ratio shall not exceed *0.38.*
Condensed silica fume is to be added to reduce the permeability of the concrete.
The cementitious content shall be 380-450 kg/m³, including 5 - 10% CSF,
with either: FA at 25 – 40%, or *GGBS at 60 – 75%.*
The cover to all reinforcement in all exposure zones shall be *75 mm.*

9 Sustainability and the role of microsilica

There are a number of things we can do to our concrete to make it sustainable. Utilising the SCMs and in particular designing triple bends gives us unique ways to tailor concrete to the best plastic and hardened properties for the work we need it to do. We can optimise the mix for:

Rheology – make it as easy as possible to handle, place and finish;

Strength – use higher strengths to allow wider spans, thinner columns, use of lightweight aggregates;

Physical Durability – improved wear and tear resistance means it will last longer, needing less repairs or rebuilds;

Chemical Durability – the better the resistance the longer lifetime, particularly in areas where repairs or rebuilds might be hazardous.

But in all these instances it is a matter of collaboration between all parties to understand what needs an has to be done to complete the project. If that happens, the results can be surprising.

9.1 South Wacker drive, Chicago

In this project the client wanted to have an innovative building, making the best use of the space and getting good value for money. The architect and the consultant worked with the ready-mix producer to ensure that if they could get a very high strength concrete, they could make design changes. The ready-mix supplier did trials, involving both the additions (microsilica) and the admixtures (superplasticisers) suppliers and said yes, this was feasible. The columns were reduced in size – so that the same dimensions were used for the whole building. High strength concrete was used for the first fourteen floors and then even with reducing strength design, the microsilica was kept in the mix to allow pumping to the top of the building - 273m. These actions saved formwork, setting up time, pumps (only one was used) and 7,650m3 of concrete and 3,000 tonnes of rebar. The environmental value of the concrete saved alone was:

Cement	3,000 tonnes
Coarse Aggregate	7,650 tonnes
Fine Aggregate	5,250 tonnes
Water	1,071,000 litres

About 1,000 truck movements / wash-downs...

However, the high performance concrete was 50% extra per m^3. This was not a problem, because the building was finished ahead of time, for less cost than the original design, and the owner got more rentable space. The cost per cubic metre for the concrete did not enter into the equation, as it was irrelevant to the final cost.

9.2 The Burj Khalifa

Currently the world's tallest building, this amazing statement of design and accomplishment has concrete at its feet and in its heart. A triple cementitious blend is the starting point – moderate sulfate resisting portland cement, fly ash and microsilica.

Despite needing high strength and pumpability, as well as high durability for the desert ground the foundations will sit in, the total cementitious content was only 450kg/m^3. Stringent durability requirements were laid down including three variations of test for water ingress. A major part of the trials was to test the pumpability of the mix – which proved successful as the concrete was transported over 600m vertically through one pump. Strength results were as expected as were the durability numbers. Consideration here was given to the fact that this first World's Tallest Building to cross the 800m mark was durable enough to last at least 100 years. The next World's Tallest Building will be the Jeddah Tower, in Saudi Arabia – which is using the same design of concrete.

Specification for the 60MPa piling (Baseline specification for all concrete).

Compressive Strengths:	45 to 80MPa	
Minimum Cement:	252 + 168 + 30 kg/m^3 (56:37:7) (MSRPC + FA + MS)	
W/C ratio:	0.34	
Flow (at site):	> 600mm	
Water Penetration	<10mm	(BS EN 12390 - 8)
Water Absorption	<1.5%	(BS 1881:122)
RCPT	<1200	(AASHTO T-277)
Water Permeability	<5mm	(Din 1048)

Results.

Compressive Strengths (150mm cubes - averages)

7 days	40.5 MPa
14 days	51.5 MPa
21 days	60.5 MPa
28 days	64.5 MPa
56 days	75.5 MPa
Water Penetration	Zero
Water Absorption	0.7%
RCP Test	590
Water Permeability	Zero

 DOI: 10.1051/matecconf/201712002011

9.3 Roadway pavement

With the enthusiastic use of de-icing salts and chemical to remove the ice and snow from our roads, it is not surprising that we get chloride attack on bridge decks and easily rutted roadways. In the United States alone there are rumoured to be a few Trillion Dollars' worth of road repairs to bring the infrastructure up to date. Hence they use cementitious blends with microsilica and microsilica with fly ash to get the durability they need.

9.4 Bridge deck - New York state DOT

The concrete used here has been designed to resist chloride ion penetration & minimize cracking – hence the triple blend: fly ash to reduce the temperature and microsilica to seal against chlorides. The basic mix design and curing requirement were:

OPC	300kg/m3
Fly Ash	80kg/m3
MS	25kg/m3
W/B	0.40
Slump	75 – 100mm
Air	6.5%

Wet Cured for 7 days.

In the UK, a major trial of high strength triple blend concrete was conducted in the early 1990s to show that a continuous concrete road was viable – with the potential for low wear and high lifetime. The trial was for Ecopave and worked well. It is interesting to note that the USA has taken on high strength pavement, with skid resistant tines 'ground' in while the UK has not. The Ecopave mix was: (kg/m^3)

Low Alkali Cement	260	
Fly Ash	55	(16%)
Microsilica	20	(6%) (335 total binder)
Sand 0-2	673	
Stone 4-8	388	
Stone 8-11	1055	
Super (Conplast 212)	4.2	
Water	87	
W/C	0.26	
Comprssive strengths:		
3 days	60 MPa	
7 days	70 MPa	
28 days	95 MPa	

9.5 Indianapolis international airport parking garage

A simple 5 floor open plan structure, and the easiest way to show that durable environmentally friendly concretes can be done without compromising the 'buildability'. The garage has 7,100 spaces. To be built in the standard fashion this would have cost $15,000 per space. By redesigning to use a concrete with opc, fly ash and microsilica, recycled aggregates and with optimised particle packing, they achieved a LEED rating and saved concrete volume and time. At the end of construction, the actual cost was only $12,000 per space.

That amounts to a saving of $21.3 million. That is the saving on construction. The long term durability could save up to $40,000 a year in maintenance costs as well.

Fig. 11. The interior and top level of Indianapolis Airport Car Parking. Note the space for vehicles and the visual light. No painting on the surface, even on the top level, just the parking bay lines.

10 An essential role

While we can use products like flay ash and ggbs to give us 'green' concrete, if we need the high performance and qualities that modern construction demands, then the addition of microsilica becomes an essential part in our concrete design.
This is because we cannot wait for 56 days or 90 days or longer for the concrete to achieve strength and resistance values needed in-situ.

Microsilica concretes have been used for more than 30 years to improve performance and sustainability in major projects across the globe. The research and the reality are there to be seen. It is our job to get things right, because we should remember that sustainability is not for us, it is for our children and grandchildren – and their grandchildren.

References.

1. Habert et al; An Environmental Evaluation of Geopolymer Based Concrete Production: Reviewing Current Research Trends – Journal of Cleaner Production · July 2011.

2. Unpublished Thesis dissertation, Loughborough University, 2016.
3. John P.H.Frearson: Microsilica Concretes-Trial Mix and Test Data -Final Three Year Report. Sandberg report number X/602/45;

Practical technical solution for clay-contaminated sands used in concrete

Pierre Estephane[1,*], and *Klaus-Alexander* Rieder[2]

[1]GCP Applied Technologies, 5006, Dubai, United Arab Emirates
[2]GCP Applied Technologies, Cambridge, Massachusetts 02140 USA

Abstract. Sand, whether natural or manufactured, shows in many instances varying degrees of high levels of clay contamination. This fact is encountered in different parts of the globe and can lead to serious problems in adjusting concrete mix proportions and requiring high water to cement ratios and/or high dosages of superplasticizers without necessarily meeting the workability requirements, even when the sand is previously washed with fresh water. In this paper, different types of sand from the Gulf Cooperation Council (GCC) region are being screened, analysed for their clay contents and consequent effects on plastic concrete quality. A technical solution is being proposed based on engineered superplasticizers. A testing protocol has been established to verify the robustness of optimized mix designs demonstrating the performance of the admixture in terms of initial and extended workability. In particular, it will be demonstrated that the customized concrete admixtures constitute by themselves a stand-alone answer to the usage of clay-contaminated sands in concrete.

1 Introduction

Without doubt, cementing mixtures and concrete existence is prominent and very important in human civilization, historically and especially throughout the last centuries. Moreover, materials scientists and engineers understand well the value of the usage of superplasticizers in modern concrete, specifically in highly performing concrete or for particular purposes. Some of these additives attribute special characteristics to concrete from initial fluidity, to workability retention and not ending by reduced viscosity, by changing the interfacing properties of the system cement-water [2].

The concrete additives, or more precisely dispersants, of the type Polycarboxylate Ethers (PCE) are the topic of our current discussion and study. These PCEs are the latest cutting-edge technology introduced in both ready-mix and precast industries [3, 14]. Their structure is mainly carboxylic groups and non-ionic side chains of polyethers covalently attached to a polymeric anionic backbone [15]. These side chains supplement the normal dispersion with a spatial or steric repulsion [2, 15, 16] that improves and leverages the performance of PCEs and widen their scope of usage when compared to other dispersants.

* Corresponding author: pierre.estephane@gcpat.com

PCEs are already widely used all over the world in the construction industry, in different types of products ranging from low to extremely high dispersing capabilities. PCEs are well known for their high quality, extended workability retention, low dosages required and not to finish with early and uniform cement hydration. But as it is naturally with almost all manufactured products, they are not impeccable, they have weaknesses and there is always an opportunity for further development. One of the main challenges in using PCEs in concrete is the clay contained in aggregates and more often in fine aggregates. Depending on the type and amount of clay, the dispersing action of PCEs is adversely affected to different magnitudes [5].

The layered mineralogical structure of clay and the tendency of PCEs' side chains to intercalate between these layers, influence the availability and effectiveness of PCEs in dispersing cement grains [2, 3, 17]. This interaction might be seen with clay impurities in aggregates and limestone [9]. These remarkable facts give cause for concern and necessitate finding a solution applicable on an industrial scale in many parts of the world, where geologic clay deposits are mainly observed.

Many regions are of interest for the current study, and the Arabian Gulf region is leading. In Qatar, clay minerals were spotted in both fluvial and aeolian sediments with straight assumptions on their presence at various zone depths [1]. In Kuwait Bay, feldspars presence was recorded in suspended sediments up to 30% of total minerals [18], as well as the abundance of clay minerals amongst other minerals in different parts of the Bay, especially the south-eastern part [11]. The coastal plain deposits in central Kuwait were found to be medium-sized sand containing clay minerals, which are having a mineralogy dominated by illite, palygorskite, mixed layer illite-montmorillonite with minor amounts of chlorite and Kaolinite [12, 22]. The environmental impact assessment in Kuwait highlights the high daily volumes of fresh water required in quarries in the process of washing out undesirable minerals, besides the wastes disposal complications accompanying the elimination of flocculated matters [13].

All above being said and then added to the scarcity of many essential raw materials in the construction industry, the complex and sometimes multi-phased washing process of the natural or manufactured sands, with all implicated transportation costs and processing costs, made it imperative to start the pursuit for a solution [1, 13]. A sustainable solution that reduces the complexity of screening out minerals and useful fine particles, helps in managing efficiently available sand reservoirs and untightens the strict specifications around fines in aggregates.

One answer to these problems is a new technology based on a specially engineered type of superplasticizer and used as a stand-alone product with clay-contaminated sand. This technical solution will be assessed and compared, at different instances, with latest generations of PCE based products before implementing the engineering changes.

Assessment will be conducted in natural clay-contaminated sand as well as in artificially doped sand with Sodium Montmorillonite Clay in both concrete and mortar. The aim of this assessment is to show and ascertain the improved performance of the newly formulated product, and to demonstrate its advantages in terms of water reduction capability, initial consistency and workability retention in clay-laden systems without adversely effecting concrete and mortar properties.

 DOI: 10.1051/matecconf/201712002012

2 Experimental

2.1 Materials

The cement used in all experiments is an ordinary Portland cement. For mortar trials, it was supplied by Holcim (CEM I 42.5R) (EN 197-1), and for concrete trials by Dubai cement (DXB). Characteristics of both cements are shown in Table 1. The particle size distribution of both cements was very similar as measured by laser diffraction, with an average particle size d50 of 18.0 µm for the CEM I 42.5R and 17.4 µm for the DXB cement.

Table 1. Phase composition (wt. %) by QXRD and BET surface area of CEM I and DXB cements.

Phase	CEM I	DXB		CEM I	DXB
Alite	58.9	60.4	Hemihydrate	2.5	4.4
Belite	14.4	12.8	Anhydrite	2.6	0.6
Brownmillerite	5.2	8.3	Arcanite	0.4	0.8
C3A_cubic	5.9	1.8	Quartz	0.1	0.2
C3A_ortho	2.3	2.7	Aphthitalite	0.4	0.2
Lime	0	0			
Periclase	0	0.2	Total SO_3	3.5	3.4
Portlandite	0.6	1			
Calcite	6.3	6.1	BET SA (m^2/g)	0.9360	1.1329
Gypsum	0.6	0.4			

The sand used in mortar tests is a CEN standard sand (EN 196-1), having naturally equant, rounded, clean and reasonably graded particles. Its lithology is mainly quartz. Table 2 and Figure 1 refer to the standard sand used.

Table 2. Sieve analysis of standard sand.

Sieve	Retained
2.36 mm	0.0%
1.18 mm	27.6%
600 µm	35.4%
300 µm	8.7%
150 µm	20.4%
75 µm	7.8%
20 µm	0.1%

Fig. 1. Macro examination and petrographic thin section of standard sand.

For concrete tests, the aggregates used were of the type gabbro and were collected from the Fujairah area in the United Arab Emirates.

In all trials where clay was intentionally introduced, it was dry sodium montmorillonite, a white fine powder with a minimum of 99% of the particles being finer than size 74 μm, having a surface area around 750 m^2/g and a specific gravity of 2.6. A typical elemental analysis would show the SiO_2 content to be around 75% and the Al_2O_3 content to be around 15%.

Two different types of superplasticizers, all being Polycarboxylate Ethers (PCE) were tested:

1. G type ADVA product, supplied by GCP Applied Technologies, and known for its robustness in concretes made with difficult sands. It will be referred to as AD1 in succeeding trials.
2. G type product based on CLARENA®, specifically developed by GCP Applied Technologies for concrete mixes with high clay contents. It will show under the code AF2 and AF3 in succeeding trials depending on the application.

2.2 Equipment and testing of engineered clay-tolerant solution

2.2.1 Determination of the clay content in sand

The clay content of sand was measured using ASTM C1777-15, Standard Test Method for Rapid Determination of the Methylene Blue Value for Fine Aggregate or Mineral Filler Using a Colorimeter. The sand is dried at 105 °C to eliminate all the moisture and to a constant mass. After impregnation in methylene blue, the absorbance of the solution is measured at specific wave length and the methylene blue value is correlated to a corrected percentage of clay in the sand in form of sodium montmorillonite.

2.2.2 Testing protocol in mortar

The objective behind these experiments is to show the improved performance of AF samples in mortar at different intensities of clay doping, in relation to extended workability or dose efficiency. Below steps were followed for all experiments:

- Weigh water (220 g) in mixing bowl.
- Weigh clay in fine powder form directly in water.
- Mix manually in order to get uniform distribution of the powder.
- Start by mixing for one minute using a mixer with a blade rotating axially in a planetary motion at a speed of 60 rpm.
- Add the EN sand (1350 g) during the next 30 seconds, continue mixing for an additional minute and a half.
- Allow the mixture to rest for 2 minutes.
- Add 700 g of cement, mix manually, add superplasticizer dispersed in 20 g of water to the mixture, and mix manually up to 1 minute.
- Continue mixing for additional 4 minutes, always at the same speed.

In total, this mixing procedure takes 10 minutes.

Mortars were kept during preparation, mixing and testing at all times at a temperature of 24.0 ± 2.0°C.

The flow of the mortar is tested using a mini-slump cone, straight after mixing, 30 minutes and 60 minutes from the first flow.

During the same time, the change in consistency and stiffening behaviour was monitored using a rotational viscometer, while ramping to 60 or 120 rpm and in both directions over time. Figure 2 shows the profile being used for this study.

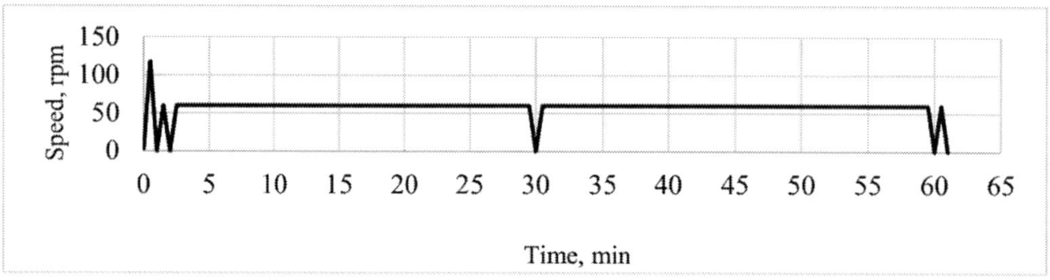

Fig. 2. Testing profile of the rotational viscometer over time.

3 Results and discussion

3.1 Screening of different sands for clay content

Many sands, natural and manufactured, washed and unwashed, were collected from different parts of the Arabian Gulf region and were tested for clay. The purpose was a better understanding of the levels of contamination we may encounter. Different levels of clay contamination were detected, depending on the location of the mining site, geology, washing and sieving processes and their efficiency. Figure 3 shows the distribution of sodium montmorillonite equivalent content for 21 samples. 14% of the samples can be classified as relatively clean, with sodium montmorillonite equivalent contents below 0.5% and little detrimental effects especially on plastic properties of concrete. 29% of the samples are moderately contaminated with sodium montmorillonite equivalent contents between 0.5% and 1.0%, while 57% are heavily contaminated above 1.0% with severe effects on the concrete properties in the plastic stage, hidden costs implications and limitations on mix designs, performance parameters and targeted applications.
These figures are highly indicative of the importance of any solution dealing with these materials and the accompanying positive impact on the construction industry.

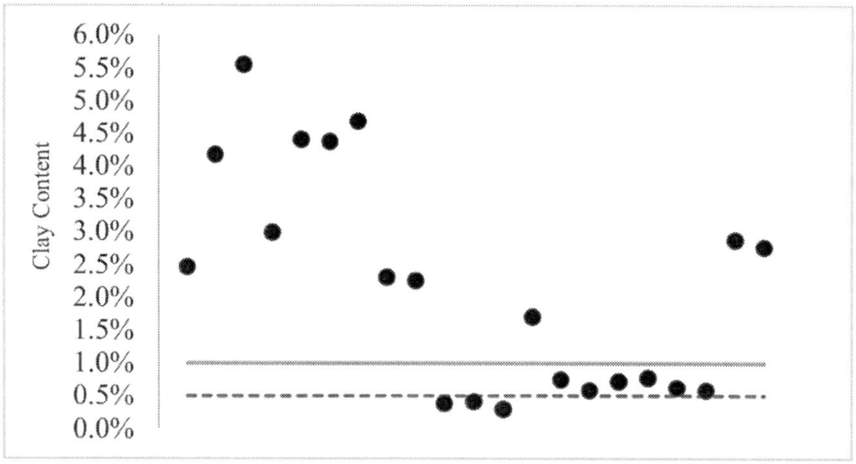

Fig. 3. Levels of corresponding clay contamination in 21 sand samples.

3.2 Mortar tests while doping sand with 0.20% sodium montmorillonite

The first tests were done with 2.73 g of sodium montmorillonite, corresponding to clay content of 0.20% by weight of sand. The mixture consisted of 240 g of water, 1350 g of EN sand and 700 g of CEM I 42.5 R cement. Table 3 represents the results of mini-slump flow obtained measured according to ASTM C1437-15. The dosages mentioned in the same table are % solids/solids corresponding to the dry solids content of the PCE material per cement content in mortar.

Table 3. Flow measured according to ASTM C1437 in mortar tests with 0.20% Sodium Montmorillonite.

Experiment	**E1**	**E2**	**E3**
Additive	AD1	AF2	AD1
Dosage, % s/s	0.186%	0.186%	0.186%
Flow, Initial, mm	220	230	220
Flow, 30 min, mm	195	200	190
Flow, 60 min, mm	160	160	150

These tests were planned for commencement aiming to get a baseline or starting point for the rest of our evaluation study and comparisons. Experiments E1 and E2 are very clear in indicating the similar performance of AD1 and AF2 at same dosage rate in clean to slightly clay contaminated sand. When sodium montmorillonite was added at a low amount, both AD1 and AF2 exhibit the same water reduction characteristics in terms of initial consistency and workability retention in mortar.

Experiment E3 was a repeat of E1 to check the variability of the test - it can be seen that the testing protocol is well controlled with minimum variability in all involved measurements.

3.3 Mortar tests while doping sand with 0.60% sodium montmorillonite

In these trials 8.1 g of sodium montmorillonite were added to the sand, corresponding to 0.60% by weight of sand. The mixture consisted of 240 g of water, 1350 g of EN sand and 700 g of CEM I 42.5 R cement. Table 3 represents the results of mini-slump flow obtained. The dosages mentioned in the same table are % solids/solids corresponding to the dry solids of PCE material per cement content in mortar.

Table 4. Flow measured according to ASTM C1437 in mortar tests with 0.60% Sodium Montmorillonite..

Experiment	**X1**	**X2**	**X3**	**X4**	**X5**	**X6**
Additive	AD1	AF2	AF2	AD1	AF2	AF2
Dosage, % s/s	0.345	0.345	0.328	0.345	0.335	0.331
Flow, Initial, mm	210	250	185	200	230	210
Flow, 30 min, mm	170	225	155	170	200	180
Flow, 60 min, mm	155	190	140	155	170	145

Looking at Table 4, one can immediately notice the considerably enhanced initial flow and extended flow retention of the additive AF2 when compared to AD1 at the same dosage in experiments X1 and X2. We are able to reduce the dosage of AF2 by more than 4%, keeping at all times a flow within a range ± 10 mm when compared to AD1, when examining experiments X5 and X6.

Fig. 4. Flow test of mortar after 60 minutes in experiments X4, X5 & X6 (from left to right)

Experiments X1 and X4 show again good repeatability of the test results.

Another way to understand the workability of above mixtures is by using a rotational viscometer (Viskomat NT) and record variations in torque as a function of time or stirring speed [19,20]. Such measures were conducted for 4 mixes as shown in Figure 4. A higher torque value indicates a mortar mix with lower workability.

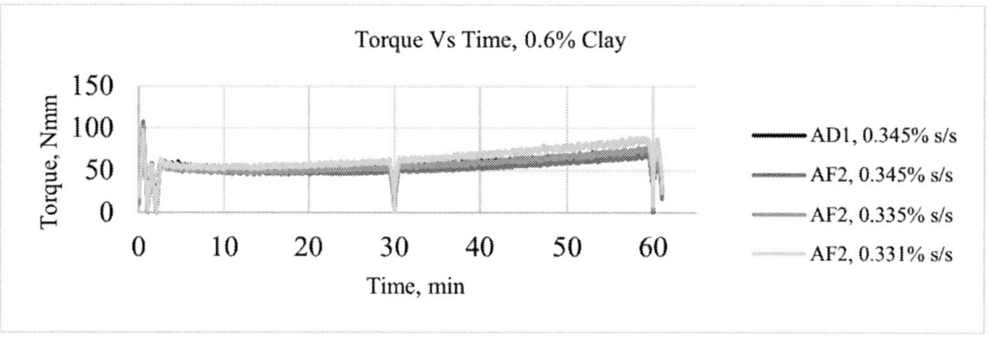

Fig. 5. Representation of the torque in relation to time of the mortar in a rotational viscometer.

The torque versus time curves presented in Figure 5 indicate that all the 4 mixes followed almost the same stiffening behavior over time, with AD1 at 0.345% s/s and AF2 at a reduced dosage rate of 0.335% s/s perfectly matching. Both are bound by AF2 at 0.345% s/s from below with more fluid consistency and higher workability, and AF2 at 0.331% s/s from top with slightly lower workability.

In the aim of having a closer look to the behavior of two similar mixtures with AD1 at 0.345% s/s and AF2 at reduced dosage 0.335% s/s, Figure 6 illustrates the torque in relation to the speed measured using Viskomat NT while ramping down from 60 rpm to 0 rpm after 60 minutes from mixing.

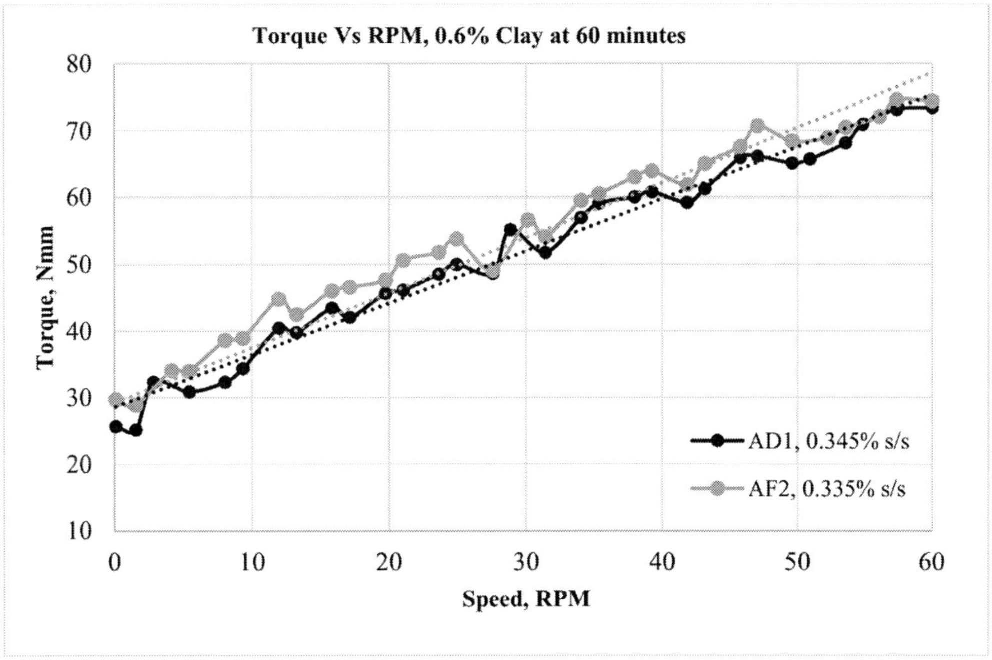

Fig. 6. Representation of the torque in relation to the rotational speed in a rotational viscometer

As a first order approximation, a linear relationship for both curves with respective equations displayed below is suggested:

AD1 at 0.345% s/s: $y = 0.780x + 28.6;\ R^2 = 0.978$

AF2 at 0.335% s/s: $y = 0.825x + 29.2;\ R^2 = 0.960$

The 2 curves are closely matching and the intercepts are almost equal.
Seeing that the 2 curves follow the model relation shear stress / shear rate, we can propose that the intercepts are proportional to the yield points of the 2 mixtures [20,21], and in this case being approximately identical.
To attain similar workability and rheology of mortar with 0.60% sodium montmorillonite, AF2 at almost 3% reduced dosage is equivalent to AD1.

3.4 Mortar tests while doping sand with 0.80% sodium montmorillonite

Another set of tests was designed by increasing the amount of sodium montmorillonite to 10.8 g, corresponding to 0.80% by weight of sand. The mixture consisted of 240 g of water, 1350 g of EN sand and 700 g of CEM I 42.5 R cement. Table 5 represents the results of the mini-slump flow obtained. The dosages mentioned in the same table are % solids/solids corresponding to the dry solids PCE material per cement content in mortar.

The flow measures of experiments P1 and P2 in Table 5 demonstrates the improved performance of AF2 when compared to AD1 at same dosage, in what concerns plastic properties up to 60 minutes. In experiment P3 the flow measures with AF2, at a dosage reduction reaching 3%, matched closely those with AD1 in P1.

Table 5. Flow measured according to ASTM C1437 in mortar tests with 0.80% Sodium Montmorillonite.

Experiment	P1	P2	P3
Additive	AD1	AF2	AF2
Dosage, % s/s	0.420%	0.420%	0.407%
Flow, Initial, mm	205	240	195
Flow, 30 min, mm	170	200	160
Flow, 60 min, mm	150	180	140

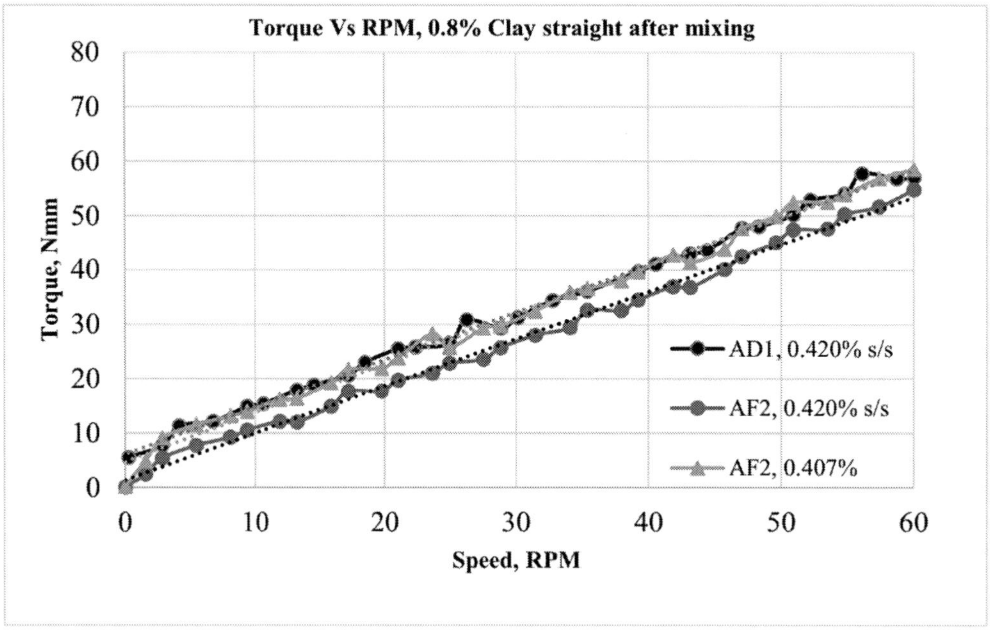

Fig. 7. Representation of the torque in relation to the rotational speed in a rotational viscometer.

Figure 7 represents the data collected from rotational viscometer straight after preparing the mortars in experiments P1, P2 and P3, in form of torque in relation to speed while ramping down from 60 rpm to 0 rpm.

In Figure 7, the data points related to AD1 at 0.42% s/s dosage and AF2 at 0.407% s/s are scattered in such a way that their representative curves are almost extending over each other.

The data points of AF2 at 0.42% s/s dosage are located lower than the other two curves, recording lower values of torque at every speed between 0 and 60 rpm.

A linear relationship for all the 3 data sets is proposed:

$$\text{AD1 at 0.420\% s/s, } y = 0.870x + 6.1;\ R^2 = 0.996$$

$$\text{AF2 at 0.407\% s/s, } y = 0.899x + 4.7;\ R^2 = 0.993$$

$$\text{AF2 at 0.420\% s/s, } y = 0.867x + 1.3;\ R^2 = 0.995$$

Examining all the above data and the similarity in yield points, we can deduce the equivalency in performance between AD1 at 0.42% s/s dosage and AF2 at a reduced dosage of 0.407% s/s. Mortar with AF2 at a higher dosage of 0.42% remains more fluid.

This is a case for a moderate sand contamination, where we witness the value of an emerging technology having a potential in solving technical problems related to cement mixtures consistency and slump retention with difficult clay contaminated sand.

3.5 Mortar tests while doping sand with 1.20% sodium montmorillonite

The last set of mortar trials was done by increasing the amount of sodium montmorillonite further to 16.2 g, corresponding to 1.20% by weight of sand. The mixture consisted of 240 g of water, 1350 g of EN sand and 700 g of CEM I 42.5 R cement. Table 6 represents the results of mini-slump flow obtained. The dosages mentioned in the same table are % solids/solids corresponding to the dry solids PCE materials per cement content in mortar.

Table 6. Flow measured according to ASTM C1437 in mortar tests with 1.20% Sodium Montmorillonite.

Experiment	R1	R2	R3	R4	R5
Additive	AD1	AF2	AF2	AF2	AF2
grams	9.03	9.03	8.58	8.4	8.22
Dosage, % s/s	0.645%	0.645%	0.613%	0.600%	0.587%
Flow, Initial, mm	190	260	220	210	195
Flow, 30 min, mm	170	250	205	195	160
Flow, 60 min, mm	160	235	180	180	150

At high levels of clay contamination the improved performance of AF2 when compared to AD1 at the same dosage rate can be clearly seen, with respect to plastic properties of mortars in experiments R1 and R2 up to 60 minutes.

To have a better understanding of the rheology of the two mixtures in experiments R1 and R5, Viskomat NT, a rotational viscometer, was used and the variation of the torque in relation to the rotational speed was measured during a ramp down from 60 rpm to 0 rpm. Figure 8 shows the two curves related to AD1 at 0.645% s/s dosage and AF2 at 0.587% s/s closely matching from 60 to 0 rpm. The torque of AF2 at 0.645% s/s dosage rate is lower than the other two curves from 60 to 0 rpm. In fact, this was seen in experiment R2 where the flow values recorded were higher and the mixture looked much more fluid.

The linear correlations for the two experiments R1 and R5 are mentioned as follows:

$$\text{AD1 at 0.645\% s/s, } y = 0.990x + 5.8;\ R^2 = 0.991$$

$$\text{AF2 at 0.587\% s/s, } y = 1.078x + 5.8;\ R^2 = 0.992$$

All above data and the full agreement between the flow measured according to ASTM C1437 and the yield points determined from experiments R1 and R5 are a key performance indicator translated in dose reduction close to 9%, while AF2 is keeping a steady efficiency comparable to AD1 within a flow range ± 10 mm extended over one hour of time.

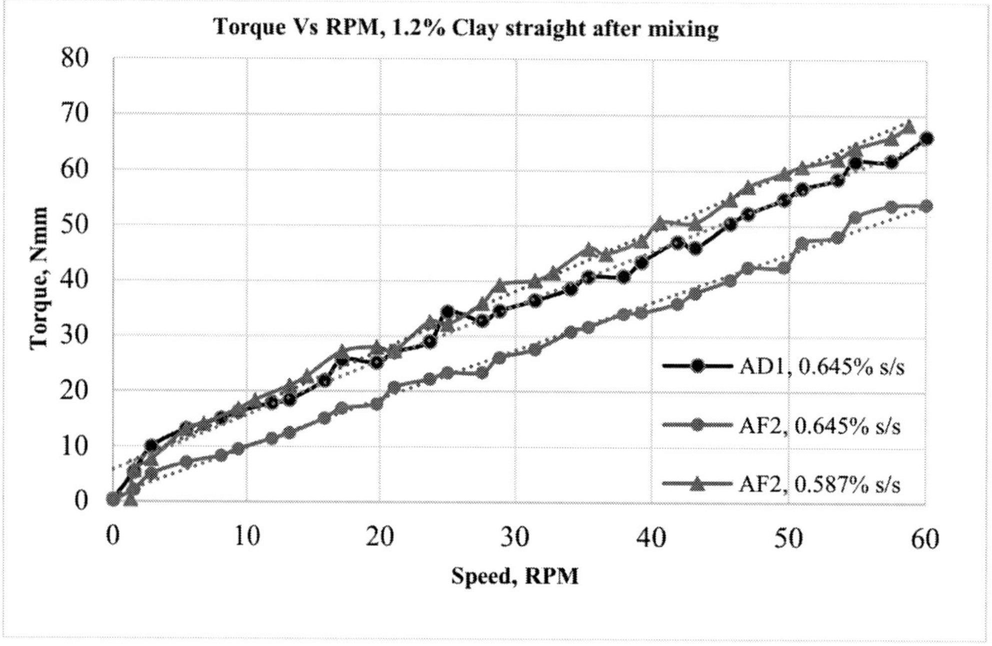

Fig. 8. Representation of the torque in relation to the rotational speed in a rotational viscometer.

At high sand contamination, the added value of the CLARENA® is clear and obvious in terms of improved plastic properties of cement mixtures or reduction of the required dose of dispersants to meet consistency parameters.

In the following section, the influence of coarse aggregates on the workability and flowability of concrete will be investigated.

3.6 Field case

This technology was applied in a practical difficult case in the field to validate the findings from the laboratory. It was a K650 concrete using 480 kg of OPC and 40 kg of silica fume at a low water to cement ratio equal to 0.28. Many different dispersants based on various technologies available in the market were tried at high dosages exceeding 10 kg/m^3, and the common performance characteristic was a quick loss of workability bringing the flow from around 600 mm flow to an almost zero slump concrete after 30 minutes from mixing. It was speculated that the poor performance was most likely linked to the clay contaminants in the sand. This triggered the idea to try AF3 in the same concrete mixture and check the capability of the new admixture.

Table 7. Measures of improved slump flow in K650 concrete

Additive	Dosage, kg/m^3	Initial Flow, mm	Flow, 30 min	Flow, 60 min
AF3	9.5	670	530	400

Table 7 shows the possibility of applying this technical solution for situations, where the problems reside in maintaining a workable concrete for a certain period of time and being pumpable. Such situations are frequently seen with higher grades concrete, where the clay content in the sand used is beyond a critical value.

4 Conclusions

The aim of the current study is to introduce and prove the efficiency of a newly engineered class of dispersants, providing a solution for workability complications, primarily concerning concrete properties in the plastic and hardened stage and specifically accompanying with the use of clay contaminated sand. This new family of products based on redesigned polymers, demonstrated an improvement in performance, with a pronounced flexibility in adjusting the initial fluidity, consistency and workability retention of concrete, a possibility of reducing the required dosage rates, as well as reducing water to cement ratios. In some difficult cases, the use of these products is imperative to overcome high clay contents in sand and to meet required technical parameters. All the preceding positive facts were assessed in mortar tests at different sodium montmorillonite doping degrees equivalent to 0.2%, 0.6%, 0.8% and 1.2% by weight of sand. Measures of slump flow and flow curves using a rotational viscometer, demonstrated the considerably improved workability of the tested solution AF at same dosages as the reference product AD. Finally, in extreme real conditions of steep loss of workability due to clay content exceeding some defined limits, CLARENA® confirmed to be the right key solution in extending slump life and providing pumpable concrete. Our proposed solution must be presented as a sustainable answer to the rigorous sand washing process practiced currently. It can help in modifying and shortening the phases of this process with all the concomitant benefits like preserving water, avoiding costs and reducing wastes. In addition, it is important to mention the beneficial roles that clay and fines can play if kept in concrete such as reduction in bleeding, segregation and formwork lateral pressure.

In future, it is a must to understand better the mode of action of these new polymers, their adsorption plus their dispersive and sacrificial effect on clay in restoring the anticipated dose-response of PCEs. Also continuous screening of sand and validation of the technology in different geographical locations is needed to broaden the scope of applications.

References

1. M. Al-Ansary, M. C. Poppelreiter, A. Al-Jabry, S. R. Iyengar, International Journal of Sustainable Built Environment **1**, 64–84 (2012)
2. C. M. Ouellet-Plamondon, G. Habert, Journal of Cleaner Production **117**, 160-168 (2016)
3. L. Lei, J. Plank, Cement and Concrete Research **42**, 1299–1306 (2102)
4. M.L. Nehdi, Construction and Building Materials **51**, 372–382 (2014)
5. Long xiong, Guangjun Zheng, Yao Bi, Chengfei Fu, Effect of typical clay upon the dispersion performance of polycarboxylate superplasticizer. *International Conference on Materials, Environmental and Biological Engineering*, (MEBE 2015)
6. N. A. Tregger, M. E. Pakula, S. P. Shah, Cement and Concrete Research **40**, 384–391 (2010)
7. M. Heikal, N.S. Ibrahim, Construction and Building Materials **112**, 19–27 (2016)

8. J. K. Norvell, J. G. Stewart, M. C. G. Juenger, and D. W. Fowler, Journal of Materials in Civil Engineering © ASCE / December / 1053 (2007)

9. A. A. Jeknavorian, L. Jardine, C. C. Ou, H. Koyata, and K. Folliard, *Seventh CANMET/ACI International Conference* **SP-217**, 1293–1316, (2003)

10. L. M. Olanitori. Mitigating the Effect of Clay Content of Sand on Concrete Strength. *31st Conference on OUR WORLD IN CONCRETE & STRUCTURES*

11. H. AlShemmari, A. M. Al-Dousari, L. Talebi, A. N. Al-Ghadban, Kuwait J. Sci. **40** (2), 159-176 (2013)

12. Elhabab and Adsani, Geochemical and Mineralogical Characters of the Coastal Plain Sediments of the Arabian Gulf, Kuwait. J Geol Geosci **3**:1 (2013)

13. MURAD D. A. Environmental Impact Assessment Case Study of the Quarries in Kuwait. *Environmental Planning and Impact Assessment Department, Environment Public Authority, Kuwait.*

14. T. Hirata, Japanese Patent JP 84, 2022 (S59-018338) (1981)

15. Flatt, R., Schober, I., Superplasticizers and the rheology of concrete. In: Roussel, N. (Ed.), *Understanding the Rheology of Concrete. Woodhead Publishing, Cambridge, UK*, 144-208 (2012)

16. H. Uchikawa, S. Hanehara, D. Sawaki, Cem. Concr. Res. **27**, 37–50 (1997)

17. C.Z. Li, N.Q. Feng, Y.D. Li, R.J. Chen, Cem. Concr. Res. **35**, 867–873 (2005)

18. Ouellet-Plamondon, C.M., Stasiak, J., Al-Tabbaa, A., Colloids Surf. A: Physicochem. Eng. Asp. **444**, 330-337 (2014)

19. Al-Ghadban, A. and A. El-Sammak, Sources, Journal of Arid Environments, **60**, 647–661 (2005)

20. P. F. G. Banfill, Magazine of Concrete Research, **43** (154), 13-2 (1991)

21. Robert J. Flatt, Domenico Larosa, Nicolas Roussel, Cement and Concrete Research **36**, 99 – 109 (2006)

22. P. F. G. Banfill, Magazine of Concrete Research, **42** (153), 213-221 (1990)

23. R.M. Quigley, F.J. Heffernan, Engineering Geology **2** (5), 351-356 (1968)

A comparative life cycle assessment (LCA) of alternative material for Australian building construction

Hangyong (Ray) Lu [1,*], *Ali* El Hanandeh [1], *Benoit* Gilbert [2], and *Henri* Bailleres[3]

[1] School of Engineering, Griffith University, 170 Kessels Road, Nathan, QLD 4111, Australia.
[2] School of Engineering, Griffith University, Parklands Drive Southport, Gold Coast, QLD, 4222, Australia.
[3] Salisbury Research Centre, Department of Agriculture, Fisheries and Forestry, Queensland Government, Australia

Abstract. The use of wood is seen as a sustainable alternative to reduce environmental impacts in building and construction sector. The low quality hardwood logs from plantation thinning are enhanced by producing engineered wood such as laminated veneer lumber (LVL). Nevertheless, engineered wood requires the use of chemicals and energy that may reduce its environmental benefits. A life cycle assessment (LCA) was conducted to compare the environmental performance of LVL produced from forestry thinning and final harvest to steel and concrete. The functional unit used in this study was a 1-m-long structural beam in a continuous beam system of 6-m-span designed according to the Australian standards. The Global Warming Potential (GWP) and embedded energy were assessed. The results indicated that LVL beam from thinned logs presented the lowest GWP impact (5.22kg-CO2-Eq). However, due to significant energy requirements for wood drying, the embedded energy in LVL was 186.78MJ which is only marginally less than steel (216.86MJ) but significantly less than concrete (352.82MJ). LVL from mature hardwood logs had slightly higher GWP than that produced from thinning; mainly due to extra energy and materials consumption in the plantation stage. Furthermore, LVL produced from mature trees had higher embedded energy than steel.

1 Introduction

The building and construction sector currently contributes approximately a quarter of the total Greenhouse gas (GHG) emissions in Australia and it is expected to increase by 110% by 2050 [1]. That is mainly attributed to the consumption of highly energy intensive materials such as concrete and steel, which leads to significant environmental impacts [2]. Wood and wood-based products are promoted as sustainable and renewable construction materials [3]; and their use has been increasing steadily over the past few decades [4].

Wood is generally identified as the most sustainable structural material and strongly recommended for substituting high-energy intensive products [4, 5]. Increasing use of wood-

* Corresponding author: hangyong.lu@griffithuni.edu.au

based material could reduce the net GHG emissions because of the relatively low energy requirement during wood products manufacturing stage compared to conventional materials [6]. In addition, wood substitution in long-life-span wood products results in accumulated carbon storage [7]. However, the benefits of using wood products will only be realised if wood is sourced from sustainable sources using environmentally responsible practices.

Nowadays, up to 40% of Australian forests are lost due to excessive harvesting. In response to growing environmental concerns about the sustainability of native forestry practices, agreements have been signed to progressively phase out logging of native forests in Australia. Plantations are now being promoted as a long-term wood supply strategy in Australia [8]. Thinning processes is a common practice, which is carried out to ensure the quality of the final harvest. Up to 70% of trees are removed by the age of 13 to 15 years [9]. The thinned logs are usually not suitable for sawn timber applications because of the high rate of defects [9]. Engineered wood products can reduce the effect from the natural defects and convert it to more uniform structural products, with higher strength and less variability in mechanical properties compared to sawn timber [10]. Engineered wood products are commonly made by forming veneers, strands and flakes through peeling, chipping or slicing processes and bonded together with adhesives to form wood panels or other shaped structural products [10, 11]. Engineered wood products have been increasingly accepted as a cost-competitive product building material [3]. Recently, engineered wood products produced from low value thinned logs were also used as substitute to replace energy-intensive materials [3, 12].

Structural Laminated Veneer Lumber (LVL) is produced by gluing layers of wood veneer sheets together with the grain of all plies running in the same direction. Compared to sawn timber, LVL has largely improved mechanical properties and used as an appropriate alternative in building structural purposes [11]. LVL can be manufactured in different sizes, such as floor beams, scaffold planking, as well as the flange material for prefabricated wood I-joists. However, many researchers mentioned that the engineered wood products may lead to much higher environmental impacts during manufacturing stage mainly due to chemicals consumption in the wood preparation phase [13, 14]. While some researchers believed that engineered wood building design offered significantly lower environmental impacts; others reported that when considering cradle-to-grave processes, the energy requirements of engineered products are nearly identical to concrete [15, 16]. Therefore, research is still needed to understand the environmental performance of engineered wood applications in building construction using comprehensive assessment.

Life cycle assessment (LCA) is a standardized method (ISO 14040:2006) which allows comprehensive assessment of products and services throughout all life stages [17]. However, LCA result can be affected by several factors such as, system boundaries, service life, manufacturing technology, energy sources, transportation distances, as well as waste management options. The existing studies could not represent the types of residential buildings commonly constructed in Australia, which makes it difficult to apply their findings to the Australian building sector. This study aims to obtain a quantitative measurement about the environmental impacts of using engineered wood in Australian residential building construction and compare it to the most commonly used material choices. The system boundaries start from hardwood plantation to the final waste disposal. Result of this study can be used to support residential building designers and builders, in making informed choices regarding the environmental impacts of construction material selection during building design and development.

2 Methods

2.1 Goal and scope

The aim of this study was to conduct a comparative life cycle assessment of using engineered wood material (LVL) manufactured from hardwood plantation logs, steel and concrete for constructing residential building in Australia. This LCA was conducted using OpenLCA software [18], following the International Organisation for Standardisation (ISO14040:2006) [17]. All the input energy and materials and output products, emissions and wastes were calculated as described in the Life Cycle Inventory Analysis section. The life cycle inventory (LCI) data of products manufacturing and energy consumption were collected from relevant literature and LCI databases. Global warming potential (GWP) and embedded energy were the two impact indictors assessed in this study. ReCiPe Midpoint (H) life cycle impact assessment (LCIA) method was used for the characterization of the emissions into the relevant impact categories.

2.1.1 Functional unit

The functional unit in this study was referenced as a 1m long section of a structural beam in a continuous beam system with 6 m spans made from engineered wood, concrete and steel following the respective Australian residential multi-stories apartment building construction codes for each material. The service life was assumed as 60 years based on Australian average building life span [19]. The specific structural details of different beams are presented in Figure 1.

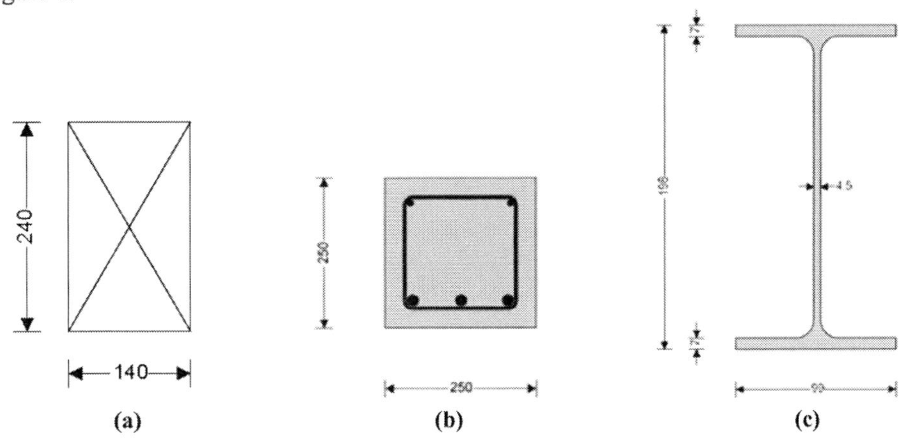

Fig. 1. (a) LVL Beam; (b) Concrete Beam Structure; (c) Steel Beam. All dimensions are in mm

2.1.2 System boundaries

The study provides a 'cradle-to-grave' LCA for laminate veneer lumber (LVL), concrete and steel beams for building construction based on resources from South-east Queensland. The system boundaries consider three life-cycle stages: raw material extraction, product manufacturing, and end-of-life treatment.

The engineered LVL beams manufactured using hardwood thinned logs from mid-rotation thinning and mature plantation hardwood from final harvest were designed as two scenarios. The system boundaries both start from hardwood plantation and up to second thinning and final harvesting, respectively. The hardwood logs are then transported to the mill for further manufacturing. The final products are transported to the building construction site using road truck. For the end-of-life treatment, the most environmentally feasible option was selected for each material. Incineration with energy recovery was the selected option for end-of-life management of LVL following the findings of Lu & El Hanandeh [20]. On the other hand, steel was assumed to be 100% recycled and concrete was sent to landfill. Furthermore, the data used in this study were only used for representing Australian current state of art.

2.2 Life Cycle Inventory (LCI) analysis

The inventory analysis involves data collection and analysis for the life cycle of the building construction material. For each stage of the product life cycle, energy and raw materials consumption, products, waste, and emissions were all determined.

2.2.1 LCI of LVL

The LVL beam was designed based on Australian Standards on Timber Structure Design AS1720.1 (2010), Timber structures - Design methods [21]. The dimensions were presented in Figure 1-a. The LCI for the LVL beam starts with the plantation stage, which includes seedling production; site preparation, tree planting, and application of herbicides and fertiliser and harvesting. Electricity is used during greenhouse operations, while diesel fuel and lubricants are required to power equipment for a series of forestry practise. The outputs are the green thinned logs from mid-rotation thinning and mature hardwood logs from final harvest. The non-merchantable slash and branches are usually treated by mechanical activities on site and were excluded in this LCA.

The LCI was conducted based on the current state of the art of hardwood plantation operations in South-east Queensland, Australia. The data was collected from published LCI for Australian forestry and wood products [22]. As thinned logs are a co-product of the forestry operation, economic allocation as reported by May et al. [22] were used to split the impacts between thinned and mature logs in the ration of 1:3. The production of 1 m^3 of hardwood from thinned logs requires 6.8L of diesel, 73.10kWh of electricity, 0.12 L of lubricant, and 0.08kg of steel. The quantity of fertiliser and herbicide consumption were also sourced.

During LVL beam manufacturing process, the hardwood logs are mechanically removed barks and cut to a certain length. The wood logs are then pre-conditioned using hot steam to ensure the logs are easier to peel and reduce veneer breakage [23]. The pre-conditioned logs are then peeled using a spindle-less veneer lathe with a high conversion rate around 70% [24]. The peeled veneers are moved to a clipper to be cut to appropriate widths. Veneer drying is an energy intensive process requiring approximately 3,800 MJ per cubic metre of hardwood veneer [25]. However, wood fuel constituted up to 80% of the total energy requirement with the remainder sourced from natural gas and electricity. The dried veneers are then glued and formed into LVL beam by Phenol-formaldehyde (PF) adhesive, which is commonly used for the engineered wood production [26]. After compressing, the LVL beam is sent to a finishing area for trimming and sanding. The air-dried density of hardwood LVL product is approximately 787 kg/m^3. Wood wastes generated during manufacturing such as veneers trim and sawdust are recycled as woody biomass fuel and combusted in boiler. LCI data of LVL manufacturing was collected based on our primary data and recently published relevant literatures [23, 25, 27], which represents common manufacturing practices in Australian mills.

The main inputs to manufacture 1 FU of structural LVL beam are: 0.056 m^3 of hardwood logs; 5.04kg of PF resin; and 0.147kg of preservative. The energy consumed during LVL manufacturing is mainly sourced from wood waste combustion in biomass boiler (Approximately 169 MJ). The other energy used includes 11.73 kWh of electricity, 0.11 litres of LPG, and 12MJ of natural gas. Transportation process required in the entire life cycle was approximately 11.22tkm. In addition, 21.5 litres of water are consumed through the entire manufacturing process. The outputs comprised 1FU of LVL beam (0.0336 m^3). Additionally, 12.7kg of the wood waste generated during manufacturing is sent to boiler for energy production.

Engineered wood waste incineration for energy recovery was considered as the most appropriate scenarios for final disposal in Queensland region [20]. This LCA assumed that LVL beam was combusted in large cogeneration plant to generate electricity, which can be distributed to the local main grid. The bottom ash was assumed to be disposed in a landfill. The LCI data of engineered wood in incineration is adopted from Lu & El Hanandeh [20]. The energy recovery through wood waste combustion was modelled as energy credit due to the avoidance of combustion of virgin feedstock according to the current Queensland electricity grid mix. It was estimated that 22 kWh of electricity could be generated per functional unit.

2.2.2 LCI of concrete structural beam

The concrete beam (Figure 1-b) was designed based on The New Australian Standard for Concrete Structures - AS3600 (2009) [28]. The LCI data on concrete and steel were sourced from the US LCI [29]. The concrete beam manufacturing process required 67.33kWh of electricity, 4.15m^3 of natural gas, 0.51L of diesel fuel and 0.22L of gasoline. The material required for 1 FU of concrete beam included 9.86kg of steel, 25.64L of water, 35.48kg of cement as well as 102.93kg of aggregate [30]. The transportation is assumed to be using B-double truck. The total transportation distance was estimated as 250km (i.e. 150km during material extraction, 50km during beam transports to building construction site, and 50km during final disposal). Concrete is demonstrated as an inert waste that is unlikely to be recycled [31]. Hence, the used concrete was assumed to be treated in landfill, meanwhile 85% of the reinforcing steel was recycled [32]. The environmental impacts of construction and closure of landfill facilities was assumed to be proportional to the mass of disposed waste beams.

2.2.3 LCI of steel structural beam

In accordance with Australian Standard for Steel Structures - AS4100 (1998) [33], steel beams are typically manufactured from hot rolled sheet steel. The dimensions of steel beam are shown in figure 1-c. The LCI data of steel beam manufacturing was sourced from published LCI data on steel production [30, 34]. The energy required for manufacturing 1FU of steel beam included 56.73 kWh of electricity, 3.62m^3 of natural gas, 0.13L of diesel and 0.32L of gasoline. The total steel required was calculated as approximately 18.33kg. The transportation processes considered in this study included transporting raw material to manufacturing facilities (150km), transporting steel beam to the building construction site (50km) as well as transporting used beams to recycle sites (50km). In final disposal stage, steel beams were assumed to be 100% recycled reported [35]. The LCI data for scrap steel recycling was sourced from published literature [30, 36].

 DOI: 10.1051/matecconf/201712002013

3 Results and discussions

To assess the overall GWP of the different beams, impact indicator values were added throughout the whole stages from material extraction to final disposal. The GHG emissions and energy consumption in each stage of the assessed options are presented in Figure 2 (a, b). Raw material extraction stage contributed significant GWP in both concrete and steel options. However, the LVL options indicated that manufacturing stage contributed the most significant impact in GWP mainly due to the PF resin consumption. In case of embodied energy, manufacturing stage is the most significant contributor to the total energy consumption in all options.

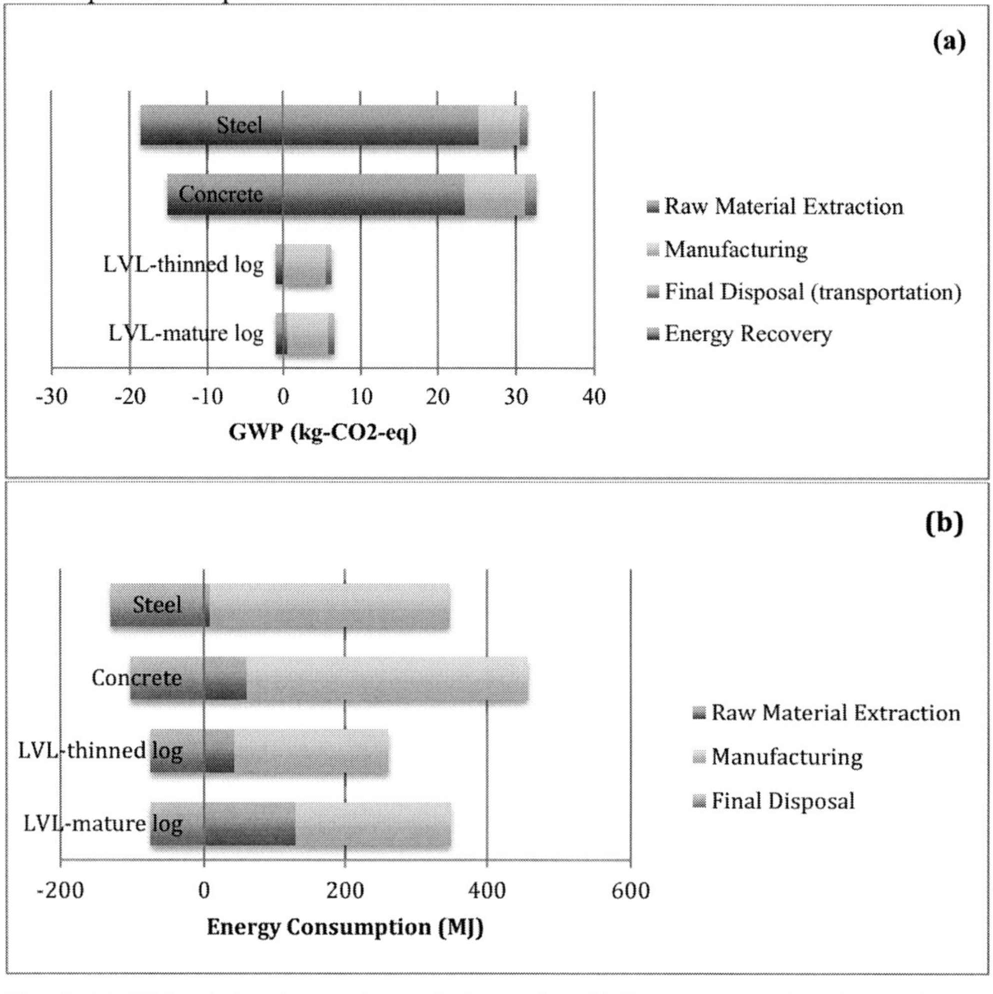

Fig. 2. (a) GHG emissions from each stage in four options (b) Energy consumptions from each stage in four options

The result indicated that the concrete had the largest environmental burden on GWP (17.43kg $CO_{2\text{-}eq}$) mainly due to the significant transportation, fossil fuel consumption, as well as use of energy intensive materials during the manufacturing process. Steel structure beam has better environmental performance that concrete on GWP (12.83kg $CO2_{\text{-}eq}$) than concrete beam. The GHG emissions released mainly from steel beam manufacturing and energy consumption. The energy required for casting concrete and steel beams were mostly sourced from fossil fuel attributed to electricity and natural gas consumption. However, the steel beam is recyclable at the end of its service life, which results in significant GHG emissions offset due to reduced energy requirements of recycled steel compared to virgin steel. Among all alternatives, LVL presented the best results on GWP mainly due to the use of significant amount of woody biomass fuel during the manufacture of LVL. According to IPCC (2007), CO_2 emissions from burning biomass are not counted towards the GWP because it has recently been sequestered from the atmosphere [37]. Furthermore, the CO_2 emissions from biomass burning were assumed to be sequestered by the next generation trees in the plantation. However, due to the significant energy requirements for wood drying, the embedded energy in LVL beam from thinned logs were about 186.78MJ which is only marginally less than steel (216.86MJ) but significantly less than concrete (352.82MJ). On the other hand, the embedded energy in the LVL beam manufactured from mature hardwood logs was calculated to be 274.2MJ/FU which is higher than steel due to energy use during the plantation stage. Furthermore, in the case of LVL beam, the materials sourced from hardwood plantation thinning had relatively less environmental impact on GWP100 (5.22kg-$CO_{2\text{-}eq}$) than that from mature hardwood logs (5.54kg-CO2-$_{eq}$), mainly due to avoided emissions during the plantation stage.

3.1 Uncertainty and sensitivity analysis

The uncertainties were caused by the variability and assumptions made in the LCA models. This section identified the most uncertain factors that may affect environmental performance. In this LCA, assumptions were made about transportation distance, energy resource consumption, as well as resin consumption, which introducing a level of uncertainty in the outcome of the LCA. Sensitivity analysis was conducted to identify the consequence by changing these uncertainties.

3.1.1 Transportation distance

Transportation distance during whole life cycle were estimated and obtained by an Internet highway directions interface [38]. However, the estimated route kilometres may not yield a real representation of actual transportation distance. A coefficient of variation (CoV) of 20% is usually assumed for the estimation of transportation distance [39]. The sensitivity analysis indicated that the changes in transportation distances had more effect on GWP than embedded energy in all scenarios. LVL beam options were slightly more sensitive to the changes in transportation distances, which caused a variation of ±16% in GWP impact indicator. Meanwhile the steel beam was the least sensitive, with GWP changing by less than 5%. However, in absolute values, concrete continued to exhibit the highest impacts. This is due to the heavy weight of concrete beam leading to higher transportation emissions due to higher consumption of diesel fuel

3.1.2 Energy consumption

The energy consumption in manufacturing and service sector fell approximately 20% during the last decade due to the improvement in technology and economy structure [40]. Therefore, in this sensitivity analysis, two cases were modelled for comparison, including (1) energy consumption saving of 20%, (2) increased energy consumption by 20%. Changes in energy consumption had proportional effect on embedded energy of all options. Nevertheless, the changes in energy consumption have slightly higher effect on the GWP of concrete and steel beams, which are attributed to the high fossil fuel energy use during the manufacturing process. A change of ±20% in energy consumption had ±9% and 8% changes in GWP for the concrete and steel beam, respectively. On the other hand, both scenarios of LVL beams showed little sensitivity to changes in energy consumption (<±4%) for GWP. This can be attributed to the high proportion of biomass energy usage in the manufacturing process of the LVL beams.

4 Conclusions and recommendations

This study compared four alternatives options of structure beam in building construction sector by using life cycle approach. Engineered LVL beams made from both hardwood forestry mid-thinning and final harvest were compared to the more mature options of steel and concrete beam. Result of this study indicated that the LVL beam from mid-rotation thinning had much better performance on GWP. Low energy intensity and utilization of low value material in the manufacturing of the LVL beams were the most significant contributors. Additionally, using biomass fuel to substitute fossil fuel as energy source during manufacturing process led to significant reduction in GHG emissions. Concrete beam was the least favourite option due to high GWP. The largest impacts of concrete beam were mainly attributed to using energy intensive materials and its heavy weight, which required more energy during manufacturing and transportation. Steel beam performed slightly better than concrete beam mainly due to its lighter weight, which led to reduction in the energy and material consumption, as well as transportation. The steel recycling contributed to significant GHG emission offset. Two scenarios were considered for the LVL beams. However, the LVL beam sourced from thinned hardwood logs was a better performer from life cycle assessment perspective as it had lower environmental impacts mainly attributed to the avoided emissions during plantation stage. Additionally, due to the significant energy requirements for wood drying, the embedded energy in thinned wood LVL beam option was only marginally less than steel but significantly less than concrete. However, LVL from mature hardwood logs presented much higher embedded energy than the LVL beam from thinned log and steel beam options.

Therefore, in order to reduce the life cycle environmental impact, the manufacture of the three types of structure beams should strive to reduce energy and raw material inputs through conservation and innovation. For instance, the raw materials should be sourced from locations close to point of manufacturing facilities, which could help to reduce the emission generation and cost due to long distance transportation. Additionally, using bio-energy products to replace fossil fuel or electricity as the main energy source could significantly reduce environmental impacts.

References

1. B. Basaglia, K. Lewis, R. Shrestha, K. Crews, in International Conference on Performance-based and Life-cycle Structural Engineering, 1433-1442 (2015)
2. L. F. Cabeza, L. Rincón, V. Vilariño, G. Pérez, A. Castell, Renew. Sust. Energ. Rev. 29, 394-416 (2014)
3. I. Z. Bribián, A. V. Capilla, A. A. Usón, Build. Environ. 46, 1133-1140 (2011)
4. L. Wang, A. Toppinen, H. Juslin, J. Clean. Prod. 65, pp. 350-361 (2014)
5. C. Knowles, C. Theodoropoulos, C. Griffin, J. Allen, Can. J. Forest Res. 41, 390-400 (2011)
6. L. Gustavsson, K. Pingoud, R. Sathre, Mitig. Adapt. Strat. Gl. 11, 667-691 (2006)
7. R. Sathre, L. Gustavsson, Appl. Energ. 86, 251-257 (2009)

8. C. J. Bradshaw, J Plant Ecol-UK 5, 109-120 (2012)
9. R. McGavin, M. Davies, J. Macgregor-Skinner, H. Bailleres, M. Armstrong, W. Atyeo, Forestry and Wood Products Australia, Melbourne, Australia (2006)
10. M. C. Barbu, R. Reh, M. Irle, Research Developments in Wood Engineering and Technology. IGI Global, 1-45 (2013)
11. F. Lam, Prog. Struct. Eng. Mat. 3, 238-245 (2001)
12. S. L. LeVan-Green, J. M. Livingston, J. Forest. 99, 25 (2001)
13. S. González-García, S. Berg, G. Feijoo, M. T. Moreira, Int. J. Life Cycle Ass. 14, 340-353 (2009)
14. B. Nebel, B. Zimmer, G. Wegener, Int. J. Life Cycle Ass. 11, 172-182 (2006)
15. J. Perez-Garcia, B. Lippke, D. Briggs, J. B. Wilson, J. Bowyer, J. Meil, Wood Fiber Sci. 37, 3-17 (2007)
16. A. B. Robertson, F. C. Lam, R. J. Cole, Buildings 2, 245-270 (2012).
17. International Standards Organisation (ISO), Environmental management–Life Cycle Assessment–Principles and Framework (ISO 14040, ed. Switzerland: second ed., 2006)
18. GreenDelta. (2014). OpenLCA. Available: http://www.openLCA.org
19. L. Aye, T. Ngo, R. Crawford, R. Gammampila, P. Mendis, Energ. Buildings 47, 159-168 (2012)
20. H. R. Lu, A. El Hanandeh, Sus. Proc. Consum. 5, 36-50 (2016)
21. Australian Standard AS 1720.1 (2010) Timber structures - Design methods (2010)
22. B. May, J. R. England, R. J. Raison, K. I. Paul, Forest Ecol. Manag. 264, 37-50 (2012)
23. E. Oneil, Cradle to Gate Life Cycle Assessment of Laminated Veneer Lumber Production from the Pacific Northwest (2013)
24. I. D. Underhill, B. P. Gilbert, H. Bailleres, R. L. McGavin, D. Patterson, Mater. Joints Timber Struct. 577-588 (2014)
25. R. D. Bergman, S. A. Bowe, Life-cycle inventory of manufacturing prefinished engineered wood flooring in the eastern United States, (University of Washington, Seattle, WA 2011)
26. D. E. Kline, Wood Fiber Sci. 37, 74-84 (2007)
27. S. Tucker, M. Syme, G. Foliente, NZ Timber Des J, 17, 3-9 (2009)
28. The New Australian Standard for Concrete Structures - AS3600 (2009)
29. National Renewable Energy Laboratory 2012 (U.S. Life Cycle Inventory Database, 2012)
30. AusLCI, The Australian National Life Cycle Inventory Database initiative (2011)
31. H. Bergsdal, R. A. Bohne, H. Brattebø, J. Ind. Ecol. 11, 27-39 (2007)
32. A. H. Buchanan, S. John, S. Love, World 15, 19 (2012)
33. Australian Standard for Steel Structures - AS4100 (1998)
34. GaBi Life Cycle Inventory Data Documentation (2016)
35. C. A. Bolin, S. T. Smith, J. Clean. Prod. 19, 630-639 (2011)
36. M. Yellishetty, G. M. Mudd, P. G. Ranjith, A. Tharumarajah, Environ. Sci. Policy 14, 650-663 (2011)
37. Intergovernmental Panel on Climate Change (IPCC), Climate Change 2007: The Physical Science Basis, Contribution of Working Group I to the Fourth Assessment Report of the Intergovernmental Panel on Climate Change (2007) Available: https://www.ipcc.ch/publications_and_data/publications_ipcc_fourth_assessment_report_wg1_report_the_physical_science_basis.htm
38. Google Maps Directions (2016)
39. G. W. Sonnemann, M. Schuhmacher, F. Castells, J. Clean. Prod. 11, 279-292 (2003)
40. S. Voigt, E. De Cian, M. Schymura, E. Verdolini, Energ. Econ. 41, 47-62 (2014)

Role of total reactive oxide ratios on strength development in activated fly ash

G.V.P. Bhagath Singh[1] , Kolluru V.L.Subramaniam[2*]

[1]Research Scholar, Department of Civil Engineering, Indian Institute of Technology Hyderabad, Hyderabad, India
[2]Professor, Department of Civil Engineering, Indian Institute of Technology Hyderabad, Hyderabad, India

Abstract. The role of individual reactive components and process variables such as molarity and temperature on alkaline activation of different low-calcium fly ash is explored. The oxide ratios in the activated system, based on the total silica (total SiO_2) in the system consisting of the reactive silica contributed by fly ash and the reactive alumina in fly ash are shown to provide consistent results for achieving the highest strength. For a given total SiO_2 content in the system, an increase in the sodium content above a certain dosage does not influence the ultimate compressive strength. An optimum (total SiO_2) to Na_2O ratio, equal to 2.66 is established for achieving maximum strength. The role of temperature within the range of 60°C-85°C is not significant when the molarity of NaOH is high. A N-A-S-H type gel with Si/Al ratio ranging between 2.5 to 3.0 and the Al/Na ratio varying between 1.30 to 0.9 is formed on decreasing the (total SiO_2)/Na_2O ratio from 6.55 to 2.66

1 Introduction

Alkali activated binders can be made from aluminosilicate source materials. The most attractive aluminosilicate source material is fly ash due to its ready availability in a large quantity. Researchers have studied the influence of several factors on the compressive strength and the structure of alkali activated fly ash-based geopolymeric binders. The influence of the process variables, such as the temperature of curing and the oxide ratios of the activating solutions on the strength of activated fly ash have been evaluated.

Most commonly used activators are combinations of sodium hydroxide (NaOH) and sodium silicate (Na_2SiO_3), which are mixed in varying proportions. Most studies have focused on the influence of molarity of NaOH and the molar ratios of SiO_2/Al_2O_3 and Na_2O/Al_2O_3 in the activating solution on the final strength of activated fly ash [1-4]. The compressive strength of activated fly ash increases with increasing concentration of NaOH in the activating solution [2]. NaOH primarily increases the molarity of the solution and also contributes the charge balancing cation, Na. The addition of Na_2SiO_3 to the NaOH solution increases the strength of the geopolymer formed after the activation. The addition of Na_2SiO_3

* Corresponding author: kvls@iith.ac.in

to the solution increases the SiO_2/Al_2O_3 and the SiO_2/Na_2O ratios, and results in the formation of more Si-O-Si bonds in the final product [5,6]. Different values of $Na_2SiO_3/NaOH$ in solution have been reported in the literature. $Na_2SiO_3/NaOH$ ratio in the solution equal to 2.5 by mass has been considered to be suitable to produce stable fly ash-based geopolymer [7]. A $Na_2SiO_3/NaOH$ ratio close to 2.0 were found to be suitable for producing the highest strength in fly ash-based geopolymer [8]. Highest strength in NaOH activated fly ash was obtained for SiO_2/Na_2O ratios in solution close to 0.69. [9]. The typical range of Na_2O/Al_2O_3 ratio in the geopolymer has been reported in the range of 0.25-0.3 and it has been shown to vary with the type of binder [10, 11]. There is a considerable variation in the findings reported in the literature since the solution ratios were arrived at considering one type of fly ash.

Fly ash contributes Al and Si to the reaction product, which is supplemented by the activators. Considering the product of the activation process, it has been shown that the strength of the material is enhanced if the Si/Al ratio is higher in the product [1, 12]. To achieve the desired elemental ratio in the end product, the activators required should consider the contribution of Si and Al contributed by fly ash. Since there is a large variability in the composition of fly ashes, the optimum solution ratios for one fly ash may not be applicable to a different fly ash since the reactive contents of the two fly ashes may be very different. Fly ashes with high reactive content would contribute higher quantities of Si and Al to the final product and may require a smaller addition of Na_2SiO_3. Activator solutions would have to be developed considering the composition and the reactive silica and alumina contents of fly ash.

Studies have shown that temperature of curing medium is another parameter that plays an important role in the geopolymerisation process [13-17]. In most of the cases researchers used the curing conditions of about 95% relative humidity and temperatures ranging from 30°C to 85°C [3, 18-20]. Along with these ratios, parameters such as water-to-geopolymer solids ratio, alkaline liquid-fly ash ratio and alkaline liquid-to-water ratio significantly influence the properties of fly ash based geopolymers [21].

An experimental study is conducted using two different low calcium siliceous fly ashes. The influence of sodium content, total reactive oxide ratios, temperature and NaOH molarity on strength development in the alkali activated fly ash is investigated. The influence of sodium is evaluated with reference to the total silica and total alumina in the activated system. The total silica in the activated system used for evaluation consists of the reactive silica contributed by fly ash and the activator solution. The influence of sodium content relative to the total reactive alumina contributed by fly ash and the total silica content of the activated system is evaluated.

2 Materials and methods

Two fly ashes, labelled B and N, were collected from a 500 MW and a 2600 MW thermal power plant located in Southern part of India. Both the fly ashes were collected directly from thermal power plants. The oxide composition of fly ashes were determined using X-ray Fluorescence spectroscopy (XRF) and are listed in Table 1. The sum may not be 100% due to the unreported oxide components which are present in trace quantities. Two fly ashes conforming to the requirements of siliceous fly ash as per the Indian Code of practice, IS 3812 [22] and Class F fly ash as per ASTM C 618 [23] with very low calcium contents.

Table 1. Oxide Composition (% by mass) of fly ashes.

Oxide Composition (% mass)	Fly ash Labels	
	B	N

CaO	2.75	3.51
SiO_2	58.14	56.92
Al_2O_3	26.06	26.97
Fe_2O_3	7.32	6.13
MgO	1.12	0.78
K_2O	2.17	2.77
TiO_2	2.02	2.80

A D2 PHASER (Bruker) benchtop automated diffractometer with Cu-Kα-radiation equipped with LYNXEYE super speed position sensitive detector was used to measure the spectrum of the fly ash. Specimens for XRD were prepared using the back loading technique to minimize preferred orientation. Incident beam with Soller 2.5° and 0.6 mm slit width and diffracted beam with Soller 4° and 8mm anti scatter slit width was used. Air scatter screen module over sample close to 2mm was used to reduce the air scattering of the sample. The readings were taken in vertical Bragg-Brentano (θ-θ) geometry between 10° and 70° at 0.02° step size with 0.6 step per second, resulting in a total measurement time about 30 minutes per scan. In order to improve powder averaging and obtain accurate intensities, the samples were rotated at 15 rpm during acquisition. The X-ray tube generator was operated at 30 kV and 10 mA. Quantitative phase analysis was performed using TOPAS 4.2 software with the. NIST SRM 676a α-Al_2O_3 external standard method approach. Crystal structure of known phases were taken from Inorganic Crystal Structure Database [24].

The reactive silica and alumina contents were determined after subtracting the crystalline forms of Silica and Alumina in the form of Quartz and Mullite from the total oxide contents. The total amorphous content determined subtracting the sum of the weight fractions of crystalline phases from unity using procedure outlined by the authors [25] and the values are listed in Table 2. It can be seen that the fly ashes used in this study have a low reactive silica content. The total reactive Alumina and silica in the two fly ashes are comparable. The reactive silica contents are lesser than the total Silica content.

Table 2. Main reactive oxide contents and Total amorphous content of fly ashes.

Composition (% mass)	Al_2O_3	SiO_2	Fe_2O_3	CaO	Total amorphous content (%)
B	15.72	18.28	5.92	2.75	48.16
N	14.82	17.72	4.23	3.51	46.73

Sodium hydroxide and sodium silicate were used as an alkali activators to activate the fly ash. Sodium silicate solution was used to increase the soluble silica content in the system and the sodium content was adjusted using sodium hydroxide. Ratios of oxides for the activated system were calculated using the reactive alumina and the total silica (referred to as (total SiO_2)) consisting of reactive silica from fly ash and silica obtained from sodium silicate solution. The (total SiO_2)/Na_2O ratio was varied without affecting the Al_2O_3/(total SiO_2) ratio. For a given fly ash, the total reactive alumina content contributed by fly ash is fixed. The ratio of the (total SiO_2) to the reactive Al_2O_3 content was kept constant with a value close to 2.0 in all mixes. The detailed experimental matrix is listed in Table 3. The molarity of NaOH was determined using its concentration in water. Increase in the sodium content is achieved decrease in the Total SiO_2/Na_2O ratio, which leads to the increasing the molarity of NaOH.

Table 3. The Solution and Total reactive oxide mix ratio of alkali activated fly ashes.

Fly ash Label	Label	Solution		(Total SiO_2)/Na_2O
		SiO_2/Na_2O	Molarity (M)	
B	a-XX	2.09	1.30	6.55
	b-XX	1.85	1.90	5.80
	c-XX	1.51	3.20	4.72
	d-XX	0.85	10.50	2.66
N	A-XX	2.23	1.00	6.55
	B-XX	1.97	1.80	5.80
	C-XX	1.61	3.10	4.72
	D-XX	0.90	10.30	2.66

XX is 60 for curing at 60 °C. XX is 85 for curing at 85°C
Total SiO_2 = Reactive SiO_2 (from fly ash) + SiO_2 (from sodium silicate solution)
SiO_2 = SiO_2 obtained from sodium silicate solution

98% reagent grade sodium hydroxide pellets supplied by the SDFCL was used to prepare the solution. Sodium silicate with the density 1.44 g/cc and with a chemical composition of 10.8% Na_2O, 30.5% SiO_2 and 41.3% H_2O supplied by kiran global chem limited was used. NaOH Solution was prepared with the required concentration of Na_2O content. After 24 hours, solution was mixed with the sodium silicate solution. Fly ash was mixed with the pre-mixed sodium silicate and sodium hydroxide solution using a pan mixer. The activator (the total mass of activating solution) to fly ash mass ratio was kept equal to 0.40. Thorough mixing was performed till a uniform consistency of the paste was obtained. After mixing, the mix was poured into cube moulds. Cubes specimens of nominal dimensions 7 cm were prepared. All the samples were prepared at room temperature (27±1°C). Immediately after casting, all specimens were cured in a sealed condition for 24 hours at room temperature before being subjected to further curing at elevated temperature and 95% relative humidity. Two different curing temperatures corresponding to 60°C and 85°C were used.

Compressive strength was measured from the cube samples at 3, 7, 14 and 28 days of age. Specimens were loaded up to failure at 2 kN/sec using a 3000 kN compression test machine. After the completion of compression test, small samples were collected from the failed cube and ground to a smaller size (<60 μm) using mortar and pestle. To stop the further reaction, samples was immersed and washed with isopropanol. The isopropanol was then evaporated at 40°C. The samples were then stored in a desiccator over silica gel in a Nitrogen atmosphere until testing.

Image acquisition and elemental analysis was performed on alkali activated fly ash samples using a bench top Phenom Pro X Scanning Electron Microscopy (SEM) coupled with Energy Dispersive X-ray spectroscopy (EDS). The samples were placed on a carbon tape and attached to stainless steel stub. The acceleration voltage 15 kv with the working distance 2 mm was used during the EDS analysis.

3 Results and discussion

The compressive strength of alkali activated fly ash B and N cured at 60°C with age is shown in Figure 1. Both the fly ashes exhibit similar tends in compressive strength at any particular age since the total (total SiO_2)/Na_2O ratio is kept same for both the fly ashes, while there is a difference in the working solution ratios. Lower (total SiO_2)/Na_2O ratio achieves higher strength than the higher the (total SiO_2)/Na_2O ratio. For (total SiO_2)/Na_2O ratio equal to 2.66, maximum compressive strength was achieved at 3 days after which there is very little improvement in compressive strength. In the mix with (total SiO_2)/Na_2O ratio of 4.72, there was a steady increment in strength till 7 days but the ultimate strength achieved is marginally higher than the (total SiO_2)/Na_2O ratio 2.66 mix. After 7 days a very little improvement in strength and the ultimate strength is almost equal to the (total SiO_2)/Na_2O ratio is 2.66 mix. The mixes with (total SiO_2)/Na_2O ratios equal to 6.55 and 5.80 show a steady improvement in strength till 28 days. The ultimate strength attained is however lower than the other two mixes. Results clearly indicates that for a given amount of total SiO_2 in the system contributed by fly ash and Na_2SiO_3, the sodium content does not influence on the strength after certain dosage in the system. There is hence an optimum (total SiO_2)/Na_2O ratio, which is close to 2.66 for achieving the maximum strength.

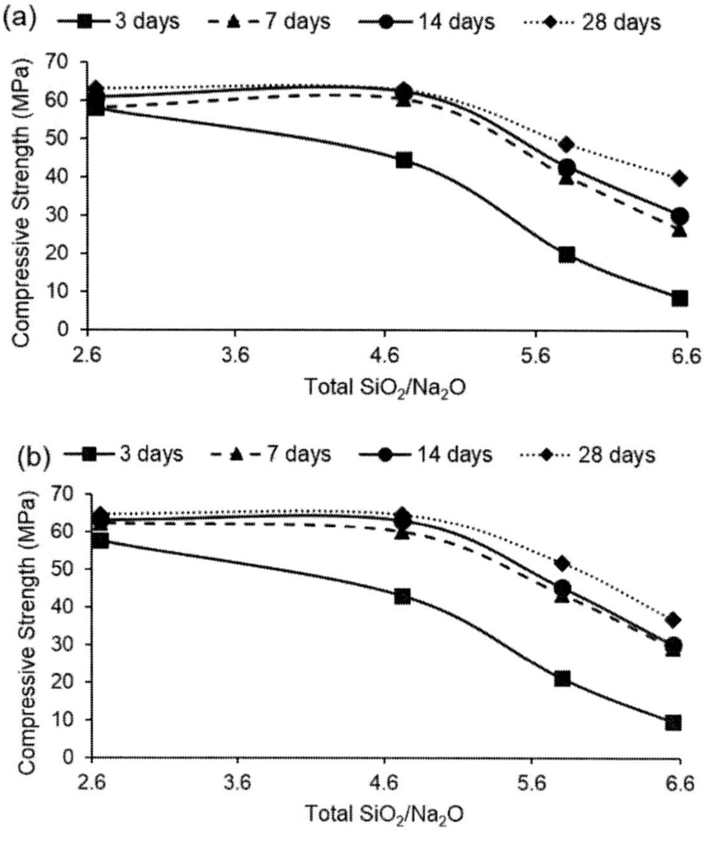

Fig. 1. Compressive strength of (a) Fly ash B and (b) fly ash N at any given age with (total SiO_2)/Na_2O ratios cured at 60°C.

The solution ratios of SiO_2/Na_2O at which (total SiO_2)/Na_2O of 4.72 and 2.66 are achieved are different for the two fly ashes. Considering the molar oxide ratios in the working solution and the total reactive oxide composition ratios (given in Table 3), the results indicate that the reactive oxide composition ratios are higher than the working solution ratios. Comparing the two fly ashes, the ratio of SiO_2/Na_2O in the working solutions varied, but the total SiO_2/ Na_2O ratio were kept identical. Nominally similar results are obtained in strength gain at identical ratio of (total SiO_2) to Na_2O, while the working solution ratios were different.

The compressive strength of the alkali activated mixes cured at 60°C and 85°C at 3 days and 28 days for both the fly ashes with age shown in Figure 2. Results indicate that both fly ashes showed nominally similar results at both curing temperatures. The compressive strength at both curing temperatures depends on the (total SiO_2)/Na_2O ratio in the system. There is higher early strength and higher ultimate strength with decreasing the (total SiO_2)/Na_2O ratio. For the mix with a (total SiO_2)/Na_2O ratio of 2.66, maximum compressive strength is achieved at the age of 3 days with very little additional improvement. Both the mixes attained identical ultimate compressive strength at both the curing temperatures. Mixes with (total SiO_2)/Na_2O ratio equal to 6.55 and 5.80 showed steady increment in compressive strength; samples cured at 85°C showed higher strength than the samples cured at 60°C at any age. There appears to be an optimum ratio of the (total SiO_2)/Na_2O in the range 4.72 to 2.66, since increasing the sodium content within this range does not appear to improve the ultimate strength.

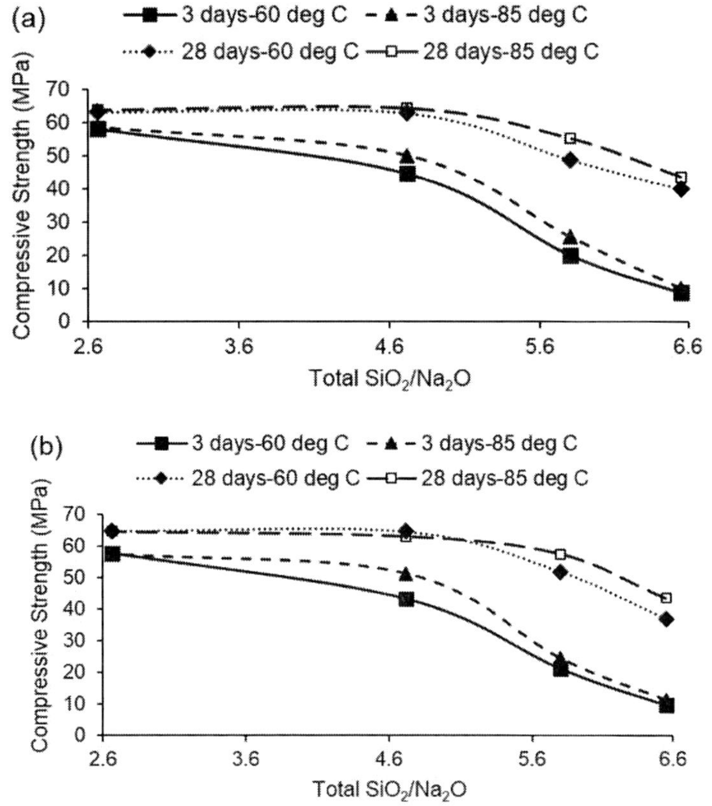

Fig. 2. Compressive strength of (a) Fly ash B and (b) fly ash N cured at 60°C.and 85°C at any given age with (total SiO_2)/Na_2O ratios

The experimental results indicate that the ratio of (total SiO_2)/Na_2O ratio close to 2.66 achieves the highest compressive strength. Lower (total SiO_2)/Na_2O ratio mixes attain a very high compressive strength by 3 days of age, and there is a small difference in the compressive strengths obtained from samples cured at the two temperatures. The results indicate that temperature has a greater influence on strength gain at higher (total SiO_2)/Na_2O ratio. Near the optimum ratio of (total SiO_2)/Na_2O, the difference observed from the two curing temperatures is not significant. The sodium content is increased at the expense of increasing the Molarity with the addition of NaOH. On increasing sodium content, the color of the

DOI: 10.1051/matecconf/201712002014

product changes from light to dark. Typical photographs of failed specimens with total SiO_2/Na_2O ratios equal to 4.72 and 2.66 are shown in Figure 3(a) and 3(b), respectively. While the ultimate strengths of the two mixes were equal, a distinctive change in color of the product is discernible. Similar behaviour was observed at both the curing temperatures and in both the fly ashes. At higher molarity of NaOH (lower SiO_2/Na_2O ratios) the influence of temperature is not significant on the rate of strength gain.

Fig. 3. Color change of the fly ash B product for (total SiO_2)/Na_2O ratio (a) 4.72 and (b) 2.66 cured at 85°C; at the age of 7 days.

Typical Backscattered image of the alkali activated (total SiO_2/Na_2O ratio equal to 4.72) fly ash B sample cured at 60°C at the age of 28 days are shown in Figure 4. The presence of partially reacted cenospheres from fly ash and crystalline phases are clearly identified. EDS-based spot and area analysis was performed on unreacted fly ash particle to determine the individual elements in the reaction product as shown in figure. The average values of Si/Al and Al/Na ratios of the all the mixes cured at 60°C for the both the fly ashes at the age of 28 days of age is listed in Table 4. Results clearly showed that the N-A-S-H type of gel with a Si/Al ratio of 2.5 and Al/Na ratio equal to 1.0 is formed in all the mixes with the variation in range.

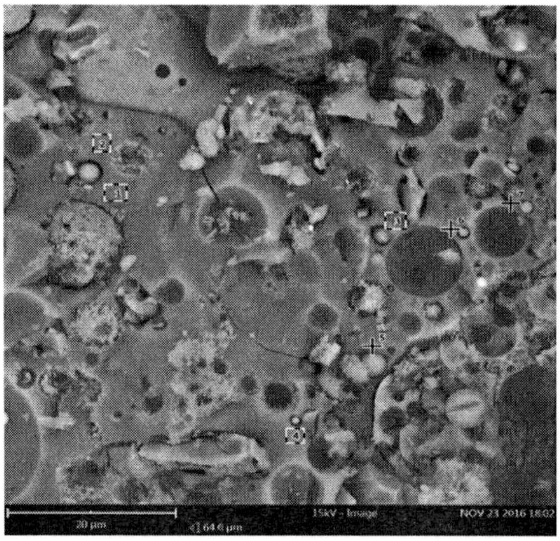

Fig. 4. SEM micrographs of Alkali activated (total SiO_2/Na_2O ratio is 4.72) fly ash B sample mix cured at 60°C at the age of 28 days.

Table 4. EDS microanalysis (atomic data) of reaction product (mostly gel) around the reacted particle after alkali activated fly ash samples cured at 60°C at the age of 28 days.

Label	Si/Al	Al/Na
a-60	2.4	1.5
b-60	2.5	1.3
c-60	2.8	1.1
d-60	3.0	0.9
A-60	2.5	1.3
B-60	2.4	1.3
C-60	2.9	1.0
D-60	3.0	0.9

4 Findings and conclusions

In this study the alkaline activation of low calcium fly ash using NaOH is investigated. Activator with varying proportions of NaOH and Na_2SiO_3 were used. The influence of oxide ratios (total SiO_2)/Na_2O and Al_2O_3/ (total SiO_2) on the compressive strength of the activated fly ashes were evaluated. In calculating oxide ratios, the reactive Alumina and reactive silica obtained from fly ash were used. SEM with EDS indicate that the reaction products contain a N-A-S-H type gel. The oxide ratios derived using the reactive component contents of fly ash in the mix are important parameters on development of compressive strength of the paste. The ratios based on (total SiO_2)/Na_2O ratios produce identical results. Highest strength are obtained for identical ratios of (total SiO_2)/Na_2O for the two fly ashes used in the study, but the solution ratios are different. The sodium content in the solution does not influence the development of compressive strength of paste beyond certain dosage. The strength of the paste increases with the decreasing ratio of the (total SiO_2)/Na_2O and the highest ultimate compressive strength is achieved for the ratio close to 2.66. The role of temperature within the range of 60°C-85°C is not significant at higher molarity. There is an insignificant difference in the compressive strength after 7 days of ages and at curing temperatures equal to 60°C and 85°C.

References

1. A. Fernandez-Jimenez, A. Palomo, Composition and microstructure of alkali activated fly ash binder: effect of the activator, Cem. Concr. Res. **35,** 1984-1992 (2005).
2. G.S. Ryu, Y.B. Lee, K.T. Koh, Y.S. Chung, The mechanical properties of fly ash-based geoploymer concrete with alkaline activators, Constr. Build. Mater.**47,** 409-18 (2013).
3. A Palomo, M.W Grutzeck, M.T Blanco, Alkali-activated fly ashes a cement for the future, Cem Concr Res **29,** 1323-9 (1999).
4. Z Xie, X.Yunping, Hardening mechanisms of an alkaline-activated class F fly ash, Cem. Concr. Res. **31**, 1245-9 (2001).
5. P. Duxson, J.L. Provis, G.C. Lukey, S.W. Mallicoat, W.M. Kriven, J.S.J. Van Deventer, Understanding the relationship between geopolymer composition, microstructure and mechanical properties, Colloids Surf. A. **269,** 47-58 (2005).
6. M. Komljenovi, Z. Bascarevic, V. Bradic, Mechanical and microstructural properties of alkali-activated fly ash geopolymers. J. Hazard. Mater. **181,** 35-42 (2010).
7. D. Hardjito, B.V. Rangan, Development and Properties of Low-Calcium Fly Ash based Geopolymer Concrete. Australia Curtin University of Technology, Perth, p.48, (2005).

8. H.Y. Leong, D. E.L. Ong, J.G. Sanjayan, A. Nazari, The effect of different Na_2O and K_2O ratios of alkali activator on compressive strength of fly ash based-geopolymer Constr. Build. Mater. **106,** 500-511 (2016).
9. M. Criado, A. Fernandez-Jimenez, A.G de la Torre, M.A.G. Aranda, A. Palomo, An XRD study of the effect of the SiO_2/Na_2O ratio on the alkali activation of fly ash, Cem. Concr. Res. **37,** 671-679 (2007).
10. R. Fletcher, K. MacKenzie, C. Nicholson, S. Shimada, The composition range of aluminosilicate geopolymers, J. Eur. Ceram. Soc. **25,** 1471-77 (2005).
11. M Rowles, B O'Connor, Chemical optimisation of the compressive strength of aluminosilicate geopolymers synthesised by sodium silicate activation of metakaolinite, J Mater. Chem. **13,** 1161 - 65 (2003).
12. A. Ferna´ndez-Jime´nez, A. Palomo, I. Sobrados, J. Sanz, The role played by the reactive alumina content in the alkaline activation of fly ashes, Micropor Mesopor Mater. **91,** 111-9 (2006).
13. A. Palomo, S. Alonso, A. Fernández-Jiménez, Alkaline activation of fly ashes: NMR study of the reaction products, J. Am. Ceram. Soc. **87,** 1141 - 45 (2004).
14. P. Sukmak, S. Horpibulsuk, S.L. Shen, P. Chindaprasirt, C. Suksiripattanapong, Factors influencing strength development in clay-fly ash geopolymer, Constr. Build. Mater. **47,** 1125-36 (2013).
15. M. Bing-hui, H. Zhu, C. Xue-min, H. Yan, G. Si-yu, Effect of curing temperature on geopolymerization of metakaolin-based geopolymers, Appl. Clay Sci. **99,** 144-148 (2014).
16. Muhammad Talha Junaid, Amar Khennane, Obada Kayali, Arezki Sadaoui, Donald Picard, Mario Fafard, Aspects of the deformational behaviour of alkali activated fly ash concrete at elevated temperatures, Cem. Concr. Res. **60**, 24-29 (2014)
17. William D.A. Rickard, Jadambaa Temuujin, Arie van Riessen, Thermal analysis of geopolymer pastes synthesised from five fly ashes of variable composition, Journal of Non-Crystalline Solids **358,** 1830-39 (2012).
18. A Ferna´ndez-Jime´nez, A. Palomo, Characterization of fly ashes. Potential reactivity as alkaline cements, Fuel **82,** 2259-65 (2003).
19. A Ferna´ndez-Jime´nez, I Garcı´a-Lodeiro, A Palomo, Durability of alkali-activated fly ash cementitious materials, J Mater Sci. **42,** 3055-3065 (2007).
20. M. Criado, A. Fernández-Jiménez, A. Palomo, Alkali activation of fly ash. Part III: Effect of curing conditions on reaction and its graphical description, Fuel **89,** 3185-3192 (2010).
21. M. Talha Junaid, Obada Kayali, Amar Khennane, Jarvis Black, A mix design procedure for low calcium alkali activated fly ash-based concretes, Constr. Build. Mater. **79**, 301-310(2015).
22. Bureau of Indian Standards, Pulverized Fuel Ash-Specification, IS 3812-part 1:2003, Manak bhavan, New Delhi.
23. ASTM C 618, Standard Specification for Fly Ash and Raw or Calcined Natural Pozzolan for Use in Concrete.
24. ICSD, Inorganic Crystal Structure Database. 2013, Fachinformationszentrum: Karlsruhe, Germany.
25. G.V.P. Bhagath Singh, K.V.L. Subramaniam, Quantitative XRD Analysis of Binary Blends of Siliceous Fly ash and Hydrated Cement, J. Mater. Civ. Eng. **28,** 04016042 (2016).

An exploratory study on the building information modeling adoption in United Arab Emirates municipal projects- current status and challenges

Senthilkumar Venkatachalam[1,*]

[1]Sustainable Construction Materials and Structural Systems Research Group, Department of Civil and Environmental Engineering, University of Sharjah, Sharjah, United Arab Emirates

Abstract. Many modern complex building projects in the public and private sectors are characterized by its poor information management which is manifested by time and cost overruns around the world. Building Information Modelling (BIM) is currently being adopted around the world with various countries mandated its implementation. In specific, the implementation of BIM is rapidly growing in developed countries such as USA, UK, Singapore and Australia, with government mandate through devising strategies and initiatives to increase the adoption. The implementation of BIM in United Arab Emirates (UAE) and other developing countries lags behind due to the existence of numerous implementation barriers. This study explores the current status of BIM implementation in the UAE municipalities, further explores the people, process and technology readiness towards BIM adoption in the UAE building projects. Further the study also explores the current challenges against the BIM adoption. The study adopts a positivists approach by collecting data through questionnaires, a sample of sixty participants across the UAE municipalities from different states were selected randomly to acquire the data. The collected data is statistically analyzed to obtain the findings. The analysis results showed that there are inadequate people and process related readiness towards BIM adoption whereas the technology readiness does exist. The same also was reflected in further analysis on BIM adoption barriers as the challenges. Further the paper also discusses the possible strategies adopted around the world through literature to address the explored challenges in the study

1. Introduction

Construction projects are complex accomplishments that require diverse purposed stakeholder's involvement to accomplish the common project outcomes [1]. The diversity on their purpose is through its components, people and processes variations and its complex

* Corresponding author: svenkat@sharjah.ac.ae

relationships and interfaces [2]. This characterization brings complexities to the project delivery. These complexities are characterized in all aspects of the projects, including people, process and technology etc., [1]. These three aspects are connected through the project information. Huge amount of information is been generated in any construction projects. These instills the complexity further and evident the collaboration related inefficiencies in their project multi-objective decision making process.

The use of IT had been so far proven to be effective in handling these huge project information, it is largely discrete until the advent of the Building Information Modeling (BIM). BIM is the process of capturing the information through the development of nD models and use of these digital models and associated tools to facilitate planning, design, construction, and operation during the project lifecycle[3]. Once modelled, it contains a wealth of data about the designed entity and it is useful for representing several views of project data including two dimensional (2D), three dimensional (3D), four dimensional (4D, schedule), five dimensional (5D, cost), six dimensional (6D, operations and maintenance) forms[4][5]. The BIM implementation in UAE municipalities is inconsistent and investigating the current status of BIM adoption, the related challenges and the possible future strategies becomes paramount important to efficiently deliver the construction projects. Hence this study aims to identify the BIM adoption status, challenges, and drivers in UAE municipal project delivery. The following sections outlines the literature review related to the study. Followed by the adopted research method. The fourth section outlines the analysis of the responses obtained followed by the discussions and conclusions.

2. Literature review

The scenario of complexity and its related collaboration inefficiencies are very much similar in the United Arab Emirates (UAE) construction industry, which hosts multiple stakeholders from all around the world to work for their common objectives in UAE construction projects [6].Building Information Modelling (BIM) is currently being adopted around the world with numerous countries mandating its implementation on public projects through legislation. This process has driven the development of BIM standards and specifications to streamline design data prior to collaboration and to meet regulatory requirements. Although BIM technologies globally are similar, many countries have developed context-specific BIM specifications to accommodate differences in their design processes, procurement processes and government policies.

BIM implementation has gained prominence in the last decade, in the United States of America, Singapore, United Kingdom, Australia and some Nordic countries[7][8].Further, many of the Asian and Middle Eastern countries started adopting the BIM on their private sector projects [9].Further, Dubai to host EXPO 2020, necessitates many more innovative and complex infrastructure projects to be built. Dubai Municipality mandated the implementation of BIM in its significant capital projects from 2014[6]. Though there are limited research to precisely measure the benefits of BIM at the project level [10], its impact in long term has been shown as huge benefits in most of the literature and further the implementation benefits also vary from project to project [11]. Some of these reported benefits to the construction companies are in terms of its improved collaboration, less lead time, less clashes, less rework and overall improvement in the project outcomes [12]. These advantages encourages more and more adoption in the construction industry around the world [13] especially in the private sectors in UAE.

BIM doesn't produce the guaranteed results always, if an appropriate modifications are made in the existing project delivery systems in terms of the three above discussed aspects such as the people, process and technology[14].Figure 1 depicts the Word cloud depicting prominence of issues studied in BIM research in the literature. There are many aspects in construction projects in which BIM implementation has a leverage. However this study focuses on some of the prominent important and basic aspects which paves the way to progressively adopt BIM.

Fig. 1. Word cloud depicting prominence of issues studied in BIM research.

People related barriers include, the reluctance to change, lack of skilled professionals, high initial cost of implementing BIM standards and specifications. This includes the cost of down time within the organization in order to change the current design workflow. It also includes the cost of continually educating and training employees in terms of the implementation of BIM standards and specifications, keeping in mind that there will be a learning curve involved for even the most proficient BIM user [15].This further will have effect on the team formation and its related organizational structure. Empowering the people with the appropriate skills and competencies become another major challenges that the construction organizations face during the implementation of the BIM. There are very limited courses taught BIM standards and specifications in the developing nations where BIM is still in the premature level of adoption. This results in BIM trainers themselves being novices. On the back of that employees in construction organizations are generally resistant and reluctant to learn new tools and work flows as they perceive this as a waste of time and a hindrance to their productivity.

The process related challenges often originates through the legal and contractual barriers of BIM standards and specifications include the uncertainty of legal liability amongst AEC stakeholders. The uncertainty on the accuracy of BIM model data may necessitates contractual clause modifications, creation and deletion of processes, roles and responsibilities among the participants [15].

Technology related barriers are otherwise tools barriers include the vague specifications about the level of details during various design stages and the amount of data, its standard format, interoperability etc., this may be nonspecific, indefinite and impractical in some projects. This may lead to adverse effect on the project data delivery and end up with the inefficiency on the final process, product and mistrust among the participants. Further, the tools barrier also include designers' BIM efficiency and the lack of BIM training and support [16]. Open interoperability may also become an issue as not all design stakeholders are utilizing the same BIM software packages in a common data environment.

Though the categorization can be specified as people, process and technology, some of the challenges under these categories may also take more than one facet. For example, one of the process related challenges associated with BIM implementation is the resistance offered by the people on the change of existing work practices and culture [14][17][18]. The categorization for the above challenge may be in terms of the people related as its behavioral oriented or it can be process oriented as the workflow dictates the people to behave in a certain manner.

Around the world there are many BIM adoption and implementation strategies introduced mainly through a TOP down (Client) approach throughout the building lifecycle [15][19][20]. Among which the important and successful ones are from the US, Nordic Singapore and more recently the UK. Local governments should play a vital role in overcoming these barriers by means of enforcing legislative BIM standards and specifications on construction projects. This enforcement can lead to the promotion of technological BIM innovation by means of driving national leadership and reform [21]. The recent Dubai Municipality's BIM mandate is an example for such pattern of BIM implementation. However, the mandate without the necessary social and infrastructure climate or BIM adoption readiness may not give the desired results. There is another approach through which the BIM is implemented in most of the pre mature adopters of BIM such as the developing nations, which is proceed with the implementation without any national or reginal standards and guidelines. Many literature have reported the pros and cons of the two approaches [15][20]. However, the top down Government involvement basically helps to facilitate a coordinated rather than fragmented approach to the adoption and implementation of BIM [19].

There have been more and more standards and guidelines developed by different countries [20][8]. These standards and the guidelines were developed to suit to their local project supply chain contexts. These standards and guidelines are developed through mainly governmental initiatives but not limited to the involvement of the research universities, educational and training institutes, professional councils and private sector collaborations.

Among all the Government led BIM implementation initiatives, the UK model is more structured and effective one[15]. The initiatives have produced more and more guidelines and standards documents to guide building information creation, sharing, storing, use, re-use as well as collaborative relationships based on BIM within the construction supply chain. These include, but not limited to:

PAS 1192-2:2013 – Specification for information management for the capital/delivery phase of construction projects using building information modelling

PAS 1192-3:2014 – Specification for information management for the operational phase of assets using building information modelling

BS 1192-4:2014 – Specification for collaborative production of information and fulfilling employer's information exchange requirements using COBie

PAS 1192-5:2015 – Specification for security-minded building information modelling, digital built environments and smart asset management
COBie (Construction Operations Building Information Exchange) – a data format for the publication of a subset of building model information focused on delivering building information not geometric modelling.

With these background the current study explores the current status of the BIM implementation in UAE municipal project delivery and capture the challenges also to propose a strategy to overcome the same. The following section explains the adopted research method in this study.

3. Research methodology

The study adopts a positivism as a research philosophy, which necessitates the deduction in the research approach. The questionnaire survey is administered as a research strategy to obtain the data from the municipal project participants. The study is a cross-sectional one and is administered towards the end of 2015 for about a month. There were 100 questionnaire distributed among the municipal project stakeholders from the different emirates. There were 60 valid responses obtained. The respondents include different stake holders which include contractors, sub-contractors, engineers, architects, project managers, government officials, facility managers, quantity surveyors and the municipal administrative authorities. Though the questionnaire was distributed proportionately to all the emirates the response spectrum shows dominance in only three emirates. However, most of the respondents do executes project in other emirates as well hence the generalizability of this exploratory study is not affected by the low respondent numbers in some of the states. Also there were only 45 respondents who would like to answer the BIM related questions the others simply not answered. There were five sections of questions were asked to the respondents through the questionnaire. The first sections captured the respondent detail which includes the stakeholder's primary job title, emirates in which he did serve, how many projects he did execute etc. The second section asks the awareness and application of the BIM in various building project life cycle stages, the third section of the questions asks the opinion of the respondents about the drivers for the BIM adoption now and in future. The fourth section of the questionnaire asks the readiness of their firms in BIM adoption in terms of people process and technology related factors. The final section lists the challenges on BIM adoption followed by some open ended questions regarding their opinion on future BIM adoption and the strategies. The questionnaire responses are analyzed using simple descriptive statistics. The obtained results are triangulated with further 3 interviews among the participant stakeholders.

4. Analysis and results

The 60 returned questionnaire responses were carefully captured in a excel sheet for further analysis. The question regarding the awareness revealed that only 48 out of 60 respondents are aware of BIM. Among the 48 only 45 were accepted to participate in further survey. Figure 2 showed the perceived level of adoption through a radar plot, the respondents who were willing to participate in the BIM related questions were asked about the potential application of BIM. The analysis showed that the intention or the adoption status that BIM are for the purposes of the 3d Modeling and walkthrough, class detection and record keeping purpose only. All other benefits are yet to be realized among the municipal project stakeholders. This reveals that the participants are lack with awareness about the BIM benefits.

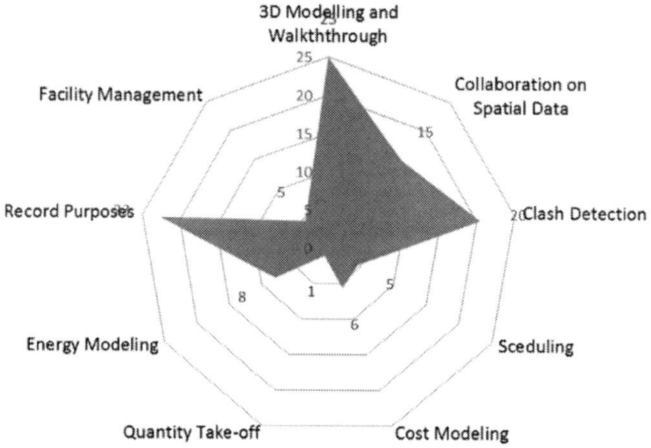

Fig.2. BIM application in Project Delivery Processes-Status

Further the questions regarding the readiness of BIM adoption reveled that there exists a lack in terms of the people and process related aspects however the technology readiness exists among the UAE municipal project stakeholders. Figure 3 showed the results in a bar chart plot. The drivers which may motivate the users to adopt BIM also been asked among the project stakeholders and the results are shown in figure 4. Overall, the drivers mentioned in figure 4 are acknowledged by most of the respondents. However, widely the respondents perceived that the BIM has advantages during the design stage of the projects. This is evident in figure 4 that many respondents agreed that they get motivated as BIM improve the design quality, reduce design time, design collaboration, reduce design rework etc.

Figure 5 showed the challenges that the participants perceived towards BIM implementation in UAE's municipal projects. It is expressed through a weighted average of the Likert scale responses. The scale is ranging between 1to 5 as strong disagreement (with the score of 1) to the strong agreement (with the score of 5).It is observed through the radar plot in figure 5 that the top three challenges which are agreed by most of the respondents are the inadequate top management support, higher cost associated with the BIM implementation and the inadequate statutory support. Though there are other important challenges which are listed as lack of BIM professionals, lack of BIM related information, lack of company policy, reluctance to use the technology etc.

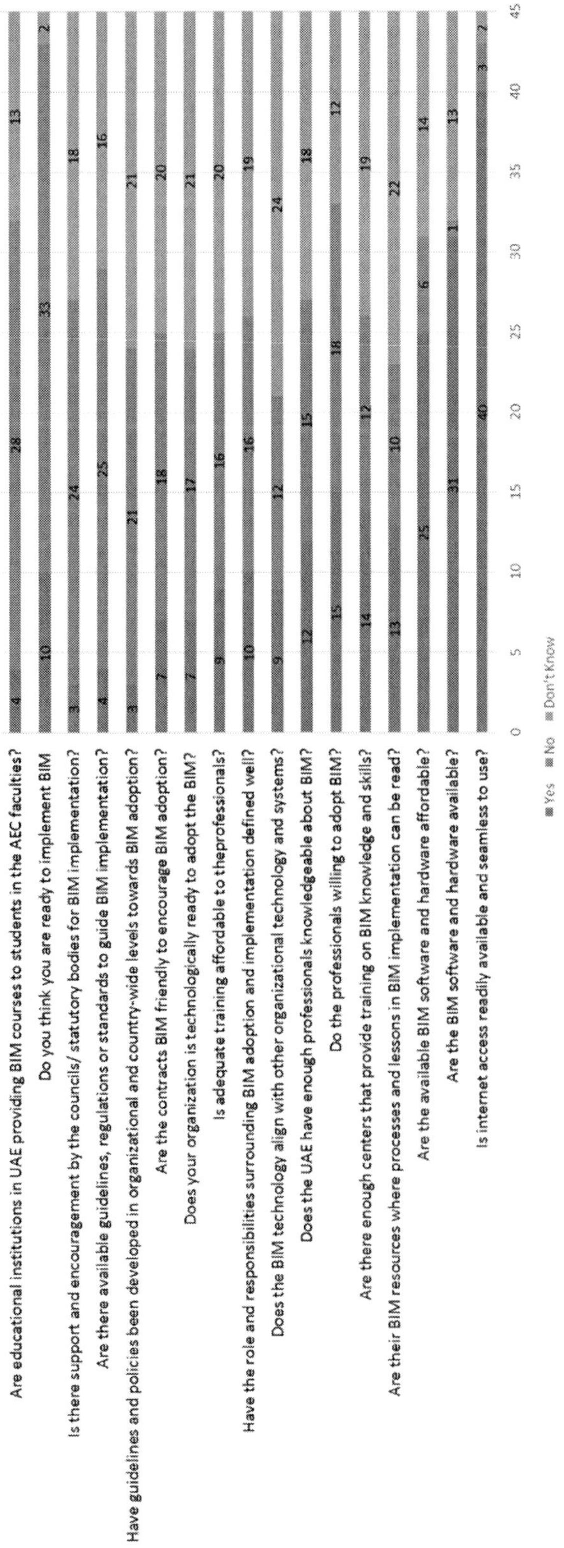

Fig.3. BIM adoption readiness in terms of people process and technology related facets.

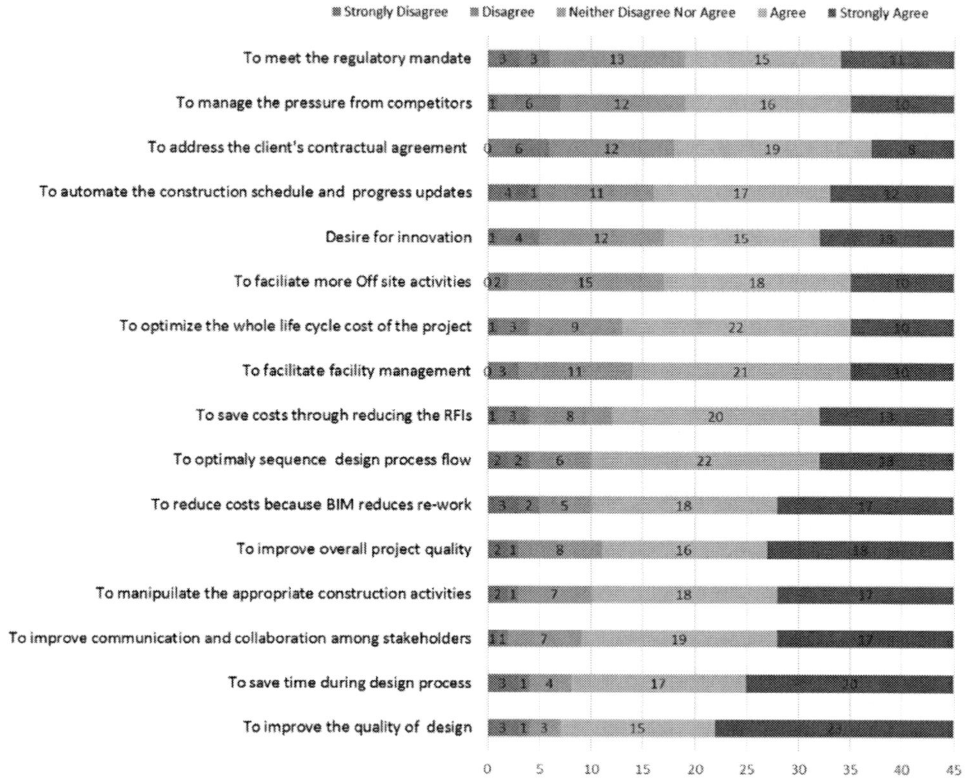

Fig.4. Perceived drivers towards BIM adoption.

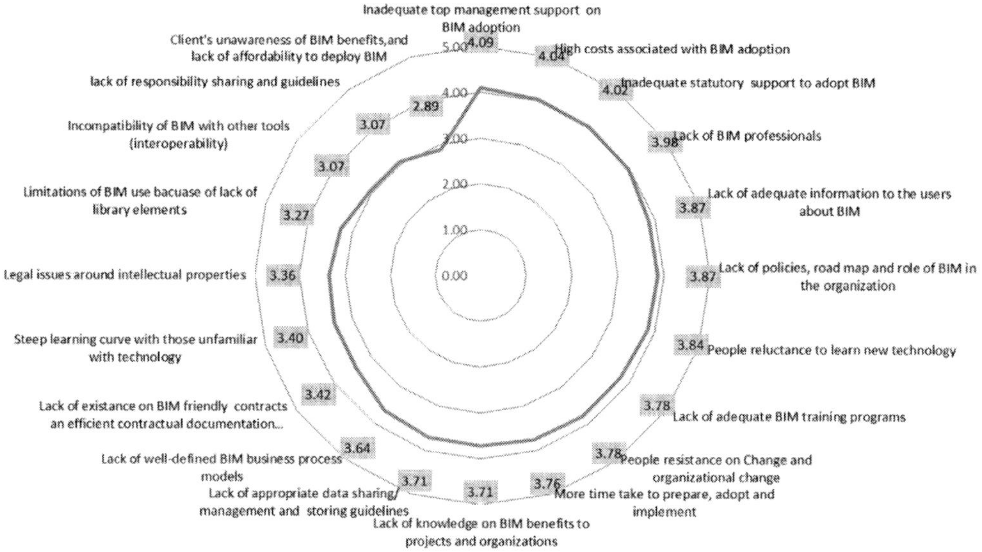

Fig.5. Perceived challenges towards BIM adoption- expressed in weighted average.

5. Discussions and conclusions

BIM is being implemented in the UAE for many years, however its implementation within the municipal projects are very low. There are initiatives such as Dubai Municipality's mandate to adopt BIM for the high capital projects. However, the adoption is still less in the projects and especially at the municipal level. It is evident from the study that the stakeholders awareness about BIM do exist however the adoption is limited. It is also evident from the survey that the perceived BIM application is mostly during the design stage. Further, it is also observed that the technological readiness exists within the study context but the people and process doesn't. To improve this end, it is necessary to have the nationwide Building Information Modeling standard and guidelines. Without this, the adoption becomes very much discrete and inconsistent in terms of its benefits realization.

Further, the respondents also perceived that the drivers which are identified in the literature is all considered and accepted to improve the BIM adoption. The top three challenges which were perceived as impediment to the BIM adoption are the inadequate top management support, higher cost associated with BIM adoption and the inadequate nationwide statutory support structure. This study results are lack with rigorous evidences because of low number of respondents, however, the findings from the study is adequate to an exploratory study. For future studies, it is recommended to have larger sample size with rigorous statistical analysis is needed to further reinforce the findings.

References

1. Fang, C., and Marle, F. Dealing with project complexity by matrix-based propagation modelling for project risk analysis.Journal of Engineering Design, **24**(4), 239–256. (2013).
2. Antoniadis, D., Edum-Fotwe, F., Thorpe, A., and McCaffer, R. Exploring Complexity in Construction Projects.1–6. (2008).
3. Azhar, S., Hein, M., and Sketo, B. "Building Information Modeling (BIM): Benefits, Risks and Challenges." Aurburn. (2008).
4. Arayici, Y., Coates, P., Koskela, L., Kagioglou, M., Usher, C., and O'Reilly, K. "BIM adoption and implementation for architectural practices." Structural Survey (2013).
5. Love, P. E. D., Edwards, D. J., and Han, S. "Bad apple theory of human error and building information modelling: A systemic model for BIM implementation." 349–354(2011a).
6. Mehran D., Exploring the Adoption of BIM in the UAE construction industry for AEC firms. Procedia Engineering 145110-1118. (2016).
7. Alabdulqader, A., Panuwatwanich, K., and Doh, J.-H. "Current use of building information modelling within Australian AEC industry." The Thirteenth East Asia-Pacific Conference on Structural Engineering and Construction (EASEC-13), C-3-1. (2013).
8. Wong, A. K. D., Wong, F. K. W., and Nadeem, A. "Attributes of Building Information Modelling Implementations in Various Countries." Architectural Engineering & Design Management, **6**(4), 288–302(2010a).
9. Cao, D., Li, H., and Wang, G. "Impacts of Isomorphic Pressures on BIM Adoption in Construction Projects." Journal of Construction Engineering and Management, **140**(12), 04014056(2014).
10. Lu, W., Peng, Y., Shen, Q., and Li, H. "Generic Model for Measuring Benefits of BIM as a Learning Tool in Construction Tasks." Journal of Construction Engineering & Management, **139**(2), 195–203(2013a).

11. Giel, B., and Issa, R. "Return on Investment Analysis of Building Information Modeling in Construction." Journal of Computing in Civil Engineering, **27**(5), 511–521. (2013).
12. Crotty, R. The impact of building information modelling: transforming construction. Routledge. (2012).
13. NBS. National BIM Report 2014. RIBA Enterprises Ltd (2014).
14. Rekola, M., Kojima, J., and Mäkeläinen, T. "Towards Integrated Design and Delivery Solutions: Pinpointed Challenges of Process Change." Architectural Engineering & Design Management, **6**(S1), 264–278(2010a).
15. Smith, P. "BIM implementation - Global strategies." 482–492(2014).
16. Davies, R., and Harty, C. "Implementing 'Site BIM': A case study of ICT innovation on a large hospital project." Automation in Construction, **30**(0), 15–24(2013).
17. Porwal, A., and Hewage, K. N. "Building Information Modeling (BIM) partnering framework for public construction projects." Automation in Construction, **31**, 204–214(2013a).
18. Jung, Y., and Joo, M. "Building information modelling (BIM) framework for practical implementation." Automation in Construction, **20**(2), 126–133(2011).
19. Atkinson, L., Amoako-Attah, J., and B-Jahromi, A. "Government's influence on the implementation of BIM." 520–527(2014).
20. Shou, W., Wang, J., Wang, X., and Chong, H. Y. "A Comparative Review of Building Information Modelling Implementation in Building and Infrastructure Industries." Archives of Computational Methods in Engineering, **22**(2), 291–308(2015).
21. Liu, Y., Zeng, N.-S., and Xu, B. "Research on the concept and framework of building information modeling ecosystem." 1799–1805. (2014)

Common practices in assessing conditions of concrete bridges

*Mohammed*Alsharqawi[1,*], *Tarek*Zayed[1], and *Saleh*Abu Dabous[2]

[1]Concordia University, Department of Building, Civil and Environmental Engineering, Montréal, Québec, Canada
[2]University of Sharjah, Department of Civil and Environmental Engineering, Sharjah, United Arab Emirates

Abstract. Bridge condition assessment is an essential step in bridge management. To ensure safety and serviceability of bridge infrastructure, accurate condition assessment is needed to provide basis for bridge Maintenance, Repair, and Replacement (MRR) decisions. In Canada and the United States, visual inspection is the common practice to evaluate a bridge condition. Meanwhile, this practice is limited to detect surface defects and external flaws. For subsurface defects, Non-Destructive Testing and Evaluation (NDT&E) technologies are being used to supplement visual inspection. This paper reviews the common practices in assessing concrete bridges' conditions and discusses the limitations of available condition assessment models. Further, this research studies six NDT&E techniques and establishes a set of selection criteria which is utilized to compare each technique in terms of providing the best inspection results. Based on the comparison, it is found that Ground Penetrating Radar (GPR) appears to be the most appropriate NDT&E techniques for inspection of concrete bridges. Thus, this paper recommends integrating GPR technology with the dominant visual inspection practice in order to establish a more accurate overall bridge condition rating system where surface and subsurface defects are assessed.

1 Introduction

Deterioration is a foremost problem in any structure during its operation. It occurs due to aging, excessive usage, environmental impacts and other factors. Bridge deterioration results in having more than 26% of Canada's bridges fall under fair, poor and very poor categories and a value of $50 billion is required to replace all of these assets [1]. The aging problem of bridge infrastructure is similar in the US as statistics show that about 28% of concrete bridge decks is either structurally deficient (SD) or functionally obsolete (FO). ASCE estimated that $20.5 billion of annual investment is needed to eliminate the country's bridge deficient backlog by 2028 [2]. The report also graded bridges' conditions as "C+" which refers to mediocre that requires attention. With a large number of aging

*Corresponding author: mohammed.alsharqawi@concordia.ca

bridges in North America and the growing problem of deterioration across the globe, condition assessment of bridges has been the subject of ongoing research.

2 Deterioration of concrete bridges

Large parts of infrastructures were constructed during the 1960's and 1970's. Accordingly, they are facing an increasingly deteriorating problem. Age is one of many factors that decreases infrastructures' robustness. Along with the effect of other factors like external environmental conditions and extreme natural hazards, infrastructure performance is reduced for users [3]. Such factors act as contributors to concrete bridge deterioration and cause vulnerability to any infrastructure. Similar to any reinforced concrete structures, bridges experience loss of integrity over time due to degradation caused by chemical (alkali–silica reaction, carbonation, corrosion, crystallization, leaching, salt and acid action), physical (temperature gradient, fatigue, overloading, shrinkage, freeze–thaw cycles, etc.) and even biological (accumulation of organic matter, living organisms, etc.) mechanisms [4]. It is important to understand different deterioration processes as each process leads to different types of defects such as cracking, scaling, spalling, concrete delamination and corrosion of reinforcing steel. It is also important to identify the basic causes of deterioration which require Maintenance, Repair, and Replacement (MRR) actions; otherwise, ineffective technique may be selected. It is evident that some mechanisms affect the reinforcement and others the concrete itself, yet all degradations mechanisms lead to structure vulnerability and thus reduce the infrastructure reliability.

Corrosion of steel reinforcement, for instance, is considered as the most leading cause of deterioration [5] and has always been viewed as the biggest problem that reasons in many consequent damages as the structure ages. Deterioration comes from the fact that rust, the final product of the electrochemical process of corrosion, occupies much larger volume than the original steel over the years. This causes internal stresses in the surrounding concrete leading to cracking. As corrosion gets more severe, those internal cracks progressively cause partial separation of concrete at the level of reinforcement known as delamination. Several delaminated areas eventually form spalling of concrete which results in structural disintegration. Cracking, delamination, and other discontinuities can occur due to overloading as well. Such defects in some cases could threaten public safety. Examples of cracking, spalling and rebar corrosion are shown in Figure 1.

3 Condition assessment of bridges

Bridge structures play a critical role in the transportation system as they serve millions of people on a daily basis. Any failure in these structures will result in both human life and economic loss. Consequently, condition assessment is performed on routine or scheduled basis to ensure public safety and prevent such catastrophic events. Bridge assessment is mainly an interpretation that identifies the appropriate Maintenance, Repair, and Replacement (MRR) activities. With the increasing number of deteriorated bridges in Canada, the US and around the globe, condition assessment techniques of concrete bridges are evolving.

3.1 Current practices and methods for condition assessment

Investigation for defects and determination of their severity level are the main objectives of bridge condition assessment. Detailed inspections are conducted regularly in order to discover serious defects and to evaluate the degree of bridge deterioration. Besides, emergency or ad hoc inspections due to specific defect are carried out after coincidences or natural disasters such as earthquakes. Concrete bridges are exposed to various forms of deteriorations. Some can be visible on the surface while others are hidden beneath it.

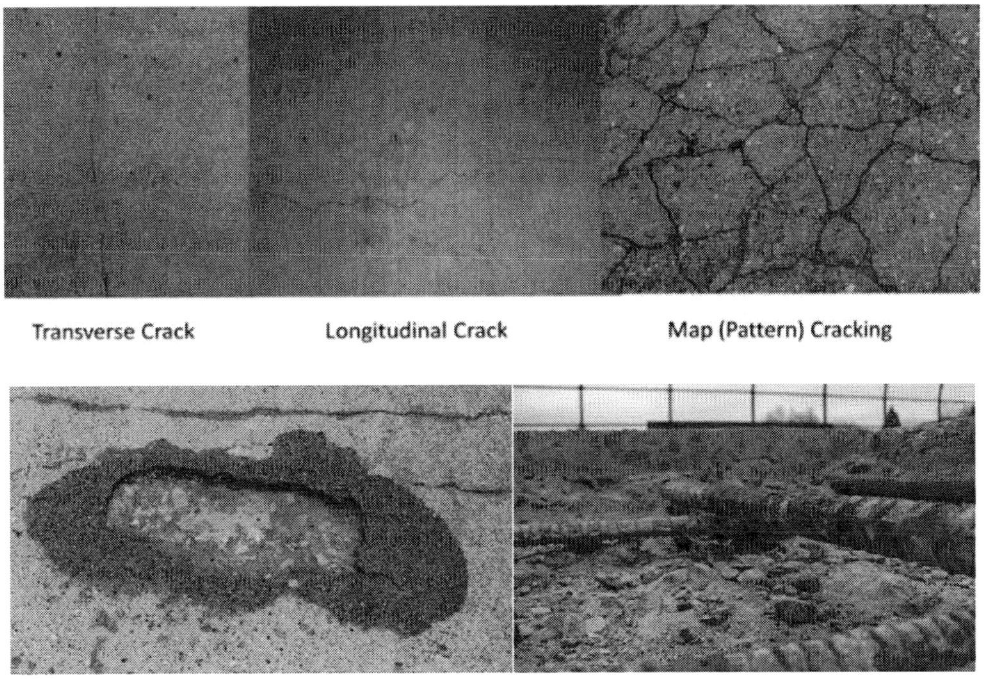

Fig. 1. (a) Classification of concrete cracks (b) Spalling [6](c) Rebar corrosion [7].

In North America, the commonly used techniques to assess bridge conditions is through visual inspection and close observation to bridge elements because it is inexpensive and requires a minimal level of experience. Visual examinations involve using specific techniques to provide valuable information on bridges physical condition. This inspection process can evaluate surface defects such as cracking and spalling. Hammer sounding and chain dragging as shown in Figure 2 are the most commonly used techniques to determine subsurface defects such as delamination where boundaries of delaminated areas within the concrete slab are measured. If severe damages are identified during the visual inspection, a more in-depth condition survey is conducted with the aid of Destructive Testing (DT) and Non-Destructive Testing and Evaluation (NDT&E) techniques. These two types of techniques are used to assess or inspect the subsurface condition of bridges as reinforcing steel corrosion. The first type is DTs; coring test is the most common test belonging to this class of inspection techniques. DTs provide direct and quick results, but they are expensive, time consuming and have a destructive nature; therefore, they cannot be applied on a regular basis. On the other hand, NDT&E techniques are less expensive; however, the results are mostly indirect and required some interpretation to get useful outputs.

In order to enhance the current inspection practices, a detailed bridge element inspection system called the "Guide Manual for Bridge Element Inspection" was published in 2011 to introduce an improvement on the widely used Commonly Recognized (CoRe) system of bridge elements. The new guide was built on the concept of element-level condition rating where in-depth assessment of bridge elements is captured. Recently, the Federal Highway Administration (FHWA) has started work on updating the widely used National Bridge Inventory (NBI) coding guide to incorporate the newly developed guide for the detailed bridge element inspection and to accurately reflect the condition and performance of highway bridges. In parallel, many transportation agencies supported the adoption and implementation of the detailed element inspection because of its ability to assist in providing sufficient Maintenance, Repair and Replacement (MRR) decisions making and for later analysis in deterioration curves and performance models.

Fig. 2. Hammer sounding and chain-drag testing [6].

3.2 Previous and related work on condition assessment

By looking at the literature, previous research works on condition assessment can be categorized into two groups. The first is based on the visual inspection where researchers try to improve the current practices for condition assessment. According to Bolar et al. [5], the condition assessment can be made more effective by determining condition of groups of elements classified based on their resilience. In his paper, he proposed a Hierarchical Evidential Reasoning (HER) framework for the condition assessment of bridge where it is classified into primary, secondary, tertiary and life safety-critical elements. One of the major advantages of using HER is that it has the capability to deal with incomplete and conflicting evidence without making strong assumption about missing data as required in other soft computing methods. Moufti et al. [8] enhanced the application of ER approach by proposing an assessment tool that can handle the prescribed uncertainties in the bridge inspection process and objectively translate the real condition of a bridge through further in-depth measurements. In his model, he hierarchy structured several levels of a concrete bridge under assessment; namely: bridge components, structural elements, and most particularly, the measured defects. Other researchers try to reduce the uncertainties associated with visual inspection and evaluation subjectivity by applying the Fuzzy Set Theory (FST). Several studies on applications of FST with condition assessment of concrete bridges have been reported. Condition evaluation of existing reinforced concrete bridges using fuzzy based analytic hierarchy approach was proposed by Sasmal and Ramanjaneyulu[9]. FST-based evaluation system for existing bridges was proposed by Chen [10] where he emphasis on the use of fuzzy sets to represent bridge condition ratings instead of precise numerical numbers. Jain and Bhattacharjee[11] applied fuzzy concepts to visual assessment of deteriorating specific distress manifestation in reinforced concrete structures. Some of the drawbacks of the above mentioned approaches are: i) the provided condition assessment depends mainly on visual inspection which can be imprecise ii) weights of bridge elements are measured by simple techniques such as Analytical Hierarchy Process (AHP) without considering interdependency and iii) some of the studies consider specific deterioration mechanisms and certain type of defects.

The other research group utilizes technologies for condition assessment. Advanced technologies have been utilized to provide information about deteriorations of bridge elements. For surface deteriorations such as cracking, a digital image processing was developed by Adhikari et al. [12] for crack quantification in such a way that it mimics the on-site visual inspections. Such developments will resolve the subjectivity problem, but still it is limited to certain type of defects. For subsurface deteriorations, a number of Non-Destructive Testing and Evaluation (NDT&E) technologies are used to evaluate subsurface conditions on bridge decks. For instance, Washer et al. [13] used Infrared thermography (IR), IR mostly used to detect voids and delamination in concrete bridge decks through measuring the radiant energy emitted from the surface. Gucunski et al. [4]used Ultrasonic Pulse-Echo (UPE) and Half-Cell Potential (HCP). UPE technology uses ultrasonic (acoustic) stress waves assess in detecting defects in concrete elements, debonding of reinforcement bars, shallow cracking, and delamination. HCP measures the potential corrosion of steel reinforcement and prestressed concrete structures. Chase [14] used Impact Echo (IE), IE detects and characterizes delamination within concrete bridge decks. Tarussov et al. [15] used Ground Penetrating Radar (GPR), GPR utilizes electromagnetic (EM) waves that have the ability to identify the subsurface defects locations for both corrosion of steel and also some concrete delamination. Table 1, summarizes the main limitations of the above mentioned technologies. In general, the main disadvantages of using NDT&E technologies are relatively the high cost and complex data interpretation in addition to detecting certain type of defects.

Table 1. Main NDT&E technologies limitations[16].

NDT&E Technology	Main Potential Limitations
Infrared thermography (IR)	-Mostly detect voids and delamination in concrete
Ultrasonic Pulse-Echo (UPE)	-Detect debonding of reinforcement, shallow cracking, and delamination
Half-Cell Potential (HCP)	-Evaluate the probable steel corrosion activity in concrete structures
Impact Echo (IE)	-Detect only delamination within bridge decks
Ground Penetrating Radar (GPR)	-Limited to steel corrosion and potential delamination

3.3 Comparison and selection of NDT&E technique

There are many NDT&E technologies that can be utilized for reinforced concrete bridge inspection. In this task, the available NDT&E technologies were studied and analysed to highlight the detection capabilities as well as investigate their limitations. The examined techniques include: Infrared Thermography, Ultrasonic, Half-Cell Potential, Impact Echo, Ground Penetrating Radar and Chain drag is added as a classical NDT&E technique. Theoretically, some or all of the techniques can be performed on the same bridge. Obviously, the bridge condition, in that case, would be better assessed as each technique will eliminate the drawbacks of the others. That solution may exist but would require more time and financial resources to inspect and interpret the results than the case if one single method is selected. In a research sponsored by the Federal Highway Administration (FHWA) in cooperation with the American Association of State Highway and Transportation Officials (AASHTO), Gucunski et al. [7] performed an independent evaluation of the capabilities and limitations of the most common NDT&E techniques to detect and characterize typical deterioration mechanisms in concrete bridge decks. Based on several performance criteria as well as field and laboratory testings, it was reported that among tested NDT&E technologies, GPR ranks first for condition assessment of concrete bridge decks.

Being aware of the fact that the best selection or combination of NDT&E technologies should be determined based on full cost-benefit analysis, however due to the time limitation, it was determined in this study that only one NDT&E technique is employed and only technical benefits are considered for its selection. These technical benefits, or selection criteria, have been identified including: (1) capable of detecting corrosion-induced defects such as (1a) corrosion and (1b) delamination; (2) does not require traffic close for bridge inspection; (3) inspection result can be reproducible; (4) usable for various bridge elements; (5) works well with asphalt overlays that is commonplace in Canada; (6) objective with minimal subjective interpretation from operators; and (7) can be used as a stand-alone technique that does not require other tests.

As shown in Table 2, consistent with the result of Gucunski et al. [7], Ground Penetrating Radar (GPR) also appears to be the most appropriate technology.

Table 2. Comparison of NDT&E techniques to selection criteria.

NDT&E Techniques	**Selection Criteria** (1a): capable of detecting corrosion (1b): capable of detecting delamination (2) : does not require traffic close (3) : inspection result can be reproducible (4) : usable for various bridge elements (5) : works well with asphalt overlays (6) : objective with minimal subjective interpretation (7) : can be used as a stand-alone technique							
	(1a)	(1b)	(2)	(3)	(4)	(5)	(6)	(7)
Infrared Thermography		x	x				x	x
Ultrasonic				x	x		x	x
Half-Cell Potential	x			x	x		x	x
Impact Echo		x		x	x	x	x	x
Ground Penetrating Radar	x	x	x	x	x	x	x	x
Chain Drag		x			x			x

4 Summary and conclusion

It has been observed from the current practices that the commonly used techniques to assess bridge conditions are mainly through visual inspection to bridge elements. Inaccurate measurements and subjective judgment are considered as a drawback of this approach. In order to improve that inspection process, many studies have been executed to account for this uncertainty and subjectivity. However, this practice is still limited to detect surface defects and external flaws. Subsurface defects are mostly measured with the aid of NDT&E techniques. Moreover, NDT&E technologies are involved in objectifying the inspection process and making it more reliable. These technologies are promising in terms of providing improvements to the traditional inspection processes; however, it is expensive and limited to certain type of defects. In this study, a selection criteria has been identified considering technical benefits of six NDT&E techniques. It is concluded that Ground Penetrating Radar (GPR) appears to be the most appropriate technology in terms of providing the best inspection results. Moreover, a current research is in progress to develop a model that integrates the dominant visual inspection practice with the selected NDT&E

technique. Hence, it is expected from this research to enhance the condition assessment of bridges by considering both surface and subsurface defects and thus establishing a more accurate condition rating.

References

1. CIRC, *Report card for Canada's infrastructure* (2016)
2. ASCE, *Report card for America's infrastructure* (2013)
3. R. Zimmerman, C. E. Restrepo, J. S. Simonoff, The age of infrastructure in a time of security and natural hazards (2009)
4. N. Gucunski, F. Romero, S. Kruschwitz, R. Feldmann, A. Abu-Hawash, M. Dunn, TRR Journal. **2201**, 34-44 (2010)
5. A. Bolar, S. Tesfamariam, R. Sadiq, StructInfrastruct E.**9**, 648-666 (2013)
6. R. Hoensheid, Evaluation of Surface Defect Detection in Reinforced Concrete Bridge Decks using Terrestrial LiDAR (2012)
7. N. Gucunski, A. Imani, F. Romero, Non-destructive testing to identify concrete bridge deck deterioration (2013)
8. S. Moufti, T. Zayed, S. Abu Dabous, TRR Journal. **2431**, 88-96 (2014)
9. S. Sasmal, K. Ramanjaneyulu, Expert Syst Appl.**35**, 1430-1443 (2008)
10. M. Chen, *CAR'09*, 269-272 (2009)
11. K. K. Jain, B. Bhattacharjee, J ConstrEngManag. **138**, 399-408 (2011)
12. R. Adhikari, O. Moselhi, A. Bagchi, Automat Constr.**39**, 80–194 (2014)
13. G. Washer, N. Bolleni and R. FenwickTRR Journal. **2201**, 27-33 (2010)
14. S. Chase, Y. Adu-Gyamfi, P. Tunuguntla, *Transportation Research Board 94th Annual Meeting*, 15 (2015)
15. A. Tarussov, M. Vandry, A. De La Haza, Constr Build Mater.**38**, 1246–1254 (2013)
16. M. Alsharqawi, S. Abu Dabous, T. Zayed, *Transportation Research Board 95th Annual Meeting*, 16-4431 (2016)

Comparison between major repair and replacement options for a bridge deck life cycle assessment: A case study

Saleh Abu Dabous[1,*], *Chaouki Ghenai*[2], *Abdallah Shanableh*[1], and *Ghadeer Al-Khayyat*[1]

[1]Departement of Civil and Environmental Engineering, University of Sharjah, Sharjah, UAE
[2]Departemnt of Sustainable and Renewable Energy Engineering, University of Sharjah, Sharjah, UAE

Abstract. Material production, manufacturing, transportation, usage, and end of lifeprocessing are usually the main contributors defining the life cycle assessment (LCA). Bridge infrastructure is important to the economy and the society. Over their life cycle, highway bridges experience several stressors that can significantly affect their structural performance and therefore require rehabilitation. This paper discusses the life cycle analysis of bridge rehabilitation decisions and demonstrates the analysis with a case study of a bridge located in Ontario, Canada. The LCA of the bridge deck is analyzed for two rehabilitation strategies: major repair and replacement. The study focuses on evaluating the different life cycle phases of the bridge deck by assessing their carbon dioxide emission, energy consumption and cost. Also, the paper presents the impact of the different elements within each phase to identify the most contributing elements. The LCA of the bridge deck is analyzed and estimated with the aid of CES EduPack 2016 software that includes a database of more than 4000 different materials and more than 200 manufacturing processes. Analysis of the case study shows that material phase causes significant life cycle impact. The study concluded that the deck replacement yields higher environmental impact and life cycle cost compared to repairing and strengthening the deck.

1 Introduction

Life cycle assessment (LCA) is a comprehensive, standardized, and internationally recognized approach used to quantify the environmental impacts in terms of resources consumption, waste production and gas emission, and any other related environmental impacts attributable to a service, asset, or a product [1, 2]. Sustainability requirements covering impacts on the environment, economy, and society can be evaluated for a certain project or product through studying its life cycle impact. Bridge management decisions, including maintenance, repair and replacement (MR&R) have direct implications on the environment, the society and the economy. Material production, transportation,

*Saleh Abu Dabous: sabudabous@sharjah.ac.ae

construction operations, maintenance and repair, and end of life phases are usually considered and assessed in bridge life cycle analysis. All the life cycle phases associated with the bridge should be identified and assessed starting from the material production phase and ending with end of life phase. The material production is generally causing the most impacts, and maintenance and repair phase of the asset comes in the second level [3]. The production of cement and the sourcing of aggregates, as the main constituent martials of concrete, have significant impact on the environment mainly in terms of the non-renewable natural resources, energy consumption and carbon dioxide emissions [4]. Every tonne of cement production releases 1.0 – 1.2 tonne of CO_2 into the environment by the time the material is put in place [5].

This research attempts to perform life cycle assessment of a bridge deck in terms of environmental impacts and cost while considering all the different phases of the deck life cycle. The LCA will be performed for bridge major repair and replacement rehabilitation options. The study aims to demonstrate the environmental impacts and cost of the different phases and attempts to address the following questions: 1) identifying the occurrence of these environmental impacts within the bridge life cycle, 2) what are the most dominant phases in terms of environmental impacts and cost throughout the bridge life cycle, and 3) what are the effects of alternative recycling options on environmental impacts. The LCA of the bridge will be demonstrated through CES EduPack 2016 software with databases that include more than 4000 different materials and more than 200 manufacturing processes. The software can analyze the LCA of a bridge deck including both cost and environmental impacts during material, manufacture, transport, use, and end of life phases.

2 Proposed assessment framework of bridge deck

The bridge life cycle starts with material production and ends with the bridge demolition. Over the bridge life cycle, MR&R decisions contribute significantly to the total impact on the environment, the society and the economy. Steele et al. (2003) reported that construction material of a bridge has significant impact on the LCA and reducing the amount of material used can significantly reduce the impact on the environment, but this target should not compromise the durability and longevity of the structure. In addition, sufficient maintenance strategies could prevent deterioration and extend the structural life. Most of studies show that bridge replacement represents higher environmental impacts than refurbishment and strengthening of the bridge [6]. The proposed framework in this paper focuses on the impact assessment of major repair and replacement decisions. Figure 1 below shows a flowchart for the rehabilitation strategies and their potential impacts to be considered in this research.

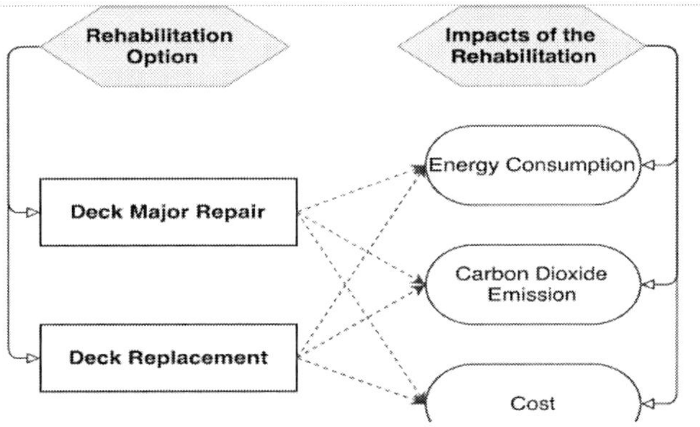

Fig. 1. Flow chart for the rehabilitation options and impacts of each.

Manufacturing of steel and concrete, which are the predominant materials for bridges construction, contribute the most to environmental impacts over the structure life cycle [6]. The bridge superstructure components include the bridge deck and its overlay and accessories, the girders, and the expansion joints. These elements require large amount of raw materials and contribute most to the environmental impacts during the material production phase. In addition, construction equipment and machineries such as earthwork excavators, compactors, cranes and many others used during construction phase contribute to the environmental impacts through energy consumption and emissions [1], [2]. The bridge construction phase accounts relatively for lower environmental impacts compared to the material production phase. It should be noted that the operation phase contributes significantly to the impacts on the environment. On the other hand, usually the transportation of construction material and equipment during the bridge different life cycle phases are of minor importance and of limited impacts on the environment [7]. The assessment of the end of life phase is associated uncertainties due to the dependency on future technologies that could take place after 50 to 100 years of a bridge service life. The environmental impacts of this stage involve the energy consumption through several activities which are mainly the demolition of the structure, sorting, transportation, and a final treatment of the generated wastes that may include reusing, recycling, or placing in landfills [1, 2]. Within the framework of the proposed analysis in this paper, the different life phases of the bridge deck and their impacts are assessed with respect to energy consumption, CO2 emission and direct cost as displayed in Figure 2. The definitions of the different phases are provided in the following sections.

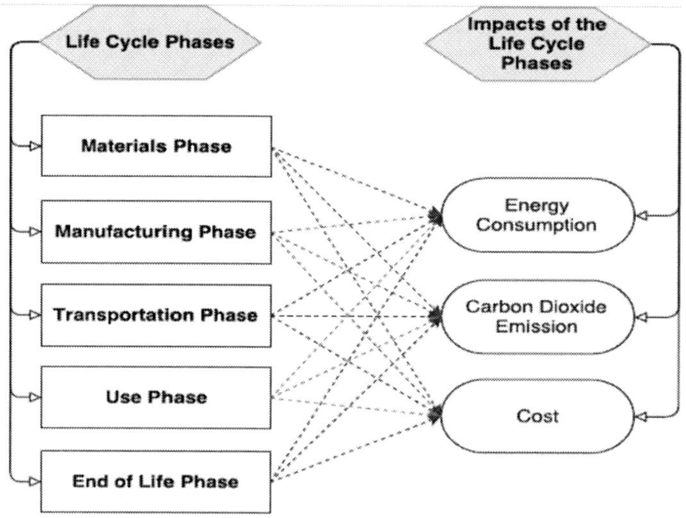

Fig. 2. Flowchart of the dzifferent life cycle phases and their impacts.

2.1 Materials phase

The material phase is considered as the most contributing phase in terms of impact on the environment. The selection of the material type has significant impact on the processes, including the consumption quantity, maintenance activities, and end of life consequences [8]. This phase captures the extraction and production of the raw materials from their original shape and source to be ready for the next phase which is the manufacturing process. Usually material production phase counts for the highest impacts on the environment [3, 7, 8].

2.2 Manufacturing and transportation phases

The manufacturing phase accounts for the different material manufacture processes. This phase starts with extracting the raw material and ends with transporting the ready products to the construction site. Transportation is the third phase in bridge life cycle assessment. This phase includes the energy consumed, carbon dioxide emitted and cost utilized during the transportation of the final products from the manufacturing plant to the construction site.

2.3 Use phase

The use phase is considered as the longest phase in bridge's life cycle and typically contains operation, repair, and routine maintenance activities [7, 8]. Different contributors are impacting the bridge structure during the use phase. Among these, traffic is the main contributor to the environmental impact. The impacts from the traffic are not included in the analysis herein due to two main reasons. First, the CO_2 emissions and the energy consumption due to the traffic will be extremely high compared to the impacts of the other phases during the expected life period of the bridge deck (25 years). However, this impact is equivalent for all of the considered rehabilitation strategies, given that the bridge serves the same number of vehicles during this phase. As a result, including the traffic impact from all the alternatives or excluding it will not affect the final results obtained from the assessment. The second reason is that the traffic is not an element of the bridge structure and therefore quantifying its environmental impacts needs to be separated from the maintenance management decision making. The analysis excludes the impact from traffic so its significant impact does not undermine the importance of the impact attained from the main elements of the bridge structure.

2.4 End of life phase

The end of life phase includes several events starting with bridge demolition, waste sorting, material recycling, and ending up with final landfill. Usually the EOL phase generate treated and recycled materials that result in benefiting the environment. To quantify the benefits out of the EOL phase for a certain project, they should be measured from the next project where the recycled material is used[8].

A detailed representation of the environmental impacts due to the deck major repair through the different life phases is shown in Figure 3 and Table 3 below. In addition, Table 4 shows the components that most contribute to the energy consumption, CO_2, and cost in the different life phases of the bridge deck major repair.

3 Bridge deck LCA: a case study

The bridge considered for this study is located in Ontario, Canada and was constructed in 1965. The deck is a three-span solid post tensioned concrete slab with a thickness of 381 mm. The five different phases considered in this study for the bridge deck life cycle assessment are: the materials phase, the manufacturing phase, the transportation phase, the use phase, and end of life phase are. The materials and their respective quantities included in this study were taken from a bill of quantity provided by Ontario department of transportation for the mentioned bridge above. The different material types were selected using the material database within the CES EduPack software. The thickness of the different materials was reasonably assumed and then the masses were calculated accordingly. The analysis was performed per unit area of the bridge deck.

To analyze the environmental impacts due to bridge's deck major repair and replacement events through CES EduPack software, three main entities need to be specified. First entity is the material,manufacture, and end of life phases. The bill of quantity of the material is an input to the software with each line representing an individual component. Quantity, component name, material, recycled content, mass, primary and secondary processes, percentage removed, and end of life are the main entry fields for the material, manufacture, and end of life phases. Second is the transport filed which accounts for the transportation of the finished products from the place of manufacturing to the construction site. The third and last entity for the analysis of the environmental impacts of the bridge deck rehabilitation event is the use filed. This field includes the product life, country electricity mix, and modes of use (static and/or mobile modes).

Once the needed information are entered into the software, the relative contributions of the different phases on the environment in terms of the energy consumption, carbon dioxide emission, and cost can be assessed. Also, the contribution of the diffident elements within each phase are compared to identify the elements with the highest impacts on the environment. The final results are demonstrated with figures and bar charts for both rehabilitation options: the major repair and the replacement of the deck. Table 1 and 2 below show the deck components, masses, primary processes, secondary processes, EOL scenarios and percentages recovered for deck major repair and replacement options respectively.

Table 1. Related information for the bridge deck major repair.

Item #	Item Description	Mass (Kg)	Primary Process	Secondary Process	EOL	% Recovered
1	Removal of asphalt wearing surface	250.0	Incl. in material value	Cutting and trimming	Landfill	0.0
2	Removal railing	2859.6	Forging		Recycle	25.0
3	Removal of concrete curbs	914.4	Incl. in material value		Landfill	0.0
4	Type A removals from top of deck (determinations only)	360.0	Incl. in material value		Landfill	0.0
5	Type B removal from deck soffit	240.0	Incl. in material value		Landfill	0.0
6	Type C removals from fascia	240.0	Incl. in material value		Landfill	0.0
7	Type C removal from deck ends (tendon anchors)	74.4	Incl. in material value		Landfill	0.0
8	Removal of existing approach slab	540.0	Incl. in material value		Recycle	5.0
9	Granular backfill	2015.0	Incl. in material value		Landfill	0.0
10	Scarify deck surface	19.2	Incl. in material value		Landfill	0.0
11	Cathodic protection	14.4	Incl. in material value		Landfill	0.0
12	Abrasive blast cleaning of rebar	1.9	Forging		Landfill	0.0
13	Abrasive blast cleaning for overlays	0.7	Incl. in material value		Landfill	0.0
14	Concrete overlay (includes padding for new crowns)	600.0	Incl. in material value		Landfill	0.0
15	Finish and cure overlay	144.0	Incl. in material value		Landfill	0.0

16	Concrete barrier wall	960.0	Incl. in material value		Landfill	0.0
17	Concrete in new deck extensions	914.4	Incl. in material value		Recycle	5.0
18	Concrete in approach slabs	540.0	Incl. in material value		Landfill	0.0
19	Stainless steel rebar (barrier wall & deck extensions)	95.6	Forging		Recycle	20.0
20	Coated rebar for overlay padding area	141.3	Forging		Recycle	15.0
21	Rebars in deck extensions	143.5	Forging		Recycle	15.0
22	Rebars in approach slabs	143.5	Forging		Recycle	15.0
23	Abutment repairs	1464.0	Incl. in material value		Recycle	5.0
24	Deck soffit repairs	360.0	Incl. in material value		Landfill	0.0
25	Bridge deck waterproofing	6.0	Polymer molding		Landfill	0.0
26	Asphalt	232.5	Incl. in material value		Recycle	15.0

Table 2. Related information for the bridge deck replacement.

Item #	Item Description	Mass (Kg)	Primary Process	Secondary Process	EOL	% Recovered
1	Removal of asphalt wearing surface	250.0	Incl. in material value	Cutting and trimming	Landfill	0.0
2	Removal railing	2859.6	Forging	Cutting and trimming	Recycle	25.0
3	Removal of existing decks including curbs	914.4	Incl. in material value	Cutting and trimming	Landfill	0.0
4	Removal of top of pier	600.0	Incl. in material value	Cutting and trimming	Recycle	10.0
5	Removal of existing approach slab	540.0	Incl. in material value	Cutting and trimming	Landfill	0.0
6	Granular backfill	2015.0	Incl. in material value	Grinding	Landfill	0.0
7	Concrete in new top of existing piers	1096.8	Incl. in material value	Cutting and trimming	Landfill	0.0
8	Prestressed members (fabrication and erection)	600.0	Incl. in material value	Cutting and trimming	Landfill	0.0
9	Concrete in barrier wall	960.0	Incl. in material value	Cutting and trimming	Landfill	0.0
10	Concrete in deck (150 mm topping)	360.0	Incl. in material value	Cutting and trimming	Recycle	5.0
11	Concrete in new deck extensions	914.4	Incl. in material value	Cutting and trimming	Recycle	5.0
12	Concrete in approach slabs	540.0	Incl. in material value	Cutting and trimming	Recycle	5.0
13	Stainless steel rebar in barrier wall	95.6	Forging	Cutting and trimming	Recycle	20.0
14	Coated rebars in deck topping	141.3	Forging	Cutting and trimming	Recycle	15.0
15	Rebars in deck extensions	143.5	Forging	Cutting and trimming	Recycle	15.0
16	Rebars in approach slabs	143.5	Forging	Cutting and trimming	Recycle	15.0
17	Bearings	425.5	Forging	Cutting and trimming	Recycle	20.0
18	Abutment repairs	1464.0	Incl. in material value	Cutting and trimming	Recycle	5.0

19	Bridge deck waterproofing	6.0	Polymer molding	Cutting and trimming	Landfill	0.0
20	Asphalt	232.5	Incl. in material value	Cutting and trimming	Recycle	15.0

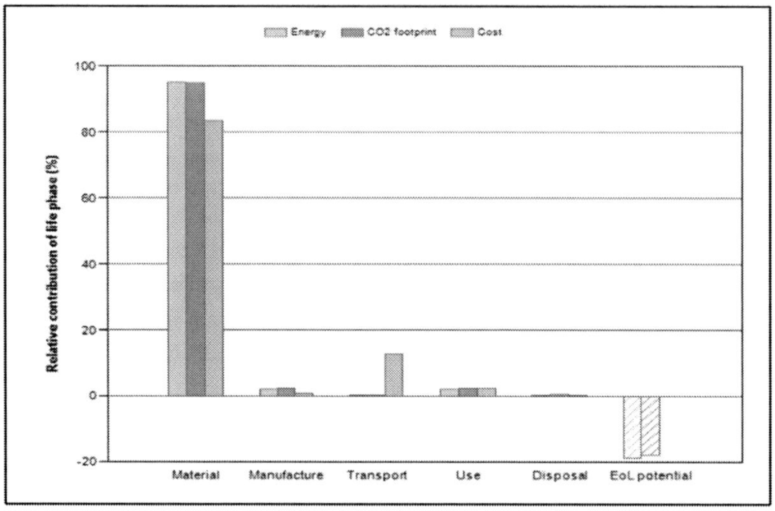

Fig. 3. Relative contribution of the different phases in terms of energy, CO2, and cost for deck major repair.

Table 3. Energy consumption, CO2 emission, and cost of the different phases for bridge deck major repair.

Phase	Energy (J)	Energy (%)	CO2 footprint (kg)	CO2 footprint (%)	Cost (USD)	Cost (%)
Material	$6.2x10^{11}$	95.1	41826.5	94.7	15823.1	83.4
Manufacture	$1.4x10^{10}$	2.2	1056.0	2.4	165.1	0.9
Transport	$1.8x10^{8}$	0.0	13.0	0.0	2415.7	12.7
Use	$1.5x10^{10}$	2.3	1060.3	2.4	483.4	2.5
Disposal	$3.1x10^{9}$	0.5	220.1	0.5	85.2	0.4
Total (for first life)	$6.5x10^{11}$	100	44175.9	100	18972.4	100
End of life potential	$-1.2x10^{11}$		-7866.6			

Table 4. Deck components with maximum contribution in the different life phases for deck major repair.

Phase	Deck component with max. contribution	Energy (J)	Energy (%)	CO2 footprint (kg)	CO2 footprint (%)	Cost (USD)	Cost (%)
Material	Removal of railing	$5.7x10^{11}$	91.6	$3.8x10^{4}$	89.8	$1.2x10^{4}$	73.2
Manufacture	Removal of railing	$1.3x10^{10}$	88.9	939.2	88.9	74.4	45.1
Transport	Granular backfill	$2.8x10^{7}$	15.2	2.0	15.2	366.7	15.2
Use (mobile mode)	Abutment repairs	$1.7x10^{9}$	11.0	116.9	11.0	53.3	11.0
Disposal	Removal of railing	$9.3x10^{8}$	29.6	65.1	29.6	18.4	21.5
End of life potential	Removal of railing	$-1.2x10^{11}$	95.6	$-7.5x10^{3}$	95.1	-	-

Figure 4 and Table 5 below present the environmental impacts due to the deck replacement through the different life phases Also, Table 6 shows the components that most contribute to the energy consumption, the CO_2 emission and cost in the different life phases of the bridge deck replacement.

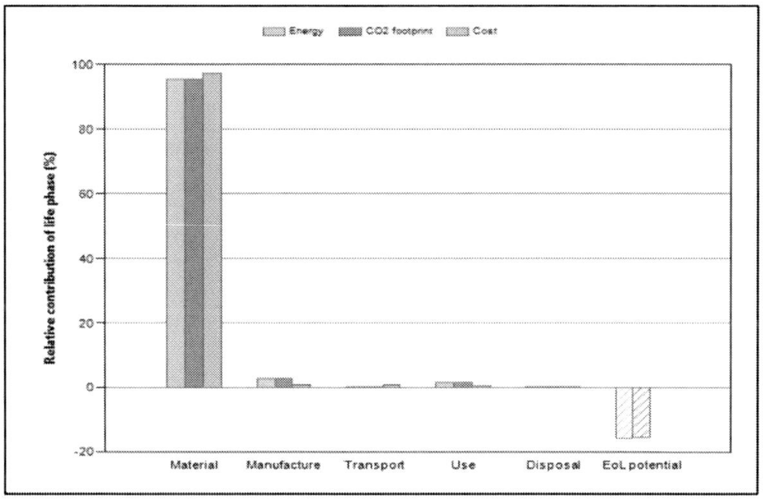

Fig. 4. Relative contribution of the different phases in terms of energy, CO2, and cost for deck replacement.

Table 5. Energy consumption, CO2 emission, and cost of the different phases for bridge deck replacement.

Phase	Energy (J)	Energy (%)	CO2 footprint (kg)	CO2 footprint (%)	Cost (USD)	Cost (%)
Material	3.6×10^{12}	95.4	254659.5	95.3	286753.9	97.5
Manufacture	1.0×10^{11}	2.7	7595.2	2.8	2962.3	1.0
Transport	6.8×10^{8}	0.0	48.6	0.0	2415.7	0.8
Use	5.6×10^{10}	1.5	3963.3	1.5	1806.9	0.6
Disposal	1.4×10^{10}	0.4	981.9	0.4	318.6	0.1
Total (for first life)	3.8×10^{12}	100	267248.4	100	294257.4	100
End of life potential	-5.9×10^{11}		-40644.5			

Table 6. Deck components with maximum contribution in the different life phases for deck replacement.

Phase	Deck component with max. contribution	Energy (J)	Energy (%)	CO2 footprint (kg)	CO2 footprint (%)	Cost (USD)	Cost (%)
Material	Bearings	3.0×10^{12}	82.7	3.8×10^{4}	89.8	2.7×10^{5}	94.5
Manufacture	Bearings	8.7×10^{10}	86.1	6.5×10^{3}	86.1	2.8×10^{3}	94.6
Transport	Bearings	4.9×10^{8}	72.0	35.0	72.0	1.7×10^{3}	72.0
Use (mobile mode)	Bearings	4.0×10^{10}	72.0	2.9×10^{3}	72.0	1.3×10^{3}	72.0
Disposal	Bearings	1.1×10^{10}	76.4	750.0	76.4	230.0	72.0
End of life potential	Bearings	- 4.7×10^{11}	79.1	-3.3×10^{4}	80.6	-	-

Table 7 shows the total energy consumption, total carbon dioxide emission, and total cost for deck major repair and replacement options. Also, Figures 5 – 7 show a graphical comparison for the environmental impacts caused by both rehabilitation options in the different life phases.

Table 7. Energy consumption, carbon dioxide emission, and cost per unit area per year for both rehabilitation options.

Rehabilitation option	Energy consumption (J/year)	CO_2 emission (Kg/year)	Cost (USD/year)
Deck major repair	$2.6x10^{10}$	1767.0	758.9
Deck replacement	$1.5x10^{11}$	10689.9	11770.3

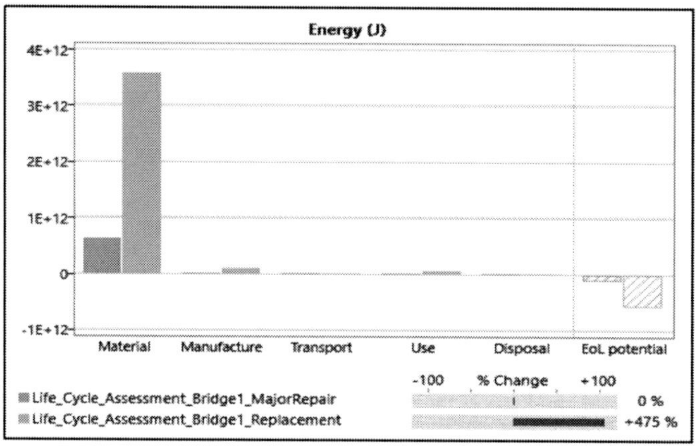

Fig. 5. Comparison of energy consumption for major repair and replacement options.

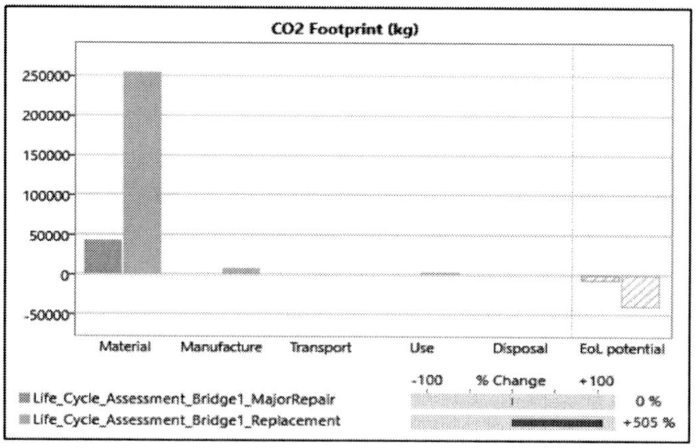

Fig. 6. Comparison of CO2 emission for major repair and replacement options.

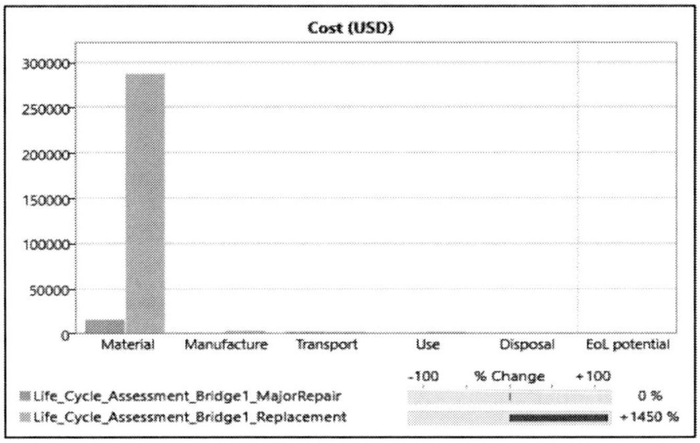

Fig. 7. Comparison of cost for major repair and replacement options.

4 Discussion of the results

A life cycle analysis was performed in this study to assess the performance of two rehabilitation strategies related to concrete bridge decks: major repair and replacement. The goal of the life cycle analysis was to determine the total energy, CO_2 emissions and the cost over the life (25 years) of the system. The total energy (J), the CO_2 emissions (kg) and the cost ($) of the main phases are related to material, manufacturing, transportation, use, and the end of life potential (Ghenai, 2012). These elements were determined using ECO Audit CES Edu pack software from Granta Design. The results of the life cycle analysis shown in Figures 3and 4 show the relative contributions of life phases (materials, manufacturing, transportation, use and end of life). The results of LCA for the bridge with the deck main repair (See Fig. 3) show that the main phase consuming more energy (95.1%), producing more carbon dioxide CO_2 (94.7%), and with the high cost (83.4%) is the material phase. The results of the Life cycle analysis for the bridge with the deck replacement (See Fig. 4) show that the material phase is the main phase that is consuming more energy (95.4%), producing more CO_2 emission (95.3%) and with the high cost (97.5%). A comparison of the total energy consumption, total carbon dioxide emission, and total cost for deck major repair and replacement options are shown in Table 7 and Figures 5-7. The total energy used and the carbon dioxide released to the atmosphere for the replacement option are respectively 5.76 and 6.05 times the one used for the bridge with the deck with major repair.

5 Conclusion

A life cycle analysis was performed in this study for the bridge deck using two rehabilitation strategies: major repair and replacement. The life cycle analysis results show clearly the benefits of using deck major repair compared to the deck replacement with respect to the life energy usage, CO_2 emisions and cost. The bridge deck replacement used 5.76 more energy and produces 6.05 more carbon dixide compared to the deck major repair. The second option with the bridge deck major repair is more sustainable, environmentally friendly and economically viable. Further future work is needed to assess the the major repair rehabilitation strategy with different materials.

References

1. G. J. Treloar, P. E. D. Love, O. O. Faniran, and U. Iyer-Raniga, "A hybrid life cycle assessment method for construction N OTE A hybrid life cycle assessment method for construction," no. January 2014, pp. 37–41, (2010).
2. G. Du, M. Safi, and L. Pettersson, "Life cycle assessment as a decision support tool for bridge procurement : environmental impact comparison among five bridge designs," pp. 1948–1964, (2014).
3. T. C. E. Dequidt, "Life Cycle Assessment of a Norwegian Bridge," (2012).
4. A. J. Martin, "Concrete bridges in sustainable development," 2004, no. December, pp. 219–230.
5. R. N. Swamy, "Holistic design : key to sustainability in concrete construction,", **4**, pp. 371–379. (2001)
6. K. Steele, G. Cole, G. Parke, B. Clarke, and J. Harding, "Highway bridges and environment — sustainable perspectives,", no. November, pp. 176–182. (2003)
7. J. Hammervold, M. Reenaas, and H. Brattebø, "Environmental Life Cycle Assessment of Bridges," no. x, pp. 153–161, 2013.
8. G. Du and R. Karoumi, "Structure and Infrastructure Engineering : Maintenance , Management , Life-Cycle Design and Performance Life cycle assessment framework for railway bridges : literature survey and critical issues," no. October, pp. 37–41, (2014).

DOI: 10.1051/matecconf/201712002018

Analysis of reasons for cost and time overrun in Indian megaprojects

Jay Mevada[1], and *Ganesh* Devkar[2*]

[1]Post graduate student in Construction Engineering and Management, Faculty of Technology, CEPT University, 380009 Ahmedabad, India
[2]Associate Professor, Faculty of Technology, CEPT University, 380009 Ahmedabad, India

Abstract. Megaprojects are becoming increasingly common in developing countries. In India, the need for and trend of rapid infrastructure development has resulted in design and implementation of projects of large scope and size, in terms of complexity and costs. As a result, megaprojects are being implemented in almost all infrastructure sectors in the country. The implementation of a megaproject attracts a lot of attention among the policy makers and researchers owing to their enormous economic, social and environmental impacts. The performance of megaprojects in terms of two key parameters – cost and time - is closely monitored by the Government of India, and this has resulted in the publication a set of reports to disclose information on project progress. Althoughquantitative analyses of the information available in these reports have been performed by the researchers, very few qualitative analysis reports are available. In this context, a qualitative research study has been undertaken in this work, to analyse the reasons for time and cost overrun in megaprojects. The analysis furthers the knowledge on megaprojects implementation by providing inference on not only key reasons for time and cost overruns but also factors and trajectories leading to the manifestation of a particular cause. The analysis indicates that delay in land acquisition, capability of contractors and contractual complexity were the major causes of time and cost overruns and these issues traverse through different layers – policy, institutional and project associated with megaprojects. The study findings are useful to policy makers to design interventions at these three layers for addressing the cause of time and cost overruns, and ensuring effective implementation of megaprojects.

1 Introduction

Infrastructure projects form the core of policy making in any country. Although priorities across different sectors, modes of financing and contractual mechanisms differ from country to country, there is an underlying and undeniable trend that has been spreading across the world –the emergence of "mega" infrastructure projects. Flyvbjerg (2015) mentions a figure of US$6 to US$9 trillion annually or 8% of the total global gross

*ganesh.devkar@cept.ac.in

domestic product (GDP)towards megaproject spending [7]. These megaprojects span different sectors including transport, airport, oil and gas, road and so on. The megaproject phenomenon is expanding into sectors such as health, education and information technology.

The conceptualization of megaprojects stretches beyond the human bounded rationality owing to far reaching economic, social, technical and policy implications resulting from them. However, the human mind perceives the concept – "mega" in terms of cost, scale, time, complexity, and money. All these concepts create a challenge to megaproject managers, owing to poor track record and history of megaprojects [3].

There are numerous policy documents, audit reports, research reports and practice notes that describe and dissect the implementation of megaprojects. These documents typically focus on two interrelated spectra,viz. the institutional mechanisms that can be created to manage the complexity, risks, finance, stakeholders and uncertainties associated with megaprojects, and the effect of these mechanisms on performance parameters like time, cost, quality, and safety. Among these performance parameters, time and cost are two key parameters that are closely monitored by the clients/ implementing agencies.

The rationale behind monitoring has been huge cost and time overruns experienced in megaprojects world over. Most megaprojects are implemented using public money. Owing to legitimacy, transparency and accountability associated with the use of these funds, the implementing agencies/audit agencies/monitoring agencies have the obligation of informing the general public about the performance of these projects. These information disclosures have been used by researchers as secondary data in a case study of megaprojects as well as archival data analysis [12].

In India, megaprojects are now increasingly being implemented and the quantum of investments in mega projects in recent years has been impressive. The Ministry of Statistics and Project Implementation (MoSPI)has been monitoring the implementation of megaprojects funded by the Central government. It periodically publishes reports providing information on the status of megaprojects on the parameters of time and cost, and brief information on reasons for overruns, if any. A literature review of Indian megaproject shows that the performance of these projects has been investigated with the case study analysis method[26,33].However, there are have not been any qualitative analysis of performance reports published by MoSPI. There are two studies that have been carried out by archival analysis; one of them does not distinguish megaprojects as a part of analysis while the other focuses on all types of projects executed with PPP mechanism, in the road sector [11]. This research study addresses this gap by conducting qualitative analysis of data published by the MoSPI pertaining to megaprojects. International literature on megaprojects includes studies using qualitative research strategies like a case study, interviews, and grounded theory[5]. However, there exists no study as yet, that has used qualitative analysis of archival data. In this context, a research study is undertaken to conduct qualitative analysis of performance reports on Indian megaprojects.

2 Megaprojects: Indian scenario

The infrastructure development has been accorded the top most priority status by the Indian government in the post liberalization era. This has resulted inmassive allocation of resources towards infrastructure development and "megaproject" has become a popular project delivery mechanism to channelize these resources for economic and social good. The Indian infrastructure sector is divided into three levels: central, state and local. Although there is a wide variety of grey literature like online magazines, newspapers, online websites and blogs that discuss different facets of megaprojects, there are no sources of precise information and there is a paucity of databases on megaprojects covering state and local level megaprojects. The megaprojects falling under the purview of the Central government are monitored by the "Infrastructure and Project Monitoring Division" of "Programme Implementation Wing" of MoSPI, and this division deals with projects that cost US$ 22.54 million or more, in terms of the parameters of time and cost overrun since inception of the project.

Based on cost, projects are categorized as two types: 1) major projects are those that cost US$ 22.54 million and above but below US$150.34 million and 2) megaprojects are those that cost US$ 150.34 million and above. The website of MoSPI contains precise information about the rationale for and date of categorization of various megaprojects. It can be seen that the monitoring of megaprojects commenced in May 2009 with publication of the 1st Report on Megaprojects in June 2009. The monthly or flash reports on megaprojects are available in the MoSPI portal; the one on January 2012 constitutes the 43rd Report on Megaprojects. Although, the 1st Megaproject flash report was published in June 2009, the subsequent reports from May 2009 to December 2011are not available in the database of MoSPI for accurate trend analysis of the number of megaprojects. The trend can however be inferred from the earliest available report – January 2012. The frequency of reports seen in the portal shows the importance afforded by the Central Government to "megaprojects" as an emerging project delivery mechanism and its close monitoring.

Figure 1 shows that in the year 2012, the number of megaprojects increased by 1.6 times from June 2009, i.e. 179 megaprojects (January 2012)compared to 112 megaprojects in June 2009. From January 2012 to September 2016, the number of megaprojects has increased 1.8 folds, i.e. from 179 (Jan 2012 to Sep 2016) to 337.

The academic literature on Indian megaprojects is very limited, although, the global research community, as well as Indian researchers, have initiated steps towards "megaprojects" research in recent years. Most of the research studies focus on urban megaprojects covering sectors such as metro, river front developments, and urban investments corridor. Siemiatycki analyzed the physical and societal implications of metro project in Delhi and concluded that the project failed to meet the transportation needs of the city [14]. The Information Technology (IT) corridor in Chennai,a prominent urban megaproject, was examined by Vijayabaskar & Babu for understanding employment creation, mobility patterns and quality of urban development [10]. Follmann conducted case analysis of two large scale river front development projects namely the Akshardham Temple Complex and the Commonwealth Games Village, and contrasted them with the slum demolitions [15]. The pressures created by megaprojects on land and environment have been discussed in a couple of papers. Raghuram stressed on technology adoption for bringing transparency and fast tracking the process of environmental impact assessment and land acquisition [26]. Pomeranz, Bandoyadhay and Perveen assessed the mega watershed projects in the Himalayas, India and river-interlinking project from the viewpoint of environmental politics and efficient use of water resources respectively [13].

Alongside the academic literature, there is ever-growing information on megaprojects in secondary data sources like newspapers, trade magazines, online construction and infrastructure industry portals, online discussion forums and social media. This discourse broadly focuses on following: 1) megaprojects investments and scope of megaprojects, 2) funding, governance and reasons behind megaprojects, 3) large scale displacements of people caused by megaprojects, 4) policy and institutional hurdles in ultra-mega power plants. Some of the megaprojects that are often referred in these platforms are Bandra-Worli Sea Link, Mumbai Metro Rail, Delhi Metro Rail, Ahmedabad Metro Rail, Gujarat International Finance-Tec City, Delhi-Mumbai Infrastructure Corridor and Navi Mumbai International Airport.

The academic and practitioner literature on megaprojects in India often whirls around the important issue of overruns in the areas of time and cost. This is not surprising, considering the global evidence on megaprojects indicating huge time and cost overruns.

MoSPIclosely monitors two parameters, i.e. cost overrun and time overrun with Indian megaprojects. Figure 1 shows the increase in number of megaprojects over the years as well as delays and cost overruns in megaprojects.

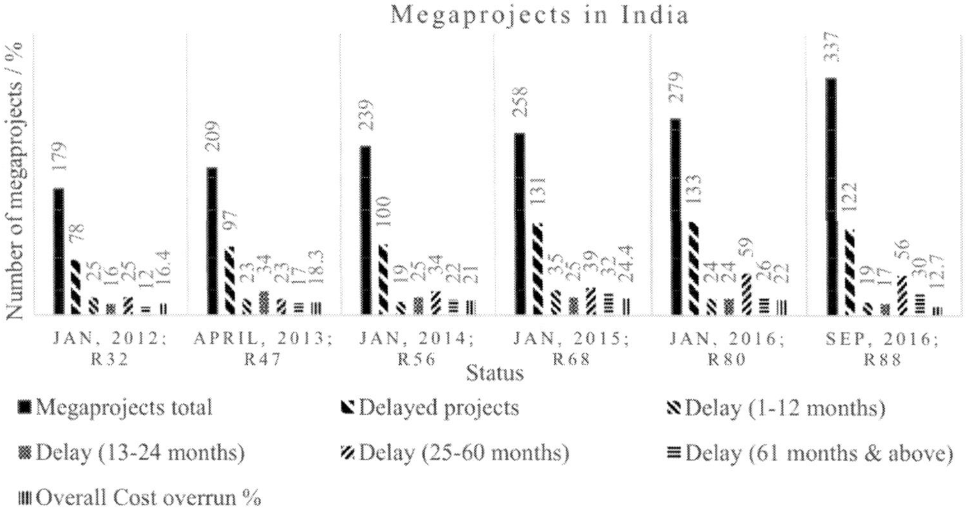

Figure 1. Megaprojects trend in India

3 Megaprojects: Complexity and Overrun

Complexity is one of distinguishing feature of megaprojects, megaproject managers and implementing agencies have been facing challenges in managing this complexity. This naturally leads to a call for focus on complexity management among academic and practitioner community in the domain of megaprojects. Hu Yi performed bibliographic analysis of construction engineering and management research to understand the status and the trends in megaproject research [5]. The analysis indicated a move towards construction megaproject management to complex project management. There is variation in the conceptualization of "complexity" associated with the megaprojects. The paper by Barrarini on concept of project complexity is one the earliest papers, which reviews literature on project complexity in the context of project management [28]. The paper reviewed different definitions of complexity and proposed the definition as "consisting of many varied interrelated parts". Further, complexities were categorized as organizational and technological. The organizational complexity refers to depth of organizational hierarchical structure (vertical differentiation) and number of formal organizational units - departments and groups, and the division of tasks (horizontal differentiation). The technological complexity refers to variety or diversity of some aspect of a task and interdependencies between tasks, teams, technologies and inputs.

Brockmann & Grimscheild defined complexity as the degree of manifoldness, interrelatedness and consequential impact of a decision field. Overall project complexity comprises three components: task (density of activities in a spatial and temporal frame), social (number and diversity of actors) and cultural complexity (diversity of the cultural software of the mind) [2]. Bosch-Rekveldthig highlighted the lack of an accepted framework to understand project complexity in the context of a large engineering project [30]. A technical, organizational and environmental framework was proposed, which addresses this gap. Some of the elements related to these complexities are as follows: technical - number of goals, goal alignment, clarity of goals, scope; organizational- project duration, size of project team / site area, resources and skills availability, financial resources; environmental: variety of stakeholders perspectives, political influence and weather conditions.

He et al (2015) developed a complexity measurement model comprising six categories - technological, organizational, goal, environmental, cultural and information complexities [32]. Some of the sub factors under these categories are as follows: 1) Technological - diversity of technology, interdependence between the technological system and the external environment. 2) Organizational complexity: number of organizational structure hierarchies and organizational units / departments, 3) goal complexity: uncertainty of goals, uncertainty of project, diversity of task, 4) Environmental complexity: multiple stakeholders, environment of changing economy, environment of changing nature, 5) cultural complexity: multiple participating countries, project team's trust, sense of cooperation, cultural differences, and 6) Information complexity: information uncertainty, level of processing information, capacity of transferring information and integration of more than one system or platform. Along the lines of Hu et al (2015), another study by Lu et al (2015) reviewed the literature on types of complexity and proposed a Task - Organizational (TO) complexity framework for measurement of complexity of large scale projects [9,32]. The task complexity factors include amount of task, pooled / sequential and reciprocal interdependence while organizational complexity factors includes amount of organizational members, degree of centralization / formalization and matrixing.

4 Research methodology

This research study follows a qualitative research strategy, which is appropriate if the research question / problem at hand is exploratory in nature. Content analysis is a qualitative research strategy that involves interpreting and coding textual material for making replicable and valid inferences. The process includes evaluation of texts such as documents, oral communication and graphics and conversion of qualitative data into quantitative data. Hesigh (1988) discussed three types of content analysis: conventional, directed and summative [22]. The procedure followed for generation of initial codes has been proposed as a key differentiating parameter among conventional, directed and summative approaches to content analysis. The codes/keywords are derived from data and theory or relevant research findings in conventional and directed content analysis, respectively, while in case of summative content analysis, it is derived from interest of researchers or review of literature.

The types of content analysis are described based on the process of analysis- inductive or deductive way. The inductive approach is adopted when there is insufficient knowledge about the phenomenon or knowledge is fragmented. On the other hand, deductive content analysis is formed on the basis of previous knowledge with the purpose of theory testing. The inductive data moves from specific to general and deductive approach moves from the general to specific, being based on earlier theory / model [25].

In the first stage, we have used summative content analysis and inductive approach as described by Hesigh 1988 and Elo & Kyngas 2007, respectively, which is followed by deductive content analysis as mentioned by Elo & Kyngas 2007 in the second stage [22-25]. We have used the NVIVO 11 software for the content analysis. The use of this software allows for analytical flexibility and enhances transparency and trustworthiness of the qualitative research process.

In the first step, we have used word frequency query in the NVIVO software, to know the number of times a particular word has been repeated in the flash reports on megaprojects published by MoSPI. Cluster analysis has been carried out in the second step to identify words falling under a specific node and Pearson's correlation has been performed between the nodes for the same word.

5 Analysis

5.1 Word frequency query

We have manually searched various progress reports on megaprojects published by the MoSPI. Although the 1stmegaproject report was published in June 2009, the earliest report available on the website of the MoSPI is January 2012. Since then, reports have been published on a monthly basis. However, a few reports are not available in the database of MoSPI. After going through the available reports, we have shortlisted 22 reports that have reported on the reasons for delays in megaprojects. These 22 reports comprise 2, 12, 8 reports published in the years 2014, 2015 and 2016, respectively. We have separated the part of this report indicating reasons for delay for specific megaprojects.

We have uploaded these reports in NVIVO software and run word frequency query for identifying top 100 reasons based on frequency of occurrences in these reports. The word query has multiple options in selecting exact match, stemmed words, synonymous words and generic words. We have referred the reasons reported and observed synonymous words are often used for reporting delays in megaprojects. Therefore, we have selected the best fit option - synonyms, which eliminate the same intent appearing again and again in different words. Many irrelevant words such as page no, year of project conceptualization, approval and so on, present in the report, were added in the stop word list of software, to avoid them from appearing in the maximum times in word frequencies. Table 1 shows the list of top 20 words related to the reasons causing delay in the megaprojects. The output of word frequency query indicates the count and length of word, and similar words that are counted under particular parent word.

Table 1. Word frequency for reasons behind delay in megaprojects.

Sr no	Word	Length	Count	Similar words
1	Land	4	408	acres, country, land, lands, shore, soil
2	Clearance	9	276	clearance, clearances
3	Fund	4	104	fund, funds
4	Performance	11	96	act, operation, operational, operations, performance
5	Contractual	11	76	contractual
6	Law	3	66	law, legal, right
7	Availability	12	61	access, availability, available, services
8	Stoppage	8	61	closure, stay, stoppage, stopped
9	Mine	4	51	excavation, mine, mines, mining
10	Renewal	7	50	rehabilitation, renewal, revival, revived
11	Equipment	9	49	equipment, equipment
12	Villagers	9	44	settlement, village, villagers
13	Rain	4	43	rain, rainfall, rains
14	Up-gradation	11	40	elevated, rising, up-gradation
15	Change	6	39	change, changed, changes, modification, shifted, switch
16	Compensation	12	37	compensation, repair, repairing, right

17	Manpower	8	37	manpower, workforce
18	Tender	6	35	raw, tender, tendered, tendering, tenders
19	Geology	7	28	geological, geology
20	Mobilization	12	28	mobilization, mobilize

'**Land** 'emerged as most frequently appearing word in the reasons behind delays in megaprojects. The construction of megaprojects often requires acquisition of large tracks of land in road, port, power and airport sector. This results in opposition of land owners to the megaprojects owing to fears of compromised livelihood opportunities, displacements, insufficient compensation and disruption in social fabric. The megaproject implementing agencies have been facing challenge in addressing concerns of these land owners in the framework of megaproject execution [12].

'**Clearance** 'was the second most frequently appearing word. The timely and speedy environmental and forest clearance has remained the bone of contestation among the approval agencies, infrastructure developers and implementing agencies [20]. The delays in clearance has led to huge time and cost overruns, and even the abandonment of projects by infrastructure developers in some cases [16]. The Central government has initiated some steps like enactment of Land Acquisition Law and release of guidelines for forest clearance to overcome hurdles in land acquisition as well as environmental and forest clearances [20,26].

The word "**performance**" has referred in the context of poor/unsatisfactory performance of contractor in executing the megaprojects. The risks, complexities and uncertainties associated with megaprojects requires matching capabilities in the contractor responsible for executing the projects. As the megaprojects are becoming a popular procurement mechanism, there is a need for up scaling of capacities of contractors in areas such as project management, financial engineering, and technology adoption. In many instances, the contractors executing the megaprojects undermine the associated complexities and uncertainties at bidding phase and it results into hurdles in project implementation [12].

The creation of megaprojects involves utilization of enormous amount of financial resources and effective utilization / management of these resources poses a challenge in front of both – the implementing agencies and contractors. The time and cost overruns severely affect the financial health of megaprojects. The implementing agencies and the contractors have to infuse financial resources, which these stakeholders often lack or face difficulties in financial resource mobilization, for bringing the project on the track. **"Fund"**, which is one of the frequently used word, reflects the scenario of financial distress faced by megaprojects. The implementing agencies are advised to closely monitor the progress of work, promptly resolve the bottlenecks and create financial buffer to tide over this situation.

The term **"contractual"** indicates that challenges faced by contracts in governing the risky and uncertain megaprojects. The construction contracts that lacks flexibility in dealing with emergent uncertainty and/or unexpected twists and turns in megaproject implementation creates hurdles. The contract designed for megaprojects should include front end mechanisms for enhancing ex-post governability and encourage collaborative and coordinated behaviour [21]. These mechanisms should be enshrined and supported by strong legal and institutional framework. This framework should provide public officials at the helm of decision making position in megaproject necessary comfort and confidence in dealing with the uncertainties.

5.2 Cluster analysis

The second stage of this study involved cluster analysis performed with NVIVO software. This exploratory technique has been used for grouping sources or nodes based on similar words or attribute values in secondary data or coded similarly by nodes. Among these three types, this research study has used clustering by word similarity. We have first created four nodes namely: overruns, technical complexity, environmental complexity and organizational complexity. This is followed by applying node - overruns to the text containing the reasons for project delay, as mentioned in the reports. This is followed by selecting a specific reason for delay and applying particular complexity code. We have conceptualized the complexities as follows:
i) Task complexity, ii) Organizational complexity and iii) Environmental complexity.

Task complexity: Megaprojects consist of multiple tasks from various fields like engineering, technology, finance and organization management. It also includes ecological protection, energy saving and social stability [9]. These tasks are interconnected in many different ways and each task is affected by a change in other tasks and leads to a corresponding change in other tasks.

Organizationalcomplexity: This complexity focuses on the project team, members experience a number of hierarchies, size and organizational structure [9,27].

Environmental complexity: This complexity refers to where the megaproject operates, such as political, regulatory environment and market conditions [27]. The environmental complexity plays a vital increasing role because environment can impact the interest and needs of project stakeholders.

The similarity between each pair of words was measured by building a table with table columns - each different word that appears in the text of the sources and table cells - number of times the column's word appears in the row's source. This is followed by calculation of a similarly index between each pair of items. We have calculated Pearson's correlation coefficient, which is shown in Table 2 below. This indicates strong correlation between overrun, and task and environmental complexity.

Table 2

Nodes A	Node B	Correlation coefficient r	Relation
Task complexity factors	Overrun	0.674	Strong
Organizational complexity factors	Overrun	0.368	Weak
Environmental complexity factors	Overrun	0.65	Strong

6. Conclusions

Megaprojects have become a popular project delivery mechanism in India. The effectiveness of this mechanism hinges on understanding and overcoming hurdles faced in the execution of these projects. This work has qualitatively analysed the reasons behind delays in megaprojects implementation. Word frequency analysis lists delay in land acquisition and environmental clearances, weak capability of contractors, rigid contracting and legal structures and fund constraints as key reasons for delay in megaprojects. Cluster analysis shows the strong correlation between overruns and task and environmental complexity.

The findings of this research study can be used by decision makers to take appropriate steps towards overcoming theses hurdles.

As for limitations, this study is based on analysis of available reports of Indian megaprojects published by the MoSPI in the public realm. Free access to all reports can improve or refine the results of this study. Further, the reports indicate the reasons for delay in megaprojects. However, it does not distinguish between reasons causing time and cost overruns. Therefore, there is an inbuilt assumption that the reasons causing the delays may lead to both time and cost overruns. In practice, this may not hold true. Finally, the quality of reporting from different implementing agencies of megaprojects cannot be ascertained. This may lead to citing generic reasons for delay rather than exact causes.

References

1. Ahn, S., Asce, A. M., Shokri, S., Lee, S., Asce, M., Haas, C. TandAsce, F. J. Manage in Eng, *Exploratory Study on the Effectiveness of Interface-Management Practices in Dealing with Project Complexity in Large-Scale Engineering and Construction Projects*, 1–12, (2016).
2. Brockmann, C., Brezinski, H., &Erbe, A.CIB World building Cong. *Innovation in Construction Megaprojects*, **1–9**(2016).
3. Davies, T.B.A. Proj. Manag. J. *The art of managing relationships in interorganizational collaboration*. **7(3)**, 47–67 (2014).
4. Hu, Y., Chan, A. P. C., &Le.J.of Management in Engineering, *Understanding the Determinants of Program Organization for Construction Megaproject Success : Case Study of the Shanghai Expo Construction*, **31**(5)(2015).
5. Hu, Y., Chan, A. P. C., Le, Y., & Jin, R. J. of Mange in Eng,*From Construction Megaproject Management to Complex Project Management : Bibliographic Analysis*, 1–11, (2015).
6. Implementation, P., Division, P. M., Bhavan, S. P., & Marg, S. 82 nd REPORT ON MEGA PROJECTS (Rs . 1000 Crore and above) *Government of India Ministry of Statistics and Programme Implementation Infrastructure and Project Monitoring Division*,**Report 82** (March, 2016).
7. Bent Flyvbjerg.*The Oxford handbook of Megaproject management* (2014).
8. LPomeranz, K. Asia Pacific J.*The Great Himalayan Watershed : Water Shortages , Mega-Projects and Environmental Politics in China , India , and Southeast Asia. ,* **7(30)**, 1–27 (2009).
9. Lu, Y., Luo, L., Wang, H., Le, Y., & Shi, Q.Int J. of Proj. Manag.*Measurement model of project complexity for large-scale projects from task and organization perspective*, **33(3)**, 610–622, (2015).
10. Vijaybaskar, M. &Babu, S.M.Economic & Political Weekly. *The Politics of Urban Mega-projects in India. Income Employment Linkages in Chennai's IT Corridor*,**lI(17)**, 85–92 (2016).
11. Rajan, T. A., Gopinath, G., &Behera, M. Jour. of Const. Eng. & Manage,*PPPs and Project Overruns : Evidence from Road Projects in India*, **140**(2004), 1–10(2014).
12. V.S.K. & Mahalingam, A. Engineering Proj.Orga. Conference.*A Framework for Post Award Project Governance of Public- Private Partnerships in Infrastructure Projects*, 1–22 (2013).
13. T Perveen, J.B.S. Economic & Political Weekly.*Interlinking of Rivers in India: Costs and Benefits*, **(Dec)**, 5307–5316 (2004).
14. Siemiatycki, M. Inter. J. of Urban and Regional Research. *Message in a Metro: Building urban rail infrastructure and image in Delhi, India*.**30(2)**, 277–292 (2006).
15. Follmann, Alexander. Habitat Inter. *Urban mega-projects for a 'world-class' riverfront - The interplay of informality, flexibility and exceptionality along the Yamuna in Delhi, India*. **45(3)**, 213-222 (2015).
16. NHAI, *Environment ministry spar over GMR project termination*.

17. Flyvbjerg, B., Bruzelius, N., and Rothengatter, W.Cambridge University Press, Cambridge, U.K, *Megaprojects and risk: An anatomy of ambition*, (2003)
18. Flyvbjerg, B. "What is a megaproject?" (May 19, 2011).
19. Flyvbjerg, B.Proj Manage J. "*From Nobel Prize to project management: Getting risks right.*"**37(3)**, 5–15 (2006).
20. Government of India. *Ministry of Environment, Forest and Climate Change Government of India.*
21. Sanderson, J.IJPM.*Risk, uncertainty and governance in megaprojects: A critical discussion of alternative explanations.***30(4)**, 432–443 (2012).
22. Helgevold, N. & Moen, V. Nordic Journal of Digital Literacy. *The use of flipped classrooms to stimulate students' participation in an academic course in initial teacher education.***2015(1)**, pp.29–42 (2015).
23. Kaefer, F., Roper, J. & Sinha, P. Kaefer Forum Qualitative Sozialforschung / Forum : *Qualitative Social A Software-Assisted Qualitative Content Analysis of News Articles : Example and Reflections Kaefer2 . Literature Review.* **16(2)**, 1–16 (2017).
24. Mayring, P. Forum Qualitative Sozialforschung / Forum : *Qualitative Social 2 . History of Content Analysis 3 . Basic Ideas of Content Analysis 4 . Procedures of Qualitative Content Analysis*, **1(2)**, pp.1–7 (2014).
25. Elo. Satu and HelviKyngas. Journal of Advanced Nursing. *The qualitative content analysis process.* **62(1)**, 107-115 (2008).
26. Raghuram; Samantha & Satyam Shivam. Brunn S, Engineering.*Mega projects in India Environmental and Land Acquisition Issues in the Road Sector,*(2009).
27. Bosch-Rekveldt, M. et al.IJPM, *Grasping project complexity in large engineering projects: The TOE (Technical, Organizational and Environmental) framework*, **29(6)**, 728–739 (2011).
28. Baccarini, D. IJPM,*The concept of project complexity—a review*, **14(4)**, 201–204 (1996).
29. Williams, T.M.IJPM.*The need for new paradigms for complex projects*, **17(5)**, pp.269–273 (1999).
30. Bosch-Rekveldt, M. et al.IJPM,*Grasping project complexity in large engineering projects: The TOE (Technical, Organizational and Environmental) framework*, **29(6)**, pp.728–739 (2011).
31. Williams, G.M.J. IJO&PM. *Now, let's make it really comlex (complicated). A systematic review of the complexities of projects,***31(9)**, pp.966–990 (2011).
32. He, Q. et al.IJPM,*Measuring the complexity of mega construction projects in China-A fuzzy analytic network process analysis*, **33(3)**, pp.549–563 (2015).
33. Pomeranz, K. The Asia Pacific Journal.*The Great Himalayan Watershed: Water Shortages, Mega-Projects and Environmental Politics in China, India and Southeast Asia.***7(30)**, pp.1–27(2009).

Mechanical and economical impacts of adding polymers into asphalt mixtures

Mena Souliman[1,*], *Waleed* Zeiada[2], and *Lubinda* Walubita[3]

[1]University of Texas at Tyler3900 University Blvd, Tyler, TX 75799
[2]Department of Civil and Environmental Engineering, University of Sharjah P.O. Box 27272, Sharjah, United Arab Emirates
[3]The Texas A&M University System, 3135 TAMU, College Station, TX 77843-3135

Abstract. Load associated fatigue cracking is one of the major distress types occurring in flexible pavements. Flexural bending beam fatigue laboratory test has been used for several decades and is considered an integral part of the Superpave advanced characterization procedure. One of the most significant solutions to sustain the fatigue life for an asphaltic mixture is to add sustainable materials such as polymers to the asphalt mixture. A laboratory testing program was performed on two gap-graded mixtures: unmodified and polymer-modified asphalt mixtures. Strain controlled fatigue tests were conducted according to the AASHTO T321 procedure. The results from the beam fatigue tests indicated that the polymer-modified gap graded mixture would have much longer fatigue life compared to the reference (unmodified) mixture. In addition, a mechanistic analysis using 3D-Move software coupled with a cost-effectiveness analysis study based on the fatigue performance on the three mixtures were performed. Overall, the analysis showed that the polymer-modified asphalt mixture exhibited significantly higher cost-effectiveness compared to unmodified HMA mixture. Although polymer-modification increases the cost of the material, the analysis showed that it is more cost effective than the unmodified mixture.

1 Introduction

In 2008, a first cooperative effort between Arizona State University (ASU) and the Swedish Road Administration (SRA) took place in testing unmodified, Asphalt rubber and polymer-modified gap-graded mixtures placed on Malmo E6 External Ring Road in Sweden. In 2009, SRA and ASU undertook another joint effort to test three types of gap-graded mixtures: unmodified, Asphalt Rubber-modified mixtures, as well as polymer-modified asphalt mixtures placed on highway E18 between the interchanges Järva Krog and Bergshamra in the Stockholm area of Sweden. This paper concentrates on the comparison between unmodified and polymer-modified gap-graded mixtures only.

---

*Corresponding author: msouliman@uttyler.edu

Rice specific gravities for the mixtures were determined. Beam specimens were prepared according to the Strategic Highway Research Program (SHRP) and the American Association of State Highway and Transportation Officials (AASHTO): SHRP M-009 and AASHTO T321-03 (equivalent European test standards are: EN12697-24 A to D). Air voids, thickness and bulk specific gravities were measured for each test specimen and the samples were stored in plastic bags until testing.

The Swedish Road Administration provided information stating that the field compaction produced air voids was around 3.0% for all mixtures. The original mix designs were done using the Marshall Mix design method. Table 1 shows the reported average aggregate gradations for both mixtures. The in-situ mixture properties of the Stockholm pavement test sections are also reported in Table 1, which include percent binder content by mass of the mix, Marshall percent void content by volume of the mixture, and maximum theoretical specific gravity of the mixes (G_{mm}) estimated at the ASU laboratory. The base bitumen used was Pen 70/100.

Table 1. Average aggregate gradations and mixture characteristics.

	Sieve Size (mm)	Reference Unmodified Mixture	Polymer-Modified Mixture
Percent Passing	22.4	100	100
	16	98	98
	11.2	65	65
	8	38	38
	4	23	23
	2	21	21
	0.063	10.5	10.5
Binder Content (%)		5.9	5.9
Air Voids (%)		2.6	2.6
G_{mm}		2.464	2.456

Previous publications showed that polymer-modified mixtures had superior fatigue resistance performance over the unmodified HMA mixture as shown in Figure 1 (1). The remaining unanswered question is: does polymer-modified mixtures represent cost-effective solutions to resist fatigue cracking compared to unmodified HMA mixtures?

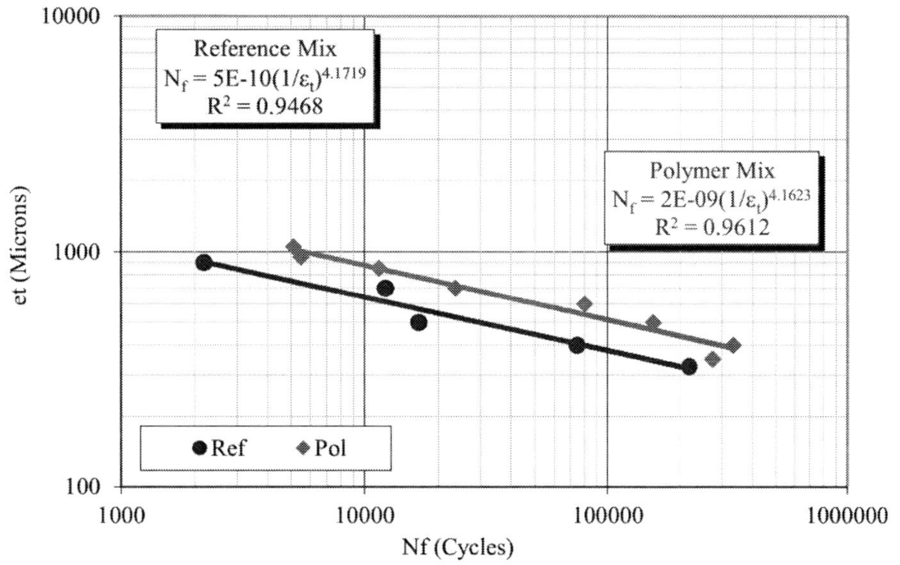

Fig. 1. Fatigue life comparison between the reference unmodified mixture and polymer-modified mixture at 21°C using N_f at 50% of the initial stiffness [1]

2 Objective

The objective of this study was to assess the cost-effectiveness of polymer modified asphalt mixtures as related to fatigue performance. The study combines the results of the mechanical fatigue laboratory evaluations with the long-term mechanistic performance as well as the cost of polymer-modified mixtures to assess the full benefits of such additives relative to their added costs.

3 Literature search

Von Quintus et al. [2] evaluated the benefits of using polymer-modified mixtures based on accelerated field pavement testing data from 36 pavement sections in the United States. It was evident that the use of polymer-modification certainly extends the service life of flexible pavements and HMA overlays. The study concluded that pavements incorporating polymer-modified mixtures showed lower amounts of fatigue cracking, transverse cracking and rutting, which extended the pavement service lives by 5 to 10 years.

Another study was conducted at the Asphalt Institute [3] to compare polymer-modified mixtures with unmodified HMA mixtures in terms of the potential to reduce occurrence of distresses and the increase in pavement life as well as conducting a life cycle cost analysis (LCCA) between the two mixtures. Eighty-four Long Term Pavement Performance (LTPP) sections were utilized in the study. Damage index (observed number of cycles to failure in the field over the predicted number of cycles to failure) was computed for the two mixtures to obtain the difference in expected service lives. It was found that polymer-modified mixtures had an extended life that ranged from one to ten years depending on the site factor (e.g., foundation, water table, drainage, etc.). In addition, the LCCA showed a potential saving ranging from 4.5% to 14% when polymer-modified mixtures are used compared to unmodified HMA mixtures.

Timm et al. [4] compared the structural responses of highly polymer-modified (HPM) asphalt on the NCAT test track with a control section. The structural behaviors of both HPM and the control section in response to temperature changes were characterized by exponential functions. At both a low temperature of 50°F and an intermediate temperature of 80°F, the HPM section experienced higher average strains compared with that of the control section (partly owing its thickness). This implies that the HPM section would have a lower strain at intermediate temperatures if the thickness of the two materials is similar. At a high temperature of 109°F, the control section experienced higher average strains compared with the HPM, despite a 1.4 in. (35 mm) thickness advantage over the HPM. The aggregate base experienced higher average vertical stresses in the HPM section than the control section at all temperatures. The accelerated field pavement testing data indicated that the HPM is effective at reducing HMA layer thickness without significant changes in performance.

The improved performance of polymer-modified pavements compared with unmodified bitumen pavements has relatively resulted from improved rheological properties of the modified asphalt binder. Modified bituminous materials can introduce measurable benefits to highway maintenance and construction divisions, in terms of better-performance and longer lasting roads, as well as cost savings in the road service life. The use of asphalt polymer modifications seems to enhance the fatigue resistance, as illustrated in a number of other studies [5,6]. What has to be ascertained in practice is the degree of modification of the asphalt mixtures that takes place and its cost effectiveness. A detailed comparison between the two mixtures (unmodified and polymer-modified) is needed to quantify the true cost effectiveness of polymer-modified asphalt mixtures as related to fatigue performance.

4 Mechanistic-empirical analysis of unmodified and polymer-modified pavements

The concept of Mechanistic-Empirical pavement design is a comprehensive approach for the design of pavement layers thickness. A mechanistic approach explains the phenomena caused by the physical action. In the pavement thickness design, the phenomena are the stresses, strains and deflections within a pavement structure and the physical causes are the loads, climatic conditions, and material properties of the pavement structure. In this part of the study, the mechanistic-empirical approach has been used to estimate the performance life of the unmodified HMA and polymer-modified pavements based on the fatigue characteristics of the mixtures.

Two pavement structures; thin and thick, and two vehicle speeds; 16 and 113 km/h (10 and 70 mph), were selected for the analyses. The thin pavement structure consisted of 100 mm (4 inches) of a surface layer over 150 mm (6 inches) of crushed aggregate base and an infinite granular subgrade, whereas the thick pavement structure consisted of 200 mm (8 inches) of surface layer over same base and subgrade as the thin pavement. The surface layer can either be a reference unmodified HMA or polymer modified.

The fatigue life (N_f) of the asphalt concrete layer is a function of the tensile strain (ε_t). The number of load cycles to fatigue failure increases exponentially with a decrease in strain level. The fatigue relationship can be modeled as follows:

$$N_f = k_1 \left(\frac{1}{\varepsilon_t} \right)^{k_2} \tag{1}$$

where k_1 and k_2 are the regression constants at 21.1°C (70°F), which is the critical temperature for fatigue cracking.

In order to predict the long-term fatigue performance of the reference unmodified HMA and the polymer-modified pavements, laboratory beam fatigue tests were conducted according to AASHTO T321 along with an analytical model were used. The analytical 3D-Move model [7, 8], which is based on a continuum-based finite-layer approach, was used in this study. The 3D-Move Analysis model can account for important pavement factors such as the moving traffic-induced complex 3D contact stress distributions (normal and shear) of any shape, vehicle speed, and viscoelastic material characterization for the asphalt concrete layer. The 3D-Move software, developed by the Western Regional Superpave Center at the University of Nevada, Reno, was used to calculate the tensile strains at the bottom of the asphalt concrete layer constructed with HMA and polymer-modified mixtures. Input parameters for the analyses include pavement layer thicknesses, materials properties, and traffic loads and speeds. Dynamic moduli measured at 5 temperatures (-10, 4.4, 21.1, 37.8, and 54.4°C) and six frequencies (0.1, 0.5, 1, 5, 10, 25Hz) for each HMA and polymer-modified mixtures [9] were used in the analysis (Figure 2). The resilient moduli of the base and the subgrade were assumed to be 193 MPa (28 ksi) and 55 MPa (8 ksi), respectively. The pavements were loaded with a standard single axle load of 80 kN (18-kips) over dual tires spaced at 304.8 mm (12 inches) with an inflation pressure of 827.4 kPa (120 psi).

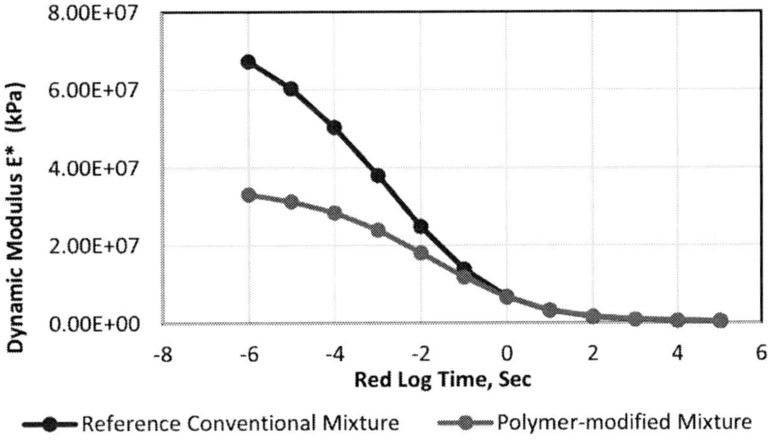

Fig. 2. Average E* master curves for HMA and polymer-modified mixtures [9].

Tensile strains at the bottom of the asphalt concrete layer were analyzed at two locations; under the center of one tire and midway between the centers of the two tires in order to identify the maximum tensile strain. A summary of the results of the maximum tensile strains is presented in Table 2. Utilizing the laboratory fatigue performance equations developed from the beam fatigue tests (Table 2), the fatigue lives for reference unmodified HMA and polymer-modified mixtures were calculated based on the maximum tensile strains determined from the 3D-Move analyses. The number of cycles calculated at the maximum tensile strain represents the fatigue life of the pavement constructed with reference unmodified HMA and polymer-modified mixtures. The ratio of fatigue life of the polymer-modified mixtures over the fatigue life of the reference unmodified HMA mixture are presented in Table 2.

Table 2. Mechanistic fatigue analysis results.

Pavement	Velocity (km/h)	Strain at the bottom of the Asphalt Layer (microns)	k_1	k_2	N_f (Cycles)*	Fatigue Ratio	Average Fatigue Ratio
Reference Unmodified Hot Mix Asphalt							
Thin	16	264	5.00E-10	4.17	416,926	1	1
	113	194			1,521,577	1	
Thick	16	135			6,852,999	1	
	113	94			30,372,021	1	
Thick							
Polymer-modified Mixture							
Thin	16	273	2.00E-09	4.16	1,338,423	3.2	2.8
	113	216			3,545,703	2.3	
Thick	16	138			22,862,988	3.3	
	113	104			74,160,270	2.4	

*Note that one cycle in the laboratory fatigue test is not equivalent to one ESAL in the field since there is a shift factor between laboratory and field conditions. The estimated N_f cycles in the laboratory is used in this paper for the purpose of comparing material alternatives only and does not affect the accuracy of calculations.

The number of repetitions to fatigue cracking was calculated using the highest strain and the appropriate k values. Summary of results that the maximum tensile strains for different combinations of pavement structures, materials, and speeds are presented in Table 2. The figure shows that, at the same thickness and vehicle speed levels, similar tensile stains at the bottom of the asphalt layer were observed among the three asphalt mixtures. The table also illustrate that, within each mixture, thicker pavement structures experience less tensile strains at the bottom of the asphalt layer compared to thinner pavements. Lastly, it was noticed that the tensile strain at the bottom of the asphalt concrete layer increases by decreasing vehicle speed from 113 km/h (70 mph) to 16 km/hr (10 mph) since the loading rate affects the modulus of the asphaltic material.

Table 2 shows the tensile strains, fatigue constants (k_1 and k_2) and number of load cycles to failure (N_f) for all combinations of pavement structures, materials, and speeds. The ratio of fatigue life of each mixture over the fatigue life of the control of HMA mixture is also shown in Table 2.

Based on the results presented in Table 2, the following observations can be made:

- Polymer-modified mixtures showed slightly larger tensile strains at the bottom of the surface layer as compared to unmodified mixtures.
- Polymer-modified mixtures has a much higher fatigue life as compared to the unmodified HMA mixtures. A fatigue ratio of 2.8 of the Polymer-modified mixtures is obtained as compared to the unmodified mixture, which implies that the pavement with polymer modifications is expected to have 2.8 times the fatigue life of the corresponding pavement with an HMA layer.
- Within each mixture, the thicker surface layer (200 mm) has significantly lower tensile strain compared to the corresponding thin asphalt concrete layer (100 mm). This resulted in higher fatigue lives for thick pavements compared to thin pavements.
- Within each mixture, the tensile strain at the bottom of the asphalt concrete layer increases by decreasing vehicle speed from 113 km/h (70 mph) to 16 km/h (10 mph). This resulted in higher fatigue lives for pavements loaded with high vehicle speed compared to low vehicle speed.

5 Cost comparison based on fatigue performance

In order to evaluate the economic values of all mixtures based on fatigue performance, a pavement section of 1.6 km (1 mile) with 4.57 m (15 feet) wide single lane was considered. Based on the assumed density of 110 lb./sq.-yd.-in, the required quantities for paving 100 and 200 mm (4 and 8 inch) thick asphalt concrete layers are as follows:

- For the 100 mm thick asphalt concrete layer: 1600 m (length) x 4.57 m (width) x 100 mm (thickness) x 59.67 kg/m2 (110 lb./square yard) = 1,936 tons.
- For the 200 mm thick asphalt concrete layer: 1600 m (length) x 4.57 m (width) x 100 mm (thickness) x 59.67 kg/m2 (110 lb./square yard) = 3,872 tons.

The cost of production of 100 tons of HMA mixture can be calculated as follows assuming typical current material costs:

- Optimum binder content in the mixture = 4.5% by total weight of mixture.
- Quantity of binder required = 4.5 tons
- Quantity of aggregates = 95.5 tons
- Total cost of binder @ $600/ton = 4.5 x 600 = $2,700
- Total cost of aggregates @ $14/ton = 95.5 x 14 = $ 1,337
- Cost of plant and equipment lump sum for 100 tons of HMA mixture = $2,500
- Total cost for the production of 100 tons of the HMA mixture =
- $2,700+$1,337+$2500 = $6,537
- Therefore, the cost of HMA per ton = $65.37

The additional cost for polymer-modified mixtures per ton was estimated to be around $0.15 per ton of asphalt mixture [10]. Therefore, the estimated cost of one ton of polymer-modified mixture is $65.37 (Cost of HMA per ton) + $0.15 = $65.52 per ton. Knowing the cost of the HMA and polymer-modified mixtures per ton, the required cost to pave 1.6 km (1 mile) of the pavement section with various mixtures were calculated as follows:

- Cost to pave one mile of 100 mm thick HMA = 1,936 tons x $65.37ton = $126,556
- Cost to pave one mile of 200 mm thick HMA = 3,872 tons x $65.37/ton = $253,113
- Cost to pave one mile of 100 mm thick polymer-modified mixture = 1,936 tons x $65.52/ton = $126,847
- Cost to pave one mile of 200 mm thick polymer-modified mixture = 3,872 tons x $65.52/ton = $253,693

Combining the cost of the pavement per mile with the fatigue lives from the mechanistic-empirical analyses, the costs of 1,000 cycles of fatigue life per pavement mile were calculated for the polymer-modified mixtures and their corresponding reference unmodified HMA mixture. This cost figure was derived by dividing the total cost of the 1-mile pavement section by the number of 1000 cycles to fatigue failure (i.e., $N_f/1000$ as summarized in Table 3). In other words, it is the cost of a 1.6 km (1-mile) pavement section for every 1,000 cycles of fatigue life were determined and are summarized in Table 3. A lower cost per mile per 1,000 cycles of fatigue life indicates more cost-effectiveness.

Table 3. Cost calculation per mile per 1,000 cycles fatigue performance.

Pavement	Velocity (km/h)	Cost ($/ton)	Cost of Pavement per 1,000 Cycles of Fatigue Life per 1.6 km (1-mile) ($)	Average Cost of Pavement per 1,000 Cycles of Fatigue Life per 1.6 km (1-mile) ($)
Reference Unmodified Hot Mix Asphalt				
Thin	16	65.37	303	108
	113		84	
Thick	16		37	
	113		8	
Polymer-modified Mixture				
Thin	16	65.52	95	36
	113		36	
Thick	16		11	
	113		3	

6 Cost-effectiveness analysis

Cost-effectiveness analysis is an economic evaluation technique for comparing the cost paid to the gained benefit for the purpose of evaluating alternatives. Several methods are available to evaluate the cost-effectiveness of alternatives [11, 12]. For the purpose of comparing the cost-effectiveness of the unmodified and polymer-modified mixture, a simple approach is used in this study by dividing the expected performance of each mixture by its cost as shown in Equation 2.

$$Cost\ Effectiveness = \frac{Expected\ Performance}{Mixture\ Unit\ Cost} \tag{2}$$

Equation 2 can be viewed as estimating the "bang for the buck" for each case. The alternative that provides the greatest ratio of the benefit to the cost is the "best." In this analysis, only the initial material's cost is used. Other costs, such as maintenance and salvage value, are ignored.

In this case, the expected performance is the predicted number of fatigue cycles to failure (Table 2), while the mixture unit cost was determined as the cost of one mile as determined earlier. Table 4 shows the coast effectiveness of different cases. For example, the cost-effectiveness for a thin (4") unmodified mixture at a vehicle velocity of 16 km/h will be the expected performance life (416,926 cycles) divided by the unit cost ($126,556 per cost of one mile), which will result in 3.3 cycles/cost of one mile.

Table 4. Cost-effectiveness of all mixture combinations.

Pavement	Velocity (km/h)	Cost of one mile ($)	N_f (Cycles)	Cost-Effectiveness (Cycles/ Cost of One Mile)	Average Cost-Effectiveness (Cycles/ cost of one mile)	Average Cost-Effectiveness Ratio
Reference Unmodified Hot Mix Asphalt						
Thin	16	126,556	416,926	3.3	41.2	1
	113	126,556	1,521,577	11.9		
Thick	16	253,113	6,852,999	27.0		
	113	253,113	30,372,021	122.4		
Polymer-modified Mixture						
Thin	16	126,847	1,338,423	10.6	105.2	2.6
	113	126,847	3,545,703	28.0		
Thick	16	253,693	22,862,988	90.1		
	113	253,693	74,160,270	292.3		

Based on the results presented in Table 4, the following observations can be made:

- The polymer-modified mixtures resulted in the higer cost-effectiveness compared to the unmodified mixtures. Cost-effectiveness values of 105.2 and 41.2 cycles per cost of one mile are obtained for the polymer-modified and the unmodified mixtures, respectively. Moreover, the average cost-effectiveness ratio for polymer-modified mixtures (calculated as the average cost-effectiveness of polymer-modified mixtures divided by average cost-effectiveness of the reference mixtures) was 2.6.
- Within each mixture, the thick surface layer (200 mm) has significantly higher cost-effectiveness compared to the corresponding thin surface layer (100 mm).
- Within each mixture, it was noticed that the cost-effectiveness increases by increasing vehicle speed from 16 km/h (10 mph) to 113 km/h (70 mph).

7 Conclusions and recommendations

Based on the results presented, the following observations are drawn:

- Polymer-modified mixtures exhibited significantly higher cost-effectiveness (larger cycles per cost of one mile) compared to reference HMA mixtures. On average, the cost-effectiveness of polymer-modified mixtures was 2.6 times higher than the unmodified mixture.
- The thicker pavement (200 mm asphalt concrete) is more cost-effective than the thinner pavements (100 mm asphalt concrete) for the same material and vehicle speed.
- The cost-effectiveness increase by increasing vehicle speed from 16 km/h to 113 km/h for the same material and pavement thickness.

Based on the results of this study, it is recommended that transportation agencies would increase the use of polymer-modified mixtures in cases where fatigue problems are common. More analyses are needed to evaluate the cost-effectiveness of these modifiers as related to other types of distresses.

References

1. Souliman, M. I., Kaloush, K. E. Assessment of the Extended Fatigue Life for Rubber and Polymer Modified Asphalt Mixtures Using Flexural Bending Beam Fatigue Test. Jordan Journal of Civil Engineering, Vo. **5**, No. 2, 2011, pp. (210-215).
2. Von Quintus, H., J. Mallela, and M. Buncher, Quantification of the Effect of Polymer Modified Asphalt on Flexible Pavement Performance. Transportation Research Record: Journal of the Transportation Research Board, No. **2001**, Transportation Research Board of the National Academies, Washington, D.C, pp. 141–154 (2007)
3. Buncher, M., Quantifying the Benefits of Polymer Modified Asphalt-Reducing Pavement Distress. Asphalt Institute : IS-215. (2009).
4. Timm, D.H., M.M. Robbins, and R. Kluttz. Full-Scale Structural Characterization of a Highly-Polymer-Modified Asphalt Pavement. Preprint CD, prepared for the 90th Annual Meeting of the Transportation Research Board, Washington, D.C., Jan. 23–27, (2011).
5. Hamed FKM. Evaluation of fatigue resistance for modified asphalt concrete mixture based on dissipate energy concept. Ph.D. thesis. Darmstadt, Germany: Technische Universität Darmstadt; (2010).
6. Soleymani HR, Zhai H, Bahia H. Role of modified binders in rheology and damage resistance behavior of asphalt mixtures. Transportation Research Record: Journal of the Transportation Research Board, No. 1875, Transportation Research Board of the National Academies, Washington, D.C. , pp.70–79.(2004)

7. Siddharthan, R.V., Yao, J., and Sebaaly, P.E., Pavement Strain from Moving Dynamic 3-D Load Distribution. Journal of Transportation Eng., ASCE, Vol. **124**(6), Nov./Dec., pp. 557-566. (1998)
8. Siddharthan, R.V., Krishnamenon, N., and Sebaaly, P.E. Pavement Response Evaluation using FiniteLayer Approach. Transportation Research Record: Journal of the Transportation Research Board, No. 1709, Transportation Research Board of the National Academies, Washington, D.C., pp. 43-49.(2000)
9. Kaloush, K., Biligiri, K., Nordgen, T., Zeiada, W., Rodezno, M., Souliman, M., Reed, J., and Stempihar, J. Laboratory Evaluation of Asphalt-Rubber Gap Graded Mixtures Constructed on Stockholm Highway in Sweden. Asphalt Rubber Conference, Munich, Germany (2012).
10. Texas Department of Transportation Estimates. (2015).
11. Lamptey, G., S. Labi, M. Ahmad, and K. Sinha. Life Cycle Cost Analysis for INDOT Pavement Design Procedures. Report FHWA/ IN/JTRP-2004/28. Joint Transportation Research Program, Indiana Department of Transportation and Purdue University, West Lafayette, IN, (2005).
12. Peshkin, D., Smith, K., Wolters, A., Krstulovich, J., Moulthrop, J., Alvarado, C. Preservation Approaches for High-Traffic-Volume Roadways. SHRP 2 Renewal Research : Report S2-R26-Rr-1. Washington, D.C. (2011).

Deformational behaviour of fly-ash based geopolymer concrete at temperatures of up to 150°c

M Talha Junaid[1]

[1]Assistant Professor, University of Sharjah, Sharjah, UAE

Abstract. The use of Geopolymer Concrete (GP*C) has been on the rise over the last few decades owing to its lower carbon emissions as compared to Ordinary Portland Cement Concrete (OPC). Recent research has also established the superior thermal properties of GPC and makes it an ideal construction material for specialized application. However, the deformational behaviour of GPC at elevated temperatures has not fully understood. If GPC is to be used as a main stream construction material for specialized applications, the exact deformational behaviour of the material under thermal loading needs to be investigated. This paper looks into the deformational characteristics of GPC (with natural crushed siliceous aggregates) when dry heated up to 150°C at near zero loading. The deformations recorded using a clip-on extensometer are used to determine the strains developed in the GPC samples due to thermal loads. Coefficient of thermal expansion (CTE) for the tested GPC samples was found and was comparable to OPC concrete at the tested temperatures. Between ambient (20°C) and 80°C the CTE for GPC was determined to be between 10.3-10.9x10^{-6}mm/mm/°C which is similar to OPC concretes. CTE for temperatures between 80°C and 150°C was determined to be 9.3-10.0x10^{-6} mm/mm/°C. First heating cycles resulted in much lower CTE which may be due to the presence of evaporable water in the samples. Like OPC, GPC is a non-homogeneous material and the variation in the materials between samples account for the slight variation in the CTE values determined.

1 Introduction

Alkali liquids (usually a soluble metal hydro-oxide and/or alkali silicate) can be used to react with silica (SiO_2) and alumina (Al_2O_3) rich natural materials, like metakaolin or with industrial by-products, like Fly Ash (FA), Silica Fume (SF), Rice Husk Ash (RHA) or Slag to produce binders [1-3]. Such binders mixed with typical coarse and fine aggregates can form concrete, usually known as Geopolymer Concrete, with mechanical and thermal properties comparable or even superior to ordinary concrete. The term "Geopolymers" was coined as the chemical reaction between the aluminosilicates and the alkaline liquid was a polymerization resulting from a source material of natural or geological origin. Although

* Corresponding author: mjunaid@sharjah.ac.ae

other researchers prefer using the term alkali activated FA (AAFA) when FA is used as a source of aluminosilicates, alkali activated Slag when Slag is the source of aluminusilicates, and so on [4].

1.1 Nature and characteristics of geopolymer concrete

Researchers agree that the basic reaction mechanism for formation of geopolymer binders is in three stages namely: dissolution of Si and Al from the source material, hydrolysis or gelation, and condensation forming a 3D network of silico-aluminates also termed as the 'geopolymer backbone' [1, 2, 4-8]. Davidovits describes geopolymerization as an exothermic reaction and has schematised it using the two stage equation (1) given below [2].

Si-Al source + Silicates + Water + Alkaline Liquid → Geopolymer Precursor

$$(Si_2O_5, Al_2O_2)n + nSiO_2 + nH_2O \xrightarrow{NaOH,\ KOH} n(OH)_3\ \text{-}Si\text{-}O\text{-}\underset{\displaystyle (OH)_2}{\overset{(-)}{Al}}\text{-}O\text{-}Si\text{-}(OH)_3$$

Geopolymer Precursor + Alkaline Ions → Geopolymer Backbone

$$n(OH)_3\ \text{-}Si\text{-}O\text{-}\underset{\displaystyle (OH)_2}{\overset{(-)}{Al}}\text{-}O\text{-}Si\text{-}(OH)_3 \xrightarrow{NaOH,\ KOH} (Na,K)^{(+)}\ \text{-}(\text{-}\underset{\displaystyle \underset{|}{O}}{\overset{|}{Si}}\text{-}O\text{-}\underset{\displaystyle \underset{|}{O}}{\overset{|(-)}{Al}}\text{-}O\text{-}\underset{\displaystyle \underset{|}{O}}{\overset{|}{Si}}\text{-}O\text{-}) + nH_2O \quad (1)$$

GPC does not require any Portland cement as a binder. The role of FA in GPC is entirely different from that when it is used as a cement replacement material in OPC concrete to enhance certain properties such as workability or to reduce the heat of hydration. In such cases FA has no pronounced effect on the strength of concrete [9] especially early strength, however, in fly ash based GPC the fly ash is the sole source of aluminosilicates for reaction with the alkaline solution to form the binder and is thus a critical factor in strength development. Coarse and fine aggregates make up 75 to 80% of the mass of GPC, as is the case with OPC concrete. Though extensive research is being conducted to determine and understand the chemical composition of GPC and the factors affecting it, the focus is now shifting to probable applications to civil and infrastructural engineering.

Currently there is sufficient evidence to suggest that geopolymer concrete has a better resistance to fire than OPC concrete [10-13]. However, most of the existing data deal with the strength or spalling of small laboratory samples [14-17]. Except for the work of Rickard et al. [12], Provis et al [18], Junaid et al. [19-22] and Pan et al. [23] on the thermal dilatation of geopolymer pastes, mortars and concrete synthesised from fly ashes of variable composition, very little is known about its deformational behaviour, creep and thermal properties of GPC. If GPC is to be used as an alternate to OPC the thermal properties of GPC need to be investigated.

The major difference between geopolymer concrete and OPC concrete is in the intrinsic chemistry of the binders. OPC attributes its strength to the hydration of cement which results in the formation of, among other compounds, C-S-H and $Ca(OH)_2$. The strength of the binder largely depends on these two compounds. Moreover, the presence of moisture and, in particular, its movements affects the thermal properties. GPC on the other hand develops its strength from the reaction of Al and Si with alkaline solutions as described above in equation (1) [24, 25]. It is therefore hypothesised that the movement of water within the geopolymer matrix will have a less significant effect on the thermal properties of GPC. This paper looks at the deformational behaviour of GPC when dry heated up to 150°C at near zero loading.

2 Experimental work

2.1 Materials

Low calcium Class F fly ash with composition as found using X-Ray Fluorescence (XRF) is tabulated in Table 1 was used in this study. Sodium silicate and sodium hydroxide were used as alkaline activators in this study. Laboratory grade D sodium silicate solution (Na2SiO3) with SiO2/Na2O between 1.95 and 2.05 was procured from IMCD Australia Limited. Sodium Hydroxide solutions (12M) were made from 98% purity flakes supplied by Redox Pyt Ltd., Australia. The silicate to hydroxide ratio was kept constant at 2.5 by weight of solutions. Locally sourced, natural crushed aggregates (nominal sizes 14, 10 and 7mm) along with river sand as fine aggregates were used in the mix.

Table 1. X-Ray Fluorescence (XRF) of Fly Ash Used.

Compound	% Composition by Weight
SiO_2	62.19
Al_2O_3	27.15
Fe_2O_3	3.23
CaO	1.97
K_2O	0.89
TiO_2	1.06
MgO	0.4
Na_2O	0.3
SO_3	0.07
L.O.I	1.75
Sum of Alumino-silicates	89.34
SiO_2/Al_2O_3	2.29

2.2 Specimen preparation

Aggregates were dry mixed with fly ash for 2 minutes. The silicates solution was then added to the dry mix and mixed for 3 minutes. Hydroxide was then added and mixed for 2 more minutes. Additional water along with any plasticizer was then added and mixed for additional 2 minutes. The mix design used in this study is summarized in Table 2. Cylinders with nominal measurements of 75mm dia and 150mm height were mixed and casted using the procedure described elsewhere [26]. The samples, after an initial rest period, were cured for 48 hours at 80°C. The samples were then placed in an environmental room with RH of 50% and temperature of 23°C till the time of testing.

Table 2. Composition and Mean Compressive Strength of GPC samples.

FA (kg/m^3)	NaOH (kg/m^3)	Na_2SiO_3 (kg/m^3)	Coarse (kg/m^3)	Fine (kg/m^3)	Water (kg/m^3)	VM^+SP^ (kg/m^3)	f_c' (MPa)
420	60	150	1190	574	32.3	8	41

^Viscosity Modifier and Superplasticizer

2.3 Test procedure

The GPC cylindrical samples were fitted with steel rings made from 316 grade steel with knife edges machined on them. A high temperature clip on extensometer was clamped on the knife edge (Figure 1a) and connected to a data acquisition software which was programmed to collect data at 10Hz.

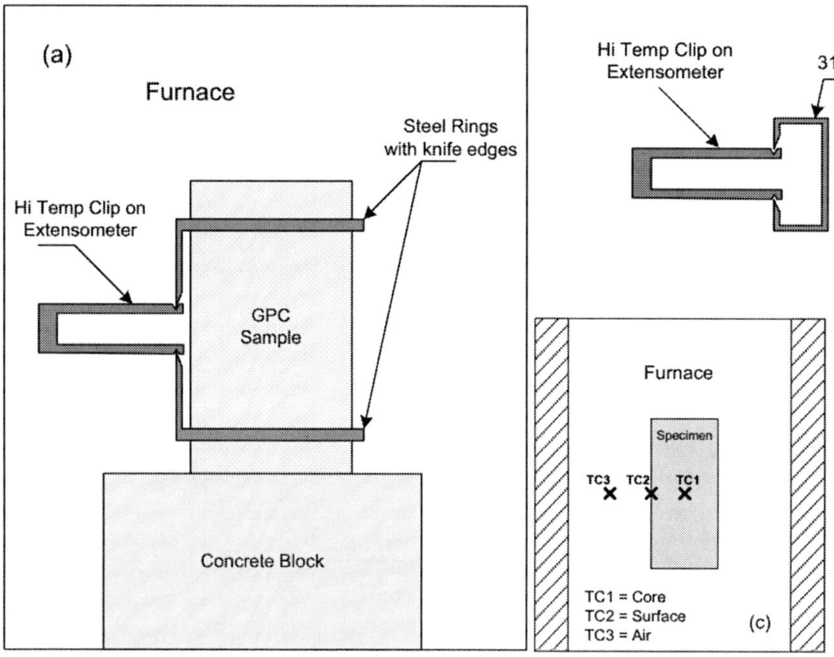

Fig. 1. Schematic representation of (a) test setup (b) strains in steel and (c) thermocouple placement in the furnace.

K-type thermocouples were placed at the core (embedded in the core), at the surface of the sample as well as within the furnace to collect temperature data with time (Figure 1c). To calibrate the clip-on extensometer, it was attached to a titanium silicate sample with a coefficient of thermal expansion (CTE) of 0.008x10^{-6} mm/mm/°C. The displacement readings obtained were used to calibrate the extensometer at the test temperatures (by subtracting these values from the test results). These values were averaged by repeating the test 5 times for each test temperature. The tests were repeated using a 316 Grade steel C-shape to establish the CTE of the steel used (Figure 1b). During calculation of the final concrete strain, the strains due to steel and the deformation caused within the high temperature extensometer were accounted for. This is further discussed in the results section. Use 170 x 250 mm paper size (W x H mm) and adjust the margins to those shown in the Table 1.

The samples were heated at 4.5°C/min until it reached the target temperatures. The temperature was then maintained for 2 hours to ensure thermal equilibrium. The furnace was then turned off and the sample allowed to cool within. Strains were measured during the ramp-up, soak, and cool down stages.

3 Results and discussion

Figure 2 represents the deformation of the different elements on the test specimen. The 'Total' displacement represents the displacement of the system as a whole including the deformation of the extensometer (Exten) computed earlier for this temperature range, the 'True' deformation which is the difference between 'Total' and 'Exten', and the expansion of the steel knife-edges (Steel). The deformation of the GPC sample (GPC) is computed by subtracting the 'Steel' extension from the 'True' deformation.

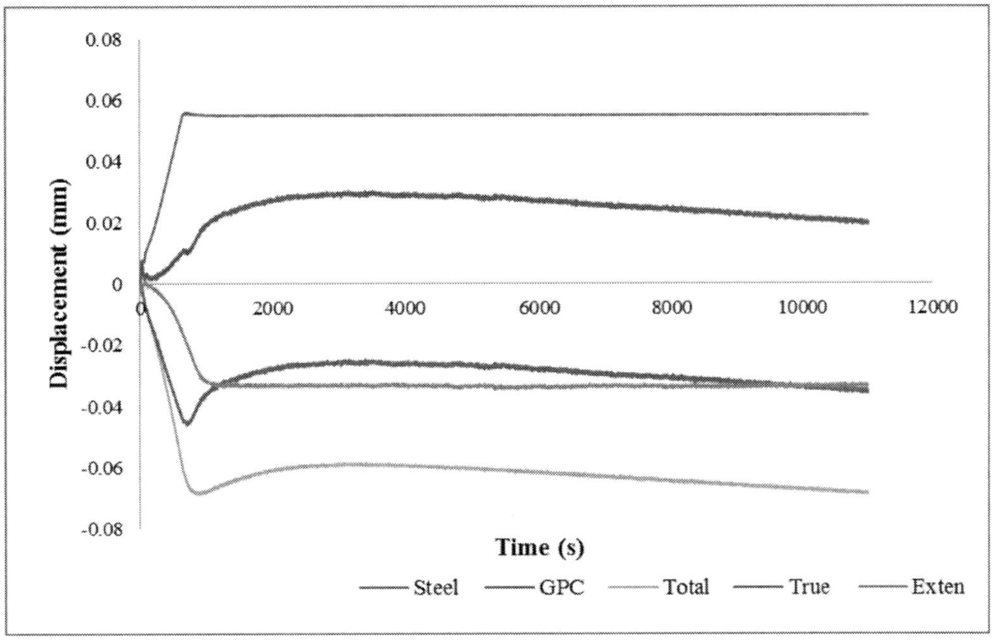

Fig. 2. Displacement readings for the test setup from Ambient to 80°C.

The displacements represented in Figure 2 were then converted to strains and are shown in Figure 3 and 4 for first, second and third heating cycles of the specimen. It can be noted that the first heating of the specimen undergoes slightly different strains as compared to the subsequent heating cycles, especially during the soak period. After the initial expansion of the GPC specimen, as the temperature is maintained, the sample starts to contract after reaching its peak strain. This can be attributed to the loss of water from the system, resulting in shrinkage. However, subsequent heating cycles reveal that the peak strains reached are slightly higher than those during first heating. This trend was observed in all samples undergoing the heating cycles. Moreover, after the peak strains are reached the strains plateaus for the duration of the heating.

 DOI: 10.1051/matecconf/201712002020

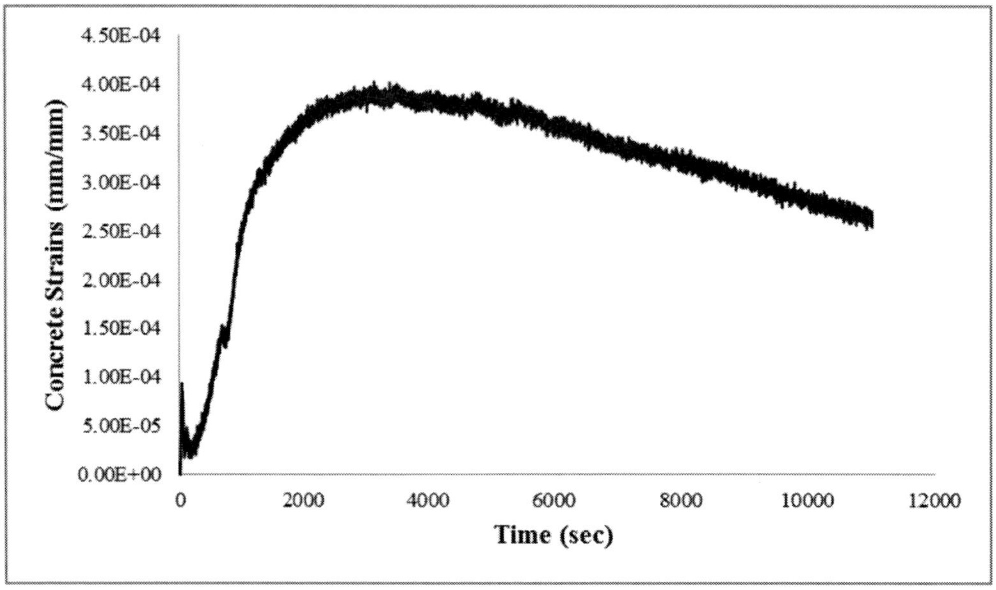

Fig. 3. Concrete Strains between Ambient and 80°C (First Heating).

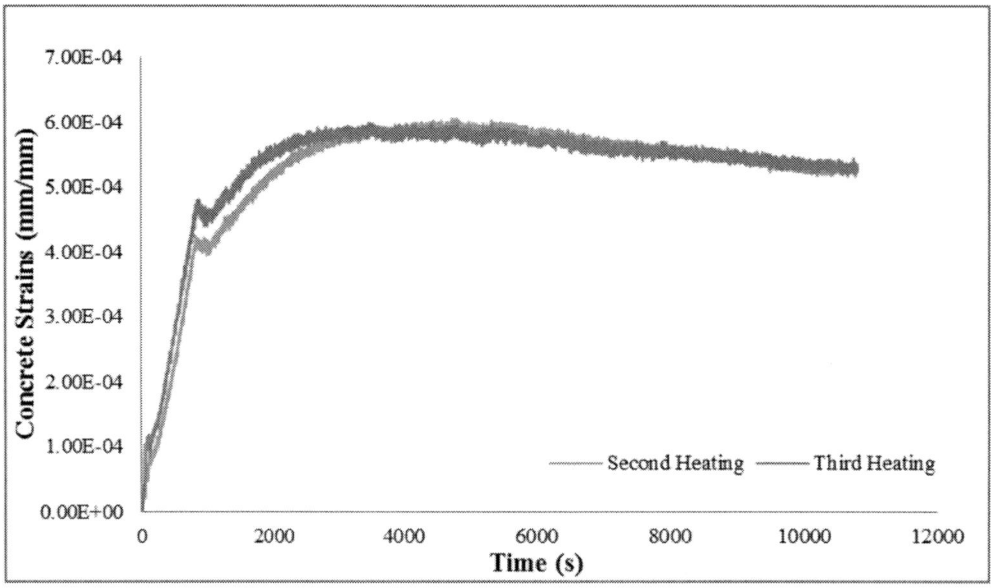

Fig. 4. Concrete Strains between Ambient and 80°C (Second and Third Heating).

The Coefficient of Thermal Expansion (CTE) was then calculated using the formula $CTE = \frac{\Delta L}{L} \times \Delta T$ where ΔL is the change in length and ΔT is the change in temperature over which this change in length was achieved. For the results presented here in Figure 3 and 4 the CTE was computed as $6.9\text{x}10^{-6}$mm/mm/°C for the first heating cycle and $10.27\text{x}10^{-6}$ and $10.31\text{x}10^{-6}$mm/mm/°C for the second and third cycles respectively. Three other samples were tested for the same temperature range with the resulting CTE between 10.3-$10.9\text{x}10^{-6}$mm/mm/°C. The results of the replicate samples are tabulated in Table 3 below.

Table 3. CTE values based on first, second and third cycle heating from ambient to 80°C.

Sample ID	Coefficient of Thermal Expansion		
	First Heating (x10⁻⁶mm/mm/°C)	*Second Heating (x10⁻⁶mm/mm/°C)*	*Third Heating (x10⁻⁶mm/mm/°C)*
Sample 1	6.9	10.27	10.31
Sample 2	6.9	10.32	10.68
Sample 3	7.1	10.74	10.91
Sample 4	7.0	10.34	10.43

Between the temperature range of 80°C and 150°C, the expansion of GPC rose steadily with temperature and became constant after thermal equilibrium was reached (Figure 5 and 6). Unlike between the ambient and 80°C temperature range, during this temperature rise there were no dips in the strains after reaching its peak. This behaviour further supports the hypothesis that the decrease in strains during first heating was due to the movement of free water in the GPC system. As the sample has undergone three cycles of heating to 80°C earlier most of the free water would have escaped. Any remaining free water would have left the system during the temperature ramp to 150°C. Therefore, at this temperature range, free water would have had a less pronounced effect on the deformational behaviour.

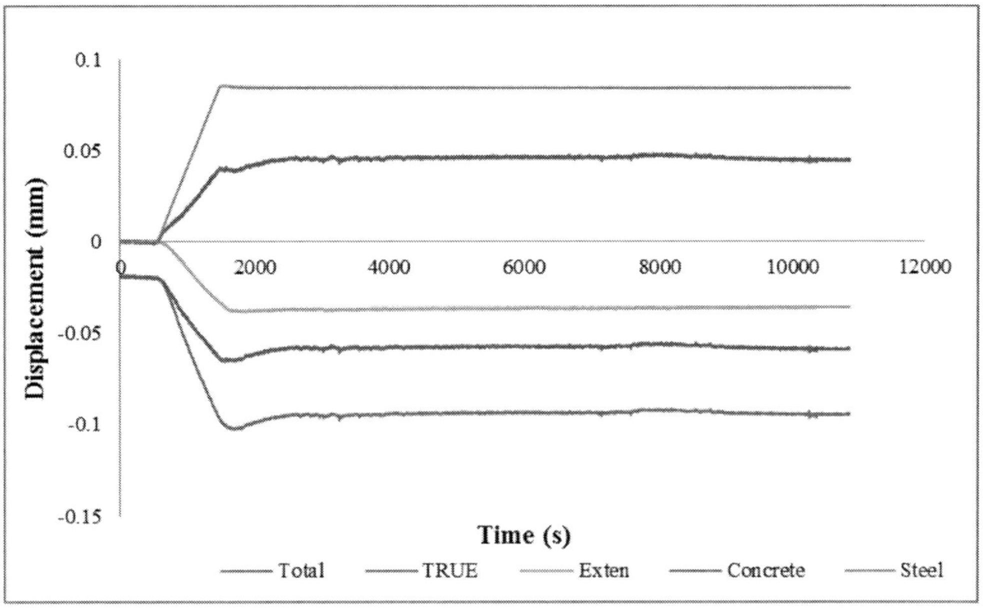

Fig. 5. Displacement readings for the test setup from 80°C to 150°C.

The CTE calculated for this sample using the first and second heating cycle results was 9.15 and 9.31x10⁻⁶mm/mm/°C, respectively. However other samples tested for this temperature ranges resulted in CTE range of 9.3 and 10.0x10⁻⁶mm/mm/°C.

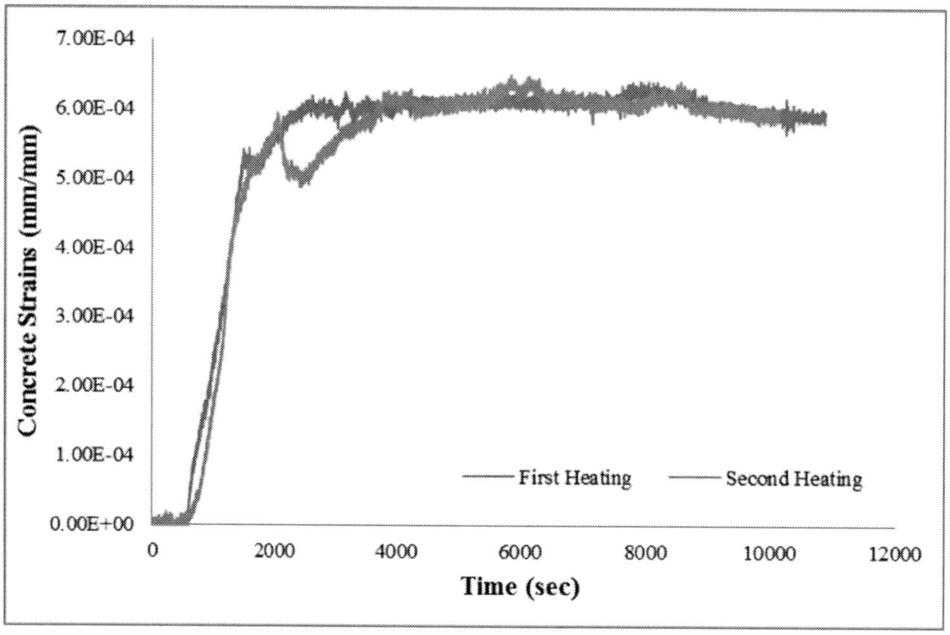

Fig. 6. Concrete Strains between 80°C and 150°C (first and second heating)

4 Conclusions

This paper presents the experimental program to determine the deformational behaviour of low calcium heat cured geopolymer concrete. The deformation of GPC is reported for first and subsequent heating cycles. The GPC samples were heated up to 80°C and then subsequently to 150°C under near zero loads. The evolution of strains and coefficient of thermal expansion are reported for the tested sample. Based on the results of tests the following conclusions can be drawn:

1. The strain development within GPC samples at near zero loads, depends on the heating cycle of the samples. During first heating the samples to 80°C, the samples developed slightly lesser peak strains than during subsequent heating cycles. During first heating to 80°C and after reaching peak strains the samples contract as the temperature is maintained. This behaviour may be attributed to the movement of free water within the GPC system resulting in shrinkage of the samples.
2. The strains developed during the subsequent heating of the GPC samples to up to 80°C results in slightly higher peak strains. During soak periods, the sample strains plateaued in contracts to the first heating cycle where they underwent contraction. This reinforces the hypothesis that the deformational behaviour of GPC under heat loading is dependent on its moisture content and the movement of moisture has a significant impact on its response.
3. The coefficient of thermal expansion of GPC samples is very similar to that of OPC. Between ambient and 80°C was determined to be between 10.3 - 10.9x10^{-6} mm/mm/°C. Between 80°C and 150°C the coefficient of thermal expansion of GPC samples was determined to be 9.3-10.0x10^{-6} mm/mm/°C, while the CTE between ambient and 100°C was determined to be 11.0x10^{-6} mm/mm/°C.

The author would like to thank his colleagues at the School of Engineering and Information Technology, UNSW, Canberra for their support during testing. The support provided by the University of Sharjah is also highly appreciated.

References

1. Davidovits, J. *Soft Mineralogy and Geopolymers.* in *Proceedings of the Geopoolymer 88 International Conference.* France.(1988)
2. Davidovits, J., *Geopolymers Inorganic polymerie new materials.* Journal of Thermal Analysis. 37: p. 1633-1656.(1991)
3. Lloyd, N. and B. Rangan, *Geopolymer Concrete with Fly Ash*, in *Second International Conference on Sustainable Construction Materials and Technologies*: Italy.(2010)
4. Fernández-Jiménez, A. and A. Palomo, *Alkaline Activation of Fly Ashes. Manufacture of Concrete Not Containing Portland Cement*, in *International RILEM Conference on the Use of Recycled Materials in Buildings and Structures.* p. 8.(2004)
5. Diaz-Loya, E.I., E. Allouche, and S. Vaidya, *Mechanical Properties of Fly-Ash-Based Geopolymer Concrete.* ACI Materials Journal. 108(3): p. 300-306.(2011)
6. Gourley, T.J. and G.B. Johnson. *Development in Geopolymer Precast Concrete.* in *International Workshop on Geopolymers and Geoplymer Concrete.* Perth, Australia.(2005)
7. Rickard, W., et al., *Assessing the suitability of three Australian fly ashes as an aluminosilicate source for geopolymers in high temperature applications.* Materials Science and Engineering: A. 528: p. 3390-3397.(2011)
8. Diaz-Loya, E.I., E. Allouche, and S. Eklund, *Factors Affecting the Suitability of Fly Ash as Source Material for Geopolymers.* Fuel. 89: p. 992-996.(2010)
9. Neville, A., *Properties of Concrete.* 4th ed: Longman Group Limited.(1995)
10. A. Ferna´ndez-Jime´nez, et al., *New Cementitious Materials Based on Alkali-Activated Fly Ash: Performance at High Temperatures.* Journal of American Ceramic Society. 91(10): p. 3308–3314.(2008)
11. Bakharev, T., *Thermal behaviour of geopolymers prepared using class F fly ash and elevated temperature curing.* Cement & Concrete Research. 36: p. 1134–1147.(2006)
12. Rickard, W., J. Temuujin, and A. van Riessen, *Thermal analysis of geopolymer pastes synthesised from five fly ashes of variable composition.* Journal of Non-Crystalline Solids. 358: p. 1830–1839.(2012)
13. Rickard, W., L. Vickers, and A. van Riessen, *Performance of fibre reinforced, low density metakaolin geopolymers under simulated fire conditions.* Applied Clay Science. 73: p. 71-77.(2013)
14. Guerrieri, M., J. Sanjayan, and F. Collins, *Residual strength properties of sodium silicate alkali activated slag paste exposed to elevated temperatures.* Journal of Materials and Structures. 43(6): p. 765-773.(2007)
15. Guerrieri, M., J. Sanjayan, and F. Collins, *Residual compressive behavior of alkali-activated concrete.* Journal of Fire and Materials. 33(1): p. 51-62.(2007)
16. Guerrieri, M. and J. Sanjayan, *Behavior of combined fly ash/slag based geopolymers when exposed to high temperatures.* Journal of Fire and Materials. 34(4): p. 163-175.(2009)
17. Provis, J., *Fire resistance of geopolymer concretes*, University of Melbourne: Melbourne.(2010)
18. Provis, J., et al., *Correlating Mechanical and Thermal Properties of Sodium Silicate-Fly Ash Geopolymers.* Colloids and Surfaces A: Physicochemical and Engineering Aspects. 336(1-3): p. 57-63.(2009)

19. Junaid, M.T., et al., *Aspects of the deformational behaviour of alkali activated fly ash concrete at elevated temperatures.* Cement & Concrete Research. 60: p. 24-29.(2014)
20. Junaid, M.T., O. Kayali, and A. Khennane, *Response of alkali activated low calcium fly-ash based geopolymer concrete under compressive load at elevated temperatures.* Materials and Structures. 50(1): p. 50.(2017)
21. Junaid, M.T., A. Khennane, and O. Kayali, *Investigation into the Effect of the Duration of Exposure on the Behavior of Geopolymer Concrete at Elevated Temperatures.* MATEC Web of Conferences. 11.(2014)
22. Junaid, M.T., A. Khennane, and O. Kayali, *Performance of fly ash based geopolymer concrete made using non-pelletized fly ash aggregates after exposure to high temperatures.* Materials and Structures.(2014)
23. Pan, Z., J. Sanjayan, and F. Collins, *Effect of transient creep on compressive strength of geopolymer concrete for elevated temperature exposure.* Cement and Concrete Research. 56: p. 182-189.(2014)
24. Davidovits, J., *Geopolymers and geopolymeric materials.* Journal of Thermal Analysis and Calorimetry. 35(2): p. 429-441.(2005)
25. Rangan, B., *Design and Manufacture of fly-ash based geopolymer concrete.* Concrete in Australia. 34(2): p. 37-43.(2010)
26. Junaid, M.T., et al., *A mix design procedure for low calcium alkali activated fly ash-based concretes.* Construction and Building Materials. 79: p. 301-310.(2015)

Challenges faced for micro-tunelling works in Jadaf, Dubai for long drive on large scale projects (case study)

Devendra Datt Bhatt[1],* *Khalil* Charif [2], and *Rabee* Rustum [3]

[1]Project Director; Gulf Petrochemical Services and Trading L.L.C (Dubai Br)
[2]Managing Director; International Foundation Group L.L.C
[3]Professor, Water and Environmental Engineering; Heriot Watt University (Dubai-campus)

Abstract. This paper presents the challenges faced during micro-tunnelling process in Emirates of Dubai. In permeable grounds, the loss of slurry increases the cutter head torque and reduces penetration rates. Overcut is formed to provide space for converging ground and enables the evenly circulation of lubricant flow on the pipe line, as such buoyancy effects of fluid lubrication act on the pipe to lessen the drag during jacking. Micro-tunnelling techniques are used to reduce penetration resistance within the pipe string by providing sufficient uniform overcut and continuously injecting polymer lubricants. It is also observed that as the Urban area in Dubai is surrounded by built structures and numerous underground services, a special consideration is required to protect these services especially DEWA ED Cables. Also, the high ground water table, fractures within the rock and proximity of creek makes things very difficult to reduce the water table level in launching / receiving Pit and have difficulties in breaking the hard ground within the Pit. At Wafi Interchange, proposed development for the extension of interchange had to be incorporated in the alignment and accordingly standard pipe fittings to be introduced in the alignment. There are services towards Sheikh Butti palace (launching pit side) and on the other side (receiving pit). Ongoing construction of Al Jaliliya children's hospital requires close coordination and safety related issues to be addressed prior to mobilization of heavy equipment at the proposed micro-tunnelling area. The receiving pit at Emarat petrol pump was also critical and special approval from RTA is required to close one lane at the heavy traffic along the service road, parallel to sheikh Rashid road. The removal of ground water was another challenge because near Sheikh Butti palace the drainage network was not completed and hence special arrangement was to be made to discharge the water near the creek, for which DM environment approval are to be sought.

* Corresponding author: [1] harshlovely@hotmail.com

1 Introduction

There are risks involved in sub-structural tunnelling that are to locate and identify the existing services to give way for proposed micro-tunelling and avoiding any interface or obstructions with existing services or as built structures. The poor geological condition (Low N Value)† and non-availability of sufficient cover can cause the surface settlement. Due to large amount of dewatering and high amount of load imposed by surrounding structures, chances of surface settlement and uneven longitudinal ground settlement increases [4]. Bore in unsuitable ground has the tendency to converse the string of the pipe including tunnelling head and the machine attached to it (known as CAN)‡. For the sub-terrain condition in hard rock ground, holes provide more suitability and are conducive ground to micro-tunnelling procedure [1, 3, 6]. The tendency of earth at forefront of micro-tunelling to move viciously towards sealed fitted in front of the pipe to be jacked and sophisticated slurry shield is used to overcome these failures.

Forces that are relevant to face load depend upon the soil condition and its nature. The front shield has the following advantages:

a) Safe working place for men to enter and work
b) Mounting for face stabilization device
c) A place to observe line and level
d) Adjusting the directions of string

Shield are controlled remotely and in most cases controlled by computer.

2 Launching and receiving shaft

In loose soil and specifically in Jadaf area where the services are quite close to launching and receiving pit, it is a challenge to make these pits. A larger amount of water is to be dewatered and special care is required to remove water only and ensure dewatering process in such a way that soil particles and other dissolved particles are minimum to avoid any future settlement, mobilization and demobilization heavy plant and equipment require special consideration and approval from government agencies to avoid any kind of disruption to the work, nearby as built structure and traffic [1, 9].

3 Sheet piling

A proper sheet piling with interlock system and vibro-hammer are used to lower the sheet pile, prior to this pre-boring is used to loosen the ground as the ground is hard after certain depth [3]. A proper designing of sheet piling is required and sufficient impeding must be done below to reduce the level, as shown in Figure 1 (a) and (b). Bracing/Wall support is to be provided per the design calculation and maximum bending moment to minimize settlement

† SPT (Standard Penetration Test) Value.

‡ A steel sleeve attached with micro-tunelling machine head to accommodate other necessary parts required for micro-tunnelling operation.

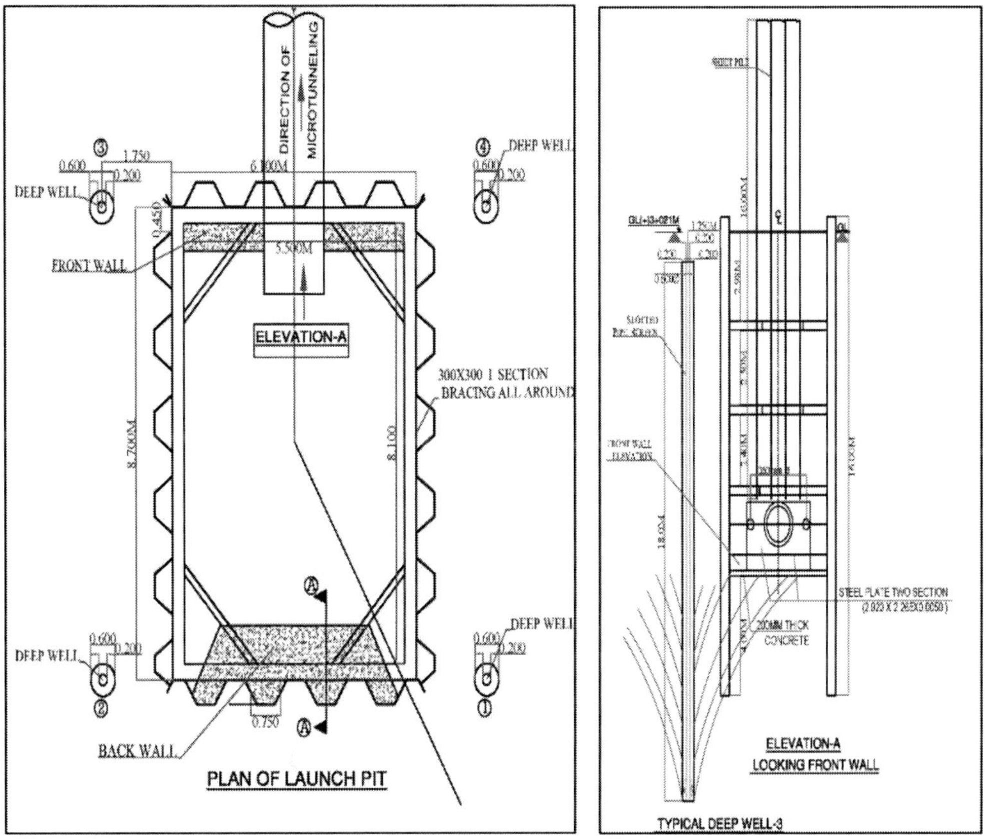

(a) (b)

Fig. 1. (a) Plan of Launch pit and location of deep well (b) Elevation A and cross-section details of deep well and sheet pile.

If proper dewatering is not in place, then due to removal of silt or dissolved particles from the existing soil, it may cause future settlements. The soil structure in Jadaf Area is soft at a depth of 4-5 meter, and thereafter SPT value increase drastically at a depth of 10-16 meter, it is then very hard and SPT value is generally more than 100 unit, in such a hard soil it is difficult to lower the sheet pile and therefore pre-drilling is a necessity to avoid any damage to the sheet pile. Such a case is found at site (receiving pit for micro-tunelling number M-01), which is described in Figure number 2. It can also be seen in the Figure that the pit is very close (around 6 meter) from the overhead bridge which is leading to the Business Bay and crossing Al Khail road. Therefore, to access work site, a special approval is required from RTA and special care is needed to monitor the vertical height limit [3-6,8,9-15]. As the area is very close to the existing creek, high amount of water has caused difficulty to lower down the water table in the receiving pit and therefore other related activities such as excavation and bracing work are affected. As the receiving pit is very close to the main road i.e. Al Khail road (refer Figure 2), it requires proper monitoring of spot level to check if any surface settlement is observed, proper designing of sheet pipe and dewatering is also required [2,7].

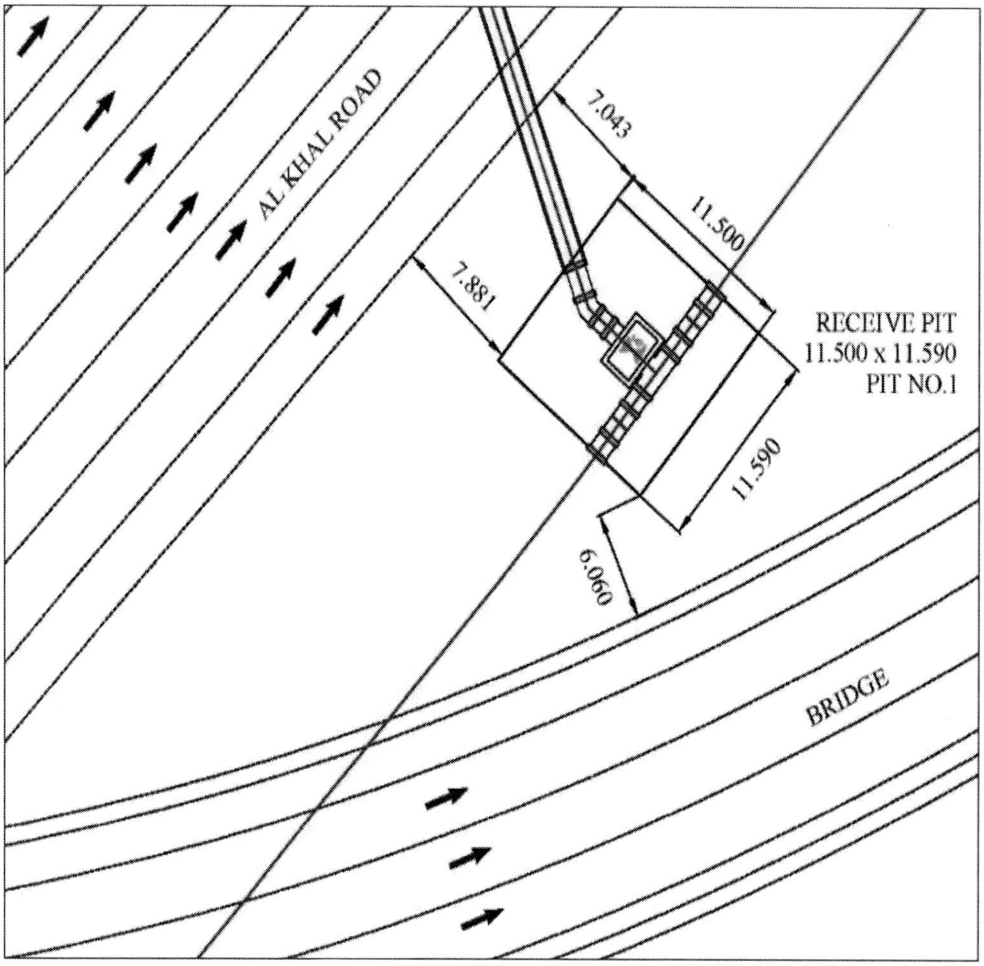

Fig. 2. Detail of Receiving pit number 1 across Al –Khail road, with surrounding fly overbridge.

4 De-watering

Place as the network in Jadaf area is not readily available at the time of construction activities, therefore a special approval is required to discharge the water into the Creek. For this purpose, a proper NOC§ is required from relevant department with water being tested to be submitted to the design team. Another difficult being faced is that as heavy equipment is being moved across the site, it can damage the de-watering discharge pipe and therefore it must be properly protected with steel sleeve. This is shown in Figure 3, and discharge point to the Creek is shown on the same drawing. Four numbers of deep well are provided for Pit Number 1 and locations of these deep well to be in a similar way as per Figure 1 (a). These four numbers are required due to high flow of water and in case of any failure as soil condition is very loose in this area. Similarly, Figure 4 provides the discharging location of de-watering (section 1-1). Special care must be considered for deep well, as particles size distribution shows that

§ No Objection Certificate.

this area has lot of fine silt content in the soil structure and to control egress silt content from soil structure and to stop them falling into deep well with water flow, surrounding aggregate** of deep well shall be of good quality and size of the aggregate to be as per the de-watering contractor's design so that silt content remains within soil structure and does not pass through de-watering system. Also, slotted pipe or vertical screen pipe (generally PVC and in special cases can be metal), as shown in Figure 1 (b) for deep well must be of good quality and as per the design [8].

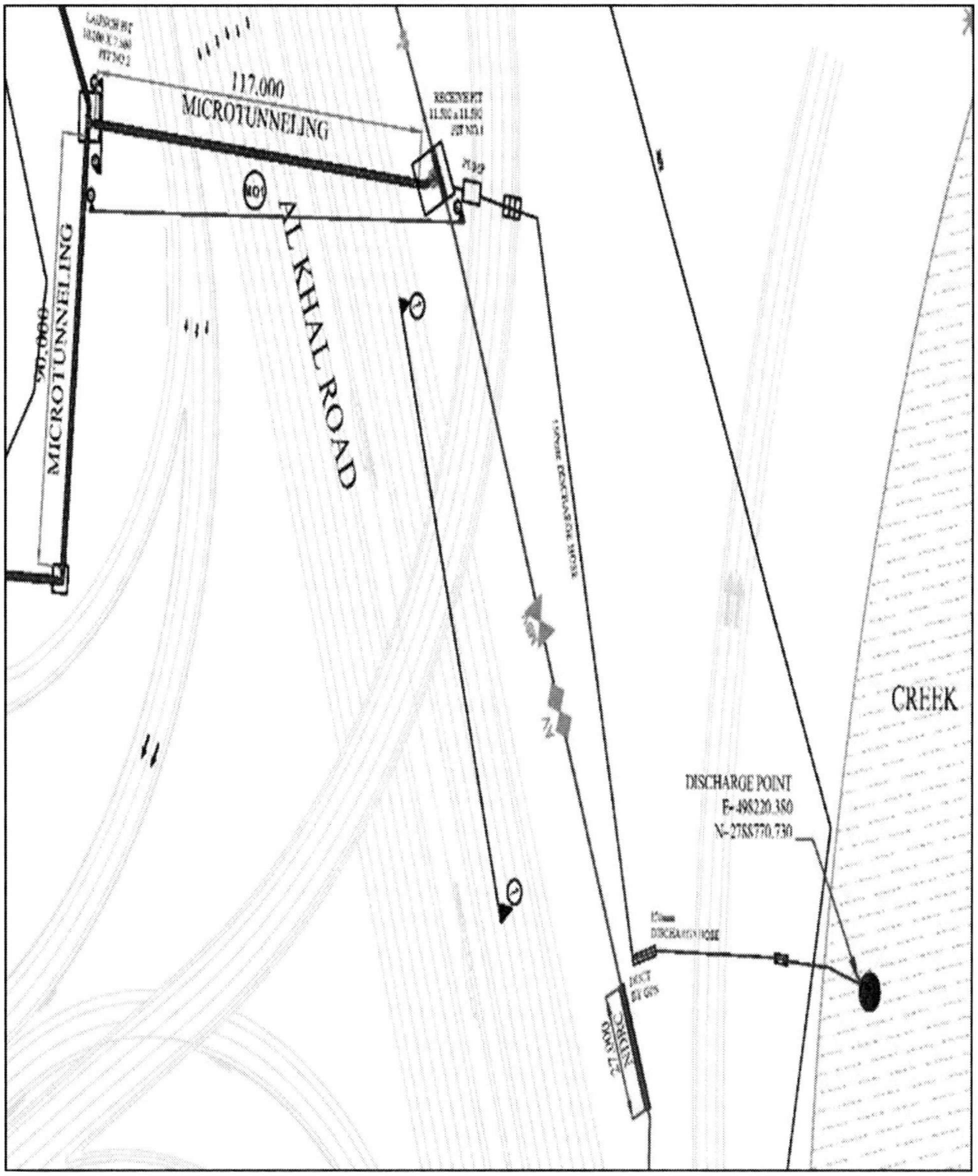

Fig. 3. Detail of de-watering network for pit no 1 and its discharge layout.

** Fine and cleaned crushed stone particle, as per de-watering sub contractor's design.

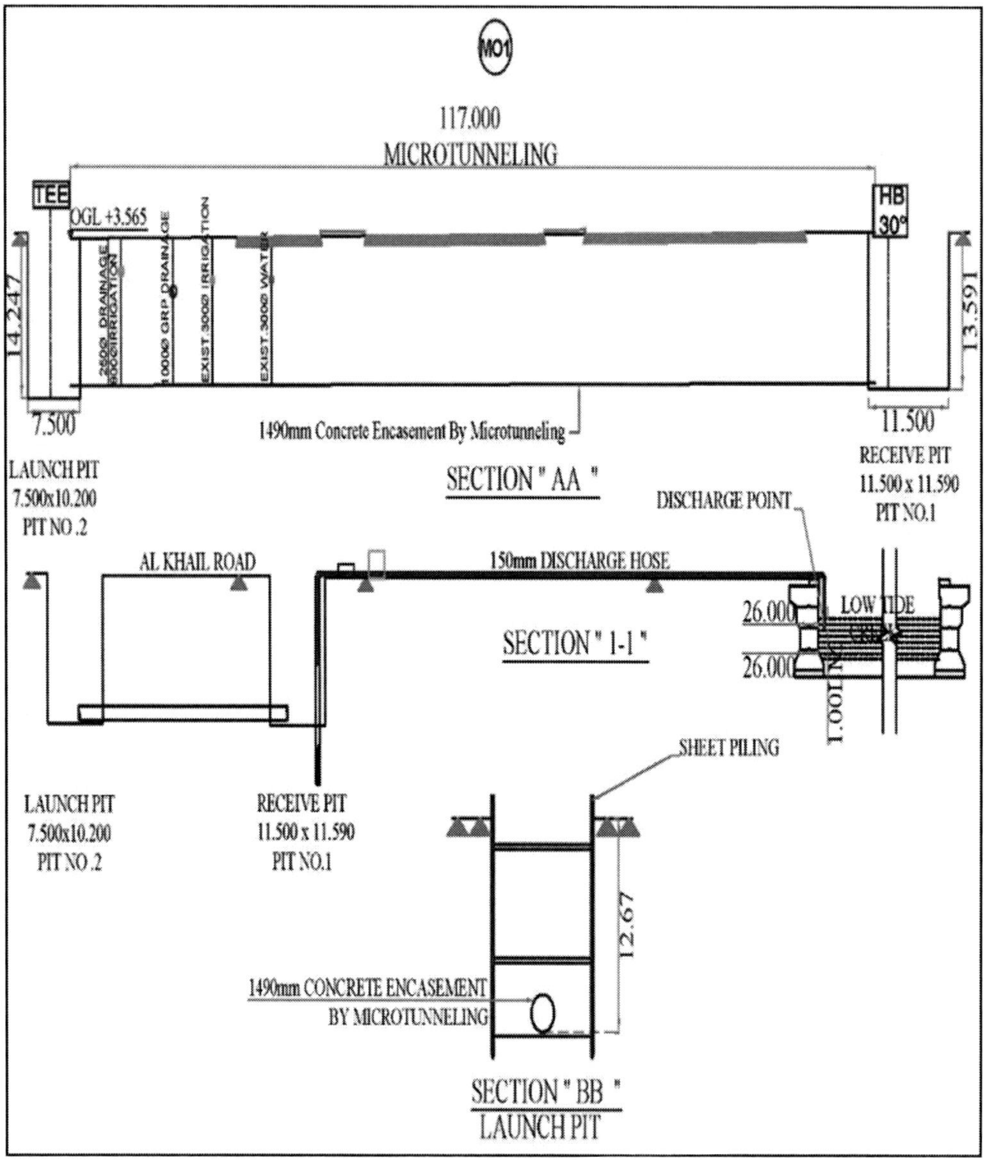

Fig. 4. Detail of de-watering network for pit number 1 and its discharge layout section 1-1, into creek.

5 Micro-tunelling

As micro-tunelling has several equipment's and requires working space, storage space to organize the equipment's such as, soil spool removal system, generator, pipes etc. In Jadaf area there are certain locations where it is very difficult to mobilize equipment and material, and due to the storage constrain it is decided that the piping materials will not be stored more than 6-7 pipes at a time. As the site area is in the city, therefore movement of heavy vehicles and noise is to be controlled accordingly. A special care is required for the summer time where cooling effect is necessary for the people working in the launching Pit. It is also necessary to control the temperature of air inside the tunnel; otherwise it can deflect laser control system and ultimately alignment of the tunnel.

In Figure 5 detail alignment of the pipeline and micro-tunnelling has been shown, to control overall movement of the traffic and other obstruction it is divided into three stages, all these are described in Figure 5.

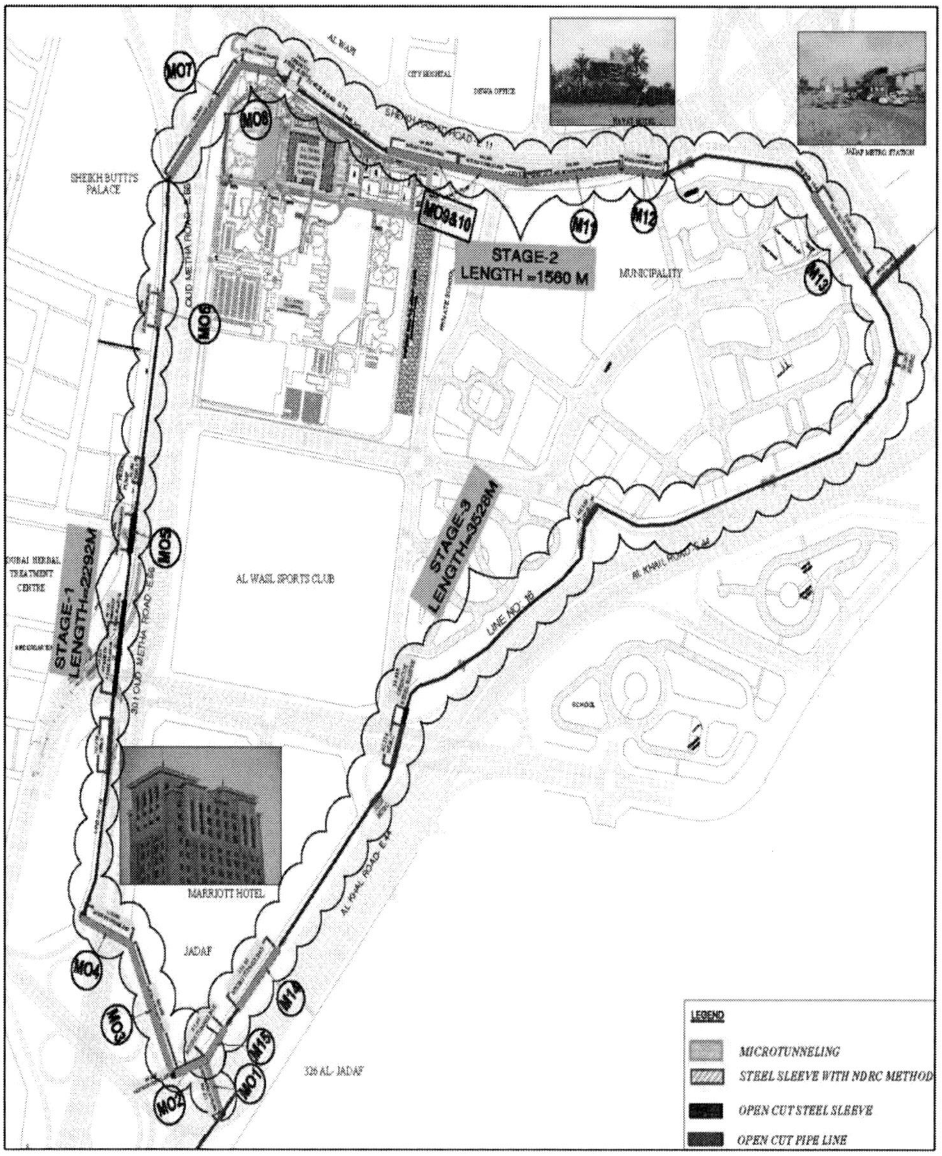

Fig. 5. Bird eye view of the study area (Jadaf- Oud Maitha – Al Khail Road).

Total micro-tunnelling works is of 2,410 meter for all the three stage for all together fifteen locations of micro-tunnelling as per Table 1.

Table 1. Micro-tunnelling details for 1200 diameter GRE water transmission Pipe.

Sl/ No.	Microtunneling Reference	Chainage		Length in meter	Stage	Line Reference No.	Location	Road Reference
		From (L.P)	To (R.P)					
1	M-01	(-)0+025.215	0+091.785	117	Stage : 1 --- 796 Meter	18	Al Khail Road Crossing	E44
2	M-02	0+097.065	0+187.065	90		18	Existing Services Crossing(Electric trough & 800mm Irrigation Pipe)	Unmade Area
3	M-03	0+493.865	0+192.865	301		18	Near Meriot Hotel	Partially unmade area & one side service road crossing
4	M-04	0+499.510	0+637.510	138		18	Oud Metha Road	E66
5	M-05	1+450.584	1+528.584	78		18	Service Road Crossing	Before EPCO petrol pump going towards Bur- Dubai
6	M-06	2+013.688	1+941.688	72		18	Service Road Crossing	After EPCO petrol pump going towards Bur -Dubai
7	M-07	(-)0+009.943	0+304.734	309	Stage : 2 --- 1,146 Meter	18A	Oud Metha road, WFI Interchange	E66
8	M-08	0+397.734	0+304.734	93		18A	Service Road Crossing	Unmade area inside Latifa hospital
9	M-09	0+761.238	0+947.238	186		18A	Service Road Crossing in front of Emirates filling station	D79
10	M-10	1+137.313	0+951.813	186		18A		
11	M-11	1+146.205	1+404.205	258		18A	Existing Services Crossing & Parallel to Sheikh Rashid Road	D79
12	M-12	1+524.468	1+410.468	114		18A	Existing Services Crossing (Partially electric trough and land scape area)	D79
13	M-13	0+458.688	0+686.688	228	Stage - 3 --- 477 Meter	16A	Service Road Crossing	Near Jadaf Metro station
14	M-14	2+308.887	2+476.887	168		16	Service Road Crossing	Near Meriot hotel along Al Khail road towards Abu Dhabi
15	M-15	2+615.195	2+534.195	81		16	Service Road Crossing	Unmade Area
TOTAL MICROTUNNELING CUMMULATIVE LENGTH				2419				

6 Site obstructions

There are many site obstructions and one such case is at micro-tunnelling number 9 & 10 where the overall length of the micro-tunelling is of 372 meter (186 meter each section joined together at mid - Pit number 5, on Figure number 6) and Receiving Pit is located at the busy service road, close to the Emarat fuel filling station. Therefore, a special permission for traffic diversion from RTA, DM†† for de-watering discharge point, Emarat fuel station for diversion of traffic, protection to their underground fuel reservoir and Horticulture for removal of grass and tree is required. Apart from this, a proper design is also required for receiving pit, movement of heavy vehicles, logistics plan for lowering and removing of sheet pile without damaging and disturbing the nearby structures, to the traffic and underground oil reservoir tank of the Emarat fuel filling station as shown in Figure 6. Similar issues were encountered near sheikh Buti palace and Al Jalilia children specialty hospital. At sheikh Buti palace where DEWA electrical services were in a close proximity to the launching pit; whereas at AL Jalila children specialty hospital, site coordination and alignment of the pipe-line and micro-tunnell to incorporate standard fitting and at the same time avoid clashes to the ongoing site activities and proposed wafi interchange additional works, such as proposed pier clashing with pipe line alignment and deep excavation of the basement of the hospital and its close proximity with launching and receiving pit for micro-tunnelling reference number M07, 08, 09 and M10.

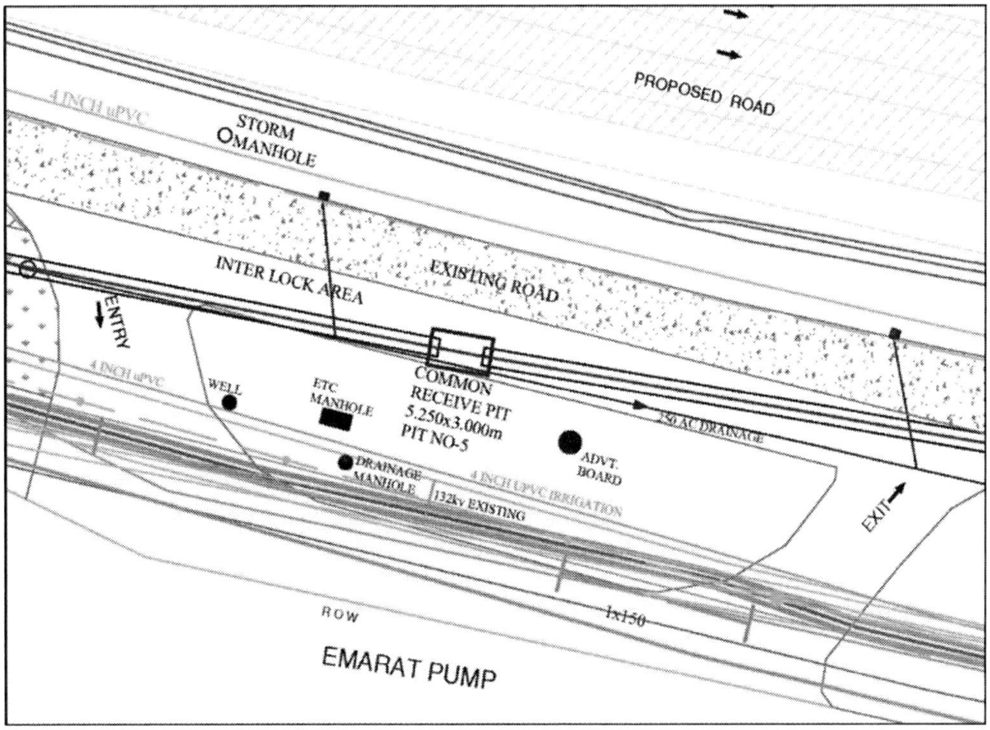

Fig. 6. Alignment of micro-tunnell 9 & 10 (meeting at common receiving pit number 5, proximity of the receiving pit to underground fuel oil tank and other services).

†† Dubai Municipality.

7 Conclusion

The present study has focused exclusively on the Jadaf area in Dubai, United Arab Emirates, in a particular pattern of the soil with specific parameter for a given O.D‡‡ (1490 mm composite pipe with concrete encashment – for I.D§§ 1200 mm*** diameter GRE††† pipe for main water transmission pipe line for DEWA) of the bore and accordingly the micro-tunnelling machine has been selected for typical soil parameter , surrounded by existing underground services and various future proposal of infrastructure. There are further regions such as crossing of water ways (canal / filled up areas etc.) and proximity of the important structure for instance hospital, hotels, educational institute etc, in a different kind of soil parameter. Similar case study is required in other areas to ascertain challenges faced during construction and the amount of surface settlement within the range predicted in theory and after effect of the micro-tunneling. It is also advisable to consider future development at designing stage itself in coordination with different agencies involved in the proposed project.

References

1. Barla, M. and Camusso, M. (2013) 'A method to design micro-tunnelling installations in randomly cemented Torino alluvial soil', Tunnelling and Underground Space Technology, **33**, pp. 73-81.
2. Beard, A. N. (2010) 'Tunnel safety, risk assessment and decision-making', Tunnelling and Underground Space Technology, **25(1)**, pp. 91-94.
3. Bergeson, W. (2014) 'Review of long drive micro-tunneling technology for use on large scale projects', Tunnelling and Underground Space Technology, **39**, pp. 66-72.
4. Camós, C. and Molins, C. (2015) '3D analytical prediction of building damage due to ground subsidence produced by tunneling', Tunnelling and Underground Space Technology, **50**, pp. 424-437.
5. Choo, C. S. and Ong, D. E. L. (October 2015) 'Back - analysis and finite element modeling of jacking forces in weathered rocks', p. 10.
6. Clarke, J. A. and Laefer, D. F. (2014) 'Evaluation of risk assessment procedures for buildings adjacent to tunnelling works', Tunnelling and Underground Space Technology, **40**, pp. 333-342.
7. Denver (2014) Wastewater Capital Projects Management, Standard Construction Specification. Denver, Colorado: Denver.
8. Fu, J., Yang, J., Zhang, X., Klapperich, H. and Abbas, S. M. (2014) 'Response of the ground and adjacent buildings due to tunnelling in completely weathered granitic soil', Tunnelling and Underground Space Technology, **43**, pp. 377-388.
9. Giardina, G., DeJong, M. J. and Mair, R. J. (2015) 'Interaction between surface structures and tunnelling in sand: Centrifuge and computational modelling', Tunnelling and Underground Space Technology, **50**, pp. 465-478.
10. Hough, C. M. and Milligan, G. W. E. (1995) Guide to best practice for the installation of pipe jacks and micro-tunnels (1 vols). Marshall Robinson Roe.
11. Knappett, J., Craig, R. F. and Craig, R. F. S. m. (2012) Craig's soil mechanics. 8th ed. edn. Abingdon, Oxon; New York: Spon Press.

‡‡ Outer diameter of the pipeline completed by micro-tunnelling method.
§§ Internal Diameter.
*** Millimeter.
††† Glass Reinforce Epoxy pipe.

12. Ong, D. E. L. and Choo, C. S. (2016) 'Back-analysis and finite element modeling of jacking forces in weathered rocks. Tunnelling and Underground Space Technology.', **51**, pp. 1-10.
13. PJA (1995) Guide to best practice for the installation of pipe jack and micro-tunnelling. London: Pipe Jacking Association.
14. Terzaghi, K. (1948) Soil Mechanics in Engineering Practice. New York: John Wiley and Sons, 1948 (1966).
15. Thomson, J. C. (1993) Pipejacking and micro-tunnelling. London; New York: Blackie Academic & Professional.

Public private partnerships for climate change mitigation – An Indian case

Dolla Tharun, and *Boeing* Laishram*

Infrastructure Engineering and Management Division, Department of Civil Engineering, Indian Institute of Technology Guwahati, Guwahati – 781 039, India

Abstract. Cities are one of the major contributors to greenhouse gas emissions. Climate change poses serious threat to urban infrastructure, quality of life, and entire urban systems. Cities need to adopt an integrated approach for improvement of city services in order to adapt to climate change and reduce their greenhouse emissions. However, the magnitude of investment required to bridge the widening infrastructure service provision demand-supply gap along with the additional investment to mitigate climate change demands the need to look for innovative financing solutions. Private investments through public private partnership (PPP) route offer an innovative mechanism for meet both the goals of infrastructure development and climate change mitigation. Private parties in PPP, however, focuses on the project economics only though they have the potential to provide innovative technical, financial and managerial solutions. The paper aims to answer the question how to integrate climate change mitigation objective in procurement process of PPP projects. The study has focused only on PPP projects in Municipal Solid Waste Management sector. The integration of climate change mitigation objective has been through design of a modified procurement protocol which promote private sector to devise project structure that fulfil both the objectives of climate change mitigation and provision of quality infrastructure services.

1 Introduction

Cities are the major contributors to greenhouse gas emissions. Most of the economic activities are concentrated in urban areas; as a result cities have a major role in climate change. Cities consume as much as 80% of energy production worldwide and account for a roughly equal share of greenhouse gas emissions. Besides, the relation between greenhouse gas emissions and how the electricity is produced and how energy is used in buildings and transit, the other urban services such as, water treatment, wastewater production, and solid waste generation are all linked among themselves and with greenhouse gas emissions as well. With development, most of the greenhouse gas emissions from cities will be driven more by the energy services for lighting, heating and cooling and less by the industrial

* Corresponding author: boeing@iitg.ernet.in

activity. In addition to this, urban population is expected to double by 2030 and the city built-up area is expected to triple during the same period. This will further increase the energy requirements.

Climate change poses serious threat to urban infrastructure, quality of life, and entire urban systems. Cities need to adopt an integrated approach for improvement of city services in order to adapt to climate change and reduce their greenhouse emissions. Cities need to build enough resilience to respond to extreme events of climate change (for example greater number and intensity of cyclones, heat waves, flooding). A few initiatives which will help the cities to deal with adaptation issues include, *inter alia*, buttressing of key infrastructure (e.g. increased robustness of water and power supply systems), greater adaptive capacity through buildings and critical infrastructure to withstand increased climate variability [1]. Reduction of greenhouse gas emissions from cities would not possible without major changes in cities including (a) increasing urban density (b) improving urban design to avoid sprawl, (c) improving city public transit, (d) changing building practices, and (e) changing sources of energy [1].

Finance to address climate change is a critical concern for cities. The current available funds for mitigation and adaptation, not limited to cities, only amounts to $9 billion per year. The maximum estimated available funding in future for climate change through the UNFCCC and other funds is $100 billion per year, which is far less than the estimated requirement of $275 billion per year for mitigation and adaptation finance [1]. In view of the extremely limited funding, cities need to find alternative sources of finance to fund the adaptation and mitigation measures. Public private partnerships have a great potential to reduce the burden on local finances and provide the necessary resources to address climate change by promoting green growth in cities.

2 Public private partnerships and climate change mitigation

In developing countries, private sector can play a significant role in infrastructure-related climate change mitigation and adaptation programmes through PPP route. In similar line, the National Action Plan on Climate Change (NAPCC) released by Prime Minister of India in June 2008, mainly looks into inclusive sustainable development and inclusion of civil society and PPP in the process [2]. PPP hold great promise for local authorities in building infrastructure for cities to diminish the causes of climate change (mitigation) and effectively protect themselves from its impacts (adaptation) in view of the budgetary constraints and lack of techno-managerial competencies to adopt green technologies. Emerging environment-friendly technologies which are applicable for adoption in urban infrastructure to meet the climate change targets demand great technical expertise and know-how, in addition to large capital availability. The private sector, who specialises in such technologies, could be main drivers for such PPP projects along with the financial institutions to provide the necessary funding to develop these projects.

Public-private partnership (PPP) has been opted by various governments as an alternative form of public procurement. Many countries have adopted PPP on account of different reasons, such as budgetary pressure, inefficient public services to infrastructure, and demand-supply gap. While, some countries have adopted PPP with the expectation of gaining operational efficiency, innovative technological solution, and getting management skills and technical expertise from private sector in delivery of public service.

PPP is a partnership between public sector and private sector for the purpose of delivering a project or service that was traditionally provided by public sector. The private sector side include investors, lenders and companies providing construction and operational services and on the public sector side include public authorities creating and implementing PPP policies as well as those procuring PPP. The parties to PPP have conflicting interests and as a result structuring PPP is quite complex as there is a need to reconcile the interests of the parties involved and relationships between them. Public organisations aim at maximizing the social welfare benefits while the private sector is concerned with maximizing their returns. An enabling environment is required for private sector to participate in infrastructure development projects which will limit the risk exposure such that the reward to them is in line with their risk appetite.

Investors to infrastructure projects, which are being implemented to meet the socio-economic changes, providing private capital undertake rigorous risk analysis to ascertain the risks which could affect the investment goals of the investors. Traditional risks which could be linked to infrastructure projects can be grouped under three broad categories: Policy, political and regulatory risks; Commercial and technical risks; and Business risks (Table 1). These risks associated with the project could affect the financial viability of the projects and these risks are required to be managed beforehand through optimal risk allocation framework, wherein the risks are allocated to the party best able to manage them at the least cost, such that the risk-return profiles are attractive to the private investors. In order to provide an attractive risk-return profile, PPPs are designed with rigid framework comprising of carefully crafted agreements wherein the rights and responsibilities are well defined in advance and mitigating measures incorporated to take care of events which the parties could foresee at the time of drafting the agreement. Security measures are incorporated in PPP arrangements to limit the exposure to wide range of risks which can come into play over the life cycle of the projects. For instance, design risk, technology risk, construction and operation-related risks are mitigated using proven technology with tested construction and operation practices by well-experienced operators. Whereas, risks such as political risk, legal risk, and planning risk are transferred to public partner. However, when climate change has to be integrated with the investment goals in the infrastructure development strategies, the investors are faced with additional challenges that limit the attractiveness of investment in low carbon climate resilient infrastructure projects, such as unsupportive environmental policy backdrop, which distorts the relative pricing of clean versus polluting projects and introduces regulatory risk and uncertainty for private investors; lack of familiarity, information, knowledge and expertise with green infrastructure projects; as well as lack of appropriately structured financing vehicles to provide the risk/return profile that the private sector expect.. These challenges, thus, introduce additional risks linked with climate change aspects of infrastructure projects besides the risks of traditional infrastructure projects (please refer Table 1).

Table 1. Risks Associated with Infrastructure Projects. [3]

Risk Category	Risk Type	Traditional risks linked to infrastructure projects	Additional risks linked to the climate change aspects of infrastructure projects
Political, policy and regulatory risks	*Policy and regulatory risk*	• Lack of political commitment / policy certainty over the long term on infrastructure planning • Tariffs regulations to increase fees with inflation fall behind schedule • High bidding costs involved in the procurement process of infrastructure projects (lack of clarity about administrative rules; risk of corrupt practices); • Fragmentation of the market and lack of policy alignment among across different levels of government.	• Lack of long term low carbon development strategies; • Trade barriers (tariff and non-tariff barriers) on green technologies and/or their inputs; • Lack of harmonized environmental regulations; • Lack of political commitment / policy certainty over the stability of specific forms of support to green investment, such as feed- in tariffs. Instability on the price of carbon, such as weak or unstable environmental regulations; • Existence of fossil fuels subsidies that make other investments more attractive to investors.
	Political and social risk	Opposition from pressure groups; corruption; Short-termism of politicians, limiting infrastructure planning and investment.	Additional forms of opposition to specific LCR technologies or processes, such as carbon capture and storage or wind farms, sea walls.
Commercial, and technical risks	*Technological risk*	Includes the risk of technology failure, obsolescence or under-performance relative to expectations.	Particularly high in the context of low carbon investments as they often involve new technologies. The level of risk will depend on the maturity of the technology.
	Construction risk	Covering delays in the completion of the project, the interface between the different contracts of subcontractors or stakeholders.	Lack of expertise in new climate mitigation and adaptation technologies.
	Operational risk	Lack of ability for management and operation of the asset, and to the decommissioning of the project.	Lack of expertise and track records in new climate mitigation and adaptation technologies.
	Environmental risk	• Unforeseen environmental hazards linked to an infrastructure project; • Climate risk, changing climate can damage the functioning of infrastructure; • Risk related to the uncertainty of climate change in infrastructure for adaptation.	
Market risks	*Business risk*	• More competitors entering the market; • Change in consumer preferences and demand.	• Technological advancement; • Lack of familiarity with new low-carbon technologies.

3 Public private partnership programme in India

The Government of India has undertaken reforms to create an enabling framework for undertaking PPP projects. The enabling framework along with the fiscal incentives has been partially successful in inviting private sector participation in various urban infrastructure sectors. The reforms have also implicitly indicated the need for adopting both proven and innovative technologies for processing, treatment, and disposal of waste and wastewater. For instance, composting, waste to energy, refuse derived fuel, incineration, pyrolysis/gasification, anaerobic digestion and biomethanation are amongst the technologies suggested for treatment of waste in India. Several PPP waste to energy and compost projects have been developed in various cities wherein the revenue, in addition to tipping fees, from selling the generated power and compost has the potential to improve the financial viability of the project and enable the project to be funded on project finance basis (off-balance sheet). However, these projects have been encountering techno-commercial challenges such as inferior quality and inadequate quantity of the waste, inappropriateness of some of the technologies in the Indian context as the technologies require high level of segregation, high capital cost and O&M costs of the technologies, and lack of market for the project outputs.

If infrastructure projects are to be developed using private investment through PPP route to meet both the goals of socio-economic development and arrest the climate change effects, then there is a need to review involvement of private sector using the current PPP models in case of climate change related infrastructure projects. Infrastructure projects with the additional goal of climate change mitigation and adaptation would mandate the need for using innovative and emerging technologies and energy efficient practices. Adoption of emerging technologies could cause conflict with the private sector interest of maximizing return and minimizing risks, and scale of infrastructure development to adapt to adverse effects of climate change could affect the project scope and size thereby affecting viability of the project. This demands not only formulation of new strategies to mitigate the risks arising from adopting green technologies but the need to design the reward mechanism so that private sector is attracted to low-carbon, climate-resilient infrastructure PPPs. The main objective of the study is to address the final research question *How to incorporate climate change mitigation measures in PPP procurement protocol?* The study has focused only on Municipal Solid Waste Management which is one of the urban infrastructure sectors with lot of potential for private sector participation through PPP route.

4 Procurement process for PPP MSW projects

The process for procurement of infrastructure projects through PPP route commences with the creation of the procurement team and the process can be considered to be complete with award of the contract to the preferred bidder. This process is undertaken over the phases, namely, Identification, Development, and Procurement stage. The typical procurement process for awarding the PPP MSW projects through the above-mentioned route is shown in Figure 1. Pre-qualification of the bidders is done based on the parameters: technical competency with reference to the experience in previous projects, financial capability with reference to the annual turnover and net worth. The evaluation of the proposals is based on different bidding parameters depending on whether the PPP project is commercially viable or not. For commercially viable projects, the bidding parameters used for proposal evaluation include: a) highest royalty per MT of MSW, b) highest authorization premium per annum to authority. While, for commercially non-viable project, the bidding parameters include: a) lowest tipping fee per MT from authority; b) lowest grant from the authority [4].

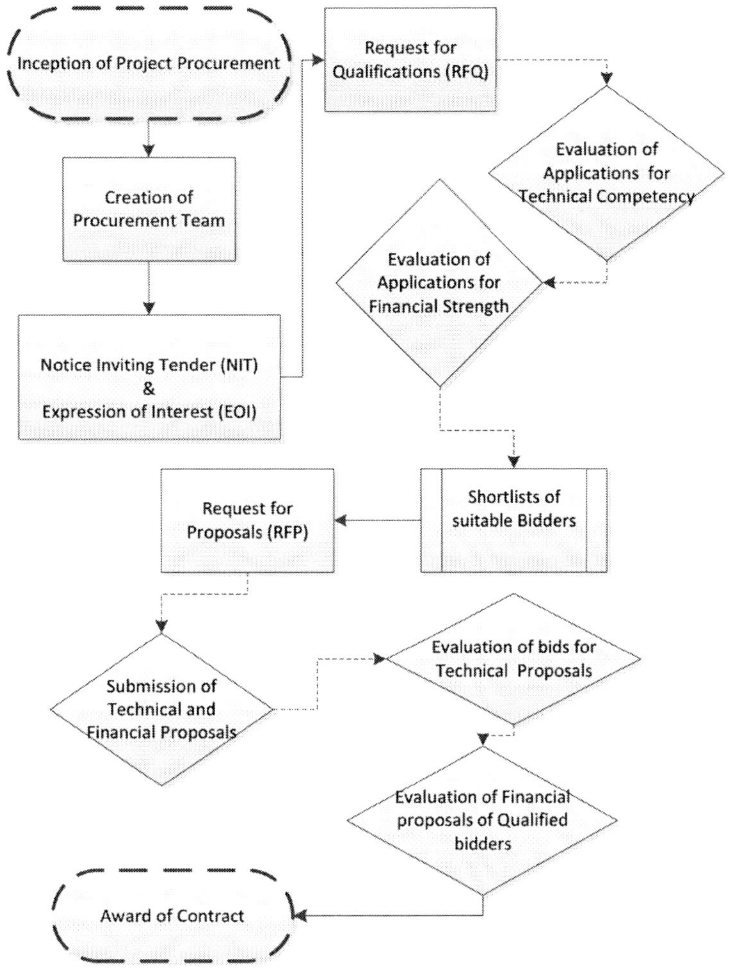

Fig. 1. Step involved in Procurement of MSW infrastructure projects.

The evaluation process of bids adopted in the PPP procurement process has also been critically reviewed to assess the focus of it on climate change (please refer Figure 2). In this model, technical proposals are evaluated in terms of the dimensions such as methodology adopted for the project appraisal, process flow and material balance, resource utilization, operation and maintenance scheme, quality assurance and quality control procedure, time schedule and environmental, health and safety policies. The existing practice of project proposal evaluation does not give too much focus on climate change. Maximum allowable scores are being pre-assigned for each and every parameter relating to the above dimensions. These maximum allowable scores can also be changed by the user depending on the perceived degree of importance of the dimension. The technical evaluation gives an output of the score a bidder gets out of maximum score of 100. A five point Likert scale, varying from 1 being poor to 5 being excellent, has been used for the performance evaluation of the technical proposals with respect to the above-mentioned dimensions.

Financial evaluation of MSW projects is done based on the assumption that the project is not financially viable. The project is assumed that it will not be able to generate enough revenues to recover the costs. Tipping fee is supported with additional capital grant, normally equaling the lowest capital grant in the bidding process. The preferred bidder is then selected in the final stage based on the composite score, which is the sum of score obtained for technical evaluation and score obtained for the financial evaluation.

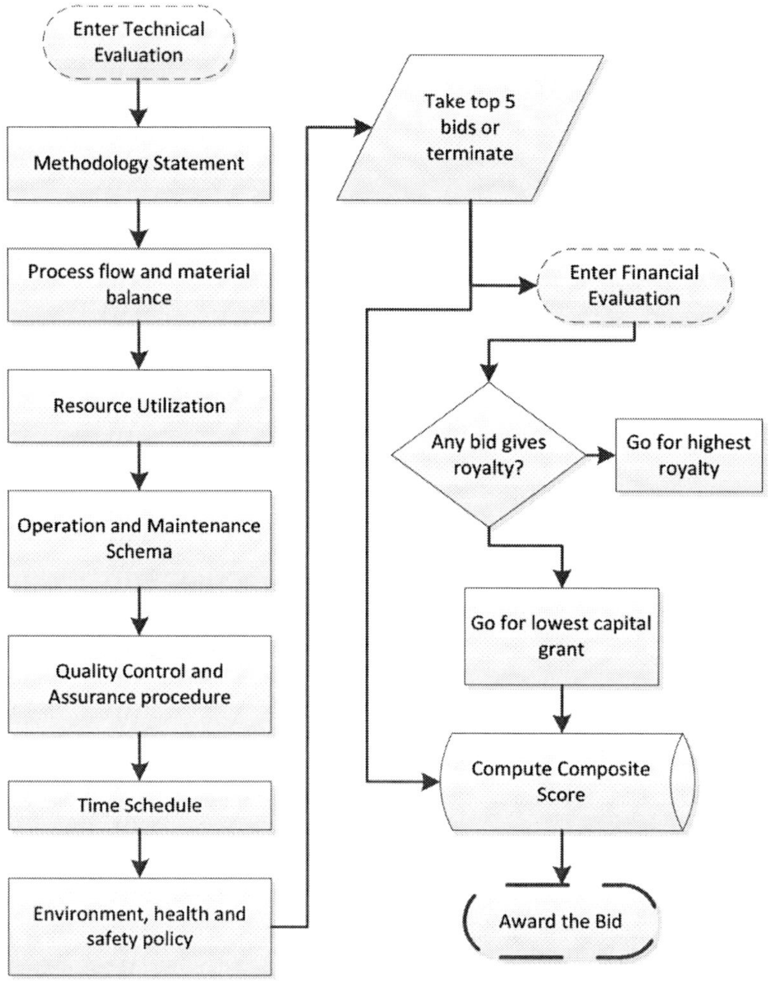

Fig. 2. Evaluation process of bids.

5 Research methodology

Case study research strategy suggested by Yin [5] had been adopted to answer the research question. Documents and interviews were the sources of evidence used in the research design of case studies. An introductory appraisal of the research is presented to the prospective interviewees and positive responses were interviewed in person. Semi structured interview template is employed to guide the interview process and to act as a reminder. The process of interview is conversational in manner and the interviewee and interviewer exercised liberty to go beyond the intended questions. Answers were hand written for transcription purposes. The interviews lasted for a minimum of 1 hour to 3 hours in a single stretch. In addition to the interview, field visits to site of operations, conversations with site in-charges, and NGOs carrying out the waste collection were carried out to gain in-depth knowledge and understanding of the project, process as well as to validate the strategies to overcome the hurdles. At least 10 hours of interviews were carried for each project included in the case studies.

Current study followed the research procedures established by [5-7]. The interview data is analyzed with a qualitative software Nvivo [Version 10] developed by QSR International [8]. Qualitative data is marked by its nature of depth of the subject the people involved. The process followed in analysis of the interview data has been drawn heavily from [6] and [7]. The case study analyses adopted for the study is based on the recommendations of Yin (2009). The interviews were verbally transcribed and were fed into Nvivo software. The same person had administered the interview and performed coding thereby eliminating the error of 'coding variance'. Analysis of the interviews have been done using 'hypothesis-coding' method in first stage of the analysis [6, 7]. Since the research didn't have a grounded theoretical propositions, 'developing case descriptions' strategy was employed in 'explanation building' analytic technique as recommended by Yin [5]. Creating codes followed a deductive approach from the literature review, hypothesis and the variables that we intend to observe. Each node holds the information about a variable or process intended to be observed. These first nodes are further refined to yield the pattern codes. The objective of pattern coding is to condense the amount of data into small and manageable units. Narrative 'rich' description is considered the most appropriate for current research objective Yin [5]. Finally, the output of the case analysis is presented in narrative and the application developed is presented in matrix displays in the culmination of all case studies.

6 Modified procurement protocol

The case study analysis has highlighted that the procurement process being used by the various local government for procurement of PPP projects in India suffers from various setbacks if the procurement process is to be designed to promote infrastructure development along with the goals of climate change mitigation. Some of the major limitations of the existing procurement protocol from climate change mitigation perspective are as follows:

1. No opportunity to innovate, as the procurement is predominantly prescriptive rather than output oriented.
2. No interlinking of payment mechanism and hence poor control on contract management.
3. No integration of incentive mechanism and promotion of private sector to optimize.
4. No emphasis on recycling.
5. The objective of climate change mitigation has not received its required utmost attention.
6. It assumes bundling of activities as a means of obtaining efficiency.

Based on the feedbacks from the case study, a modified procurement protocol has been devised. The modified procurement protocol is shown in Figure 3. The principles and objectives governing the modified procurement protocol to promote both infrastructure development and climate change mitigation other than what is already existing are outlined as follows:

1. Promotion of low carbon technology for effective reduction of GHG emissions from all the process involved in the project is of main emphasis.

2. Best value in this context of infrastructure not only mean financial sustainability but also project which meets the need by providing service which is both sustainable and utilized the low carbon technology.
3. Modified procurement protocol facilitates, promotes, and incentivizes the private sector for emission reduction in appropriate manner.
4. Payment mechanism should be robust enough to capture the necessity of service levels of the project as well as high remission reductions.

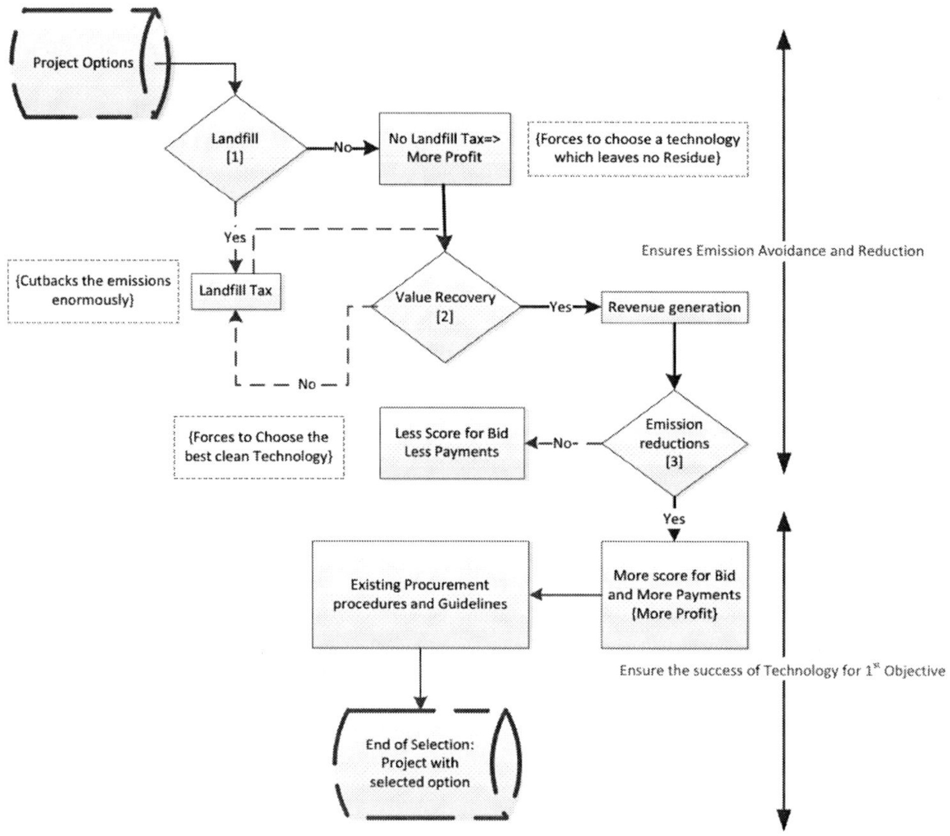

Fig. 3. Modified Procurement Process for PPP MSW Projects.

Performance based payments are recommended and the performance is measured according to the emission reductions. Any changes occurring during the asset life because of technological advancements, adaptation or modified usage can be incorporated to facilitate emissions reduction. However, the incentive is only limited to the performance payments. The rational of the new protocol is to promote the usage of clean technology and in mitigating GHG emissions and the rationale for the modified framework is based on the following concepts:

Landfill, material recycling, and emission reductions: The importance of each category can be viewed in terms of all the dimensions of sustainability. Carbon pricing comprises of two components: (i) carbon tax, and (ii) emission trading schemes [9]. From sustainability perspective, economic sustainability of landfill can be looked in terms of the value of the land consumed instead of some other opportunity value, environmental sustainability with respect to the methane emission that are caused from the material going into it, whatever insignificant quantity may be, and social sustainability with respect to the social discomfort caused by the site it at all any. The desired direction of procurement process could be as follows: A landfill tax on the private entity for dumping the waste into the sanitary landfill could force him to innovate and build facilities from time to time that will force him to divert the waste from landfill. In such case, a no tax scenario itself is the profitable scenario so that the process brings legitimacy. In spite of the agreement of respondents on the philosophy of landfill tax, the applicability of such methodology is being questioned owing to the factors like large number of small-scale projects or distributed project in a city. Implementation of landfill tax requires unbundling of the package 'construction of sanitary landfill and operation' amongst the various PPP players. Though such bundling might be possible, the responsibility of landfill care is vested with the public sector in US, where this model is widely used.

Similarly, material recovery has an unseen, unharnessed potential, which is often less communicated or underestimated. The more we recycle the items with high emission potential, lesser will be the impact on emission of the whole project. Even though the direct emission reduction is low for steel and aluminum, huge reduction in GHG is possible if the virgin products are replaced with recycled products. This is referred as downstream emissions or indirect emissions. Task force setup by GOI strongly recommends the diversion of at least 75% of waste from landfill using appropriate technologies [10]. It should be noted that recycling and source segregation is the key while technology can provide additional support for diversion rates.

Furthermore, quantitative emission reduction measurement is of utmost importance and it is the crux of low carbon pathways. This helps in setting the benchmark and investing in much lower technologies in future. However, the critical concern here is that unless the benefit or reward that a private sector can accumulate over years is greater than the investment it may require to bring in much more efficient systems, the private sector may settle with the business-as-usual scenario in comparison to the opportunity cost. This demands a rewarding mechanism based on the objective emission reduction the project is achieving in a given financial year. A payment mechanism in terms of rupees per tCO_2e reduced from the project scenario will help and encourage the private sector to always innovate and bring new potential measures even in the contract management phase. This, paradigm of performance-based payments is entirely different from the payment mechanisms being used in conventional project scenario. However, owing to the high competency levels required in public sector in spite of wide range of transient competency enhancement strategies used in India like transaction advisors, these kind of dynamic payment mechanisms require higher competency levels in public sector organizations also. Much rigorous studies are further needed to guide the ULBs and PPP coordinating agencies in order to implement such kind of payment mechanism.

7 Conclusions

Development of urban infrastructure using private sector participation has been widely adopted by various cities to create a sustainable built environment. Private sector participation has been sought by the governments to supplement the scare budgetary resources in the light of increasing infrastructure demand-supply gaps. Furthermore, the rationale for widespread private sector participation in urban infrastructure development is based on the efficiency gains brought by their participation. Through well-designed procurement protocol, public private partnership approach has the potential to promote innovative solutions for infrastructure development.

Public private partnerships have huge potential for incorporation of climate change mitigation measures besides providing quality infrastructure services. The current procurement process being used by Indian Government for private sector participation in development of MSW projects has been studied through case study research approach to study the extent to which it promotes climate change mitigation measures besides infrastructure development goals fulfilment. The current procurement process has limitations in promoting this objective and to overcome this a modified procurement protocol has been formulated based on the inputs from the case study research. The modified protocol incorporates criteria that will encourage them to incorporate measures for climate change mitigation while structuring MSW projects to be developed through PPP route. This procurement protocol, however, needs to be studied in depth to examine its applicability.

The authors appreciate the support received from HUDCO's Human Settlement Management Institute (Research and Training Wing of HUDCO), New Delhi for carrying this research. The views expressed herein are those of the author and not necessarily those of the HSMI, HUDCO.

References

1. World Bank, *Cities and climate change - An urgent agenda*, in *Urban Development Series - Knowledge papers*. 2010, The World Bank: Washington DC.
2. Government of India. *National Action Plan on Climate Change*. 2008 [Accessed on 2012 August 18]; Available from: pmindia.nic.in/Climate%20Change.doc.
3. Kennedy, C. and J. Corfee-Morlot, *Mobilising investment in low carbon, climate resilient infrastructure*. 2012, OECD: Paris.
4. MoUD, *Model RFP for ISWM on PPP-BOT : Volume 1 Instruction to Bidders*. 2010, Ministry of Urban Development, Government of India: New Delhi, India.
5. Yin, R.K., *Case study Research*. 2009, Thousand Oaks, California: SAGE Publications, Inc.
6. Miles, M.B., M, H., & Saldana, J. (2014). *Qualitative Data Analysis: An Expanded Sourcebook* 2014, Thousand Oaks, California: SAGE Publications, Inc.
7. Saldaña, J., *The Coding Manual for Qualitative Researchers*. 2011, Thousand Oaks, California: SAGE Publications, Inc.
8. QSR, *Nvivo 10*. 2012, QSR International Pty. Ltd.: Melbourne.
9. ADB, *Policies and practices low- carbon green growth in Asia*. 2012, Asian Development Bank: Phillippines.
10. Planning Commission, *Report of the Task Force on Waste to Energy*. 2014, Planning Commission, Government of India: New Delhi.

Factors influencing cost over-run in Indian construction projects

A. *Cindrela Devi*[1,*] , and K. *Ananthanarayanan*[2]

[1]Doctoral student, Civil Engineering Department, Indian Institute of Technology Madras, India
[2]Professor, Civil Engineering Department, Indian Institute of Technology Madras, India

Abstract. Construction cost is the most important criteria of project success and hence the construction project performance is generally expressed in terms of cost and its variance from the budget. In spite of having extant literature, cost estimation methods, cost indices etc., construction projects rarely meet the budgeted cost. This research study focuses on the construction cost overrun and to identify the various factors that affects the construction cost performance. Based on an extensive literature review and input from industry experts, sixty eight factors that causes cost overrun were identified for investigation. Further, a structured questionnaire survey was conducted among the industry experts and the collected data has been analysed statistically. It is concluded that the factors namely scope creep, construction delays, rework and practise of awarding the contract to the lowest bidder are most significant factors for construction cost overrun in non-infrastructural Indian projects. The relative importance of the listed factors used to guide the project team in addressing the cost related risks involved in the projects. The findings are expected to bridge the gap in the current construction cost management practices.

1 Introduction

The construction project performance is generally expressed in terms of time and cost variance against its baseline. Out of the four fundamental constraints namely scope, cost, time and quality, cost performance is the most essential and common issue in the global construction industry [1]. It is important to measure the cost variance in construction to understand the performance of the project and thereby to understand financial risks involved in the project execution. The cost variance, resulting as project cost overrun is denoted as a negative impact on economy and the profitability. Several perspectives on cost overrun are available in extant literature and are proven valid. However, it is not adequately explained on why the cost overrun keeps occurring though sufficient knowledge on cost overrun has been largely shared.

The causes of inaccuracy in forecasts are different for different projects. The change of governance structures for forecasting the project development as an external factor also threat the project planning and execution. It was found that projects do not perform as forecasted,

* Corresponding author: cindreladevi@gmail.com

in terms of costs: almost 9 out of 10 projects fall victim to significant cost overrun. The underlying causes of such inaccurate cost projections were investigated and summarized [2].

The research is proposed to improve current construction practices of determining the construction cost and controlling the cost as the project progresses. The aim is to propose a framework for managing the risk factors involved in cost performance of non-infrastructural construction projects and thereby developing construction cost assurance. This will be achieved on proper understanding of various risk factors which are the causes of cost overrun and understanding their impact on the project execution. To obtain this, the causes of cost overrun are to be identified and classified according to the different stages and progress of the project

2 Motivation and need for study

Cost overruns in construction projects around the world are very common. It represents the performance of the construction project and indicates the economic losses of the project. Cost overrun study was conducted across 20 nations in five continents and found that the average cost overrun for megascale projects could range from 20.4% to 44.7% [3]. Over the past 70 years, there have been no systematic improvements in cost overrun of the projects [4]. There is no established pattern for the occurrence of cost overrun [5]. In fact, cost overrun have the same magnitude as they had 30 years before, no improvements exists [6]. Hence the prediction of cost overrun, cause, its pattern, its frequency, impact on the progress are need to be explored to have a successful project completion.

The construction progress of projects are directly linked with the cash-flow and hence any unlikely performance of construction progress will directly hit the financial commitments. Traditionally, cost-benefit analysis is practiced for the investment purpose. It represents a most likely value of the project rather than the actual value. The input variables are carefully considered in 'what-if' scenarios or sensitivity tests for the analysis. However, the assessment of construction projects increasingly requires a greater understanding of its complexity on project progress and corresponding cash-flow. Hence, the number of 'what-if' scenario combinations were generated which makes the decision-making process more difficult. Therefore cost related risk factors are analysed in this research which will give strong base rather than having analysis on mythical or impractical situations.

Referring to the literature, it is obvious that there are number of causes of overruns and the impact on the construction costs are almost in similar patterns. In a study, eighty six explanatory factors had been listed out for the cost overrun [7]. It will be practically useful and have logical meaning for developing classification of causes based on their effect on the overruns of infrastructure projects [8]. However, in order to understand how and why cost overrun occur, listing down the causes of cost overrun can contribute minimum. Hence it is required to establish a framework for the cost performance of construction project in general and cost overrun in detail.

3 Review of literature

Cost is noted as one of the most important parameters and key driving force of project success among other considerations [9]. The measures such as total budgeted cost, cumulative budgeted cost, cumulative actual cost, cumulative earned value and estimated cost at completion are useful in evaluating the cost performance of the project [10, 11]. Poor cost performance has been a major concern for all stakeholders in the construction project. Despite the large number of reported cases of poor performance, it seems that construction projects have increasingly experienced significant cost overruns on all projects ranging from the simplest to more complex projects such as nuclear plants, transportation systems and oil & gas platforms [12]. As a result, many construction organizations have faced successive financial burdens, which often have led to insolvency and bankruptcy especially private construction sector.

Extant literature on cost performance is available for the government funding construction projects such as rail and road projects. However, non-infrastructural projects and profit oriented projects like residential, commercial constructions have not been studied on cost performance as the market is of huge competitors and the information on costing is being kept confidential. The challenge in the non-infrastructural construction project is sourcing of funds, cash-flow management, and completion of project on time. The performance of project is directly related to the cash-flow of the project, which in turn looped to the payment to the contractors that affects the construction progress [13].

The cost performance of the project highly rely on cost overrun, which is referred as an increase in the budget, cost or any growth in the cost. It has been calculated as the difference between the actual cost and the forecasted cost of the project [14]. If the percentage of cost overrun for the same project is compared by different planners may be different. The contradiction in the results of these studies is because of the point of reference based on which the cost overrun is measured. The reference point for estimating the cost overrun should be derived at the detailed planning stage where design, scope, specification and final cost are developed [15]. An alternative point of reference for determining cost overrun is the award of contract, which is the final negotiated and agreed price at the construction commencement [16, 17]. The relationship between project size and cost overruns is notable as the larger the construction project is greater in the percentage cost overrun. The factors that contribute cost overrun are price escalation, poor design and implementation, inadequate financial plans, administrative uncertainty and the lack of coordination between enterprises [18]. Apart from these factors mentioned, the completion time or the duration of the project will also play a vital role.

Too many complexities in construction projects allow more probable cost overrun causes to emerge, and so risks and uncertainties increases. Hence, there might be greater variance in the cost and time schedule which leads to unsuccessful projects [19]. There are four types of explanations had been categorized for forecasting inaccuracy namely technical, economical, psychological and political explanations [20].

Two main causes of cost overruns because of the inaccuracies in the cost estimates are optimism bias and strategic misrepresentations [21]. Optimism bias can be explained as the cognitive disposition to evaluate future events in fairer light than they might actually be in reality [22]. The strategic misrepresentation is accounted for the systematic underestimation of the true costs and overestimation of benefits. The political and organizational pressures are very high for strategic misrepresentation as it is committed intentionally. Human judgments are generally optimistic due to overconfidence and incomplete information [23]. Thus people will underestimate the costs, and overestimate the benefits of the same action. This behaviour is called as planning fallacy. To get the true costs of the project and to have control over the cost overrun, better understanding is required on its pervasiveness of inaccuracy and risk in decision-making.

The other critical problem is project finance that obtains funds to bridge the time between making expenditures and obtaining revenues. Based on the concept design, the budget and the construction plan, the cash flow for a project can be estimated. Normally, this cash flow will involve expenditures in early periods. The challenge is covering this negative cash balance in the most beneficial or cost effective way. [24].

The above findings of various research have been integrated to develop a cost control mechanism with respect to non-infrastructural construction project in order to achieve project success (i.e) meeting the budgeted cost, scheduled time with the acceptable quality

4 Research objectives and methodology

The research is proposed to improve current construction practices of determining the construction cost and controlling the cost of non-infrastructural project. The objectives of the research are as follows

- To identify the causes of cost overrun and their impact on project performance at various stages of the project
- To identify cost performance risk indicators and its frequency of occurrence in the project life cycle and thereby accurate cost can be predicted

The proposed methodology includes;

- Studying the literature on construction cost performance, cost overrun and its causes.
- Doing the case/ethnographical studies to observe the cost performance and to collect the data. Empirical evidence of cost performance risk factors, cost overrun and its causes were studied
- Questionnaire surveys and interviews were conducted with the stakeholders and professionals in Indian construction. The questionnaire was designed based on review of related literature and ethnographic studies. Statistical analysis was employed to discover the linkages between the cost overrun, risks involved and their impact on project performance.
- Identifying and categorizing the causes and impact of overrun in different stages of the project so as to develop a theoretical framework that portrays the occurrence of cost overrun at different stages of the project execution. It will demonstrate the relationships between types of cost overrun, its causes, and corresponding management approaches. Thus, the understanding of the costs and cost related risks, risks and risk related costs with respect to the project cost performance will be emphasized. The framework will sort the cost overrun causes and their occurrence with respect to the project progress. It will recognize and describe patterns of relationships among causes, overrun and corresponding management methodologies. This can be used as an indicator to predict the risk involved in the project execution at various stages and for the stakeholders.

5 Questionnaire survey

5.1 Questionnaire design

The questionnaire has been designed in such a way that it considers the objective of the study with the aim to answer the research questions. Based on an extensive literature review and input from industry experts, sixty eight factors that causes cost overrun were identified as shown in the Table.1. To validate the listed factors with respect to Indian construction sector, few discussions and feedback sessions were conducted with the industry professionals.

5.2 Contents of the questionnaire

The questionnaire has two sections comprising general information and the causes of cost overrun. The major causes of cost over-runs from the literature have been listed and the respondents were asked to state the frequency of causes of cost over-run happening in their past or current projects. The reply has been noted in terms of five point likert scale for each of the listed causes that was organised in the form of frequency scaling (1 = never, 2 = seldom, 3 = sometimes, 4 = often , 5= always).The questionnaire was distributed to planning managers and cost consultants of 15 mega sized non-infrastructural projects like residential, commercial, hotel etc. The objective of the questionnaire survey is

- To confirm the findings of cost over-run from the literature, in lieu with industry practices
- To understand the existence of cost overrun / risk factors in Indian construction (non-infrastructure) projects
- To establish the relative importance of the various factors responsible for cost overruns
- To understand the linkages between the causes of overrun

5.3 Data Analysis

In order to establish the relative importance of the various factors that are responsible for cost over-run, the score for each factor is calculated. This is done by summing up the scores given by all respondents. The relative importance weight is derived using the following equation.

$$\text{Relative Importance Weight (RIW)} = \frac{\sum_{i=1}^{5} a_i.n_i}{\sum_{j=1}^{N} x_j} x\ 100 \qquad (1)$$

where i, is score of the factor ranging from 'never' to 'always' as shown below, N is total no of factors, x_j is the sum of the jth factor, j is the factor from 1 to N, a_i is the weight expressing for the ith response, n_i is the frequency of the ith response from all respondents

Using the equation (1), the Relative importance weight is calculated for the cost over-run factors describing for a response of 'never' $a_1 = 1$, for a response of 'seldom' $a_2 = 2$, for a response of 'sometimes' $a_3 = 3$, for a response of 'often' $a_4 = 4$, for a response of 'always' $a_5 = 5$. The RIW is a useful measure of relative positioning of particular factor. Though this score does not represent the statistically significant measurement, it is a good indicator of occurrence of cost overrun in projects

Table 2. The causes of cost over-run with perceived ranks.

S.No	Causes of overrun	RIW	S.No	Causes of overrun	RIW
1	Increase in material prices	1.77	35	Changes in the specifications	2.10
2	Inflation	1.70	36	Design error	1.54
3	Increase in wages	1.67	37	High interest charged by the Bankers on loan	1.17
4	Price fluctuations	1.74	38	Incomplete drawings	1.90
5	Labour cost increased due to	1.24	39	Inadequate specifications	1.67
6	Financial difficulties of	0.90	40	Waste on site	1.80
7	slow or delayed payment of	1.40	41	Equipment availability and failure	1.17
8	Market conditions	1.64	42	Delay in decision making	1.84
9	Fluctuation in money	1.14	43	Poor financial control at site	1.24
10	Deficiencies in cost estimate	1.34	44	Many works being done at the same time	1.67
11	deficiencies in the	1.14	45	Lack of constructability	1.04
12	Insurance problems	0.74	46	Complicated design	1.34
13	Problems related to work	1.00	47	Changes in the scope of the project	2.07
14	Problems related to workers'	0.97	48	Insufficient equipment and technology	1.20
15	Additional works	2.21	49	Optimism bias	1.07
16	Inaccurate quantity take-off	1.24	50	Fraudulent practices	0.84
17	Lack of experience in project	1.00	51	Disputes on site	1.47
18	Lack of communication	1.60	52	Owner interferences	1.30
19	Lack of experience of project	1.14	53	Lack of experience of local regulations	1.30
20	Contractor's poor site	1.54	54	Change of regulations by Govt	1.60
21	Inadequate contractor	1.30	55	Political complexities	1.57
22	Shortage of site workers	1.84	56	Practice of assigning the contract to the lowest	2.04
23	Unrealistic contract duration	1.40	57	strategic misrepresentation	1.14
24	Mistakes during construction	1.50	58	Change of Laws and regulatory framework	1.54
25	Inaccurate site investigation	1.24	59	Unpredictable weather conditions	1.90
26	Lack of coordination between	1.90	60	Unforeseen site conditions	1.57
27	Rework	2.00	61	Deficiencies in the social structure	1.04
28	Unexpected subsoil	1.70	62	Heritage materials being discovered	0.63
29	poor technical performance	1.10	63	Construction delays	2.34
30	Design changes	2.14	64	Inaccurate estimates	1.27
31	Incorrect planning and	1.80	65	Cash flow and financial difficulties faced by	1.94
32	delay in material procurement	1.84	66	Insufficient time to prepare estimate	1.00
33	Delays in design	1.70	67	Shortage of materials	1.37
34	Late delivery of materials and	1.50	68	Lack of experience in contract.	1.24

In order to measure the degree of agreement among the respondents on the construction cost over-run on the cost performance of the projects, two hypotheses were developed, as follows:.

- Null hypothesis H_0: Disagreement among respondents in occurrence of the cost overrun in the construction projects
- Alternative Hypothesis H_1: Agreement among respondents in occurrence of the cost overrun in the construction projects

If there is significant agreement among respondents on the cost overrun issues on construction projects, the contribution made in this research are highly relevant in achieving the targeted cost performance and thereby achieving success in the project. To test the commonality among the respondents, Kendall's coefficient of concordance (W) is considered as relevant for the study and with respect to the attribute ranking. Since the ranks have more ties in the factors listed, the Kendall's coefficient has been calculated considering the ties between the factors using the below formula.

$$\text{Kendall's Coefficient of Concordance (W)} = \frac{12s^2 - 3m^2k(k+1)^2}{m^2(k^3-k) - mT} \quad (2)$$

where k is total no of factors, m is no of respondents, T is ties in the scores, S is sum of scores. The Calculated W value is 0.56. As the number of factors (N) is too large to check on the critical values of Kendall's, chi-square approximation of sampling distribution of W is calculated with the below equation.

$$\text{Chi square } X^2 = k(m-1)W \quad (3)$$

The calculate Chi-square value is 567.72 is greater than the critical value of chi-square from the table for the degree of freedom = 67 and significant level = 0.05. Hence, the null hypothesis is rejected and alternative hypothesis is accepted. Therefore, it is concluded that there is significant degree of agreement among the respondents for the occurrence of cost – overrun in the non-infrastructural Indian construction projects.

5.4 Survey results and conclusions

A summary of all the factors causing the cost overrun is been listed and as seen in Table 2, the overall ranking reveals that various factors influence the cost performance in projects. The most significant factors that cause construction cost overrun are shown in Table 3 based on the Eq (1) with the appropriate occurrence on the project life cycle and classified as internal and external issues

Table 3. Most significant factors of causes of cost over-run.

Project stage	Internal Issues	External Issues
Pre-planning	Financial difficulties	Delay in regulatory approvals
Planning & Design	Scope creep Design / specification changes Poor selection of contractors Ineffective procurement planning	Changes of laws and regulatory framework Political complexities Unforeseen site conditions
Execution and Monitoring	Rework Construction Delays Ineffective planning & monitoring Delay in decision making Contractual disputes	Inflation / price fluctuations

Considering all the identified factors along with their potential occurrence in the project life cycle, Fig.1 shows the percentage of frequency of cost overrun in the project performance. The study says that the cost over-run occurs in detailed design stage about 68% as often and always and about 37% at execution stage of the project.

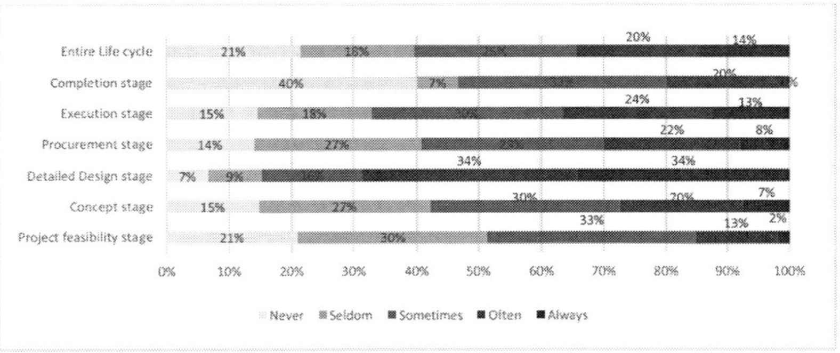

Fig. 1. Cost overrun occurs during the phases of construction.

This study has been focused on assessing cost overrun problems and their causative factors in the non-infrastructural Indian construction projects. The structured questionnaire was used to acquire information on the relative importance of cost overrun factors. A descriptive statistical analysis was carried out and the following findings were discovered

- Cost overrun is a major issue in project performance and agreed by 97% of the respondents
- The cost baseline is been fixed at different stages of project for various projects and is been revised many times in the project
- The most significant cause of cost overrun includes scope creep, construction delays, rework and practise of awarding the contract to the lowest bidder
- the changes in design and specifications, changes in scope, additional works, design errors, incomplete design, delay in design delivery reflects in delayed construction, excess contractual claims, disputes at site and poor project management
- price fluctuations, inflation, increase in wages and material prices could be related to the contractual clauses to avoid cost overrun during execution of the project
- poor coordination at site, poor communication are the internal factors that reflects the delay in construction and contractual disputes
- proper procurement strategy to be executed for awarding the contracts / consultants / suppliers as the result shown as the practice of boarding lowest bidder reflect cost overrun and extra claims at the project closure
- delayed construction contributes poor project performance in terms of cost also. It results in poor scheduling of project, complexity of constructions not considered in scheduling of the project. Hence appropriate risk analysis tool and scheduling methods to be used to mitigate.
- the internal factors mentioned in the Table 3 can be mitigated through proper training to the planning team / managers
- the external factors mentioned in the Table 3 can be mitigated through appropriate risk analysis technique before project inception

6 Recommendation

The objective of this study has been to investigate the statistical relationship between the cost overrun factors and the causes. The data has revealed that the most significant cost overrun factors that occur in various stages of the project. With overwhelming statistical significance, it is evident that there is significant agreement among the respondents on the occurrence of cost overrun in the non-infrastructural Indian projects. From the findings of this study, project managers can effectively mitigate risks of significant overruns by executing corrective and preventive action plans. Few other recommendations for mitigating the cost over-run risks are as follows:

- Decision making is important in all stages on project life cycle cost
- To include cost specific (eg escalation) clauses as a part of contract agreement
- To develop cost effective project planning & design by minimizing waste, having fixed price contract for key services, procuring critical and volatile materials early, having corporate alliance with key vendors / contractors and having long term supply agreements / item rate contracts
- Grouping of causes -- with respect to the responsibilities of stakeholders rather than explaining reasons for the overruns
- Identifying where the cost overrun could have happened in the project timeline against project progress and how could have been eradicated

References

1. Donald Towey, *Cost Management of Construction Projects* (2013)
2. Flyvbjerg, Bent, Mette K. Skamris Holm, and Søren L. Buhl. J. American Planning Asso, 68, **3**, 279-295 (2002)
3. Flyvbjerg, B., N. Bruzelius and W. Rothengatter. *Megaprojects and Risk: An Anatomy of Ambition* (2003a).
4. Bruzelius, N., B. Flyvbjerg, W. Rothengatter. Transport Policy, 9, **2**, 143–154 (2002).
5. Lundman P. *Cost Management for Underground Infrastructure Projects: A Case Studyon Cost Increase and its Causes*, Luleå University of Technology, report. (2011)
6. Flyvbjerg, Bent, Mette K. Skamris Holm, and Søren L. Buhl. Transport Reviews, 24, **1**, 3-18 (2004)
7. Memon, A. H., Rahman, I. A. & Aziz, A. A. A. Int. J. Real Estate Studies, 7 (2012)
8. Cantarelli C, Flyvbjerg B, Molin E J E, & Van Wee B. European J. Trans. and Infra Research, 10, **1**, 5-18 (2010)
9. Azhar, N., Rizwan U. Farooqui & Ahmed, S.M. Advancing and Integrating Construction Education, Research & Practice, 499-508 (2008).
10. Gido, J. and Clements, J. P. *Successful Project Management*, 2nd edition. (2003)
11. Suqrat Babar, Muhammad Thaheem, J. Const. Engg. Man. 143, **3** (2017)
12. Al-Bahar, J.F. and Crandall, K.C. J. Const. Engg. Man., 116, **3**, 533- 546 (1990).
13. Wenhua Hou. J. Convergence Information Technology, 7, **18**, 29-35 (2013)
14. Merrow A H, Rahman I A & Azis A. Int J. Sustainable Construction Engineering & Technology, 2, **1**, 57-71 (2011)
15. James Odeck. Transport policy, 11, **1**, 43-53 (2004)
16. Ganuza, J. J.of Industrial Eco, 55, **4**, 633-660. (2007)
17. Love P E D, Wang X, Sing C & Tiong R. J. Const. Engg. Man. 139, 321-330 (2013)
18. Cantarelli C C, Molin E J E, van Wee B & Flyvbjerg, B. Transport Policy, **22**, 49–56 . (2012)
19. Abdelgawad, M., Fayek, A.R. J. Const. Engg. Man. 137, 294–302 (2010).
20. Flyvbjerg, Bent, Mette K. Skamris Holm, and Søren L. Buhl. J.American Planning Association, 71, **2**, 131-146 (2005)
21. Flyvbjerg, Bent and Cowi. *Procedures for Dealing with Optimism Bias in Transport Planning,* Guidance Document, UK Department for Transport. (2004)
22. Lovallo, Dan and Daniel Kahneman. *Delusions of Success: How Optimism Undermines Executives' Decisions,* Harvard Business Review, 7, 56-63. (2003)
23. Kahneman, D. and Tversky, A. Econometrica, 47, 313-327. (1979)
24. Jing Du, Byung-Cheol Kim, Dong Zhao, J. Const. Engg. Man. 142, **6** (2016)

A review of hot climate concreting, and the appropriate procedures for ordinary jobsites in developing countries

Nabil Bella[1,*], *Ilham Aguida* Bella[1] , and *Aissa* Asroun[2]

[1] FIMAS Laboratory of reliability of materials and structures in south, Tahri Mohamed University, B.P. 417, 08000, Béchar, Algeria
[2] LGCE Laboratory of civil engineering and environment, Djillali Liabes University, BP 89, 22000. Sidi Bel Abbés, Algérie

Abstract. Hot weather concreting involves some procedures to reduce negative effects caused principally by excessive water evaporation from the concrete surface. Potential problems for fresh concrete are: increased demand for water, increased the tendency the rate of slump loss corresponding to add water on job-site, an increased in execution rate, increased tendency for plastic shrinkage cracking and increased difficulty in controlling occluded air. Potential problems for hardened concrete may include: reduction of resistance at 28 days and long-term resulting of higher water demand and/or higher temperature of concrete, decreased durability resulting from cracking. Most developing countries have hot climate, ordinary jobsites in developing countries are characterised by reduced of human resources, equipment and infrastructures. This paper briefly reviews hot climate concreting procedures, especially the latest research in developing countries, and discusses the most appropriate in developing countries.

1 Introduction

Hot climate concrete is when the concrete is mixed, transporting, and pured in an elevated ambient temperature, low humidity and under solar radiation or exposed to the wind; Or any combination of the above-mentioned climatic conditions, which tends to deteriorate the qualities of the hardened or hardened concrete by the acceleration of the rate of water loss or the rate of hydration of the cement, which requires precautions To be taken, is defined the concreting in the hot climate by the ACI 305. Other standards, in order to avoid hot climate problems, impose a maximum limit of temperature. For example, BS 8110: Part 1: 1985 (British standard), at placing time no part of the concrete must have a temperature exceeding 30 °C, and the desirable value is generally 27°C. ASTM C 94 (AASHTO-M157) for ready-mix concrete, also give a limit of the temperature of the concrete when casting between 29 ° C and 32 ° C.

* Corresponding author: bella5dz@yahoo.fr

ACI 305 (hot climate concreting) does not impose any limit for temperature of concrete, it considers that hot-weather concreting is more complex to be simplified at concrete placing temperature or delivery temperature, because durability is a general term and difficult to be quantified, but which can be perceived by the sense of the resistance of the concrete to the climatic actions.

In some works in southwestern USA concreting is prohibited in extremely hot weather in the summer months. In Arabian golf countries, where the annual amber temperature is generally between 10 °C and 50 °C. In the period between June and September, the permissible maximum concrete temperature is usually between 35 ° C and 38 ° C. Some authors [2, 3] consider the climate in Arab golf as an extremely hot climate.

This article focus on the effects of the hot climate on the concrete characteristics and the procedures of hot weather concreting based on important standards on this field: ACI, ASTM and AASHTO, and other important references; finally, conclusion of recommendations to improve ordinary construction practices in developed countries will be done.

2 Effects of hot climate concreting

2.1 Water demand and slump loss

The loss of workability is due to double action of evaporation of water and heat of hydration. This can lead to a tendency to compensate the lack of workability by the addition of water with detrimental consequences to the resistance [5, 6] and durability [6]. Other research shows that increasing the E/C ratio with an adequate cure improves resistance [7]. Increasing the mixing water also results in increased drying shrinkage.

2.2 Excessive evaporation

ACI Committee 305 indicate that precautions to remedy plastic shrinkage should be taken if the evaporation rate approaches 1 $kg/m^2/h$, this rate is estimated from the ACI 305 nomograph (Fig. 1).

Although this diagram has been criticized [8] and new formulas more precise and adequate to concrete have been proposed [9, 10], because it connects to the evaporation of an open surface of water rather than to a concrete surface, other researcher [11] ensures that the nomograph gives reasonable evaporation rates when the climatic data are taken with precise from jobsite.

2.3 Shorter time set

Due to the acceleration of the hydration reactions of hot climate concreting, setting time will be reduced by the action of high temperatures. Climatic conditions influence setting time, and wind is most influential parameter compared to the action of temperature, relative humidity and adjuvants (water reducer and retarder) [12, 2]. Which, shortens the time available for transport, placing and finishing.

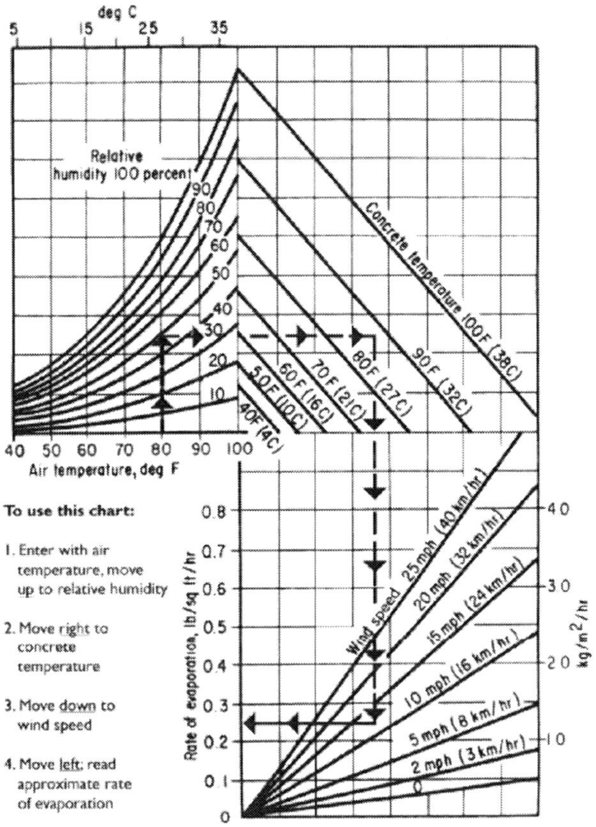

Fig. 1. ACI nomograph for estimating rate of evaporation of surface moisture from concrete [1].

2.3 Plastic shrinkage

If the climatic conditions (temperature, relative humidity and wind) are severe, such as the hot and arid climate and the form of the structure has a large free surface (such as slabs and floors), water will evaporates from the surface more rapidly than bleeding rate (i.e. the rate of rise of water), which will create capillaries pores between the solid particles (Fig. 2.); and as a result, tensile stress will act to bring the walls of the capillaries pores closer together (Fig. 3).

The hydrostatic tensile stress results from the concavity of the meniscus in the water-air interface. This hydrostatic stress increases when the diameter of the meniscus decreases. If this surface tension created by the evaporation of water from the concrete in the plastic state is greater than the tensile strength of the concrete, concrete will crack due to plastic shrinkage.

This cracks can have a length from 1 to 2 cm to 1 to 2 m, with a depth that can penetrate the slabs, with depths greater than 23 cm, and thicknesses of 0.1 to 3 mm [13]. It is clear that, plastic shrinkage cracks weaken the structure, allow the penetration of water, making rebar vulnerable to corrosion and reduce the durability.

DOI: 10.1051/matecconf/201712002024

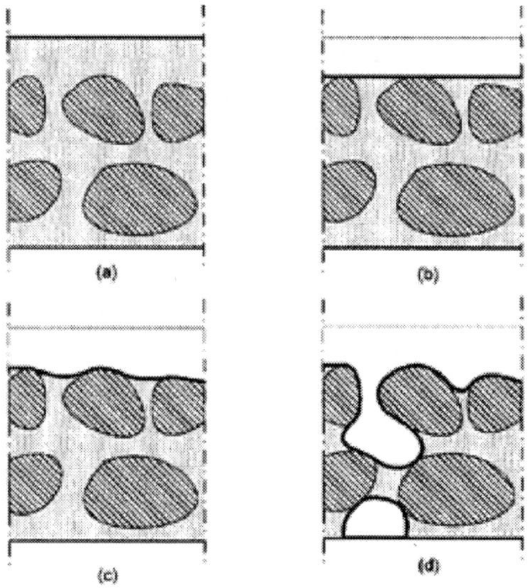

Fig. 2. Diagram of capillary stress in a cement matrix, (a) initial state, (b) loss of water due to evaporation, (c) formation of water menisci on the surface of the concrete, and (d) The opening pressure due to excessive evaporation.

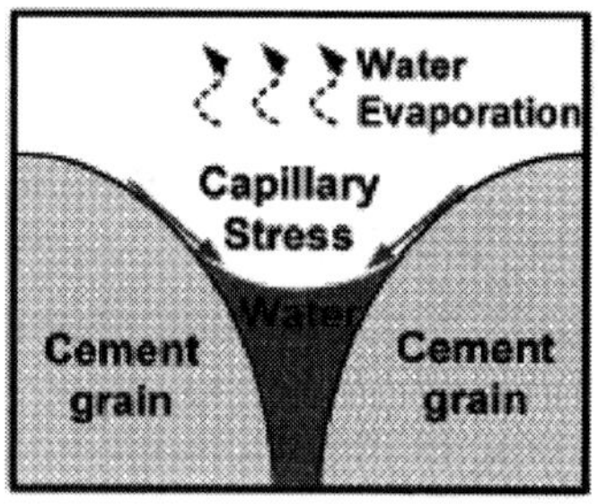

Fig. 3. Graphic representation of the liquid between two spherical particles.

2.5 Rise of temperature and thermal cracking

After demoulding of a structural concrete elements its surface will cool more quickly than its interior. Due to thermal contraction, the surfaces will be under traction and the inside will be compressed. If the tensile stresses on the surfaces exceed the tensile strength, the concrete cracks [14]. Moreover, the high temperature of the interior of the concrete also favours self-desiccation cracking [6].

2.6 Strength

The hydration reactions of cement are thermally activated; i.e., they are accelerated by temperature. Which normally give great resistance, but this is true only at early age (<2 days). Some researches show [15, 16, 17] that long-term strength (28 days) is reduced by accelerated hardening. Probably because the hydration product of concrete which has accelerated hardening is less dense than the hydration products of the concrete in normal temperature (20 ± 5 ° C).

2.7 Durability

The high temperature of concrete affects the initial workability and the rate of loss of workability and consequently the tendency to add more water to the mixer in jobsite. This would lead directly to concrete that is more vulnerable to freeze-thaw, climatic conditions, sulphate attack and penetration of carbon dioxide and chloride solutions leading to reinforcement corrosion. The low workability resulting from high temperature can lead to low compaction which also leaves the concrete more vulnerable to deterioration [8]. Plastic shrinkage and thermal cracking at early ages are also classed as deterioration mechanisms for concrete because they allow moisture, carbon dioxide, oxygen or chlorides to easily access to concrete reinforcements.

3 Verification of curing conditions at jobsite

Conditions (such as: temperature, relative humidity and wind velocity) during the curing period should be estimated before construction. Activities during construction focus on verifying actual conditions and making adjustments to jobsite procedures.

3.1 Evaporation conditions

From climatic data, evaporation rates can be calculated to determine if critical evaporotion rates exist, using the ACI chart 305 shown in Figure 1.

In general, evaporation rates greater than 0.3 kg/m2 /h will present problems. However, the exact thresholds depend on particular conditions of concrete bleeding at jobsite.

3.2 Calculation of setting time

Using the concrete placement temperature, setting start time can be estimated, as described in the following equation:

$$TOS = TOS_{std} \cdot e^{R\left(\frac{1}{CT} - \frac{1}{StdTemp}\right)} \tag{1}$$

TOS: setting time in jobsite temperature with the same unit as the standardized test
TOS_{std}: set time under standard conditions, with any unit
CT: temperature of the concrete at jobsite [K]
StdTemp: temperature of the concrete in laboratory [K]
R: constant.

4 Hot climate concreting procedures

4.1 Cooling of components

The amount of heat contained in a body or mass of material is the product of its specific mass, heat and temperature. The various ingredients in a concrete composition are present in different masses and they have a large difference in specific heat. The temperature of the fresh concrete can be approximated to:

$$T = \frac{0.22(T_a W_a + T_c W_c) + T_w W_w + T_a W_{wa}}{0.22\ (W_a W_c) + W_w + W_{wa}} \tag{2}$$

Where:
T: the temperature of the fresh concrete,
Ta, Tc, Tw: the temperature of the granulates, cement and mixing water respectively,
Wa, Wc, Ww, Wwa: weight of aggregates, cement, mixing water and free water on granulates in kg/m3 respectively.

Cooling of granulates with naturally available cold water is often the least costly way to reduce the temperature of the concrete; And using artificially cooled water for mixing concrete and cooling granulates can sometimes provide as much or more cooling than crushed ice used as part of mixing water.

On large construction jobsite, it may be necessary to install two cement silos so that the time between delivery and use is extended to reduce the temperature of the cement. In order to reduce the temperature of the cement. The silo cladding should, whatever happens, be painted white to minimize the temperature due to solar rays.

4.2 Production and deliverance

Temperature reductions can be achieved by painting the concrete mixing plant and the mixing trucks in white. As an example, concrete in a clean white body can be approximately 1.5 °C cooler than concrete in a red body based on a one-hour delivery time. The period between mixing and delivery should be kept to a minimum. The shipping of the mixing plant trucks should be carefully coordinated with the rate at which the concrete is placed to avoid prolonged stop at the jobsite before the trucks are unloaded. Extremely low temperatures can be achieved nitrogen [18], but the process is expensive.

4.3 Placement

The planning of the transport, placement, vibration and finishing of the concrete must aim at the fastest possible speed. It can be considerably advantageous to concreting at night where the components are probably at their lowest temperature. The temperature of the moulds, reinforcement and previous cast concrete can be reduced by shading it before the concreting operation. Fresh concrete should be protected by windbreaks especially in hot, arid climates. Freshly placed concrete should also be shaded from direct sun rays.

4.4 Cure

4.4.1 Initial Cure

The initial curing period is defined in ACI 308 R as the period between the placement of the concrete and the application of the final cure. The appropriate time for application of the final cure is approximately the start time of setting.

The initial cure is achieved through the use of evaporation reducers [19] which are relatively new developed products specifically designed for excessive evaporation conditions (they reduce evaporation by up to 80% in the shade by windy weather and about 40% of the direct sunlight), generally are based on alcohol a recent research proposes a silicone-based evaporation retarder [20].

The procedures is to apply evaporative reducers (just after finishing) in sufficient quantity and frequency so that the concrete does not lose critical amounts of water by evaporation. Using equipment similar to that used to apply the curing products. The relatively common practice is to apply part or of the curing product very early after placement to serve as an effective evaporative reducing agent.

4.4.2 Final cure

The final cure period is defined as the time interval between the application of the initial cure and the end of the programmed cure. Final cure methods can be classified into three types:

4.4.2.1 Curing product method

Curing product method is normally the most economical method for curing large surfaces because the relatively low labour costs. After the application is accomplished satisfactorily, it requires little or no additional attention. The negative side to the use of curing product methods is the relatively complicated issues of choice and conformity specification that are frequently required, and the skill required to properly application.

The characteristics of the curing products are covered by ASTM C 309 and ASTM C 1315. Selection criteria for curing products include: water retention, colorants, drying time, type and quantity of solids, Volatiles Organic Compounds (VOC), compatibility with coatings. Each of these criteria is discussed below.

Water retention is the main property of curing products. It is measured according to ASTM C 156, which measures the loss of water after a fixed period (usually 72 hours) of exposure to standard drying conditions. The standardized water retention limit (ASTM C 309) is a moisture loss of 0.55 kg/m2 (maximum). ACI 305 R recommends reducing the limit to 0.39 kg/m2 for hot weather concreting.

The standardized drying time in ASTM C 150 is 4 hours under prescribed laboratory drying conditions. Laboratory conditions represent an evaporation rate of approximately 0.43 $kg/m^2/h$. The following empirical equation is useful for estimating the required drying time under drying conditions.

$$\text{Drying time} = \text{ER-0.67} \quad (3)$$

ER: the evaporation rate (in $kg/m^2/h$) estimated using the ICA 308 nomograph.
The white dye reflects sunlight and helps control of temperature in hot weather. In addition to these functional properties, the dye is a very strong indicator of the amount and uniformity of the application.

The curing agents give good results if applied after the setting time. The guides recommend the application after the disappearance of the surface brightness of the concrete, this practice can lead to poor performance. The application of the curing product in this way slows or stops the evaporation of the bleeding water, which accumulates under the membrane or dilutes the curing product. In both cases the membrane will be damaged and this performance may be reduced during the final cure period. In some cases this damage is visible as cracks or tears in the membrane.

There is no method to check the uniformity of uncolored curing product, but two approaches are potentially practical for white-colored curing products. Visual checking is the easiest practice. The controller looks for less whitish surfaces. The products of curing of whitish coloring conforms to ASTM C 309 and applied at a rate equal to 5 m^2/L, have a very whitish appearance. Any trace of gray is an indication of serious under-dosage. Portable reflectometers are marketed to evaluate the quality of the paint. These devices are effective for measuring the application of the whitish coloring product.

4.4.2.2 Water curing method

The water cure method includes immersion, vaporization and burlap or other wet absorbent materials. The water curing method is generally considered to be the only effective cure method for avoiding internal drying cracking for concretes with low E/C ratios (which becomes critical when E/C is less than 0.4) .

ASTM C 94 contains conditions on the mixing water and the curing water. The conditions on the burlap are more complicated.

Fogging can be applied at any time after placing as long as it is not as heavy to prevent runoff. Watering can not be tolerated before the setting time due to the danger of washing the cement fines. Absorbent materials should not normally be applied before to setting start time because of the danger of physically damaging the surface.

Water absorbent materials can be covered with plastic sheet to eliminate evaporation and to reduce the amount of water needed for an effective cure. The burlap can be impregnated on one side with an impermeable layer.

Extreme hot weather conditions can be a problem when using the water curing method. Laboratory research has shown that a concrete exposed to an evaporation rate greater than 1.4 kg/m2/h may develop a cooling gradient above a depth of 50 millimeters above 13 ° C. Covering wet absorbent materials with plastic overlay will significantly prevent evaporative cooling.

Typical advice is that any concrete is inspected at least once a day or more often if conditions permit, and if the found concrete is dry, the situation is corrected and an additional day is added to the required cure.

4.4.2.3 Curing with plastic sheet cover

This method is simple and relatively free of compliance with standards. This method is probably impractical for large surfaces and/or windy conditions, but can be very practical for smaller surfaces. Some customers do not allow the use of plastic cover directly against the concrete surface due to the non-regular coloring that sometimes develops.

ASTM C 171 describes the specifications of the plastic sheet cover, and contains relatively simple requirements. Particular attention must be paid to the overlap of the sheets (50 mm).

Single sheets of polyethylene work well for water conservation, but the use of this type of material can result in non-homogeneous colorization of the surface of the concrete. For this reason, some organisms do not allow curing with plastic sheets. Laminated plastic sheets/burlap are also manufactured to avoid this problem. White colored covers are available, which help to control the temperature by reflecting sunlight.

The plastic sheets must be applied after the start time of setting. As in the water cure method, daily visual inspection is the normal method of verification. If dry areas are found, the situation is corrected and an additional day is added to the required cure.

5 Recommendations

Developing countries are generally the hottest areas in the world, and the concrete industry develop very fast, because of the need of more infrastructures and buildings, this lead to hot climate concreting, understand hot climate negative effect on fresh and hardened concrete and hot climate procedures is primordial for realizing durable concrete, for that following recommendations can be drawn:

- Most important problems of hot climate concreting for fresh concrete, are plastic shrinkage, excessive evaporation and early stiffing.
- Most important problems of hot climate concreting for hardened concrete, are strength drop on long term, rebar corrosion and reduced durability.
- Estimate and control of climatic conditions on jobsite is primordial.
- When evaporation rate is critical use hot climate concreting procedures.
- Keep cool as possible concrete components.
- Paint with all equipment to reduce its temperatures.
- Place concrete in the coolest period of the day.
- Use initial cure by using evaporation retarder or use curing product.
- Use adequate final cure method according to the characteristics the element.
- If any lack of cure is observed add another day to the cure period.

References

1. Murdock L. J., Brook K. M. and Dewar J. D. Concrete Materials and practices; Edward Arnold edition; London (1991)
2. El Rayyes M. S. Remedies to rapid setting in hot-weather concreting; in: Admixtures for concrete. Improvement of properties symposium RILEM; Barcelona (1990)
3. Awad K. W., MazenH. and FakihR. Quality Control and Innovations in materials/ Hot-Weather Concreting. Proceedings, ACI Fifth International Conference, Canun, Mexico (2002)
4. Al hozaimy A.M. Effect of retempering on the compressive strength of ready-mixed concrete in hot-dry environments, Cement & Concrete Composites, **29**: 124–127 (2007)
5. Al-KubaisyM.A. and PalanjianA.S.K. Retempering Studies Of Concrete In Hot Weather, RILEM Proceeding: Properties of Fresh Concrete, Hanover, West Germany (1990)
6. Kay E.A. Hot and cold weather concreting, in: Advanced Concrete Technology, Concrete Properties Edited by J. Newman, B. S. Choo, Elsevier (2003)
7. Ait-Aider H., Hannachi N.E., Mouret M. Importance of W/C ratio on compressive strength of concrete in hot climate conditions; Building and Environment **42**: 2461–2465 (2007)
8. BerhaneZ. (1984), Evaporation of Water from Fresh Mortar and Concrete at Different Environments, ACI Journal, November-December.
9. Uno P. J. Plastic Shrinkage Cracking and Evaporation Formulas, ACI Materials Journal; July-August: 365-375 (1998)
10. Al-Fadhala, M., and Hover, K. C. Rapid Evaporation from Freshly Cast Concrete and the Gulf Environment, Construction and Building Materials, V. **15**, No. 1, Jan., pp. 1-7 (2001)
11. Kenneth C. Hover. Evaporation of Water from Concrete Surfaces, ACI Materials Journal/September-October (2006)
12. Ahmadi B. H. Initial and final setting time of concrete in hot weather, Materials and Structures/Matériaux et Constructions, Vol. **33**, October, pp. 511-514 (2000)
13. Neville AM. Propriétés des bétons, Eyrolles (2000)
14. Chatterji S. Probable mechanisms of crack formation et early ages of concretes: A literature survey, Cement and Cconcrete Research. Vol. **12**, pp. 371-376,

USA (1982)
15. Mamillan M. Recherches expérimentales sur l'accélération du durcissement du béton par le chauffage. Cement and Concrete Research. Vol. **1**, pp. 211-223 (1971)
16. Mouret M., BascoulA. and Escadeillas G. Drops in concrete strength in summer related to the aggregate temperature. Cement and Concrete Research, Vol. **27**, No. 3, pp. 345-357, 1997. Pergamon, Elsevier Science USA (1997)
17. Bella N. Traitement thermique de béton de sable. Mémoire de magistère, USTO-MB, Oran (2004)
18. Aitcine P., Shipping J., Byung G. K., Pierre C., Nkinama B., et Petrov N. L'interaction ciment /superplastifiant cas des PNS, Bulletin LCPC N°/ 223, juillet/ Août (2001)
19. William A. Cordon &Derle Thorpe. Controle of rapid drying of fresh concrete by evaporation controle, Journal of the American Concrete Institute, August, pp. 977-985 (1965)
20. Bella Nabil, AsrounAissa and Bella IlhamAguida, The Use of a New Type of Evaporation Retardant Based on Silicone in Hot Climate Concreting, XIII-ICCC-International Congress on the Chemistry of Cement, Madrid, Spain, 3-8 July, (2011)
21. Bella Nabil, Asroun Aissa and Bella Ilham Aguida, Use of a New Approach (Design of Experiments Method) to Study Different Procedures to Avoid Plastic Shrinkage Cracking of Concrete in Hot Climates, Journal of Advanced Concrete Technology, **9**(2) 149-157 (2011)

MATEC Web of Conferences **120**, 02025 (2017) DOI: 10.1051/matecconf/201712002025
ASCMCES-17

A comparative study of physical and chemical properties of different pozzolanic materials used for roller compacted concrete RCC dams

Abdallah I. Husein Malkawi [1,*], *Ehab* Shatnawi [2], and *Dima* A. Husein Malkawi [3]

[1]Prof. and Chancellor, Fahad Bin Sultan University, Kingdom of Saudi Arabia
On Leave from Jordan University of Science and Technology, Irbid-Jordan
[2]Project Geotechnical Engineer, CH2M,VA 20171, USA
[3]Master Student, Civil Engineering Department, The University of Akron, Akron, OH, USA

Abstract. This paper addresses the feasibility and the efficiency of using Natural Pozzolan and/or Rock flour in Roller Compacted Concrete (RCC) gravity dams. For this purpose, five identical mortar trial mixes were prepared using five different supplementary materials, i.e., fly ash produced in South Africa (proven to be effective in RCC construction), fly ash produced in Turkey, Jordanian natural pozzolan, Saudi natural pozzolan, and rock flour from Mujib Dam basalt quarry. The physical and chemical properties of these pozzolanic materials were determined. The effectiveness of each one of these mineral admixtures used as a cement replacement material in controlling alkali silica reaction are studied and analyzed. Correlations were made between the mechanical properties for the five proposed mixes and a control mix using the Jordanian Portland Cement. The results demonstrate that the performance of Natural Pozzolana and/or rock flour as compared with that of fly ash and other pozzolanic material is very satisfactory and can be effectively used in RCC construction.

1 Introduction

Roller-compacted concrete (RCC) is defined as "concrete compacted by rolling compaction which, in its unhardened state, will support a roller while being compacted" (American Concrete Institute ACI). RCC dams consist of concrete placed at a lower water-to-cement ratio as compared to conventional concrete with the aid of compaction equipment and methodologies normally employed for earth-fill placement. RCC has gained worldwide acceptance as an alternative to conventional concrete in dam construction due to the construction advantages and proved performance [1]. According to Dunstan (2002) RCC dams continue their development worldwide and by the end of (2001) there were (232) complete large RCC dams and a further 31 were under construction, by the end of 2011 over 500 large RCC dams were developed in the world [2]. Recently there are more than 550 RCC dams in the world and approximately 45% are located in Asia. "During the meeting entitled

* Corresponding author: mhusein@just.edu.jo

The Practicalities and Economics of RCC by Malcolm Dunstan, held at the British Dams Society in London on 20 January 2014". Forbes, et al. (1999, 2001) and Forbes (2008 and 2012) in his authoritative papers stated clearly that more significant innovations, improvement, efficient construction and at less cost have taken place since the first development of RCC dams in early 1990's [3,4,5,6]. Factors that affect the material properties of RCC are the same as those that affect the properties of conventional concrete, which are water-to-cement ratio, quality of grading in mixing, and degree of compaction and curing. The objective of RCC proportioning is to provide a dense and stable mass that meets the strength, durability, and permeability requirements for its application. Materials used for RCC include cementitious materials (Portland cement and pozzolans such as fly ash), aggregates, water, and admixtures. A wide range of materials have been used successfully to produce RCC mixtures. It is common practice in proportioning concrete mixtures for dams to utilize large aggregates in order to decrease cementitious material requirements and lower the costs and heat of hydration. RCC can be made with any of the basic types of cement or combination of cement and pozzolanic material. Selection of cementitious materials to resist the chemical attack of sulfate and potential alkali reactivity with certain aggregates should follow the same standard procedures adopted for CVC. The compressive strength of concrete is less dependent on the aggregate bond than tensile strength. Thus, it is apparent that the relationship between the tensile strength and compressive strength of concrete not only varies with the method of test, but also varies with the type and maximum size of aggregate. The strength of RCC is primarily dependent upon, the quality of the aggregate, degree of compaction, properties of cement, pozzolanic material, water, and admixtures if used, and grading of aggregate. Alkali aggregate reaction, as defined in the ACI Committee (1988), is a chemical reaction in mortar or concrete between alkalis (sodium and potassium) from Portland cement or other sources and certain constituents of some aggregates; under certain conditions, deleterious expansion of the concrete or mortar may result. Classic research by [7] correctly diagnosed the failure as being due to expansions caused by a chemical reaction between the alkalis contained in the cement paste and certain reactive forms of silica within the aggregate. Stanton found opal and chert to be common forms of reactive silica. [8] addressed the possibility and the efficiency of using Jordanian natural pozzolana in RCC gravity dams. For this purpose, two identical trial mix properties were proposed and undertaken using two different supplementary cementitious materials, i.e., fly ash brought from South Africa and Jordanian natural pozzolana. The study showed that Jordanian natural pozzolana produced satisfactory results. [9] studied the potential of alkali aggregate reaction of different types of Jordanian aggregate. The results indicate that the main types of Jordanian aggregates could be considered as innocuous aggregates, with no expected expansion resulting from alkali aggregate reaction, except some types of minerals such as opal and some types of sands and cherts. [10] states that all good RCC has a paste content of about 19% to 21% by volume, regardless of the cement and pozzolan or fly ash content. The paste includes "all material finer than 75 microns – cement, slag, pozzolan (fly ash), aggregate fines, admixtures, water, and air". Mixes with less paste are harsh and tend to segregate, whereas mixes with excess paste tend to produce less strength per kilogram of cementation material. Therefore, low cementitious content RCC requires aggregate fines in order to provide adequate paste without excessive water, and high cementitious content mixes require cleaner aggregates.

2 Objectives

The main objective of this study is to address the feasibility and the efficiency of using natural pozzolan and/or basalt rock flour in roller compacted (RCC) gravity dams. To achieve this, five different supplementary cementitious materials were used in preparing the mortar sample mixes, these are, fly ash produced in South Africa (proven to be effective in RCC construction), fly ash produced in Turkey, Jordanian natural pozzolan, Saudi natural pozzolan, and rock flour from Mujib Dam basalt quarry. The physical and chemical properties for each one of these cementitious materials will be determined. The effectiveness of each one of these mineral admixtures used as a cement replacement material in controlling alkali silica reaction are to be studied and analyzed. Correlations will be made between the mechanical properties for the five proposed mixes and the control mix using Jordanian Portland Cement.

3 Materials properties

3.1 Pozzolanic materials

The selection of pozzolanic material suitable for RCC should be based on its conformance with applicable standards (ASTM C-618). This standard defines pozzolan as "siliceous or siliceous and aluminous materials which in themselves possess little or no cementitious value but will, in finely divided form in the presence of moisture, chemically react with calcium hydroxide at ordinary temperatures to form compounds possessing cementitious properties". The type of pozzolanic material can include natural pozzolans; diatomaceous earth; Industrial waste material as fly ash or silica fume. The use of a pozzolanic material in RCC serves some purposes such as partial replacement for cement to reduce heat generation, increase workability and compressive strength at great ages, if the material has high pozzolanic activity with cement, increase durability, and reduce the effects of Alkali Aggregate Reaction and to reduce cost and, as a mineral addition to the mixture, to provide fines to improve workability.

3.2 South Africa fly ash

Fly ash is a residue of the combustion of the finely ground coal used in electric power generation. The fly ash class F material was not available in local markets and an attempt was made to import enough quantities from South Africa in order to be used in RCC mixtures. The fly ash material was tested in South Africa by Ash Resources.

3.3 Turkey fly ash

Fly ash was imported from Isken Sugozu Power Plant, located in the South East of Turkey. Chemical analysis is shown in Table 1.

3.4 Jordanian natural pozzolan

The natural pozzolan, which is exposed in large deposits throughout north-eastern Jordan at Tel-Alremah is used in the RCC mixture. The pozzolana was extracted, crushed, grinded and tested by the Jordan cement factory.

3.5 Saudi natural pozzolan

The Saudi natural pozzolan is available in large deposit throughout the western part of the Kingdom of the Saudi Arabia, and in the outskirts of Al-Madinah. The pozzolan was brought, crushed, grinded by the Pozzolan Product Factory (Burkani) located in the Jeddah Industrial Area. Table 1 shows the chemical analysis of the Saudi natural pozzolan.

3.6 Basalt rock flour

The rock flour used in this study was crushed basalt rocks came from a basalt quarry, located on the south rim of Wadi Mujib in the southern part of Jordan.

3.7 Portland cement

Portland cement Type II (low heat) was used in these mixes. The cement is manufactured by the Jordan Cement Factory. The chemical properties of the cement were periodically tested by the Royal Scientific Society Laboratories and the Department of Quality Control.

Table 1. Chemical analysis for the pozzolanic material.

Determinant	South Africa fly ash	Turkey fly Ash	Jordanian pozzolan	Saudi pozzolan	Rock Flour
Free Water (%)	0.1	0.0	0.1	0.1	0.1
R_density (g/cm^3)	2.20	2.10	2.86	2.8	2.85
LOI (%)	0.81	1.10	2.56	1.26	2.31
Water Reqr. (%)	90	102	100	100	105
Blain	2380	3670	4610	3200	5870
SO_3 (%)	0.42	0.82	0.35	0.1	-
CaO (%)	4.18	3.02	9.75	8.53	11.02
SiO_2 (%)	51.06	56.54	42.56	48.44	49.25
Al_2O_3 (%)	32.23	18.68	13.55	15.97	13.36
Fe_2O_3 (%)	3.42	10.83	13.24	13.2	11.3
MgO (%)	1.06	4.45	11.15	7.89	8.76
TiO_2 (%)	1.70	0.83	2.54	-	1.57
K_2O (%)	0.58	1.46	1.17	1.37	0.51
Mn_2O_3 (%)	0.01	0.16	0.14	-	0.190
Na_2O(%)	-	-	-	-	0.51

3.8 Aggregates

The selection of a suitable aggregate depends on a number of factors, including local availability, economics, size of the project, strength and durability requirements, and workability. The selected aggregates were crushed basalt which came from a basalt quarry, located high above the left abutment to Wadi al-Mujib, in Southern part of Jordan.
The Cilas Granulometre 715 device was used to determine the grain size distribution of the four cementitious materials used in this study. Tables 1 & 2 show the physical and chemical analysis for the five types of pozzolanic materials and the Jordanian OPC. Also, Figure 1 shows the grain size distribution of the five selected cementitious materials.

Table 2. Chemical analysis for the Portland cement.

Item	Result (%)
SiO_2	19.92
Al_2O_3	5.00

Fe_2O_3	3.31
CaO	63.54
MgO	3.36
SO_3	3.05
Na_2O	0.19
K_2O	0.53
L.O.I	1.49

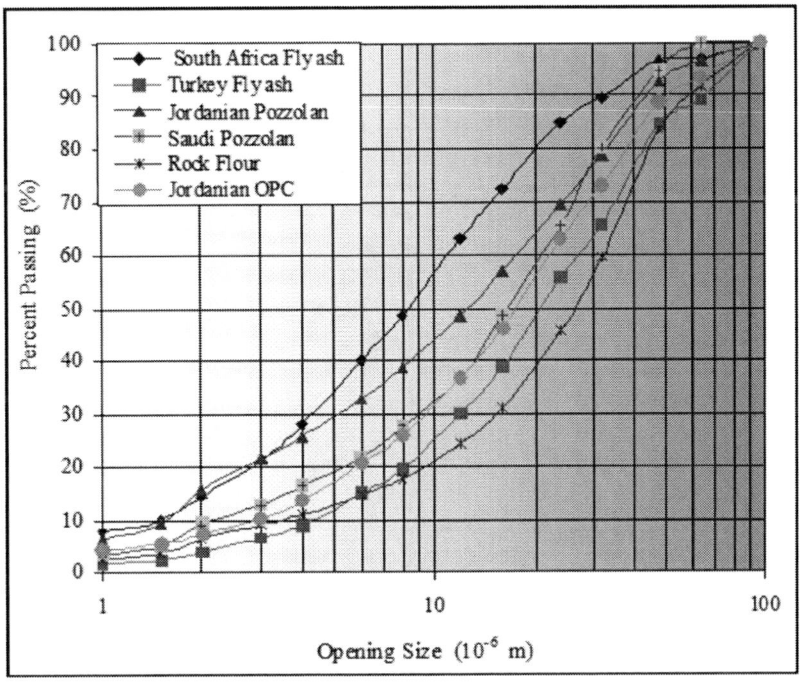

Fig. 1. Grain Size Distribution.

4 Laboratory work

4.1 Compressive strength

This investigation was carried out under laboratory conditions using procedures as described in ASTM C 109. The materials include (a) ordinary cement with class of strength 42.5 MPa produces at Jordanian Cement Factory. (b) South Africa fly ash, Turkey fly ash, Jordanian pozzolana, Saudi natural pozzolan and basalt rock flour and basalt aggregate crushed to sand size. The test specimens from a control mixture and from a test mixture were molded in accordance with ASTM Test Method C 109/C 109M. In the test mixture, six percent with 10, 15, 20, 25, 30, and 40 % of the mass of the cement used in the control mixture were replaced by the same mass of mineral admixture. The following weights were used to prepare the (50mm) cubic samples:

1- Control Mixtures:

500 g of Portland cement

1375 g of graded standard sand (graded as ASTM C 778).

242 ml of water

2- Test Mixtures:

450, 425, 400, 350, and 300 g of Portland cement.

50, 75, 100, 150, and 200 g of mineral admixtures.

1375 g of graded standard sand

242 ml of water.

The graded sand used was the basalt aggregate that was crushed to the sizes and amounts shown below:

Table 3. Seive analysis.

Sieve	Amount used,%
No. 16 - No. 30	5
No. 30 - No. 40	30
No. 40 - No. 50	42
No. 50 - No. 100	23

After the aggregate was separated into the various sieve sizes, each size was washed with water over the sieve to remove adhering dust and fine particles from the aggregate. The portions retained on the various sieves were dried and stored individually in a clean container. The dry materials for the test mortar were prepared using 1 part of cement to 2.75 parts of graded aggregate by mass to cast 50 x 50 x 50 mm cubes. Water to cement ratio equal to 0.485 by mass was used. After the mix proportions were prepared, the specimens were mechanically mixed in accordance with the procedure given in ASTM C 305. The specimens were molded immediately after the mixing of the mortar batch was finished, placing and tamping as required by ASTM C 109. Immediately upon completion of molding, the test specimens were placed in the moist room at 23.0 ± 2.0°C for 24 h, then the cubes were removed from the moist room and stored in storage water tanks as specified in test method C 109/C 109M. The compressive strength was determined, as specified in Test Method C 109/C 109M, of two specimens of the control mixture and two specimens of the test mixture at ages of 7, 28, 90, 180, and 360 days.

4.2 Alkali silica reactions

This investigation was carried out under laboratory conditions using procedures as described in ASTM C 1260. The materials include ordinary OPC cement with class strength of 42.5 MPa produced at JCF, mineral admixtures of South Africa fly ash, Turkey flyash, Jordanian natural pozzolan, Saudi natural pozzolan, and rock flour, and aggregate composed of basalt aggregate crushed to sand size.

4.3 Strength activity index with portland cement

Tests specimens from a control mixture and from a test mixture were molded in accordance with Test Method C 109/C 109M. In the test mixture, 20 % of the mass of the amount of cement used in the control mixture was replaced by the same mass of the test sample. ASTM C-311 suggested that six-cube batches to be made as follows:
1- Control Mixtures:
500 g of Portland cement
1375 g of graded standard sand (Ottawa sand, ASTM C 778).
242 ml of water
2- Test Mixtures:
400 g of Portland cement.

100 g of test sample.
1375 g of graded standard sand
242 ml of water
Although Specification C 618 specifies that "meeting the 7 day or 28 day Strength Activity Index will indicate specification compliance", 4 cubes of each batch were used for 7 and 28 day testing. After molding, the specimens were placed in the moist room at 23.0 ± 2.0°C for 24 h, then the cubes were removed from the moist room and stored the in saturated lime water as specified in Test Method C 109/C 109M. The compressive strength was determined, as specified in Test Method C 109/C 109M, of two specimens of the control mixture and two specimens of the test mixture at ages of 7 days, 28 days, 90 days and 180 days (see Figs 2-5).

4.4 Heat of hydration

The heat of hydration is the quantity of heat, in joules per gram of dehydrated cement, developed upon complete hydration at a given temperature. The most common method of determining the heat of hydration is by measuring the heat of the solution of dehydrated and hydrated cement in a mixture of nitric and hydrofluoric acids, the differences between the two values represent the heat of hydration. This method is described in BS 4550: Part 3: Section 3.8: 1978, and ASTM Standard C 186-78.

5 Test results and discussion

5.1 Compressive strength

Detailed data summarize the results of the average compressive strength at age 7, 28, 90 and 180 days are shown in Figures 2 to Figure 5. It can be seen that the average peek compressive strength of RCC increases from 6 MPa to more than 12 MPa at the age of 180 days.

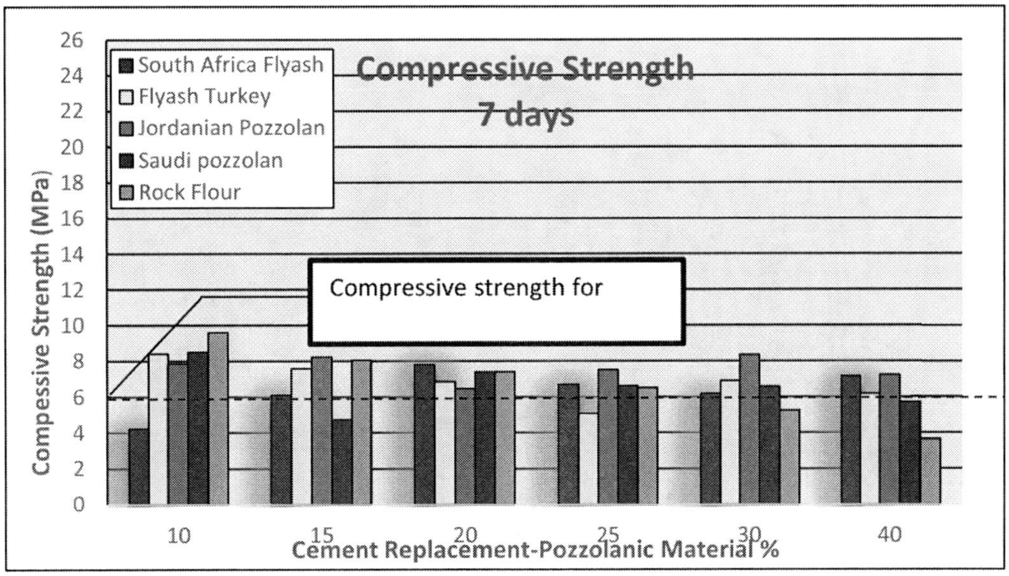

Fig. 1. Compressive Strength with Different Percent of Pozzolanic at 7 days.

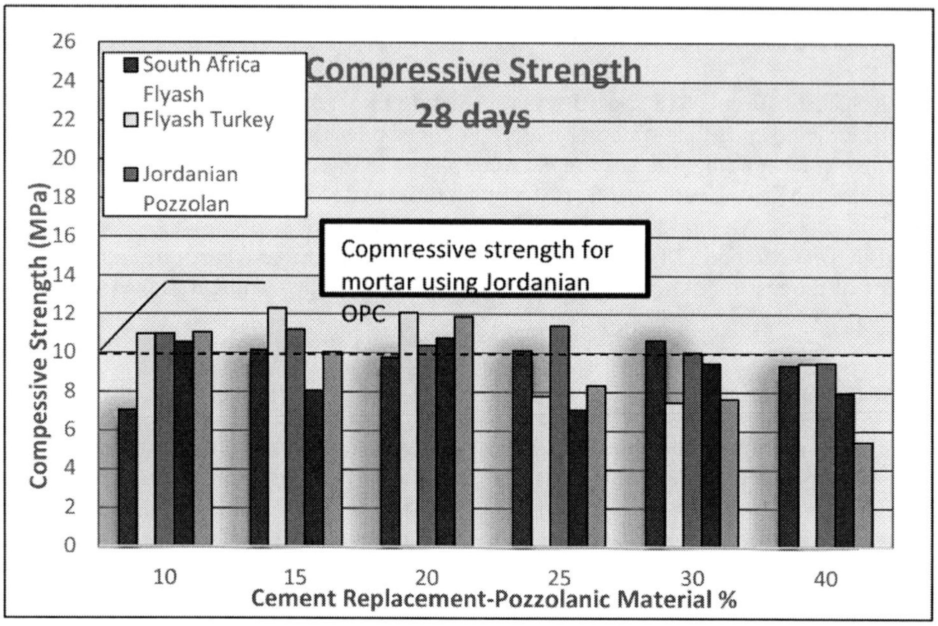

Fig. 2. Compressive Strength with Different Percent of Pozzolanic Materials at 28 days.

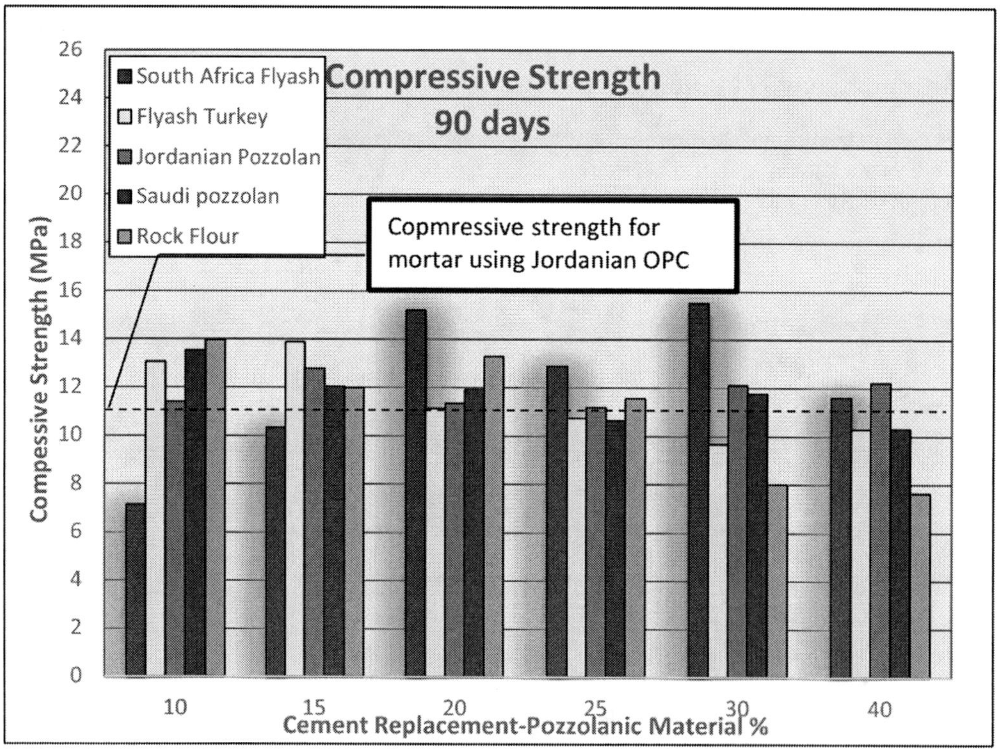

Fig. 4. Compressive Strength with Different Percent of Pozzolanic Materials at 90 days.

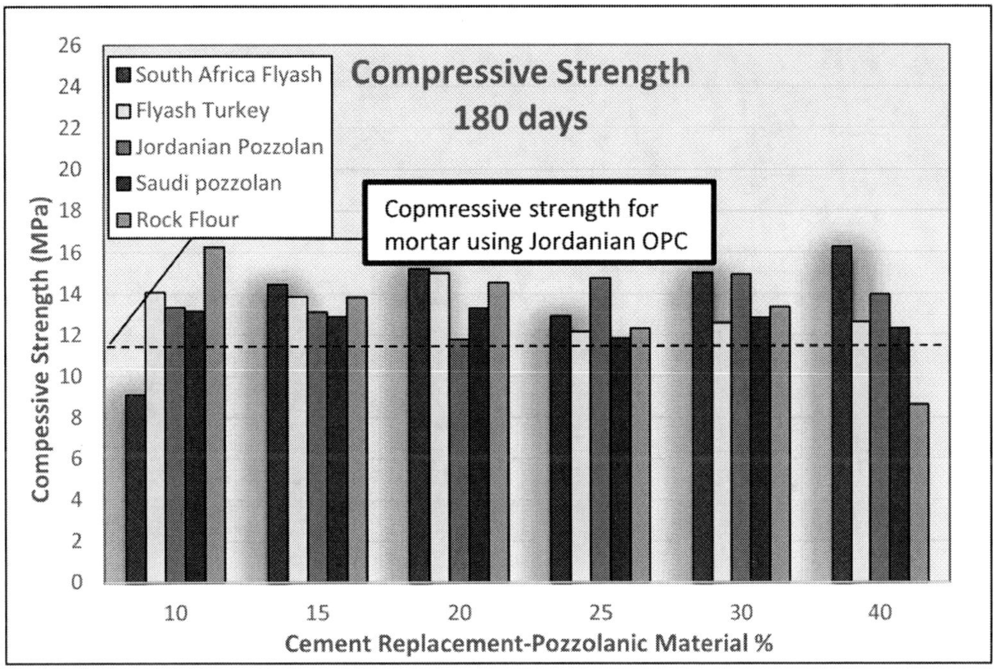

Fig. 5. Compressive Strength with Different Percent of Pozzolanic Materials at 180 days.

5.2 Activity Index

The pozzolanic activity index is calculated as follow:
Strength activity index with Portland cement = (A/B)x100
Where:
A= average compressive strength of test mixture cubes.
B= average compressive strength of control mix cubes.
The activity index for the different tested pozzolanic materials, is shown in Figure 6

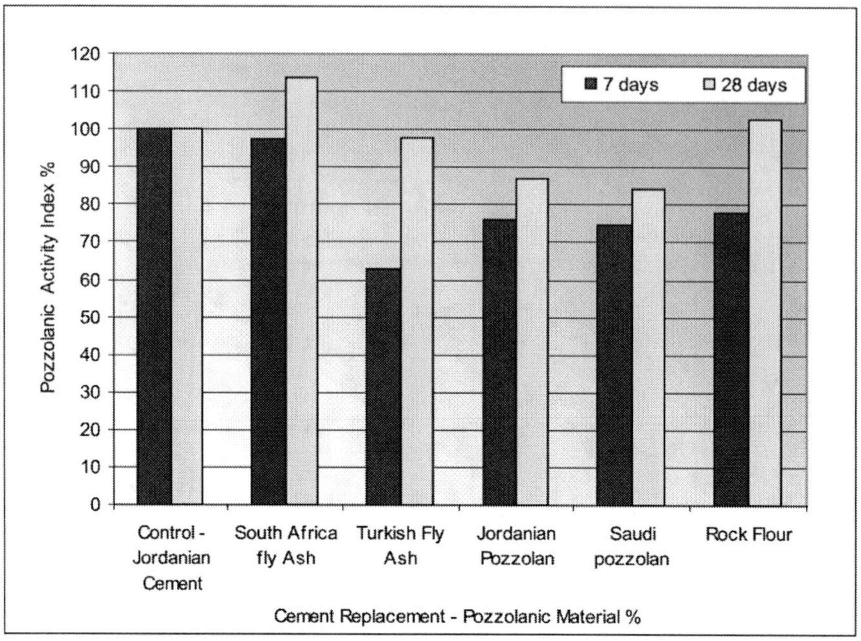

Fig. 6. Strength Activity Index with Portland cement

5.3 Alkali silica reaction

A 14-day expansion greater than 0.20% indicated a potentially reactive aggregate in the field, while a 14-day expansion smaller than 0.10% indicated a non-reactive (innocuous) aggregate. Recent specifications have mentioned that 14-day expansions higher than 0.10% should be used to classify aggregates as reactive. Figure 7, shows the test results for evaluating effectiveness of mineral admixture in controlling Alkali Silica Reactions.

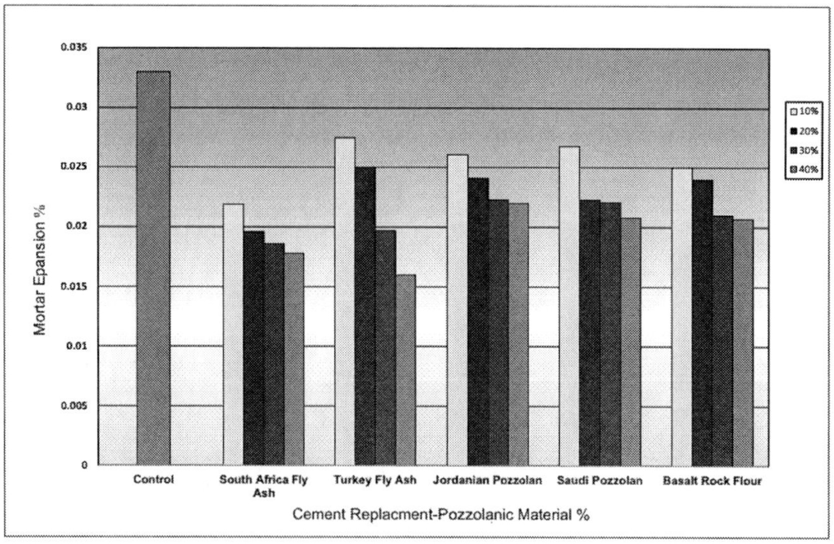

Fig. 7. 14-day expansions for mortar prepared by using different types of pozzolanic admixture

5.4 Heat of hydration

Using the procedures listed above, the heat of hydration was calculated for the five different mixtures, that is the with fly ash from South Africa, fly ash from Turkey, Saudi natural pozzolan, Jordanian natural pozzolan and Rock flour. The results at age 7 days are shown in Figure 9, the heat of hydration for the zero percent of the pozzolanic material reflects the heat of hydration for Jordanian OPC cement.

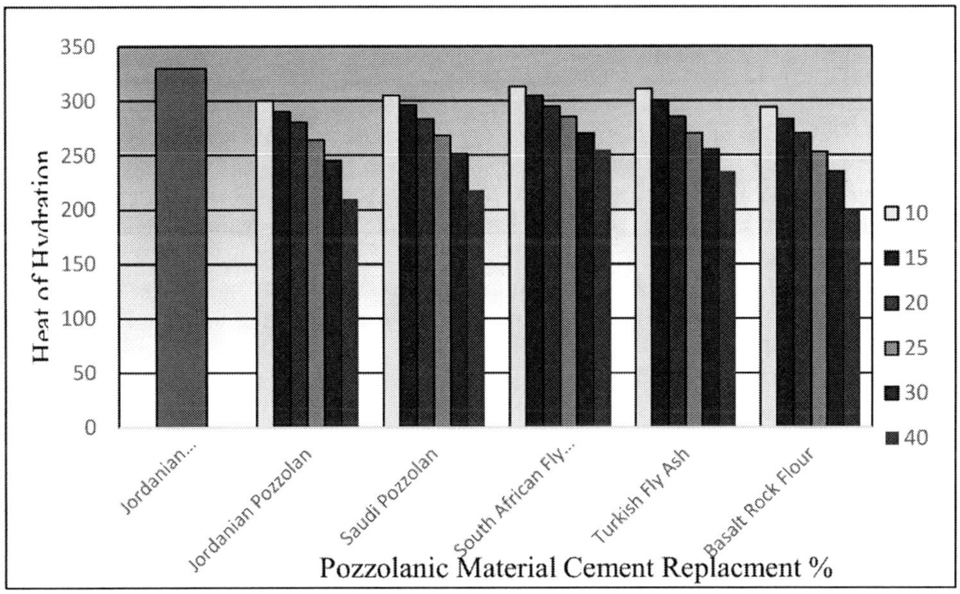

Fig. 8. The Accumulative Heat of Hydration after 7 days

6 Conclusion

Based on this study, the experimental and laboratory work performed in this investigation, indicate the following:

1. High strength of RCC mixtures used was obtained through specimens tests. It is clear that the contribution of fly ash and pozzolan can exhibit excellent strength of RCC especially over long time.
2. The results demonstrated that the performance of Natural Pozzolana as compared with that of fly ash and other pozzolanic material is very satisfactory and can be effectively used in RCC construction.
3. Natural pozzolana showed clearly its effectiveness in controlling alkali silica reaction
4. Using the pozzolanic materials in mixtures gives lower heat generation and water requirements that are improved workability in RCC mixtures used. This can be accomplished at cost savings despite higher cost of fly ash due to the water reduction and additional pozzolanic strength contribution of the ash.
5. Basaltic fines of about 20% replacement seems to be ideal in terms of achieving long term compressive strength, and less accumulative heat of hydration, compared to other cementitious materials.

References

1. Luna R. and Wu Y. (2000) "Simulation of Temperature and Stress Field during RCC Dam Construction". Journal of Construction Engineering and Management, September/October.
2. The International Journal on Hydropower & Dams – World Atlas, 2011.
3. Forbes, B. A. (2012) "Innovations of Significance and their Development on Some Recent RCC Dams" Invited Lecture, 6th International Symposium on Roller Compacted Concrete (RCC) Dams, Zaragoza, 23-25 October.
4. Forbes, B.A., Yang, L., Tang, G., and Yang, K. (1999) "Jiangya Dam China, Some Interesting Techniques Developed for High Quality RCC Construction". Proc. Of the International Symposium on Roller Compacted Concrete Dams. Chengdu, China, April.
5. Forbes, B.A. (2008) "RCC-New Developments and Innovations". Proc. Of the 50th Brazilian Congress on Concrete-CBC2008, 1st RCC Symposium. Salvador, Brazil, September.
6. Forbes, B.A., Iskander, M.M., and Husein Malkawi, A.I, (2001) "High RCC Standards Achieved at Jordan's Tannur Dam". International Journal on Hydropower & Dams, Issue Three.
7. Stanton, D. E. (1940) "Expansion of concrete through reaction between cement and aggregate". Proc. ASCE 66: 1781-1811.
8. Husein Malkawi, A.I.; Shaia, H.A.; Mutasher, S.A. and Aridah, M.F. (2003) "A Comparative Study of Mechanical Properties of RCC Trial Mix Using two Different Cementitious Materials (Fly Ash and Natural Pozzolan)". 4th International Symposium on Roller Compacted Concrete (RCC) Dams, 17-19 November, Spain.
9. Husein Malkawi, A.I.; Abed, A. and Rababa, S. (2003) "Behavior of the reaction between certain Jordanian rock-aggregate and cement in roller compacted concrete (RCC) dams". 4th International Symposium on Roller Compacted Concrete (RCC) Dams, 17-19 November, Spain.
10. Schrader, E. (2002) "Experience and Lessons Learn in 30 Years of Design, Testing, Construction, and Performance of RCC Dams". In Husein Malkawi et al. (ed.), *Roller Compacted Concrete Dam Construction in the Middle East*; Proc. intern. Conf., Irbid-Jordan, 7-10 April 2002. Jordan University of Science and Technology & Technische Universität München.
11. Dunstan, M.R. (2002) "The state of art of RCC dams at the end of 2001". In Husein Malkawi et al. (ed), *Roller compacted concrete dam construction in Middle East;* Proc. Int. Conf., Irbid- Jordan, 7-10 April 2002. Jordan University of Science and Technology & Technische Universität München.
12. Hobbs, D.W. (1988) Alkali-silica reaction in concrete. London: Thomas Telford.
13. Lane, D.S and Ozyildirum, H. (1990) "Evaluation of the effect of Portland cement alkali content, fly ash, ground slag, and silica fume on alkali-silica reactivity". Cement, Concrete, and Aggregates Journal 21(2): 126-140.
14. Mantuani, D.L (1983). Handbook of concrete aggregates. New Jersy: Noyes Publications.
15. Thomas, M.D. & Innis, F.A. 1998. Effect of slag on expansion due to alkali-aggregate reaction in concrete. ACI Materials Journal 95 (6): 716-721.
16. Wang, H. & Gillott, J.E. (1993) "Effect of three natural Zeolite-containing Pozzolanic materials on alkali-silica reaction". Cement, Concrete, and Aggregates Journal 15: 24-44.

Clay bricks prepared with sugarcane bagasse and rice husk ash – A sustainable solution

Muhammad Azhar Saleem[1,*], *Syed* Minhaj Saleem Kazmi[2], and *Safeer* Abbas[3]

[1]Associate Professor, Department of Civil Engineering, University of Engineering and Technology, Lahore, Pakistan
[2]Junior Lecturer, Department of Civil Engineering, Mirpur University of Science and Technology, Mirpur, AJK, Pakistan
[3]Assistant Professor, Department of Civil Engineering, University of Engineering and Technology, Lahore, Pakistan

Abstract: This study aims to characterize the clay bricks produced by the addition of the two agricultural waste materials i.e. sugarcane bagasse and rice husk ash. Disposing off these waste materials is a very challenging task and is a hazard to environment. The sugarcane bagasse and rice husk ash were collected locally from the cities of Peshawar and Wazirabad, respectively. These were mixed with the clay for brick manufacturing in three different proportions i.e. 5, 10 and 15% by weight of clay. Mechanical i.e. compressive strength and modulus of rupture and durability properties i.e. water absorption; freeze-thaw and sulphate resistance of these bricks were evaluated. Test results indicated that the sulphate attack resistance and efflorescence of clay bricks incorporating sugarcane bagasse and rice husk ash have been increased significantly. However, no significant effect on mechanical properties was observed. Furthermore, the additions of wastes have reduced the unit weight of bricks which decrease the overall weight of the structure leading to economical construction. Therefore, it can be concluded that the addition of waste materials in brick manufacturing can minimize the environmental burden leading towards more economical and sustainable construction.

1 Introduction

Clay products (burnt clay bricks or blocks) are one of the oldest materials used in the construction activities. Bricks/blocks are prepared after burning the clay in a kiln. Raw materials and production method play an important role in developing brick properties. Bond development between the clay particles occur through firing process. Silica (SiO_2) is considered as the main constituent of clay that helps in bond development. Generally different fluxing agents are added in clay to lower down the melting temperature. Fluxing agent helps to achieve required brick properties at lower temperature, leading to fuel saving.

*Corresponding author: msale005@fiu.edu

Production and consumption of burnt clay bricks is still common in many parts of the world. China is considered as the top country in brick production [1]. Similarly, India with annual 180 billion tons of brick production is the second largest brick producing country [2]. Extensive utilization of natural resources like clay has caused an alarming situation [3-5]. Moreover, day by day increasing waste generation is also causing environmental problems. Therefore, researchers are focused to develop new construction materials by utilizing waste products in construction. Waste material consumption will not only reduce the environmental burden but also lead towards an economical and sustainable solution.

Different waste materials have been investigated in the production of bricks with improved performance. Waste glass, fly ash, agricultural and industrial wastes was used effectively in the production of burnt clay bricks [6]. Utilization of waste glass in clay bricks can improve the strength and porosity [7-8]. Similarly, utilization of fly ash in brick production was very effective. Improved strength and water absorption was observed in case of fly ash bricks. Furthermore, utilization of fly ash is environment friendly leading towards green construction [9-10]. Utilization of marble powder in burnt clay bricks can be very helpful in controlling the environment pollution and air borne diseases. Marble powder replacement 15-20% by weight of clay was found effective in brick production [11]

Pakistan is an agricultural country producing approximately one million rice husk annually [12]. This husk is used as a fuel source in brick and paper industry. As a result of combustion, rice husk ash (RHA) is obtained. Similarly, Pakistan produces around 50 million tons of sugarcane annually [13]. Bagasse is obtained after utilization sugarcane in sugar production. Bagasse is also consumed as fuel source and as a result sugarcane bagasse ash (SBA) is produced. There is no proper mechanism for the disposal of these ashes. As a result, environmental pollution is increasing by these wastes. Different researchers have studied the utilization of RHA and SBA in clay bricks [1, 14-17]. Lighter bricks with improved thermal conductivity can be prepared by using agro wastes in brick production [18-19].

In this study clay bricks have been prepared after incorporating RHA and SBA in different proportion (5%, 10%, 15% by weight of clay). Burnt clay bricks were prepared in brick kiln on a massive scale. The utilization of these waste will be helpful in saving the natural resources along with economical and sustainable construction.

2 Materials and methods

Clay was attained from a brick kiln in Pakistan. Similarly, RHA and SBA were obtained from a sugar milland industrial brick kiln located in Pakistan respectively. Firstly, RHA and SBA were added and mixed manually (Fig. 1(a)) with clay in different proportions presented in Table 1. Afterwards, mixture was prepared by adding water (Fig. 1(b)) and allowed for 3 hours to attain homogeneity. Molds (228x114x76 mm) were then filled with clay lumps to prepare the bricks. Bricks were first dried in the sun for 10 days (Fig. 2(a)) and then burnt in a kiln for 36 hours at approximately 800°C. Brick specimens were removed after 45 days from the kiln (Fig. 2(b)).

Raw materials' properties were studied by chemical analysis, particle size distribution and specific gravity. X-ray fluorescence (XRF) and X-ray diffraction analysis (XRD) were performed to determine the chemical composition. Particle size distribution and specific gravity of raw materials were determined in accordance with ASTM D422 (Standard test method for particle-size analysis of soils) and ASTM D854 (Standard test methods for specific gravity of soil solids by water pycnometer), respectively.

Mechanical and durability properties like unit weight, compressive strength, modulus of rupture, water absorption, initial rate of absorption and efflorescence were determined in accordance with ASTM C67 (Standard test methods for sampling and testing brick and structural clay tile).However, ASTM C20 (Standard test methods for apparent porosity, water absorption, apparent specific gravity, and bulk density of burned refractory brick and shapes by boiling water) was used to determine the apparent porosity of brick specimens.

Table 1. Mixture proportions of raw materials.

Brick series	Clay (%)	RHA (%)	SBA (%)
C	100	-	-
RHA5	95	5	-
RHA10	90	10	-
RHA15	85	15	-
SBA5	95	-	5
SBA10	90	-	10
SBA15	85	-	15

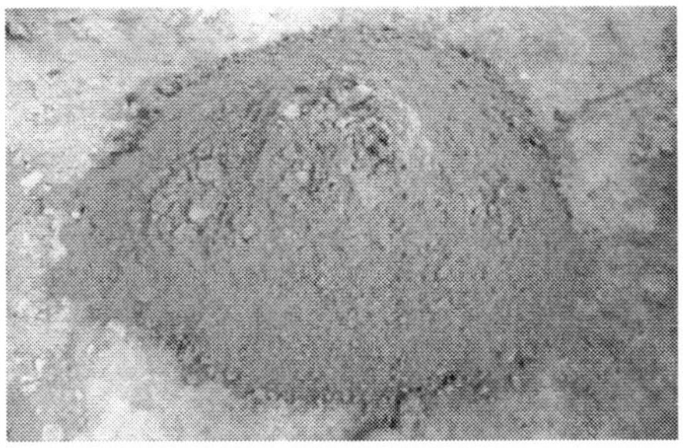

a) Soil placed with dry ash

b) Mixture of soil with ash

Fig. 1. Preparation of mixture for brick production.

1. Bricks placed for firing in kiln

2. Bricks placed for sun drying

Fig. 2. Drying and burning process of brick production.

3 Results and discussion

3.1 Properties of raw materials

Table 2 shows the chemical composition of raw materials. It was observed that silica and alumina were the main skeleton components of clay. Presence of silica within the range of 50-60% makes it suitable for brick production [6]. Moreover, the amount of fluxing agents (K_2O, CaO, MgO, Fe_2O_3 and TiO_2) and calcium oxide in clay was greater than 6% and 9%. Therefore, it can be classified as low refractory calcareous clay [20]. Silica was also found as the main chemical component of RHA and SBA. Other oxides like alumina, iron oxide, calcium oxide and magnesium oxide were found in small amount.

Table 2. Chemical and physical properties of the raw materials.

Components	Clay	RHA	SBA
SiO_2 (%)	57.05	77.21	87.97
Al_2O_3 (%)	11.91	6.87	1.84
Fe_2O_3 (%)	4.96	4.69	2.65
CaO (%)	8.98	3.65	2.65
MgO (%)	2.52	1.45	0.72
TiO_2 (%)	0.68	-	-
P_2O_5 (%)	0.14	-	-
SO_3 (%)	-	0.37	0.15
MnO (%)	0.08	-	-
Na_2O (%)	1.86	1.24	0.28
K_2O (%)	2.21	2.59	0.32
LOI (%)	9.59	4.71	10.45
pH	8.55	-	-
Unit Weight (kg/m^3)	1120	550.54	258.6
Specific Gravity	2.24	2.11	1.96

Loss on ignition (LOI) was observed higher for SBA (10.45%)as compared to clay (9.59%) and RHA (4.71%). It may be related to the combustion of carbonates and unburnt particles.

Figure 3-5 shows the x-ray diffraction (XRD) scans of clay, RHA and SBA Quartz was observed as the major mineral present in the clay along with cinnabar, halite, alumina, calcite, fluorite and hematite minerals in minor amount. Furthermore, RHA and SBA were also found comprised of the quartz (SiO_2). Other minerals present in RHA were hematite, calcite and fluorite. Whereas, SBA comprised of calcite, corundum, hematite, fluorite, halite and bornite minerals in minor amount.

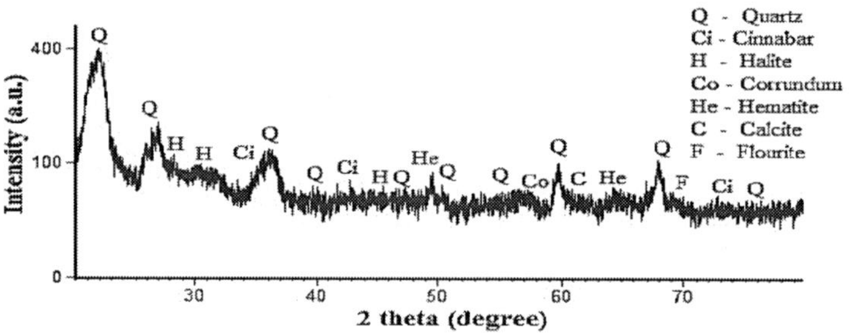

Fig. 3. XRD pattern of clay.

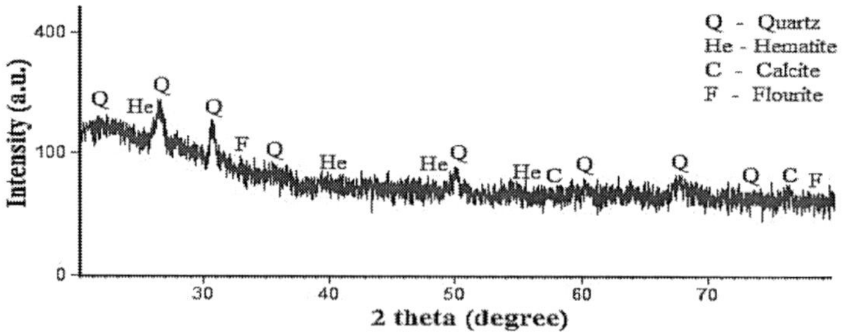

Fig. 4. XRD pattern of RHA.

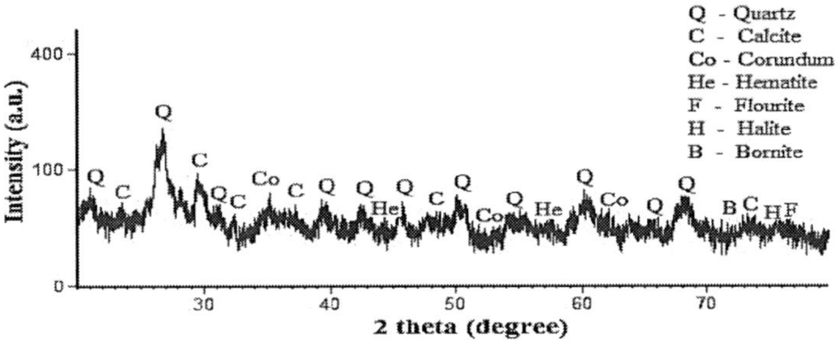

Fig. 5. XRD pattern of SBA.

Figure 6 shows the particle size distribution curves of raw materials. It was observed that a wide range of sizes were present in the raw materials. Major amount of clay particle sizes was observed within the sand and silt size. However, particles of RHA and SBA were observed mostly within the sand and silt sizes, respectively. As far as, specific gravity is concerned. Clay (2.24%) showed a higher specific gravity than RHA (2.11%) and SBA (1.96%), which indicated the production of lighter bricks by incorporating these wastes.

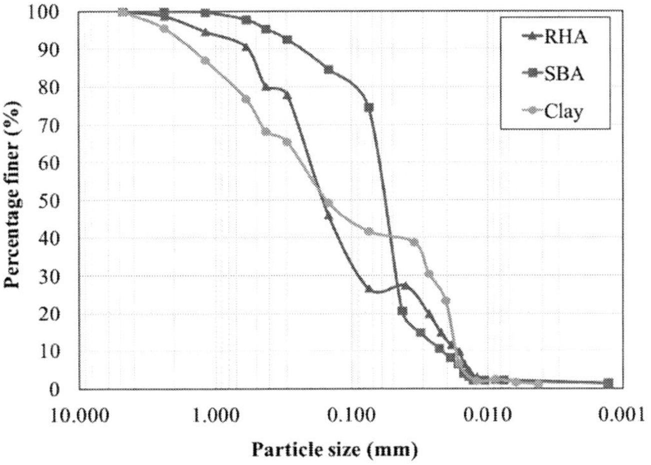

Fig. 6. Particle size distribution for RHA, SBA and clay.

Table 3. Particle size distribution analysis of raw materials used.

Particle content	Raw materials		
	Soil (%)	RHA (%)	SBA (%)
Gravel	0	0	0
Sand	58.29	73.38	25.47
Silt	40.46	25.12	73.03
Clay	1.25	1.50	1.50

3.2 Mechanical and durability properties

3.2.1 Weight per unit area

Results of weight per unit area are shown in figure 7. It was observed that lighter bricks can be produced after incorporating RHA and SBA in clay bricks. Lighter bricks up to 15% and

4% were prepared after incorporating 15% of SBA and RHA in replacement of clay, respectively. Lesser unit weight of RHA (550.54 Kg/m^3) and SBA (258.6 kg/m^3) as compared to clay (1120 kg/m^3) was considered responsible for lighter weight bricks. Light weight bricks can be very helpful in earthquake affected areas and can be economical by reducing the transportation cost.

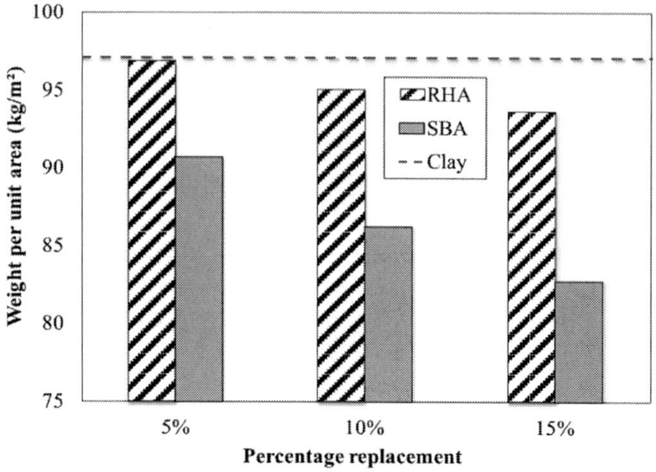

Fig. 7. Effect of RHA and SBA wastes on weight per unit area of bricks.

3.2.2 Compressive strength

Figure 8 showed the compressive strength results of clay bricks incorporating RHA and SBA. It was observed that with waste addition strength of burnt clay bricks were reduced. For instance, decrease of 14% in compressive strength was observed after incorporating 5% of SBA by clay weight in brick specimens. 50% reduction in compressive strength was observed with 15% SBA replacement with clay. Addition of RHA also showed a similar reduction in strength. Increase in porosity was considered as the main reason of strength reduction. Brick specimens showed compressive strength of 6.62 and 7.18 MPa after incorporating 5% of RHA and SBA, respectively, which satisfied the Building Code of Pakistan 2007requirement for minimum compressive strength.

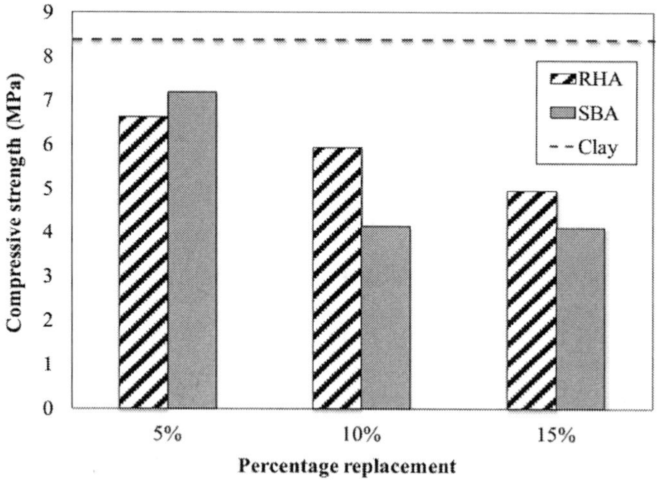

Fig. 8. Compressive strength of bricks incorporating RHA and SBA.

3.2.3 Modulus of rupture

Figure 9showed that the modulus of rupture reduced after addition of SBA and RHA in burnt clay bricks. 15% replacement of SBA with clay showed minimum flexural strength of burnt clay brick i.e., 0.67 MPa. Similar, decrease in strength was also observed with RHA addition. Reduction in strength was related to increased porosity with waste. However, all the brick specimens satisfied the minimum permissible limit for modulus of rupture i.e., 0.65 MPa according to ASTM C 67 [10].

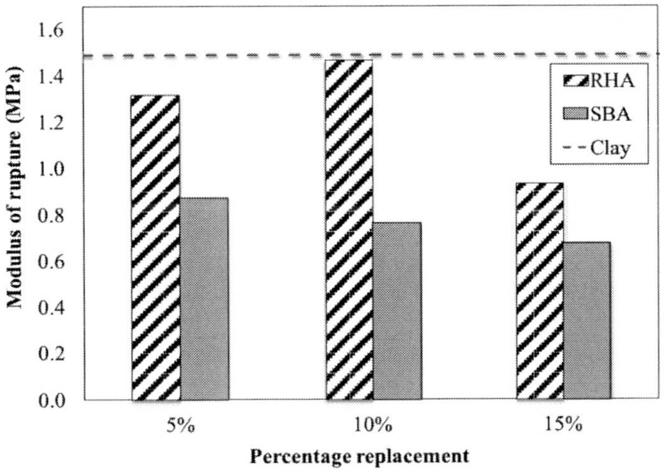

Fig. 9. Effect of RHA and SBA wastes on modulus of rupture of bricks

3.2.4 Apparent porosity

Porosity is the property of burnt clay bricks, which is related to the durability [21]. Apparent porosity results of burnt clay bricks incorporating RHA and SBA are shown in figure 10. It was observed that porosity of brick specimens increased with waste addition. Least porosity of 36% was observed for control specimens. However, higher porosity of 44% was observed after incorporating 15% of SBA in burnt clay bricks. Similarly, 4% increase in porosity was observed, when RHA addition was increased from 5% to 15% by weight of clay in the mixture. Change in size and number of pores could be the possible reason of increased porosity [22]. Bricks with higher porosity are usually considered lighter having good thermal properties [23].

3.2.5 Water absorption

Water absorption of burnt clay bricks incorporating RHA and SBA was presented in Figure 11. It was observed that with addition of SBA and RHA wastes, water absorption of burnt clay bricks increased. For instance, 24% of absorption was observed after incorporating 10% of SBA, however brick specimens without SBA showed 17% of absorption. Increased absorption with waste addition could be related to increased porosity. Lesser absorption was observed for bricks incorporating RHA as compared to bricks with SBA. Porous nature of SBA was considered responsible for such a behaviour [24]. Water absorption results of burnt clay bricks are also consistent with porosity results. Small amount of waste addition (i.e., 5%) showed water absorption less than 22%. Therefore, these bricks can be used in moderate weathering resistant environment effectively according to ASTM C62 (Standard specification for building brick solid masonry units made from clay or shale).

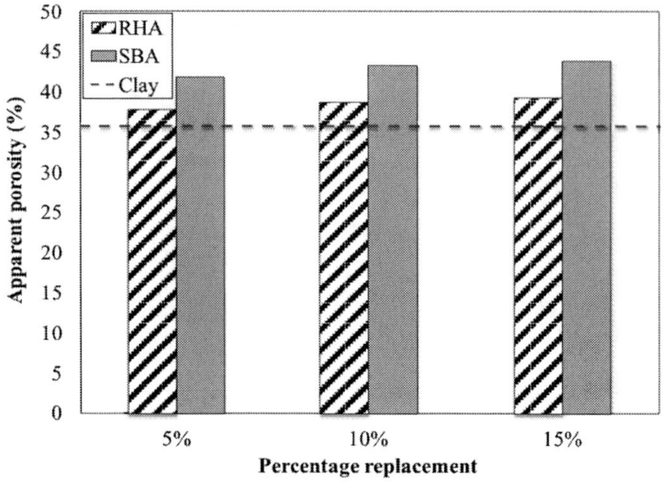

Fig. 10. Effect of RHA and SBA wastes on apparent porosity results of bricks.

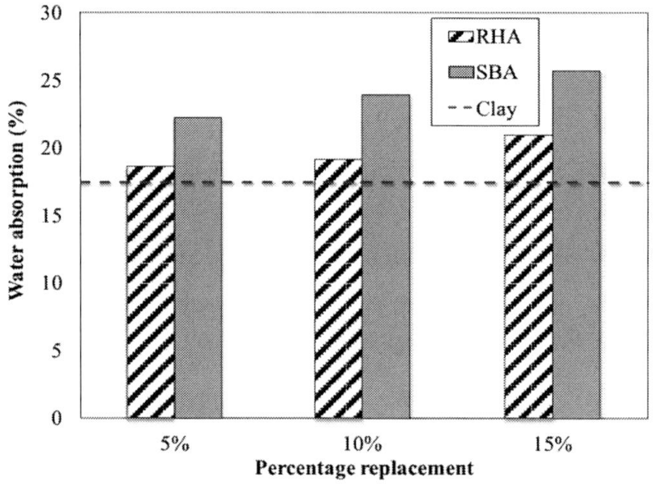

Fig. 11. Water absorption of brick specimens incorporating RHA and SBA.

3.2.6 Initial rate of absorption

Water absorbed during one minute through the brick surface is termed as initial rate of absorption (IRA) (ASTM C67 2003). Bond between brick and mortar is greatly affected by IRA. Results of IRA are shown in figure 12. It was observed that IRA increased with incorporation of RHA and SBA in burnt clay bricks. For instance, IRA value of 0.78 g/cm^2/min was observed with mixture incorporating 15% of SBA by clay weight. However, control brick showed 0.46 g/cm^2/min of absorption. Higher IRA was observed for SBA addition as compared to RHA, which is related to the porosity [25]. All the results of IRA were observed higher than 0.15 g/cm^2/min, therefore these bricks should be wetted before masonry works for proper bond with mortar (ASTM C62 2013).

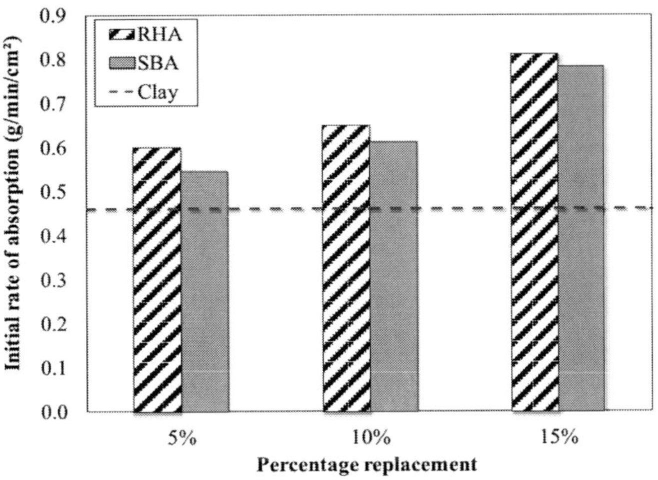

Fig. 12. Initial rate of absorption of brick specimens incorporating waste.

3.2.7 Efflorescence

Control specimens were observed with slight efflorescence i.e., 5% of the surface area after 7 days. However, brick specimens incorporating RHA and SBA showed no efflorescence. To study the efflorescence for a longer time, brick specimens were kept in efflorescence conditions for 45 days. Control specimens showed minor efflorescence i.e., 10% of the surface area, however brick specimens incorporating RHA and SBA showed no efflorescence after 45 days. Generally, amount of calcium oxide (CaO) is considered responsible for causing efflorescence [20]. Addition of RHA and SBA being silica rich waste materials, reduced the quantity of calcium oxide in burnt clay bricks, therefore improved the efflorescence resistance of bricks.

4 Conclusions

Agriculture waste (RHA and SBA) utilization in burnt clay bricks is an effective way of disposal of waste materials leading to sustainable construction. Lighter bricks can be produced after addition of RHA and SBA in burnt clay bricks. Lighter bricks are helpful in achieving economy during construction. Brick specimens after incorporating RHA and SBA showed less compressive and flexural strength. An addition of 5% of waste in burnt clay bricks satisfied the Building Code of Pakistan requirement for minimum compressive strength. Similarly, all the bricks fulfilled ASTM C 67 requirement for flexural strength.

Furthermore, porosity, water absorption and initial rate of absorption was increased with the addition of waste in burnt clay bricks. High porosity is usually related with good insulation properties. Burnt clay bricks with 5% waste addition can be used in moderate weather according to water absorption results. Efflorescence results were also encouraging. Based on the study, it can be concluded that bricks incorporating RHA and SBA up to 5% can be effectively used for construction purposes leading to sustainable construction.

References

1. A.S. More, A. Tarade, A. Anant, Int. J. Sci. Res. Publ. **4(7)**, 1-6 (2014).
2. A. Pawar, D. Garud, Int. J. Res. Eng. Technol. **3(9)**, 75-80 (2014).

3. L. Zhang, Const. Build. Mater. **47**, 643-665 (2013).
4. L. Henry, B. Shankha, J. William, S. Melissa, *Proceedings of World of Coal Ash* (Covington, Kentucky, USA, 2007).
5. A.A. Kadir, N. Maasom, Int. J. Zero Waste Generation **1(1)**, 21-26 (2013).
6. P. Velasco, M. Ortiz, M. Giro, L. Velasco, Const. Build. Mater. **63**, 97-107 (2014).
7. S.M.S. Kazmi, S. Abbas, M.L. Nehdi, M.A. Saleem, M.J. Munir, J. Mate. Civil Eng. (to be published).
8. S.E. Chidiac, L.M. Federico, Can. J. Civil Eng. **34**, 1458-1466 (2007).
9. E.R. Kumar, N. Hooda, Int. J. Res. Aeronaut. Mech. Eng. **2(9)**, 56-67 (2014).
10. A. Shakir, S. Naganathan, K. Mustapha, Const. Build. Mater. **41**, 131-138 (2013).
11. F. Saboya, G. Xavier, J. Alexandre, Const. Build. Mater. **21**, 1950-1960 (2007).
12. M.J. Munir, A.U. Qazi, S.M.S. Kazmi, A. Khitab, S.Z. Ashiq, I. Ahmed, Pak. J. Sci. **68(1)**, 53-62 (2016).
13. S. Abbas, S.M.S. Kazmi, M.J. Munir, Const. Build. Mater. **132**, 61-70 (2017).
14. S.M.S. Kazmi, S. Abbas, M.J. Munir, A. Khitab, J. Build. Eng. **7**, 372-378 (2016).
15. S.M.S. Kazmi, S. Abbas, M.A. Saleem, M.J. Munir, A. Khitab, Const. Build. Mater. **120**, 29-41 (2016).
16. S.M.S. Kazmi, M.J. Munir, S. Abbas, M.A. Saleem, A. Khitab. *International Conference on Advanced Materials and Emerging Technologies* (University of Engineering and Technology, Lahore, 2016).
17. M.V. Madurwar, S.A. Mandavgane, R.V. Ralegaonkar, J. Energy Eng. **141(3)**, (2014).
18. K.C.P. Faria, R.F. Gurgel, J.N.F. Holanda, J. Environ. Manage. **101**, 7-12 (2012).
19. D. Tonnayopas, P. Tekasakul, S. Jaritgnam, *Proceedings of Technology and Innovation for Sustainable Development Conference* (Thailand, 2008)
20. I. Netinger, M. Vracevic, J. Ranogajec, S. Vucetic, Gradevinar **66(3)**, 197-209 (2014).
21. K. Elert, G. Cultrone, C.R. Navarro, E.S. Pardo, J. Cult. Herit. **4(2)**, 91-99 (2003).
22. C.A. Garcia-Ubaque, G. Liliana, C.M. Juan, Afinidad LXX **561**, 61-66 (2013).
23. G. Majkrzak, J. Watson, M. Bryant, K. Clayton, *Proceedings of World of Coal Ash* (Covington, Kentuck, USA, 2007).
24. M.V. Madurwar, S.A. Mandavgane, R.V. Ralegaonkar, Curr. Sci. **107(6)**, 1044-1051 (2014).
25. T. Banu, M.M. Billah, F. Gulshan, A.S.W. Kurney, Am. J. Mater. Eng. Technol. **1(3)**, 35-40 (201

Use of recycled tires in non-structural concrete

Zamzam Al Rawahi[1,*], and *Muhammad* Bilal Waris[2]

[1] Sultan Qaboos University, Department of Civil and Arch. Engineering, P.O Box: 33, Al-Khoud 123, Muscat, Sultanate of Oman
[2] Sultan Qaboos University, Department of Civil and Arch. Engineering, P.O Box: 33, Al-Khoud 123, Muscat, Sultanate of Oman

Abstract. This research addresses the issue of tire waste management and natural aggregate resource depletion. It investigates use of commercially produced recycled tire rubber as replacement for fine and coarse aggregate in non-structural concrete. Two replacement levels of 10% and 20% were considered for fine aggregate with 0% or 10% of coarse aggregate. The study employed a mix proportion of 1:5:4 (cement: fine aggregate: coarse aggregate) with a water-to-cement ratio of 0.25, which is normally utilized in concrete block manufacturing in Oman. The mixes were tested for their thermal conductivity, water absorption and compressive strength. The behavior of mixes exposed to 100 and 200°C was also studied and the samples were later tested for compressive strength. The results showed improvements in compressive strength after exposure to heat. Thermal conductivity was reduced as the percentage replacement increased for both fine and coarse aggregate. During heat exposure, the temperature rise was faster in rubberized mixes, and the compressive strength of all mixes improved after the exposure to heat. Water absorption and void content increased with increase in replacement percentage. The compressive strength did not show a clear trend with the replacement and this is due to the sensitivity of the stiff mix used in the study and its inherent lean nature. The results indicate that the lean nature of the mix makes it insensitive to small replacement investigated in this research.

1 Introduction

Tire waste represents serious health and environmental concern in many ways. Waste tires from vehicles are often improperly stored and disposed in Oman by either disposal into stockpiles or illegal dumping. The lack of proper management of these waste tires presents a potential threat to human health and environment [1]. On the other hand, saving natural resources utilized in form of aggregate is a concern. Concrete could be one of the possible solutions for the utilization of this tire waste and conserve the natural resources.

Several studies have been conducted to investigate use of tire rubber waste in concrete. The studies where carried out by replacing fine and coarse aggregate with fine and coarse

* Corresponding author: zamzam7793@gmail.com

tire rubber in different replacements percentage. According to previous researchers, it is recommended to replace no more than 30% of fine aggregate. With super fine rubber (500-600 micron) [2], the crump rubber enhances sound insulation and thermal conductivity properties of the resulting concrete [3]. Also, the larger the rubber particle size, the higher the effect on compressive strength [4] and as the percentage replacement increased the compressive strength decrease [1-2, 5-6]. Water absorption increases as the percentage replacement increase [1, 5-6]. In addition, the use of tire replacements in structural concrete is likely to face loss in strength, hence, its application to non- structural concrete is more suitable [2].

Therefore, this paper shall investigate the use of tire waste rubber in non-structural concrete as fine and coarse aggregate replacements with different replacement proportions, and study physical properties of the rubberized concrete including water absorption, thermal conductivity and compressive strength. Also, it shall verify the conformance of the rubberized concrete with the specified standards for non- structural concrete.

2 Experimental program

2.1 Specifications

This research investigates the performance of rubberized concrete mix to be used in non-structural concrete that follow two standards, Oman Concrete Products (OCP) and ASTM Standards. Oman Concrete Products [7] is one of the leading companies in Oman in the field of infrastructure construction. It has ISO 9001-2000 certification. Materials produced by OCP are designed based on a combination of several international standards and also environmental, economical, and technical factors specific to construction in Oman [2]. This research uses a mix proportion of 1:5:4 (cement: fine aggregate: coarse aggregate) and w/c ratio of 0.25 according to OCP.

2.2 Materials

Materials used in this study include Ordinary Portland cement (Type 1/ ASTM C150 [8]), 10 mm natural course aggregate, fine aggregate, tap water, crumb tires (<20 mm as coarse aggregate replacement and 0.8-4 mm as fine aggregate replacement). Specific gravity and water absorption for coarse and fine aggregate were tested according to ASTMC127 [9] and ASTM C128 [10] respectively. Water absorption for tire rubber was measured and calculated according to ASTMD570 [11]. Physical properties of aggregate and tire rubber are shown in Table 1. Figure 1(a) shows the gradation curve of fine aggregate and fine tire rubber. It is clear that approximately 60 % of tire crumps are within the size range 1.18 - 2.36 mm. Figure 1(b) shows the gradation curve of coarse aggregate and coarse rubber. Coarse tire crumps varied between 2.36 and 14 mm in size.

Table 1. Physical properties of aggregate and tire rubber.

Material	%Absorption	Specific Gravity
Coarse Aggregate	3.63	2.77
Fine Aggregate	3.34	2.68
Fine Crumped Rubber	3.48	1.05
Coarse Crumped Rubber	5.57	1.05

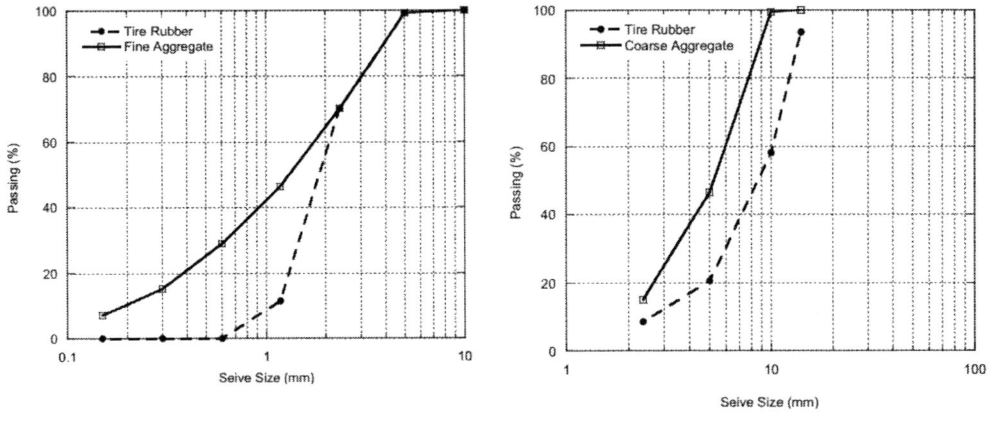

(a) Fine aggregate and tire rubber (0.8-4 mm)

(b)Coarse aggregate and tire rubber.

Fig. 1. Particle size distribution of natural aggregate and tire rubber.

A control mix (CM) will be produced with 2 sets of mixes. Set 1 has 0% coarse aggregate replacement (CAR) and set 2 has 10% coarse aggregate replacement. Each set will have 10% and 20% fine aggregate replacements (FAR). The experimental test scheme is summarized in Table 2.

Table 2. Experimental test scheme.

	Control mix	**Set 1**		**Set 2**	
Coarse aggregate replacement (CAR)	0%	0%		10%	
Fine aggregate replacement (FAR)	0%	10%	20%	10%	20%

2.3 Specimen preparation and testing

Concrete was casted according ASTM C192 [12] using a vibration table. Despite the stiff mix produced, no segregation was observed as it can be observed in Figure 2. All the samples were cured for 28 days at room temperature.

Fig. 2. Cross section of 100 mm cube with CAR-0%, FAR-10%.

DOI: 10.1051/matecconf/201712003002

Four tests were carried out: thermal conductivity (Figure 3), water absorption, compressive strength and heating test (Figure 4). Heating test is a non-standard test where concrete is exposed to heat at 2 different temperatures, 100 and 200°C. The temperature was measured using thermocouples, 1 inside each cube and one in the oven to indicate the temperature at the surface of the cube. The cubes were then inserted in the furnace and the thermocouples were plugged into the data logger to take a reading of the temperature every 10 seconds. Table 3 lists the investigated mechanical and physical properties of hardened concrete along with the standards used and cubes size. Three specimens were tested for each mix and the property is reported as the average of the three values.

Fig. 3. Setup for thermal conductivity test.

(a) Cubes with thermocouples inside for heating test

(b) Apparatus used for heating test.

Fig. 4. Setup for investigating behavior under exposure to heat.

Table 3. Tests on hardened concrete

Property Test	Sample Size	Test Standard
Thermal Conductivity	100 mm Cube	ISO 22007-2 [13]
Water Absorption	100 mm Cube	ASTM C 642 [14]
Compressive Strength	150 mm Cube	BS EN 12390 [15]
Heating Test	150 mm Cube	--

3 Results and analysis

3.1 Thermal conductivity test

Figure 5 presents the results of thermal conductivity, which decreased as the percentage replacement increased. Thermal conductivity of the control mix was found to be 1.0174 W/mK, while for those with FAR-10% and FAR-20% of CAR-0% mix were found to be 0.9505 and 0.8848 W/mK respectively. For CAR-10% mixes, thermal conductivity of FAR-10% was found at 0.9692 W/mK, while it was 0.8890 W/mk for FAR-20%. Low values of thermal conductivity for mixes with tire rubber indicate that rubberized concrete has a better insulation than plain concrete and this is partly be due to the low density of rubberized concrete samples.

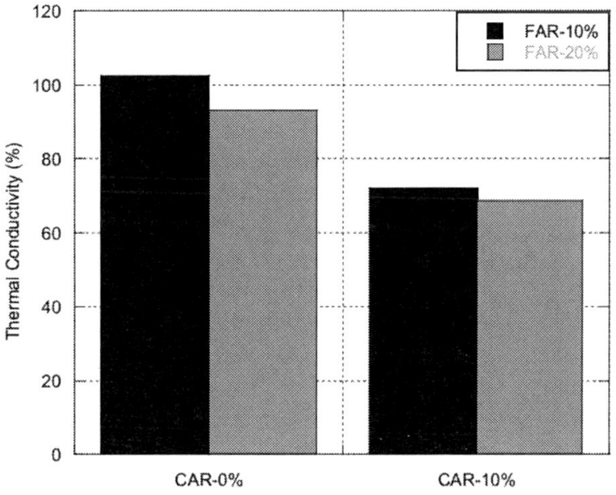

Fig. 5. Thermal conductivity.

3.2 Water absorption test

Table 4 shows the results for water absorption, bulk density and percentage voids in the mixes. As shown in Table 4, the rubberized concrete exhibited lighter density than plain concrete. The reduction in the bulk density is mainly due to the replacement of a heavier material (aggregate) by a lighter material (tire rubber). Replacement of fine and coarse aggregate by rubber influence the porosity of concrete which affects the water absorption capacity of concrete. It can be observed from Table 4 that water absorption percentage did not exceed the limit specified by OCP based on BS 6073 [16], which is 15%. The water absorption of rubberized concrete is, therefore, still acceptable. Samples with no CAR has a higher water absorption percentage than samples with CAR-10%. The reason behind this inverse relationship may be due to the sensitivity of the mixes to variation in compaction due improvement in workability with addition of rubber as reported in previous studies [1, 2].

Table 4. Results of water absorption test.

Set	CM	1		2	
Coarse Aggregate Replacement (CAR)	**0%**	**0%**		**10%**	
Fine Aggregate Replacement (FAR)	**0%**	**10%**	**20%**	**10%**	**20%**
Weight (kg)	2.13	2.03	1.93	1.96	1.96
Bulk Density (kg/m^3)	1990	1920	1930	1960	1950
Absorption (%)	12.93	13.25	12.05	9.32	8.88
Voids (%)	30.75	30.52	27.77	22.48	21.19

3.3 Compressive strength test

The weights and compressive strength values of 28 days samples are presented in Table 5. Samples with CAR-0% have a higher compressive strength value than samples with CAR-10% as it can be observed from Figure 6. Also, as the fine aggregate replacement increased, the compressive strength decreased and this can be clearly observed at CAR-10% mixes. These results indicate that as the percentage replacement increased, the compressive strength decreased. However, the effect of fine aggregate replacement is reversed in CAR-0% mixes, as the percentage replacement increased, the compressive strength increased. This may be due to variation in the compaction effort since it is a manual process as previously explained.

Table 5. Summary of weights and compressive strength values.

Set	CM	1		2	
CAR	**0%**	**0%**		**10%**	
FAR	**0%**	**10%**	**20%**	**10%**	**20%**
Weight (kg)	7.03	6.82	6.75	6.63	6.48
Compressive Strength (MPa)	2.95	3.48	4.74	3.20	3.02

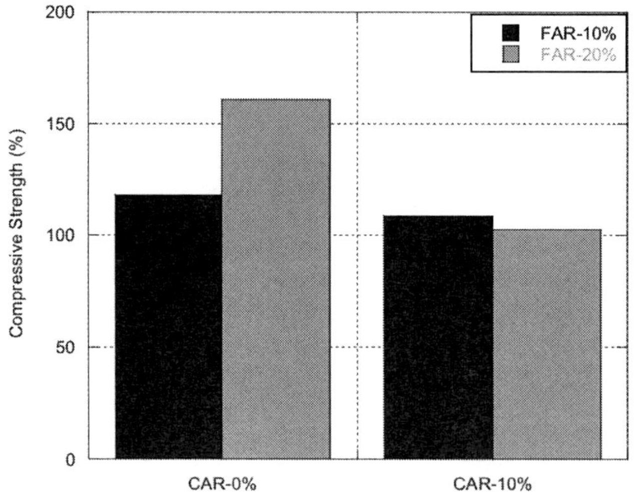

Fig. 6. Compressive strength test results as percentage of control mix.

3.4 Heating test

The temperature gradient at the core of 150 mm cube during exposure to 100°C for 3 hours is shown in Figure 7. Figure 8 shows the temperature gradient at 200°C for 6 hours. The data was collected at 10 sec intervals but an average of one minute is reported in the graphs. For 100°C, all mixes of CAR-0% reached a temperature of approximately 60°C after 3 hours, while for the CAR-10%, the FAR-20% reached a temperature of nearly 70°C. For 200°C, in CAR-0% all mix shows near similar behavior as the control mix with slight variations. For the CAR-10%, the FAR-10% shows almost identical behavior has the control mix while the FAR-20% clear steep temperature gradient. Relating the results to the thermal conductivity test, there was heat dissipation in the thermal conductivity test through the sides exposed to the environment while in case of the oven heating all the sides were exposed to heat and hence it showed a low thermal insulation. For the 200°C, a plateau is observed in all mixes at 100°C, this could be a result of thermo-chemical changes in the concrete matrix due to complete evaporation of free and bonded water.

The heat exposure was extended until the core reached thermal equilibrium. The compressive strength test was carried out for these cubes to investigate the effect of exposure to high temperature. It can be observed from Figure 9 that the compressive strength of all mixes has improved after heat exposure. When the concrete samples are heated (<300°C), free water evaporates first, followed by capillary water, and finally by physically bound water. The evaporated water will provide a moist environment inside the oven so hydration reactions may take place. In these conditions, hydration process is accelerated that increases strength of concrete [17].

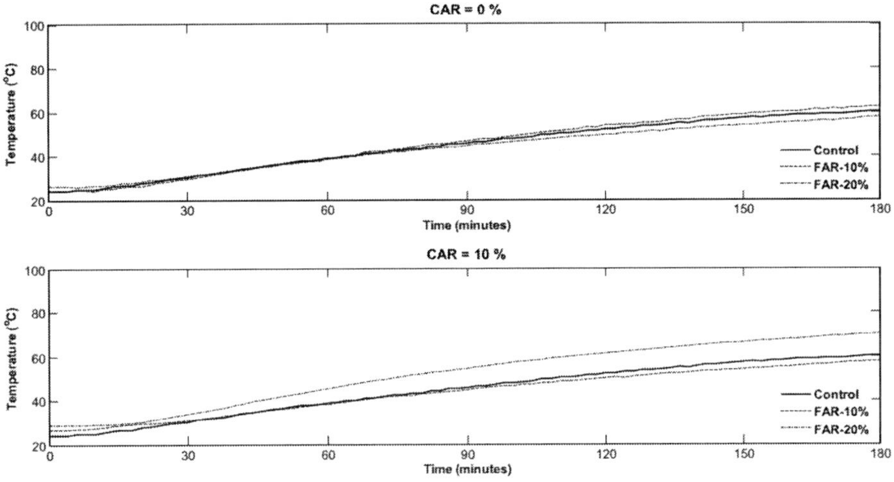

Fig. 7. Temperature gradient for all mixes at 100°C.

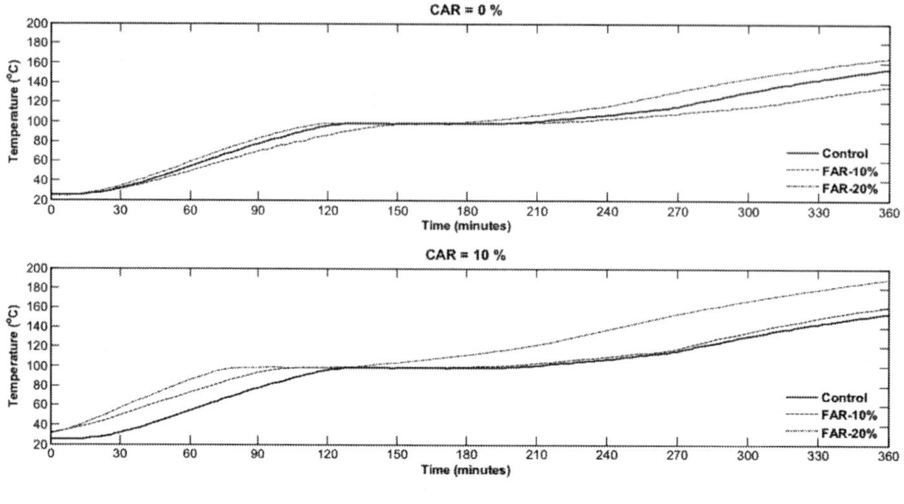

Fig.8. Temperature gradient for all mixes at 200°C.

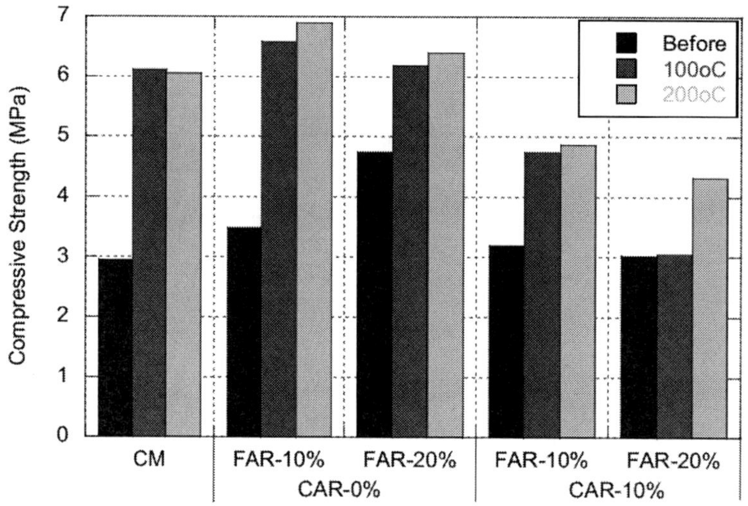

Fig. 9. Compressive strength of samples before and after exposure to 100 and 200°C.

4 Conclusion

Five concrete mixes have been studied in this research with two levels of replacement for fine aggregate and one for coarse aggregate. Mix proportion and w/c ratio were kept constant as 1:5:4 (cement: fine aggregate: coarse aggregate) and 0.25 respectively. The samples were tested for their thermal conductivity, water absorption, compressive strength and heat exposure. Rubberized concrete provides better thermal insulation properties as seen by decreasing thermal conductivity values as compared to those with no or less rubber content. Water absorption of rubberized concrete mixes was not greatly affected by the replacement. All water absorption values did not exceed the maximum limit specified by OCP which is 15%. Voids content in rubberized concrete can be reduced by a better compaction effort or by adding superplasticizer to improve the workability and hence reduce the void content. Compression strength is relatively insensitive to the replacement due to the lean nature of the mix. Since the void content was found to be slightly higher in rubberized concrete compared to those with no or less tire replacement, the heat penetrates easily and rapidly into the rubberized concrete samples at exposure to heat. The compressive strength of the samples has improved after heat exposure at 100°C and 200°C. The overall performance of concrete under replacement was good and shows evident improvements in the the compressive strength after heating and in thermal insulation.

References

1. Ali, N. 'Utilization of Recycled Tires for Aggregate Replacement in Concrete', Research Project Report, Sultan Qaboos University, Oman (2013).
2. Najwani, H . ' Use of Recycled Rubber in Non-Structural Concrete', Research Project Report, Sultan Qaboos University, Oman (2015).
3. Sukontasukkul, P. and Wiwatpattanapong, S. 'Lightweight Concrete Mixed with Superfine Crumb Rubber Powder Part 1: Insulation Properties', *KMUTNB*, **19**(3), p. 2552 (2009).
4. Lijuan, L. Shenghua, R. Lan, Z. 'Mechanical properties and constitutive equations of concrete containing a low volume of tire rubber particles', Construction and Building Materials, **70**, pp. 291–308 (2014).
5. Gupta, T., Chaudhary, S .and Sharma, R. 'Assessment of mechanical and durability properties of concrete containing waste rubber tire as fine aggregate', Construction and Building Materials, **73**, pp. 562–574 (2014).
6. Sandeep, V., Kumar, N. and Sudharani, Ch. 'Using Tires Wastes as AggregatesInConcreteToFormRubcrete–MixForEngineeringApplications',IJRET: International Journal of Research in Engineering and Technology, **3**, pp. 2321-7308 (2014).
7. Oman Concrete Products, www.ocpblock.com
8. ASTM C150, Standard Specification for Portland Cement.
9. ASTM C127-15, Standard Test Method for Relative Density (Specific Gravity) and Absorption of Coarse Aggregate.
10. ASTM C128-15, Standard Test Method for Relative Density (Specific Gravity) and Absorption of Fine Aggregate.
11. ASTM D570-98, Standard Test Method for Water Absorption of Plastics.
12. ASTM C192, Standard Practice for Making and Curing Concrete Test Specimens in the Laboratory.
13. ISO 22007-2, Determination of Thermal Conductivity and Thermal Diffusivity - Part 2: Transient Plane Heat Source (Hot Disc) Method.
14. ASTM C642-13, Standard Test Method for Density Absorption and Voids in Hardened Concrete.
15. BS EN12390, Testing Hardened Concrete, Part 3: Compressive Strength of Test Specimens.
16. BS 6073, Precast Concrete Masonry Units. Guide For Specifying Precast Concrete Masonry Units.
17. Al-Jabri, K., Waris, M. and Al-Saidy, A., 'Effect of aggregate and water to cement ratio on concrete properties at elevated temperature', Fire and Materials; **40**:913–925 (2015).

Performing self-compacting concrete with electric arc-furnace slag as aggregates

Amaia Santamaría[1,*] , *José Tomás San-José*[1], *Iñigo Vegas*[2], *Vanesa Ortega-López*[3], *Javier Jesús González*[1]

[1]UPV/EHU, Dept. of Engineering Materials, Alameda Urquijo, s/n, 48013 Bilbao, Spain
[2]TECNALIA. C/Geldo, Parque Tecnológico de Bizkaia, Edificio 700, 48160 Derio, Spain
[3]University of Burgos, EPS, Dept. of Civil Engineering, C/Villadiego s/n, 09001 Burgos, Spain

Abstract. The electric steelmaking industry is of great importance to the economy of the Basque Country (Spain). In all, 600,000 tons of electric arc-furnace (EAF) slag are produced every year; a by-product that this research group believes can be transformed into a useful resource. One of the uses of this material is as an aggregate in hydraulic mixes. Many studies have demonstrated that hydraulic mixes manufactured with EAF slag have at least the same mechanical behaviour and durability as ordinary concrete. However, their weaknesses are their higher density and poorer workability. In this paper, the aim is to demonstrate that manufacturing slag concrete to an acknowledged standard of workability is possible; so the objective is to manufacture self-compacting concrete using EAF slag in partial substitution of aggregates. Our analysis of the successful manufacture of three different self-compacting mixes, their properties in the fresh state and their mechanical behaviour yielded very encouraging results.

1 Introduction

Increased global steel production is mainly due to unrestricted growth in the Asian market. The economic crisis and the Asian market are the factors that have influenced European production of steel over the past years. There was a decrease of 16% in EU production between 2000 and 2010, being almost constant this production over the past 5 years.

Our interest centres on the production of steel in electric arc-furnaces (EAF). In all, 15% of the global production of this product is in the European Union, of which most of it is produced in Italy, followed by Germany and in third position, Spain, with 15% of all EAF steel produced in the EU.

* Corresponding author: amaia.santamaria@ehu.eus

Table 1. EAF steel production [1,2].

Per year	EAF steel (Million tons)	Land extension ($x10^3km^2$)	t EAF steel/ km^2 land	Population (million)	t EAF steel/ person
World	430	148,940	2.88	7,376.5	0.058
European Union	66	4,442	14.85	508.45	0.13
Spain	10	506	19.76	46.77	0.21
Basque Country	4	7.23	552.94	2.19	1.83

In the Basque Country, a small territory in the north of Spain where this research was carried out, all steel production use this type of electric furnace. The data presented in Table 1 show that the production per square kilometre is 200 times higher when compared with global production, 36 times higher when compared with Europe, and 28 times higher in comparison with Spain. If we compare these figures against population density, the results are similar [1,2]. These data imply that the sector generates large amounts of steelmaking waste, which have to be stockpiled within a small territory.

Nowadays, the European Union is trying to support the transition towards a more circular economy, where the waste generated in one industry becomes a secondary raw material for another industry. So, in line with European policy, in the Basque Country 600,000 tn. of EAF slag are generated every year, which this research group is convinced can be turned into a useful by-product, investing it with added value. Due to its physical and chemical properties, many researchers have proposed its use as an aggregate in hydraulic and bituminous mixes [3-6].

This research group has been studying the use of EAF slag in concrete since the late 1990s. Since then, many published works from these and other authors have concluded that concrete manufactured with EAF slag has at least the same mechanical properties as concretes manufactured with natural materials [7,8], and of similar durability [9,10]. The disadvantages of using this raw material in the aforementioned way are its higher specific weight and the lower workability of its mixes.

In the opinion of this research group the higher density of slag concretes can be compensated by its gain in strength, making this EAF slag concrete suitable for manufacturing structural elements [11]. Its poor workability can be solved through slight modifications to the traditional mix design. The objective of this research is to demonstrate that it is possible to perform EAF slag concrete with good workability, even by manufacturing a self-compacting concrete, with electric arc furnace slag in substitution of a percentage of the conventional aggregate.

In the 1980s, Okamura [12] communicated the technique of manufacturing self-compacting concrete to the world, a concrete that can fill a formwork with no need for vibration and with no segregation of the coarse aggregate. His procedure to produce these kinds of mixtures can today be considered as "the traditional or conventional way to make self-compacting mixes".

That traditional or conventional method for manufacturing self-compacting concrete in brief consists in slightly decreasing the coarse aggregate content, so as to obtain good matrix flowability [13]. This matrix phase must be of sufficient deformability to be able to compact by its own weight, but also it must be highly viscous so that the coarse aggregate fraction is efficiently transported.

Some months ago, this research team successfully prepared self-compacting mortars [14] that satisfy the specifications of current standards [15]. In the present work, the manufacture of self-compacting concrete is the relevant objective.

2 Material and methods

2.1 Materials

Two types of cement were used: first, a Portland cement type I 52.5 R; second, a Portland cement type IV/B-V 32.5-N; both in accordance with UNE-EN 197-1 standard. Water from the urban mains supply of the city of Bilbao was used containing no compounds that could affect the hydraulic mixes.

A commercial crushed natural limestone qualified as fine aggregate (maximum size 4.75 mm, fineness modulus 2.9 units, bulk density 2.6 Mg/m^3), and a medium-size limestone aggregate (sized 5-12 mm, fineness modulus 6 units) were used partially or totally in the mixes; the main mineral component of both aggregates was calcite (95%) and they can be considered classical or conventional components of concrete.

The same material mentioned as fine aggregate was also used in the mixes as an ultrafine fraction after sieving through 1.18 mm (passing through ASTM sieve Nº 16), the fineness modulus of which was 1.5 units. The grading of all the above-mentioned fractions is shown in Figure 1.

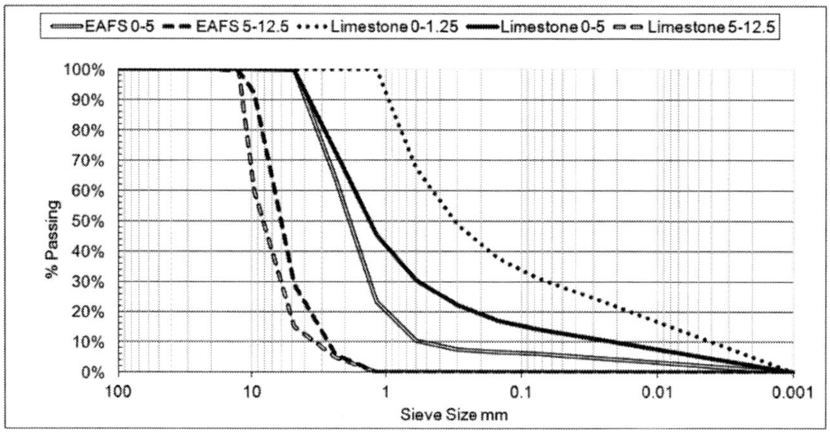

Fig. 1. Grading of the aggregates.

Crushed Electric Arc-Furnace slag (EAFS) in two size fractions (fine <4.75 mm and medium 4.75-12.5 mm), was used in this research, supplied by the company Hormor-Zestoa. Its main chemical compounds are Fe_2O_3, CaO, SiO_2 Al_2O_3, having also minor presence of MgO MnO, Cr_2O_3 and other oxides. Its main crystalline components, obtained after X-ray diffraction test, were Wüstite, Ghelenite and Kirsteinite. Its physical properties are related to water absorption of 1.2%, and specific gravity of 3.42 Mg/m^3. Their grading is also shown in Figure 1, and it is a crushed aggregate whose particles are sharp and rough form.

2.2 Design of the mixes

Several attempts took place to manufacture self-compacting concretes, using the method proposed by Okamura. The results were not as expected, obtaining concretes with good workability but they did not fulfil the requirements of a self-compacting concrete. Two indispensable factors had to be changed to obtain self-compacting concretes.

The first one was to find a suitable admixture, plasticizer and viscosity conditioner. Some of the commercial superplasticizers had effects on the concretes manufactured with natural aggregates that differed from the effects it had on the concretes manufactured with EAF slag. This observation means that in spite of the supposedly unique interaction between the superplasticizers and the cement particles, the aggregate also influences this chemical process.

Table 2. Mix proportions.

Mix Design (kg/m³)		SCC1	SCC2	SCC3
CEM I 52.5 R		300	300	210
Fly ash				140
Limestone	Sieved fraction <1.18 mm	670	900	900
	Fine aggregate <5 mm	500		
	Medium aggregate 5-12 mm	670		
EAFS	<5 mm		535	535
	5-12 mm		685	685
Water		195	195	185
Superplasticizer		2%	2%	2%

The second indispensable factor was the need to use a suitable proportion of fine aggregate of a size below 1.18mm (Figure1). EAF slag has insufficient fine fractions to behave as a conventional aggregate and it is necessary to add a selected fraction of natural fine aggregate. In addition, due to the rough shape of the coarse EAF slag, the fine fraction must be increased comparing with the quantity needed for manufacturing self-compacting concretes with natural aggregates.

With this adjustment, three different self-compacting mixes were manufactured with the proportions shown in Table 2 and the grading shown in Figure 2. The first mix SCC1, the reference mix, is manufactured with natural limestone aggregate. Both SCC2 and SCC3, manufactured with EAF slag aggregate, are of similar grading. The only difference between them is the cement type: cement type I and type IV were used in the manufacture of mixes SCC2 and SCC3, respectively.

The components of concrete were put in a vertical-axis concrete mixer, and were kneaded over ten minutes. These mixtures were poured in cubic moulds of 100x100x100 mm after performing the flowability tests; they were demoulded after 24 hours and kept immersed in water until their mechanical testing.

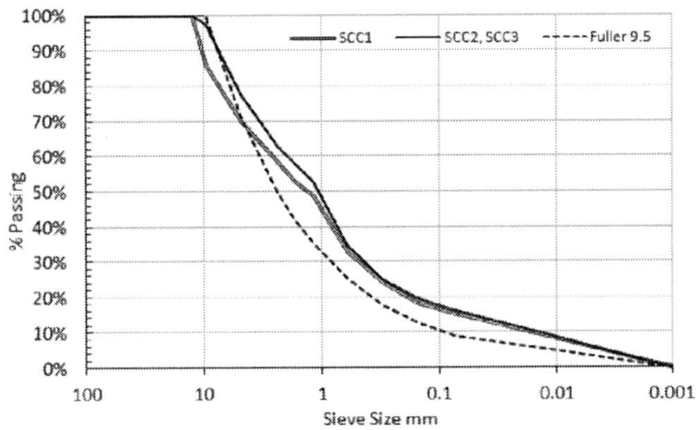

Fig. 2. Grading of mixes.

3 Fresh properties

The slump flow test and the L-box test [15] were respectively used to evaluate the flowability of the concretes, and their the passing ability through the reinforcing bars in formworks. The results of these tests are shown in Table 3. Remarkably, none of the mixes showed evidence of either segregation of coarse aggregate or blocking.
A self-compacting concrete is considered so if it has the minimum required value of 550mm in the slump flow test. Three different classes of self-compactablity are stabilized in the EFNARC [15] according to the flowability of the concretes:

- SF1 550-650 mm.
- SF2 660-750 mm.
- SF3 760-850 mm.

Table 3. Fresh properties test result.

Mixture	Slump-flow in mm	Passing ability L-box
SCC1	580 (SF1 class)	0.85 (PA2 class)
SCC2	680 (SF2 class)	0.9 (PA2 class)
SCC3	560 (SF1 class)	0.8 (PA2 class)

In this research, the aim was to reach a minimal standard of the SF1 class. Both SCC1 and SCC3 met the SF1 standard, even though SCC2 obtained a flowability class of SF2, all which were nevertheless really encouraging results.
Taking into account the passing ability, two different classes of self-compacting concrete also showed differences in the EFNARC [15]:

- PA1: ≥ 0.80 with 2 rebars.
- PA2 : ≥ 0.80 with 3 rebars.

In this case, the L-box used to evaluate the passing ability had three rebars, so all the concretes were shown to have a passing ability class of PA2. Due to the classical loss of workability in mixes using electric arc furnace slag as aggregate, it could be logic to use a L-box with two rebars, class PA1.
In spite of some results at the limit of fulfilling the requirements for self-compacting concretes, it can be said that the challenge of mixing a self-compacting concrete using EAF slag aggregate was achieved in relation to its fresh properties.

4 Hardened properties

4.1 Density

The dry and fresh densities of concretes are shown in Table 4; the correlation between both is good, as may be expected. The dry density was obtained after drying the samples in an oven at 60°C for seven days, after a curing period of 28 days following their casting.

Table 4. Density results.

Mixture	Fresh density [Mg/m^3]	Dry density [Mg/m^3]
SCC1	2.36	2.34
SCC2	2.62	2.53
SCC3	2.67	2.63

 DOI: 10.1051/matecconf/201712003003

4.2 Strength

Even if the aim of this research was focused on obtaining concretes that in the fresh state fulfil the requirements to be considered self-compacting concrete, a minimum value of 40 MPa at 28 days was considered sufficient threshold for the use of these concretes in structural applications. The aim of this research team was to obtain this threshold without increasing the cement content, 300 kg per cubic meter, in its attempts to manufacture a more sustainable concrete. After 7, 28, 90, and 180 days immersed in water, the specimens were removed and their compressive strength was tested. The results are shown in Figure 3

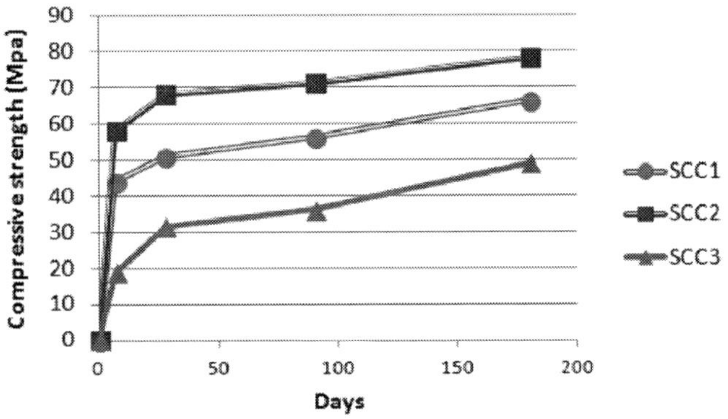

Fig. 3. Compressive strength.

The reference concrete reached 50MPa at 28 days, a higher than expected value, and at 180 days this mix reached a value 65 MPa. Mix SCC2 obtained the highest strength of all three, due to the influence of the EAF slag, reaching a value of 68MPa at 28 days and 76MPa at 180 days.

Mix SCC3 reached a value of 32 MPa at 28 days, exceeding the required strength of 40 MPa after 180 days.

5 Discussion

Enhancement of the mix flowability of SCC2 compared with mix SCC1 was achieved by over a 32% increase in the limestone fine fraction. It was expected due to the higher density and more irregular shape of the EAF slag to be worse the workability of SCC2. But the increase in the fine fraction allows achieving this good value, without increasing the cement content.

Even though mix SCC3 had the same mix proportions used to manufacture mix SCC2, the change of cement type entailed a negative effect on flowability. This effect is probably due to the dispersive action of the admixtures that were only effective on the clinker particles, without exerting repulsive forces between the fly ash particles.

The average density of the slag concretes may be seen to increase by about 12% with respect to those of concrete made with conventional aggregates (the values shown in the SCC1 mix were 2.35 Mg/m^3 and close to 2.65 Mg/m^3 in EAFS slag mixes). This is an expected value; explain with the higher density of the EAF slag in comparison with the density of natural aggregates (2.6 Mg/m3 vs 2.42 Mg/m^3). This issue is advantageous in the constructions where the concrete works by gravity. Contrary, in structural elements it seems to be a problem, but this issue is normally compensated by the increase in the strength and stiffness of EAF slag concretes.

As it has been stated in the previous section the concrete that reaches the higher compression strength value was the one manufacture with EAF slag and manufactured with Cement I. It has enhanced strength 32% higher than the natural aggregate concrete. As stated in a previous work [16] from this research group, the ITZ between EAF slag aggregates and the cement matrix will often be of better quality than the ITZ of natural aggregates and cement matrix with a consequent increase in its macroscopic compressive strength.

These values are encouraging taking into account that cement Type IV/B 32.5 N, according EN-1015, containing a high proportion of fly ash (see Table 2) was used to manufacture this mix rather than cement I 52.5 R that was used in the manufacture of the other mixes.

A full analysis of all the data confirms the successful manufacture of a self-compacting structural concrete containing EAF slag in partial substitution of conventional aggregate in terms of its mechanical strength.

6 Conclusions

The conclusions of this study are as follows:

- Manufacturing self-compacting concrete with electric arc furnace slag as aggregate is feasible.
- A suitable increase of the fine aggregate fraction is advisable when EAF slag is used.
- Finding a compatible admixture is essential, as not all admixtures will work appropriately with the EAF slag aggregate;.
- The mechanical strength tests on these concretes showed encouraging results.
- In the absence of performing durability tests, it could be said that well-performed self-compacting concrete with EAF as aggregate have been manufactured.

Our thanks to the Basque Regional Government (IT781-13 Research Group), to the Vice-Rectorate of Investigation of the University of the Basque Country (UPV/EHU) for grant PIF 2013, to the FEDER program, and to the Spanish Ministry (MINECO) (project BIA2014-55576-C2-2-R), for funding this research work. Materials were supplied by Hormor-Zestoa, Arcelor Mittal and CHRYSO Admixtures.

References

1. *Steel Statistical Year book.* International iron and steel institute (2005)
2. *Steel Statistical Year book.* International iron and steel institute (2015)
3. M. Pasetto, N. Baldo, Mix design and performance analysis of asphalt concretes with electric arc furnace slag, Construction and Building Materials **25**(8), p 3458-68 (2011).
4. F. Faleschini, M. A. Fernández-Ruiz, M.A. Zanini, K. Brunelli, C. Pellegrino, E. Hernández-Montes, High performance concrete with electric arc furnace slag as aggregate: Mechanical and durability properties, Construction and Building Materials **101** 113-121 (2015).
5. K. Morino, E. Iwatsuki, Utilization of electric arc furnace oxidizing slag as concrete aggregate, Minerals, Metals and Materials Society/AIME, REWAS'99: *Global Symposium on Recycling, Waste Treatment and Clean Technology*. pp. 521-30 (1999).
6. I. Papayianni, E. Anastasiou, Concrete incorporating high-calcium fly ash and EAF slag aggregates, Magazine of Concrete Research **63**(8) 597-604 (2011).
7. J.M. Manso, J.J. Gonzalez, J.A. Polanco, Electric arc furnace slag in concrete, Journal of materials in civil engineering **16**(6) 639-645 (2004).
8. J.T. San-José, I. Vegas, I. Arribas, I. Marcos, The performance of steel-making slag concretes in the hardened state, Materials & Design **60** 612-619 (2014).

9. I. Arribas, I. Vegas, J.T. San-José, J.M. Manso, Durability studies on steelmaking slag concretes, Materials & Design **63** 168-176 (2014).
10. I. Arribas, I. Vegas, J.T. San-José, JM. Manso. Durability studies on steelmaking slag concretes. Mater Des.;**63**:168-76 (2014).
11. C. Pellegrino, F. Faleschini. Experimental behavior of reinforced concrete beams with electric arc furnace slag as recycled aggregate. ACI Mater J.;**110**(2):197-205 (2013).
12. H. Okamura, Self-compacting high performance concrete, Concrete International **19**(7) 50-54 (1997).
13. H. Okamura, M. Ouchi, Self-compacting concrete, Journal of advanced concrete technology **1**(1) 5-15 (2003).
14. A. Santamaría, J.J. Gonzálezlez, J.T. San-José, I. Vegas, I. Arribas, E. Rojí.Performance of self-Compacting concrete containing EAF slag as aggregate. *8th European slag conference*. 21-23 October (2015).
15. EFNARC, Specification, Guidelines for Self-Compacting Concrete, European Federation of Producers and Applicators of Specialist Products for Structures (2002).
16. I. Arribas, A. Santamaría, E. Ruiz, V. Ortega-López, J.M. Manso, Electric arc furnace slag and its use in hydraulic concrete, Construction and Building Materials 90 68-79 (2015).

Effect of GGBS and curing temperature on microstructure characteristics of lightweight geopolymer concrete

Hilal El-Hassan[1,*], *Najif* Ismail[1], *Sara* Al Hinaii[1], *Asma* Alshehhi[1], *Noor* Al Ashkar[1]

[1]Department of Civil and Environmental Engineering, United Arab Emirates University, Al Ain, United Arab Emirates, P.O. Box 15551

Abstract. Cement replacement by supplementary cementitious materials has been gaining momentum as a sustainable mechanism to reduce greenhouse gas emissions while also recycling industrial by-products. This paper presents the development and microstructure characterization of fly ash-based lightweight geopolymer concrete incorporating ground granulated blast furnace slag (GGBS). Concrete samples were prepared with 0%, 25% and 50% GGBS replacement and cured at 30°C, 60°C, and ambient temperature. While dune sand and lightweight expanded clay were used as aggregates, a mixture of sodium silicate and sodium hydroxide served as the alkaline activation solution. Microstructure evaluation was carried out at 7 and 28 days employing scanning electron microscopy (SEM), Fourier transform infrared spectroscopy (FTIR), and differential scanning calorimetry (DSC). Residual fly ash and GGBS were identified in the concrete and bonded to geopolymeric reaction products. The microstructure highlighted the formation and coexistence of aluminosilicate hydrate and aluminum-rich calcium silicate hydrate with traces of sodium. Subsequent polymerization was also verified by an increase in FTIR and DSC peaks.

1 Introduction

The global production of concrete is approximately one cubic meter per capita, making it one of the largely produced materials on earth [1]. It is a composite material mainly comprising cement and coarse and fine aggregates. Portland cement is manufactured by mixing limestone and clay in specific proportions at elevated temperatures. The process consumes 1.6 kg of raw material, requires 3.7 MJ of energy, and emits 1 kg of CO_2 per kg of cement produced [2, 3]. In fact, the cement industry alone is accountable for 5-7% of the global CO_2 emissions [4, 5], leading to an increase in the concentration of CO_2 in the atmosphere [6, 7]. As a result, cement production is becoming an increasing global pressing issue from an ecological, social, and environmental standpoint.

To alleviate consumption of natural resources and emission of CO_2 from cement manufacturing, scientists and environmentalists propose the incorporation of supplementary cementitious material (SCMs) in concrete mixes. In most cases, these SCMs are pozzolanic

[*] Corresponding author: helhassan@uaeu.ac.ae

and non-pozzolanic industrial solid waste. Fly ash, a residual by-product from the combustion of coal, is highlighted as a primary replacement to ordinary Portland cement (OPC) in concrete due to its pozzolanic properties and global abundance. In one technique, it served as the main binder in an inorganic geopolymer concrete (GPC). The aluminosilicate material was activated through mixing with an alkaline solution of sodium silicate (Na_2SiO_3) and potassium or sodium hydroxide (KOH or NaOH). When compared to OPC, the geopolymer technology could reduce carbon dioxide emissions up to 80-90% [8]. In addition, fly ash-based GPC experienced improved early strength, low permeability, excellent resistance to chemical attacks, good freezing-thawing cycles and enhanced thermal durability in high temperature environments [9-14]. Further work investigated the incorporation of other SCMs, as ground granulated blast-furnace slag (GGBS), in geopolymer concrete [15-18]. The microstructure provided evidence of the coexistence of aluminosilicate gel and calcium silicate hydrate (C-S-H) gel [16, 19]. The geopolymeric matrix also highlighted the presence of unreacted fly ash as a pore-filling material, leading to a good bond between the geopolymer paste and aggregate at the interfacial transition zone (ITZ) [20, 21]. In addition, higher GGBS replacement presented a more compact and less porous concrete [15].

Despite the research efforts in studying the microstructure, the results were somewhat lacking. The combined effect of curing temperature, multiple aluminosilicate sources, dune sand addition, and lightweight aggregate use on the microstructure of geopolymer concrete has not been investigated yet. As such, this research aims to study the effect of 25% and 50% fly ash replacement by GGBS and curing temperatures of 30°C, 60°C, and outdoor ambient conditions on the microstructure of lightweight GPC. To achieve a sustainable concrete mix design, locally-available lightweight expanded clay aggregates (LECA) and dune sand were used as aggregates. The morphological changes and reaction products were characterized in GPC by employing scanning electron microscopy (SEM), Fourier transform infrared spectroscopy (FTIR), and differential scanning calorimetry (DSC). The proposed work could provide valuable information on the behavior of GPC to validate their use in the building and construction industry.

2 Materials and methods

2.1 Concrete sample preparation

Geopolymer concrete mixes were prepared with locally-available material including GGBS, fly ash, dune sand, and expanded clay aggregates. The chemical composition of the as-received materials were determined using X-ray fluorescence (XRF) and are shown in Table 1. The morphology of constituent materials is presented as SEM micrographs in Fig. 1. The alkaline activator solution was prepared two hours prior to incorporation in the GPC mix by mixing sodium hydroxide (concentration = 18M) with sodium silicate at solid and solution ratios of 1:1 and 1:1.5, respectively. Table 2 presents the chemical composition of each solution.

For this study, 100-mm concrete cubes were cast. The mixture proportions, shown in Table 3, were designed following a procedure for geopolymer concrete [22]. Three mixes were prepared while varying the percent replacement of fly ash by GGBS (0%, 25%, and 50%). By mass, the ratio of dune sand-to-LECA was fixed to 1.73, while that of total aggregate-to-cementitious material was 3.27. The alkali solution-to-cementitious material ratio was set to 0.4 for all mixes. Water was added to the mixes to improve the workability of fresh geopolymer concrete. The expanded clay aggregates were used at surface-saturated dry (SSD) condition to minimize water or alkali solution absorption during mixing. The raw materials were mixed in a drum mixer, and samples were compact formed on a vibrating table to simulate industry practice. Samples were then demolded and cured for 24 hours at outside ambient temperature (A), 30°C, and 60°C. Later, all concrete specimens were left outside to cure until testing. Samples were labeled N-T based on the mix number (1, 2, 3) and curing temperature (A, 30°C, or 60°C). It is worth noting that curing condition A, outside ambient temperature, was proposed to study the possibility of preparing GPC without the need of a controlled environment or furnace.

Table 1. Chemical composition of as-received materials.

Material	Constituent (%)							
	SiO_2	Al_2O_3	Fe_2O_3	CaO	MgO	Na_2	K_2	LOI
Fly ash	48.0	23.1	12.5	3.3	1.5	-	-	1.1
GGBS	34.7	14.4	0.8	42.0	6.9	-	-	1.1
Dune sand	63.9	3.0	0.7	14.1	1.3	0.4	1.1	-

Table 2. Chemical composition of alkali solutions.

Solution	Density (kg/m^3)	Solids (%)	H_2O (%)	Na_2O (%)	SiO_2 (%)
NaOH	1380	37.5	62.5	8.9	28.6
Na_2SiO_3	1610	51.8	48.2	40.2	0
NaOH + Na_2SiO_3	1518	43.3	56.7	21.5	17.1

Table 3. Mixture proportions of geopolymer concrete mixes.

Mix #	Component (kg/m^3)							
	Fly Ash	GGBS	Dune Sand	LECA	NaOH	Na_2SiO_3	Added Water	Total
1	302	0	628	364	48	73	5	1420
2	227	74	628	364	48	73	12	1426
3	151	148	628	364	48	73	20	1432

Fig. 1. SEM of as-received dune sand (a), fly ash (b), GGBS (c), and LECA (d).

2.2 Microstructure evaluation

2.2.1 Scanning electron microscopy

The morphology of produced geopolymer concrete specimens was investigated after 7 and 28 days using scanning electron microscopy (SEM). Test samples were extracted from untested concrete with typical flat surface and coated with a gold layer to ensure electron conductivity. High-vacuum SEM mode and energy-dispersive X-ray (EDX) analysis were employed (JEOL-JSM 6390A) to examine the changes in concrete microstructure and morphology.

2.2.2 Fourier transform infrared spectroscopy

Fourier transform infrared (FTIR) spectroscopy was employed to evaluate the degree of polymerization reaction and identify the reaction products after 7 and 28-day subsequent curing. Specimens were prepared by mixing potassium bromide (KBr) with finely-ground geopolymer concrete at a 2:1 ratio. The FTIR spectra were obtained in transmittance mode from 400 to 4000 cm^{-1} at a resolution of 1 cm^{-1} using a Varian 3100 FT-IR spectrometer.

2.2.3 Differential scanning calorimetry

Powdered samples were examined by differential scanning calorimetry (DSC) using a DSC Q200. Specimens, on the range of 10 mg, were packed and sealed into aluminum containers in preparation for testing. DSC curves were obtained by heating the samples from 20°C to 550°C at a heating rate of 10°C/min under a constant flow of nitrogen gas (50 ml/min). Resulting curves showed the heat flow between the sample and reference pan as a function of temperature.

3 Results and discussion

3.1 SEM Analysis

The effect of replacing cement by GGBS and alternating the curing temperature on the morphology of GPC was examined using SEM imaging. Fig. 2 shows the micrograph of mix 1 after 7-day curing. Aluminosilicate glass represented in unreacted fly ash spheres are heavily dispersed throughout the microstructure of concretes regardless of curing temperature. The ambient-cured sample (1-A) presented some microcracks on the order of 3 μm. The aggregate-geopolymer paste interface of concrete mix 2 appeared to be weakly bonded, evidenced by the crack propagation along the interfacial transition zone. Upon the replacement of 25% fly ash with GGBS, concrete micrographs of Fig. 3 showed agglomerated angular slag particles with some fly ash spheres attached to the reaction products. It seemed that while fly ash particles did not dissolve in the mix, the geopolymerization reaction occurred on the surface or external shell only [23]. Further incorporation of GGBS produced an intermix of source materials in the paste within a diverse matrix as shown in Fig. 4. It is clear that the addition of water to improve workability formed a more porous concrete microstructure.

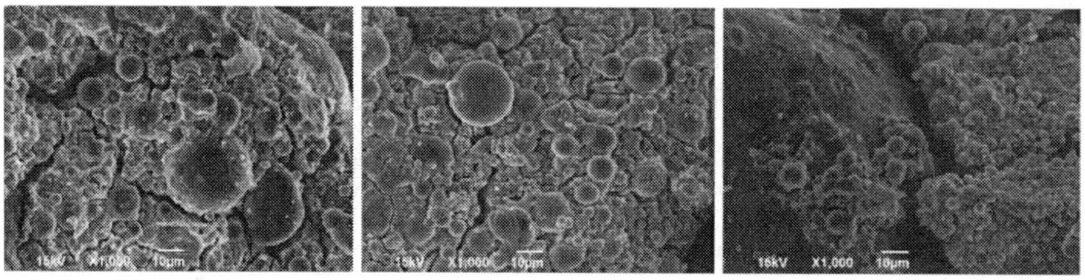

(a) (b) (c)

Fig. 2. Micrograph of mix 1 after 7 days curing at (a) ambient conditions, (b) 30°C, (c) 60°C.

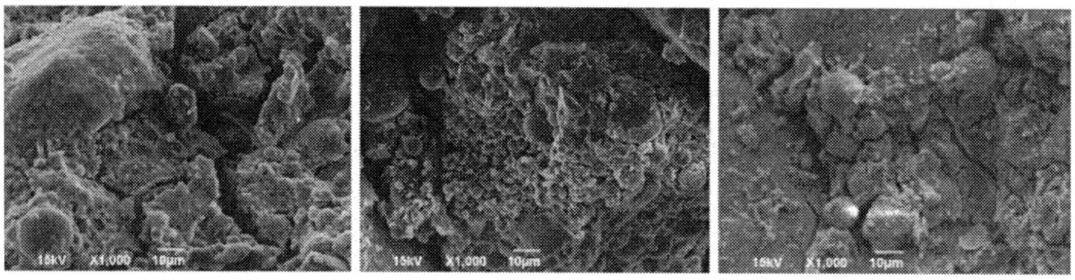

(a) (b) (c)

Fig. 3. Micrograph of mix 2 after 7 days curing at (a) ambient conditions, (b) 30°C, and (c) 60°C.

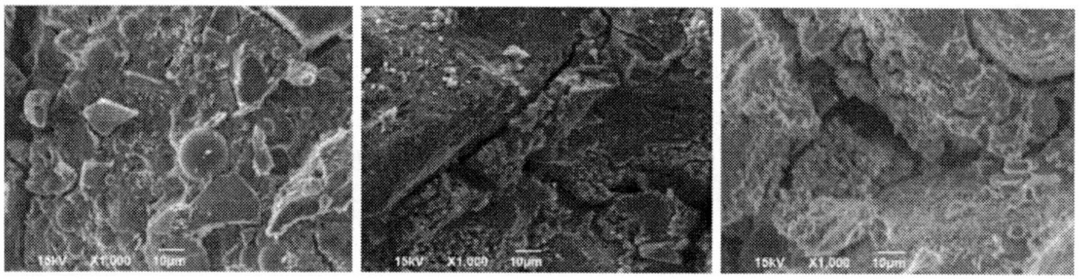

(a) (b) (c)

Fig. 4. Micrograph of mix 3 after 7 days curing at (a) ambient conditions, (b) 30°C, (c) 60°C.

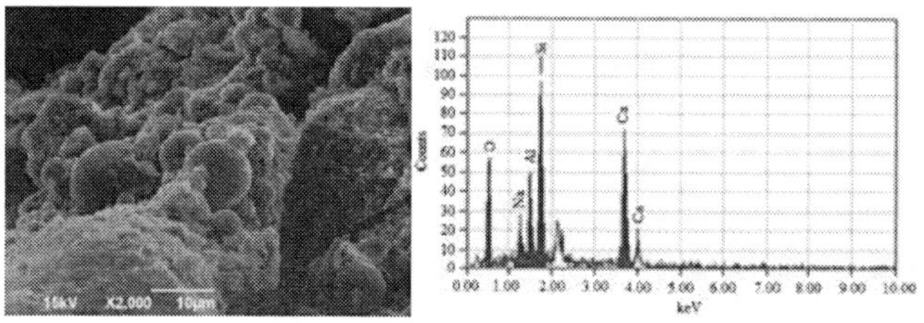

Fig. 5. SEM micrograph (a) and EDX spectrum (b) of specimen 3-60°C after 7 days curing.

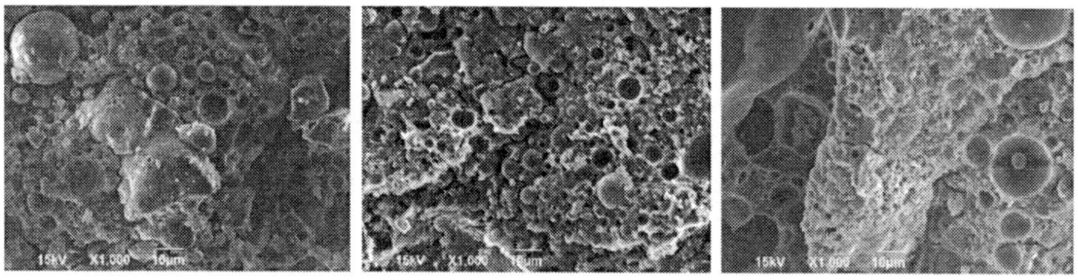

(a) (b) (c)

Fig. 6. Micrograph of mix 1 after 28 days curing at (a) ambient conditions, (b) 30°C, and (c) 60°C.

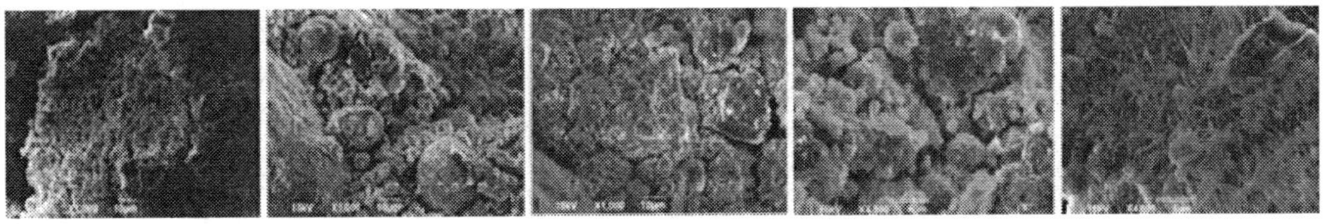

(a) (b) (c) (d) (e)

Fig. 7. Micrograph of mix 2 after 28 days curing at at (a) ambient conditions, (b) 30°C (Mag. 1000), (c) 60°C (Mag. 1000), (d) 30°C (Mag. 4000), (e) 60°C (Mag. 4000).

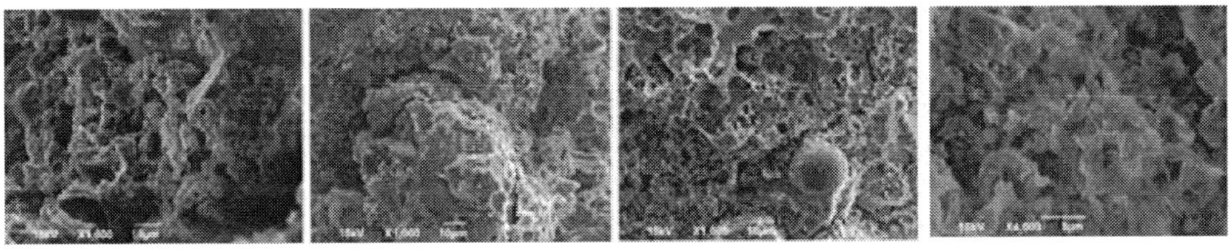

(a) (b) (c) (d)

Fig. 8. Micrograph of mix 3 after 28 days curing at (a) ambient conditions, (b) 30°C (Mag. 1000), (c) 60°C, (d) 30°C (Mag. 4000)

EDX analysis was employed over a GPC surface area to characterize the reaction products. Fig. 5(a) and 5(b) depict the SEM micrograph and corresponding EDX spectrum of concrete 3-60. It highlights the presence of calcium (Ca), silicon (Si), sodium (Na), aluminum (Al), and oxygen (O), indicating the coexistence of an aluminium-modified C–S–H gel with N–A–S–H gel [15]. It was however difficult to distinguish C–S–H and N–A–S–H gels through their morphology.

Micrographs of Figures 6(a) to 6(c) study the morphology of mix 1 at the age of 28 days. With further integration of fly ash spheres into the geopolymer paste, it is clear that the geopolymerization reaction developed during subsequent curing. Microcracks, although still present, have decreased to the range of 1 μm. A more grainy microstructure was observed in Fig. 7 for concretes with 25% GGBS replacement. Fly ash spheres were coated with glassy reaction products. Ettringite was identified at higher magnifications as shown in Fig. 7(d) and 7(e). As it was also detected in mix 3 (Fig. 8(c)), it was confirmed to be a reaction product that formed in concretes cured at 30°C and 60°C during subsequent curing.

3.2 FTIR analysis

The FTIR spectra of three geopolymer concrete mixes at the age of 7 days is shown in Fig. 9. The stretching vibration mode of Si–O–T (T being Si or Al) on the range of 820 cm^{-1} was noted at different curing temperatures. However, a shift in the wave number of concrete 1-30 towards a lower wavelength was noticed, signifying a lower degree of silica polymerization [24]. However, with the incorporation of GGBS in mixes 2 and 3, the intensity of this vibration decreased due to the formation of C–S–H and N–A–S–H gels, associated with a reduction in the amount of Al [25]. When samples were cured in ambient conditions, a band at 1420 cm^{-1} was detected. It suggested the carbonation of unreacted NaOH, resulting in Na_2CO_3 [26, 27]. The stretching vibrations associated to –OH bending vibrations and H–O–H stretching vibrations (1600–3600 cm^{-1}) were consistent throughout the mixes. These bands represented water and provided evidence of an alkali-activation reaction in geopolymer concrete [28].

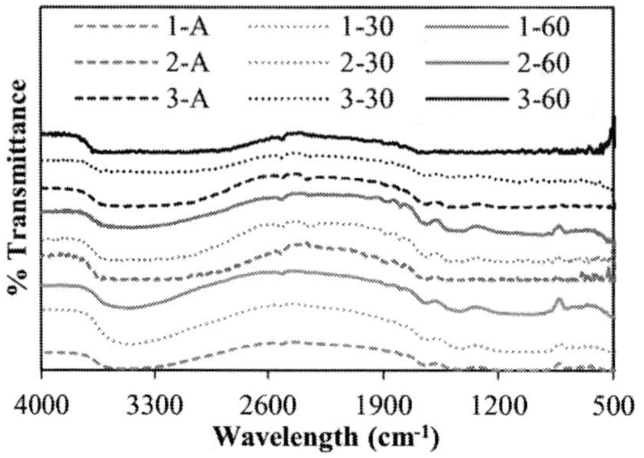

Fig. 9. FTIR spectra for geopolymer concretes after 7-day curing.

The spectra of 28-day geopolymer concrete specimens are shown in Fig. 10. An increase in H–O–H stretching vibrations at 3500 cm^{-1} was observed in comparison to 7-day specimens. This provided evidence that the geopolymerization reaction further developed and produced more polymerization products during subsequent curing. It is also worth noting that the intensity of these bands had increased in mix 3 with higher slag content.

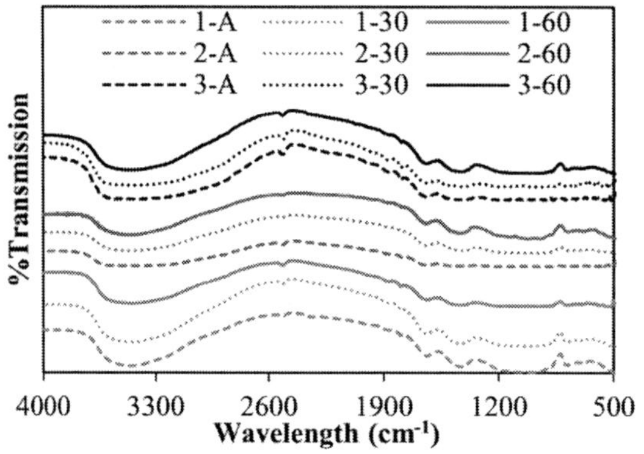

Fig. 10. FTIR spectra for 28-day geopolymer concretes.

3.3 DSC analysis

Fig. 11 depicts the DSC thermograms of geopolymer concrete mixes after 7 days subsequent curing. Results show an endotherm between 110°C and 130°C. While some work reported it as evidence of the decomposition of combined "non-evaporable" water in the form of aluminosilicate gel [29], others work associated it to the evaporation interstitial water [30]. However, it was noticed the endothermic peak shifted towards higher temperatures (170-180°C) in mixes 2 and 3. The increase in GGBS content contributed to the degree of reactivity and reduced the setting time by developing aluminum-modified C-S-H (or C-(A)-S-H) gel [25, 27].

The DSC thermograms of 28-day geopolymer concretes are presented in Fig. 12. The increase in intensity of the endotherm at low temperatures indicated further geopolymerization reaction. In mix 3, an additional endotherm was detected around 100°C, possibly due to the formation of ettringite as reported in SEM analysis. C-S-H was also identified as the broad endothermic shoulder up to 350°C [31]. The geopolymerization reaction products were highlighted as an intermix of calcium silicate hydrate and aluminosilicate hydrate in the form of a calcium aluminosilicate hydrate (C-(A)-S-H) gel.

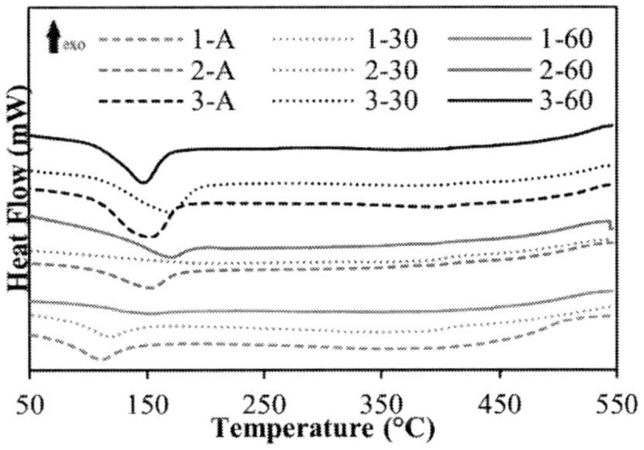

Fig. 11. DSC thermograms for 7-day geopolymer concretes.

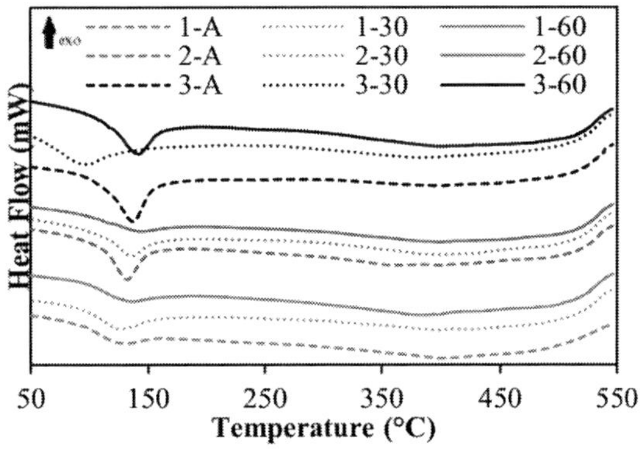

Fig. 12. DSC thermograms of geopolymer concrete after 28-day subsequent curing.

4 Conclusion

The microstructure of lightweight geopolymer concretes exposed to different curing temperatures has been investigated. The geopolymer concretes were prepared using GGBS, fly ash and an alkaline activation solution. The reaction products resulting from polymerization were identified. Based on the findings of this study, the following conclusions may be drawn:

1. Microstructure evaluation through SEM highlighted an amorphous geopolymeric matrix. At 7 days, the polymerization reaction was incomplete with much unreacted fly ash and slag particles, producing a weak bond at the aggregate-paste interface. The degree of reactivity however increased after 28 days with reaction products forming on the external shell of fly ash spherical particles. The addition of GGBS to concrete mixes cured at 30°C and 60°C resulted in the formation of ettringite.
2. DSC curves highlighted a major endotherm occurring in the range of 110-130°C. It was associated to the evaporation of combined "non-evaporable" water. The water could be either interstitial or in the form of aluminosilicate hydrate gel. After 28 days subsequent curing, ettringite was detected in mix 2 and 3. Calcium silicate hydrate was detected in a broad endotherm after 350°C.

3. The main FTIR peaks at 820 cm^{-1} have been attributed to the asymmetric stretching vibration of Si–O–Si or Si–O–Al bonds. Unreacted NaOH could carbonate in the presence of atmospheric CO_2 to produce Na_2CO_3, evidenced by the peak at 1420 cm^{-1}. Subsequent curing led to an increase in H–O–H stretching vibrations, indicating higher degree of reaction and formation of more hydration products.
4. The coexistence of aluminum-modified C-S-H gel and N-A-S-H gel was noted, evidenced by lower intensity of vibrations at 820 cm^{-1} using FTIR and endothermic peaks on the range of 170°C using DSC. Morphologically, the two reaction products could not be easily distinguished. In fact, they were intermixed in the form of a calcium aluminosilicate hydrate (C-A-S-H) gel.

Financial support for this study was provided by the United Arab Emirates University under the grants 31N249 and 31N241. Moustafa Mansour, Mohammed Al-Mawri, Muath Bassam, Abdalla El-Hashmi, and Faisal Ali assisted with specimen preparation. Abdelrahman Alsallamin and Ehab Shehab helped in performing the microstructure analysis. Ashtech International supplied the fly ash. Emirates Cement Factory provided the ground granulated blast furnace slag and aided with chemical analysis of constituent materials.

References

1. US Geological Survey. Minerals commodity summary - cement USA: USGS; (2016) [cited 2016 October 6].
2. Afkhami B, Akbarian B, Beheshti A N, Kakaee AH, Shabani B. Energy consumption assessment in a cement production plant. Sustainable Energy Technol and Assess.**10**:84-9 (2015).
3. Thakur RN, Wu Z. Development of High-Performance Blended Cements. USA: The University of Wisconsin; (2000).
4. Davidovits J. High-Alkali Cements for 21st Century Concretes. Special Publication.**144** (1994).
5. Benhelal E, Zahedi G, Shamsaei E, Bahadori A. Global strategies and potentials to curb CO2 emissions in cement industry. Journal of Cleaner Production.**51**:142-61 (2013).
6. Herzog HJ, Eliasson B, Kaarstad O. Capturing greenhouse gases. Scientific American.**282**:72-9 (2000).
7. Earth System Research Laboratory. Trends in Atmospheric Carbon Dioxide. (2013).
8. Davidovits J, editor Chemistry of Geopolymeric Systems, Terminology. 2nd International Conference on Geopolymer; Saint Qunentin; (1999).
9. Yang K-H, Song J-K, Lee J-S. Properties of alkali-activated mortar and concrete using lightweight aggregates. Materials and Structures.**43**(3):403-16 (2010).
10. Li Z, Ding Z, Zhang Y, editors. Development of Sustainable Cementitious Materials. International Workshop on Sustainable Development and Concrete Technology; Beijing, China; (2004).
11. Wallah SE, Rangan BV. Low-Calcium Fly Ash Based Geopolymer Concrete: Long-Term Properties. Perth, Australia: Curtin University of Technology, (2006).
12. Gourley J, Johnson G, editors. Developments in Geopolymer Precast Concrete. Fourth World Congress Geopolymer; Saint-Quentin, France; (2005).
13. Abdullah MMAB, Jamaludin L, Hussin K, Bnhussain M, Ghazali CMR, Ahmad MI. Fly Ash Porous Material using Geopolymerization Process for High Temperature Exposure. International Journal of Molecular Sciences.**13**(4):4388 (2012).

14. Abdullah MMAB, Hussin K, Bnhussain M, Ismail KN, Yahya Z, Abdul Razak R. Fly Ash-based Geopolymer Lightweight Concrete Using Foaming Agent. International Journal of Molecular Sciences.**13**(6):7186 (2012).
15. Nath P, Sarker PK. Effect of GGBFS on setting, workability and early strength properties of fly ash geopolymer concrete cured in ambient condition. Constr Build Mater.**66**:163-71 (2014).
16. Puertas F, Fernandez-Jimenez A. Mineralogical and microstructural characterisation of alkali-activated fly ash/slag pastes. Cem Concr Compos.**25**:287-92 (2003).
17. Puertas F, Martiez-Ramirez S, Alonso S, Vazquez T. Alkali-activated fly ash/slag cement strength behaviour and hydration products. Cem Concr Res.**30**:1625-32 (2000).
18. Akçaözoğlu S, Atiş CD. Effect of Granulated Blast Furnace Slag and fly ash addition on the strength properties of lightweight mortars containing waste PET aggregates. Constr Build Mater.**25**(10):4052-8 (2011).
19. Wang SD, Scrivener KL. 29Si and 27Al NMR study of alkali-activated slag. Cem Concr Res.**33**:769-74 (2003).
20. Das S, Yang P, Singh SS, Mertens JCE, Xiao X, Chawla N, et al. Effective properties of a fly ash geopolymer: Synergistic application of X-ray synchrotron tomography, nanoindentation, and homogenization models. Cem Concr Res.**78**, Part B:252-62 (2015).
21. Abdulkareem OA, Albakri MM, Hussin K, Ismail KN, Binhussain M. Mechanical and Microstructural Evaluations of Lightweight Aggregate Geopolymer Concrete before and after Exposed to Elevated Temperatures. Materials.**6**:4450-61 (2013).
22. Lee NK, Lee HK. Setting and mechanical properties of alkali-activated fly ash/slag concrete manufactured at room temperature. Constr Build Mater.**47**:1201-9 (2013).
23. Izquierdo M, Querol X, Davidovits J, Antenucci D, Nugteren H, Fernández-Pereira C. Coal fly ash-slag-based geopolymers: Microstructure and metal leaching. Journal of Hazardous Materials.**166**(1):561-6 (2009).
24. Mollah MYA, Yu W, Schennach R, Cocke DL. A Fourier transform infrared spectroscopic investigation of the early hydration of Portland cement and the influence of sodium lignosulfonate. Cem Concr Res.**30**(2):267-73 (2000).
25. Al-Majidi MH, Lampropoulos A, Cundy A, Meikle S. Development of geopolymer mortar under ambient temperature for in situ applications. Constr Build Mater.**120**:198-211 (2016).
26. Ferone C, Colangelo F, Roviello G, Asprone D, Menna C, Balsamo A, et al. Application-Oriented Chemical Optimization of a Metakaolin Based Geopolymer. Materials.**6**(5):1920 (2013).
27. Nath SK, Kumar S. Influence of iron making slags on strength and microstructure of fly ash geopolymer. Constr Build Mater.**38**:924-30 (2013).
28. Somna K, Jaturapitakkul C, Kajitvichyanukul P, Chindaprasirt P. NaOH-activated ground fly ash geopolymer cured at ambient temperature. Fuel.**90**(6):2118-24 (2011).
29. Colella C. Use of Thermal Analysis in Zeolite Research and Application. In: Rincon JM, Romero M, editors. Characterization Techniques of Glasses and Ceramics. Berlin, Heidelberg. p. 112-37: Springer Berlin Heidelberg; (1999).
30. Perera DS, Vance ER, Finnie KS, Blackford MG, Hanna JV, Cassidy DJ. Disposition of Water in Metakaolinite Based Geopolymers. Advances in Ceramic Matrix Composites XI. p. 225-36: John Wiley & Sons, Inc.; (2006).
31. Wang K, Shah SP, Mishulovich A. Effects of curing temperature and NaOH addition on hydration and strength development of clinker-free CKD-fly ash binders. Cem Concr Res.**34**(2):299-309 (2004).

 DOI: 10.1051/matecconf/201712003005

Development of a low-cost cement free polymer concrete using industrial by-products and dune sand

Najif Ismail[1,*], *Moustafa* Mansour[1] , and *Hilal* El-Hassan[1]

[1]Department of Civil and Environmental Engineering, United Arab Emirates University, PO Box 15551, Al Ain, United Arab Emirates

Abstract. Alkali-activated polymer concrete (APC) can potentially reduce CO_2 emissions associated to concrete production by 84%. The binder in APC herein was synthesized using a combined sodium silicate-sodium hydroxide solution (i.e., alkali activator), alumino-silicate rich precursor (fly ash) and slag. Light weight expanded clay and desert dune sand were used as aggregates. An overview of an experimental program was presented, which involved evaluation of fresh and mechanical properties of the produced APC and counterpart mortar (APM). Variables investigated were the fly ash to slag ratio and curing conditions. The curing regimes adopted herein included 24 hours of curing at ambient conditions, 30oC, and 60°C. The experimental program was undertaken in two stages, of these the first stage involved physical and chemical testing of constituent materials and the second stage involved testing or produced APM/APC. Reported were the setting times, workability, compression strength, strength development, flexural strength, tensile splitting strength, and plastic shrinkage strains. Relationship between strength results were investigated and effectiveness of codified predictive equations was evaluated.

1 Introduction

Global demand for cement and concrete has risen noticeably over the past few decades, driven by increasing urbanisation that reflects continuing growth in construction activity across the globe. Incidentally, cement is produced using an energy intensive process, with each ton of cement production emitting almost equal amount of carbon dioxide (CO_2) to the atmosphere [1]. To this end, substantial efforts have been invested over the past few decades to find an alternate for cement in concrete.

Alkali-activated polymer concrete (APC) is deemed a possible strategy to reduce these CO_2 emissions and to put abundant industrial waste to good use, which would otherwise be accumulating in landfills and polluting the environment. APC can be produced using minerals rich in alumino-silicate (e.g. metakaoline, clay, etc.) or industrial by-products such as fly ash and blast furnace slag. The increasing cement usage rates in developing countries

* Corresponding author: najif@uaeu.ac.ae

DOI: 10.1051/matecconf/201712003005

and the Gulf region, supports furtherance in research associated to APC. This being the motivation to undertake this study, an experimental program was undertaken. The produced APC mixtures are appropriate for construction of façade wall panels and masonry blocks, which could reduce seismic mass of structure and offer superior thermal performance when introduced in building envelope.

APC is synthesised from a source material in the presence of a strong alkali activator solution. The key characteristic desired of a source material is the presence of a high percentage of alumina and silica. A blend of two very common and abundant industrial by-products were used as source materials in this study, which were pulverized fly ash and ground granulated blast-furnace slag (referred to slag hereafter). Past research has shown that a combination of sodium hydroxide (NaOH) and sodium silicate (Na_2SiO_3) performs the best, where the addition of Na_2SiO_3 in alkali increases the rate of hardening [2].

APC mix design requires consideration of several variables, which include alkali-to-source material ratio, the type and concentration of alkali, NaOH-to-Na_2SiO_3 ratio, silica-to-alumina ratio, alkali oxide-to-silica ratio, and water-to-alkali oxide ratio. Amongst others are the water quantity, proportion of binder and aggregates, resting period, curing temperature, and curing time [3, 4]. Figure 1 shows a synopsis of APC mixture variables reported in literature.

2 Material properties

Results of unit weight testing and XRF analysis are given in Table 1. Sieve analysis was performed on dune sand (DS), river sand, and expanded clay aggregate (ECA) in accordance with ASTM C136 [5]. The chemical composition of fly ash, slag, and dune sand was determined by performing XRF analysis. The fly ash used herein was characterised as low calcium class F as per ASTM C168 [6]. Oven-dried bulk densities/unit weights of all dry constituent materials were determined using ASTM C29 [7].

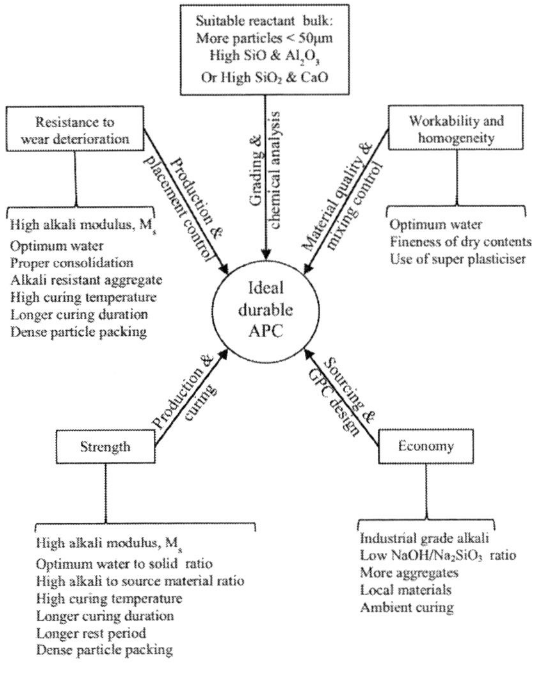

Fig. 1. Factors affecting the fresh and mechanical properties of APC.

DOI: 10.1051/matecconf/201712003005

Figure 2a and 2b show the particle size distribution and photos of as-received constituent materials. The as-received fly ash and slag powders were also subjected to X-ray diffraction (XRD) analysis with a Cu-Kα radiation at room temperature. The powder specimens were scanned at 2θ between 10° and 80°, with each step being 0.02° and 0.5 seconds. The experimentally obtained XRD patterns are presented in Figure 3. The diffraction spectrum for fly ash showed presence of amorphous phases at 18-38° (2θ), whereas Quartz, Mullite and Hematite were the main crystalline phases identified. The XRD pattern for slag was typical of a vitreous material, with a large halo located between 25° and 35° (2θ). The lower intensity diffraction lines were attributed to Gehlenite.

The alkali-activator solution used to synthesise APM/APC in the current study was prepared by mixing industrial grade Na_2SiO_3 solution and 18M concentration NaOH solution. A pre-weighed mass (40×18=576g) of 98% pure NaOH flakes was dissolved in distilled (de-ionised) water to prepare an 18M solution. The Na_2SiO_3 solution was purchased locally that had a concentration of 37.5% and a modulus of 3.2. The two solutions were combined with a fixed mass ratio of 1:1.5 ($NaOH$:Na_2SiO_3) to yield an alkali-activator solution with a solution modulus (M_s, molar SiO_2/Na_2O ratio) of 1.

Table 1. Chemical composition of constituent materials determined using XRF analysis.

Constituent	Density (kg/m³)	SiO_2 %	Al_2O_3 %	Fe_2O_3 %	CaO %	MgO %	LOI %
Fly ash	1262	48.04	23.14	12.46	3.25	1.53	1.07
Slag	1236	34.70	14.38	0.80	41.95	6.85	1.08
Dune Sand	1693	64.94	3.01	0.71	14.07	1.28	0.52

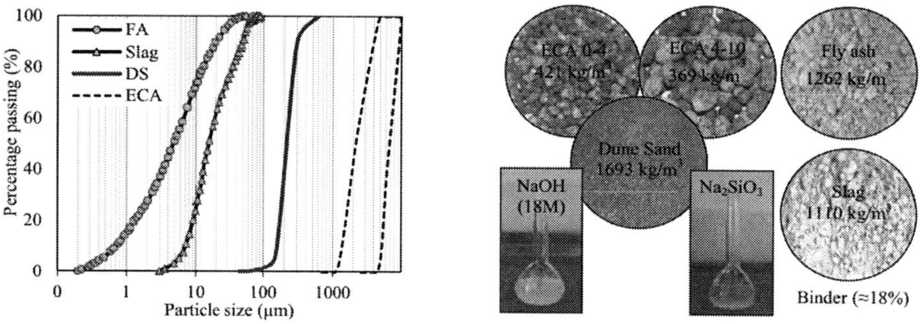

(a). Grading curves (b). Photos and densities

Fig. 2. APC constituent materials.

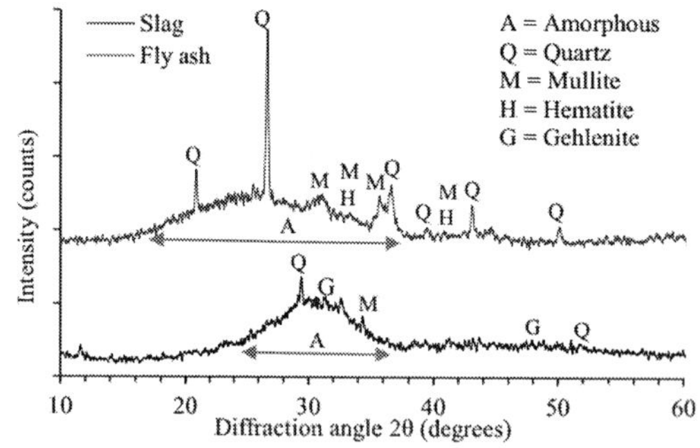

Fig. 3. XRD results for fly ash and slag.

Table 2. Alkali-activated polymer mortar (APM) and concrete (APC) mixture proportions.

Material	Density (kg/m^3)	Constituent (kg) per batch of APM			Constituent (kg) per batch of APC		
		APM1	APM2	APM3	APC1	APC2	APC3
Fly ash (1)	1262	433	325	216	302	227	151
Slag (2)	1236	-	95	190	-	74	148
Dune sand	1693	900	900	900	628	628	628
ECA*	494	-	-	-	364	364	364
NaOH	1589	70	70	70	48	48	48
Na_2SiO_3	1392	104	104	104	73	73	73
Water (3)	1000	-	-	41	-	10	20
W/SM	-	0.21	0.23	0.25	0.21	0.23	0.25

Where: ECA = expanded clay aggregate; SM = total mass of source material (1+2); and W = total water (water in alkali solution+3). *combined mass of the two grades of ECA

3 Experimental program

A testing program was undertaken to develop a suitable APC for use in building construction. The mixture proportions adopted to prepare test specimens in this study are given in Table 2. Three different mixture designs were used, with each of these mixtures varying in fly ash-to-slag (F/S) ratio with a fixed alkali-to-source material mass ratio of 0.4. The F/S ratios of 100/0, 75/25, and 50/50 were used in mixes 1 to 3. Test specimens were left for one day to rest at 22°C inside the laboratory and were then cured in ambient conditions, 30°C, and 60°C.

3.1 Testing procedures

Figure 4 shows photos of testing setups used. Initial and final setting times of each APM batch were measured using a Vicat needle apparatus at a temperature of 22°C in accordance with ASTM C191 [8], and slump was measured in accordance with ASTM C143 [9]. From each test set, 27 APM cubes (50×50×50 mm3) were tested for compression strength in accordance with ASTM C109 [10] after 3, 7 and 28 days, in sets of three.

Sets of three replicate APC cylinders (100×200 mm^2) from each test set were tested at the ages of 3, 7, and 28 days for compression strength in accordance with ASTM C39 [11] and for 28-days split tensile strength using standard test procedure detailed in ASTM C496 [12]. Three APC prisms (100×100×500 mm^3) from each test set were tested for 28-days flexural strength in accordance with ASTM C78 [13]. Shrinkage strains on ambient cured APC specimens were monitored in accordance with ASTM C175 [14] using 75×75×300 mm^3 prisms.

(a). f'_c (b). f_{ct} (c). f_r (d). ε_{sh}

Fig. 4. Photos of APC testing setups.

4 Experimental results and discussion

4.1 Slump and setting times

Slump values for mixes APC1-3 were 248, 241, and 156 mm respectively. The initial setting time ranged between 19-31 minutes and final setting time ranged between 89-117 minutes. Preliminary tests showed that slag addition increased water demand due to water consumed in the hydration of free lime and the APC mixtures became dry. Accordingly, extra water was added to mixture APC2 and APC3. Furthermore, all mixes started to set within 10-15 minutes after mixing when the APC was produced at an ambient temperature of about 40°C in summer. Generally, the addition of slag to the mixture reduced the workability and accelerated hardening.

4.2 Compression behaviour

Peak recorded compression stress during testing of APM cubes and APC cylinders after 7 and 28 days were determined and reported as compression strength (see Figure 5a and 5b). The APM compressive strength (f'_j) increased as the curing temperature was increased, with specimens cured at 60°C exhibiting the highest strength, which evidenced occurrence of polymerization with a higher degree. Of all the APM mixtures, APM2 (F/S = 75/25) was the strongest, with specimens cured at 60°C exhibiting f'_j of 78.2 MPa after 7 days and 111.6 MPa after 28 days. The strength development over time was significant for APM2, the increase in strength from 7 to 28 days was noted to be 39%, 51.4% and 42.7% for ambient, 30°C and 60°C curing temperatures respectively. Test results showed that 25% fly ash replacement with slag is the optimum percentage of the percentages examined in this study. APM3 samples (F/S = 50/50) were not significantly affected by curing temperature, with similar measured f'_j for all three curing conditions observed. This similarity in measured f'_j is attributed to the presence of more slag content because calcium oxide present in slag hydrated faster and promoted early-age strength development even prior to heat curing application.

The compressive strength of APC cylinders (f'_c) is much lower than that of APM cubes (see Figs 5c and 5d), which ranged between 6.3 MPa and 7.8 MPa after 7 days and between 6.5 MPa and 8.5 MPa after 28 days. This is due to the presence of weak ECA aggregates, occupying 55% volume in the mixture. Nevertheless, APC specimens results followed a trend similar to that observed in APM cube testing results in terms of the relation to F/S proportion and curing temperature.

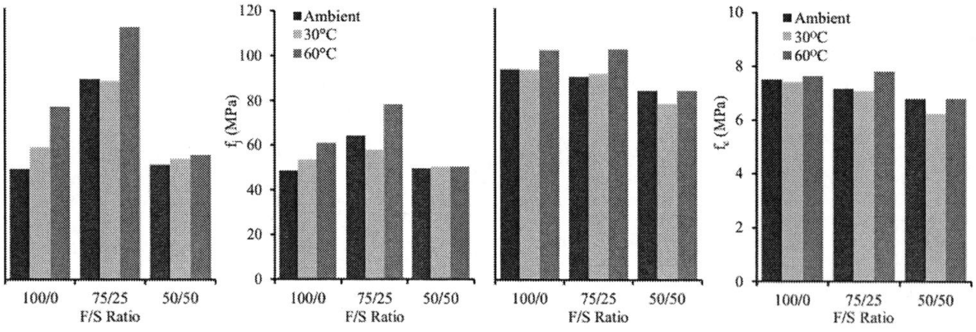

(a). APM-7 days (b). APM- 28 days (c). APC- 7 days (d). APC- 28 days

Fig. 5. Compression strength test results.

The compressive strength of ambient cured specimens of APC1 and APC2 had higher strength than their counterpart APC cylinders cured at 30°C. This is because the ambient temperature reached up to 45°C during the testing duration. Similar strength was recorded for ambient and 60°C cured cylinders of APC3, whilst the strength dropped by approximately 8% for those cured at 30°C.

4.2 Tensile and flexural behaviour

Experimental flexural strength (f_r) and the split tensile strength (f_{ct}) were calculated from recorded maximum applied loads and measured test specimen dimensions using Equations 1 and 2 respectively, where F_r is maximum force from bending test, F_{ct} is maximum applied splitting force, b is test prism width, h is test prism height, d is test cylinder diameter, and l is test specimen length. ACI 318 [15] propose Equation 3 and AS 3600 [16] propose Equation 4 to predict the APC tensile strength (f_{ct}) from compression strength (f'_c).

$$f_r = F_r\, l / b\, h^2 \quad (1)$$

$$f_r = 2\, F_{ct}\, l / \pi\, d\, l \quad (2)$$

$$f_{ct} = 0.36\, (f'_c)^{1/2} \quad (3)$$

$$f_{ct} = 0.56\, (f'_c)^{1/2} \quad (4)$$

It can be seen in Figures 6a and 6b that the f_r slightly decreased with increasing slag content. Likewise, the splitting tensile strength reduced when 50% of fly ash was replaced but increased by 11.8% to 6.3% when 25% fly ash replaced by slag when compared to their counterpart fly ash only APC specimens. It was noted that the strength of some specimens cured at 30°C resulted in a higher strength than that of ambient cured specimens, which was possibly due to variation in material properties and/or production quality. Curing temperature of 60°C yielded the highest strength in all cases.

In Figure 6c, a correlation between the compressive strength and the flexural/ tensile strength was noted. The flexural/tensile strength was found to be proportional to the measured compressive strength, with measured tensile strength lying in the middle zone between the values predicted using the two equations, but closer to the prediction made using the equation recommended by Australian concrete standards. This also suggested that the APC behaved differently than cement concrete and strength models developed for the latter may not be applied to APC.

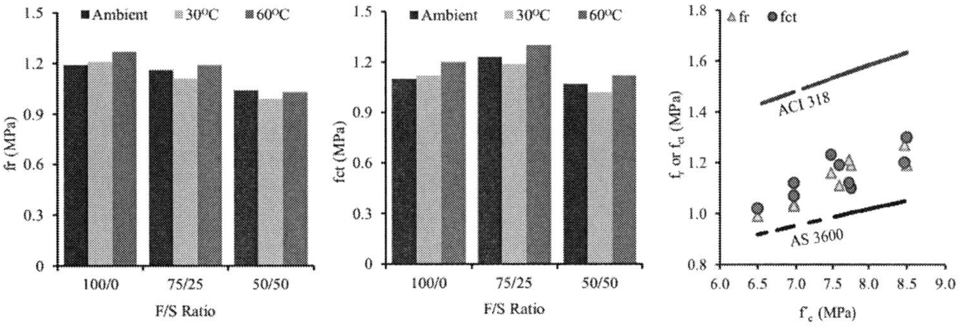

(a). Flexural strength (b). Split tensile strength (c). Correlation with f'_c

Fig. 6. Flexural and tensile strength test results.

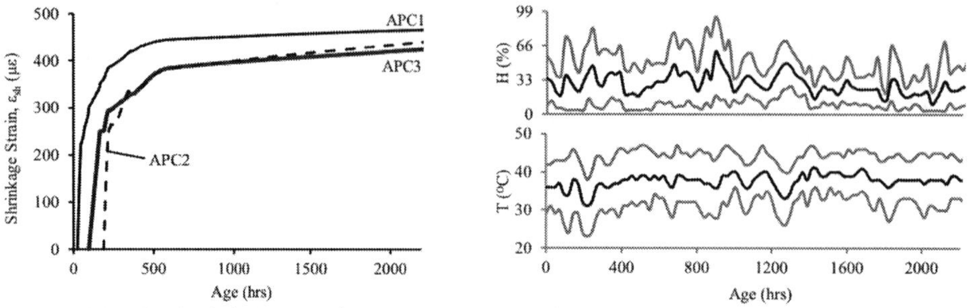

(a). Plastic shrinkage strains (b). Humidity (H) and temperature (T)

Fig. 7. Shrinkage testing results.

4.3 Plastic shrinkage strains

The shrinkage strains, ambient temperature, and ambient relative humidity were monitored over a period of 2207 hours from mixing time. Figure 7a shows the shrinkage values versus time elapsed since mixing and Figure 7b shows the recorded ambient temperature and relative humidity values. It was noted that the volume changed mainly in the first four days due to moisture movement from APC to the surrounding environment. This followed occurrence of shrinkage strains at a gradually reducing rate between 4 and 25 days, with these strain rates approaching zero after 25 days. Observed shrinkage trends of all mixtures were similar, however higher shrinkage strains were recorded in prisms produced using APC1 mixture. This indicated that plastic shrinkage strains reduced when fly ash was replaced with slag. This observation was similar to those reported in a previous study [17]. After 2207 hours since mixing, the shrinkage strain values reached 466, 424 and 440 micro-strain ($\mu\varepsilon$) for APC1, APC2 and APC3 respectively.

5 Conclusions and recommendations

The effects of fly ash replacement by slag and curing conditions on properties of fresh and hardened APM and APC were studied experimentally. The addition of slag reduced workability and setting time due to water consumed for hydration of free-lime present in slag. The compression strength of APM2/APC2 mixture was the highest, and high curing temperature promoted polymerization and, consequently, resulted in high strength. That makes 25% fly ash replacement by slag the optimum percentage among the examined. Also, slag addition to APC positively influenced plastic shrinkage by reducing the shrinkage strains. The measured APC strength was noted to directly relate with curing temperature, with 60°C curing temperature to result in the best mechanical properties. Ambient curing in UAE summer conditions had better or comparable results to those observed for specimens cured at 30°C. The concrete strength was highly associated to quality of ECA, whilst used ECA with low tensile strength produced APC with low mechanical strength. Based on experimental results, further studies are recommended to be undertaken to optimise the APC mixture and/or to determine the effects of other variables, which can be used to develop realistic strength prediction/ behavioural models.

Financial support for this study was provided by the United Arab Emirates University under the grant 31N249. Assistance of Robin Debeer, Mohammed Al-Mawri, Muath Bassam, Abdalla El-Hashmi, Faisal Ali, and Abdelrahman Sallamin are gratefully acknowledged. Ashtech International supplied fly ash. Emirates Cement Factory provided slag and assisted with x-ray fluorescence testing.

References

1. Olivier JGJ, Maenhout GJ, Muntean M, and Peters J. Trends in global CO2 emissions: 2015 Report. Ispra, Italy: European Commission Joint Research Centre Institute for Environment and Sustainability; 77 (2015)
2. de-Vargas AS, Dal Molin DCC, Vilela ACF, da-Silva FJ, Pavao B, and Veit H. The effects of Na2O/SiO2 molar ratio, curing temperature and age on compressive strength, morphology and microstructure of alkali-activated fly ash-based geopolymers. Cement and Concrete Composites. **33**(6), p. 653-60 (2011)
3. Hardjito D, Wallah SE, Sumajouw DMJ, and Rangan BV. On the development of fly ash-based geopolymer concrete. ACI Materials Journal. **101**(6), 467-72 (2004)
4. Junaid MT, Kayali O, Khennane A, and Black J. A mix design procedure for low calcium alkali activated fly ash-based concretes. Construction and Building Materials **79**, 301-310 (2015)
5. ASTM C136-14: Standard test method for sieve analysis of fine and coarse aggregates. West Conshohocken, PA: American Society for Testing and Materials International (2014)
6. ASTM C168-15: Standard specification for coal fly ash and raw or calcined natural pozzolan for use in concrete. West Conshohocken, PA: American Society for Testing and Materials International (2015)
7. ASTM C29-16: Standard test method for bulk density ("unit weight") and voids in aggregate. West Conshohocken, PA: American Society for Testing and Materials International (2016)
8. ASTM C191-13: Standard test methods for time of setting of hydraulic cement by Vicat needle. West Conshohocken, PA: American Society for Testing and Materials International (2013)
9. ASTM C143-15: Standard test method for slump of hydraulic-cement concrete. West Conshohocken, PA: American Society for Testing and Materials International (2015)
10. ASTM C109-16: Standard test method for compressive strength of hydraulic cement mortars (using 50 mm cube specimens). West Conshohocken, PA: American Society for Testing and Materials International (2016)
11. ASTM C39-15: Standard test method for compressive strength of cylindrical concrete specimens. West Conshohocken, PA: American Society for Testing and Materials International (2015)
12. ASTM C496-11: Standard test method for splitting tensile strength of cylindrical concrete specimens. West Conshohocken, PA: American Society for Testing and Materials International (2011)
13. ASTM C78-16: Standard test method for flexural strength of concrete (using simple beam with third-point loading). West Conshohocken, PA: American Society for Testing and Materials International (2016)
14. ASTM C490-11: Standard practice for use of apparatus for the determination of length change of hardened cement paste, mortar, and concrete. West Conshohocken, PA: American Society for Testing and Materials International (2011)
15. ACI 318-08: Building code requirements for structural concrete. Farmington Hills, MI: American Concrete Institute (2008)
16. AS 3600-09: Concrete structures. Sydney, NSW, Australia: Standards Australia International (2009)
17. Deb PS, Nath P, Sarker PK, Drying shrinkage of slag blended fly ash geopolymer concrete cured at room temperature, Procedia Engineering, **125**, 594-600 (2015)

Development of building blocks using vegetable oil and recycled aggregate

Mohamed I. Attia[1], *Salah* Zoorob[2], *Khaled* Hassan[1*], *Hozaifa* El-husseini[1], *James* M. Reid[2] and *Mohamed* S. Al kuwari[3]

[1]TRL-QSTP-B, Qatar Science and Technology Park, Qatar.
[2]TRL, Crowthorne House, Nine Mile Ride, Crowthorne, Berkshire, UK.
[3]Ministry of Municipality and Environment, Qatar.

Abstract. The primary objective of this research was to contribute towards greater sustainability of the construction industry in the Qatar by proposing methods to reduce its dependency on primary imported materials. In this investigation, recycled and secondary aggregates (RSA) were combined with non-traditional binders to develop a unique method of manufacturing construction and building blocks. Following an extensive phase of laboratory trials and experimentation, it was realised that many types of graded mineral aggregates, when mixed with vegetable oils (virgin or waste) at optimal proportions, then compacted and thermally cured at elevated temperatures can readily generate hardened composites that have the mechanical characteristics of conventional building blocks. The resultant blocks have been named "Vegeblocks" and are viewed as viable alternatives to conventional concrete blocks. Furthermore, the research has demonstrated the feasibility of producing Vegeblocks composed of 100% recycled aggregate and discarded waste cooking oil. Based on physical and mineralogical properties, each type of aggregate has an optimum oil content for maximum compressive strength, beyond which, any additional oil will result in reduction in mechanical properties. Acceptable compressive strength values were achieved by thermally curing Vegeblocks at of 170 °C for 24 hours.

1 Introduction

Qatar has limited supply of high quality aggregate and the construction industry depends heavily on imported gabbro [1]. Ongoing research in Qatar focused on evaluating the suitability of exploiting recycled and secondary aggregate (RSA) in various construction applications including concrete, concrete blocks, and granular subbase for roads [1, 2].

The concept of utilizing alternative binders to Portland cement for construction blocks has the advantages of reducing costs with more impact on the environment. Lutfi presented a patent for using vegetable oil as a binding agent for construction material and for soil stabilization [3]. In 2006 a team of researchers from Leeds and Nottingham Universities

* Corresponding author: khassan@trl.co.uk

continued the research on using vegetable oils as binders in the development of building blocks, and named this class of product Vegeblocks [4, 5]. It was reported that vegetable oil could be mixed with aggregates at ambient temperatures to produce a workable, compactable product, which when thermally cured produced an oxy-polymerized composite of high strength named Vegeblock. This work was later presented in a patent [6].

Forth and Zoorob's patent [6] focused on curing the mixed aggregate and oil in an oven, resulting in oxidation of the vegetable oil, which in turn caused a significant increase in block strength. One of the main differences between the two patents is that Forth and Zoorob [6] focused on curing the Vegeblock sample after extraction from the compaction mould, while Lutfi [3] proposed curing the sample whilst still in the mould. Further details on possible oxy-polymerization processes that take effect in vegetable oil and cause it to harden and act as a rigid binding agent was presented in [5, 7].

2 Objectives

The aims of the research were to improve the sustainability of the construction industry in Qatar by reducing CO_2 emissions from manufacture of cement and import of high quality primary aggregates. The specific objectives were:

- Use non-traditional and environmental friendly binders for developing construction blocks.
- Using recycled and secondary aggregate (RSA) for developing construction blocks.
- Develop an understanding of the factors affecting the strength of the developed blocks.
- Optimize the development of novel building blocks that utilize vegetable oil as the binding agent instead of cement. Furthermore, investigate the potential for using 100% recycled and secondary aggregate (RSA) in the development of Vegeblocks.
- Evaluate the suitability of producing mid-size blocks based on the optimized formulations as a step towards full scale mechanical production of Vegeblocks.
- Initial evaluation of the effect of direct flame on Vegeblocks.

3 Experimental program for optimization of vegeblocks production

3.1 Materials

3.1.1 Aggregate

- Limestone from excavation waste (EW);
- A blend of EW and incinerator bottom ash (IBA) (50/50 by weight);
- A blend of EW and unwashed sand (50/50 by weight), simply referred to as sand in this paper.

The research focused on using locally available materials in Qatar. Qatar is underlain by geologically young limestone that is highly variable in strength and hence is not used as aggregate for concrete or concrete blocks [1]. There has been a major boom in construction in Qatar over the last decade, with many major infrastructure projects and new buildings ahead of the 2022 World Cup. This has resulted in large quantities of limestone being generated in Qatar during excavation for building and infrastructure construction. Most of this material is dumped in landfills in the desert, though some is now being recycled as aggregate for use in construction [8].

IBA is generated from an Energy-from-Waste plant that opened in 2012 to handle municipal waste from Doha, the capital of Qatar. The IBA is stockpiled in a landfill next to the Energy-from-Waste plant. The suitability of this IBA for use in construction was evaluated earlier [1].

In the south west of Qatar the limestone is overlain by occasional deposits of alluvial gravelly sand. This material has been widely exploited for use as fine aggregate in concrete. However, the sand normally has high sulphate and clay contents, so it has to be washed to become acceptable for use in concrete. However, much of the sand has sulphate content that is too high to be treated by current technology. This unwashed sand was used in the trials reported here.

All the aggregates listed above were sieved through a 5 mm sieve. The gradation of all three aggregate types is presented in Fig 1. Material was classified in accordance with the unified soil classification system (USCS) [9]. EW+sand was found to be poorly graded sand (SP), EW is well graded sand with clay (SW-SC), the EW+IBA was classified as well graded sand, but the liquid limit and plastic limit were not well defined, thus it was not possible to accurately define the type of fines (silty or clay), nonetheless the material lies between (SW-SM and SW-SC).

3.1.2 Vegetable oil

Two vegetable oil types were trialled in this investigation, namely virgin palm oil (VPO) and post-consumer or used palm oil (UPO). Palm oil is classified as non-drying oil, with iodine value ranging from 43 to 58. The iodine value is a measure of the average degree of unsaturation of a lipid; the higher the iodine value, the greater the number of C=C double bonds and the higher the potential for polymerisation [10]. Non-drying oil does not harden when it is exposed to air at ambient temperatures, though it will oxidize by heating at higher temperatures, and it is readily available in Qatar.

3.2 Sample preparation

800 grams of aggregate was mixed manually with a predetermined amount of oil, which was enough to produce three compacted cubes 50×50×50 mm of each mixture type, as presented in Figure 2. The oil content was defined as a percentage of aggregate weight.

The samples were compacted and tested following ASTM C 109 [11] for 50mm mortar cube specimens. Compaction was conducted manually using a tamping rod, 32 blows each layer for 2 layers. The compacted samples were carefully extruded whilst being kept supported on a flat aluminium base plate and gently transferred into a pre-heated oven whilst resting on the base plates. Curing of compacted samples was carried out in a ventilated oven pre-set to a temperature in the range of 170 to 190 °C.

Fig 2 shows some of the samples prepared during this phase. It shows an example of the change in colour caused by varying the oil contents; as oil content increases, the samples become darker. The experimental programme was conducted in three stages.

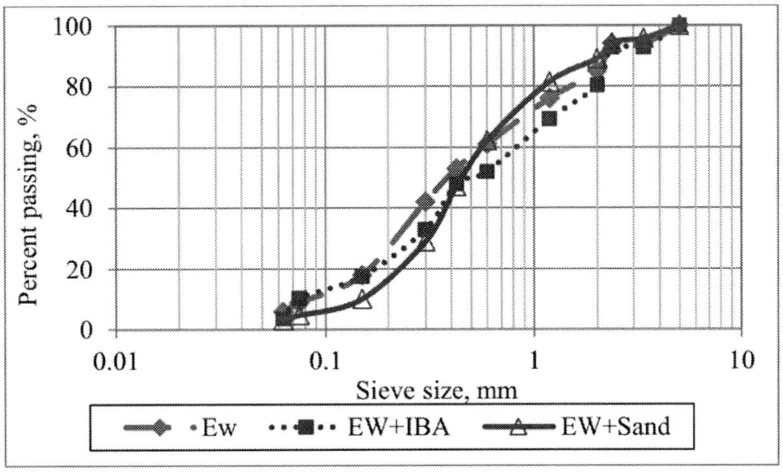

Fig. 1. Gradation of the aggregate materials.

3.2.1 Stage 1

Determine the optimum oil content (OOC) for each aggregate and oil type. The remaining variables were held constant at:

- Curing duration: 24 hours
- Curing temperature: 170 °C

This combination of curing duration and temperature had been proven in a preliminary stage of the research to result in acceptable strength for Vegeblocks composed of sand and palm oil. This combination was also reported to be acceptable production parameters for samples composed of limestone and rapeseed oil [5].

Preliminary testing had shown that curing temperatures lower than 160 °C resulted in samples that were not adequately cured within 24 hours, while temperatures higher than 200 °C resulted in a reduction in the compressive strength of the cured sample.

3.2.2 Stage 2

To understand the effect of curing duration on samples made of recycled materials. For this stage all samples were prepared at the OOC values as defined in stage one. The samples were all cured at 170 °C for durations of 12, 24 and 48 hours.

a-Sample preparation (EW to the left, sand to the right and Mixed aggregate with the oil in the middle)

b- Cured samples

Fig. 2. Samples of Vegeblocks during preparation and after curing.

3.2.3 Stage 3

Define an optimum curing temperature. In this stage all samples were prepared at the optimum oil content, and cured for 24 hours. The duration of 24 hours was selected because it was the shortest period where samples started to show a good level of curing and developed the target strength, as will be presented in the results section. The curing temperatures were 170, 180, and 190 °C. Figure 3 presents a summary of the experimental programme.

3.3 Testing

There are two main requirements for non-load bearing blocks in Qatar according to the Qatar construction specification (QCS 2014), Section 13, Part 4[12]; mainly compressive strength and water absorption.

All cubes samples were tested (after curing in the oven) for compressive strength following ASTM C 109 [11], and similarly, density was determined for all samples. The target was to produce non-load bearing blocks that fulfil the QCS 2014 [12] minimum strength requirements of 4.0 MPa.

The optimum conditions were defined as the lowest curing temperature and shortest curing duration that achieved the strength requirements at the optimum oil content.

Water absorption was evaluated for samples prepared at optimum conditions. In this test procedure, the dry weight of each sample was first measured, then each test specimen was immersed in water for 30 minutes, and the saturated weight was re-measured in air. This test procedure was according to QCS 2014 [12], test method outlined in [13]. Water absorption is defined in Equation 1:

$$\%\ Water\ Absorption = [(W_s - W_d) / W_d] \times 100 \tag{1}$$

where:
W_s = Weight of specimen after immersion in grams (Saturated surface dry);
W_d = Dry weight of specimen in grams.

For each material composition and curing condition, at least 3 samples were prepared and tested. All the results presented in this paper are for the average of three samples at each mixing and curing condition. A flowchart of the experimental programme is shown in Fig 3.

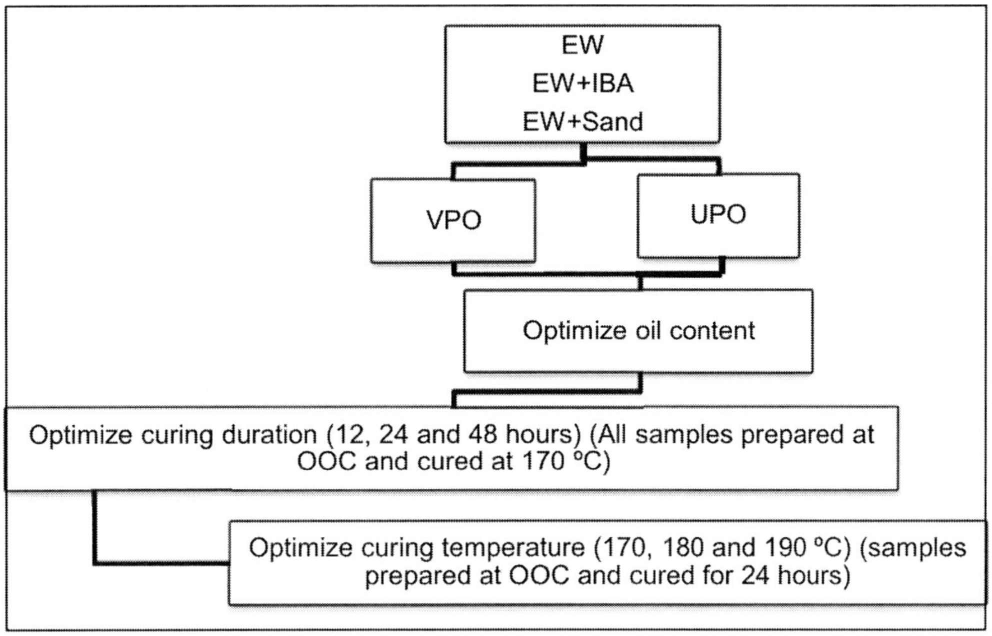

Fig. 3. Overview of optimization experimental programme.

4 Optimization results and analysis

4.1 Effect of oil content and type on compressive strength

Figure 4 (a and b) shows that increasing the oil content improves the strength of the Vegeblock up to a certain limit (optimum), after which the strength begins to decline. The aggregate type has a clear effect on the strength of the developed strength. The (EW + IBA) mixture blended with VPO and cured for 24 hours had the lowest compressive strength results, barely achieving the 4.0 MPa minimum strength requirements for non-load bearing blocks. The EW oil blends had acceptable compressive strength results, achieving 5.0 MPa using VPO and 8.0 MPa using UPO. As can be observed from Fig. 4, the highest compressive strength was achieved with the (EW + sand) mixture.

The results also show that used palm oil gave slightly higher strength compared to the virgin palm oil. This is clearer with the EW mixture exhibiting approximately 35% higher strength for the used oil neat optimum content. At oil content higher than the optimum both virgin and used oil gave similar results and trends for the different aggregate types.

The optimum oil content values for the evaluated materials are presented in Table 1. EW + IBA had the highest optimum oil content of 16 %, this was followed by 12 % for EW, then EW+ sand had OOC of 8 to 10% depending on oil type. These oil contents are considered high compared to the 5% oil content used in the earlier research by [5] and the 4% to 4.5% OOC obtained in preliminary trials using sand and vegetable oil. This can be attributed to the finer gradation and higher absorption value of RSA compared to the sand used in the earlier trials. EW and IBA also had relatively high fine contents compared to the sand, which would require more oil to coat all the particles with an adequately thick layer of oil that could bind the particles together.

The water absorption values of the raw material are also presented in Table 1. EW+IBA had high water absorption compared to other materials. Higher water absorption of the aggregate may have resulted in higher optimum oil content. This high water absorption would also affect the available oil (effective oil) that can actually bind the particles. The strength gain is dependent on the oil content. Samples with low oil content had low compressive strength. It was possible to break some samples with bare hands after curing at low oil contents (such as 4 and 6 % for EW). These samples were also difficult to handle before heat curing (fragile samples).

It was not possible to test some samples that had high oil content (for example, 16% oil content for a mixture of sand + EW). In these cases, the samples became stuck to the base plates following curing making it impossible to separate the samples from their supports without damaging the sample. It is likely that oil content was too high and resulted in binder drainage from the sample during curing.

Table 1. Optimum oil content for different aggregate types.

Aggregate type	Oil type	Optimum oil content	Water absorption of aggregate (ASTM C128), %
EW	VPO	12	4.70
EW	UPO	12	4.70
EW+IBA	VPO	16	6.50
EW+IBA	UPO	16	6.50
EW+Sand	VPO	10	2.90
EW+Sand	UPO	8	2.90

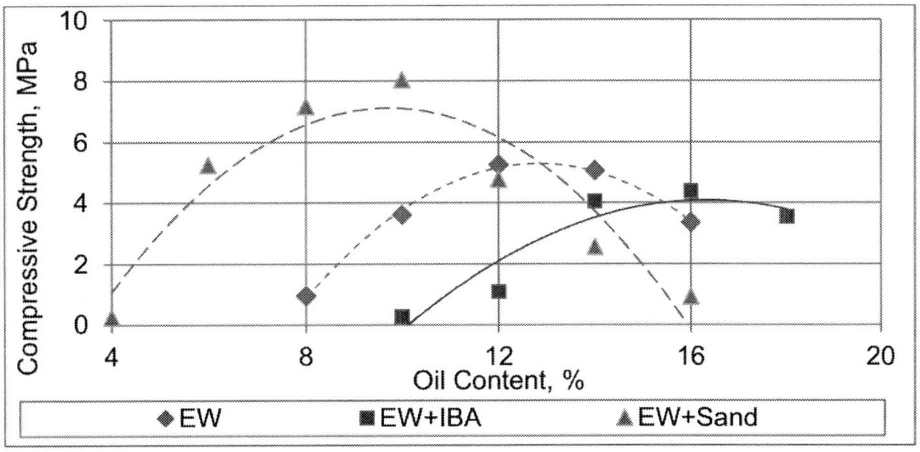

a - RSA aggregate + VPO

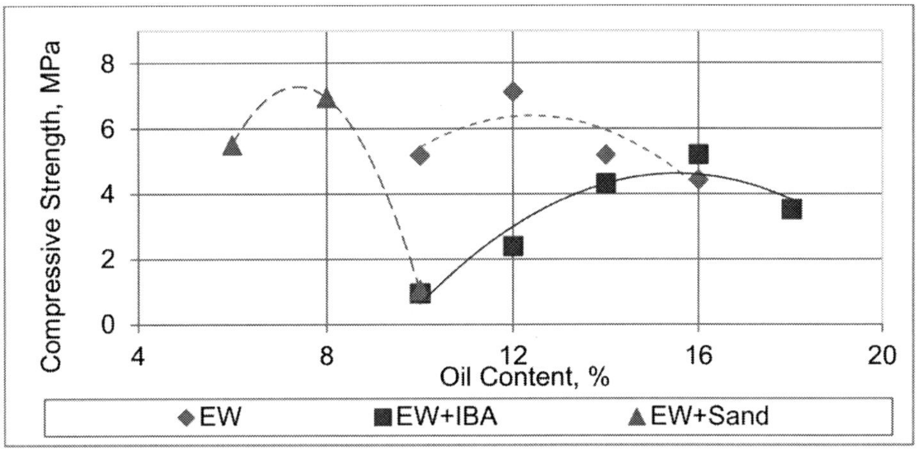

b- RSA aggregate + UPO

Fig. 4. Effect of oil content on compressive strength of RSA aggregate blends.

4.2 Effect of curing duration on compressive strength

It is expected that the oxy-polymerization reactions that are dominant during thermal curing of vegetable oils will be time dependent. It is important from an environmental and economic point of view to determine the minimum curing duration necessary to develop the required strength of the Vegeblocks. The effect of curing duration was evaluated at three levels (12, 24 and 48 hours). Compressive strengths for all materials investigated are presented in Fig 5. For all materials, increasing curing duration caused a clear increase in compressive strength. The 12 hour period was too short for curing the samples; as many samples showed a friable bottom at this short period; and after compressive strength testing it could clearly be seen that the core of the sample was not well cured. Comparing the results for 24 to 48 hours indicated improvement in the strength by 4 to 41% for different materials. The EW + sand with UPO had the lowest improvement in compressive strength when curing duration was extended from 24 to 48 hours, while EW + sand with VPO had the highest.

It is expected that a longer curing duration can improve the strength; however, the shorter period (24 hours), was adequate for achieving the target compressive strength of 4.0 MPa.

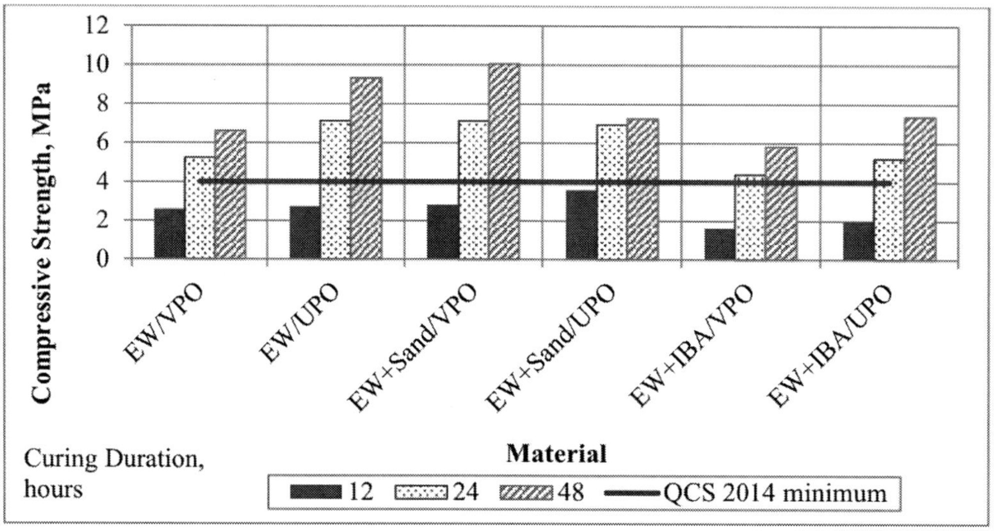

Fig. 5 Effect of curing duration on compressive strength.

4.3 Effect of curing temperature on compressive strength

The effect of curing temperature on the strength of Vegeblocks is presented in Fig 6. Increasing the curing temperature from 170 to 190 °C improved the compressive strength of the various blends by an amount ranging from 27% up to 58%. The highest strength improvement was achieved for (EW + IBA) mixed with VPO whereas (EW + sand) mixed with UPO exhibited the least improvement in strength. At short times of curing, the higher temperature can result in a faster rate of oxy-polymerization reaction, which in turn will produce a higher compressive strength. The 170 °C was enough to achieve the target strength with 24 hours of curing.

4.4 Water absorption

The water absorption was evaluated for EW and EW+ sand mixtures only. The EW absorbed 1.5% while EW + sand absorbed 1.9% of water. This is far below the QCS 2014 requirement of 7% maximum water absorption. These values looks small compared to the water absorption of 4 to 11% reported after 24 hours soaking in water [5]. It is also smaller than particles water absorption reported in Table 1, reflecting that oil had coated the aggregate particles. However, it must be noted that the low water absorption is based on soaking in water for only 30 minutes, which is the current Qatar construction specification requirement [12].

 DOI: 10.1051/matecconf/201712003006

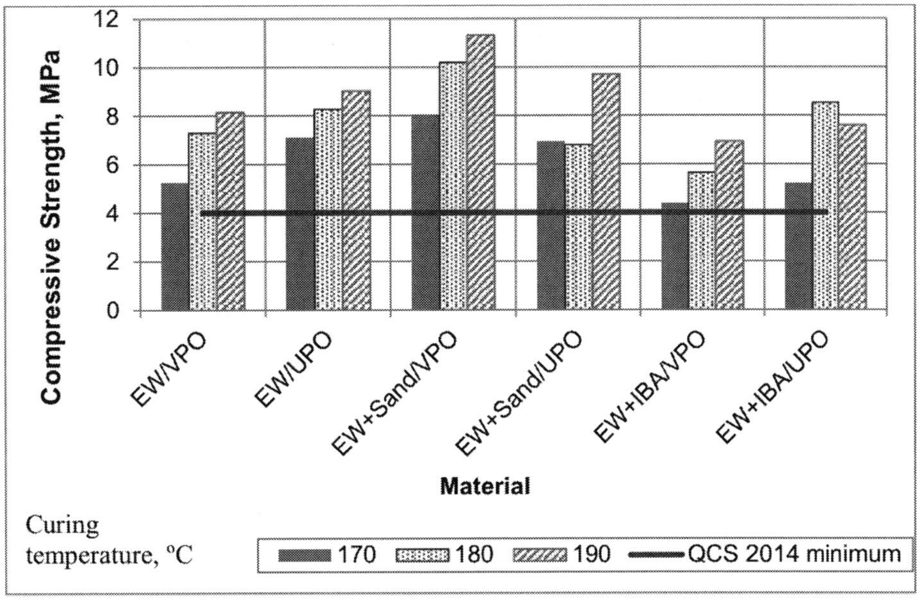

Fig. 6. Effect of curing temperature on compressive strength.

5 Manual block production (mid-size blocks)

This step aimed to give a better insight of the suitability of producing real size construction blocks and prove that the Vegeblock concept can be taken further for full scale production. The concern was that the Vegeblock would be too weak while in its green (uncured) condition, to the level that it would not be possible to produce real size construction blocks.

A mould was designed to produce significantly larger sized VegeBlocks than the 50 mm cubes. The block was designed to be a 200*100*50 mm hollow block. The mould and produced block are presented in Fig 7. The height of the block was chosen to be 50 mm to avoid long walls and to simulate the initial cubes that were manufactured in the previous stage of the research. The thickness of the web was similar to that of regular produced concrete blocks (40 mm). This thickness was also considered acceptable as it would facilitate the curing process.

The material was poured inside the mould and compacted over two layers. Each layer was subject to 25 hits by the manual compaction tool shown in Fig 7. The compaction process was conducted while the sample was supported by a steel plate underneath it in order to be carried to the oven. After de-moulding the sample, the Vegeblocks were transferred carefully to the oven and cured for 24 hours at 170 °C (optimum conditions as defined for small cubes).

At this stage samples were prepared at the optimum oil content (OOC) only. Used palm oil (UPO) was used to produce the Vegeblocks. The produced Vegeblocks were made of the following materials:

- Excavation Waste (EW) with OOC of 12%;
- EW+ unwashed sand with OOC of 8%;
- EW+ incinerator bottom ash (IBA), with OOC of 16%.

This experiment proved to be successful, and the Vegeblock were compacted and cured. This presented the first step towards full scale/mechanical production of the Vegeblock.

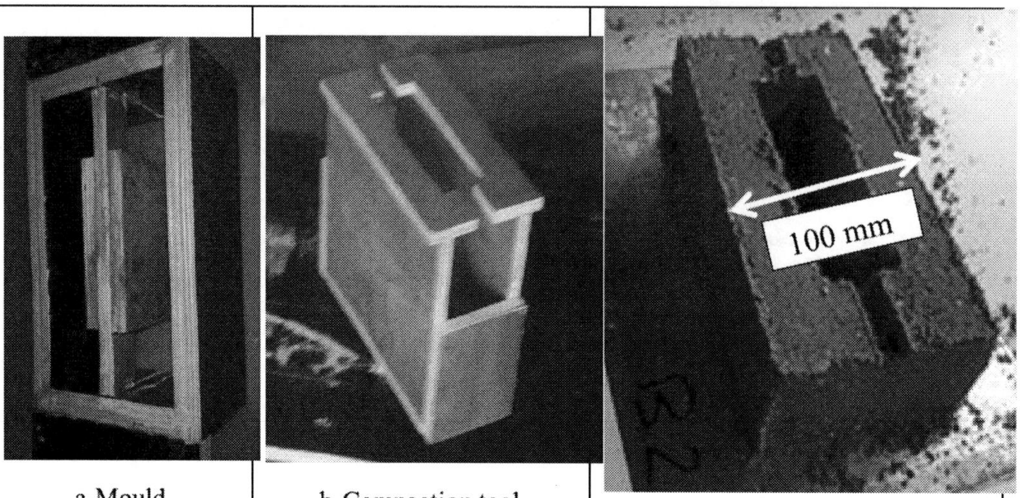

a-Mould | b-Compaction tool | C- Cured blcok

Fig. 7. Mould for Manual block production, and the produced Vegeblock.

6 Effect of direct flame on vegeblocks

There was a concern that Vegeblock would catch fire quickly due to the existence of the vegetable oil as a binding agent in the blocks. Initial fire resistance testing was conducted in a third party laboratory in Qatar. The aim of this test was to evaluate the fire hazardous of the Vegeblock, and whether it would catch fire once exposed to direct flame from a gas burner. The building blocks were exposed to a direct flame source two inches away from one face of the sample for two hours. The temperature was measured at the front and back sides of the block using thermocouples. The block was visually investigated before and after the test to evaluate the development of any failures. Fig 8 shows the setup of the experiment a few minutes after the Vegeblocks were exposed to the fire. It was not possible to run this test on the small 50 mm cubes. The test was conducted on mid-size block samples made of EW and EW+ sand.

The Vegeblock did not catch fire during the test. For Vegeblocks made of EW, a crack appeared after 70 minute. The temperature at the point of cracking had reached 483°C on the surface of the VegeBlock and 78.5°C on the back side. This was the only time the EW VegeBlock was damaged and thereafter until the 120 minute mark, the block had remained only with that crack. The EW+ sand specimen did not incur any physical damage. The surface colour in both materials changed very quickly.

The temperature profiles of the Vegeblock were recorded and compared to another test results for a conventional concrete hollow block, as presented in Fig 9. The comparison shows that Vegeblock specimens had an initial rapid rise of surface temperature within 30 minutes compared to the linear increase shown by the conventional concrete block, then the Vegeblock's surface temperature increases gradually until about 70 minutes. Over the last 30 minutes of the test, the surface temperature on both VegeBlock specimens increased by only 3-4°C every ten minutes. In contrast, the conventional hollow block's surface temperature continued rising in surface temperature at a fast rate throughout the entire two hours reaching 781°C whereas the VegeBlock reached 505°C in the first 60 minutes and reached a maximum of 528°C after 120 minutes. The temperature of the back side for

Vegeblocks was similar to conventional concrete blocks and the maximum temperature was about 100 °C at the end of the experiment.

The experiment proved that the oil in the Vegeblock had polymerized to an extent that it is not a fire hazard, and would not catch fire if it is exposed to flame for two hours and for temperature up to 500 °C.

A- Fire Test setup	b- EW after 2 hours	c- EW+sand after 2 hours

Fig. 8. Fire test setup and samples after the test.

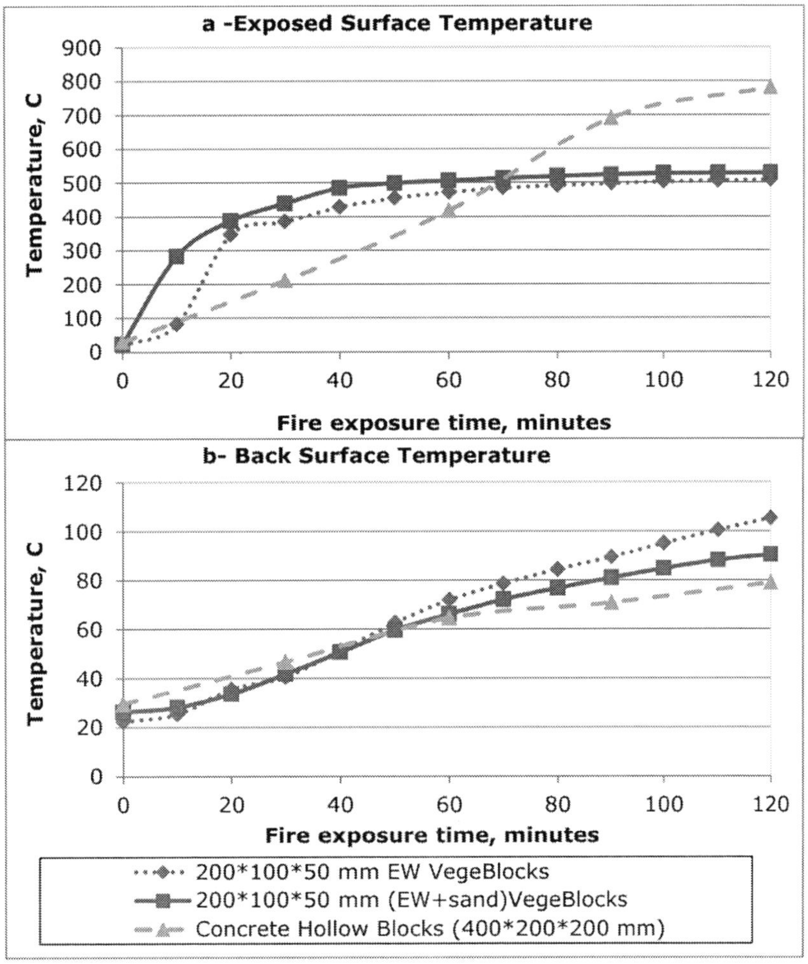

Fig. 9. Temperature profile of tested samples.

7 Summary and conclusions

This paper presented part of the research aimed at evaluating the viability of producing building blocks using recycled aggregate and vegetable oil as a binding agent. Different types of locally available recycled aggregate were evaluated including:

- Limestone excavation waste (EW)
- EW plus incineration bottom ash (IBA) (50/50 by weight)
- EW plus unwashed sand (50/50 by weight)

Two types of oils were used in the production of the Vegeblocks namely:

- Virgin palm oil (VPO)
- Used palm oil (UPO)

The research developed an understanding of the effect of oil content, curing duration and curing temperature on the compressive strength of the Vegeblocks.

Based on the work carried out in the investigation, the following conclusions were made:

- Graded mineral aggregates, when mixed with vegetable oil (virgin or waste) at optimal proportions, then compacted and cured at elevated temperatures can result in blocks (Vegeblock) that have the mechanical characteristics of a conventional building block and can be a viable alternative to concrete blocks.
- Furthermore, the research demonstrated the possibility of producing blocks out of 100% recycled material (i.e. recycled aggregate and waste oil).
- Excavation waste (EW) and IBA mixtures (which was the weakest material evaluated), achieved the minimum strength required by the QCS 2014 for non-load bearing blocks. Vegeblocks could be a sustainable application for such waste materials.
- For each type of aggregate, there is an optimum oil content (OOC) to achieve the target strength. Increasing the oil content beyond the optimum reduces the strength of the Vegeblocks. The OOC varied from 8% for (EW + sand) up to 16% for blends composed of (EW + IBA).
- A curing duration of 24 hours gave acceptable levels of strength for Vegeblocks at the optimum oil content. Increasing the curing duration beyond 24 hours improved the strength.
- A high temperature is needed for the Vegeblocks to be adequately cured and to achieve the target strengths. The recommended minimum temperature for curing was found to be 170 °C and increasing the curing temperature up to 190 °C resulted in even higher strengths.
- The very delicate and careful operation required to remove and handle the freshly compacted uncured samples is a slight drawback in this process as it may create a challenge when producing samples at full-scale production.
- Mid-size hollow blocks (200*100*50 mm) were successfully produced as a step towards full scale production of Vegeblocks.
- The experiment proved that the oil in the Vegeblock had polymerized to an extent that it is not a fire hazard, and would not catch fire if it is exposed to flame for two hours and for temperature up to 500 °C.

The work presented in this paper is part of ongoing research for evaluating and developing a deeper understanding of Vegeblock performance. The Authors are grateful to the Qatar National Research Fund, National Priority Research Program (NPRP 6–1472– 2–600) for funding the project.

References

1. K.E. Hassan, J.M. Reid and M.S. Al-Kuwari. PPR736, TRL (2015).
2. K.E. Hassan, M.S. Al-Kuwari and J. M. Reid. PEAT, **15**, 2 (2014).
3. R. Lutfi. Intern. Pub. Number WO 01/74948 A1. World International Property Organization (2001).
4. J.P. Forth and S. Zoorob. *Factor 10 Engineering for Sustainable Cities,* IABSE, UK, (2006).
5. S.E. Zoorob, J.P. Forth and H.K. Bailey. *Twenty-First Int. Conf. Solid Waste Tech. and Management*, USA, 511-520, (2006).
6. J.P. Forth and S. Zoorob. Patent No. US 8,298,330 B2, (2012).
7. T. Heaton, J. Fisher and J.P. Forth. J. of the Am. Oil Chem. Soc., **89**, 6, 1101-1111, (2012).
8. J.M. Reid, K.E. Hassan and M. S. Al-Kuwari (2016). *Proceedings of the Institution of Civil Engineers - Waste and Resource Management.* **169,** 1, 21-29, (2016).
9. ASTM D2487. Am. Soc. Testing & Mat., USA (2011).
10. C. L. Alsberg and A. E. Taylor. http://journeytoforever.org/biofuel_library/fatsoils/fatsoils1.html, accessed May (2009).
11. ASTM C109/C109M. Am. Soc. Testing & Mat., USA (2013).
12. QCS, Sec. 13, Part 4, Qatar Standards – Ministry of Environment (2014).
13. State of Qatar. Ministry of Municipal Affairs & Agric. CML Test Method 9-97 (1997).

Overview of studies on the effect of recycled aggregates sourced from tested cylinders on concrete material and structural properties

Hamad Bilal[1*] and *Dawi* Ali[2]

[1]American University of Beirut, P.O. Box: 11-0236, Riad El-Solh, 1107 2020, Beirut, Lebanon
[2]CODE Consultants & Designers sarl, Mina El Hoson, Phoenicia St., Sfeir Bldg., Beirut, Lebanon

Abstract. The paper presents an overview of a two-phase research program that was designed at the American University of Beirut (AUB) to investigate the effect of replacing different percentages of natural coarse aggregates (NCA) with recycled coarse aggregates (RCA) on the properties of the produced concrete. The source of RCA was tested cylinders in batching plants which would help recycling and reusing portion of the waste products of the concrete industry. In the first phase, the fresh and hardened mechanical properties of the produced concrete mix were investigated. The variables were the concrete strength (28 or 60 MPa) and the percentage replacement of NCA with RCA from crushed tested cylinders [0 (control), 20, 40, 60, 80, or 100%). Normal strength tested cylinders were used as source of the recycled aggregates for the normal strength concrete (NSC) mix and high strength tested cylinders were used for the high strength concrete (HSC) mix. Tests included plastic state slump and hardened state mechanical properties including cylinder compressive strength, cylinder splitting tensile strength, modulus of elasticity, and standard beams flexural strength. The results indicated no significant effect on the slump and around 10% average reduction in the hardened mechanical properties for both investigated levels of concrete compressive strength. In the second phase, the structural behavior of normal strength concrete (NSC) reinforced concrete beams prepared by replacing different percentages of NCA with RCA from tested concrete cylinders was studied. For each of three modes of failure (flexural, shear, or bond splitting), three beams with different percentages replacement [0 (control), 40, or 100 percent] were tested. One replicate was prepared for each beam to validate the test results. Results indicated no significant difference in the ultimate load reached or load-deflection behavior that could be related to the percentage replacement of NCA with RCA.

* Corresponding author: bhamad@aub.edu.lb

1 Introduction

Construction and demolition wastes (CDW) are produced worldwide due to the development of cities and/or due to emergencies such as earthquakes and wars. CDW can be defined as a mixture of surplus materials generated during new construction, renovation, and demolition of buildings, roads, bridges, and other structures. the boost in construction activities that led to the annual production of 500,000 ton of CDW in the city of Beirut by the demolition of a large number of existing old facilities either because of the limited land for new development or due to the fact the existing buildings are structurally defected. Another source of cementitious based CDW comes from quality control procedures. Common practices require the sampling and laboratory testing of standard cylinders. These tested cylinders constitute a significant portion of the dumped CDW. A typical batching plant generates between 20 and 80 tons of waste per month.

The accumulation of such CDWs has detrimental impact on the environment especially in developing countries that lack plans and methods to manage, handle, dispose, recycle, and reuse dismantled waste properly. Therefore, the incorporation of the cementitious portions of CDW as recycled concrete aggregates (RCA) in concrete mixes has become a commonly proposed solution that would help the preservation of natural resources, alleviation of construction waste disposal problem with its negative environmental impact, and production of a sustainable and green concrete material.

In 2004, the Federal Highway Association (FHWA) in USA reported that 38 States recycle concrete as an aggregate base for roadways and 11 States incorporate recycled aggregates into newly produced PCC mixes. States that use RAC have reported that concrete with RCA equally performs as the conventional natural aggregate concrete (NAC) prepared with crushed limestone aggregates.

The American Concrete Institute (ACI) Committee 555 report stated that the recycling of concrete is a relatively simple process [1]. After the removal of non-cementitious component, RCA can be produced by breaking and crushing existing concrete into a material with a specified size and quality. The quality of RCA is very dependent on the quality of the recycled material used.

2 Previous research

2.1 Research on mechanical properties of RAC

Many papers are reported in the literature on research conducted on the mechanical properties of RAC. Nagataki et al. investigated the effect of several physical properties of recycled aggregates including the structure and amount of adhered mortar on the performance of recycled aggregate concrete [2]. The results indicated that all possible variations in the recycled aggregate properties should be taken into account in designing durable recycled aggregate concretes.

In 2007, Rahal compared several mechanical properties of RAC and NAC made with the same mix proportions to achieve a nominal concrete strength between 25 and 30 MPa [3]. Rahal concluded that at 28 days, the cube and cylinder compressive strength and the indirect shear strength of RAC were on the average 10% less than those of NAC. The reduction in the modulus of elasticity was only 3%.

Etxeberria et al. investigated RAC produced using four different percentage replacements of NCA with RCA used in the wet unsaturated condition: 0, 25, 50 and 100% [4]. Results indicated lower modulus of elasticity values when recycled coarse aggregates were used.

Yang et al. studied the influence of the type and percentage replacement of NCA with RCA on concrete properties [5]. Test results showed that the fresh and hardened properties of concrete containing recycled aggregates were dependent on the relative water absorption of aggregates. In addition, the modulus of rupture and modulus of elasticity of RAC were lower than the design equations specified in ACI Building Code [6] when the relative water absorption of aggregates was above 2.5%.

Recently in 2012, Ahmed reported the effect of introducing RCA originating from CDW in West Australia on the properties of concrete [7]. The percentage replacements by weight of NCA with RCA were 25, 50, 75 and 100%. The results indicated increase in the mechanical properties of concrete made with 25% replacement recycled aggregates. RAC containing 40% replacement of cement with fly ash had greater compressive strength at old ages and lower indirect tensile and flexural strengths at all ages.

2.2 Research on structural behaviour of RAC

In 2001, Han et al. reported on tests of the shear capacity of twelve simply supported beams cast with different percentage replacements of NCA with RCA [8]. Variables in the test included the span to depth ratio (1.5, 2, 3, and 4); the type of aggregates that included natural, washed recycled and non-washed recycled aggregates; and the amount of shear reinforcement used in the specimens (0, 0.089, 0.507, and 0.823 percent). It was found that the ACI Building Code equations overestimated the shear capacity of RAC beams with span/depth ratio greater than 3.

In 2009, Gholamreza el al. published the results of flexural strength tests of RAC beams [9]. Variables included the percentage replacement of NCA with RCA, and the flexural reinforcement percentage. The results indicated comparable flexural strength when using recycled aggregates and proportioning them according to the EMV method. It was also found that ACI Building Code equations over-estimated the modulus of rupture.

Sun-Woo Kim et al. tested 144 pullout specimens to investigate the influence of RCA on the bond behavior of deformed bars in concrete [10]. The variables included: the aggregate size, percentage replacement of NCA with RCA, the reinforcing bar direction (vertical and horizontal), and the reinforcing bar location [75 and 225 mm (2.95 and 8.86 in.) from the bottom]. Results showed that for the same percentage replacement of NCA with RCA, the bond strength increased while decreasing the maximum recycled concrete aggregate size. Results also showed that the increase in the percentage replacement had a negative impact on the compressive strength. The top-bar effect was very significant with bottom cast bars performing better than top cast bars at all ages.

3 Research significance

This paper gives an overview of an extensive two-phase research program that was conducted at the American university of Beirut (AUB). The objective of the first phase was to investigate the effect of replacing different percentages NCA with RCA on the plastic and hardened mechanical properties of the produced concrete, whereas the objective of the second phase was to study the effect on the structural behavior of reinforced concrete members. The significance of the research is two-fold. First is the source of RCA which was tested cylinders in batching plants thus helping the recycling and reusing of portion of the waste products of the concrete industry and mitigating its negative environmental impact. Another reason for significance is the investigation of the difference in flexure, shear, and bond splitting structural behavior between NAC and RAC beam elements. No single previous study covered all three structural modes in one research project.

4 Experimental program and test results

4.1 Phase 1

In the first phase, six normal strength concrete (NSC) and six high strength concrete (HSC) mixes were prepared. The nominal intended concrete compressive strength of the mixes was 28 MPa for the NSC mixes and 60 MPa for the HSC mixes. Normal strength tested cylinders were used as source of the recycled aggregates for the NSC mixes and high strength tested cylinders were used for the HSC mixes. The six mixes of each strength type were identical except for the percentage replacement of NCA with RCA. Replacement values of 0, 20, 40, 60, 80, and 100 percent were tested. The mixes are referred to as R0 (Control mix), R20, R40, R60, R80, and R100, respectively.

Natural sand with particle sizes ranging from 0.075 to 1.18 mm was used in all NSC and HSC mixes. All used natural coarse aggregates (NCA) were from the same source. Also all recycled coarse aggregates were the waste product of regular tests conducted at a specific batching plant on NSC and HSC mixes prepared and delivered to the different construction sites. All coarse aggregates, natural and recycled, were washed from dust material prior to oven drying. The mix design was performed in accordance with Absolute Volume Method stipulated in the ACI Committee 211 Standard: ACI 211.1-91 [11]. The mixes were designed and batched to meet the consistency values of 100 mm for NSC and 200 mm for HSC. The water/cement ratio prior to water correction due to absorption was 0.57 for NSC and 0.33 for HSC. No admixtures were used in the NSC mixes, however a high performance concrete admixture based on Polycarboxylate Ether was used in the HSC mixes.

The actual slump test results indicated that replacement of NCA with different percentages of RCA had no negative impact on consistency of the concrete mix. This could be related to the fact that all aggregates were batched at SSD conditions regardless of their type. The slump values ranged between 90 and 160 mm for the NSC mixes and between 180 and 210 mm for the HSC mixes.

For the NSC mixes, the replacement of different percentages of NCA with RCA led to an average reduction of 9.8% in the ultimate compressive strength. The trends of compressive strength gain over age were similar for all six mixes regardless of the percentage replacement of NCA with RCA. The average reduction in the splitting tensile strength for all six mixes was 13.5% and was similar for all percentage replacements of NCA with RCA. Moreover, the average reduction in the modulus of elasticity was 11%. Whereas the reduction was only 3.52% for the 20 percent replacement mix, close to what was reported by Rahal [3], it jumped above 10% for the other larger replacement percentages mixes. The average reduction in the flexural strength for the six mixes relative to the control mix with no RCA was 9.2%, and similar to the splitting tensile strength tests the reduction in flexural strength relative to the control mix was almost consistent for all percentage replacement values.

For the six HSC mixes, the average reduction in the concrete compressive strength relative to the control mix was 11%, similar to the value for the NSC mixes which was 9.8%. Similar to the NSC mixes, the trends of compressive strength gain over age were similar for all six mixes. Whereas the 28-days strength of the control mix (58 MPa) was very close to the targeted design value of 60 MPa, the compressive strength values of the mixes with different percentage replacement of NCA with RCA ranged from 49.7 to 52.7 MPa. This proves that a compressive strength value above 50 MPa is achievable for RAC.

The average reductions in the splitting tensile strength and the flexural strength relative to the control mix were 10.3 and 10.8%. For the NSC mixes the value for the splitting tensile strength was slightly higher (13.5%) but the value for the flexural strength (9.2%)

was slightly lower. Similar to the NSC mixes, the reductions in the splitting tensile strength and the flexural strength were comparable for all percentage replacements of NCA with RCA. Finally, the average reduction in the modulus of elasticity was 11.3%. The reduction was only 5.01% for the 20 percent replacement mix, close to what was reported by Rahal for NSC mixes [3], but ranged from 9.09 to 14.66% for the other larger replacement percentages mixes.

4.2 Phase 2

Based on the results of the first phase, 18 reinforced concrete beams were tested in three series each with different percentage replacement of NCA with RCA sourced from tested and crushed concrete cylinders [0(control), 40, or 100%]. The intended concrete strength used was 28 MPa (4,000 psi). In each series, two replicate beams were tested to fail in flexure (Flexure beams), two to fail in shear (Shear beams), and two beams were tested to fail in bond splitting of the concrete cover in the splice region (Bond beams). Replicates were used to validate the test results.

Section details of the different types of beams are shown in Figures 1, 2, and 3. Plastic spacers between the steel cages and the form sides were used to guarantee side and bottom covers of 30 mm. In the shear beams, the spacing of the stirrups in the shear region was large enough to ensure shear failure. In the bond beams, the bottom reinforcing bars were spliced 305 mm at the middle of the beam. This splice length was designed to develop a steel stress less than yield to ensure bond splitting mode of failure. The beams were tested in positive bending. The loading system was designed to produce a constant moment region in the middle of the beam specimen. A view of the test setup is shown in Figure 4. The loads were applied monotonically in increments of 10 kN; i.e. the testing protocol was load control and not displacement control.

The beams are identified by a four-part notation system (refer to Table 1). The first part indicates the concrete strength of the mix (N for normal strength). The second part specifies the mode of failure (F for flexure, S for shear, and B for bond splitting). The third part designates the type of concrete mix used in casting the beam specimen (N for natural aggregates with 0 percentage replacement, 40 for 40 percent replacement of NCA with RCA, and 100 for 100 percent replacement), whereas the fourth part designates the listing number of the two replicates (1 or 2).

Test results of all 18 beams are presented in Table 1. The listed data includes the ultimate load reached and the corresponding mid-span deflection. Values of ultimate loads of replicate beams are within 10%. Also listed in Table 1 are the concrete compressive strength values at the day of testing. The latter indicate that relative to the batch with no RCA, mixes with 40% and 100% replacement percentages of NCA with RCA had reductions in the compressive strength amounting to 9 and 11.7%, respectively. The average ultimate loads of replicate beams in each series of a given mode of failure are listed in Table 2 along with ratios of ultimate loads relative to the control beam with no replacement of NCA with RCA. The average ultimate loads are also shown in Figure 5.

Although the concrete compressive strength values of the 40% and 100% replacement flexural beams were less than that of the control mix, however the maximum flexural loads of these specimens were slightly larger than that of the control beam. Also, the load-deflection histories were almost identical. Thus, the replacement of even 100% of NCA with RCA did not adversely affect the flexural strength of the beam specimen.

DOI: 10.1051/matecconf/201712003007

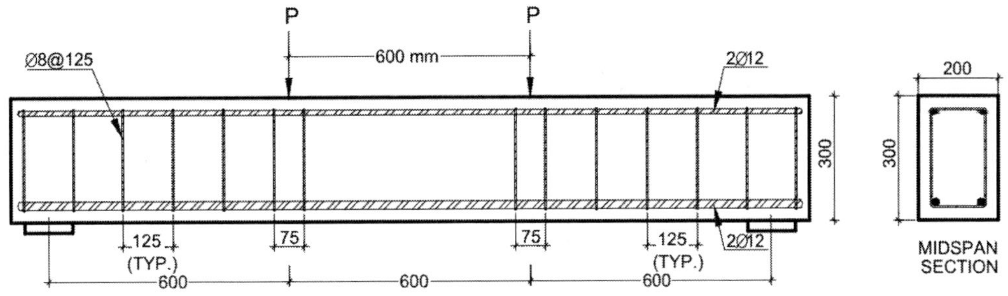

Fig. 1. Flexural beams; all dimensions are in mm.

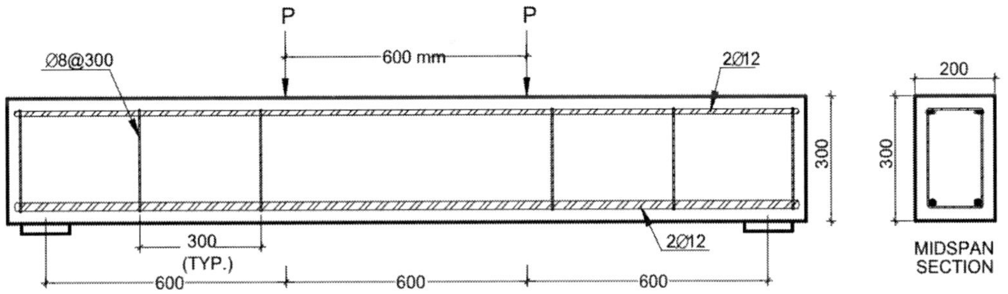

Fig. 2. Shear beams; all dimensions are in mm.

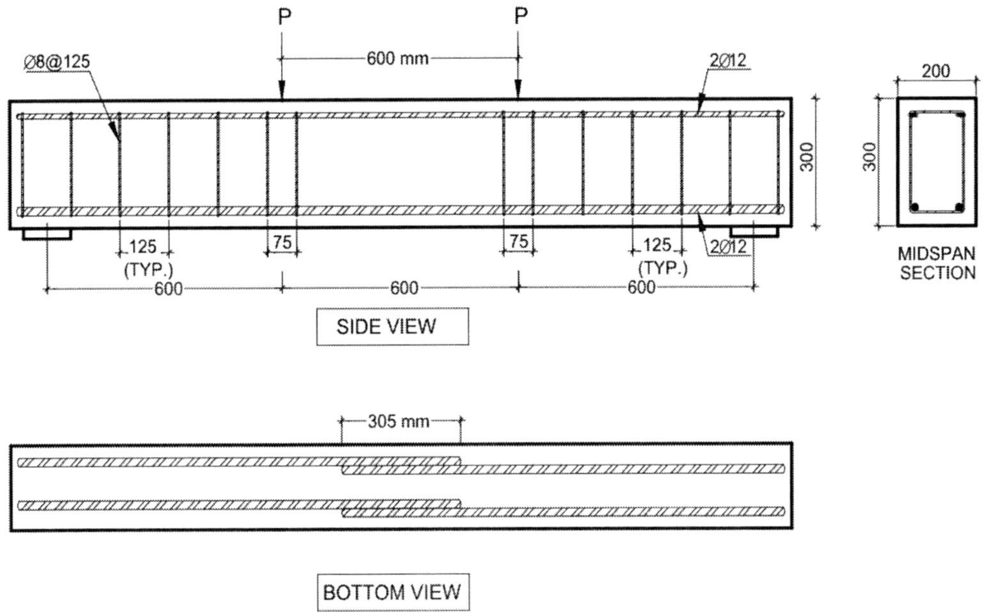

Fig. 3. Bond beams; all dimensions are in mm.

 DOI: 10.1051/matecconf/201712003007

Fig. 4. View of the test setup.

Table 1. Ultimate loads and corresponding centerline deflections.

Series	Beam Notation	f'c At day of testing (MPa)	Pmax (kN)	Mid-span deflection At Pmax (mm)
Series 1 Flexural Beams	NFN-1*	33.4	199.2	7.1
	NFN-2	33.4	194.3	8.2
	NF40-1	30.7	204.2	9.1
	NF40-2	30.7	223.1	9.4
	NF100-1	29.5	240.8	11.5
	NF100-2	29.5	213.6	12.1
Series 2 Shear Beams	NSN-1	33.4	155.2	6.1
	NSN-2	33.4	170.5	7.5
	NS40-1	30.7	159.1	7.3
	NS40-2+	30.7	166.6	6.8
	NS100-1	29.5	158.4	8.5
	NS100-2	29.5	160.2	7.8
Series 3 Bond Beams	NBN-1	33.4	118.1	4.3
	NBN-2	33.4	120.2	4.5
	NB40-1	30.7	140.6	3.8
	NB40-2	30.7	135.2	4.2
	NB100-1++	29.5	120.2	5.2
	NB100-2	29.5	122.2	5.3

* NFN-1: The first "N" stands for Normal strength concrete, "F" stands for flexural mode of failure; the second "N" stands for natural coarse aggregate or 0% replacement of NCA with RCA; and "1" stands for the first of two identical replicates.

\+ NS40-2: "N" stands for Normal strength concrete, "S" stands for shear mode of failure; "40" stands for 40% replacement of NCA with RCA; and "2" stands for the second of two identical replicates.

++ NB100-1: "N" stands for Normal strength concrete, "B" stands for bond splitting mode of failure; "100" stands for 100% replacement of NCA with RCA; and "1" stands for the first of two identical replicates.

Table 2. Average ultimate loads for the replicate beams in the three series.

Series	Beam Notation	Average ultimate load (kN)*	Ratio relative to the control beam**
Series 1 Flexural Beams	NFN	196.8	-
	NF40	213.7	1.09
	NF100	227.2	1.15
Series 2 Shear Beams	NSN	162.9	-
	NS40	162.8	0.99
	NS100	159.3	0.97
Series 3 Bond Beams	NBN	119.2	-
	NB40	137.9	1.16
	NB100	121.2	1.02

* This is the average value of two replicate beams.
** This is the ratio of the ultimate load of beams with different percentage replacements of NCA with RCA relative to the control beam with 0 replacement in each series.

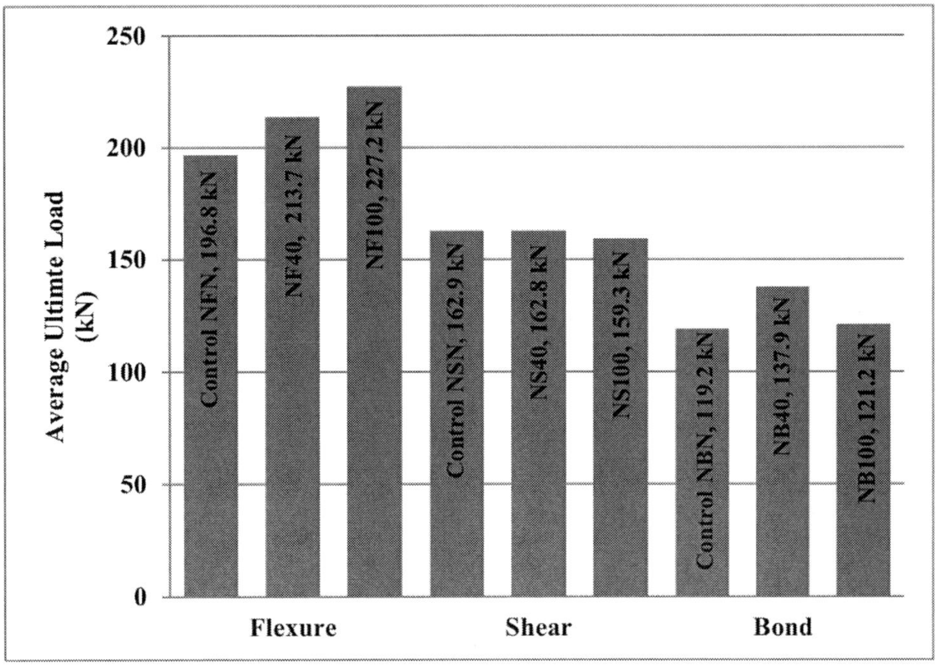

Fig. 5. Average ultimate loads for the beams in each series.

The ultimate loads of the 40 and the 100% replacement shear beam specimens were comparable to that of the control beam (Table 2 and Figure 5). Thus, the shear capacity was not affected by any percentage replacement of NCA with RCA. Also, the load-deflection histories of the shear beams with different percentage replacements were very similar with sharp drops in load after ultimate indicating brittle behavior.

Failure of all bond beams occurred along the mid-span where the provided 305 mm splice length was less than the required design length. Bond splitting cracks propagated along the splice length on the side and on the bottom tension face of the beam. The average maximum loads were 119.2 kN for the control shear beams, 137.9 kN for the 40% replacement beams (16% greater than the control beam), and 121.2 kN for the 100% beams (2% greater than the control beam). Thus, the bond capacity was not negatively affected by any percentage replacement of NCA with RCA.

5 Conclusion

The paper reports on a two-phase research program conducted at the American University of Beirut.

The objective of the first phase was to study the effect of replacing part or all of the natural crushed limestone coarse aggregates (NCA) with recycled coarse aggregates (RCA) resulting from tested and crushed concrete cylinders in batching plants, on the plastic and hardened mechanical properties of the produced normal and high strength concrete mixes (NSC and HSC). Results indicated that for both types of concrete mixes with nominal strength values of 20 and 60 MPa, the replacement of NCA with different percentages of RCA (0, 20, 40, 60, 80, or 100%) did not have a negative effect on the consistency of fresh concrete. As for the NSC hardened concrete properties, average reductions of 9.8, 13.5, 11, and 9.2% were recorded relative to the control mix in the compressive strength, splitting tensile strength, modulus of elasticity, and flexural strength values, respectively. The average reductions in HSC were 11, 11.3, 10.3, and 10.8%, respectively. The reductions were not significantly affected by the percentage replacement of NCA with RCA or by the level of concrete strength.

For each of the three modes of failure, three beams with different percentages replacement of NCA with RCA [0 (control), 40, and 100 percent] were tested in the second phase. One replicate was prepared for each beam to validate the test results. Thus a total of 18 beams were tested with 6 for each mode of failure. The loading system produced a constant moment region at mid-span of the beam. Cylinder compressive strength tests indicated that mixes with 40% and 100% replacement percentages of NCA with RCA had reductions in the compressive strength relative to the mix with 0% replacement amounting to 9 and 11.7%, respectively. However, the latter reduction in cylinder compressive strength did not have a significant impact on the mode of failure or ultimate strength or load-deflection behavior of reinforced concrete beams tested to fail in flexural or shear or bond splitting of the concrete cover in the splice region that could be related to the percentage replacement of NCA with RCA.

Results of the research program have a positive impact on the usage of recycled coarse aggregates produced by crushing tested concrete cylinders in the concrete construction industry. However, further research is needed to check the effect of recycled aggregate concrete sourced from tested concrete cylinders on other types of structural members such as columns and slabs.

References

1. ACI Report 555R-01. Removal and reuse of hardened concrete. ACI Committee 555, American Concrete Institute, Farmington Hills, Michigan (2010)
2. S. Nagataki, A. Gokce, T. Saeki, Effects of recycled aggregate characteristics on performance parameters of recycled aggregate concrete. *ACI SP 192*, American Concrete Institute, Farmington Hills, Michigan. 53-72 (2000)

3. K. Rahal, Mechanical properties of concrete with recycled coarse aggregate. Build. & Env. **42**(1), 407–415 (2007)
4. M. Etxeberria, E. Vázquez, A. Marí, M. Barra. Influence of amount of recycled coarse aggregates and production process on properties of recycled aggregate concrete. Cem. & Concr. Res. **37**(5), 735–742 (2007)
5. K.H. Yang, H.S. Chung, A.F. Ashour. Influence of type and replacement level of recycled aggregates on concrete properties. ACI Mat. J. **105**(3), 289-296 (2008)
6. ACI Committee 318. Building Code Requirements for Reinforced Concrete and Commentary (ACI-318-11/ACI-318R-11), American Concrete Institute, Farmington Hills, Michigan (2011)
7. S.F.U. Ahmed. Properties of concrete containing construction and demolition wastes and fly ash. ASCE J. of Mat. in Civ. Eng. **25**(12), 1864-1870 (2012)
8. B.C. Han, H.D. Yun, S.Y. Chung. Shear capacity of reinforced concrete beams made with recycled-aggregate. *ACI SP 200*, American Concrete Institute, Farmington Hills, Michigan. 503-516 (2001)
9. F. Gholamreza, A.G. Razaqpur, O.B. Isgor, A. Abbas, B. Fournier, S. Foo. Flexural performance of steel-reinforced recycled concrete beam. ACI Str. J. **106**(6), 858-867 (2009)
10. S.W. Kim, H.D. Yun. Influence of recycled coarse aggregates on the bond behavior of deformed bars in concrete. Eng. Str. **48**, 133-143 (2012)
11. ACI Report 211.1-91. Standard practice for selecting proportions for normal, heavyweight, and mass concrete. ACI Committee 211, American Concrete Institute, Farmington Hills, Michigan (1991)

Feasibility of coal fly ash based bricks and roof tiles as construction materials: a review

M.N. Akhtar[1]* *O.* Hattamleh[2] *and J.N.* Akhtar[3]

[1] Department of Civil Engineering Fahad Bin Sultan University, Tabuk, Saudi Arabia
2 Department of Civil Engineering Hashemite University, Zarqa 13133, Jordan
3University Polytechnic Civil Engineering section, A.M.U Aligarh

Abstract. The aim of present study is to investigate about the potential use of coal fly ash along with other natural and solid wastes for the production of coal fly ash based bricks and roof tiles. The study is based on the comprehensive reviews available from the previous experimental data on coal fly ash based bricks and roof tiles. The study intendeds to provide the essential technical information and data for the use of fly ash mix with other solid wastes and reveal their suitability as construction materials. It has been found that samples were non-hazardous in nature and vigorously used as an additional construction materials and their compositions are perfectly fit to make the strong composite material for bricks and tiles. The three past studies have been demonstrated that, fly ash based bricks and roof tiles provides a sustainable supplement to the traditional clay bricks and roof tiles, that not only increases the efficiency of traditional bricks and roof tiles but also helps significantly to reduce the environmental issues associated with the disposal of these waste materials. In addition to this study highlights the potential use of fly ash for producing sustainable construction materials.

1 Introduction

Fly ash (F.A) particles typically solidify while they are still in suspension in exhaust gases and thus are generally spherical in shape. F.A is composed primarily of silica (SiO_2), alumina (Al_2O_3), and iron oxide (Fe_2O_3). Physical and chemical requirements for F.A usually vary depending on its intended use. Accordingly, specific requirements for use of F.A in concrete or soil stabilization are described in [1]. Standard specification for coal F.A and Raw or Calcined Natural Pozzolan for use in Concrete and [2], standard specification for F.A and other Pozzolans use with Lime for Soil stabilization, respectively. The plasticity index of mixture of fly ash and clay decreases on drastically with increase of replacing ratio of F.A was to be determined [3]. The traditional construction materials such as concrete, bricks, hollow blocks, solid blocks, pavement blocks and tiles are being produced from the existing natural resources. This damages the environment due to continuous exploration and depletion of natural resources. Moreover, various toxic substances such as high concentration of carbon monoxide, oxides of sulfur, oxides of nitrogen, and suspended particulate matters are invariably emitted to the atmosphere during the manufacturing process of construction materials. The emission of toxic matters contaminates air, water, soil, flora, fauna and aquatic life, and thus influences human health as well as their living standards. Therefore, the issues related to environmental conservation have gained great importance in our society in recent years [4]. Municipal solid waste incineration (MSWI) ash is used in part as raw materials for cement clinker production by taking advantage of the high contents of SiO_2, Al_2O_3, and CaO. It is necessary for environmental reasons to establish a material utilization system for the incineration waste ash residue instead of disposing these ashes into landfill [5]. The utilization of solid wastes in construction materials is one of such

* Corresponding author : nakhtar@fbsu.edu.sa

innovative efforts. The cost of construction materials is increasing day by day because of high demand, scarcity of raw materials, and high price of energy. From the standpoint of energy saving and conservation of natural resources, the use of alternative constituents in construction materials is now a global concern.

For this, the extensive research and development works towards exploring new ingredients are required for producing sustainable and environment friendly construction materials. The present study investigates about the potential use of fly ash based bricks and tiles as construction material resulting in either full or partial replacement of clay bricks and tiles.

2 Chemical analysis of F.A and R.S.D

The procedures used in the chemical analysis of fly ash to determine the oxides of silica, aluminum, iron, calcium, magnesium, total sodium and total potassium is as follows heat the sample prior to fusing as indicated for material previously ashed in ASTM Designation: D 4326. Weigh 0.50 g of the ashed fly ash sample and 6.5 g of fluxing material into the crucible. Add 400 µL of liquid lithium bromide to the sample before fusing. Determine loss on ignition, available sodium oxide, available potassium oxide and available alkali in accordance with ASTM Designation: C 311. 3. Determine sulfur trioxide with the carbon sulfur determinator. The apparatus used in the study are XRF (X-ray Fluorescence) X-ray diffraction (XRD), and wavelength dispersive spectroscopy (microprobe WDS). X-ray microanalysis (EDS) techniques are being used in the chemical analysis of fly ash (F.A) and radish stone dust (R.S.D). In the construction industry, the combination of SEM and EDS are being used particularly for quality control and verification of material composition. The VEGA3 SEM, which is fully compatible with any EDS system, is a powerful tool for quality control and development in the construction industry. Locally available T.F.A and R.S.D samples were examined in the present study. After examination of the samples it has been found that both samples were non-hazardous in nature and vigorously used as an additional construction materials. EDS analysis was performed using Bruker's Quantax system to determine the chemical composition of the T.F.A and R.S.D particles. The chemical composition of T.F.A and R.S.D particles and the element distribution over the sample surface was determined by EDS. Exact chemical composition is then summarized in Table 1 and 2.

Table 1. List of Elements and Concentrations of T.F.A.

Element	Weight%	Atomic%
C	14.42	21.05
O	54.59	59.83
Al	11.96	7.77
Si	16.47	10.29
K	0.73	0.33
Ca	0.94	0.41
Ti	0.88	0.32
Totals	100.00	

Table 2. List of Elements and Concentrations of R.S.D.

Element	Weight%	Atomic%
O	59.96	75.21
Al	11.11	8.26
Si	19.53	13.95
K	2.17	1.11
Fe	2.81	1.01
Au	4.42	0.45
Totals	100.00	

Apart from oxygen, silicon and aluminum, from which the fly ash is usually composed; many other elements were found as well. In R.S.D samples iron and many other elements were also found with the combination of oxygen, silicon and aluminium. The presence of diverse elements in fly ash particles might be attributed to the type of coal being used. The Quantitative results from point analysis of T.F.A and R.S.D are being used in the study are also shown in Figure.1 and Figure.2.

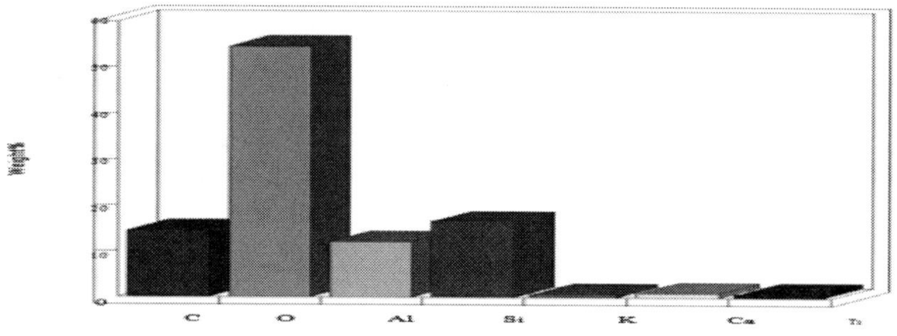

Fig. 1. Quantitative results from point analysis of TFA.

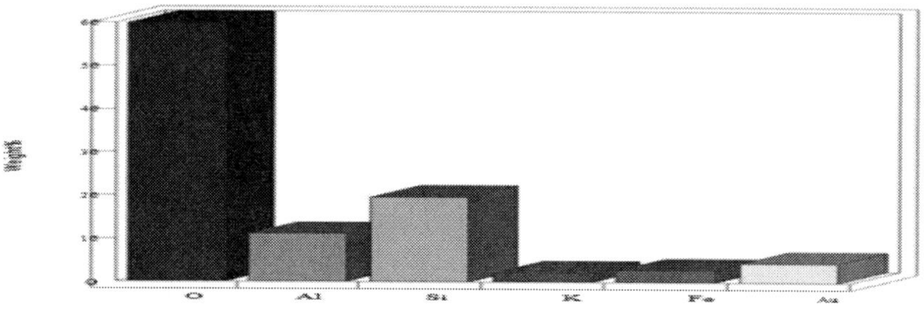

Fig. 2. Quantitative results from point analysis of RSD.

The fly ash consists of fine powdery particles that are predominantly spherical in shape, either solid or hollow and mostly glassy in nature. The particles size distribution of most bituminous coal fly ashes is generally similar to that of silt (< 0.075mm). Although, Sub-bituminous coal fly ashes or class C fly ash is generally slightly coarser than bituminous coal ranges between 0.002 mm - 0.1 mm. Figure. 3 shows the variation of particles in XRF, T.F.A at 3000x Magnification and Fig. 4 shows SE image of T.F.A at 6000x Magnification were collected from Harduaganj Thermal Power Station U.P (India).

Fig. 3. Spectrum from point analysis of TFA particles at 3000x.

Fig. 4. SE image of TFA at 6000x.

Construction material such as Radish Stone Dust (R.S.D) admixtures are mixed with cement particles and improved the particle packing of cement paste, thus reducing permeability (k). Figure. 5 and 6 shows the XRF, R.S.D particles at 3000 x magnification and SE image of R.S.D at 6000x magnification was collected locally.

Fig. 5. Spectrum from point analysis of RSD particles at 3000x

Fig. 6. SE image of RSD 6000x

3 Comparison of past experimental results

Only few studies have been conducted on fly ash based bricks and tiles. Since fly ash based bricks and tiles is an assemblage of fly ash and other materials.The comprasion is based on three recent past stidies. It is generally believed that the strength and durability of fly ash based bricks and tiles would lie somewhere between that of bond of fly ash and other mixed materials. It may be true in cases when one combination of mould, i.e., cement, fly ash, sand and stone dust, etc is substantially weaker and softer than other, for example variation of these materials play a significant role to change their strength as reported by [6]. Series of tests were conducted on different combinations with partial replacement of clay for making fly ash bricks and it was observed that the combination of materials play a significant role to change the compressive strength. The combination of fibre reinforced fly ash lime stone dust brick (25FRFALSDB9') was found to have highest compressive strength (9.155 MPa) with 50% Treated fly ash combination at 10% cement. It was found that when the amount of treated fly ash changes with respect to the stone dust and sand replacement, the compressive strength increased and was found to be maximum 25% stone dust-sand combination with 50% treated fly ash. The variation of strength is shown in Figure. 7(a). Based on experimental data [7] series of tests were conducted on fly ash based brick tiles. It was observed that the compressive strength of these tiles increases with increase in bond strength of the composite material along with other factors. It has been also observed that the permeability reached at lower value at the same whereas compressive strength was reached at maximum. After experimental investigation it has been carried out that the tiles were failed on lower compressive strength as compare to the conventional clay brick tiles. The variation of compressive strength is shown in Figure. 7(b). The highest compressive strength reached (6.896 MPa) for the combination of 15TFASDBT. In other study was carried out on fly ash based roof tiles with waste polythene fibres [8]. The compressive strength has been increased by 30.65% compared to the past study carried out by [7]. The studied compressive strength behavior has been shown in Figure. 7(c) and (d). The detailed experimental data of the past investigations are given in Table 3, shows the compressive strength behavior of the studies.

Table 3. Comparison of Past Experimental Results of Fly ash Based Bricks and Roof Tiles.

J.Alam et al. (2013)		J.N Akhtar et al. (2010)				Akhtar M.A et al. (2016)			
Mould Designations	Compressive Strength (MPA)	Mould Designations	Compressive Strength (MPA)	Mould Desig-ations	Compress-ve Strength (MPA)	Mould Designations	Compr-ssive Strength (MPA)	Mould Designat-ons	Compressive Strength (MPA)
00FABT	3.636	00FAB	3.348	00FALSD B1	3.666	00 TFASDRT 70	3.56	10TFASDR T15	6.04
05FABT	3.688	05FAB	3.69	05FALSD B2	4.235	05 TFASDRT 70	3.84	15TFASDR T15	6.54
10FABT	3.755	10FAB	4.39	10FALSD B3	7.156	10 TFASDRT 70	4.25	20TFASDR T15	6.98
15FABT	3.94	15FAB	5.385	15FALSD B4	6.569	15 TFASDRT 70	4.19	25TFASDR T15	6.35
20FABT	3.69	20FAB	4.45	20FALSD B5	4.79	20 TFASDRT 70	3.89	30TFASDR T15	6.3
00TFABT	4.055	00FRFAB	3.748	00FRFAL SDB1'	3.666	00 TFASDRT 60	3.74	10TFASDR T20	8.52
05TFABT	4.1	05FRFAB	3.99	05FRFAL SDB2'	4.235	05 TFASDRT 60	4.06	15TFASDR T20	8.75
10TFABT	4.31	10FRFAB	4.79	10FRFAL SDB3'	8.156	10 TFASDRT 60	4.47	20TFASDR T20	9.01
15TFABT	4.535	15FRFAB	6.35	15FRFAL SDB4'	6.569	15 TFASDRT 60	4.35	25TFASDR T20	8.95
20TFABT	4.315	20FRFAB	4.75	20FRFAL SDB5'	4.79	20 TFASDRT 60	4.15	30TFASDR T20	7.99
00TFASDBT	4.845	00FALB	3.685	10FALSD B6	8.16	00 TFASDRT 50	4.05		
05TFASDBT	5.3	05FALB	4.065	15FALSD B7	8.33	05 TFASDRT 50	4.35		
10TFASDBT	5.48	10FALB	4.865	20FALSD B8	8.520	10 TFASDRT 50	4.61		
15TFASDBT	5.9	15FALB	6.599	25FALSD B9	8.890	15 TFASDRT 50	4.53		
20TFASDBT	5.714	20FALB	4.596	30FALSD B10	7.480	20 TFASDRT 50	4.41		
10TFASDBT	4.69	00FRFALB	4.185	10FRFAL SDB6'	8.315	10 TFASDRT 10	4.61		
15TFASDBT	6.896	05FRFLAB	5.465	15FRFAL SDB7'	8.545	15 TFASDRT 10	4.53		
20TFASDBT	6.24	10FRFALB	5.85	20FRFAL SDB8'	8.625	20 TFASDRT 10	4.41		
25TFASDBT	5.77	15FRFALB	7.859	25FRFAL SDB9'	9.155	25 TFASDRT 10	4.38		
30TFASDBT	5.39	20FRFALB	5.46	30FRFALS DB10'	7.755	30 TFASDRT 10	4.25		

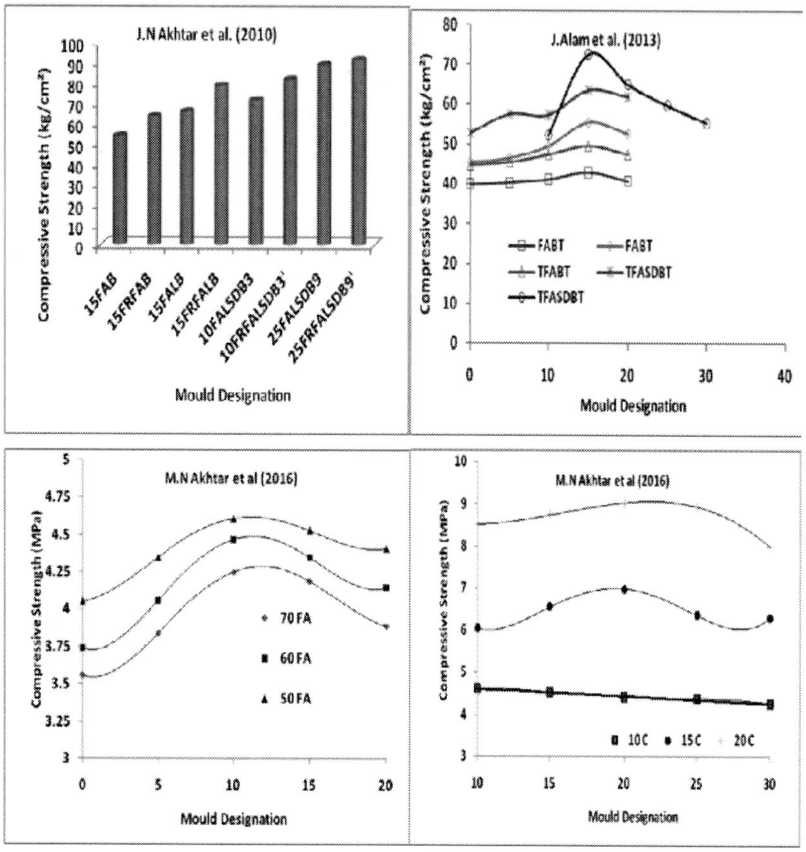

Fig. 7. (a),(b),(c), and (d) work of other researches with experimental curves obtained in several published studies.

4 Stress-strain behavior of fly ash based bricks and tiles

The tests were performed in accordance with [9], and [10]. The stress-strain curves for the standard clay tiles obtained by averaging the stress strain data from ten samples of each type of clay tiles. The tiles were found to be behaving linearly up to about one-third of the ultimate failure load after that the behavior became highly nonlinear. An average stress-strain curve for all the tiles types used in the study is also shown in Figure. 8 [8]. The stress-strain curves for fly ash based tiles for six grades of mould obtained by averaging the data from ten specimens of each grade. Initial straight portion of the stress-strain curve (up to about one-third of mould strength) is followed by a nonlinear curve almost same as stress-strain curves for standard clay brick tiles (SCBT) discussed above which extends well beyond the strain limits corresponding to the tiles samples. Strain readings on the falling branch of the stress-strain curve could not be recorded for the very weak moulds because of its brittle and explosive crushing failure after reaching the ultimate strength. Figure. 9(a) shows that the stress-strain curve comparison for six different combination grades of moulds TFASDRT. Data points represent experimental results and solid lines represent corresponding trend lines. A comparison of stress-stain curves generated by the TFASDRT and SCBT obtained in the present experimental study is shown in Figure. 9(b), which show a very close match between the TFASDRT and SCBT curves [8]. The stress strain curves of different test series of FAB, FRFAB, FALB, FRFALB, FALSDB and FRFALSDB have been plotted and shown in Figure 10 [6] . Each stress-strain curve is the average of three bricks pair [6].

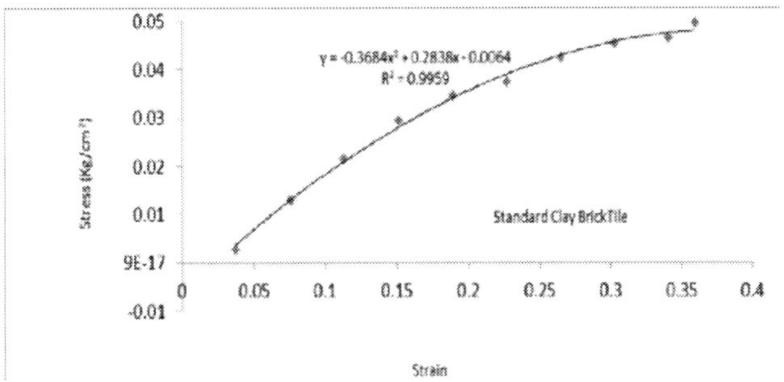

Fig. 8 Stress-strain curve behavior for standard clay brick tiles

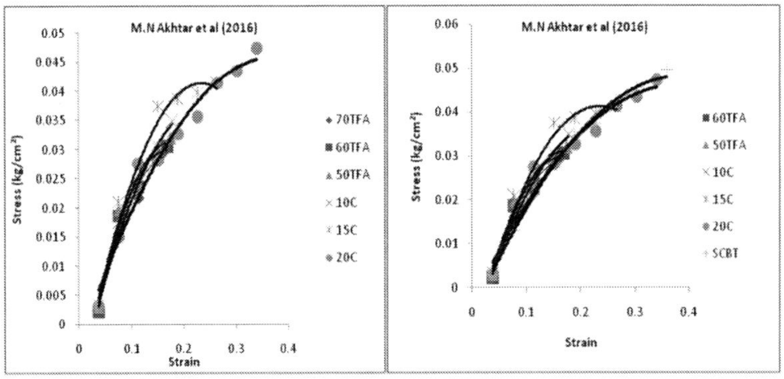

Fig. 9 (a,b). Stress-strain curve behavior of TFASDRT and Comparison of stress-strain curves of TFASDRT and SCBT.

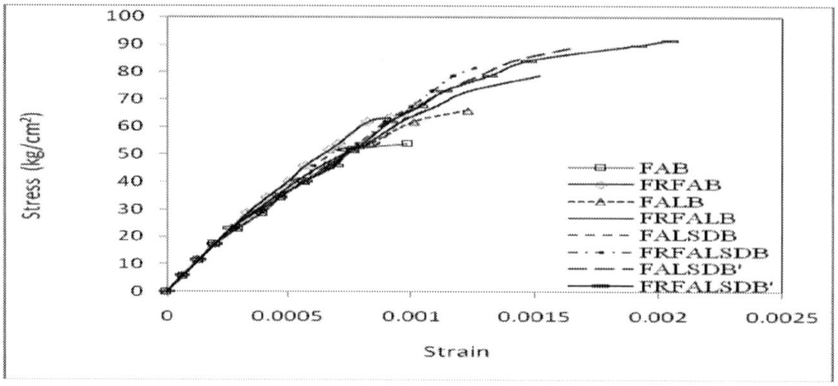

Fig. 10. Stress vs. Strain variation of fly ash based bricks.

5 Benefits of fly ash based composite material with cement

Cement is a binding material, which is widely used in construction industry. It is made from limestone, clay and various additives, such as powered gypsum and fly ash. Gypsum ($CaSO_4.2H_2O$) is a common and soft mineral. It is used as a set retarder in the early stage of hydration. It keeps the mortar of the concrete in a plastic stage for longer time and therefore it prolongs concrete workability. The combination of T.F.A and R.S.D with cement discloses the various important benefits of the composite material. The replacement of cement by fly ash reduces water demand and the decreased water demand has a little or no effect on drying shrinkage/cracking. So, the reduction in free lime and the resulting increase in cementitious compounds combined with the reduction in permeability and enhance the mortar durability. The additional binder produced by the fly ash reaction with available lime allows fly ash based composite material to continue to gain strength over time. The failure modes of tested samples are shown in Figure 11. Weak mortar was found to be very weak and soft as compared to other three i.e., intermediate, strong and very strong. Intermediate mortar perform stable as compared to weak and strength increased about (29.41%), then Strong mortar performed well in terms of strength and ductility as compared to other two, strong mortars was found (51.40%), higher than intermediate mortar, in comparison of strong mortar very strong mortar was reached the highest strength about (29.08%) more than the strong [8].

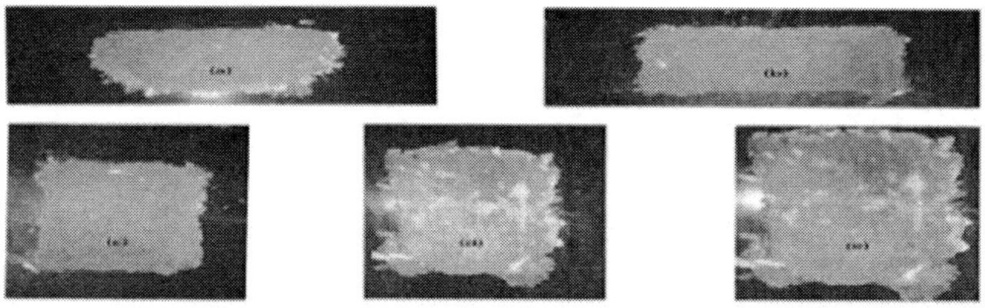

Fig. 11. Typical failure modes of: (a) Weak, 00TFASDRT (70%TFA); (b) Intermediate, 10TFASDRT (60%TFA); (c) Intermediate, 10TFASDRT (50%TFA); (d) Strong, 20TFASDRT (15% C) (e) Very strong, 20TFASDRT (20% C).

The spherical shaped particles of fly ash act as miniature ball bearings within the concrete mix,thus providing a lubricant effects. This same effect also improves concerete pumpability by reducing frictional lisses during the pumping process and flat work finish ability [11]. The replacement of cement by fly ash reduces the water demand for a given slump. When fly ash is used at about 20 percent of the total cementious, water demand is reduced by approximately 10 percent. Higher fly ash contents contain will yield higher water reductions. The decreased water demand has little or no effects on drying shriankage / cracking. Some fly ash is known to reduce drying shrinkage in certain situations [11]. Replacing cement with the same amount of fly ash can reduce the heat of hydration of concrete. This reduction in heat of hydration does not sacrifice long-term strength gain or durability. The reduced heat of hydration lessens heat rise problems in mass concrete placements [11]. The additional binder produced by the fly ash reaction with available lime allows fly ash concrete to continue to gain strength over time. Mixtures designed to produce equivalent strength at early ages (less than 90 days) will ultimately exceed the strength of straight cement concrete mixes [11].

Control point defining the characteristics of the study

Based on the past studies five control points were indentified, which correspond to the experimentally observed data and representing compressive behavior, stress-strain characteristics and safe utilization of fly ash as bricks and roof tiles in civil engineering construction industry.

The control points identified during the studies are as follows:

1. The chemical analysis identified that the waste materials are non-hazardous in nature.
2. The compressive strength of fly ash based bricks and tiles was found very closed to the standard strength of clay bricks and tiles.
3. The compressive strength [8], has been increased by 30.65% compared to the past study carried out by [7].
4. Addition of waste Polythene Fibre (W.P.F) provides more strength in the tiles as compared to the other past studies refer (Figure. 11 Typical failure modes).
5. Stress-strain curves have been generated by past experimental data and compared with the experimental curve obtained by standard clay brick and tile is fall almost on the same pattern.

Conclusions

Fly ash have become an important raw material for various industrial and construction applications. The objective of this paper was to investigate that the fly ash based bricks and roof tiles fits as construction materials and it can replace the conventional clay bricks and roof tiles. As shown, in the past studies the compressive behavior and stress-strain characteristics is almost as same as the standard clay bricks and roof tiles. The available tests mentioned in the past studies fouced only on the strength of the bricks and tiles and the strength based criteria is not enough to replace any building material in the construction industry. Design needs durability the construction materials fly ash based bricks and tiles must fulfill the acceptance criteria for durability tests. Measuring the durability factor of bricks and tiles to determine its longevity, and ability to withstand local weathering and use, is difficult to predetermine. Absorption and permeability test should be considered the main test to evaluate where bricks and tiles is subjected to such as basements, water reservoirs, building roofs, tunnels etc.In addition, the ample scope in the continuation of this study is for the investigation of the mechanical properties of the same samples.

References

[1] ASTM, "American Standards for Testing and Materials", Standard C618-05 "Standard specification for coal fly ash and raw or calcined natural pozzolan for use in concrete". (2005).
[2] ASTM, "American Standards for Testing and Materials". Standard C593-06 "Standard specification for fly ash and other pozzolans for use with lime for soil stabilization". (2006).
[3] X, Lingling, Guo Wei, Wang Tao and Yang Nanru Study on fired bricks with replacing clay by fly ash in high volume ratio Construction and Building Materials 19 (2005) 243–247.
[4] Y, Xue et al.,. Utilization of municipal solid waste incineration ash instone mastic asphalt mixture: Pavement performance and environmental impact. Construction and Building Materials Vol. **23**, Issue 2, pp. 989–996 (2009)
[5] Charles Hoi King Lam et al.,. Utilization of municipal solid waste incineration ash in Portland cement clinker. Clean Technologies and Environmental Policy. Springer Vol. **13**, Issue 4, pp. 607–615. (2011)

[6] J.N Akhtar et al.. An Experimental study on fibre-reinforced fly ash based lime bricks. International Journal of physical Sciences Vol.**5** (11), pp.1688-1695, ISSN 1992-1950 (2010)

[7] J. Alam et al. Flay ash Based Brick Tiles: An Experimental Study. International journal of emerging Trends in Engineering and development. Issue 3, Vol.**6** ISSN 2249-6149 (2013).

[8] M.A Akhtar et al. Sustainable fly Ash Based Roof Tiles Waste Polythene Fibre: An Experimental Study. Open Journal of Civil Engineering, Vol- **6**, Issue 2 pp.314-327 (2016). **DOI:** 10.4236/ojce.2016.62026

[9] ASTM, "Standard test methods for sampling and testing brick and structural clay tile." Masonry test methods and specifications for the building industry, ASTM-C67-00, 4th Ed., Philadelphia. (2001c).

[10] Indian Standards Indian standard methods of test of burn clay building bricks—Part 1: Determination of compressive strength, IS 3495, 3rd Rev., Bureau of Indian Standards, New Delhi, India. (IS.1992a).

[11] Newsletter CBRI ENVIS Centre on fly ash (CBRI), Roorkee Vol.**7** Issue(2) pp.1-8ISSN No.0974-2026.

DOI: 10.1051/matecconf/201712003009

Utilizing waste materials to enhance mechanical and durability characteristics of concrete incorporated with silica fume

Ali Hamza[1] *, *Shahram* Derogar[2], and *Ceren* Ince[1]

[1]Civil Engineering Department, Middle East Technical University, Northern Cyprus Campus, Kalkanlı, Güzelyurt, North Cyprus via Mersin 10, Turkey
[2]Civil Engineering Department, European University of Lefke, Lefke, North Cyprus via Mersin 10, Turkey

Abstract. Construction and demolition wastes are increasing significantly due to augmented boom of modern construction. Although the partial cement replacement materials do promote the idea of sustainable construction, the use of construction and demolition waste can also be considered to be viable option to advance the sustainability in modern construction practices. This paper investigates the use of industrial waste materials namely marble dust and crushed bricks as replacement of natural fine aggregates along with the use of silica fume as a partial cement replacement on the mechanical properties and durability characteristics of concrete. Partial replacement levels of waste materials were 10 and 20 percent by volume while the partial replacement level of silica fume was kept to 20 percent at all concrete samples. The results reported in this paper show that the use of marble dust as a replacement material to the natural fine aggregates resulted in an increase in the mechanical properties of concrete. However, the use of crushed bricks did not substantially contribute in the development of strength. Water permeability of concrete incorporated with both silica fume and waste materials (marble dust and crushed bricks) decreased significantly. The decrease in water permeability of concrete was attributed to the pozzolanic reaction of silica fume with calcium hydroxide of cement and the filler effect of the waste materials of marble dust and crushed bricks. The use of waste materials also enhance the freeze and thaw resistance of concrete. Authors strongly suggest that the pozzolanic reaction and the development of the microstructure of the concrete through the use of waste materials are largely responsible from the advances in the durability of concrete.

1 Introduction

Construction and Demolition Waste (CDW) is considered to be an alarming issue for our modern world. Significant portions of waste production are likely to be seen from

* Corresponding author: ali.hamza@metu.edu.tr

manufacturing processes, industries and municipal waste [1]. With the increasing shortage of natural resources and ever increasing cost, waste utilization has become a great alternative for disposal [1]. Utilizing CDW seems practical; it promotes the development of sustainable construction practices [2]. In the literature, the use of supplementary cementitious materials (SCMs) especially silica fume contributes various improvements of mechanical properties and durability [3]. However, there is still a need to develop the sustainability in concrete manufacturing by reducing the burden on natural resources. Around 80% of volume of concrete is made of aggregates, which is normally derived from natural resources [2]. As the natural aggregates are continued to eventually receive their expiration date in various parts of the world, the concrete industry needs to seek for other possible alternatives [2].

Construction and demolition waste includes various types of wastes; rubbles, stones, recycled/crushed aggregates, industrial wastes such as marble dust, crushed bricks, saw dust etc. However, with respect to concrete, the common aspect among all of these waste materials is that they may contribute as a filler material in the concrete manufacturing [4]. Many research studies have highlighted the effects of filler materials on the properties of concrete manufacturing. Based on their conclusions, the benefit of filler material usage results in improved workability with reduced cement content [5, 6]. Furthermore, due to their filler ability, pores between aggregates can be fulfilled improving the permeability and porosity of the concrete manufactured [7]. These waste materials can be pozzolanic or inert, depending on the physical and physio-chemical properties which may further affects the performance of concrete [8].

Marble, being commonly used building material contributes to a by-product environmental waste i.e. marble dust [9]. However, marble dust exhibits the properties of filler material due to its limestone origin [10], the possibility of use of marble dust as a filler material is likely to happen. According to [10, 11], use of limestone dust as a filler material showed quite good performance in concrete manufacturing. Brick and tile manufacturing industry produces quite a high number of rejected fired products due to being off-standard i.e. various bricks might be broken, underburned, overburned or distorted. Since bricks are porous and do possess the filler ability, they can be recycled in concrete as filler material [12, 13].

Therefore, the aim of this paper is to offer an integrative approach of using marble dust and crushed bricks as partial replacement of fine aggregates as filler material. Silica fume will also replace Portland cement up to 20% by volume. The idea is to investigate the possible role of marble dust and crushed bricks used as a partial replacement to fine aggregates along with incorporation of silica fume on the strength development and durability characteristics. Strength development includes compressive strength test as a key parameter being investigated. Tests such as water permeability and freezing/thawing resistance for determining the pore structure of concrete are included in the durability characteristics. The possible effect of addition of silica fume along with these filler replacement materials is also highlighted addressing its significant role on strength development and durability characteristics of the concrete.

2 Materials and mix design

This experimental study was directed on freshly-mixed and harden state concrete utilizing Portland cement and Silica Fume as binders. Type V Portland cement was attained from Adana Cimento Ltd. and conforms to [14]. Silica Fume was procured from Cyprus Environmental Enterprises Ltd. (CEE) and conforms to [15]. Waste materials being marble dust and crushed bricks will replace proportions of fine aggregates. Marble dust has been obtained from Izmer Ltd. whereas crushed bricks are obtained from Gurdag Madencilik Ltd. Apart from these materials, different components of concrete incorporates fine aggregates, coarse aggregates and water. Particle size distribution of marble dust and crushed bricks was necessary since these materials are to replace fine aggregates, the procedure is carried out conforming [16]. Various replacement levels of silica fume incorporated with marble dust and crushed bricks were used to investigate the significant role on the fresh and harden state properties of Portland cement concrete. Mix design calculations includes the assumption of slump value kept constant to 100 mm. Hence, the water/binder (w/b) ratio have been maintained with respect to the slump value. The masses of binders and waste materials required to produce the mix proportions by volume were measured from precisely determined values of relative density. For fine and coarse aggregates, specific gravity was determined conforming [17, 18].

3 Experimental procedures

3.1 Measurement of slump

The slump of fresh concrete produced having different volume substitution levels of silica fume, marble dust and crushed bricks was measured utilizing the standard slump test apparatus [19]. The slump value has been kept consistent to 100 mm. By keeping such consistency of slump value, the water/binder proportion has been kept between the limits of 0.50 to 0.70.

3.2 Measurement of compressive strength

Compressive Strength testing was led on different harden substitution levels of concrete including silica fume, marble dust and crushed bricks collectively. The testing has been led from 28 days to 180 days on 150 mm standard blocks. Specimens from every substitution level were tested utilizing a compressive testing machine. The loading rate has been kept constant to 0.6 MPa/sec. An average compressive strength has been taken among the substituted levels of concrete created as per [20].

3.3 Measurement of water permeability

Standard cubic molds of 150 mm were utilized for deciding the water permeability for every substitution level of concrete created. Every specimen was settled into the permeability testing machine and water has been infiltrated by capillary action from the base of the specimen. Adjustment of the specimen is critical on the permeability machine to eliminate the risk of water spillage from the sides of the specimen. All specimens were subjected to 1 atmospheric pressure (0.1 MPa). It is significant to address that the mass measurements are needed to be taken before and after the test to ensure each specimen's water penetration limit. By following 72 hours, the test is completed. Every specimen was released from the permeability machine and split into two equivalent parts for deciding the water infiltration/penetration depth. Average depth was taken from four equidistant spots along the base of the specimen from where the water pressure has been applied. The system followed has been directed as per [21].

3.4 Measurement of freeze and thaw resistance

For freeze and thaw cycles, cubic molds of 150mm have been set up in freeze and thaw cabinet. Similar methodology has been applied as reported by [22]. It incorporates freezing and thawing for the cured specimens of different substitution levels at – 30oC to 30oC. Two cycles have been kept up once a day. For 12 h every day, the specimens have been encountering freezing and thawing consequently when the temperature comes to - 30oC or 30oC. The cycle continues going twice every day for next 180 days. The relative humidity is kept up within the limits from 85% to 95%. It is significant to mention that the surface of the specimens have not been secured during the freezing and thawing, so that the specimen is completely exposed to moisture variations.

4 Experimental results

4.1 Effect of waste materials on the water content of concrete

The effect of waste materials on the water content of concrete is investigated in this section. Crushed bricks and marble dust are used as sand replacement materials whereas silica fume is used as a cement replacement material for concrete. Figure 1 shows the slump (mm) versus the mix proportions of concrete. Water: binder is also shown in Figure 1.

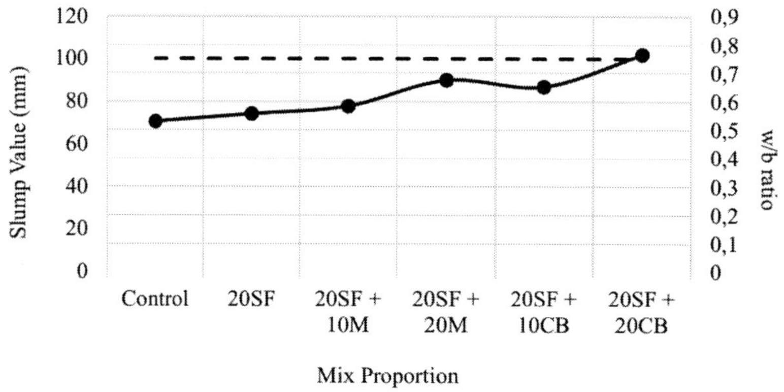

Fig. 1. Slump (mm) and water: binder ratios for all the mix proportions of concrete; slump values (mm), ------; water: binder ratio; SF: Silica Fume; CB: Crushed Brick and M: Marble Dust.

It is shown in Figure 1 that the use of silica fume as a cement replacement material resulted in an increase water: binder ratio to achieve the same consistency of control specimen. It is also shown in Figure 1 that the increase in the sand replacement materials such as marble dust and crushed bricks resulted in an increase in water: binder ratio to attain 100 mm of slump value. The increase in the water demand of concrete with increased replacement level of marble dust and crushed bricks are attributed to the finer particle size of the waste materials compared to sand. Finer particle size of waste materials used as a replacement to sand resulted in an increase water demand to attain constant slump value of 100 mm.

4.2 Effect of waste materials on the compressive strength of concrete

Marble dust and crushed bricks are used as sand replacement materials and their replacement levels were 10 and 20%. Silica fume is used as a cement replacement material and replacement level was kept constant to 20%. Figure 2 shows the effect of the use of marble dust on the compressive strength of concrete incorporated with silica fume. The results show that the increase in the replacement level of marble dust resulted in a slight decrease in compressive strength of concrete compared to concrete that incorporated with only silica fume. It must however be noted that each replacement level of marble dust produced higher compressive strength both at 28, 90 and 180 days compared to the concrete control sample.

Influence of the use of crushed bricks on the compressive strength of concrete incorporated with silica fume is shown in Figure 3. It is shown in Figure 3 that the increase replacement level of crushed bricks resulted in a greater decrease in compressive strength of concrete compared to the compressive strength of concrete with replacement levels of marble dust. Although concrete with 10% replacement level of crushed bricks provide higher compressive strength than the concrete control specimen, compressive strength of concrete with 20% replacement level of crushed bricks fall below the compressive strength of control specimen.

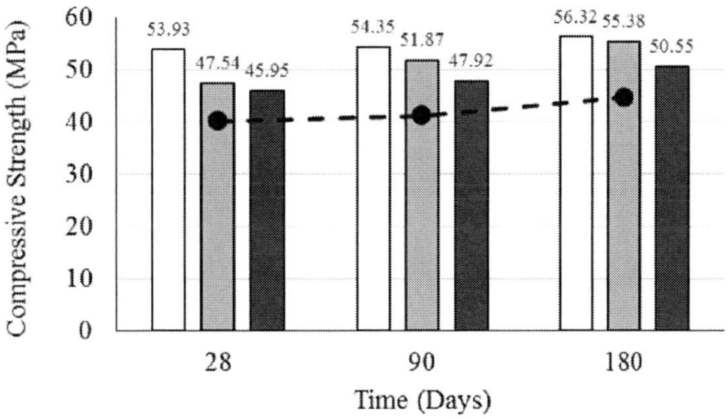

Fig. 2. Compressive strength of concrete incorporated with silica fume; %20 SF; 20% SF with 10% marble dust; 20% SF with 20% marble dust; concrete control specimen.

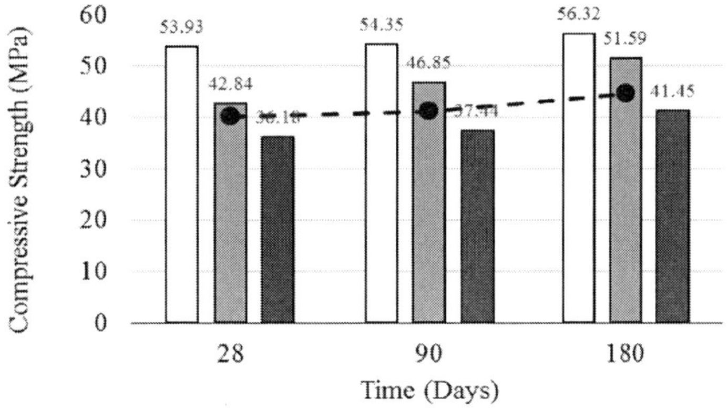

Fig. 3.
Compressive strength of concrete incorporated with silica fume; %20 SF; 20% SF with 10% crushed bricks; 20% SF with 20% crushed bricks; concrete control specimen.

4.2 The effect of waste materials on the water permeability of concrete

The influence of the use of marble dust and crushed bricks as sand replacement material on the water permeability of concrete is studied in this section. Figure 4 show the water penetration depths of concrete specimens prepared using 20% silica fume and 20% silica fume with 20% marble dust and 20% crushed bricks.

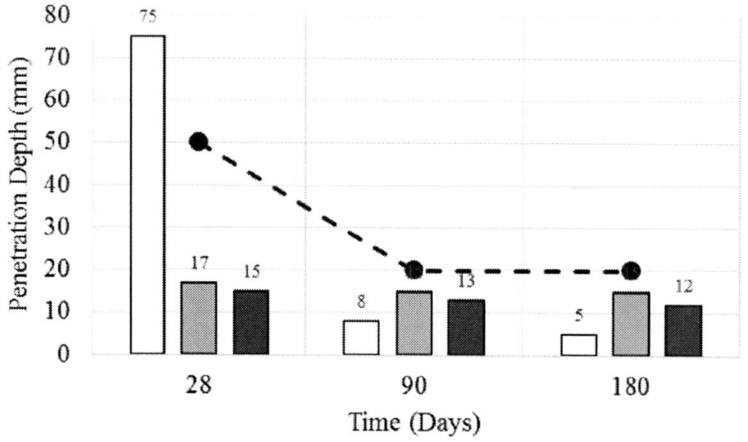

Fig. 4. Water penetration depth of concrete; %20 SF; 20% SF with 20% marble dust; 20% SF with 20% crushed bricks; concrete control specimen.

It is shown in Figure 4 that the use of 20% silica fume resulted in a great reduction in water penetration depth of concrete at 90 and 180 days. It must also be noted that the use of both 20% marble dust and 20% crushed bricks resulted in a dramatic decrease in the water penetration depth of concrete at 28th days however, the water penetration of these samples remained constant at 90 and 180 days. It must be underlined here that both replacement materials resulted in a reduced water penetration of concrete at all test days and therefore enhanced the water permeability of concrete incorporated with silica fume.

4.4 The effect of waste materials on the resistance to freeze and thaw of concrete

The influence of the use of marble dust and crushed bricks as sand replacement material on the freeze and thaw resistance of concrete is studied in this section. Figure 5 shows the compressive strength of silica fume concrete prepared using 10 and 20% replacement level of marble dust that were subjected to freeze and thaw cycles for 28, 90 and 180 days.

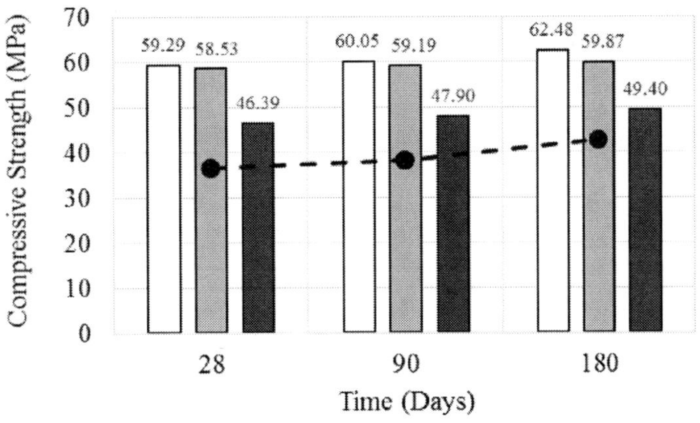

Fig. 5. Compressive strength of concrete subjected to freeze and thaw cycles; %20 SF; 20% SF with 10% marble dust; 20% SF with 20% marble dust; concrete control specimen.

The results shown in Figure 5 indicate that the increase replacement levels of marble dust resulted in a decrease in freeze and thaw resistance compared to the concrete incorporated with silica fume only. It must however be underlined that the compressive strength concrete at all replacement levels of marble dust produced higher compressive strength compared to the control specimen at all test days. The results show that the use of marble dust in combination with a pozzolanic replacement material of silica fume enhances the freeze and thaw resistance compared to the control specimen.

Furthermore, Figure 6 shows the compressive strength of silica fume concrete prepared using 10 and 20% replacement level of crushed bricks that were subjected to freeze and thaw cycles for 28, 90 and 180 days.

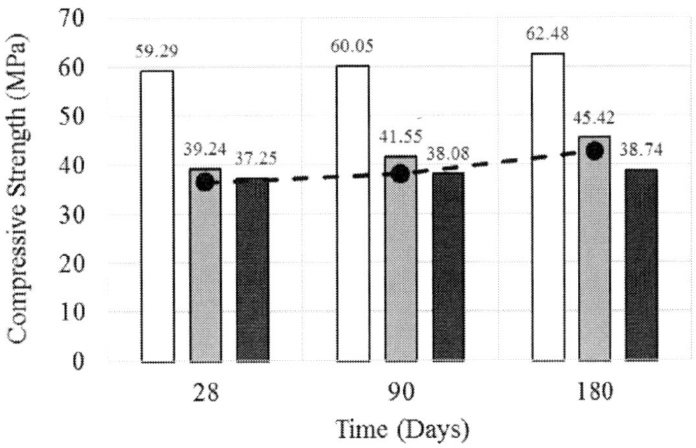

Fig. 6. Compressive strength of concrete subjected to freeze and thaw cycles; %20 SF; 20% SF with 10% crushed bricks; 20% SF with 20% crushed bricks; concrete control specimen.

It is shown in Figure 6 that the use crushed bricks as a replacement material to sand resulted in a great decrease in the compressive strength compared to compressive strength of concrete incorporated with silica fume only. While the compressive strength of concrete with 10 % crushed bricks were almost equivalent to the compressive strength of control specimen, compressive strength of concrete with 20% replacement level of crushed bricks were less than that of the control. The results suggest that the use of crushed bricks are less effective in enhancing the freeze and thaw resistance of concrete than the marble dust. In both cases 10% replacement level provided a better resistance to freeze and thaw cycles of concrete.

5 Conclusion

The paper investigates the use of waste materials namely marble dust and crushed bricks on the strength development and durability of concrete. Marble dust and crushed bricks are used as sand replacement materials and 20% silica fume is also used as a cement replacement material for the preparation of concrete specimens. Replacement levels of the waste materials were 10 and 20%. The results indicated that the increase in the waste replacement materials resulted in an increase water: binder ratio to attain the constant slump for concrete. The finer particle size of the waste materials were attributed for the increase in water content in order to attain the consistency of concrete samples constant. It must be noted that the use of both marble dust and crushed bricks enhanced the compressive strength compared to the control concrete. It must also be underlined that compressive strength of concrete incorporated with silica fume is decreased with the increase replacement levels of waste materials. Although the use of waste materials decreased the water penetration depth compared to the control specimen, the better resistance to water permeability is attain in concrete samples that were incorporated with silica fume. The results indicate the use of marble dust enhanced the freeze and thaw resistance compared to the control concrete however, no significant improvement is achieved with the replacement materials of crushed bricks. It must also be noted that the incorporation of silica fume alone, resulted in a greater resistance to freeze and thaw resistance of concrete. Therefore, these waste materials such as marble dust and crushed bricks do provide positive impact by reducing the burden of construction and demolition waste as they can be utilized in concrete manufacturing. Furthermore, the significance of improved mechanical properties and durability characteristics by using these waste materials along with silica fume contributes towards the field of sustainable construction.

Authors would like to thank Mr. Arsen Ancı, Director of Izmer Enterprises Ltd. and Mr. Necat from Gürdağ Madencilik Ltd. for material supply of marble dust and crushed bricks necessary for the experimental work of this study.

References

1. Vaidya, D., Performance of Marble Waste as Partially Replacement of Sand and Aggregate in Concrete. Imperial Journal of Interdisciplinary Research, **2**(7) (2016)
2. Bektas, F., Wang, K. and Ceylan, H., Effects of crushed clay brick aggregate on mortar durability. Construction and Building Materials, **23**(5), 1909-1914 (2009)
3. Behnood, A. and Ziari, H., Effects of silica fume addition and water to cement ratio on the properties of high-strength concrete after exposure to high temperatures. Cement and Concrete Composites, **30**(2), 106-112 (2008)
4. Topcu, I.B., Bilir, T. and Uygunoğlu, T., Effect of waste marble dust content as filler on properties of self-compacting concrete. Construction and Building Materials, **23**(5), 1947-1953 (2009)
5. Ye, G., Liu, X., De Schutter, G., Poppe, A.M. and Taerwe, L., Influence of limestone powder used as filler in SCC on hydration and microstructure of cement pastes. Cement and Concrete Composites, **29**(2), 94-102 (2007)

6. Poppe, A.M. and De Schutter, G., Cement hydration in the presence of high filler contents. Cement and Concrete Research, **35**(12), 2290-2299 (2005)
7. Assie, S., Escadeillas, G. and Waller, V., Estimates of self-compacting concrete 'potential'durability. Construction and Building Materials, **21**(10), 1909-1917 (2007)
8. Domone, P.L., Self-compacting concrete: An analysis of 11 years of case studies. Cement and Concrete Composites, **28**(2), 197-208 (2006)
9. Corinaldesi, V., Moriconi, G. and Naik, T.R., Characterization of marble powder for its use in mortar and concrete. Construction and Building Materials, **24**(1), 113-117 (2010)
10. Bonavetti, V., Donza, H., Menendez, G., Cabrera, O. and Irassar, E.F., Limestone filler cement in low w/c concrete: a rational use of energy. Cement and Concrete Research, **33**(6), 865-871 (2003)
11. Yahia, A., Tanimura, M. and Shimoyama, Y., Rheological properties of highly flowable mortar containing limestone filler-effect of powder content and W/C ratio. Cement and concrete Research, **35**(3), 532-539 (2005)
12. Mansur, M.A., Wee, T.H. and Lee, S.C., Crushed bricks as coarse aggregate for concrete. Materials Journal, **96**(4), 478-484 (1999)
13. Cavalline, T.L. and Weggel, D.C., 2013. Recycled brick masonry aggregate concrete: Use of brick masonry from construction and demolition waste as recycled aggregate in concrete. Structural Survey, **31**(3), 160-180 (2013)
14. ASTM C150/C150M-16e1 Standard Specification for Portland Cement, ASTM International, West Conshohocken, PA (2016)
15. ASTM C1240-15 Standard Specification for Silica Fume Used in Cementitious Mixtures, ASTM International, West Conshohocken, PA (2015)
16. ASTM C33/C33M-16e1 Standard Specification for Concrete Aggregates, ASTM International, West Conshohocken, PA (2016)
17. ASTM C127-15 Standard Test Method for Relative Density (Specific Gravity) and Absorption of Coarse Aggregate, ASTM International, West Conshohocken, PA (2015)
18. ASTM C128-15 Standard Test Method for Relative Density (Specific Gravity) and Absorption of Fine Aggregate, ASTM International, West Conshohocken, PA (2015)
19. ASTM C143/C143M-15a Standard Test Method for Slump of Hydraulic-Cement Concrete, ASTM International, West Conshohocken, PA (2015)
20. BS EN 12390-3. Testing hardened concrete. Compressive strength of test specimens. London: BSI (2002)
21. BS EN 12390-8. Testing hardened concrete. Depth of penetration of water under pressure. London: BSI (2009)
22. Khoury, N.N. and Zaman, M.M., Environmental effects on durability of aggregates stabilized with cementitious materials. Journal of Materials in civil Engineering, **19**(1), 41-48 (2007)

The use of a concrete additive to eliminate returned concrete waste volumes

Johannes Bester[1,*], *Deon* Kruger[2] , and *Broadley* Miller[3]

[1]University of Johannesburg, PO Box 524, Auckland Park, 2006, Johannesburg, South Africa [2]University of Johannesburg, PO Box 524, Auckland Park, 2006, Johannesburg, South Africa
[3]University of Johannesburg, PO Box 524, Auckland Park, 2006, Johannesburg, South Africa

Abstract. This paper investigates the effects of the use of a recently developed two-component powdered product made from polymers and inorganic compounds that can be mechanically mixed into returned fresh ready-mix concrete to allow for the separation of the concrete into fine and coarse aggregates. This allows for the re-use of the returned concrete as aggregates in the manufacturing of new concrete. The returned concrete waste can therefore be eliminated, thus reducing virgin aggregate usage, as well as reducing the environmental impact of returned concrete. In this study, the treated recycled fresh concrete was separated into fine and coarse aggregates, and then used at replacement levels of 0%, 25%, 50%, 75% and 100%. The effect of the product on the material classification, and on important fresh and hardened properties of the concrete for the above-mentioned replacement values was tested. For the fine aggregate, the results indicate minimal changes in both the fresh and hardened properties. For the coarse aggregate, the results show a marked improvement of flexural strength with an increase in replacement value when coarse aggregates are used. Very high replacement levels may be used with very little effect on the quality of the new concrete.

1 Introduction

A study conducted by Obla et al for the RMC Research & Education Foundation found that there are four ways in which returned concrete can be used, of which one way is to discharge the concrete at the concrete plant, allow the concrete to harden, and then crushed. The crushed material can then be used as fill or as base for pavements. The report further states that the fines proofs to be a significant challenge for the producer to dispose off [1]. This leaves the question whether returned concrete, in the fresh state, could be utilized, rather than allowing the returned concrete to harden.

Returned concrete from construction sites have for long been a problem in South Africa (SA). This product have typically been send of to landfill sites, or has been treated in settling ponds. Both of these methods are not environmentally desirable. In addition, these methods are not cost effective, and are in conflict with municipal environmental regulations [2]. The

* Corresponding author: jannesb@uj.ac.za

triple bottom line theory of sustainable development consists of three parts, namely enonomic, environment- and social factors, as shown in Figure 1 [3]. It is important that a balance is achieved between these three factors, to achieve an increase in sustainability. This having been said, any small contribution from any of these factors will have an effect on sustainability.

A new two-component powdered product made from special polymers and inorganic compounds has entered the market in South Africa (SA). This product allows concrete returned to mixing plants, at the end of the day, to be completely and quickly recovered directly from truck mixers. The product, when added to concrete whilst still in its plastic state, first absorbs the water in the concrete mix and rapidly sets it, to transform the concrete into a granular material with excellent mechanical characteristics which, may be used to partially replace natural aggregates.

Fig. 1. Triple bottom line [4].

Returned concrete (Ret. Con.) is the unused portion left over in the ready-mix truck after delivery and use by the client. This material, when still in the fresh concrete state, is currently not being utilized by most ready-mix concrete suppliers and causes major waste streams for proper disposal of large amounts of water used to wash out the agitator and aggregate that is merely landfilled or used as fill or base material. The water after washing needs to be properly processed to remove the hazardous components before it can be discharged from the site. This waste stream therefor has great potential to reduce operating costs of ready-mix suppliers if the material could be more efficiently utilised as well as offering them the opportunity to reduce their environmental impact on the surrounding environment.

The objective of this research project is to prove that returned concrete, a common wasted material, can be successfully recycled to be used in concrete to replace or partially substitute the virgin aggregate needed in the production of new concrete, thus reducing the need for as much virgin material and also reducing the waste stream destined for landfilling by finding an alternate use for the waste stream generated in the construction industry.

2 Experimental design

The aggregate was produced from the treatment of returned concrete from a ready-mix vehicle, to test the effects of its use as specified by the manufacturer. After treatment and discharge of the material from the ready-mix vehicle, the aggregate was air-dried until use as aggregate replacement in the test concrete mixes.

 DOI: 10.1051/matecconf/201712003010

A control mix is designed and made using the procedure set out in the Fulton's Concrete Technology handbook, [5], using a specified slump, water-cement ratio and readily available crushed stone and sand aggregate used in typical ready-mix concretes mixes supplied by the Afrisam Roodepoort Plant. The control strength aimed for, and achieved, was 20MPa.

3 Results

3.1 Fresh properties

3.1.1 Slump values

The slump values of the treated returned concrete, where the sand portion was used in the new mixes, for all the sand replacement values remained mainly unchanged when compared to the control mix, as can be seen in Figure 2. This could be that the properties of the more round shape of the particles counteract the higher water absorption rates expected from the returned concrete aggregates. The same was observed for the stone portion, as can be seen in Figure 3.

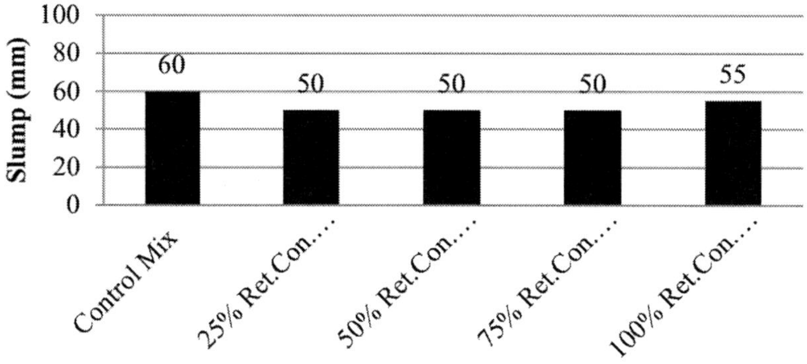

Fig. 2. Slump values for the returned concrete sand aggregate concrete.

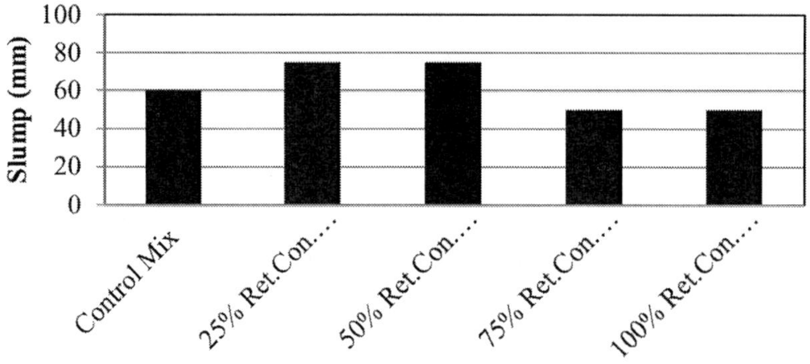

Fig. 3. Slump values for the returned concrete stone aggregate concrete.

This was surprising, as the manufacturer of the product stated that concrete mixes containing the recycled material as aggregate at a 50% replacement level would require around 10ℓ per 1 m³ concrete to yield the same workability as a concrete mix containing only virgin aggregates, yet the workability slightly increased at a 50% replacement with the same water content.

3.1.2 Slump retention

The slump retention results of the mixes containing both the returned concrete sand and stone material, shown in Figure 4 and Figure 5, had much the same slopes as that of the reference mix. This again was interesting as the material was expected to have higher water absorption from the lower relative density than that of the reference material, which should have caused a reduction in the slump retention experienced. Therefore, as with returned concrete sand and stone material, it can be assumed that the use of the returned concrete sand material as aggregate in concrete would have little effect on the slump retention properties of the concrete made therewith.

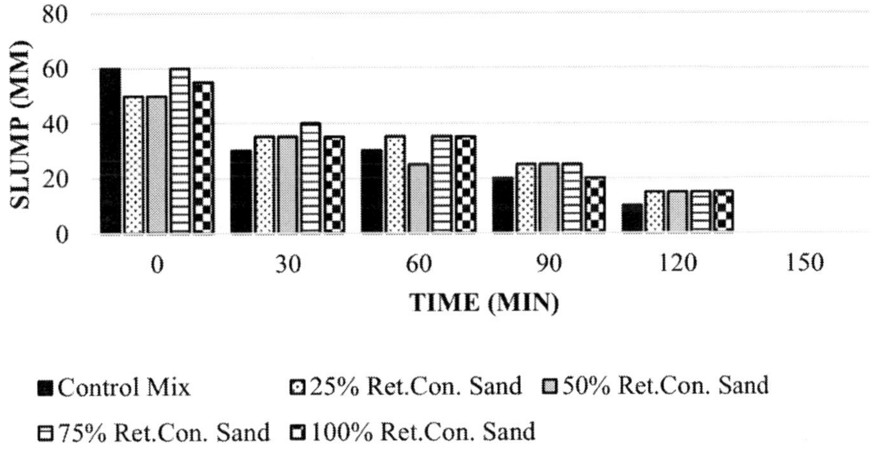

Fig. 4. Slump retention for the returned concrete sand aggregate concrete.

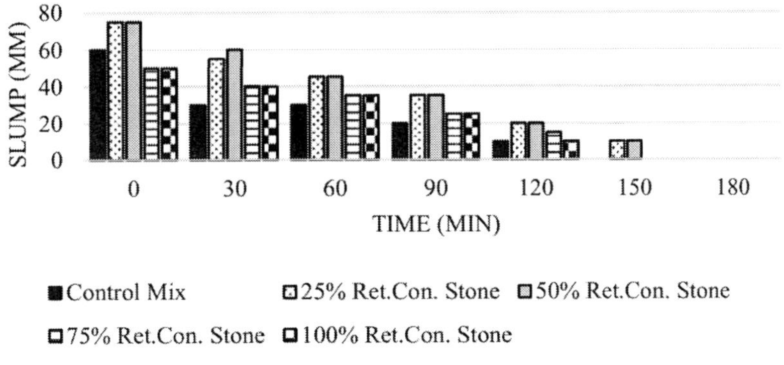

Fig. 5. Slump retention results for the returned concrete stone aggregate concrete.

3.1.3 Setting time

The returned concrete stone material setting time results, seen in Figure 6, showed a slight increase in the setting times at the 25% replacement mix and thereafter a trend of decreasing setting times as the replacement value increased. The setting times of the 50% replacement concrete however could be presumed to be a slight outlier as the setting times achieved by this mix is slightly lower than expected if the trend is to be believed. The setting times of the returned concrete sand material (Figure 7), showed a marked reduction in the setting times achieved by the mixes containing the returned concrete sand material compared to that of the control material. What was interesting to note was that, although the initial setting time showed a trend of reduced set setting times as the level of replacement increased, the final set time seemed to remain unaffected by the level of replacement. This was unexpected as the material was thought to have a large degree of un-hydrated cement particles present that should have caused the setting phase time to decrease.

The reduction in the initial setting times is presumed to be as a result of the higher water absorption properties of the returned concrete sand material taking water out of the concrete mix, effectively reducing the workability, however, this seems to conflict with the slump retention results shown earlier, that revealed that the concretes containing the returned concrete sand material as aggregate had much the same slump retention properties as that of the control mix.

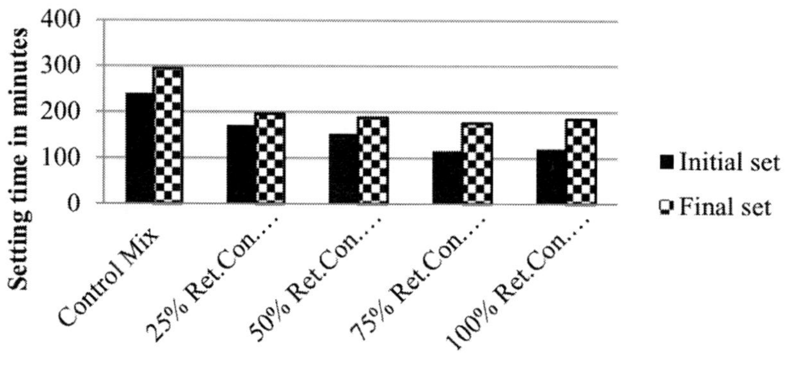

Fig. 6. Setting time results for the returned concrete sand aggregate concrete.

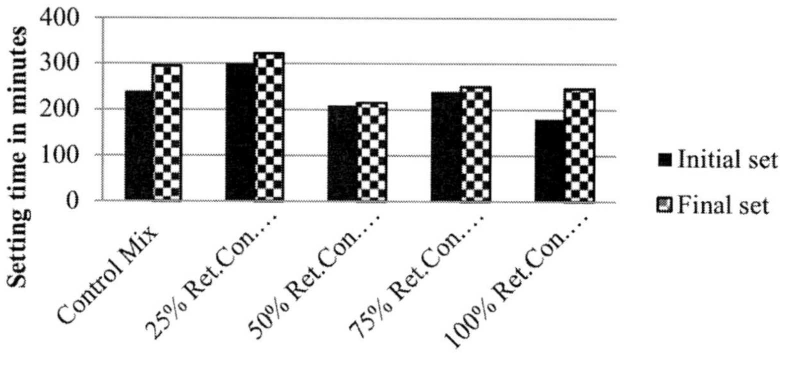

Fig. 7. Setting time results for the returned concrete stone aggregate concrete.

3.1.4 Bleeding

It is important to note that the bleed volume results in this study were only the bleed volumes measured on the surface of the concrete after the 280 minute test period, and not the total bleed volume of the concrete. This is because weight measurements of the test samples after being oven-dried were not done and therefore, the total bleed volumes could not be calculated.

The bleed volumes achieved by the concrete mixes containing the returned concrete stone material, shown in Figure 8, yielded much lower bleed volumes than that of the reference mix. The mixes also show a slight trend toward lower bleed volumes as the level of replacement increased. The reason expected for this large difference in bleed volumes experienced is due to the higher water absorption of the material absorbing free water in the mix.

The concrete sand material, Figure 9, had an even larger effect on the bleeding than that of the returned concrete stone material which was unexpected as the lower amount of very fine particles present in the returned concrete sand material compared with that of the reference material was expected to increase the bleeding experienced. However, it is believed that the recycled material's higher absorption caused it to absorb the bleed water from the mix, thus effectively reducing the bleed amounts on the surface of the concretes.

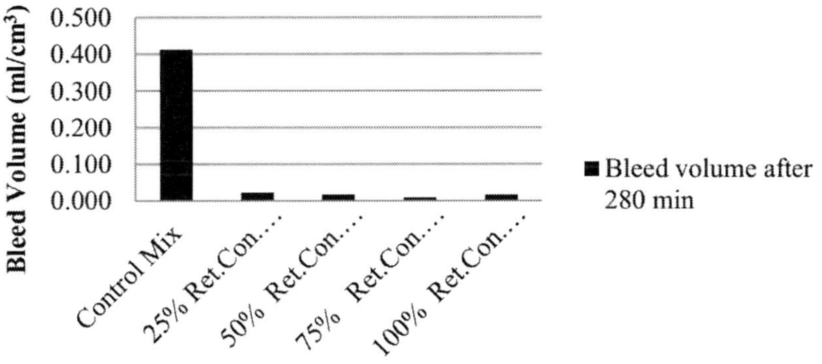

Fig. 8. Bleed volume results for the returned concrete sand aggregate concrete.

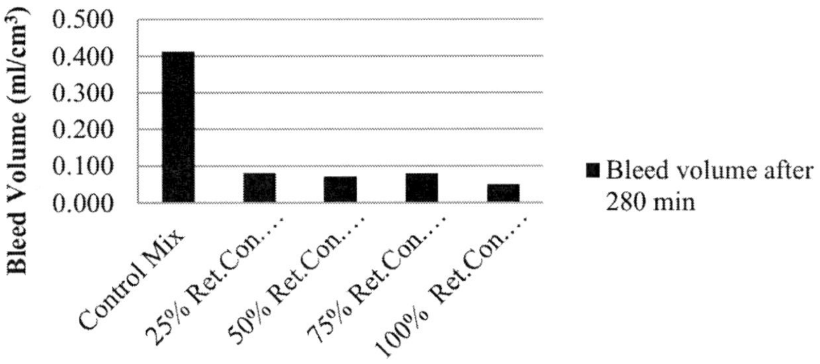

Fig. 9. Bleed volume results for the returned concrete stone aggregate concrete.

3.2.1 Compressive strength

The compressive strengths achieved by both the returned concrete sand and stone material, as shown in Figures 10 and 11, were marginally lower than that of the reference concrete, yielding less than 10% loss in compressive strength at a 25% replacement value and the material continues the trend with RCA of a decreasing compressive strength as the level of replacement increased.

The loss in compressive strength is at odds with the manufacturer of 'Product A' claims that states although it retards the compressive strength gain initially, in that the material actually yields a marginally higher compressive strength after 28 days than that of a reference mix containing only virgin aggregates. The reason for this loss in compressive strength could be as a result of the original concrete's low compressive strength, as it was a specified 20 MPa concrete, as stated earlier. This lower strength original concrete could have caused the attached cement mortar present on the returned concrete stone material to be of a lower strength, thus negatively affecting the strength capabilities if the concretes made therewith.

The concrete's loss in strength seemed to be much less reactive to the increase in the level of replacement, than that of the lab waste RCA and almost seems at odds with the manufacturer's recommendations not allowing replacement values of more than 50%.

Although the compressive strength of the concretes decreased with the replacement of virgin aggregate with that of the returned material, the reduction in compressive strengths achieved at low levels of replacement is quite small, i.e. 25% replacement having a loss of less than 10% of its compressive strength.

The almost perfect distribution of results could also be an indication that the material is consistent in its properties. Therefore, it is expected that the effects of the lab waste material on concretes made therewith, can be controlled and therefor mitigated with proper design considerations.

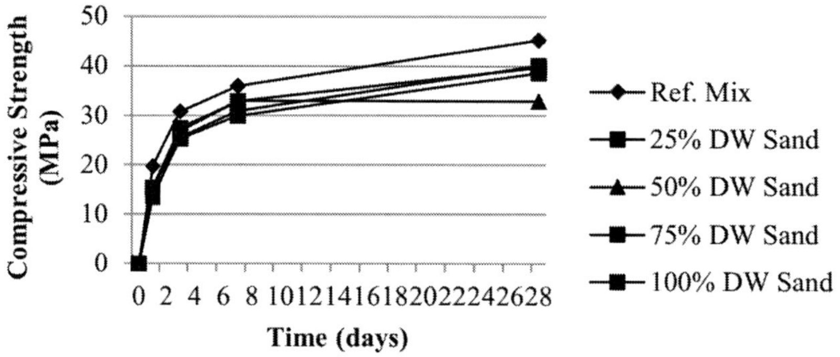

Fig. 10. Compressive strength results for the returned concrete sand aggregate concrete.

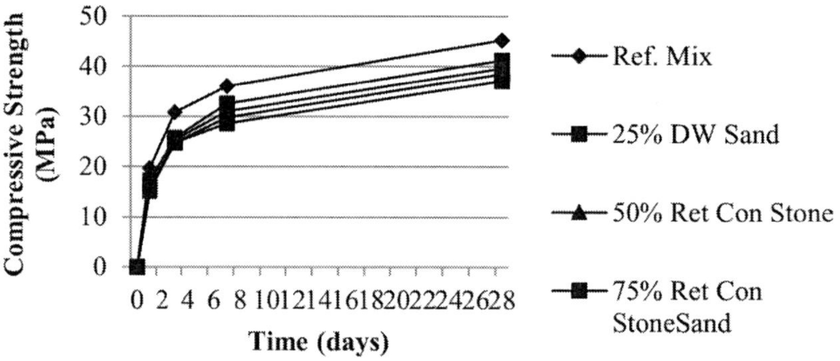

Fig. 11. Compressive strength results for the returned concrete stone aggregate concrete.

3.2.2 Flexural strength

The flexural strengths achieved by the returned concrete stone material, shown in Figure 12, impressed the most, consistently outperforming the reference mix at any replacement level. The returned concrete stone material again showed a trend of decreasing flexural strengths achieved as the level of replacement increased, as expected when the compressive strengths achieved are taken into account.

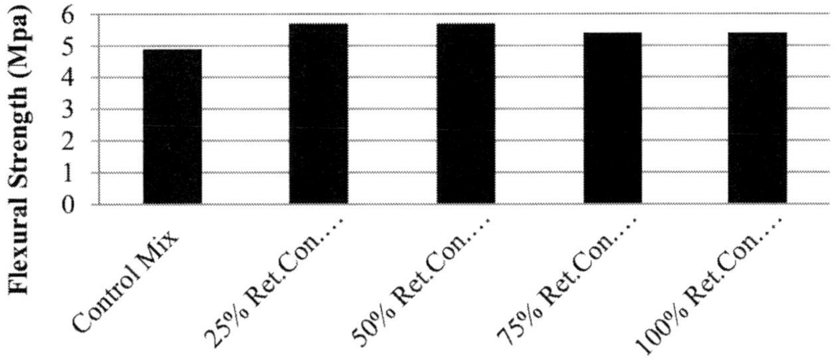

Fig. 12. Flexural strength results for the returned concrete stone aggregate concrete.

The reference concrete achieved a flexural strength to compressive strength ratio of around 1:10, as expected from most concretes containing virgin aggregates in SA. However, the mixes containing the returned concrete stone RCA's yielded ratios of around 1.4:10. This higher flexural strength is due to the improved interfacial bond strength between the aggregate and the cement mortar paste, caused by the rougher texture of the RCA compared with that of the virgin aggregate.

The flexural strength results achieved by the returned concrete sand material, shown in Figure 13, were marginally lower than that of the reference concrete and surprisingly showed a trend of increasing flexural strengths as the level of replacement increased, possibly caused by the better interfacial bond strength of the recycled aggregates to the new cement mortar due to the rougher texture.

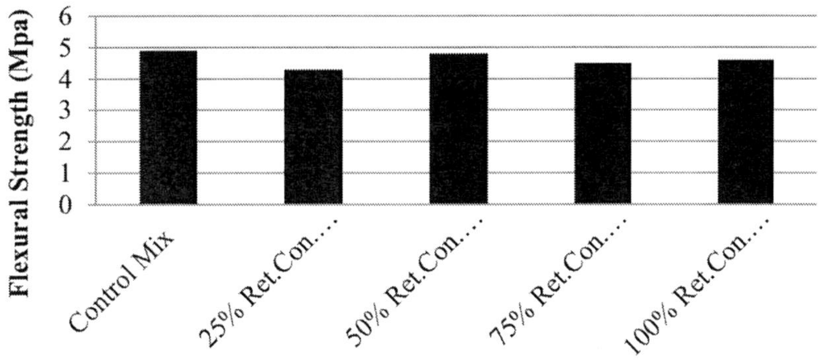

Fig. 13. Flexural strength results for the returned concrete sand aggregate concrete.

The 50% replacement mix again tested better than expected, as with the compressive strengths showed earlier, yielding a higher flexural strength than the other replacement levels. However, when compared with the trend produced by the other mixes this much higher value could be considered as an outlier.

The ratios of flexural strength to compressive strength tested almost exactly the same as the reference mix with values around 1.1:10 and the 100% replacement level concrete achieving the best result at almost 1.2:10. This means that the returned concrete sand material reacts in much the same way as the reference aggregate and thus would not adversely affect the concrete flexural properties if it were used as aggregate.

4 Conclusions

From the results obtained in this research project, it can be seen that the slump for both the sand and stone replacement, at any replacement values, was not affected. The slump retention was also largely not affected. The initial set and final set was shorter than for the control mix. There is a massive reduction in the bleed volume for both the sand and stone replacement mixes, at all the replacement values. This could cause the fresh concrete to undergo plastic shrinkage cracking, and should be taken into consideration in the mix design, as well as in site practice.

In addition, the hardened properties of the concrete when the returned concrete was used, showed a slight decrease when both the stone and sand fraction was used. This is evident in all the replacement values. When the flexural strength is considered, there is an increase in the flexural strength of the concrete that was made using the stone fraction of the treated concrete, but a very slight decrease in the sand fraction.

The results from this study indicates that the two-component powdered product may be used on returned concrete, in order to allow the returned concrete to be used in new concrete, with minimal changes to the mix design.

References

1. K. Abla, H. Kim, C. Lobo, *Crushed returned concrete as aggregates for new concrete.* Final report to the RMC Research & Education Foundation. Project 05-13. (2007)
2. http://www.leadingarchitecture.co.za/recover-returned-concrete-within-minutes. Date accessed: 2016/12/07.
3. G. Owens, (Editor). *Fundamentals of Concrete.* Midrand. The Concrete Institute, pp.8. (2013).
4. http://www.theconcreteinstitute.org.za/in-perspective. Date accessed: 2016/12/15.
5. B. Addis, J. Goodman. *Concrete mix design. Chapter 11 in Fulton's concrete technology.* Midrand: The Cement & Concrete Institute, pp. 219-224.

Enhancement of concrete brick with marble waste

Zainelabdin Khalid Zainelabdin Elfaki*

Mining House Company, General management, Riyadh, Saudi Arabia

Abstract -This experimental study of the reinforcement of concrete brick with marble waste aims to show the influence of partial replacement of cement with marble waste on the workability and compressive strength of concrete. Three replacement percentages of 10%, 20% and 30% were used with different curing times of 7, 14and 28 days. For the optimum results and in order to reduce the effect of human errors on the study accuracy, two specimens were prepared for each percentage and curing time. For this study, marble powder from Mining House Company in Saudi Arabia was used. After doing all the experiment works it was observed that the replacement of Concrete brick with marble waste enhanced the loading strength and the workability of prepared concrete bricks. It was also observed that the 20% replacement of cement with marble waste is the best replacement percentage compared to the other replacement percentages. Marble waste from manufacturing companies either in powder, sludge or slabs form has serious bad effect on the environment. Furthermore, the usage of marble waste in such industries has many other benefits such as economical benefit of selling the marble wastes and saving the natural resources.

1 Introduction

Concrete is a vital ingredient in infrastructure development with its versatile and extensive applications. Concrete is the second largest material consumed by human beings after food and water as per WHO [1] (World Health Organization). As Cement is one of the concrete basic materials, today, utilization of blended cements is usually preferred due to their economical and Technical benefits and indirect advantages such as their ability of decreasing CO2 emissions by reducing clinker production in plants [2]. The world cement production has been increasing constantly since the early 1950s, especially in developing countries. World cement production was about 2100 million tons in 2004, with a yearly growth of about 7–8%; it was doubled in less than 20 years. Asia has been the driving force with China and India currently the main cement producers (respectively: 930 million and 128 million tones or 44% and 6%, respectively, of world production) [3].

Leaving the waste materials to the environment directly can cause environmental problem. Hence, many countries have been working on how to reuse the waste material so that they reduced hazards to the environment). The developed countries have strict rules to protect the environment whereas many developing countries have almost no rules to protect the environment against wastes. However, wastes can be used to produce new products or

* Corresponding author: zain998@yahoo.com

can be used as admixtures so that natural sources are used more efficiently and the environment is protected from waste deposits. During the cutting process about 25% marble is resulted in dust [4].

One of the logical means for reduction of the waste marble masses calls for utilizing them in building industry itself. Some attempts have been made to find and assess the possibilities of using waste marble powder in mortars and concretes and results about strength and workability were compared with control samples of conventional cement-sand mortar/concrete. When marble powder is partially replaced in cement by weight, there is a marked reduction in compressive strength values of mortar mix with increasing marble powder content when compared with control sample at each curing age. Surprisingly, on increasing marble waste fine aggregate ratio i.e. when marble waste / granules are partially replaced in fine aggregate by weight then there is increase in compressive strength values of marble waste mortar at each curing age [5].

2 Experimental procedure

2.1 Materials

For this study, three replacement percentages of 10%, 20% and 30% were used with different curing times of 7, 14and 28 days. For the optimum results and in order to reduce the effect of human errors on the study accuracy, two specimens were prepared for each percentage and curing time.

2.1.1 Aggregates

Fine aggregates
Natural fine aggregates easily available, cheap price and widely used in construction were used.
Maximum size of 4 mm fine aggregates with 2775 kg/m3 density was used for all specimen preparation. Fine aggregates take 2/7 or 28.57 % of the concrete weight as we are using the 1, 2 and 4 mixing design.
Course aggregates
Natural course aggregates or Natural gravels (NG) commonly used for building applications were used.
Maximum size of 16 mm with 2733 kg/m3 density Course aggregates was used. 4/7 or 57.14 % of concrete prepared is Course aggregates.

2.1.2 Cement

Normal grade Ordinary Portland cement (OPC) was used for all samples preparation. Table 1 shows the physical and chemical composition of OPC.

2.1.3 Marble waste

Dry powder marble waste was used from Mining House Company Riyadh, Saudi Arabia. Figure 1 shows the marble waste generated at marble factories.

Fig. 1: Marble waste in factory.

Table 1: Physical properties and chemical composition of OPC.

Test	Unit	British Standard	Result
Chemical Composition			
Lime Saturation factor (LSF)	%	0.66-1.2	0.94
Insoluble Residue (IR)	%	5 max.	0.2
Magnesium Oxide (MgO)	%	5 max.	1.5
Sulphate Contest (SO_3)	%	3.5max.	3
Loss In Ignition (LOI)	%	5max.	2
Chloride (Cl)	%	0.1 max.	0.01
Physical Properties			
Specific Surface Area	M^2/kg	225 min.	347
Soundness	mm	10 max.	0.7
Setting time- Initial	Mins.	45 min.	135
Setting time- Final	Mins.	600 max.	195
Compressive Strength for Concrete Cubes (1:2:4:0.6) 3 days	N/mm^2	Not less than 11.5	21.8
Compressive Strength for Concrete Cubes (1:2:4:0.6) 28 days	N/mm^2	Not less than 26	43.3

2.2 Methods

Marble waste chemical composition

The chemical composition of marble waste samples used was studied by using XRD (X-Ray Diffraction) analyzing device. Elements that marble waste consists of determines its chemical and physical properties which effects the mechanical

Compressive Strength

Compressive strength of concrete is the most important characteristic property which also affects other concrete properties. Water to Cement ratio and curing age are the main important factors influencing Compressive strength besides compaction. By measuring the load of concrete specimens, we can calculate the value of compressive strength either in Megapascal (MPa) or Pounds per square inch (psi).

Workability

Workability is considered to be that property of concrete that determines its capacity to be placed and consolidated properly and to be finished without harmful segregation. It embodies such concepts as moldability, cohesiveness, and compactability.
Workability is affected by: the grading, particle shape, and proportions of aggregate; the amount and qualities of cement and other cementitious materials; the presence of entrained air and chemical admixtures; and the consistency of the mixture. Procedures in this Standard Practice permit these factors to be taken into account to achieve satisfactory placeability economically.

2.2.1 Preparation of Specimens

A replacement of the Cement quantity by 10, 20 and 30 % of marble waste and testing the effect of this replacement to the mechanical properties of the concrete made. Table 2 shows the specimens prepared with relevant details

Table 2: All specimens prepared with relevant details.

	Specimen Number	Replacement percentage (%)	To be cured for (days)	W/C
1	S-10-7-1	10	7	0.55
2	S-10-7-2	10	7	0.55
3	S-10-14-1	10	14	0.55
4	S-10-14-2	10	14	0.55
5	S-10-28-1	10	28	0.55
6	S-10-28-2	10	28	0.55
7	S-20-7-1	20	7	0.55
8	S-20-7-2	20	7	0.55
9	S-20-14-1	20	14	0.55
10	S-20-14-2	20	14	0.55
11	S-20-28-1	20	28	0.55
12	S-20-28-2	20	28	0.55
13	S-30-7-1	30	7	0.55
14	S-30-7-2	30	7	0.55
15	S-30-14-1	30	14	0.55
16	S-30-14-2	30	14	0.55
17	S-30-28-1	30	28	0.55
18	S-30-28-2	30	28	0.55

3 Result and discussion

3.1 Chemical Composition

The XRD analysis shows that marble waste mainly composed of Oxide based elements such as: Silica (SiO_2), Calcium (CaO), Phosphorus (P_2O_5), Magnesium (MgO), Potassium (K_2O), Sodium (Na_2O), Copper (CuO), Manganese (MnO), Zinc (ZnO),Iron (Fe_2O_3), Aluminum (Al_2O_3), Titanium (TiO_2), and Sulphur trioxide (SO_3). The next figures show the results of the analysis by XRD analyzing device.

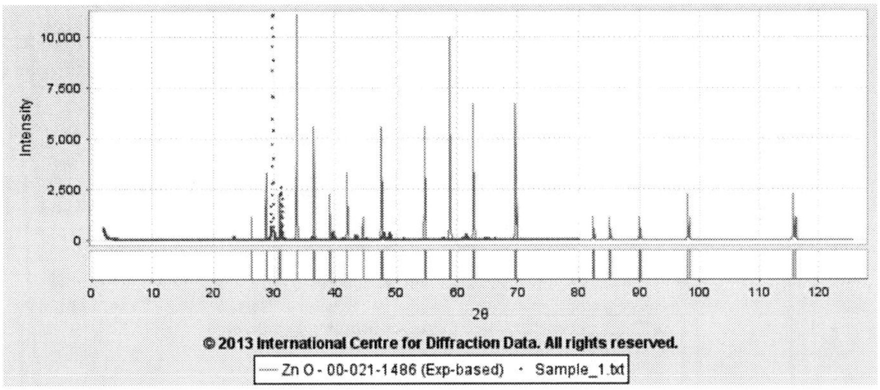

Fig. 2: XRD analysis for ZnO in marble waste sample.

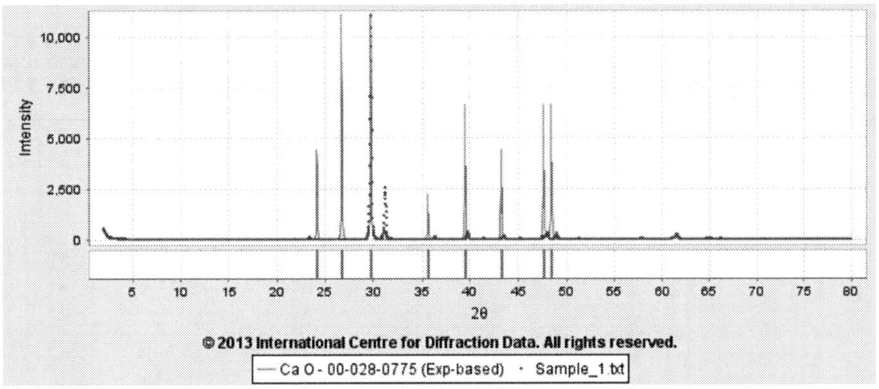

Fig. 3: XRD analysis for CaO in marble waste sample.

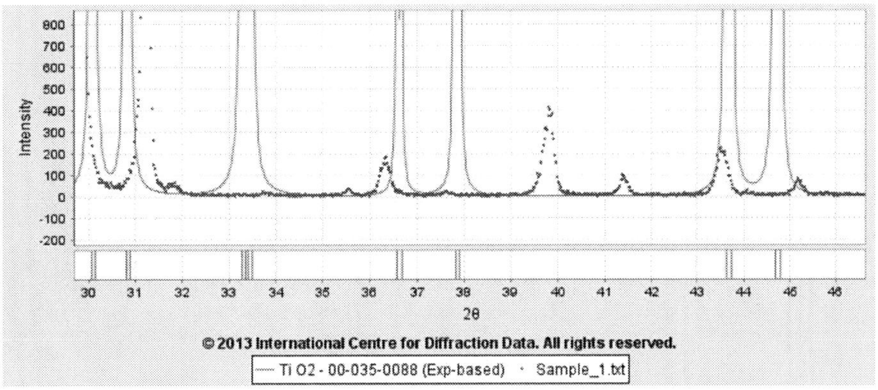

Fig. 4: XRD analysis for TiO_2 in marble waste sample.

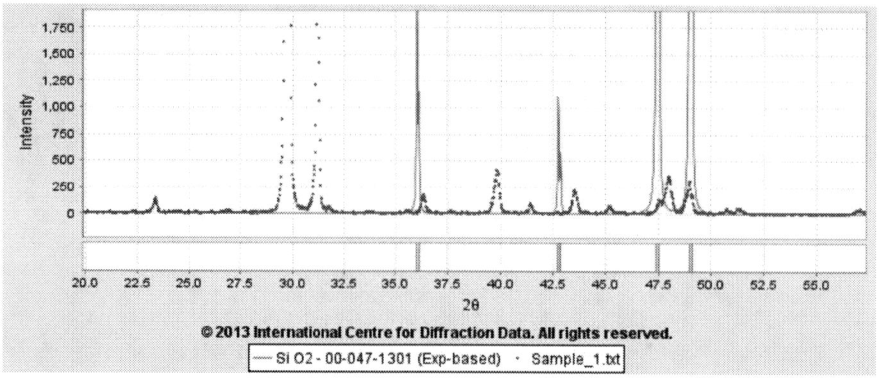

Fig. 5: XRD analysis for SiO_2 in marble waste sample.

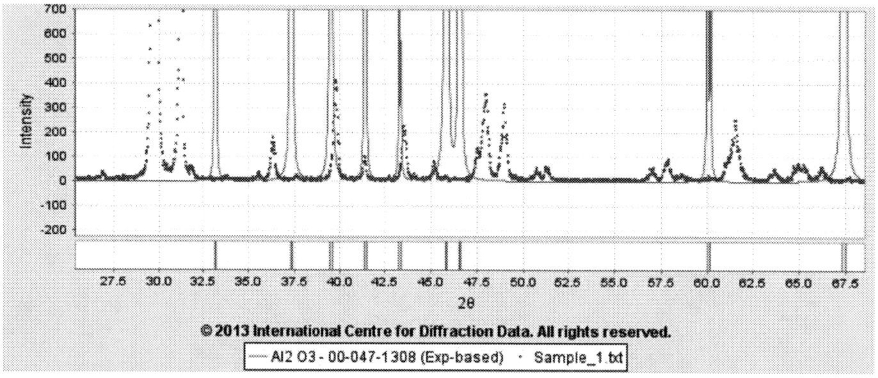

Fig. 6: XRD analysis for Al_2O_3 in marble waste sample.

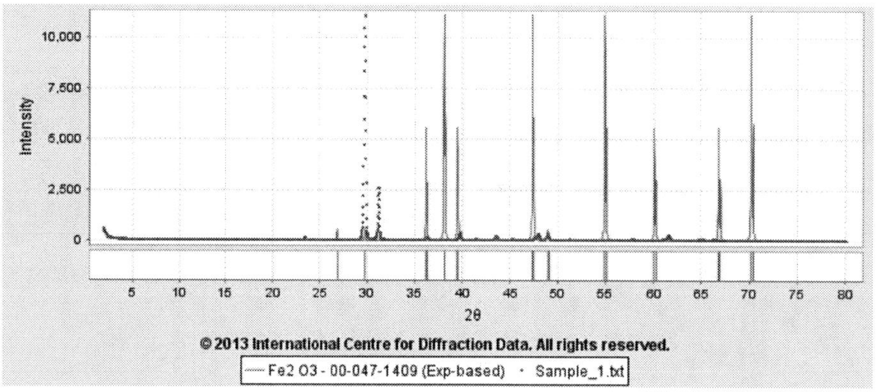

Fig. 7: XRD analysis for $Fe2O_3$ in marble waste sample.

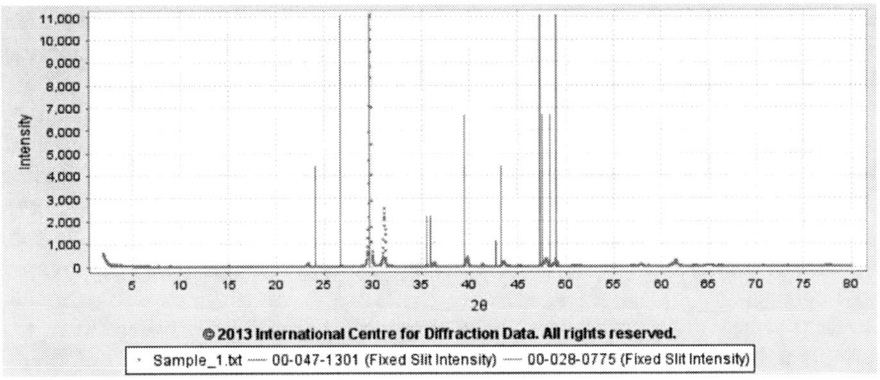

Fig. 8: XRD analysis for CaO and SiO_2 in marble waste sample.

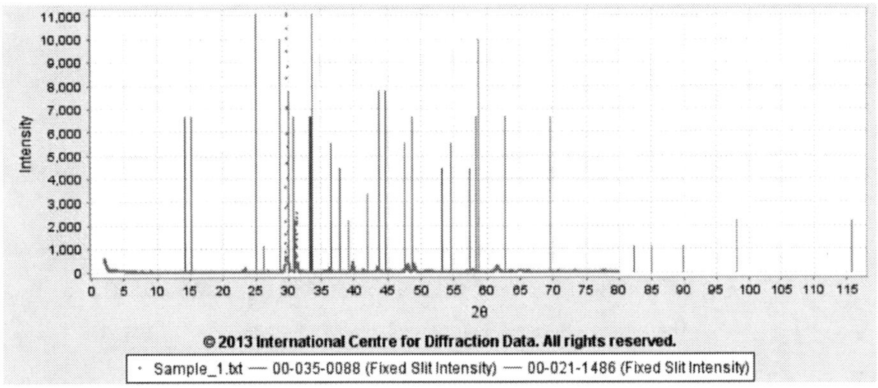

Fig. 9: XRD analysis for TiO_2 and ZnO in marble waste sample..

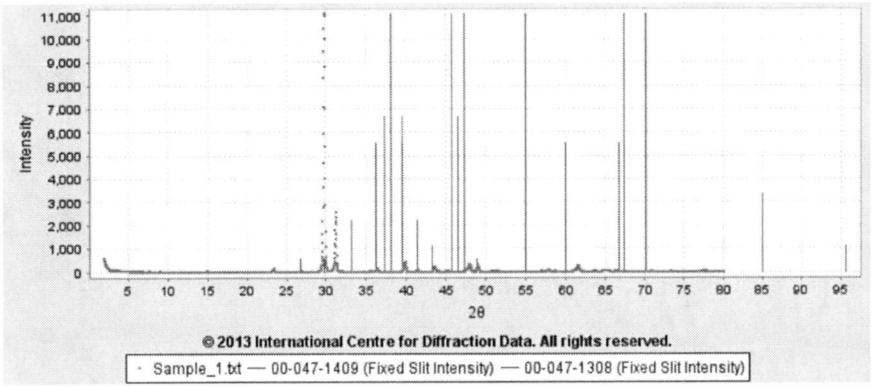

Fig. 10: XRD analysis for Al_2O_3 and Fe_2O_3 in marble waste sample.

3.2 Sieve Analysis of Marble Waste

Importance of sieving in chemical engineering can observed in produces consistent results and reliable. Despite the minimalist design, test sieves are capable for producing repeatable results regardless of when and where to implement each test. Also, because of their simple design, and its very effective data, this type of particle-size analyzer plays an important role in many production facilities, including those found in food processing, pharmaceuticals, cosmetics and chemicals, agricultural and manufacturing industries.

On the other hand, in the sedimentation which is based on the terminal velocity on the liquid particles suspended since the time of deposition is the longest of the minute particles, which means this technique determines the size of the particles as a function of settling.

After experiment have been carried out to analysis the particle size distribution in sieving process. In order to obtain desire product, different aperture has been used in sieving process. It shows that as aperture size of sieve decrease, the cumulative undersize also decrease based on graph obtained.

All eight samples show that porosity is extremely high range between 20 % to 40%. Porosity of the study area changes from the first measured as 0 m to the last point measured as 80 m. as shown in (figure:1.0) it is assumed that porosity of the study area is stable and the lowest porosity is at the middle of the study area which is at 40 m distance.

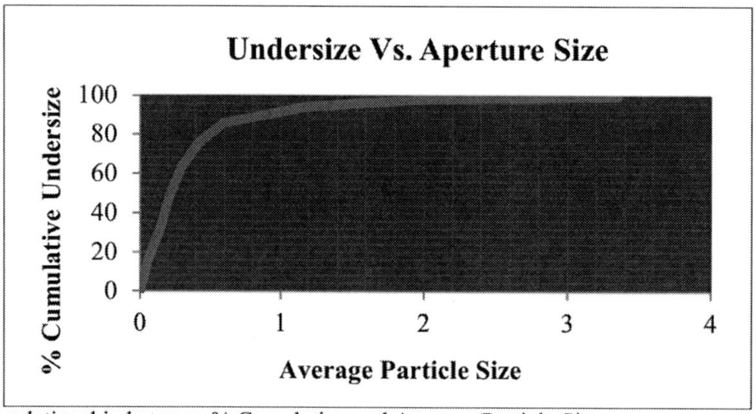

Fig. 11: the relationship between % Cumulative and Average Particle Size

Table 3: Slump test results for fresh concrete made with marble waste replacement

Sample date	Marble Waste replacement %	Course Aggregates (kg)	Fine Aggregates (kg)	Cement (kg)	Water (kg)	W/C	Marble waste (kg)	Slump difference (mm)
13-05-2013	10	30.548	15.274	6.8734	4.2	0.55	0.7637	35
28-05-2013	20	30.548	15.274	6.1095	4.2	0.55	1.527	30
11-06-2013	30	30.548	15.274	5.346	4.2	0.55	2.2911	25

3.3 Slump test

As the amount of cement decreased by the incensement in marble waste quantity, the Slump decreased as shown in Figure 4. The reason behind this is that cement reacts with water more than marble waste does. Furthermore, the size of cement particles is much lower than marble waste which affects the packing density of cement. Table 3 shows the slump test results for fresh concrete made with marble waste replacement.

3.4 Compaction Factor

Compaction factor, measures the degree of compaction achieved when a concrete mixture is subjected to a standard amount of work. The compacting factor of concrete of is used to investigate the Workability of Concrete at low, medium and high workability.

A rising in compaction factor is found as the cement replacement increases. This is due to the reduction in cement amount where the cement has better reactivity with water essentially and also with aggregates.

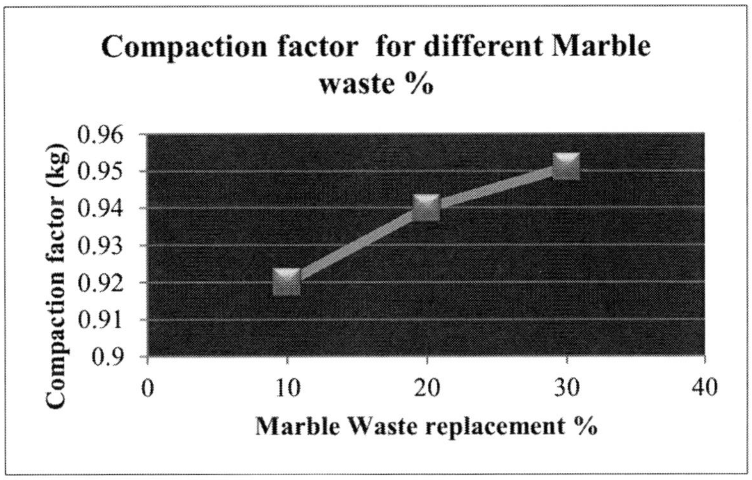

Fig. 12: Compaction factor for different marble waste replacement percentages.

3.5 Compressive Strength

Compressive strength of concrete is the most important characteristic property which also affects other concrete properties. Water to Cement ratio and curing age are the main important factors influencing Compressive strength besides compaction. By measuring the load of concrete specimens, we can calculate the value of compressive strength either in Megapascal (MPa) or Pounds per square inch (psi).

For 10 % replacement:

As the substitution percentage is low, no much difference is observed to the results compared to concrete with no marble waste replacement. However, for 14 and 28 days we found almost similar data due to curing conditions. Table 4 shows all measured data while figure 7 present results for easier observation.

For 20 % replacement:

High Compressive strength is observed even with 20% replacement. This result encourages future works to be done for industry usage. As marble waste quantity increased in concrete mixture it enhances the concrete hardness and this results in good loading properties and thus, such cubes meet the construction requirements.

Table 4: Compressive strength data for 10% marble waste.

Specimen	Curing days	sample Curing start date	sample Curing end date	Load (KN)	Compressive Strength (Mpa)
S-10-7-1	7	14/5/2013	21/5/2013	375	16.667
S-10-7-2	7	14/5/2013	21/5/2013	350	15.556
S-10-14-1	14	13/6/2013	27/6/2013	350	15.556
S-10-14-2	14	13/6/2013	27/6/2013	550	24.444
S-10-28-1	28	14/5/2013	11/6/2013	440	19.556
S-10-28-1	28	14/5/2013	11/6/2013	460	20.444

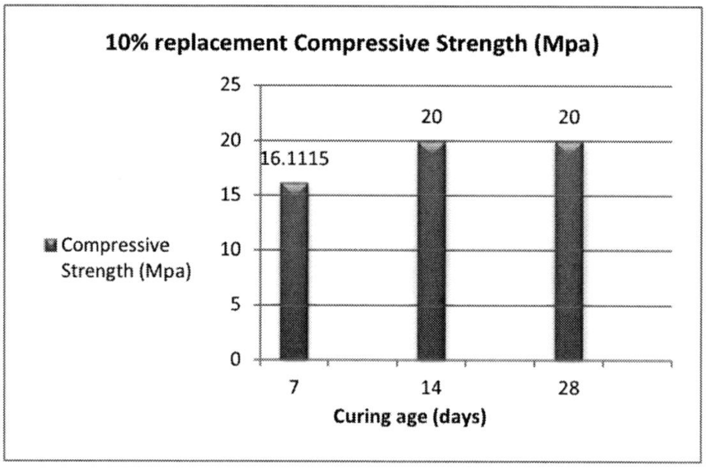

Fig. 7: Compressive strength for Marble waste with 10 % replacement.

Table 5: Compressive strength data for 20% marble waste.

Specimen	Curing days	sample Curing start date	sample Curing end date	Load (KN)	Compressive Strength (Mpa)
S-20-7-1	7	29/05/2013	5/6/2013	320	14.222
S-20-7-2	7	29/05/2013	5/6/2013	350	15.556
S-20-14-1	14	29/05/2013	12/6/2013	420	18.667
S-20-14-2	14	13/06/2013	27/06/2013	375	16.667
S-20-28-1	28	29/05/2013	26/06/2013	480	21.333
S-20-28-2	28	29/05/2013	26/06/2013	430	19.111

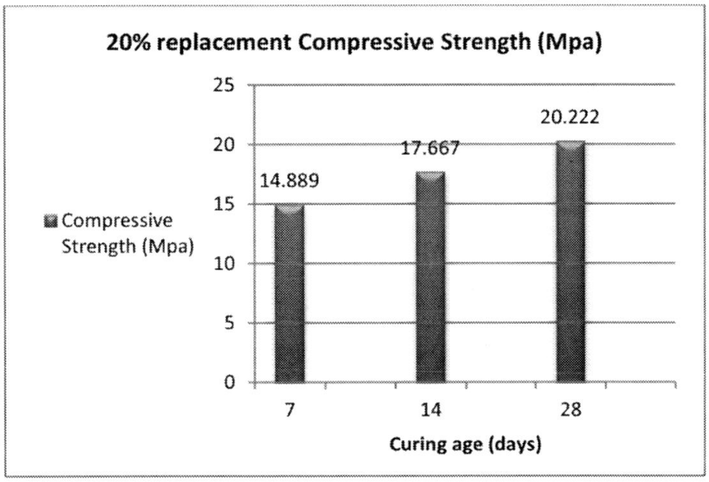

Fig. 8: Compressive strength for Marble waste with 20 % replacement.

For 30 % replacement:

As 30% of cement is replaced by marble waste, this will result in low loading and compressive strength. The reduction on loading properties is due to the lowering in cementations materials quantity as cement has higher reactivity and water absorption compared to marble waste in such big replacement percentage.

Compressive strength is the most important property not only as a mechanical point of view but in concrete making overall.

Concrete loading is the only way to predict and calculate the compressive strength and by doing so, it is revealed from the above tables and graphs that 20% replacement is the optimum percentage as it gives around 20 Mpa compared to 10% and 30%.

For 10% replacement, there were no much difference between obtained results and fully cementitious concrete and this is due to the slight change in the cement amount compared to 20% and 30% replacement.

Compressive strength increases with curing days rising, fair enough, but for 30% a reduction of the compressive strength value is observed and this is due to the reduction on cement quantity as cement gives the strength to concrete.

Table 6: Compressive strength data for 30% marble waste.

Specimen	Curing days	sample Curing start date	sample Curing end date	Load (KN)	Compressive Strength (Mpa)
S-30-7-1	7	6/6/2013	13/06/2013	390	17.333
S-30-7-2	7	13/06/2013	13/06/2013	370	16.444
S-30-14-1	14	11/6/2013	25/06/2013	240	10.667
S-30-14-2	14	11/6/2013	25/06/2013	230	10.222
S-30-28-1	28	11/6/2013	9/7/2013	220	9.778
S-30-28-2	28	11/6/2013	9/7/2013	220	9.778

4 Conclusion

Recycling of wastes is one of the best solutions now days for environmental protection and for economical and sustainable use of resources. Day by day, the increasing in the consumption of Marble waste can be intuitively observed. This will result in a warning signs to the environment.

The possibility of using marble waste as a partial substitutive to cement in concrete industry has been studied and the effect of this replacement to the mechanical properties of concrete.

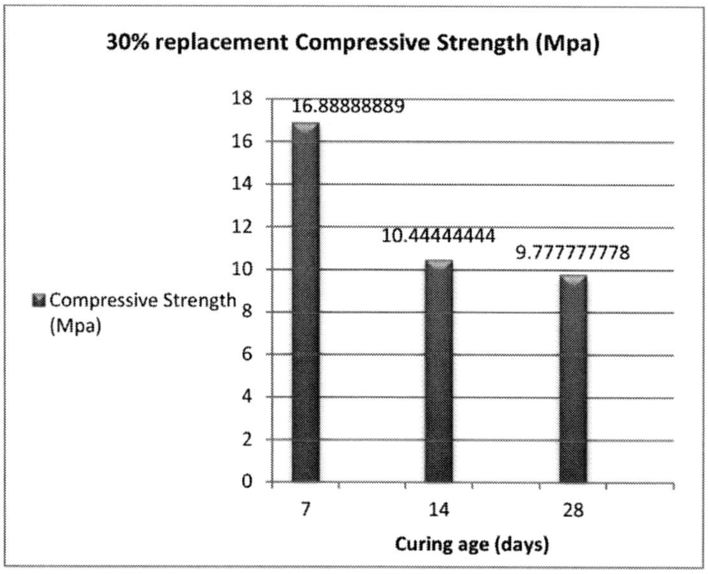

Fig. 9: Compressive strength for Marble waste with 30 % replacement.

The density of concrete which its cement has been replaced partially increase with the rising in the replacement percentage.

From all 18 samples prepared, the 20% replacement of marble waste is the optimum replacement percentage, and this give a successful of the project.

The XRD (X-Ray diffraction) results it is clear that marble waste is mainly consisting of oxide based components such as:

Silica (SiO2), Calcium (CaO), Phosphorus (P2O5), Magnesium (MgO), Potassium (K2O), Sodium (Na2O), Copper (CuO), Manganese (MnO), Zinc (ZnO), Iron (Fe2O3), Aluminum (Al2O3), Titanium (TiO2) and Sulphur trioxide (SO3).

Experimental works revealed that replacing of Ordinary Portland Cement (OPC) results in reduction in the Slump of the concrete made. This is true for high percentage replacement due to the reduction amount of cement because cement has higher reactivity compared to marble waste.

The study revealed the reduction of Compaction factor with the rising in replacement percentage again this is due to the reduction on the cement.

The study justifies the possibility of the replacement of Cement by marble waste up to 20% and this could result in reduction of the building cost and the prevention of the environment by the usage of marble waste instead of throwing it out of the city.

References

1. N.Krishna Murthy, A.V.Narasimha Rao, M.Vijaya Sekhar Reddy and P.Pamesh (2012), "The Influence of Metakaolin on the Modulus of Elasticity of Concrete", IOSR Journal of Engineering (IOSRJEN), Volume 2, Issue 11 (November 2012), PP 18-23
2. Kamile TOSUN (2009), "Portland Limestone Cement Part I - Preparation of Cements", Digest 2009, December 2009 1337-1355
3. T. W. Bremner (2001), "Environmental aspects of concrete problems and solutions", the 1st All-Russian Conference on Concrete and Reinforced Concrete, 9-14 September 2001.
4. M. Karaşahin, S.Terzi, "Pavement Management System, Past, Present, Future, 4.Asfalt Symposium,, Ankara, 2004.
5. Hanifi Binici1 (2007), "Influence of marble and limestone dusts as additives on some mechanical properties of concrete", Scientific Research and Essay Vol. 2 (9), pp. 372-379, September 2007[6] Baboo Rai1 (2011), "Influence of Marble powder/granules in Concrete mix", International journal of civil and structural engineering, Volume 1, No 4, 2011
6. ACI 211.1. (1993). "Standard practice for selecting proportions for normal, heavyweight and mass concrete". ACI Manual of Concrete Practice, 38 p.
7. Mustafa Yavuz C- elika,_, Eyu p Sabah(2007), "Geological and technical characterisation of Iscehisar (Afyon-Turkey) marble deposits and the impact of marble waste on environmental pollution", Journal of Environmental Management 87 (2008) 106–116
8. BS 1881-102:1983Part 2: Slump test , British Standards Institute, London, U.K
9. BS 1881-103:1993, Part 103: Method for determination of compacting factor, British Standards Institute, London, U.K
10. Kryžanowski, A., Šušteršič, J. (2003). "Performance of concrete exposed to long-term underwater abrasion loading", 21st Congress of International Commission on Large Dams ICOLD Montreal, Proceedings, Q.82-R.13, p. 207-218
11. Shamsad Ahmad (2005), "A study of Permeability and Tortuosity of Concrete", Our world in Concrete and Structure Conference: 23 – 24 August 2005, Singapore
12. Tommy E. Nantung (1998), "Determination of Water-to-Cement Ratio In Fresh Concrete Using Microwave Oven", Indiana Department of Transportation Research
13. Harrison, T. A., Early-Age Thermal Crack Control in Concrete, Report 91, Construction Industry Research and Information Association, London, 1981
14. ACI E701. (2007). "Aggregates for Concrete". ACI Manual of Concrete Practice, Materials for Concrete Construction
15. ASTM, C 150-04, "Standard Specification for Portland Cement", American Society for
16. Testing Materials", American Society for Testing Materials,USA, 1994.
17. ASTM, C 188-09, "Standard Test Method for Density of Hydraulic Cement", American Society for
18. Testing Materials", American Society for Testing Materials,USA, 1994.
19. ASTM, C 187-11, "Standard Test Method for Amount of Water Required for Normal Consistency of Hydraulic Cement Paste", American Society for Testing Materials", American Society for Testing Materials,USA, 1994.
20. ASTM, C 204-11, "Standard Test Methods for Fineness of Hydraulic Cement by Air-Permeability Apparatus", American Society for Testing Materials", American Society for Testing Materials,USA, 1994.
21. Jorge F.Carvalho _, PauloHenriques,Patrı´cia Fale´ , GabrielLuı´s (2008), "Decision criteria for the exploration of ornamental-stone deposits: Application to the marbles of

the Portuguese Estremoz Anticline", International Journal of Rock Mechanics & Mining Sciences 45 (2008) 1306–1319

22. Valeria Corinaldesi a, Giacomo Moriconi a, Tarun R. Naik (2009), "Characterization of marble powder for its use in mortar and concrete", Construction and Building Materials 24 (2010) 113

Performance of concrete blended with pozzolanic materials in marine environment

Asad-ur-Rehman Khan[1,*]

[1]NED University of Engineering and Technology, Karachi-75270, Pakistan

Abstract. Reinforced concretes structures located at or near the coast line needs to be repaired more frequently when compared to structures located elsewhere. This study is continuation of previous studies carried out at the Department of Civil Engineering, NED University of Engineering and Technology, Karachi, Pakistan to study the performance of concrete made up of cements blended by pozzolonic materials. Different pozzolanic materials (blast furnace slag, fly ash and silica fume) were used in the study. Tests conducted during the study to compare the performance of samples cast from concrete of different mix designs were Compressive Strength Test (ASTM C 39), Flexural Strength Test (ASTM C 293), Rapid Migration Test (NT Build 492), Absorptivity of the oven-dried samples (ASTM C 642) and Half Cell Potential (ASTM C 876). Use of cements blended with pozzolanic materials, used during the study, proved to be effective in enhancing the performance of the concrete exposed to marine environment. Use of pozzolans in concrete not only provides a sustainable and feasible solution to the durability problems in coastal areas, it also helps in conservation of natural resources and reduction of pollution and energy leading to a green environment.

1 Introduction

Deterioration of reinforced concrete (RC) structures in coastal areas has been one of the major concerns for civil engineers. Huge amount of money is being spent annually in rehabilitation and repair of these deteriorated RC structures. Failure of construction materials to sustain their due performance against environmental aggressions is largely responsible of these unwanted distresses. Role of the oceanic aggressive agents become highly undermining in marine environment. The deterioration of RC in marine environment, that may be shore decks, bridges, piles, piers or any other RC structure built in coastal areas, is predominantly because of exposure to sea water. The cost of repairing replacing deteriorated structures has become a major liability for users and different agencies.

Most of the onshore structures are constructed with normal carbon steel reinforcement due to it being economical compared to stainless steel or galvanic protection alternatives. Before the actual degradation of concrete corrosion of reinforcement bar is driven up majorly due to carbonation and chloride attack [1]. This results in the loss of alkalinity in concrete around the reinforcement bars leading to the initiation of actual corrosion.

* Corresponding author: asadkhan@neduet.edu.pk

For most of the structures exposed to marine environment the major extent of chlorides in concrete originates from the external sources. The penetration of this chloride occurs through various transport mechanisms depending upon the exposure conditions. In the case seawater exposure particularly, the chloride poses a much, in fact the sole, threat to steel in concrete than sulfate do to concrete as calcium sulpho-aluminate or ettringite (the expansive product of sulfate and tri-calcium aluminate in the cement) is more soluble in the presence of chloride and hence it does not initiate a considerable disruptive expansion.

A significant function for durability indexes is the water to cement ratio, lesser the pores, denser the microstructure, lesser penetration of water and ionic aggression and in turns lesser chance of deterioration. Another renowned way of mitigating the degradation of concrete, under the attack of ions, is the replacement of OPC by supplementary cementitious materials like fly ash, slag and silica fume as explained by Philip et al. [2]. These fine particles of pozzolans help in densely packing up the pores leading to permeation and also give rise to formation of excess ultra-stable calcium-silicate-hydrate (C-S-H) through pozzolanic reaction.

Corrosion inhibitors like calcium nitrate can be used in reinforced concrete which does not actually stop corrosion, but delays its initiation. Air-entrainers reduce the probability of segregation and bleeding, ensuring a uniform paste of concrete and improving durability. External waterproofing membranes can also be used to hinder the penetration and absorption of water to a great extent. Membrane system can be categorized as asphalt-impregnated fabric, cementitious, polymer, elastomeric and asphalt-laminated. The initiation of corrosion of steel reinforcement can also be delayed by coating the reinforcement bar with the referenced anodic zinc. Zinc has more electronegative potential than steel that means it is more sensitive to corrosion.

This study is continuation of previous studies carried out at the Department of Civil Engineering, NED University of Engineering and Technology, Karachi, Pakistan ([3-5]) to study the performance of concrete made up of cements blended by pozzolonic materials along with calcium nitrate dosing and external waterproofing applications, exposed to marine environment. Performance of different mix designs was determined by conducting Compressive Strength Test (ASTM C 39 [6]), Flexural Strength Test (ASTM C 293 [7]), Rapid Migration Test (NT Build 492 [8]), Absorptivity of the oven-dried samples (ASTM C 642 [9]) and Half Cell Potential (ASTM C 876 [10]). Results showed that in almost all cases, use of cements blended with pozzolanic materials resulted in an enhanced performance of the concrete exposed to marine environment.

2 Experimental programme

Nine different mix designs were used in the study in order to fulfill the scope of studying the performance of marine concrete enhanced by the addition of pozzolans like fly ash, silica fume and slag, along with the use of corrosion inhibitor (calcium nitrate). Details of mix designs are shown in Table 1. The water to cement ratio was kept fixed as 0.45. Proportions of the ingredients were designed as per ACI 211.1 [11], except for mix design A. Mix design A contains the ingredients in 1:2:4 ratio (M15 nominal design), and serves as control mix design as it represents general practice in local construction industry for the construction of general residential buildings. Air entrainer was used in all the mix designs except mix design A. Optimum 4% dosage of calcium nitrate was added in a mix design. Moreover, for workability and ensuring proper compaction with w/c ratio of 0.45, high-range water reducing plasticizer was used. For each mix design 0.2 cubic yard concrete was prepared. Minimum cover of 3$''$ was maintained for the reinforcement as per specification of ACI 357 [12].

Supplementary cementitious materials used in the study as a replacement of Portland cement in concrete were locally available slag cement containing 30 % slag, Fly ash (Class-C) with the activity index up to 88.72 %, and silica fume from BASF Chemicals & Polymers, MasterLife SF 100. Low water to cement ratio, aimed to make concrete denser and more impermeable, was achieved by the help of High-Range Water Reducer (Type-F) from BASF Chemicals & Polymers, MasterPozzolith LD10 and Super Plasticizer was used additionally for increasing workability. Maximum size of coarse aggregate used was ¾ inch, sand passing #4 sieve was used as fine aggregate. Air entrainer (MasterAir 720) and waterproofing membrane (MasterSeal 550) from BASF Chemicals & Polymers, were also used in the study.

Table 1. Mix Designs

Mix Design	**A**	**B**	**C**	**D**	**E**	**F**	**G**	**H**	**I**
Water/Cement Ratio	0.45	0.45	0.45	0.45	0.45	0.45	0.45	0.45	0.45
Cement (lb/c.yrd)	500	450	460	350	400	360	225	500	500
Slag (Replacement by %)				30%			27%		
Slag (lb/c.yrd)				150			135		
Fly Ash (Replacement by %)					20%	20%	20%		
Fly Ash (lb/c.yrd)					100	100	100		
Silica Fume (Replacement by %)		10%	8%			8%	8%		
Silica Fume (lb/c.yrd)		50	40			40	40		
Water (lb/c.yrd)	225	202.5	207	157.5	180	162	101.3	225	225
Fine Aggregate (lb/c.yrd)	1000	1000	1000	1000	1000	1000	1000	1000	1000
Coarse Aggregate (lb/c.yrd)	2000	2000	2000	2000	2000	2000	2000	2000	2000
Calcium Nitrate (oz)									500
Micro Air Entrainer (oz)		12	12	12	12	12	12	12	12

2.1 Compressive and flexural strength

Strength has always been the most important characteristic of concrete along with durability. Reduction in compressive and flexural strengths of all mix designs used in the study was determined by exposing respective specimens to marine environment for 100 days after 28 days of curing in normal conditions. An actual splash zone near the construction site of a Quay wall at the harbor port, by China Harbor Engineering Company Group, was selected as marine environment and strength reduction was determined after 100 days. Concentration of different ions in the marine water samples from China Harbor port is shown in Table 2.

Compressive strength was determined by testing 6 inch diameter cylinders as per ASTM C 39, while flexural strength was determined by testing 24 inch long beams with 6 inch square cross-section subjected to three-point bending, in accordance with ASTM C 293. Three (03) samples of each type were cast for testing. A set of samples was maintained in controlled non-aggressive environment, a set was coated with waterproofing membrane layer for exposure in marine environment while a set of sample were exposed to the marine environment without water proofing membrane.

Table 1. Ionic Concentrations present in Sea Water Samples

Ions	China Harbor Port
Sulfate	27.7 gm/10 ltr
Sodium	91.96 gm/10 ltr
Potassium	11.65 gm/10 ltr
Calcium	12 gm/10 ltr
Magnesium	15.32 gm/10 ltr
Chloride	254.92 gm/10 ltr

2.2 Chloride penetration (R.M.T)

Chloride induced corrosion in marine reinforced structures is a key mechanism directly affecting the durability of the structure. Thus, it can be used as an effective quantifiable durability index of concrete. Electrically-accelerated chloride penetration testing method, known as Rapid Migration Test (RMT) or Rapid Chloride Migration Test (RCMT), was used in the present study, which quantify the ingression of chloride ions. This test was standardized in 1999 by Nordic Council of Ministers as NT Build 492 (1999).

Specimen of 2 inch thickness and 4 inch diameter were used as defined in NT Build 492. The average depth of the chloride penetration was used for further calculation of Chloride-ion Migration Coefficient or Non-Steady State Migration Coefficient ($\boldsymbol{D_{nssm}}$). The diffusion coefficient was calculated using the simplified formula proposed by NT Build 492 for the Non-Steady State Migration Coefficient (D_{nssm}) defined in Eq. 1.

$$D_{nssm} = \frac{0.0239(273+T)L}{(U-2)t}\left(x_d - 0.0238\sqrt{\frac{(273+T)Lx_d}{(U-2)}}\right) \tag{1}$$

Where;
D_{nssm} = Non-steady state migration coefficient, $\times 10^{-12}$ m²/s;
U = absolute value of the applied voltage, V
T = average value of the initial and final temperature in the anolyte solution, C
L = thickness of the specimen, mm
x_d = average value of the penetration depths, mm
t = test duration, hour

2.3 Absorption capacity test

Absorptivity of samples cast from all the mix designs was determined in accordance with ASTM C642. Specimens were extracted from 4 inches diameter cylindrical samples by slicing them with the help of a water cooled diamond saw. Two (02) inches thick slices were obtained for the test. The water absorbed is simply represented in percentage of the sample's dry weight and calculated by Eq. 2 stated below.

$$\% \, water \, absorption \, = \frac{(W_2 - W_1)}{W_1} \, x \, 100 \qquad (2)$$

Where,
W_1 = Oven dry weight of sample;
W_2 = Wet weight of sample after it reaches a constant weight.

2.4 Corrosion rate rest

The corrosion potential was measured by the help of copper/copper sulfate electrode. The setup was prepared in accordance with the specifications of ASTM C-876. The specimens were subjected to a chlorine environment of 5% NaCl solution. The corrosion potential readings were obtained weekly. The corrosion activity is evaluated in accordance with ASTM C-876. Evaluation criteria suggested by ASTM C876 when Copper/Copper Sulfate electrode is utilized is as follows:

1. If potentials over an area are greater than -0.20 volt, there is a probability greater than 90 percent that no steel corrosion is occurring in the area at the time of measurement.
2. If potentials are in the range of -0.20 to -0.35 volt, corrosion activity of the steel in the area is uncertain
3. If potentials are less than -0.35 volt, there is a probability greater than 90 percent that steel corrosion is occurring in the area at the time of measurement.

3 Results and discussion

3.1 Compressive strength

Compressive strengths of all mix designs before and after exposure to marine environment are shown in Figure 1. Targeted compressive strength for the mix designs was 5000 psi which was achieved by most of the mix designs except for mix designs involving fly ash, with or without silica fume and slag (E, F and G). This may be attributed to quality of fly ash used in the study, and more experimentations and testing may be needed to account for for the poor performance of mix designs involving fly ash. Enhanced strength of concrete can be noted in mix designs in which cement was replaced by silica fume, B and C, and blast furnace slag, D. Addition of calcium nitrate in mix design I and air entrainer in mix designs I and H seems to have no effect on strength as the strengths similar to control mix design were achieved by mix designs H and I. All the samples showed reduction in compressive strength after exposure of 100 days to marine environment. Performance of mix designs B,C, H and I was better than rest of the mix designs in terms of strength reduction, followed by mix designs E, F, A, D and G. Highest loss of strength was found in Mix design G with all the three pozzolans. Use of Air entrainer appears to have improved the performance in the marine environment as strength reduction in almost all mix designs is smaller than mix design A except D and G.

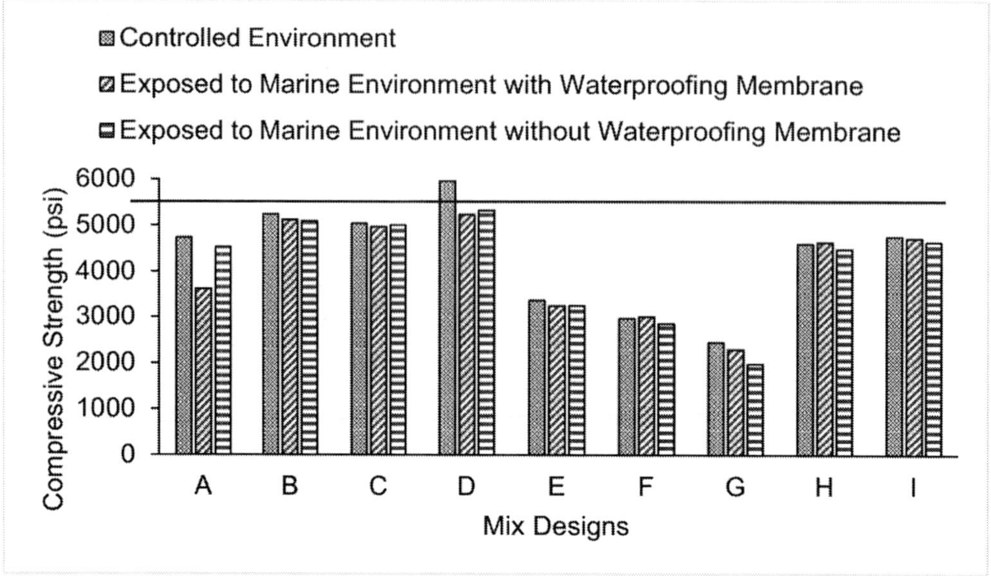

Fig. 1. Compressive Strength Results of Different Mix Designs

3.2 Flexural strength

Flexural strengths of all mix designs before and after exposure to marine environment are shown in Figure 2. A trend similar to compressive strength can be seen in Figure 2 that silica fume and slag attributed to the increase in strength while fly ash influenced a considerable loss in flexural strength as well. Mix design D containing 30% slag showed highest value of strength and least strength degradation in marine exposure. Maximum strength reduction is noted in mix design A. Performance of mix designs H and I was almost similar. Use of air entrainer and water proofing membrane seems to have contributed in minimizing the loss of flexural strength when exposed to marine environment.

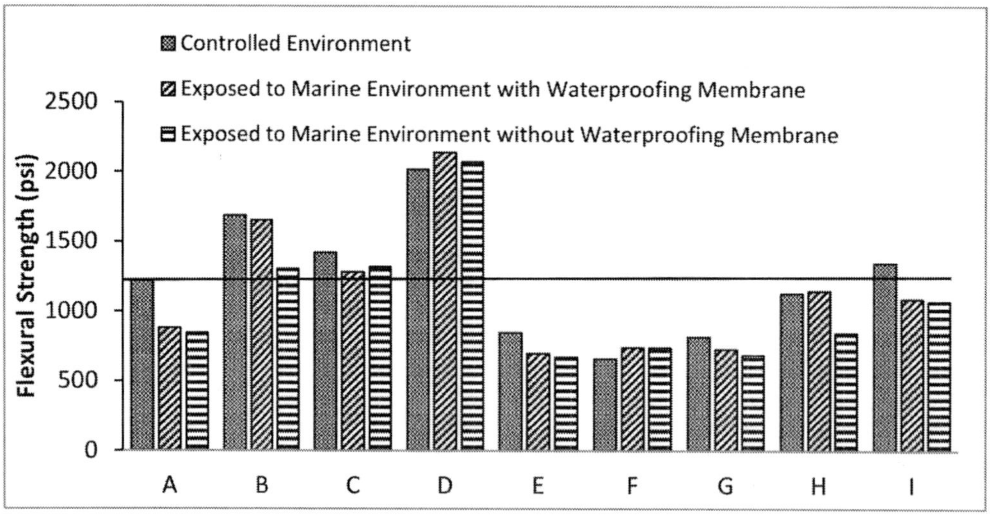

Fig. 2. Flexural Strength Results of Different Mix Designs

3.3 Chloride penetration (R.M.T)

Chloride-ion migration test results of different mix designs are compared in Figure 3. Authenticity and reliability of the results can be justified by the fact that diffusion coefficient of the same mix design did not differ by a wide scale. The slag cement mixes G, D and F, showed a much lower value of diffusion coefficient, than the mixes A, H and I containing only OPC. Fly ash and silica fume also contributed to the reduction in the diffusion coefficient's value for B, C and E mix designs. Performance of the mix design G, with fly ash, silica fume and slag, was found to be best against chloride ingression and showed least chloride penetration depth as well. Dosage of calcium nitrate and air entrainer seems to have no effect on providing resistance against chloride ions induced with a potential difference.

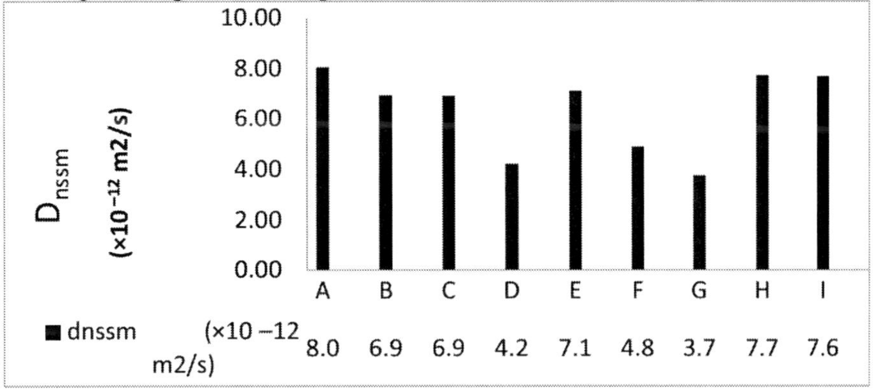

Fig. 3. Chloride Rapid Migration Coefficient Results for Different Mix Designs

3.4 Absorption capacity test

Percentage of water absorption of dry specimens by weight as per ASTM C-642 of all the mix designs is shown in Figure 4. Mix Design E shows the maximum water absorption as compared to the other mix designs followed by mix designs A, F, G, H and I. Silica fume on the other hand proved to be an effective means of reducing absorptivity, despite its small portion replacement of cement (8% to 10%) in mix designs B and C. Slag, that is effective when it replaces the cement in a large proportion, also showed a considerable decrease in absorptivity. Calcium Nitrate did not seem to affect the absorption property of concrete. Use of air entrainer contributed in reducing the absorption capacity when exposed to marine environment as can be seen in mix designs H and I as compared to the control mix design A.

3.5 Corrosion rate test

The half-cell potential was noted with the half-cell submerged setup, developed in accordance to ASTM C876. The corrosion potential of steel embedded in concrete, exposed to a chloride aggressive environment of 5% NaCl for a period of 35 days, showed the mutually comparable rate of corrosion of steel in reinforced concrete due to salt attack, presented in Figure 5. It is clearly visible that the corrosion potential, that represents the corrosion activity in concrete specimen, decreased adequately with the use of pozzolans. Silica fume alone and in combination with fly ash reduced the potential by a wider margin as can be noticed for mix designs C and F. Use of calcium nitrate as corrosion inhibitor helped in delaying the corrosion process, a great deal. A significant low corrosion activity was witnessed in the concrete samples containing calcium nitrate.

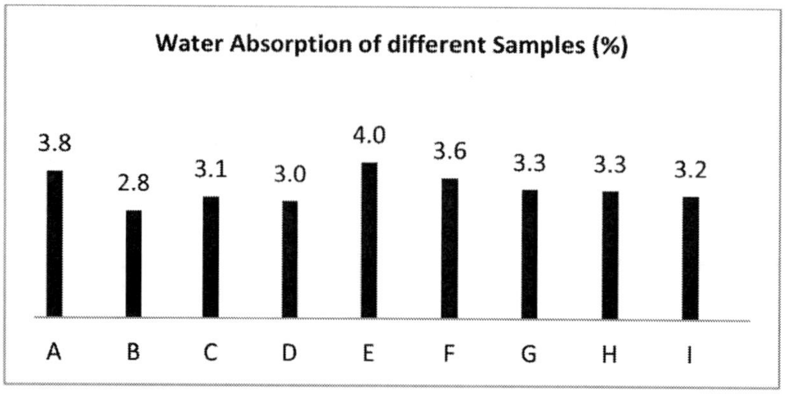

Fig. 4. Absorptivity for Different Mix Design

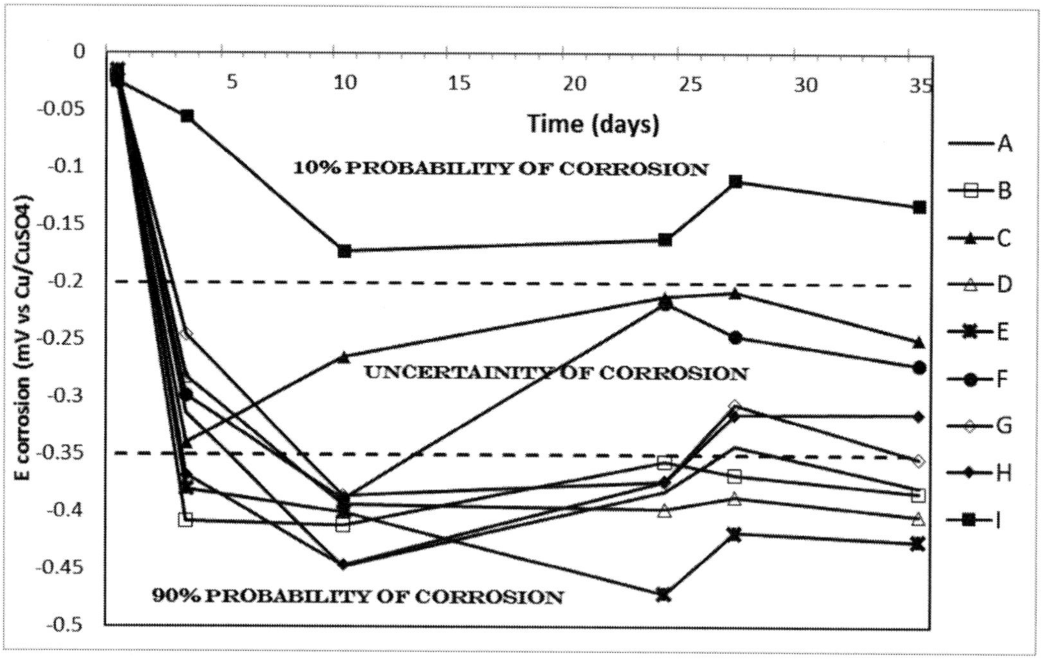

Fig. 5. Variation of Half-Cell Corrosion Potential of Reinforcement versus Time of exposure to Chloride Attack (5% NaCl)

4 Conclusions

Following conclusions can be drawn from the study:

1. Performance of all the mix designs in terms of strength gain were better or comparable to control mix design except for mix designs E, F and G where fly ash was used separately or with silica fume and slag. This may be attributed to slow strength gain of fly ash which also affected the performance of silica fume and slag.
2. Resistance to chloride migration was best observed for the mix design G followed by D and F indicating that slag only or combination of slag, silica fume and fly ash can produce dense, impermeable and durable concrete.

3. Least absorption capacity was observed for the mix designs B and D followed by C and I reflecting better performance of Slag and silica fume as compared to fly ash which again shows highest absorption capacity which is even higher than the control mix design A.
4. Performance of mix design I, with calcium nitrate as corrosion inhibiter, was the best in Half Cell corrosion rate measurements followed by mix designs C, F, H and G. Performance of fly ash alone was again not comparable to mix designs with slag and silica fume.
5. As per observed durability results, use of silica fume in combination with slag would be a preferred option in marine concrete within their optimal ranges as defined in the literature. Addition of calcium nitrate as corrosion inhibitor is expected to produce better performance in resisting corrosion.

The authors are indebted to the Department of Civil Engineering at NED University of Engineering & Technology, Karachi, Pakistan and the University itself, in the pursuit of this work. Authors also gratefully acknowledge the assistance and support provided by the Shahzeb Baloch, Afnan Waseem, Syed Ashad Ali, Osama Hussain, Haris Ali Abbasi, Abdur Rafae.

References

1. G.J. Verbeck, *ACI SP-49* (American Concrete Institute, Detroit, United States, 1975)
2. S.Z. Philip, *Offshore Mediterranean Conference and Exhibition*, (Ravenna, Italy, 2007)
3. A.R. Khan, T. Zahra, *Civil Engineering for Sustainability and Resilience International Conference, CESARE '14* (Amman, Jordan, 2014)
4. A.R. Khan, M.J. Anis, M.S. Ahmed, *Second International Conference on Performance-based and Life-cycle Structural Engineering, PLSE 2015* (Brisbane, Australia, 2015)
5. A.R. Khan, M.S. Ahmed, *fib Symposium 2016, Performance-based Approaches for Concrete Structures* (Cape Town, South Africa, 2016)
6. ASTM C 39 (American Society of Testing and Materials, 2004)
7. ASTM C 293 (American Society of Testing and Materials, 2002)
8. NT Build 492 (Nordic Council of Ministers, Espoo, Finland, 1999)
9. ASTM C 642 (American Society of Testing and Materials, 2006)
10. ASTM C 876 (American Society of Testing and Materials, 1987)
11. ACI 211.1 (American Concrete Institute, 1991)
12. ACI-357 (American Concrete Institute, 1997)

Production of durable and cost effective sewer pipes using petroleum and industrial waste by-products

Maisa El Gamal[1]*, *Amr* El-Dieb[2], *Khaled* El-Sawy[3], *Amr* Sweedan[2] *and Abdel-Mohsen* Mohamed[4]

[1] Assistant Professor, University College, Zayed University, Abu Dhabi, P.O. Box, 144534, Abu Dhabi, UAE
[2] Professor, Civil and Environmental Engineering Department – United Arab Emirates University, P.O. Box 15551, Al Ain, UAE
[3] Professor, Architecture Department - Effat University P.O. Box 34689, Jeddah, KSA
[4] Associate Provost and Chief Academic Officer, Zayed University P.O. Box, 144534, Abu Dhabi, UAE

Abstract. The Arabian Gulf environment has an adverse impact on concrete structures because of the high ambient temperature, humidity, salt contaminated dust, sea water and underground salts. As an innovative solution, reinforced modified sulfur concrete (MSC) pipes are recognized as a durable and economical solution for deterioration of pipelines. This work describes the manufacture and characterization of new MSC based on a cost effective sulfur modification process. Bitumen, a by-product from crude oil refining process was used to modify elemental sulfur and enhance its physical, mechanical properties, and mostly to increase its corrosive resistivity. The study has focused on optimizing the proportions of an offered MSC mixes that are composed of modified sulfur (MS) as a binder, crushed sand, dune sand and ladle furnace (LF) slag as aggregates, and ground granulated blast furnace slag (GGFBS) as a filler. To maximize the physical and mechanical properties of MSC sewer pipes, different mixtures were prepared and investigated. The optimum mix of MSC has a maximum compressive strength of 64 MPa, maximum splitting tensile strength of 4.5 MPa, maximum flexural strength of 21 MP, with a high corrosion resistance in acidic and salty environments.

1 Introduction

These Improving concrete durability in sewerage systems against attack by strong mineral acids and salts can be accomplished by using modified sulfur concrete (MSC). The MSC is a thermoplastic composite concrete consisting of chemically modified sulfur polymer and aggregates [1]. The product is a building material with a smooth surface that is easy to design and sets much faster than normal concrete. The MSC produced from waste materials is corrosion resistant to harsh environments and impervious to moisture infiltration [2]. MSC

*Corresponding author: maisa.elgamal@zu.ac.ae

has many industrial applications such as sidewalks and road blocks, construction of drainage and sewage facilities, basic coverage foundation, making the acid tanks, manufacturing and repairing concrete platforms, etc. [3, 4]. Research and development of MSC are in advance to discover possible usage of elemental sulfur present in many nations as a by-product of the oil and gas industry. The MSC produced from waste materials is corrosion-resistant and impervious to moisture infiltration.

Process of the sulfur concrete manufacturing is based on the "hot" technology in which all the mixed components are heated until 140 -150°C. The sulfur used in sulfur concrete production can be mixed with any type of traditional aggregate. Quantity should be optimized according to practical criteria as well as the mechanical properties [3]. When sulfur and aggregate are hot-mixed and cooled to produce sulfur concrete, the liquid sulfur binder initially crystallizes to monoclinic sulfur (Sb). As it continues to cool, the material goes through a solid phase transition to orthorhombic sulfur (Sa), which causes the material to shrink in volume [5]. This reduction in volume creates internal stress and causes durability problems, especially when exposed to freeze thaw cycling. The modification of elemental sulfur suppresses the crystalline phase change that occurs upon sulfur cooling below 95.6°C from monoclinic to orthorhombic forms with a significant increase in density and internal stresses. Such residual stresses contribute to failure of elemental sulfur concrete products by impact or thermal change. Mixing elemental sulfur with a suitable thermoplastic material preferably a polyolefin prevents the residual stresses build-up between sulfur crystals, which causes the elemental sulfur concrete material to crack. While the phase transformation still occurs during sulfur cooling, the polymer acts as an acquiescent layer between sulfur crystals, thereby overcoming the drawback of elemental sulfur concrete products. [6, 7].

At present, a great deal of research and innovation are being conducted in many countries including United Arab Emirates [8], USA, Canada, Poland and Russia where building materials such as sulfur concrete and sulfur asphalt have been used because these materials are environmental friendly and cost effective [9]. Studying the effect of different proportions of MSC on physical and mechanical properties of MSC is relatively difficult due to the complexity and number of variables in this system. To this end, in the current study, different MSC mixtures were manufactured to be used in sewer systems. Experimental mixture design was used taking into consideration the properties desired for its specific use with the following characteristics; high mechanical properties equivalent to or better than those of PCC, low porosity and hydraulic conductivity and improved resistance to attack by most acids and/or salt solutions.

2 Materials and Methods

2.1 Raw materials

Sulfur: granular elemental sulfur, with purity of 99.9%, was obtained from Al Ruwais refinery, United Arab Emirates (UAE). Bitumen: is an olefin hydrocarbon polymeric material, which is black oily, viscous, with specific gravity of 1.0289 g/cm3, kinematics viscosity of 431 cSt at 135°C and softening point of 48.8°C produced from the distillation of crude oil. Crushed sand from Ras Al Khaimah with particles size 38 μm - 4 mm. Dune sand: obtained from Al Ain area, UAE. The most common constituent of sand is silica, with specific gravity of 2.58 g/cm3. Ladle Furnace (LF) slag was produced by the Electric Arc Furnace process from Emirates Steel in Abu Dhabi, UAE. Ground Granulated Blast Furnace Slag (GGBFS) was obtained from Al Sharjah cement factory, UAE.

2.2 Sulfur modification

Sulfur is modified by mixing elemental sulfur and the chemical stabilizer (2.5-5% bitumen) to form a sulfur-containing polymer. To facilitate the compatibility between elemental sulfur and the bitumen, surfactant agent (Triton X-100) was used as emulsifying agent. The mixing process was maintained at about 140°C for a period of 45-60 minutes, (Figure 1). Detailed description of the preparation can be found in [10]. The development of the reaction is followed by monitoring the changes in viscosity and homogeneity of the mixture. The final product is a sulfur-containing polymer, which on cooling possess glass like properties.

Fig. 1. Sulfur modification.

2.3 Production of sulfur concrete

MSC was prepared according to the procedure described in ACI 548.2R-93 for mixing and placing sulfur concrete [11]. The preheated coarse, fine aggregates and filler were mixed with molten modified sulfur cement at controlled rate of 130-140°C. The temperature was maintained throughout the mixing process. Finally, the hot mixture was poured in pre-heated cubic and cylindrical moulds and settled on a vibrating table for 1 minute. After curing at 40°C for 24 hours, the samples were kept at room temperature.

2.4 Mixing protocol and measured responses

To determine the mix proportions of MSC, five constituents were used; sulfur cement (X1), crushed sand (X2), Dune sand (X3), LF Slag (X4) and GGBFS (X5). The proportions of the mixture components were selected in terms of mass fraction, based on preliminary tests. The lower and upper limits of the proportions were chosen based on pre-laboratory experimental results, Table 1.

Table 1. Lower and Upper limits of different MSC ingredients mass fraction.

Constituent	ID	Lower limit by mass fraction	Upper limit by mass fraction
Sulfur cement	X1	0.18	0.28
Crushed sand	X2	0.11	0.40
Dune sand	X3	0.13	0.36
LF slag	X4	0.05	0.22
GGBFS	X5	0.05	0.12

MINITAB17 software was used to propose various recommended proportions to be used in the design and statistical evaluation of experimental results. MINITAB17 proposed 31 different mixtures in which the ingredients are varied as represented in Table 2. Also, in order to achieve more representative results of the components' effect, the 31 suggested mixtures were duplicated. Additional mixtures were conducted to test the developed mixture models. The successful optimization of the MSC composition requires identifying the best proportions of the materials to reach optimum mechanical response. A generalized linear model is basically a regression model. Like all regression models, it is made up of a random component and a function of the design factors (the x's) and some unknown parameters. In a standard normal-theory the response (y) in linear regression model as Eq.1

$$y = \beta_0 + \beta_1 X_1 + \beta_2 X_2 + \cdots + \beta_i X_i + \in \cdots \ldots \ldots \ldots \ldots \ldots \ldots (1)$$

Where: Y: is the response; compressive strength, splitting tensile strength and flexural strength; X1, X2, ... Xi are the variables; sulfur cement (X1), crushed sand (X2), Dune sand (X3), LF Slag (X4) and GGBFS (X5); and β1, β2, ... coefficient

This statistical design enables comparing the different mathematical regression models based on their best-fit to the experimental results. In addition, it helps identifying the significant terms of a selected mathematical model so that other insignificant terms are eliminated allowing for a simpler model. In particular, it was hypothesized that the special cubic model correlated the composition of the MSC mixture with the required characteristics. As the statistical analysis was described in details for three mechanical responses; compressive strength, splitting tensile strength and flexural strength. For each of the three responses, the model was fitted using linear (Eq. 2), quadratic (Eq.3) and special cubic (Eq. 4), respectively [12].

$$E(y) = \sum_{i=1}^{p} \beta ixi \ldots \ldots \ldots \ldots \ldots \ldots \ldots \ldots \ldots \ldots \ldots \ldots \ldots (2)$$

$$E(y) = \sum_{i=1}^{p} \beta ixi + \sum_{i<j}^{p}\sum \beta ijxi\, xj \ldots \ldots \ldots \ldots \ldots \ldots \ldots \ldots \ldots (3)$$

$$E(y) = \sum_{i=1}^{p} \beta ixi + \sum_{i<j}^{p}\sum \beta ijxi\, xj + \sum_{i<j<k}\sum\sum \beta ijkxixj\, xk \ldots \ldots (4)$$

Where: Y represents the response; the variables are; sulfur cement (X1), crushed sand (X2), Dune sand (X3), LF Slag (X4) and GGBFS (X5).

Table 2. Proportions of MSC mixtures.

Mix No.	Sulfur cement	Crushed sand	Dune sand	LF Slag	GGBFS
1	0.22	0.40	0.23	0.05	0.10
2	0.27	0.11	0.33	0.22	0.07
3	0.22	0.40	0.18	0.15	0.05
4	0.18	0.40	0.17	0.15	0.10
5	0.22	0.40	0.13	0.15	0.10
6	0.22	0.38	0.25	0.05	0.10
7	0.18	0.37	0.25	0.15	0.05
8	0.22	0.33	0.25	0.15	0.05
9	0.18	0.32	0.25	0.15	0.10
10	0.26	0.15	0.30	0.18	0.12
11	0.21	0.32	0.28	0.14	0.06
12	0.22	0.40	0.25	0.05	0.08
13	0.22	0.40	0.25	0.08	0.05
14	0.20	0.40	0.25	0.05	0.10
15	0.18	0.40	0.25	0.07	0.10
16	0.22	0.34	0.24	0.12	0.08
17	0.21	0.39	0.23	0.08	0.09
18	0.22	0.30	0.26	0.15	0.07
19	0.21	0.39	0.20	0.13	0.06
20	0.24	0.22	0.25	0.17	0.11
21	0.23	0.35	0.19	0.14	0.10
22	0.21	0.38	0.24	0.08	0.09
23	0.22	0.28	0.28	0.15	0.07
24	0.21	0.35	0.24	0.13	0.06
25	0.28	0.11	0.36	0.19	0.07
26	0.24	0.24	0.27	0.15	0.10
27	0.19	0.39	0.24	0.12	0.06
28	0.21	0.39	0.24	0.08	0.08
29	0.22	0.35	0.25	0.10	0.07
30	0.20	0.39	0.24	0.08	0.09
31	0.22	0.31	0.27	0.10	0.10

2.5 Durability studies of MSC

Specimens from the optimum MSC mixture were immersed in different solutions; water at room temperature, water at 60°C, 20% sulfuric acid solution, and 5% sodium chloride solution tank for 18 months. The reduction in compressive strength with respect to the air (un-exposed) specimens was measured.

3 Results and Discussion

3.1 Sulfur Modification

In the modification process, bitumen, a by-product from crude oil refining process was used to modify elemental sulfur with the aid of a non-ionic surfactant agent to facilitate the compatibility between the sulfur and the bitumen. Addition of bitumen in amounts of 2.5–5% wt. to elemental sulfur initiates chemical reactions are dependent on the bitumen content, heating temperature and reaction time. Some competing reactions such as bitumen incorporation into sulfur molecules and/or de-hydrogenation with liberation of hydrogen sulfide could occur. The reaction of sulfur with bitumen at 140°C produces a mixture of polysulfide and free elemental sulfur, which held as monoclinic sulfur crystal. It is supposed that the growth of large crystals may be restricted because bitumen particles serve as nucleation sites that cause the formation of many small crystals instead of fewer large orthorhombic sulfur crystals. The resulting modified sulfur leads to sulfur solidified products that are not highly stressed and have good durability and corrosion resistance [13].

3.2 Production of MSC

The modified sulfur is a thermoplastic material that becomes liquid when heated above its melting point and can be mixed with aggregates. Crystallization of sulfur in this case is controlled by the relative percentage and distribution of space between aggregates and sulfur binder. The effect of modification seems to be related to the increase of degree of sulfur polymerization. Sulfur itself possesses a tendency to polymerize; the modified sulfur would thus increase this tendency or maintain it for a longer time. The quality and quantity of the aggregates have substantial effect on the properties and durability of the finished MSC. No chemical reaction is required for setting as Portland cement concrete (PCC), thus minimizing incompatibilities between binder and aggregate. The MSC offers many benefits as an alternative construction material compared to PCC.

3.3 Collective models and analysis of variance

In the statistical mix design, compressive strength, splitting tensile strength, flexure strength and stability are greatly influenced by the proportions of crushed sand, dune sand, LF slag, GGBFS and exist of optimum proportions of sulfur cement. These design allow us to estimate linear, quadratic (the linear + the interaction terms) and special cubic interaction effects of the components, then compare the model. Table 3 shows that, the special cubic model provides an adequate fit to MSC responses; compressive strength, splitting tensile strength and flexural strength as represented in equation No. 4. The good agreement between the predicted and experimental values verified the validity of the model along with the existence of the optimal point.

3.3 Mechanical properties of MSC

Mechanisms involved in preparation MSC are different from those used in PCC. MSC is a thermoplastic material that is mixed and cast hot. On solidification and cooling to ambient

Table 3. Measured compressive, splitting tensile and flexural strength values of MSC mixes.

Mix No.	Compressive Strength (MPa)		Splitting Tensile Strength (MPa)		Flexure Strength (MPa)	
	Experimental	Prediction	Experimental	Prediction	Experimental	Prediction
1	54.5	55.0	3.66	3.68	17.08	17.30
2	40.0	40.0	2.78	2.78	12.66	12.64
3	50.0	49.3	3.36	3.31	16.76	16.53
4	39.1	39.2	2.82	2.82	14.77	14.78
5	57.0	55.6	3.91	3.84	17.92	17.59
6	63.7	66.3	4.45	4.58	19.00	19.09
7	55.7	55.9	3.60	3.59	16.70	16.78
8	50.0	47.9	4.31	4.29	18.60	18.96
9	50.0	50.0	3.62	3.62	13.47	13.48
10	38.7	39.4	3.57	3.61	16.07	16.27
11	56.2	55.7	3.64	3.67	17.04	16.90
12	61.0	59.1	4.37	4.28	20.66	20.37
13	61.3	62.7	3.91	4.03	17.05	17.11
14	45.0	45.5	3.39	3.41	14.06	14.32
15	47.2	45.9	3.60	3.54	18.86	18.43
16	61.2	63.4	3.95	3.81	19.62	19.64
17	61.7	59.9	3.95	3.85	20.30	19.98
18	41.7	44.5	2.95	3.02	19.49	18.98
19	53.0	55.8	3.36	3.47	16.50	16.47
20	42.2	41.0	3.00	2.93	13.61	13.35
21	54.6	59.4	3.53	3.57	17.61	18.05
22	60.5	59.0	3.94	3.99	20.74	19.93
23	47.8	50.2	3.44	3.44	16.61	17.59
24	54.7	53.5	3.63	3.52	21.03	19.93
25	39.0	38.9	3.13	3.12	13.07	13.06
26	53.5	51.5	3.86	3.77	16.62	16.08
27	53.3	53.3	3.56	3.56	17.78	17.88
28	55.0	56.7	3.80	3.78	17.83	18.35
29	47.4	48.3	3.24	3.30	20.85	21.49
30	49.0	50.9	3.15	3.14	18.34	19.27
31	44.7	46.0	3.13	3.19	18.08	18.35

temperatures, the liquid sulfur binds the aggregate and forms a rigid MSC. The PCC is prepared at ambient temperatures and relies on chemical hydration reaction to form the concrete. Table 4 shows a comparison between mechanical properties of the prepared MSC and a typical 40 MPa PCC. In MSC, the majority of the matrix is composed of sulfur coated aggregates and sulfur accumulating in the voids between particles leading to the reduction of the overall porosity of the MSC. The molten modified sulfur acts as binder material for these aggregates and filler. It can penetrate deeply into the gaps between coarse and fine aggregates and filler producing a strong and dense structure. Moreover, MSC solidifies and achieves its high strength because of modified sulphur, as an encapsulating agent. The fast curing of MSC contributes to shortening the construction period. MSC has high flexural and tensile strength because of the thermoplastic properties of sulfur cement. It is estimated that crushed sand and LF slag develop very strong interlocking properties leading to improved mechanical properties. GGBFS increased the density by filling the pores and better packing the particles. Sand dune enables the molten sulfur to adhere more easily on the surface of the aggregates.

Table 4. Comparison between mechanical properties of manufactured MSC and 40 MPa PCC.

Property (MPa)	Compressive Strength	Splitting Tensile	Flexure Strength
MSC*	64	4.5	21
PCC**	40	3.2	10

* Average measured values after 3 days.
**Average measured values after 28 days.

3.4 Statistical model for prediction of MSC strength

The different strength values of MSC, as predicted by the special cubic model along with the corresponding experimental measurements are presented in Fig. 2a-c. The comparison between these values indicates an excellent agreement between the predicted and experimental data [14]. The maximum differences between the measured and predicted compressive strength, splitting tensile and flexural strength of the mixes are 4%, 3.4% and 5.2%, respectively.

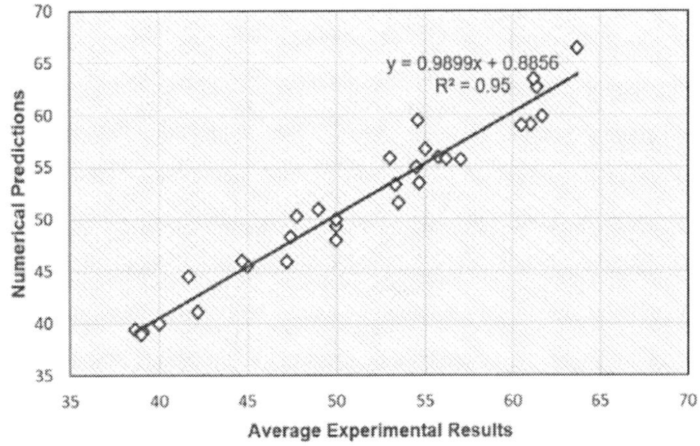

Fig. 2a. Correlation between experimental and predicted values of compressive strength for MSC.

This covenant between the predicted and experimental values validates the accuracy of the model and confirms the existence of optimal mixture. In general, the higher the R-squared, the better the model fits of data. The adjusted R^2 is a useful tool for comparing the explanatory power of models with different numbers of predictors. The coefficient of determination R^2 of the compressive strength, splitting tensile and flexural strength are 0.95, 0.98and 0.96, respectively, using a 5% significance level as represented in Figs. 2a, 2b and 2c, respectively.

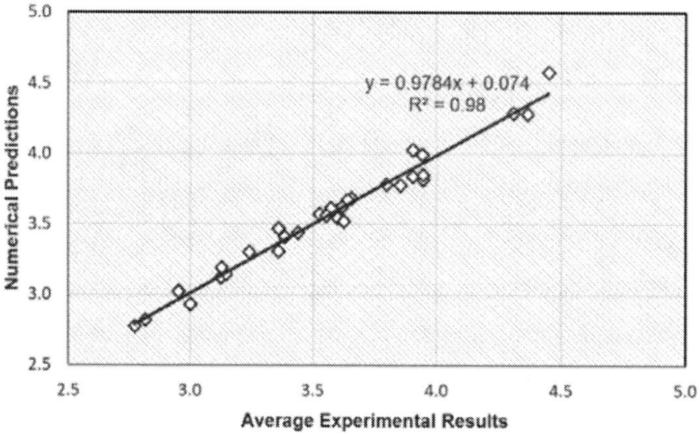

Fig. 2b. Correlation between experimental and predicted values of splitting tensile strength for MSC.

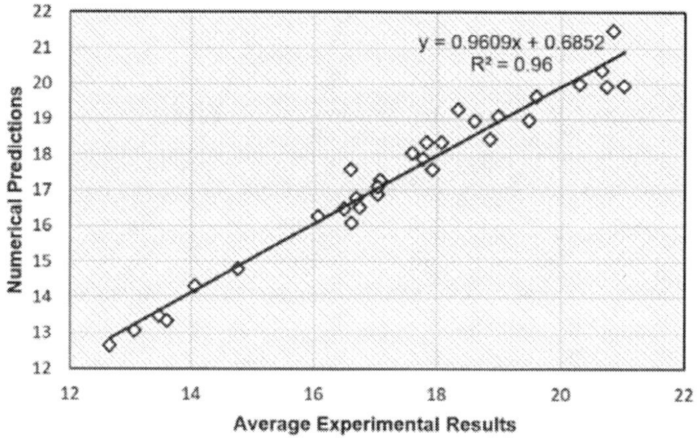

Fig. 2c. Correlation between experimental and predicted values of flexural strength for MSC.

3.5 Durability of MSC

Figure 3 shows the reduction in compressive strength of the optimized MSC after exposure for 18 months to different solutions; water at room temperature (23±3°C), water at 60°C, 20% sulfuric acid solution, and 5% sodium chloride solution. Durability results indicated that the manufactured MSC has high resistance to neutral, acidic, and saline environments.

DOI: 10.1051/matecconf/201712003013

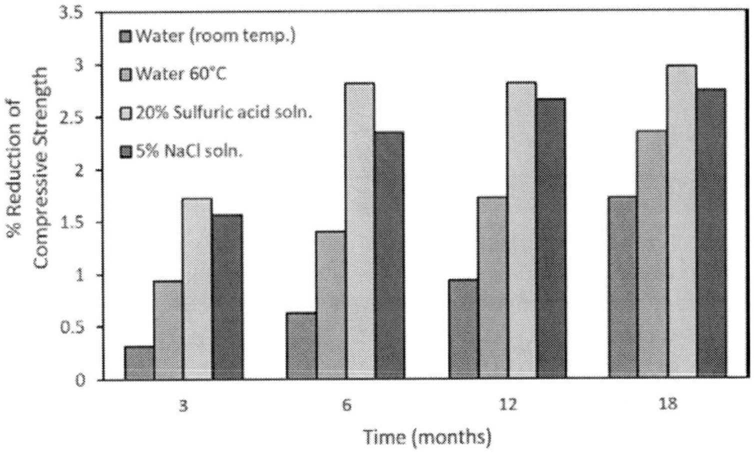

Fig. 3. Compressive strength reductions of MSC after exposure 18 months to different environments.

The least reduction in the compressive strength was observed when the MSC was exposed to distilled water at room temperature. As the water temperature increased the strength reduction also increased. Exposure to 20% sulfuric acid solution produced the highest compressive strength reduction after 3months of exposure time, then it was stabilized after 6 months and almost no further reduction occurred, which indicated a very good performance under these aggressive environments. This behaviour can be attributed to the structure of MSC where the majority of the matrixes are composed of sulfur-coated materials (aggregates and fillers). Figure 3 shows that there was a little reduction in compressive strength of MSC tested in 5% saline solution, which can be attributed to the formation of expansive salts (e.g., Gypsum, Ettringite and Thaumasite) that resulted in loss of bonding properties [8].

3.6 Testing of MSC Pipe

The optimized MSC mixture No. 22, Table 2 (modified sulfur: 0.21, crushed sand: 0.38, dune sand: 0.24, LF slag: 0.08 and GGBFS:0.09) was used to cast a reinforced pipe segment with outer diameter of 360 mm, inner diameter of 300 mm and length of 600 mm.

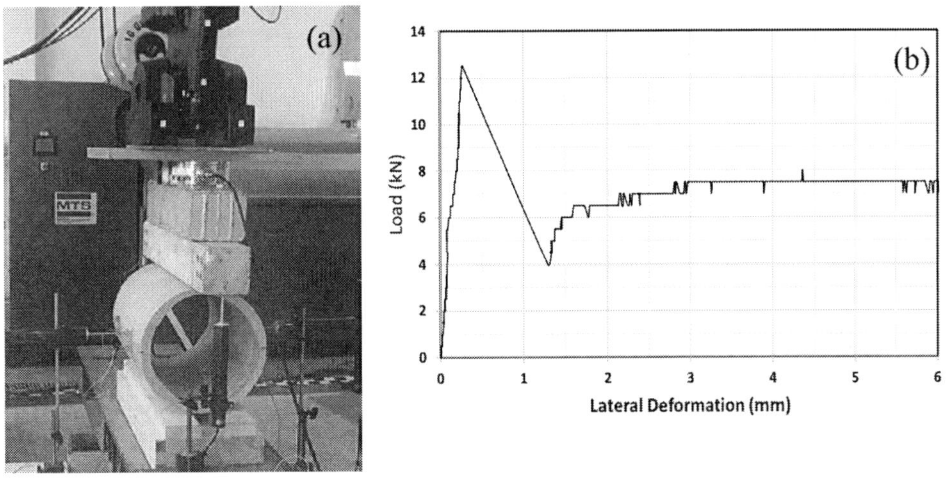

Fig. 4. (a) Testing setup and (b) load-lateral deformation curve of the produced MSC pipe segment.

The pipe segment was tested under vertical line load and three-edge bearing standard test as per ASTM C497, as shown in Figure 4a. The reinforced MSC pipe segment tested failed under a load of 12.5 kN and the average lateral deformation at maximum load was 0.25 mm as shown in Figure 4b.

4 Conclusion

This work describes the modification of elemental sulfur using an olefin hydrocarbon polymeric material (bitumen) which resulted in controlling crystallization of sulfur and forming a new structure with fine grain sizes. The manufacture of a MSC consisting of modified sulfur, crushed sand, dune sand, LF slag, and GGBFS was conducted using statistical experiment-based mixture design to optimize proportions of MSC in order to maximize its mechanical properties. The developed model showed good correlation between predicted and measured properties. The optimum MSC mixture showed improved resistance to harsh environments over 18 months, particularly in water at different temperatures, 20% sulfuric acid solution and 5% sodium chloride solution. The optimized MSC mixture was used to cast a reinforced pipe segments to be used in sewer system. The pipe segment failed under a load of 12.5kN and lateral deformation of 0.25 mm. The presented MSC has good potentials to be used in sewer pipes.

References

1. J. Moon, P. D. Kalb, L. Milian, P. A. Northrup, Cement and Concrete Composites, **67** (2016)
2. A.M.O. Mohamed, M. El Gamal, Sulfur Concrete for the Construction Industry: A Sustainable Development Approach: J, Ross Publishing, (2010)
3. V. Gracia, E. Vàzquez, S. Carmona, International RILEM Conference on the Use of Recycled Materials in Buildings and Structures. Nov. 8-11, Barcelona (2004).
4. P. Naufal Rizwan, Dhiwagar S. Maria Jesuraj, V. Vipin, V. Anbuvel, S. Prabhakaran IJERT, 3, **8**, (2014)
5. S.L. Lin, J.S. Lai, E.S.K. Chian, 1995. Waste Management 15 (5/6), (1995)
6. A.M.O. Mohamed, M. El Gamal, US Patent App. 14/480,351(2014)
7. A.M.O. Mohamed, M. El Gamal, Environ. Geol. 53, **1** (2007)
8. A.M.O. Mohamed, A. El-Dieb, K. El-Sawy, and M. El Gamal, Environmental Geotechnics EG2 (2015)
9. N. Ciak, J. Harasymiuk, Technical Sciences, 16, **4** (2013)
10. A.M.O. Mohamed, M. El Gamal (2015) US Patent App. 20150027345 (2015)
11. ACI Committee 548, American Concrete Institute, Farmington Hills, Mich (1993).
12. D. C. Montgomery, design and analysis of experiments 8th edition john Wiley & sons, Inc. (2012)
13. A.M.O. Mohamed, M. El Gamal, GCC patent 0003089 (2014)
14. A. A. Jaafer, American Journal of Civil Engineering and Architecture, Vol. 1, **4** (2013)

DOI: 10.1051/matecconf/201712003014

Use of alternative waste materials in producing ultra-high performance concrete

Shamsad Ahmad*

Civil & Environmental Engineering Department, King Fahd University of Petroleum & Minerals, Dhahran-31261, Saudi Arabia

Abstract. In a corrosive environment similar to that of the Arabian Gulf, use of high-performance concrete is one of the options to ensure a target service life of concrete structures. However, in absence of good quality coarse aggregates, it is a challenging task to produce high-performance concrete. Recently, the possibility of producing ultra-high-performance concrete (UHPC) has been widely reported in the literature. UHPC is produced without coarse aggregates at very low water to cementitious materials ratio, high amounts of cement, mineral admixtures, and superplasticizer along with fine quartz sand as aggregate, quartz powder as micro-filler, a nd steel fibres for fracture toughness. In the present work, an effort was made to utilize local waste materials as alternative mineral admixtures and local dune sand as aggregate in producing different UHPC mixtures without addition of quartz powder. The mechanical properties, shrinkage, and durability characteristics of the UHPC mixtures were studied. Test results indicate that it is possible to produce UHPC mixtures using alternative waste materials, which would have targeted flow, strength, toughness, and resistance against reinforcement corrosion. The information presented in the paper would help in optimum selection of a mixture of UHPC considering the availability of local materials, exposure conditions and structural requirements.

1 Introduction

In recent years, the possibility of producing ultra-high-performance concrete (UHPC) has been widely reported in the literature. UHPC is produced without coarse aggregates at very low water to cementitious materials ratio, high cementitious materials content, and quartz fine sand, quartz powder as micro-filler, silica fume or any other suitable mineral admixture, steel or polymer fibres, and high dosage of superplasticizer. UHPC is reported to possess very high strength, elastic modulus, ductility (i.e., fracture toughness), and excellent durability characteristics. The mechanical properties of UHPC typically reported in literature are: compressive strength > 150 MPa, flexural tensile strength > 25 MPa, modulus of elasticity > 50 GPa, and fracture energy > 20000 J/m^2 [1-4].

* Corresponding author: shamsad@kfupm.edu.sa

The mixtures of UHPC are developed based on four principles: (i) reduction in porosity through close packing of raw materials and reduction in water/cementitious materials ratio with the help of superplasticizer; (ii) improvement in microstructure through close packing of solid materials, facilitating hydration and pozzolanic reactions, and improving the interfacial transition zone between aggregates and cementitious products; (iii) enhancement in homogeneity by eliminating coarse aggregate and using fine raw materials; (iv) increase in toughness by using fibres. Besides selection of suitable raw materials and mixture design, the mixing methods and curing regimes also significantly affect the quality of UHPC [5].

The specifications regarding ingredients, mix design, mixing methods, and curing of UHPC, as reported in literature, are briefly described in the following sub-sections along with the objectives and significance of the present work.

1.1 Ingredients

Fine quartz sand having particles in the size range of 150 to 600 μm is used as the only aggregate because coarse aggregate is eliminated from UHPC [2, 3, 5]. Quartz powder having particle size distribution in the range of 0.1 to 100 μm is used as micro-filler [2, 3, 5]. If very fine quartz sand is available, the UHPC can be produced without the quartz powder [6]. Ordinary Portland cement with low C_3A content, low alkali content, and low to medium fineness is used as primary binder [2, 3, 5]. Silica fume, a byproduct from ferro-silicium industries with particle size distribution ranging from 0.1 to 1.0 μm is generally used as mineral admixture [2, 3, 5]. Polycarboxylates are used as most effective superplasticizer for UHPC to achieve the desirable fluidity [7]. Steel fibers having length of 13 mm and diameter of 0.2 mm are often used in UHPC mixtures [5].

In addition to cement, silica fume is primarily used as supplementary cementitious materials in UHPC. However, the use of silica fume may not be economical, particularly if it is not available locally at a low price. The partial replacement of cement by locally available cementitious materials would be economical and environment-friendly. The silica fume may be partially or fully replaced and cement may be partially replaced (without compromising with the quality of UHPC) by locally available supplementary cementitious materials such as ground granulated blast furnace slag, fly ash, metakaolin, limestone powder, rice husk, steel slag powder, cement kiln dust, nanomaterials (nano-SiO_2, nano-$CaCO_3$, nano-Fe_2O_3, nano-TiO_2), etc. [5, 8-14].

1.2 Design of UHPC mixtures

Particle size distribution, cement content, supplementary cementitious material content, water/cementitious materials ratio, dosage of superplasticizer, and fiber content are the parameters that are optimally selected to satisfy the requirements of UHPC in fresh and hardened states [1, 15-16].

The particle size ranges of cement, supplementary cementitious materials (such as silica fume), fine quartz sand and quartz powder should be selected in such a way that the grading curve for the mixture of granular materials would match with the optimum grading curves, reported in literature, for example optimum grading curve given by Schmidt and Fehling [17] and particle size distribution model for optimization of the composition of mixture of granular materials used for UHPC [18-19].

The values of the mix design parameters can be selected within their ranges as reported in literature [20], as follows: water to total binder ratio 0.15-0.24, cement content 800-1100 kg/m^3, silica fume content 150-300 kg/m^3, silica fume to cement ratio 0.15-0.35, cement and silica fume content 950-1400 kg/m^3, quartz and sand 1000-1400 kg/m^3, steel fiber content 190-250 kg/m^3.

In order to obtain an optimum mixture design satisfying the intended fresh and hardened properties of UHPC, various trial mixtures should be considered by choosing different sets of mixture design parameters within their reported ranges satisfying the absolute volume equation, i.e. the sum of absolute volumes of all the ingredients including air content should be equal to the total volume of UHPC considered for mixture design for example 1 m^3. After mixture design, the UHPC mixtures should be prepared and tested for flow. The dosage of superplasticizer should be adjusted to keep the flow as 200 ± 20 mm. Once the required flow for each of the trial mixtures is achieved, the mixtures should be tested for their hardened properties. An optimum mixture of UHPC, out of various trial mixtures considered, should be finally selected based on the high performance at low consumption of cement. UHPC with lower cement content would be economical and environment-friendly because of reduced carbon dioxide footprint.

1.3 Preparation and curing of UHPC mixtures

UHPC mixtures are prepared by mixing the ingredients in a proper sequence. First cement, supplementary cementitious materials, sand and filling powder are mixed in dry state for about 10 minutes and then the mixture of water and superplasticizer is added in steps over a period of about 5 to 10 minutes till appropriate flow ability of the mortar matrix is achieved. Finally, the fibers are added [5]. The porosity of an UHPC mixture can be reduced enhancing its hardened properties by using vacuum mixing [21-22]. Various curing regimes reported for producing UHPC include standard room temperature curing for 28 days, heat curing under atmospheric pressure, and autoclave curing [5].

1.4 Objective and significance of the present work

The main objective of the present work was to explore the possibility of producing UHPC mixtures utilizing dune sand and several industrial waste materials such as micro silica, natural pozzolana, fly ash, limestone powder, cement kiln dust, and pulverized steel slag, locally available in Saudi Arabia. The outcomes of the present work would be useful in producing UHPC mixtures by concrete industries in Saudi Arabia with a wider range of options regarding the use of readily available raw materials and meeting out the intended strength and durability requirements.

2 Experimental work

2.1 Materials

The ingredients used to prepare different mixtures of UHPC are listed in Table 1 along with their specific gravities. Locally available dune sand having fine quartz particles was used in its naturally-graded form in all the mixtures. The particle size distribution of the dune sand used in this study is given in Table 2. Due to very fine size range of the dune sand, the quartz powder was not used. A liquid superplasticizer (commercial name: Glenium 51) was used to obtain the desired flow. Glenium 51 is a polycarboxylic ether (PCE) based superplasticizer with 65% water content by weight that does not contain chlorides and complies with ASTM C494 Types A and F. Varying dosage of this superplasticizer was used to obtain a flow of 200 ± 20 mm for all the mixtures. Micro copper-coated steel fibres of 0.22 mm diameter and 13 mm length with an aspect ratio, l/d, of 59 were utilized. Fibres are made up of high strength steel (tensile strength greater than 2850 MPa). Locally available supplementary cementitious materials namely micro silica, natural pozzolana, limestone powder, cement kiln dust, and pulverized steel slag were used. The CaO and SiO_2 contents of all cementitious materials used in the present study are presented in Table 3.

Table 1. List of ingredients used for producing UHPC mixtures.

Ingredient	Specific gravity
Basic ingredients of UHPC	
Water	1.00
Cement (Type I conforming to ASTM C150)	3.15
Dune sand	2.53
Superplasticizer (Glenium® 51)	1.09
Steel fibers	7.85
Micro silica	2.25
Waste materials used to partially replace micro silica and sand	
Natural pozzolana	3.00
Fly ash	2.25
Limestone powder	1.60
Cement kiln dust	2.79
Pulverized steel slag	3.75

Table 2. Particle size distribution of dune sand used.

Sieve opening, mm	Cumulative % retained
4.75	0
2.4	0
1.2	0
0.6	3.8
0.3	38.6
0.15	78.1
0.075	99.0

Table 3. CaO and SiO_2 contents of the cementitious materials used.

Cementitious material	CaO content (%)	SiO_2 content (%)
Cement (Type I)	64.35	22.00
Micro silica	0.48	92.50
Natural pozzolana	8.06	42.13
Fly ash	8.38	45.30
Limestone powder	45.70	11.79
Cement kiln dust	49.30	17.10
Pulverized steel slag	40.80	16.47

2.2 Design of mixtures

Design of several trial mixtures of UHPC was carried out considering various design variables, such as water to cementitious material (w/cm) ratio, micro silica content, sand content, and fiber content. Each trial mixtures of UHPC were prepared and tested for flow before casting to test for compressive strength after 28-days of moist curing. Based on a maximum flow of 230 mm and a maximum 28-day compressive strength of 161 MPa, a basic optimum mixture of UHPC (designated as M0) having mixture proportion as follows: w/cm ratio = 0.145 (by mass); cement content = 900 kg/m^3; micro silica content = 220 kg/m^3; sand content = 1005 kg/m^3; water content = 162.4 kg/m^3; superplasticizer content = 40.3 kg/m^3 (3.6% of the mass of cementitious materials); and steel fiber content = 157 kg/m^3, was selected.

Taking the basic mixture of UHPC (M0) as a benchmark, various trial mixtures of UHPC were considered by partially replacing micro silica and dune sand using different permutations and combinations of the five waste materials, listed in Table 1, keeping everything same as of mixture M0. These trial mixtures were prepared and tested for flow and 28-day compressive strength and based on adequate flow and maximum 28-day compressive strength, a total of seven optimum mixtures of UHPC with partial replacement of silica fume and dune sand were selected. The details of eight optimally selected UHPC mixtures for detailed study (i.e., M0 without replacement by waste materials and M1 to M7 with replacement by waste materials) are presented in Table 4.

Table 4. Details of eight optimally selected UHPC mixtures for detailed study

Optimally fixed parameters for all 8 mixtures: w/cm ratio = 0.145 (by mass); Cement content = 900 kg/m^3; Mineral admixture content = 220 kg/m^3; Water (w) content = 162.4 kg/m^3; Superplasticizer content = 40.3 kg/m^3 (3.6% of the mass of cementitious materials); Steel fiber content = 157 kg/m^3								
Mixture ID	**Optimum use of waste materials in addition to MS**	**MS (kg/m^3)**	**Waste materials (kg/m^3)**					**Sand (kg/m^3)**
			NP	**FA**	**LSP**	**CKD**	**PSS**	
M0	Only MS	220	-	-	-	-	-	1005
M1	40% MS and 60% NP	88	132	-	-	-	-	1042
M2	40% MS and 60% FA	88	-	132	-	-	-	1005
M3	80% MS and 20% LSP	176	-	-	44	-	-	985
M4	80% MS and 20% CKD	176	-	-	-	44	-	1015
M5	10% sand replaced by LSP	220	-	-	87	-	-	868
M6	5% sand replaced by CKD	220	-	-	-	48	-	962
M7	5% sand replaced by PSS	220	-	-	-	-	48	972
MS: Micro Silica; NP: Natural Pozzolana; FA: Fly Ash; LSP: Lime Stone Powder; CKD: Cement Kiln Dust; PSS: Pulverized Steel Slag								

2.3 Tests

Flow test was conducted on each of the eight selected mixtures of UHPC in accordance to ASTM C1437 and subsequently test specimens were prepared to conduct tests for determining their compressive strength, flexural strength, fracture toughness, drying shrinkage, water absorption, DIN water penetration depth, chloride permeability, electrical resistivity, and reinforcement corrosion rate. The details of specimens and test methods used for determining various hardened properties of UHPC mixtures are given in Table 5.

3 Results and discussion

The summary of test results are presented in Table 6. It can be seen from Table 6 that almost all mixtures have satisfied the flow requirements (required range of flow is 180 mm to 220 mm).

All eight mixtures possess compressive strength above 150 MPa. The compressive and flexural tensile strengths and fracture toughness of the UHPC mixtures developed under this work are comparable to that of a patented UHPC mixture with a proprietary name Ductal® [24].

Except few mixtures, the shrinkage is in the same range as that for the normal concrete mixtures.

The water absorption for these UHPC mixtures is found to be around one-tenth of the permissible limits for normal concrete. Water penetration depth is zero in all cases indicating negligible water permeability. Chloride permeability and electrical resistivity values for all eight mixtures of UHPC are indicating "very low" to "negligible" risk of reinforcement corrosion. Very low corrosion current densities measured for all the UHPC mixtures (around one-tenth of the threshold value) have confirmed that they are highly corrosion- resistant.

Table 5. Details of specimens and test methods used to determine hardened properties.

Property	Specimen shape and size	Specimens per mixture	Test method
1. Compressive strength	50 mm cube	3	ASTM C 109
2. Flexural tensile strength (using 3 point bend test)	40 × 40 × 160 mm prism	3	ASTM C 78
3. Fracture Toughness	40 × 40 × 160 mm prisms (notched)	3	Jenq and Shah [23]
4. Drying shrinkage	25 × 25 × 275 mm prism	3	ASTM C 356
5. Water absorption	75 × 150 mm cylinder	3	ASTM C 642
6. Water penetration depth	100 mm cube	3	DIN 1048
7. Rapid chloride permeability	100 × 200 mm cylinder	3	ASTM C 1202
8. Electrical resistivity	75 × 150 mm cylinder	3	Two probe Wenner method
9. Reinforcement corrosion rate	A centrally embedded rebar in 75 × 150 mm concrete cylinder	3	Linear polarization resistance method

Table 6. Summary of test results of eight optimally selected UHPC mixtures

Mixture ID	Flow (mm)	28-d Compressive strength (MPa)	28-d flexural strength (MPa)	28-d fracture toughness (MPa √m)	Drying shrinkage strain after around 6 months (microns)	Water absorption (%)	Water penetration depth (mm)	Rapid chloride permeability (Coulombs)	Electrical resistivity (kΩ-cm)	Corrosion rate after exposure to chloride for six months (μA/cm²)
M0	230	161	31	8.0	450	0.43	0	9	179.6	0.038
M1	195	154	29	5.2	845	0.62	0	66	205.8	0.038
M2	210	158	32	6.5	446	0.40	0	75	110.6	0.017
M3	255	152	31	4.0	594	0.40	0	37	245.1	0.020
M4	220	153	25	4.4	585	0.35	0	78	182.1	0.033
M5	215	163	29	4.0	867	0.32	0	35	313.0	0.029
M6	180	153	26	4.0	1147	0.40	0	75	176.8	0.039
M7	210	161	24	4.5	718	0.41	0	49	215.4	0.026

4 Conclusions

Based on the results of the present study, following conclusions can be drawn:

1. The local dune sand and several waste materials available in abundance in Saudi Arabia have potential to produce UHPC mixtures.
2. The local dune sand was found to be naturally well graded cutting the need for quartz powder. Arranging quartz powder for producing UHPC is a big hassle and consumes a lot of time, energy and money.
3. The developed mixtures of UHPC having strength and durability several times more than the normal concrete could be utilized alternatively in special structural members, such as long span bridge girders, large-diameter concrete pipes, heavy-duty tanks, high-rise buildings, precast concrete members etc., and for repairing the deteriorated concrete structures.

The author gratefully acknowledges the financial support received from King Abdulaziz City for Science and Technology (KACST), Saudi Arabia, through King Fahd University of Petroleum & Minerals (KFUPM), Saudi Arabia, for carrying out this project under the National Science, Technology and Innovation Plan (NSTIP) funding (Project No. AM-08-2009, Code Number: 09-ADV751-04). The technical support received from the Civil & Environmental Engineering Department and the Center for Engineering Research at the Research Institute, KFUPM, are also acknowledged.

References

1. F. Larrard, T. Sedran, Cem. Concr. Res., **24**(6), 997 (1994)
2. P. Richard, M. Cheyrezy, Cem. Concr. Res., **25**(7), 1501 (1995)
3. J. Ma, H. Schneider, Lacer, **7**, 25 (2002)
4. G. Moriconi, *Proceedings of Jean Pera Symposium on Specialty Cements and Sustainability Issues*, 21 (2007)
5. C. Shi, Z. Wu, J. Xiao, D. Wang, Z. Huang, Z. Fang, Constr. Build. Mater., **101**, 741 (2015)
6. S. Ahmad, I. Hakeem, M. Maslehuddin, Eur. J. Environ. Civ. En., **20**(9), 1106 (2016)
7. Ch. Schröfl, M. Gruber, J. Plank, Cem. Concr. Res., **42**(11), 1401 (2012)
8. C. Shi, D. Wang, L. Wu, Z. Wu, Cem. Concr. Compos., **61**, 44 (2015)
9. H. Yazıcı, M.Y. Yardımcı, H. Yiğiter, S. Aydın, S. Türkel, Cem. Concr. Compos., **32**(8), 639 (2010).
10. H. Yazıcı, H. Yiğiter, A.S. Karabulut, B. Baradan, Fuel, **87**(12), 2401 (2008)
11. Yu, R., Spiesz, P. and Brouwers, H.J.H., Cem. Concr. Compos., **55**, 383 (2015)
12. Z. Wu, C. Shi, K.H. Khayat, S. Wan, Cem. Concr. Compos., **70**, 24 (2016)
13. M.S.M. Norhasri, M.S. Hamidah, A.M. Fadzil, O. Megawati, Constr. Build. Mater., **127**, 167 (2016)
14. A. Tafraoui, G. Escadeillas, T. Vidal, Constr. Build. Mater., **112**, 980 (2016)
15. K. Sobolev, Cem. Concr. Compos., **26**, 901 (2004)
16. C.M. Tam, V.W.Y. Tam, K.M. Ng, Mag. Concr. Res., **62**(10), 701 (2010)
17. M. Schmidt, E. Fehling, *Proceedings of Seventh International Symposium on the Utilization of High-Strength/High-Performance Concrete*, Washington, DC, USA, ACI SP 228, 51 (2005)
18. R. Yu, P. Spiesz, H.J.H. Brouwers, Cem. Concr. Res., **56**, 29 (2014)
19. Z.Y. Peng, *Ph.D. Thesis*, Wuhan University of Technology, China (2009)
20. A. Zubair, *M.S. Thesis*, King Fahd University of Petroleum & Minerals, Saudi Arabia (2012)

21. J. Dils, G.De. Schutter, V. Boel, Mater. Struct., **45**(11), 1673 (2012)
22. J. Dils, V. Boel, G.De. Schutter, Mater. Struct., **48**(11), 3485 (2015)
23. Y. Jenq, S.P. Shah, J. Eng. Mech., **111**, pp. 1227 (1985)
24. Ductal®, Chapter 2: Materials and Production. FHWA-HRT-13-060. Virginia, Federal Highway Administration (2013)

Technical and environmental characterisation of recycled aggregate for reuse in bricks

Sabrina Sorlini[1,*], *Luca* Rondi[1], *Nicola* Bettini[1], *Carlo* Collivignarelli[1], *Giovanni* Plizzari[1]

[a]Department of Civil Engineering, Architecture, Land, Environment and of Mathematics, University of Brescia, Via Branze 43, Brescia, 25123, Italy

Abstract. Waste mud coming from an aggregate washing plant was formerly used as filling material for a pond, aimed at the recovery of an abandoned quarry. Once completed the filling capacity of the pond, the need for identifying a possible reuse of mud produced by the plant arose in order to avoid landfill disposal. Therefore, mud has been geometrically, physically and chemically characterised for its recovery as construction material. A variety of tests was carried out on mud samples as required by EN technical specifications and by Italian environmental standards, focusing particularly on leaching behaviour. The tested material showed satisfactory physical and chemical properties and a release of pollutants below the limits set by the Italian code. Many mix-designs for the production of unfired bricks made of waste mud, sand and straw, stabilised and non-stabilised with lime, gypsum or cement, were developed. The bricks were tested in order to evaluate mechanical properties and leaching behaviour. Mud bricks provided remarkable compressive strength, even if not suitable for structural elements. The use as interior design to minimise humidity changes and to facilitate a thermal insulation is fostered, thus strengthening the so-called green building economy.

1 Introduction

The construction industry is a main issue for environmental sustainability because it produces large amounts of carbon dioxide, exploits natural resources and produces but rarely reuses/recovers wastes. A growing spread of the green building economy appears to be the only solution able to ensure sustainable development. Although in recent years the green building economy and the popular environmental awareness were mainly focused on energy, material (i.e. natural resources) saving and waste recycling have probably the most significance for green economic transformation. Following this idea, it will be possible to decrease quarrying activities (considering that natural aggregates are not a renewable resource) and amount of wastes to dispose of in landfills, thus avoiding environment degradation and pollution. For instance, in Italy, quarrying activities still represent the main source for aggregate supply in the construction sector, with an amount of natural resources extracted that was stated at about 120 million m3 in 2012 [1]. In the late years, industrial

* Corresponding author. sabrina.sorlini@unibs.it

waste management became an increasing problem, due to the restrictions in landfill disposal prescribed by international regulations. Therefore, waste recovery is considered as an increasingly necessary option, especially the reuse as recycled aggregate in construction can reduce the exploitation of natural quarries. Several types of waste can be recovered in this field, like steel slags, municipal solid waste incineration (MSWI) slags, construction and demolition (C&D) waste, mud from natural aggregate washing etc. Italian regulation on waste recovery allows the use of these materials both directly in the environment (road and railway basements, excavation filling etc.), and indirectly, through the production of concrete and bitumen [2, 3].

Several experimental researches investigated the opportunity to use waste mud or dredged sediments specifically for the production of bricks [4, 5, 6, 7]. Other investigations of mud properties covered the recovery as material for geopolymers production, as containment barrier or protection of groundwater and as wastewater treatment material [8, 9, 10].

This study aims at investigating the possible recovery of waste mud obtained from an aggregate washing plant as unfired bricks. An extensive experimental campaign was carried out in order to determine geometrical, physical and chemical characteristics of mud, according to [11]. Leaching behaviour was also carefully investigated. Bricks produced using only mud and fibre/lightening or stabilised with lime, gypsum or cement were designed and tested in order to determine mechanical properties and to verify the environmental suitability.

2 Experimental program

2.1 Recycled mud

The material employed in this research was mud obtained from an aggregate washing plant located in northern Italy (Bergamo, Italy). The plant treats, by means of crushing and sieving systems, inert material coming from quarries, construction and demolition waste and excavation of construction sites. The product obtained from this plant is an aggregate used in the construction sector (Fig. 1).

In order to optimize the aggregate granular separation, the raw materials are regularly washed with water in the sieving devices. The residue resulting from washing operations consists of wet fine aggregates, which is about 2-10% (w/w) with respect to the raw material treated in the plant. This mixture of fine aggregate and water is then treated by means of a sedimentation and a filter press system. In the settler, an anionic polyelectrolyte DRYFLOC PCP 1® based on polyacrylamide, soluble in water and insoluble in solvents, is added in order to facilitate the liquid-solid separation. The final fine aggregate (named "recycled mud") is mainly composed of silt, and in slight percentages of sand and clay.

Originally the mud was used as filling material for a pond, aimed at the recovery of an abandoned quarry. This study was aimed at researching an alternative solution for mud reuse in the construction sector in order to avoid landfill disposal.

A comprehensive characterisation was carried out on different waste mud samples, hereinafter named "S1", "S2", "S3", and "S4".

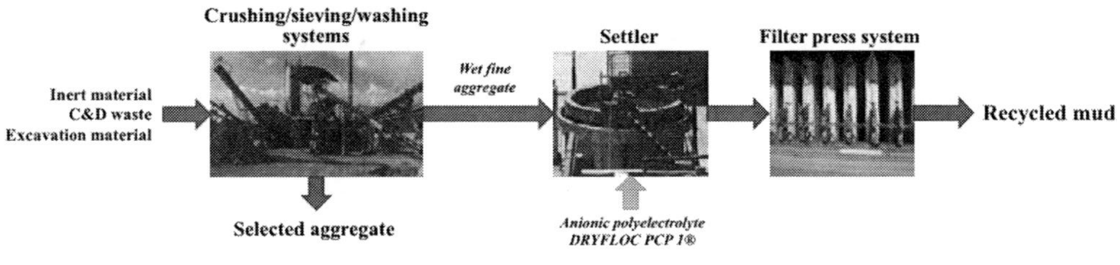

Fig. 1. Scheme of the aggregate washing plant producing the mud investigated in this research.

2.2 Recycled mud characterisation

Technical properties of mud samples (geometrical, physical and chemical) were determined in accordance with the UNI Norm for "Aggregates for mortar" [11] and its Italian accomplishment [12]. The latter provides, for several requirements, limit values in accordance with the different final destinations. The environmental characterisation of mud (chemical composition and leaching behaviour) was carried out in accordance with the Italian Norm [2], which rules over the waste characterisation, reuse and recovery, and the Italian Ministerial Decree that is the specific reference for the leaching test [3].

2.2.1 Geometrical and physical properties

The geometrical properties investigated in this study were the particle size distribution and the quality of fines. The granulometric composition was determined through the testing method [13] specifically addressed to aggregates characterised by particles with grain size lower than 75 μm. The quality of fine particles was carried out by means of two tests: sand equivalent and methylene blue value.

Regarding physical properties, the Atterberg limits were firstly determined for the assessment of mud plasticity behaviour, since materials mainly composed by fine particles (as silt and clay) are strongly influenced by the interaction between grains (and their mineralogical composition) and water. Particle density and water content tests were also carried out in order to acquire important values for the optimal design of final mixtures. The alkali-silica reactivity was investigated in compliance with [14], in order to evaluate the possible decay of mud bricks (in terms of cracks).

Each testing method is defined by [11] and refers to specific technical standards.

2.2.2 Chemical properties

The chemical characterisation of mud was carried out according to [11]. Water-soluble constituents, water-soluble chloride, water-soluble and acid-soluble sulphate, total sulphur, presence of humus and lightweight contaminants were the main properties investigated. Moreover, the organic contaminant content was determined (in compliance with [15]), in order to verify the presence of organic matter able to influence mechanical performances of mud bricks. Mortar samples of 40 x 40 x 160 mm with non-treated and pre-treated (by an incineration process at 480°C for 4 hours) mud samples were designed and tested to measure initial setting time and compressive strength.

Regarding the environmental suitability of mud samples, oxides content, chemical composition and release of pollutants tests were carried out. The X-Ray Diffraction analysis (XRD) was applied to determine the content of oxides, in compliance with [16]. The leaching test was carried out according to [17] based on contaminants extraction in a single stage in demineralised water (liquid/solid ratio, L/S = 10) for a period of 24 h [3].

2.3 Mud bricks production and characterisation

Several mixtures based on waste mud (dried at 105°C for 3 hours), binder, lightening and fibre (saturated with water in order to hinder its absorption during the mixture realisation) were produced and tested to identify a final product with mechanical properties suitable for the intended use. Table 1 lists all the mixtures performed.

Table 1. Mixtures performed for the mud brick production.

Mixture	Binder (w/w)	Lightening (v/v)	Fibre (v/v)
RM	-	-	-
RM-L6	L 6%	-	-
RM-L5-G5	L 5% + 5% G	-	-
RM-C6	C 6%	-	-
RM-L6-RH1:1	L 6%	-	RH 1:1
RM-L6-EV1:1	L 6%	EV 1:1	-
RM-L6-SD1:1	L 6%	-	SD 1:1
RM-L6-S1:1	L 6%	-	S 1:1
RM-L6-S1:0.5	L 6%	-	S 1:0.5
RM-L10-S1:0.5	L 10%	-	S 1:0.5
RM-G6-RH1:1	G 6%	-	RH 1:1
RM-G6-EV1:1	G 6%	EV 1:1	-
RM-G6-SD1:1	G 6%	-	SD 1:1
RM-C6-RH1:1	C 6%	-	RH 1:1
RM-C6-EV1:1	C 6%	EV 1:1	-
RM-C6-S1:1	C 6%	-	SD 1:1
RM-SA1:1-S1:1	-	S 1:1	-
RM-SA1:1-S1:1-EP1:0.5	-	EP 1:0.5	S 1:1
RM-SA1:1-S1:1-EP1:1	-	EP 1:1	S 1:1
RM-SA1:1-S1:1-SD1:1	-	SD 1:1	S 1:1
RM-SA1:1-L6-S1:1	L 6%	-	S 1:1
RM-SA1:1-L10-S1:1-EP1:1	L 10%	EP 1:1	S 1:1
RM-SA1:1-C3-S1:1	C 3%	-	S 1:1
RM-SA1:1-C6-S1:1	C 6%	-	S 1:1
RM-SA1:1-C6-S1:1.5	C 6%	-	S 1:1.5
RM-SA1:1-C10-S1:1	C 10%	-	S 1:1
RM-SA1:1-C10-S1:1-EP1:1	C 10%	EP 1:1	S 1:1

Legend: Recycled Mud (RM), Sand (SA), Lime (L), Gypsum (G), Cement (C), Expanded Vermiculite (EV), Expanded Perlite (EP), Rice Husk (RH), Saw Dust (SD), Straw (S).

In order to increase mechanical performances, lime (L), gypsum (G) and Portland cement 42.5 R (C) in different percentages were adopted as binder materials, whereas expanded vermiculite (EV) and expanded perlite (EP) were used as lightening to decrease the specific weight of mixtures. Finally, rice husk (RH), sawdust (SD) and straw (S) were adopted as fibre to increase strength, improve the lightening of the final mixtures and drain the moisture during the drying stage. Some mixtures were designed adding sand (in a weight ratio 1:1 compared to mud) in order to improve strength and geometric stability of bricks. Binders were added in different percentages relative to dried mud weight (w/w), whilst lightening and fibre relative to dried mud volume (v/v).

Mixtures without sand where prepared in cubes of 10 x 10 x 10 cm (handmade), whilst mixtures containing sand both in cubes of 10 x 10 x 10 cm (handmade) and in bricks with 5 x 10 cm cross section and variable length (extruded with a machine developed specifically to produce mud bricks [18].

Moisture, specific weight and compressive strength were determined for the different mixtures, then the best performing ones were subjected to a deeper investigation focused on direct tensile strength [19] and leaching behaviour [17].

3. Results and discussion

3.1 Recycled mud properties

3.1.1 Geometrical requirements

The results of the grain size distribution analyses are provided in Fig. 2. The tested material was composed by 80% silt, about 10% sand and 10% clay. Significant differences in the granulometric distribution amongst S1 and S2 samples were not highlighted, whilst a greater discrepancy could be noted relative to the ideal distribution curves for rammed earth and brick production [20]. The low sand content suggested the need for a particle size correction during the production of mud bricks, in order to guarantee a higher compressive strength.

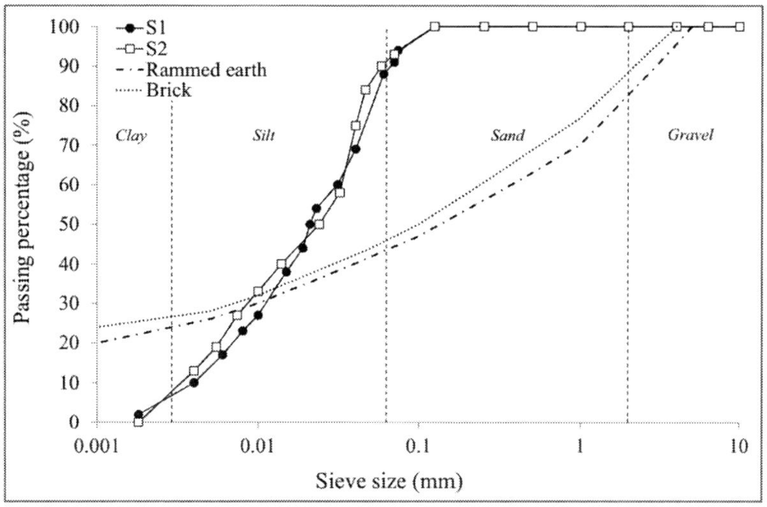

Fig. 2. Grain size distribution curves of recycled mud samples, in compliance with ASTM D 422-63, and ideal curves for rammed earth and brick production.

Test results for the assessment of fines quality (conducted on sample S1) indicated, as expected, the suitability of the material to adsorb ions or other elements in solution. Methylene blue test provided a high value of adsorption (3.1 mg/kg) considering that for the use of aggregates in concrete, the specific standard [21] fixes a value lower than 1.2 mg/kg. Sand equivalent resulted equal to 0.09% (the reference value for aggregates in concrete is set higher than 70%), meaning that the amount of ultrafine particles represented all the material analysed, whilst the sand component was almost no measurable.

 DOI: 10.1051/matecconf/201712003015

3.1.2 Physical requirements

Mud particle density was measured to be 2,728 kg/m^3. This value seemed slightly high, but suitable if typical values of bulk density (in the range 1,600-2,100 kg/m^3) are considered. As specified in [11], an aggregate with these characteristics (particle density higher than 2,400 kg/m^3) does not provide durability problems in mortars when subjected to phenomena of freezing and thawing. Water content was evaluated according to the testing method [22] and provided a value of 12% for S1 and S2 sample. These results were comparable to the ones obtained in a previous characterisation made on samples S3 and S4, where water contents were equal to 12 and 14% respectively. Even if water content strictly depends on the compression rate set in the filter press, these values permit a suitable recovery of waste mud in unfired bricks, since they are close to the optimum water content range (between 15 and 25%) that maximises plasticity and workability, as confirmed by [23].

Mud samples were analysed in order to determine the Atterberg limits, which provide important information on plasticity behaviour. Results indicate that waste mud samples were characterised by low plasticity (Fig. 3), thus a poor ability to permanently deform under stress conditions, and low compressibility, thus a poor attitude to volume reduction if subjected to compression. S1, S2 and S3 did not show an appropriate behaviour in terms of optimal extrusion (since they are located outside the grey reference area pointed out in Fig. 3), probably owing to the low clay fraction content. The presence of clay minerals, in fact, makes possible for particles to slide over each other due to the water retained in the interstitial spaces, thus favouring the extrusion [24].

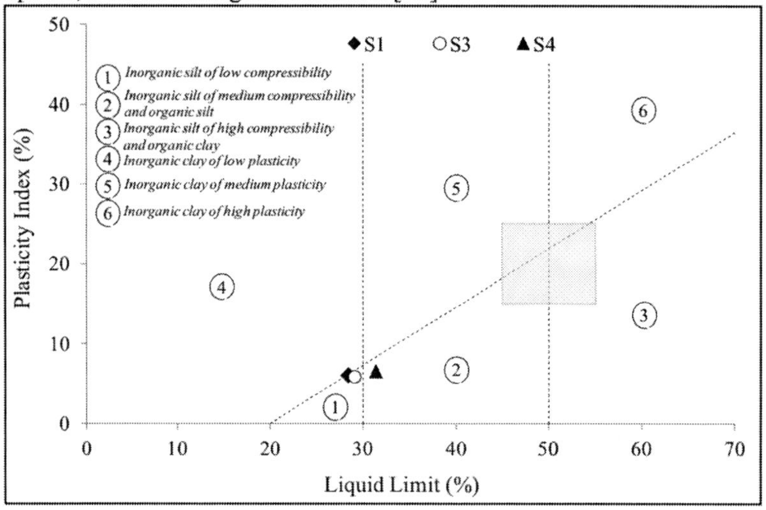

Fig. 3. Representation of mud samples in Casagrande's diagram. Grey rectangular area marks the optimal field for extrusion moulding, as highlighted in [25].

Alkali-silica reactivity was investigated in compliance with accelerated test proposed by [14]. After 14 days testing, the expansion of mud mortar samples was equal to 0.072%. This value was lower than the reference limit of 0.1%, thus avoiding the long-term expansion test and assuring the absence of cracking phenomena when waste mud is recovered in unfired bricks.

3.1.3 Chemical requirements

The mineralogical composition of mud sample S1 was investigated through the XRD analysis. Table 2 shows that the material was mainly composed of quartz and muscovite.

High concentrations of sulphates, chlorides, barium, total chromium, arsenic and zinc were found in the chemical composition of mud samples (Table 3).

Table 2. Mineralogical composition of sample S1 carried out by means of XRD test [16].

Parameter	Unit	S1
Quartz	% SiO_2	30
Muscovite	% $K_2Al_4(Si, Al)_8O_{20}(OH)_4$	27
Plagioclase	% $CaAl_2Si_2O_8$-$NaAlSi_3O_8$	14
Dolomite	% $CaMg(CO_3)_2$	13
Calcite	% $CaCO_3$	13
Chlorite	% $(Mg,Fe,Al)_6(Si, Al)_4O_{10}(OH)_8$	3

Table 3. Waste mud chemical composition and percentage of oxides content.

Parameter	Unit	Waste mud sample				Limits [2]
		S1	S2	S3	S4	
Sulphates	mg/kg	55.7	-	-	-	-
Chlorides	mg/kg	57.4	-	-	-	-
Antimony	mg/kg	5.8	< 5.0	-	-	10.0
Arsenic	mg/kg	35.6	18.2	13.9	-	20.0
Barium	mg/kg	237	237	-	-	-
Total chromium	mg/kg	16.8	28.4	5.8	35.2	20.0
Cobalt	mg/kg	7.6	-	-	-	150.0
Nickel	mg/kg	16.7	-	14.2	13.0	120.0
Lead	mg/kg	24.3	44.7	15.2	11.4	100.0
Copper	mg/kg	24.3	21.9	17.8	14.3	120.0
Vanadium	mg/kg	31.1	< 0.5	27.9	-	90.0
Zinc	mg/kg	128	279	88	50	150
Total silicon (SiO_2)	%	-	52.3	40.0	48.0	-
Iron oxide (FeO)	%	-	2.3	-	-	-
Calcium oxide (CaO)	%	-	< 1.0	-	-	-
Aluminium oxide (Al_2O_3)	%	-	2.7	-	-	-
Magnesium oxide (MgO)	%	-	5.9	-	-	-
Manganese oxide (MnO)	%	-	0.1	-	-	-

Note: in all the samples tested, the following pollutants had concentrations under the detection limit: F-, CN-, Be, Cd, Cr VI, Hg, Tl, Se, Sn, C<12, C>12, asbestos.

Regarding the latter three contaminants, in particular S1, S2 and S4 samples provided concentrations above the limit set by [3] that states a list of pollutant concentration limits above which a soil in a green residential, public and private area needs to be reclaimed. This means that these mud samples are not suitable for reuse in soil with this type of use. A slightly high variability in arsenic, total chromium, lead, vanadium and zinc content was highlighted amongst samples. This was probably due to the variability of the raw material fed and treated in the plant (construction and demolition waste, excavated soil and quarried aggregate).

The content of oxides characterising the inert materials was also evaluated (Table 3). S2 sample was characterised by a higher content of total silicon, as compared to sample S1 (analysed by means of XRD test). The comparison between values obtained in this study and results from other researches, where dredged sediments or sludge from aggregate washing plant were analysed [25, 5, 7], highlighted the same content of total silicon (average value of 55%), a lower content of aluminium oxide (average value of 13%) and a higher content of magnesium oxide (average value of 1.5%).

Chemical properties were determined [11, 13]. Table 4 provides the results of each parameter investigated and the related more restrictive limits (since different limits are proposed by the standards, in function of the different final destinations which the use of the mortar can be addressed to).

Table 4. Chemical properties of waste mud samples, in compliance with [15].

Parameter	Unit	S1	S2	Limits of UNI EN 13139 and UNI 11320
Water-soluble constituents	%	0.10	0.12	< 1.00
Water-soluble chloride	%	< 0.01	< 0.01	< 0.06[a]
Water-soluble sulphate	%	< 0.01	< 0.01	-
Acid-soluble sulphate	%	< 0.01	< 0.01	< 0.20[a]
Total sulphur	%	0.03	0.07	< 0.10[a]
Presence of humus	-	Colourless	Colourless	Colourless
Lightweight contaminants	%	0.02	0.03	< 0.25[a]

[a]: more restrictive limit

Acid-soluble and water-soluble sulphate content, water-soluble chloride content, water-soluble constituents and lightweight contaminants were in compliance with the reference limits and, in some cases, even under the detection limit for both samples. Total sulphur content test provided values below the limit, even if S2 sample (0.07%) was quite close to the reference limit (0.10%). Presence of humus was also detected, in order to verify if organic matter can influence the mechanical behaviour of mud when recovered in unfired bricks. The analysis carried out on samples S1 and S2 provided negative results, thus avoiding to further look for organic matter by means of fulvo acid content test, as specified by the reference standards.

Mortar samples produced with Portland cement (CEM I) and untreated or pre-treated (by an incineration process at 480°C for 4 hours) S1 specimen were analysed after 28 days of curing, in order to evaluate the possible influence of organic contaminants on compressive strength and initial setting time. This latter property was not evaluated owing to the low plasticity of silt that did not permit to correctly carry out the test. Compressive strength resulted to be equal to 16.0 MPa and 17.3 MPa for mortar samples, respectively made by using untreated and pre-treated waste mud, thus strengthening the slight influence of organic matter on mortar mechanical resistance.

Table 5 summarises results of leaching tests carried out in this research (samples S1 and S2) and in the previous characterisations of mud (samples S3 and S4).

Table 5. Concentrations of the pollutants in the leachate, obtained in compliance with [17].

Parameter	Unit	Waste mud sample				Limits [3]
		S1	S2	S3	S4	
Fluorides	mg/L	0.22	0.31	0.14	0.24	1.50
Sulphates	mg/L	4.88	4.58	4.65	3.48	250.00
Chlorides	mg/L	3.64	4.47	2.41	1.86	100.00

Arsenic	µg/L	< 5	< 1	35	< 5	50
Barium	mg/L	0.01	0.01	< 0.02	0.06	1.00
Beryllium	µg/L	< 5.00	< 5.00	6.55	< 5.00	10.00
C >12	mg/L	-	< 0.1	-	-	-
C <12	mg/L	-	< 0.1	-	-	-
PAHs	mg/L	-	< 0.05	-	-	-
DOC	mg/L	17.2	23.1	12.9	3.2	30.0
pH	-	6.5	7.4	-	-	5.5-12.0

Note: in all the samples tested, the following pollutants had concentrations under the detection limit: NO_3^-, CN^-, Cd, Co, Cr, Hg, Ni, Pb, Cu, Se, V, Zn, asbestos.

All the parameters analysed provided concentrations below the limits indicated by [2]. A quite significant variability (from a minimum value of 3.2 to a maximum of 23.1 mg/L) was observed for the release of DOC, comparing the results obtained in this and in the previous experimental campaign. In S2 sample, the release of Polycyclic Aromatic Hydrocarbons (PAHs) and heavy (C >12) and light (C <12) hydrocarbons was also investigated even if not required by [2]. The aim of these additional analyses was to verify the release of substances potentially present in excavated soils or construction and demolition waste treated inside the plant.

3.2 Unfired bricks performances

3.2.1 Mechanical strength

The production of brick characterised by different mix-design aimed at providing the best mechanical performances and the most eco-friendly product. The objective of the research was to define a mix-design with the higher compressive strength (in the range 1-2 MPa) accompanied by a low weight (specific weight from 1200 to 1500 kg/m^3).

The first mixture was composed only of waste mud. Then, mud brick composition was varied each time using different amounts and types of binder, lightening and fibre. The inadequacy of the waste mud particle size distribution required a correction of the mixtures with the addition of sand, in order to improve mud brick geometrical stability, reduce shrinkage and increase compressive strength. Table 6 shows results concerning physical-mechanical properties of different mud brick mixtures.

Mixtures prepared only with recycled mud and binder (RM, RM-L6, RM-L5-G5, RM-C6) demonstrated interesting compressive strength values (from a minimum of 0.95 to a maximum of 4.16 MPa), but related to high specific weight (from a minimum of 1570 to a maximum of 1710 kg/m^3).

For this reason, lightening and fibre were added in order to decrease final density, even if these elements brought to decrease mechanical resistance under the 1 MPa originally set as research goal. Amongst this second mixture category, indeed, only RM-C6-EV1:1 provided suitable results regarding both specific weight (1.42 kg/dm^3) and compressive strength (1.76 MPa).

Table 6. Values of moisture, specific weight and compressive strength for the different mixtures.

Mixture	Moisture (%)	Specific weight (kg/dm^3)	Compressive strength (MPa)
RM	29.6	1.71	1.61
RM-L6	44.1	1.65	1.57
RM-L5-G5	34.6	1.57	0.95
RM-C6	35.3	1.71	4.16

RM-L6-RH1:1	67.5	1.11	0.40
RM-L6-EV1:1	89.9	0.97	0.24
RM-L6-SD1:1	60.4	1.41	0.67
RM-L6-S1:1	58.9	1.17	0.48
RM-L6-S1:0.5	50.4	1.33	0.55
RM-L10-S1:0.5	54.6	1.34	0.79
RM-G6-RH1:1	50.6	1.51	0.56
RM-G6-EV1:1	55.6	1.33	0.69
RM-G6-SD1:1	56.9	1.64	0.71
RM-C6-RH1:1	56.5	1.41	0.42
RM-C6-EV1:1	52.4	1.42	1.76
RM-C6-S1:1	57.0	1.47	0.87
RM-SA1:1-S1:1	25.4	1.50 (CU) - 1.58 (BR)	0.48 (CU) - 2.08 (BR)
RM-SA1:1-S1:1-EP1:0.5	44.0	1.44 (CU) - 1.32 (BR)	0.57 (CU) - 2.17 (BR)
RM-SA1:1-S1:1-EP1:1	n.a.	1.38 (CU) - 1.22 (BR)	0.47 (CU) - 1.45 (BR)
RM-SA1:1-S1:1-SD1:1	41.0	1.37 (CU) - 1.33 (BR)	0.80 (CU) - 2.06 (BR)
RM-SA1:1-L6-S1:1	49.0	1.42 (CU) - 1.37 (BR)	0.67 (CU) - 1.47 (BR)
RM-SA1:1-L10-S1:1-EP1:1	n.a.	1.35 (CU) - 1.32 (BR)	0.82 (CU) - 1.29 (BR)
RM-SA1:1-C3-S1:1	29.0	1.50 (CU)	0.43 (CU)
RM-SA1:1-C6-S1:1	n.a.	1.52 (CU)	1.01 (CU)
RM-SA1:1-C6-S1:1.5	40.0	1.48 (CU)	0.99 (CU)
RM-SA1:1-C10-S1:1	49.0	1.51 (CU) - 1.43 (BR)	1.81 (CU) - 3.86 (BR)
RM-SA1:1-C10-S1:1-EP1:1	n.a.	1.50 (CU)	1.85 (CU)

Legend: Recycled Mud (RM), Sand (SA), Lime (L), Cement (C), Gypsum (G), Rice Husk (RH), SawDust (SD), Straw (S), Expanded Vermiculite (EV), Expanded Perlite (EP), Cubes (CU), Bricks (BR), data not available (n.a.).

The addition of sand was the key element in order to gain the performance objectives. Moreover, the production of both cubes and bricks highlighted that the latter ones had better properties in terms of compressive strength and specific weight, as expected, due to a favourable aspect ratio and to the production via mechanical extrusion. Comparing same mixtures, in fact, extruded bricks showed a higher compressive resistance (up to about 4) with respect to handmade cubes, owing to the lower slenderness and to the reduction of eigenstresses during curing.

Adding a more eco-friendly binder as lime, mechanical properties of mixtures did not significantly vary. Longer curing in wet (or with high relative humidity) environment of specimen containing lime should however be considered to obtain adequate performances in terms of compressive strength.

Finally, the dosage of cement clearly increased the compressive resistance (on average equal to 1.42 MPa for cubes compared to the previous 0.61 MPa for mixtures without cement), as expected, but with a subsequent increase of specific weight (from an average value of 1.42 kg/dm^3 of previous mixtures to 1.50 kg/dm^3 of the latter ones containing cement).

Three out of all the mixtures tested were then considered for a deeper investigation. Table 7 shows the composition of mud bricks specifically produced for further analyses.

Table 7. Composition of mud bricks produced for deeper characterisation.

Mud brick	Silt and sand (g)	Straw (g)	Water (g)	Cement (g)
A (stabilised)	7,800	224	1,800	0
B1 (non-stabilised)	8,000	226	2,300	400
B2 (non-stabilised)	8,000	227	2,500	800

Mixture A was not stabilised with any binder, whereas mixtures B1 and B2 were stabilised with cement in percentages of 5 and 10% respectively. All of these three new mixtures were designed in bricks of 40 x 40 x 160 mm side, according to the requirements set by [19]. Table 8 exhibits the obtained results, indicating that the three proposed mixtures might be considered a good basis for the realisation of non-structural masonry elements.

Table 8. Mechanical properties of the three new mixtures.

Mud brick	Specific weight (kg/dm^3)	Compressive strength (MPa)	Direct tensile strength (MPa)
A	1.65	1.20	0.31
B1	1.62	1.36	0.57
B2	1.63	1.54	1.03

The non-stabilised mud brick, owing to the lower values of compressive and tensile strength, can be used for specific architectural solutions (if subject to limited stress), in combination, for example, with load bearing timber structures. In addition, it can be useful as hygrometric stabiliser in environments exposed to significant moisture variations and as soundproofing element. Conversely, stabilised solutions can be used for the same purposes mentioned above, but even for outdoors and structural elements subjected to higher loads.

3.2.2 Environmental suitability

Stabilised (B2) and non-stabilised (A) mud brick samples were analysed by means of leaching test, in order to verify the environmental suitability.

The release of pollutants from the experimental bricks [17] was below the limits [3] for almost all samples (Table 9). Unexpected concentrations above the limits were provided by copper (0.09 with a limit of 0.05 mg/L), nickel (66 and 17, respectively for stabilised and non-stabilised mud bricks, with a limit of 10 μg/L), total chromium (71 with a limit of 50 μg/L) and DOC (156 and 65, respectively for B2 and A mud brick specimens, with a limit of 30 mg/L). Since waste mud samples did not provide releases of those pollutants higher than the reference limits (as reported in Table 5), the exceedance of the maximum concentrations allowed [2] was probably due to the materials mixed to the recycled mud for unfired brick production. In particular, the high concentration of DOC can be traced back to straw, whilst metals to cement. Nevertheless, limits set by the Italian legislation are referred to aggregates (i.e. waste mud) and not to final products (i.e. unfired bricks).

Table 9. Major concentrations of the pollutants released through the leaching test, in compliance with

Parameter	Unit	Mud brick		Limits [3]
		A	B2	
Fluorides	mg/L	< 0.2	0.9	1.5
Sulphates	mg/L	15.7	31.1	250.0
Chlorides	mg/L	8.9	15.3	100.0
Cyanide	μg/L	< 0.02	7.90	50.00
Arsenic	μg/L	14.8	< 5.0	50.0
Barium	mg/L	0.07	0.10	1.00
Copper	mg/L	0.02	0.09	0.05
Zink	mg/L	0.010	< 0.005	3.000
Nickel	μg/L	17	66	10
Total chromium	μg/L	< 5	71	50
DOC	mg/L	65	156	30
pH	-	8.9	11.5	5.5-12.0

Note: in all the samples tested, the following pollutants had concentrations under the detection limit: NO^{-}_{3}, Be, Co, V, Cd, Pb, Se, Hg, C<12, C>12, PAHs, asbestos.

4. Concluding remarks

The possibility of making unfired bricks using recycled mud coming from an aggregate washing plant was investigated in this study. The conclusions derived from this research are as follows:

- The geometrical characterisation of recycled mud highlighted, as expected, a lack of sand in the grain size distribution analysis. In order to guarantee higher compressive strength to mud bricks, a particle size correction adding a coarser fraction (i.e. sand) is therefore required.
- The evaluation of waste mud physical requirements provided suitable results. Mud particle density was higher than 2,400 kg/m3, thus guaranteeing (according to EN 13139 standard) mortars durability if subjected to freeze-thaw phenomena. The determination of Atterberg limits indicated that the recycled mud was characterised by low plasticity and low compressibility. Finally, alkali-silica reactivity was not detected, thus assuring the absence of cracking phenomena in unfired bricks.
- The evaluation of chemical requirements [11] provided, for all waste mud samples and parameters investigated, values below the more restrictive limits set by the reference standard.
- Concerning chemical properties, recycled mud samples highlighted a slightly high variability in chemical composition (mostly in arsenic, total chromium, lead, vanadium and zinc content), probably due to the variability of the material treated in the aggregate washing plant and for which an accurate selection should be carried out. The leaching test provided concentrations well below the limits [3] for all the parameters.
- Unfired bricks made by using recycled mud provided remarkable compressive resistances, even if not suitable for load-bearing structural elements. For being suitable for structural elements, mud bricks should however be designed adding lime or cement.
- Non-stabilised mud bricks can be used for plastic and creative architectures, featuring natural colours and by an inherent sense of warmth. The high hygroscopic adjustment

capacity makes mud bricks suitable for the use in areas subjected to high daily variations in moisture. Finally, they can be even employed as soundproofing elements.

- Stabilised (using lime or cement) mud bricks can be used for the same purposes mentioned above, but even for outdoors and structural elements subjected to higher loads. These mud bricks, however, lack in hygroscopic adjustment capacity, compared to the non-stabilised ones.

This study was conducted under the research agreement "Recovery of mud coming from quarry activities in the construction sector" between the University of Brescia and the company "Cava al Tiro srl" (Martinengo, Bergamo, Italy).

The Authors wish to thank all the "Cava al Tiro" staff and the students (J. Donneschi, V. Andreoli, F. Lodrini, C. Zenari and F. Beretta) for the collaboration in the research project.

Sabrina Sorlini planned and supervised the research activities and the paper drafting; Luca Rondi did data collection and analysis on mud characterisation, and was responsible for the paper drafting; Nicola Bettini did data collection and analysis on unfired brick characterisation; Carlo Collivignarelli and Giovanni Plizzari were responsible for the agreement between University of Brescia and "Cava al Tiro srl".

References

1. Legambiente, Rapporto Cave 2014 - I numeri, il quadro normativo, il punto sull'impatto economico e ambientale dell'attività estrattiva nel territorio italiano. Ufficio Urbanistica di Legambiente. Available at: http://www.legambiente.it/sites/default/files/docs/rapporto_cave.pdf (2014).
2. Legislative Decree 04/03/2006 n. 152. Norme in materia ambientale. Published on Gazzetta Ufficiale n. 88 dated 04/14/2006 - Supplemento Ordinario n. 96.
3. Ministerial Decree 04/05/2006 n.186. Regolamento recante modifiche al Decreto Ministeriale 5 Febbraio 1998 "Individuazione dei rifiuti non pericolosi sottoposti alle procedure semplificate di recupero, ai sensi degli articoli 31 e 33 del Decreto Legislativo 5 Febbraio 1997 n. 22". Published on Gazzetta Ufficiale n. 115 dated 05/19/2006.
4. H. Binici, O. Aksogan, T. Shah, Constr. Build. Mater. **19**, 313-318 (2005).
5. K. Hamer, V. Karius, Waste Manage. **22**, 521-530 (2002).
6. H. Liu, H. Yang, Adv. Mater. Res. **641-642,** 492-495 (2013).
7. A. Mezencevova, N.N. Yeboah, S.E. Burns, L.F. Kahn, K.E. Kurtis, J. Environ. Manage. 113, 128-136 (2012).
8. G. De Los Cobos, A. Parriaux, Bull. Int. Assoc. Eng. Geol. **52**, 53-58 (1995).
9. C. Lampris, R. Lupo, C.R. Cheeseman, Waste Manage. **29**, 368-373 (2009).
10. D.S. Parker, G.A. Horstkotte, G.A. Carthew, J. Env. Eng. Div. ASCE **101**, 985-1004 (1975).
11. UNI EN 13139. Aggregates for mortar (2003).
12. UNI 11320. Aggregates for mortar – Additional provision, for the application of EN 13139 (2009).
13. American Society for Testing and Materials, ASTM D 422-63, 2007. Standard Test Method for Particle-Size Analysis of Soils.
14. UNI 8520-22. Aggregates for use in concrete – Determination of potential reactivity of alkali in aggregates (2002).
15. UNI EN 1744-1. Tests for chemical properties of aggregates - Part 1: Chemical analysis (2010).

16. UNI EN 13925-2. Non-destructive testing - X-ray diffraction from polycrystalline and amorphous materials - Part 2: Procedures (2003).
17. UNI EN 12457-2. Leaching: Compliance test for leaching of granular waste materials and sludges - Part 2: One stage batch test at a liquid to solid ratio of 10 L/kg for materials with particle size below 4 mm (2004).
18. J.P. Temga, A. Mazzù, J.P. Nguetnkam, D. Palazzini, R. Ndjouenkeu, F. Vitali, Int. J. Sust. Eng. (2013).
19. UNI EN 1015-11. Methods of test for mortar for masonry - Part 11: Determination of flexural and compressive strength of hardened mortar (2007).
20. A. Gottfried, *Quaderni del Manuale di progettazione edilizia*, Hoepli, Milano (Italy), ISBN 88-203-3220-5 (2003).
21. UNI 12620. Aggregates for concrete (2013).
22. CNR UNI 10008. Aggregate natural water content (1963).
23. M. Achenza, U. Sanna, I manuali del recupero dei centri storici della Sardegna – Il manuale tematico della terra cruda. ITACA (2006). Available at http://www.sardegnadigitallibrary.it/mmt/fullsize/2009022319494200003.pdf
24. M.J. Ribeiro, J.M. Ferreira, J.A. Labrincha, Ceram. Int. **31**, 515-519 (2005).
25. B. González-Corrochano, J. Alonso-Azcárate, M. Rodas, J. Environ. Manage. **90**, 2801-2812 (2009).

MATEC Web of Conferences **120**, 04001 (2017) DOI: 10.1051/matecconf/201712004001
ASCMCES-17

Performance of steel-making slag concrete reinforced with fibers

Vanesa Ortega-López[1,*], *José Antonio Fuente-Alonso*[2], *Marta Skaf*[2], *Amaia Santamaría*[3], *Ángel Aragón*[1], *and Juan Manuel Manso*[1]

[1] Department of Civil Engineering/University of Burgos. C/Villadiego s/n. 09001 Burgos, Spain.
[2] Department of Construction/University of Burgos. C/Villadiego s/n. 09001 Burgos, Spain.
[3] Department of Mining and Metallurgical Engineering/University of Basque Country. Alameda Urquijo s/n. 48013 Bilbao, Spain.

Abstract. In this research, the possibility of making concrete reinforced with fibers and manufactured with recycled aggregates from carbon steel production was explored. Electric arc furnace slag (EAFS) was used as coarse and medium aggregate, and part of the sand sizes. Metallic and synthetic fibers were added in different amounts. Initially, the properties of EAFS and their suitability to be used in the manufacture fiber reinforced concrete were analysed. Then, a series of fiber reinforced concrete mixtures were developed incorporating EAFS, and they were compared with the reference mixtures, made with conventional components plus fibers and made with EAFS without fibers. A series of tests were performed, including concepts such as consistency, compressive strength, flexural strength, splitting tensile strength, resistance to water penetration or toughness. The results show that it is possible to make a suitable steel-slag concrete reinforced with fibers, complying with the standard requirements for it use in pavements and slab, and improving their proprieties respect to the control mixtures.

1 Introduction

Electric Arc Furnace Slag (EAFS) is a by-product of the steel electric industry, produced following the melting of scrap steel. It is usually presented as gravel sized aggregate.

In Spain, approximately 70% per cent of total steel production (14 millions of tons) is from steel electric furnaces, together with 1MT of EAFS [1].

The steelmaking industry is mainly concentrated in the north of Spain. Therefore, it is so important for us to reuse this by-product in order to reduce slags landfilling and contribute to global sustainability.

In general, steelmaking slags are products which have certain risk of expansion, mainly due to a potential presence of free lime and periclase. So, an outdoor weathering and some additional cautions are necessary to prevent the occurrence of any expansive phenomena [2-

* Corresponding author: vortega@ubu.es

6]. In the last decades, important works on EAFS [7-10] and LFS characterization [11-13] have been published; it has been demonstrated the suitability of both slags in different applications, such as bituminous mixtures [14-18], concrete and mortars [19-34], self-compacting concrete [35-37], making clinker [38], soil stabilization [39-41] and others [42, 43]. Despite these recycling possibilities, in Spain, approximately 23% of the electric slag is accumulated in landfill sites.

On the other hand, the construction sector is an important consumer of natural aggregates. Total consumption of natural aggregates in Spain is around 90 MT per year, resulting in a consumption of natural resources and exploitation of quarries.

With this work, we tried to convert the EAFS in steel aggregate for manufacturing steel-slag concrete.

The aim of the research carried out to outline this paper, was based on the reuse of EAF slags in the construction of concrete pavements and concrete slabs reinforced with fibers, in order to avoid the spread of the superficial cracks during the initial drying of the concrete, to improve its flexural and splitting tensile strength and to increase its capacity to absorb rupture energy.

For that, the slag was used as coarse and medium aggregate in the concrete mixture, and the sizes of sand were completed with 50% of slag and 50% of siliceous sand; which is mean approximately 75% of EAFS and 25% of natural aggregate in the finale concrete mixture. The siliceous sand, with rounded morphology, counteracted the effect of the surface irregularity of the EAFS, improving the fluidity throughout the concrete mass [44]. Furthermore, different dosages and types of fibers (metallic and synthetic) were studied in order to determine which are the most suitable for the aim proposed.

2 Materials

The following materials were used in this research:

- Ordinary Portland Cement (OPC) CEM I 42,5R.
- Water from an urban water supply
- Natural siliceous aggregates: It has been provided in three granulometric sizes: 0/4, 4/12 and 12/25 mm; the size limestone aggregate employed in the EAFS concrete was the sand 0/4 mm, with fine content according with UNE-EN-933-1of [45, 46]1.58% and sand equivalent according with UNE-EN 933-8 of 89.
- Superplasticizers: Policarboxilato modified with water. Density of 1.08 g/cm3, pH of 5 and solid content of 36%.
- EAFS aggregates: It has been provided in three granulometric sizes: 0/4, 4/10 and 10/20 mm (Figure 1). The lack of fines is appreciated on the sand slag. EAFS is heavy, with density over 3500 kg/m^3 and a very resistant aggregate (loss Angeles Loss: 24%, Flakiness index: 3%). The main compounds of the EAFS are shown in Table 1.
- Metallic fibers (RL-45/50-BN): hard drawn steel filaments with formed ends to improve adhesion; Synthetic fibers (M-48): polyolefins curly monofilament (Figure 2).

Table 1. Main chemical composition of the EAFS.

Component	CaO	SiO2	MgO	Al_2O_3	Fe_2O_3	MnO
EAFS wt.-%	27.7	19.1	2.5	13.7	26.8	5.3

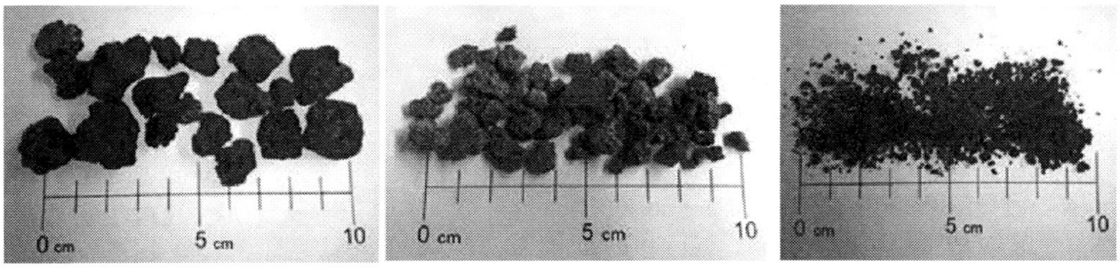

Fig. 1. EAFS granulometric sizes.

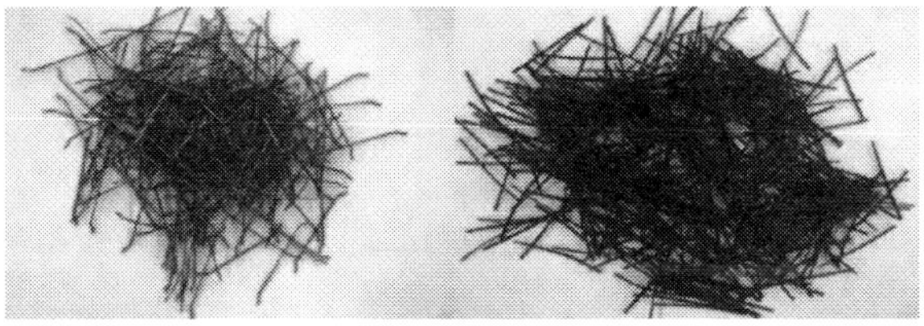

Fig. 2. Metallic and Synthetic fibers.

3 Mix design

Some dosages were used for the reference mixtures, which had previously been endorsed by other authors in bibliography:

- Reference mixture **P**, was a mixture with natural aggregates and 30 kg of metallic fibers per m^3 of concrete, according with Turmo et al [47].
- Reference mixture **E**, was a mixture with 75% of EAFS aggregates and 25% of siliceous sand without fibers, according with Manso et al [30].

Other mixes under study in this research (**EM1**, **EM2**, **ES1**, **ES2**) had the same dosage as the reference mix E, but including two types of fiber reinforcements (metallic and synthetic) in different quantities, within the range of 0.4-0.6% of fibers by unitary volume of concrete. Table 2 shows the compositions of the mixes under study.

For all mixtures, the following parameters were constant: amount of cement: 363 kg/ m^3; relation water/cement (w/c): 0.5; relation coarse aggregate/fine aggregate/cement: 3/3/1.

Table 2. Mix proportioning of components in manufactured concretes.

Mix design (kg/m^3)		P	E	EM1	EM2	ES1	ES2
Cement		360	360	360	360	360	360
Water		180	180	180	180	180	180
Siliceous aggregates	Size 0/4 mm	800	500	500	500	500	500
	Size 4/12 mm	575	-	-	-	-	-
	Size 12/20 mm	465	-	-	-	-	-
EAFS aggregates	Size 0/4 mm	-	515	515	515	515	515
	Size 4/10 mm	-	670	670	670	670	670
	Size 10/20 mm	-	550	550	550	550	550
Plasticizer (1-1.5% wt. of cement)		3.63	5.44	5.44	5.44	5.44	5.44
Metallic Fibers		30	-	30	45	-	-
Synthetic Fibers		-	-	-	-	3.5	5

DOI: 10.1051/matecconf/201712004001

4 Results and discussion

4.1 Consistency of freshly mixtures

The consistency of freshly mixed concrete was measured with Abrams Cone, according to UNE 12350-2, before and after including the fibers. The results show that the mixture with natural aggregates and fibers had soft/plastic consistency and the rest of the mixtures manufactured with EAFS or EAFS plus metallic/synthetic fibers had dry consistency.

4.2 Compressive strength

The compressive strength was measured on cylindrical concrete specimen, with 150 mm of diameter and 300 mm of height, cured during 28 days in moist chamber at 20+/-2 °C and 95% of moisture. The assay was performed in triplicate. According to UNE 83507, the results obtained are shown in Table 3.

Results show higher compressive strength in the concretes with steel-slag aggregates than the concretes with natural aggregates, even when those contained reinforcing fibers. The mixture E (with EAFS without fibers) had an increase of 43% of compressive strength respect to the mixtures P (with natural aggregates and fibers).

EM2 and ES1 are the mixtures with better results in this test, with an increase of 10-12% of compressive strength with respect to the same concrete without fibers. These results are according with other authors who said that fiber volumes added to concrete mixes at 0.5%, 1.0% and 1.5% by volume of concrete improve the compressive strength between 4% and 19% [48].

Table 3. Compressive strength, Flexural strength and Splitting tensile strength.

Mixture	Compressive Strength 28 days (MPa)	Flexural Strength 28 days (MPa)	Splitting Tensile Strength 28 days (MPa)
P	46.30	5.15	4.33
E	66.05	6.80	4.20
EM1	54.00	5.95	5.01
EM2	72.62	7.00	5.45
ES1	74.04	7.13	5.23
ES2	66.50	6.88	4.59

4.3 Flexural strength

The flexural strength was measured on concrete specimen with dimensions 150x150x600 mm and 100x100x400 mm, cured in moist chamber during 28 days. According to the UNE 83509, the load was applied in two spaced points between them 1/3 of the length of the specimen (Figure 3). The results obtained are shown in Table 3.

All the steel-slag concretes had better results than the conventional one, even with fibers. The average of the flexural strength for steel-slag concretes with fibers had an increase upper than 30% respect to the reference mixture P.

It was well known that the steel-slag concrete improves the flexural strength [6] and that the fibers provide better deformation performance by flexion [49] according to the results obtained for concrete manufactured with EAFS and fibers.

 DOI: 10.1051/matecconf/201712004001

The results were similar to the previous ones of compressive strength, reaching the best results the mixtures EM2 and ES1.

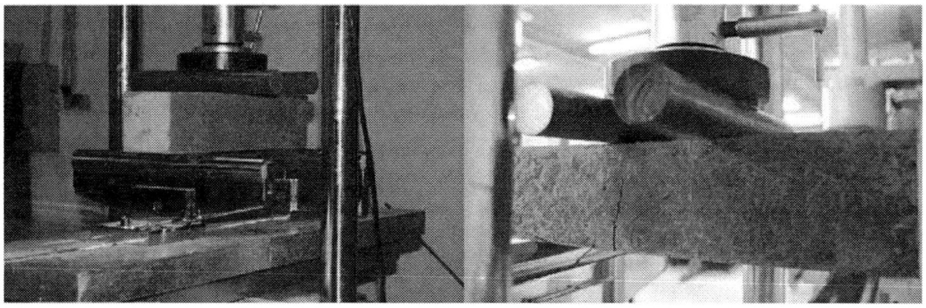

Fig. 3. Flexural strength test.

4.4 Splitting tensile strength

The splitting tensile strength was measured on cylindrical concrete specimen with diameter of 150 mm and height of 300 mm, cured in moist chamber during 28 days. According to the UNE 12390-6, the load was applied over two opposite lines until failure. The results provide the maximum load that the pavement is able to support (Figure 4). The results obtained were also shown in Table 3.

All the mixtures with fibers had better performance at splitting tensile strength than the mixture E (without fibers), especially the mixtures with EAFS and fibers, EM2 and ES1. It is possible that higher amount of synthetic fibers in mixture ES2 than in mixture ES1 damages the strength characteristics of the hardened concrete, well by its problematic mix process or by the low hydration of the concrete components.

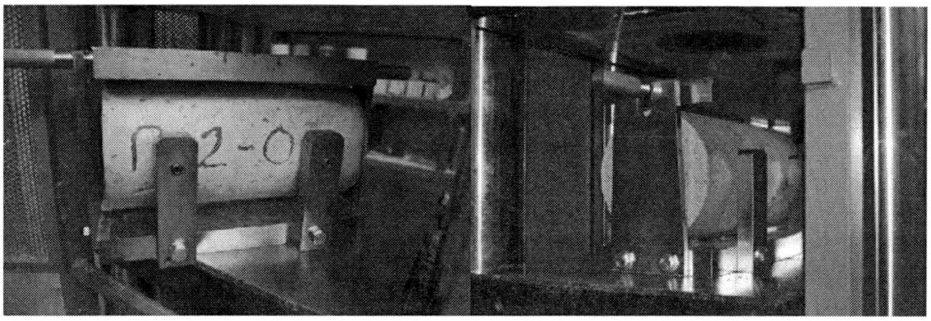

Fig. 4. Splitting tensile strength test

4.5 Resistance to water penetration

The resistance to water penetration under pressure was measured on cylindrical concrete specimen with diameter of 150 mm and height of 300 mm, cured in moist chamber during 28 days. According to the UNE EN 12390-8, the water was applied under pressure of 500+/-50 kPa during 72 hours in one of the specimen faces. After that, the specimen was broken and the depth of water penetration was measured. The results of the maximum depth of water penetration are shown in Figure 5.

All the mixtures with EAFS had better resistance to water penetration than the conventional concrete with natural aggregates. However, in this test, the fibers in the concretes manufactured with EAFS did not improve the results respect to concrete without them.

Even so, all the mixtures fulfilled the requirements of the EHE standard [50], which specifies, for the worst environmental exposure, the value of maximum depth of water penetration under 30 mm. Therefore, these concretes were considered impermeable enough for its use in pavements.

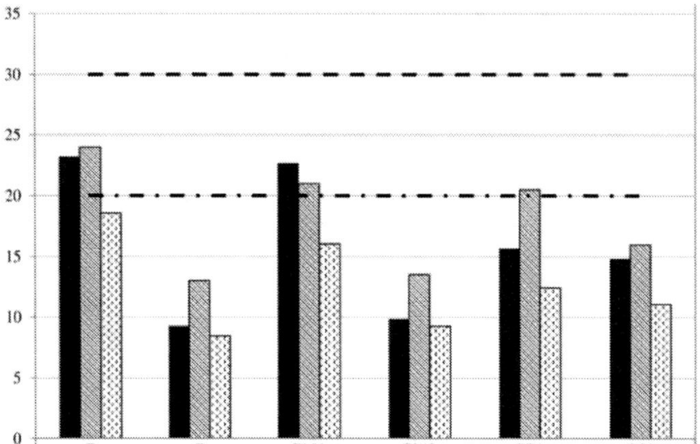

	P	E	EM1	EM2	ES1	ES2
Water penetration area (cm^2)	23.21	9.28	22.70	9.85	15.67	14.79
Maximum depth (mm)	24.00	13.00	21.00	13.50	20.50	16.00
Average depth (mm)	18.57	8.44	16.04	9.28	12.42	11.07
Maximum depth in EHE (mm)	30	30	30	30	30	30
Average depth in EHE (mm)	20	20	20	20	20	20

Fig. 5. Depth of water penetration under pressure.

4.6 Toughness by compression

The toughness is the energy required for a prespecified total deterioration or breakage of the material. It is one of the most important characteristics that the fibers provide to the material.

The toughness by compression was measured on cylindrical concrete specimen with diameter of 150 mm and height of 300 mm. According to the UNE 83508, the results provide the area bounded by the load-deformation curve from the origin 0, and the ordinate corresponding to a deformation of 1.125 mm.

The mixture EM2 (toughness of 1949300 N.mm) had better results than the mixture ES1 (toughness of 1887900 N.mm), which means that metallic fibers provided greater post-cracking strength and higher increase of concrete ductility.

4.7 Impact strength

Another important characteristic that the fibers bring to the concrete, once again directly related with toughness, is increased concrete strength against impacts, collapse, and other dynamic loading. A drop weight impact test was applied to cylindrical concrete specimens, in accordance with the UNE 83514 [45] standard. The test results, shown in Table 4, record the number of hits until the first crack appears and the number of hits up until breakage of the specimen.

In view of test results, the fiber-reinforced concretes (i.e EM2, ES1, P) clearly appear to have much better impact strength than the equivalent not-reinforced concrete (E, P without fibers), both against the first signs of cracking and especially against breakage. Additionally, the concrete mixes with electric arc furnace slag (i.e E, EM2, ES1) showed better impact strength than the corresponding mixes made with conventional aggregates (P not fibers, P).

Table 4. Impact and abrasion resistance.

Property		E	EM2	ES1	P	$P_{not\ fibers}$
Impact	Number of hits until first crack	11	46	24	19	6
Strength	Number of hits until breakage	13	155	88	50	8

5 Conclusions

The conclusions from this work can be summarized as follows:

- Steel-slag concretes reinforced with fibers, metallic or synthetic, at around 0.4-0.6% by volume of concrete, provided concretes with suitable mechanical behavior.
- The mixture reinforced with metallic fibers provided slightly better results of Toughness and Impact Strength than the concrete reinforced with synthetic fibers.
- Due to the poor workability and docility of the fiber reinforced concretes manufactured with EAFS, it is recommended a water/cement ratio over 0.5 and the use of plasticizer additive, as well as the incorporation of siliceous sand in order to compensate the lack of fines of the EAFS.

The authors wish to express their gratitude to the Junta de Castilla y León (Regional Government), to the Spanish Ministry (MINECO) for FEDER Funds through project BIA2014-55576-C2-1-R, which contribute to financing this research work. It could be not have gone ahead without the materials supplied by Hormor-Zestoa, Syka admixtures and beaker and Cementos Alfa.

References

1. Steel Statistical Yearbook : World Steel Association (2014).
2. Wang, G., Y. Wang, and Z. Gao, Use of steel slag as a granular material: volume expansion prediction and usability criteria. Journal of Hazardous Materials, **184** p. 555-560 (2010).
3. Pellegrino, C., et al., Properties of concretes with black/oxidizing electric arc furnace slag aggregate. Cement and Concrete Composites, **37**(1): p. 232-240 (2013).
4. Juckes, L.M., The volume stability of modern steelmaking slags. Transactions of the Institution of Mining and Metallurgy, Section C: Mineral Processing and Extractive Metallurgy, **112**(3 DEC.): p. 177-197 (2003).
5. Goldring, D.C. and L.M. Juckes, Petrology and stability of steel slags. Ironmaking and Steelmaking, **24**(6): p. 447-456 (1997).
6. Manso, J.M., J.J. Gonzalez, and J.A. Polanco, Electric arc furnace slag in concrete. Journal of Materials in Civil Engineering, **16**(6): p. 639-645 (2004).
7. López, F.A., A. López-Delgado, and N. Balcázar, Physico-chemical and mineralogical properties of EAF and AOD Slags. Afinidad, **53**(461): p. 39-46 (1996).
8. Luxán, M.P., et al., Characteristics of the slags produced in the fusion of scrap steel by EAF. Cement and concrete Research, **30**(4): p. 517-519 (2000).
9. Vazquez Ramonich, E. and M. Barra, Reactivity and expansion of electric arc furnace slag in their application in construction. Materiales de Construcción, **51**(263-264): p. 137-148 (2001).
10. Frías Rojas, M., M.I. Sánchez, and A. Uria, Study of the inestability of black slags from EAF steel industry. Materiales de Construcción, **52**(267): p. 79-83 (2002).

11. Shi, C., Characteristics and cementitious properties of ladle slag fines from steel production. Cement and concrete Research, 2002. **32**(3): p. 459-462 (2002).
12. Posch, W., H. Presslinger, and H. Hiebler, Mineralogical evaluation of ladle slags at voestalpine stahl GmbH. Ironmaking and Steelmaking, **29**(4): p. 308-312.
13. Setién, J., D. Hernández, and J.J. González, Characterization of ladle furnace basic slag for use as a construction material. Construction and Building Materials, **23**(5): p. 1788-1794 (2009).
14. Pasetto, M. and N. Baldo, Mix design and performance analysis of asphalt concretes with electric arc furnace slag. Construction and Building Materials, **25**(8): p. 3458-3468 (2011).
15. San José, J.T. and A. Uría, Escorias de horno de arco eléctrico en mezclas bituminosas. Arte y Cemento, 1905: p. 122-125 (2001).
16. Skaf, M., et al., Ladle furnace slag in asphalt mixes. Construction and Building Materials, **122**: p. 488-495 (2016).
17. Pasetto, M. and N. Baldo, Fatigue performance of asphalt concretes made with steel slags and modified bituminous binders. International Journal of Pavement Research and Technology, **6**(4): p. 294-303 (2013).
18. Pasetto, M. and N. Baldo. Resistance to permanent deformation of base courses asphalt concretes made with RAP aggregate and steel slag. in 12th International Conference on Asphalt Pavements, ISAP 2014. Raleigh, NC: Taylor and Francis - Balkema (2014).
19. Manso, J.M., et al., Ladle Furnace Slag in Construction. J Mater Civil Eng., **17**: p. 513-518 (2005).
20. Bosela, P., et al., Fresh and hardened properties of paving concrete with steel slag aggregate. Propiedades para firmes del hormigón fabricado con áridos siderúrgicos. Carreteras: Revista técnica de la Asociación Española de la Carretera, **4**(166): p. 55-66 (2009).
21. Abu-Eishah, S.I., A.S. El-Dieb, and M.S. Bedir, Performance of concrete mixtures made with electric arc furnace (EAF) steel slag aggregate produced in the Arabian Gulf region. Construct Build Mater, **34**: p. 249-256 (2012).
22. Kim, S.W., Y.J. Lee, and K.H. Kim, Bond behavior of RC beams with electric arc furnace oxidizing slag aggregates. J Asian Architect Build Eng., **11**(2): p. 359-366 (2012).
23. Pellegrino, C. and F. Faleschini, Experimental behavior of reinforced concrete beams with electric arc furnace slag as recycled aggregate. ACI Materials Journal, **110**(2): p. 197-205 (2013).
24. Anastasiou, E., K. Georgiadis Filikas, and M. Stefanidou, Utilization of fine recycled aggregates in concrete with fly ash and steel slag. Constr Build Mater, **50**: p. 154-161 (2014).
25. Manso, J.M. and J. Setién, Investigación de nuevos usos de las escorias de horno eléctrico de arco (EAF): la oportunidad de los hormigones. Hormigón y Acero, **241**: p. 51-57 (2006).
26. Santamaría, A., et al., The use of steelmaking slags and fly ash in structural mortars. Construction and Building Materials, **106**: p. 364-373 (2016).
27. Arribas, I., et al., Electric arc furnace slag and its use in hydraulic concrete. Construction and Building Materials, **90**: p. 68-79 (2015).
28. Arribas, I., et al. Application of steel slag concrete in the foundation slab and basement wall of the Tecnalia kubik building. in 6th European Slag Conference Proceedings. Madrid, Euroslag (2010).
29. Etxeberria, M., et al., Properties of concrete using metallurgical industrial by-products as aggregates. Construction and Building Materials. **24**(9): p. 1594-1600 (2010).
30. Manso, J.M., et al., Design and elaboration of concrete mixtures using steelmaking slags. ACI Materials Journal, **108**(6): p. 673-681 (2011).

31. Papayianni, I. and E. Anastasiou, Concrete incorporating highcalcium fly ash and EAF slag aggregates. Magazine of Concrete Research, **63**(8): p. 597-604 (2011).
32. Polanco, J.A., et al., Strength and durability of concrete made with electric steelmaking slag. ACI Materials Journal, **108**(2): p. 196-203 (2011).
33. San-José, J.T., et al., The performance of steel-making slag concretes in the hardened state. Materials and Design, **60**: p. 612-619 (2014).
34. Arribas, I., et al., Durability studies on steelmaking slag concretes. Materials and Design, **63**: p. 168-176 (2014).
35. Anastasiou, E.K., I. Papayianni, and M. Papachristoforou, Behavior of self compacting concrete containing ladle furnace slag and steel fiber reinforcement. Materials and Design, **59**: p. 454-460 (2014).
36. Tomasiello, S. and M. Felitti, EAF slag in self-compacting concretes. Facta universitatis - series: Architecture and Civil Engineering, **8**(1): p. 13-21 (2010).
37. Santamaría, A., et al., Self-compactin concrete incorporating electric arc-furnace steelmaking slag as aggregate. Materials and Design, **115**: p. 179-193 (2017).
38. Akin Altun, I. and I. Yilmaz, Study on steel furnace slags with high MgO as additive in Portland cement. Cement and Concrete Research, **32**(8): p. 1247-1249 (2002).
39. Ortega-López, V., et al., The long-term accelerated expansion of various ladle-furnace basic slags and their soil-stabilization applications. Construction and Building Materials, **68**: p. 455-464 (2014).
40. Manso, J.M., et al., The use of ladle furnace slag in soil stabilization. Construction and Building Materials, **40**: p. 126-134 (2013).
41. Montenegro, J., et al., Ladle furnace slag in the construction of embankments: expansive behavior. J Mater Civil Eng., **25**(8): p. 972-979 (2013).
42. Bäverman, C., et al., Serial batch test performed on municipal solid waste incineration bottom ash and electric arc furnace slag, in combination with computer modelling. Waste management and Research, **15**(1): p. 55-71 (1997).
43. Sasamoto, H., et al., Development of fishing block using EAF refining slag. Tetsu-To-Hagane/Journal of the Iron and Steel Institute of Japan, **89**(4): p. 461-465 (2003).
44. Pellegrino, C. and V. Gaddo, Mechanical and durability characteristics of concrete containing EAF slag as aggregate. Cement and Concrete Composites, **31**(9): p. 663-671 (2009).
45. EN Euronorm. European Committee for Standardization: Rue de Stassart, 36. Belgium–1050 Brussels.
46. CEN. European Committee for Standardization. rue de Stassart, 36. Brussels B-1050.
47. Turmo, J., et al., Study of the shear behaviour of fiber reinforced concrete beams. Materiales de Construcción, **58**(292): p. 5-13 (2008).
48. Yazici, S., G. Inan, and V. Tabak, Effect of aspect ratio and volume fraction of steel fiber on the mechanical properties of SFRC. Construction and Building Materials, **21**(6): p. 1250-1253 (2007).
49. Bemal, S., et al., Mechanical behaviour of steel fibre-reinforced alkali activated slag concrete. Materiales de Construccion, **59**(293): p. 53-62 (2009).
50. Instrucción para realización de obras de hormigón estructural, M.d. Fomento, Editor., Comisión Permanente de Hormigón: Madrid (2008).

Bond behaviour of GFRP reinforced geopolymer cement concrete

Biruk Hailu Tekle[1], *Amar* Khennane[1*], *Obada* Kayali[1]

[1]SEIT, UNSW Canberra, Northcott Drive, Canberra ACT 2600, Australia

Abstract. Bond plays a key role in the performance of reinforced concrete structures. Glass fibre reinforced polymer (GFRP) reinforcing bar and Geopolymer cement (GPC) concrete are promising alternative construction materials for steel bars and Ordinary Portland Cement (OPC) concrete respectively. In this study, the bond behaviour between these two materials is investigated by using beam-end specimen tests. The bond behaviour of 15.9 mm diameter sand-coated GFRP bar was investigated. An embedment length of six and nine times the bar diameter were used. The free end and the loaded end bond-slip-relationships, the bond failure mode and the average bond stress were used to analyse each of the specimens. Additionally, the distribution of tensile and bond stress along the embedment length was investigated by installing strain gauges along the embedment length in some of the specimens. Test results indicate that a significant difference exists between the free end and loaded end bond-slip curves, which is due to the lower elastic modulus of the GFRP bars. Furthermore, it was found that the tensile and bond stress distribution along the embedment length is nonlinear and the nonlinearity changes with the load.

1 Introduction

Corrosion of reinforcing steel bars and environmental problems associated with Ordinary Portland cement (OPC) concrete have lead researchers to look for new construction materials. These materials ideally are to be noncorrosive and environmentally friendly while providing the required mechanical performances. Glass fiber reinforced polymer (GFRP) reinforced Geopolymer cement (GPC) concrete stand out to be good alternatives with these desired properties. GFRP bars are characterized by high tensile strength, high durability, light weight, and electromagnetic permeability [1], while GPC concrete exhibits rapid a rate of strength development, resistance to sulphate attack, acid resistance, little drying shrinkage, low creep, improved resistance to fire, and prolonged handling time [2, 3, 4].

For GFRP bar and Geopolymer concrete to be used together, they should show a good composite action. The composite action is developed through the bond between the concrete and the surface of the bar. In reinforced concrete structures, bond achieves the transfer of load from the concrete to the reinforcing bar. In such a way, the weak tensile strength of concrete is assisted by the high tensile strength of the bar. There are many studies carried out

* Corresponding author: a.khennane@adfa.edu.au

on the bond properties of GFRP bars with the normal OPC as well as with GPC concretes [5-9, 10-11]. However, with the exception of some recent works [12, 13], not much has been done on the bond properties of GFRP reinforced GPC concrete. Both Tekle et al [12] and Maranan et al [13] studies used pullout tests to compare the bond performace of GFRP reinoforced GPC concrete with GFRP reinforced OPC concrete and steel reinforced GPC concrete, respectively. These studies revealed the promising performance of the GFRP reinforced GPC concrete. Nonetheless, further research is still needed for a better understanding of the bond mechanisms between these two materials. The tensile and bond stress distributions along the embedment length of GFRP reinforced GPC concrete are two areas, which have not been studied yet. A nonlinear tensile stress and bond stress distribution have been reported for both GFRP and steel with OPC concrete [6]. As there are inherent differences in the surface deformations and mechanical conditions of GFRP bars from steel bars, design guidelines such as ACI 318-08 and ACI 408-03 developed for steel cannot be directly used for GFRP bars [14, 15]. The difference in the properties of GPC and OPC concrete may also affect the bond properties, which again makes it inconvenient to use models and guidelines developed for GFRP reinforced OPC concrete such as ACI 440.1R-03 [16]. Thus, the aim of this study is to provide more understanding of the bond between GFRP and GPC concrete using a more reliable test method (beam-end specimen) according to ASTM A944 [17].

2 Experimental program

2.1 Materials

Fly ash based geopolymer concrete with an average compressive strength of 35 MPa is used. Table 1 shows the mix proportion of each component of the geopolymer concrete. Class F fly ash complying with ASTM C618 standard was used [18]. A combination of sodium hydroxide and sodium silicate solutions was used as the alkaline liquid. The sodium hydroxide solution had a concentration of 12 M and was prepared by dissolving 98% caustic soda in distilled water. SiO_2/Na_2O ratio of 1.95-2.05 was used for the sodium silicate solution. Coarse aggregate with a nominal maximum aggregate size of 14 mm and natural sand fine aggregate both satisfying AS 2758.1:2014 were used [19]. OPC concrete with an average compressive strength of 35 MPa was also prepared for a comparison as a control specimen. The OPC concrete mix proportion is also shown in Table 1.

Table 1. Mix proportions.

Ingredient	**GPC concrete**	**OPC concrete**
Cement (kg/m^3)	-	357
Fly Ash (kg/m^3)	420	-
Coarse aggregate [a] (kg/m^3)	1090	944
Fine Aggregate (kg/m^3)	630	814
12M NaOH Solution(kg/m^3)	60	-
Na_2SiO_3 solution (kg/m^3)	150	-
Water (kg/m^3)	35-45	225
SP (kg/m^3)	4	-
VM (kg/m^3)	4	-

[a] maximum aggregate size of 14 mm

High-modulus straight GFRP bars with sand-coated surface and 15.9 mm nominal diameter are used. The bar is composed of glass fibres in a vinyl ester matrix. The fibre and resin contents are 65% and 35% by volume respectively. The mechanical and physical properties of the bar are given in Table 2.

Table 2. Mechanical and physical properties of GFRP bars.

Property	**#5 GFRP**	
	Specification	**Experimental**
Tensile strength (MPa)	1184*	1505
Bar diameter (mm)	15.9*	18.9
Elastic modulus in tension (GPa)	62.6±2.5*	62.6
Tensile strain at failure (%)	1.89	2.40
Poisson's ratio	0.25	-
Coefficient of longitudinal thermal expansion (/°C)	6.2×10^{-6}	-
Coefficient of transverse thermal expansion (/°C)	23.8×10^{-6}	-
Weight (g/m)	558	576

*Nominal values

2.2 Specimens Preparation

Six beam-end specimens were cast and tested. All the beam-end test specimens were prepared according to ASTM A944 recommendation. A schematic diagram of the beam-end specimens is shown in Figure 1. The specimens had a length of 600 mm, a height of 450 mm, and a width of 220 mm. Shear reinforcement were provided with 12 mm diameter steel bars. These reinforcements are placed in the specimens so as not to intercept longitudinal splitting cracks that occur at bond failure [15]. GFRP bars with an embedment length of six and nine times the bar dimeter were used. The rest of the bar inside the concrete was passed through a PVC tube to avoid contact between the concrete and the bar. Plywood moulds were used for casting the concrete. To avoid premature failure of the GFRP bars from the conventional grips, a special anchor designed according to ACI 440.3R is attached at the loaded end of the GFRP bar [16].

The mixing and curing of all the GPC concrete specimens were carried out according to previous research [20]. The beam-end specimens were cast and vibrated with a poker vibrator. The compressive strength of each beam-end specimen is tested from three cylinders casted from the same mix. The specimens were then allowed to rest for 24 hours before placing them into the oven where they were cured for 24 hours at 80°C. After demolding, the specimens were placed in a temperature and humidity controlled room (23°C and 50% relative humidity) until the time of testing, i.e. 7 days for GPC concrete specimens. The OPC specimens on the other hand were stored in a fog room until the test day, which was 28 days.

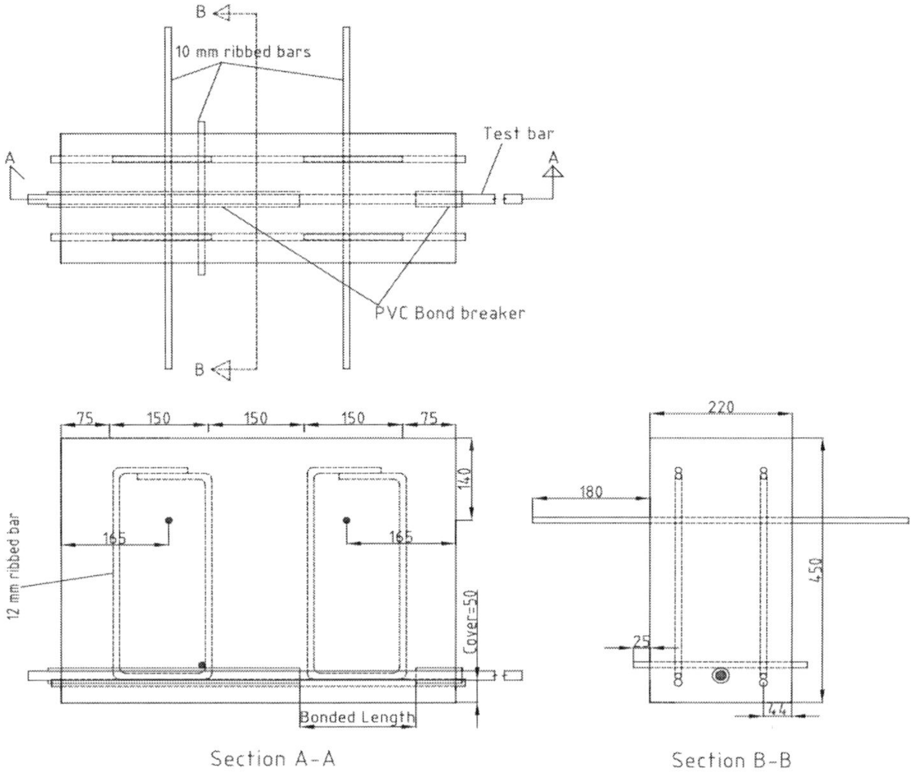

Fig. 1. Schematic of beam-end specimen.

2.3 Test setup and instrumentation

The pullout tests were performed using a universal testing machine with a capacity of 100 kN. The set-up of the experiment is as shown in Figure 2.

Fig. 2. Beam-end specimen under test.

Some of the GFRP bars were strain gauged along the embedment length. The sand coating of the bars was removed from the part of the bar surface where the strain gauges are glued. To minimize the reduction of bond area, a small area (6×6 mm^2) was used for bonding the strain gauges. Four linear variable displacement transducers (LVDTs) were attached at the free and loaded ends of the GFRP bar (two at each ends). The load was applied in displacement control mode at a rate of 1 mm/min. The strain, displacement and forces were recorded using an automatic data acquisition system.

3 Results and discussions

3.1 Failure mode and bond strength

The experimental results are given in Table 3. The specimens are identified by the type of concrete used (G for GPC concrete and O for OPC concrete), the diameter of the bar used, the embedment length as a multiple of the bar diameter (6d and 9d) and specimen number. The maximum pull out load was used to calculate the average bond stress by assuming a uniform bond stress distribution along the embedment length as per Eq. (1). The corresponding slips at this load are also reported for both free and loaded end:

$$\tau = \frac{P}{\pi d l} \tag{1}$$

where, τ is average bond strength in MPa; P is applied pull out load in N; d is diameter of the rebar in mm; and l is bonded/embedded length in mm.

Table 3. Summary of experimental results.

Specimen	Load P_{max} (kN)	Average bond stress τ (MPa)	Free end slip (mm)	loaded end slip (mm)	Failure mode
G16-6d-1	67.9	14.2	265	498	S
G16-6d-2	75.8	15.8	355	603	S
G16-9d-1	81.3	11.3	580	824	T
G16-9d-2	82.1	11.4	474	765	T
G16-9d-3	88.6	12.2	448	562	T
O16-9d-1	78.0	10.8	252	750	T

S = splitting failure; T = splitting of the concrete in the transverse direction

As shown in Figure 3, transverse tensile cracking caused the failure of the 9d specimens. The applied tensile load on the bar is transferred to the concrete via the bond. The concrete around the bonded region applies tensile stress on the concrete on the other side of the bonded region. This results in the tensile cracking at the end of the embedment length. Tekle et al. [12] reported the splitting failure of specimens with a similar embedment length in case of pullout cylinder specimens. In both the beam-end specimens of the present study and pullout specimens reported in [12], the tensile failure of the concrete causes a premature failure before the bond reaches its maximum strength. Nonetheless, the types of concrete failure observed in the pullout specimens [12], and the present beam-end specimens are different. This can be explained by considering the differences in the test specimens and their setup. The concrete in the pullout specimens are under compression due to the support condition. Furthermore, compressive struts are formed between support points for the concrete and the surface of the reinforcing bar placing the reinforcement surface in compression [15]. Unlike the pullout specimens, the beam-end specimens are in tension around the bonded region, hence more prone to tensile failure by cracking of the concrete in the transverse direction than in the pullout specimens.

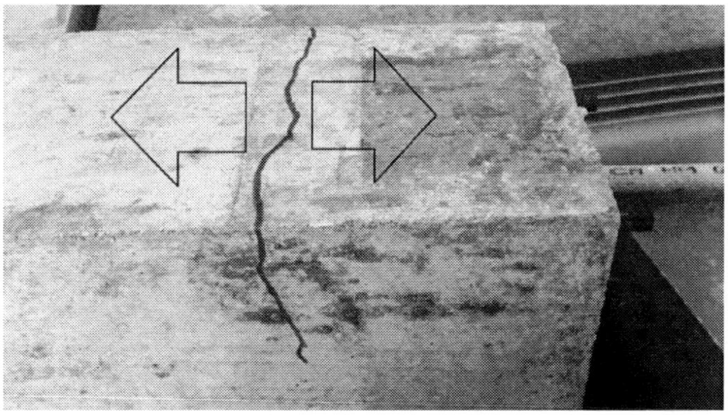

Fig. 3. Failure mode of O16-9d-1 specimen.

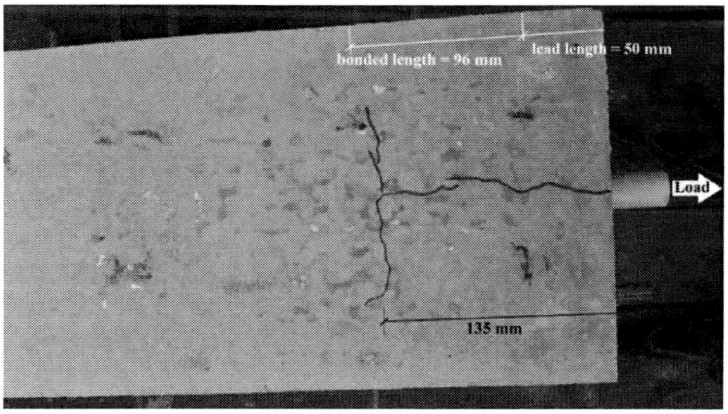

Fig. 4. Failure mode of G16-6d-1 specimen

A splitting failure was observed for the 6d specimens. This type failure was also observed by Tekle et al [12] on their GFRP reinforced GPC concrete pullout specimens for a similar embedment length. However, contrary to the results reported in [12], the splitting failure in this experimental work was accompanied with a transverse tensile crack, which again is due to the tensile stress generated in the beam-end specimens. Figure 4 shows these cracks on the top surface of the beam-end specimen.

The difference in failure in 6d specimens and 9d specimens can be explained by considering the bond stress and the generated radial stress. It is known that as the embedment length increases, bond stress decreases in turn causing the radial stress to decrease [7, 8, 12]. Thus, the lower radial stress delayed the splitting failure of the 9d specimens due to their longer embedment length. Before the occurrence of the splitting cracks in the longitudinal direction, the concrete cracked in the transverse direction due to the tensile stress. On the other hand, and because of the shorter embedment length, the 6d specimens generate higher radial stresses, which manage to split the concrete. Transverse cracks were also observed in 6d specimens. However, these cracks were not fully developed like in the 9d specimens.

3.2 Bond-slip relationships

A typical average bond stress-slip relationship of some of the specimens is shown in Figures 5 and 6. The bond-slip curve starts with a high initial slope due to the undamaged bond between the concrete and the bar. The bond stiffness gradually decreases as slippage of the bar occurs. Tekle et al. [12] described two distinct regions in case of splitting failure; i.e.: the linear and nonlinear regions. As can be seen in Figure 5 and 6, a similar trend is also observed.

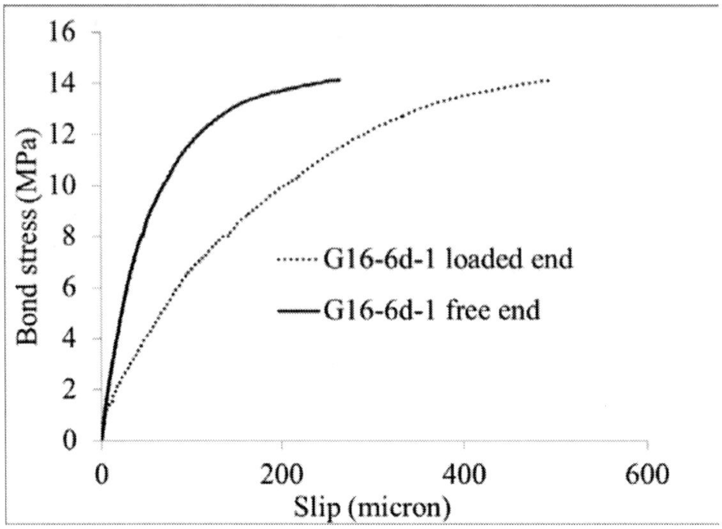

Fig.5. Bond-slip curve of G16-6d-1 specimen.

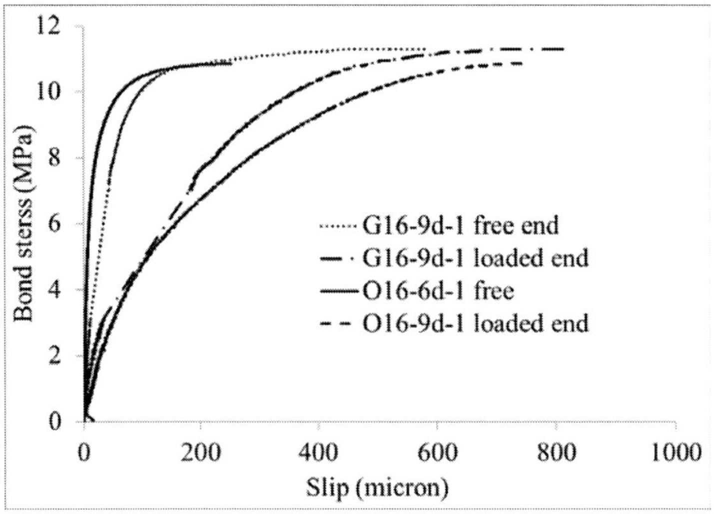

Fig. 6. Bond-slip curve of G16-9d-1 and O16-9d-1 specimens.

Comparison of the free end and the loaded end slips shows that there is a significant difference between them. This behavior is typical of GFRP bars and is due to their low elastic modulus. Steel bars with a similar condition showed a negligible amount of slip difference between free end and loaded end due to their higher elastic modulus [5, 9].

3.3 Bond distribution

Figures 7 and 8 show respectively the tensile and bond stress distribution of G16-9d-3 specimen. Each curve corresponds to a specific load represented as a percentage of the maximum pullout load. The tensile and bond stress were determined from strain gauges placed along the embedment length of the bar. A nonlinear tensile stress distribution is observed, and is more pronounced at lower load levels. Benmokrane et al. [6] also observed this nonlinearity of tensile stress along the embedment length for GFRP reinforced OPC concrete. The tensile stress on the bar increases rapidly from the free end to the loaded end.

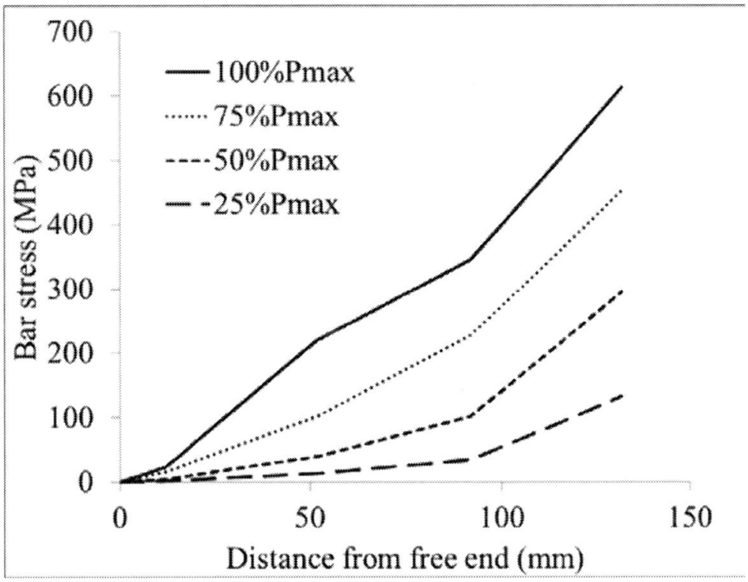

Fig. 7. Tensile stress distribution along embedment length for G16-9d-3.

At low load levels, a low value of the bond stress was observed at the free end. However as the load increases, the bond stress in the vicinity of the free-end increases, and the difference in bond stress along the embedment length becomes less pronounced. This can be explained by the redistribution of the bond stress along the embedment length as the load increases. This nonlinearity of the bond stress along the embedment length results in a lower bond stress value for longer embedment lengths when calculating average bond stress using the assumption of uniform bond stress.

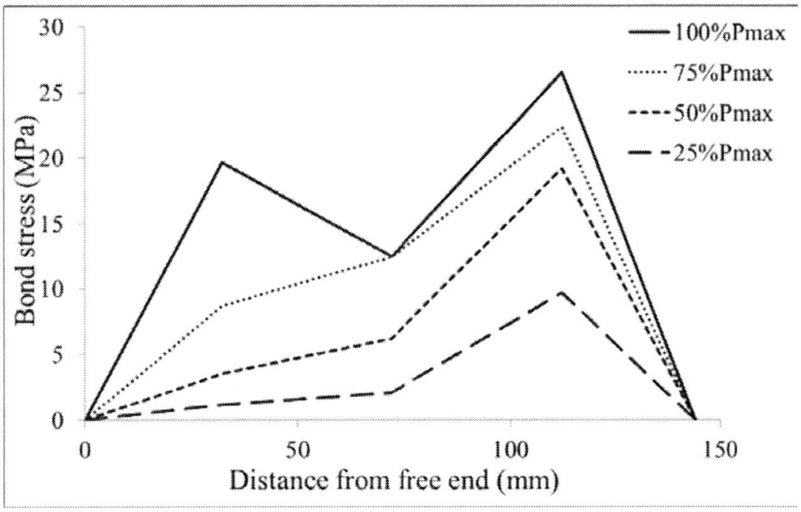

Fig. 8. Bond stress distribution along embedment length for G16-9d-3.

4 Conclusion

In this study the bond property between GFRP bars and fly ash based GPC was investigated using ASTM A944 beam-end specimens. For each specimen, the specimen failure mode, the average bond stress at the point of failure, the slip at the loaded and free end and the bond stress-slip relationship curves were determined. The results of this study can be summarised as follows:

1. The bond slip behaviour of sand coated GFRP reinforced GPC is similar to that of GFRP reinforced OPC concrete. Both concrete OPC and GPC concretes with GFRP bar display similar stages consisting of initial stiffening and nonlinear behaviour prior to peak stress.
2. No bond failure was observed in all the specimens studied in this paper. The specimens failed due to tensile cracking of the concrete. Longitudinal and lateral tensile cracking were observed. The type of tensile cracking was found to be dependent on the embedment length of the bar.
3. Distribution of tensile and bond stress along the embedment length of the bars is nonlinear. As the load increases, the bond stress at the vicinity of the free-end increases due to redistribution of bond stress along the embedment length.

References

1. L.C. Bank, Composites for Construction: Structural Design with FRP Materials. John Wiley & Sons, Inc. (2006).
2. D.Hardjito, and B. V. Rangan, "Development and Properties of Low Calcium Fly Ash-Based Geopolymer Concrete." Research Report GC 1, Curtin University of Technology, Perth, Australia, (2005).
3. M.T. Junaid, A. Khennane, O. Kayali, A. Sadaoui, D. Picard, and M. Fafard, "Aspects of the deformational behaviour of alkali activated fly ash concrete at elevated temperatures." Cem. Concr. Res., **60**, 24-29. (2014).
4. Junaid, M.T., Khennane, A., and Kayali, O. (2015). "Performance of fly ash based geopolymer concrete made using non-pelletized fly ash aggregates after exposure to high temperature." Mater. Struct. **48**(10), 3357-3365
5. M. Pecce, G. Manfredi, R. Realfonzo, and E. Cosenza, "Experimental and Analytical Evaluation of Bond Properties of GFRP Bars." J. Mater. Civ. Eng., **13**(4), 282-290 (2001).
6. B. Benmokrane, B. Tighiouart, and O. Chaallal, "Bond Strength and Load Distribution of Composite GFRP Reinforcing Bars in Concrete." ACI Mater. J., **93**(3), 246-252 (1996).
7. E. Cosezna, G. Manfredi and R. Realfonzo, "Behaviour and Modelling of Bond of FRP Rebars to Concrete." J. Compos. Constr., **1**(2), 40–51, (1997).
8. R. Okelo, and R.L. Yuan, "Bond strength of fiber reinforced polymer rebars in normal strength concrete." J. Comp. Constr., **9**(3), 203-213 (2005).
9. F. Focacci, A. Nanni, and C. E. Bakis, "Local Bond-Slip Relationship for FRP Reinforcement in Concrete." J. Compos. Constr., **4**(1), 24-31(2000).
10. Y.Cui, and O. Kayali, "Bond Performance of Steel Bar and Fly Ash based Geopolymer Concrete." Concrete Institute of Australia Biennial National Conference: Understanding Concrete, Gold Coast, Australia (2013).
11. Ee Hui Chang et Al, Bond Behaviour of Reinforced Fly Ash-Based Geopolymer Concrete Beams, The 24th Biennial Conference of the Concrete Institute of Australia,. Luna Park, Sydney Sep 17, (2009).

12. B. H. Tekle, A. Khennane, O. Kayali "Bond properties of sand-coated GFRP bars with fly ash based geopolymer concrete." J. Compos. constr. **20**(5), (2016).
13. G., M. A. Maranan, "Bond Stress-Slip Behavior: Case of GFRP Bars in Geopolymer Concrete." ASCE J. Mater. Civ. Eng., **27**(1), 1-9. (2015).
14. American Concrete Institute (ACI). "Building Code Requirements for Structural Concrete and Commentary." ACI 318M-08, Farmington Hills, Mich.
15. American Concrete Institute (ACI). "Bond and Development of Straight Reinforcing Bars in Tension." ACI 408R-03, Farmington Hills, Mich.
16. ACI (American Concrete Institute). "Guide for the design and construction of structural concrete reinforced with FRP bars." ACI 440.1R-06, Farmington Hills, MI. (2006).
17. American Society for Testing and Materials (ASTM). "Standard test method for comparing bond strength of steel reinforcing bars to concrete using beam-end specimens" ASTM A944 - 10, West Conshohocken, PA. (2010).
18. American Society for Testing and Materials (ASTM). "Standard Specification for Coal fly ash and raw or calcined natural pozzolan for use in concrete." ASTM C618 - 12a, West Conshohocken, PA. (2012).
19. Australian Standard (AS). "Aggregates and rock for engineering purposes - Concrete aggregates." AS 2758.1:2014. Australia. (2014).
20. M.T. Junaid, O. Kayali, A. Khennane and J. Black, "A mix design procedure for low calcium alkali activated fly ash-based concretes." Constr. Build. Mater., **79**, 301-310, (2015).

Shear behavior of steel fiber reinforced concrete using full-field displacements from digital image correlation

Sahith Gali[1], and *Kolluru.V.L.*Subramaniam[2,*]

[1]Indian Institute of Technology Hyderabad, Hyderabad, TS 502285, India.
[2]Indian Institute of Technology Hyderabad, Hyderabad, TS 502285, India.

Abstract. Reinforced concrete beams with discrete hooked-end steel fibers at 0.5% volume fraction are tested with a shear span to depth (a/d) ratio equal to 1.8. Digital image correlation (DIC) technique was used to obtain the full-filed displacements from the beam. The formation and propagation of a shear crack which directly influences the load response and the peak load in the load response of the beam is monitored using the displacement information available from DIC. There is a continuous increase in slip across the crack faces with increasing load, which produces an increase in the crack opening. The dilatant behavior indicated by the proportion of crack opening to slip displacement obtained from the control and the SFRC beams is identical. Failure in control beams is brittle and was produced by the opening of dominant shear crack within the shear span. At the peak load, the shear crack pattern in fiber reinforced concrete is identical to the crack pattern in the control beam. The fiber reinforced concrete beams exhibit post peak load carrying capacity with continued slip of the dominant shear crack. The crack bridging stress provided by the fibers results in a significant increase shear transfer across the crack which provides significant post-peak load carrying capacity with increasing slip of the shear crack.

1 Introduction

Often in shear critical members, the requirement of shear reinforcement leads to close spacing of stirrups. This problem is compounded while considering the ductile detailing requirements to meet the concerns of seismic design, which requires provision of extra stirrup. The extra stirrups, often lead to heavy congestion of steel. Methods of reducing steel congestion, which provide the required capacity and ductility have often involved the use of fiber reinforcement in concrete. A survey of literature indicates that there is an increase in the load carrying capacity and ductility with the addition of steel fibers [1-12]. ACI 318 (2008) [13] allows the usage of steel fibers as minimum shear reinforcement. Most of the data in the literature has been obtained for slender beams with shear span to depth (a/d) ratios greater than 2.5 in which beam action governs the failure. There is very limited data available from beams with low a/d ratio i.e short beams for which arch action governs the failure. The

[*] Corresponding author: kvls@iith.ac.in

role of fibers in beams with low a/d ratio is still not clear. An a/d ratio less than 2.0 was selected to contribute additional data to the limited available data set. Further, for small a/d ratio, failure is shown to be produced by a shear crack within a small region [2, 3, 9], which allows for studying the behavior of the shear crack.

Many researchers have proposed empirical equations for the shear strength of fiber reinforced concrete by modifying the existent equations available for the reinforced concrete beams. In all the codes the shear strength equation for reinforced concrete is given as two separate components those are the contribution coming for concrete and stirrup contribution [14-16]. Researchers have modified the concrete contribution term to propose the shear strength expression for fiber reinforced concrete. These proposed equations takes into account major factors responsible for shear behavior which includes shear span to depth ratio, percentage of longitudinal reinforcement, and other factors include volume of fibers, fibers aspect ratio, fiber geometry, fiber tensile behavior and also concrete matrix properties [1-12] [17]. All the proposed equations are empirical relations estimated from the experiments on steel fiber reinforced concrete beams and few on other types of fibers. The major effect of fiber parameters is in improving stress transfer across the crack due to the bridging action which in turn effects the post cracking behavior of concrete beam.

Addition of fibers increases the post cracking stress transfer ability in concrete, which primarily contributes to improving the ductility [18-19]. To assess the true potential of fibers the post cracking stress transfer across crack has to be accounted for. This is accounted for by including an additional term contributed by steel fibers to the total shear resistance. The post cracking stress transfer is particularly important for shear where shear transfer by aggregate interlock provides significantly to shear capacity of RC elements [20]. The slip displacement along rough cracks also results in dilatancy under shear and torsion [21-24]. The crack bridging provided by fibers can potentially provide for increased mobilization of aggregate interlock, thereby providing shear load capacity. An understanding of the influence of fibers on the post-cracking shear stress transfer across rough cracks surface and its influence on the shear capacity of reinforced concrete beams needs to be understood to develop design provisions which consider the influence of fibers on shear capacity.

Motivation for the study comes from the need to understand the role of discrete steel fiber reinforcement in contributing to the shear behavior of reinforced concrete beams. The overall objective of this study is to investigate the influence of steel fibers on the shear response of reinforced concrete. In particular, it is proposed to investigate the influence of fibers on shear behavior of beams with potential arch action. Beams with shear span to depth ratio of 1.8 was considered in the present study. The influence of the steel fibers on the shear transfer across rough cracks in concrete and its influence on the shear behavior of reinforced concrete is investigated. An approach which combines experimental and digital image analysis is presented to develop an understanding of shear response of reinforced concrete with steel fibers.

2 Experimental program

Reinforced concrete beams of dimensions 125mm width, 250mm depth and 1500mm length were used in the test program. In the test program, both control and fiber reinforced concrete beams were cast for testing. These beams include three beams each of control and fiber reinforced concrete with fiber volume fractions ($v_f = 0.5\%$). Dramix^{RM} 3D hooked end and glued non-balling steel fibers of aspect ratio 80 (length 60mm and diameter of 0.75mm) manufactured by Bekaert industries were used in this study. The fibers have a modulus of elasticity of 210 GPa and tensile strength equal to 1225 MPa.

All beams were reinforced with two reinforcing bars of 16 mm diameter placed with a clear cover of 20 mm as shown in Figure 1. The reinforcing bars confirmed to the requirements of IS 1786 (2008) [25] and had a prescribed yield stress defined at 0.2% offset strain equal to 500 MPa, Elastic modulus equal to 200 GPa and an ultimate strain equal to 0.16. One half of the beam was reinforced with 8 mm closed shear stirrups at a spacing of 150 mm. The stirrups were provided in one half of the beam to reinforce it against shear failure and to ensure that shear failure occurred only on one side of the beam. The part of the beam without shear reinforcement allowed for a closer examination of the shear crack patterns and to evaluate the mechanisms contributing to shear resistance. Closely placed stirrups with a spacing equal to 100 mm were also provided at the supports of beam to avoid possible anchorage failure.

All the beams were tested with a shear span to depth ratio equal to 1.8. The testing was performed using a servo-controlled test machine where the net deflection of the center of the beam was measured using LVDT. The test was performed in displacement control where the deflection of the beam was increased in a controlled manner. The loading arrangement used in this study confirmed to the requirements of ASTM 1609, which ensured that no torsion is generated in the beam due to eccentricity. The loads were applied using hardened steel rollers, which ensured perfect line loading at precisely marked locations of the beam. Each roller was attached to the actuator through individual swivel head which allowed for reducing the influence of torsion in the loading arrangement.

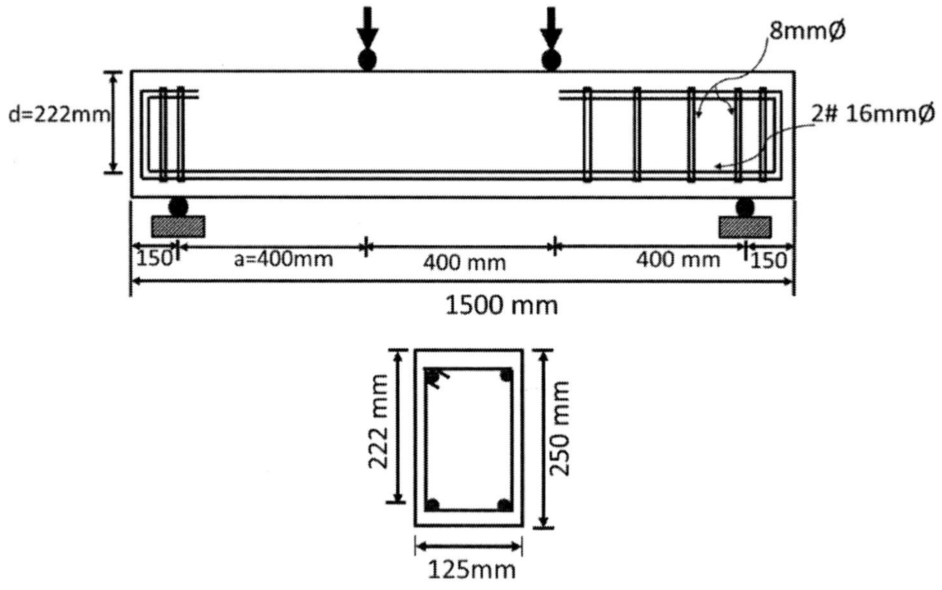

Fig. 1. The specimen geometry and the loading arragement used for shear tests.

The surface displacements, strains and development of cracks during the entire load response were determined using a digital image analysis procedure called digital image correlation. Digital image correlation (DIC) is a full-field optical technique which provides spatially continuous measurement of displacements across the surface of the specimen. DIC relies on correlation of the random pattern of speckles between images of the deformed and reference (undeformed) configurations of the specimen within small neighborhoods called subsets [27-28]. A sprayed-on speckle pattern was created on the surface of the beam without the shear stirrups. A uniform coat of white paint and after the white paint dried, the sprayed-on speckle was created by spraying a black mist.

During a test, digital images of the specimen were captured for use in digital image correlation using a high resolution camera (5 mega pixel), which was controlled by a computer. Uniform light intensity was ensured across the surface of the beam using normal white light. The camera was fitted with a 50 mm lens and was placed at a distance of 1 m from the specimen surface. Calibration for the pixel size was performed using a graduated rule placed in front of the specimen. From this measurement, the physical calibration was established. Typical calibration was in the range of 7-10 pixels per mm. A reference image was captured in the undeformed state prior to the initiation of loading program. Images of the specimen were captured at regular intervals during the test. A photograph of the test setup is shown in Fig. 2(a) and a schematic figure of the DIC test setup is shown in Fig. 2(b). For the setup used in this study, the random error in the measured displacement is in the range of 0.02 pixels and a conservative estimate of the resolution in strain obtained from the digital correlation was 17 $\mu\varepsilon$. [29-30]

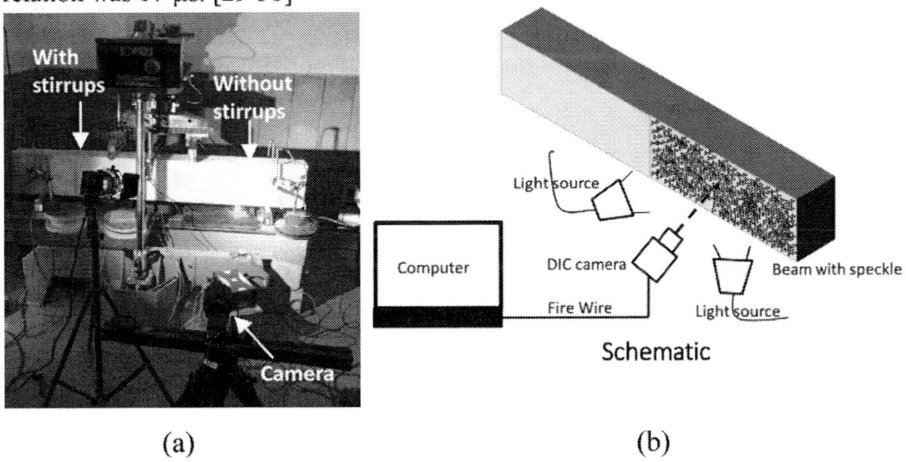

Fig. 2. (a) Photograph of the Test setup; (b) schematic of the test setup.

3 Experimental results

The load-deflection responses of the control and hooked end steel fiber reinforced concrete (SFRC) beams are shown in Fig. 3. Failure in control specimens was sudden, immediately after attaining the peak load and could not be controlled even in displacement control. The load response of the SFRC beams is nominally similar to the load response obtained from the control specimens up to the peak load of the control beams. In the 0.5% SFRC beams (containing 0.5% volume fraction steel fibers), there is a small increase in the peak load when compared to the control beams. After the peak load there is a continuous decrease in the load carrying capacity with increasing deflection following which, the load carrying capacity essentially levels to a constant value with increasing deflection.

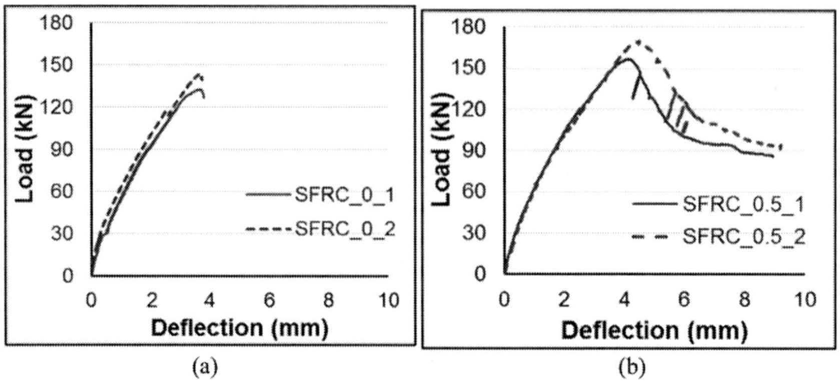

Fig. 3. Load-Deflection Response of beams: (a) Control beams; and (b) SFRC beams.

The crack patterns on the side of the beam without shear stirrups for both control and FRC beams are shown in Fig. 4. The crack patterns shown in the figure were identified visually and marked on the surface of the beam. These crack patterns were marked on the back surface, which did not have the speckle pattern. The numbers in the figure indicate the applied load. With the application of load, initially, flexural cracks were formed in the constant moment region. With an increase in load, flexural cracks increased in length and newer flexural cracks were formed along the span. In control beam (labelled SFRC_0_1), at an applied load in the range of 95-100 kN a shear crack emanating from the flexural crack located approximately 215 mm from the support up to the mid-height of the beam. The crack angle was in the range 41 to 43 degrees with respect to the horizontal. The crack was located at a distance approximately equal to the depth of the beam, in the range of 230-250 mm from the support. With increasing load, this shear crack continued to propagate up to the top of the beam. Ultimate failure was by a crack which propagated along the rebar towards support producing splitting along the rebar. While a flexure-shear crack was also formed, it was arrested in the compression region. In 0.5% SFRC beams (labelled SFRC_0.5_1), a shear crack was also initiated from a flexural crack located approximately 210 mm from the support at a comparable load in the range of 100-110 kN. The location of the shear crack and the angle of the crack were also nominally similar to those observed in the control beams. At loads corresponding to the peak load in the control beams, the shear crack was fully established between the level of the flexural reinforcement and the compression region close to the top of the beam. The sudden failure by splitting along the flexural reinforcement was not observed. There was a noticeable decrease in the flexural crack spacing in SFRC beams compared to the control beams. A secondary crack was observed to form from the support close to the peak load and it joined with the main shear crack. The shear crack continued to open with increasing deflection after the peak load.

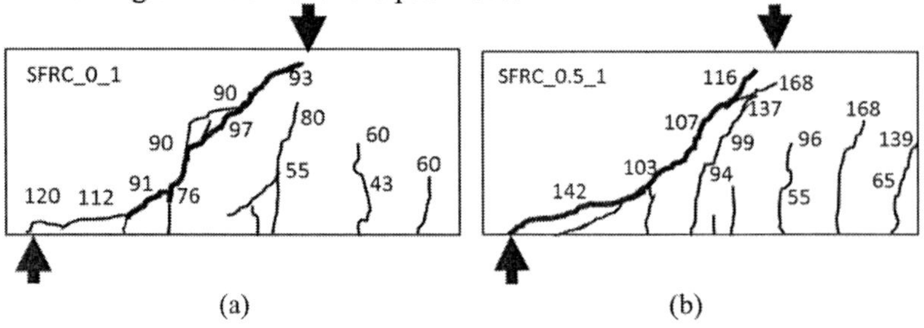

Fig. 4. Crack pattern in beams (a) Control beam; and (b) SFRC beam.

In all beams, the shear crack was formed at distance of 250 mm from the support at the mid-height location of the beam. The crack angles within the shear region extending over the middle third of the beam are also nominally similar in all specimens, between 41 and 43 degrees with respect to the horizontal. The load responses of the SFRC beams are nominally similar to the load responses obtained from the control specimens, up to peak load of the control. The shear crack extends between the level of the rebar and the top of the beam at the peak load. Cracking in the matrix therefore does not appear to be influenced by the presence of discrete steel fibers. This indicates that the internal mechanism of resistance is essentially identical in both SFRC and control beams up to peak load. Failure in control specimens was very brittle, with very visible opening of the shear crack. The failure may therefore be attributed to the failure of the internal stress transfer across the shear crack. At corresponding loads, the crack opening in SFRC specimens are significantly smaller when compared with the control beams. The shear crack continued to open in a controlled manner in the 0.5% SFRC past peak load.

5 Digital image analysis of shear cracks

During the loading process, digital images were taken at regular intervals of load. These images were processed using the VIC 2D^{TM} software to obtain the full-field surface displacements. Typical contours of displacement in the X-direction at two distinct points in the load response of a control beam are shown in Fig. 6. The formation of cracks are identified by a break in the contours. The use of DIC allows for accurate identification of cracks, which may not be delineated visually. The crack patterns identified using DIC were nominally similar to those obtained from visual examination. The use of DIC allows for tracking changes in the material response across a crack after the shear crack is identified. From a record of images, the location of the crack can be identified and the opening and shear displacements across the shear crack can be studied from a load point before the initiation of the crack. To understand the shear crack opening behavior in the beams, guage points identified with two sets of boxes (labelled G1 through G6) about 80X80 pixels were considered at locations on either side of the crack such that the line joining their centers is perpendicular to the crack direction as shown in the Fig. 6(b). The exact location of the guage points was decided considering the load points close to failure, where the shear crack was fully established. The guage points are identified with the center of the box and were located in the shear region within the middle third portion of the beam. The location of the guage points were kept fixed in all images for analysis. The average values of displacements were obtained from within the area of each box from each location; the average horizontal and vertical displacements were obtained from the values of displacements of all the pixels within each box.

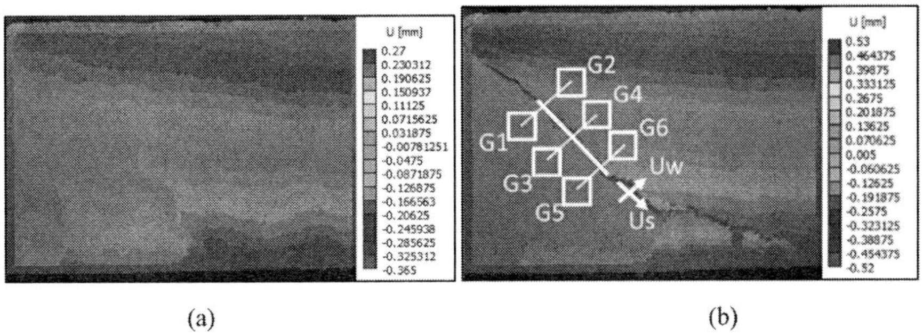

(a) (b)

Fig. 5. Contour plot of displacement in x-direction from the SFRC_0_1 beam at different points on the load response: **(a)** at 95 KN and **(b)** at 132KN.

Typical horizontal and vertical displacements from two pairs of points across the shear crack, located within the shear region (middle third of the beam) of the beam, were used to calculate the components of relative displacements in the directions which are perpendicular and parallel to the crack. The component of relative displacement which is parallel to the crack gives the slip across the crack (U_s) and the component which is perpendicular to the crack gives the crack opening displacement across the crack (U_w). The U_w and U_s obtained from the two points located at the mid-height of the beam, G3 and G4 are shown in Fig. 6. The U_w and U_s of the two guage points are initially identical with increasing load. The formation of the shear crack can be identified by the divergence of the displacements measured from the two guage points. With increasing load there is an increasing divergence in the displacements of the pair of guage points. The average slip and average crack opening displacement across the crack was determined as the average of the difference in the relative displacement from pairs of points G1 and G2, G3 and G4, G5 and G6.

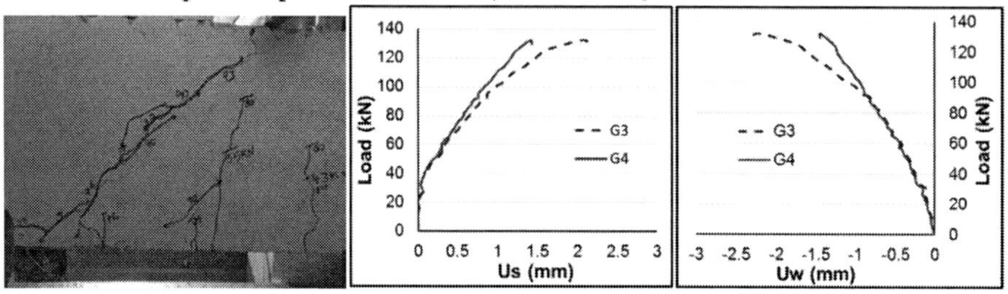

(a) (b) (c)

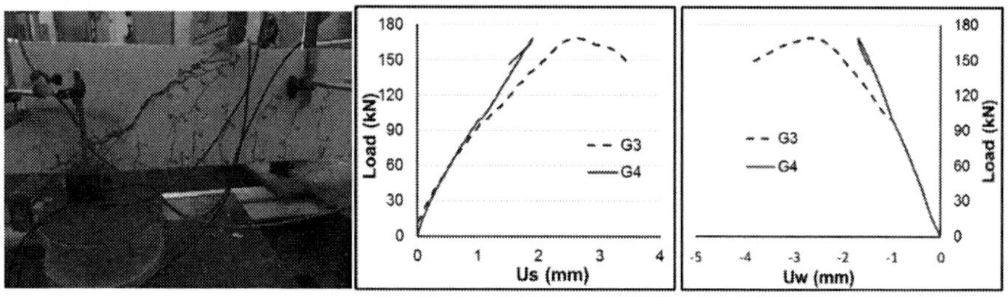

(d) (e) (f)

Fig. 6. A photograph showing crack pattern of **(a)** SFRC_0_1 **(d)** SFRC_0.5_1 beams; At G3 and G4, the variation of vertical displacement of **(b)** SFRC_0_1 **(e)** SFRC_0.5_1; horizontal displacements of **(c)** SFRC_0_1 **(f)** SFRC_0.5_1 beams.

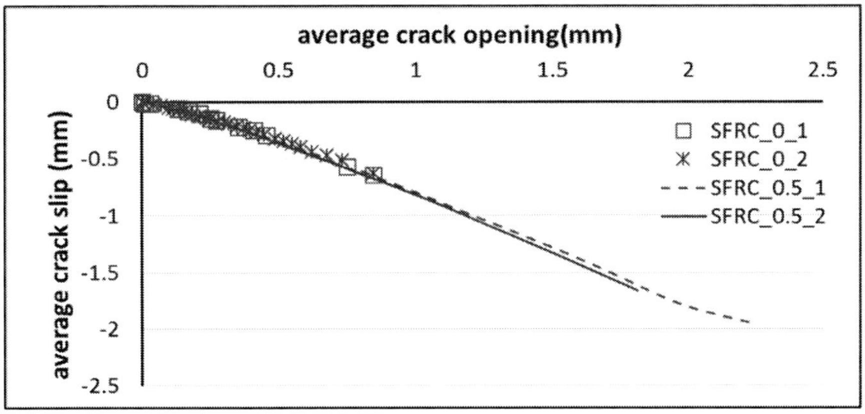

Fig. 7. Crack opening-slip response of control and SFRC beams.

The average crack slip as a function of the average crack opening across the primary shear crack formed in the pre-peak load response obtained from both control and SFRC beams is shown in Fig. 7. The resulting behavior obtained from the shear region indicates a dilatant behavior. There is an increase in crack opening associated with an increase in slip between the rough crack surfaces. With the application of load, after the onset of the shear crack, the crack opening increases nominally in linear manner with an increase in the average slip. The dilatant behavior (crack opening – slip response) exhibited by both control and the SFRC beams is identical. This suggests that the relationship between slip and crack opening is of the same origin. In the 0.5% SFRC beams, the crack slip and the crack opening continue to increase past the peak load for the control specimens.

A comparative analysis of the control and 0.5% SFRC beams is performed using the measured responses from each. The shear dilatant behaviors recorded from a control specimen (SFRC_0_1) and a 0.5% SFRC beam (SFRC_0.5_1) are shown in Fig. 8(b) and Fig. 8(c), respectively. The load responses of the two beams are plotted in Fig. 8(a) for reference. The crack opening and the crack slip measurements at a load corresponding to the peak load in the control beam is marked in the dilatant behavior. At the peak load of the control specimen, both the crack opening and the crack slip in the SFRC beam are smaller than the corresponding values in the control specimen. The crack opening and the crack slip values recorded at the peak load in the response of the SFRC beam are comparable to the values obtained from the control specimen at peak load. The dilatant behavior, the general trend in the crack opening and crack slip relationship is not altered after the peak load.

 DOI: 10.1051/matecconf/201712004003

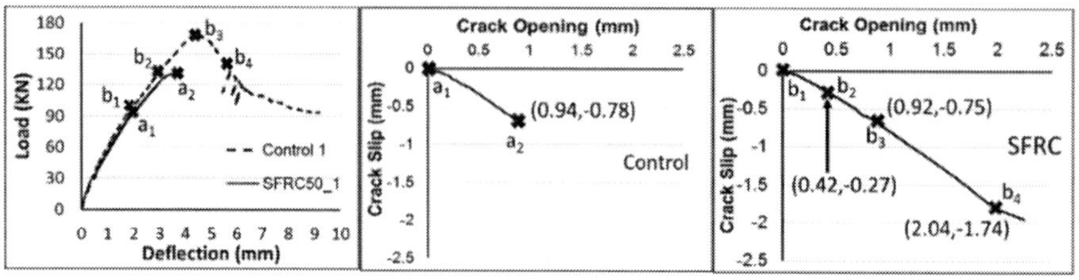

(a) (b) (c)

Fig. 8. (a) Load-response; Crack opening-slip response of **(b)** control_1 and **(c)** SFRC50_1 beams.

6 Analysis of shear failure of beams

The free body diagram of a part of beam in shear span for which one end is the crack plane is shown in Figure 9. Considering the diagonal crack at the end from experimental tests is essentially at 45^0 to the horizontal (thick line shown in Fig 9a), the crack plane is idealized as a straight line inclined at 45^0 as shown in the free-body diagram. The applied forces and the internal forces of resistance from concrete and reinforcement are as shown in the Fig 7b. The coordinate system for evaluating forces is taken in directions which are aligned parallel and perpendicular to the crack plane. The internal forces of resistance include, the resultant normal compressive force on the crack plane (C) from concrete acting at a distance z from the top fiber of the beam, shear resistance across the crack coming from the aggregate interlock and other factors (S), the tensile force in the reinforcement (T), and the dowel force from the reinforcement (D). The resultant compressive force (C) is the vectorial sum of the normal forces acting on the crack plane. S represents the shear demand on the inclined crack plane resulting from the requirement of equilibrium of forces in a direction parallel to the crack. The shear resistance generated across the crack face would depend upon the level of compression (C) and the may take as μ*C where μ is the constant of proportionality, which depends on the roughness of the crack surface and the contact area of the rough crack.

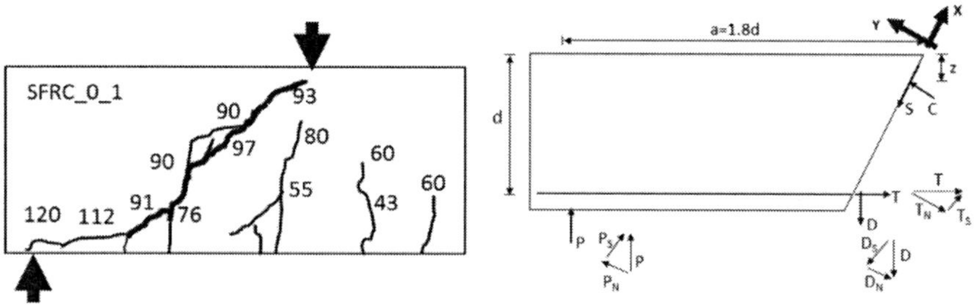

(a) (b)

Fig. 9. (a) crack pattern of control beam; **(b)** Free body diagram of the part of the control beam in shear span.

By taking components of forces along the crack plane as shown in the Figure 9,

$$T_N = T_S = T/\sqrt{2} \qquad \text{--(1)}$$

$$P_N = P_s = P/\sqrt{2} \qquad \text{--(2)}$$

$$D_N = D_s = D/\sqrt{2} \qquad \text{--(3)}$$

Taking moment about top right most point of the beam and assuming z is small, the moment contribution from C is insignificant,

$$T*d = P*(1.8)d - (D*d) \qquad \text{--(4)}$$

By considering the force equilibrium along Y-axis,

$$\Sigma F_y = P_N - T_N + C - D_N = 0$$

$$C = T_N - P_N + D_N \qquad \text{--(5)}$$

By considering the force equilibrium kalong X-axis,

$$\Sigma F_x = P_S + T_S - S - D_s = 0$$

$$S = T_S + P_S - D_s \qquad \text{--(6)}$$

Considering Equation (5), the tensile stress obtained from the strain gauge data at the crack location.

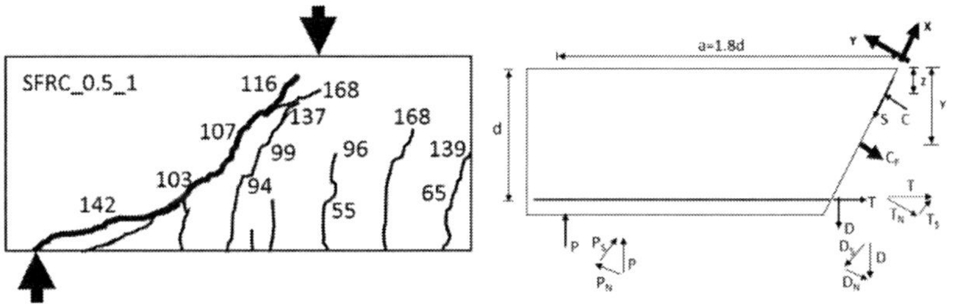

(a) (b)

Fig. 10. (a) Crack pattern of beam with 0.5% fibers, (b) Free body diagram of the part of the SFRC beam in shear span

In fiber reinforced concrete beams, due to the presence of fibers, there will be an additional average normal resistant force across the crack associated with opening of the crack. The additional the crack closing stresses are identified with C_F in Fig 10.The fiber force (C_F) is shown acting at a distance y from the top of the beam (the Fig 10).
Taking moment about point A shown in the figure,

$$\Sigma M = T*d - P*(1.8d) + C_F*(\sqrt{2}\,y) + (D*d) = 0 \qquad \text{--(7)}$$

Considering force equilibrium along Y axis,

$$\Sigma F_y = P_N - T_N + C - C_F - D_N = 0$$

$$C = T_N - P_N + C_F + D_N \qquad \text{--(8)}$$

Considering force equilibrium along X axis,

$$\Sigma F_x = P_S + T_S - D_s - S = 0$$

$$S = T_S + P_S - D_s \qquad \text{--(9)}$$

The resultant compressive force (C) on the crack face now contains an additional contribution of the fiber force and is higher than the corresponding value in control beam. After peak, the value of C depends on the crack closing stresses available as C_F. Even at the higher displacement which results in larger crack opening, the fibers provide crack bridging stresses, which ensures contact compressive stress across the compression along the crack plane. This allows the shear carrying capacity across the crack face and the beam continues to carry load. From this it can be established that fibers do help in transfer of shear across the shear crack with the additional compression coming across the crack with presence of fibers.

7 Summary and conclusions

Addition of fibers do not influence shear cracking load or the crack angle. There is continuous increase in slip across the crack faces of the dominant shear crack with increasing load. The crack exhibits a dilatant response, where the increase in slip displacement produces an increase in crack opening. Shear slip to crack opening relationship of a rough shear crack up to peak is not altered by the fibers. In control beams, when the shear crack reaches the level of the flexural reinforcement, it grows along the rebar up to the support point and the beam fails by splitting along this crack as the resistance coming from the aggregate interlock is lost with increased crack width. Fibers are effective after the formation of crack and sustain the shear transfer ability across the rough crack. This helps sustain shear load carrying capacity of the beam. Sudden splitting along the shear crack is suppressed by the fibers bridging the crack leading to a continuous resistance to load for a larger deflection of the beams. With the addition of steel fibers failure mode changes from brittle to ductile.

Analysis of shear response of reinforced concrete beam shows that control specimens failed when compression generated by rebar is insufficient to sustain aggregate interlock after yielding of the flexural reinforcement. In fiber reinforced beams, the additional compression across the crack provided by the fibers helps sustain the shear transfer across the crack.

References

1. M. A. Mansur, K. C. G. Ong, & P. Paramasivam., Shear strength of fibrous concrete beams without stirrups. *Journal of Structural Engineering*, **112(9)**, 2066-2079, (1986).
2. A.K. Sharma., Shear strength of steel fiber reinforced concrete beams. In *ACI Journal Proceedings*, Vol. **83**, No. 4. (1986).
3. R. Narayanan, & I. Y. S. Darwish, (1987)., Use of steel fibers as shear reinforcement. *ACI Structural Journal*, Vol. **84(3)**.
4. S. A. Ashour, G. S. Hasanain, & F. F. Wafa., Shear behavior of high-strength fiber reinforced concrete beams. *ACI Structural Journal*, Vol. **89**(2), (1992).
5. V. C. Li, R. Ward, & A. M. Hmaza, (1992). Steel and synthetic fibers as shear reinforcement. *ACI Materials Journal*, Vol. **89(5)**, (1992).
6. K. H. Tan, K. Murugappan, & P. Paramasivam., Shear behavior of steel fiber reinforced concrete beams. *Structural Journal*, Vol. **90(1)**, 3-11, (1993).
7. M. Imam, L. Vandewalle, F. Mortelmans, & D. Van Gemert., Shear domain of fibre-reinforced high-strength concrete beams. *Engineering structures*, Vol. **19(9)**, 738-747, (1997).
8. Y. K. Kwak, M. O. Eberhard, W. S. Kim, & J. Kim., Shear strength of steel fiber-reinforced concrete beams without stirrups. *ACI Structural Journal*, Vol. **99(4)**, (2002).
9. S. H. Cho, & Y. I. Kim., Effects of steel fibers on short beams loaded in shear. *Structural journal*, vol. **100(6)**, 765-774, (2003).
10. C. Cucchiara, L. La Mendola, & M. Papia., Effectiveness of stirrups and steel fibres as shear reinforcement. *Cement and Concrete Composites*, Vol. **26(7)**, 777-786, (2004).

11. D. R. Sahoo, & N. Kumar., Monotonic behavior of large-scale SFRC beams without stirrups. *Engineering Structures*, Vol. **92**, 46-54, (2015).
12. D. R. Sahoo, S. Bhagat, & T. C. V. Reddy., Experimental study on shear-span to effective-depth ratio of steel fiber reinforced concrete T-beams. *Materials and Structures*, Vol. **49(9)**, 3815-3830, (2016).
13. ACI 318-08, Building code requirements for structural concrete (ACI 318-08) and commentary, American Concrete Institute, Farmington Hills, MI, (2008)
14. R. Park, & T. Paulay., Reinforced concrete structures. John Wiley & Sons, (1975).
15. J. G. MacGregor, J. K. Wight, S. Teng, & P. Irawan., Reinforced concrete: mechanics and design (Vol. **3**). Upper Saddle River, NJ: Prentice Hall, (1997).
16. G. Winter, & A. Nilson., Design of concrete structure, 9th ed., McGraw Hill, New York, (1986).
17. M. Khuntia, B. Stojadinovic, & S. C. Goel., Shear strength of normal and high-strength fiber reinforced concrete beams without stirrups. *Structural Journal*, Vol **96(2)**, 282-289, (1999).
18. L. Vandewalle., Design method for steel fiber reinforced concrete proposed by RILEM TC 162-TDF. In P. Rossi, & G. Chanvillard (Eds.), *Fifth International RILEM Symposium on Fibre-Reinforced Concrete (FRC)* (pp. 51-64). RILEM Publications SARL, (2000).
19. L. Vandewalle., RILEM TC162-TDF: Test and Design methods for Steel Fibre Reinforced Concrete: Bending test (final recommendation). *Materials and Structures*, Vol. **35**, 579-582, (2002).
20. M.P. Collins, E. C. Bentz, E. G. Sherwood, & J. K. WIGHT., Where is Shear Reinforcement Required? Review of Research Results and Design Procedures. *ACI structural journal*, Vol. **106(4)**, 556-559, (2009).
21. J. C. Walraven., *Aggregate interlock: a theoretical and experimental analysis* (Doctoral dissertation, TU Delft, Delft University of Technology), (1980).
22. J. C. Walraven, & H. W. Reinhardt., Theory and experiments on the mechanical behaviour of cracks in plain and reinforced concrete subjected to shear loading. *HERON,* Vol. **26** *(1A),* (1981).
23. J. C. Walraven, Fundamental analysis of aggregate interlock. *Journal of the Structural Division*, ASCE, Vol. **107(11)**, 2245-2270, (1981).
24. K. V. Subramaniam, J. Popovics, and S. P. Shah., Testing Concrete in torsion: Instability analysis and Experiments, *Journal of Engineering Mechanics, American Society of Civil Engineers* (ASCE), Vol. **124 (11)**, pp. 1258-1268, (1998).
25. IS 1786, High strength deformed steel bars and wires for concrete reinforcement, *Bureau of Indian Standards*. New Delhi, (2008).
26. ASTM, C. 1609/C 1609M-06. *Standard Test Method for Flexural Performance of Fiber–Reinforced Concrete (Using Beam with Third-Point Loading), ASTM International, USA*, (2006).
27. M. A. Sutton, W. J. Wolters, W. H. Peters, W. F. Ranson, & S. R. McNeill., Determination of displacements using an improved digital correlation method. *Image and vision computing*, Vol. **1(3)**, 133-139. (1983).
28. M. A. Sutton, S. R. McNeill, J. Jang, & M. Babai., Effects of subpixel image restoration on digital correlation error estimates. *Journal of Optical Engineering*, Vol. **27(10)**, 870-877, (1988).
29. C. Carloni, & K. V. Subramaniam., Direct determination of cohesive stress transfer during debonding of FRP from concrete. *Composite Structures*, Vol. **93(1)**, 184-192, (2010).

30. K. C. Reddy, and K. V. Subramaniam, "Crack Propagation and Post cracking Behavior of Macrosynthetic Fiber Reinforced Concrete," accepted for publication in *Magazine of Concrete Research*, (2017).

Behaviour of fibre reinforced concrete using steel slag coarse aggregate produced in Qatar

Wael Alnahhal[1,*]

[1]Assistant Professor, Department of Civil and Architectural Engineering, Qatar University, Qatar.

Abstract. The state of Qatar suffers from the shortage of natural resources needed for concrete production. Therefore, it is essential to investigate the feasibility of using by-product recycled materials as aggregates to maintain the concrete construction industry. Several types of recyclable materials are currently used in concrete. One of the potential resources of recycled concrete is steel slag. Knowing that Steel slag is the most significant solid waste generated by Qatar Steel Company in Qatar, replacing of natural coarse aggregate with steel slag aggregate will have a significant environmental and economic impact to the state of Qatar. This paper presents the compression and flexural test results of different concrete mixes made of steel slag coarse aggregate combined with a newly developed basalt chopped fibres. The parameters investigated included the volume fraction of the fibre used and the type of coarse aggregates (natural aggregates "Gabbro" and steel slag aggregates). Plain concrete specimens containing natural coarse aggregates and steel slag aggregates with no fibres added were also tested to serve as control. Test results showed that adding the basalt chopped fibres to the concrete mixes enhanced their flexural tensile strengths at different percentages. In addition, the compressive strength of concrete made with steel slag aggregate was higher than that made with natural gabbro aggregate. Test results clearly showed that steel slag aggregates can be used as sustainable and eco-friendly alternative materials in concrete structures.

1 Introduction

The State of Qatar has experienced tremendous economic development in the last decades. As a result, massive quantities of building materials have been used to accommodate the great pace in construction. The state is anticipated to spend USD 17 billion on its infrastructure as part of its preparation to host the Soccer World Cup in 2022. Most of these expenses will go towards the construction of new hotels, stadiums, facilities, and tourism's attractions (from Qatar 2022 bidding committee official website).

Concrete is the main material used in construction in Qatar and in most of the Gulf Cooperation Council (GCC) countries. However, most of these countries suffer from the shortage of natural resources required for concrete production. For example, the quantities

* Corresponding autour: wael.alnahhal@qu.edu.qa

of imported aggregates in the State of Qatar have increased from 9.5 million tons in 2006 to 21.5 million tons in 2008 (data from the Qatar Customs Database). Therefore, it is essential to investigate the feasibility of using recycled materials in concrete production.

Several types of recyclable materials are currently used in concrete. One of the potential resources of recycled concrete is steel slag, knowing that steel slag is the most significant solid waste generated by Qatar Steel Company in Qatar [1]. Several researchers have shown that steel slag can be used as an aggregate in structural concrete [2-6]. The use of steel slag in concrete applications has major economic and environmental benefits to Qatar. Steel slag aggregate (SSA) can partially replace the relatively expensive imported Gabbro aggregates commonly used in Qatar, thus saving on the total cost of the concrete projects. Recycling the steel waste will also have a significant environmental impact due to the reduction in demand of virgin aggregates. Moreover, using waste materials would partially solve a significant environmental concern from land-filling the large quantities of waste materials produced every year in Qatar.

There has been always a general belief about the undesirable impact of using SSA in concrete production [2]. However, a number of publications have contradicted this idea [3-6]. Recently, a number of researchers studied the possibility of using steel slag in concrete mixes as a coarse aggregate to make sustainable concrete [3-4]. They reported that the use of the SSA improved the flexural and compressive strength of the concrete. According to a study by Krishnasami et al., it was reported that the flexural and compressive strength of self-compacted concrete made with coarse SSA is higher than that of the concrete with natural aggregate (NA) [5].

A number of studies focused on determining the optimal replacement percentage of coarse and fine aggregates by steel slag to produce a concrete that would accommodate a higher strength than the normal concrete. Devi et al. found that 40% and 30% individual replacement for fine and coarse aggregates by steel slag respectively resulted in improved compressive and flexural strengths [6]. In addition, Qasrawi found that the optimum percentage of slag aggregate replacement is 30 % of coarse aggregate content [3]. Therefore, a proper mix design is necessary to obtain the required quality of concrete made of steel slag aggregate.

Previous investigations have shown that using discrete steel fibres in concrete increases its ductility due to the large compressive strains exhibited at failure [7-9]. However, among the disadvantages of using steel fibres is corrosion, especially in the harsh environment that characterizes the Arabian Gulf area, and the added weight to the concrete structure. Therefore, chopped basalt fibres were proposed in this study. Chopped basalt fibres are non-corrosive macro fibres made from basalt fibre-reinforced polymer bars. Chopped basalt fibres possess higher tensile strength and higher stiffness compared to conventional synthetic fibres [10]. The fibres act as a proactive reinforcement that provides the immediate tensile load carrying capacity when micro cracks develop in concrete. Bajaj [11] reported that the crack width that develops in plain concrete is larger than the crack width in polypropylene fibre concrete, which is larger than the crack width obtained in reinforced concrete having chopped basalt fibres in the mix.

This paper presents a part of a large research program that aims at investigating the feasibility of using SSA as a low-cost alternative sustainable material in fibre reinforced concrete (FRC). The program also aims at investigating the use of basalt chopped fibres in combination with SSA to produce sustainable concrete. The results of preliminary tests that have been conducted to determine the properties of the obtained FRC are presented.

2 Experimental program

The aim of this experimental program is to characterize the properties of hardened FRC made of SSA in combination with chopped basalt fibres. The results of compressive and flexural tests are presented. Three different volume fractions of chopped basalt fibres were investigated. The properties of plain concrete containing 100% natural coarse aggregate with no fibres added to the mix were also tested to serve as benchmark for the FRC. The maximum aggregate size used in all concrete mixes in this study was 25 mm.

2.1 Materials and mixture proportions

Type I Portland cement was used in this study. Washed sand was used as fine aggregate. The two types of coarse aggregate used were the Gabbro and the SSA. Gabbro is a natural coarse aggregate that is not available in Qatar but imported from the United Arab Emirates. On the other hand, the steel slag aggregates used in this study have been produced in Qatar as a by-product of steel manufacturing. Test results for the specific gravity and absorption ratio for both types of aggregates are shown in Table 1. As anticipated, the SSA showed higher specific gravity and unit weight and relatively higher absorption than the Gabbro aggregates. Figure 1 and Figure 2 show the sieve analysis obtained for both types of aggregates with respect to the upper and lower ASTM limits. It can be noticed that the SSA curve was close to the upper ASTM limit whereas that of the Gabbro aggregates was between the ASTM limits.

Six concrete mixes containing either natural or steel slag aggregates were studied and compared. The first mix was a control mix and did not contain any chopped basalt fibres. For other mixes, two volume fractions of chopped basalt fibres were used namely, 0.25% and 1%. All mixes had a water/cement ratio of 0.51. The weight proportions of all ingredients are shown in Table 2. The target compressive strength of all mixes was 40 MPa.

Table 1: Specific Gravity and Absorption Ratio for Gabbro and SSA.

Aggregate Type	Bulk Specific Gravity (Dry)	Bulk Specific Gravity (SSD*)	Bulk Specific Gravity (APP**)	Absorption (%)
Gabbro	2.886	2.905	2.942	0.654
SSA	3.242	3.311	3.484	2.145

* (SSD= Saturated Surface Dry)

** (APP= Apparent)

Table 2: Weight Proportions of the Ingredients.

Ingredients	Weight Proportion
Cement	1.00
Water	0.51
Coarse Aggregate	2.6
Fine Aggregate	2.5

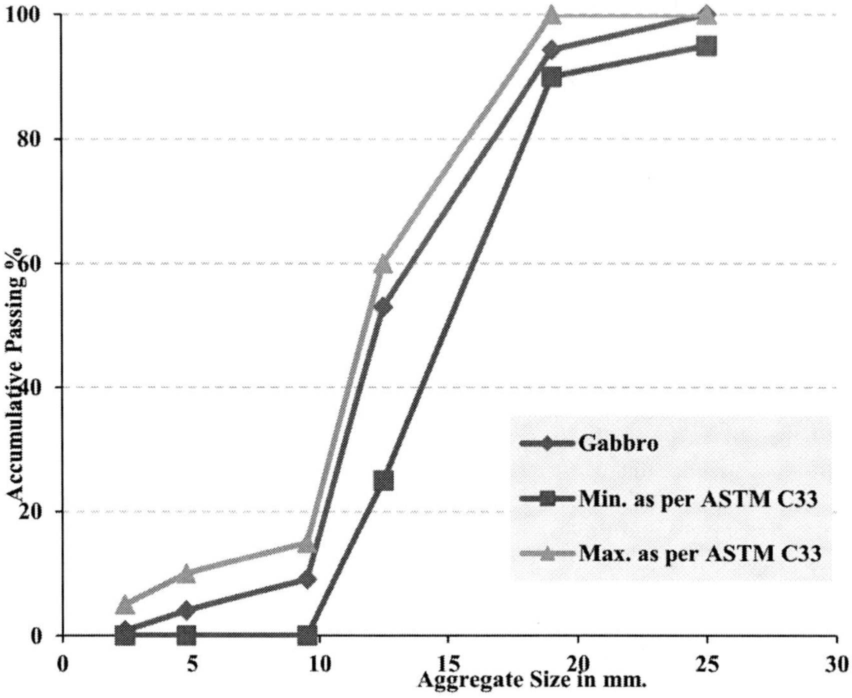

Fig. 1: Sieve Analysis for Gabbro.

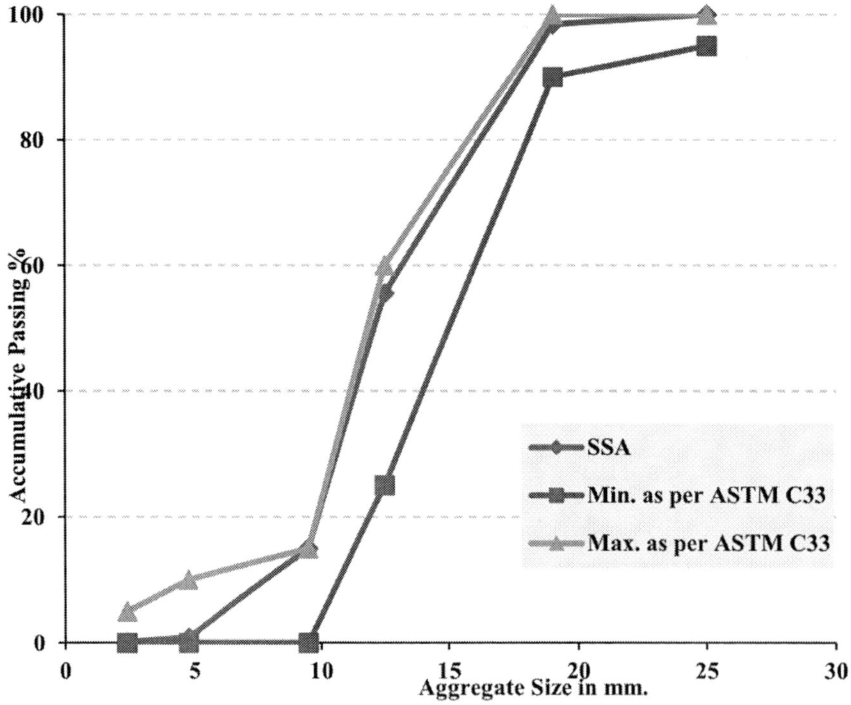

Fig. 2: Sieve Analysis for SSA.

2.2 Compression and flexural tests

The compressive strength of all concrete mixes was measured using standard cylinders of 150x300 mm according to the provisions of ASTM C192/C192M-14. The cylinders were moist-cured for 28 days after mixing. Five test cylinders were performed for each mix after 28 days according to ASTM C39-14C. Three flexural strength tests were conducted for each mix using standard prisms of 150x150x500 mm according to ASTM C78-07. The aim of these tests was to study the effect of the fibres on the flexural performance of the hardened concrete. The prisms were tested under four-point loads until failure.

3 Results and discussion

The measured average compressive strengths are shown in Table 3. It was noticed that the average compressive strength of the concrete mixes made of Gabbro aggregate (G-1, G2, and G-3) were lower than that of the concrete mixes made of SSA (SSA-1, SSA-2, and SSA-3). As shown in Table 3, the compressive strength results indicated better performance of the fibre reinforced concrete over the control mix (without fibres) for steel slag aggregates. However, there is no correlation between the compressive strength and the percentage of fibre for gabbro aggregates, since increasing the percentage of fibre did not influence the concrete compressive strength.

Figure 3 showed the compression test results. It can be observed that adding a volume fraction of 0.25% to the Gabbro concrete resulted in an increase in the compressive strength of 17.4%. However, the compressive strength increased by 15.4 % when the volume fraction of the chopped fibres increased to 1%. For concrete mixes with steel slag aggregates, adding volume fractions of 0.25% and 1% of the fibres to the mix resulted in an increase in the compressive strengths of 3.8% and 15.8%, respectively. The increase in compressive strength in SSA mixes was obvious when fibres were added.

It is clearly observed that the compressive strength of concrete cast using SSA aggregate had better results than the concrete made with normal Gabbro aggregate. The average increase in the strength of SSA concrete was 15.2% compared with Gabbro concrete. This can be attributed to the high physical properties of the SSA compared to that of the Gabbro aggregates.

Table 3: Average Compressive Strength of Concrete Cylinders.

Specimen	Aggregate Type	Chopped Basalt Fiber Dosage % by Volume	Average Compressive Strength, f'_c (MPa)
G-1	Gabbro	0 %	39.6
G-2	Gabbro	0.25 %	46.5
G-3	Gabbro	1 %	45.7
SSA-1	Steel Slag	0 %	47.4
SSA-2	Steel Slag	0.25 %	49.2
SSA-3	Steel Slag	1 %	54.9

DOI: 10.1051/matecconf/201712004004

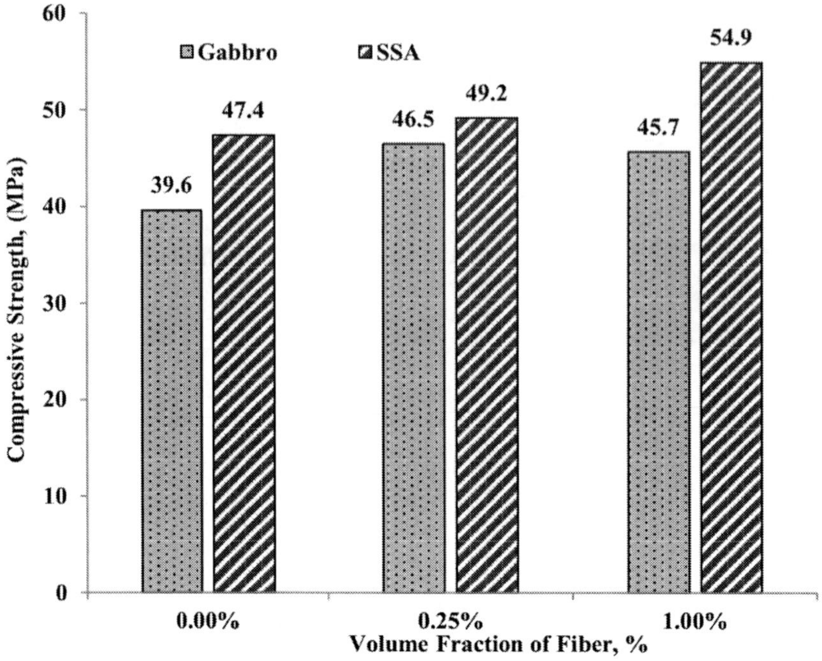

Fig. 3: Compression test results.

The measured average flexural strengths are shown in Table 4. There was an obvious better performance of the fibre reinforced concrete over the control mix (without fibres) for both types of aggregates used. Moreover, the effect of adding fibres to the concrete mix was more effective if compared to the compression test results. This is demonstrated in Figure 4 that compares the flexure test results of both types of concrete mixes. It can be observed that adding a volume fraction of 0.25% to the Gabbro concrete resulted in an increase of 9.6% in the flexural strength. The gain in flexural strength further increased to 23.3% when the volume fraction of the chopped fibres increased to 1%. For concrete mixes with steel slag aggregates, adding fibres with volume fractions of 0.25% and 1% to the mix resulted in an increase in the flexural strength of 14.1% and 29.8%, respectively.

It is clearly observed that the tensile strength of concrete cast using SSA aggregate had slightly better results than the concrete made with normal Gabbro aggregate. The average increase in the strength of SSA concrete was 6.7% compared with Gabbro concrete.

Table 3: Average Flexural Tensile Strength of Concrete Prisms.

Specimen	Aggregate Type	Chopped Basalt Fiber Dosage % by Volume	Average Tensile Strength, f_r (MPa)
G-1	Gabbro	0 %	6.04
G-2	Gabbro	0.25 %	6.62
G-3	Gabbro	1 %	7.45
SSA-1	Steel Slag	0 %	6.25
SSA-1	Steel Slag	0.25 %	7.13
SSA-1	Steel Slag	1 %	8.11

DOI: 10.1051/matecconf/201712004004

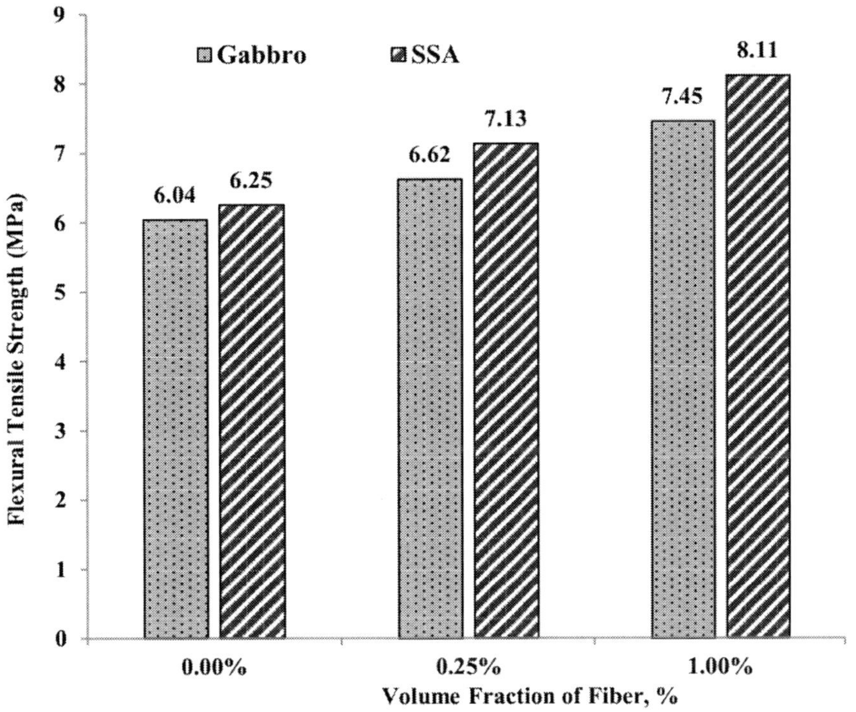

Fig. 4: Flexural testing results.

4 Conclusions

Preliminary test results on the use of steel slag aggregates and chopped basalt fibres in concrete mixes were presented. Compression tests on standard cylinders and flexural tests on standard prisms were conducted. The following conclusions are withdrawn from these tests:

- The compressive and flexural strengths of concrete cast using steel slag aggregate had better results than the concrete made normal Gabbro aggregate.
- The addition of chopped basalt fibre at volume fractions of 0.25% and 1% to the Gabbro concrete did not show a clear correlation between the two parameters. However, adding of chopped basalt fibre at volume fractions of 0.25% and 1% to the Gabbro resulted in an increase in the flexural strength were 9.6% and 23%, respectively.
- The addition of chopped basalt fibre at volume fractions of 0.25% and 1% to the SSA concrete resulted in an increase in the compressive strength of 3.8% and 15.8%, respectively. The corresponding gains in the flexural strength were 14.1% and 29.8%, respectively.
- Both compressive strength and flexural tensile strength of concrete made with steel slag aggregate were higher than those made with natural gabbro aggregates. Test results clearly showed that steel slag aggregate can be used as sustainable and eco-friendly alternative materials in concrete structures.

The authors show their gratitude to Qatar University for their financial support through the internal research grant QUUG-CENG-CAE-14/15-5. The authors are grateful to the Slag Aggregates Producer. W.L.L (SAP) for providing the SSA. In particular, special thanks are due to Mr. Devassy Baby who was very supportive of the research proposal.

References

1. Qatar Steel Sustainability Report (2013): http://www.qatarsteel.com.qa/Publications/ SitePages /sustainreport.aspx.
2. I. Netinger, D. Bjegovic, and G. Vrhovac (2011). "Utilisation of steel slag as an aggregate in concrete," Materials and Structures, **44**, 1565–1575.
3. H. Qasrawi (2012), "Use of Relatively High Fe2O3 Steel Slag as Coarse Aggregate in Concrete," ACI Material Journal V. **109**, 2012, pp. 471-478.
4. S. Tarawneh, E. Gharaibeh, & F. Saraireh (2014). "Effect of using steel slag aggregate on mechanical properties of concrete," American Journal of Applied Sciences, 11(5), 700.
5. R. Krishnasami, R. Malathy (2013) "Significance of Blast Furnace Slag as Coarse Aggregate in Self Compacting Concrete", Applied Mechanics and Materials Vols. **357-360**, 2013, pp. 829-833.
6. V. Devi, B. Gnanavel (2014). "Properties of Concrete Manufactured using Steel Slag," 12th Global Congress on Manufacturing and Management, GCMM-2014. Procedia Engineering 97, pp 95 – 104.
7. K. Holschemacher, T. Mueller, and Y. Ribakov (2010). "Effect of steel fibers on mechanical properties of high-strength concrete," Materials and Design, 31(5), 2604-2615.
8. Y. Mohammadi, R. Carkon-Azad, S.P. Singh, and S. Kaushik (2009). "Impact resistance of steel fibrous concrete containing fibers of mixed aspect ratio," Construction and Building Materials, 23(1), 183-189.
9. J. Katzer and J. Domski (2012). "Quality and mechanical properties of engineered steel fibers used as reinforcement for concrete," Construction and Building Materials, 34, 243-248.
10. S. Adhikari (2013). "Mechanical and structural characterization of mini-bar reinforced concrete beams," Ph. D. Dissertation, University of Akron, USA
11. S. Bajaj (2012). "Effect of corrosion on physical and mechanical properties of reinforced concrete," M. Sc. Thesis, University of Akron, USA.

AUTHOR INDEX

AUTHOR INDEX

AUTHOR INDEX

EDP Sciences – Web of Conferences
17, Avenue du Hoggar
Parc d'Activité de Courtabouf
BP 112
F-91944 Les Ulis Cedex A
France

ISBN 978-1-5108-4704-0

CURRAN ASSOCIATES INC.
proceedings.com

International Conference on Advances in Sustainable Construction Materials & Civil Engineering Systems (ASCMCES-17)

MATEC Web of Conferences Volume 120 (2017)

Sharjah, United Arab Emirates
18-20 April 2017

Part 2 of 2

edp sciences

International Conference on Advances in Sustainable Construction Materials & Civil Engineering Systems (ASCMCES-17)

MATEC Web of Conferences Volume 120 (2017)

Sharjah, United Arab Emirates
18-20 April 2017

Part 2 of 2

Editors:

A. Shanableh
S. Barakat
S. Al-Toubat
K. Hamad
M. Maalej
M. Omar
R. Al-Ruzouq

ISBN: 978-1-5108-4704-0

Printed from e-media with permission by:

Curran Associates, Inc.
57 Morehouse Lane
Red Hook, NY 12571

Some format issues inherent in the e-media version may also appear in this print version.

Printed by Curran Associates, Inc. (2017)

For additional information, please contact EDP Sciences – Web of Conferences at the address below.

EDP Sciences – Web of Conferences
17, Avenue du Hoggar
Parc d'Activité de Courtabœuf
BP 112
F-91944 Les Ulis Cedex A
France

Phone: +33 (0) 1 69 18 75 75
Fax: +33 (0) 1 69 28 84 91

contact-edps@webofconferences.org

Additional copies of this publication are available from:

Curran Associates, Inc.
57 Morehouse Lane
Red Hook, NY 12571 USA
Phone: 845-758-0400
Fax: 845-758-2633
Email: curran@proceedings.com
Web: www.proceedings.com

TABLE OF CONTENTS

PART 1

PART 2

Numerical simulation of concrete slabs strengthened with PTFE sheets subjected to blast load

Rawy Youssef[1,*], *Walid* Attia[1], and *Mohamed* Laissy[2]

[1]Faculty of Engineering, Cairo University, Giza, Egypt
[2]Faculty of Engineering, Prince Mugrin Bin Abdulaziz University, Medina, Saudi Arabia

Abstract. Due to the noticeable increase in bombing activities, the understanding of the structures response against blast loads has become one of the major concerns in structural engineering. This present study investigates the blast loading hazards that can take place in concrete slabs. Concrete slabs strengthened using PolyTetraFluoroEthylene (PTFE) are used to investigate the control of the hazards. The thickness of the slabs is 200 mm. The models are exposed to a blast load equivalent to 20 Kg of TNT explosives with a standoff distance equals to 500 mm. The concrete slabs are strengthened using different thicknesses and different number of sheets. In order to investigate the effect of PTFE sheets, AUTODYN software is used to create the numerical models. Concrete is modelled using Lagrange solver while the air and TNT explosives are modelled using Euler solver. The results of the study show that PTFE sheets control concrete fragmentation and debris propagation. The use of PTFE can improve concrete resistance to one of the blast components because of its stability at high temperatures, thermal insulation, and resistance to weathering. PTFE has a wide working temperature range from -100°F to +400°F (-73°C to 204°C) in which its properties remain functional.

1 Introduction

An explosion can lead to many hazards that can jeopardize human safety. Some of those hazards are blast pressure, heat, structural collision, vibration, and fragments. Human tolerance to those hazards can vary widely [1]. Von Ramin et al. [2] studied the fragments distribution and energy from debris of a one way span masonry wall to predict the minimum safety ranges. Lin et al. [3] conducted a parametric study to investigate the responses of FRP-strengthened RC panels under blast loads. They found that increasing the thickness of the GFRP sheet decreases the maximum and residual deflections of RC panels. Tai et al. [4] studied the effect of meshing and steel reinforcement in concrete slabs. The study of the dynamic response of RC slabs showed that the mesh size is very sensitive to the shock wave propagation and should be as fine as possible. Also, the reinforcement ratio

* Corresponding author: rawy.adly@yahoo.com

DOI: 10.1051/matecconf/201712004005

in the concrete slab affects damage location. Weerheijm et al. [5] studied the failure of a one way concrete slab under internal explosion.

The objective of this research is to study the control of hazards in concrete slabs using PTFE sheets. A parametric study is conducted using AUTODYN to compare the behaviour of the slabs with different number and thickness of PTFE sheets.

PTFE $(C_2F_4)_n$ is an artificial highly crystalline fluoropolymer consisting of carbon and fluorine, it is best known commercially as Teflon. PTFE is chemically inert and resistant to attacks from corrosive chemicals. PTFE is highly dense and has a high melting temperature which makes it highly stable at high temperatures and has high impact strength [6, 7].

1.1 Blast wave and debris propagation

Figure 1 illustrates the idealization of a blast wave pressure-time history. The curve can be divided into two phases: positive and negative phases. At time t_A, detonation takes place. A sudden increase in pressure occurs, it is equal to (P_P - P_O) and called overpressure. From t_A to t_B, the positive phase represents the instantaneous rise of pressure and decrease of pressure to ambient condition. The negative phase starts at time t_B and ends with the decay of the wave. The negative phase is longer than the positive phase and the pressure turns into a very small value of suction [8].

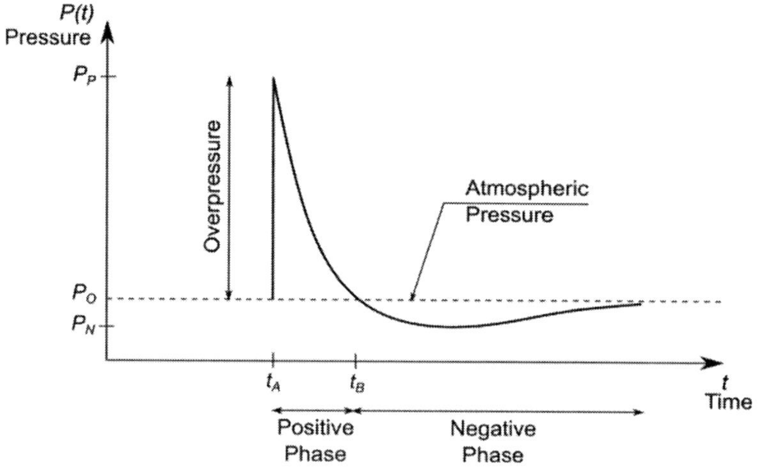

Fig. 1. Blast wave pressure-time history [8].

During an explosion, fragments may be formed from any object in the path of the blast wave. Fragments can travel for long distances thus increasing the blast radius and human injuries. The distance travelled by a fragment is affected by initial direction, size, shape, and explosion yield [8].

1.2 Structural response to shock waves

Analysis of structures under blast loads can be a complex process. In order to simplify this process, it is widely accepted to model the structure as a single degree of freedom SDOF.

Figure 2 illustrates the SDOF system for analysis. The structure is idealized as a lumped mass M having a weightless spring with stiffness K and an idealized triangular pulse as shown in Equation (1). Equation of motion for the un-damped SDOF is shown in Equation (2). Design for blast loads is affected mainly by the first cycle response, thus damping is neglected. [1, 9]

 DOI: 10.1051/matecconf/201712004005

Fig. 2. (a) Un-damped SDOF system, (b) Blast load

$$F(t) = F_m\left(1 - \frac{t}{t_d}\right) \quad (1)$$

$$Ma + Ky = F_m\left(1 - \frac{t}{t_d}\right) \quad (2)$$

Due to the dynamic analysis simplification, transformation factors for load, mass, and resistance are used to convert the required system for analysis into a SDOF.

1.3 Concrete behaviour at high strain rates

Concrete behaviour is different under dynamic and static loading. In dynamic loading, for the same strain value concrete reaches a higher value of stress. Strain rates for blast loads are 10^{11} times the static strain rates. According to Bischoff et al. the relative increase in concrete ultimate strength in uniaxial compression can reach a value of 2 [10] and according to Malvar et al. ultimate tensile strength can increase up to 7 times [11]. To account for this behaviour, codes of practice introduced the dynamic increase factor DIF which is defined as the ratio of the dynamic to static strength [1].

2 Numerical simulation

In order to conduct this study, accurate modelling of materials, loading, boundaries, meshing, and gauges are carried out. Approaches and techniques developed by AUTODYN software are used [12]. Setup for the analysis is 20Kg of TNT explosives placed at a standoff distance of 500mm from the concrete slab as shown in Figure 3.

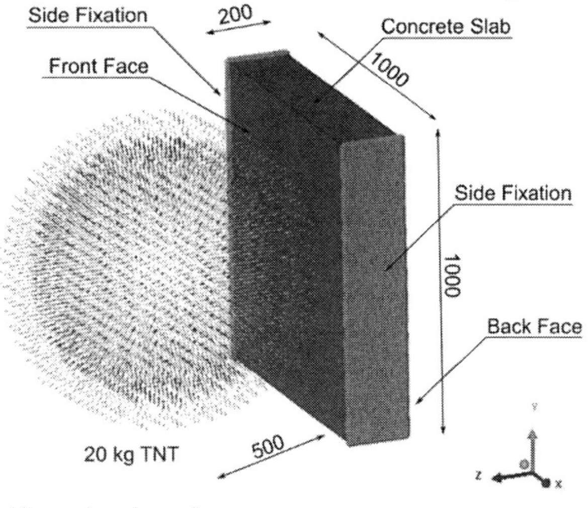

Fig. 3. Models setup. (dimensions in mm)

2.1 Material description

The materials used to simulate the concrete slab, PTFE sheets, and the surrounding medium are CONC-25MPA, TEFLON, and AIR respectively. The materials are taken from the AUTODYN materials library and adjusted to the conditions of the study. Equation of state, strength model, failure model, and properties for concrete, Teflon, and air are shown in Table 1.

Table 1. Materials models and properties.

Material	Equation of State	Strength Model	Failure Model	Erosion Control	Erosion Strain	Reference Density (g/cm^3)	Shear Modulus (kPa)
CONC-25MPA	P alpha	RHT concrete	RHT concrete	Instantaneous geometric strain	0.7	2.5	1.67x10^7
TEFLON	Shock	Von Mises	Hydro (Pmin)	Instantaneous geometric strain	0.5	2.16	2.33x10^6
AIR	Ideal Gas	-	-	-	-	1.225x10^{-3}	-

2.2 Geometry and mesh description

For this study 24 models are carried out, all test specimens are 1000x1000x200 mm concrete slabs defined using Lagrange solver with mesh size 10x10x5 mm. PTFE sheets are modelled by filling the concrete slab part using the TEFLON material. The air and explosives are defined using Euler solver with mesh size 40x40x60 mm. All slabs are modelled as one way slabs using side fixation as boundary condition. Geometric details for models are illustrated in Figure 4 and Table 2. Figure 5 shows a sample from the models.

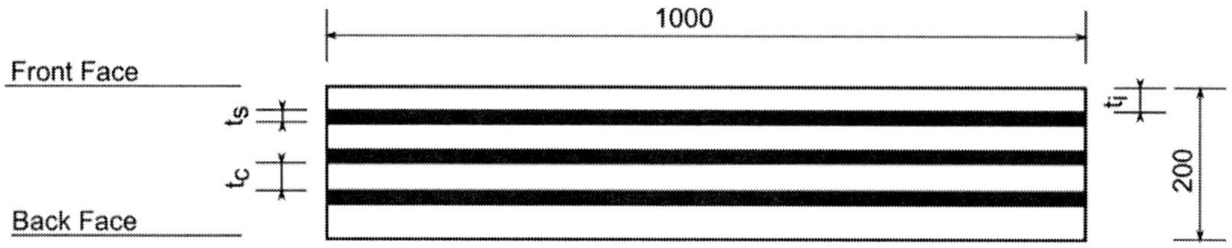

Fig. 4. Models schematic. (dimensions in mm)

Table 2. Sheets configurations.

#	*ts* mm	*ns*	*tc* mm	*ti* mm	#	*ts* mm	*ns*	*tc* mm	*ti* mm
S1	-	-	-	-	S13	30	3	50	10
S2	5	1	-	10	S14	5	1	-	195
S3	10	1	-	10	S15	10	1	-	190
S4	20	1	-	10	S16	20	1	-	180
S5	30	1	-	10	S17	30	1	-	170
S6	5	2	50	10	S18	5	2	50	140
S7	10	2	50	10	S19	10	2	50	130
S8	20	2	50	10	S20	20	2	50	110
S9	30	2	50	10	S21	30	2	50	90
S10	5	3	50	10	S22	5	3	50	85
S11	10	3	50	10	S23	10	3	50	70
S12	20	3	50	10	S24	20	3	50	40

ts is the PTFE sheet thickness in mm.
ns is the number of PTFE sheets.
tc is the concrete thickness between PTFE sheets in mm.
ti is the initial concrete thickness at the front face in mm.

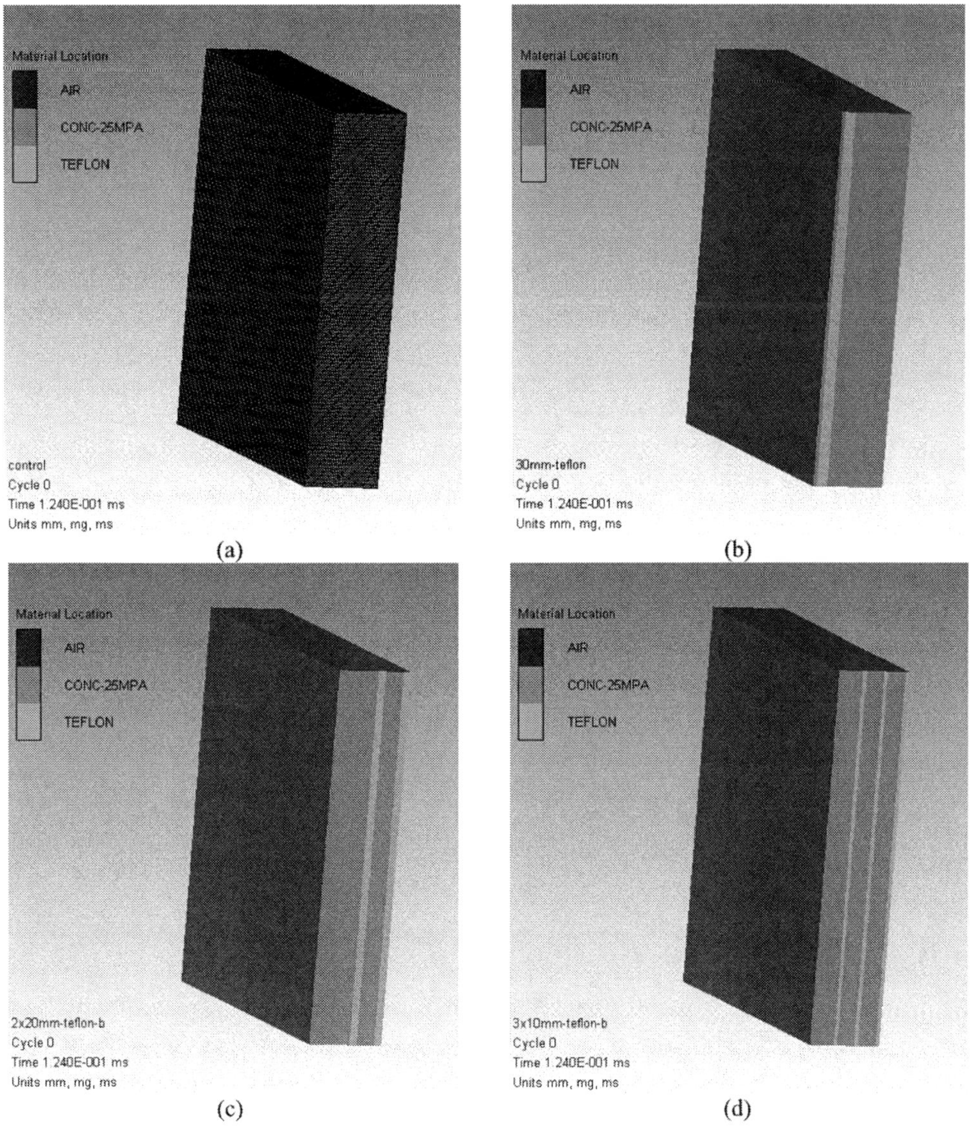

Fig. 5. Sample of models (a) S1 model, no PTFE sheets and showing the mesh grid, (b) S5 model, 30mm sheet at the front face, (c) S20 model, two 20mm sheets at the back face, and (d) S23 model, three 10 mm sheets at the back face.

2.3 Explosion Simulation Validation

In order to validate the TNT charge used in the simulation, overpressure from manual calculations and the simulation are compared. Overpressure is calculated manually from Kinney and Graham [8] overpressure formulation described in Equation (3) and UFC 3-

 DOI: 10.1051/matecconf/201712004005

340-02 technical manual [1]. Scaled distance of the equivalent TNT explosives is calculated using Equation (4).

$$P_{SO} = P_o \frac{808\left[1+\left(\frac{Z}{4.5}\right)^2\right]}{\sqrt{\left[1+\left(\frac{Z}{0.048}\right)^2\right]\left[1+\left(\frac{Z}{0.32}\right)^2\right]\left[1+\left(\frac{Z}{1.35}\right)^2\right]}} \quad (3)$$

$$Z = \frac{R}{\sqrt[3]{W}} \quad (4)$$

R is the distance between the centre of the explosion and the gauge in m
W is the weight of the TNT explosives in kg.
Z is the scaled distance in m.
P_{SO} is the overpressure in kPa.
P_O is the ambient pressure equal to 101.325 kPa.

Pressure-time history for the 20 kg explosives in the AUTODYN simulation is illustrated in Figure 6. Overpressure is taken as the peak value from the pressure-time history and compared with manual calculations in Table 3. Comparison shows that AUTODYN overpressure value is relatively close to the overpressure values manually calculated.

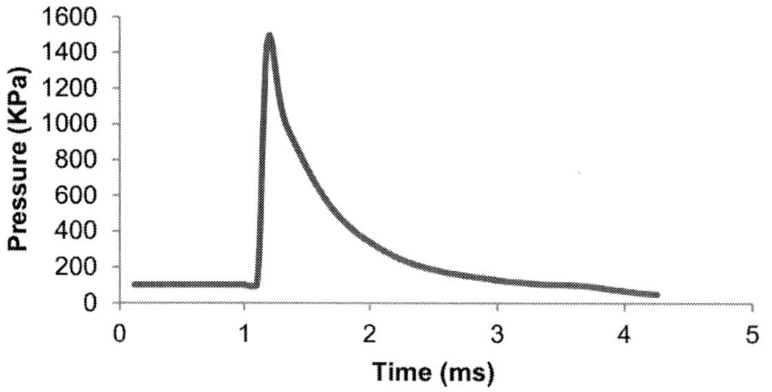

Fig. 6. Pressure-time history from the AUTODYN Model for R = 2.5m and W = 20kg.

Table 3. Overpressure Calculations.

R	*W*	*Z*	*Pso* (kPa)		
m	**kg**	**m/kg$^{1/3}$**	**Kinney & Graham**	**UFC 3-340-02**	**AUTODYN Model**
2.5	20	0.921	1204	1380	1497

3 Results

From this study, we can notice that the maximum deflection is found in slabs at midspan. Concrete experiences mostly flexural failure at midspan and a combination of flexural and shear failure at the side fixation. The use of PTFE sheets reduces the debris and fragmentation up to 90%, thus decreasing the blast radius and human injuries. Table 4 shows results for maximum fragment propagation (*FP*), approximate fragmentation percentage (*F*), and pressure behind the slabs (*P*). Figures 7 to 13 show the numerical models.

Table 4. Numerical simulation results.

#	*FP* mm	*F* %	*P* kPa	#	*FP* mm	*F* %	*P* kPa
S1	768.35	100	106.69	S13	143.22	1	105.83
S2	838.37	100	106.43	S14	604.48	75	106.71
S3	844.84	100	106.44	S15	686.92	65	106.65
S4	785.5	100	106.65	S16	552.8	60	106.42
S5	873.58	100	106.85	S17	569.46	50	106.24
S6	757.04	65	106.51	S18	667.2	80	106.62
S7	911.03	60	106.41	S19	576.49	70	106.39
S8	685.73	45.5	106.47	S20	296.81	10	106.04
S9	705.5	40	106.22	S21	168.39	5	105.94
S10	694.46	37.5	106.43	S22	635.32	60	106.53
S11	920.69	30	106.34	S23	378.18	5	106.28
S12	957.92	15	106.22	S24	175.3	1	106.07

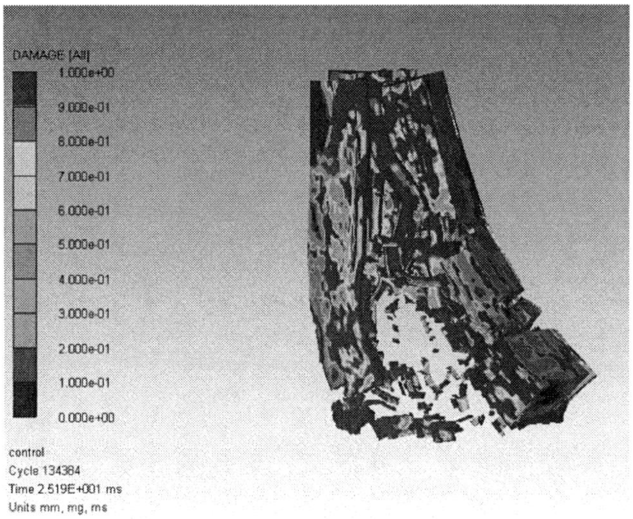

Fig. 7. S1 model, no PTFE sheets and showing the damage.

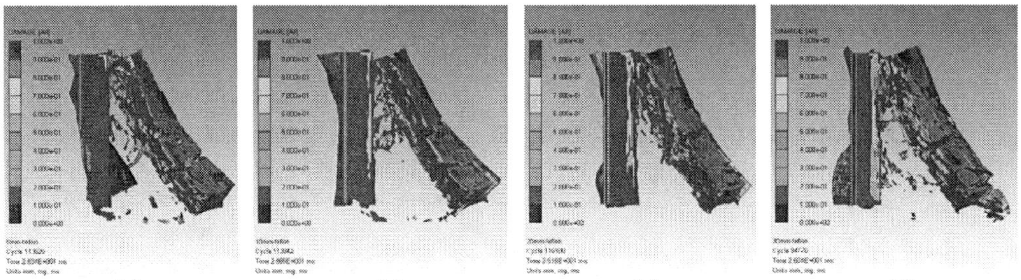

(a) (b) (c) (d)

Fig. 8. One PTFE sheet at the front face models showing the damage, (a) S2 model, (b) S3 model, (c) S4 model, (d) S5 model.

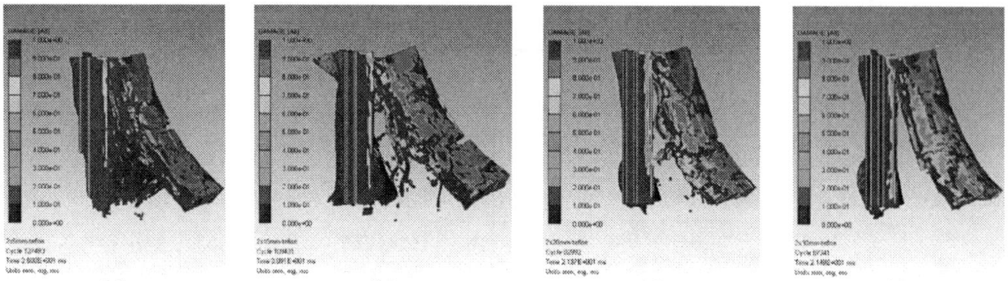

(a) (b) (c) (d)

Fig. 9. Two PTFE sheets at the front face models showing the damage, (a) S6 model, (b) S7 model, (c) S8 model, (d) S9 model.

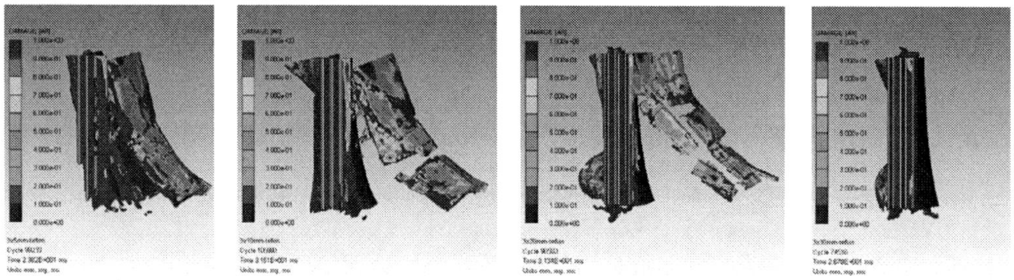

(a) (b) (c) (d)

Fig. 10. Three PTFE sheets at the front face models showing the damage, (a) S10 model, (b) S11 model, (c) S12 model, (d) S13 model.

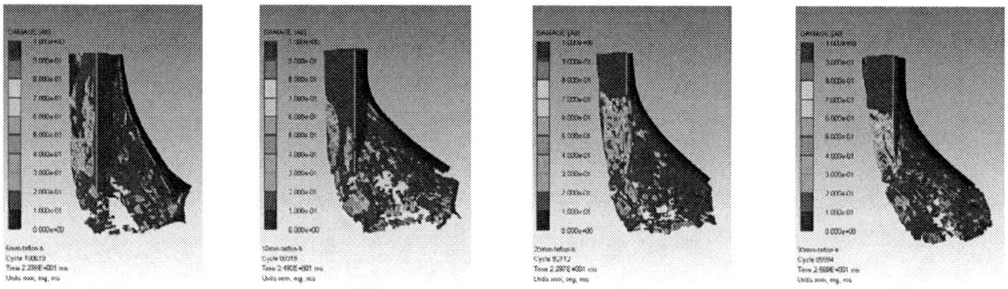

(a) (b) (c) (d)

Fig. 11. One PTFE sheet at the back face models showing the damage, (a) S14 model, (b) S15 model, (c) S16 model, (d) S17 model.

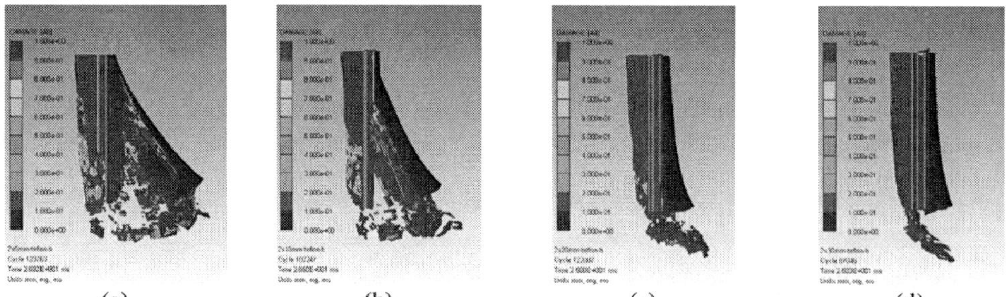

(a) (b) (c) (d)

Fig. 12. Two PTFE sheets at the back face models showing the damage, (a) S18 model, (b) S19 model, (c) S20 model, (d) S21 model.

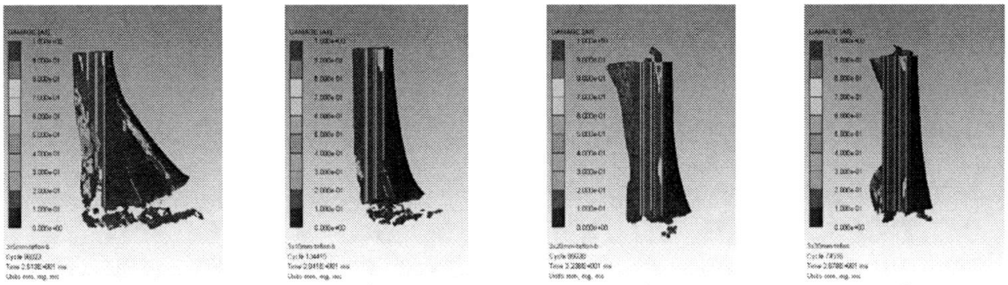

(a) (b) (c) (d)

Fig. 13. Three PTFE sheets at the back face models showing the damage, (a) S22 model, (b) S23 model, (c) S24 model, (d) S13 model.

4 Conclusion

From this study, we can conclude the following:

- Numerical simulation of the blast wave propagation in ANSYS AUTODYN is satisfactory.
- Optimum placement of PTFE sheets is at the back face of the slab using the sandwich configuration.
- Increasing thickness and number of PTFE sheets decreases fragmentation by up to 90%.
- PTFE sheets placed at the front face did not diminish the blast wave pressure.
- PTFE should be used for blast fragmentation protection purposes only and not in the structural design.
- PTFE sheets can be added to new structures or used to protect existing structures.
- Experimental testing should be conducted to verify the results.

References

1. *UFC-3-340-02 Structures to Resist the Effects of Accidental Explosions. US Army Corps of Engineers, Naval Facilities Engineering Command, Air Force Civil Engineer Support Agency* (2008)
2. M. Von Ramin, A. Stolz, Chem. Eng. Trans. **48**, 85-90 (2016)
3. X. Lin, Y. X. Zhang, *Proceedings of The International Conference on Computational Methods* **2**, 86-93 (2015)
4. Y. S. Tai, T. L. Chu, H. T. Hu, J. Y. Wu, Theor. Appl. Fract. Mec. **56**, 140-147 (2011)
5. J. Weerheijm, H. S. Lim, *Proceedings of FraMCoS-6 Conference*, (2007)
6. R. O. Ebewele, *Polymer Science and Technology* (CRC Press, New York, 2000).
7. C.A. Sperati, H. W. StarkweatherJr, *Fluorine-containing polymers. II. Polytetrafluoroethylene*, 465-495 (Springer-Verlag Berlin Heidelberg, 1961)
8. G. F. Kinney, K. J. Graham, *Explosive Shocks in Air* (Springer-Verlag Berlin Heidelberg, New York, 1985)
9. J. M. Biggs, *Introduction to Structural Dynamics* (McGraw-Hill, 1964)
10. P.H. Bischoff, S.H. Perry, Mater. Struct. **24**, 425-450 (1991)
11. L.J. Malvar, C.A. Ross, ACI Mater. J. **95**, 735-739 (1998)
12. *Century Dynamics Company, Inc. AUTODYN Interactive Non-Linear Dynamic Analysis software. Theory Manual,* (2005)

Contribution to optimize decision parameters in activated-sludge process using ANFIS model

Radia. Maachou[1], *Abdelouahab Lefkir*[2*], *Tarek Merabtene*[3], *Amel Hamriche*[4], *Abdemalek Bermad*[4]

[1]USTHB, laboratory of faculty of chemistry, electrochemistry-corrosion, metallurgy and mineral chemistry, BP 32 El-Allia, pc 16111, Algeria.
[2]ENSTP, TPiTE Laboratory, Bp 32 cité sidi Garidi, Kouba, Algers, Algeria.
[4]ENP, Civil engineering, Material and Environment Laboratory, Bp 32 Hassen-Badi 16000, Algeria.

Abstract. The monitoring of activated sludge processes is difficult because of their slow dynamics, the complexity of their behavior. To obtain the desired level of performance in an activated sludge system, a proper balance must be maintained between aeration (energy consumption) and the quantity of sludge recirculation. The objective of this study is to determine the necessary amount of recirculated sludge in activated sludge process in the aim to reduce the energy consumption with respect of environmental standards. To achieving this objective, a comparative study between three models was performed. In order to reduce the complexity of activated-sludge process, the first is a linear model based on principal component analysis method (PCA), the second is an Adaptive Neural Fuzzy Inference System model (ANFIS), while the third is based on coupling of those two approaches. The best values of validation criterion obtained with the last method show the high performance of the hybrid model proposed...

1 Introduction

Awareness of environmental problems caused by discharges from wastewater treatment plants and the will to improve and preserve the quality of receiving waters has led the adoption of higher regulatory standards, compliance with discharge standards, on the hand, and energy consumption consequent [1], on the other hand, are the key issue for managers of wastewater treatment plants.

In recent years, cost minimization has become increasingly important in the control and operation of wastewater treatment plants. In order to run plant economically efficient, operational costs such as pumping energy, aeration energy and dosage of different chemicals should be minimized. At the same time, the discharges to the recipient should be kept at a low level. Of course, minimizing the operational costs and at the same time treat the wastewater properly may lead to a conflict of interest that must somehow be solved [2].

The ASP is the most used in Algeria in the wastewater treatment process. Although the purification performance and reliability of this method are approved, several malfunctions

* Corresponding author: a_lefkir06@yahoo.fr

may occur due to the qualitative and quantitative changes of water to be treated, the aeration time in the aeration tanks, the amount of recirculated sludge [3].

In this context, the cost of the energy consumption during the aeration process is important. In general, approximately one third of the total operating cost of wastewater treatment plant (WWTP) is due to the energy requirements, and the energy consumed for the aeration represents approximately 60-65% of the total energy consumption [4,5,6].

The aim of this work is focused on reduction of energy consumption on predicting the amount of recycle sludge necessary.

2 Materials and methods

2.1 Activated sludge process

The activated sludge process (ASP) is one of a number of unit processes used for the treatment of wastewaters. The basic feature of the process is the decomposition of complex dissolved and suspended organic substrates into simple end-products such as carbon dioxide and water. These systems involve inoculation with floc forming bacteria, which oxidize the organic matter, stabilizing the wastewater under aerobic conditions. A settling tank is used to separate the biomass constituents according to their settling abilities [7]. Decomposition is achieved by a heterogeneous culture of micro-organisms the activated sludge, which in part utilize the waste organic substrates in the synthesis of their own biological cell material [8].

Many different processes happen simultaneously in wastewater treatment plants (WWTP), which were originally designed to reduce the biological oxygen demand, total suspended solids (SS) and nitrogen and phosphorus pollution [9]. The treatment plant generally includes a pre-treatment phase, during which the large elements (unrefined) are removed, followed by secondary treatment (or biological) to remove the organic material and dissolved. Finally, tertiary treatment is applied to destroy microorganisms.

After pre-treatment, the pre-treated water is directed to the aeration basins that are mixed with an aerated biomass and kept in suspension. The mixed liquor produced is composed of flocculated sludge and treated water directed to the clarifiers. The clarifier is an integral part of the activated sludge system. It has two main functions: it separates the biomass from the water in order to produce a good quality effluent free from settle able solids and it also thickens the biomass. In output thereof, the biomass is separated by decantation; part of the biomass is recycled (recycle sludge) to the basin because it contains numerous microorganisms which are needed for the reaction [10] to maintain an appropriate biomass concentration. The excess biomass is removed from the system and constitutes the secondary sludge that is drained and will be treated in a further sludge treatment process. Recycle sludge (RS) is a critical control variable as it redistributes the sludge between the secondary clarifier and the aeration tank, such that the healthy population of biomass is maintained in the aeration basin.

Thus, the operator must maintain a continuous return of activated sludge to the aeration tank or the process will show drastic decrease in performance. To obtain the desired level of performance in an activated sludge system, a proper balance must be maintained between aeration (energy consumption) and the quantity of sludge recirculation (pumped from the secondary clarifier back to the aeration tank), and the amount of excess sludge withdrawn from the system (usually pumped from the secondary clarifier towards sludge treatment).

2.2 Constraint of activated sludge process

The dynamic behaviour of ASP is a result of many complex mechanisms [11] due to the specific growth rate of bacteria in biological process which varies with time and is influenced by many factors such as the substrate concentration, temperature, pH, dissolved oxygen concentration, amount of inflow [12], imprecision and uncertainty of several parameters which are difficult to identify and strong nonlinearities. These changes may be the result of metabolic changes in biomass or of unpredictable genetic changes and unobservable from a macroscopic point of view [13]. Given all these factors, it is very difficult to construct a mathematical model for control of the process. In order to meet these demands, the use of advanced modelling methods by fuzzy logic model is required [14]. The basic components of Fuzzy Neuro model and techniques that have been used for their development will be discussed in the following.

2.3 Fuzzy logic and neural networks

Recently, soft computing method was proposed as a useful tool to model complex systems for process industries. In the last decades, the research on soft computing methods such as Neural network techniques and fuzzy logic fuzzy [15].

Artificial neural network technique is powerful because it can learn to represent complicated data pattern or the data relationship between input and output variables of the system being studied.

Fuzzy neural (FN) are hybrid systems that combine the theories of fuzzy logic and neural networks, thus can make effective use of easy interpretability of fuzzy logic, as well as superior learning ability and adaptive capability of neural networks.

A general fuzzy system has basically three components: fuzzification, fuzzy inference and defuzzification

1)The fuzzification consists in splitting in several terms the parameters which compose the characteristic vector. Each decomposition or term gets closer to a word of the natural language. For instance, the concentration can be "high", "medium" or "low". There exit two ways to set the fuzzification: a fuzzification with an equal distribution of the terms, and a fuzzification with a repartition adapted to the training data set.

The fuzzification of the input variables, which convert digital inputs available in fuzzy parts. It is then possible to associate variables of the membership the coefficients of the fuzzy subsets taking values in the range [0, 1].

2)The fuzzy inference is the process of formulating the mapping from given input(s) to output(s) using fuzzy logic. The mapping them provides a basis from which decisions can be made. The process of fuzzy inference involves all the topic such as fuzzification, defuzzification, implication and aggregation. It has found successful applications in a wide variety of fields, such as automatic control, data classification, decision analysis, expert systems, time series prediction, robotics, and pattern recognition. Because of its multidisciplinary nature, the fuzzy inference system is known by numerous others names, such as fuzzy-rule-based system, fuzzy expert system, fuzzy model, fuzzy associative memory, fuzzy logic controller and simply (and ambiguously) fuzzy system.

The basic structure of a fuzzy inference system consists of three conceptual components:

A rule base, which contains a selection of fuzzy rules;

A database (or dictionary), which defines the membership functions used in the fuzzy rules;

And a reasoning mechanism, which performs the inference procedure upon the rules and given facts to derive a reasonable output or conclusion.

We conclude that fuzzy inference is consisted by the basic rules and the database. The combination of inputs with the fuzzy rules allows drawing conclusions. (Produced the image of fuzzy part stemming the fuzzification, by a fuzzy relation usually built from rules.

For example:

If ((Y_{BOD} is high) **and** (Y_{COD} is high) **and** (Y_{SS} is high) **and** (Y_{SS} is high)) **then** (Ren is high)

If ((Y_{BOD} is low) **and** (Y_{COD} is low) **and** (Y_{SS} is low) **and** (Y_{SS} is low)) **then** (Ren is low)

if *((RS is high)* **and** *(Y_{COD} is high)* **and** *(Y_{SS} is high))* **then** *(Ren is high)*

3)Defuzzification: The inference methods provide a resulting membership function for the output variable, so it is a fuzzy information. By this step, is then made the return in the real sizes of output. It is for this purpose calculated from the degrees of membership in all fuzzy sets of the output variable, the abscissa corresponding to the value of this output, so it converts the blurs on outputs inference mechanism in digital outputs. In other words, where the linguistic fuzzy output has to be translated into a numerical value such as the necessary recycle sludge.

ANFISs were first proposed by Jang in 1992. These are adaptive networks whose function is similar to FISs. The aim of an ANFIS is to integrate the best features of fuzzy systems and NNs. The advantages of an ANFIS over an FIS are as follows:

•An ANFIS can simulate and analyze the mapping relation between input and output data through a learning algorithm to optimize the parameters of a given FIS.

•An FIS is fully dependent on its membership functions. An ANFIS involves networks which combine nodes and directional links, and some learning rules are associated with these networks. The networks learn relationships between inputs and outputs.

•An ANFIS can be trained without any need for the expert knowledge usually required for the standard fuzzy logic design [16] (Tyagi K et al, 2014).

3 Results and discussions

Database collection of the downstream and upstream from the station was conducted. The series of data was collected daily from January 2006 until December 2013. In addition, 250 daily data yields describing the pollution control were collected during a weekly measurement.

ANFIS model is applied to reduce the energy consumption. The data of Recycle sludge were separated into two subsamples: the training data and validation data. The training data were chosen among filtered data where elimination of NO3- observed, and the validation data were chosen among unfiltered data of NO3 where excess of NO3 observed. The input parameters such as the removal yield of Suspended Solids (RSS), Biological oxygen demand (YBOD), Chemical oxygen Demand (YCOD) were selected to represent the pollution and RS was selected to present the amount of recycle sludge, this last is divided on the maximum value. The resulting values are normalized (between 0 and 1). Decision parameters is energy ratio.

3.1 The membership functions

Development of the fuzzy neural network model involves using both fuzzy logic and neural networks. Here, the fuzzy logic is really a matter of the fuzzy set theory particularly used to deal with process imprecise infonnation with a changed membership function. This fuzzy logic technique uses non-conventional membership function that is different than the classical membership function that takes only two values: one, when an element belongs to the set, and zero, when it does not. Because fuzzy sets describe vague concepts based on the premise that the elements used are not numbers but belong to words or the value of a linguistic variable, an element of a fuzzy set naturally belongs to the set with membership values from the interval [0,1]. A fuzzy membership function associated with a given fuzzy set maps an input value to its appropriate membership value [17].

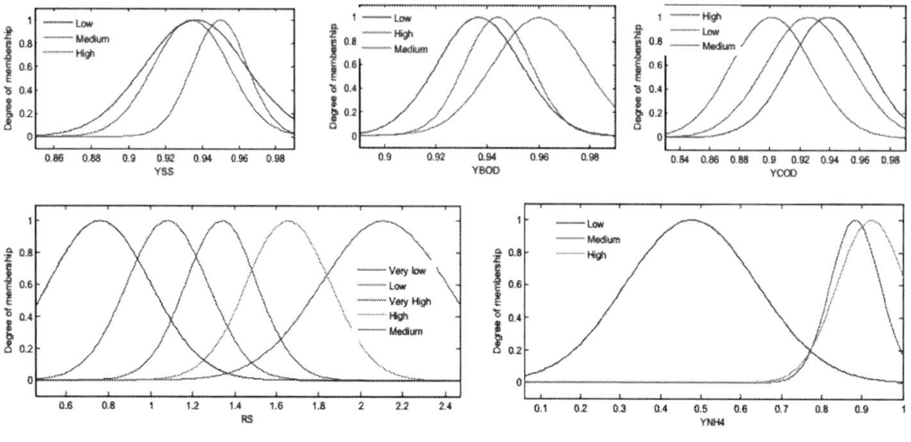

Fig 1. Example of membership functions for the input and output variables.

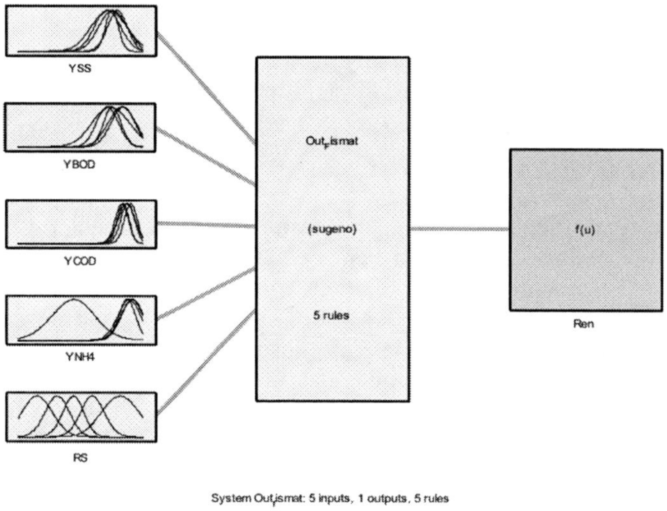

Fig 2. Sugeno fuzzy inference system (FIS) system.

3.3 Simulation results

In the first part, ANFIS model has been proposed and applied to simulate recycle sludge (RS) in activated sludge process in objective to optimize the energy consumption given by the fig3.

We symbolize Ratio energy by Ren/ Ren [kWh / m3] = energy consumption [kWh / day] / input flow [m3 / day].

In order to evaluate the predicting performance of ANFIS, the Root Mean Square Error criterion (RMSE) is used to calculate the difference between the simulated and observed values and the correlation coefficient R [14] are given in table 1:

The correlation coefficient allows assessment of the degree of relationship while the RMSE criterion used to measure the distortion between the observed and the simulated values. We concluded that the coupling of the two criteria is necessary for a proper assessment model.

Simulation and prediction have shown that the resulting model can adequately predict the process with satisfactory results. These last were obtained during the learning and validation periods, revealing the advantages of fuzzy reasoning and justifying the predictive power of the model developed to simulate the amount of the recycle sludge in activated sludge process [14].

Table 1. Validation criteria of learning and validation periods.

	validation criteria	Values
Learning period	RMSE	11,2
	R(%)	83,37

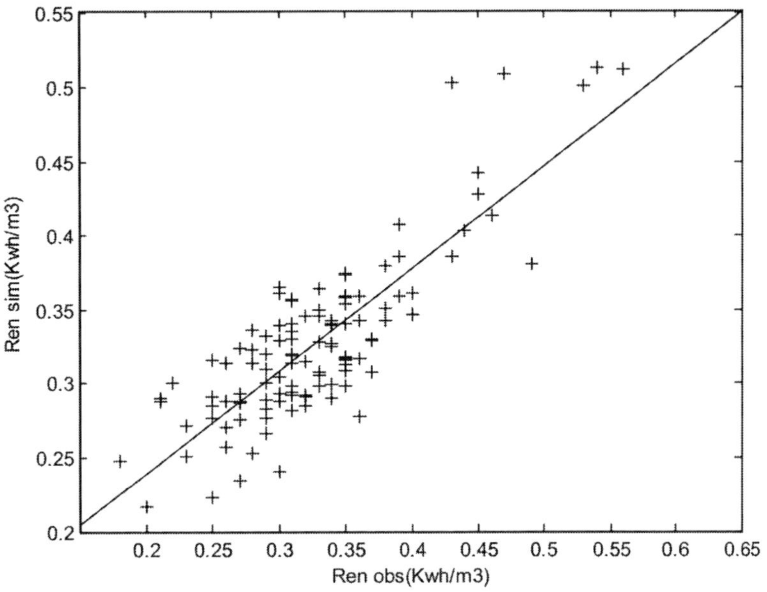

Fig 3. Correlation line between Ren (Obs) and Ren (Sim)

In the second part, the reduction of energy was given by the fig2

The variation of the observed and simulated energy ratio based on the learning sample is given by a histogram shown in fig 4.

According to the histogram plot, we find significant differences between the observed values and simulated values of Ren, this confirms that there is energy consumption.

The optimized Ren percentage equal to 14.34%

DOI: 10.1051/matecconf/201712005001

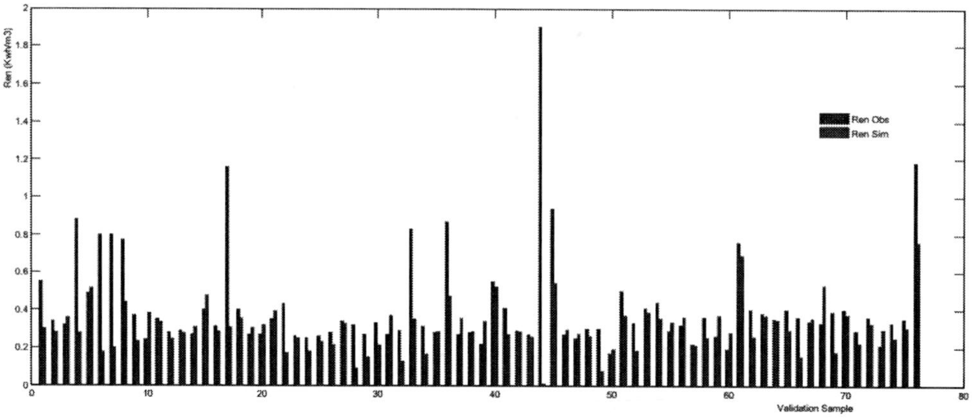

Fig 4. Comparison between energy ratio observed & optimized for sample validation.

4 Conclusions

The present study concludes that the application of ANFIS can improve Sludge recycling system, providing the optimal yields. ANFIS indicates the high ability of modelling a more complicated process such as the activated sludge processes, for improving the performance of wastewater treatment plant and can lead to a better understanding of the system. However, the complexity and uncertainty in the process make the task somewhat complicated using traditional deterministic models.

If the economy is only a few percent, it can be very significant in terms of kWh per year per decade.

References

1. Holmberg A., Modelling of the activated sludge process for microprocessor-based state estimation and control, water research **16**, 1233-1246, (1982).
2. P. Samuelsson, B. Halvarsson, Bengt Carlsson, cost efficient operation of a denitrifying activated sludge process, Water Research., **41**, P2325 – 2332, (2007)
3. Fernandez FJ., Castro MC., Rodrigo M.A., Canizares P., Reduction of aeration costs by tuning a multi-set point on/off controller: A case study, Control Engineering Practice **19**, 1231–1237 (2011).
4. Rieger L., Alex J., Gujer W., & Siegrist H., Modelling of aeration systems at wastewater treatment plants. Water Science & Technology, **53**, 439–447 (2006)
5. Duchène Ph., Cotteux E., & Capela S., Applying fine bubble aeration to small aeration tanks. Water Science & Technology, **44**(2–3), 203–210 (2001).
6. Rustom R, Modelling Activated Sludge Wastewater Treatment Plants Using Artificial Intelligence Techniques (Fuzzy Logic and Neural Networks), thèse doctorat, Heriot Watt University, UK (2009).
7. Mafalda C., Sarraguça et al., Quantitative monitoring of an activated sludge reactor using on-line UV-visible and near-infrared spectroscopy, Anal Bioanal Chem.,**395**,1159–1166, (2009).
8. Tong R. M, Beck M. B., Latten A.,, fuzzy control of the activated sludge wastewater treatment process, aulomatica, vol. **16**, pp. 659 701 pergamon press, (1989)
9. Lefkir A., Maachou R., Bermad A., Khouider A., Factorization of physicochemical parameters of activated sludge process using the principal component analysis, Desalination and Water Treatment, 1–6 (2015).

10. Chen Y, Zhiqiang C, Abraham K, application of fuzzy reasoning to the control of an activated sludge plant*, fuzzy sets and systems **38** 1-14 (1990)
11. Ruano M.V., J. Ribes ., Sin G., Seco A., Ferrer J., A systematic approach for fine-tuning of fuzzy controllers applied to WWTPs, Environmental Modelling & Software **25**, 670–676 (2010)
12. Bououden S., Chadli M.,.Karimi H.R. 2015, Control of uncertain highly nonlinear biological process based on Takagi–Sugeno fuzzy models, Signal Processing, 108-205 Mafalda et al., (2009).
13. Grizale P V H, Modélisation et commande floues de type Takagi-Sugeno appliquées à un bioprocédé de traitement des eaux usées, thèse d'état, Automatic. Université Paul Sabatier, Toulouse III, (2007)
14. Maachou R., Lefkir A., Khouider A. And Bermad A., control of recycle sludge in activated sludge process using adaptive neuro-fuzzy logic controller (anfis) proceedings of the 14th international conference on environmental science and technology rhodes, greece, 3-5 september, (2015)
15. Qiao J, Li W, Han H, Soft Computing of Biochemical Oxygen Demand Using an Improved T–S Fuzzy Neural Network Chinese Journal of Chemical Engineering **22** (2014) 1254–1259, (2014).
16. Tyagi K, Arun Sharma., An adaptive neuro fuzzy model for estimating the reliability of component-based software systems , 38-51, Applied Computing and informatics, (2014)

Desalinated and blended water in Saudi Arabia: human exposure and risk analysis from disinfection byproducts

Imran Rahman Chowdhury[1], and *Shakhawat* Chowdhury[2*]

[1]King Fahd University of Petroleum and Minerals, Dhahran, Dhahran 31261, Saudi Arabia
[2]King Fahd University of Petroleum and Minerals, Dhahran, Dhahran 31261, Saudi Arabia

Abstract. Saudi Arabia produces the largest amount of desalinated water as a single country. The desalinated water is typically blended with treated groundwater, pH adjusted and chlorinated prior to supply to the communities. The desalinated seawater and/or blended water contains various types of disinfection byproducts (DBPs), some of which may induce cancer risks to human through lifetime exposure. In this study, occurrences of trihalomethanes (THMs) in desalinated and blended water in Saudi Arabia were investigated and their exposure and risks were predicted. The chronic daily intakes of $CHCl_3$, BDCM, DBCM and $CHBr_3$ were estimated to be 8.38×10^{-5}, 7.57×10^{-5}, 2.54×10^{-5} and 4.32×10^{-4} mg/kg-day respectively. The overall cancer risk was 1.78×10^{-5} with the range of $7.40\times10^{-7} – 9.26\times10^{-5}$ and the average hazard index was 3.49×10^{-2} with the range of $1.20\times10^{-3} – 2.34\times10^{-1}$. The loss of disability adjusted life years (DALY) were estimated to be 25.1 per year and the average cancer risk had 8.48×10^{-7} DALY per person per year. The financial burden was estimated to be US$2.72 million with the range of US$2.52–2.91 million. The findings may assist in better understanding and reducing cancer risks from DBPs in desalinated and blended water.

1 Introduction

Desalination of seawater has been in practice for satisfying domestic water demands in many countries in the Middle East, the Mediterranean Basin, as well as in Australia and USA. In the Arabian Gulf, total production of desalinated water was reported to be approximately 11 million m^3/day while Saudi Arabia produced approximately 4.5 million m^3/day [1, 2], which has made Saudi Arabia as the largest producer of desalinated water as a single country [1, 2]. For desalination, two main processes: thermal (MSF: multi-stage flash; MED: multi effect distillation); and membrane (RO: reverse osmosis) are in practice, and the dominant process is MSF, due mainly to low energy cost in this region. In addition, large-scale plants with advanced technologies are being established in these countries indicating the increase in desalinated water supplies in future.

* Corresponding author: schowdhury@kfupm.edu.sa

The desalination plants receive feed water mainly from the open seawater intakes while some plants receive feed water via different intakes and coastal locations. The pretreatment is performed using continuous or intermittent chlorination with initial doses of 0.5 – 2.0 mg/L to achieve the target residuals of 0.25 – 0.5 mg/L. Free chlorine is also applied in different stages of desalination process as well as in final disinfection. Consequently, different types of disinfection byproducts (DBPs) are formed in the desalinated seawater, which can pose public health risks. The average concentration of bromide in seawater from the Arabian Gulf was reported to be 64,000 µg/L [3]. Past studies have reported that seawater could contain bromide and iodide in the ranges of 50,000 – 80,000 µg/L and 21 – 60 µg/L respectively [3-5], which can increase the formation of brominated and iodinated DBPs. The brominated and iodinated DBPs are much more cytotoxic and genotoxic than the chlorinated DBPs [6].

In the past, many studies have focused on human exposure and risk from DBPs in freshwater sourced drinking water. Despite the public health relevance, not much is known about the DBPs in desalinated and blended water, their exposure and risk. In Saudi Arabia, desalinated water supplies more than 60% (~1620 million m^3/year) of domestic water demand [2], which is likely to be increased in future. The desalinated water is blended with the treated brackish groundwater, pH adjusted and chlorinated prior to supplying to the consumers. The blended water stays in the regional pipelines, storage tanks, blending stations and WDS for several days. In addition, maximum daily temperature during summer often exceeds 40°C, which can increase the reaction rates by several folds. The water has the environments and constituents conducive to increased formation of DBPs, in particular, the brominated and iodinated DBPs [7]. There is a need to understand DBPs concentrations in desalinated and blended water, and the consequent exposure and risk. In this study, formation and distribution of THMs in desalinated and blended water from different desalination plants in Saudi Arabia were investigated. Exposure to THMs were estimated through ingestion, inhalation and dermal absorption pathways. Cancer and non-cancer risks were predicted which were used to predict the societal burden in terms of disability adjusted life year (DALY) applying the concept of YLL (years of life lost due to premature mortality) and YLD (years of life lost due to disability).

2 Methodology

2.1 Data Collection

Occurrences of THMs (chloroform: $CHCl_3$; bromodichloromethane: BDCM; dibromochloromethane: DBCM; and bromoform: $CHBr_3$) and water quality parameters were investigated in desalinated seawater and blended water (mixture of desalinated seawater and treated groundwater in the ratio of 40-60%) in the WDS in Al-Khobar and Dhahran, Saudi Arabia for a period of one year (Feb, 2014 – Jan 2015). The samples were collected on bi-weekly basis and analyzed in duplicates following the standard methods [8, 9].

2.2 Exposure and Risk Assessment

Exposure to THMs occurs through multiple pathways, including ingestion of drinking water, and inhalation and dermal contact during showering, bathing, house cleaning and swimming [10]. To incorporate uncertainty, 5000 random data are generated for each parameter following the statistical distributions. Exposure to THMs through ingestion is predicted using the THMs in cold water. THMs exposure through inhalation pathway is predicted using THMs in the shower air, which is estimated using the partition coefficients and mass-balance equations. In assessing dermal exposure during showering, showering events are divided into the unsteady and steady states and exposure. Moreover, to incorporate the higher level of exposure during early life adjustment factor was used (i.e. adjustment factor of 10 fold and 3 fold for the age group <2 years and 2 – 16 years, respectively; no adjustment factor for the age group > 16 years) [11]. Further details are summarized below.

2.1.1 Ingestion pathway

The chronic daily intakes (CDI) of THMs through ingestion of drinking water is computed following USEPA [7, 12-14] as:

$$CDI_{ing} = \frac{C_w \times EF \times IR \times ED \times CF}{BW \times AT} \tag{1}$$

Where, CDI_{ing} = CDI via ingestion (mg/kg-day); C_w = THMs concentrations in drinking water (μg/L); EF = frequency of exposure (days/year); IR = rate of drinking water ingestion (L/day); ED = duration of exposure (year); BW = weight of the body (kg); AT = averaging time (days); and CF = 0.001 (conversion factor of mass from μg to mg).

2.1.2 Inhalation pathway

The CDI of THMs through inhalation pathway is predicted following USEPA [7, 12-14] as:

$$CDI_{inh} = \frac{E_r \times C_a \times t \times R \times F \times EF \times ED \times CF}{BW \times AT} \tag{2}$$

Where, CDI_{inh} = CDI via inhalation (mg/kg-day); E_r = THMs absorption rate in respiratory system; t = shower duration (min/shower); R = breathing rate (m^3/min); F = shower frequency (shower/day); CF = 0.001 (conversion factor of mass from μg to mg); and C_a = concentrations of THM in shower air (μg/m^3). C_a depends on various factors, including water flow rate of shower, shower stall volume, THMs in cold water, mass transfer rate, shower duration and exchange rate of shower air. C_a is be modeled following Chowdhury (2013) [7].

2.1.3 Dermal contact pathway

THMs can be absorbed through human skin during showering. Human skin is a complex organ, which acts as a barrier to chemical intrusion through dermal pathway. Depending on shower duration, dermal exposure often needs unsteady and steady state exposure assessments while unsteady state exposure can be different from the steady state assessments [7]. However, diffusion of chemicals before reaching steady state can be significantly different from the steady state values [15]. To achieve steady state between the chemicals in water attached to the skin surface (i.e. upper layer of stratum corneum) and lower layer of stratum corneum, there is a need of lag time, which was reported to be in the range of 9.8 - 391.2 minutes for different DBPs [7]. The lag time prior to achieving steady state condition, diffusion through skin, octanol-air partition coefficient, CDI during unsteady and steady states are estimated following past studies [7, 16]. Upon estimation of CDI through the unsteady and steady states of exposure, the total CDI through dermal route is calculated as:

$$CDI_{derm} = CDI_{derm\text{-}ust} + CDI_{derm\text{-}ss} \tag{3}$$

Where, $CDI_{derm-ust}$ = CDI of THMs during unsteady-state period (mg/kg-day); $CDI_{derm-ss}$ = CDI of THMs through dermal route (mg/kg-day) during steady-state condition; $CDI_{derm-ss}$ = total CDI through dermal route. Upon estimation of route specific CDI, lifetime cancer and non-cancer risks can be estimated as:

$$CR = \sum_{i=1}^{m} \sum_{j=1}^{n} CDI_{ij} \times SF_{ij} \tag{4}$$

$$HI = \sum_{i=1}^{m} \frac{CDI_i}{R_f D} \tag{5}$$

Where, i = 1,2,3…m represent different THMs (i.e. $CHCl_3$, BDCM, DBCM and $CHBr_3$); j = 1,2,3…n representing different routes of exposure (i.e. ingestion, inhalation and dermal contact); CR = cancer risk; SF = slope factor ($[mg/kg/day]^{-1}$) for specific route; R_fD = reference dose (mg/kg/day). The oral SF for BDCM, DBCM and $CHBr_3$ are 0.062, 0.084 and 0.0079 per mg/kg-day respectively [17], while for the other routes, SF data are not available. In this study, oral SF from USEPA were used to estimate human health risk. The R_fD for $CHCl_3$, BDCM, DBCM and $CHBr_3$ are 0.01, 0.02, 0.02 and 0.02 mg/kg-day respectively [17].

2.1.3 DALY analysis

The disability adjusted life year (DALY) estimates the number of years lost due to premature death and disability because of a disease or injury [18]. To asses DALY, three possible scenarios are considered: (i) a proportion of affected people will die from cancer; (ii) a proportion will be cured from cancer; and (iii) rest of them will live with cancer sequelaes. The healthy years of life lost in a group of populations is calculated following Soerjomataram et al. (2012) [19] as:

$$DALY = YLL + YLD \tag{6}$$

Where, DALY = total disability adjusted years of life lost; YLL = total years of life lost due to premature death from cancer; YLD = total years of life lost due to disability. YLL is calculated by multiplying the number of death with the years of life expected to live at the time of death. YLD is calculated by multiplying the incidence rate with the disable years caused by the disease and severity of the disease (e.g., disability weight). YLL and YLD were estimated following Pan et al., (2014) [20] as:

$$YLD = \sum_{x} n_x P_x \left(1 - S_x\right)\left(e_x - T_D\right) \tag{7}$$

$$YLD = \sum_{x} n_x P_x \left[\left(1 - S_x\right) DW_y L_y + S_x \left\{DW_y L_y + P_{seq} DW_{seq} \left(e_x - T_D\right)\right\}\right] \tag{8}$$

Where, n_x = number of population for each age group; e_x = standard life expectancy, as reported in the Global Burden of Disease Studies (GBDS) [18]; T_D = median time to die (e.g., 2.20 years for bladder cancer) [21]. DW = adjusted disability weight (range: 0.0 - 1.0) obtained from GBDS [18]; L = duration of disability, which is divided into three stages: time for diagnosis and treatment (L_D) times for pre-terminal stage (L_M) and times for terminal stage (L_T). The average values of L_D, L_M and L_T were reported to be 4, 3 and 1 months respectively [19]. P_{seq} = proportion of the various sequelaes; DW_{seq} = disability weight of various sequelaes; T_c = median time to cure for cancer (average: 4 years for bladder cancer) [21]; P_x = age-specific cancer risk; S_x = survival rate. P_x and S_x were estimated following Pan et al., (2014) [20].

3 Results and discussion

3.1 Data statistics

Concentrations of THMs and water quality parameters for the desalinated and blended water are shown in Table 1. Concentrations of THMs in blended water were significantly higher than that in the desalinated water (Table 1).

Table 1. Summary of water quality parameters and THMs in desalinated and blended water.

Parameter	Desalinated water		Blended water	
	Average	Range	Average	Range
DOC (mg/L)	0.78 (0.33)	0.39–1.68	1.98 (0.41)	1.27–3.14
TC (mg/L)	0.55 (0.09)	0.33–0.72	1.1 (0.23)	0.69–1.54
FRC (mg/L)	0.44 (0.16)	0.24–0.72	0.82 (0.23)	0.07–1.29
Bromide (mg/L)	0.28 (0.13)	0.10 – 0.64	0.30 (0.21)	0.18 – 0.76
Water temperature (°C)	26.3 (5.8)	20–39	25.4 (6.4)	19–37
pH	6.9 (0.49)	6.6–7.7	7.13 (0.12)	6.7 – 8.1
Turbidity (NTU)	0.21 (0.09)	0.13–0.38	0.33 (0.15)	0.14–0.46
UV_{254} (/cm)	0.02 (0.01)	0.01–0.04	0.04 (0.01)	0.02–0.058
Conductivity (μS/cm)	230.4 (43.9)	149-348.7	415.3 (72.4)	266.4-552.5
THMs (μg/L)	10.08 (2.1)	0.1-33.6	19.2 (4.7)	2.1-52.4

DOC: Dissolved organic matter; TC: total chlorine; FRC: Free residual chlorine; Values within brackets are standard deviations

Table 2. Concentration of THMs (μg/L) in desalinated and blended water in gulf countries.

Country	THMs	Average	Std. dev.	Min	Max
Saudi Arabia	$CHCl_3$	0.71		0	9.3
	BDCM	0.86		0	7.77
	DBCM	1.46		0	13.73
	$CHBr_3$	15.74		0	62.42
Qatar	$CHCl_3$	0.6		0.01	4.96
	BDCM	0.21		0.01	2.66
	DBCM	0.52		0.01	2.74
	$CHBr_3$	20.89	15.2	3.99	72.95
Kuwait	$CHCl_3$	0.81	1.13	0	5.87
	BDCM	2.44	0.84	0.99	7.87
	DBCM	6.63	2.11	2.74	13.73
	$CHBr_3$	30.16	19.6	3.38	77.42
Bahrain	$CHCl_3$				
	BDCM	0.6	0.12		
	DBCM	0.64	0.05		
	$CHBr_3$	5.9	0.15		

In this study, average concentrations of THMs in the desalinated and blended water were 10.08 and 19.2 μg/L respectively with the ranges of 0.1 – 33.6 and 2.1 – 52.4 μg/L respectively (Table 1). Concentrations of THMs in desalinated and blended water from few other plants in Saudi Arabia, and some Arabian Gulf countries are summarized in Table 2.

In the major cities of Saudi Arabia, THMs in desalinated water were in the range of 0.1 – 41.7 μg/L [22]. In blended water, THMs were in the ranges of 0.1 – 66.7 μg/L [5, 23]. THMs in the desalinated and blended water from Bahrain, Kuwait, Qatar and UAE were in the ranges of 0.27 – 6.4, 7.11 – 104.89, 4.02 – 83.31 and 7.0 – 15.0 μg/L respectively (Table 2). Figure 1 shows the concentration variability of THMs in Saudi Arabia. On average, $CHCl_3$, BDCM, DBCM and $CHBr_3$ were approximately 5.2, 6.2, 9.5 and 79.1% respectively.

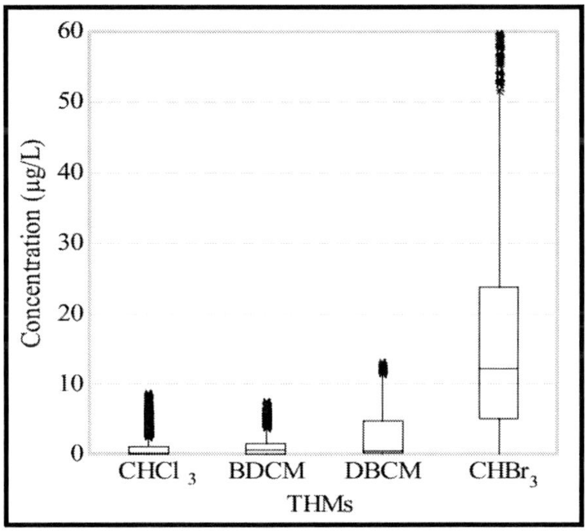

Fig. 1. Concentration of THMs in desalinated and blended water in Saudi Arabia. The bottom of the box and top of the box are the first quartile (Q1) – 25% of the data values and the third quartile (Q3) – 75% of the data values, respectively; the upper whisker and lower whisker extend to the highest data value within the upper limit: Q3 + 1.5*(Q3 – Q1) and to the lowest value within the lower limit: Q1 – 1.5*(Q3 – Q1), respectively; Values beyond the whiskers are outliers; the horizontal bar in the middle of box is the median of the data values.

The THMs database obtained in this study was expanded using additional data from the desalinated and blended waters in different desalination plants and WDS in the major cities in Saudi Arabia [5, 7, 16, 22-25]. Using the combined data, the best-fit statistical distributions were developed for THMs. In this study, $CHCl_3$ and BDCM followed triangular distribution while DBCM and $CHBr_3$ followed the Gamma distribution (Table 3).

Table 3. THMs concentration (μg/L) in desalinated water in Saudi Arabia.

THMs	Average	Range	Distribution
$CHCl_3$	0.71	0.0 – 9.30	T(0.0, 0.76, 9.3)
BDCM	0.86	0.0 – 7.87	T(0.0, 0.98, 7.87)
DBCM	1.46	0.0 – 13.73	Gamma(0.4626, 2.028, 0.0111)
$CHBr_3$	15.74	0.0 – 62.42	Gamma(1.043, 13.77)

Table 4. Proportion and disability weight (DW) for bladder cancer.

Sequelae	Proportion (%)	DW
Incontinence	5	0.157
Impotence	10	0.195
Primary infertility	16	0.18

Secondary infertility	16	0.1

The proportion of various sequelaes and disability weight of various sequelaes are shown in Table 4. The other relevant parameters for estimating exposure and risk assessment are obtained from previous studies [7, 13-17, 20, 26].

3.2 Chronic daily intake of THMs

The CDI for $CHCl_3$, DBCM, BDCM and $CHBr_3$ contributed 13.6, 12.3, 4.1 and 70.0% of total CDI respectively. On average, ingestion, inhalation and dermal routes contributed approximately 63.4, 22.3 and 14.3% of total CDI, respectively. The average CDI of $CHCl_3$, BDCM, DBCM and $CHBr_3$ through all routes were 8.38×10^{-5}, 7.57×10^{-5}, 2.54×10^{-5} and 4.32×10^{-4} mg/kg-day respectively while their ranges were $6.22\times10^{-7} - 3.97\times10^{-4}$, $1.14\times10^{-6} - 3.57\times10^{-4}$, $1.20\times10^{-7} - 5.33\times10^{-4}$ and $7.62\times10^{-8} - 4.92\times10^{-3}$ mg/kg-day respectively. The frequency distribution of CDI through all routes are shown in Figure 2. The CDI of $CHCl_3$ and BDCM followed lognormal distribution whereas, the CDI of BDCM and $CHBr_3$ followed Gamma distribution (Figure 2). The frequency distributions for $CHCl_3$, BDCM, DBCM and $CHBr_3$ showed skewed distributions with right sided long tail, indicating that these fractions of data might pose elevated risk to human.

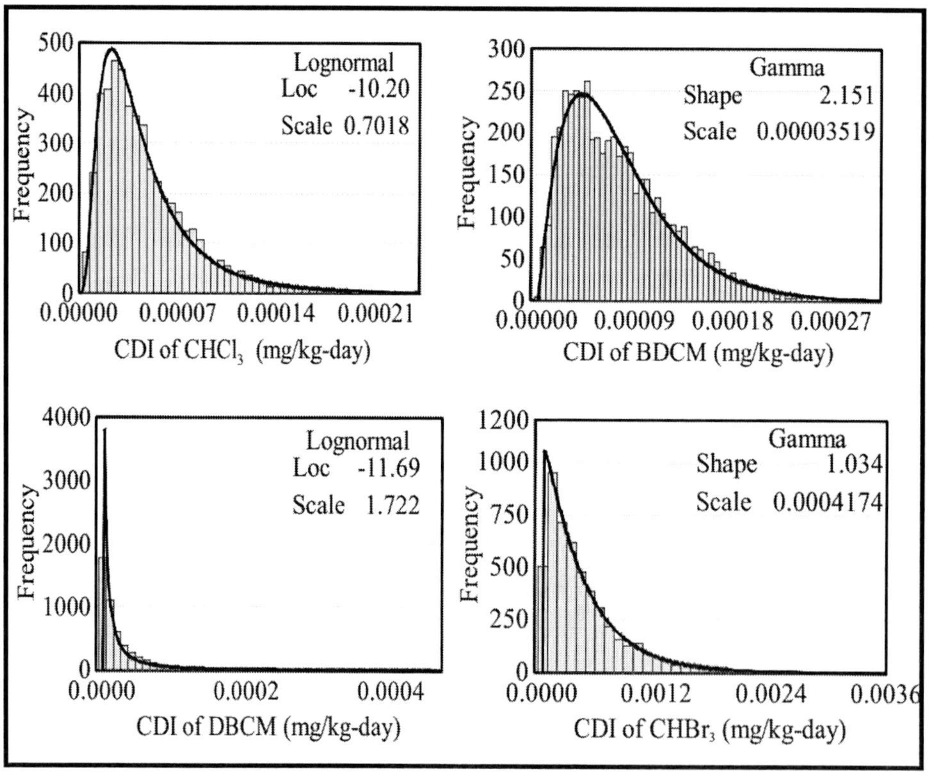

Fig. 2. Frequency distribution and fits of CDI of THMs.

3.3 Risks of THMs

The lifetime excess cancer risks and hazard indices are shown in Table 5. The average cancer risk considering all age groups was predicted to be 1.78×10^{-5} with range of $7.40\times10^{-7} - 9.26\times10^{-5}$. The ingestion, inhalation and dermal routes contributed approximately 65.4, 23.5

and 11.1% of overall cancer risks. With respect to age groups, average cancer risks during <2 years, 2 – 16 years and >16 years were 4.5×10^{-6}, 5.2×10^{-6} and 8.0×10^{-6} respectively. These three age groups contributed approximately 25.4, 29.3 and 45.3% of total cancer risks respectively. In the early life, despite the exposure was only for 2 years, it contributed 25.4% of the overall risk, indicating that appropriate protection during this period (birth to < 2 years) might reduce the overall risk significantly. The overall hazard index was estimated to be 3.49×10^{-2} with the range of 1.20×10^{-3} – 2.34×10^{-1}. Hazard indices through ingestion, inhalation and dermal routes were estimated to be 2.23×10^{-2} (range: 1.05×10^{-3} – 1.45×10^{-1}), 7.89×10^{-3} (range: 1.21×10^{-4} – 8.18×10^{-2}) and 4.77×10^{-3} (range: 3.84×10^{-5} – 4.53×10^{-2}) respectively (Table 5).

Table 5. Cancer risks and hazard indices for exposure to THMs.

Pathways	Cancer Risk				Hazard Index			
	Average	Std. Dev.	Minimum	Maximum	Average	Std. Dev.	Minimum	Maximum
Ingestion	1.16×10^{-5}	6.53×10^{-6}	5.76×10^{-7}	5.26×10^{-5}	2.23×10^{-2}	1.45×10^{-2}	1.05×10^{-3}	1.45×10^{-1}
Inhalation	4.17×10^{-6}	3.31×10^{-6}	1.11×10^{-7}	3.54×10^{-5}	7.89×10^{-3}	6.70×10^{-3}	1.21×10^{-4}	8.18×10^{-2}
Dermal Contact	1.97×10^{-6}	1.59×10^{-6}	1.31×10^{-8}	1.35×10^{-5}	4.77×10^{-3}	3.90×10^{-3}	3.84×10^{-5}	4.53×10^{-2}
Total	1.78×10^{-5}	1.02×10^{-5}	7.40×10^{-7}	9.26×10^{-5}	3.49×10^{-2}	2.33×10^{-2}	1.20×10^{-3}	2.34×10^{-1}

The overall DALY was estimated to be 25.1 with the range of 23.2 – 26.8, meaning that on average 25.1 years are likely to be lost due to cancer risks from exposure to THMs. Among the DALY, the YLL and YLD were estimated to be 18.6 (17.1 – 19.9) and 6.5 (6.0 – 6.9) respectively. The YLL and YLD contributed approximately 74 and 26% of total DALY. DALY for different age groups are shown in Figure 3. The largest DALY (5.6) were estimated for 15-39 age group (Figure 3). The cancer risk in terms of DALY was obtained as 8.48×10^{-7} per person per year (PPPY), which was lower than the USEPA guideline value of 1.0×10^{-6} [27]. One DALY was associated with a loss of US$ 108,600 [28], which means one year of healthy human life costs US$ 108,600, indicating that the total financial burden for Saudi Arabia was US$ 2.72 million with the range of US$ 2.52 – 2.91 million per year.

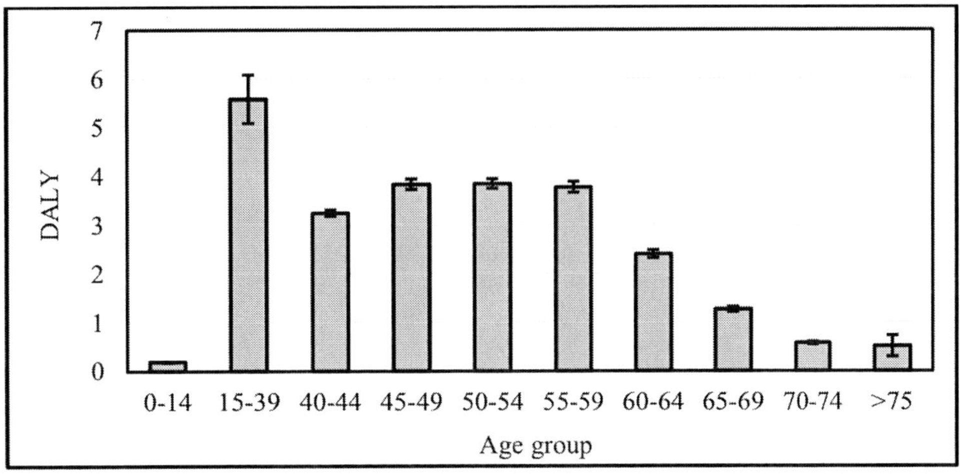

Fig. 3. DALYs for different age groups. Error bars represent standard deviation.

4 Conclusion

This study presented and applied the methodologies of estimating human health risk and DALY from exposure to DBPs in desalinated and blended water in Saudi Arabia. The cancer risks of DBPs in desalinated and blended water was predicted to be 1.78×10^{-5}, which was higher than the minimum risk recommended by the USEPA. Cancer risks through ingestion route was highest (65.4%) followed by dermal (23.5%) and inhalation (11.1%) routes. With respect to age groups, the highest contributor group was >16 years (45.3%), which had the largest lifespan (16+ to death). In contrast, the exposure during the early life (birth to < 2 years) contributed approximately 25.4% of the overall risks. Control of early life exposure can reduce the risk significantly. The cancer risks in terms of DALY was estimated to be 8.48×10^{-7} per person per year, which is below the reference risk level as recommended by the USEPA. The DALY can be used for analyzing financial burden from cancer risks. It can also be used for cost effectiveness analysis, which may help to improve quality of desalinated water and may help to prioritize the hazardous material present in desalinated water. In this study, THMs were used for analyses, which were a fraction of various DBPs present in desalinated water. Few other DBPs with possible cancer risks are HAAs, iodo-THMs, bromate, NDMA while sufficient information on these DBPs are not available to date to conduct risk analysis for desalinated and blended water. Future study may further look into these DBPs and their risks.

The authors would like to acknowledge the support of research group funding for the Water Research Group at the King Fahd University of Petroleum and Minerals (KFUPM), Dhahran 31261, Saudi Arabia.

References

1. M. Al-Zahrani, A. Musa, S. Chowdhury, S. Environ Dev Sustain, **18**, 3, 777-798 (2016)
2. Ministry of Water and Electricity, *The Ministry of Water and Electricity annual report, 2013, Saudi Arabia* (2013)
3. A. Abdel-Wahab, A. Khodary, N. Bensalah, J. Environ. Prot., **1**, 4, 456-465 (2010)
4. World Health Organization, *Guidelines for drinking water quality* (2010)
5. D. Kim, G. Amy, T. Karanfil, Water Res., **81**, 343–355 (2015)
6. S.D. Richardson et al., Environ. Sci. Technol., **42**, 22, 8330-8338 (2008)
7. S. Chowdhury, Sci Total Environ., **463-464**, 922-930 (2013)
8. United States Environmental Protection Agency, *Method 551.1 determination of chlorination disinfection byproducts, chlorinated solvents, and halogenated pesticides/herbicides in drinking water by liquid-liquid extraction and gas chromatography with electron-capture detection* (1995)
9. American Public Health Association, *Standard methods for the examination of water and wastewater* (1998)
10. D.A. Savitz et al., Am J Epidemiol, **164**, 11, 1043-1051 (2006)
11. United States Environmental Protection Agency, *Supplemental guidance for assessing susceptibility from early-life exposure to carcinogens* (2005a)
12. S. Chowdhury, P. Champagne, P.J. McLellan, Sci Total Environ., **407**, 14, 4189-4206 (2009)
13. United States Environmental Protection Agency, *Human health risk assessment protocol* (1998)
14. United States Environmental Protection Agency, *Human health risk assessment protocol for hazardous waste combustion facilities* (2005b)
15. T.E. McKone, R.A. Howd, Risk Anal, **12**, 4, 543-556 (1992)
16. S. Chowdhury, J Water Health, **14**, 3, 489-503 (2016)
17. United States Environmental Protection Agency, *Integrated risk information system: Online database*, Available at: https://cfpub.epa.gov/ncea/iris2/atoz.cfm
18. C.J.L. Murray, A.D. Lopez, *Harvard University Press*, 1, 1-35 (1996)
19. I. Soerjomataram et al., BMC Medical Research Methodology, **12**, 125 (2012)

20. S. Pan et al., J. Hazard. Mater., **280**, 288–294 (2014)
21. M. Smaastuen et al., *Long-term cancer survival: patterns and trends in Norway 1965-2007, Cancer Registry of Norway* (2008)
22. N.M. Fayad, J Am Water Works Assoc, **85**, 1, 46-50 (1993)
23. J. Le Roux et al., Desalination, **359**, 141–148 (2015)
24. M. Ahmad, A.S. Bajahlan, Environ Monit Assess, 159, 1-14 (2009)
25. P.C.M. Kutty, A.A. Nomani, T.S. Thankachan, *Analyses of water samples from Jeddah seawater RO / MSF plants for organic pollutants, Issued as Technical Report No. SWCC (RDC)-14* (1991)
26. United States Environmental Protection Agency, *Exposure factors handbook* (2011)
27. United States Environmental Protection Agency, *Risk Assessment Guidance for Superfund. Volume I: Human Health Evaluation Manual (Part A)* (1989)
28. C.P. Lee, G.M. Chertow, S.A. Zenios, Value Health, **12**, 1, 80-87 (2009)

Analyzing trend and developing an empirical formula for estimating rainfall intensity: a case study Eskisehir city, Turkey.

Sinan Jasim Hadi[1,*], and *Mustafa* TOMBUL[1]

[1]Department of Civil Engineering, Faculty of Engineering, Anadolu University Eskisehir, Turkey

Abstract. Trend and stationarity analysis conducted for the city center of Eskisehir province, Turkey. The Intensity- Duration – Frequency (IDF) curves and formulas for the estimation of the design rainfall intensity have been developed. Through the application of four different frequency methods, the IDF has been developed for a record of 64 years. The functions employed are; Normal, Log – Normal, Gumbel, and Log –Pearson Type III. A formula for each function developed by determining the parameters with the help of a non-linear regression method. The most appropriate fit of probability distribution determined by goodness of fit test i.e. Chi – Square. The results of the study revealed no evidence of significant trend or nonstationarity in any duration. With an exception to the Normal function which exceeded the critical region in several durations, all the functions found to show high fitting. LPT III identified as the best fit. The coefficient of correlation for the developed formulas found high in all the functions except for LPT III. The study endorsed that the estimation of rainfall intensity for any storm duration and return period can be done for Eskisehir using LPT III and Gumbel formulas as they are evaluated as best fitting.

1 Introduction

The frequency and intensity of extreme rainfall events are expected to increase in the future for the regions that already have intense and frequent events, whereas for the regions having less events are expected to witnessed clear decreases in the amount of rainfall [1]. Stationarity and trend analyses are considered as good tools for investigating the historical change in the time series data especially in hydrological observations. Stationarity has been taken into consideration in many studies as a result of it being one of the main assumptions of analysing the frequency of extreme rainfalls [2]. Therefore, conforming stationarity or detecting nonstationarity is crucial for extreme rainfall observations [3, 4]. There are several studies which have been conducted around the world in order to examine stationarity/nonstationarity and trend analysis of extreme rainfall data [5-7].

Management and planning of water resources includes the determination of required discharge capacity of channels, pumping station capacity and planning out the design and

* Corresponding author: sinan.jasim@yahoo.com

building of sewage and storm systems. This management can be eased by using statistical methods that use extreme rainfall data for carrying out the assessment of water resource management. These methods, particularly the IDF curves, can play an important role in: reducing the loss of property and life by judging and assessing hazards, the damage that can occur, and the preventive methods that need to be implemented [8]. The definition of Intensity – Duration – Frequency (IDF) curve of rainfall is given by [9, 10] as the illustration of the amount of rainfall falling on a watershed for a given period of time in the form of a graph.

The mathematical relationship among Return period T (which is also known as frequency referring to the annual frequency of exceedance f), Rainfall intensity I and Storm duration d can easily be found from the IDF curves [9, 11, 12]. Numerous studies have been conducted around the world to develop formulas from the designed IDF curves such as [8, 9, 13-15]. Several studies of developing such a formula have been, also, conducted in several cities in Turkey. For example, [16-21].

The presence of a huge mountain range, different surface shapes and three seas in the surrounding of Turkey has resulted in a varying climate all over the country. In order to get a clear understanding of that climate variation, each of the provinces must be studied. Eskisehir is one of these provinces where neither the historical changes in the extreme rainfall events nor the construction of IDF have been studied. The objective of this study to analyze the trend of the annual maximum rainfall of Eskisehir and develop IDF curves using several distribution functions followed by deriving formula can be used for estimating the rainfall intensity for any return period and duration.

2 Study area and data collection

Eskisehir is a north-western province of Turkey lying in between Istanbul (cultural and touristic capital of Turkey) and Ankara (administrative capital of Turkey); it is 233 km away from Ankara and 324 km from Istanbul. The location of the study area, shown in Figure 1, is between 39°44'00"N and 39°49'00"N latitudes and 30°27'00"E and 30°36'00"E longitudes. The study area has an area of 190 km2 and is inhabited by 755,427 people. Only the centre of Eskisehir city which can be represented by the used rainfall station is chosen.

Data was collected for the chosen metrological station located in the centre of Eskisehir from General Department of Meteorology - Ministry of Forest and Water Affairs, Tukrey. Data from 1940 to 2010 except for the period between 1950 and 1956 (7 years' duration) was recorded at the station. The recorded extreme rainfall storms have the duration of 5, 10, 15, 30, 60, 120, 180, 240, 300, 360, 480, 720, 1080, 1440 min.

3 Trend analysis and nonstationarity

It is important to detect the trend in any time series data in order to identify whether there is an increase or decrease during the historical observations. In this study, three trend tests were implemented: Mann-Kendall test (MK), Cox and Stuart test, and Sen's Slope test. Null hypothesis (H0) of MK test is that no monotonic trend is present whereas the alternative hypothesis (Ha) is that Monotonic trend is present. P value for every duration was calculated and compared with the confidence level. In the same manner, Cox and Stuart test has the same hypotheses (i.e. H0: No monotonic trend, and Ha: Monotonic trend is present) and the same P value comparison. On the contrast, the statistic test, was calculated for Sen's slope test in order to be compared with the confidence interval.

Trend analysis is implemented to detect whether there is a general increase or decrease in the time series data. However, this increase or decrease not always ensures the existence of nonstationarity in the series. Therefore, further analysis was conducted to detect whether nonstationarity is present. In this study, three nonstationarity tests were implemented: Kwiatkowski–Phillips–Schmidt–Shin (KPSS) test, Phillips–Perron (PP) test, and Augmented Dickey–Fuller (ADF) test. Null hypothesis (H0) of KPSS test is that data are stationary and the alternative hypothesis (Ha) is that nonstationarity is present. On the other hand, the hypotheses of PP and ADF tests are the opposite (i.e. H0: Nonstationary, and Ha: Stationary).

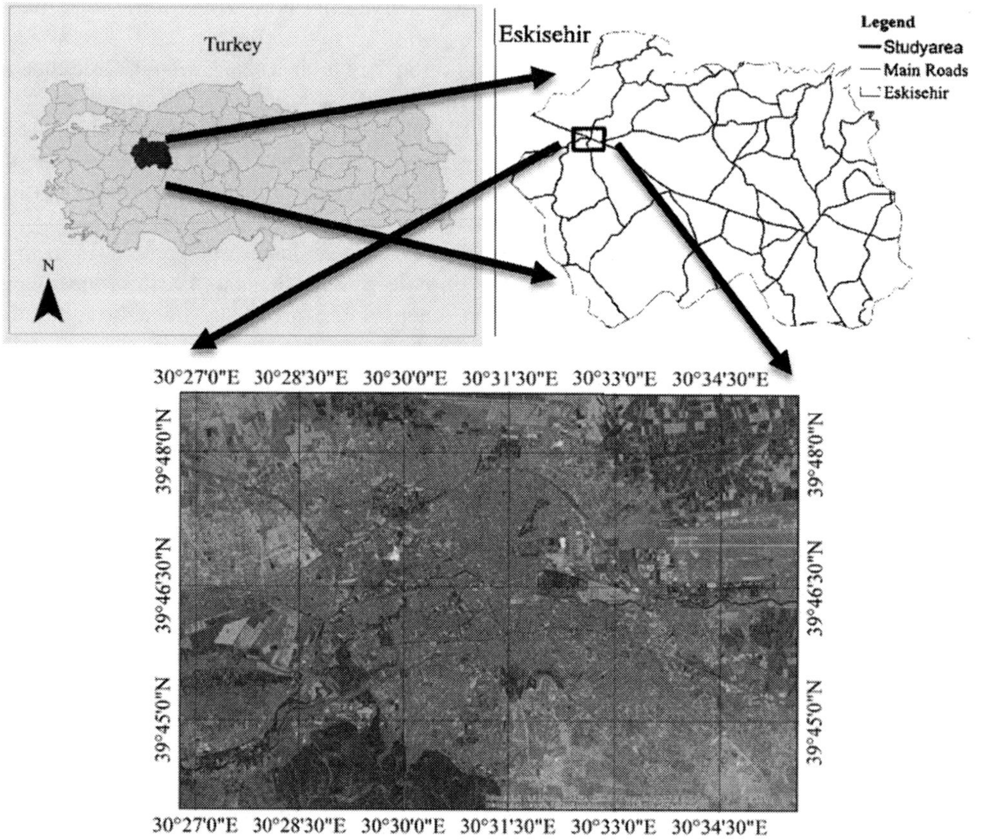

Fig. 1. Map of the study area (i.e. city centre of Eskisehir province represented by one rainfall station, Turkey) extracted by ArcGIS 10 using administrative and satellite imagery base maps and the main roads digitized from the StreetMap of ArcGIS online [22].

4 Development of IDF curves

The IDF curves are developed through a three steps procedure [23, 24]. Firstly, the records are fit into any of the probability distribution functions for a specific storm duration, for instance 10 minutes, 20 minutes, 30 minutes, 1 hour, … 24 hours. Then, the probability distribution function selected in the earlier step is used for determining rainfall intensities for a specific storm duration and one of the return periods that have already been defined (e.g. 2, 5, 10, 50, 100 years, etc.). Lastly, two approaches can be used: (a) an equation or mathematical relation is developed relating rainfall intensities, return periods and storm durations, (b) a graphical illustration of the rainfall intensities determined earlier for specific storm durations and each of the defined return periods. IDFs for four functions; Normal, Log-Normal, Gumbel, and Log-Pearson; were developed and shown on Figure 2.

5 Fitting of probability distribution function

There are some of the theoretical probability distribution functions are used commonly and were applied in various areas around the globe: GEV (Generalized Extreme Value Distribution), Gumbel (Extreme Value Distribution type I), Normal Distribution, Log-Normal Distribution, Pearson, Log Pearson Type III Distribution, etc. See [8, 9, 17, 23, 25]. In this study, four common probability distribution functions were used for the purpose of developing the IDF curves. These functions are: Normal, Log-Normal, Gumbel, and Log-Pearson. The goodness of fit test; Chi-Square test was used for testing the best fitting function.

6 Derivation of empirical IDF formula

The maximum rainfall intensities are calculated as dependent variable and the return period and the storm duration as independent variables using any of the four IDF curves developed earlier [8, 9, 12, 25]. A power-law relation as in Eq. (1) can be used for this purpose due to the advantage of having the intensity I_T being dependent on the return period T_r and storm duration t_d separately.

$$\mathbf{I_T} = \frac{fn\,(T_r)}{fn\,(t_d)} \quad (1)$$

The function $\mathrm{fn}(T_r)$, according to [8, 23, 24] can be given as:

$$\mathbf{fn(T_r)} = a\,(T_r)^m \quad (2)$$

According to [23, 25], the storm duration $\mathrm{fn}(t_d)$ can be given by:

$$\mathbf{fn(t_d)} = (t_d + b)^e \quad (3)$$

Placing Eq. (2) and (3) in this Eq. (1), following power-law equation is obtained:

$$\mathbf{I_T} = \frac{a\,(T_r)^m}{(t_d+b)^e} \quad (4)$$

In the equation, a, m, b and e represent empirical parameters and their values are to be determined. To calculate these parameters, the least squares method used and an equation for every distribution function was developed.

7 Results and discussion

P values of MK and Cox and Stuart tests for all the durations that shown in Table 1 are higher than any of the confidence levels (0.01, 0.05, and 0.1) which means we fail to reject the null hypothesis that data has no trend. Also, as can be seen in Table 1, all the Sen's slope test values are within the confidence interval which, hence, leads to the same decision that we fail to reject the null hypothesis that no monotonic trend is present. Therefore, there is no evidence that there is a significant trend in the observed extreme rainfall events.

Nonstationarity analyses results are shown in Table 2. The critical values for every test and confidence level are shown, also, in order to ease the inference. The result showed that even if the test values compared with the 0.1 confidence level value which is 0.347, they still all inside the confidence interval. So, we fail to reject the null hypothesis that the series are stationary. PP and ADF tests are left sided test. In specific, the least critical value correspond to 0.01 confidence level for PP test is -4.138 and for ADF test is -4.040. By comparing the test statistics for each duration with these values, it can be obviously seen that all the values are less which means that for both tests the null hypothesis is rejected. Due to the rejection of the null hypothesis and as the null hypothesis in these two tests is that series are nonstationary, it is concluded that there is no evidence the data is nonstationary.

Table 1. Test statistics values for each duration of three Trend analysis tests: Mann-Kendall (MK), Cox and Stuart, and Sen's Slope.

Test	5 m	10 m	15 m	30 m	1 h	2 h	3 h	4 h	5 h	6 h	8 h	12 h	18 h	24 h
MK Test	0.17	0.14	0.13	0.23	0.38	0.32	0.27	0.16	0.29	0.39	0.16	0.19	0.51	0.94
Cox and Stuart	0.99	0.97	0.99	0.97	0.78	0.88	0.5	0.88	0.78	0.78	0.78	0.78	0.5	0.22
Sen's Slope	**0.04**	**0.06**	**0.09**	**0.09**	**0.06**	**0.04**	**0.06**	**0.07**	**0.05**	**0.05**	**0.07**	**0.06**	**0.04**	**0.003**
Sen's Slope Test	0.02	0.02	0.02	0.06	0.08	0.07	0.06	0.05	0.06	0.08	0.05	0.05	0.10	0.16
Confidence Interval (Upper and lower value)	**0.09**	**0.14**	**0.19**	**0.24**	**0.22**	**0.18**	**0.17**	**0.18**	**0.17**	**0.18**	**0.18**	**0.17**	**0.17**	**0.14**

Confidence levels for MK and Cox and Stuart test are: 0.1, 0.05, 0.01
Confidence Level for Sen's Slope is: 0.05
Bold numbers are negative values

Table 2. Test statistics values for each duration of three Nonstationarity analysis tests.

Test	5 m	10 m	15 m	30 m	1 h	2 h	3 h	4 h	5 h	6 h	8 h	12 h	18 h	24 h
KPSS Test	0.25	0.31	0.32	0.2	0.09	0.06	0.09	0.05	0.05	0.04	0.03	0.03	0.04	0.09
PP Test	**5.98**	**6.98**	**6.72**	**6.99**	**7.41**	**7.75**	**7.51**	**8.13**	**7.93**	**7.92**	**7.17**	**7.11**	**7.84**	**7.06**
ADF Test	**5.92**	**6.95**	**6.71**	**6.99**	**7.41**	**7.75**	**7.51**	**8.13**	**7.93**	**7.92**	**7.17**	**7.11**	**7.84**	**7.06**
C. Level	KPSS	0.1	0.05	0.01	PP Test	0.1	0.05	0.01	ADF Test	0.1	0.05	0.01		
Critical value		0.35	0.46	0.74		**3.17**	**3.49**	**4.14**		**3.15**	**3.45**	**4.04**		

Bold numbers are negative values

Kwiatkowski–Phillips–Schmidt–Shin (KPSS), Phillips–Perron (PP), and Augmented Dickey–Fuller (ADF).

According to the IDFs seen in Figure 2, the intensity and return period in all the functions are directly proportional to one another, i.e. if the return period increases the intensity was also seen increasing for a specific storm duration. It was also seen that rainfall intensity and storm duration are inversely proportional for the four functions. The intensity was seen changing dramatically for a certain return period until duration reaches 8-10 hours and after that it starts changing gradually. Also, the figure show that Log-Normal function has the highest intensity values for high return periods and values deduced from rest of the functions were close to each other.

To identify which function is the best fitting among the used distribution functions, the goodness of fit tests must be implemented. In this study, Chi-Square test was utilized. Two significance levels (i.e. $\alpha = 0.01$ and $\alpha = 0.05$) were used. The results of Chi-Square test are shown in Table 3. Based on the result in the table, all the distributions were considered to be well fitting the data as the test values are not significant in all the storm durations with an exception of Normal function which has significant result in several durations: 3, 6, 8 hours at the significance level $\alpha = 0.05$ and 12 hours at $\alpha = 0.01$.

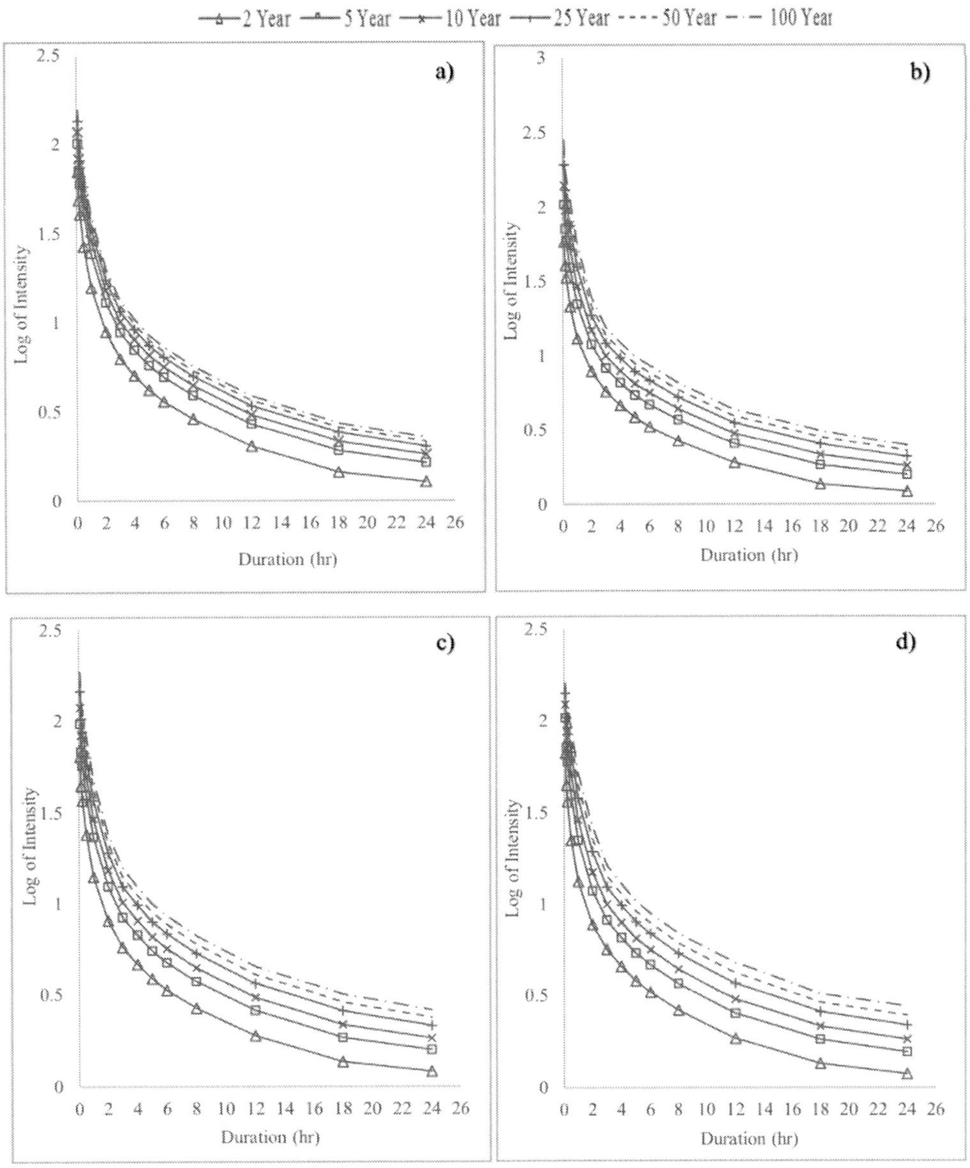

Fig. 2. Fitted Intensity- Duration – Frequency (IDF) Curve using extreme rainfall events for 14 storm durations (i.e. 5, 10, 15, 30 minutes and 1, 2, 3, 4, 5, 6, 8, 12, 18, 24 hour) and 6 return periods (i.e. 2, 5, 10, 50, 100 years) for four functions: a) Normal, b) Log-Normal, c) Gumbel, and d) Log-Pearson.

Table 3. the result of goodness of fit test (Chi – Square) of annual maximum rainfall.

Duration	Normal	LogNormal	Gumbel	Log - Pearson III
5 m	9.28	3.22	8.34	4.83
10 m	1.06	8.25	4.92	2.94
15 m	3.98	3.74	7.57	2.15
30 m	3.49	2.6	5.36	5.36
1 h	7.39	4.25	3.97	4.01
2 h	9.61	4.02	4.66	5.05

3 h	11.89**	4.15	8.56	2.45
4 h	8.91	6.83	10.8	7.8
5 h	9.62	8.7	10.3	1.85
6 h	12.14**	6.8	5.4	8.86
8 h	12.35 **	5.02	6.67	3.97
12 h	15.49 *	4.93	6.87	7.16
18 h	7.08	2.63	2.71	3.77
24 h	4.57	2.27	1.44	2.25
When $\alpha = 0.01$, degree of freedom = 5, the critical region is > 15.086				
When $\alpha = 0.05$, degree of freedom = 5, the critical region is > 11.07				
* X^2 significant at $\alpha = 0.01$				
** X^2 significant at $\alpha = 0.05$				

Table 4 lists out the parameters along with their obtained values and the developed formulas. The parameters set of each function formula was worked out using least squares method which implies the obtainment of accurate parameters. Determined parameters were substituted in Eq. (4) in order to obtain a formula for every function that can estimate the intensity for any return period and storm duration.

In the same table, coefficients of correlation were also listed for all the formulas of the four functions. This coefficient indicates the relation between the actual values and predicted values that were obtained by the developed formula. The highest correlation values obtained belong to the formulas of Log-Normal and Gumbel functions whereas the lowest value is for the formula of LPT III function.

Table 4. calculated parameters and coefficient of correlation for every function used in the derived formulas.

Function/Parameter	a	b	e	m	Derived Equation	Correlation
Normal	19.10	0.14	0.87	0.18	$I_T = \frac{19.10\,(T_r)^{0.18}}{(t_d + 0.14)^{0.87}}$	0.995
Log - Normal	13.70	0.14	0.97	0.34	$I_T = \frac{13.70\,(T_r)^{0.34}}{(t_d + 0.14)^{0.97}}$	0.996
Gumbel	16.93	0.14	0.88	0.24	$I_T = \frac{16.93\,(T_r)^{0.24}}{(t_d + 0.14)^{0.88}}$	0.996
Log - Pearson	22.83	0.28	1.01	0.21	$I_T = \frac{22.83\,(T_r)^{0.21}}{(t_d + 0.28)^{1.01}}$	0.992

8 Conclusion

Eskisehir is one of Turkey provinces requires its climate to be studied. The trend of extreme rainfall events for Eskisehir was investigated for the period between 1940 and 2010 (with an exception of the years between 1950 and 1956) using three tests: Mann-Kendall test (MK), Cox and Stuart test, and Sen's Slope test. There has been no significant trend found throughout the studied period. According to these results, the argument that areas with intense and frequent extreme storms are more likely to witness an increase while on the other hand, the areas which have less intense and frequent storms are expected to witness a decrease is proved, as Eskisehir considered as an area that has moderate extreme rainfall events. This could be due to the location of the study area in the middle of Turkey where the effect of the Mediterranean and Black sea climate is not present and therefore concluding that the precipitation of Eskisehir has not been affected by the climate change.

Stationarity of the extreme rainfall events for all the 14 storm durations examined using several tests: Kwiatkowski–Phillips–Schmidt–Shin (KPSS), Phillips–Perron (PP), and

Augmented Dickey–Fuller (ADF) tests. Log – Normal, Log –Pearson Type III, Normal, and Gumbel functions used to fit IDF curves with 6 return periods (i.e. 2, 5, 10, 50, 100 years) and 14 storm durations. Formulas representing these curves developed for the reason that can be used for estimating rainfall intensity rather than the curves. Therefore, it can be stated that the developed formulas for the study area of the Log-Pearson III and the Gumbel function, with preference given to the Gumbel function, can be used in order to estimate the rainfall intensity for any return period and storm duration.

In general, the use of formula proved to be a good alternative for getting the values from a graph i.e. IDF curve in order to obtain design storm. In this study, only the city center of Eskisehir (i.e. the urban area) with one rainfall station was studied while the consideration of studying the area with a regional effect by using many stations is highly recommended. Also, the effect of the climate change along with effect of the seas and mountains, surrounding Turkey and Eskisehir respectively, on the study area must be studied in detail.

The authors would like to thank (DMİ) Devlet Meteroloji İşleri, Türkiye (General Department of Meteorology - Ministry of Forest and Water Affairs, Turkey) for providing the data to complete this study

References

1. IPCC, *Climate Change 2007: The Physical Science Basis. Contribution ofWorking Group I to the Fourth Assessment Report of the Intergovernmental Panel on Climate Change*. Cambridge and NewYork: Cambridge University Press, (2007).
2. A. G. Yilmaz, M. A. Imteaz, and B. J. C. Perera, Investigation of non-stationarity of extreme rainfalls and spatial variability of rainfall intensity–frequency–duration relationships: a case study of Victoria, Australia, Int. J. Climatol., (2016).
3. E. A. Rosenberg, P. W. Keys, D. B. Booth, D. Hartley, J. Burkey, A. C. Steinemann, and D. P. Lettenmaier, Precipitation extremes and the impacts of climate change on stormwater infrastructure in Washington State, Clim. Change, **102**, 1–2, pp. 319–349, (2010).
4. a. G. Yilmaz and B. J. C. Perera, Extreme Rainfall Non-Stationarity Investigation and Intensity-Frequency-Duration Relationship, J. Hydrol. Eng., (2013).
5. S. Shahid, Trends in extreme rainfall events of Bangladesh, Theor. Appl. Climatol., **104**, 3, pp. 489–499, (2011).
6. H. Shang, J. Yan, M. Gebremichael, and S. M. Ayalew, Trend analysis of extreme precipitation in the Northwestern Highlands of Ethiopia with a case study of Debre Markos, Hydrol. Earth Syst. Sci., **15**, 6, pp. 1937–1944, (2011).
7. E. M. Douglas, M. Asce, and C. A. Fairbank, Is Precipitation in Northern New England Becoming More Extreme ? Statistical Analysis of Extreme Rainfall in Massachusetts , New Hampshire , and Maine and Updated Estimates of the 100-Year Storm, Most, **16**, 3, pp. 203–218, (2011).
8. S. A. AlHassoun, Developing an empirical formulae to estimate rainfall intensity in Riyadh region, J. King Saud Univ. - Eng. Sci., **23**, 2, pp. 81–88, (2011).
9. I. H. Elsebaie, Developing rainfall intensity–duration–frequency relationship for two regions in Saudi Arabia, J. King Saud Univ. - Eng. Sci., **24**, 2, pp. 131–140, (2012).
10. B. S. Dupont and D. L. Allen, Revision of the rainfall intensity duration curves for the commonwealth of kentucky., USA, (2000).
11. D. Koutsoyiannis, *On the appropriateness of the gumbel distribution in modelling extreme rainfall*, in *Proceedings of the ESF LESC Exploratory Workshop, European Science Foundation, National Resea*, (2003), pp. 303–319.
12. D. Koutsoyiannis, D. Kozonis, and A. Manetas, A mathematical framework for studying rainfall intensity-duration-frequency relationships, J. Hydrol., **206**, 1–2, pp.

118–135, (1998).
13. V. R. Baghirathan and E. M. Shaw, Rainfall depth-duration-frequency studies for Sri Lanka, J. Hydrol., **37**, 3–4, pp. 223–239, (1978).
14. A. Gert, D. J. Wall, E. L. White, and C. N. Dunn, Regional Rainfall intensity-duration-function curves for Pennsylvania, Water Resour. Bull, **23**, pp. 479–486, (1987).
15. C. Chen, Rainfall Intensity-Duration-Frequency Formulas, ASCE J. Hydraul. Eng., **109**, 12, pp. 1603–1621, (1983).
16. R. Acar, S. Celik, and S. Senocak, Rainfall Intensity-Duration - Frequency Model using an artificial neural network approach, J. Sci. Ind. Res., **67**, pp. 198–202, (2008).
17. R. Acar and S. Senocak, *Modelling of Short Duration Rainfall (SDR) Intensity Equations for Ankara, Turkey*, in *Balwois 2008*, (2008), May, pp. 1–9.
18. S. Senocak and R. Acar, Modelling of Short Duration Rainfall (SDR) Intensity Equations for Erzurum, Turkey, J. Eng. Sci., **13**, 1, pp. 75–80, (2007).
19. S. Senocak and R. Acar, Modelling of short-duration rainfall intensity equations for the Agean region of Turkey, FRESENIUS Environ. Bull., **16**, 9b, (2007).
20. O. L. Asikoglu and Ertugrul Benzeden, Simple generalization approach for intensity–duration– frequency relationships, Hydrol. Process., **28**, pp. 1114–1123, (2014).
21. T. Haktanir, M. Cobaner, and O. Kisi, Frequency analyses of annual extreme rainfall series from 5 min to 24 h, Hydrol. Process., **24**, 24, pp. 3574–3588, (2010).
22. ESRI, ArcGIS online, 2016. [Online]. Available: https://www.arcgis.com/home/index.html. [Accessed: 12-Jan-2016].
23. W. T. Chow, *Handbook of Applied Hydrology*. McGraw-Hill, (1988).
24. V. P. Singh, *Elementary hydrology*. New Jersey: Prentice Hall, (1992).
25. L. Nhat, Y. Tachikawa, and K. Takara, Establishment of intensity-duration-frequency curves for precipitation in the monsoon area of Vietnam, Annu. Dis. Prev. Res. Inst, 49B, pp. 93–103, (2006).

Removal of nitrate, ammonia and phosphate from aqueous solutions in packed bed filter using biochar augmented sand media

Ali El Hanandeh[1,*], *Aswin* Bhuvaneswaran[1], *Philiphi* de Rozari[1,2]

[1]School of Engineering, Griffith University, Nathan, QLD Australia
[2]Nusa Cendana University, Kupang, Indonesia

Abstract. Nutrients from wastewater are a major source of pollution because they can cause significant impact on the ecosystem. Accordingly, it is important that the nutrient concentrations are kept to admissible levels to the receiving environment. Often regulatory limits are set on the maximum allowable concentrations in the effluent. Therefore, wastewater must be treated to meet safe levels of discharge. In this study, laboratory investigation of the efficiency of packed bed filters to remove nitrate, ammonium and phosphate from aqueous solutions were conducted. Sand and sand augmented with hydrochloric acid treated biochar (SBC) were used as packing media. Synthetic wastewater solution was prepared with PO_4^{3-}, NO_3^-, NH_4^+ concentrations 20, 10, 50 mg/L, respectively. Each experiment ran for a period of five days; samples from the effluent were collected on alternate days. All experiments were duplicated. Over the experiment period, the average removal efficiency of PO_4^{3-}, NO_3^-, NH_4^+ were 99.2%, 72.9%, 96.7% in the sand packed columns and 99.2%, 82.3%, 97.4% in the SBC packed columns, respectively. Although, the presence of biochar in the packing media had little effect on phosphate and ammonium removal, it significantly improved nitrate removal.

1 Introduction

Nutrient removal from wastewater before discharge to natural water bodies is essential because excessive nutrient loads can lead to eutrophication. Traditional wastewater treatment methods such as activated sludge are not efficient in removing nutrients. Therefore, it is important to find cost effective technologies to reduce the nutrient concentrations to admissible levels to the receiving water bodies. Filtration is a well-established technology for wastewater treatment. The removal efficiency of nutrients in filters depends on the filtration media used. The use of activated carbon (AC) can enhance the performance of filter media

* Corresponding author: a.elhanandeh@griffityh.edu.au

because of its high surface area and adsorption capacity. Nevertheless, AC is very expensive which makes it impractical for treatment of large volumes of wastewater [1].

In recent years, the interest in using biochar in wastewater treatment has been increasing due to its low cost of production, relatively high surface area and active surface functional groups [2, 3]. A number of researchers reported that biochar can be used to remove nutrients from aqueous solutions in batch and column experiments [4-6]. Furthermore, Gupta et al. [7] reported that augmenting media with biochar in constructed laboratory scale constructed wetland improved nitrogen and phosphorus removal from wastewater. On the other hand, Yao et al. [6] found that the effects of the addition of biochar to sand media were not universal and it depended on the biomass origin of the biochar. Other researchers also reported that biochar preparation conditions, such as pyrolysis method, pre-treatment, heating rate and time and final pyrolysis temperature also affected the adsorption characteristics of the biochar [8, 9]. Therefore, the aim of this study was to investigate the suitability of commercially produced biochar derived from eucalyptus woody biomass prepared under slow pyrolysis conditions at pyrolysis temperature of 500 °C as a media amendment in sand filters for the removal of phosphate, nitrate and ammonium from aqueous solutions.

2 Material and methods

The effect of the addition of biochar to sand media on nutrient removal from aqueous solutions was investigated using laboratory scale column experiments. All experiments were duplicated. In total four (4) plexi-glass columns with internal diameter of 56 mm and 200 mm height were used (Figure 1).

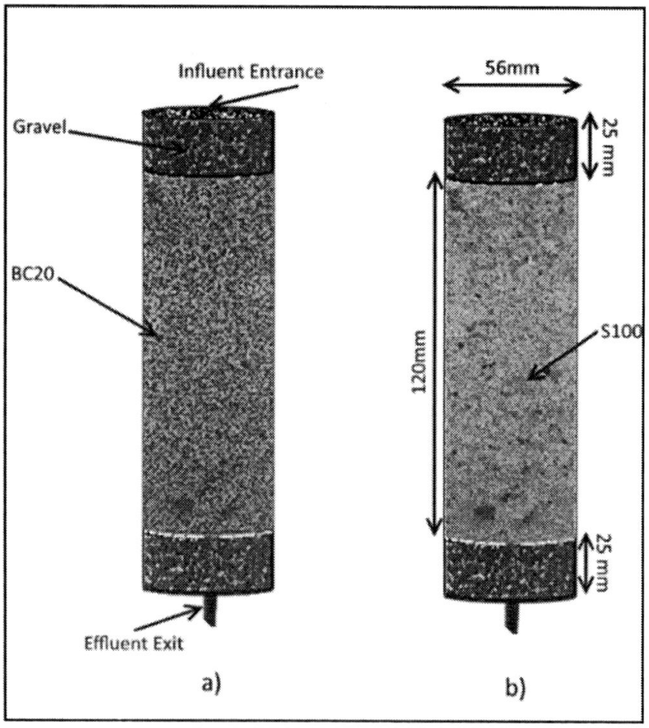

Fig. 1. Filter design (a) sand and (b) SBC media.

2.1 Material procurement characterisation and preparation

Sand was purchased from a local hardware store. Biochar was purchased from a local biochar supplier. The biochar was prepared from eucalyptus woody biomass using slow pyrolysis method at 500 °C.

Two treatments were prepared: sand alone (control) and sand mixed with 20% biochar by volume, henceforth referred to as SBC. Both media types were washed with distilled deionised water and 0.1M HCl acid to remove impurities. Then the media was rinsed with deionised distilled water until the pH of the effluent was neutral. The basic characteristics of both media are given in Table 1.

Table 1. Physical characteristics of media.

Property	Sand	SBC
Bulk density (g/cm^3)		
• Dry	• 1.46	• 1.22
• Wet	• 1.80	• 1.58
Porosity (%)	29.8	39.2
Particle Grain Size (μm)	125<d<710	125<d<710

2.2 Column packing

First gravel layer was added to a depth of 25 mm. This layer acts as drainage layer. The sand or SBC were added in three layers to total height of 120 mm. After adding each layer, the layer was compacted by dropping a 50 g weight 10 times from 100 mm height. Finally another layer of gravel (25 mm) was add onto to ensure equal distribution of water, prevent scouring of the filter media and prevent floatation of biochar. To prevent media mixing, plastic fabric (mesh) were laid between grave and sand media. All media were wet packed in the columns. Figure 1 shows a schematic diagram of the packed columns.

2.3 Chemicals

All chemical were laboratory grade chemicals. Solution containing phosphate, nitrate and ammonium with concentrations of 20, 10 and 50 mg/L, respectively was prepared by adding 176.5 mg of KH_2PO_4, 145.35 mg of KNO_3 and 148.58 mg of NH_4Cl to a 2L measuring flask and then filling it to the 2L mark with deionized water. The pH of the solution was 5.53. The stock was stored in the fridge at 4 °C.

2.4 Loading of the columns, sample collection and analysis

Fifty millilitres (50 mL) of the stock solution was added daily to each column. The effluent was collected in a beaker, the effluent was discarded at the end of each day. Samples were collected on alternate days (1, 3 and 5) from the beaker. Samples were frozen until analysed.

Phosphate in the effluent were analysed using Spectrophotometry and the Molybdenum Blue Method. Nitrate and ammonium were analysed using Discrete Chemistry Analyser (Westco Smartchem 200). MS Excel™ with data analysis Toolpack was used to perform all statistical tests and data analysis.

3 Results and discussion

The concentration of different nitrogen compounds in the effluent as well as the removal efficiency of the corresponding compounds are shown in figures 2-4. The NO_3^-

concentration were in the range 2.3-3.4 mg/L-N in the effluent from the sand media and 1.5-2.2 mg/L-N in the SBC effluent. The NH_4^+ concentration ranged from 0.2 to 4.1 mg/L-N in the effluent from sand and SBC media. In the case of phosphate, the concentration in the effluent from the sand media ranged from 0.12 to 0.24 mg/L-P and 0.08 to 0.16 mg/L-P in the effluent from the SBC media.

3.1 Nitrate removal

Figure 2 shows the concentration of the nitrate ion in the effluent from the sand and SBC media as well as the removal efficiency of nitrate by each media. As expected, Figure 2 shows that the concentration of NO3- in the effluent increased over time as active adsorption sites were being occupied. Nevertheless, the NO3- concentration was reduced significantly from 20 mg/L-N in the influent to a 5-day average of 2.77±0.43 and 1.81±0.23 mg/L-N in the effluent from the sand and SBC media, respectively. This reduction represents an average removal efficiency of 72.3±4.7% and 81.9±2.5% in the sand and SBC media, respectively. Figure 2 also shows that the NO_3^- concentration in the effluent from the SBC media was consistently lower that in the sand media which indicates that the addition of biochar had positive effect on the capacity of the media to remove nitrate. On average, the SBC media retained 13.3% more nitrate than sand alone. A t-test confirmed that improved removal rate was significant at α=0.05 (p=0.0007). Yao, et al. [6] tested 13 biochars and reported that the majority of them had little or no ability to remove nitrate. Nevertheless, they further reported that sandy soil augmented with 2% biochar derived from peanut hull or Brazilian pepperwood released approximately 34% less nitrate than soil alone. Our results suggest that biochar derived from eucalyptus is one of the few biochars that has the capacity to improve nitrate removal. This is important in Australian context as well as other countries were eucalyptus is abundant.

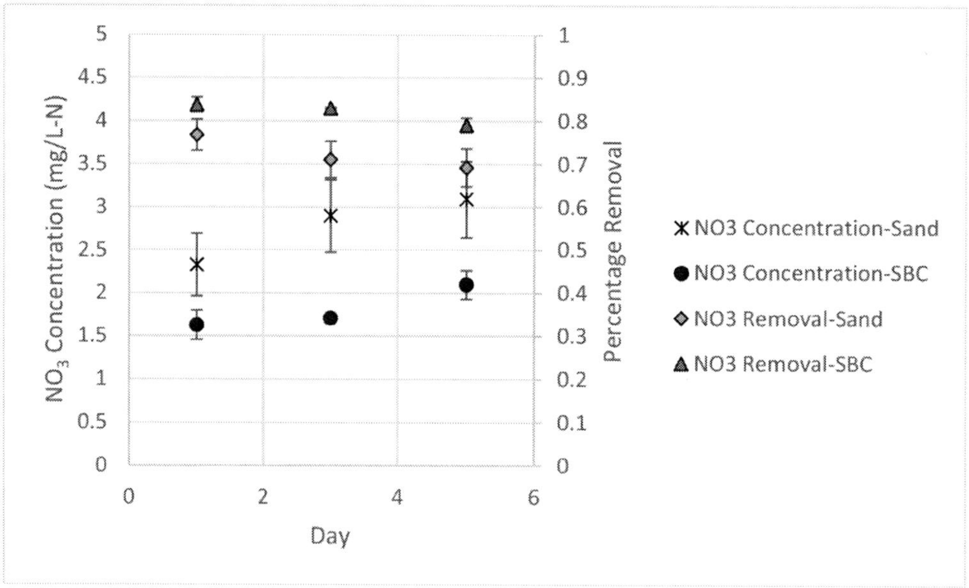

Fig. 2. Nitrate concentration in the effluent and removal efficiency.

3.2 Ammonium removal

The concentration of ammonium in the effluent as well as the removal efficiency of ammonium in each media are shown Figure 3. The results show that the ammonium concentration in the effluent from both media were similar. Suggesting that the addition of biochar did not have adverse effect on the media. In the earlier mentioned study, Yao, et al. [6] reported that nine out of the 13 biochars had some capacity to remove ammonium with varying efficiencies. Hale, et al. [10] who studied the adsorption of ammonium by biochars prepared from corn cob and cacao shell suggested that ammonium is weakly adsorbed on the surface of the biochar through electrostatic exchanges with other cationic species. This is consistent with the findings of Taghizadeh-Toosi, et al. [11] who reported that ammonium adsorbed by biochar was readily available to plants.

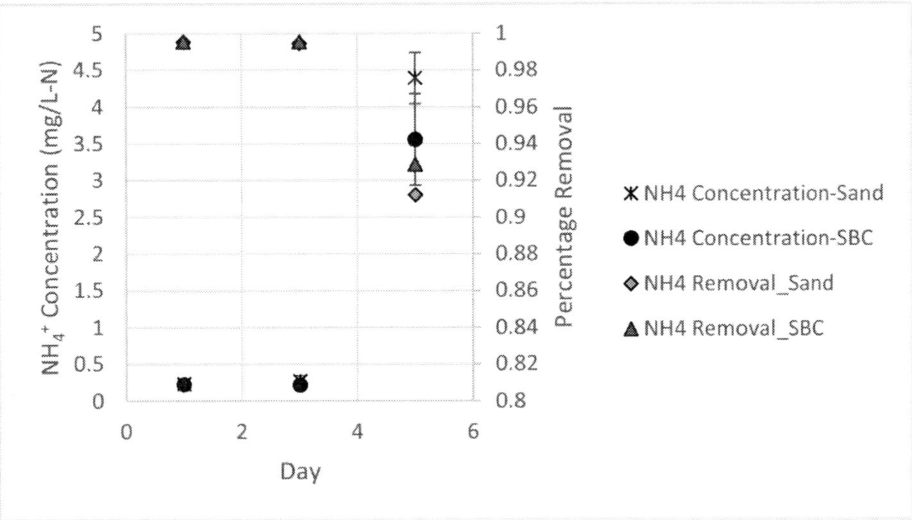

Fig. 3. Ammonium concentration in the effluent and removal efficiency.

Upon close inspection of the results, a trend was observed which indicated that the addition of biochar may enhance the NH_4^+ removal efficiency overtime. Unfortunately, the experiment was terminated after 5 days, thus being unable to establish the long term effect. Therefore, we suggest that the effect of biochar addition should be studied over a longer period of time.

3.3 Phosphate removal

Figure 4 shows the phosphate concentration in the effluent from sand and SBC filter media. The removal rate is also shown in Figure 4. Effluent from sand consistently had higher phosphate concentration than that from the SBC media, thus suggesting that the addition of biochar had positive effect on phosphate removal. The removal efficiency of phosphate has remained constant over the period of the experiment. This is unlike the case for nitrate and ammonium which showed a decreasing trend overtime.

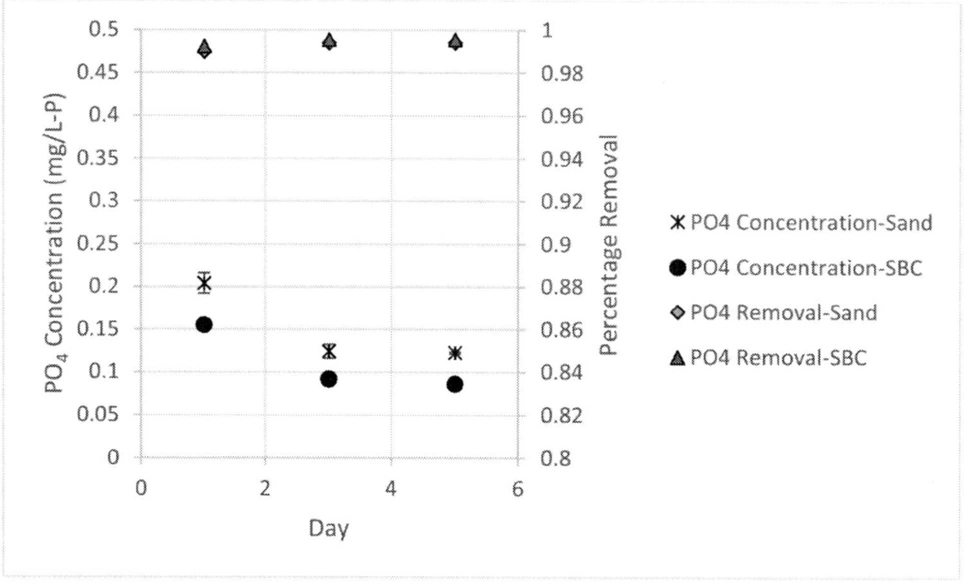

Fig. 4. Phosphate concentration in the effluent and removal efficiency.

The removal efficiency of phosphate in sand was extremely high (>99%); as a result, the effect of biochar addition on the removal efficiency (in relative terms) was insignificant. A number of researchers reported that biochar had little capacity for PO_4^{3-} removal [6, 12]. On the other hand, others reported that biochar has the capacity to remove or enhance phosphorus removal [4, 5, 7, 13]. The apparently conflicting findings can be explained by the fact that the characteristics of biochar depend on the parent biomass and preparation conditions which affect their adsorption capacity of different chemicals [9, 14].

4 Conclusion

Biochar prepared from eucalyptus woody biomass was used as augmenting material for sand filters in laboratory scale column experiments. The effect of biochar addition on nutrients removal (NH_4^+, NO_3^-, PO_4^{3-}) were investigated. The results showed that sand media was very effective in removing nutrients from aqueous solutions. The results further revealed that the addition of biochar had enhanced the removal efficiency of nutrients. The addition of biochar had the greatest effect on the removal of nitrate which increased by 13.3%. In the cases of phosphate and ammonium the improvement was marginal. Nevertheless, a developing trend of enhanced performance over time was detected. It is recommended that the long term effect of biochar addition be further investigated.

References

1. D. Mohan, A. Sarswat, Y. S. Ok, and C. U. Pittman Jr, Bioresource Technol. **160,** 191 (2014).
2. M. Ahmad, A. U. Rajapaksha, J. E. Lim, M. Zhang, N. Bolan, D. Mohan, et al., Chemosphere **99**, 19 (2014).
3. H. N. Tran, S.-J. You, and H.-P. Chao, Waste manage. Res. **34**, 129 (2016).
4. T. A. Ghezzehei, D. V. Sarkhot, and A. A. Berhe, Solid Earth **5**, 953 (2014).
5. Y. Yao, B. Gao, M. Inyang, A. R. Zimmerman, X. Cao, P. Pullammanappallil, et al., J. Hazard. Mater. **190**, 501 (2011).

6. Y. Yao, B. Gao, M. Zhang, M. Inyang, and A. R. Zimmerman, Chemosphere, **89**, 1467 (2012).
7. P. Gupta, T.-w. Ann, and S.-M. Lee, Environ. Eng. Res. **21**, 36 (2016).
8. C. A. Takaya, L. A. Fletcher, S. Singh, K. U. Anyikude, and A. B. Ross, Chemosphere **145**, 518 (2016).
9. Z. Mahdi, A. El Hanandeh, and Q. Yu, Waste Biomass Valor. (2016).
10. S. E. Hale, V. Alling, V. Martinsen, J. Mulder, G. D. Breedveld, and G. Cornelissen, Chemosphere **91**, 1612 (2013).
11. A. Taghizadeh-Toosi, T. J. Clough, R. R. Sherlock, and L. M. Condron, Plant and Soil **350**, 57 (2012).
12. P. de Rozari, M. Greenway, and A. El Hanandeh, Sci. Total Environ. **569–570**, 123 (2016).
13. J. D. Streubel, H. P. Collins, J. M. Tarara, and R. L. Cochran, J. of Environ. Qual. **41**, (2012).
14. S. Yavari, A. Malakahmad, and N. B. Sapari, Environ. Sci. Pollut. R., 1 (2016).

DOI: 10.1051/matecconf/201712005005

Date seed derived biochar for Ni(II) removal from aqueous solutions

Zainab Mahdi[1,*], *Ali* El Hanandeh[1], and *Qiming* Yu[1]

[1]School of Engineering, Griffith University, 170 Kessels Road, Brisbane, Australia

Abstract. The purpose of this study was to investigate the adsorption characteristic of biochars derived from date seed for Ni^{2+} ions removal from aqueous solutions. Two biochars were prepared by slow pyrolysis of date seed for 3 h at 450 °C (DSB450) and 550 °C (DSB550). The adsorption of Ni^{2+} was carried out by batch experiments at room temperature. The effects of pyrolysis temperature, contact time, initial metal concentration, and solution pH were investigated. The results showed that biochar prepared at higher temperature (DSB550) had higher adsorption capacity of Ni^{2+} from aqueous solution than biochar prepared at lower temperature (DBS450). Adsorption efficiency of Ni^{2+} ions was pH dependent and the maximum adsorption was found to occur at pH around 6.0. To describe the equilibrium isotherms, the experimental results were analyzed by the Langmuir, and Freundlich isotherms. The adsorption isotherm for Ni^{2+} by DSB550 was best fit to Langmuir isotherm with (R^2 = 0.94). The maximum adsorption capacity of Ni^{2+} of DSB550 biochar was 0.609 mmol g^{-1}. Pseudo-first order, pseudo-second order, and intraparticle diffusion models were used to model the kinetic parameters and mechanism of adsorption process. The results showed that the adsorption kinetics of these biochars are well described by a pseudo-second order kinetic model with correlation coefficient (R^2 = 0.99). The results of the study indicated that biochar derived from date seed biomass is a suitable material for adsorption of Ni^{2+} ion from aqueous solution.

1 Introduction

Heavy metal pollution poses significant threat to humans and the environment. Heavy metal can bio-accumulate in living tissues due to their non – biodegradability [1]. The majority of metallic pollutants in aquatic systems are due to the discharge of various untreated industrial effluents such as battery manufacturing, tanneries, electroplating, and chemical manufacturing. Nickel (Ni^{2+}) is a highly toxic ion which can induce embryotoxic and nephrotoxic effects, allergic reactions, and contact dermatitis [2].

Therefore, it is important to treat wastewater containing Ni^{2+} to safe levels before it is discharged to the environment. Various treatment methods including chemical precipitation, oxidation/reduction, and electrochemical treatment, ion exchange, and membrane technologies are conventionally used for heavy metal removal. Adsorption using

[*] Corresponding author: a.elhanandeh@griffith.edu.au

activated carbon (AC) has also proven to be an effective method. However, the high cost of the conventional methods may place a barrier to their utilization on large scale [3].

Biochar has been promoted as a low cost adsorbent for the removal of various contaminants from wastewater. It is a carbon rich solid material created by thermal decomposition of biomass under limited oxygen conditions [5]. Several studies reported that biochar has high adsorption capacity towards heavy metal ion due to the abundance of surface functional groups, and its amorphous structural arrangement [5]. Biochar can be prepared from a variety of organic materials, such as woody biomass and agricultural residue such as date stone. The total world production of date is 7,166,695 tons per year [6]. Date stone represents approximately 10% of the date weight; as a result, large quantities of waste are discarded as unwanted material [7].

Several studies investigated the use of date stones for the preparation of activated carbon either chemical or physical activation. Alhamed & Bamufleh [8] prepared activated carbon from date palm waste using $ZnCl_2$ as activating agent. Girgis and El-Hendawy [9] used date pits for activated carbon production after chemical activation using H_3PO_4 acid. Bouchelta et al. [10] used date seeds for AC preparation using physical activation with water vapor. However, AC is costly material due to high energy and chemicals requirements. Biochar, on the other hand, is cheaper to produce and has shown good heavy metal removal capacity. Therefore, the aim of this work was to investigate the ability of biochar prepared from date stone biomass for Ni^{2+} ion removal from aqueous solution.

2 Materials and methods

Biochars were prepared in the lab from date stones using slow pyrolysis process at terminal temperatures of 450 and 550 °C for 3 h. Detailed description of biochar preparation can be found in our previous works [11, 12]. Adsorption studies were carried out using batch adsorption experiments at room temperature. The effects of pH solution, contact time, and initial metal ion concentration on Ni^{2+} removal efficiency were studied. The kinetic and equilibrium batch experiments were carried out to study the adsorption mechanism of the Ni^{+} from aqueous solution.

2.1 Chemicals

All the chemicals used in the experiments were of analytical reagent grade. Deionized water was used in preparing the solutions throughout this study. Stock solution of 5.0 mM of metal ion was prepared by dissolving 1.454 g $Ni(NO_3)_2.6H_2O$ supplied by (Scharlau Chemical) in 1 L of deionized water. Then dilute solutions of different concentrations ranging from 1.0 to 4.0 mM were prepared by successive dilutions to the required working concentrations. The standard solutions for atomic absorption spectroscopy analysis were freshly prepared before use. The initial pH of the solutions was adjusted from 2.0 to 6.0 using 0.1M NaOH and 0.1M HCl.

2.2 Batch adsorption experiment

Batch adsorption studies for Ni^{2+} removal was carried out by mixing 0.1 g of the biochar with 10 mL of Ni^{2+} solution (initial Ni^{2+} concentration 1.0 mM) the solution was kept at pH = 6. All the experiments were carried out at room temperature using 50 ml glass container (Hybex) soaked with 0.1 M HCl and washed with deionized water before use. The mixture was agitated for 24 h using rotary shaker. After 24 h, the mixture of biochar and metal solution was separated using Millex® syringe filter pore size 0.45 μm connected to a glass syringe (10 mL). The final concentration of Ni^{2+} ion was determined using atomic absorption spectrophotometer (Avanta, GBC-SDS) and the amount of Ni^{2+} ions adsorbed on the biochar q_e (mmol g^{-1}) and removal efficiency (R) were calculated using mass balance equation:

$$q_e = \frac{(C_o - C_e)\,V}{W} \tag{1}$$

$$R = \frac{(C_o - C_e)}{C_o} \times 100 \tag{2}$$

where, q_e is the quantity of metal ion adsorbed onto the biochar (mmol g^{-1}), C_o is the initial metal ion concentration (mM), C_e is the final concentration of metal ion at equilibrium (mM), V is the solution volume (L) and W is the biochar dosage (g).

Blank run adsorption experiments (without biochar) were carried out in identical manner to find the effect of glassware on Ni^{2+} ion removal. The investigations showed that the quantity of metal ions adhering on the container walls were negligible (2.36± 0.65 %). Adsorption performance of date seed biomass in their native form was also investigated and compared to the biochar. The effect of pH of the solution was investigated over pH range from 2 to 6 by mixing 0.1 g of biochar with 10 mL of Ni^{2+} solution at initial metal concentration of 1.0 mM. The pH of the solution was adjusted by using 0.1 M NaOH or HCl as required. The adsorption experiments were conducted in triplicate under similar conditions and the average value was reported.

2.3 Adsorption isotherms studies

The equilibrium isotherm is used to describe adsorption performance relation between adsorbate phase (metal solution, C_e mM) and adsorbent phase (biochar, q_e mmol g^{-1}) at equilibrium [13]. The isotherms experiment was conducted at room temperature (23 ± 2 °C) by mixing 0.1 g of the biochar with 10 mL of Ni^{2+} solution with initial Ni^{2+} concentrations range of (1.0-4.0 mM). The solution was kept at pH 6.0. The experimental results were fitted to Langmuir and Freundlich isotherm models.

2.4 Adsorption kinetics studies

The kinetics experiments were carried out by conducting 1.0 g of the biochar with 100 mL of Ni^{2+} solution at initial Ni^{2+} concentration of 1.5 mM. The pH of the solution was kept constant at 6.0. The content in the container was shaken using rotary shaker for the duration of the experiment. Samples were withdrawn at time intervals and analyzed to find the amount of Ni^{2+} adsorbed at time as shown below:

$$q_t = \frac{(C_O - C_t)V}{W} \tag{3}$$

where, q_t is the quantity of Ni^{2+} adsorbed onto the biochar (mmol g^{-1}) at given time, C_o is the initial Ni^{2+}concentration (mM), C_t is the Ni^{2+} concentration at given time (mM), V is the solution volume (L) and W is the biochar dosage (g).

3 Results and discussion

3.1 Effect of pyrolysis temperature on Ni^{2+} adsorption

Adsorption equilibrium uptake is the main parameter used to compare adsorption performance of different adsorbents towards adsorbate. Adsorption performance of date seed biomass in their native form as well as biochars was investigated and compared as shown in Figure 1. It can be seen from Figure 1 that DSB550 biochar had higher adsorption capacity for Ni^{2+} ions than the original biomass. This can be attributed to the influence of pyrolysis on the chemical and physical biochar characteristics. DSB450 biochar had marginally lower adsorption capacity than DS550. However, successful implementation of heavy metal ion adsorption from aqueous solution depends largely on the total cost of adsorption process including cost of original biomass, energy, handling, transportation, drying, disposal, ...etc. [14]. After adsorption, laden biochar can be classified as hazardous material which can be expensive to transport, handle, treat and dispose of due to the high degree of caution requirement and environmental regulations. Therefore, DSB550 was chosen for further investigation in this study.

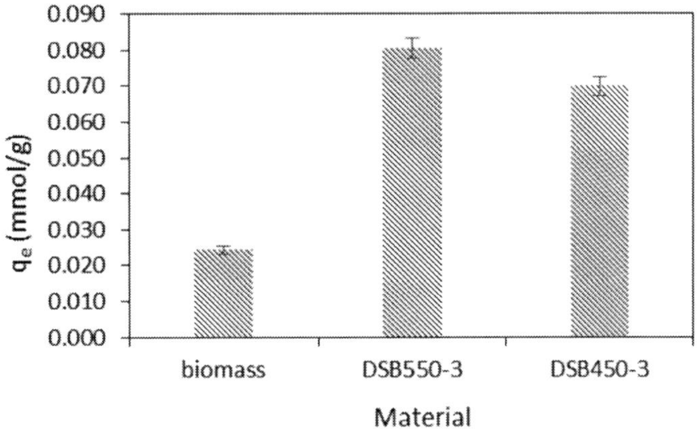

Fig. 1. Adsorption performance of biochar as well as original biomass for Ni^{2+} ion removal.

3.2 Effect of Solution pH on Ni^{2+} Adsorption

The effect of solution pH on Ni^{2+} adsorption was investigated over the pH range 2 to 6 and keeping all other parameters constant as shown in Figure 2. It can be seen from Figure 2 that the adsorption amount of Ni^{2+} increased as pH increased from 2 to 6. This can be attributed to the effect of pH on affinity between adsorbent (biochar) and adsorbate (metal ion solution) and the speciation of metals and dissociation of active functional sites on the adsorbent [15]. At low pH, the surface functional groups (mainly oxygen-containing groups) linked to the H^+ ion making competition between Ni^{2+} ions and H^+ ions which resulted in a lower electrostatic repulsion [16]. Similar findings were observed Kılıç et al. [17] for almond shell biochar who reported that maximum adsorption of Ni^{2+} occurred around pH 7. In this study, pH values higher than 6 were not used to avoid the precipitation of nickel ion in the form of $Ni(OH)_2$.

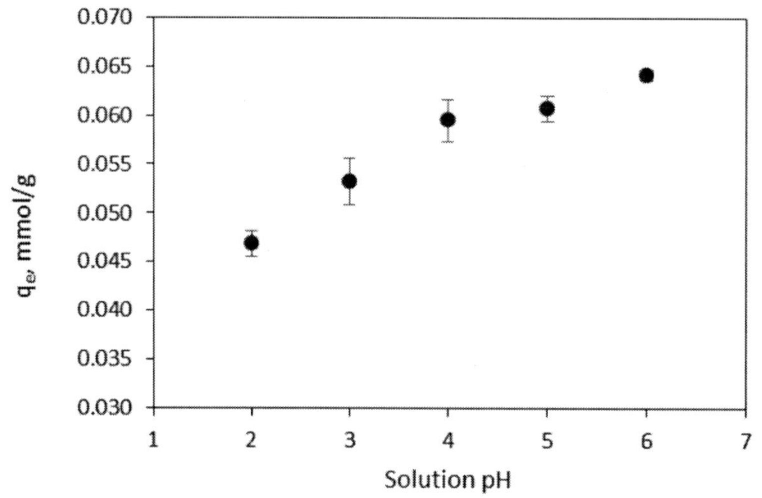

Fig. 2. The effect of pH solution on Ni^{2+} ion removal onto DSB550 biochar .

3.3 Kinetics of Ni^{2+} adsorption

Figure 3 shows the influence of contact time on Ni^{2+} adsorption onto DSB550 biochar. A two stage adsorption mechanism was observed: the first was rapid adsorption and occurred within the first hour followed by slower stage until the reaction reached equilibrium. This can be explained by the availability of a large number of vacant sites on the adsorbent surface at the beginning, as time passed the remaining vacant sites were difficult to access [18]. Adsorption kinetics of Ni^{2+} onto the biochar DSB550 was fitted using pseudo-first order, pseudo second-order models, and intraparticle diffusion model. The Pseudo first-order rate of Largergren (1898) is expressed as follows:

$$\frac{dq_t}{dt} = k_1(q_e - q_t) \tag{4}$$

The linear form of the above equation can be written as below:

$$\ln(q_e - q_t) = \ln q_e - k_1 t \tag{5}$$

where, q_e and q_t are the amount of Ni^{2+} adsorbed per unit mass (mmol g^{-1}) at equilibrium and at time, t respectively. K_1 (min^{-1}) is the pseudo first order adsorption rate constant.

A linear form of the Pseudo second order kinetic model is expressed as following equation:

$$\frac{t}{q_t} = \frac{1}{k_2 q_e^2} + \frac{t}{q_t} \tag{6}$$

The second-order rate constant k_2 and q_e are obtained from the intercept and the slope of the plot obtained by plotting t/q against time t .

The Weber–Morris intraparticle diffusion model is described as follows [19]:

$$q_t = k_p t^{0.5} + C \tag{7}$$

where k_p is the intraparticle diffusion rate constant (mmol g^{-1} min$^{-0.5}$), t is the time (min), and C is the intercept (mmolg^{-1}), which represents the thickness of the boundary layer effects.

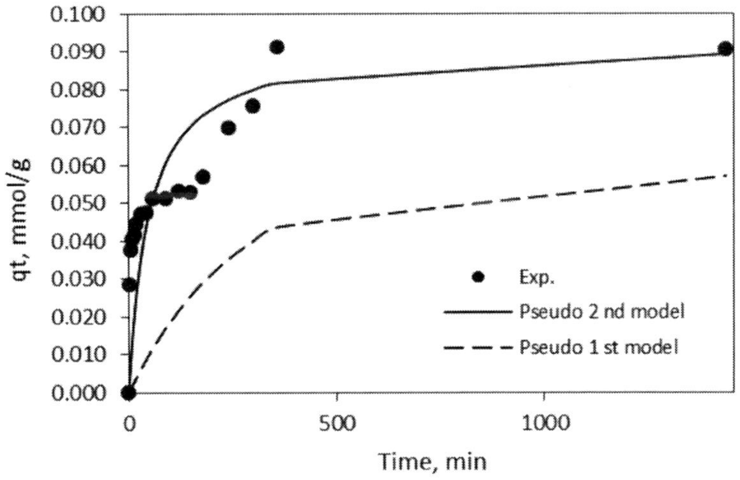

Fig. 3. Effect of contact time on Ni^{2+} adsorption.

The kinetics parameters of adsorption kinetics models were presented in Table 1. The $q_{e,calc}$ from the pseudo second order model showed good agreement with the experimental values with $q_{e,exp}$ with correlation coefficient (R^2) of 0.92.

Table 1. Adsorption kinetics parameters of DSB550 biochar.

Pseudo first order model		**Pseudo second order model**		**Intraparticle diffusion model**	
k_1	0.004	k_2	0.234	k_p	0.003
q_{exp}, mmol/g	0.090	q_{exp}, mmol/g	0.090	C	0.03
q_{calc}, mmol/g	0.057	q_{calc}, mmol/g	0.089	q_{calc}, mmol/g	0.087
R^2	0.84	R^2	0.99	R^2	0.87

3.4 Equilibrium isotherms for Ni^{2+} adsorption

The adsorption performance of an adsorbent is a function of the initial metal ion concentration [17]. The amount of Ni^{2+} adsorbed onto DSB550 biochar increased with increasing initial Ni^{2+} concentrations and finally reached to an upper adsorption limit as shown in Figure 4. This can be attributed to the availability of vacant sorption sites on the surfaces of the biochar [17]. With increasing metal concentration, the adsorption of Ni^{2+} decreased because of the saturation of the sites.

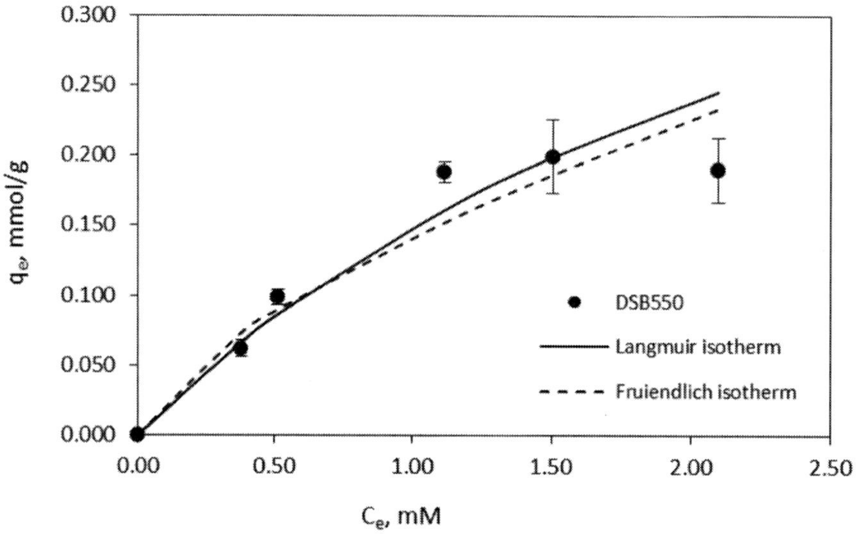

Fig. 4. Equilibrium isotherms of Ni^{2+} adsorption onto DSB550 biochar

The adsorption isotherms were characterized using Langmuir and Freundlich isotherms models. Langmuir isotherm (1918) model assumes that the surface of the adsorbent is homogeneous with monolayer sorption onto surface and has finite number of identical sites. Langmuir isotherm is given by equation [20]:

$$q_e = \frac{q_m K_L C_e}{1 + K_L C_e} \tag{8}$$

This can be arranged to the linear form as shown below:

$$\frac{1}{q_e} = \frac{1}{q_m} + \frac{1}{K_L q_m C_e} \tag{9}$$

where, q_e is the amount of Ni^{2+} adsorbed (mmol g^{-1}), C_e is the equilibrium concentration of the adsorbate (mM), q_m is the maximum adsorption capacity (mmol g^{-1}), and K_L is the adsorption equilibrium constant (L mmol $^{-1}$). The empirical Freundlich isotherm model (1906) assumes the surface of adsorbent is heterogeneous with multilayers and it is expressed as shown in equation below:

$$q_e = K_F C_e^{\frac{1}{n}} \tag{10}$$

A linear form of the Freundlich model is given as:

$$\ln q_e = \ln K_F + \frac{1}{n} \ln C_e \tag{11}$$

where, K_F (mmol g^{-1}) and n (g L^{-1}) are the Freundlich constants related to adsorption capacity and adsorption intensity, respectively. The adsorption isotherm parameters of Langmuir and Freundlich isotherms are given in Table 2. The dimensionless separation factor (R_L) of Langmuir is used to predict the affinity between the metal ion solution and the biochar. R_L can be calculated by the following equation:

$$R_L = \frac{1}{1 + K_L C_o} \tag{12}$$

where, C_o is the highest initial metal concentration in the solution (mM). This is often used to judge whether a sorption process is a thermodynamically favorable process or not: $R_L > 1$ it is an unfavorable adsorption; $R_L = 1$, indicates a linear adsorption; when $0 < RL < 1$, adsorption is favorable; when $R_L = 0$, it is an irreversible adsorption [21]. As shown in Table 2 the value of RL for this study was 0.437 which indicates a favorable adsorption. The adsorption isotherm of Ni^{2+} was adequately fitted to Langmuir isotherm with R^2 equal to 0.94. According to Freundlich model, the value of adsorption intensity ($1/n = 1.47$) which was higher than 1.0 thus confirming a favorable adsorption and Ni^{2+} adsorption is more of chemisorption than physical adsorption [22].

Table 2. Calculated parameters of the adsorption isotherms models

Langmuir isotherm model		**Freundlich isotherm model**	
k_L L mmol $^{-1}$	0.322	k_F mmol g^{-1}	0.141
R^2	0.94	R^2	0.88
q_{max}, mmol/g	0.609	1/n	1.47

The adsorption uptake capacities for Ni^{2+} ions on various adsorbent were shown in Table 3. The adsorption capacity of biochar derived from date stone was higher than that reported by Kılıç et al., [17] and Vijayaraghavan et al., [23]. However, the comparison with other reported adsorbent is difficult due to the varying experimental conditions employed in those studies such as initial concentrations and the adsorbent doses.

Table 3. Adsorption capacity toward Ni^{2+} from aqueous solution

Adsorbent	**pH**	**q_{max} mg g^{-1} q_{max} (mmol g^{-1})**	**References**
Seaweeds	4.5-4	20.63 (0.351)	[23]
Waste apricot	5	101.01 (1.72)	[24]
Almond shell biochar	7	22.22 (0.378)	[17]
Hickory wood biochar Modified hickory wood biochar	-	0.24 (0.004) 0.89 (0.015)	[25]
Date seed biochar	6.0	35.74 (0.609)	This study

4 Conclusion

In this study, biochar prepared from date seed biomass was used for Ni^{2+} adsorption from aqueous solutions. Biochar prepared at higher pyrolysis temperatures (DSB550) showed higher adsorption capacity for Ni^{2+}. The amount of Ni^{2+} ion adsorbed was found to vary significantly with process parameters such as initial Ni^{2+} concentration, solution pH, and contact time. The maximum adsorption capacity was 0.609 mmol/g. The Langmuir isotherm model showed a good fit ($R^2 = 0.94$). Adsorption kinetics showed good agreement of the experimental data with the pseudo-second order kinetic reaction model ($R^2=0.99$)

indicating a favorable adsorption and that Ni^{2+} adsorption is more of chemisorption than physical adsorption. The high adsorption capacity of biochar towards Ni^{2+} ions suggest that biochar has the potential to be an economical and efficient sorbent for treatment of wastewaters contaminated with Ni^{2+} ion.

The authors would like to thank Miss Jasmine Rossler for her assistance in conducting the experiments.

References

1. L. Järup, Br Med Bull, **68**,167-182, (2003)
2. Y.C. Sharma, *A guide to the economic removal of metals from aqueous solutions* (John Wiley & Sons, 2012)
3. D. Mohan, A. Sarswat, Y.S. Ok, C.U. Pittman, Bioresource Technol., **160**, 191-202, (2014)
4. J. Lehmann, M.C. Rillig, J. Thies, C.A. Masiello, W.C. Hockaday, D. Crowley, Soil Biol Biochem., **43**, 1812-1836, (2011)
5. X. Chen, G. Chen, L. Chen, Y. Chen, J. Lehmann, M.B. McBride, A.G. Hay, Bioresource Technol, **102**, 8877-8884, (2011)
6. Food and Agriculture Organization of the United Nations, http://www.fao.org/statistics/en/ (2012)
7. T. Ahmad, M. Danish, M. Rafatullah, A. Ghazali, O. Sulaiman, R. Hashim, M.N.M. Ibrahim, Environ. Sci. Pollut. Res., **19**, 1464-1484, (2012)
8. Y.A. Alhamed, H.S. Bamufleh, Fuel, **88**, 87-94, (2009)
9. B.S. Girgis, A.N.A. El-Hendawy, Microporous Mesoporous Mater, 52, 105-117, (2002)
10. C. Bouchelta, M.S. Medjram, O. Bertrand, J.P. Bellat, J Anal. Appl. Pyrol., 82, 70-77, (2008)
11. Z. Mahdi, A. El Hanandeh, J.Y. Qimimg, *EPPM2015,* (2015)
12. Z. Mahdi, A. El Hanandeh, J.Y. Qiming, Waste Biomass Valor, 1-13, 2016)
13. S. Wang, (Doctoral dissertation, UNIVERSITY OF FLORIDA, 2014)
14. R. M. Ali, H.A. Hamad, M.M. Hussein, G.F. Malash, Ecol. Eng., **91**, 317-332, (2016)
15. F.M. Pellera, A. Giannis, D. Kalderis, K. Anastasiadou, R. Stegmann, J.Y. Wang, E. Gidarakos, J. Environ. Manage, **96**, 35-42, (2012)
16. N.T. Abdel-Ghani, M. Hefny, G.A. El-Chaghaby, Int. J. Environ Sci. & Tech., **4**, 67-73, (2007)
17. M. Kılıç, Ç. Kırbıyık, Ö. Çepelioğullar, A.E. Pütün, Appl. Sur. Sci., **283**, 856-862, (2013)
18. Q. Cheng, Q. Huang, S. Khan, Y. Liu, Z. Liao, G. Li, Y.S. Ok, Ecol. Eng., **87**, 240-245, (2016)
19. L. Sun, D. Chen, S. Wan, Z. Yu, Bioresource Technol., **198**, 300-308, (2015)
20. M. Zhang, Chem. Eng. J, **172**, 361-368, (2011)
21. F.C. Wu, R.L. Tseng, R.S. Juang, J. of hazard. Mater., **69**, 287-302, (1999)
22. A. El Hanandeh, R.A. Abu-Zuryk, I. Hamadneh, A.H. Al-Dujaili, Water Sci. Technol. **74**, 1899-1910, (2016)
23. K. Vijayaraghavan, J. Jegan, K. Palanivelu, M.Velan, Sep. Purif. Technol., **44**, 53-59, (2005)
24. S. Erdoğan, Y. Önal, C. Akmil-Başar, S. Bilmez-Erdemoğlu, Ç. Sarıcı-Özdemir, E. Köseoğlu, G. Icduygu, Appl. Surf. Sci., **252**, 1324-1331, (2005)
25. Z. Ding, X. Hu, Y. Wan, S. Wang, B. Gao, J. Ind. Eng. Chem., **33**, 239-245, (2016)

Comparative assessment between GR model and tank model for rainfall-runoff analysis using Kalman filter- application to Algerian basins-.

Mohamed Amireche[1,*], *Tarek* Merabtene[2], *Abdelmalek* Bermad[3], and *Djamel* Boutoutaou[4]

[1]University of Oum El Bouaghi, Department of hydraulic, 04000 Oum El Bouaghi, Algeria
[2]University of Sharjah, Department of Civil and Environmental engineering, Sharjah, UAE
[3]ENPolytechnic , Department of hydraulic,10 street Hassan.Badi- El harrach-Algeirs, Algeria
[4]University of Ouargla, Department of civil engineering, 30000 Ouargla, Algeria.

Abstract. Modeling the rainfall-runoff relationship with conceptual models has always been a fascinating subject for hydrologists in view of its practical importance and complexity. This study presents a comparative assessment of the performance of two well established rainfall-runoff conceptual models. A first model called: Model 'Genie Rural' (i.e. Agricultural Engineering) and abbreviated GR, developed by Cemagref has been extensively tested in the Mediterranean watersheds and some basins in African countries. When applied to Algerian basins, the different version of the GR model gave satisfactory results, particularly for long time steps (monthly and annual data). In this work, the tank Model by Sugawara using Kalman filter for adaptive calibration is developed and tested for the first time to assess rainfall-runoff in Algerian basins. The results appear to be very prominent and far better than those given by the GR models including daily time steps. Indeed, a comparison between the two models established for daily and monthly data was performed on the three (03) Algerian Basins (i.e. Isser, Zardezas basin and Cheffia). Calibration of the Tank model parameters was performed by Kalman filter. Furthermore, the structure of tank model (i.e: number of tanks, number of outlets in each tank, and their location) was determinated for the studied basins.

1 Introduction

The rainfall-runoff relationship is undeniably useful in the field of hydrology and water projects. The design of structures and dam management can't be achieved without the use of rain and runoff parameters. Conceptual models have taken a large part in modeling this relationship. The "tank model" Sugawara [1,2] is one of these methods. Indeed, the major difficulty in this model is the calibration of parameters in order to achieve a good estimate of the observed data.

*Corresponding author: amiremoh@hotmail.com

2 Formulation

2.1 Tank model

Since its introduction in early 1950's by Sugawara [3], the tank model has found growing and wide applications in the field of hydrology including flood control and groundwater modeling. Figure 1 introduces the general structure of a four-stage tank model widely applied in catchment characterization. The tank model is a conceptual rainfall-runoff model characterized by several interconnected storage units, called tanks, to account for surface, intermediate, and groundwater runoff components.

The first tank usually consists of two side outlets and infiltration outlet, while the tanks of intermediate and groundwater zones are equipped with one side and one infiltration outlet, respectively. The runoff components from the outlets are usually expressed as a linear function of the storage amount, i.e., water level in a tank. Whenever a linear relation does not generate satisfactory result a simple square relation can also be used (Parsad, 1967; Jermar, 1987 , [4,5]).

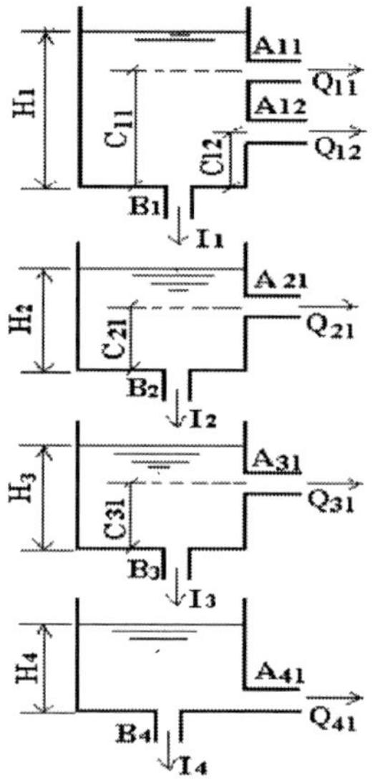

Fig. 1- Four-stage tank model.

The discharge, i.e., runoff component, from a side outlet is defined by:

$$Q_{ij}(t) = A_{ij}\,(H_i(t) - C_{ij})^{\gamma}\; U(H_i(t) - C_{ij}) \tag{1}$$

where γ=1 for linear approximation, Qij the runoff height component from side-outlet j of tank i (unit of length/time), Aij the runoff coefficient of side-outlet j of tank i (unit of 1/time), and Cij the height of side-outlet j of tank i (unit of length). As for the bottom tank, the value of Cij (i.e., C41) usually equals 0, whereas the side-outlet represents the continuous base flow.

The infiltration to the lower tank, or deep percolation in case of the bottom tank expressed by:

$$I_i = B_i\ H_i \tag{2}$$

where *Ii* is the infiltration from tank i to lower tank i+1 in height (unit length/time), and *Bi* the infiltration outlet coefficient.

The total discharge *QT(t)*, affected by a given observation noise *w(t)* for each time period *t*, is given by:

$$QT(t) = \sum_{i=1}^{N} \sum_{j=1}^{J_i} Q_{ij} + w(t) \quad (3)$$

In control theory Eq. (3) is termed the observation equation, where *i* is the tank index *(i=1,...,n=4)*, and *j* the side-outlet index in tank *i (j=1,..,Ji)*. According to the schematic structure of the four-stage model given in Figure 1, *Ji*=2 if *i*=1, and *Ji*=1 otherwise (*i*=2,...,4).

The system equation of the model is formulated by the equation system representing the changing of the water level in each tank by considering a given system noise *u(t)*:

$$\frac{dH_1}{dt} = \left(Rain(t) - Evt(t) - Q_{11}(t) - Q_{12}(t) - I_1(t)\right) + u_1(t) \quad \text{for i=1; j=1.}$$
$$\frac{dH_i}{dt} = \left(I_{i-1}(t) - Q_{ij}(t) - I_i(t) + u_i(t)\right) \quad \text{for i=2,3 or 4th tank, j=1.} \quad (4)$$

where *Rain(t)* and *Evt(t)* is the rainfall and evapotranspiration in [length/time], respectively. The parameters *Aij, Bi, Cij,* and water depth *Hi* are the unknown variables of the tank model to be calibrated. The developed methodology integrates Kalman filtering technique to optimize the model parameters in adaptive mode as discussed later.

2.2 Kalman Filter

Kalman filtering is the most commonly applied method for estimating the state of a linear system with known system dynamics and known statistics distribution of the assumed Gaussian error (i.e., noise), and it was developed by Kalman in early 1960's (Athans et al., 1968; Balakrishnan, 1984) [6,7] .

The system state and observation equations of the Kalman filtering are expressed in discrete form as:

System state equation: $$X(k+1) = \Phi(k)X(k) + \alpha(k) + u(k) \quad (5)$$

Observation equation: $$Y(k) = H(k)X(k) + \beta(k) + w(k) \quad (6)$$

Where, at each time step *k*, *X(k)* is the system vector to be estimated, *H(k)* and *Φ(k)* the known transition matrices, *α(k)* the known constant vector, *u(k)* the white Gaussian system noise vector, *Y(k)* the observed vector, *β(k)* known constant vector, and *w(k)* the white Gaussian observation noise vector. Details on the Kalman filtering procedure can be found in, e.g., Kitanadis and Bras (1980) [8], Anderson and Burt (1985) [9], and Bras and Rodriguez-Iturbe (1985) [10] .

2.3 Formulation of tank model by Kalman filtering:

In application to the tank model [11,12], the system state equation and observation equation (Eqs. 5 and 6) are given by Eqs. (3) and (4). The state system vector X for the four-stage tank (**Figure 1**) model is given by:

$$\mathbf{X} = [x_1 \ x_2 \ \ldots \ x_{17}]^T$$
$$= [H_1 \ H_2 \ H_3 \ H_4 \ A_{11} \ A_{12} \ A_{21} \ A_{31} \ A_{41} \ B_1 \ B_2 \ B_3 \ B_4 \ C_{11} \ C_{12} \ C_{31} \ C_{41}]^T$$

(7)

If we define by $F(X)$ a vector function of the state system vector X given by:

$$\mathbf{F} = [f_1, f_2, \ldots, f_{17}]^T \tag{8}$$

the transition equation $\frac{dX}{dt}$ representing the system dynamics of X may be expressed as:

$$\frac{dX}{dt} = F(X) \tag{9}$$

In the specification of the four-stage tank model system state equation (Eq.4), the discrete form of (Eq. 5) is given by:

$$\begin{aligned}
f_1(X) &= Rain(t) - Evt(t) - x_5(x_1 - x_{14})U(x_1 - x_{14}) - x_6(x_1 - x_{15})U(x_1 - x_{15}) - x_{10}x_1 + u_1(t) \\
f_2(X) &= x_{10}x_1 - x_7(x_2 - x_{16})U(x_2 - x_{16}) - x_{11}x_2 + u_2(t) \\
f_2(X) &= x_{11}x_2 - x_8(x_3 - x_{17})U(x_3 - x_{17}) - x_{12}x_3 + u_3(t) \\
f_4(X) &= x_{12}x_3 - x_9x_4 - x_{13}x_4 + u_4(t) \\
f_i(X) &= 0 \quad (i=5, 6, \ldots, 17)
\end{aligned} \tag{10}$$

2.4 Formulation of GR model:

2.4.1 GR2M model:

The GR2M model (i.e :2-parameters model of Rural Engineering) is a global model Rainfall-runoff with two parameters for monthly scale. Its development was initiated at Cemagref in the late 1980s, with application targets in the field of water resources and low flows periods.

This model has known several versions, successively proposed by Kabouya (1990), Kabouya and Michel (1991), Makhlouf (1994), Makhlouf and Michel (1994),Mouelhi (2003), Mouelhi and al. (2006b), which have gradually improved the performance of the model.

Its structure, although empirical, class it in the group of conceptual tanks models with a procedure for monitoring the moisture state of the basin which appears to be the best for taking into account the previous conditions and to ensure continuous operation of the model. Its structure combines a production tank with a tank of Routing as well as an opening on the outside other than the atmospheric environment. These Three functions simulate the hydrological behavior of the basin [13].

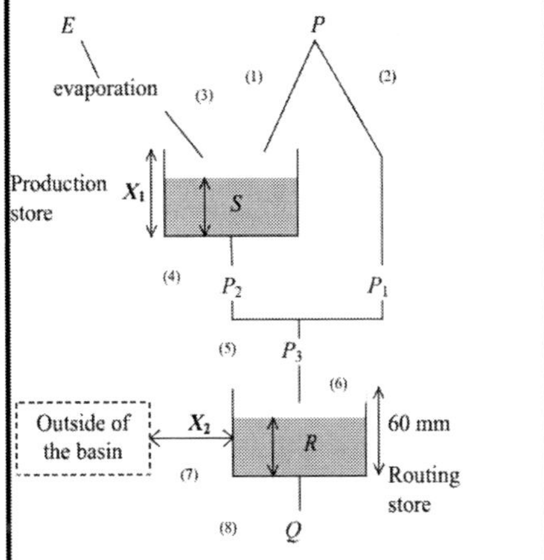

(1) $$S_1 = \frac{S + X_1\varphi}{1 + \varphi\frac{S}{X_1}} \quad \text{with} \quad \varphi = \tanh\left(\frac{P}{X_1}\right)$$

(2) $$P_1 = P + S - S_1$$

(3) $$S_2 = \frac{S_1(1-\psi)}{1 + \psi\left(1 - \frac{S_1}{X_1}\right)} \quad \text{with} \quad \psi = \tanh\left(\frac{E}{X_1}\right)$$

(4) $$S = \frac{S_2}{\left[1 + \left(\frac{S_2}{X_1}\right)^3\right]^{1/3}} \qquad P_2 = S_2 - S$$

(5) $$P_3 = P_1 + P_2$$

(6) $$R_1 = R + P_3$$

(7) $$R_2 = X_5.R_1$$

(8) $$Q = \frac{R_2^2}{R_2 + 60} \qquad R = R_2 - Q$$

Fig. 2 - Architecture of GR2M model.

The model has two optimizable parameters :
X1 : capacity of tank production (mm)
X2 : exchange underground coefficient (-).
On a large sample of watersheds, the values given in Table 1.

Table 1 - Values of GR2M parameters obtained on a large sample of watersheds.

Parameter	Median	At confidence interval (90%)
X1(mm)	380	140- 2640
X2(-)	0.92	0.21 -- 1.31

2.4.2 GR4J Model:

The GR4J model (i.e : 4 parameters Rural Engineering model for daily scale) is a global model Rainfall-runoff with four parameters. Its development was initiated at Cemagref In the early 1980s, with a view to developing a Robust and reliable rainfall-runoff simulation for applications for Management of water resources and engineering (dimensioning of works, forecasting of floods and low flows, management of tanks, impact detection, etc.).

This model has had several versions, successively proposed by Edijatno and Michel (1989), Edijatno (1991), Nascimento (1995), Edijatno and al. (1999), Perrin (2000), Perrin (2002) and Perrin and al. (2003) which have improved the performance of the model [13].

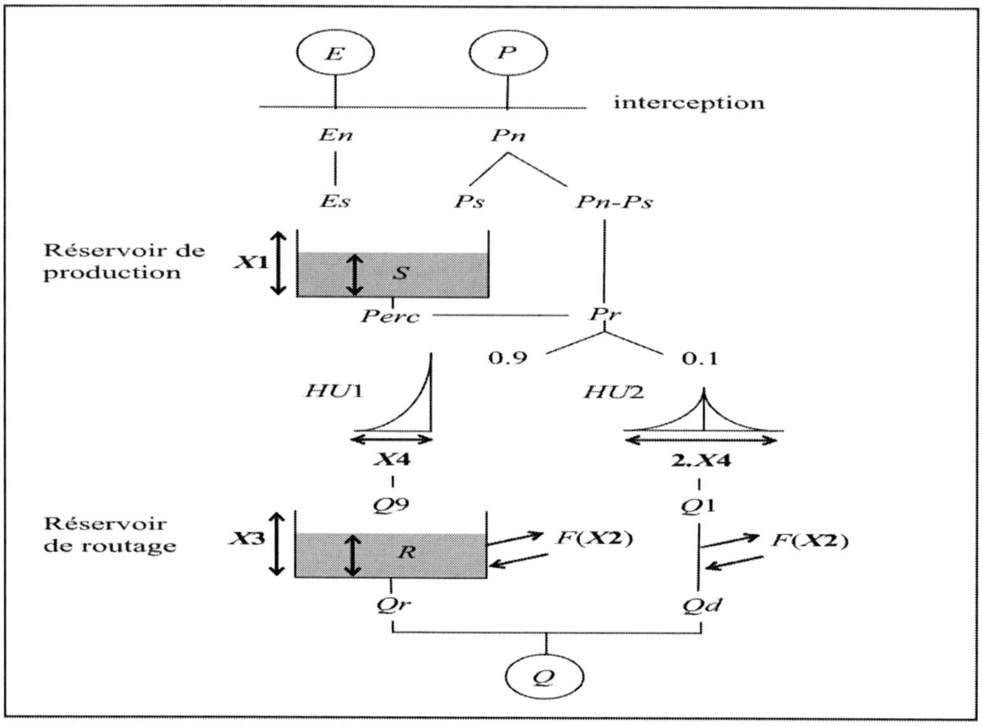

Fig. 3 - Architecture of GR4J model.

GR4J model has four parameters to be calibrated :
*X*1: capacity of production tank (mm). *X*2: exchange underground coefficient (mm)
*X*3: daily capacity of routing tank (mm)
*X*4: base time of hydrograph unit *HU*1 (day)
On a large sample of watersheds, the values given in Table 2, [13] :

Table 2 - Values of GR4J parameters obtained on a large sample of watersheds.

Parameter	Median	At confidence interval (80%)
X1(mm)	350	100 -- 1200
X2(mm)	0	-5--3
X3(mm)	90	20--300
X4(day)	1.7	1.1--2.9

3 Studied basins presentation

The input data used are the daily data of rainfall and concomitant hydrometric and evapotranspiration data (ETP). The data of rainfall and hydrometric issued by ANRH (the National Agency of Water Resources) and the ANBT (National Agency of Dams and Transfers).

In the basin of the Cheffia, we will use the data for daily evapotranspiration from the city of Annaba, calculated by the method of Penmann. For Zardezas basin they are are calculated by a Colorado tray (coefficient K = 0.7) [14]. For Isser catchement we used the formula Oudin (2004), [15].

Table 3. Morphological characteristics of the studied watersheds.

Basin	Cheffia (03-15)	Isser (09-04)	Zardezas (03-09)
Area (km2)	575	1044	345
Coefficient Kc	1.20	1.25	1.26
Average Altitude (m)	420	665	628
Roche slope index	0.11	0.11	0.17

Table 4- hydrometric stations studied.

Stations (code)	Basin	Lambert coordinates	Installation date	Area (km^2)
Cheffia(031501)	Cheffia	X=977.70 -Y=381.32	1978	575 km^2
Lakhdaria (090501)	Isser	X=579.10 - Y=368.75	1954/65	1044 km^2
Zardezas (030902)	Zardezas	X=878.75 - Y=370.67	1989	345 km^2

Table 5 Period of observations used in this study.

Rainfull stations(code)	Hydrometric Stations	Daily observations Periods	Calibration/ validation -Daily	Calibration/ validation -Monthly
Chefia (031501)	Cheffia(031501)	1986-1990	03 years / 2 years	1978-1986/ 1987-1991
Isser(090403)+ (090413)+(090419)	Isser (090501)	1992-1994	02 years / 1 year	1980-1992/ 1993-1996
Zardezas (030903)	Zardezas (030902)	1990-1995	04 years / 02 years	1990-1993/ 1994-1996

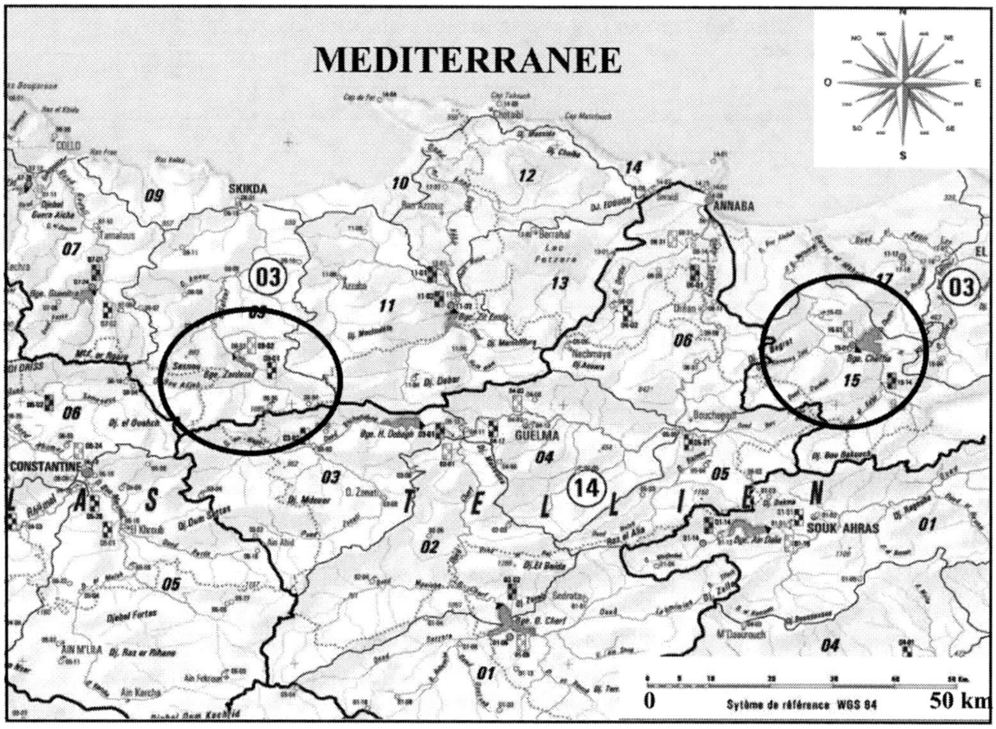

Fig. 4- Position of Zardezas (code :03 09) and Cheffia (code:03 15) Algerian basins. .

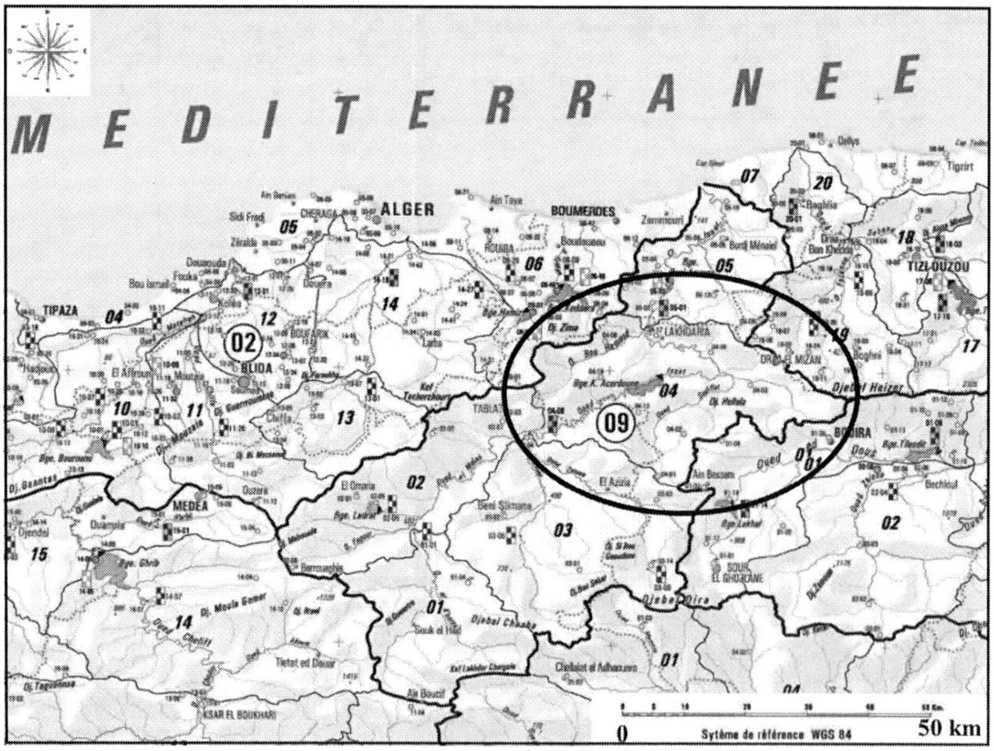

Fig. 5- Position of Isser (code: 09 04) Algerian basin.

4 Results and comments

4.1 Definition of validation criteria

To assess the quality of our results we calculated the following numbers: Nash Sutcliffe *(N.S.E)*, R^2 . These parameters are defined in the following: Considering that:. qiobs: observed flow, qisim: simulated flow, q0obs: the average of the observed flows , q0sim : the average of the simulated or estimated flows [16,17] .

- *N.S.E : Nash Sutcliffe number:* $$N.S.E = 1 - \left[\frac{\sum_n^1 (qiobs - qisim)^2}{\sum_n^1 (qiobs - q0obs)^2}\right] \quad (11)$$

- Determination coefficient: $$R^2 = \left[\frac{\sum_n^1 (qisim - q0sim)(qiobs - q0obs)}{\sqrt{\sum_n^1 (qisim - q0sim)^2} \cdot \sqrt{\sum_n^1 (qiobs - q0obs)^2}}\right]^2 \quad (12)$$

4.2 Results and interpretations

4.2.1 Not-updating approach

In this approach the structural parameters of the tank are fixed, only the dynamic parameters change, in our case the water levels in the tanks are (H1, H2, H3 and H4).

4.2.2 Results Summary

Table 6 - Summary results of all models studied.

Basin	Step	GR4J		Tank by Kalman filter	
		Nash	R^2	Nash	R^2
Isser	Calibration	0.72	0.73	0.97	0.97
	Validation	0.43	0.44	0.91	0.91
Cheffia	Calibration	0.71	0.72	0.98	0.98
	Validation	0.78	0.82	0.99	0.99
Zardezas	Calibration	0.81	0.81	0.998	0.998
	Validation	0.74	0.82	0.99	0.99
Basin	Step	GR2M		Tank by Kalman filter	
		Nash	R^2	Nash	R^2
Isser	Calibration	0.28	0.37	0.82	0.83
	Validation	0.11	0.20	0.97	0.97
Cheffia	Calibration	0.75	0.76	0.96	0.96
	Validation	0.74	0.88	0.96	0.96
Zardezas	Calibration	0.83	0.82	0.78	0.80
	Validation	0.58	0.83	0.95	0.95

DOI: 10.1051/matecconf/201712005006

4.2.3 Daily results

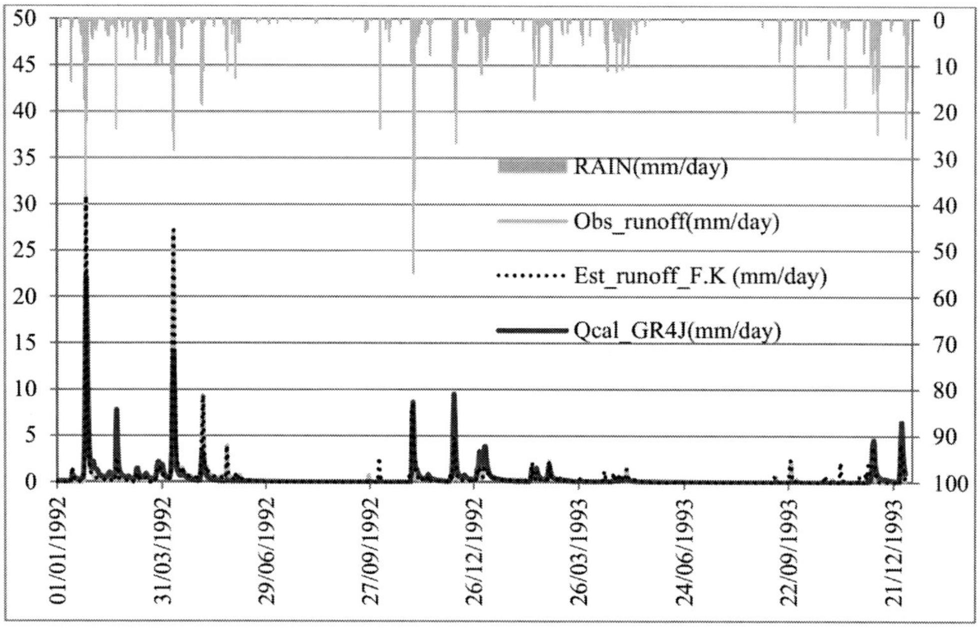

Fig. 6- Runoff with tank model by Kalman filtering Model and GR4J -Calibration - Isser basin.

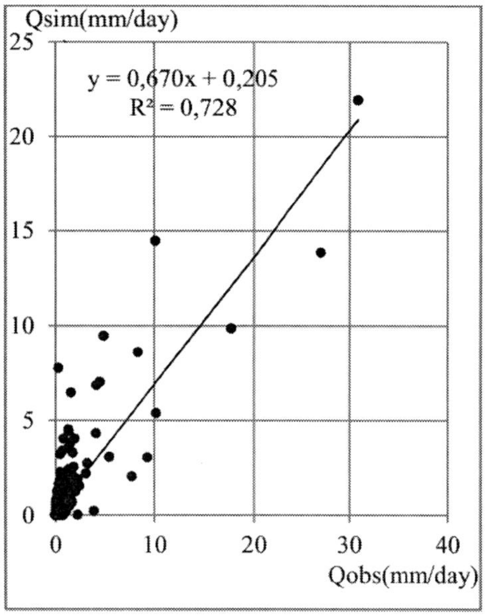

Fig. 7- Simulated runoff as a function of observed runoff with GR4J model-Calibration of Isser basin.

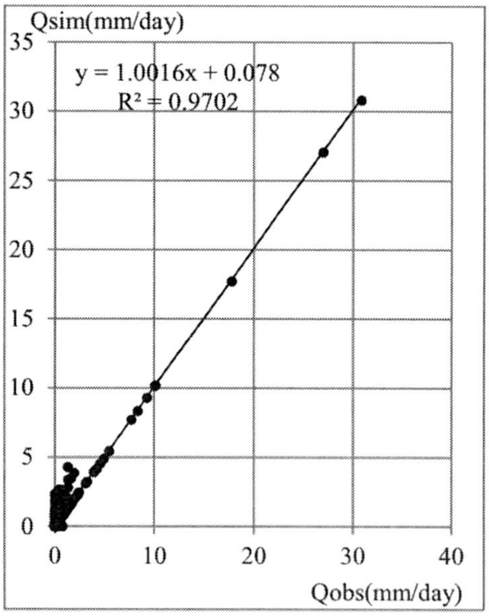

Fig. 8- Estimated runoff as a function of observed runoff with tank model by Kalman filtering- Calibration of Isser basin.

DOI: 10.1051/matecconf/201712005006

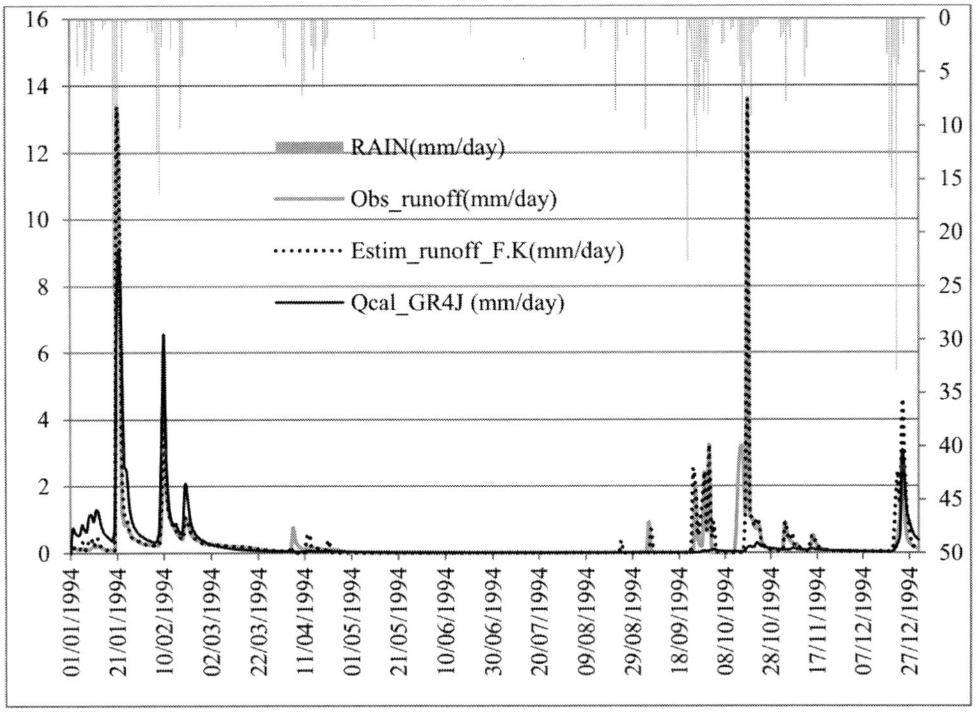

Fig. 9- Runoff with tank model by Kalman filtering Model and GR4J -Validation- Isser basin.

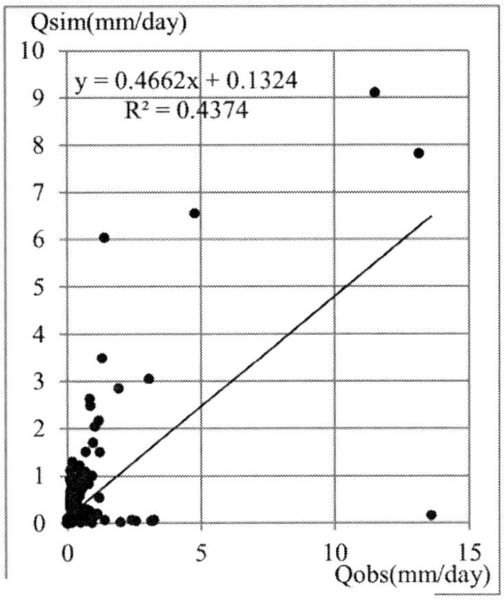

Fig. 10- Simulated runoff as a function of observed runoff with GR4J model- Validation of Isser basin.

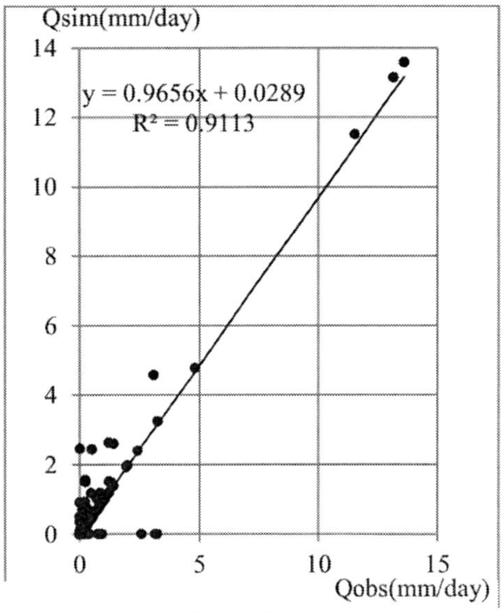

Fig. 11- Estimated runoff as a function of observed runoff with tank model by Kalman filtering- Validation of Isser basin.

4.2.4 Monthly results

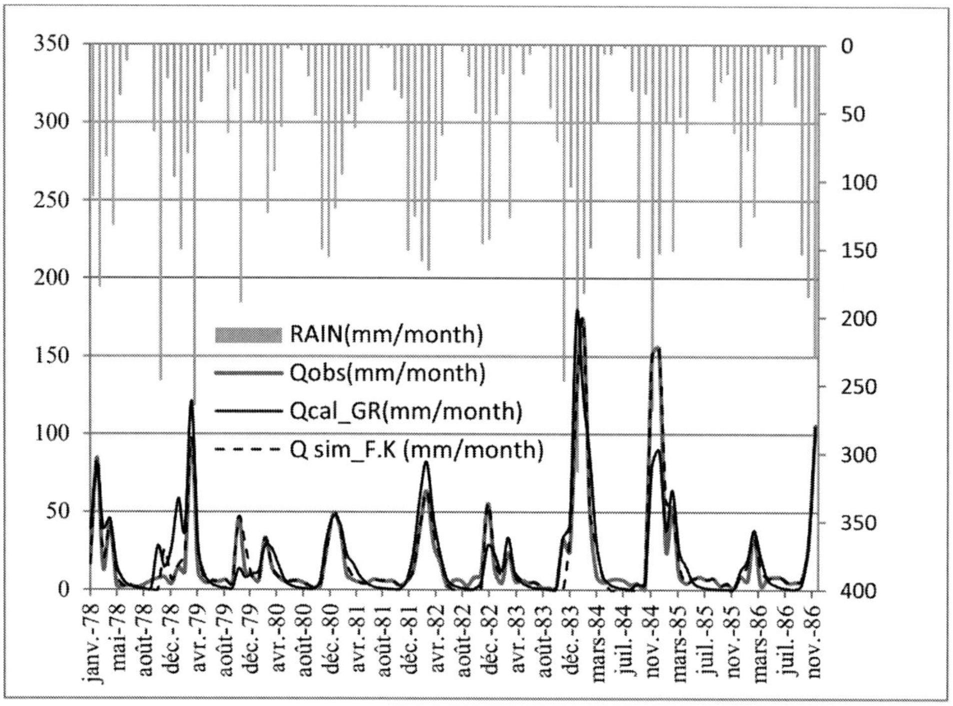

Fig. 12- Runoff with tank model by Kalman filtering Model and GR2M -Calibration - Cheffia basin.

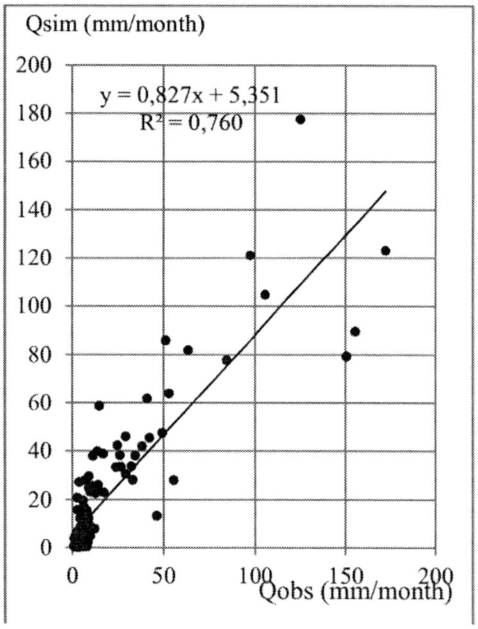

Fig. 13- Simulated runoff as a function of observed runoff with GR2M model-Calibration of Cheffia basin.

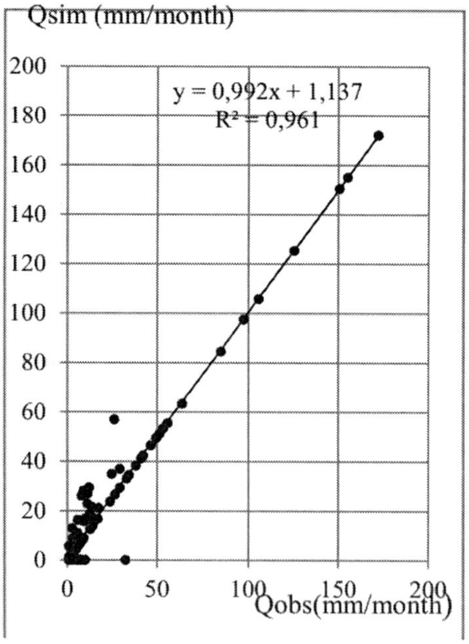

Fig. 14- Estimated runoff as a function of observed runoff with tank model by Kalman filtering- Calibration of Cheffia basin.

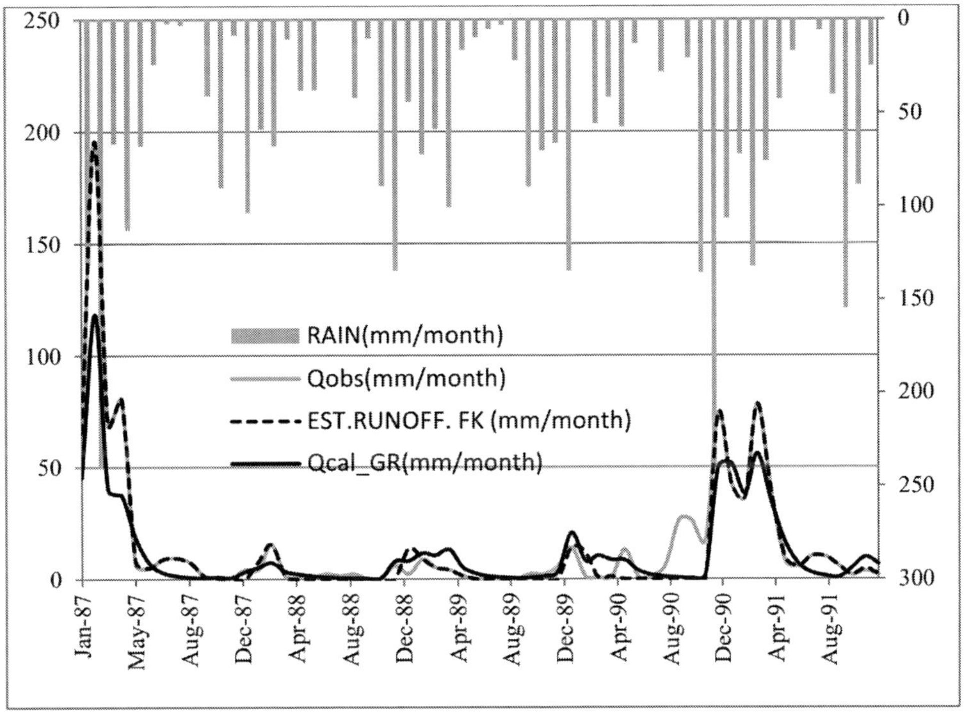

Fig.15- Runoff with tank model by Kalman filtering Model and GR2M-Validation- Cheffia basin.

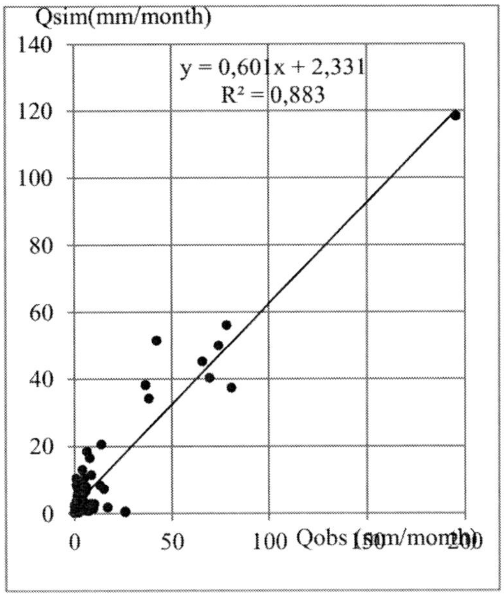

Fig. 16- Simulated runoff as a function of observed runoff with GR2M model-Validation of Cheffia basin.

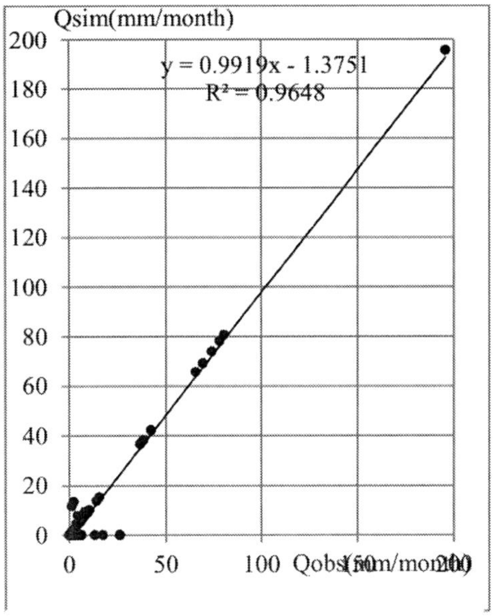

Fig. 17- Estimated runoff as a function of observed runoff with tank model by Kalman filtering- Validation of Cheffia basin.

5 Adopted models

In previous results after , the tank model by Kalman filter with the approach "Not_updating" seems best for modeling our three catchments. We have thus reduced the number of tank to 2 upper tanks which reduces the number of parameters to 10: (A11, A12, C11, C12, A21, C21, B1, B2, H1, H2).

The first 08 parameters are fixed and present the structural parameters of the Tank, and for a watershed given the variable parameters are only reduced to levels : H1 and H2, this greatly simplifies the problem and is in itself a very useful results.

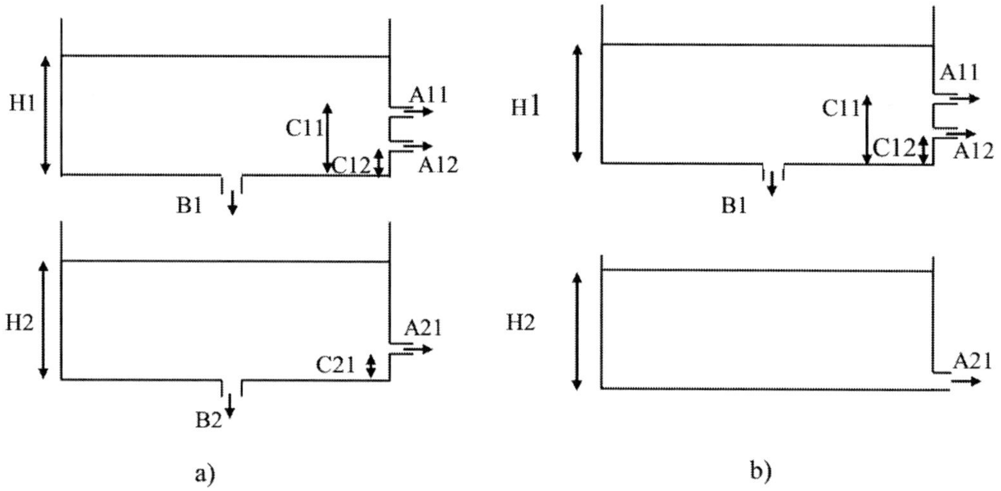

Fig. 18 - Adopted simplified structures in modeling of the three basins.

a)-Daily model ; b) Monthly model

We found a similarity between the tanks of the GR model and those of the adopted tank model from the present study. This similarity appears in the number of tanks as well as in the meaning of outlet's discharge.

6 Conclusion

Through this study we were able to model three (03) basins by GR and tank model using Kalman filter for two different time scales: daily and monthly. The graphical analysis and that of the validation parameters (Nash and R^2) showed that the modeling results by the tank model using Kalman filter are better than those given by the GR models on these Algerian basins.

In view of their simplicity and practicality, especially for manual calibration, GR models have been widely used in North Africa and particularly in Algeria. The excellence of the results given by tank model using Kalman filter leads us to take more interest in these models in this region of the world, especially, for the prediction, simulation, filling and correction of hydrometric series.

References

1. Sugawara, (1961): An analysis of runoff structure about several Japanese rivers", J. of Geophys., JSCE, 2 (1961).

2. Sugawara,M : Automatic calibration of the tank model. Hydrological Sciences, **24**(3), pp375-388 (1979).
3. Sugawara, M: Runoff analysis, Kyoritsu Edition (1974).
4. Prasad, R: A non-linear hydrologic system response model, J. of the Hydr. Div., ASCE, **4**, pp. 201-221, (1967).
5. Jermar M.K: Water resources and water management, Development in water science, **28**,pp. 33-40, (1987).
6. Athans M., Wishner R.P. and Bertolini A. : Suboptimal state estimation for continuous-time nonlinear systems from discrete noisy measurement , IEEE Transaction on automatic control, **13(5)**, (1968).
7. Balakrishnan, N., and Joshi ,P.CProduct moments of order statitistics from doubly truncated exponential distribution, Naval research Logistics Quarterly , **31**, pp 27-31. (1984). .
8. Kitanidis, P.K., and Bras, R.L.,: Real-time forecasting with a conceptual hydrologic model, 1.Analysis of uncertainty, Water Resour. Res., **16(6)**, pp. 1025-1033(1980).
9. Anderson, M.G., and Brut, T.P. Hydrological forecasting, Wiley-Interscience Pub., (1985).
10. Bras, R.L., and Rodrigueze-Iturbe, I.,Random Functions and Hydrology, Addison-Wesley Pub (1985).
11. Merabtene, T., Jinno, K., Kawamura, A., and Matsunaga, T., Interactive user interface for rainfall-runoff analysis by tank model, Memoirs of the Faculty of Engineering, Kyushu University, **57(3)**, pp. 107-120 (1997a).
12. Merabtene, T., Jinno, K., and Kawamura, A.,: Decision support system for the water resources management, Proc. of the 5th Symposium on Water Resources of the Japan Society of Civil Engineers, Tokyo, pp. 461-466 (1997b).
13. C. Perrin, C.Michel, V.Andréassian: Modèles hydrologiques du Génie rural, Cemagref, pp3-16 (2007).
14. Benkaci Ali Tarik: Modélisation pluie-débit mensuelle et journalière par les modèles conceptuels et les systèmes neuro-flous (Application aux bassins algériens), thèse de doctorat, pp1-207, (2006).
15. Ludovic Oudin: Recherche d'un modèle d'evapotranspiration potentielle pertinent comme entrée d'un modèle pluie-débit global, Mémoire de doctorat de l'Ecole Nationale du Génie rural, des eaux et forêts, Centre de : Paris , pp 1-495(2004) .
16. Nash, J.E.& Sutcliffe, J.V: Riverflow forecasting through conceptual model,J.Hydrol **10**,pp. 282-290.(1970).
17. P.Krause, D.P.Boyle2, and F.Base :Comparison of different efficiency criteria for hydrological model assessment, Advances in Géosciences,**5**, pp. 89-97(2005).

MATEC Web of Conferences **120**, 05007 (2017) DOI: 10.1051/matecconf/201712005007
ASCMCES-17

Conversion of CruTS 3.23 data and evaluation of precipitation and temperature variables in a local scale

Sinan Jasim Hadi[1,*], and *Mustafa* TOMBUL[1]

[1]Department of Civil Engineering, Faculty of Engineering, Anadolu University Eskisehir, Turkey.

Abstract. The precipitation and temperature data obtained from the last version 3.23 of Climate Research Unit Time Series (CruTS) data are evaluated in a local scale. The data obtained for the period 1901-2014 consists of ten climate variables: Precipitation, Mean temperature, Diurnal Temperature Range, Wet-day Frequency, Vapour Pressure, Cloud Cover, Maximum temperatures, Minimum temperatures, frost day frequency, and potential evapotranspiration. These data converted to ASCII format which is usable in GIS environments. Daily precipitation and temperature data of 9 station located in Seyhan basin in Turkey for the period 1973-2000 used for the assessment after converting them to monthly total precipitation and average temperature. The assessment implemented using several statistical measurements: correlation, RMSE, mean difference, median difference, and Mann-Whitney test. The study found that CruTS has high agreement with the observed data in the meaning of the correlation and low RMSE in all stations. The result of mean, median, and Mann-Whitney tests indicate no significant difference between CruTS and the observed data in all station with the exception of two stations; Tomarza station has significant values in the temperature values in all tests; Goksun station has significance difference in the mean test in the two climate variables.

1 Introduction

Climate Research Unit Time Series (CruTS) was named based on the name of the producer of these data i.e. the Climate Research Unit, University of East Anglia, Norwich, UK. Initially these data developed by [1] for the period 1901- 1996, updated by [2] for the period 1901 - 2002, and finally updated one more time by [3] for the period 1901-2009. The data are still being added to the collection and corrected.

The last version 3.23 of CruTS datasets consists of 0.5° latitude-longitude grid cells covering the global with the exception of Antarctica region. The datasets provide monthly values of ten climate variables. Six of these variables are independent: Precipitation (pre), Mean temperature (tmp), Diurnal Temperature Range (dtr), Wet-day Frequency (wet), Vapour Pressure (vap), and Cloud Cover (cld). The remain four (i.e. Maximum

*Correspondingauthor: sinan.jasim@yahoo.com

temperatures (tmx), Minimum temperatures (tmn), frost day frequency (frs), and potential evapotranspiration (pet)) are calculated and estimated using the independent variables [4].

CruTS is routinely updated using several resources: CLIMAT monthly data for about 2400 station, for the last 4-5 years, produced by the collaboration between World Meteorological Organization (WMO) and the US National Oceanographic and Atmospheric Administration (NOAA), Monthly Climatic Data for the World (MCDW) produced by UK National Climatic Data Center (NCDC) for incorporating in WMO by 200 stations, World Weather Records (WWR) decadal data for about 1700 stations during 1991-2000 published as exchange between the archive center of NCDC and the National Meteorological Services (NMSs). The aforementioned sources provide a systematic incorporation in the updating process.

The importance of CruTS datasets has been increasing in Geosciences researches with a focus on the climate change studies due to two reasons: the long period that covering from 1901 – 2014 that give the ability for studying the trend and the change in the climate, and the high spatial resolution which helps in local not only regional studies. CruTS with its different versions have been widely used in many researches and in many applications such as trend analysis, climate change, desertification and many others. For example, but not limited to [5-10].

In order to use this data for applications, they need to be validated in different scales. [3] created the CruTS 3.10 data using Climate anomaly method (CAM) [11] and validated them in a sub-continental scales by using: temperature dataset developed by University of Delaware (UDEL) based on GHCN-M [12] for the period 1901-2008, and precipitation data developed by Global Precipitation Climatology Center (GPCC) for the period 1901-2009. They found a tight agreement between the temperature data of CruTS and UDEL.CruTS and GPCC precipitation data have high agreement also, but not as tight as temperature data.

The main objective of this study is to evaluate the last version of CruTS 3.23 released 1 September 2015 in a local scale.The dataset to be compared with the observed monthly total precipitation and mean temperature data of 9 stations located in Seyhan basin, Turkey for the period 1973 – 2000. Additionally, the CruTS 3.23 dataset for the period 1901 – 2014 for all the variables converted to GIS-based ASCII files to make them usable in GIS environments.

2 Data collection

CruTS data are originally available at Centre for Environmental Data Archival of the British Atmospheric Data Center (BADC) which can be reached by (http://www.badc.nerc.ac.uk). The data provided in two format: ASCII and NetCDF. In the same time, a temporal or spatial subset of the datasets can be obtained with CSV file format from CEDA OGC WEB SERVICES/ Web Processing Service reached by (http://wps-web1.ceda.ac.uk/).

The Consortium for Spatial Information, CSI, which is the CGIAR spatial scientists community (http://www.cgiar-csi.org/) processed the CruTS 3.10 data which covers the period 1901 to 2009 and converted them to ASCII in order to facilitate their use in GIS environment.

In this study, CruTS 3.23 data for the period 1901 – 2014 released 1 September 2015 evaluated by comparing them with the observed data. The data set collected from CEDA OGC WEB SERVICES/ Web Processing Service with CSV format. The collected data set converted to ASCII format to provide a full set contains the entire available variables and period.

Observations of 9 stations were collected from DMI (DevletMeteorolojiIsleri); the Tukrish General Directory of Meteorology. These stations located at three sub-basins called (coded): 1822, 1801, and 1805 as part of Seyhan basin located in the southwest of Turkey. The area is taking part from four Turkey's provinces: Adana, Kayseri, Kahramanmaras, and Sivas, as seen in Figure 1.

The data were collected for the period from 1/1/1973 to 31/12/2000. For every station, the collected data were total daily precipitation and average daily temperature. These data converted to monthly total precipitation and monthly average temperature to be compared, after that, with the CruTS data.

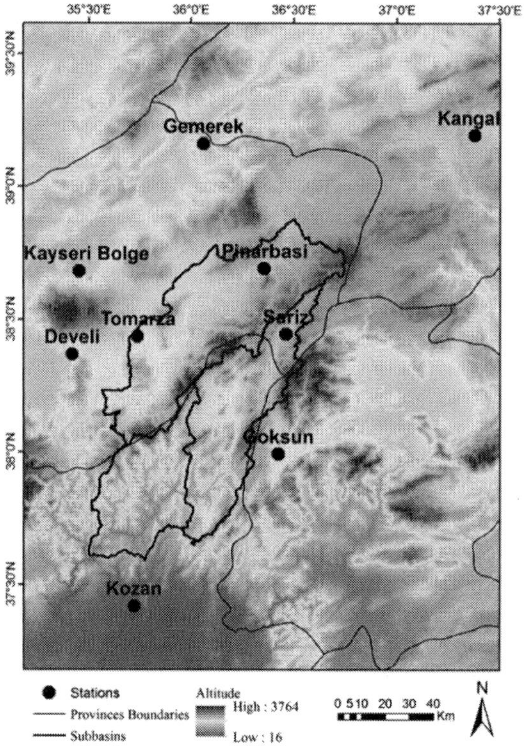

Fig. 1. Study area with the 9 stations.

3 Methodology

The obtained CruTS data of ten climate variables for the period 1901 -2014 have csv file format as one file for every month. MATLAB was used for reading files and converting them to ASCII after adding the required text to those files. The obtained ASCII files can be then easily used in any GIS software for further analysis.

The obtained files of the precipitation and temperature were then imported to R (i.e. Statistical Programming Language) for implementing the main objective of this study. CruTS data has a period from 1901 -2014 while the observed data 1973 -2000, therefore subsets of the CruTS data were extracted. The values of the CruTS ASCII files cells where the stations are located were extracted that 9 values obtained for each month which represented by one ASCII file.

The strength of the relationship between the values extracted from CruTS data and the observed data were evaluated by calculating the correlation coefficient which is given as:

$$r_{X,Y} = \frac{\sum_{i=1}^{N}(x_i - \bar{x})(y_i - \bar{y})}{\sqrt{\sum_{i=1}^{N}(x_i - \bar{x})^2}\sqrt{\sum_{i=1}^{N}(y_i - \bar{y})^2}} \quad (1)$$

Where x_i is the value of the CruTS data for month i, $y_i$is the observed value of the same month, $\bar{x}$ is the mean of the CruTS data, $\bar{y}$ is the mean of the observed data, and N the number of data which represents months in this case. The result of the correlation for every station and for the two climate parameters; precipitation and temperature shown in table 1.

In order to measure the error between the two datasets, Root Mean Squared Error (RMSE) was calculated for each station using:

$$RMSE = \sqrt{\frac{\sum_{i=1}^{N}(x_i - y_i)^2}{N}} \quad (2)$$

The result of RMSE listed in table 1.

The significance of the difference between the mean of the CruTS data μx and the mean of observed data μy was tested using a hypothesis test:

Ho: $\mu x = \mu y$

Ha: $\mu x \neq \mu y$

The test conducted for each station using t student's test:

$$t = \frac{\bar{x} - \bar{y}}{SE} \quad (3)$$

The standard error SE is calculated by:

$$SE = \sqrt{\frac{{s_x}^2}{N_x} + \frac{{s_y}^2}{N_y}} \quad (4)$$

Where s_x and s_y are the standard deviation of the CruTS and Observed data respectively [13, 14].

After calculating the t statistic, P – value calculated for every station in order to compare them with the significance level.

Some arguments saying that mean should not be used in case of the data are not normally distributed unless the number of observations large enoughwhich is the case here in the precipitation data. Although the samples used here can be considered as sufficiently large, the median test was implemented.

In addition, Non-Parametric test named Mann-Whitney (known as Rank-Sum test and Wilcoxon Rank-Sum test) was used. This test is used for checking whether one group has higher or lower values than a second group. The two-sided hypothesis of this test is:

Ho: Prob[$x > y$] = 0.5

Ha: Prob[$x > y$] $\neq$ 0.5

Where x the first group data (i.e. CruTS data) and y the second group data (i.e. the observed data). In words, if the null hypothesis rejected it means that the data of two groups are different and vice versa. To save some space, the details of median and Mann-Whitney test are not mentioned here; for detailed description see [14-17]. The P-values of mean, median, and Mann-Whitney test are shown in table 2.

4 Results and discussion

The main objective of this study is to evaluate the CruTS 3.23 data by comparing them with the observed data of 9 stations located at the southwest of Turkey. To achieve this objective and to make the CruTS data (i.e. obtained with CSV file) usable in several software, especially GIS softwares, the files of the monthly 10 climate variables mentioned earlier were converted to ASCII. An example of the obtained files (i.e. the first month January of 1973 of precipitation) was added to ArcGIS and symbolized shown in figure 2.

In order to ease the access, the ASCII files of the ten climate variables for the period 1901-2014 are available for download as a compressed file for every variable from (https://goo.gl/vFGNdp) or all the variables in one compressed file from (https://goo.gl/TXfxZm).

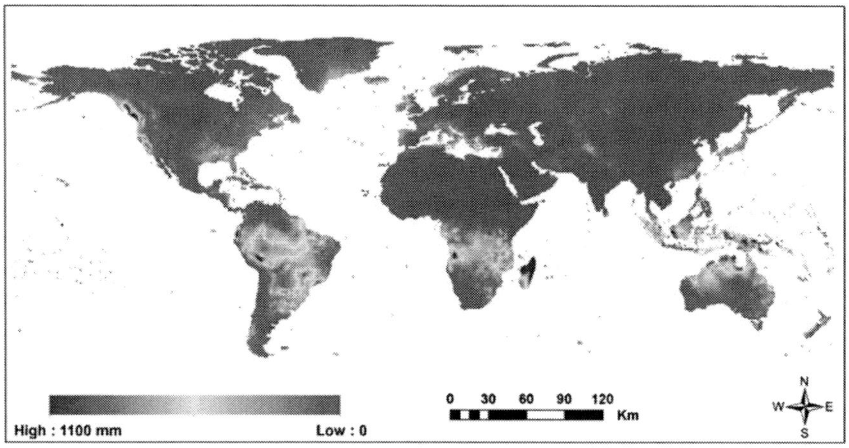

Fig. 2. Global Precipitation map of January 1973 obtained from CruTS 3.23 data.

Precipitation and Temperature variables for the period 1973-2000 were extracted from the CruTS data in order to have the same monthly data collected from the DMI for the 9 station. By having subsetting the original data, consistent data are obtained that can be used for comparison.

Box plots were created for precipitation and temperature (i.e. figure 3 and 4 respectively) illustrating the data of every station from the two sources CruTS and observation side by side. Looking at figure 3, Kozan station has the highest values of the total monthly precipitation while other stations are close to each other. In all stations, data coming from CruTS and observed data are close to each other in terms of the 0.25, 0.5 (i.e. median), and 0.75 quantiles with an exception of Goksun station in which the observed data has obvious higher values than CruTS data. Tomarza station shows a significant difference in the median between the two data sources. Based on the time series plot of the precipitation (i.e. figure 5), there is an obvious mismatch between the two data sources; highest in Tomarza station, a bit lower in Goksun station while other stations matching better.

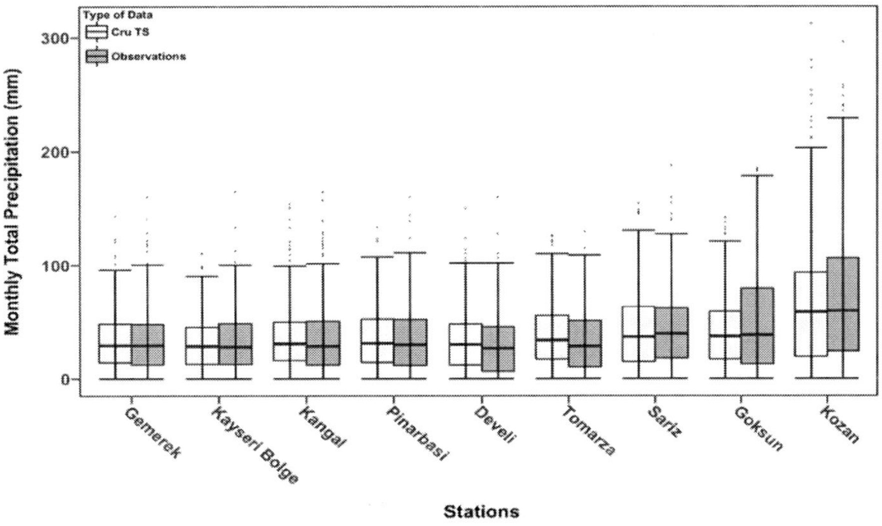

Fig. 3. CruTS and observed precipitation data boxplot.

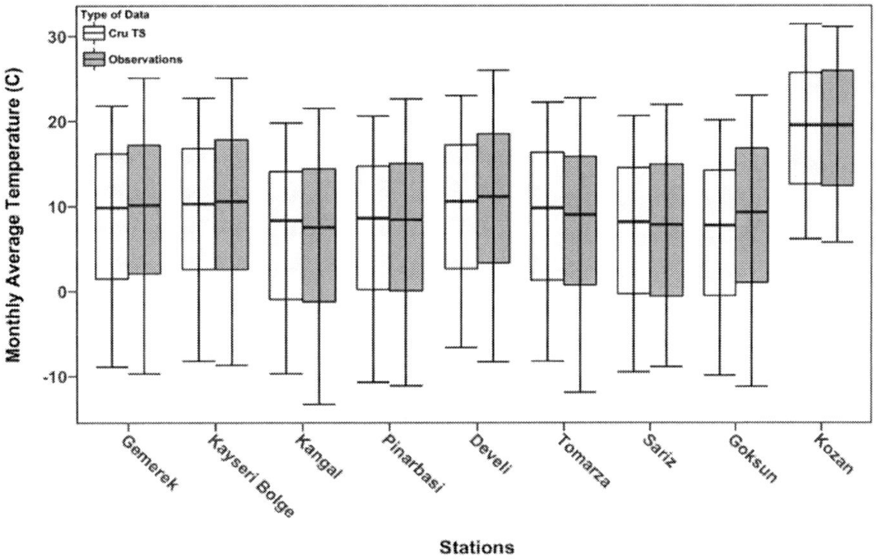

Fig. 4. CruTS and observed temperature data boxplot.

Based on the temperature box plot figure 4, Kozan station has, also, the highest temperature values in comparison with other stations which are not much different from each other. Similarly, to the precipitation, Goksun station has the highest difference between CruTS and observed data where as other stations have slight close differences. According to the temperature time series plot figure 6, the temperature in comparison with precipitation has changed, generally, smoothly. In addition, all stations show a high matching except Goksun station where there is a slight mismatch between CruTS and observed data. According to figure 1 which shows the Digital Elevation Model (DEM), Kozan station is located at a low altitude in comparison with other station which are located

DOI: 10.1051/matecconf/201712005007

in higher altitude and that could be the reason that this station has highest values in both precipitation and temperature.

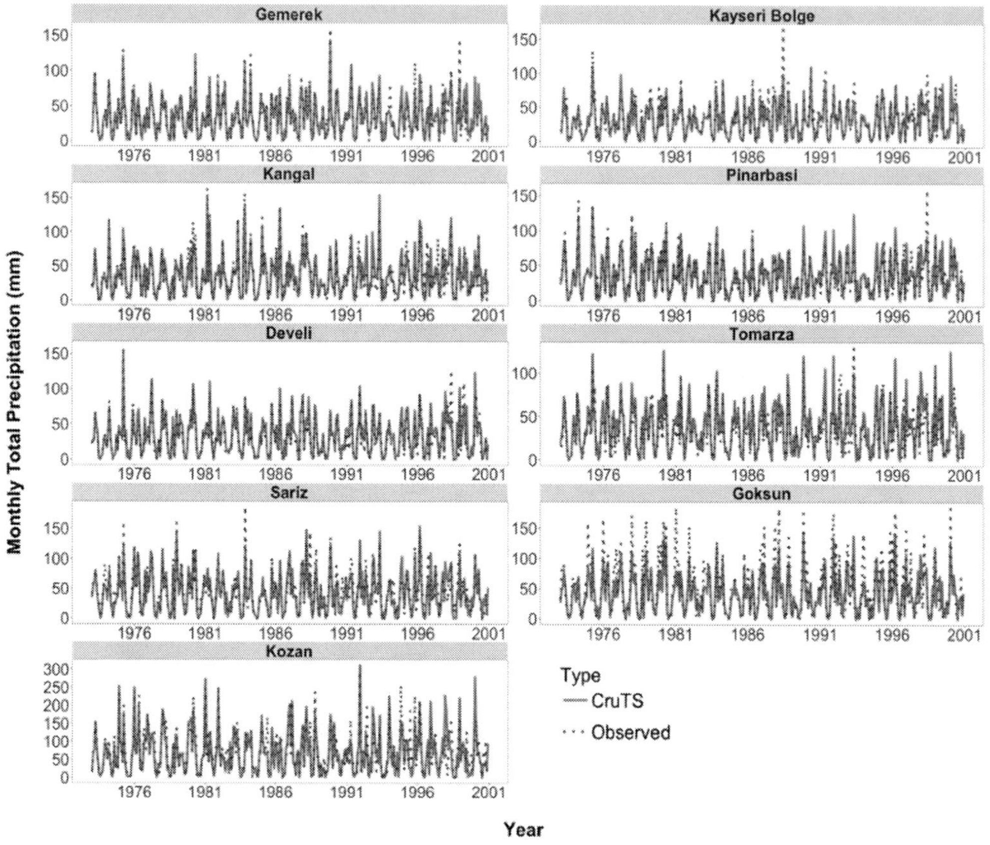

Fig. 5. Time series plot of the monthly total precipitationobtainedfromCruTS and Observed data for the period 1973 - 2000.

Observed precipitation and temperature data were used for the comparison with the CruTS 3.23 in the meaning of several statistical measurements. Correlations listed in table 1 showed very high agreement for all stations in the temperature data ranges from 0.989 for Kangal station to 0.996 for KayseriBolge station. From the same table, RMSE which measures the error between the two datasets, illustrates that Kozan station has the lowest while Goksun station has the highest error between the CruTS 3.23 and observed temperature data which agrees with what have been resulted from figure 4.

In general, the agreement between the precipitation data was lower than temperature data. The precipitation correlations lowest value is 0.87 for Kozan station while the highest value is 0.942 for Gemerek station. RMSE results for precipitation, also, showed higher error values. The highest error belongs to Kozan station while Gemerek station has the lowest. This lower agreement in the precipitation could be resulted from the nature of creating this data as these data created by interpolating a number of stations. The temperature data has more homogeneous distribution than precipitation around the interpolated area regardless the used method.

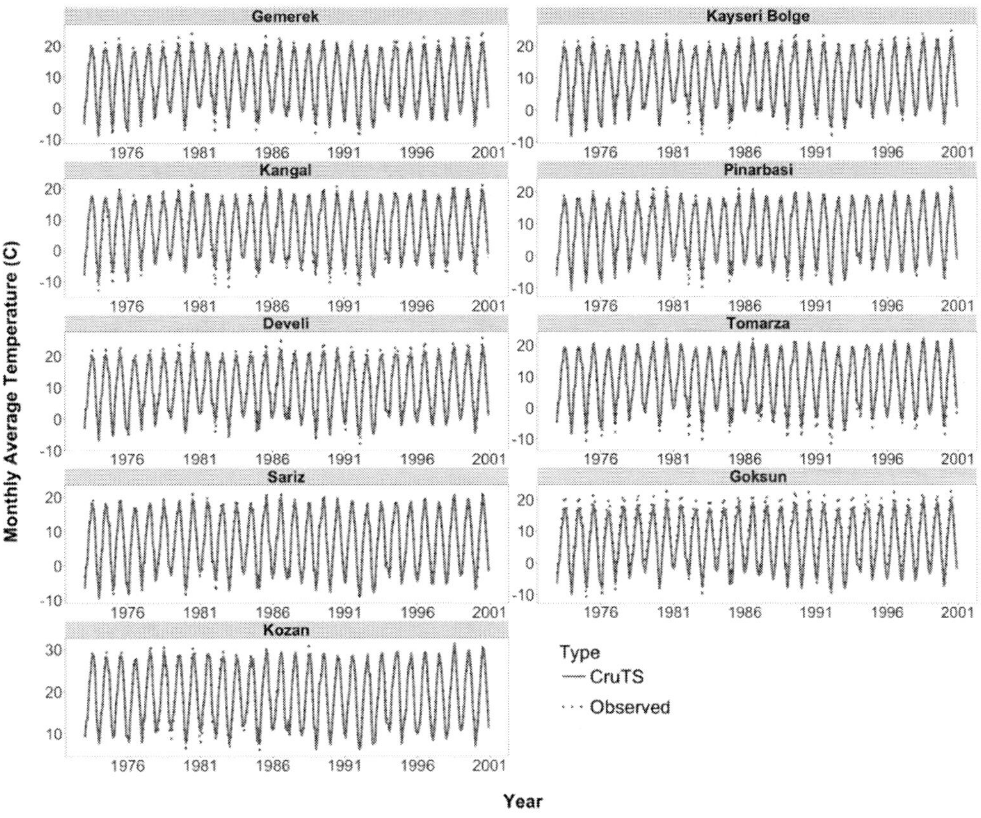

Fig. 6. Time series plot of the monthly average temperature obtained from CruTS and Observed data for the period 1973 - 2000.

The P values of testing the difference between the means of two sources data for two variables and every station listed in table 2. The result shows no significant values in most of the stations which means fail to reject the null hypothesis that the there is no difference between the means with several exceptions; Tomarza station has a significant precipitation mean difference at the significance level $\alpha = 0.05$ and not significant at$\alpha = 0.01$ where P values is 0.018, Goksun station is significant in both precipitation and temperature at both significance levels with values 0.005 and 0.003 respectively.

The median P value results listed in the same table indicate that two variables in all stations have no significant difference in the mean with the exception of the precipitation variable in Tomarza station which is significant at the significance level $\alpha = 0.05$ and not significant at$\alpha = 0.01$ with the value 0.029. Surprisingly, temperature variable in Kozan station has a P value = 1 which indicates the median are same in the two data sources and that agrees with very high value of this station's mean result which is 0.949.

The Non – Parametric Mann-Whitney test p values are also listed in table 2. These values indicate fail to reject the null hypothesis (i.e., no difference between the two groups) as no significance difference for the two variables in all station except two stations i.e., Tomarza and Goksun stations. Precipitation variable is significant in Tomarza station at the significance levels $\alpha = 0.05$ only which agrees with the two mean and median tests. Goksun station has a significant value in the temperature variable at the two significance levels $\alpha =$ 0.01 and 0.05 which matches the mean test.

Table 1. The correlation and MSRE between CruTS and observed data for precipitation and Temperature.

Station Name	Precipitation		Temperature	
	Correlation	RMSE(mm)	Correlation	RMSE (C)
Gemerek	0.947	9.007	0.992	1.383
Kayseri Bolge	0.916	10.866	0.996	1.178
Kangal	0.924	11.694	0.989	1.562
Pinarbasi	0.920	11.288	0.991	1.217
Develi	0.929	10.618	0.995	1.409
Tomarza	0.891	13.663	0.992	1.506
Sariz	0.885	16.320	0.994	0.948
Goksun	0.870	25.387	0.992	2.414
Kozan	0.870	30.326	0.992	0.930

Table 2. The P value of the mean difference, median difference and Mann-Whitney tests.

Station Name	Mean		Median		Mann-Whitney test	
	Precipitation	Temperature	Precipitation	Temperature	Precipitation	Temperature
	P value					
Gemerek	0.582	0.324	0.977	0.693	0.343	0.222
Kayseri Bolge	0.397	0.452	0.862	0.762	0.793	0.338
Kangal	0.524	0.580	0.464	0.340	0.257	0.718
Pinarbasi	0.518	0.813	0.462	0.873	0.355	0.721
Develi	0.130	0.161	0.083	0.507	0.060	0.102
Tomarza	0.018*	0.265	0.029*	0.490	0.021*	0.365
Sariz	0.548	0.743	0.289	0.806	0.594	0.677
Goksun	0.005**	0.003**	0.823	0.139	0.291	0.001**
Kozan	0.614	0.949	0.911	1.000	0.618	0.980

* significant at $\alpha = 0.05$.
* *significant at $\alpha = 0.01$.

In general, CruTS 3.23 data has a tight agreement with the observed data. Most of the used 9 stations has; high correlations; low errors; and no significance in the differences tests. The result agrees with the result of [3] who found close-fitting matches in a sub-continental scale during their updating of CruTS 3.10 data. As an exception, Goksun station has a statistical significance in the difference in several test which agrees with the low correlation and high error in comparison with other stations. However, although Kozan station has the lowest correlation (sharing the same value with Goksun) and highest error in the precipitation variable, it has no significance in any of the tests.

5 Conclusion

This study aims to evaluate the CruTS 3.23 data for only precipitation and temperature variables in a local scale. To have suitable format of the data, files of monthly data of ten climate variables; precipitation, mean temperature, diurnal temperature range, minimum and maximum temperature, vapour pressure, cloud cover, rain days, frost days and potential evapotranspiration for the period January 1901 to December 2014 was converted to GIS-based ASCII formats.

The local scale comparison results indicate that the two sources of data have a tight agreement in both precipitation and temperature, however, the agreement of temperature variable is higher.

The three tests (i.e., Mean difference, Median difference, and Mann-Whitney) results show no significance difference between the two data sets in all stations with an exception of two stations; Tomarza and Goksun stations. Tomarza station has significant difference only in the precipitation variable in the three test while Goksun station has significant difference only in the mean test of the two variables. Kozan station shows the opposite of these two stations as, for example, it has a difference in the median reaches to zero. This difference among stations originated from the nature of the CruTS data that the different versions were created by interpolating a number of station around the world.

This research was funded by Anadolu University under the following project: BAP-000261. The authors would like to thank (DMİ) DevletMeterolojiİşleri (General Department of Meteorology - Ministry of Forest and Water Affairs) for providing the data to complete this study.

References

1. M. New, M. Hulme, and P. Jones, "Representing twentieth-century space-time climate variability. Part II: Development of 1901-96 monthly grids of terrestrial surface climate," J Climate. **13**, 13, pp. 2217-2238, (2000).
2. T. D. Mitchell and P. D. Jones, "An improved method of constructing a database of monthly climate observations and associated high-resolution grids," Int J Climatol. **25**, 6, pp. 693-712, (2005).
3. I. Harris, P. D. Jones, T. J. Osborn, and D. H. Lister, "Updated high-resolution grids of monthly climatic observations - the CRUTS3.10 Dataset," Int J Climatol. **34**, 3, pp. 623-642, (2014).
4. P. D. Jones and I. Harris. (2008, 10/08/2016). *Climatic Research Unit (CRU) time-series datasets of variations in climate with variations in other phenomena*. Available: http://catalogue.ceda.ac.uk/uuid/3f8944800cc48e1cbc29a5ee12d8542d
5. R. A. Garcia, M. Cabeza, C. Rahbek, and M. B. Araújo, "Multiple dimensions of climate change and their implications for biodiversity," Science. **344**, 6183, (2014).
6. U. Hellden and C. Tottrup, "Regional desertification: A global synthesis," Global Planet Change. **64**, 3-4, pp. 169-176, (2008).
7. H. Hoff, M. Falkenmark, D. Gerten, L. Gordon, L. Karlberg, and J. Rockstrom, "Greening the global water system," Journal of Hydrology. **384**, 3-4, pp. 177-186, (2010).
8. M. Kottek, J. Grieser, C. Beck, B. Rudolf, and F. Rubel, "World map of the Köppen-Geiger climate classification updated," Meteorologische Zeitschrift. **15**, 3, (2006).
9. B. Poulter, D. Frank, P. Ciais, R. B. Myneni, N. Andela, J. Bi, *et al.*, "Contribution of semi-arid ecosystems to interannual variability of the global carbon cycle," Nature. **509**, 7502, (2014).
10. A. K. Taxak, A. R. Murumkar, and D. S. Arya, "Long term spatial and temporal rainfall trends and homogeneity analysis in Wainganga basin, Central India," Weather and Climate Extremes. **4**, pp. 50-61, (2014).
11. T. C. Peterson, T. R. Karl, P. F. Jamason, R. Knight, and D. R. Easterling, "First difference method: Maximizing station density for the calculation of long-term global temperature change," Journal of Geophysical Research: Atmospheres. **103**, D20, pp. 25967-25974, (1998).
12. T. C. Peterson, R. Vose, R. Schmoyer, and V. Razuvaëv, "Global historical climatology network (GHCN) quality control of monthly temperature data," Int J Climatol. **18**, 11, pp. 1169-1179, (1998).
13. D. M. Diez, C. D. Barr, and M. C. etinkaya-Rundel, *OpenIntro Statistics*: Duke University, (2015).

14. D. R. Helsel and R. M. Hirsch, "Statistical Methods in Water Resources," in *Techniques of Water Resources Investigations*. vol. Book 4, ed: U.S. Geological Survey, p. 522, (2002).
15. M. Hollander, D. A. Wolfe, and E. Chicken, *Nonparametric Statistical Methods*, Third Edition ed.: Wiley Online, (2014).
16. J. Hájek, Z. Šidák, and P. K. Sen, *Theory of Rank Tests*, Second Edition ed.: Academic Press, (1999).
17. R. H. McCuen, *Modeling Hydrologic Change: Statistical Methods*, 1 edition ed.: CRC Press, (2002).

Semi-2D modeling of river morphological changes caused by exceptional flooding

Tew-Fik Mahdi*

Polytchnique de Montreal, Department of Civil, Geological and Mining.C.P.6097, Succ. Centre-Ville, Montréal, Québec, H3C 3A7. Canada

Abstract. This paper presents the application of the model, MHYSER 1.0 (Model for the HYdraulics of SEdiments in Rivers, version 1.0), to study river morphological changes due to exceptional flooding. MHYSER 1.0 is a semi-two-dimensional model using the stream tubes concept to achieve lateral variations of velocity, flow stresses and sediment transport rate. Each stream tube has the same conveyance as the other ones. In MHYSER 1.0, the uncoupled approach is used to solve the set of conservation equations'. After the backwater calculation, the river is divided in a finite number of stream tubes of equal conveyances. The sediment routing and bed adjustments calculations are accomplished separately along each stream tube taking into account lateral mass exchanges. MHYSER 1.0 is applied to the Ha! Ha! River (Quebec, Canada), a tributary of the Saguenay River, for the 1996 "downpour". The results show good correlation of with collected data. This application shows the capabilities of this model and predicts its promising role in solving complex real engineering cases.

1 Introduction

Precipitation-runoff floods and dam failure floods result in an unusual rapid water surface elevation, and high velocity outflow through the downstream river. The inundation of river banks may cause significant erosions, important landslides and creates potentially unstable embankments, as those observed in the aftermath of the Saguenay floods in 1996 [1].

This paper presents the model, MHYSER 1.0, Model for the HYdraulics of SEdiments in Rivers, version 1.0, [2, 3, 4]. Following the presentation of MHYSER 1.0, a reach of the Ha! Ha! River (Quebec, Canada), a tributary of the Saguenay River, is described and used as a test case to demonstrate the capabilities of the MHYSER 1.0.

The parameters used in this research, especially Manning coefficient, are based on literature equations. Since the exercise is to demonstrate the capabilities of MHYSER 1.0 to capture morphological changes, the calibration of the model was not performed. Please note that an automated tool, OPTIMHYSER, is available to calibrate automatically any river model based on MHYSER 1.0 [5].

* Corresponding author: Tewfik.Mahdi@polymtl.ca

2 Overview of MHYSER 1.0

MHYSER 1.0 is a quasi-steady model in which the water discharge hydrograph is approximated by bursts of constant discharge. MHYSER 1.0 performs basically five groups of operations which are attached to water phase, stream tubes, sediment phase, riverbank stability analysis and stream power minimization respectively.

MHYSER 1.0 uses the continuity equation and the energy equation, when there are no changes in the flow regime, while the momentum equation is used with the continuity equation when there are changes from supercritical to subcritical flows, or vice-versa. For backwater computations, the standard step method is used [6] and the friction losses are computed by a uniform flow formula as it is generally admitted [7]. Irregular cross-sections can be handled regardless of whether the river reach is of a single channel or multiple channels. For the latter case, the variables related to the cross-sectional geometry are computed for each sub-channel and are summed to obtain the total values. Moreover, internal conditions such as weirs, falls and sluices are modeled by rating curves.

After the backwater profiles are calculated, the cross sections are divided into sections of equal conveyance or stream tubes. These stream tubes are used as if they were conventional one-dimensional channels with known hydraulic properties, and sediment routing can be carried out within each stream tube almost as if they were independent channels. Once the top widths are determined, the velocities of the stream tubes are calculated giving a crosswise velocity distribution at every cross section.

Stream tube locations are allowed to vary with time. Therefore, although no material is allowed to cross stream tube boundaries during a time step, lateral movement of sediment is described by lateral variation of the stream tube boundaries. For bed load sediment transport, MHYSER 1.0 uses the transport functions of DuBoys [8], Meyer-Peter and Müller [9] and Parker [10]. For bed-material load, the used equations are: Laursen [11], Toffaleti [12], Engelund and Hansen [13], Ackers and White [14], Yang [15, 16, 17]. Both the space and temporal derivatives are approximated by first order finite difference operators [18]. Transport potential of cohesive sediments is computed using the approach of Partheniades [19] and Mehta et al. [20].

To compute the bed changes, the sediment transport is computed for each individual sediment size fraction within each stream tube. The bed changes are computed as a sum of the bed change for each particle size. To maintain numerical stability the time step is determined by a Courant-Friedrichs-Lewy (CFL) condition [21]. Since the kinematic wave speed of the bed changes is not easily quantified, numerical experimentation is required to determine suitable time step to be used for a simulation.

MHYSER 1.0 uses the method of Bennett and Nordin's method [22] for bed composition accounting procedure by dividing the bed into conceptual layers. The top layer, or active layer, contains the bed material available for transport. Beneath the active layer, the storage layer or inactive layer exists. Below these two layers there is the undisturbed bed. The active layer is the most important layer in this procedure. Erosion of a particular size class of bed material is limited by the amount of sediments of that size class presented in the active layer. At the end of each time step, bed material is calculated in each stream tube. At the beginning of the next time step, after the new locations of the stream tube boundaries are determined, these values are used to compute the new layer thickness and bed composition.

The determination of depth and width adjustments is based on the minimum stream power theory [23] which states that a river will minimize its stream power subject to hydrologic, hydraulic, sediment, geometric, geologic, and man-made constraints. Choosing the direction for channel adjustments is made by minimizing total stream power at different stations. If alteration of the channel widths results in lower total stream power than raising or lowering of the channel's bed, then channel adjustments progress in the lateral direction. Otherwise, the adjustments are made in the vertical direction.

3 Application: the 1996 lake Ha!Ha! flood

3.1 Site description

From July 18 to 21, 1996, unusually heavy rains affected the Saguenay region of Québec, Canada, between Lake St. Jean and the St. Lawrence River (figure 1). This led to widespread flooding and damage including the destruction of many run-of-the-river dams on rivers discharging into the Saguenay River and Saguenay Fjord and course adjustments of the rivers in the region. The most severely affected during the 1996 floods was the Ha! Ha! River. This river drains a catchment of 610 km^2. The Ha! Ha! River links Lake Ha! Ha! to the Ha! Ha! Bay, an arm of the Saguenay Fjord.

The Ha! Ha! Lake is impounded by a concrete dam which suffered little damage during the flood. Lake breaching nevertheless occurred when a small earthen dyke situated to the south of the dam was overtopped by the rising water level. Incision of a new outlet channel resulted, bypassing the concrete dam and leading to rapid drainage of the main lake. Because of the incision, the lake level dropped from a level of 381m to a new level of 370m (average mean sea level). Brooks and Lawrence [24] give a detailed description of this flood and the corresponding damages. They estimate the peak outflow to be in the range 1080–1260m^3/s at a surveyed cross-section 27 km downstream of the dyke. In this paper, the study area extends from the dyke to the falls located 9.8 km downstream.

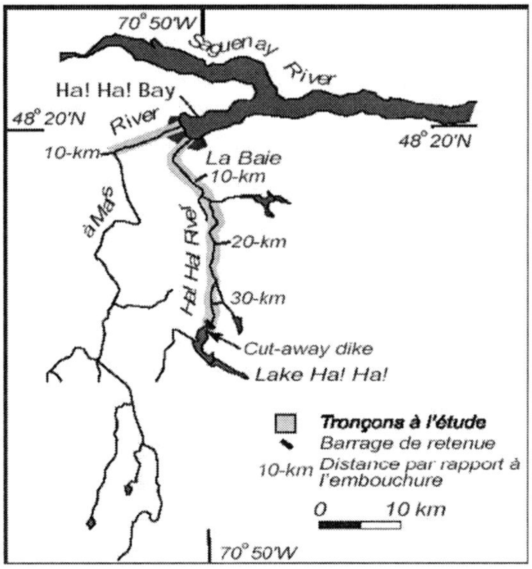

Fig. 1. Ha! Ha! River. [25]

3.2 Input data

The outflow discharge hydrograph (figure 2), the transversal cross sections and the sediment particle size distribution curve (figure 3) have been used by Mahdi and Marche [26]. For the simulations, a time step of 6 minutes and 98 transversal cross sections are used.

DOI: 10.1051/matecconf/201712005008

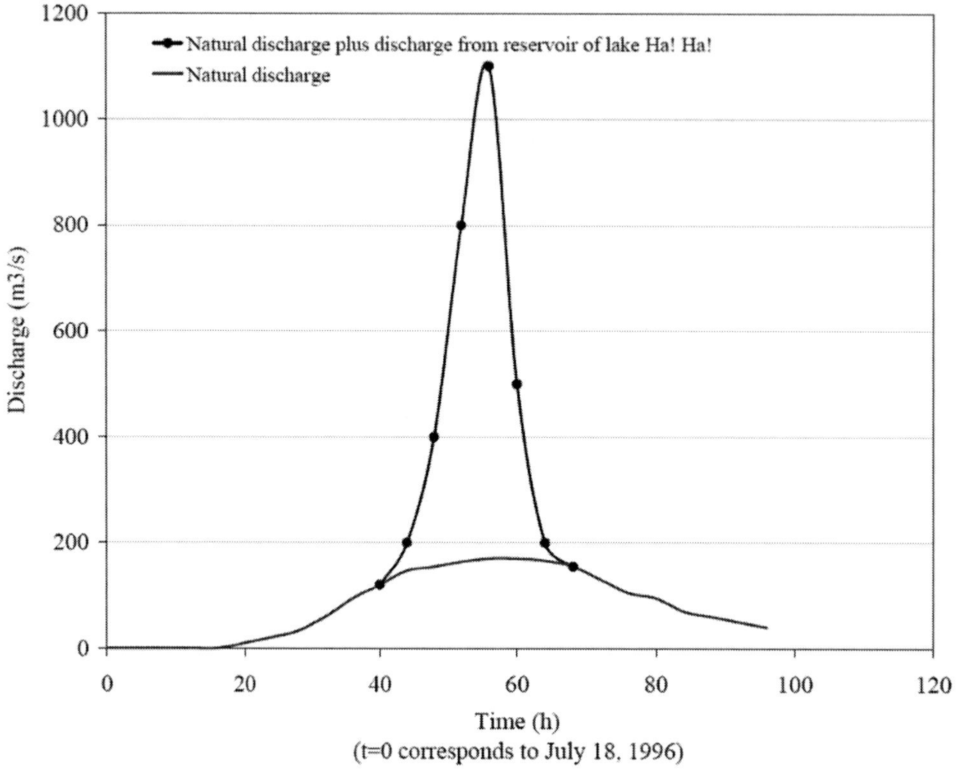

Fig. 2. Outflow discharge hydrograph.

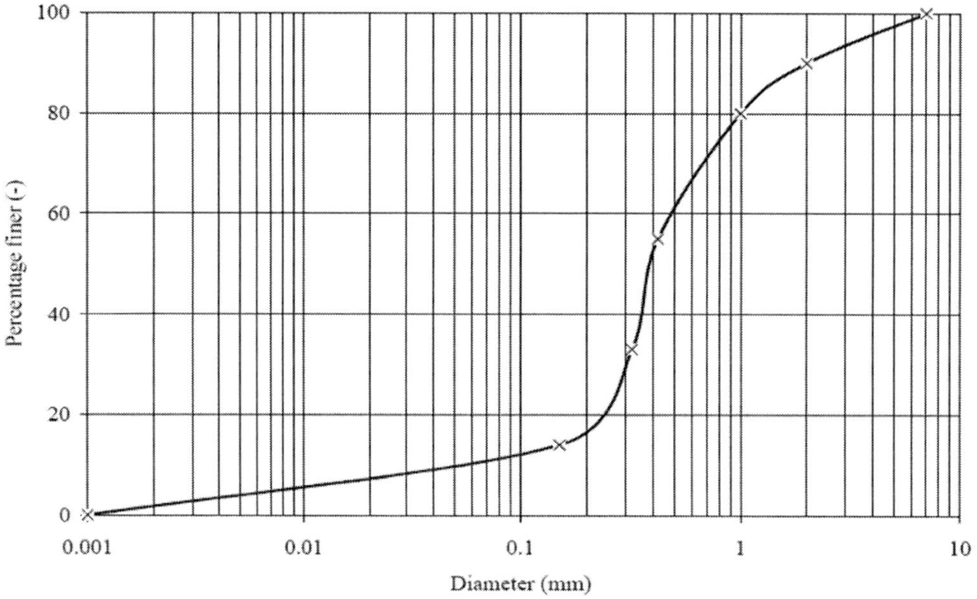

Fig. 3. Sediment particle size distribution used in the simulations.

3.3 Results

3.3.1 Longitudinal profile's evolution

Figure 4 demonstrates the initial and final measured longitudinal profiles with the final simulated one. MHYSER 1.0 simulates well the erosion and deposition tendencies even if for the upstream cross sections, the predicted thalweg is 3m (22%) deeper than the observed one. The erosion is overestimated, while the deposition is underestimated as shown in figure 4.

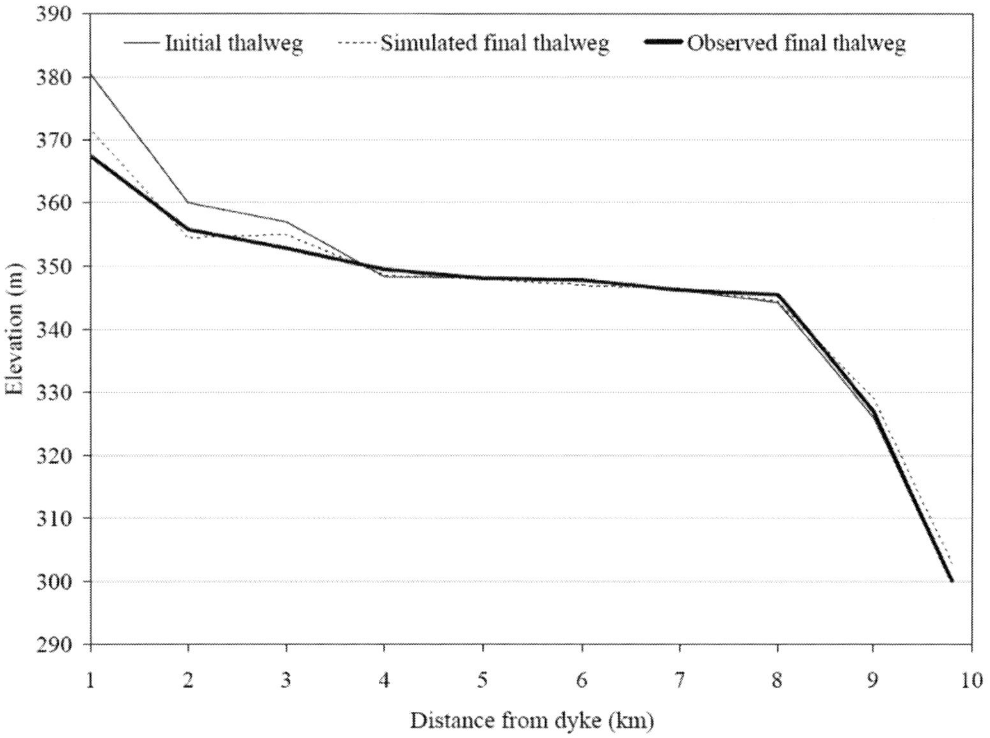

Fig. 4. Plot of the initial, final measured and simulated thalwegs.

3.3.2 River banks' retreat

Figure 5 shows the initial and final measured river's widths superposed to the final simulated ones. MHYSER 1.0 captures well the trend of riverbank's retreat.

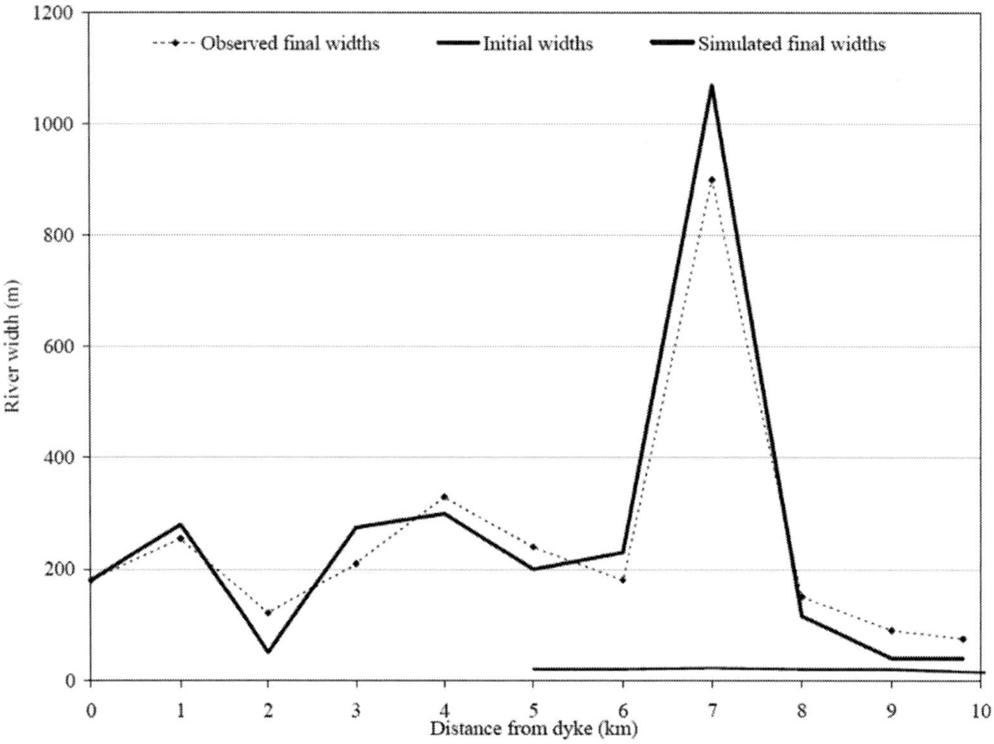

Fig. 5. Plot of the initial, final measured and simulated river widths.

3.3.3 Cross-sections' evolution

Figure 6 and 7 shows examples of simulated and measured cross-sections after the flood passage. Using MHYSER 1.0, some sections were well simulated but, others were not as illustrated by these figures.

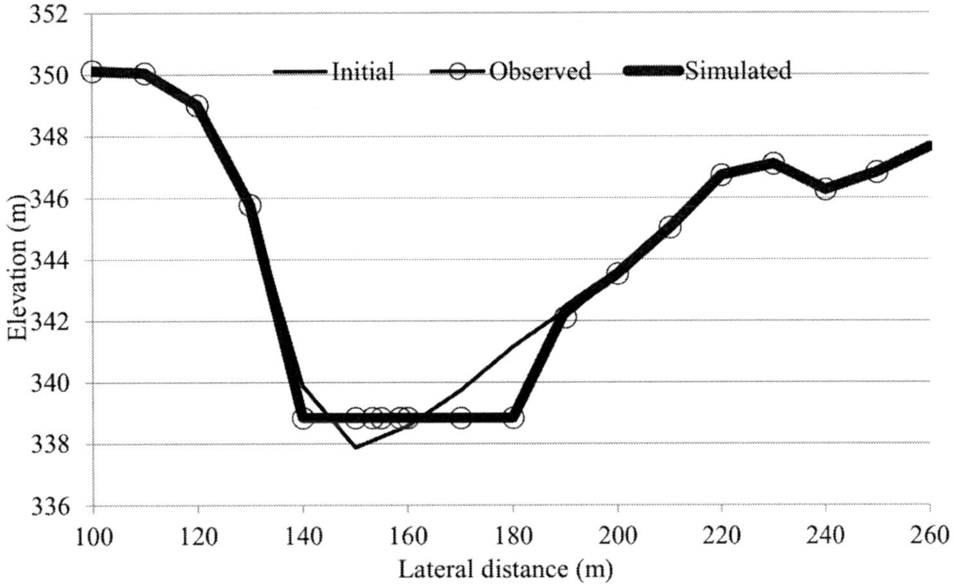

Fig. 6. Example of a good prediction of cross-section's evolution by MHYSER 1.0.

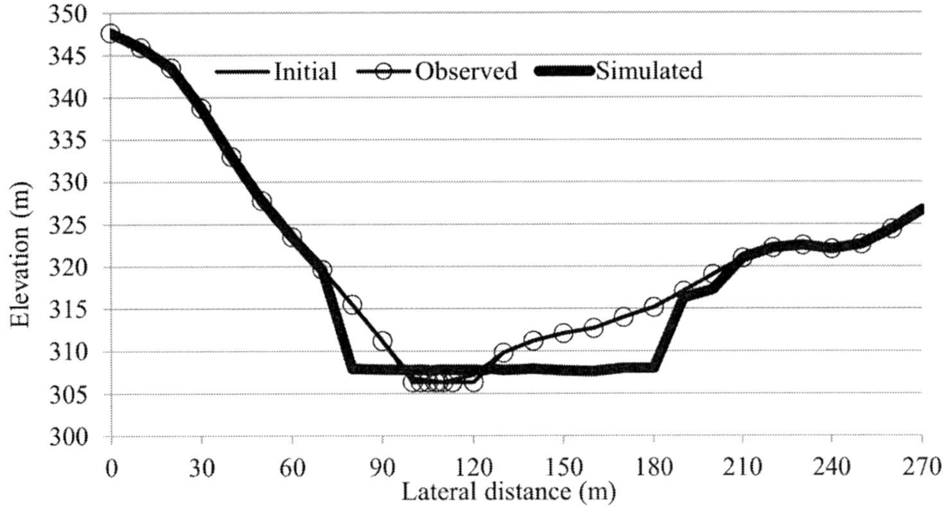

Fig. 7. Example of a poor prediction of cross-section's evolution by MHYSER 1.0.

4 Conclusion

Several river systems could suffer extensive catastrophic floods in the event of a dam break. This paper presents MHYSER 1.0, a semi-two-dimensional model, using the stream tubes concept to achieve lateral variations of velocity, flow stresses and sediment transport rate. Applied to the Ha! Ha! River (Quebec, Canada) during the 1996 "downpour", MHYSER 1.0 predicts well the river's changes. Please note that MHYSER1.0 cannot be used to predict exact changes of a river reach. MHYSER 1.0 is well suited to predict the evolution trend of a river reach undergoing a flood. This application shows the capabilities of this model and predicts its promising role in real engineering cases.

This research was supported in part by a National Science and Engineering Research Council (NSERC) Discovery Grant, application No: RGPIN-2016-06413.

References

1. M. Lapointe, S. Driscoll, N. Bergeron, Y. Secretan, M. Leclerc. Water Ress. Res. **34(9)** 2383-2392 (1998)
2. T. Mahdi. Nat. Hazards **49(3)** 590-603 (2009)
3. T. Mahdi. Nat. Hazards **49(3)** 577-589 (2009)
4. T. Mahdi. Nat. Hazards **49(3)** 565-576 (2009)
5. J. McKibbon, T. Mahdi. Nat. Hazards **54(3)** 879-899 (2010)
6. F. M. Henderson. *Open channel flow*. MacMillan Book Company, NY (1966)
7. S. C. Jain *Open channel flow*. Wiley, Chichester (2000)
8. M. P. Du Boys. Annales des Ponts et Chaussées **18(5)** 141-195 (1879)
9. E. Meyer-Peter, R. Müller. 2nd IAHR Congress Stockholm, 39-64 (1948)
10. G. Parker. J. Hydr. Res. **28(4)** 417-436 (1990)
11. E. Laursen. J. Hydr. Div. ASCE **84(HY1)** 1-36 (1958)
12. F. Toffaleti. J. Hydr. Div. ASCE **95(HY1)** 225-246 (1968)
13. F. Engelund, E. Hansen. *A monograph on sediment transport in alluvial streams*. Teknish Forlag, Technical Press, Copenhagen, Denmark (1972).
14. P. Ackers, W. White. J.Hydr. Div. ASCE **99(HY11)** 2041-2060 (1973)
15. C. T. Yang. J.Hydr. Div. ASCE **99(HY10)** 1679-1704 (1973)
16. C. T. Yang. J. Hydrology **40** 123-138 (1979)
17. C. T. Yang. Div. ASCE 110**(HY12)** 1783-1797 (1984)
18. C. Hirsh. *Numerical computation of internal and external flows*: fundamentals of numerical discretization. John Wiley, NY (1990)
19. E. Partheniades. 3rd Int. Symp. River Sedimentation, Jackson, MS, 3-25 (1986)
20. A. J. Mehta, E. J. Hayter, W. R. Parker, R. B. Krone, A. M. Teeter. J. Hydr. Eng. ASCE **115(8)** 1076–1093 (1989)
21. J. A. Cunge, F. M. Holly, A. Verwey. *Practical aspects of computational river hydraulics*. Pitman, London (1980)
22. J. Bennet, C. Nordin. Hydrological Sci. Bull. **22** 555-569 (1977)
23. H. H. Chang, J. C. Hill. *Applying Research to Hydraulic Practice*. Proc. ASCE Hydr. Div. Conf. Ed. Peter E. Smith (1982)
24. G. R. Brooks, D. E. Lawrence. Geomorphology **28** 141–168 (1999)
25. Couture, R. and Evans, S.G. 2000. *Achievements and Prospects*. Proc. 53rd Canadian Geotechnical Conference. Ed. Geological Survey Can., Ottawa, ON (2000)
26. T. Mahdi, C. Marche. *Can. J. Civil Eng.* **30** (3) 568-579 (2003)

Extraction solvent's effect on biogas production from mixtures of date seed and wastewater sludge

Wameed Radeef[1,*], *Abdallah* Shanableh[1,2], *Mohamed* Semreen[3], *and Ismail* Saadoun[4]

[1]Department of Civil & Environmental Engineering, University of Sharjah, 27272 Sharjah, United Arab Emirates
[2]Research Institute of Sciences and Engineering, University of Sharjah, 27272 Sharjah, United Arab Emirates
[3]College of Pharmacy, University of Sharjah, 27272 Sharjah, United Arab Emirates
[4]Department of Applied Biology, University of Sharjah, 27272 Sharjah, United Arab Emirates

Abstract. The objective of this study was to evaluate the impact of extraction solvent on biogas production from mixtures of date (Phoenix dactylifera) seed and wastewater treatment sludge. Date palm seed is a locally produced waste which is used in different useful purposes. Date seed is rich in proteins, carbohydrates and lipids and have the potential for biogas enhancement. In this study, three extraction solvents, hexane, ethyl acetate and Ethanol:water (1:1), were used on two types of date seeds (locally known as Khalas and Khudari). Date seed powder of size 0.425 - 0.6 mm after extraction was mixed with wastewater treatment sludge at dry date seed/dry sludge solids ratios of 0%, 10%, 20%, 40%, 60%, and 80%. The study showed a direct relationship between biogas production and ratio of date seeds/wastewater sludge in the mixture. Furthermore, the results confirmed that the quantity of biogas produced from samples containing raw date seed was almost same as the quantity produced from samples containing date seed after extraction. The results revealed that the used extraction solvents had no effect on biogas production

1 Introduction

Biogas produced from anaerobic digestion (AD) is a promising source of renewable energy that can address the global energy needs [1]. Over the last decades, bio-fuel production from waste products has been a subject of intensive research efforts. Global warming impact and fuel availability have risen many concerns over the use of fossil fuel [2]. Many researchers suggest that bio-fuel production from waste products is a good option for energy recovery [3]. Although there are some limitation in producing bio-fuel from organic wastes, researchers recommend that it is the right time to develop suitable renewable energy sources.

* Corresponding author: wradeef@sharjah.ac.ae

Biogas is considered as one of the most environmentally friendly and energy-efficient fuel over other types of biofuel [4]. The production of bio-fuel in the form of methane biogas requires readily available and biodegradable organic feedstocks [5-8]. There are generally two sources of organic matter that are suitable for bio-fuel production, which are: energy crops; and organic waste products.

In this study, wastewater treatment sludge and date palm seeds were co-digested to produce biogas. Both raw and after extraction date seeds were prepared and then co-digested with wastewater sludge at different dry weight ratios. Date palm, scientifically called Phoenix dactylifera, seeds are rich in biodegradable organic matter, including Protein (5-5.5%), Carbohydrates (81-83%) and Lipids (10-12%) [9]. Date palm seeds provide a good source of organic matter to the microorganisms involved in AD process and eventually can potentially contribute to and enhance biogas production from wastewater sludge.

The main focus of this paper is to study the effect of extraction solvents, used on date seeds, on biogas production from mixtures of wastewater sludge and date seeds.

2 Materials and methods

Wastewater primary sludge was brought from Dubai Wastewater Treatment Plant. The sludge was obtained from the bottom of the primary sedimentation tank. The total solids (TS) of the sludge was modified to approximately 2%. Then, the sludge was distributed in 750 ml plastic containers and stored in freezer to be used in the future as necessary.

Khalas and khudari date palm seeds were used for biogas production. The date seeds were isolated and thoroughly washed in water to remove date flesh remains. Then, they were dried in the oven at 40 degree for 24 hours. The size of the date seed was adjusted to 0.425 - 0.6 mm by using a centrifugal mill with a rotational speed of 10,000 rpm as shown in Figure 1.

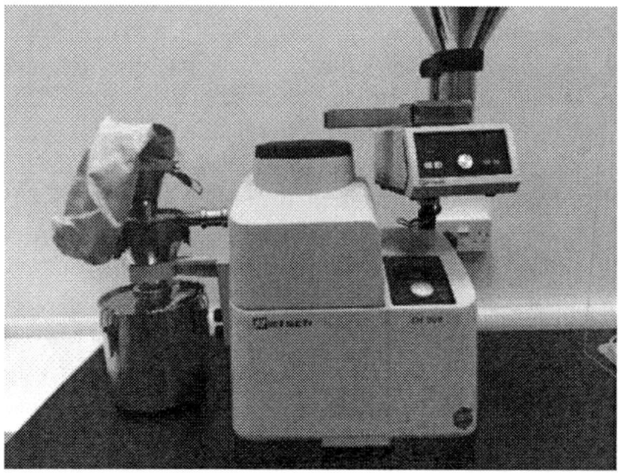

Fig. 1. Ultra Centrifugal. Mill ZM 200

Three extraction solvents, hexane, ethyl acetate and ethanol:water (1:1), were used on two types of date seeds khalas and khudari. Samples of sludge and date palm seeds mixed at different date seed/wastewater total solids dry weight ratios (0%, 10%, 20%, 40%, 60%, and 80%) were prepared and filled in 50 ml serum bottles. Then, the serum bottles were capped by using rubber stoppers and aluminium sealing rings and were incubated under mesophilic temperature (35°C) as shown in Figure 2.

DOI: 10.1051/matecconf/201712005009

Fig. 2. Samples of Date Seed Co-digested with Wasterwater Treatment Sludge.

Four groups of samples were prepared for each type of the two mentioned types of date seeds (khalas and khudari). The four groups of samples are: wastewater sludge and raw date seed, wastewater sludge and date seed after extraction by hexane, wastewater sludge and date seed after extraction by ethyl acetate, wastewater sludge and date seed after extraction by thanol:water (1:1). All samples were co-digested in duplicates to ensure accuracy.

3 Results and discussion

The cumulative biogas production from raw khalas date seed/sludge mixtures after 9 weeks of incubation is shown in Figure 3. The maximum biogas production was achieved from the date seed/sludge samples containing 80% date seed and the minimum biogas production was from samples containing sludge alone. The volume of the produced biogas increased as the quantity of the added date seed increased. The order of biogas production in terms of date seed ratio in the seed/sludge mixtures was as follows: 80%>60%>40%>20%>10%>0%.

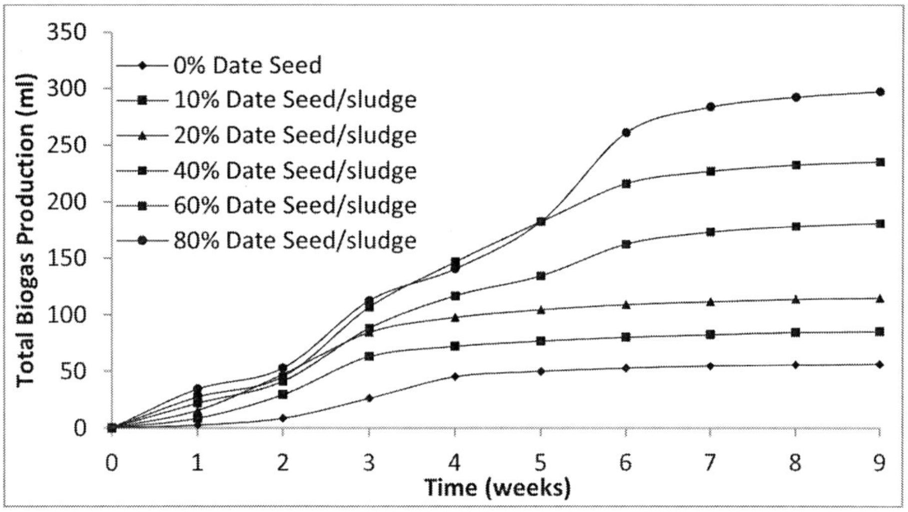

Fig. 3. Cumulative Biogas Production from Mixtures of Raw Khalas Date Seed and Wastewater Sludge.

Samples containing 80% khalas date seed produced about 527% biogas compared with 100% produced by the samples containing 0% date seed after 9 weeks of incubation. The results indicate big enhancement in biogas production. Likewise, the samples containing 80% khudari date seed produced 539% biogas compared with 100% produced by the sludge alone as shown in Figure 4. Both date seed types, khalas and khudari, did almost the same enhancement to the produced biogas when they were added to the wastewater sludge samples.

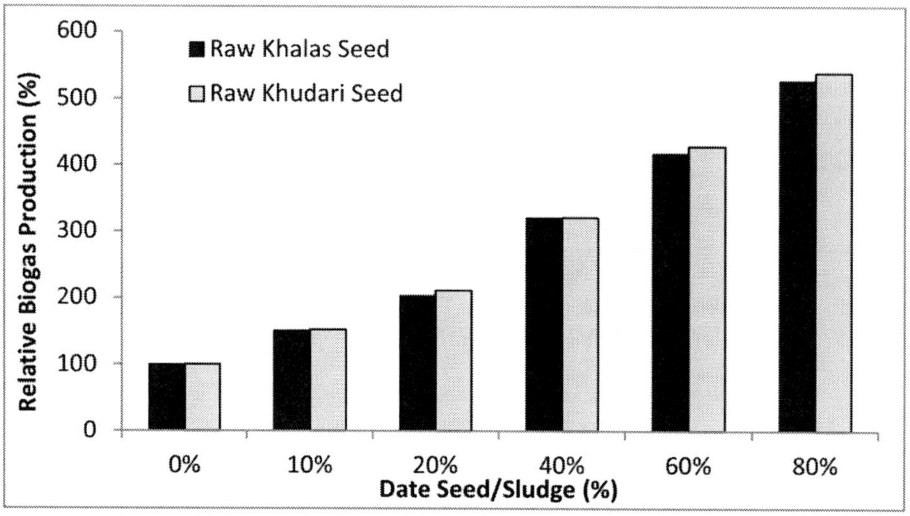

Fig. 4. Comparison of Relative Biogas Production from Mixtures Containing Raw Khalas and Raw Khudari Date Seeds.

The data in Figures 5 and 6 show that raw date seed and date seed after extraction by different solvents had almost the same effect on biogas production. However, raw date seed produced a little bit more biogas than extracted date seed. The three used solvents, hexane, ethyl acetate and ethanol:water (1:1), had no effect on biogas production.

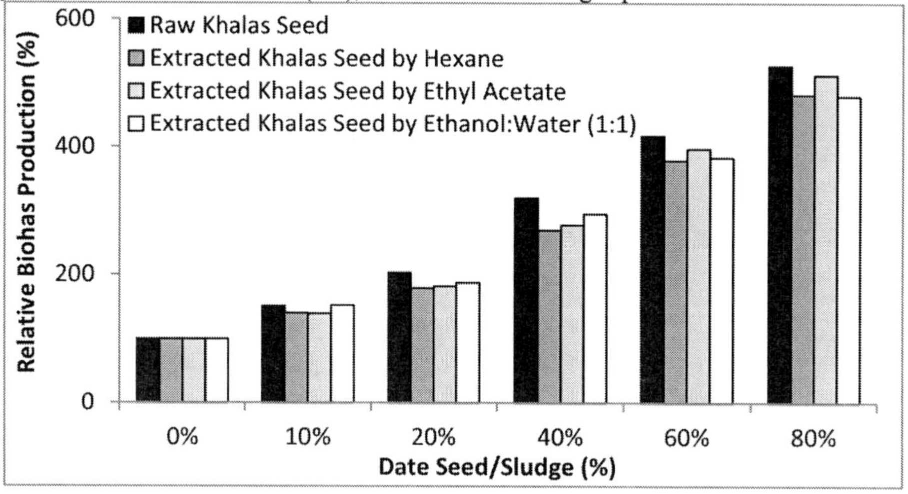

Fig. 5. Comparison of Relative Biogas Production from Mixtures Containing Raw and Extracted Khalas Date Seeds.

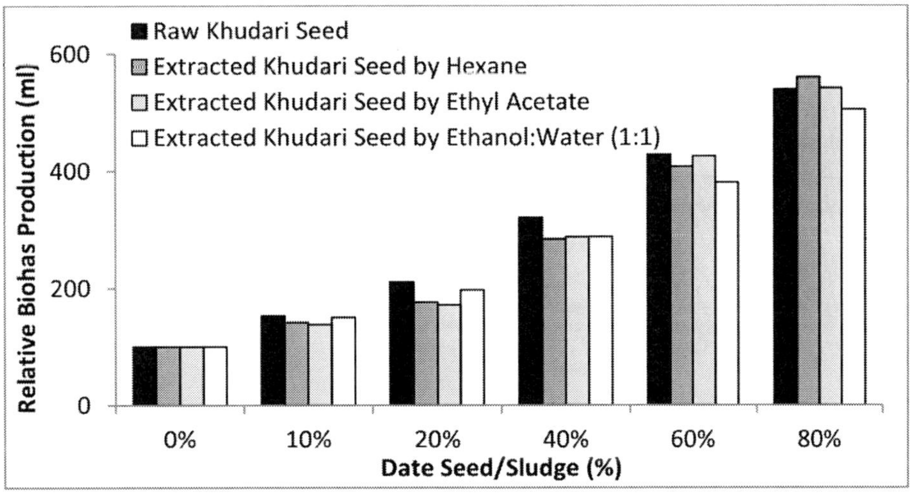

Fig. 6. Comparison of Relative Biogas Production from Mixtures Containing Raw and Extracted Khudari Date Seeds.

3 Conclusions

Both date seed types, khalas and khudari, are rich in organic nutrients and had almost the same effect on biogas production. The production of biogas increased as the ratio of date seeds in the mixture increased. The maximum biogas production after 9 weeks of incubation was produced from the date seed/sludge samples containing 80% date seed and the order of biogas production, in terms of date seed ratios, was 80% > 60% > 40% > 20% > 10% > 0%. Samples containing 80% khalas date seeds produced 527% biogas and the samples containing 80% khudari date seeds produced 539% biogas compared with 100% produced by the samples containing 0% date seed. Solvents used in the date seed extraction process had no effect on biogas production.

This study was conducted with support from the Research Institute for Sciences and Engineering (RISE), University of Sharjah, UAE (Project number: 15020401002-P).

References

1. A. Cuéllar, M. Webber, "Cow power: the energy and emissions benefits of converting manure to biogas", Environ Res Lett **3**, 034002 (2008)
2. A.C., Hansen, Q. Zhang, P.W.L. Lyne, Ethanol-diesel fuel blends – a review, Bioresour. Technol **96**, 277–285 (2005)
3. Y. Kalogo, and H. Monteith, "Energy and resource recovery from sludge", Global Water Research Coalition, 2008
4. H. Fehrenbach, J. Giegrich, G. Reinhardt, U. Sayer, M. Gretz, K. Lanje, J. Schmitz, "Kriterien einer nachhaltigen Bioenergienutzung im globalen Maßstab", UBA-Forschungsbericht **206**, 41–112 (2008)
5. A.I. Al-Turki, Y.M. El-Hadidi, and S.A. Al-Yahya, "Utilization of date pits as a potential source of biogas and organic fertilizer", Food, Agriculture and Environment **2**, 369-374 (2004)
6. G.D. Zupančič, and V. Grilc, "Anaerobic treatment and biogas production from organic waste", Institute for Environmental Protection and Sensors, Slovenia

7. O.L. Momoh, and L. I. Nwaogazie, "Effect of waste paper on biogas production from co-digestion of cow dung and water hyacinth in batch reactors", J. Appl. Sci. Environ. Manage **12**, 95-98 (2008)
8. T.D. Yavini, U.H. Taura, N. Mohammed, and J.M. Namo, "Comparative study of mesophilic biogas production potentials of selected agro-wastes", Int. J. of Eng. and Sci. **3**, 1-6 (2014)
9. S. Besbes, C. Blecker, C. Deroanne, G. Lognay, N.E. Drira, N.E., and H. Attia, "Quality characteristics and oxidative stability of date seed oil during storage", Food Sci. Tech. **10**, 333-338 (2004)

MATEC Web of Conferences **120**, 05010 (2017) DOI: 10.1051/matecconf/201712005010
ASCMCES-17

Bridging the gap between sustainability, the regulatory, and the recycling of construction and demolition waste in the UAE, with specific reference to the Emirate of Sharjah

Marta Bialko[1,*]

[1]University of Sharjah, Architectural Engineering Department, Sharjah, UAE

Abstract. Waste management is currently an inherent part of sustainable construction and, thus, sustainable development. Construction and demolition waste is an issue continuously intensifying in every country. The primary object of this paper is to present issues associated with a gap between sustainable building regulations in the United Arab Emirates, and the recycling facilities such as the Bee'ah facility in Sharjah. Existing construction and demolition waste management has been investigated in terms of available literature as well as for construction sites and the precast concrete factory located in Sharjah.

1 Introduction

The objective of the research is to review the sustainability dimension of building regulations in the United Arab Emirates (UAE) with respect to current practices implemented by contractors to manage construction and demolition (C&D) waste, along with the recycling process as the last step of sustainable waste management. The findings are based on interviews with contractors, and project managers as well as a literature review. The UAE legislation was assessed for the three emirates of Abu Dhabi, Dubai, and Sharjah each and compared with other regulatory bodies.

1.1 Definition of Sustainable development

As we know, the planet has limited natural resources for energy and materials and this is why it is important to conduct urban development while protecting the environment so future generations can use those natural resources. This is the Brundtland Report [1] definition of sustainability that says: 'Sustainable development is development that meets the needs of the present without compromising the ability of future generations to meet their own needs'. Sustainable development has undergone numerous transformations since the first principles were adopted by the United Nations at a conference in Stockholm in 1972, and when the last amendments were made on 25-27 September 2015 in New York as

* Corresponding author: mbialko@sharjah.ac.ae

the document “Transforming our world: the 2030 Agenda for Sustainable Development” (Agenda 2030). This document contains 17 goals relating sustainable development to the society, the environment, and the economy [2]. Table 1 shows selected goals of sustainable development in Agenda 2030 as they relate to sustainable development in the construction industry.

Table 1. Selected goals of the sustainable development in Agenda 2030, as they relate to sustainable development in the construction industry. [2]

Goal 3.	Ensure healthy lives and promote well-being for all at all ages
Goal 6.	Ensure availability and sustainable management of water and sanitation for all
Goal 7.	Ensure access to affordable, reliable, sustainable and modern energy for all
Goal 8.	Promote sustained, inclusive and sustainable economic growth, full and productive employment and decent work for all
Goal 9.	Build resilient infrastructure, promote inclusive and sustainable industrialization and foster innovation
Goal 11.	Make cities and human settlements inclusive, safe, resilient and sustainable
Goal 12.	Ensure sustainable consumption and production patterns
Goal 13.	Take urgent action to combat climate change and its impacts
Goal 15.	Protect, restore and promote sustainable use of terrestrial ecosystems, sustainably manage forests, combat desertification, and halt and reverse land degradation and halt biodiversity loss

Sustainable development has been already implemented into building regulations in many countries around the world. It is included in the Regulation (EU) No 305/2011 of the European Parliament and of the Council of the European Union of 9 March 2011 [3] which introduced sustainable construction as a new, seventh basic requirement: “Sustainable use of natural resources. The construction works must be designed, built and demolished in such a way that the use of natural resources is sustainable and in particular ensure the following:

a) reuse or recyclability of the construction works, their materials and parts after demolition;
b) durability of the construction works;
c) use of environmentally compatible raw and secondary materials in the construction works.”

The Directive 2008/98/EC of the European Parliament and of the Council [4] on waste sets a waste management hierarchy with prevention, reuse, recycle and recover as the most preferable since they produce the least waste, to disposal as the least preferable.

Adapting a building to the principles of sustainable development requires much effort, change, and wise planning over the full life cycle of a building. The principles of sustainable development in the construction industry pay special attention to the use of building materials that can be reused or recycled.

1.2 Sustainable regulations in the UAE

Many countries around the world have their own systems and standards to protect the environment and regulate work in the construction industry. The most popular standards used throughout the world are Leadership in Energy and Environmental Design (LEED) and BRE Environmental Assessment Method (BREEAM). The oldest system is BREEAM that has been created in the UK in 1990 [5] and the LEED system generated in the United States in 1998 [6]. The UAE has only relatively recently joined the list of countries with their own system with the Emirate of Abu Dhabi establishing Estidama in 2010 [7] and the Emirate of Dubai creating the 'Green Building Regulations and Specifications in the Emirate of Dubai' in 2011 that has been mandatory for all new buildings since 2014 [8].

LEED, BREEAM, Estidama and the Green Building Regulations and Specifications in the Emirate of Dubai all guide and assess buildings according to similar principles based on goals of the sustainable development written in Agenda 2030 (Table 2).

Table 2. Comparison of systems BREEAM, LEED, Estidama and the Green Building Regulations and Specifications in the Emirate of Dubai with goals of the sustainable development written in Agenda 2030.

Goals of the sustainable development, Agenda 2030 [2]	BREEAM [5]	LEED [6]	Estidama [7]	Green Building Regulations and Specifications in the Emirate of Dubai [8]
Goal 3 and 11	Health and Wellbeing	Indoor Environmental Quality	Livable Buildings	Building Vitality
Goal 6	Water	Water Efficiency	Precious Water	Resource Effectiveness: Water
Goal 7	Energy	Energy and Atmosphere	Resourceful Energy	Resource Effectiveness: Energy
Goal 8	Management	Integrative Process	Integrative Development Process	-
Goal 9	Transport	Location and Transportation	Livable Buildings (Outdoors)	Ecology & Planning
	Innovation	Innovation	Innovating Practice	-
Goal 12	Materials and Waste	Materials and Resources	Stewarding Materials	Resource Effectiveness: Materials and Waste
Goal 13	Pollution	Sustainable Sites	Natural Systems	Ecology & Planning
Goal 15	Land Use and Ecology	Regional Priority		

As shown in Table 2, all statements in BREEAM, LEED and Estidama fully cover the sustainable development objectives of Agenda 2030 but the Green Building Regulations and Specifications in the Emirate of Dubai needs to improve in two aspects in order to do the same. The objectives of sustainable development as established by the organization of the United Nations generally dictate the law protecting the environment and the principles of ecological systems in the UAE. It also indicates how sustainability has been incorporated into UAE legislation.

1.3 Definition of construction and demolition (C&D) waste

C&D waste can be defined in many ways, one of which is "the waste material produced in the process of production, construction, renovation, or demolition of structures" [9]. As per the Building Research Establishment (BRE) (1978) it can be also defined as "the difference between materials ordered and those placed for fixing on building projects". A more detailed definition was provided by A. Dania [10] who said "construction and demolition waste is a complex waste stream, made up of a wide variety of materials which are in the form of building debris, rubble, earth, concrete, steel, timber, and mixed site clearance materials, arising from various construction activities including land excavation or formation, civil and building construction site, clearance, demolition activities, road-work, and building renovation." The definitions of C&D waste can be a subject of a separate study.

1.4 Classification of sources of C&D waste in a full life cycle of a building

Many studies have been conducted to specify sources of the C&D waste throughout the entire life cycle of a building. The main four stages named by Ekanayake and Ofori [11, 12] are (1) design, (2) procurement, (3) handling of materials, and (4) operation. They confirmed that the maximum numbers of causes of waste are already created at the design stage [11]. A survey conducted by Saunders and Wynn (2004) also proved that wrong design is the main reason for excessive off-cuts and the production of waste [12]. Research by Bossink and Brouwers [13] and by Gavilan and Bernold [14] specified the six main causes of waste as design, procurement, material handling, operation, residual, and others. Lingard classified the four causes of [15], procurement, handling, operation, and a cultural cause that added a behavioural motive.

C&D waste was also classified by amounts produced annually. Studies conducted in Kuwait [16] reveal that the main sources of C&D waste are (in decreasing order from the most produced) "demolition of old buildings, leftovers from new construction, building repair and maintenance, and manufacturing debris". The U.S. Department of Energy [17] waste generation statistics specify19.5 kg/m2 as a minimum for new construction and 757 kg/m2 as a minimum for building demolition. In summary, C&D waste is produced over the full life cycle of a building, beginning at the design stage and continuing throughout construction and use of a building (e.g. renovations and repairs), but the largest amount of waste is generated during demolition.

2 Case studies

Three case studies including two construction sites at different stages and with different functions, and one concrete precast factory were visited to study the general production of C&D waste and sustainable approaches. Conclusions were compared with the literature findings.

2.1 C&D waste generated in the production of construction materials & components – case study A

The C&D waste generated during the production of construction materials has been studied in general using the example of the Sharjah factory Emirates Stone that produces concrete components, glass reinforced concrete components (GRC), and glass reinforced plastic (GRP) components. The wide range of items produced includes internal walls, external walls with insulation, external elevation panels, foundations, and concrete columns, slabs, and fences. It also makes many types of GRC and GRP decorative and free-standing elements. All are used in all types of buildings up to eight floors in height. External walls have from 20 to 40 mm of Styrofoam insulation and slabs are produced in widths of 1200mm and lengths of 120 meters and cut as per project requirements.

Elements with simple shapes such as foundations, columns, slabs, and unornamented walls are cast in steel moulds that can be used at least 100 times before being cleaned, repaired and polished ready for re-use. Ornamented walls are cast in wooden moulds that include rubber forms for the details. Wooden formwork made from MDF can be used from five to eight times.

C&D waste was produced in the manufacture of components of all types. The following are the results of my own observations as well as interviews with factory representatives.

2.1.1 Produced type of waste:

- Wood – old wooden moulds are discarded along with the steel ones.
- Steel – two types of steel waste were observed:
  - Steel from moulds – parts that are already worn out and cannot be renewed or reused are sent to a landfill at the factory.
  - Cut-offs from steel reinforcing bars is also directed to the factory landfill.
- Concrete and reinforced concrete – defective / broken during production, or old or worn-out components (e.g. walls, slabs, or fences) brought from construction sites are send to the factory landfill.
- Plastic – this type of waste is also placed together with other materials in the factory landfill.
- GRC and GRP – defective / broken during production, or old worn out components (e.g. decoration panels, ornaments, benches, or fountains) brought from construction sites are sent to the factory landfill.

2.1.2 Conclusions

As representatives of the precast factory admit, the waste is not segregated. The factory landfill contains different types of C&D waste stored along with general waste (e.g. packaging, soil, or even organic waste generated by employees). The factory produces large amounts of valuable C&D waste suitable for recycling (i.e. concrete, reinforced concrete, MDF, wood, and steel) and large amounts of non-recyclable waste (i.e. GRC and GRP). Due to the mixing of waste, many of these valuable materials are wasted in the landfill. Furthermore, the factory does not have a waste management plan and as representatives declare the waste is not collected regularly. In this case, the problem is a lack of a waste management plan and a lack of awareness of the impact of waste on the environment. Additionally, the management is unaware of the financial benefits that can result from C&D waste management and recycling, that results in a lack of support from their side to train the staff.

2.2 C&D waste generated during the process of construction – case study B and C

Two different construction sites at various stages of completion and with different functions were chosen as examples of generating construction waste, and were visited. Both construction sites were for buildings of precast concrete components supplied by the factory described in case study A.

The first project (case study B) is a two-storey building that will be dormitories for the Medical College of the University of Sharjah (Sharjah, the UAE), where interior works were in progress. At the construction site, preliminary segregation of construction waste such as cardboard, aluminium, plastic pipes, or dirty metal waste was seen. However, the containers were small and clean recyclable waste (e.g. aluminium pipes) was mixed with dirty waste (e.g. metal covered in oil). A skip with various types of unsorted general waste e.g., plastic packaging, soil, and organic waste, that also contained recyclable materials such as cardboard, or wooden pallets was observed during site walks. A project manager admits that the same garbage truck irregularly collects both segregated and unsegregated waste. Only cardboard is collected separately in accordance with recycling factory requirements. According to the project manager, waste is segregated at the recycling plant (Bee'ah). This however is not the case since Bee'ah confirms that any recyclable waste mixed with prohibited waste is sent to landfill. Valuable materials are therefore still being wasted even though there is basic segregation of construction waste onsite. The reason is a lack of training of the staff and thus a lack of knowledge about the C&D waste management plan. The main problem is a lack of environmental awareness.

The second project is the Arabic College Language Building (case study C) in Sharjah, the UAE that is single storey and was visited during an early construction stage where the foundations were finalized and the ground floor was being constructed. During this phase, very little construction waste was observed. According to a project manager, waste on the site is not segregated and there is no plan to. Furthermore, all concrete waste (i.e. defective or broken during construction or transport, concrete components) is send back to the precast factory. In this case the problem is more complex because there is no waste management plan and the project manager does not consider it as necessary to segregate waste. The waste management issue is passed back to the factory producing the precast concrete components. There is therefore a lack of environmental awareness and professional training. At the site, the improper storing of concrete elements that can damage them and generate more waste was also observed.

Table 3. The summary of the main causes of C&D waste observed during the study.

	Case study A	Case study B	Case study C
Type	Precast concrete factory	Residential	Educational
Stage	Production	Finishing	Structure
Type of C&D waste	- Wood after repeated use - Steel (after repeated use and off-cuts) - Concrete and reinforced concrete - Plastics - GRC and GRP	- packaging (cardboard, and wooden) - aluminium cut-offs - plastics - dirty metal waste - soil	- wooden packaging - precast concrete
Possible causes	- a lack of the waste management plan - a lack of awareness	- a lack of the waste management plan - a lack of awareness	- improper storing - lack of a waste management plan

	- a lack of management's support - a lack of training	- a lack of training	- a lack of awareness - a lack of training

Most of the causes including lack of C&D waste management plan are connected to a lack of awareness and can be grouped under 'culture', according to Lingard [15]. Only improper storing can be categorized [15] as a handling issue. The study performed in Jordan [18] confirms that 'poor strategy for waste minimization' is one of the major reasons of material waste in construction, and lists it in a ranking of causes of wastage as fifth out of 25. Additionally, the same ranking lists 'wrong and lack of storage of materials' as fourth. Findings in the research done in the UAE [12], confirm that 'lack of workers' awareness' is the main cause of material wastage. This is a general issue in the UAE construction industry. Moreover, the same ranking [12] lists 'inappropriate storage' as sixth out of 13.

3 The required percentage of diverted C&D waste from landfill as per selected international regulations

In the recycling process waste materials are converted into new reusable materials and objects. A high percentage of recycled building materials depend on the type of material used originally, building technology, and the waste management. The required percentages of diverted C&D waste from landfill are compared in Table 3 for BREEAM, LEED, Estidama and the Green Building Regulations and Specifications in the Emirate of Dubai. The comparison clearly shows that the respective requirements are similar to the each other and converge at around 70%. However, the LEED and Estidama have the basic mandatory rates consecutively 50% and 30% where the non-standard requirements are 70% for both of them. The situation is slightly different for BREEAM which lists separate percentages for construction and demolition waste that range between 60% and 75%.

The comparison of the requirements of the above standards shows that in the United States, Europe, and the UAE, the required percentages of diverted C&D waste from landfill are very similar. This shows that since 2010 [12] the sustainable side of building regulations including C&D waste management, in the UAE has improved and is continuing to do so.

Table 4. The comparison of required percentage of diverted C&D waste from landfill for BREEAM, LEED, Estidama and the Green Building Regulations and Specifications in the Emirate of Dubai.

The required percentage of diverted C&D waste from landfill as per international standards. *			
BREEAM [5] % - as shown below	LEED [6] % by volume or weight	Estidama [7] % by volume or weight	Green Building Regulations and Specifications in the Emirate of Dubai [8] % by volume or weight
Construction waste: • ≥ 60% (by weight) or ≥50% (by volume) (1credit) • ≥ 75 (by weight) or ≥65% (by volume) (Exemplary level, 2 credits) Demolition waste: • ≥ 70% (by weight) or ≥ 60% (by volume) (1 credit) • ≥ 75% (by weight) or ≥ 65% (by volume) (Exemplary level, 2 credits)	50 % (1 point) 70 % (2 points)	30 % (mandatory credit) 50 % (1 optional credit point) 70 % (2 optional credit points) **	Minimum 50 %

* As per each listed in the table standards diverted C&D waste from landfill must be recycled or reused.

** As per Estidama rating system, all new projects in Abu Dhabi (after 2010) must meet all mandatory credit requirements to obtain a 1 Pearl rating. To reach a higher Pearl rating, a project must meet all mandatory credit requirements along with optional credit points requirements [7].

4 Recycling of C&D waste in the UAE, with specific reference to the Emirate of Sharjah

In the UAE, the most recent government statistics reveal that almost 26 million tonnes of waste were collected across the country in 2014 [19]. Statistics are not available for C&D waste collected across all emirates as each emirate has different companies to collect waste and recycle it. Recycling factories in Abu Dhabi and Al Ain have C&D waste capacity of approximately 1,825,000 tons per emirate yearly [20]. This number is not related to C&D waste collected in total but only to recycled bulk. As per a 2011 Dubai Municipality report 6,638,471 tons of C&D waste was generated for only Dubai, where the C&D Recycling Facility can process approximately 1,635,200 tons annually [21].

In Sharjah, Bee'ah Construction & Demolition Waste Recycling Facility processes more than 2,190,000 tons of C&D waste per year. Bee'ah's Material Recovery Facility is the largest in the Middle East and the third largest in the world [22]. Tables 4 and 5 show C&D waste (in tonnes) for the three waste types of C&D (mineral), Mix C&D, and Asphalt processed by Bee'ah, for 2015 and 2016. In Bee'ah the recovery rate of C&D waste in 2015 is 97.8 % and in 2016 (up to the end of April) is 98.1 % [22].

Table 5. Tonnages received of each category of C&D waste in 2015 and in 2016 up to end of April. (As per statistics provided by Bee'ah) [22]

Types of C&D waste material as per Bee'ah categorization	2015	2016 up to the end of April*
C&D (mineral)	135,789.74 ton	66,404.04 ton
Asphalt	7,637.76 ton	3,090.81 ton
Mix C&D	39,261.02 ton	41,642.77 ton
Total	182,688.52 ton	111,137.62 ton

Bee'ah uses a combination of manual and mechanical sorting of waste to recover materials such as paper, mixed paper, cardboard, aluminum cans, steel cans, and many others. The same technique is used in the recycling of C&D waste [22].

The full C&D waste recycling process is shown in Fig. 1. General waste is given a first examination at the first inspection point by employees. If it is classified as C&D waste it is sent to the main gate where the waste is weighted, a license is checked and a payment received. If the waste is prohibited because it includes gypsum, powder, marble, asbestos, green landscaping waste, sand with oil or it is sewage or hazardous material then it is sent to landfill. Waste recognized as recyclable is directed to the second checkpoint where a detailed inspection is performed. Processing begins only if the waste doesn't contain asbestos. Presorting is performed first and large items removed. Next, material with a maximum size of 350 mm is sent to the 'jaw crusher' machine and its size reduced to less than 120 mm. Steel is then extracted by a magnetic separator and then a screen machine sifts C&D material according to size. Material between 0 and 50 mm is directed to a machine that sifts into the three groups of 0 - 5 mm, 5 - 14 mm, and 14 - 50 mm. Material sifted is stored before being sold. Material of 50 to 75 mm, and 75 mm and above is sent to two separate sorting chambers where waste that cannot be converted into aggregate is removed. After that, a final size reduction is performed by impact crusher and the remainder of the steel is sorted. Finally, the sifting machine sorts material into their three sizes of 0 - 10 mm, 10 - 40 mm, and 40 - 75 mm that are finally sent to the sales department [22].

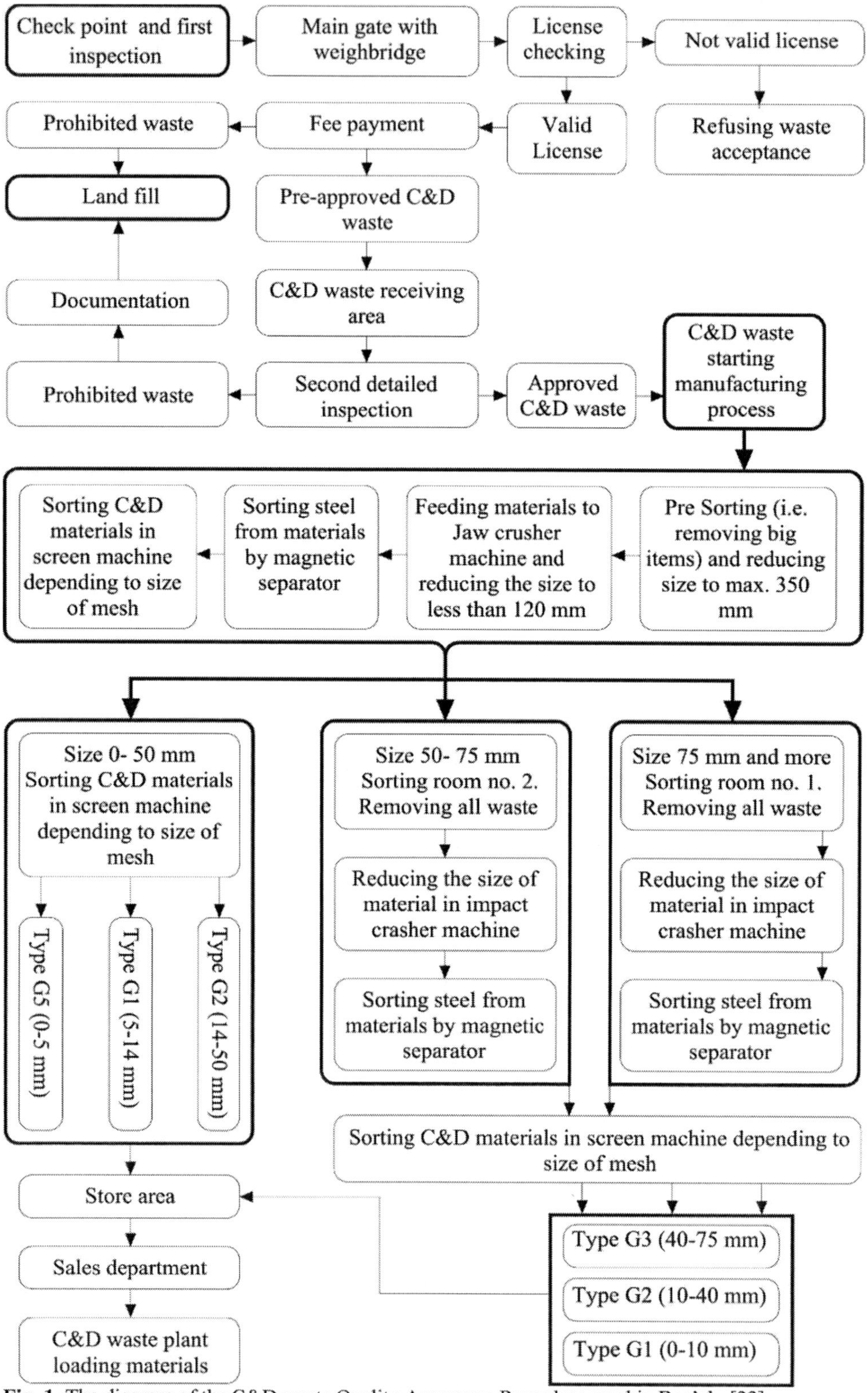

Fig. 1. The diagram of the C&D waste Quality Assurance Procedure used in Bee'ah. [22]

Manufactured Recycled Crushed Aggregate (RCA) consists of clean, hard, durable, angular fragments of rock, concrete and sand fragments of uniform quality. The RCA produced from C&D waste is delivered by Bee'ah in six sizes (Fig. 2) and is guaranteed by the Abu Dhabi Quality and Conformity Council, established by His Highness Sheikh Khalifa Bin Zayed Al Nahyan, President of the United Arab Emirates and ruler of Abu Dhabi, by a Certificate of Conformity. Moreover, it is authorized to hold the Abu Dhabi Government's Trustmark for Environmental Performance [23].

Type of RCA by Bee'ah		
Grade 1: size 0 mm – 10 mm	Grade 2: size 10 mm – 40 mm	Grade 3: size 40 mm – 85 mm

Grade 1: size 5 mm – 14 mm	Grade 2: size 14 mm – 50 mm	Grade 5: size 0 mm – 5 mm

Fig. 2. Recycled Crushed Aggregate (RCA) from C&D waste by Bee'ah. [22]

The use of RCA products has many environmental benefits including limiting the consumption of natural resources and diverting C&D waste from disposal to landfill [23]. Moreover, the price of materials produced from C&D waste is very attractive for investors. Based on studies done in Kuwait [16] "a required amount of recycled aggregate annually it's 5% for reinforced concrete work, 50% for plain concrete and 100% for road works and backfilling" and this suggests large savings for investors.

5 Summary and conclusions

This study compares different sustainable regulations including regulations in the UAE, with goals adopted by the United Nations in Agenda 2030 [2]. Various sources of C&D waste have been surveyed and compared with literature findings. Recycling facilities and the recycling process have been also investigated using the emirate of Sharjah [22] as an example.

The conclusions are as follows:

1. The sustainability of regulatory in the UAE, including C&D waste management, have improved since the previous research conducted in 2010 [12]. The only existing sustainable regulations are established in the emirates of Abu Dhabi and Dubai even though Dubai regulations need improvement in two sectors pointed out in Table 2. The

other five emirates lack clear and easily accessible public regulations to control C&D waste management.
2. The main source of the C&D waste seen in the presented case studies from Sharjah is caused by a lack of environmental awareness and professional training of staff. This confirms the results of studies conducted in Abu Dhabi and Dubai [12].
3. The current implementation of C&D waste management is the weakest point in the chain of regulatory – practice – recycling.
4. Recycling facilities are there and ready to process large amounts of C&D waste. However, the contractors' and workers' awareness of ecological and financial benefits that come from recycling is what needs to be improved.
The following actions are recommended:
1. Clear and easily accessible public regulations created by emirates of Sharjah, Ajman, Ras Al-Khaimah, Fujairah, and Umm al-Quwain. These regulations should consider sustainable goals.
2. Governmental bodies should encourage investors, consultants, and contractors to prevent and reduce C&D waste by commencing a program promoting awareness of the ecological and financial benefits that result from recycling.
3. More C&D waste studies should be conducted to support amending local regulations to current implementation practices.
4. More studies and testing should be undertaken to produce new materials from recycled C&D waste.
In summary, regulatory and recycling facilities in the UAE have improved although particular areas still need addressing. Moreover, as observed and proven, significant improvements in implementation are needed.

References

1. UN, Report of the World Commission on Environment and Development, *Our Common Future* (1987)
2. UN, Resolution adopted by the General Assembly on 25 September 2015, *Transforming our world: the 2030 Agenda for Sustainable Development* (2015)
3. Official J. of the EU, *Regulation (EU) No 305/2011 of the European Parliament and of the Council of 9 March 2011 laying down harmonised conditions for the marketing of construction products*, L **88/5**, (2011)
4. Official J. of the EU, *Directive 2008/98/EC of the European Parliament and of the Council of 19 November 2008 on waste*, L **312/3** (2008)
5. BRE Global, *BREEAM International New Construction 2016*, SD233, 1.0 (2016)
6. USGBC, *LEED v4 Building Design and Construction* (2016)
7. Department of Municipal Affairs, Abu Dhabi, *The Pearl Rating System for Estidama, Building Rating System: Design & Construction* (2010)
8. Dubai Municipality, Dubai Electricity & Water Authority, *Green Building Regulations and Specifications in the Emirate of Dubai* (2011)
9. M. Yeheyis, K. Hewage, M. Shahria Alam, C. Eskicioglu, R. Sadiq, Clean Techn. and Env. Policy, **15(1)**, 81-91 (2013)
10. A. Dania, J. Kehinde, K. Bala, *Proceedings of the 3rd Scottish Conference for Postgraduate Research of the Built and Natural Environment*, 121–129 (Glasgow, 2007)
11. L. Ekanayake, G. Ofori, *Proceedings: Strategies for a Sustainable Built Environment*, (Pretoria, 2000)
12. A. Al-Hajj, K. Hamani, Arch. Eng. and Des. Man., **7.4**, 221-235 (2011)
13. B. A. G. Bossink, H. J. H. Brouwers, J. of Con. Eng. and Man., **122(1)**, 55–60 (1996)

14. R. M. Gavilan, L. E. Bernold, J. of Con. Eng. and Man., **120(3)**, 536–552 (1994)
15. H. Lingard, P. Graham, G. Smithers, Cons. Man. and Ec., **18(4)**, 383–393 (2000)
16. N. Kartam, N. Al-Mutairi, I. Al-Ghusain, J. Al-Humoud, Waste Man., Department of Construction Engineering, Kuwait University, **24**, 1049-1059 (2004)
17. U.S. Department of Energy, Greening Fed. Facil., An Energy, Environmental, and Economic Resource Guide for Federal Facility Managers and Designers, 2nd ed, 170–171 (2001)
18. G. A. Bekr, J. of Con. Eng., **2014**, Art. ID 283298 (2014)
19. United Arab Emirates Data Portal. Socio-Economic Data. Collected waste. http://bayanat.ae/vndyxhe/education-statistic (retrieved: 05.11.2016)
20. Tadweer The center of Waste Management – Abu Dhabi, http://www.cwm.ae/ (retrieved: 15.11.2016)
21. A. A. Saifaie, Envirocities eMagazine, **4**, 4-7 (2013)
22. Sharjah Environment Company (Bee'ah) http://beeah.ae/ (retrieved: 15.11.2016)
23. Abu Dhabi Quality and Conformity Council, Abu Dhabi Certification Scheme for Recycled Crushed Aggregate. Issue 1.0 (2013)

Preparation and characterization of modified bentonite nano composites for phosphorus removal

Ghadeer Al-Khayyat[1,*], *Karin* Schitzenhofer[2], *and* Abdallah Shanableh[3]

[1]Departement of Civil and Environmental Engineering, University of Sharjah, Sharjah, UAE
[2]Institute of Sanitary Engineering and Water Pollution Control, University of Natural Resources and Life Sciences, Vienna, Austria
[3]Research Institute of Sciences and Engineering, University of Sharjah, Sharjah, UAE

Abstract. Due to the restrictions of phosphorus uptake in commonly applied treatment systems such as biological treatment and precipitation, it has been focused to find improved removal technologies. The application of natural nanomaterials as adsorbents is favorable due to high adsorption capacity, low environmental impact and reasonable cost. Specially modified nanoclay presents an efficient and cheap alternative with good recovery abilities. However, the material has to be characterized in detail to understand its behavior. For the characterization, modified bentonite (BNT) was produced through exchanging the cations naturally bound to the bentonite surface with metal polycations. In this study, solutions containing hydroxy-polycations of aluminum (Al-BNT), iron (Fe-BNT), and a combination of aluminum and iron (Al-Fe-BNT) were used to prepare three samples of BNT for phosphorus adsorption. To obtain BNT in an accurate surface – volume ratio, it has to be grinded and sieved. To gain information on the influence of grinding on particle size of modified BNT, the samples were characterized through X-ray fluorescence (XRF), X-ray powder diffraction (XRD), and scanning electron microscopy (SEM). A clear determnation of particle size using SEM was difficult due to agglomeration of samples as a result of milling process. The need of ultrasonication or another separation method increases the production effort of nanosized modified BNT. As due to the improved surface - volume ratio, BNT with smaller mean particle size is expected to perform better, leading to a higher adsorption capacity and adsorption rate and therefore, this additional effort might pay off.

[*]Corresponding author : U15100751@sharjah.ac.ae

1 Introduction

The United Arab Emirates (UAE) experiences a dramatic increase in population that result in an increase in water demand. The UAE is highly dependent on desalination and purification of the available groundwater resources to help meeting the demand for water. The wastewater treatment plants in the UAE are using conventional systems to treat the wastewater without nitrogen or phosphorus removal [1] and therefore, there is a need for further treatment for the removal of such nutrients.

Phosphorus is an essential nutrient for the growth of organisms which may lead to eutrophication, therefore its removal is a major environmental concern specially when it reaches the surface freshwater bodies [2], [3]. Adsorption methods are widely implemented for phosphorus removal from wastewater because they are cheap, efficient, and allow for recovery of adsorbed phosphorus which can be used as fertilizer [1]. Several natural minerals, commercial products and by-products such as zeolite, bentonite, magnetite, alum sludge, fly ash, red mud, gibbsite and boehmite work as phosphorus adsorbents [4]–[9]. Due to their low cost, intrinsic properties and wide availability, clays are favorable in comparison with the other available adsorbents [1], [10], [11].

Clay particles modified with polycations of aluminum, iron, lanthanum or a combination of them can significantly increase the adsorption capacity. The polycations work as pillars between clay layers and therefore increase the surface area, pore volume and permanent porosity [1], [12]. Recently, nano-materials are attracting the attention of researchers to investigate their ability in the removal of pollutants from wastewater. The smaller size of nano-particles compared to micro-sized particles increases the surface area and therefore enhances the adsorption capacity of material against several contaminants [13], [14]. Reducing the size of the clay particles to the nano-level can significantly enhance their ability as adsorbents due to the extensive surface area in addition to the special charge characterization of the clay material [10]. Several studies were performed for the removal of phosphorus and some other anions from wastewater using iron based nano-particles, zinc oxide nano-particles, magnesium oxide nano-particles, copper oxide nano-particles, carbon nano-tubes, and graphene based nano-particles [15]–[18].

The study herein was focused on synthesis and characterization of nano-sized modified clay adsorbents for the removal of phosphorus from water. The modified bentonite adsorbents were prepared using solutions containing hydroxy-polycations of aluminum (Al-BNT), iron (Fe-BNT), and mixtures of aluminum and iron (Al-Fe-BNT). The nano-sized level of particles was reached through ball milling and the samples were characterized through x-ray fluorescence (XRF), x-ray powder diffraction (XRD), and scanning electron microscopy (SEM).

2 Materials and methods

The bentonite particles used in this study are mostly containing sodium (Na-montmorillonite) and it was used in previous study [1]. The bentonite was obtained from a local supplier in Dubai city, United Arab Emirates (Poudrszan Industrial and Mineral Group). The bentonite swell index was observed to be 20 ml/g which is an indication of highly expansive bentonite [1]. The modified bentonite adsorbents were prepared based on previously documented methods [1], [12], [19]. The bentonite was prepared with three types of stock solutions and two types of pillaring solutions (Table 1) to obtain three types of modified bentonite possessing different Al/Fe ratios. The stock solutions consisted of 1) 1 mol/L sodium carbonate through dissolving 105.99 g of dry sodium carbonate in 1 L of distilled water, 2) 0.5 mol/L aluminium chloride through dissolving 120.7 g of aluminium chloride

($AlCl_3.6H_2O$) in 1 L of distilled water, and 3) 0.2 mol/L iron chloride through dissolving 54.1 g iron chloride ($FeCl_3.6H_2O$) in 1 L of distilled water. The Al pillaring solution (Solution A) was obtained by adding 300 mL of stock solution 1 as drops into 500 mL of stock solution 2 to reach a final ratio of [OH]/[Al]=2.4. The procedure was conducted with a magnetic stirrer at 60 °C and then the solution was aged for a week. The Fe pillaring solution (Solution B) was prepared by slowly adding 100 mL of stock solution 1 to 500 mL of stock solution 3 to reach a ratio of [OH]/[Fe]=2. Similarly, the procedure was conducted with a magnetic stirrer at 60 °C and the solution was aged for a week.

Table 1. Preparation conditions of BNT adsorbents [4].

Bentonite sample	Bentonite (g)	Distilled Water (mL)	Solution A (mL)	Solution B (mL)	Ratio mmol (Al + Fe) / g BNT
Blank BNT	10	1000	-	-	-
$Al\text{-}BNT_{2.4}$	10	745	255	-	8
$Fe\text{-}BNT_2$	10	520	-	480	8
$Al_4\text{-}Fe_4\text{-}BNT$	10	632	128	240	8
Annotations: Solution A was prepared by mixing 300 mL of 1.0 M Na_2CO_3 with 500 mL of 0.5 M $AlCl_3.6H_2O$; Solution B was prepared by mixing 100 mL of 1.0 M Na_2CO_3 with 500 mL of 0.2 M $FeCl_3.6H_2O$.					

All clay preparations were produced at a final volume of 1 L. After several washings with distilled water to get rid of chloride in the supernatant which was tested by silver nitrate, the clay was separated by filtration and dried at 105 °C for 24 hours. The dried clay was grinded for varying time intervals to obtain different sizes in micro and nano levels. For getting micro size samples, the original bentonite was grinded and sieved, and a size of 75 μm to 150 μm was obtained. Then, the particels size were further reduced through ball milling method. Ball milling is a widely-used technique for particle size reduction but the problem of agglomeration usually occures whenever it's applied in a dry powder form. A combination of ball milling and ultrasonication can effectively reduce the particle size and minimize agglomeration. The particles were grinded in ball milling machine (Type PM100) using Teflon containers of size 25 ml with eight Teflon balls of 12 mm diameter for 1 hour, 2 hours, and 3 hours under dry conditions at 450 rpm. The particles were then characterized through x-ray powder diffraction (XRD) and scanning electron microscopy (SEM) to ensure reaching the nano-size level and recording the particles sizes.

3 Results and discussion

A knowledge of the material properties and particle characterization is important specially when the material is to be implemented in an application. In this research, the bentonite adsorbents were characterized by XRF, XRD, and SEM. XRF was performed using (HORIBA, XGT-7200V) with a beam diameter of 1.2 mm and for a duration of 200 seconds. Herein, the chemical composition of the modified bentonite samples was determined through XRF. Table 2 and Fig. 1 below show XRF results with silicone as the most dominant element in bentonite.

Table 2. Chemical composition results of bentonite samples using XRF.

Bentonite sample	Element								
	Mg	Al	Si	S	Cl	K	Ca	Ti	Fe
Fe-BNT	2.54	5.68	30.31	0.16	2.45	0.22	0.54	0.20	57.89
Al-BNT	3.71	21.43	61.40	0.48	0.29	0.83	2.27	0.79	8.80
Al4-Fe4-BNT	3.24	11.29	42.80	0.16	0.90	0.42	1.26	0.40	39.52

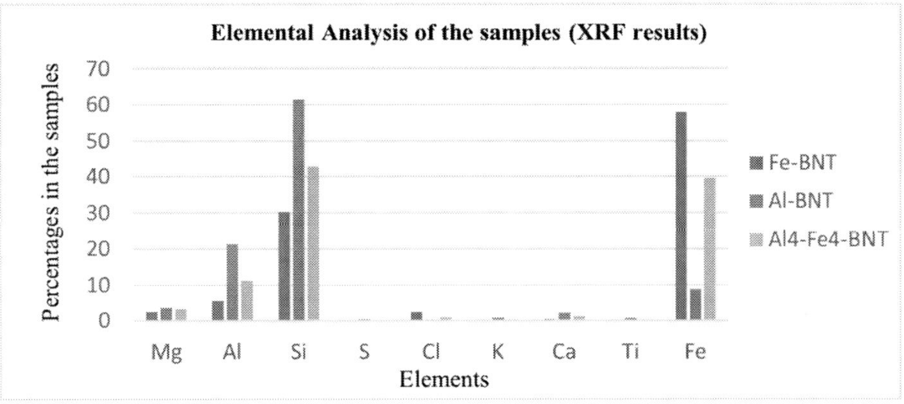

Fig. 1. Percentages of the chemical elements in each of the three modified bentonite samples.

The 1, 2 and 3 hours' ball milled samples of BNT, Al-BNT, Fe-BNT, and Al_4-Fe_4-BNT were analyzed through X-ray powder diffraction (XRD) and shown in Fig. 2 below. XRD test was performed using (Bruker D8 advance) with 40 kv applied voltage and 50 mA current. The step size was 0.2 degree with 0.2 seconds per step for angles between 5 and 85 degrees. XRD test was carried to ensure that structure of particles was not damaged with modifying original bentonite or with reducing particle size. It was found that as the particle size decreased, the count value increased. Also, chemical composition of the different minerals presented in the samples were determined and shown in Fig. 3 below. Cristobalite, quartz, kyanite, dolomite, and aerinite were the dominent minerals in bentonite samples.

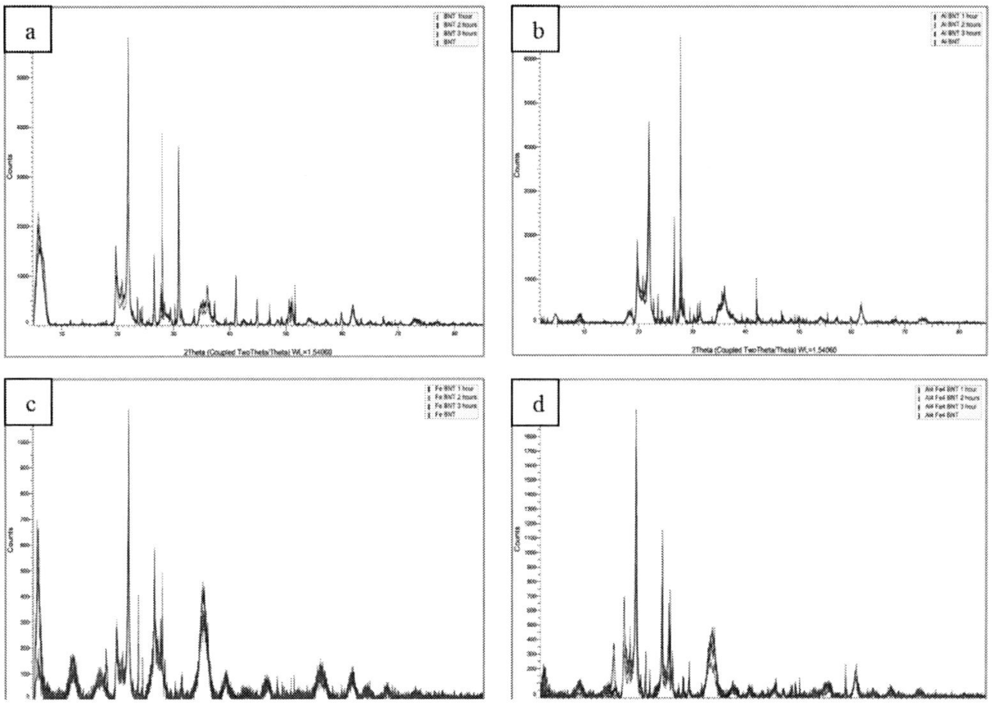

Fig. 2. XRD results of 1, 2, and 3 hours' ball milled samples. (a) BNT sample, (b) Al-BNT sample, (c) Fe-BNT sample, and (d) Al_4-Fe_4-BNTsample.

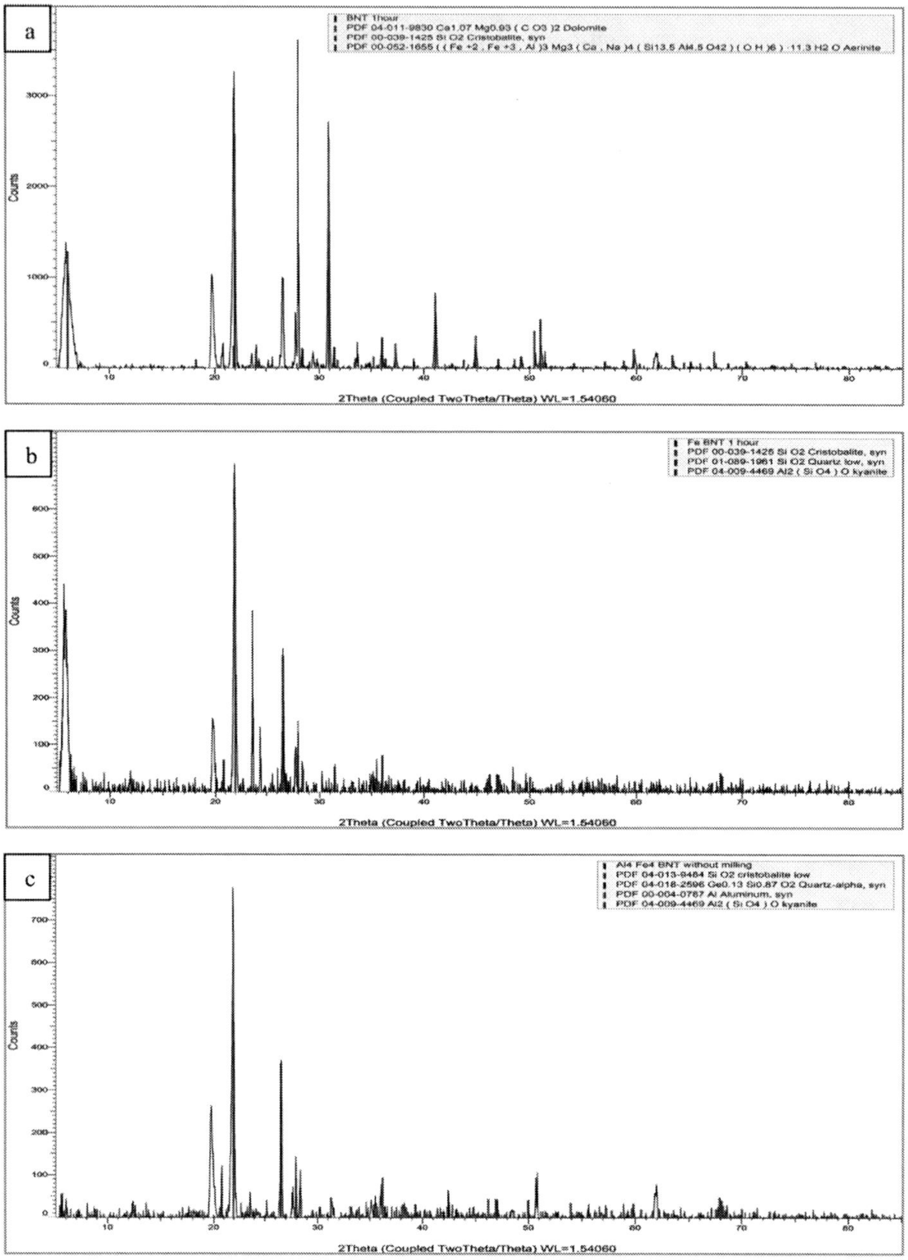

Fig. 3. XRD results for chemical composition of minerals in the samples. (a) BNT sample, (b) Fe-BNT sample, and (c) Al_4-Fe_4-BNT sample.

Prior to characterization of bentonite samples through scanning electron microscopy (SEM) and to reduce particles agglomeration, the particles were dispersed in an ultrasonication path at 24° C for approximately 1 hour. Then, the samples were coated with carbon layer of 20 nm thickness for a duration of 2 minutes using carbon evaborater device (Quorum-CA7625). SEM was perfromed to determine the particle size of bentonite samples using (Tescan, environmental SEM, source: LaB 6) with 30 kv voltage. Fig. 4 below shows

SEM results after 1 hour of ball milling. Furtehr sonication is needed to minimize the presence of agglomerartion as shown in SEM images (Fig. 4).

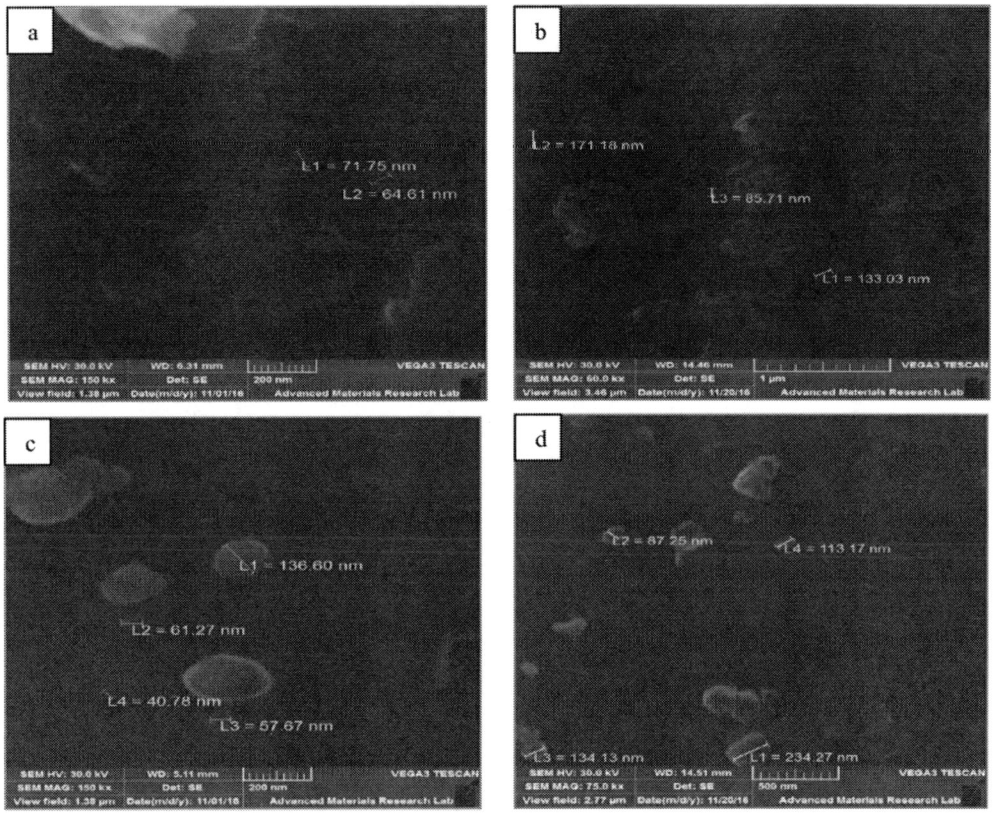

Fig. 4. SEM results of 1 hour ball milled bentonite samples. (a) BNT sample, (b) Al-BNT sample, (c) Fe-BNT sample, and (d) Al_4-Fe_4-BNTsample.

To study the impact of sample particle size against its adsorption capacity for phosphorus with time, an adsorption kinetics experiment was performed. A stock solution of phosphates (50 mg/l of phosphorous) was prepared by dissolving 0.221 gm of KH_2PO_4, dried at 120° C for 2 hours in 1 liter distilled water. The size of the phosphorus stock solution was 50 ml, and the amount of adsorbent added to the solution each time was 0.1 g. The sample pH was adjusted to 5 by using HCL or NaOH, as it's the most appropriate value for phosphorus adsorption. The adsorption experiment was performed with initial concentration of 25 mg P/l. The test was carried on magnetic stirrers at 200 rpm and phosphorus concentration was measured at intermediate intervals (15, 30, 45, 60, 90, 120, 180, minutes … etc.) and up to 5 hours. For adsorbent separation from the solution, a centrifugation technique was applied at 6,000 rpm for approximately 20 minutes. The PO_4-P concentration was measured according to the ascorbic acid standard method with a UV-Vis spectrophotometer.

Table 3 and Fig. 5 below show the results of adsorption experiment using bentonite samples. The results showed significant increase in adsorption capacity of modified bentonite compared to the blank samples. Fe-BNT in the micro size level reached an adsorption capacity of 9.8 mg P/l which is the highest among other adsorbents. Adsorption capacities of samples in the nano size level were not as expected and that's due to the agglomeration of the particles before and during the experiment. Agglomerated particles reduced the surface area for phosphorus to be adsorbed on bentonite samples.

Table 3. Adsorption capacity of modified bentonite against phosphorus.

Bentonite sample	Size Level	Adsorption Capacity (mg P / g BNT)								
		15	30	45	60	90	120	180	240	300
BNT	Micro	0	0	0.2	0	0	0	0.2	0	0
	Nano	0.4	0.2	0.4	0.2	0.2	0.2	0.4	0.2	0.2
Al-BNT	Micro	1.4	1.4	1.4	1.8	2.6	4.0	1.0	3.0	5.4
	Nano	0.8	0.8	1.8	2.4	1.8	3.4	0.6	2.2	4.8
Fe-BNT	Micro	4.2	7	6.8	7.4	7.6	8	8.6	8.2	9.8
	Nano	4.6	5.2	5.6	6.2	5.4	5.6	7.2	7.8	7.8
Al_4-Fe_4-BNT	Micro	0.8	3.4	3.2	5.4	4.4	4.6	4.2	3.6	7.8
	Nano	2.4	3.4	3.6	3.8	3.8	5.6	5.4	5.4	6.2

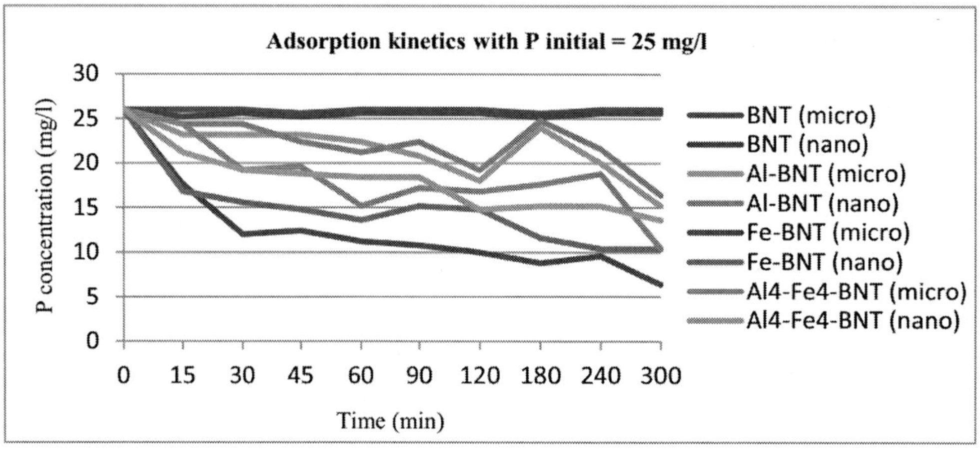

Fig. 5. Adsorption kinetics results of modified bentonite samples against phosphorus.

4 Conclusion

This study was performed for the synthesis and characterization of modified nano bentonite samples for phosphorus adsorption. Samples characterization was done through XRD, XRF, and SEM tests. Size of bentonite particles was around 70 nm after 1 hour of ball milling. The adsorption kinetics experiment showed that modified bentonite samples in micro size level performed better than nano size level. Agglomeration of nano bentonite samples prevented maximum adsorption of phosphorus by bentonite particles. Further treatment for nano samples should be considered before running phosphorus adsorption experiment.

Support from Advanced Materials Research Lab and Environmental Engineering Lab at the University of Sharjah are both acknowledged.

References

1. A. M. Shanableh and M. M. Elsergany, "Removal of phosphate from water using six Al-, Fe-, and Al-Fe-modified bentonite adsorbents.," *J. Environ. Sci. Health. A. Tox. Hazard. Subst. Environ. Eng.*, vol. 48, no. 2, pp. 223–31. (2013)
2. L. Zeng, X. Li, and J. Liu, "Adsorptive removal of phosphate from aqueous solutions using iron oxide tailings," *Water Res.*, vol. 38, pp. 1318–1326. (2004)
3. A. Genz, A. Kornmu, and M. Jekel, "Advanced phosphorus removal from membrane filtrates by adsorption on activated aluminium oxide and granulated ferric hydroxide," vol. 38, pp. 3523–3530. (2004)
4. E. W. SHIN, J. S. HAN, M. JANG, S.-H. MIN, J. K. PARK, and R. M. ROWELL, "Phosphate Adsorption on Aluminum-Impregnated Mesoporous Silicates : Surface Structure and Behavior of Adsorbents," *Environ. Sci. Technol.*, vol. 38, no. 3, pp. 912–917. (2004)
5. E. López, B. Soto, M. Arias, A. Núñez, D. Rubinos, and M. T. Barral, "Adsorbent properties of red mud and its use for wastewater treatment," *Water Res.*, vol. 32, no. 4, pp. 1314–1322. (1998)
6. J. Edzwald, "Phosphate adsorption reactions with clay minerals," *Environ. Sci. Technol.*, vol. 10, no. 5, pp. 485–490. (1976)
7. E. Oguz, "Removal of phosphate from aqueous solution with blast furnace slag," *J. Hazard. Mater.*, vol. 114, pp. 131–137. (2004)
8. E. Oguz, "Thermodynamic and kinetic investigations of PO3-4 adsorption on blast furnace slag.," *J. Colloid Interface Sci.*, vol. 281, no. 1, pp. 62–7. (2005)
9. J. Chen, H. Kong, D. Wu, X. Chen, D. Zhang, and Z. Sun, "Phosphate immobilization from aqueous solution by fly ashes in relation to their composition," *J. Hazard. Mater.*, vol. 139, pp. 293–300. (2007)
10. G. Yuan and L. Wu, "Allophane nanoclay for the removal of phosphorus in water and wastewater," *Sci. Technol. Adv. Mater.*, vol. 8, no. 1–2, pp. 60–62. (2007)
11. M. L. Schlegel, A. Manceau, D. Chateigner, and L. Charlet, "Sorption of Metal Ions on Clay Minerals: I. Polarized EXAFS Evidence for the Adsorption of Co on the Edges of Hectorite Particles," *J. Colloid Interface Sci.*, vol. 215, pp. 140–158. (1999)
12. S. Tian, P. Jiang, P. Ning, and Y. Su, "Enhanced adsorption removal of phosphate from water by mixed lanthanum / aluminum pillared montmorillonite," *Chem. Eng. J.*, vol. 151, pp. 141–148. (2009)
13. M. Anjum, R. Miandad, M. Waqas, F. Gehany, and M. A. Barakat, "Remediation of wastewater using various nano- materials," *Arab. J. Chem.* (2016)
14. S. P. Gubin, Y. A. Koksharov, G. B. Khomutov, and G. Y. Yurkov, "Magnetic nanoparticles : preparation , structure and properties," (2005)
15. S. Mahdavi and D. Akhzari, "The removal of phosphate from aqueous solutions using two nano-structures: Copper oxide and carbon tubes," *Clean Technol. Environ. Policy*, vol. 18, no. 3, pp. 817–827. (2016)
16. B. D. Martin, L. De Kock, T. Stephenson, S. A. Parsons, and B. Jefferson, "The impact of contactor scale on a ferric nanoparticle adsorbent process for the removal of phosphorus from municipal wastewater," *Chem. Eng. J.*, vol. 215–216, pp. 209–215. (2013)
17. B. D. Martin, "Removal and recovery of phosphorus from municipal wastewaters using a ferric nanoparticle adsorbent,". (2010)
18. T. Shahwan, C., Üzüm, A. E. Eroğlu, and I. Lieberwirth, "Synthesis and characterization of bentonite/iron nanoparticles and their application as adsorbent of cobalt ions," *Appl. Clay Sci.*, vol. 47, no. 3–4, pp. 257–262. (2010)
19. L. Zhu and R. Zhu, "Simultaneous sorption of organic compounds and phosphate to inorganic – organic bentonites from water," vol. 54, pp. 71–76. (2007)

Flood inundation mapping for the Gwadar city

Karamat Hussain Barlaas[1] , *Habib* ur Rehman[2] , *Mohsin* Siddique[3*], *Abdul Sattar* Shakir[2], and *Muhammad* Naeem Akhtar[2]

[1]Public Works Department, AJ&K Muzaffarabad, Pakistan
[2]Department of Civil Engineering, UET, Lahore, Pakistan
[3]Department of Civil and Environmental Engineering, University of Sharjah, UAE

Abstract. Floods wreak havoc with human society, infrastructure and the environment. Accuracy and reliability of the forecasts concerning the resultant inundation depths and extents of these floods are primarily dependent upon the employed hydrologic modeling approach. In the present study, a more precise distributed hydrological modeling approach has been exploited with grid-based spatial discretization for the simulation of flood inundation in Akara river watershed which contains a proposed metropolitan-level developed coastal city of Gwadar in the 'Balochistan' province of Pakistan. Institute of Industrial Sciences Distributed Hydrological Model, a physically based distributed hydrological model, has been employed. This model, through fully implicit finite difference schemes, utilizes diffusive wave approximations of 1D and 2D Saint Venant's governing equations for the river and over-land flow components, respectively. While the model is calibrated against the cyclonic flood event of June 2007, to earmark the potential hazardous areas inside as well as outside the Master Plan boundary the flood inundation simulation for the worst precipitation events of 50 and 100 years return periods are performed. The simulated flood inundation depths at various known locations showed good agreement with the observed ones. Moreover, the simulations depict that the simulated flood inundation depths at the same locations increase by 1.07 to 1.87 times and 1.21 to 1.89 times respectively due to 50 and 100 years return period rainfalls.

1 Introduction

Proposed mega scale development in the Gwadar City would increase the land use in Akara river watershed and consequently the area would become more vulnerable to flood inundations in the wake of any likely worst event of precipitation. Relatively more precise appraisal of this inundation is, therefore, inevitable. The effectiveness of a hydrologic forecast about timings, depths and extents of the inundation is indisputably highly dependent upon the degree of precision and accuracy of the techniques involved in the quantification of the surface and river flows in response to a precipitation event.

* Corresponding author: msiddique@sharjaha.ac.ae

In the abovementioned context, the application of remote sensing (RS) and geographic information system (GIS) techniques have become popular in Pakistan to study and predict the various aspects of flooding in the rivers. Many researchers have been applying these RS and GIS techniques either standalone or coupled with additional rainfall-runoff models for the past many years. For example, Haq et al. applied RS and GIS to study the flood monitoring and damage assessment in Sindh province of Pakistan [1] ; Hussain et al. mapped Pakistan 2010 flood using RS data [2]; Khan et al. used RS techniques to conclude that RS data integrated from multispectral and microwave sensors could be used to supplement stream gauges in sparsely gauged large basins to monitor and detect floods [3] and Shakeel et al. performed pre and post flood analysis of 2014 flood along the river Chenab using RS and GIS technique [4].

The present study is, therefore, aimed to use GIS techniques in a more precise physically based distributed hydrological model named as Institute of Industrial Sciences Distributed Hydrological Model (IISDHM) to perform flood inundation in Akara river watershed and to prepare flood inundation maps for the Gwadar city in the event of 50 and 100 years return period rainfalls.

1.1 Location, extent, demography and climate of the Gwadar city

Gwadar district lies in the extreme south of 'Balochistan' province of Pakistan. On the north, it is adjoined by other two 'Balochistan' districts; Turbat and Awaran. On its west, it bounds Pakistan with Iran while on the south it constitutes more than 600 km long coast of Pakistan with Arabian Sea. In the south, its coast embraces the Gulf of Oman, across which lies the oil rich Middle East. On its eastern side lies the Lasbela, a bounding district of Balochistan with Sindh province. Geographically the Gwadar district is located in-between longitude 60° E – 66° E & latitude 25° N – 27° N (Figure 1). The old town of Gwadar lies on the Arabian Sea coast about 48 km to the east of the Gwadar bay. Based on the census of 1981, population of the Gwadar town area is about 43,250 persons with density of 7.4 persons per km^2 [5].The newly proposed Gwadar city by Gwadar Development Authority stretches across the newly named bays as East Padi Zar (East Bay) and West Padi Zar (West bay).

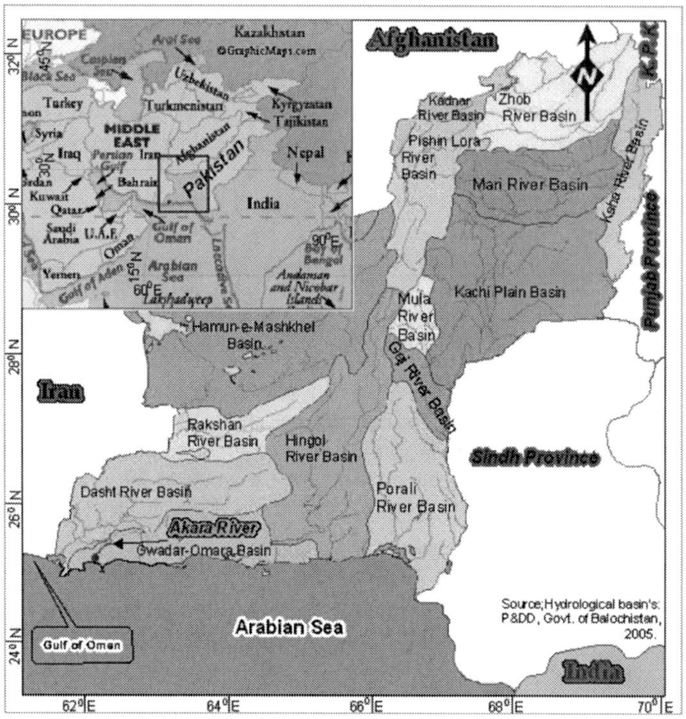

Fig. 1. Geographic location and hydrological context of Gwadar city.

The climate of Gwadar may be classified as arid with warm summer and mild winter. The annual rainfall varies from 75 to 100 mm [6]. The winter rain decreases with the increase in longitude. Rainfall is highly variable in intensity and duration. It is also highly unpredictable, because it occurs inconsistently and occasionally implying that there is no fixed cycle of the rainfall. Actually, downpours are mostly due to local atmospheric disturbances in the sea [7]. Hydrologically the Gwadar city is located in the 'Akara' river basin. It is one of the many sub basins of the major Gwadar-Omara basin of the Balochistan as shown in Figure 1.

2 Institute of industrial sciences distributed hydrological model (IISDHM)

IISDHM is a physically based distributed hydrological model comprising of major processes of hydrological cycle, i.e., interception, evapo-transpiration, river flow, surface flow, and unsaturated and saturated zone flows. A schematic demonstrating of all of the components of IISDHM model is shown in Figure 2. IISDHM is capable of modeling both the overland and sub-surface flow ([8], [9], [10], [11], [12] & [13]). For the subsurface flow, it makes use of 3D Richards Equations. For modeling the two dimensional flow, the model operates on square grid spatial resolution. With the resolution fineness, the model has certain processing constraints owing to its data structure.

For the present study, the surface and river components of the IISDHM are used for the flood inundation simulation. The model makes use of implicit finite difference scheme for solving its diffusive wave form governing equations. In this model, one-dimensional

Saint-Venant Equations (1 & 2) simulate the river flow while the two-dimensional Saint-Venant Equations (3, 4, & 5), simulate the surface flow routing.
The One-dimensional Saint-Venant Equations for river flow can be written as:

$$\frac{\partial Q}{\partial x}+\frac{\partial A}{\partial t}=q \tag{1}$$

$$\frac{\partial Q}{\partial t}+\frac{\partial}{\partial x}[\frac{Q^2}{A}]=-g(\frac{\partial Z}{\partial x}+S_f) \tag{2}$$

Where, Q is the discharge through the cross-sectional area A, x is the distance along longitudinal direction of flow, t is the time, q is the lateral inflow or out flow distributed along the x-axis of the waterway, Z is the water surface elevation with respect to datum, g is gravitational acceleration constant and S_f is frictional slope.

The 2D Saint–Venant's governing equations for the overland flow used in present IISDHM can be written as:

$$\frac{\partial uh}{\partial x}+\frac{\partial vh}{\partial y}+\frac{\partial h}{\partial t}=q \tag{3}$$

$$u\frac{\partial u}{\partial x}+v\frac{\partial u}{\partial y}+\frac{\partial u}{\partial t}=-g(\frac{\partial z}{\partial x}+S_{fx}) \tag{4}$$

$$u\frac{\partial v}{\partial x}+v\frac{\partial v}{\partial y}+\frac{\partial v}{\partial t}=-g(\frac{\partial z}{\partial x}+S_{fy}) \tag{5}$$

Where, u & v are velocities of flow in x and y-directions and S_{fx} and S_{fy} are friction slopes in x and y-directions, respectively.

For river and surface flow routing, diffusive approximations of 1D and 2D Saint-Venant's equations are used based on method of diffusion analogy [14]. It is done by solving implicitly the respective equations through finite difference schemes with a uniform network of square grids. The exchange of flow between the channel network and floodplains is simulated using the floodplain compartment concept. The floodplain compartments are surface grids along the river channel network, which are considered as boundary condition in overland flow routing. Flow transfer between flood plain compartments and river is assumed to occur along the respective Δx reaches adjoining the river and floodplain compartments. Further, this flow is assumed as flow over broad crested weir with submergence correction.

Now, the flow can be either away from the river or into the river, depending upon the relative water surface elevations of the river grid and the flood plain compartment. The river elevations are computed using 1D diffusive wave model equations for channel network and the flood plain elevations are computed by 2D diffusive wave model equations for overland flow. The exchange of flow between floodplain compartment and river reach in each time step Δt is computed by a simple storage routing relation:

$$V_l^{t+1}=V_l^t+(I^{t+1}-O)\Delta t \tag{6}$$

Where, V_l is the volume of water in the flood plain compartment at time $t+1$ or t depending on the water elevation, I is the inflow from the river grids to adjacent floodplain compartments, and O is the outflow from the floodplain compartments to adjacent river grids [11].

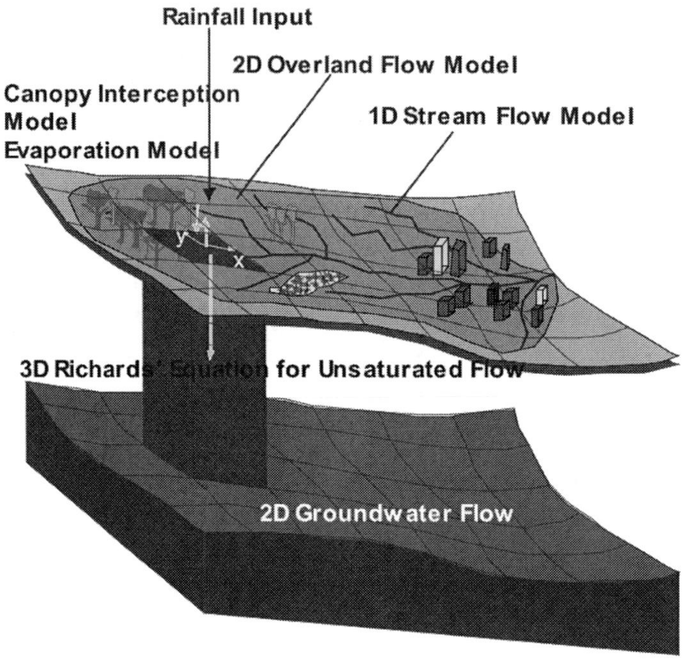

Fig. 2. Schematic representation of IISDHM.

3 Methodology

Methodology in computational modeling is most of the times subservient to the type of the input file formats of the employed models. In the present work, IISDHM dictates its two types of data sets (i.e., spatial and temporal) whose collections from various sources and subsequent formatting in line with the input requirement of the model instantiate the methodology of the present research. Transformation of the output in the graphical format is another feature of the methodology. Schematic diagram of the flood inundation mapping through hydrological modeling, shown in Figure 3, is the best illustrative of the methodology. Spatial and temporal data sets formatted in appropriate form serve as input to model. IISDHM after execution of its respective modules produces output in ASCII format files. These files are, further, processed through ARCGIS version 9.2 in grid format.

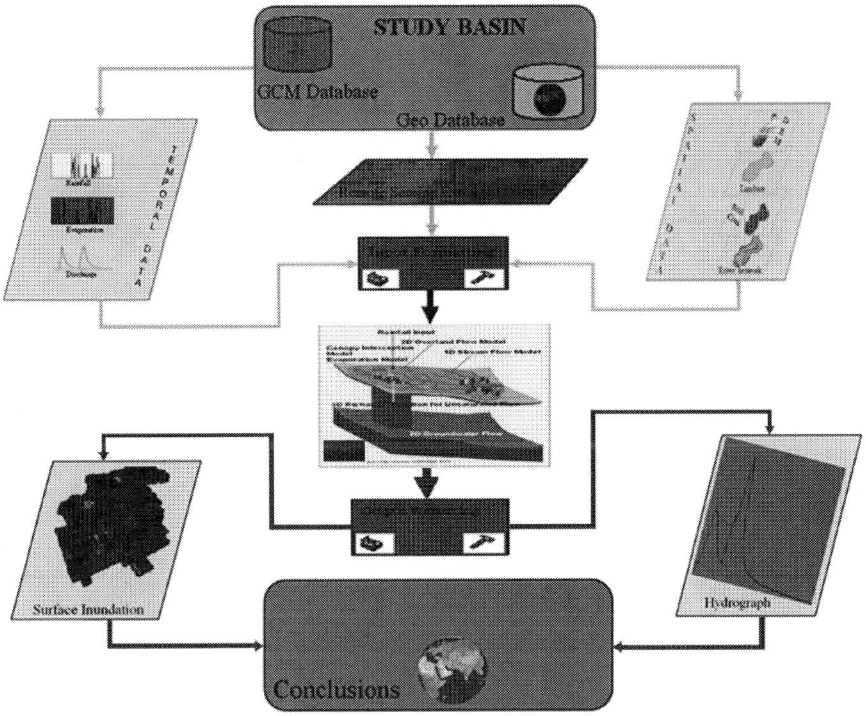

Fig. 3. Schematics of flood inundation mapping.

4 Data collection and model settings for analysis

4.1 Spatial data

Digital Elevation Model (DEM) provides the most important data set that governs the overall flow pattern in the catchment. Resolution of the DEM plays a very crucial role in determining the accuracy and robustness of the results of a model. However, different computational and data access constraints permitted the use of 1 km grid resolution. DEM from GTOPO Official website of U.S Geological Survey (USGS) was accessed for downloading DEM of concerned basin as shown in Figure 4(i). Land use map of the study area was downloaded from the Food and Agriculture Organization (FAO) official website. Its grid resolution was 919 m and the grid size was, therefore, rescaled to 1 km according to the requirement of DEM resolution. Watershed extracted from DEM and its various phases of processing are shown in Figure 4.

Owing to the scope of the present study, the arbitrary values of parameters corresponding to sub-surface flow were utilized during the run of the model. Moreover, Gwadar city's proposed Master Plan was obtained from the Gwadar Development Authority (shown in Figure 4(ii)). It required many adjustments regarding its resolution and spatial reference, which were achieved by intensive geo- processing in ARCGIS. Land use map shown in Figure 4(iii) was also tailored accordingly. The study specific clipped tile was processed for Akara river raster delineation as shown in Figure 4(iv). It was then overlaid by the city Master Plan shown in Figure 4(v). Projected 3D view of the fit-in Master Plan is shown in Figure 5.

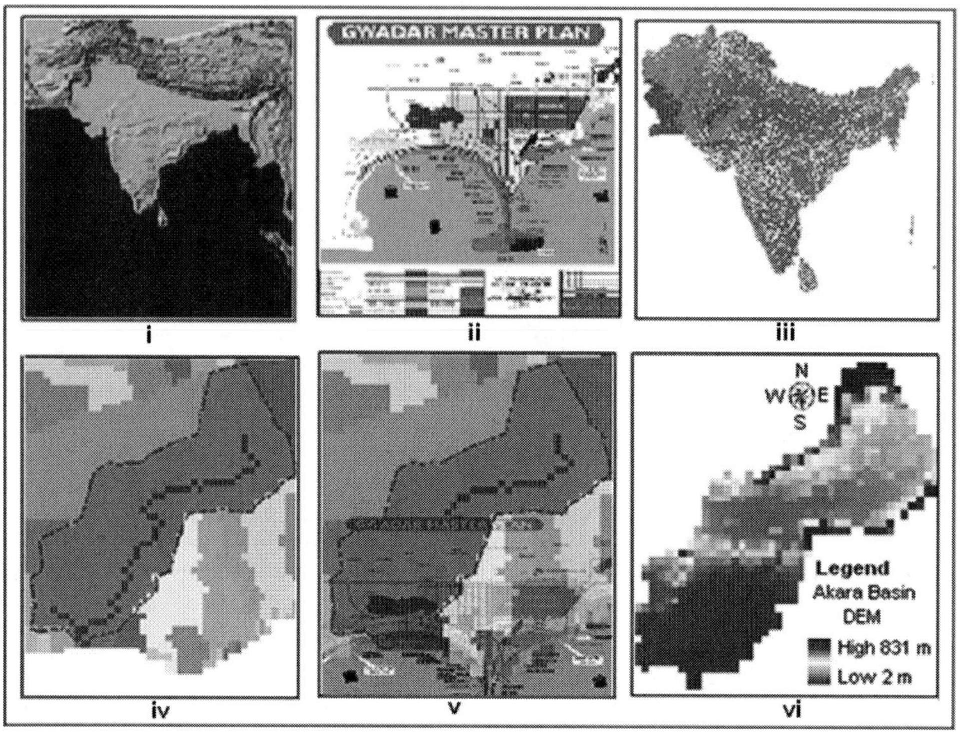

Fig. 4. i) Downloaded DEM, ii) Gwadar Master Plan, iii) Land use map of South Asia, iv) Master Plan specific clipped and processed watersheds, v) watersheds in clipped tile overlaid by the city Master Plan, vi) Akara river basin DEM (processed).

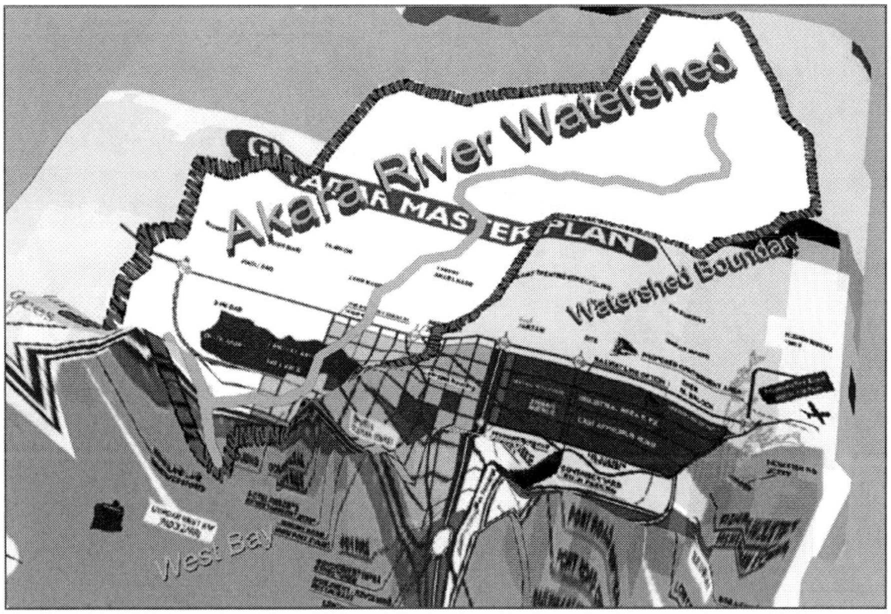

Fig. 5. 3D view of the fit-in Master Plan

4.2 Temporal data

Temporal data comprises of precipitation, evaporation and stream flow data sets. Stream gauging data corresponding to precipitation data of duration 2002-2007 could not be available. Instead the maximum surface inundation mark that resulted from the past cyclonic precipitation of June 2007 was utilized. Precipitation data for the duration 2002-2007 regarding the Gwadar city is tabulated below in Table 1. Precipitation frequency curve through log normal distribution concerning this data is plotted in Figure 6. Projected precipitations for 50 and 100 years return periods were computed graphically for their ultimate use in simulations. The hourly distributions of 50 and 100 years return periods are given in Table 2.

Table 1. Precipitation for Gwadar city (2002-2007).

Year	Monthly Precipitation (mm)												Total mm	Mean mm	Max mm	Min mm
	JAN	FEB	MAR	APR	MAY	JUN	JUL	AUG	SEP	OCT	NOV	DEC				
2002	0	0	0	0	0	-	-	-	-	-	-	-	0	0.00	0.00	0.00
2003	18.5	0	0	0	0	0	12.9	7.8	0	0	0	9.6	48.8	4.07	18.50	0.00
2004	38.3	0	0	0	0	0	0	0.5	0	0	0	18	56.8	4.73	38.30	0.00
2005	56	17.9	15	0	1	0	0.4	4	0	0	0	0	94.3	7.86	56.00	0.00
2006	5.5	8.7	24.2	0	0	0	0	0	0	0	0	102	140.4	11.70	102.00	0.00
2007	0	5.2	33.5	0	0	51	0	1	0	0	1	0	91.7	7.64	51.00	0.00
Mean	19.72	5.30	12.12	0	0.17	10.20	2.66	2.66	0	0	0.20	25.92				
Max	56.00	17.90	33.50	0.00	1.00	51.00	12.90	7.80	0.00	0.00	1.00	102.00				
Min	0.00	0.00	0.00	0.00	0.00	0.00	0.00	0.00	0.00	0.00	0.00	0.00				

The frequency curve is indicative of the interesting result about the return period of precipitation amount of 91.7 mm comprising mainly the precipitation of the 'YEMYIN' cyclone of 2007. The return period corresponding to this amount is about 3 years which has had been naturally validated by 'PEHT' cyclone of June 4, 2010.

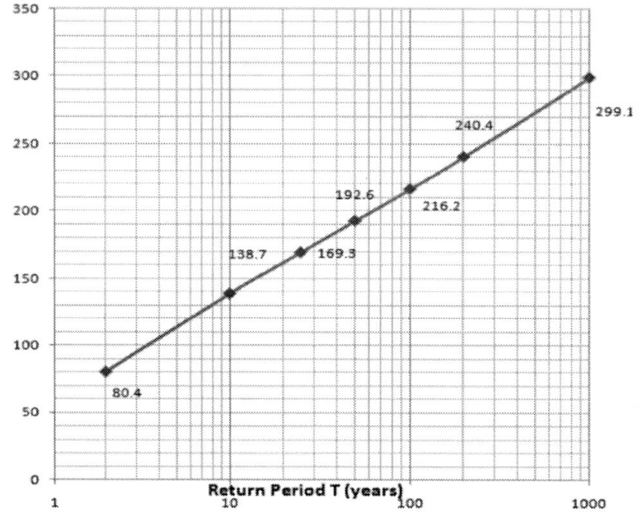

Fig. 6. Precipitation frequency curve (log normal distribution).

Table 2. Distributions of the simulated precipitation.

SN	Day of June	Hour Of Pptn.	Precipitation (pptn) Regarding the			SN	Day of June	Hour Of Pptn.	Precipitation (pptn) Regarding the		
			Base Year	Return Periods Of					Base Year	Return Periods Of	
			2007	50 Years	100 Years				2007	50 Years	100 Years
			mm	mm	mm				mm	mm	mm
1	25	16:00:00	1.0	2.0	2.3	16	26	07:00:00	0.0	0.0	0.0
2	25	17:00:00	1.1	2.3	2.6	17	26	08:00:00	0.0	0.0	0.0
3	25	18:00:00	2.1	4.5	5.0	18	26	09:00:00	0.0	0.0	0.0
4	25	19:00:00	0.0	0.0	0.0	19	26	10:00:00	2.6	5.5	6.2
5	25	20:00:00	0.0	0.0	0.0	20	26	11:00:00	3.0	6.2	6.9
6	25	21:00:00	0.0	0.0	0.0	21	26	12:00:00	5.7	12.0	13.5
7	25	22:00:00	0.0	0.0	0.0	22	26	13:00:00	3.0	6.2	6.8
8	25	23:00:00	0.0	0.0	0.0	23	26	14:00:00	3.3	6.9	7.8
9	25	24:00:00	0.0	0.0	0.0	24	26	15:00:00	6.3	13.4	15.1
10	26	01:00:00	0.0	0.0	0.0	25	26	16:00:00	5.6	11.5	13.3
11	26	02:00:00	0.0	0.0	0.0	26	26	17:00:00	6.2	13.2	14.8
12	26	03:00:00	0.0	0.0	0.0	27	26	18:00:00	12.1	25.4	28.5
13	26	04:00:00	0.0	0.0	0.0	28	26	19:00:00	0.5	1.1	1.2
14	26	05:00:00	0.0	0.0	0.0	29	26	20:00:00	0.6	1.2	1.4
15	26	06:00:00	0.0	0.0	0.0	30	26	21:00:00	1.1	2.3	2.6
						Total	2	30	54.2	113.8	128.0

4.3 Model settings for analysis

Model set up of IISDHM generally includes preparation of input data i.e. spatial data, temporal data, calibration of the model and application of the model. Spatial data includes files related to watershed boundary, river network, DEM, land use and soil map of the catchment. Temporal data consists of hourly rainfall data and evaporation data. In the present study, model set up comprises of 1 km spatial resolution and 1 hour temporal resolution. Spatial resolution is achieved by processing the downloaded DEM of 990 m resolution to 1 km. Respective DEM was obtained from official website of USGS and is known as GTOPO DEM 1 km. Land-use map of the respective catchment obtained from official database of FAO was also converted to 1 km resolution. Respective Manning roughness coefficients were assigned according to the land-use of each Grid. For the channel network, Akara River along with its Chatt tributary was specified. Gauging records of it however could not be available; the boundary conditions were therefore assigned accordingly.

Model calibration in abstract models is always a pivotal phase before the application of the models. Repeated model execution was carried out by tuning the respective parameters, such as manning roughness coefficient, run-off coefficients, until the simulated results were found satisfactorily close to the observed inundation values of the specific past records. The maximum inundation caused by precipitation event of June 2007 was the only record utilized for this purpose. In the absence of an explicit evidence of maximum inundation depth record inside the study watershed, the general maximum inundation depth of 14 feet (420 cm) quoted David Darg [15] Assistant Director in Operation Blessing International, was utilized in the calibration process. This maximum inundation depth was supposedly referred to a location inside the study watershed. The cell corresponding to maximum inundation was mapped and its reference x and y coordinates in Universal Transverse Mercurator (UTM) Projection Coordinate System were found to be 439045.587537 m and 2818038.18951 m respectively and hence it was declared as the model calibration reference in the present study. This point lies in the extreme north of the watershed and is shown in the simulation results of the next section.

After calibration, the model is applied against two prospective worst precipitation events of 50 and 100 years return periods. The model application results are produced both graphically and in tabular formats and various inferences are made about the inundation mapping pattern and drainage characteristics of the area under consideration. The results, discussions and conclusions are described in the following section.

5 Results and discussions

IISDHM with regards to its overland flow model designates mainly the roughness coefficient and the runoff coefficients as the major calibration parameters. The evaporation parameters in the current scenario of diffusive flow routing environment are not much effective compared with above mention parameters. Flooding event of cyclonic precipitation of 25th and 26th of June 2007 was utilized and the repeated runs were carried out. The appropriate matching of the simulated and observed amounts by adjusting the parameters was achieved. The results of this simulation are shown in Figure 7.

According to simulated results, maximum inundation of 420 cm occurred outside the Master Plan Area in a cell located at extreme north of the watershed. This depth takes place in the 68th hour of simulation after the start hour of precipitation that was the 16th hour of June 25, 2007. Against the same hour, maximum inundation inside the Master Plan Area is

102 cm at 2 km southwest of Shahdad area as shown by the city specific enlarged map in Figure 8.

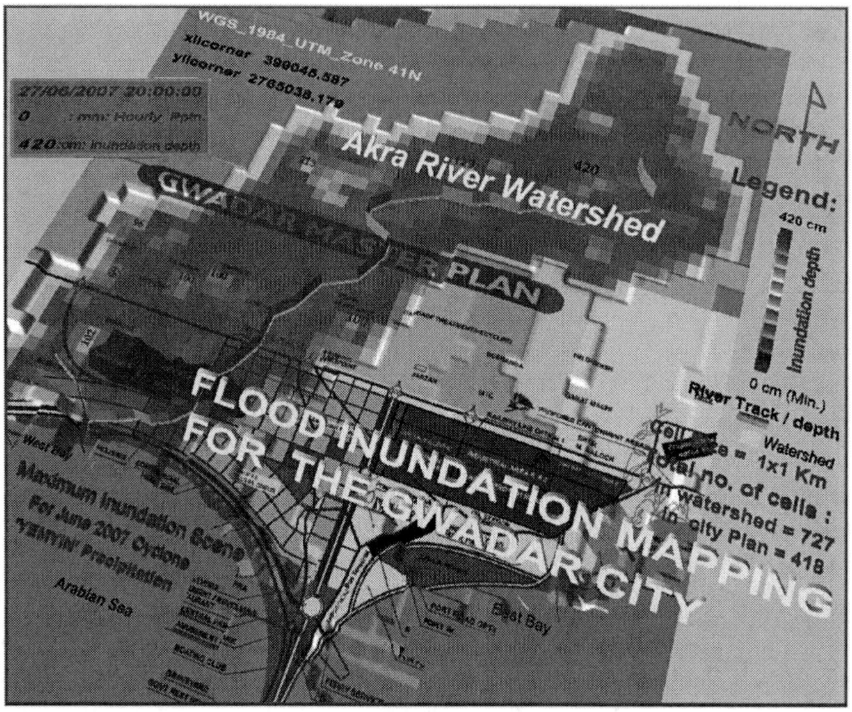

Fig. 7. Simulated view for the cyclonic event of June 2007.

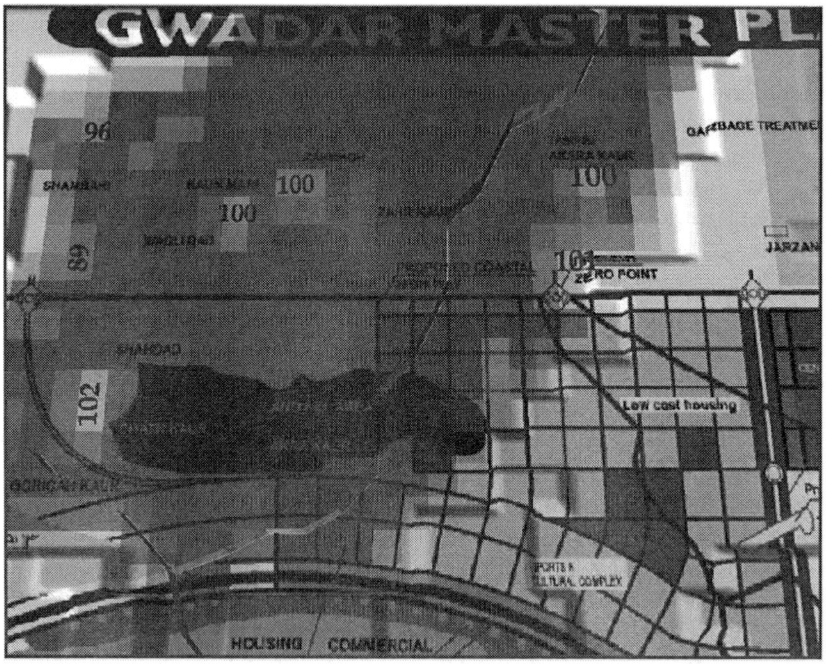

Fig. 8. Enlarged inundation view of June 2007 inundation.

5.1 Model application for 50 years return period

Model application for the 50 years return period precipitation (Figure 9) shows the maximum inundation of 517 cm outside the Master Plan Area at the location corresponding to the observed inundation depth of 420 cm during the event of cyclonic precipitation of June 2007. The simulated maximum inundation inside the Master Plan Area is 180 cm at the northern part of Shambari area in contrast to the maximum inundation of 101 cm in the locality of Shahdad area during the cyclonic event of 2007. Besides this contrast of location, there is another temporal discrepancy regarding the occurrence hour of the maximum inundation i.e. in the previous case it is the 68th hour while in this case it is the 43rd hour of precipitation.

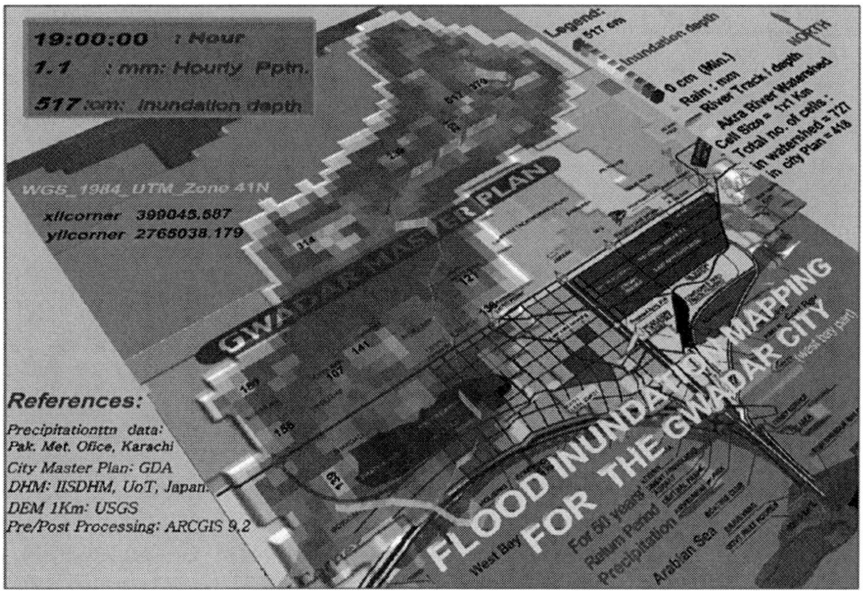

Figure 9. Inundation view for 50 years return period

These spatial and temporal contrasts of inundation are indicative of the varying drainage characteristics of the corresponding watershed under different precipitation intensities that explicitly reflect the temporal scale of the inundation. The other contrasts in inundation depths are due to relatively centrally located cells/clusters of the Master Plan Area. These refer to the locations of Kaun Mari, Zahrkoh and Akara Kaur locations. Corresponding to event of 2007, these depths are 100 cm for each cell while in the case of inundation for 50 years return period these depths rise to 107 cm, 141 cm and 121 cm, respectively. Figure 10 shows the enlarged view of the Master Plan area.

Fig. 10. Enlarged inundation view for 50 years return period.

Another comparison that is noteworthy with regards to inundation is the minimum inundation observed in the context of labeled cells. This minimum simulated inundation depth is observed as 89 cm in the Shambari locality cell while in the 2nd corresponding case the minimum depth takes place in the Kaun Mari locality cell. This spatial shift regarding the minimum depth in both of the cases may also be considered as another characteristic of the drainage of the watershed.

5.2 Model application for 100 years return period

Results from the application of the model for 100 years return period precipitation are shown in the Figure 11. The results are almost similar to those of 50 years return period precipitation. The only temporal difference of one hour earlier (i.e., 42nd hour of simulation) occurrence of maximum inundation depth is observed in this simulation.

The location of maximum inundation like the previous cases is the same. The inundation is exceeded by 4 cm and 2 cm outside and inside the Master Plan Area, respectively, in comparison to the case of 50 years return period (i.e. maximum depth reached outside Master Plan Area in the current scenario is 521 cm as compared to 517 cm in case of 50 years return period). Similarly maximum depth arrived inside the Master Plan Area at North of Shambari is 182 cm as compared to 180 cm in the case of 50 years return period.

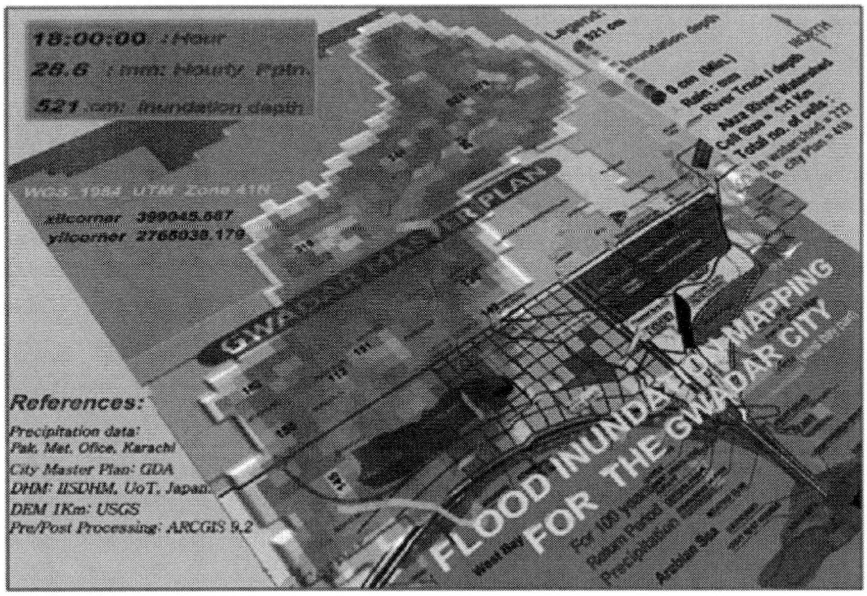

Fig. 11. Inundation view for 100 years return period.

However, the minimum depth in the Master Plan Area is exceeded by 5 cm as compared to the minimum depth simulated in 50 years return period case. The location of minimum depth is also the same (i.e. KaunMari). Figure 12 shows the enlarged view of inundation in Master Plan Area.

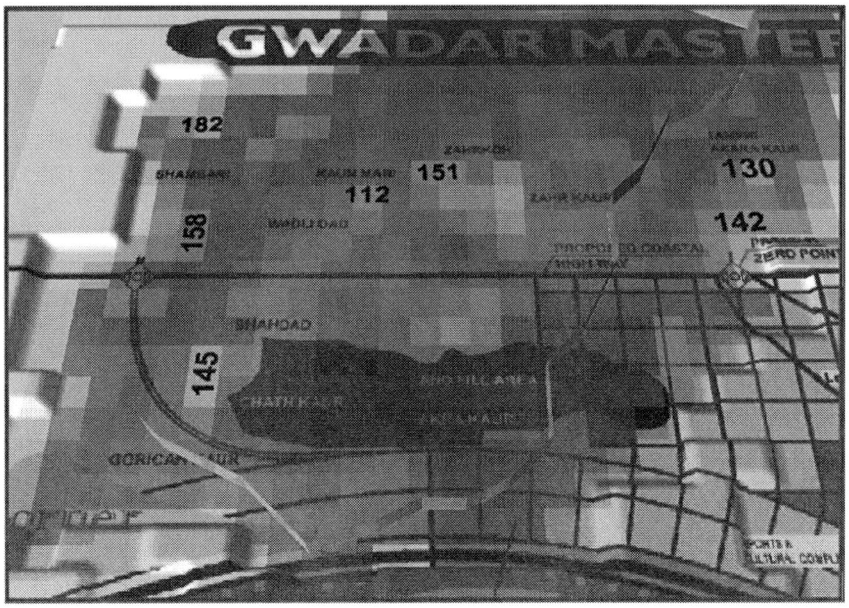

Fig. 12. Enlarged Inundation view for 100 years return period.

Summarized results of both simulations are given in Table 3 along with location coordinates.

Table 3. Summarized results.

Cell / Cluster ID	X Co-ordinate of the cell/ cluster	Y Co-ordinate of the cell/ cluster	Name of City/ Town Area in the Vicinity Of the Cell	No of Cells in the cluster	Model Calibration On the basis of		Model Application For Projected precipitations of			
					Inundation from precipitation of Cyclone 'Yemyin' in Balochistan On June 25, 26 & 27, 2007		50 Years' Return Period		100 Years' Return Period	
					Surface Inundation Depth. (Max.)	Hour Of Max. Inund.	Surface Inundation Depth (Max.)	Hour Of Max. Inund.	Surface Inund. Depth (Max.)	Hour Of Max. Inund.
Index	m	m	Location	-	cm	hour	cm	hour	cm	hour
*R_1	413045.58	2795038.1	Shahdad	2	102	68	139	43	145	42
R_{2a}	412045.58	2800038.1	Shambari	2	89	68	158	43	158	42
R_{2b}	427045.58	2800038.1	ZeroPoint	1	101	68	136	43	142	42
R_3	417045.58	2801038.1	KaunMari	1	100	68	107	43	112	42
R_{4a}	419045.58	2802038.1	Zahrkoh	2	100	68	141	43	151	42
R_{4b}	428045.58	2802038.1	Akra-Kaur	1	100	68	121	43	130	42
R_5	412045.58	2804038.1	North of Shambari	1	96	68	180	43	182	42
R_6	[illegible]	[illegible]	Outside ***MPA	[illegible]	420= Simulated **420=Observed	68	517	43	521	42

**R1, R2, R3, R4 and R5 are the indices of respective cell-containing rows with subscripts number indicating increase of rows form bottom to top in Watershed, while the subscripts 'a' and 'b' represent centre left and centre right cells in the same numbered row, respectively.*
*** David Darg, July 30, 2007 [15].*
****Master Plan Area (MPA)*

6 Conclusions and recommendations

Based on the results and discussions, the following conclusions can be drawn:

1. Spatial analysis result illustrates that about half of the Master Plan Area of Gwadar city lies in the study watershed, which implies that it is relatively the largest watershed concerning the city Master Plan. Hence, the detailed hydrologic analysis of this watershed is of prime importance in the context of future planning, design or development activity for the Gwadar city.
2. The temporal data analysis through precipitation frequency analysis enabled the estimation of various return periods concerning potential worst precipitations. Based on the frequency curve, the return period of annual precipitation in 2007, comprising mainly that of 'Yemyin' Cyclone, is about 3 years which is really alarming particularly with reference to a coastal city like Gwadar, that is undergoing a mega scale developmental.
3. During the calibration of the model, the simulated results showed a very good agreement with the observed past inundation results.
4. The application of model for 50 and 100 years return period clearly earmarked the potentially hazardous areas both inside and outside the Master Plan Area by depicting the location, extent and depth of inundation. Simulated results, further, depict that the flood inundation depths at the same locations increase by 1.07 to 1.87 times and 1.21

to 1.89 times, respectively, due to 50 and 100 years return period rainfalls. The concerned authorities are, therefore, recommended to consider this inundation variation spectrum in conjunction with any other prospective similar study of the corresponding area.

5. The comparative statement of results enabled the preparation of the flood vulnerably spectrum of the potentially hazardous areas such as northern locations of Shambari where the maximum inundation of 180 cm is observed for 50 years return period precipitation. Moreover, outside of Master Plan Area, the potential hazardous location lies at extreme north with maximum inundation depth of 517 cm.

References

1. M. Haq, M. Akhtar, S. Muhammad, S. Paras, J., Rahmatullah, The Egy. J. RS and Space Sci, Vol. **15**, pp. 135-141, (2012).
2. E. Hussain, S. Urala, A. Malik, J. Shana, In Proc. of the ASPRS ann. conf., Milwaukee, WI, USA, 1–5 May 2011, (2011).
3. S. I. Khan , Y. Hong. J.J. Gourley , M. U. Khattak and T. D. Groeve., RS., Vol, **6**, pp. 2393-2407, (2014).
4. M. Shakeel, K. Munir, K. Hina, R. Saeed, Q. Arshad., J Earth Sci Clim Change. Vol. **6**, pp. 318, (2015).
5. Population Census Organisation (PCO), Population Census -1981, Yearly book, Statistics Division, Ministry of Economic Affairs and Statistics, 69-E, Adeel Plaza, Blue Area, Islamabad, (1981).
6. Directorate of Agriculture (DOA), Balochistan, 1999-2000, Quetta, Pakistan, (2000).
7. U. R. Habib, B. Rabin., *Proc. International Symposium*, AIT, Bangkok, (2005).
8. R. Jha, S. Herath, and K. Musiake., *Proc. of congress of the IAHR*, pp 627-632, (1997).
9. R. Jha, S. Herath, and K. Musiake,. A. J. of Hyd. Eng. Vol., **42**, (1998).
10. D. Dutta, S. Herath and K. Musiake., Distributed hydrological model for flood inundation simulation, A. J. of Hyd. Eng., Vol. **43**, pp 25-30 (1997).
11. D. Dutta, N. Shulian, and K. S. Rajan., Modelling and socio-economic impact analysis of floods in coastal cities under sea-level rising scenarios in Bangkok, Thailand, *Proc. International Symposium,* AIT, Bangkok, (2005).
12. U. R. Habib, and B. Rabin., Modeling of floods in coastal cities under sea level rising scenarios: a case study of Karachi, Pakistan, *International Symposium on Floods in Coastal Cities under Climate Change Conditions, Thailand* (2005).
13. H. A. Bakhsh, U. R. Habib, A.S. Shakir, and M. K. Noor., Flood inundation modeling for Malir watershed of Karachi considering future mean sea level rise, Pakistan *J. of Eng. & App. Sci, UET, Lahore,* Pakistan, Vol. **9**, pp. 34-47, (2011).
14. J. George, *Theoretical Physics*, 2nd ed., Hanfer publishing company, New York, pp. 590-594, (1950).
15. D. David., http://www.alertnet.org/db/blogs/34417/2007/05/27-102352-1.htm, accessed in Jan 2014, (2007).

MATEC Web of Conferences **120**, 06001 (2017) DOI: 10.1051/matecconf/201712006001
ASCMCES-17

Design and analysis of single piles under lateral loads in cohesionless soils based on the standard penetration test (SPT)

Abdesselem Laouedj[1], and *Ali* Bouafia[2,*]

[1]University Mouloud Mâamery, Faculty of Construction, Department of Civil Engineering, Tizi-Ouzou, Algeria
[2]University Saâd Dahleb, Faculty of Technology, Department of Civil Engineering, Blida, Algeria

Abstract. The Standard Penetration Test (SPT) is nowadays widely used in geotechnical projects to characterize cohesionless soils due to its simplicity and cost effectiveness. This test is appropriate to estimate the resistance and density of cohesionless soils and is largely practised in the Arabian Gulf region within the scope of foundation projects. Since this test was internationally standardized, it may be considered as a practical tool of soil characterization as well as of pile foundation design. This paper is aimed at presenting a practical method of constructing the P-Y curves on the basis of the SPT test derived from a thorough analysis of a worldwide database of pile loading tests in sandy soils. The two main parameters of the P-Y curve, namely the lateral subgrade reaction modulus and the lateral soil resistance, were correlated by a back-analysis procedure to the N_{spt}.

1 Introduction

The design of pile foundation submitted to lateral loads based on serviceability limit states as well as ultimate limit states requires the analysis of lateral load-deflection behaviour of a single pile. However, the response of a pile under lateral loads is rather a complex problem of soil/pile interaction due to the three-dimensional character of the lateral soil reaction, the diversity of the mechanical and geometrical parameters involved and the inherent nonlinearity of the soil behaviour. A pragmatic approach consists of analysing the pile deflections by calibrating the P-Y curves based on the experimental load-head deflections data obtained from full-scale lateral loading tests carried out within the scope of geotechnical projects.

The P-Y curve describes a local relationship at a given depth along the pile between the lateral soil reaction P undertaken by a spring at the pile/soil interface and the lateral pile deflection Y at the same depth. As shown in Figure 1, the P-Y curve has a non-linear shape and is characterized by an initial slope denoted E_{ti} and called the lateral reaction modulus, and a horizontal asymptote P_u corresponding to the lateral soil resistance [1].

* Corresponding author: ali.bouafia@univ-blida.dz

In spite of the large number of applications of the Standard Penetration Test (SPT) in foundation engineering, there is seemingly no publication highlighting the contribution of SPT test to the definition of the parameters of P-Y curves.

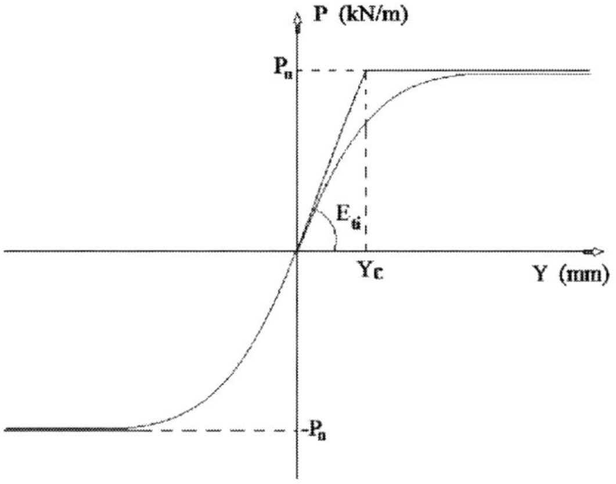

Fig. 1. Schematic shape of the P-Y curve [1].

This paper aims at presenting a practical method of construction of the P-Y curves for piles in sand on the basis of the SPT test, derived from a detailed interpretation of full-scale lateral loading tests on single piles carried out in sandy soils.

2 Methodology of analysis of the P-Y curves

As illustrated in figure 1, the P-Y curves were usually described as hyperbolic shaped functions, which show an inherent non-linear load-deflection response. In many research works, the experimental P-Y curves were fitted by the following hyperbolic function [2]:

$$P = \frac{y}{\frac{1}{E_{ti}} + \frac{y}{P_u}} \tag{1}$$

At a given depth z, the P-Y curve parameters E_{ti} and P_u may be correlated to the N_{spt} and the pile characteristics, according to the following general equation:

$$f(E_{ti}, P_u, N_{spt}, D, B, E_pI_p, \sigma_{v0}) = 0 \tag{2}$$

E_pI_p, D and B are respectively the pile flexural stiffness, the pile embedded length and the pile diameter (or the dimension perpendicular to the lateral load direction). By using the Vashy-Bukingham's theorem of dimensional analysis this equation is transformed into an equivalent equation described by (N-k) dimensionless parameters noted π, k being the number of fundamental units in equation (2). Since N=7 and k=3, the equivalent equation is written as follows:

$$g(\pi_1, \pi_2, \pi_3, \pi_4) = 0 \tag{3}$$

$\pi_1 = \frac{E_{ti}}{N_{spt}\sigma_{v0}}$ is noted hereafter K_E and called the modulus number,

$\pi_2 = \frac{P_u}{\sigma_{v0} B}$ is called the lateral resistance factor and noted by K_N,

$\pi_3 = \frac{D}{B}$ is the pile slenderness ratio,

$\pi_4 = \frac{E_p I_p}{\sigma_{v0} N_{spt} D^4}$ is the lateral pile/soil stiffness ratio noted hereafter by K_R.

The terms K_E and K_N should be independent since the first one corresponds to the small displacements of the pile, whereas the second one corresponds to the large displacements. Consequently, the equation (3) may by uncoupled to the two following equations:

$$K_E = \frac{E_{ti}}{N_{spt}\sigma_{v0}} = h(N_{spt}, K_R, \frac{D}{B}) \tag{4}$$

$$K_N = \frac{P_u}{\sigma_{v0} B} = j(N_{spt}, K_R, \frac{D}{B}) \tag{5}$$

Deriving the functions h and j allows a practical formulation of the P-Y curves parameters as follows:

$$E_{ti} = K_E N_{spt} \sigma_{v0} \tag{6}$$

$$P_u = K_N B \sigma_{v0} \tag{7}$$

The lateral pile/soil stiffness K_R described by the term π_4, will hereafter be calculated on the basis of a "Characteristic soil modulus" along the pile, noted E_c, as follows:

$$K_R = \frac{E_p I_p}{E_c D^4} \tag{8}$$

$E_c$is defined as the weighted average value of the product ($N_{spt}(z).\sigma_{v0}$ (z)) along the effective length D_e:

$$E_c = \frac{1}{D_e} \int_0^{D_e} N_{spt}(z) \sigma_{v0}(z) dz \tag{9}$$

D_e is the effective embedded length of the pile, beyond which the pile segments do not deflect, and is computed as follows [2]:

$$D_e = min\ (D, \pi L_0) \tag{10}$$

The elastic length (or the transfer length) $L_0$is given by:

$$L_0 = \sqrt[4]{\frac{4 E_p I_p}{E_{ti}^c}} \tag{11}$$

E_{ti}^c is called "Characteristic subgrade reaction modulus" (or the average subgrade reaction modulus) of the equivalent homogeneous soil given by:

$$E_{ti}^{c} = \frac{1}{D_e} \int_{0}^{D_e} E_{ti}(z)dz \tag{12}$$

The back-analysis procedure was undertaken for each pile-loading test by seeking the values of K_E and K_N giving the best calibration of the experimental loading curve at the pile top. In other words, it is aimed at finding for each pile loading test, on the basis of the least squares technique, the couple (K_E, K_N) giving the best value of the regression coefficient R. The function h and j will be then derived by fitting respectively the values of K_E and K_N as functions of the variables D/B, K_R and N_{spt} appearing in equations (4) and (5).

The tool of the back-analysis was the P-Y curves-based software SPULL (**S**ingle **P**ile **U**nder **L**ateral **L**aods), developed at the University of Blida [21], [22]. During this process, sets of hyperbolic P-Y curves along the pile, described by the equation (1), were input and the output was the pile head deflection for a given lateral load, which provides a predicted load-deflection curve, as illustrated in figure 2, for a couple of parameters (K_E, K_N) and the corresponding regression coefficient R computed. The trial-and-error process continues until an excellent agreement is found between the experimental deflections and the predicted ones. Consequently, the coefficient R should be as close as possible to 100% with a threshold of acceptance fixed at a level of 95%.

3 Description of the database of pile loading tests

A lot of technical documents and reports describing full-scale lateral loading of piles are available in the pile foundations literature. References [3-20] were used to build a medium sized database of pile loading tests. 44 loading tests were carried out in 15 sites where the SPT test was carried out within the scope of the geotechnical investigation.

The test piles were made from reinforced or prestressed concrete, steel pipe, HP pile or composite materials, and installed into the soil according to many techniques: boring, driving, cast-in-drilled hole, and cast-in-steel shell.

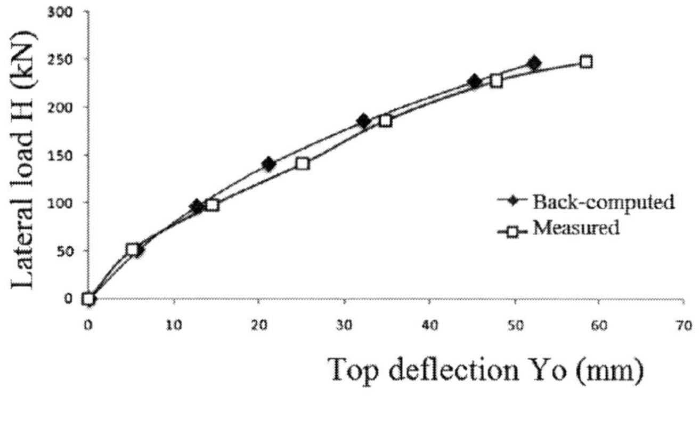

Fig. 2. Illustrative example of back-computation

The margin of slenderness ratio D/B and the pile/soil stiffness ratio K_R for the piles analysed are respectively: 10-60, and 10^{-4}-$2x10^{-1}$. It is to be noted that 70% of the test piles are characterised by K_R less than 10^{-2}, which usually corresponds to flexible piles. Figure 3 illustrates a typical experimental arrangement of a pile loading test.

Lithology of the experimental sites is mainly composed of multi-layered deposits of silty and/or clayey sand whose N_{spt} profile is not homogeneous with depth. In some sites, a ground water table was found.

4 Presentation of the results

Back-computation of the experimental piles led to conclude that the slenderness ratio D/B has a negligible effect on both the parameters K_E and K_N. However, it was found all the piles analysed were flexibles ($D > \pi L_0$) which limits the proposed values of K_E and K_N to the flexible piles. Moreover, K_E and K_N depend on the position of the P-Y curve with respect to the ground water level. Table 1 summarizes the values of these parameters, which allows directly defining the P-Y curves on the basis of the equations (1), (6) and (7). In figure 4 are shown typical histograms of the modulus number and the factor of lateral resistance.

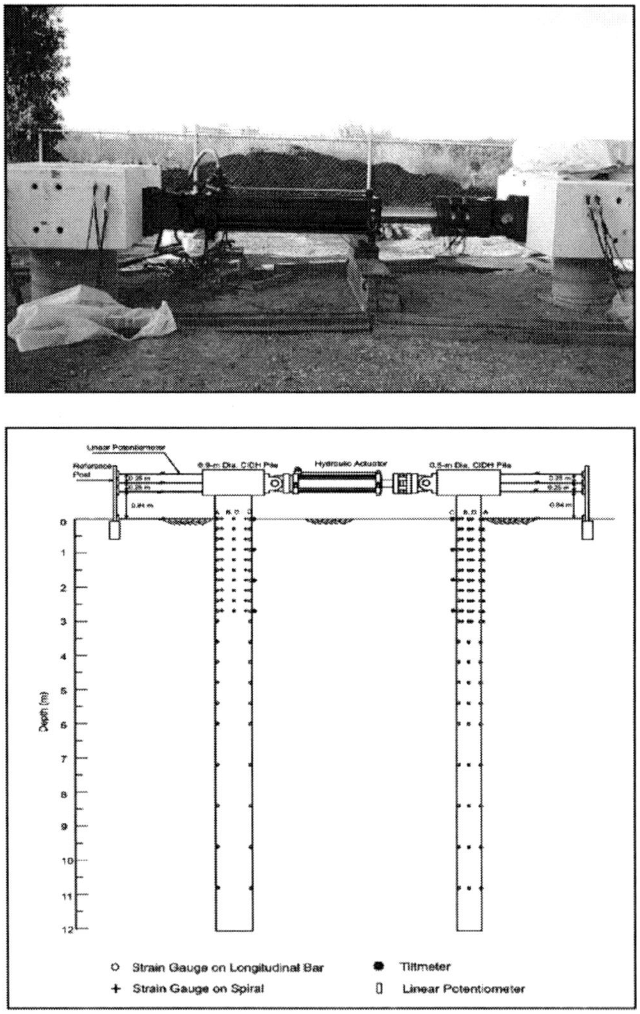

Fig. 3. Typical lateral load test set-up and locations of instruments [3].

According to table 1, submerged layers of sand exhibit lesser values of K_E and K_N than those above the water table.

Table 1. Values of the parameters K_E and K_N

	Flexible piles (D > πL_0)	
Above water table	K_E=318.0	K_N=25.3
Below water table	K_E=171.4	K_N=16.0

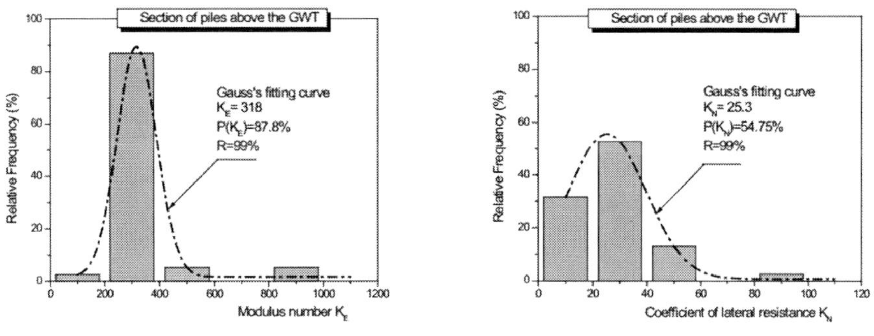

Fig. 4. Typical histograms of the parameters K_E and K_N

In all the geotechnical design using the SPT test (bearing capacity of shallow foundations and piles, liquefaction analysis, etc) a normalized value N_{spt}^1 is used rather than the gross measured N_{spt} value. According to ASTM standard, the coefficient of normalization C_N is given by [23]:

$$N_{spt}^1(z) = N_{spt}(z)C_N = N_{spt}(z)\sqrt{\frac{\sigma_{ref}}{\sigma_{v0}(z)}} \tag{13}$$

σ_{ref} is a stress reference conventionally equal to 100 kPa, and the coefficient C_N should be limited to the margin 0.5-2. It is suggested to reformulate the equation (6) in order to take into consideration N_{spt}^1 rather than N_{spt} as follows:

$$E_{ti}(z) = K_E N_{spt}^1(z)\sqrt{\frac{\sigma_{v0}^3(z)}{\sigma_{ref}}} \tag{14}$$

5 Conclusions

This paper presents a new practical method of defining the parameters of P-Y curves for single piles under lateral loading in sand on the basis of the Standard Penetration Test (SPT) derived from a thorough analysis of a worldwide database of pile loading tests in sandy soils, which presents a contribution to the use of the SPT in the design of laterally loaded piles.

The two parameters of the P-Y curve, namely the subgrade reaction modulus and the lateral soil resistance were correlated to the N_{spt} depending on the position of the curve with respect to the ground water table as well as on the pile/soil stiffness ratio.

Further works will focus on the predictive capability of such a method of construction of P-Y curves by undertaking a validation process and analysing many case studies of pile loading tests.

References

1. A. Bouafia, *P-Y curves from the prebored pressuremetertest for laterally loaded single piles*, Proceedings of the 18th International Conference on Soil Mechanics and Geotechnical Engineering, Paris, 2695-2698 (2013).
2. A. Bouafia, *Single piles under horizontal loads in sand Determination of P-Y curves from the prebored pressuremeter test.* SpringerVerlag, J. Geotechnical and Geoenvir. Eng., **25**, 283-301 (2007).
3. S. A. Ashford, T. Juirnarongrit, *Evaluation of Pile Diameter Effect on Initial Modulus of Subgrade Reaction*, ASCE, J. Geotechnical and Geoenvir. Eng., **129(3)**, 234-242 (2003).
4. J. L. Briaud, B. H. Moore, G. B. Mitchell, *Analysis of Pile Loading Tests at Lock and Dam 26*, ASCE, Geotechnical Special Publication **22(2)**, 925-942 (1989b).
5. D. A. Brown, C. Morrison, L. C. Reese, *Lateral load behaviour of pile group in sand*, ASCE, J. Geotechnical and Geoenvir. Eng., **114(11)**, 1261-1276 (1988).
6. S. Fernandez, M. Pando, M. Ashour, *Case study of the influence of flexural stiffness on the developed soil reactions of three laterally loaded piles*, ASCE, GeoFlorida, 1543-1552 (2010).
7. M. Georgiadis, C. Anagnastopoulos, S. Saflekou, *Centrifuge testing of laterally loaded piles*, Canadian Geotechnical Journal, **259**, 208-216 (1992).
8. D. N. Jennings, S. J. Thurston, F. D. Edmonds, *Static and dynamic lateral loading of two piles*, Proceedings of the 8th WCEE, San Francisco, CA., **3**, 561–568 (1985).
9. T. Juirnarongrit, S. A. Ashford, *Lateral load behaviour of cast-in-drilled-hole piles in weakly cemented sand*, J. Transportation Research Board, 190-198 (2004).
10. S. Kumar, M. Alizadeh, L. Lalvani, *Lateral load deflection response of single piles in sand*, J. Geotech. Eng., Southern Illinois University, (2000).
11. R. L. Little, J. L. Briaud, *Full-scale cyclic lateral load tests on six single piles in sand*, Final report, Miscellaneous paper GL-88-27. Texas: Geotechnical Division, A&M University, Texas, (1988).
12. B. J. Meyer, L. C. Reese, *Analysis of single piles under lateral loading*, Research report 244–1, Project 3-5-78-244, Centre for transportation research, University of Texas, 1–182 (1979).
13. M. W. O'Neill, J. M. Murchison, *An evaluation of p-y relationships in sands*, A.P.I. report PRAC 82-41-1, University of Texas, 174 (1983).
14. H. G. Poulos, *Load deflection prediction for laterally loaded piles*, Australian Geomechanics Journal, **G3(1)**, 1-8 (1973).
15. L. C. Reese, W. R. Cox, F. D. Koop, *Analysis of laterally loaded piles in sand*, Proceedings of 6th offshore technology conference, Houston, **2080**, 473–83 (1974).
16. L. C. Reese, W. F. Van Impe, *Single piles and pile groups under lateralloading.* A.A. Balkema, Rotterdam, Netherlands, 463 (2001).
17. T. J. Weaver, S. A. Ashford, K. M. Rollins, 2005. *Lateral Resistance of a 0.6 m Drilled Shaft in Liquefied Sand*, ASCE, J. Geotechnical and Geoenvir. Eng., **131(1)**, 94–102 (2005).
18. K. M. Rollins,J. D. Lane, T. M. Gerber, *Measured and computed lateral response of a pile group in sand*, ASCE, J. Geotechnical and Geoenvir. Eng., **131(1)**, 103–114 (2005).

19. K. M. Rollins, T. M. Gerber, J. D. Lane, S. A. Ashford, *Lateral Resistance of a Full-Scale Pile Group in Liquefied Sand*, ASCE, J. Geotechnical and Geoenvir. Eng., **131(1)**, 115–125 (2005).
20. P. F. Ruesta, F. C. Townsend, *Evaluation of a laterally loaded pile group at Roosevelt Bridge*, ASCE, J. Geotechnical and Geoenvir. Eng., **123(12)**, 1153–1161 (1997).
21. A. Lachenani, *Contribution à l'étude des pieux chargés latéralement - Modélisation par éléments finis et analyse d'une base de données*, (in French), M.Sc. Civ. Eng. University of Blida, Algeria. (2003).
22. A. Laouedj, *Pieux sous charges horizontales - Développement d'un programme de calcul et étude comparative des méthodes de calcul à la base des courbes P-Y*, (in French), M.Sc. Civ. Eng., University of Blida, Algeria, 210 (2011).
23. ASTM, *Standard Practice for Determining the Normalized Penetration Resistanceof Sands for Evaluation of Liquefaction Potential*, Standard D 6066 – 96, 16 (1998).

Comparative study of four consolidation settlement estimation methods of a railway embankment.

Ahcène Benamghar[1,*] , and *Mohamed* Boudjellal[2]

[1]LTPiTE Laboratory, ENSTP, 1 Rue Garidi Kouba, Algiers, Algeria
[2]ENSTP, 1 Rue Garidi Kouba, Algiers, Algeria

Abstract. Embankment is one of the oldest and most common constructions of public works due to its low cost, simplicity and good performance. This is why it is often required for economic reasons even if underlying soil is of poor quality. Meanwhile, this is not without risk, since we notice sometimes significant damages such as intolerable settlements, landslides and erosion. These risks are particularly dangerous in the case of railways where tolerances are mainly severe. In this work we propose a comparative study of four theoretical methods for estimating the consolidation settlement of a railway embankment. These methods are Terzaghi (1925), Asaoka (1978), Tan (1995) and Chunlin (2014). We applied each method to estimate the final settlement of the 15m high embankment at PK245+000 of the railway under construction Boughezoul-M'sila (Algeria).

1 Introduction

Algeria has recently initiated an ambitious program concerning basic infrastructure equipment like motorways and railways. So far, the biggest project in this grandiose program is unquestionably the east-west highway. This nerve center, with a length of 1216km, connects the Algerian-Tunisian border to the Algerian-Moroccan border through the capital, Algiers, and the main cities of the north .

Similarly, several railway lines were also finished and put into service linking many cities especially in the highlands region. Meanwhile other lines are still under construction in this region in order to strengthen the existing network, upgrade and make it more environmentally friendly. Indeed, the network will be fully electric and will respect the environment in a very rigorous set of specifications. Altogether there are almost 13000km of rail lines to modernize or achieve which have been entrusted to the domestic firm Infrafer.

The rail line concerned in this study is situated in the highlands region of Algeria, connecting the city of Boughezoul, south of the capital Algiers, to the city of M'sila in

* Corresponding author: a.benamghar@enstp.edu.dz

the east (151km). This line is a part of a longer one called the "Highlands Ring" intended to link the main cities in this fast growing part of the country (Fig. 1).

The Highlands region, also known as High plains, is located in the Atlas Mountains in northern Algeria. A steppe-like natural area stretching more than 500km in an east-west direction from the Moroccan border to the Aures Mountains in the eastern part of the country

Topographically, the area is between 400m of elevation in the west to 1300m in the east. It mainly consists of undulating alluvial plains resulting from ancient sedimentary processes affecting the Tell and Saharan Atlas Mountains.

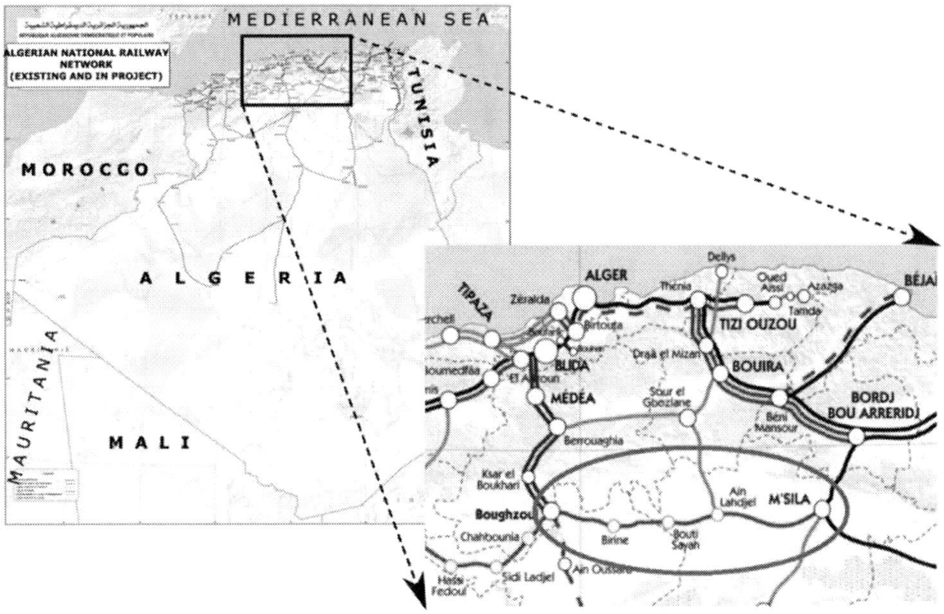

Fig. 1. Algerian railway network with a zoom on the study area. The Boughezoul-M'sila railway project is shown circled by an ellipse (bottom right) (original figures from ANESRIF).

Despite favorable topography for rail construction, there are places on the route of the project that pose challenges for engineers. One of these challenging problems is compressible soils which are very frequent in recent quaternary soils spreading in the region. The problem becomes particularly difficult when these soils must accommodate high embankments to fill topographic depressions.

The most important backfill on the route of the Boughezoul-M'sila line is located at PK245+000. It must have an exceptional height of 15m on a compressible soil (wet and saturated gypsum clay). The purpose of this work was to study the final settlement of that soil in the future. This work was particularly interested in the consolidation settlement through a comparative study of four different theoretical methods. These methods are Terzaghi [1], Asaoka [2], Tan [4] and Chunlin [6].

DOI: 10.1051/matecconf/201712006002

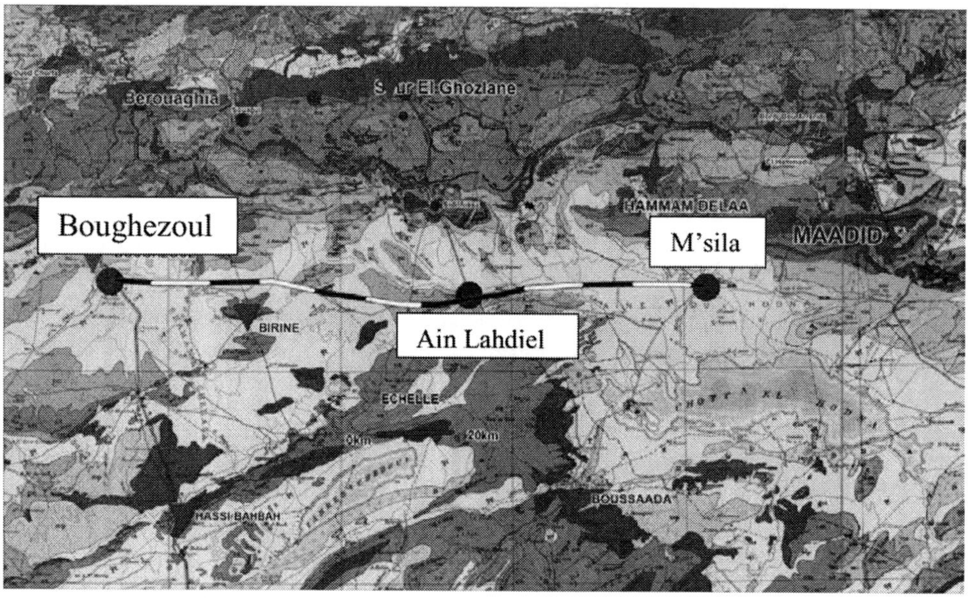

Fig. 2. Drawing of the Boughezoul-M'sila railway line. Extract from the geological map of Algeria [7] (1:500.000).

2 Theory and method

The compressibility phenomenon is often very slow and may extend over time, well after the establishment of the embankment. It can last for decades, mainly depending on the thickness of the compressible layer and its permeability. These vertical down deformations are known as "settlement".

2.1 Magnitude of settlement

According to Terzaghi's theory, the total settlement (S_∞) of a soil consists of primary settlement (S_p), secondary settlement (S_s) and settlement induced by lateral displacements (S_{lat}). The primary settlement has two components: an immediate settlement (S_i) and a deferred settlement associated with consolidation (S_c). Hence the overall formulae [8, 9]:

$$S_\infty = S_p + S_s + S_{lat} = S_i + S_c + S_s + S_{lat} \tag{1}$$

Where:
S_i : immediate settlement,
S_c : consolidation settlement,
S_s : secondary settlement,
S_{lat}: lateral displacement induced settlement

In order to estimate the final settlement, one should calculate separately each one of the four parameters of Eq. (1). Among all these parameters, only the consolidation settlement is the most difficult to estimate because it depends on the time. Each one of the four methods, used in this study, has its own approach to estimate this important variable.

The three other parameters (S_i , S_s , S_{lat}) are the same for the four methods; we give in the following section the formulas to calculate them.

2.2 Immediate Settlement S_i

According to Terzaghi this settlement is given by [6]:

$$S_i = \frac{q}{E} \cdot B \cdot I \quad (2)$$

Where: q: vertical stress applied to the foundation soil (KPa) [10]
B: width of the loaded area (m),
I: influence coefficient depending on the embankment geometry and the compressible soil thickness,
E: Young's modulus, estimated in our case using the oedometer modulus E_{oed}: given by: $E = \frac{2}{3} E_{oed}$, (KPa).

2.3 Secondary Settlement S_s

In the PhD thesis of Akou [11] or Leroueil et al. [12], we can learn that the secondary settlement S_s mainly occurs at the end of consolidation for a time t greater than t_{100} which gives the following equation:

$$S_s = C_\alpha \cdot H \cdot log(\frac{t}{t_{100}}) \quad (3)$$

Where: t_{100}: time corresponding to the end of primary consolidation,
C_α: coefficient of creep,
H: thickness of the compressible layer (m).
$t = 2 \cdot t_{100}$

2.4 Lateral displacement induced settlement S_{lat}

According to Akou [11], there is no way to calculate this settlement, but using Bourges and Mieussens [9] study concerning lateral displacement under embankments which allows to convert the volume corresponding to these movements into supplementary settlement:

$$S_{lat} = 0.11 \frac{D}{B} S_{oed} \quad (4)$$

Where: D: total height of the compressible layer (m)
B: half-width of the embankment (m)
S_{oed}: oedometric settlement (cm)

3 Methods compared in this work

This paper compared four theoretical methods in computing consolidation settlement. These are the well-known Terzaghi method [1], the Asaoka method [2], the Tan method [4] and the recently proposed Chunlin method [6]. In the following section are presented the main principles of each one. The related computation results can be found in the case study section, below.

3.1 Terzaghi method [1]

In the two-dimensional case, consolidation settlement S_c is calculated for the virgin compression zone. According to the stress-strain curve used, S_c is expressed by the following formulae [8]:

- Normally consolidated soil:

$$S_c=C_{ci}\frac{H_i}{1+e_{0i}}log\left[\frac{\sigma'_{v0i}+\Delta\sigma_i}{\sigma'_{pi}}\right] \tag{5}$$

- Over consolidated soil, where we can face two possibilities:

  - <u>Fist case</u> : $\boldsymbol{\sigma'_{v0} + \Delta\sigma_i < \sigma'_p}$

$$s_c= c_{si}\frac{H_i}{1+e_{0i}}log\left[\frac{\sigma'_{pi}}{\sigma'_{v0i}+\Delta\sigma_i}\right] \tag{6}$$

  - <u>Second case</u> : $\sigma'_{v0} + \Delta\sigma_i > \sigma'_p$

$$S_c= C_s\frac{H_i}{1+e_{0i}}log\left[\frac{\sigma'_{pi}}{\sigma'_{v0i}}\right]+C_{ci}\frac{H_i}{1+e_{0i}}log\left[\frac{\sigma'_{v0i}+\Delta\sigma_i}{\sigma'_{pi}}\right] \tag{7}$$

Where: $\Delta\sigma_i$ the stress due to embankment and road overload in the ith layer on embankment axis,

σ'_{v0}: Effective stress due to the weight of land in the ith layer on the axis of embankment,

σ'_{pi} : Pre-consolidation stress in the layer.

3.2 Asaoka method [2]

This method is based on the one-dimensional settlements δ_0, δ_1, δ_2...etc. at times 0, Δt, $2\Delta t$...etc. which can be expressed as a first order approximation that represents a line written as δ_{n-1} = f (δ_n), known as the Asaoka graph [3, 6] (Fig. 7):

$$S_{n+1}=\beta_0+\beta_1\ S_n \tag{8}$$

Where: β_0 represents the intercept and β_1 the slope of the line (Fig. 7).

When the primary ultimate settlement is reached, $\delta_n = \delta_{n-1} = \delta_{ult}$, it is written as:

$$S_{ult}=\beta_0/(1-\beta_1) \tag{9}$$

And the settlement $S(t)$ at time t can be computed as:

$$S(\mathrm{t}) = \left(\frac{\beta_0}{1-\beta_1}\right) - \left[\frac{\beta_0}{1-\beta_1} - S_0\right]\beta_1^{\mathrm{t}} \tag{10}$$

Where: S_0 is the initial settlement.

In Eq. (10), S_0 must be determined first. The different values of S_0 lead to different values of S(t). This is why the accuracy strongly depends on the initial time choice.

3.3 Tan method [4]

Tan and Chew [3] have shown that in-situ measurements of soil settlement as a function of time, expressed as t/δ=f(t) (Fig. 3, right), show the same shape as the theoretical curve of Terzaghi (Tv/U=f(Tv), Fig. 3, left). This is why these authors proposed to extrapolate this property to identify levels at 60% and 90% of consolidation for any settlement dataset.

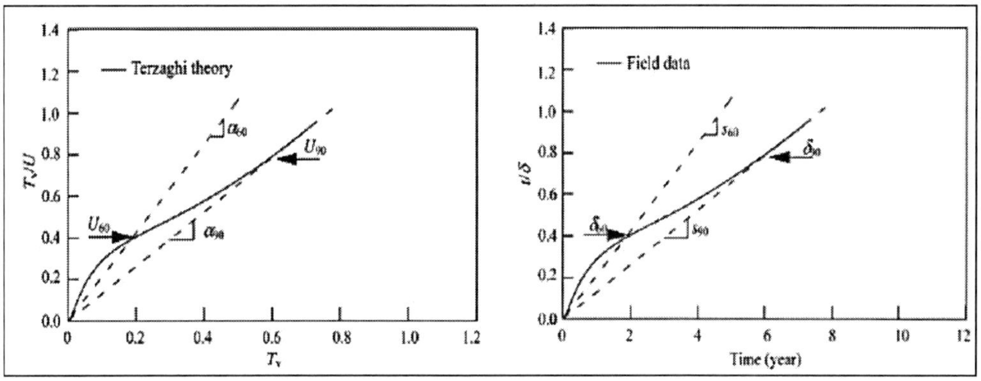

Fig. 3. Hyperbolic curve of Terzaghi (left) and in-situ measurements as a function of time (right) [3].

The S_{60} and S_{90} slopes can be found by:

$$S_{60} = S_i \frac{\propto_{60}}{\propto_{\mathrm{i}}} \tag{11}$$

$$S_{90} = S_i \frac{\propto_{90}}{\propto_{\mathrm{i}}} \tag{12}$$

Where: S_{i} is the slope of the initial linear portion of the graph t/δ=f(t) and α_i the slope of the theoretical graph Tv/U=f(Tv) (Fig. 3, left).

Finally, Tan and Chew [3] showed that the primary final settlement can be obtained using one of the following three ratios:

$$\delta_f = \frac{\propto_{\mathrm{i}}}{S_i} = \frac{\delta_{60}}{0.6} = \frac{\delta_{90}}{0.9} \tag{13}$$

And the settlement $\delta(t)$ can be computed as follow:

DOI: 10.1051/matecconf/201712006002

$$\frac{t}{\delta(t)}=\left(\frac{1}{6}\right)\frac{S_i}{\propto_i} \tag{14}$$

However, according to Chunlin [6] the Tan method limitation resides mainly, also, in the determination of the initial time.

3.4 Chunlin method [6]

Chunlin [6] blames Asaoka and Tan methods [2, 4] to include parameters difficult to determine accurately, particularly the choice of the initial time. Chunlin [6] begins with the calculation of settling time from the one-dimensional consolidation equation of Terzaghi, by writing:

$$S_t=S_\infty\left(1-\frac{8}{\pi^2}e^{bt}\right) \tag{15}$$

Where: S_t: settlement at time t
S_∞: final settlement,
b: parameter given by in situ data or by Asaoka method.

To simplify calculations, Chunlin [6] introduces a new variable called "potential settlement" by writing:

$$S_p=S_\infty\frac{8}{\pi^2}e^{bt}=S_\infty-S_t \tag{16}$$

Where: S_p is the potential settlement to happen in the future.

In Eq. (16), S_∞ comes from the Terzaghi's theory. The parameter b represents the drainage conditions in the Terzaghi's consolidation equation. It may be obtained using the C_v consolidation coefficient and the length of the drainage under various types of drainage. These two parameters must be determined first [6].

From Eq. (16) we can deduce the logarithm $ln\left[S_p\pi^2/(8*S_\infty)\right]$=b·t which is a straight line equation of slope b (Fig. 9).

4 Case study

4.1 Geological framework

A geological and geotechnical investigation program was carried out in order to adequately characterize the geological formations and to obtain the respective geotechnical parameters of the railway [7].

In lithostratigraphic terms, the main geological formations crossed in the section between the Pk235+100 and the Pk250+100 of the railway line are characterized by ancient formations of the Miocene and Pliocene, covered by quaternary formations (Fig. **2**). In summary, the Miocene (Mi) is characterized by formations of green marls while the Pliocene (Pc) is represented by thick formations of encrusted conglomerates of limestone and gypsum. Finally the Quaternary (Qt) is characterized by formations of calcareous and calcaro-gypsum crusts [7].

4.2 Geotechnical framework

In situ investigations and laboratory tests were carried out all along the route of the railway line. The results for the studied area in PK245 zone showed the following facts [7].

In summary, two layers of soil were found. A 5m layer of wet sandy clay followed by a 15m saturated layer of gypsum clay (Fig. 4 and Fig. 5).

Table 1. Physical characteristics of the wet sandy clay layer (mean values) [7].

	Granulometry									Classification	
Parameter	D_{max}	0.2		0.8	0.02		0.002			**GTR**	**UIC**
Unit	mm	%		%	%		%				
Values	10	95		62	54		0				
Parameter	ω	γ_h	γ_d	vbs	E	w_l	w_p	I_p	I_c	A_3	Q_{s0}
Unit	%	g/cm^3	g/cm^3	g/100g	KPa	%	%	%	%		
Values	15	1.78	1.55	3.73	0.55	54	27	27	1.5		

Table 2. Physical characteristics of the saturated of gypsum clay layer (mean values) [7].

	Granulometry									Classification	
Parameter	D_{max}	0.2		0.8	0.02		0.002			**GTR**	**UIC**
Unit	%	%		%	%		%				
Values	30	90		80	76		62				
Parameter	ω	γ_h	γ_d	vbs	E	w_l	w_p	I_p	I_c	A_4	Q_{s0}
Unit	%	g/cm^3	g/cm^3	g/100g	KPa	%	%	%	%		
Values	30	1.95	1.5	5.14	0.67	83	33	50	1.06		

From Table 1 and Table 2 we can deduce the following summaries:

- $\gamma_d = 15.5\ KN/m^3 \leq 16\ KN/m^3$ ➔ low density soil,
- Plasticity index $I_P > 40$ ➔ very plastic,
- Consistency index $Ic > 1$ ➔ hard clay,
- Methylene blue value vbs > 2.5 ➔ sensitive to water.

Table 3. Compressibility characteristics of wet and saturated clay layers [7].

Parameter	**Compressibility**				200KPa
	pc (KPa)	C_c	C_g	C_s	$C_v\ (m^2/s)*10^7$
	Wet clay layer				
Values	520	17.50	11.02	0	6.49
	Saturated clay layer				
Values	213	18.00	9.07	0	6.49

From Table 3 we can deduce the following summaries:

- The pre-consolidation stress is 520 KPa for the wet clay layer and 213 KPa for the saturated clay layer,
- $C_c = 0.17 \sim 0.18 \leq 0.2$: moderately compressible clay,
- $C_g = 0.09 \sim 0.1$: moderately swelling clay.

4.3 Application to consolidation settlement Sc calculation at PK245+000

For this case study we have selected the worst case in the Boughezoul-M'sila railway line. Namely, a portion of 15km long (PK245+000) where the height of embankment arrives up to 15 m, reposing on a moderately compressible soil (Fig. 4). In this section we focus only on consolidation settlement *Sc* calculation.

The other settlements (immediate, secondary and lateral) are straightforward and easy to obtain. They are taken from the master thesis work of Boudjellal [13].

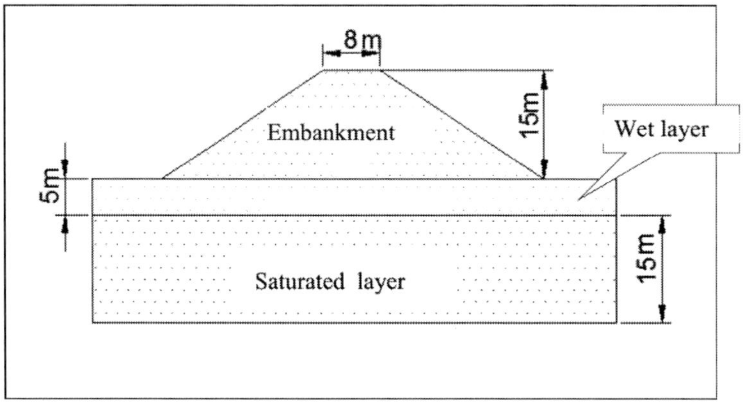

Fig. 4. Cross section with 15m height embankment at PK245+000.

4.3.1 S_c calculation using Terzaghi's method

We have an over-consolidated clay with H/B~0.3 then ➔ μ= 0.6

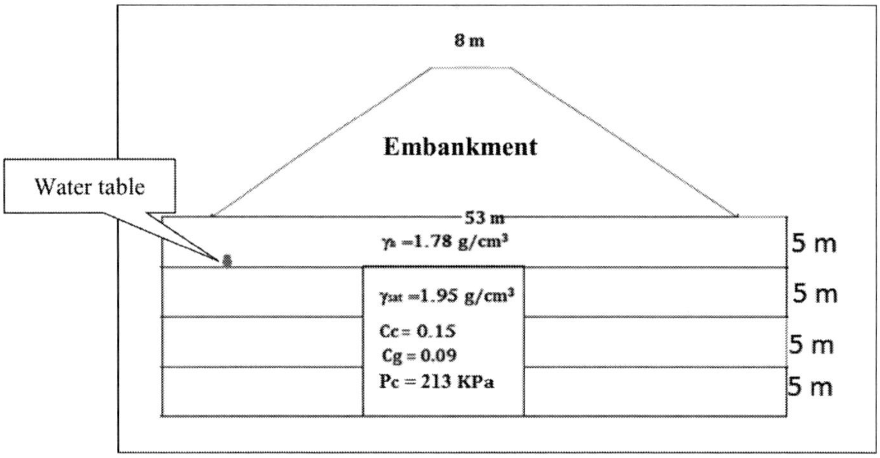

Fig. 5. Embankment profile at PK245+000.

The stress due to the embankment is given by Eq. (7), where:

$\Delta\sigma_i = 2 * I * q$ and $q = \gamma_{emb} * h_{emb} = 18.50 * 15 = 277.5$ KPa. (given that γ_{emb} and h_{emb} are the specific weight and the height of the embankment respectively)

Table 4 Computation of the stresses due to the embankment, Boudjellal [13].

Hi	z	a/z	b/z	I	$\Delta\sigma_i$ (KPa)
5	-	-	-	-	-
5	7.5	0.53	3.53	0.5	277.5
5	12.5	0.32	2.12	0.48	266.5
5	17.5	0.23	1.51	0.43	238.5

Table 5 Calculation of consolidation settlement for embankment, Boudjellal [13].

Hi (m)	e	Cc	Cg	σ'vo (KPa)	σ'p (KPa)	Δσi (KPa)	σ'v (KPa)	S_{oed} (m)
5	-	-	-	-	-	-	-	-
5	0.67	0.21	0.1	112.75	213	277.5	390.25	0.09
5	0.67	0.21	0.1	160.25	213	266.5	420.75	0.12
5	0.67	0.21	0.1	207.75	213	238.5	446.25	0.17
							Σ	0.38

Taking into account the correction of Skempton and Bjerrum [10] with μ =0.6

$$S_c = \mu \cdot S_{oed} = 0.6 \cdot 38.0 = 22.8 \text{ cm}$$

4.3.2 Speed and consolidation time

The relationship between the time factor *Tv* and the degree of consolidation *Uv* makes it possible to calculate the settlement corresponding to a consolidation time *t*. Conversely, in order to know the time required to reach a certain value of the settlement, the corresponding degree of consolidation is calculated and then the curve expressing *Uv* is used as a function of the time factor *Tv* according to the form of the initial distribution of the interstitial overpressure to determine *Tv* value and then the required consolidation time. We have:

$$T_v = \frac{C_v}{H'^2} t \tag{17}$$

Where: H' drainage distance H' = H/2 and H the height of the compressible layer.

Table 6 Speed and settlement time calculation, Boudjellal [13].

H(m)	*Cv* 10^{-7}(m²/s)	*Tv*=0.033 *Uv*=20%	*Tv*=0.197 *Uv*=50%	*Tv*=0.844 *Uv*=90%	*Tv*=2 *Uv*=99.4%
		t (years)	*t* (years)	*t* (years)	*t* (years)
20	6.49	0.16	0.96	4.14	9.92

Calculation results (Table 6) give a settlement time of 9 years, 11 months and 14 days:

Table 7 Evolution of settlements over time, Boudjellal [13]

Time (years)	*Tv*	*Uv*	Settlement δ (cm)
0.099	0.02	0.16	3.648
0.298	0.06	0.276	6.2928
0.496	0.1	0.356	8.1168
0.744	0.15	0.437	9.9636
0.992	0.2	0.504	11.4912
1.488	0.3	0.613	13.9764
1.984	0.4	0.697	15.8916
2.481	0.5	0.764	17.4192
2.974	0.6	0.816	18.6048
3.473	0.7	0.856	19.5168
3.969	0.8	0.887	20.2236
4.466	0.9	0.912	20.7936
4.962	1	0.931	21.2268
9.924	2	0.994	22.6632

Using Table 7 we obtained the hyperbolic curve in Fig. 6.

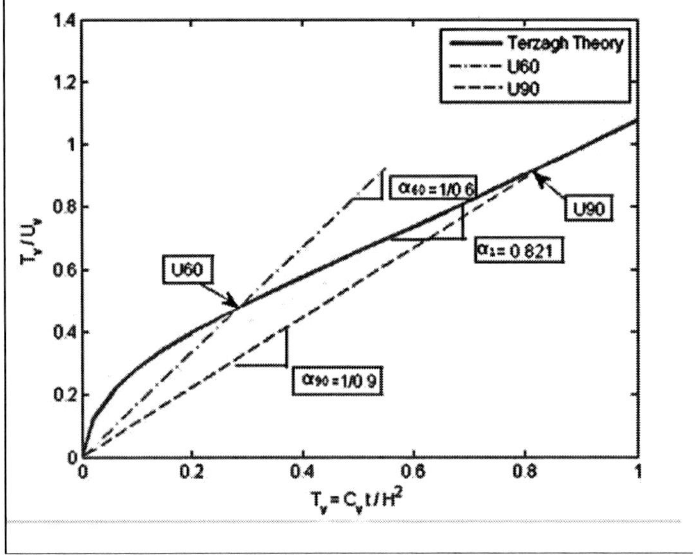

Fig. 6. Hyperbolic curve of settlements, at PK245+000, according to Terzaghi's theory.

4.3.3 S_c calculation using Asaoka method

The Asaoka method [2] considers only successive settlements. Table 8 shows the successive settlements S_n and S_{n+1} obtained from the settlement table as a function of time (Table 7).

Table 8 Successive settlements obtained using Table 7, Boudjellal [13].

S_n (cm)	S_{n+1}(cm)	S_n (cm)	S_{n+1}(cm)
3.648	6.2928	15.8916	17.4192
6.2928	8.1168	17.4192	18.6048
8.1168	9.9636	18.6048	19.5168
9.9636	11.4912	19.5168	20.2236
11.4912	13.9764	20.2236	20.7936
13.9764	15.8916	20.7936	21.2268

Graphical determination of settlement (Fig. 7):

$$S_{n+1} = \beta_0 + \beta_1 S_n \quad (18)$$

Where: $\beta_0 = \frac{\Delta(S_{n+1})}{\Delta S_n} = 0.7772$ and $\beta_1 = 5.0646$

Which gives: $S_{n+1} = 5.0646 + 0.7772\, S_n$

And finally $S_c = \beta_0 \,/\, (1 - \beta_1) = 22.73\text{cm}$

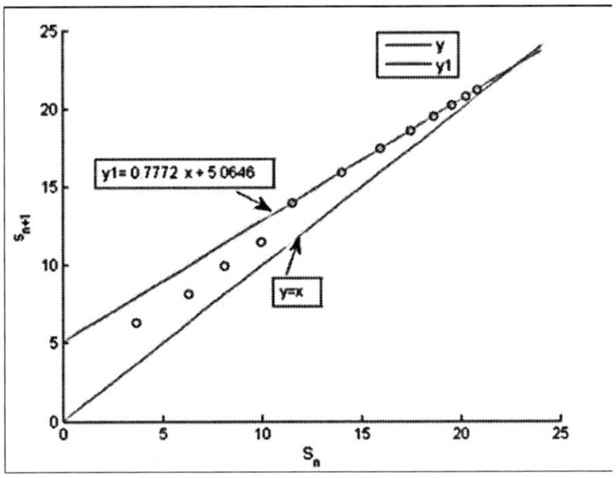

Fig. 7. Asaoka curve for PK245+000 railway embankment, Boudjellal [13].

4.3.4 S_c calculation using Tan hyperbolic method

Using the *Cv* value obtained previously (Table 6, *Cv* = 6.49 10^{-7} m²/s) we plotted the t=f(t/δ) curve from the calculation data in following Table 9.

Using the curve in Fig. 8, Tan method [4] emphasizes on the linear part of the graph (between δ_{60} and δ_{90}) which is written as:

$$\frac{t}{\delta} = \alpha \cdot \delta + \beta$$

Where α and β are determined graphically: $\alpha = \frac{\Delta(\frac{t}{\delta})}{\Delta t} = 0.0405$ and $\beta = 0.0349$

Then: $\frac{t}{\delta}$=0.0405*δ+0.0349

We obtain $S_c = \frac{\alpha_i}{s_i} = \frac{0.82}{0.0405} = 20.25\text{cm}$

Table 9 Settlement according to time/settlement ratio obtained from Table7, Boudjellal [13].

t (years)	t/δ (years/cm)	t (years)	t/δ (years/cm)
0,099	0,027	2,481	0,142
0,297	0,047	2,977	0,160
0,496	0,061	3,473	0,177
0,744	0,074	3,969	0,196
0,992	0,086	4,466	0,214
1,488	0,106	4,962	0,233
1,984	0,124	9,924	0,437

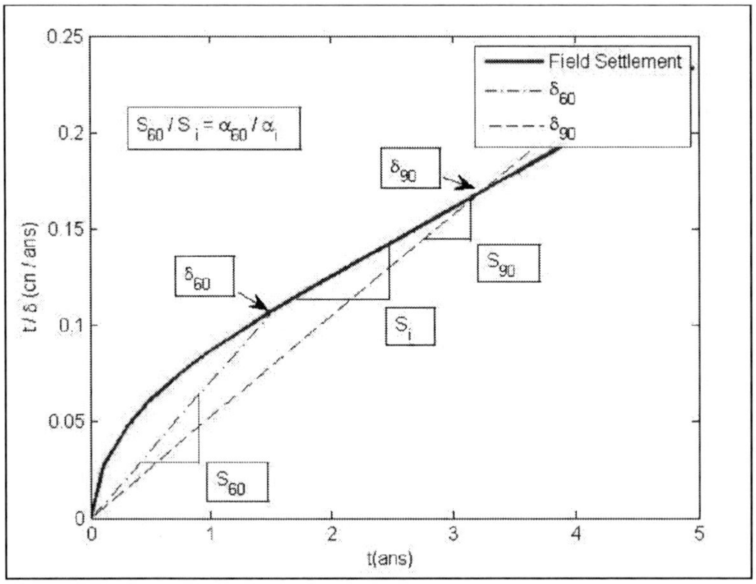

Fig. 8. Tan Hyperbolic curve for embankment settlement at Pk245+000.

4.3.5 S_c calculation using Chunlin method

The graph of Fig. 9 gives the slope of the line b = -0.0014 and according to the hypothesis of Chunlin [6] the final settlement S_∞ = 59.08cm (Table 11) is the same as the Terzaghi's method.

Table 10 Calculations using Chunlin method [6] at PK245+000.

t (years)	Ln $((Sp *\pi^2)/(8 *S_\infty))$	*t* (years)	Ln $((Sp *\pi^2)/(8 *S_\infty))$
36.225	0.036	905.639	-1.234
108.672	-0.113	1086.767	-1.483
181.127	-0.23	1267.895	-1.728
271.691	-0.364	1449.023	-1.97
362.255	-0.491	1630.151	-2.22
543.383	-0.739	1811.279	-2.464
724.511	-0.984	3622.558	-4.906

Then from Eq. (16), we obtain S_c at $t = 3622.5$ days by:

$$S_c = 22.8^* \left(1 - \frac{8}{\pi^2} e^{-0.0014*3622.5}\right) = 22.78 \text{ cm}$$

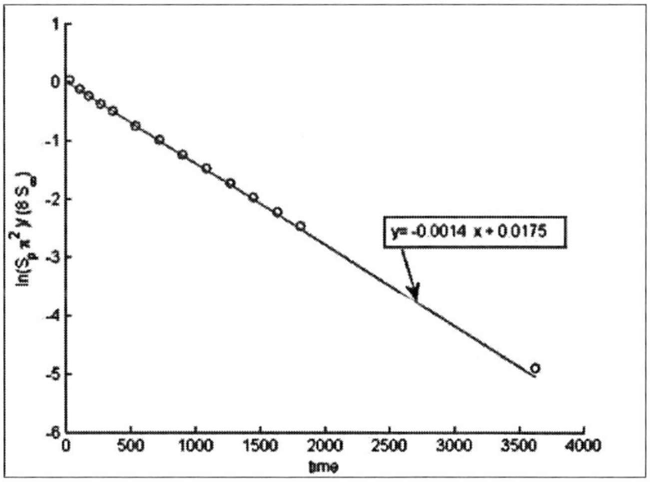

Fig. 9. Chunlin curve for railway embankment at PK245+000.

5 Results and discussion

The calculation results of consolidation settlement, S_c at PK245+000, according to the four methods studied in this work are summarized in Table 11 below. To make the comparison more interesting we also presented the other types of settlements (initial, secondary and lateral) as well as final settlement S_∞ as calculated by Boudjellal [13] for the same embankment of the Boughezoul-M'sila railway project.

These results show a nearly identical final settlement S_∞ regardless of the method. This is predictable since all these methods are based, in their calculations, on the final settlement of Terzaghi. But regarding consolidation settlement S_c, which is the subject of this study, we note that the results obtained are almost, also, similar between the different methods.

However, the methods of Chunlin [6] and Asaoka [2] are the most interesting because they are more experimental (graph of successive settlements), simple, easy and free of subjective parameters. On the other hand, the method of Terzaghi [1] remains rather long and empirical whereas the Tan method [4] suffers mostly from the choice of the initial time which can distort the results if inaccurate. Moreover, these results confirm those obtained by Chunlin [6] in his research, particularly at the end of the settlement period.

Table 11. Result of Sc settlement calculation at PK245+000 according to the four methods.

Method	**Different settlements (cm)**				**Final settlement after 9 years S_∞(cm)**
	S_i	S_c	S_s	S_{lat}	
Terzaghi	26.83	22.80	6.3	3.15	59.08
Asaoka		22.73			59.01
Tan		20.75			57.03
Chunlin		22.78			59.06

6 Conclusion

Knowing that the case study we have dealt with is a high embankment (15m) on a railway track, where the tolerances to settlement errors are of the order of mm; this work has been particularly interesting. Therefore, we applied the four methods, Terzaghi [1], Asaoka [2], Tan [4] and Chunlin [6], to estimate consolidation settlement of the embankment in question. All of these methods showed very close results, within a few tenths of a millimetre.

We can conclude that for the estimation of an intermediate phase settlement, which could prove crucial in certain projects such as railways, the new Chunlin method [6] is particularly attractive by its simplicity and precision. This method, together with Asaoka method [2], is even more precise for intermediate settlement calculation, as shown by Boudjellal [13]. Meanwhile for the final settlement all four methods are comparable. This is explained by the fact that they all are theoretically based on Terzaghi's assumptions and use the same S_∞ value. However, in the intermediate phase the precision of the Asaoka or Chunlin methods [2, 6] is explained by the fact that these two methods are much more based on the experimental side than the empirical one, as opposed to the two other methods.

References

1. K. Terzaghi, Principles of soil mechanics, Settlement and consolidation of clay. *Eng. N. R.,* **95**, pp874–878 (1925)
2. A. Asaoka, Observational procedure of settlement prediction. *S. and F.,* **18** pp 87–101 (1978)
3. S. Tan, S. Chew, Comparison of the hyperbolic and Asaoka observation method of monitoring consolidation with vertical drains. *S. F.,* **36**, pp31-42 (1996)
4. S. Tan, Validation of hyperbolic method for settlement in clays vertical drains. *S. F.,* **35**, pp101-113 (1995)
5. S. Tan, Hyperbolic method for settlements in clays with vertical drains. *Can. Geo. J.* **31**, pp125-131 (1994)
6. L. Chunlin, A simplified method for prediction of embankment settlement in clays. *J. R. M. Geo.* China, pp61-66 (2014)
7. SETIRAIL, Rapport géotechnique projet ferroviaire Boughezoul–M'sila, lot 1, dossier 302, Algeria (2011)
8. G. Costet, J. Sanglerat, Cours de mécanique des sols 1, Plasticité et calcul des tassements. Troisième édition, Dunod, 277p, Paris (1981)
9. F. Bourges, C. Mieussens, Déplacements latéraux à proximité des remblais sur sols compressibles. Méthodes de prévision. *Bull. L. L.P.C.* **101**, pp73-100, Paris (1979)
10. A. W. Skempton, L. Bjerrum, A contribution to the settlement analysis of foundations on clay. *Géotech.* **7**, pp168-178 (1957)
11. Y. Akou, Etude expérimentale et modélisation de l'élargissement des remblais sur sols. PhD thesis, ENPC, 254p, France (1995)
12. S. Leroueil, J. P. Magnan, F. Tavenas, Remblais sur argiles molles. Technique et documentation - Lavoisier, 342p, Paris (1985)
13. M. Boudjellal, Etude comparative des méthodes d'estimation des tassements des remblais de voie ferrée de grandes hauteurs. Master Thesis. ENSTP, Algeria (2015)

Geologic environments for nuclear waste repositories

Evan K. Paleologos[1,*], *Abdel-Mohsen O.* Mohamed[2], and *Kosmas* Pavlopoulos[3]

[1]Dept. of Civil Engineering, Abu Dhabi University, Abu Dhabi, United Arab Emirates
[2]Assoc. Provost & Chief Academic Officer, Zayed University, Abu Dhabi, United Arab Emirates [3]Dept. of Geography and Planning, Paris Sorbonne University-Abu Dhabi, United Arab Emirates

Abstract. High-level radioactive waste (HLW) results from spent reactor fuel and reprocessed nuclear material. Since 1957 the scientific consensus is that deep geologic disposal constitutes the safest means for isolating HLW for long timescales. Nuclear power is becoming significant for the Arab Gulf countries as a way to diversify energy sources and drive economic developments. Hence, it is of interest to the UAE to examine the geologic environments currently considered internationally to guide site selection. Sweden and Finland are proceeding with deep underground repositories mined in bedrock at depths of 500m, and 400m, respectively. Equally, Canada's proposals are deep burial in the plutonic rock masses of the Canadian Shield. Denmark and Switzerland are considering disposal of their relative small quantities of HLW into crystalline basement rocks through boreholes at depths of 5,000m. In USA, the potential repository at Yucca Mountain, Nevada lies at a depth of 300m in unsaturated layers of welded volcanic tuffs. Disposal of low and intermediate-level radioactive wastes, as well as the German HLW repository favour structurally-sound layered salt stata and domes. Our article provides a comprehensive review of the current concepts regarding HLW disposal together with some preliminary analysis of potentially appropriate geologic environments in the UAE.

1 Introduction

Two lines of defense are always used for land disposal of radioactive waste. The first and most important is the geologic setting, which is selected so that it would facilitate the operation of the disposal facility and in case the engineering measures fail it will contain or retard the spread of radioactive waste. Site selection includes low seismicity and low volcanic activity, geologic environments that would exclude faults or extensive fracturing, geochemical conditions that would hinder or retard the corrosion of containers, hydrologic setups of low precipitation and deep water table that would minimize water infiltration and fluid transport of the waste in case of failure, etc. More general considerations also include

* Corresponding author: evan.paleologos@adu.ac.ae

distances from population centers, a simple legal framework of the repository area (preferably to be owned directly by the national government, rather than by a state or a municipality), low political representation, relative ease of access, etc.

Engineered barrier systems are constructed in order to: (a) isolate the waste from the biosphere for extremely long periods of time, (b) ensure that in case of failure contaminants migrating from the waste canisters will be of low concentration compared to natural background levels, and (c) provide assurance that any risk from inadvertent human intrusion would be minimal. Ensuring that engineered barrier systems will perform according to design criteria requires integration of site characterization data, waste and engineering properties' characteristics, in situ and laboratory testing, and modeling [1].

2 The international experience on HLW repositories

High-level radioactive waste results in two ways:

a. Spent reactor fuel, which is the fuel used in commercial nuclear reactors after it is no longer efficient in producing electricity, because the fission process has slowed down,
b. Reprocessed nuclear material for defense purposes. This waste in United States exists in Hanford, Washington, and Savannah River Site, South Carolina.

Despite ideas about HLW disposal in ocean trenches and the marine environment [2], or to the outer space, the scientific consensus of the last sixty years is that deep geologic disposal constitutes the safest means of isolating this type of waste for long timescales [3]. Efforts to identify locations for the permanent sitting of HLW have been in progress, among others, in USA, Sweden, Belgium, China, Finland, France, Germany, Japan, Russia, Switzerland, Canada, Italy, Mexico, the Netherlands, Taiwan, South Korea, Spain, and the UK. It must be kept in mind that, for any site selected, there will always exist significant uncertainty in a repository's performance since HLW containment and repository safety is mandated for tens of thousands of years (in the case of USA for one million years) and scientific predictions at such time scales are highly speculative.

Different types of geologic environments have been studied in various countries [27]. Sweden and Finland are proceeding with construction of deep underground repositories mined in bedrock at depths of 500m, and 400m respectively [4, 5]. Countries, such as Denmark and Switzerland, having smaller quantities of HLW are considering disposal into crystalline basement rocks through deep boreholes that would reach depths of about 5,000m [6]. The initial 1957 proposal by the National Research Council of the US National Academy of Sciences favored disposal in structurally-sound layered salt strata and domes, or abandoned salt mines because very little water passes through salt, and fractures are self-healing due to the creep of salt, which is plastic [3]. Such locations have been in use in many countries for the disposal of low- (LLW) and intermediate-level (ILW) radioactive wastes, and this is the environment of the Waste Isolation Pilot Plant (WIPP) in USA for the disposal of defense transuranic wastes (long lived ILW), 650m below ground level. The German national center under investigation for the disposal of HLW lies in a salt dome in Lower Saxony with the final decision to be reached by 2019 [7].

Belgium has its candidate site in a deep clay formation (referred to as Boom clay), where experiments have been conducted in an underground research laboratory (HADES) that includes access shafts and two galleries at a depth of 250m [8]. France accelerated its efforts for HLW disposal after the passage of its 1991 Waste Act that specified the creation of underground laboratories, and in 1998 the French government approved construction of a repository in clay formations in eastern France. The French radioactive waste disposal agency, ANDRA, was expected to apply for construction and operating license in 2014 [7].

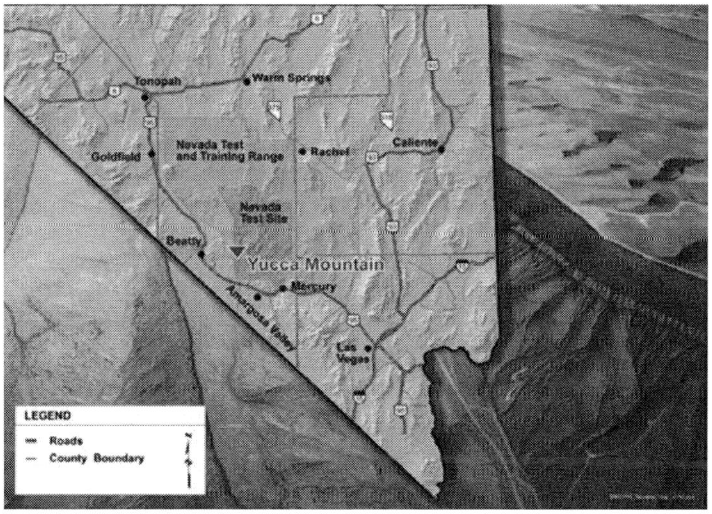

Fig. 1. Location of the potential US site for disposal of HLW at Yucca Mountain, Nevada (US DOE, http://energy.gov/photos/yucca-mountain).

In USA the Nuclear Waste Policy Act was amended in 1987 to designate the Yucca Mountain, Nevada (Fig. 1 and 2) as the sole site to be investigated as a US permanent repository for the country's 70,000 metric tons of HLW [7]. The potential repository would lie 300m below ground surface and another 300m above the underlying water table in unsaturated layers of welded volcanic tuff [9]. The site was selected because of its remoteness, its low-precipitation desert environment, its prior use as an underground testing facility of nuclear weapons, and because its land, part of the Nevada Test Site, belongs directly to the Federal government. The US Department of Energy's progress report to the US Congress on the study of the Yucca Mountain, as a potential HLW disposal site, is contained in the about 1,500 pages report "Viability Assessment of a Repository at Yucca Mountain," while supporting detailed scientific analyses are included in numerous reports and summarized in the even longer "Yucca Mountain Site Description" report [10, 11].

Fig 2. First curve of the main tunnel of the exploratory studies facility at Yucca Mountain, Nevada, October 1995. (US DOE, http://energy.gov/photos/yucca-mountain).

 DOI: 10.1051/matecconf/201712006003

The Yucca Mountain Project has met with delays and even cancellation in 2009 [12, 13]. However, in August 2013 the Federal Appeals Court ordered the US Nuclear Regulatory Commission (NRC) to resume its review of the DOE's application for a license to construct and operate the Yucca Mountain repository. In October 2014, the 781-page NRC report concluded that the 2008 license application by the U.S. Department of Energy (DOE) included "multiple barriers to isolate radioactivity from the environment" for hundreds of thousands of years, and hence the "repository design meets the requirements that apply after the repository is permanently closed" [14]. The US DOE's overall repository safety strategy lies in four tenets: (1) limited water contact with the waste packages, (2) long waste-package lifetime, (3) low radionuclide release rate from breached waste packages, and (4) reduction of radionuclide concentration during transport from breached waste packages. All four are meant to serve the US DOE's overall objective "to contain and isolate the radioactive wastes so that the dose impact to humans is attenuated to a relatively benign level" for a period of one million years [10].

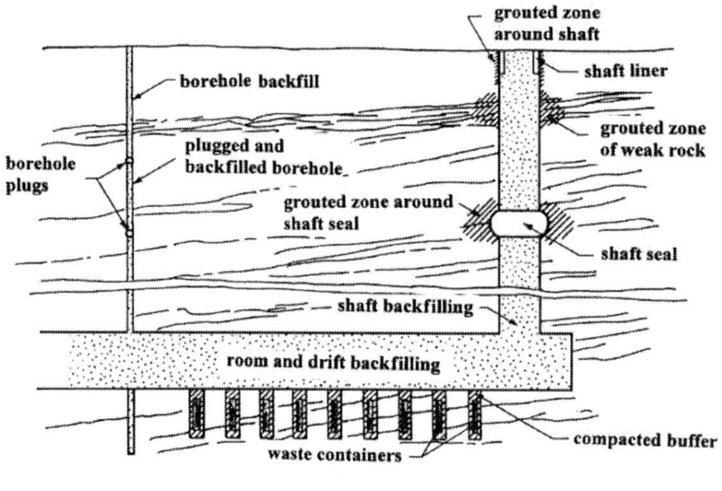

Fig. 3. Conceptual engineered barrier system in a deep rock repository in Canada [27].

The Canadian proposals, for the disposal of HLW concentrate on deep burial in the plutonic rock masses of the Canadian Shield [15]. Canadian concepts for deep disposal facilities identify several important components, intended to serve as both engineered and natural geologic barriers to long-term radionuclide migration. The vault system is constructed to a specific shape in a rock mass with known physical, mechanical, chemical, hydrogeological and tectonic characteristics. The rock mass serves as the primary natural, geologic barrier for radionuclide migration. The primary engineered barrier involves the sealed cylindrical containers that house the heat-emitting radioactive waste. The primary engineered geological barrier is a highly dense mixture of bentonite clay and crushed quartz sand within which the waste containers are emplaced. This engineered clay barrier, which in its compacted state is also referred to as a ''buffer'' is either compacted in situ, in emplacement boreholes that are drilled into the base of the galleries of the waste disposal vault system, or placed around the waste containers in pre-compacted units.

The use of bentonite clays as engineered geological barriers is advocated in many concepts put forward for deep disposal of HLW. The potential of bentonite clays to act as geochemical filters for the sorption of radionuclides is an important factor in their selection as an engineered geological barrier. It is expected that, no matter the precautions, radionuclide migration will occur eventually, as a result of the natural disintegration of the waste containers. Bentonite clay also acts as a buffer and can protect the canisters containing the spent fuel from corrosion, as well as from minor bedrock movement. The clay will gradually absorb moisture and by swelling and filling the space around it is expected to fill any fractures in the hosting geologic environment. Finally, if a canister develops cracks the clay buffer will prevent water from entering the canister [16-23].

The secondary engineered geological barrier is the backfill material that fills the major part of the waste disposal vault system. In addition, other engineered barriers, such as the backfilling of the access shafts and boreholes, the use of bulkheads for the shaft and borehole sealing, etc., form the complete system of engineered barriers intended to minimize the migration of radionuclides from the repository to the biosphere throughout the life of the radioactive waste. A conceptual description of the Canadian engineered barrier system and its components are shown in Fig.3.

3 The geologic environment of the UAE

One of the first steps for the determination and characterization of geologic formations as appropriate host environments for HLW disposal is the understanding of their origin and their classification in terms of their desirable properties for such a role.

3.1 Structural geologic history

The Arabian Peninsula consists of two main geologic provinces: (i) the Arabian Shield, as a paleo-geographic area to the west, composed mostly of Precambrian igneous and metamorphic rocks, which has remained a largely positive region since the Cambrian; and (ii) the Arabian Platform (that includes Qatar and the UAE) as a vast area to the east of the Arabian Shield, which has undergone periodic subsidence, and which has accumulated a sequence of sedimentary rocks ranging in age from Cambrian to recent geologic times.

The UAE is situated on the north-eastern margin of the Arabian plate. During the Early Mesozoic the area was situated near the southern margin of the large Tethys Ocean, which separated Eurasia to the north from the paleo-continent of Gondwana (Africa, Arabia, Australia and India) to the south. In the Triassic period rifting affected the paleo-continent of Gondwana, leading to the development of an ocean basin called Neo-Tethys. The majority of the rocks of the UAE were formatted within and on the margins of this basin. Closure of the Neo-Tethys occurred during the Late Cretaceous to Paleogene as Africa and Arabia moved northwards colliding with Eurasia. This event created the distribution of the rock units that are seen in the UAE, as well as many major structural features [24]. The geologic formations of the UAE can be divided into five main units [28]:

a. The Middle Permian–Lower Cretaceous carbonate platform sequence (Hajar Supergroup) is exposed in the northern part of the UAE, where it forms the high mountains of the Musandam peninsula.
b. The Late Triassic-Upper Cretaceous sequence of fine- to coarse-grained limestone turbidites and deep water sediments, dominated by mudstones and cherts with minor volcanic rocks and olistostrome deposits (Masafi-Hatta). They were folded and thrusted during the obduction of the ophiolites during the Upper Cretaceous [25].
c. The sequence of metamorphic rocks consisting of green schists, amphibolite schists, and meta-volcanic and meta-igneous formations (Masafi-Ismah-Bani Hamid).
d. The Upper Cretaceous UAE-Oman ophiolite that extends for over than 500 Km through the mountains of northern Oman and the UAE (Hajar Mountains). The ophiolite was thrusted or obducted onto the margin of the Arabian continent during the closure of the Neo-Tethys in Upper Cretaceous.

e. The Upper Cretaceous-Neogene formations that are exposed along the western edge of the Hajar Mountains and in outliers in the desert to the west.
f. The Quaternary formations that cover the above formations. Fluvial gravel depositions are well developed in the wadis passing into large-coalesced alluvial fans or bahadas along the mountains piedmonts. Sand dunes of various types overlap the oldest geological formations. Along the UAE coastlines a lot of coastal barriers, spits, lagoons have developed, with the locally existence of mangrove stands, bounded by sabkha flats.

3.2 Tectonics and lithology

Two main tectonic features had a deep influence on the structural evolution of Abu Dhabi: the central Arabian arch and the Rub al Khali basin. These two structures plunge gently towards the North east towards the Arabian Gulf. General thickening of sedimentation is observed in this direction indicating an increasing rate of subsidence.

Gentle and simple folds are common characteristics. The main fold axes directions in Abu Dhabi near the top of Thamama formations are N-S, E-W/NE-SW, and NW-SE. The "Arabian folds" aligned N-S is the most dominant and probably related to deep-seated basement tectonics. Diapiric salts are believed to underlie most of the structures of Abu Dhabi. Furthermore, in the western Abu Dhabi regions and islands there exist salt pierced structures, surrounded by Tertiary and Quaternary deposits associated with Cambrian sediments carried up by salt intrusions.

The Hajar Mountains, as we know them today, are the result of oogenesis during late Oligocene to early Miocene (ca. 30–20 million years ago), followed by uplift and erosion under generally arid or semi-arid conditions, continuing to the present day. This deformation and uplift are presumed to be related to the collision of the Eurasian plate with the Afro-Arabian plate and other marginal plates, coupled with the opening of the Red Sea. The Hajar Mountains, however, are situated relatively far from the relevant plate boundaries and no conceptual model has yet been proposed which fits this mid-Tertiary phase of activity into our evolving understanding of plate tectonics and its relationship to mountain building.

The surface geologic formations of UAE are covered by sand fields. Outcrops are confined only to Ras Al Khaima, Al Mleha and Al Madam area, and to the Al-Ain area to the East, where the Hajar Mountains form part of that region. Sand dune ridges reaching heights of 150m inland and plain gravel areas exist into the inter-dune areas. Evaporate flats (sabkha) dominate the coastal plains, which extend more than 80km southwards into the sand deserts. Consequently the geology of this area is based exclusively on subsurface information. Data from below the Permian is quite scarce, as only 3 wells have penetrated the Permian sequences of Abu Dhabi. Offshore Abu Dhabi the sedimentary section from the Upper-Permian to recent times has a maximum thickness of more than 5500m [26].

3.3 Potentially appropriate environments

In the Emirate of Abu Dhabi six geologic environments have been identified:
a. Hard fractured rock to surface, which includes carbonate rocks, schists, and the ophiolite complex;
b. Hard fractured rock overlain by relatively high-permeability sedimentary rocks in which advective transport dominates, as limestones;
c. Hard fractured rock overlain by a sedimentary rock sequence containing at least one significant low-permeability formation in which diffusion dominates solute transport;

DOI: 10.1051/matecconf/201712006003

d. Evaporites;
e. The Carbonate host rock; and
f. Non-evaporitic rock with hypersaline groundwater.

According to the stratigraphic correlation of Abu Dhabi onshore wells from the Paleozoic to the recent era a number of groups have been identified consisting of carbonate (limestone-dolomite), shale limestone, and anhydrite. Two main groups below the deeper aquifer system are important for further investigation for radioactive waste geological repositories. Both are below the shallow (160-460m) aquifer system and the deepest one (from 700 to 1600m) are represented in the onshore structural cross section of ADCO's fields shown in Fig. 4.

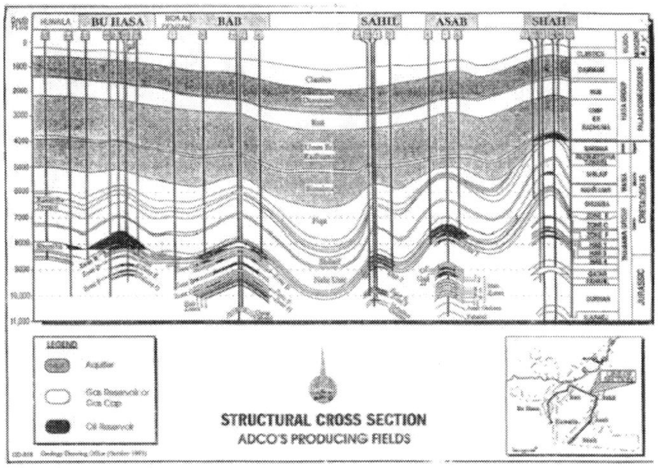

Fig. 4. Structural cross section ADCO's producing fields [26].

The first one is part of the deeper formation of the Bab-Agrab- Nahr Umr group of Middle Cretaceous era. This formation thickens gradually from about 75m in the north to a maximum of 200m in the south and southwest. It thins to approximately 100m in the extreme east of the offshore area towards the Northern Emirates, in various depths between 2600-3000m approximately. The Nahr Umr formation consists of a sequence of shales with some rare sand lenses, glauconitic silts, and occasional beds of limestone, mostly packstones and wackestone.

The second one belongs to the Ruwais-Fiqa formation of the Upper-Cretaceous. It is a deep open marine-shelf environment deposit that ranges in thickness from 120m to more than 360m in offshore Abu Dhabi, in depths between 2000-2500m approximately. Its lower part consists of soft marl, calcareous shales, and argillaceous limestone and grades upwards into mainly argillaceous lime-wackestone.

4 Summary and conclusions

In this paper we have highlighted the different geologic environments and methods adopted by various countries for the disposal of high-level radioactive waste. Further, we have discussed the nature of the geological environment in UAE to outline the issues associated with the geologic formations in the country. Due to the nature of fractured rocks of the geologic formations, more investigations are required to be able to identify the suitability of sites. For example, cross-hole investigation can be used to develop and test geophysical and hydraulic methods and instruments for the purpose of detecting and mapping fracture zones. Also, electromagnetic (radar), seismic, and hydraulic (sinusoidal) techniques are well developed and tested, and could be used for such investigations. It is worth noting that a disturbance, such as a fracture zone, will affect the chemical and hydraulic transmission properties that would impact on the overall safety of the designed engineered barrier systems for the containment of radioactive wastes.

References

1. T. Avanidou, E. K. Paleologos, Water Resour. Research, **38**, 11 (2002)
2. U.S. Academy of Sciences-National Research Council, *Radioactive waste disposal into Atlantic and Gulf coastal waters; a report from a working group of the committee on oceanography of the National Academy of Sciences-National Research Council*, **48** (1959)
3. National Research Council, *The disposal of radioactive waste on land*, 146 (1957).
4. SKB (Swedish Nuclear Fuel and Waste Management Company), *Long-term safety for the final repository for spent nuclear fuel at Forsmark*, TR-11-01, 3 vol. (SKB, Sweden, (2011)
5. POSIVA OY, *Safety case for the disposal of spent nuclear fuel at Olkiluoto—Performance assessment 2012* (POSIVA 2012-04, Finland, 2013)
6. P. Brady, B. Arnold, S. Altman, P. Vaughn, *Sandia Report SAND2012-7789*, 30 (Sandia, September 2012)
7. National Research Council, *Disposition of high-level waste and spent nuclear fuel: The Continuing societal and technical challenges, Committee on disposition of high-level radioactive waste through geological isolation, Board on Radioactive Waste Management*, 212 (US National Academies Press, 2001)
8. ONDRAF/NIRAS (Belgian Agency for Radioactive Waste and Enriched Fissile Materials), *Waste plan for the long-term management of conditioned high-level and/or long-lived radioactive waste and overview of related issues*, NIROND 2011-02, (ONDRAF/NIRAS, Belgium, 2011)
9. US DOE, *Assessment of disposal options for DOE-managed high-level radioactive waste and spent nuclear fuel*, 32 (US DOE, October 2014)
10. US DOE, *Viability assessment of a repository at Yucca Mountain*, 5 Volumes and Overview (US DOE, Office of Civilian Radioactive Waste Management, 1998)
11. US DOE, *Yucca Mountain Site Description, Yucca Site Characterization Project, draft A*, 3 books (US DOE, Office of Civilian Radioactive Waste Management, Apr. 1998)
12. US DOE, *Analysis of the total system life cycle cost of the Civilian Radioactive Waste Management Program, Fiscal Year 2007 (July 2008)*, DOE/RW-0591 (US DOE, Office of Civilian Radioactive Waste Management, 2008)
13. US DOE, *Yucca Mountain repository license application*, DOE/RW-0573, Rev.1. (US DOE, Office of Civilian Radioactive Waste Management, 2008)
14. US NRC (United States Nuclear Regulatory Commission), *Safety evaluation report related to disposal to high-level radioactive wastes in a geologic repository at Yucca Mountain, Nevada. Volume 3: repository safety after permanent closure*, NUREG 1949-Vol. 3, 781 (US NRC, October 2014)
15. L.H. Johnson, D.M. LeNeveau, D.W. Shoesmith, D.W. Oscarson, M.N. Gray, R.J. Lemire, C. Garisto, *The disposal of Canada's nuclear fuel waste: The vault model for post closure assessment*, AECL-1071, COG-93-4 (AECL Research Report, 1994)
16. R.N. Yong, A.M.O. Mohamed, D.M. Xu, Eng. Geol., **28**, 315-324 (1990)
17. R.N. Yong, A.M.O. Mohamed, I. Shooshpasha, C. Onofrei, Eng. Geol., **47**, 351-365 (1997)

18. A.M.O. Mohamed, R.N. Yong, B.H. Kjartanson, C. Onofrei, Geotech. Test. J., GTJODJ, **19**, 2, 155-163 (1996)
19. J. Graham, N.A. Chandler, D.A. Dixon, P.J. Roach, T. To, A.W.L. Wan, *The buffer/container experiment: Volume 4-results, synthesis, issues*, Review Report (Fuel Waste Technology, AECL, Whiteshell Labs, Canada, 1996)
20. J. Graham, J.M. Oswell, M.N. Gray, Can. Geotech. J., **29**, 1033-1043 (1992)
21. J. Graham, F. Saadat, M.N. Gray, Eng. Geol., **28**, 391– 403, (1990)
22. A.P.S. Selvadurai, Eng. Geol. **41**, 219– 238 (1996)
23. Selvadurai, A.P.S., Special Issue of Eng. Geol., **47**, 311– 317 (1997)
24. A.H.F. Robertson, M.P. Searle, in the *Geology and Tectonics of the Oman Region*, **49**, 3-25 (Geological Society of London, Special Publication, 1990)
25. A.H.F. Robertson, A.E.D. Kemp, D.C. Rex, C.D. Blome, in the *Geology and Tectonics of the Oman Region*, **49**, 285-305 (Geological Society of London, Special Publication, 1990)
26. F.N. Irani, *Geology of the UAE*, 143 (Schlumberger Drilling and Measurements, 2001)
27. A.M.O. Mohamed, E.K. Paleologos, D.N. Singh, V.G. Silvestre Rodrigues, *Fundamentals of geo-environmental engineering*, (Elsevier, to be published, 2017)
28. A.S. Alsharhan, A.E.M. Nairn, *Sedimentary basins and petroleum geology of the Middle East*, 843 and appendices (Elsevier, 2003)

Amelioration physical and chemical parameters of marl miocene doped with the pozzolan

Laila Mesrar [1,*] *Ahmed* Banamar[1], *and Raouf* Jabrane [2]

[1] Normandie University, UNIHAVRE, CNRS, LOMC, 76600 Le Havre, France
[2] Department of Earth Sciences, Laboratory of Georesources and Environment, Faculty of Science and Technology of Fez, Morocco.

Abstract. Given the increasing exploitation of clay in Morocco for ceramic industry and for the purpose of the benefit reuse of this resource, a study is conducted in order to carry out the identification and characterization of Miocene marl after doping with pozzolan. The objective of this work is to improve the physicochemical characteristics of Miocene marl of Fez region, by doping it pozzolan at different contents: 5, 10, 15 and 20%. Physical and chemical characterization of this material was carried out before and after doping with pozzolan following various analytical technics, such as the X-ray fluorescence and geotechnical investigation. The result of geochemical analysis showed a higher rate of aluminum oxide (Al_2O_3), iron oxide (Fe_2O_3) and titanium oxide (TiO_2). Further characterization of Atterberg limits have shown that the plasticity of doped materials increases. Similarly the mechanical behaviour of marls doped with pozzolan have significantly improved, providing a mechanical strength of obtained materials up to 670N when firing at 1000 ° C, compared to the raw marl.

1 Introduction

The use of clay resources by human, mainly in the manufacture of building materials, dates from antiquity. Clay is a cheap and abundant material. However its low mechanical strength and non -uniformity colour after cooking limits its added value. The addition of an organic reinforcement in a mineral matrix is often a solution to improve the implemented mechanical elements [1].

Economic issues related to the rising costs of these methods (long chemical processes), and their complications, therefore, leads to explore other simple and low cost manufacturing methods such as doping with external material. In this work, we evaluate the influence of doping with pozzolan the Miocene marl of the region of Fez (Morocco).

However, the estimation of these resources to develop new materials, most environmentally friendly has become a concern of both scientific and industrial aims. The activity of extraction rocks and minerals, whose product is mainly export-oriented, experienced the past three decades a significant decline. The recent global economic crisis

* Corresponding author: mesrarlaila@gmail.com

has further complicated the situation because of the very significant decline in investment in the field of mineral exploration. It has resulted in an alarming decline in exports and even a reversal of the situation: Morocco became an importer of some very expensive minerals but nonetheless vital to its industry. It becomes capital and urgent to draw up an inventory (usage and needs) of many minerals available in Morocco to find alternative solutions (domestic substitute products that are competitive in the point of view of quality but also the cost). Therefore, selection and recovery of industrial rocks and minerals require the involvement of different disciplines in particular, recognition, geological mapping and mineralogical identification and characterization of technological specifications.

In particular, The Miocene marl substrates, which are abundant in the region of Fez, remain insufficiently appreciated. However, as regard to the new demands of the society in terms of sustainable development and energy saving, recovery of the clay and its use in building materials may be more environmentally sustained and become of scientific and industrial concern. It was also investigated from different points of view in literature [2, 3, 4, 5].

In this perspective, the doping of the marl with pozzolan rocks could increase the added value of the clay used in the technology and building industry. A major problems encountered by users of these materials is their fragility and high shrinkage induced by cooking and causing defects in the bricks [6], and can go until it bursts. In order to improve the material performance, the efficiency and the quality in the manufacture of this marl, we opted for doping by dry method this material by the pozzolan, at different percentages of 5, 10, 15 and 20%. The results of this study allow us to conceive and suggest new improvements of the physicochemical and mechanical parameters for clay so as to rise the benefit use of this low cost material.

2 Materials and methods

2.1 Material

The marl Miocene of Fez presents two distinct facies by their colour and lithology: blue marls and yellow marls (Figure.1). These marls flush to the surface especially at the top of the hill of Koudiat Ben Jallik (Figure 2).

The basic raw material is a marl Miocene of the region of Fez (Figure 3) which is an outcrop in the eastern area of Fez (Figure 2) at the Ben Jallik quarry. This region is known for its expertise in the field of pottery and tiles. The importance of clay-based materials in the ceramic industry is due to their plastic properties [7] allowing final ceramic products to retain their shapes after firing. The process is complex because usually the shaped product depends on the mineralogical composition of the raw material and the nature and quantity of impurities in clay.

The grey or blue marls of Fez, attributed to the Vindobonian, form a powerful monotone series of great lithological homogeneity reaching in Fez almost at 1000m elevation (Figure 4).

The base of the series is constituted by transgressive facies (sand more or less consolidated passing to compact sandstones, locally crystalline sandy limestone, conglomerates) that can pass laterally or vertically to blue marls of the Tortonian, normally located above. Blue marls are very compact and show traces of stratification (Figure 5). They are often affected by at least three families of vertical, oblique and subhorizontal diaclases.

These marls constitute the regional substratum of the whole region and appear locally in the city center and in the valleys of the wadis Zitiune and Boufekranne.

This marl is distinguished by its gray green or blue colour. The superficial layers of the latter are generally altered and become yellowish (Picture 2).

The rocks pozzolan selected for this study come from an open pit quarry, available in large quantities, the career of pozzolan is located at 40 Km from the city of Fez.

Fig. 1. A forehead of the career of Koudiat Ben Jallik.

Fig. 2. Finely compacted marl of vertical diaclase.

Fig. 3. MNT of the study area Koudiat Ben-Jalik.

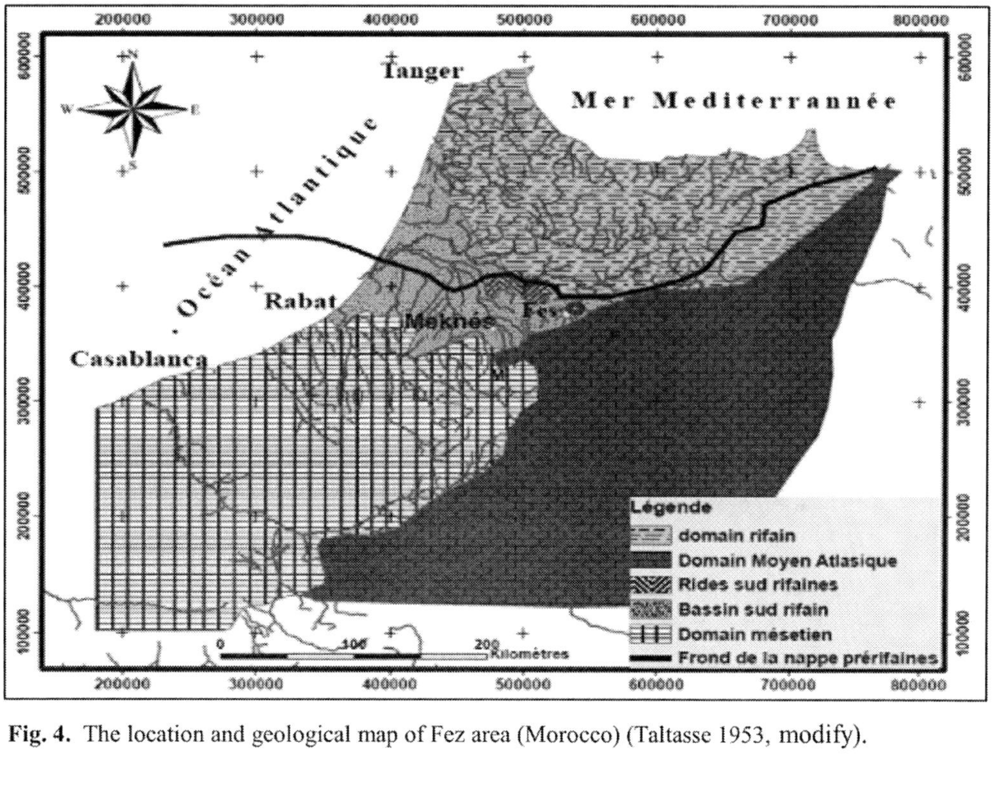

Fig. 4. The location and geological map of Fez area (Morocco) (Taltasse 1953, modify).

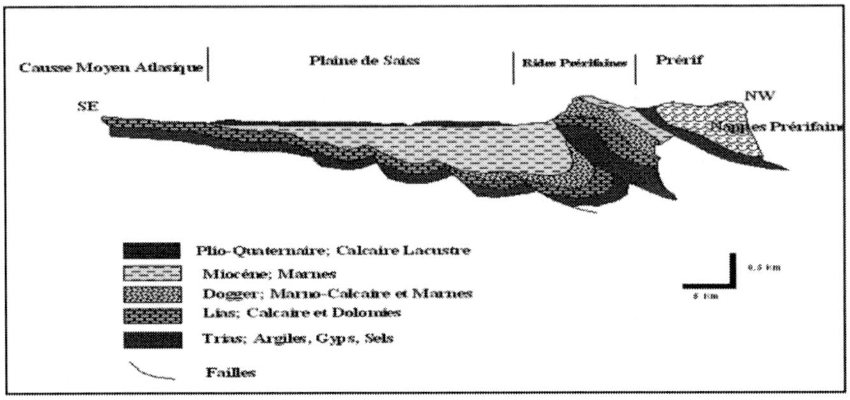

Fig. 5. Geological cross section of South Rif (DRH et ONI, 1971-1977).

2.2 Preparation of composite marls / Pozzolan rocks

Many methods have been developed for the synthesis of powders, which are widely used in laboratory research and industry [8].

The method used for mixing rocks pozzolan and marls involved three steps: mixing of precursors, reaction in solid and grinding. In the present study, we used the marl Miocene and doped by pozzolan at different percentages (5, 10, 15 and 20%).

2.3 Methods

In order to follow the homogeneity and the degree of pozzolan incorporation in the clay sheets several analysis/ techniques were used.

Chemical analysis: the quantitative chemical analysis of clays was performed by X-ray fluorescence spectrometry (WD-XRF) used spectrometer type Axios sequential brand Panalytical. This analysis was made after grinding the material and taking 10g of each sample, and placed them in a hydraulic press, in order to make the pellets. The limits of detection of each element and error consideration for analysis are given in Table 1.

Mineralogical analysis: X-ray diffraction DRX study was done with the Philips diffractometer XPERT-PRO" PW 3064, with the Kα1, 2 radiation of copper. X-ray diffraction (XRD) data were obtained using oriented samples prepared on glass slides in the 2θ range of 2–40° at a scanning speed of 2° min−1.

The investigation of geotechnical properties of material was performed in the laboratory of geosciences and environment (university of Fez). The evaluation of plasticity was performed by the Casagrande method. The Atterberg limits are conventional physical constants indicating the clay transition from plastic to liquid state (WL) and from the plastic to solid state (WP). The plasticity index, which represents the difference between the liquid limit and plastic limit, defines the extent of the plastic field of clay.

Bricks production is carried out using a standard metal mold, provided for the operation. It has a length of and a width of 10mm. Bricks are previously dried in the open air until the stabilization of the withdrawal. They are fired in electric oven programmable to the temperatures of 900°C, 1000°C and 1100°C, with a bake cycle of 30 minutes and a bearing of one hour and a half to the maximum firing temperature. The rate temperature of firing rise 10°C/min. The parameters resulting from the bricks firing are observed as the materials shrinkage along drying and firing, the ignition loss, and the mechanical resistance to compression.

3 Results and discussion

3.1 Chemical analysis

The chemical composition of the samples were given in Table 1. The most abundant oxides in the samples were SiO_2, Al_2O_3 and Fe_2O_3, whereas MgO, K_2O, Na_2O and TiO_2 presented only in small quantities. Iron oxide (Fe_2O_3) is the main colorant component, contents of Iron oxide being sensitive to firing and so lead to produce unexpected results in colour and texture of the fired clays [2]. The colour caused by iron oxide (Fe_2O_3) depends on the changes in the ratio of CaO. This might explain how high CaO contents may turn the colour to yellowish and pinkish shades [5]. Carbonates in clay samples are present in relatively higher content of CaO with was an indication of expressive calcite and dolomite. The low content of alkaline oxides (K_2O, Na_2O) was a consequence of kaolinite clay that naturally have low flux materials content. The easiest way to determine the influence of a doping is to make measurements of tested material before and after doping.

Thus, analysis of the chemical composition by XRF marl doped with pozzolan reveals a significant enrichment in alumina (Al_2O_3), iron oxide (Fe_2O_3) and titanium oxide (TiO_2). Their concentration increases as the doping level is increasing.

However, doping causes a slight decrease in some major element remains like CaO, which is varying quantitatively.

Table 1. Percentage of major elements of the crude marl (Z1) and doping with Pozzolan.

Sample	SiO_2	Al_2O_3	CaO	Fe_2O_3	MgO	K_2O	Na_2O	TiO_2
Z1 Brut	42.3 0.020	13.1 0.010	9.18 0.010	7.63 0.005	3.2 0.003	1.02 0.003	1 0.017	0.33 0.002
Pozzolan Brut	37.8 0.020	17 0.010	3.88 0.010	17.3 0.005	6.2 0.003	0.33 0.003	0.82 0.017	2.3 0.002
Z1 doped with 5%	42.5 0.020	13.3 0.010	8.77 0.010	7.63 0.005	3.22 0.003	1.01 0.003	0.98 0.017	0.35 0.002
Z1 doped with 10%	42.3 0.020	13.4 0.010	8.75 0.010	7.93 0.005	3.27 0.003	1.06 0.003	1.01 0.017	0.43 0.002
Z1 doped with 15%	42.2 0.020	13.5 0.010	8.70 0.010	8.51 0.005	3.4 0.003	1.06 0.003	0.97 0.017	0.44 0.002
Z1 doped with 20%	42.5 0.020	13.4 0.010	8.92 0.010	8.86 0.005	3.4 0.003	1.06 0.003	0.97 0.017	0.5 0.002

3.2 Mineralogical analysis

The mineralogical compositions of marl crude and doped showed significant differences, which reflects the change in the mineralogical structure of composites marl / pozzolan. The mineralogical compositions are reported in Figure 6, which refers to the relative abundance non-clayey minerals (quartz, calcite), and in small quantity the phyllite minerals (kaolinite, illite).

In all X-Ray diffraction of doped samples, we noted the appearance a hematite peak may be explained by the exchange between the pozzolan minerals and marl material.

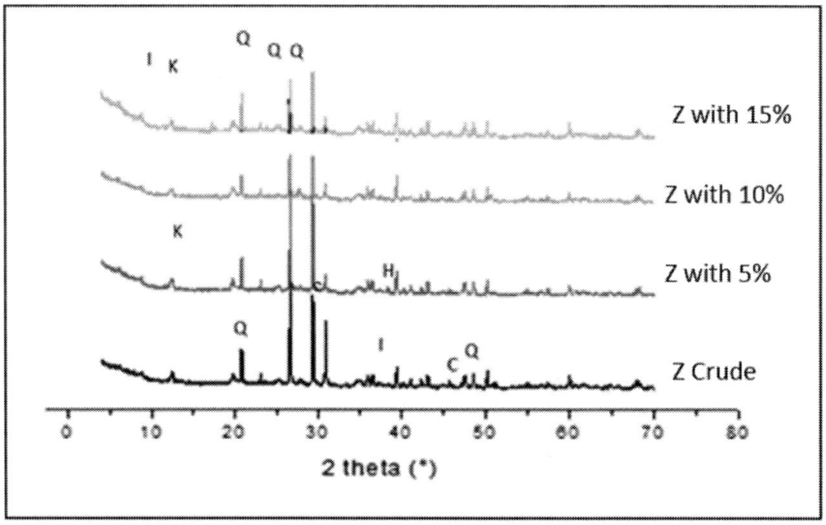

Fig. 6. Diffractograms RX of the raw sample z1 and doped with pozzolan at 5, 10 and 15. Q: quartz; C: calcite; I: illite; and K: kaolinite; H: Hematite.

3.3 Geotechnical characterization (Atterberg limits)

Based on the results of Table 2, we note that the raw material Z1 has an average plasticity that increases with the level of doping. However, the greater plasticity is observed in the doped marl with 20% pozzolan. Therefore we can deduce that the doping of this clay with pozzolan increases the plasticity of these materials. Previously this phenomenon was observed during handling.

Table 2. The results of plasticity and liquidity tests in the Samples.

Sample	Liquid Limit (Wl)	Plastic Limit (Wp)
Z1 Brut	43.2	19.40
Z1 doped with 5%	43.2	27.40
Z1 doped with 10%	43.5	28.30
Z1 doped with 15%	43.3	29.40
Z1 doped with 20%	43.3	30

As regards the Atterberg limits, we note that the calcareous crude clays (Z1) show an intermediate plasticity. The clay doped with 5, 10 et 15% pozzolan rocks has a high plasticity and is more suitable for the use in bricks and pottery because it accounted for acceptable extrusion (Figure 7) .

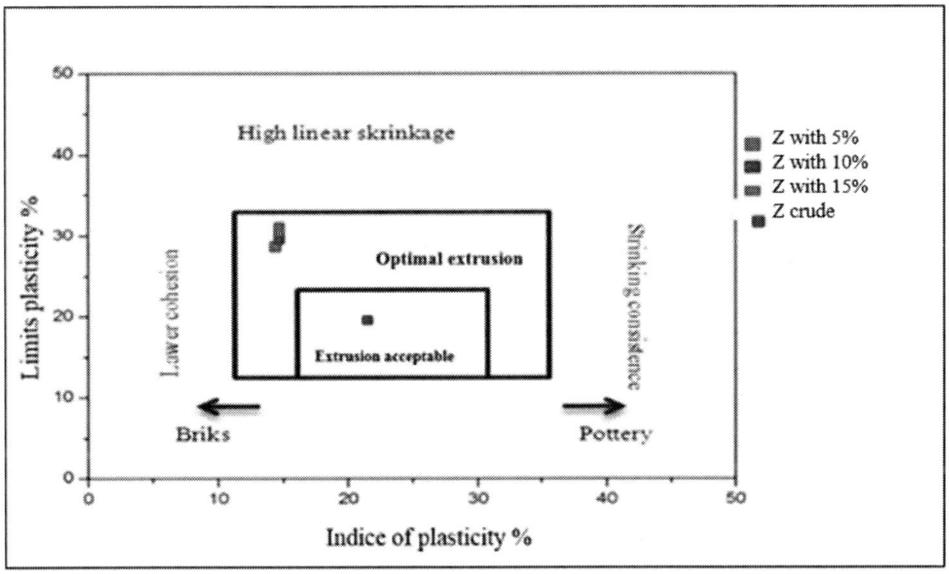

Fig. 7. Casagrande diagram with domains showing the optimum and acceptable domain for clay Shaping by extrusion (Highly, 1978 in [4].

3.4 Effect of firing

During the rise of firing temperature, the physicochemical transformations of the phase's marls affect the material (Figure 8). Each mineral in the mixture behaves as if it was alone. The decomposition products or new structural varieties, which take place under the influence of heat, react together to form new compounds [9].

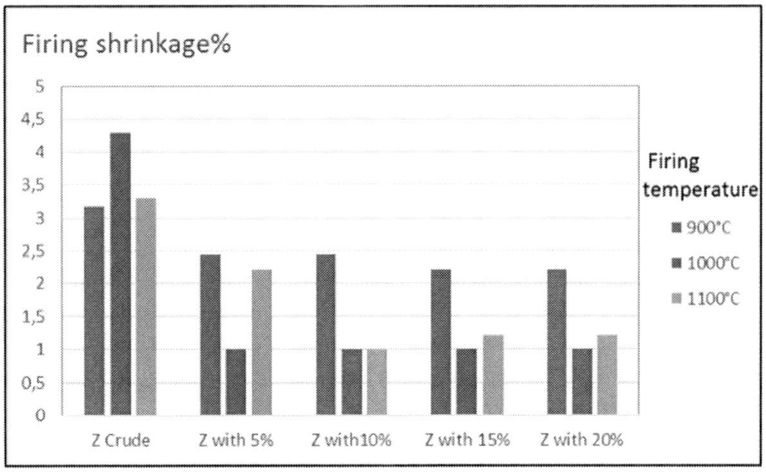

Fig. 8. Influence of firing temperature on the variation of linear shrinkage.

However, upon baking, the initial crystalline minerals disappeared and make space in varying proportions to an amorphous phase and new minerals [10].Shrinkage accompanies this solid reaction, and is associated with changes in porosity and particle size.

The proportions of the clay phase and / or associated minerals play an important role in these changes as well as the particle size [11]. Removing charged in the raw material is very important. This is due to the expansion case by new new-formed minerals [12].

In this study, all doped materials attended a shrinkage reduction until its cancellation. Therefore, being concluded the doping plays a role in limiting internal shrinkage. So we can conclude that the pozzolan plays a degreasing agent limiting withdrawal.

3.5 Mechanical properties

The bricks of marl doped at 10% of pozzolan have a relatively high resistance up to 525 N compared with bricks doped at 5% (Figure 9). However, the increase in percentage of pozzolan is not indefinitely correlated positively with mechanical strength.

At 1000 ° C, the bending strength increased from 900 °C in all the bricks and peaked its maximum (670 N) in the marl doped with 10% of the pozzolan. The explanation of the increase in resistance is linked to the modification of the crystalline structure of our marls [1], because at this temperature the vitrification begins and induces a stronger resistance.

At 1100°C, overall, the flexural strength near a slight decrease estimated approximately 30N this decrease can be attributed to a beginning of alvealing [7]. However, the marl doped with 10% displays a maximum resistant which that of 640N.

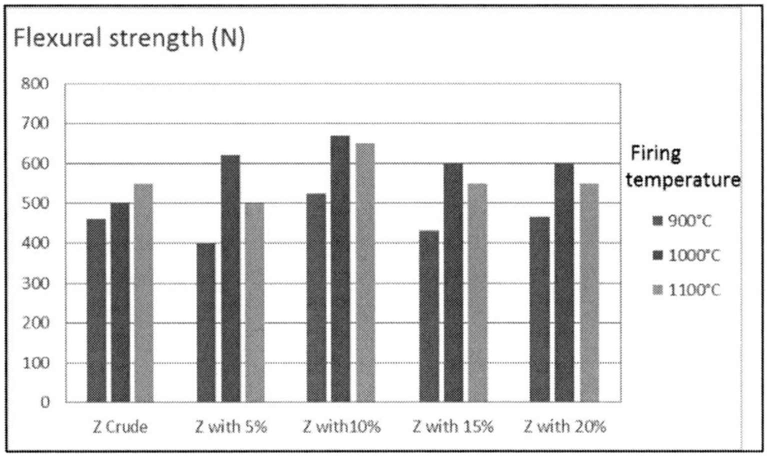

Fig. 9. The flexural strength of the bricks firing at different temperatures.

4 Conclusion

The results chemical analysis of marl doped with pozzolan reveals a significant enrichment in alumina (Al_2O_3), iron oxide (Fe_2O_3) and titanium oxide (TiO_2). The mineralogical analysis indicate the presence of Hematite peaks in all samples doping. Technological tests showed suitable industrial properties of these marls, their doping with pozzolan induced an increase in their plasticity. The pozzolan acts as a degreaser and decreases shrinkage.

The mechanical resistance of the doped bricks marl is dependent on quantity pozzolan doping and the temperature of firing. The marl doped with 10% has a relatively high resistance compared with bricks doped with 5, 15 and 20%. Therefore, additional content of doping increases mechanical strength of bricks up to a certain amount less than 20%.

This resistance is in fact related to the decrease of the porosity induced by the increase of the content of the doping. This causal relationship is in full agreement with the results obtained in the study of the porosity of the bricks. The bricks of marls fired at 1000°C and doped with 10% gives better mechanical strength of up to 670 N. So, suggesting the use of this composite as eco-geo-materials in construction work, tiles and in pottery, is the main recommendation from the present study.

References

1. L. Mesrar, and R.Jabrane. Characterization of Clays of the Sillon South Rif (Morocco), Clay Research, Vol. 33, No. **2**, p 65-73 (2016).
2. L. Mesrar , O. El Aroussi, M. Lakrim, A. Lahrach, A. Chaouni, and R. Jabrane, Technological valorisation of the Miocene clay in the region of Fez (Morocco) : Characterisation and exploitation possibilities. PESD, VOL. 7, no.**1**, p 311-317(2013).
3. F. El Fgaier, Z. Lafhaj, C. Chapiseau. Use of clay bricks incorporating treated river sediments in a demonstrative building: case study. Constr. Build. Mater, no **48**, p 160 - 165 (2013).
4. C. Fiori, B. Fabbri, G. Donati, I. Venturi. Mineralogical composition of the clay bodies, Appl. Clay Sci, **4**, p 461 –473 (1989).
5. M. Bondi. Clay materials for ceramic tiles from the Sassuolo District Northern Apennines, Italy. Geology, composition and technological properties, Appl. Clay Sci, **15**, p 337-366 (1999).
6. L. Mesrar, M. Akdim, M. Lakrim, I. Akhrif, and R. Jabrane. The physico-
7. Mechanical characteristics of the clays in and after doping with metal oxide Al_2O_3 in the region of fez (Morocco). PESD, vol. 9, no. **1**, p 103-110 (2015).
8. L.Mesrar. Caractérisation chimique, mineralogique, technologique nanotechnologique des marnes du sillon sud riffain: Etude et valorisation. Thèse de Doctorat, soutenu à l'université de Fez (Maroc), p 200 (2013).
9. A.Baccour, A. Zghal, M. Medhioub, and T. Mhiri. Caractérisation Physicochimique et mécanique de matériaux céramiques obtenus à partir des argiles Tunisiennes, Verres, Céramiques/Composites, Vol.1. N°**2**, p 25-33 (2011).
10. C.A Jouenne. Traité de céramiques et de matériaux minéraux. *Ed. Septima, Paris*, p 620 (1984).
11. M.J. Trimdade. Transformations of calcareous rich clays with firing: a comparative study between calcite and dolomite, Appl. Clay Sci .**42**, p 345 -355 (2009).
12. M. El ouahabi. Valorisation industrielle et artisanale des argiles du Maroc. Thèse de Doctorat, soutenu à l'université de liège, p 202 (2013).
13. S. Boussen, D. Sghaier, F. Chaabani, B. Jamoussi, A. Bennour. Characteristics and industrial application of the Lower Cretaceous clay deposits (Bouhedma Formation), Southeast Tunisia: Potential use for the manufacturing of ceramic tiles and bricks. Appl. Clay Sci. **123**, p 210–221 (2016).

Stress transfer from pile group in saturated and unsaturated soil using theoretical and experimental approaches

Raid R. al-Omari[1,*], *Mohammed Y.* Fattah[2], and *Abbas M.* Kallawi[3]

[1] Professor, Civil Engineering Department, College of Engineering, Al-Nahrain University, Baghdad, Iraq. Email: tosharaid@yahoo.com.
[2] Professor, Building and Construction Engineering Department, University of Technology, Baghdad, Iraq. Email: myf_1968@yahoo.com.
[3] Lecturer, University of Baghdad, Baghdad, Iraq. Email: mabbas153@yahoo.com.

Abstract. Piles are often used in groups, and the behavior of pile groups under the applied loads is generally different from that of single pile due to the interaction of neighboring piles, therefore, one of the main objectives of this paper is to investigate the influence of pile group (bearing capacity, load transfer sharing for pile shaft and tip) in comparison to that of single piles. Determination of the influence of load transfer from the pile group to the surrounding soil and the mechanism of this transfer with increasing the load increment on the tip and pile shaft for the soil in saturated and unsaturated state (when there is a negative pore water pressure). Different basic properties are used that is ($S = 90\%$, $\gamma_d = 15$ kN / m^3, $S = 90\%$, $\gamma_d = 17$ kN / m^3 and $S = 60\%$, $\gamma_d = 15$ kN / m^3). Seven model piles were tested, these was: single pile (compression and pull out test), 2x1, 3x1, 2x2, 3x2 and 3x3 group. The stress was measured with 5 cm diameter soil pressure transducer positioned at a depth of 5 cm below the pile tip for all pile groups. The measured stresses below the pile tip using a soil pressure transducer positioned at a depth of 0.25L (where L is the pile length) below the pile tip are compared with those calculated using theoretical and conventional approaches. These methods are: the conventional 2V:1H method and the method used the theory of elasticity. The results showed that the method of measuring the soil stresses with soil pressure transducer adopted in this study, gives in general, good results of stress transfer compared with the results obtained from the theoretical and conventional approaches.

1 Introduction

The bearing capacity of unsaturated soils can be viewed as an extension of saturated soil mechanics. The unsaturated soil can be visualized as having a cohesion consisting of two components. One component is the effective cohesion, and the other component is due to matric suction which is usually ignored in the conventional bearing capacity analysis, with this concept, the conventional bearing capacity theory is applicable to unsaturated soils [1].

*Correspondingauthor: myf_1968@yahoo.com

The stresses on the underlying strata produced by piles are difficult to estimate for several reasons [2]:

1. Influence of pile cap especially when it is in direct contact with the ground.
2. Unknown friction and end bearing effect on pile.
3. Overlap of stresses from adjacent piles, which is difficult to evaluate.
4. Influence of driving piles on the adjacent soil.
5. Time dependent effect such as consolidation, thixotropy, varying load and change in ground water table.

The typical stresses/soil pressures produced from shaft friction or end bearing of a single pile and a group of piles are shown in Figure 1. If piles are used in groups, there may be an overlap of stresses, as shown in Figure 1b, d and e, if the spacing is too close. If the overlap is large, the soil may fail in shear or settlement. Though, the overlapping zone of stresses obviously decreases with increased pile spacing, it may not be feasible since the pile cap size becomes too large and hence expensive [3].

For these reasons, it is common practice to simplify the stress computation, the load is assumed to spread from fictitious rigid footing located at a depth of L/3 from the bottom of the pile. Using the theory of elasticity as presented in Bowles (1996) [2], for three cases of stress distribution, illustrated in Figure 2, these are:

Case1: end bearing, **Case 2**: constant skin friction, and **Case 3**: linear variation of skin friction. The stress is obtained as:

$$\sigma_z = \frac{P}{L_p^2} K_z \tag{1}$$

where: σ_z = vertical stress, P = applied load, L_p = embedment pile length and K_z = stress coefficient.

The stress coefficient K_z is obtained from the equations presented for every case mentioned or directly from tables presented in the same reference.

Vanapalli and Taylan (2011, 2012), suggested more experimental and numerical studies in progress to check the validity of the modified α, b and λ methods for different coarse and fine-grained soils [4-5]. These studies will be useful to better understand the influence of matric suction on the load carrying capacity of the piles. They also suggested different techniques of testing to alleviate some experimental problems. For instance, instead of drilling a hole into the initially compacted soil, they suggested compacting the soil around the pile using a specially designed compactor, such a technique would reduce disturbance during pile installation for tests and likely provide better correlation results as it eliminates the problems associated with the soil disturbance. In addition, this technique also provides better contact between the pile and the soil.

Mohsen (2012) [6] and Fattah et al., (2014) [7] investigated the behavior of a single pile in unsaturated soils using a pile foundation with (0.6 m) diameter and (12 m) length, embedded in fully saturated and partially saturated Iraqi soils, within Baghdad city. The analysis was carried out using the finite element programs SIGMA/W and SEEP/W.

The main objectives of the research are to investigate the influence of the following factors on the behavior of single and pile group foundation: Investigating the behavior of single pile and pile group foundation system in saturated and unsaturated soils. And studying the stress distribution within the soil system and below the pile tip and negative pore water pressure variation with water content.

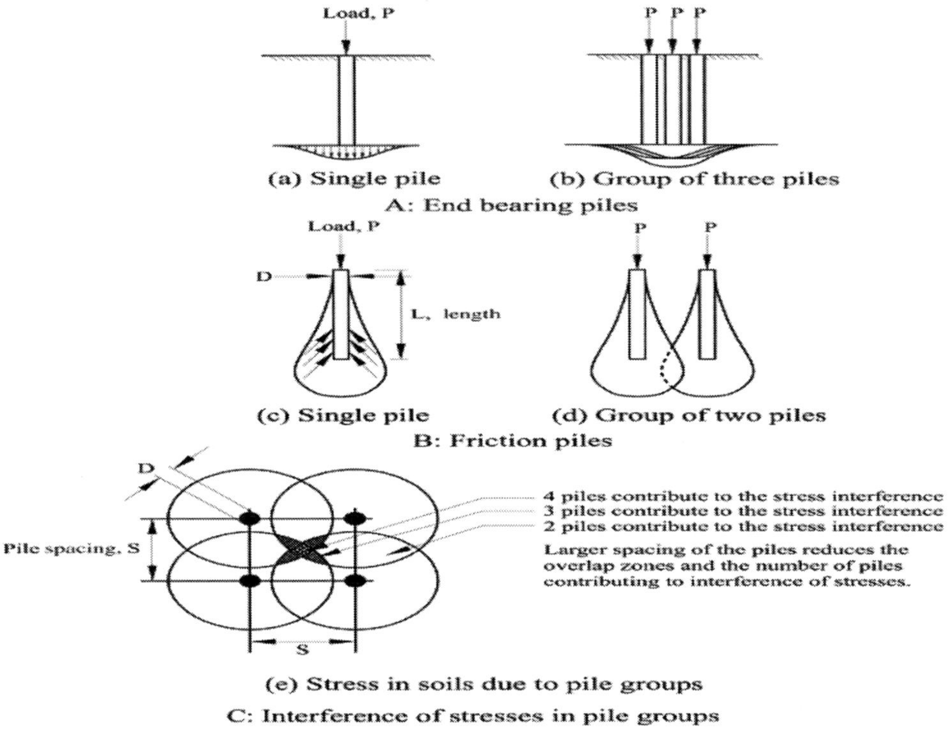

Fig. 1: Stresses in soil due to loads on single pile and pile groups [3].

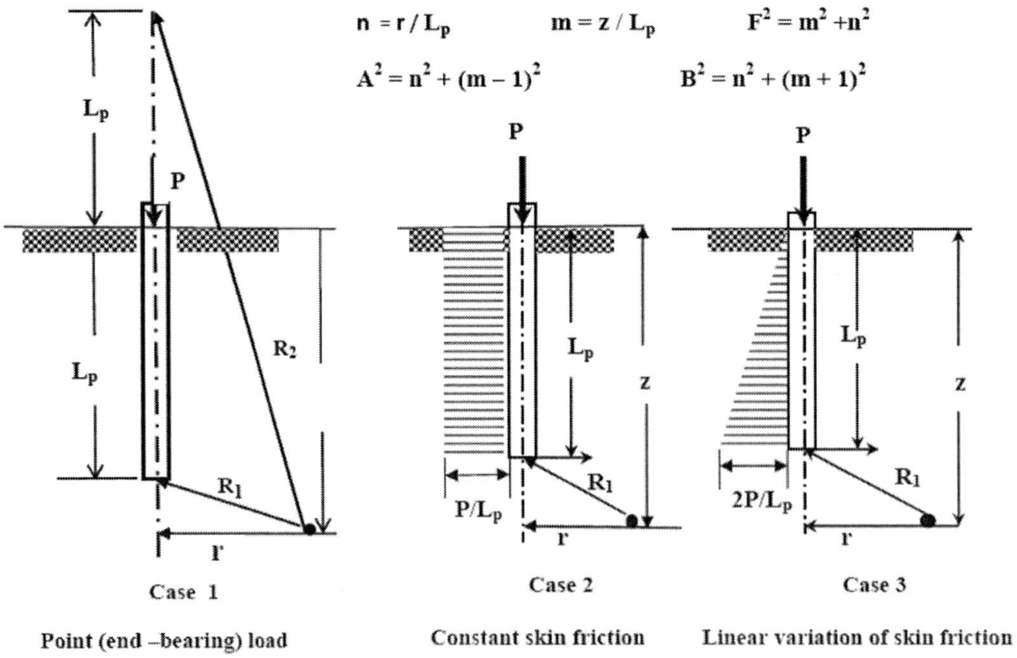

Fig. 2: Pile – soil system for the evaluation of the soil stresses using Mindlin solution [2].

2 Experimental Work

Two sets of containers were used in the experimental work, These containers are designed to serve the saturation process and can be subdivided into three types according to the pile group tested as follows: Container 1 with internal dimensions of 20 cm length, 20 cm width and 35 cm height, this container was used for single pile test (compression and pull out test). Containers 2 with internal dimensions of 30 cm length, 30 cm width and 35 cm height, this container was used for pile of (2x1) and (2x2) group configuration. Container 3 with internal dimensions of 40 cm length, 30 cm width and 50 cm height, this container was used for pile groups of (3x1), (3x2) and (3x3) configuration.

It should be noted that the spacing between piles is the same for all pile groups s = 3 (i.e. 6 cm).

In saturated tests, the water was supplied to the container through a main water supply plastic tube of 10 mm diameter, this tube is branched into two pipes of the same diameter on every side of the container to ensure gradual and uniform saturation process from the base of the container to the surface (saturation was progressed from the bottom to the top of the soil surface in order to expel any entrapped air during saturation process). The physical and chemical properties of the soil used are illustrated in Table 1. According to the Unified Soil Classification System (USCS), the soil can be classified as (CL –ML) with 38% sand, 28% silt and 34% clay. Fattah et al. (2013) concluded that the maximum suction is measured for soils of low consistency (liquid limit < 40%) [8].

Six model pile groups were tested, these are single pile, 2x1, 3x1, 2x2, 3x2 and 3x3 groups. The model piles used were made of aluminum metal with a square cross section (20 mm x 20 mm) and 1.2 mm thickness. The pile model was closed at the toe with an aluminum plate (closed end pile), and attached to the pile cap with two small aluminum angles (20 mm x 15 mm), 2 mm thick by two bolts and nuts as shown in Figure 3.

Table 1: Physical and chemical properties of soil used.

Index property	Index value
Optimum moisture content (OMC)%	13.3
Liquid limit %(LL)	25
Plastic limit %(PL)	19
Plasticity index%(PI)	6
Activity (Ac)	0.19
Specific gravity (Gs)	2.68
Gravel (larger than 4.76 mm)%	0
Sand (4.76 -0.074 mm)%	38
Silt (0.074 - 0.005)%	28
Clay (less than 0.005 mm)%	34
Gypsum content %	0.8
Total dissolved salts TDS%	1.3
SO_3 content %	0.41
pH value	8.73
$CaCO_3$ content %	30.7
Soil symbols (U.S.C.S)	CL-ML

Fig. 3 Connecting the pile to pile cap.

All piles have a length of 200 mm to give an L/D ratio = 10, with the same spacing for all groups s = 3. To facilitate and minimize the disturbance of the soil during pile inserting, another aluminum pile of square cross section (15 mm x 15 mm) was used which is slightly smaller in cross section in comparison with the used pile. This pile was inserted prior to the pile installation to facilitate the pile jacking and minimize as possible the soil disturbance. After that, the pile group was jacked to the soil model slowly and checked within the jacking process to be in the horizontal alignment.

The cap used in the test was also modeled using aluminum plate with smooth surface and thickness of 6 mm. It was connected to the pile group with bolts and nuts. The stresses below the pile groups are measured by soil pressure transducer type (KDE-200 kPa) located at 0.5L (5o mm) down the piles (center piles). The transducer is connected to a digital data logger (TC-32K) through a five channel switching box (CSW 5A).

` The soil was prepared in accordance with the type of test (saturated or unsaturated), and the amount of soil according to the testing container adopted.

The saturated soil was prepared according to the following steps:

1. The sieved soil is prepared at a water content equal to 12.5% which is in the range of optimum moisture content.
2. Each 10 kgs of conditioned soil is mixed gradually and thoroughly with an amount of water corresponding approximately to the water content of 12.5% by hand for a period of 10 minutes to give a uniform distribution of moisture content.
3. After mixing with water, the wet soil is kept in double tightened polythene sheets and left for about 2-4 days as a curing period. Two samples of the soil are extracted to check the actual water content, after that the pre-calculated amount of soil according to the box selected is spread to occupy 50 mm depth of the box. This quantity of soil is compacted to the desired depth using a special tamping steel hammer of (100 mm x 100 mm) in size. The soil is tamped according to the calculated dry density. Before placing the second layer using the same procedure, the surface of the first layer is scratched to give a good interaction with the next layer.
4. A pressure transducer device 50 mm diameter type (KDE-200kPa) is located at 0.5L (5cm) down the piles (center piles). The transducer is connected to a digital data

logger (TC-32K) through a five channel switching box (CSW 5A). The transducer is placed at a depth of 0.25L (5 cm) below the pile tip in all tests for stress measurements below the pile tip.

The same procedure for the saturated soil is used here except that the water content was pre calculated according to the selected degree of saturation.

Figure 4 shows details of the equipment that was designed to serve the objective of the proposed testing program. The loading frame was constructed using two steel channels (150 mm x 50 mm x 8 mm), and 2 m length as supporting columns. They are connected with two other steel channels of the same dimensions, 1.2 m length. One of these channels was to fix the column channel against movement during the load application, while another channel was allowed to move vertically for about 500 mm according to the container length used during the test—full details are given by Khallawi, 2015 [9].

The vertical load was applied on the model piles by means of a hydraulic jack, the applied load was measured using a calibrated proving ring with 2 kN capacity. A dial gauge with 0.01 mm accuracy was used for measuring settlements of the pile group model. The load is applied incrementally with a period of 5 minutes for every load increment as adopted in some pile load tests—U. S. Army Corps of Engineers, 1991 [10].

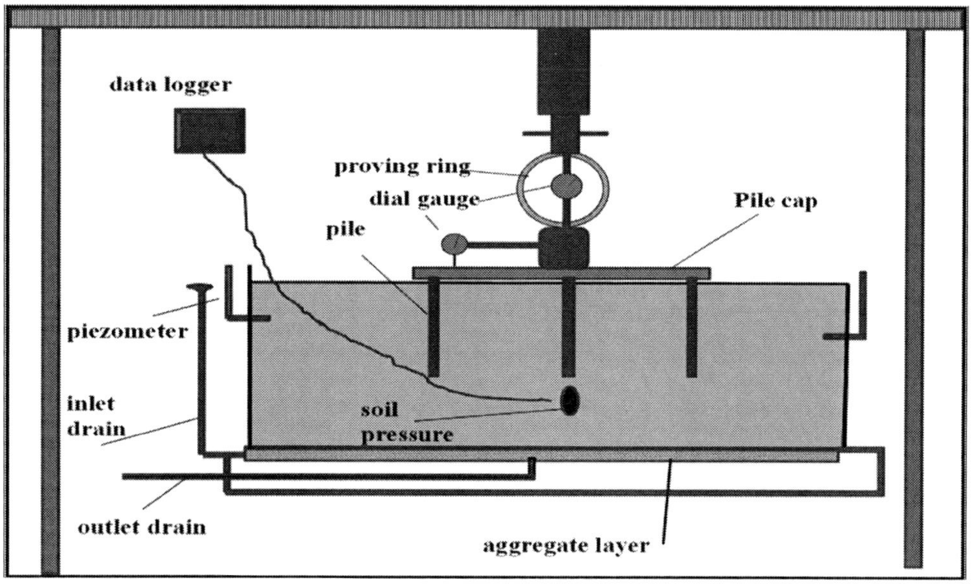

Fig. 4 Diagram of setup configuration.

3 Results and discussion

The results of comparison of stress transfer in saturated soil with properties (S = 90%, γ_d =15 kN/m^3), (S = 90%, γ_d=17 kN/m^3) and unsaturated soil with (S = 60%, γ_d=15 kN/m^3) are presented in Figures 5 to 7 for single, 2x1, 3x1, 2x2, 3x2 and 3x3 pile groups, respectively.

These figures show a good agreement between the results of experimental and the theatrical approach, it can be noticed that the vertical pressure in experimental work is not varying linearly with the applied load, while the linear line is obtained for theoretical analysis. That may cause the stress transfer below the pile tip measured from experimental work is more reasonable than that of the theoretical and conventional 2V:H methods because the experimental results show that the tip resistance increases rapidly before pile failure, while the other two methods consider that the tip resistance is constant throughout the test.

Further, for theoretical analysis, the stress is transferred to the pile tip and shaft by a linear variation and the percentage of load shared by the shaft and tip is calculated according to final results of load shared between the shaft and tip obtained from the experimental work for all pile groups. These values are also presented in Tables 2, 3 and 4. Further The Poisson's ratio used in the theoretical analysis is 0.4 is used for saturated soil, while a value of 0.2 is used for unsaturated soil as suggested by Bowles (1996) [2].

Table 2: Percentage of stress values used in the analysis of soil (S = 90%, γ_d=15 kN/m^3.

Pile group	Single	2x1	3x1	2x2	3x2	3x3
% shaft	50	75	90	90	90	95
% tip	50	25	10	10	10	5

Table 3: Percentage of stress values used in the analysis of soil (S = 90%, γ_d=17 kN / m^3).

Pile group	Single	2x1	3x1	2x2	3x2	3x3
% Shaft	50	75	85	90	90	95
% Tip	50	25	15	10	10	5

Table 4: Percentage of stress values used in the analysis of soil (S = 60%, γ_d=15 kN/m^3.

Pile group	Single	2x1	3x1	2x2	3x2	3x3
% Shaft	0	70	80	88	90	92
% Tip	100	30	20	12	10	8

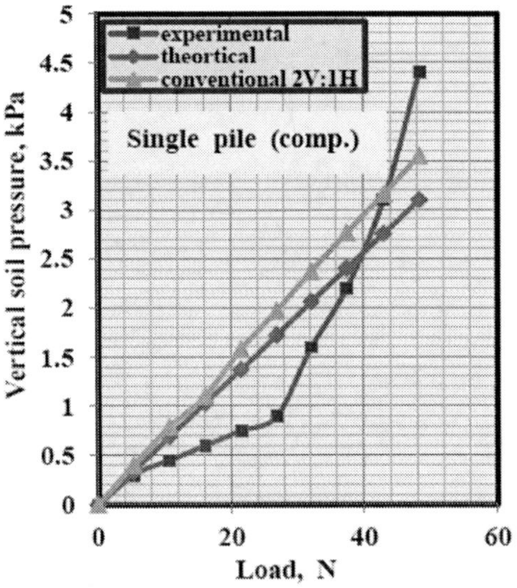

a. Single pile.

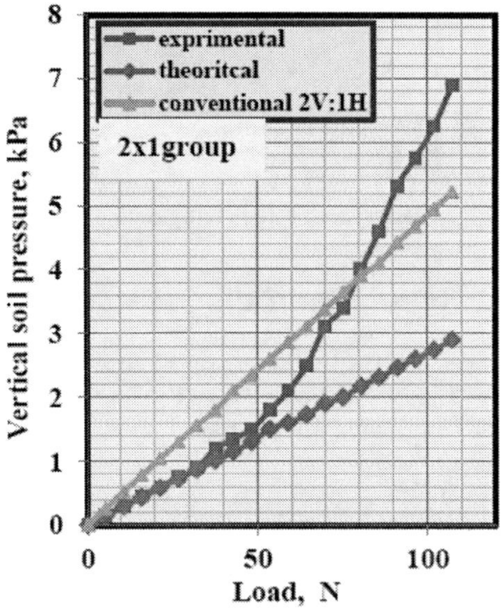

b. 2x1 group.

DOI: 10.1051/matecconf/201712006005

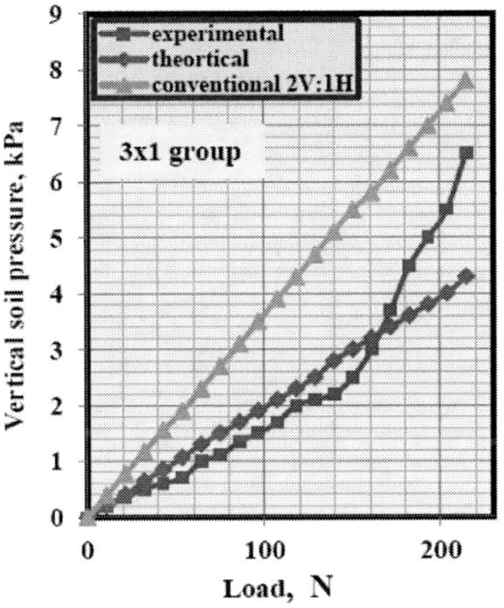

c. 3x1 group.

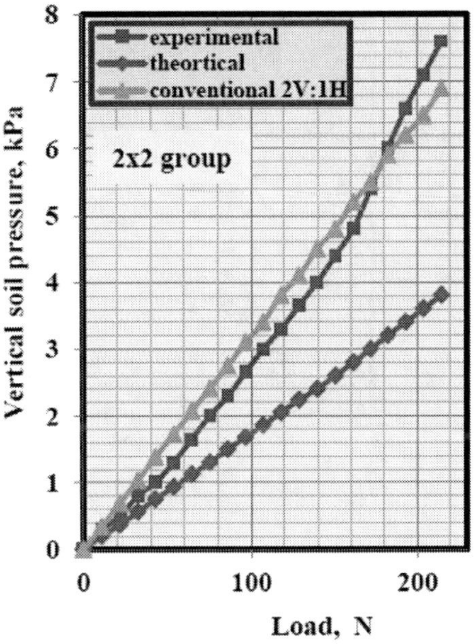

d. 2x2 group.

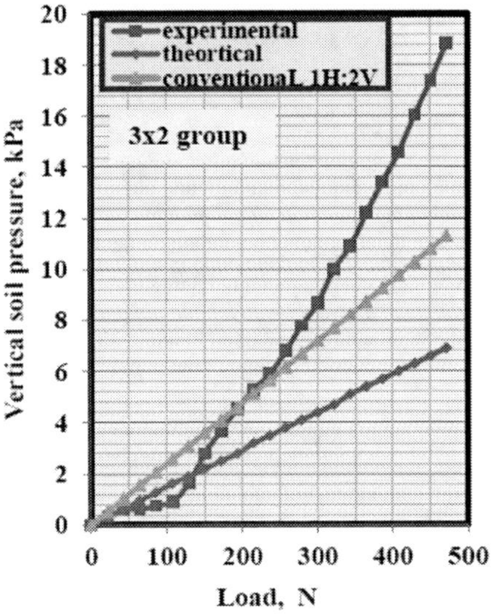

e. 3x2 group.

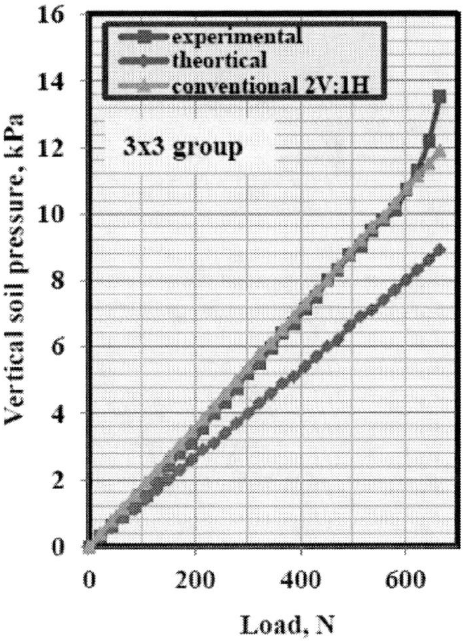

f. 3x3 group.

Fig. 5 Stress transfer below the pile tip during the load application to single and pile group pile in saturated soil (S = 90%, γ_d=15 kN/m^3) compared with that of theoretical and conventional 2V:1H approaches.

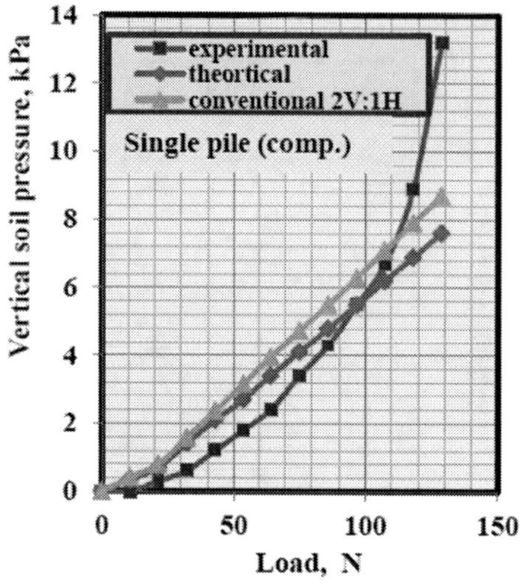

a. Single pile.

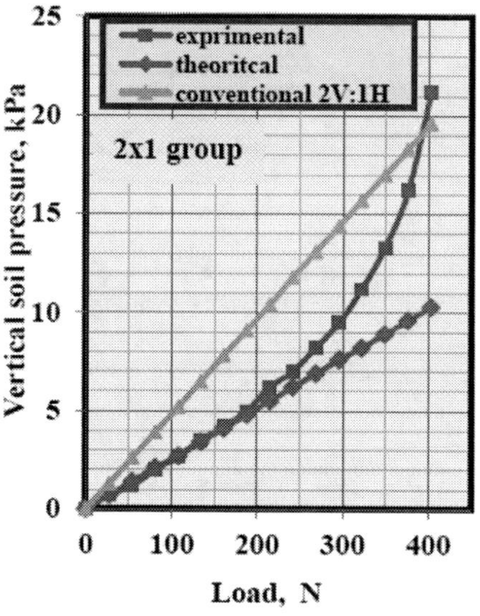

b. 2x1 group.

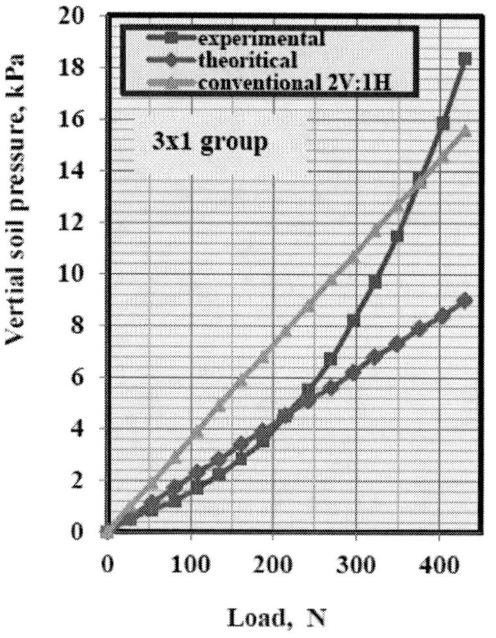

c. 3x1 group

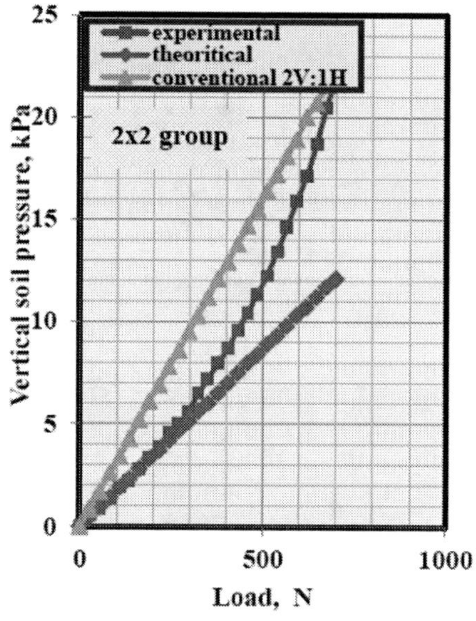

d. 2x2 group.

Fig. 6 Stress transfer below the pile tip during the load application to pile group in unsaturated soil (S = 60%, γ_d=17 kN/m^3) compared with that of theoretical and conventional 2V: 1H approaches.

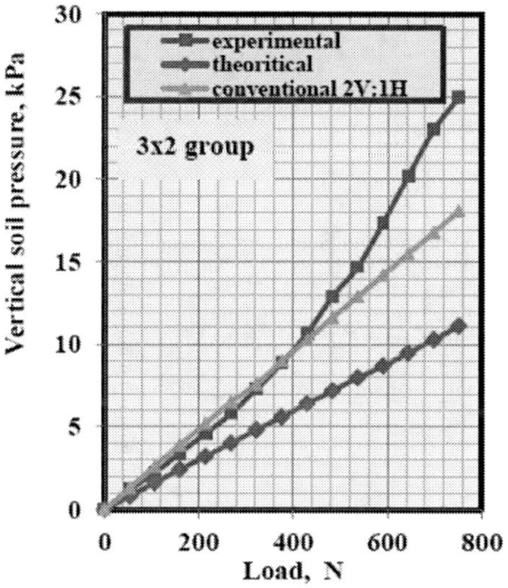

e. 3x2 group.

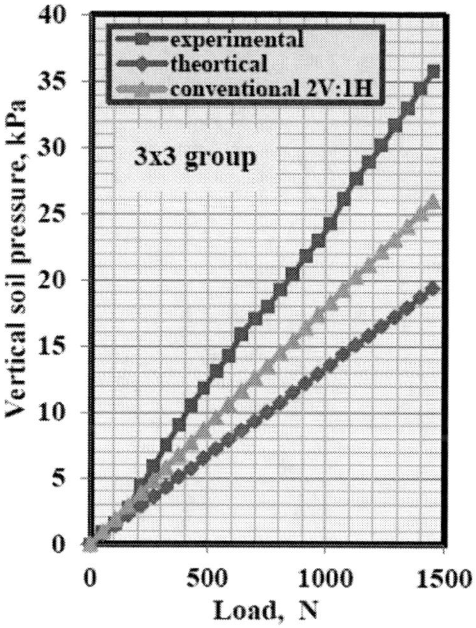

f. 3x3 group.

Fig. 6: Continued.

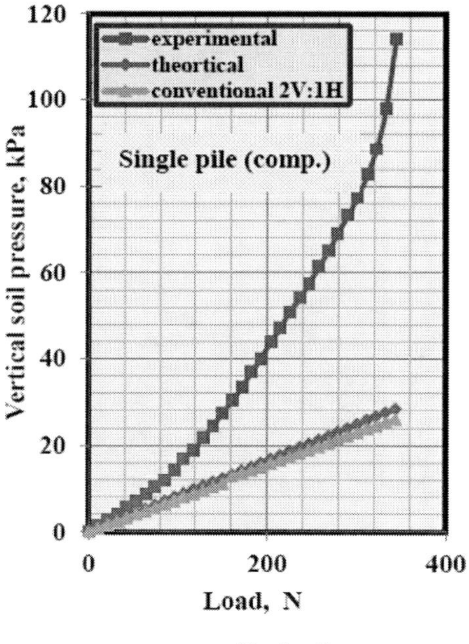

a. Single pile.

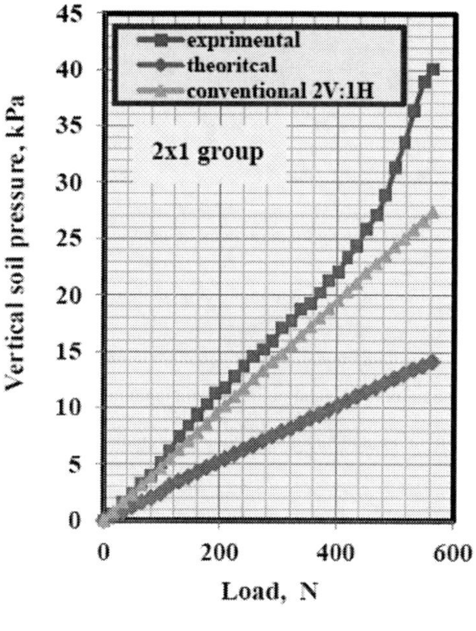

b. 2x1 group.

Fig. 7 Stress transfer below the pile tip during the load application to pile group in unsaturated soil (S = 60%, γ_d=15 kN/m^3) compared with that of theoretical and conventional 2V: 1H approaches.

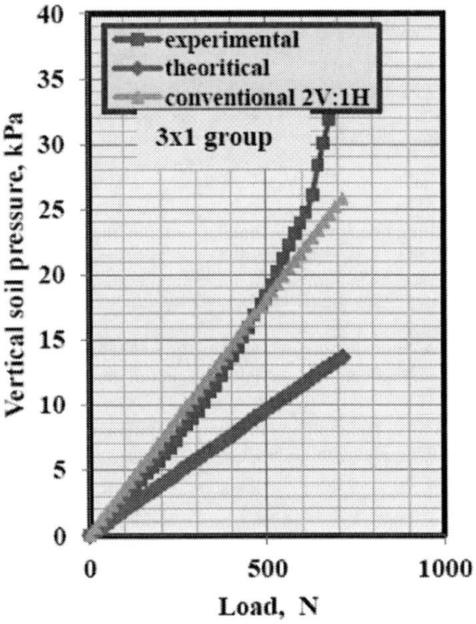

C. 3x1 group.

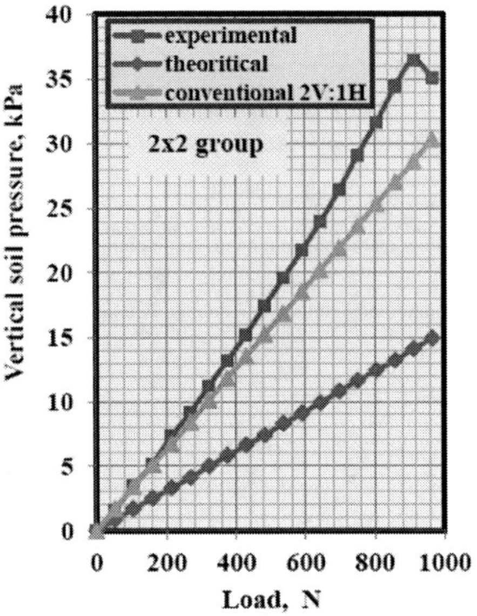

d. 2x2 group.

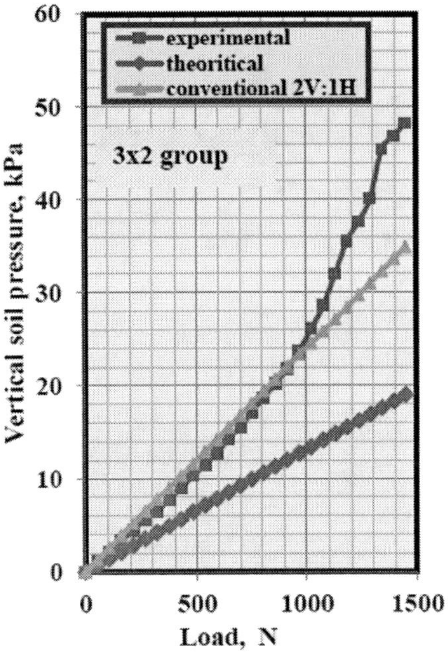

e. 3x2 group.

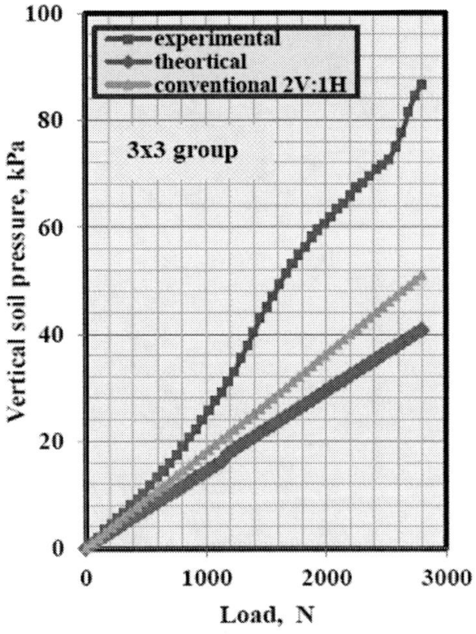

f. 3x3 group.

Fig. 7 Continued.

The stress transferred is important in the analysis of the pile foundations, since it provides the amount of stress transferred to the underlying soil below and around the pile groups, and consequently, the capacity of the soil to carry these stresses safely.

However, there is a complexity associated with studying these stresses due to the difficulty accompanying the method for the stresses to be transferred to underlying soil, since this phenomenon is controlled by many factors reviewed in the introduction.

4 Conclusions

From the presented figures reviewed, it can be said that the analysis of stress transferred from theoretical approach depends on sharing stresses transferred between the pile and tip and these percent's on most occasions depend on engineering judgment, while the other is the conventional 2V:1H method. These two methods give a linear stress transferred results, however, the theoretical method take into account only one parameter of soil property (Poisson's ratio). In contrast, the direct measuring of stress transferred with a soil pressure transducer positioned at 0.25L below pile tip through experimental tests, gives a non- linear response for all piles tested with different soil properties, as shown in the figures.

In brief, it can be concluded that the method of measuring the soil stresses with soil pressure transducer adopted in this study, compared with the results obtained from the theoretical and conventional approaches, gives in general, good results of stress transfer, However the theoretical analysis takes only one soil parameter, Poisson's ratio, into account.

References

1. Fredlund, D.G. and Rahardjo, H., (1993), "Soil Mechanics for Unsaturated Soils". John Wiley, New York.
2. Bowles, J. E., (1996), "Foundation Analysis and Design". McGraw-Hill International Book Company.
3. Rao, N. S. V., (2011), "Foundation Design, Theory and Practice". John Wiley & Sons (Asia) Pte Ltd, 2 Clementi Loop, 02-01, Singapore.
4. Vanapalli, S. K, and Oh, W.T., (2011), "Interpretation of the Bearing Capacity of an Unsaturated Soils Extending the Effective and Total Stress Approaches", in Proc. 5th Int. Conf. on UNSAT, 2010, pp: 1223-1229.
5. Vanapalli, S. K. and Taylan, Z. N., (2011), "Estimation of the shaft capacity of model piles in a compacted fine-grained unsaturated soil". Proc. 14th Pan-Am Conf. on Soil Mechanics and Geotechnical Engineering & 64th Canadian Geotechnical Conf.
6. Mohsen, I. M., (2012), "Behavior of Single Pile in Unsaturated Clay Soils". M.Sc. Thesis, Building and Construction Engineering Department, University of Technology, Iraq.
7. Fattah. M. Y., Salim, N. M., Mohsin, I. M., (2014), "Behavior of Single Pile in Unsaturated Clayey Soils". Engineering Technology Journal, Vol.32, Part (A), No.3. pp: 763-787.
8. Fattah, M. Y., Yahya, A. Y., Al-Hadidi, M. Th., Ahmed, B. A., (2013), "Effect of Salt Content on Total and Matric Suction of Unsaturated Soils", European Scientific Journal, Vol. 9, No. 9, pp. 228-245.
9. Khallawi, A.M. (2015), "Load Transfer of Pile Group in Saturated and Unsaturated Clayey Soil". Ph.D. Thesis, Al-Nahrain University, Baghdad, Iraq.
10. U S Army Corps of Engineers, (1991), "Design of Pile Foundation". Engineering Manual.

MATEC Web of Conferences 120, 06006 (2017) DOI: 10.1051/matecconf/201712006006
ASCMCES-17

Dynamic response of plates on elastic foundation under eccentric impact load

Mohammed Y. Fattah[1,*] *Mohammed* J. Hamood[1], and *Sura* A. Abbas[1]

[1]Building and Construction Engineering Department, University of Technology, Baghdad, Iraq

Abstract. In this study, nonlinear three-dimensional finite element analysis has been used to conduct a numerical investigation of the effect of applied impact load on the foundation based on sandy soil using the finite element method by ANSYS (Version 11) computer program. The 8-node brick elements are used to model the concrete of foundation and the soil under the foundation which are denoted by Solid 65 for concrete and Solid 45 for the soil while the interface is modeled by using three-dimensional surface-to-surface (Target 170 and Contact 174) contact elements connected with concrete and soil. As a case study, a square concrete foundation with dimensions of 3×3×0.3 m placed on the foundation soil 15 m deep and 9 m away from the edge of plate is subjected to impact load. A parametric study is carried out to investigate the effect of several parameters including: foundation thickness, load eccentricity and amplitude of impact load. It was concluded that the load eccentricity increases the displacement and decreases the stress at the foundation center. This is attributed to non-uniformly distributed stresses on the loaded area, where the loads are concentrated locally within the loaded area. The presence of damping leads to a considerable decrease in the foundation displacements and stresses. The increase in the damping ratio reduces the vertical displacement of the foundation at the same time at all damping ratios.

1 Introduction

Plates on elastic foundation are often used in civil engineering problems, such as building infrastructures, tanks or silo foundations, aerospace engineering etc. [1].

Plates are commonly used structural elements and are subjected to wide variety of static and dynamic loads. Broadly, the dynamic loads can be classified into three major categories:

- Steady state harmonic loads.
- Random loads as in the case of seismic excitation on tall buildings.
- Impulsive or shock loads as in the case of hammer foundations, missile impact, blast loads on structures, or an aircraft crashing against a civil engineering

*Corresponding author: myf_1968@yahoo.com

DOI: 10.1051/matecconf/201712006006

structure. The term "impact" covers a wide range of topics that are of great interest to do the research work.

2 Previous studies on plates resting on soil

The literature generally refers to three distinguished methods, these are:
1) Analytical methods:
The earliest formulation of the foundation model was due to Winkler (1867) [2]. This model represents a continuous elastic foundation by a set of closely spaced independent linear springs. Since the foundation is characterized only by springs, the Winkler type model is also called one-parameter model. The vertical displacement at the foundation surface, w is related to the intensity of applied load (q) or to the foundation reaction (p) by the modulus of the foundation and is defined as the reaction of the foundation per unit surface area per unit deflection [3]. In order to eliminate the deficiency of Winkler model, improved theories have been introduced on refinement of Winkler's model, by visualising various types of interconnections such as shear layers and beams along the Winkler springs [4-7]. These theories have been attempted to find an applicable and simple model of representation of foundation medium [8].

Levinton (1949) introduced a simplified method to analyze beams on elastic foundations, in which the contact pressure is represented by a number of redundant reactions [9]. Another foundation model was proposed by Pasternak in 1954 [6], which acquires shear interaction between springs by connecting the ends of the springs to a layer consisting of incompressible vertical elements which deform only by transverse shearing. This class of mathematical models has another constant parameter which characterizes the interaction implied between springs and hence is called two-parameter models. Two-parameter foundation models are more accurate than the one parameter (e.g. Winkler) foundation model.

Wang and How (2001) analyzed the rectangular thick raft resting on a homogeneous elastic half space [10]. They used the Mindlin plate theory to model the raft in order to allow for the effect of transverse shear deformation in the raft as it bends under transverse loading. By approximating the displacement function using complete two-dimensional polynomials of sufficient degree, the Ritz method was shown to be an accurate technique in solving this class of the raft-soil interaction problems. They compared the settlement and the bending moment of the raft for various loading conditions.
2) Experimental methods:
Al-Azawi (1984) tested experimentally simply supported reinforced concrete slabs subjected to both impact loading and static loading [11]. The impact load was produced by a freely falling steel mass which impacted a (50 mm diameter x 500 mm long) cylindrical steel bar placed in contact with the slab at its center. The adopted variables were the falling mass, the boundary conditions of the slabs, and the slab thickness. A theoretical analysis was carried out to determine the deflection-time history at the center of a reinforced concrete slab. The central transient deflection of the slabs was calculated and found in a good agreement with the experimental transient deflection.

Abbasi et al. (1992) analyzed reinforced concrete beams and plates under impact loading [12]. Three-dimensional nonlinear finite element analysis of the reinforced concrete has been modeled. Experiments were conducted in the laboratory involving drop hammer loading on circular concrete slabs. The load cell was fixed in position so that it could not displace off its position when the penetrating mass impinges on the specimen. The impact velocity was measured with help of a simple photo gate type arrangement. The results showed good agreement of the predicted values with their experiments.

Baidya (2004) presented an experimental study on the natural frequency of a foundation on a layered soil system subjected to dynamic loading [13]. It was observed that the natural frequency of the system decreases due to presence of the soft layer at the top, whereas it increases due to presence of the stiff layer at the top. It was also observed that presence of a soft layer at top is more dangerous than at depth.

Fujikake and Soeun (2009) studied the impact responses of simply supported reinforced concrete beams through an experimental study and presented an analytical model to predict the maximum midspan deflection and maximum impact load, to evaluate the damage levels of reinforced concrete (RC) beams when subjected to impact loadings [14]. The experimental impact responses of the RC beams were simulated with a two degree-of-freedom mass- spring-damper system model, in which the loading rate effects were duly considered.

3) Numerical and finite element methods:

Emrich et al. (1982) used finite element method to investigate the behavior of reinforced concrete beams subjected to impact load. Nonlinear constitutive laws for concrete and steel were assumed [15]. Newmark implicit time scheme was used to solve nonlinear dynamic equations. A consistent mass matrix was used; neither structural nor material damping was considered. Results were verified by results from experiments and were in good agreement.

Miyamoto et al. (1991) used Dracker-Prager to model concrete when they investigate the analytical and numerical failure modes of reinforced concrete slabs subjected to impulse loads [16]. They examined the rate of dynamic load, the maximum deflection in the center, the impulse load-midspan deflection curves from zero to ultimate load, the propagation of cracks through the cross section, and the pattern at failure. The numerical failure mode was determined as well as the load rating in failure mode and the distribution of cracking while they carried out a comprehensive study in this field.

Tameroglu (1996) studied a different solution technique for free vibrations of rectangular plates with clamped boundaries resting on elastic foundations and subjected to uniform and constant compressive, unidirectional forces in the mid- plane [17]. The method was based on the use of a non-orthogonal series expansion consisting of some specially chosen trigonometric functions for the deflection surface (w) of the plate. The orthogonalization of the series and other calculations were performed using Fourier expansion of Bernoulli polynomials under some realistic approximations for the limiting values of the boundary conditions. It was concluded that by this method, one does not need to use the solution of the differential equation of the problem.

Tee (2005) studied the dynamic response of a finite circular plate resting on sand by using ABAQUS program [18]. In the analysis, a free-drop impact system was considered to generate the dynamic loading on the plate free surface. Two finite element models were built, one with a slide line underneath the target plate and another one without the slide line. The numerical results of the finite element method for the radial strain at the bottom of the target plate were compared with the experimental measurement. The numerical results showed good agreement with the experimental results.

Borgerhoff et al. (2011) presented the analyses of impact tests of reinforced concrete slabs with dominating punching behavior and confirmed that the layered shell element of the program SOFiSTiK was suitable for a reliable numerical simulation of the problem, provided that the slabs had a minimum of transverse reinforcement, which was sufficient to assure that the ultimate limit state with respect to punching is not exceeded [19]. The requirement was the implicitly fulfilled in the design of a structure which has to resist a specified impact load. In the case of checking reinforced concrete plates without shear reinforcement (or with significant exceedance of punching resistance), empirical perforation formulae or analyses by use of simple two-mass vibration models are applicable.

Mahmoudpour et al. (2011) presented a coupled scaled boundary finite element model to examine the dynamic response of a structure considering the soil-structure interaction [20]. The analysis was performed in time domain. The material behavior of soil and structure was assumed to be linear. They reported that when the system was subjected to sine excitation, the reduction in displacement and base shear was more significant when the loading frequency was close to natural frequencies of the structure.

Based on the literature reviewed, it is found that many literatures are reported on nonlinear behavior of soil and concrete, and considerable amount of literatures are reported on the three-dimensional nonlinear analysis of plate on elastic foundation, but very few studies adopted the dynamic response of soil-structure interaction problems. The present research aims to analyze the foundation by three-dimensional nonlinear finite element method using ANAYS 11 finite element software, in which both the structure and its supporting soil are modeled as continua. The foundation and soil are discretized by eight nodded brick elements. The soil is modeled as elasto-plastic material. The finite element procedure is used to analyze the foundation subjected to concentrated impact load.

3 Description of the problem

The present finite element formulations and solution procedures are applied to the transient dynamic analysis of a plate on elastic foundation with material nonlinearities. A concrete foundation with dimensions of 3×3×0.3 m is placed on the foundation soil 15 m deep and 9 m away from the edge of plate. The boundary conditions are chosen such that all the lateral boundaries and the base of the problem are restricted both horizontally and vertically. The geometry, loading conditions and all details of the problem are shown in Figure (1).

The foundation is subjected to the transient dynamic load described by the following equation:

$$f(t) = 0.5 - 0.5\cos(2\pi t/T_o) \qquad (1)$$

$$p(t) = p_o f(t) \qquad (2)$$

where:

t: time for each point,

T_o: impact duration, and

p_o: the peak amplitude of the load.

The loading function is shown in Figure 2. The duration of this dynamic load is 0.5 seconds with a time step (Δt = 0.01 sec.). The properties of the material for concrete and soil are summarized in Table 1. The soil foundation is cohesionless dense sand with an angle of internal friction, f equals 40°. The nonlinear behavior for both soil and concrete is considered. The Drucker-Prager model is used to model the soil elasto-plastic behavior, while the linear elastic model is used for modeling the behavior of concrete. The 8-node brick elements in **ANSYS** are used to represent the concrete of foundation and the soil under the foundation which are denoted by Solid 65 for concrete and Solid 45 for the soil, and the interface between the concrete and supporting soil is modeled by using three-dimensional surface-to-surface (Target 170 and Contact 174) contact elements connected with concrete and soil [21].

 DOI: 10.1051/matecconf/201712006006

The load is applied at the foundation center for eccentric loading and at distances of (e/B = 0.1, 0.2, 0.3, where B is the width of the foundation) away from the center for concentric loading. No damping has been considered in this problem.

4 Parametric study

A parametric study is performed to investigate the influence of several important parameters on the behavior of the foundation on elastio-plastic soil subjected to impact loads. The parameters studied are foundation thickness, load eccentricity and amplitude.

The analysis is carried out by the program (ANSYS V11) [22]. The parameters used in the analysis are summarized in Table 2.

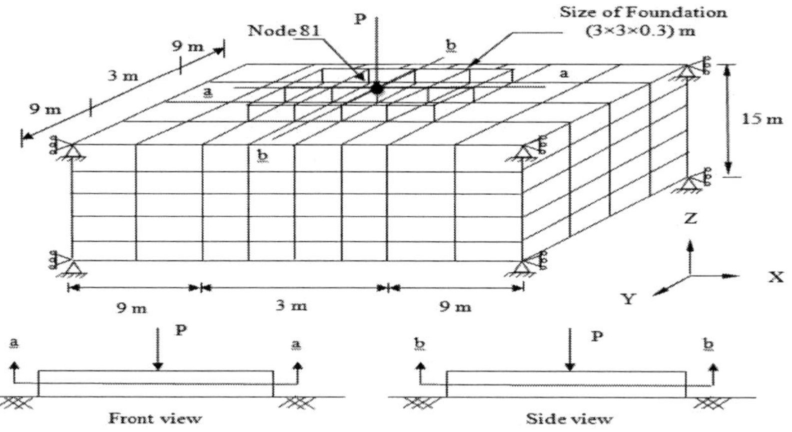

Fig. 1: Typical finite element mesh.

Table 1: Material properties for concrete and soil in the basic problem.

	Symbol	Definition	Value
Concrete	f'_c	Compressive strength (MPa)	25
	E_c	Young's modulus (MPa)	23500
	f_t	Tensile strength (MPa)	3.1
	ν	Poisson's ratio	0.15*
	ρ_c	Density (kg/m^3)	2400
Interface	μ	Coefficient of friction	0.6*
Soil	Es	Young's modulus (MPa)	50
	c_u	Cohesion	0
	ϕ	Frication angle	40°
	ν	Poisson's ratio	0.3*
	ρ_s	Density (kg/m^3)	1800

Notes:*Assumed value, $E_c = 4700\sqrt{f'c}$ and $f_t = 0.62\sqrt{f'c}$.

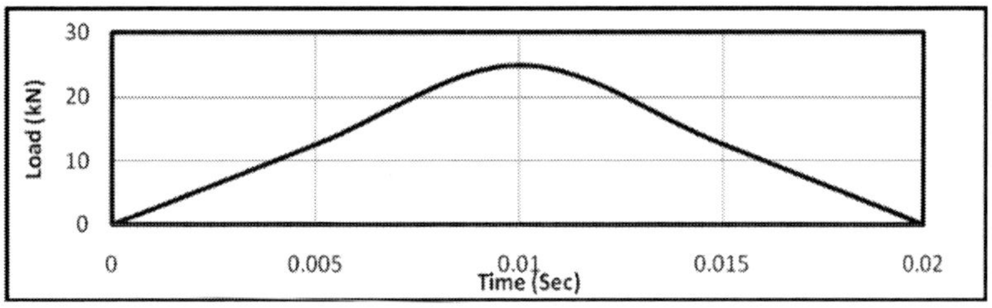

Fig. 2: The loading function.

Table 2: Parameters of the main problem.

parameter	Range of value	Unit
Foundation thickness	0.3, 0.5, 0.75	m
Foundation length	3, 4, 5, 6	m
Load amplitude, p_o	25, 75, 100	kN

5 Results and discussion

The eccentricity of the foundation (e) is defined as the distance between point load and the center of foundation. Different values of load amplitude (25, 75 and 100) kN are applied at e/B (0.1, 0.2 and 0.3) from center of foundation with three values of foundation thickness (0.3, 0.5, 0.75) m.

The eccentricity of the foundation is increased to (e/B = 0.1, 0.2, and 0.3) to study its effect on the distribution of displacements and stresses. The results are summarized in Figures 3 to 17. From Figures 4 to 11, it is noticed that the settlement takes place on one side of foundation and upward displacement on the other side with increasing the eccentricity to (0.2 and 0.3) and this is due to the vertical displacement which acts to push the soil under loading causing an uplift at the farther side. It is also observed that in these figures, with the presence of eccentricity, the maximum vertical displacement will turn aside from center of foundation because the maximum vertical displacement will happen under the point of load application.

From Figures 12 to 17, it noticed that the larger thickness with eccentricity (e/B =0.3) causes a stiff behavior (more rigidity), and the oscillation of stress becomes smaller.

From Figure 12, it is noticed that the oscillation of stress for the farther points from point load increases greater than the nearer points. This is because the foundation length on the other side increases making the foundation behave as a long cantilever beam.

From Figures 3 a to 5 a, it can be noticed that when the load eccentricity increases from (e/B = 0.1 to 0.2 and 0.3) for amplitude loads (p = 25, 75 and 100) kN and in case of t = 0.3 m, the maximum vertical displacement at node (81) in the center of foundation increases by about (0.05, 0.7 and 20.5) %, respectively.

On the other hand, from Figures 12 to 14, it can be seen that when the load eccentricity increases from (e/B = 0.1 to 0.2 and 0.3) for amplitude loads (p = 25, 75 and 100) kN and in case of t = 0.3 m, the maximum vertical stress at node (81) decreases by about (22.7, 28.2 and 29.7) %, respectively.

From Figures 6 a to 8 a, it can be concluded that when the load eccentricity increases from (e/B = 0.1 to 0.2 and 0.3) for amplitude loads (p = 25, 75 and 100) kN and in case of t = 0.5 m, the ratio of maximum vertical displacement at e/B of (0.1, 0.2 and 0.3) to maximum vertical displacement at no load eccentricity (e/B = 0) increases by about (1.4, 32.4 and 87) %, respectively. On the other hand, from Figures 15 a to 17 a, it can be seen that when the load eccentricity increases from (e/B = 0.1 to 0.2 and 0.3) for amplitude loads (p = 25, 75 and 100) kN and in case of t = 0.5 m, the maximum vertical stress at node (81) decreases by about (4, 9.6 and 10.8) %, respectively.

From Figures 9 a to 11 a, it can be noticed that when the load eccentricity increases from (e/B = 0.1 to 0.2 and 0.3) for amplitude loads (p = 25, 75 and 100) kN and in case of t = 0.75 m, the ratio of maximum vertical displacement at e/B of (0.1, 0.2 and 0.3) to maximum vertical displacement at no load eccentricity (e/B = 0) increases by about (11.6, 52.3 and 111.5) %, respectively.

On the other hand for , it can be noted that when the load eccentricity increases from (e/B = 0.1 to 0.2 and 0.3) for amplitude loads (p = 25, 75 and 100) kN and in case of t = 0.75 m, the maximum vertical stress at node (81) decreases by about (2.3, 4.3 and 10.6) %, respectively.

From Figure 3 a, the maximum vertical displacement takes place at time (0.24 sec.) for (e/B=0.1), thickness (0.3 m) and amplitude loads (25, 75 and 100) kN, while the maximum vertical displacement occurs at time (0.45 sec.) for (e/B=0.1), thickness (0.5 m) under the same loads, Figure 6 a.

From Figures 7 a and 8 a, the maximum vertical displacement takes place at time (0.49 sec.) for (e/B=0.1 and e/B=0.2), thickness (0.75 m) and amplitude loads (25, 75 and100) kN. Also, it can be seen that the maximum vertical displacement takes place at time (0.5 sec.) for (e/B=0.2) and the thickness (0.3and 0.5) m and loads (25, 75 and 100) kN, Figures 4 a and 7 a.

On the other hand, the maximum vertical displacement occurs at time (0.5 sec.) for (e/B=0.3) and thickness (0.3, 0.5 and 0.75) m and loads (25, 75 and 100) kN, Figures 5 a, 8 a and 11 a, respectively. From Figure 12 a, the maximum vertical stress takes place at time (0.23 sec.) for (e/B=0.1), thickness (0.3 m) and amplitude loads (25, 75 and 100) kN, while the maximum vertical stress occurs at time (0.26 sec.) for (e/B=0.2), thickness (0.3 m) and the same loads, Figure 13 a. Also, it can be seen that the maximum vertical stress occurs at time (0.16 sec.) for (e/B=0.3), thickness (0.3 m) and the same loads, Figure 14 a. From Figure 15 a, the maximum vertical stress takes place at time (0.37 sec.) for (e/B=0.1), thickness (0.5 m) and amplitude loads (25, 75 and 100) kN, while the maximum vertical stress occurs at time (0.14 sec.) for (e/B=0.2), thickness (0.5 m) and the same loads, Figure 16 a. Also, it can be noticed that the maximum vertical stress occurs at time (0.37 sec.) for (e/B=0.3), thickness (0.5 m) and the same loads, Figure 17 a.

These figures reveal that increasing the load eccentricity from e/B = 0.1 to 0.2 and 0.3 results in increasing the maximum displacement and reducing the maximum stress. This is due to non-uniformly distributed stresses on the loaded area, where the loads are locally concentrated within the loaded area, and this leads to increase the stresses in this area and decrease it in other area, this difference in concentration of loads leads to decrease the vertical stress [23-24].

DOI: 10.1051/matecconf/201712006006

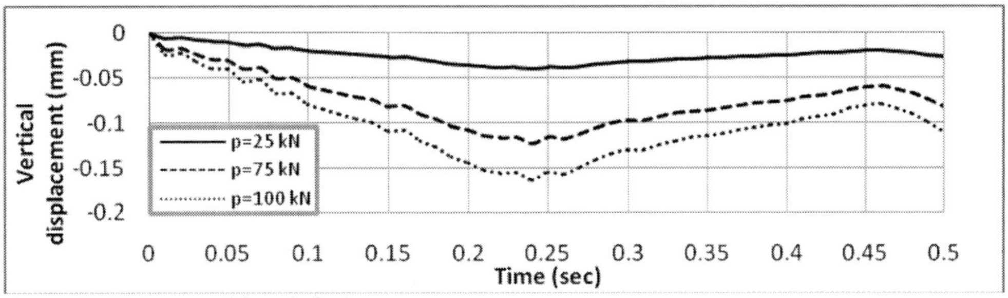

a) at node 81 (center of foundation)

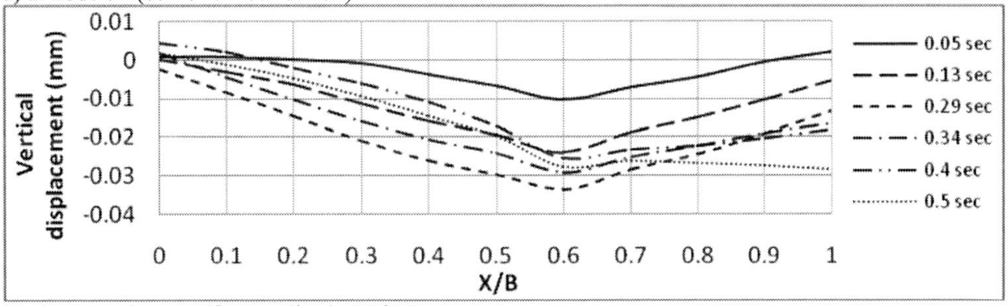

b) along section a-a (for amplitude 25 kN)

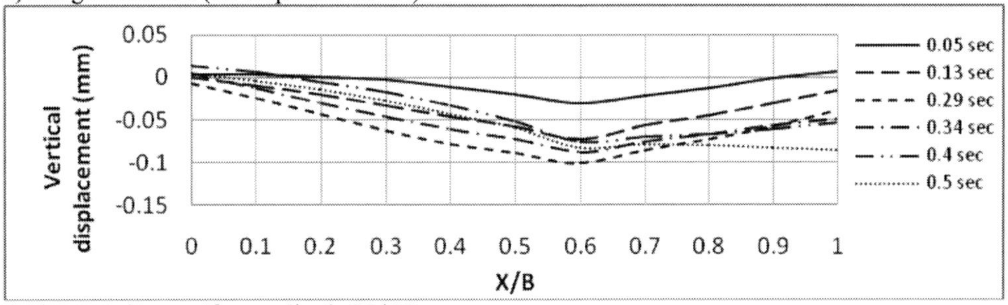

c) along section a-a (for amplitude 75 kN)

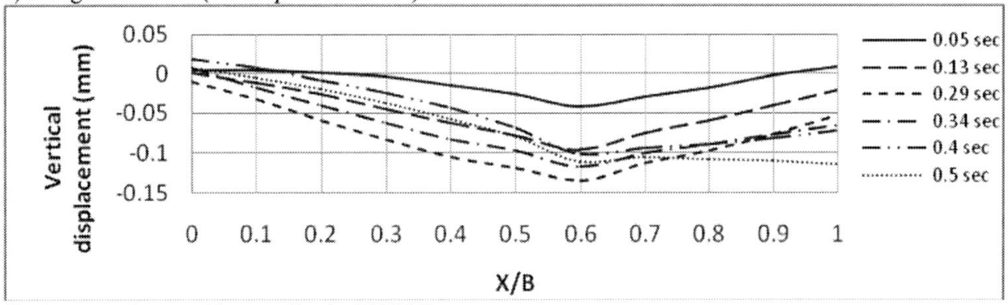

d) along section a-a (for amplitude 100 kN)

Fig. 3: Dynamic response of the foundation to impact load, vertical displacement, t=0.3 m, B=3 m, L=3 m, e/B=0.1, D=0.

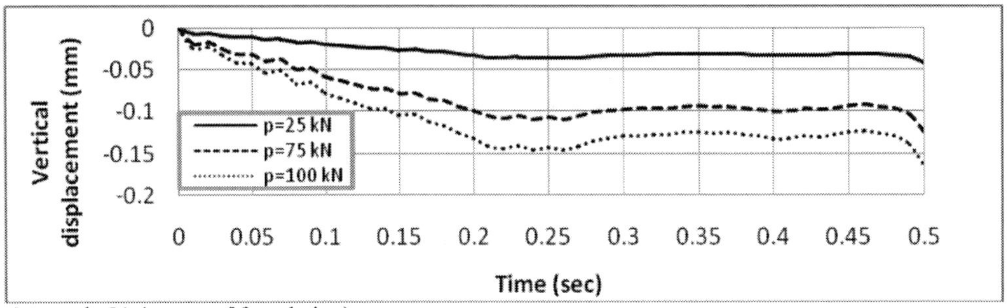

a) at node 81 (center of foundation)

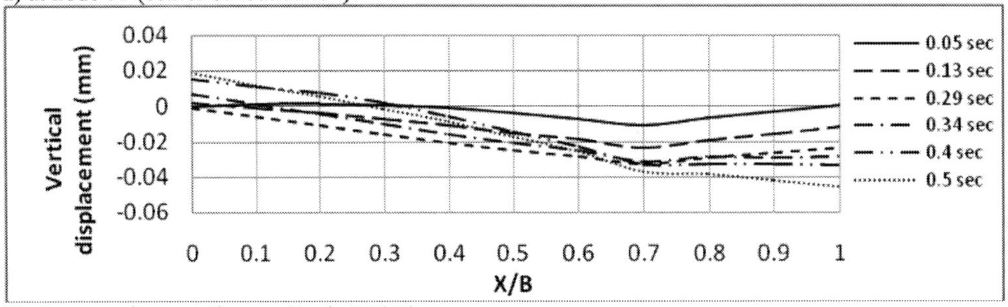

b) along section a-a (for amplitude 25 kN)

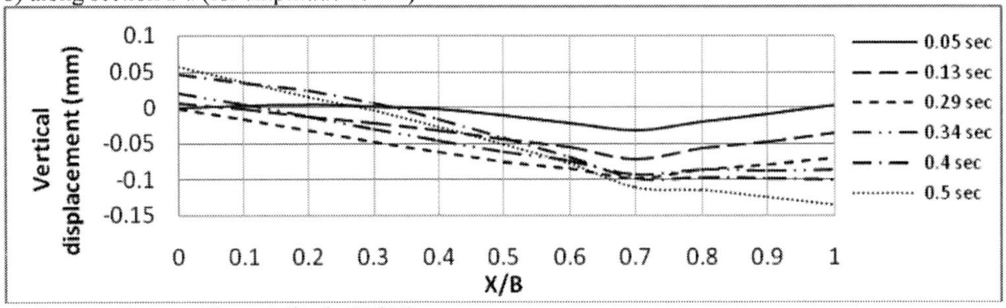

c) along section a-a (for amplitude 75 kN)

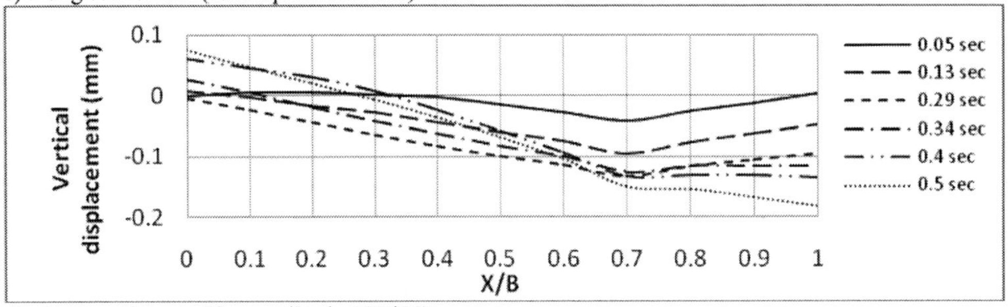

d) along section a-a (for amplitude 100 kN)

Fig. 4: Dynamic response of the foundation to impact load, vertical displacement, t=0.3 m, B=3 m, L=3 m, e/B=0.2, D=0.

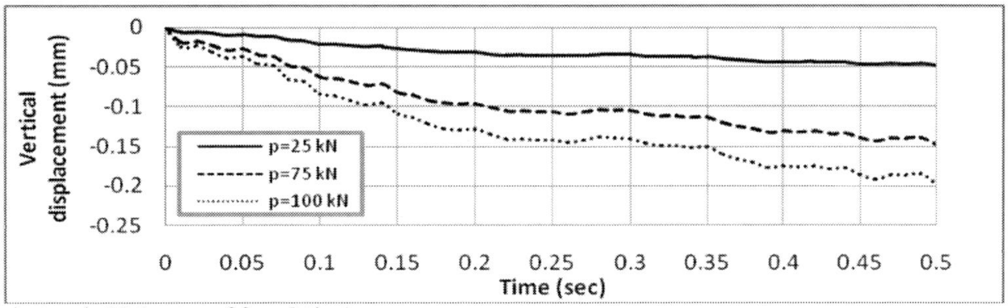

a) at node 81 (center of foundation)

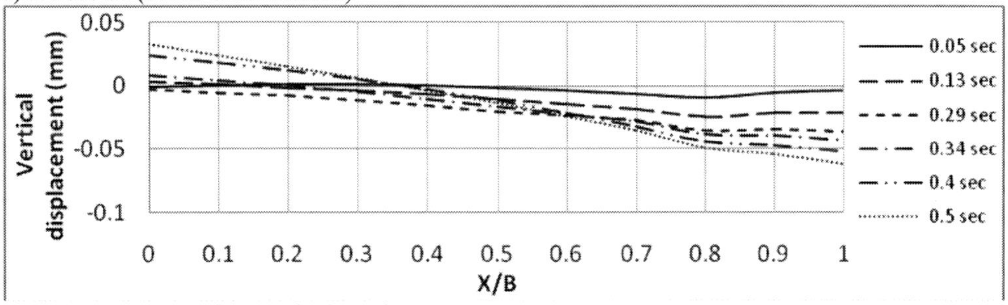

b) along section a-a (for amplitude 25 kN)

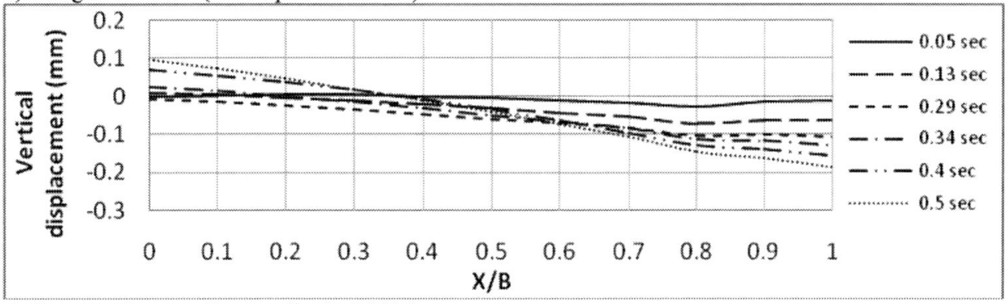

c) along section a-a (for amplitude 75 kN)

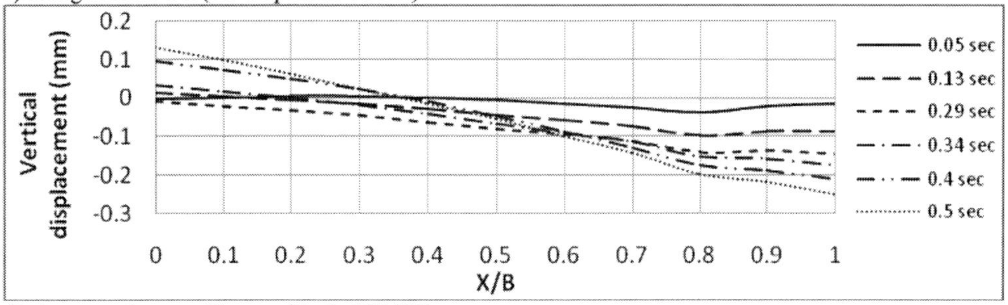

d) along section a-a (for amplitude 100 kN)

Fig. 5: Dynamic response of the foundation to impact load, vertical displacement, t=0.3 m, B=3 m, L=3 m, e/B=0.3, D=0.

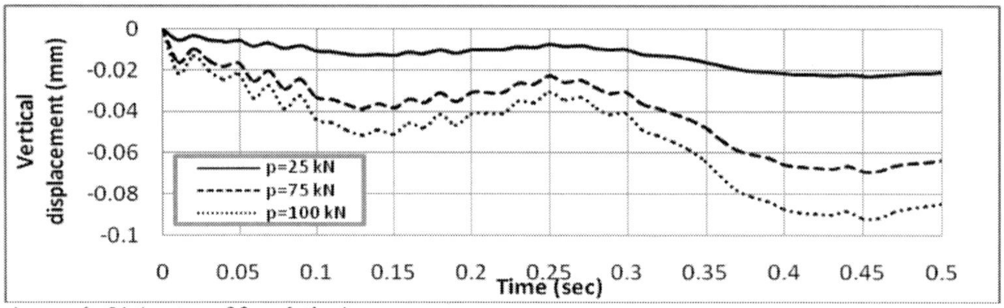

a) at node 81 (center of foundation)

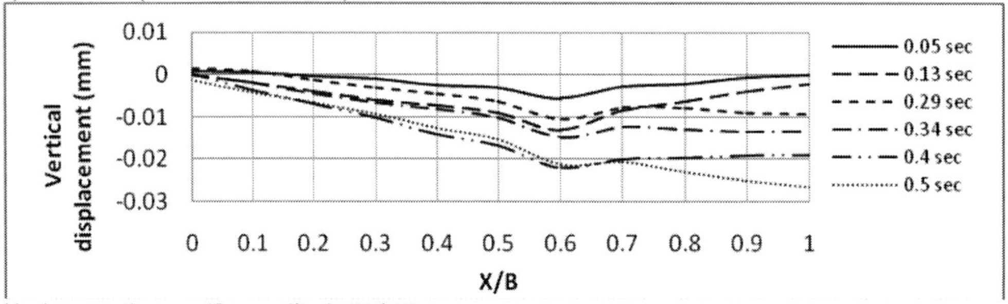

b) along section a-a (for amplitude 25 kN)

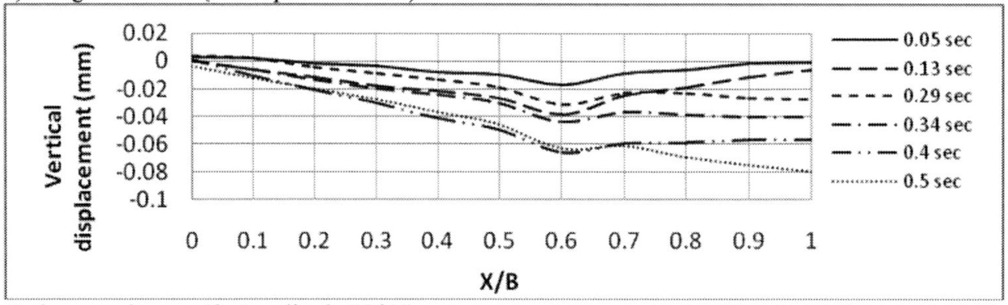

c) along section a-a (for amplitude 75 kN)

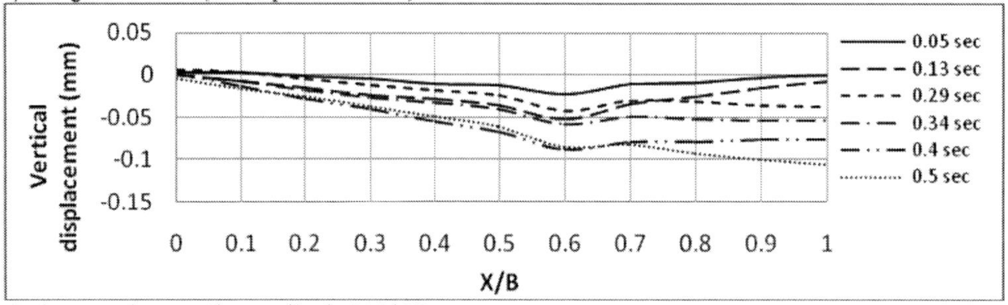

d) along section a-a (for amplitude 100 kN)

Fig. 6: Dynamic response of the foundation to impact load, vertical displacement, t=0.5 m, B=3 m, L=3 m, e/B=0.1, D=0.

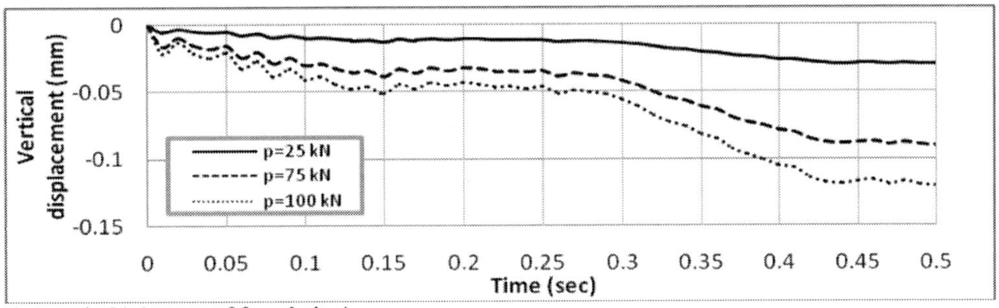

a) at node 81 (center of foundation)

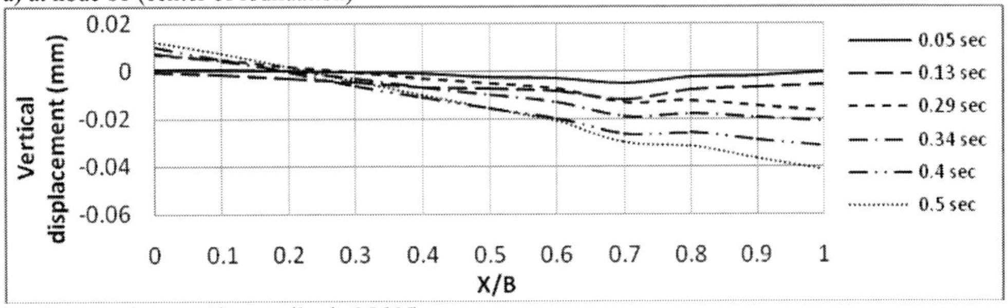

b) along section a-a (for amplitude 25 kN)

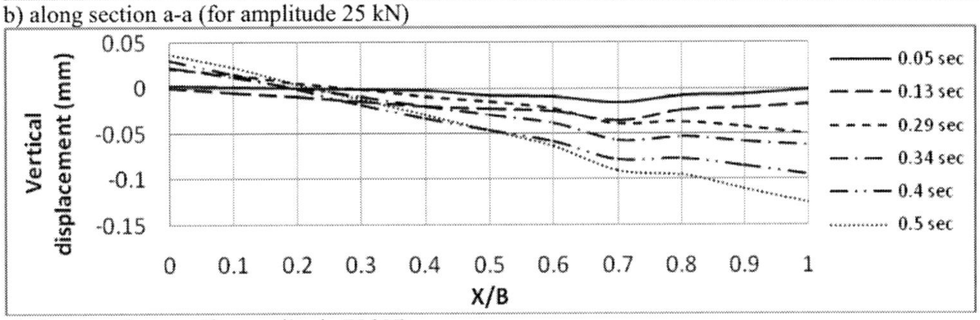

c) along section a-a (for amplitude 75 kN)

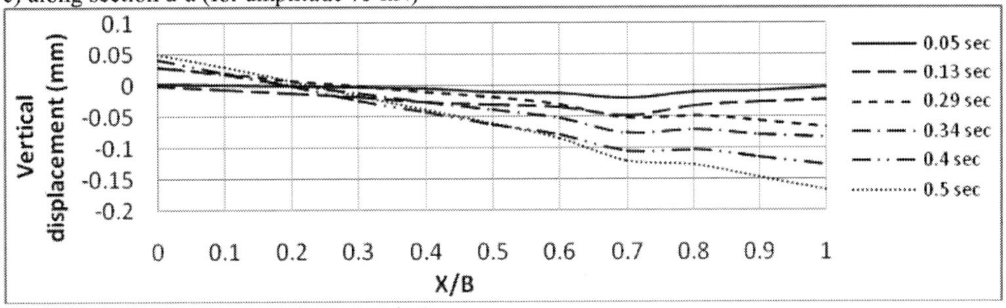

d) along section a-a (for amplitude 100 kN)

Fig. 7: Dynamic response of the foundation to impact load, vertical displacement, t=0.5 m, B=3 m, L=3 m, e/B=0.2, D=0.

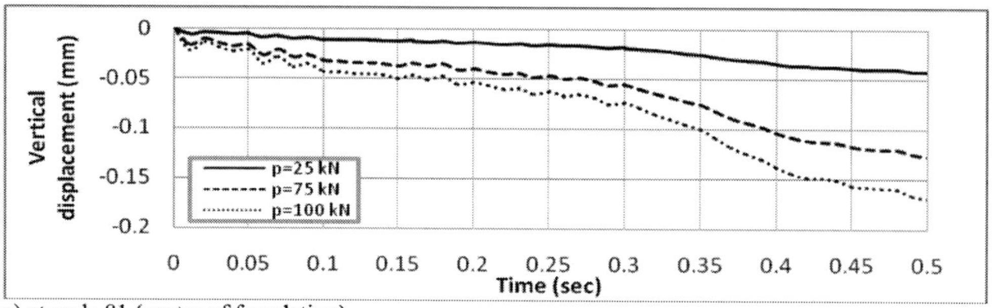

a) at node 81 (center of foundation)

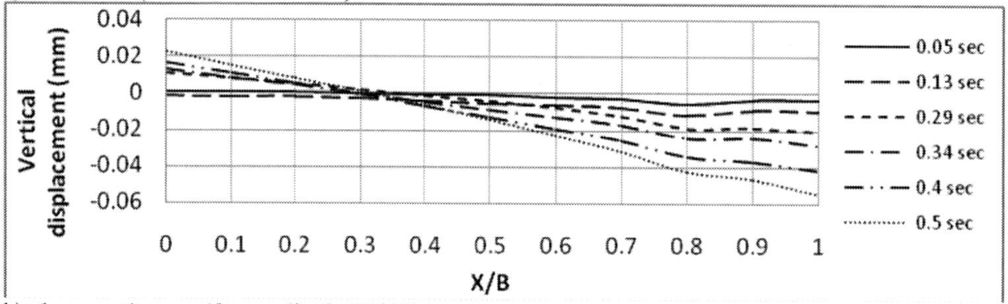

b) along section a-a (for amplitude 25 kN)

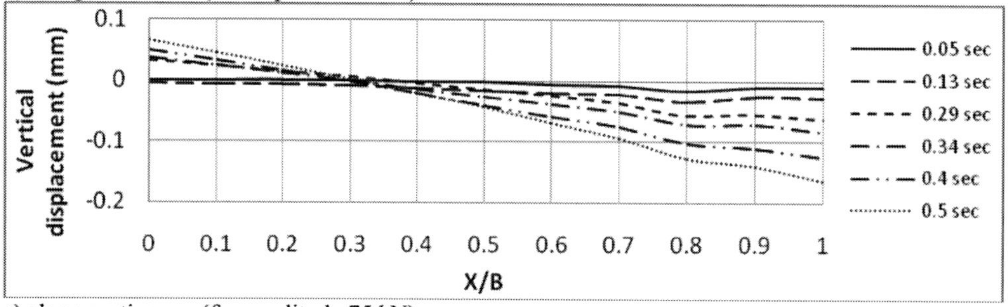

c) along section a-a (for amplitude 75 kN)

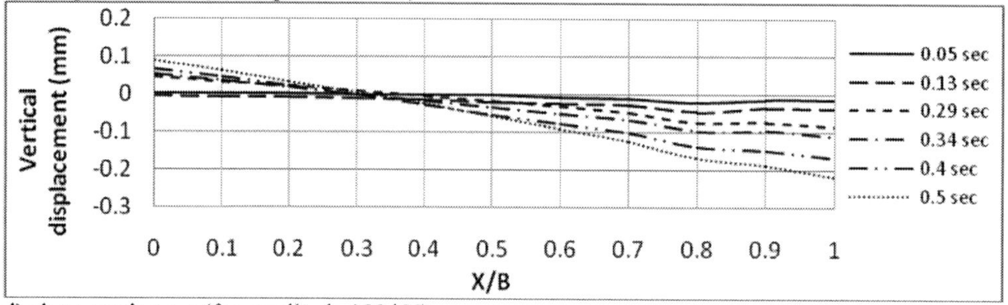

d) along section a-a (for amplitude 100 kN)

Fig. 8: Dynamic response of the foundation to impact load, vertical displacement, t=0.5 m, B=3 m, L=3 m, e/B=0.3, D=0.

DOI: 10.1051/matecconf/201712006006

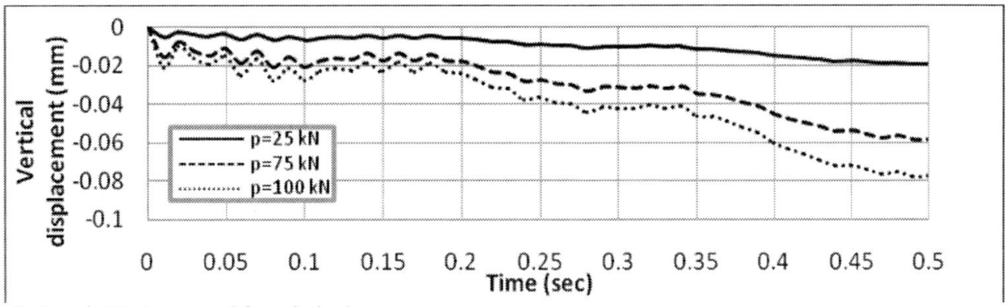

a) at node 81 (center of foundation)

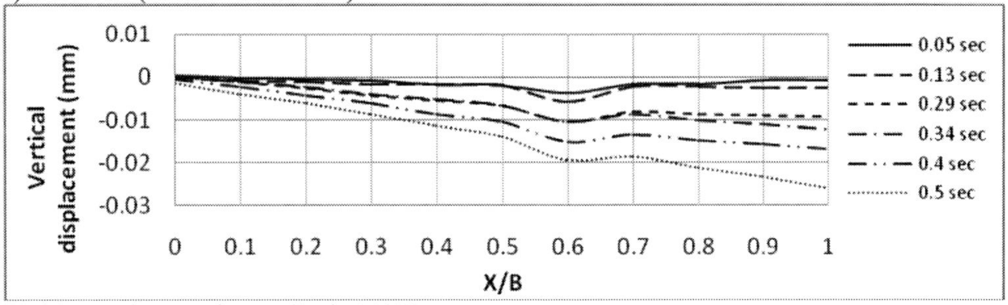

b) along section a-a (for amplitude 25 kN)

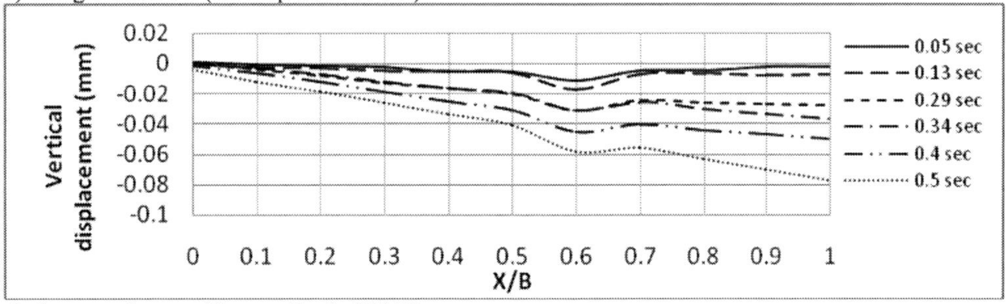

c) along section a-a (for amplitude 75 kN)

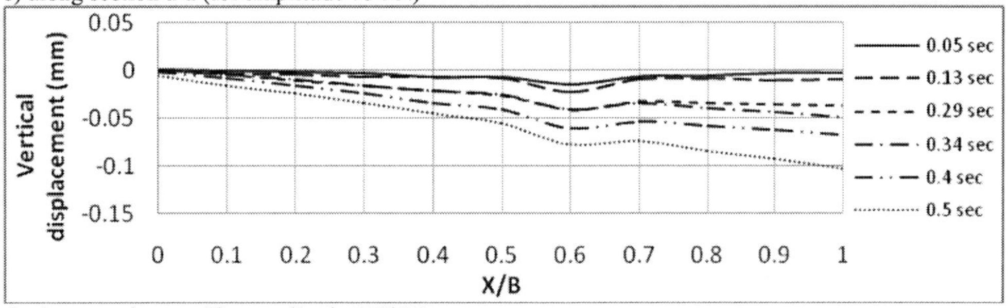

d) along section a-a (for amplitude 100kN)

Fig. 9: Dynamic response of the foundation to impact load, vertical displacement, t=0.75 m, B=3 m, L=3 m, e/B=0.1, D=0.

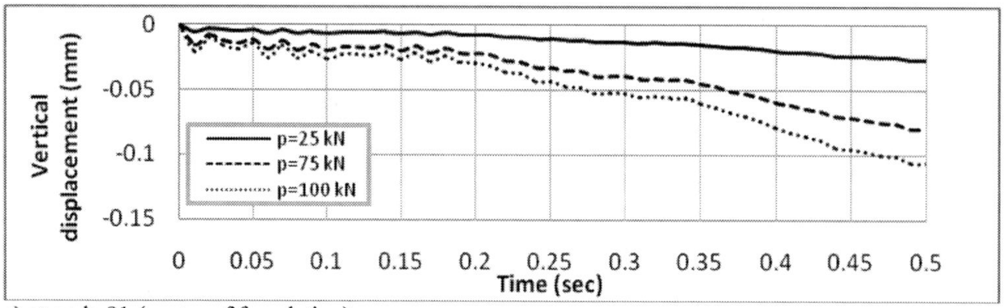

a) at node 81 (center of foundation)

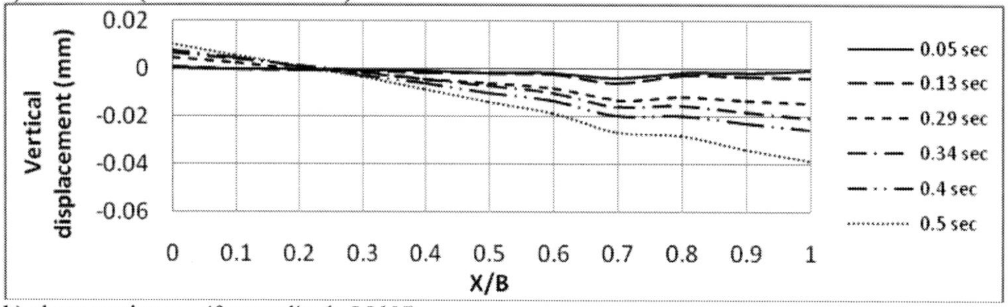

b) along section a-a (for amplitude 25 kN)

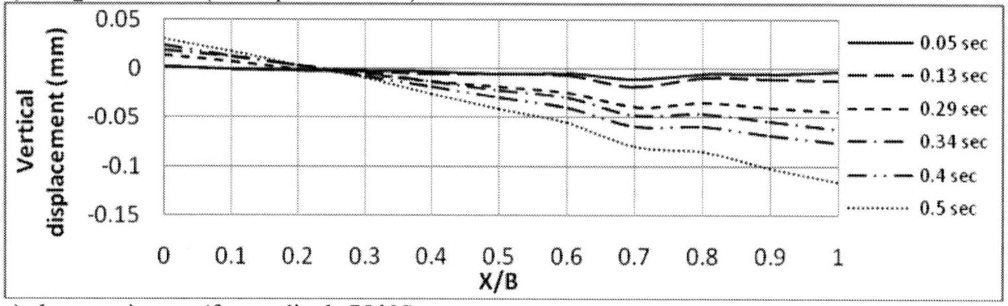

c) along section a-a (for amplitude 75 kN)

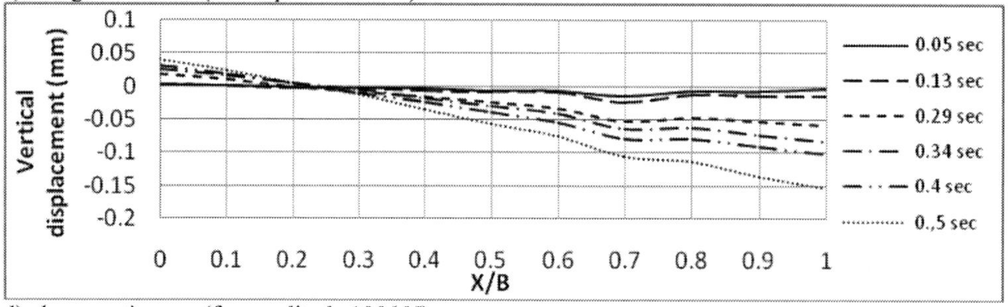

d) along section a-a (for amplitude 100 kN)

Fig. 10: Dynamic response of the foundation to impact load, vertical displacement, t=0.75 m, B=3 m, L=3 m, e/B=0.2, D=0.

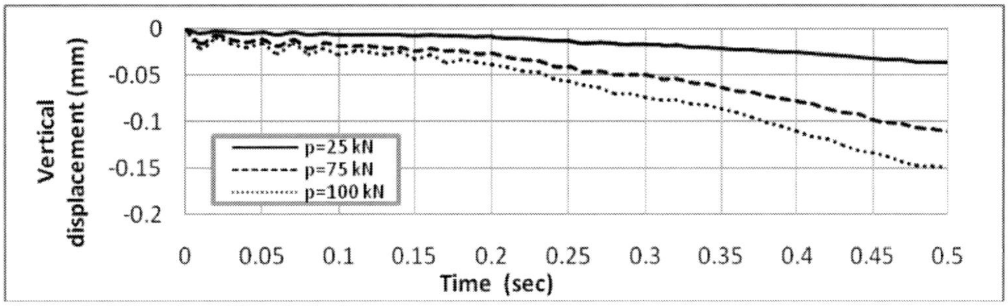

a) at node 81(center of foundation)

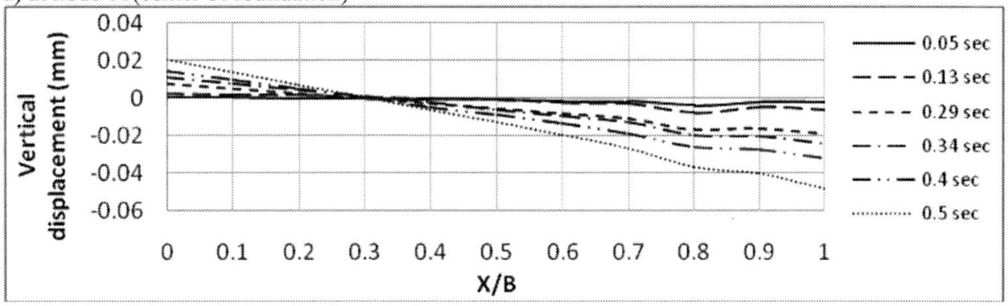

b) along section a-a (for amplitude 25 kN)

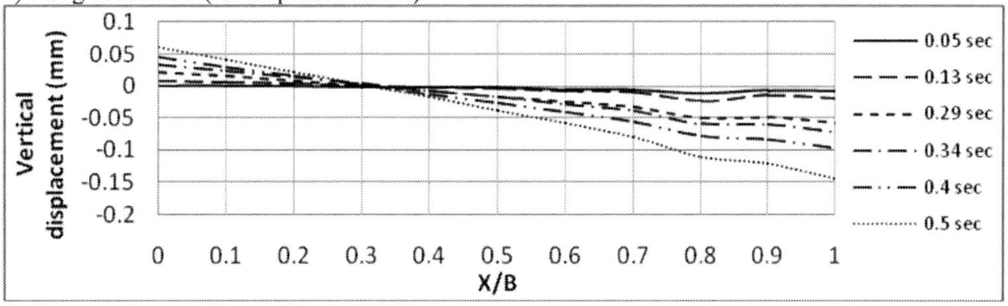

c) along section a-a (for amplitude 75 kN)

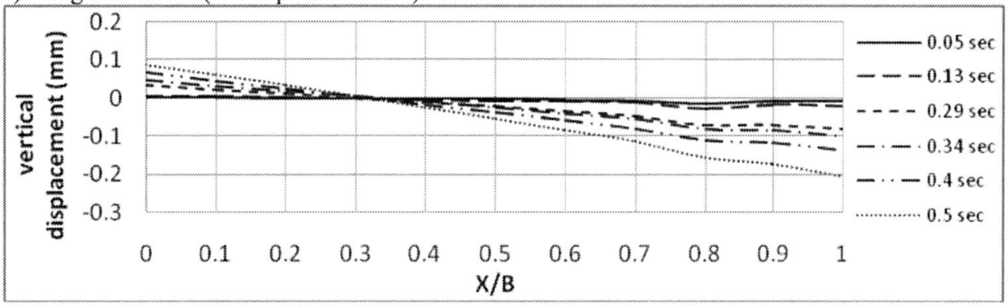

d) along section a-a (for amplitude 100 kN)

Fig. 11: Dynamic response of the foundation to impact load, vertical displacement, t=0.75 m, B=3 m, L=3 m, e/B=0.3, D=0.

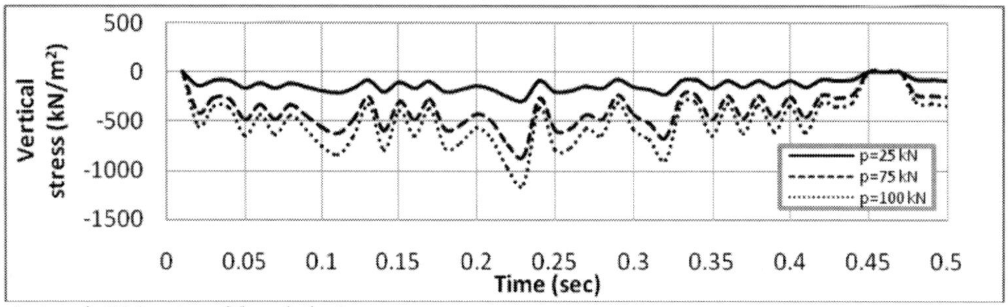

a) at node 81(center of foundation)

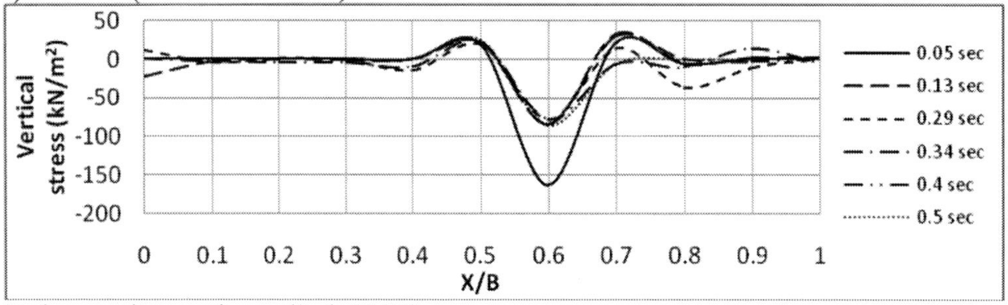

b) along section a-a (for amplitude 25 kN)

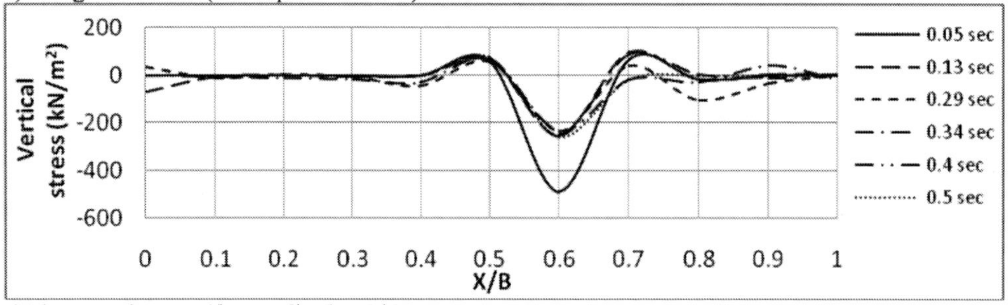

c) along section a-a (for amplitude 75 kN)

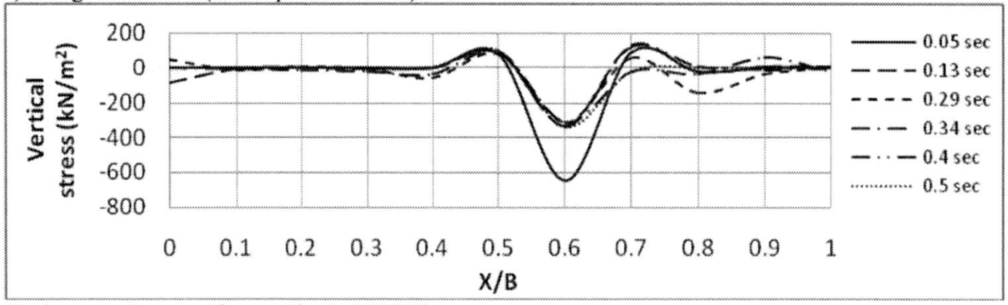

d) along section a-a (for amplitude 100 kN)

Fig. 12: Dynamic response of the foundation to impact load, vertical stress, t=0.3 m, B=3 m, L=3 m, e/B=0.1, D=0.

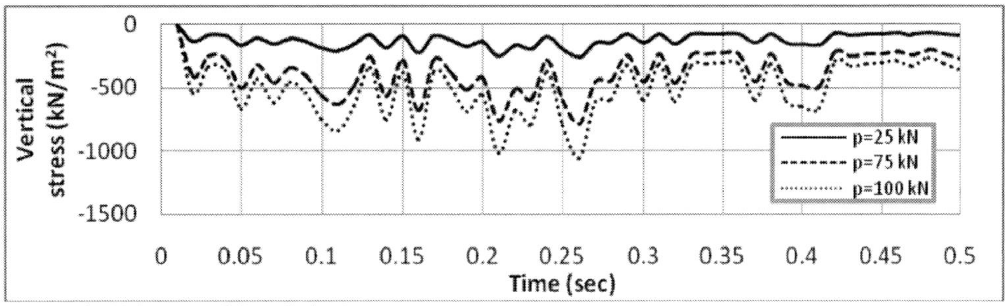

a) at node 81(center of foundation)

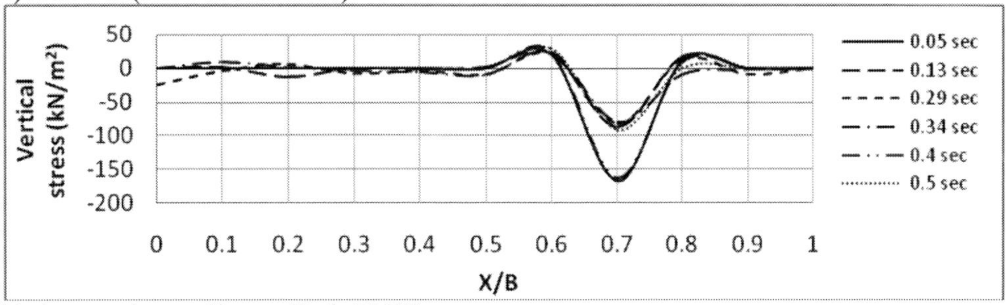

b) along section a-a (for amplitude 25 kN)

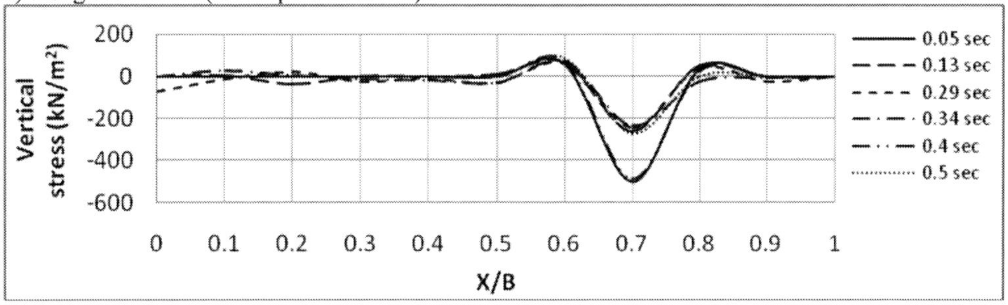

c) along section a-a (for amplitude 75 kN)

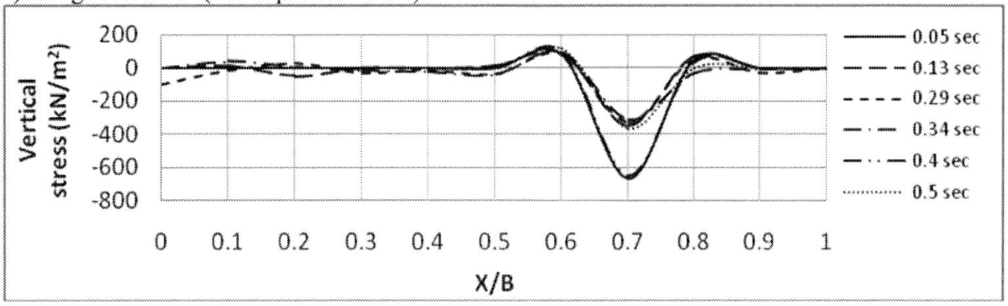

d) along section a-a (for amplitude 100 kN)

Fig. 13: Dynamic response of the foundation to impact load, vertical stress, t=0.3 m, B=3 m, L=3 m, e/B=0.2, D=0.

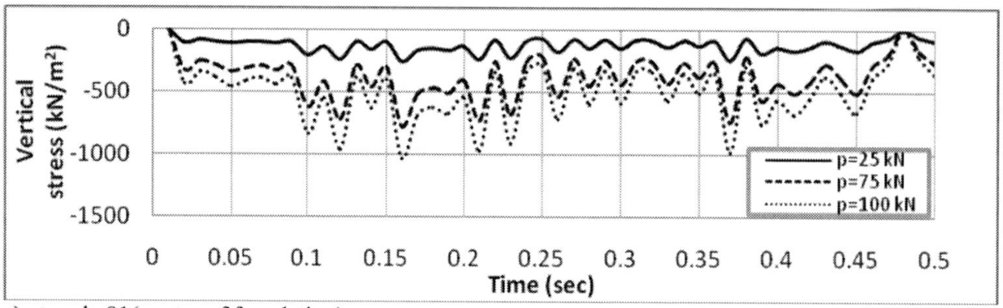

a) at node 81(center of foundation)

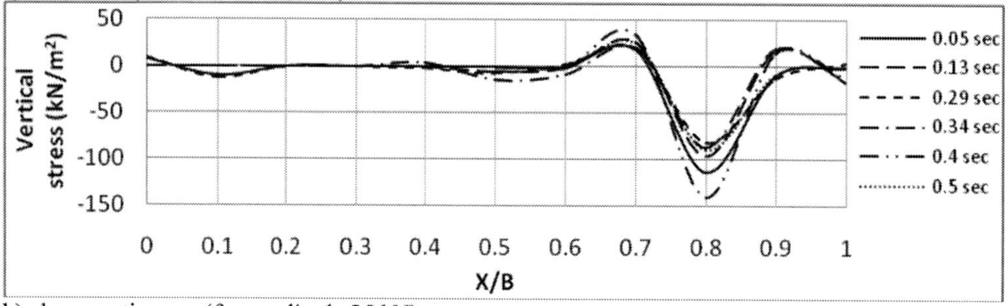

b) along section a-a (for amplitude 25 kN)

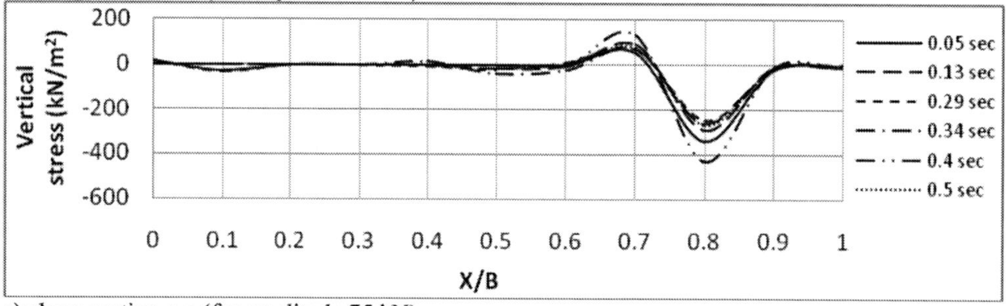

c) along section a-a (for amplitude 75 kN)

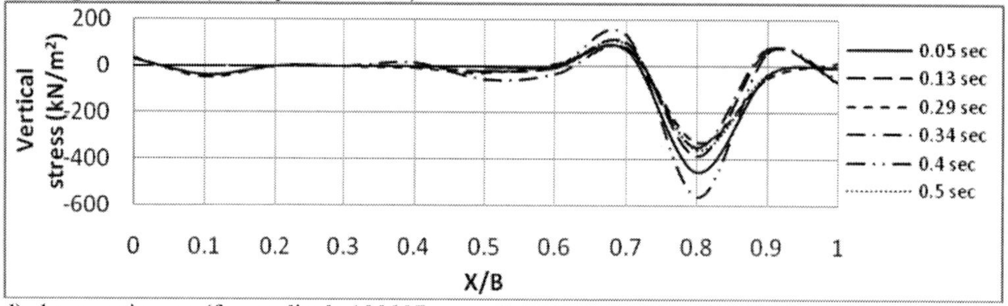

d) along section a-a (for amplitude 100 kN)

Fig. 14: Dynamic response of the foundation to impact load, vertical stress, t=0.3 m, B=3 m, L=3 m, e/B=0.3, D=0.

 DOI: 10.1051/matecconf/201712006006

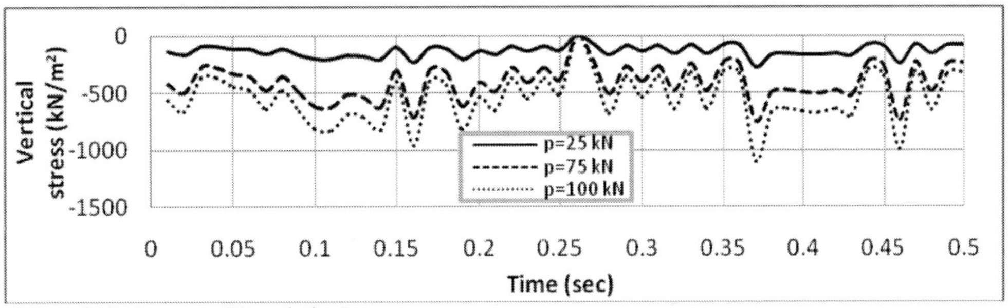

a) at node 81(center of foundation)

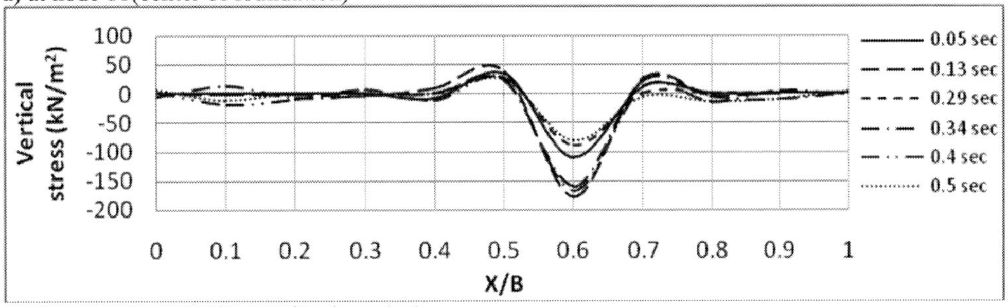

b) along section a-a (for amplitude 25 kN)

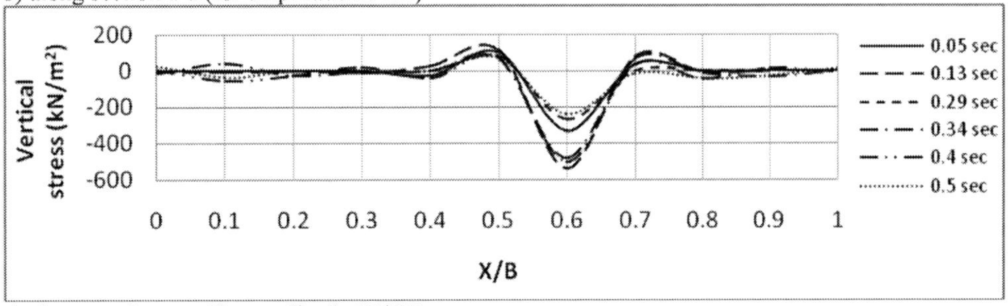

c) along section a-a (for amplitude 75 kN)

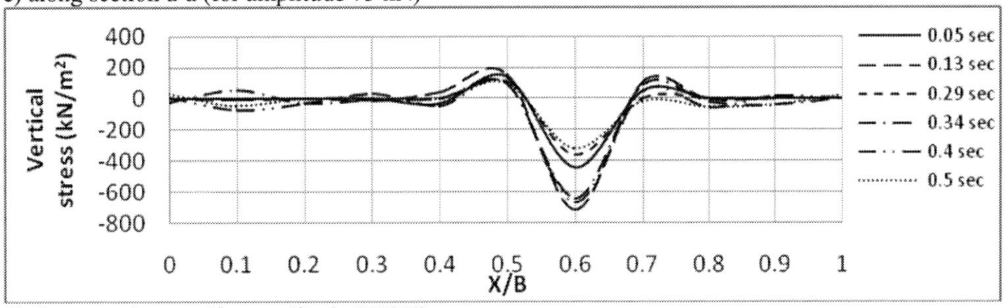

d) along section a-a (for amplitude 100kN)

Fig. 15: Dynamic response of the foundation to impact load, vertical stress, t=0.5 m, B=3 m, L=3 m, e/B=0.1, D=0.

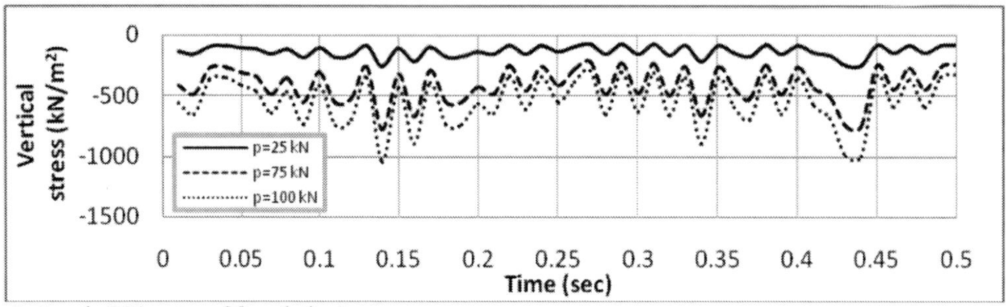

a) at node 81(center of foundation)

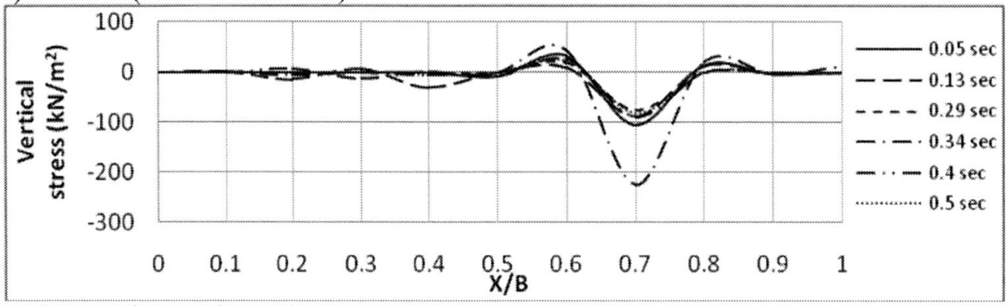

b) along section a-a (for amplitude 25 kN)

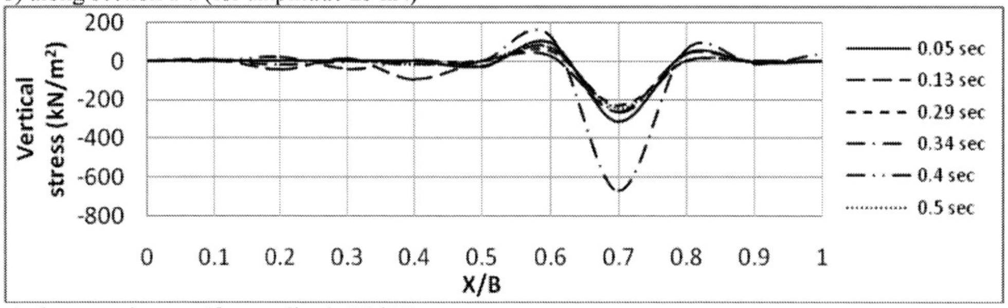

c) along section a-a (for amplitude 75 kN)

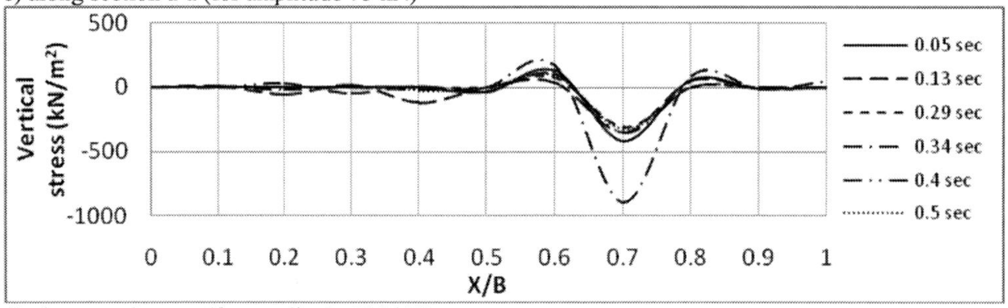

d) along section a-a (for amplitude 100 kN)

Fig. 16: Dynamic response of the foundation to impact load, vertical stress, t=0.5 m, B=3 m, L=3 m, e/B=0.2, D=0.

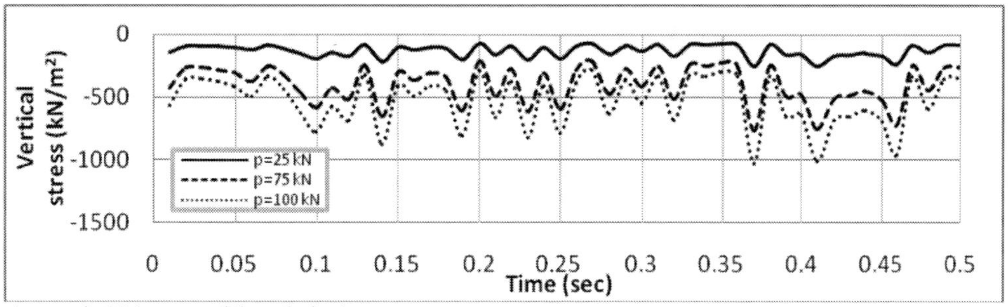

a) at node 81(center of foundation)

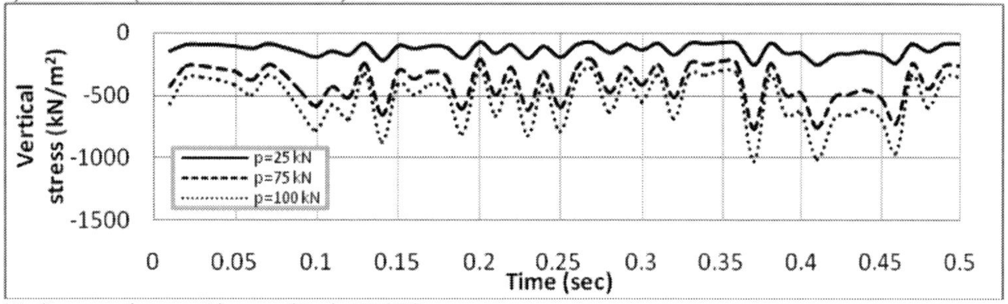

b) along section a-a (for amplitude 25 kN)

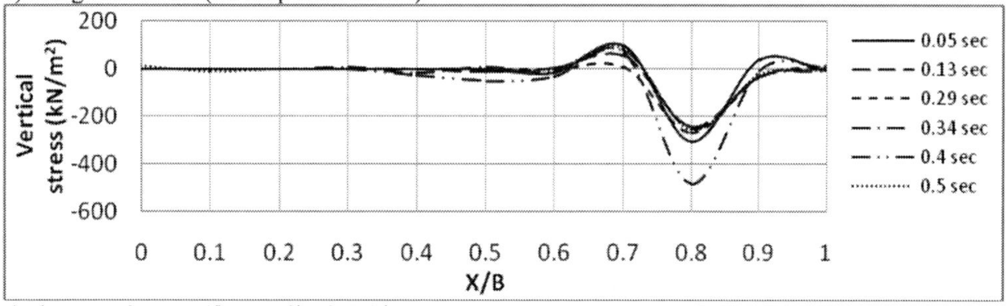

c) along section a-a (for amplitude 75 kN)

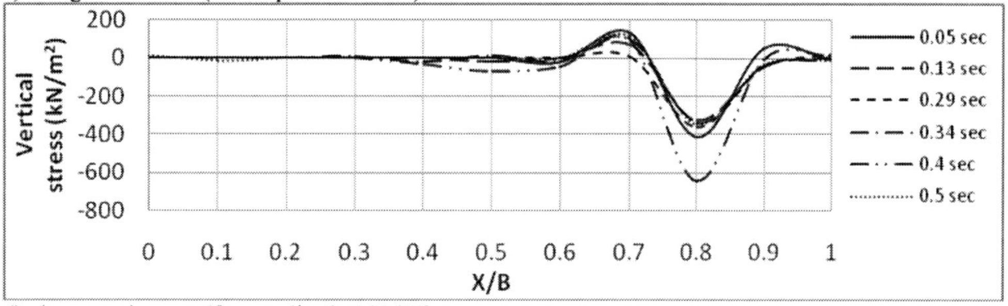

d) along section a-a (for amplitude 100 kN)

Fig. 17: Dynamic response of the foundation to impact load, vertical stress, t=0.5 m, B=3 m, L=3 m, e/B=0.3, D=0.

6 Conclusions

As a result of the finite element analysis carried out in this study, the following conclusions could be made:

1) As the foundation thickness increases, the time for maximum displacement to take place increases due to the geometrical damping induced by the foundation. When the thickness of foundation increases from (t = 0.3 to 0.5 and 0.75) m for amplitude load (po = 25, 75 and 100) kN, the maximum

vertical displacement at the center of foundation decreases by about 45 and 58 %, and the maximum vertical stress decreases by about 22 and 50 %, respectively.

2) The load eccentricity increases the displacement and decreases the stress at the foundation center. This is attributed to non-uniformly distributed stresses on the loaded area, where the loads are concentrated locally within the loaded area. The oscillation of stress for the farther points from the point load increases more than the nearer points. This is because the foundation length on the other side increases making the foundation behaves as a long cantilever beam.

References

[1] Pavlou, D.G., Bancila, G., Lucaci, R., Belic, F., Dan, D., Pavlou, M. G.,Tirtea, A., Gruin, A. and Baera, C., (2007), "An Exact Solution of the Plate on Elastic Foundation under Impact Loading", 3rd Wseas International Conference on Applied and Theoretical Mechanics, Spain, December 14-16, pp. 79-83.

[2] Winkler, E., (1867), „Die Lehre von der elasticitaet und festigkeit“, Prag Dominicus, Berlin, p.p. 182.

[3] Rao, N. S. V. (2011), “Foundation Design: Theory and Practice”, John Wiley & Sons (Asia), First Edition, Pte Ltd.

[4] Filonenko-Borodich, M. M. (1940), “Some Approximate Theories of the Elastic Foundation (in Russian)”, UchenyeZapiski MGU, 46, 3-18.

[5] Hetenyi, M., (1946), “Beams on Elastic Foundations”, Ann Arbor, University of Michigan Press.

[6] Pasternak, P. L., (1954), “Fundamental of a New Method of Analysis of an Elastic Foundation by Means of Two Foundation Constants (in Russian)”, Gos. Izd.Lit.poStrait i Arkh.,Moscow, USSR.

[7] Kerr, Arnold D (1964). "Elastic and viscoelastic foundation models." Journal of Applied Mechanics 31.3: 491-498.

[8] Teodoru, B. and Musat, V., (2008), “Beam Elements on Linear Variable Two-Parameter Elastic Foundation”, University of Gheorghe Asachi, Vol. 3.

[9] Levinton, Z., (1949), “Elastic Foundation Analyzed by the Method of Redundant Reactions”, Trans., ASCE, Vol. 114, pp. 40- 52.

[10] Wang, C.M. and How, Y.C., (2001), “Analysis of Rectangular Thick Rafts on an Elastic Half Space”, Computers and Geotechnics, Vol. 28, Issue 3, pp. 161-184.

[11] Al-Azawi, T. K., (1984), “Impact of Reinforced Concrete Slabs”, Ph.D. Thesis, University of Sheffield, England.

[12] Abbasi, M.S.A., Baluch, M. H., Azad, A. K. and Abdel-Rahman, H.H., (1992),"Nonlinear Finite Element Modeling of Failure Modes in Reinforced Concrete Slabs”, Journal of Computers and Structures, Vol. 42, No. 5, pp. 815-823.

[13] Baidya, E. A., (2004), "Dynamic Response of Footings Resting on a Sand Layer of Finite Thickness", Journal of Geotechnical and Geoenvironmental Engineering, ASCE, Vol. 130, No. 6, pp. 651-655.

[14] Fujikake, K., Li, B. and Soeun, S., (2009), "Impact Response of Reinforced Concrete Beam and its Analytical Evaluation", Journal of Structural Engineering, ASCE, Vol. 135, No. 8, pp. 938-950.

[15] Emrich, F., Herter, J. and Puffer, G., (1982), "Nonlinear Finite Element Analysis of Reinforced Concrete Beams under Impact Load in Comparison with Experimental Results”, Computers and Structures, BAM, Berlin (West), Vol. 4, No. 2, pp. 455-471.

[16] Miyamoto, A., King, M. W. and Fuji, M., (1991),"Nonlinear Dynamic Analysis of Reinforced Concrete Slabs under Impulsive Loads", ACI Structural Journal, July-August, Vol. 88, No. 4, pp. 411-419.
[17] Tameroglu, S. S., (1996), "Vibrations of Clamped Rectangular Plates on Elastic Foundations Subjected to Uniform Compressive Forces", Journal of Engineering Mechanics, ASCE, Vol. 122, No. 8, pp. 714-718.
[18] Tee, C. H., (2005), "Dynamic Response of Plates and Buried Structures", M.Sc. Thesis, University of West Virginia.
[19] Borgerhoff, M., Stangenberg, F. and Zinn, R., (2011), "Numerical Simulation of Impact Tests of Reinforced Concrete Slabs with Dominating Punching", Transactions, SMiRT 21, New Delhi, India, Vol. 774, pp. 1-8, November 6-11.
[20] Mahmoudpour, S., Attamejad, R. and Behnia, C., (2011), "Dynamic Analysis of Partially Embedded Structures Considering Soil-Structure Interaction in Time Domain", Mathematical Problems in Engineering, Volume 2011, Article ID 534968, 23 pages, doi:l0.1 155/2011/534968, Hindawi Publishing Corporation.
[21] Fattah, M. Y., Hamood, M. J., Abbas, S. A., (2013), "Simulation of Behavior of Plate on Elastic Foundation under Impact Load by the Finite Element Method", Engineering and Technology Journal, Special issue, 1st International Conference on Geotechnical and Transportation Engineering, Vol. 31 Part (A), No. 19, pp. 44-58.
[22] ANSYS Manual, Version. 11, (2007): SAS IP, Inc.
[23] Lenczner, D., (1972), "The Elements of Load Bearing Brickwork", 1st Edition, pp. 11-23.
[24] Fattah, M. Y., Hamood, M. J., Abbas, S. A., (2014), "Behavior of Plate on Elastic Foundation under Impact Load", Engineering and Technology Journal, Vol. 32 Part (A), No. 4, pp. 1007-1027.
[25] Teodoru, B., (2009), "Beam on Elastic Foundation the Simplified Continuum Approach", University of Gheorghe Asachi, Vol. 4.

Tsunami hazard assessment along Diba-Oman and Diba-Al-Emirates coasts

Issa El-Hussain[1,*], *Rachid* Omira[2], *Khalid* Al-Bulushi[3], *Ahmed* Deif[1,4], *Zaid Al*-Habsi[1], *Ghazi* Al-Rawas[1], *Adel* Mohamad[1,4], *Khalifa* Al-Jabri[1], *Maria Ana* Baptista[5]

[1]Earthquake Monitoring Center, Sultan Qaboos University, P.O. Box 50, Al-Khoudh, PC 123, Muscat, Oman
elhussain@squ.edu.om, Tel. +(968) 24142642 Fax +(968) 24413137
[2]Instituto Português do Mar e da Atmosfera, Rua C do Aeroporto, 1749-077 Lisboa, Portugal
[3]Department of Earth Science, United Arab Emirates University
[4]National Research Institute of Astronomy and Geophysics, Helwan, Egypt
[5]Instituto Superior de Engenharia de Lisboa, Instituto Politécnico de Lisboa, R. Conselheiro Emidio Navarro, 1, 1959-007 Lisboa, Portugal

Abstract. Tsunami is among the most devastating natural hazards phenomenon responsible for significant loss of life and property throughout history. The Sultanate of Oman and United Arab Emirates are among the Indian Ocean countries that were subjected to one confirmed tsunami in November 27, 1945 due to an Mw 8.1 earthquake in Makran Subduction Zone. In this study, we present preliminary deterministic tsunami hazard assessment for the coasts of Diba Oman and Diba Al-Emirates, which are located on the western coast of the Oman Sea. The tsunami vulnerability of these cities increases due to the construction of many critical infrastructures and urban concentration along their coasts. Therefore, tsunami hazard assessment is necessary to mitigate the risk on the socio-economic system and sustainable developments. The major known source of tsunamis able to impact both coasts of Oman and United Arab Emirates is the Makran Subduction Zone (MSZ) which extends for approximately 900 km. The deterministic approach uses specific scenarios considering the maximum credible earthquakes occurring in the MSZ and computes the ensuing tsunami impact in the coasts of the study area. The maximum wave height graphs and inundation maps are obtained for tsunami scenarios caused by 8.8 earthquake magnitude in eastern MSZ and 8.2 magnitude from western MSZ. The Mw8.8 eastern MSZ causes a maximum inundation distance of 447 meters and a maximum flow depth of 1.37 meter. Maximum inundation distance larger than 420 meters occurs due to the Mw8.2 western MSZ scenario. For this scenario, numerical simulations show a maximum flow depth of about 2.34 meters.

* Corresponding author: elhussain@squ.edu.om

DOI: 10.1051/matecconf/201712006007

1 Introduction

A tsunami is a series of water waves that propagate from the location of the tsunamigenic source toward the shore. The term tsunami can be referred to the water waves caused by rapid vertical displacement of the water column in open oceans or inland water bodies.Tsunamis can be severely destructive to the infrastructure, human life, and the economy of the coastal areas. They may inundate large onshore areas based upon the local bathymetry and topography. The hydrostatic and hydrodynamic forces associated with the tsunamis may damage structures in such coastal areas. The most frequent cause of tsunamis is large submarine earthquakes. Less frequently, tsunamis are generated by submarine landslides, by pyroclastic flows and caldera collapses during volcanic eruptions, by meteorite impacts, and by snow avalanches. These long-period tsunami waves can travel very far from the source without much attenuation. The waves decrease in velocity and increase in amplitude as they reach shallow water zones near the shore. The amplification of the wave depends on the local, near-shore bathymetry. The maximum ground elevation that the tsunami waves reach above a datum is defined as run-up.

Tsunamis generated in the Indian Ocean pose a great threat to all the countries of the region. Oman and United Arab Emirates (UAE) are among the affected countries. The main objective of this research study is to assess the tsunami hazard posed by the Probable Maximum Tsunami (PMT) at the coastal areas of Diba Oman and Diba UAE (Fig. 1) and to determine whether any protective measures are required.

2 The study area and tsunamigenic source

Tsunami hazard along the coast of Diba, Oman and UAE, is mainly due to the seismic activity at the Makran subduction zone, located circa 150 to 600 km east of the Diba coast. The tsunamigenic source region that has the potential to affect the coasts of Diba Oman-UAE is delineated on the basis of historical earthquake records, tsunami data, and sea bathymetric data. The source mechanism and characteristics of the Probable Maximum Tsunami (PMT) are evaluated based on the maximum magnitude earthquake estimated by [1]. The established source parameters are then used to numerically model the tsunami caused by the PMT scenarios. The potential coastal inundation and run-ups maps are estimated for each PMT scenario. The potential tsunami impact on Diba Oman-UAE coasts (Fig. 1) is conducted utilizing Deterministic Tsunami Hazard Assessment approach.

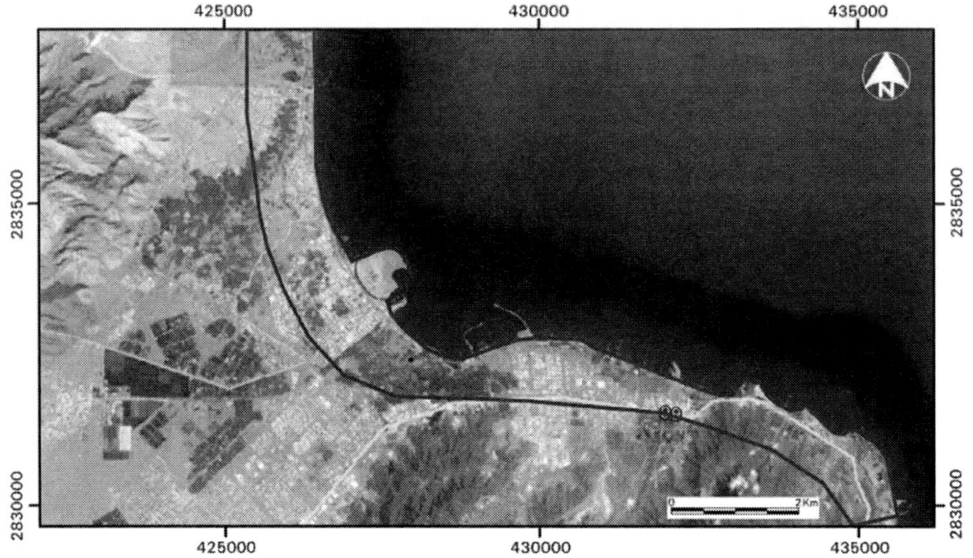

Fig. 1. Map show the Diba Oman and Diba Al Emirates coastal areas along the Sea of Oman.

3 Historical tsunami records

The primary tsunami threat to Diba Oman-UAE coastal area is from the MSZ located east-northeast of the UAE coast. Although large earthquakes along the MSZ are infrequent, they have the potential to generate destructive tsunamis in the Northern Arabian Sea. Quick survey of the tsunami history in the Oman-UAE indicates that at least one confirmed tsunami might have struck their coasts in 1945. A great earthquake with moment magnitude 8.1 occurred off Makran coast of Pakistan, about 100 km south of Karachi in 1945. The earthquake generated a very destructive tsunami, which is reported to affect Pakistan, the western coast of India, Iran, UAE and Oman, killing hundreds of people and causing great destruction. A maximum run-up of 13 m is reported in Pakistan [2, 3]. The tsunami is recorded in Muscat but no details of damage were documented. The tsunami was generated along an active Makran subduction zone, which marks the boundary between the Arabian Plate and the stable Eurasian Plate. Therefore, it is anticipated that future tsunamis can be generated from large earthquakes along MSZ. All published articles in relation to tsunamis from the Makran fault are reviewed [1-12].

4 Data collection and quality assurance

Tsunami simulation results are strongly dependent on the accuracy and resolution of the Digital Elevation Model (DEM), and must ensure that the onshore and off-shore components of the potential flooding area are properly described in terms of vertical data. This is particularly important because the vertical references of topographic maps made on-shore and nautical charts are often different. Therefore, the first step is the establishment and validation of a DEM from all available bathymetric datasets: local bathymetry around the coasts and navigation channel, regional bathymetry from available digital charts, and its validation, particularly with the direct inspection of the most sensitive locations.

Tsunami simulation models use grids of different resolution for different stages of tsunami propagation because the onshore propagation (inundation stage) requires much higher resolution, to be able to correctly describe the target area (Fig. 2).

The original bathymetric set of data is prepared to be used in the numerical simulation by specifying: the region of interest, the desired grid resolution (i.e. the density of nodes present in the resulting grid). The compiled grid was checked and corrected for errors that may appear due to the interpolation process. For deep ocean tsunami simulation, GEBCO 30 second arc resolution gridded data were used [13].

Digitization of bathymetric regional charts and generation of near-shore bathymetric grid including the study area was performed. The fusion of high-resolution depth and height data requires a common definition for the vertical reference surface. The zero height contour line is, in general, obtained from the topographic data and is used as the boundary for land-sea transition (Fig. 3). Once this procedure is completed, depth and height data are merged subjected to the zero line constraint and carefully edited on the onshore-offshore envelope.

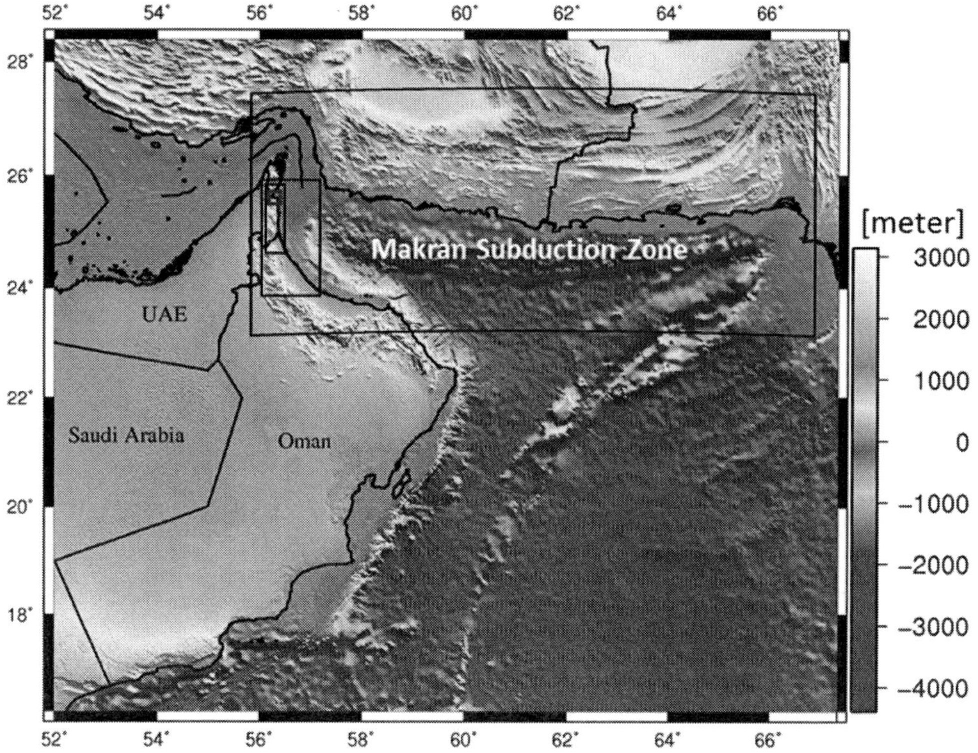

Fig. 2. DTM showing bathymetry and topography grid with four nesting grid levels used for Diba tsunami simulation with grid spacing of 3750m, 750m, 150m and 30m respectively.

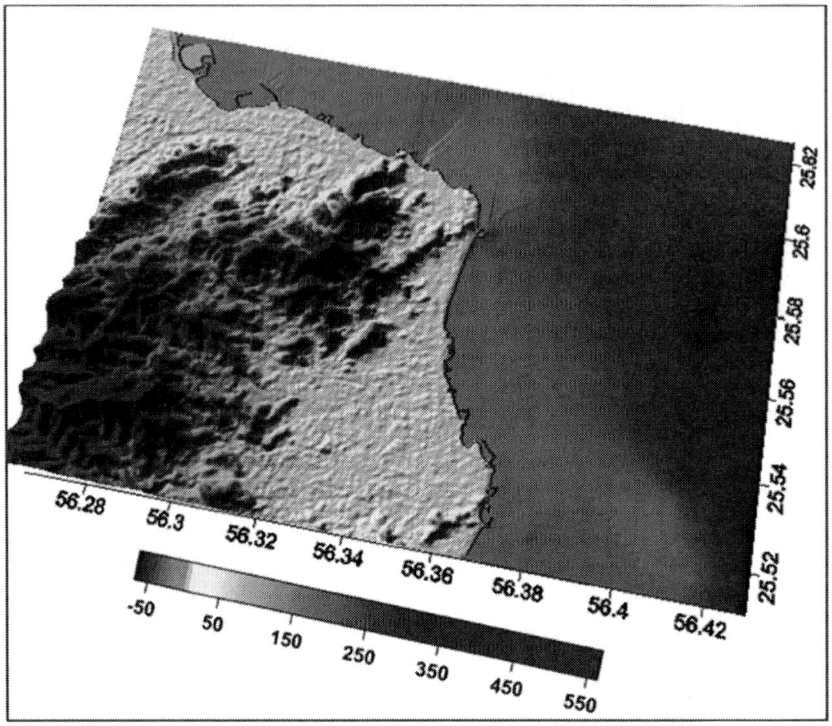

Fig. 3. A 3D view of level 4 grid 30m resolution for Diba UAE.

5 Tsunami model implementation

The numerical model used to simulate tsunami propagation and run-up is based on the Non-Linear Shallow Water equations, solved infinite differences scheme. Two of the implementations that have been widely used in previous studies are MOST [14] and COMCOT [e.g. 15]. Both models are able to simulate the generation, propagation, inundation, run-up and run-down, and have been benchmarked.

6 Deterministic tsunami hazard assessment

The methodology called Deterministic Tsunami Hazard Assessment (DTHA) is based upon the definition of tsunami scenarios. In this case, the different hazard maps result from a set of individual sources that may affect the target area. Based on available seismic information, two earthquakes, with several hundred years return period [1, 9] representing worst case tsunami scenarios for Diba coasts are simulated Table 1. The impact of each individual scenario is presented in terms of maximum wave height, tsunami travel time, and maximum inundation distance.

Table 1. Makran Subduction Zone Tsunamigenic Earthquake Scenarios.

Makran Subduction Zone Scenarios								
Scenario	**MSZ**	**Length(km)**	**Width(km)**	**Slip(m)**	**Dip(°)**	**Strike(°)**	**Rake(°)**	M_w
One	Eastern	461	110	11.1	7	263	90	8.8
Two	Western	201	82	6	7	281	90	8.2

7 Results

Numerical simulations of the tsunami propagation from the source area to the target area with onshore propagation (inundation) are conducted. The outputs of this task are grids and maps of tsunami travel time (Fig. 4), synthetic waveform tsunamis at specific points (virtual tide gage) along the target area (Fig. 5a and b), and maximum wave height (Fig. 6). Fig 4 is coarse grid map of bathymetry and topography with 30 second arch resolution with contours representing the tsunami travel times (TTT) in hours following the occurrence of Mw 8.8 earthquake event in Eastern Makran Subduction Zone. Fig 6 is the Tsunami hazard map showing the maximum inundation distances of 40 and 447m, the maximum flow depths of 1.37 and 1.26m, and the runup of 1.05 and 1.16m for Diba Oman and Diba Al Emirates respectively, due to the Mw 8.8 scenario from eastern Makran Subduction Zone.

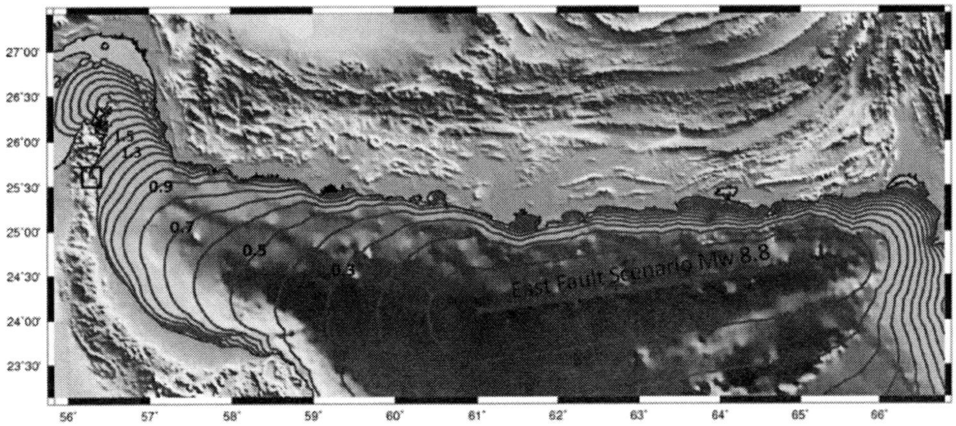

Fig. 4. Tsunami travel time in hours (red contours) for scenario one M_W 8.8, eastern MSZ.

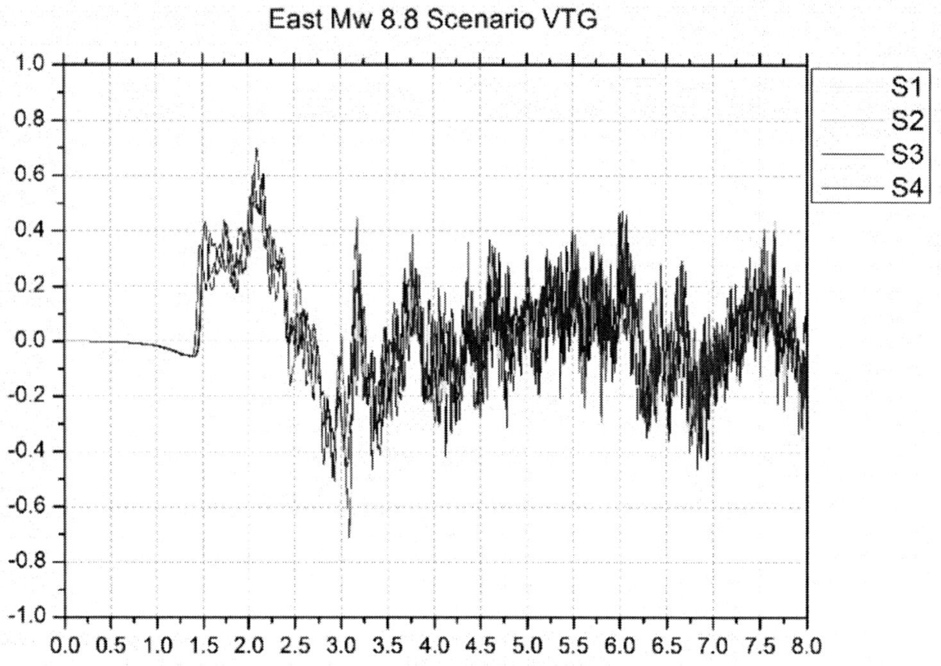

Fig. 5a. Synthetic waveforms at VTG from S1 to S4 (locations in Figure 6), maximum wave height of 0.8 m.

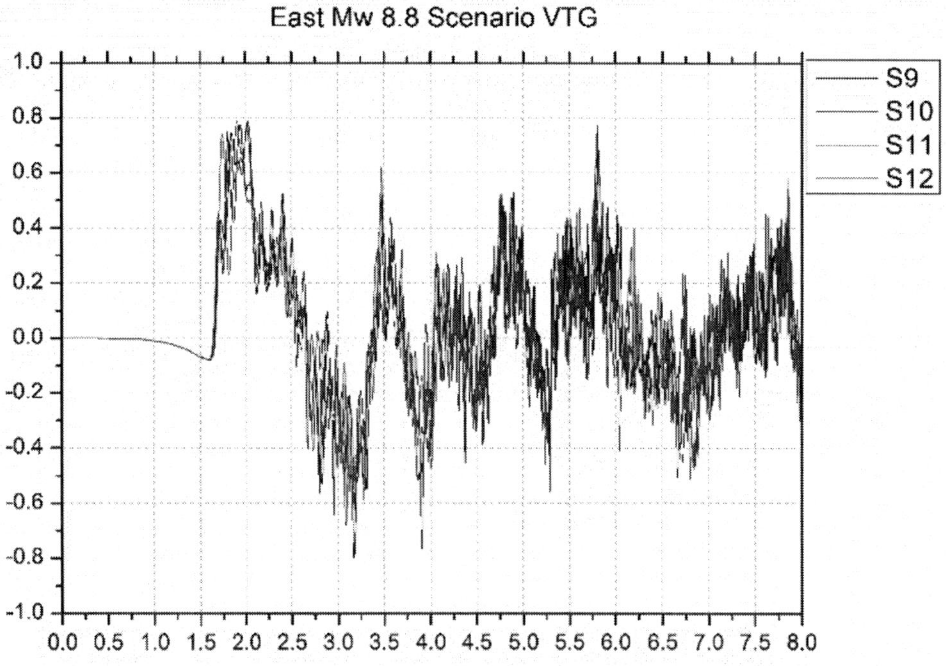

Fig. 5b. Synthetic waveforms at VTG from S9 to S12 (locations in Figure 6), maximum wave height of 0.8 m.

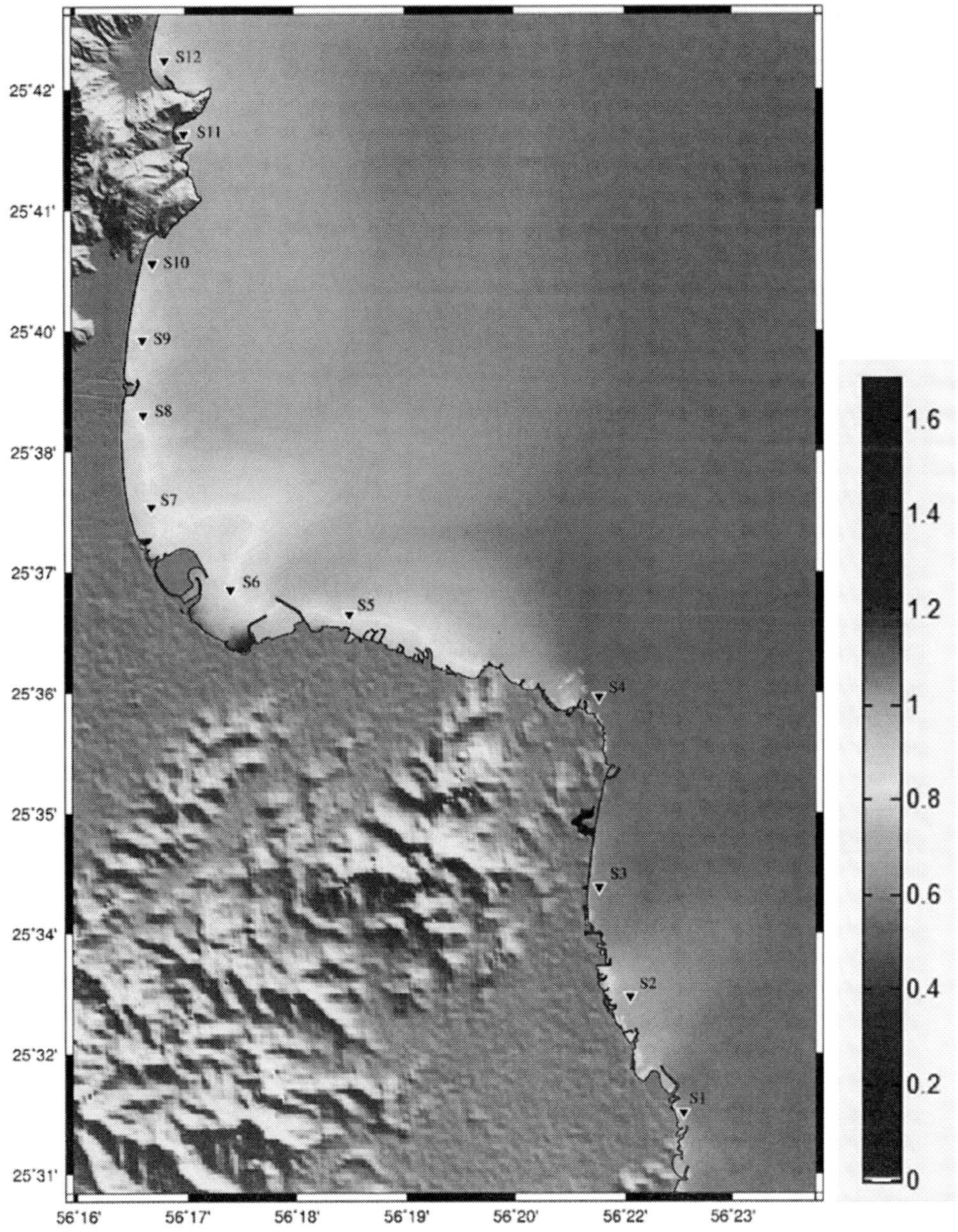

Fig. 6. Tsunami hazard map for Diba Oman with 5m grid spacing representing maximum wave height and inundation in meters for Scenario One M_w 8.8.

Fig 7 is coarse grid map of bathymetry and topography with 30-second arc resolution with contours representing TTT in hours following the occurrence of Mw 8.2 earthquake event in western Makran Subduction Zone. Figs 8a and b show the synthetic waveform of the tsunami at specific points (virtual tide gage) along the target area. Fig 9 is the tsunami hazard map showing the maximum inundation distances of 150 and 420m, the maximum flow depths of 2.18 and 2.34m, and runup of 2.42 and 2.57m for Diba Oman and Diba Al Emirates respectively, due to the Mw 8.2 scenario from western Makran Subduction Zone.

DOI: 10.1051/matecconf/201712006007

MCES-17

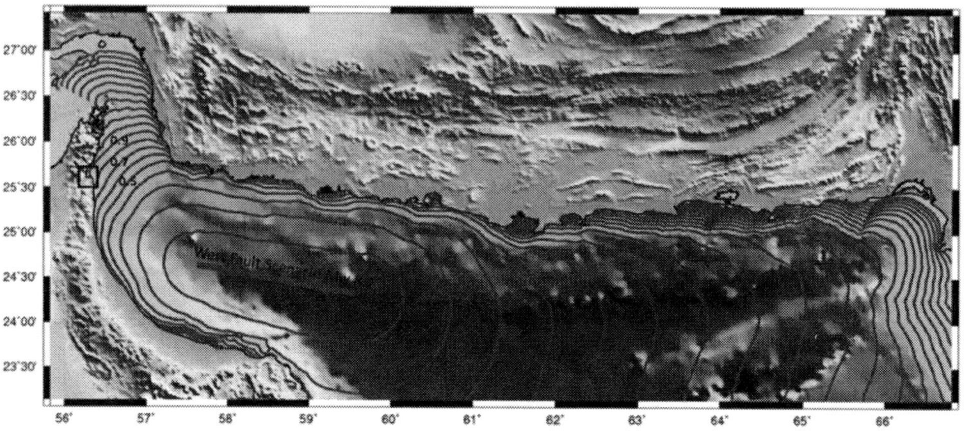

Fig. 7. Tsunami travel time in hours (red contours) for scenario two M_w 8.2, Western MSZ.

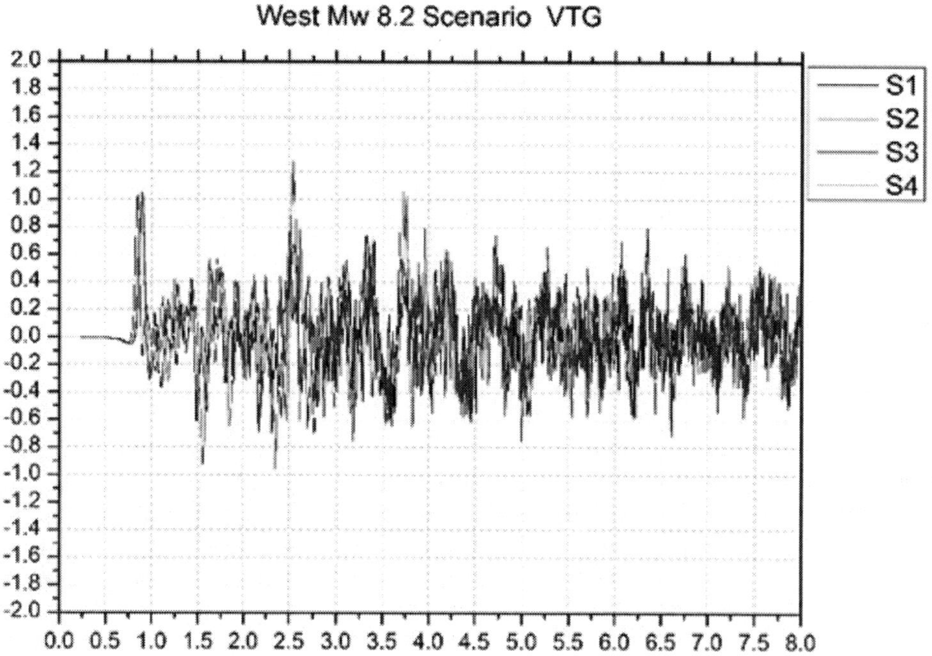

Fig. 8a. Synthetic waveforms at VTG from S1 to S4 (locations in Figure 9), maximum wave height of 1.2m.

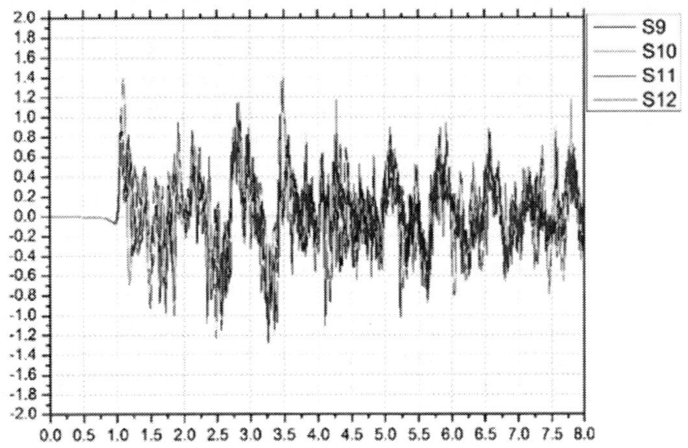

Fig. 8b. Synthetic waveforms at VTG from S9 to S12 (locations in Figure 9), maximum wave height of 1.4m.

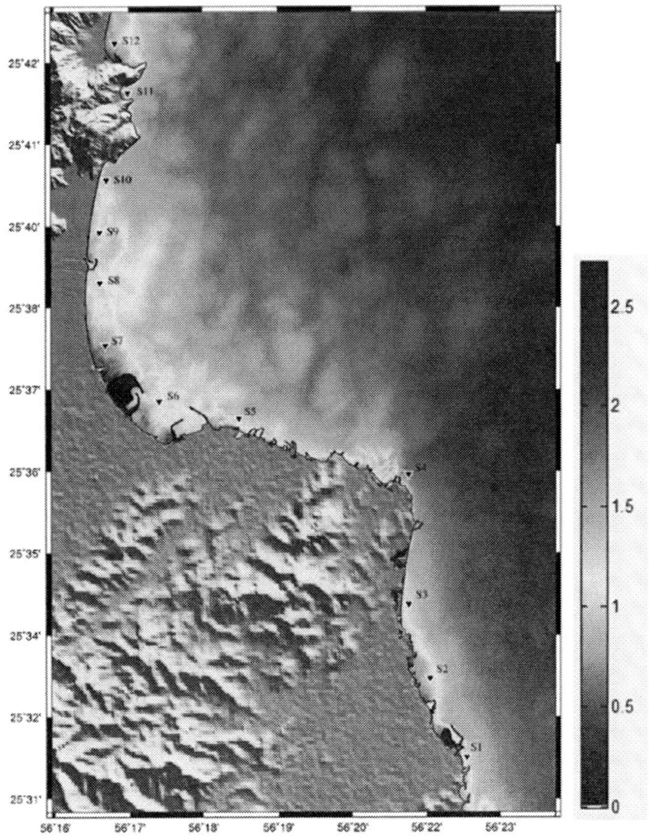

Fig. 9. Tsunami hazard map for Diba Oman with 5m grid spacing representing maximum wave height and inundation in meters for Scenario two M_w 8.2., also showing the locations of 12 virtual tide gauges (VTG).

8 Conclusions

Deterministic tsunami hazard assessment for Diba coast shows a maximum run-up of 1.16m and maximum flow depth of 1.37m due to an earthquake of Mw 8.8 from eastern Makran Subduction Zone. The run-up would be 2.57m and the flow depth would be 2.34m due to an earthquake scenario of Mw 8.2 from western Makran Subduction zone. The hazard from western Makran subduction would be higher than that of the eastern Makran subduction for even a lower magnitude due to close proximity to Diba coasts.

References

1. I.El-Hussain,A.Deif, k. Al-Jabri, S. Al-Hashmi, K. Al-Tobi, Y. Al-shigbi, M. Al-Saifi, 2012. Nat. Hazard, **64**:173-210(2012)
2. M. Heidarzadeh, M.D. Pirooz, N.H. Zaker, C.E. Synolakis. Pure Appl. Geophys. **165**(11-12): 2045-2058(2008)
3. M. Mokhtari, I.A. Fard, K. Hessami , Natural Hazards **47**(2):185-99. doi: 10/1007s11069-007-9208 (2008)
4. D. R. Burbidge, P.R. Cummins, R. Mleczko, H. Latief, M. Mokhtari, D. Natawidjaja, C.P. Rajendran, C. Thomas, Geoscience Australia professional opinion. No: 2009/11 (2009)
5. D.E. Byrne, L.R. Sykes, S.M. Davis, J Geophys Res**97**-B1: 449–478,(1992)
6. A. Deif, I. El-Hussain, J Geophys Eng **9**: 773-783(2012)
7. A. Deif, I. El-Hussain, K. Al-Jabri, N. Toksoz, S. El-Hady, S. Al-Hashmi, K. Al-Tobi, Y. Al-shijbi, M. Al-Saifi, Arabian J. Geosciences, **6**, 4947–4960(2013)
8. I. El-Hussain, R. Omira, A. Deif, Z. Al-Habsi, G. Al-Rawas, A.M.E. Mohamed, K. Al-Jabri, M. A. Baptista, Arabian J. Geosciences, **9**: 1-14 (2016)
9. G. Hoffmann, K. Reicherter, T. Wiatr , C. Gru¨tzner , T. Rausch, Nat Hazards, (2013)
10. S. Neetu, I. Suresh, R. Shankar, B. Nagarajan, R. Sharma, S. Shenoi, A. Unnikrishnan, D. Sundar, Natural hazard, published online may (2011)
11. E. Okal, H.M. Fritz, C.E. Synolakis, P.E. Raad, Y. Al-Shijbi, M. Al-Saifi, Earthquake Spectra,22, S203-S218(2012)
12. M. Rodriguez, N. Chamo-Rooke, H. Hebert, M. Fournier, P. Puchon, Natural Hazards and Earth System Sciences, **13**, 416-424 (2013)
13. GEBCO http://www.gebco.net/data_and_products/gridded_bathymetry_data/gebco_30_second_grid/(2014)
14. C. E. Synolakis, E.N. Bernard, V. V. Titov, U. Kânoğlu, F.I. Gonzalez, Pure and Applied Geophysics, **165**(11-12), 2197-2228(2008)
15. R. Omira, L.M. Matias, M.A. Baptista, Pure App. Geophys.doi:10.1007/s00024-016-1367-z(2016a)

MATEC Web of Conferences **120**, 06008 (2017) DOI: 10.1051/matecconf/201712006008
ASCMCES-17

Effects of pore fluid salinity on the collapsibility of low plasticity clay

Omar Al Hattamleh[1],* and *M.N.* Akhtar [2]

[1]Civil Engineering Department, College of Engineering Hashemite University, Zarqa, Jordan P.O. Box 150459, Zarqa 13115, Jordan
[2]Civil Engineering Department, College of Engineering, Fahad Bin Sultan University, Tabuk, Kingdom of Saudi Arabia.

Abstract. This paper presents the results on the influence of pore fluid salinity on the Collapsibility of low plastic clay from Hashemite town, Jordan. Double odometer tests were carried out on natural specimens using pore fluids with different salt concentrations. The results indicate a considerable amount of collapse takes place for the sample with increasing the initial applied pressure on the sample regardless of the initial molding water content. However, the dry side of optimum sample shows larger collapse than that on wet side of optimum sample and vice versa. At low applied initial stresses, the higher the salts percentage in the sample, the larger the collapse occurs. Moreover, the amount of collapse tends to increase with increasing the applied pressure to reach a certain value for a given salts percentage, then the amount of collapse asymptotes to its maximum value or level off with increasing the pressure on the sample.

1 Introduction

Collapsibility of soils which defined as large reduction of volume in soil deposits upon saturation or wetting. This process may or may not be a result of the application of external applied pressure. The collapsible soils appear in strong and stable dry natural state. Upon loading or wetting huge settlement of soil layer occurs and may yield a catastrophic consequences for structures built on it. Such soils are often termed collapsible or metastable soils (among others; Benatti and Miguel, 2013; Mouria et al. 2014; Shalaby, 2014; Al Hattamleh et al. 2015).

Collapse settlement is observed in unsaturated soils of either manmade fills due to poor compaction or in natural soils deposits. The process tends to modify metastable soils and converts it to stable non-collapsible soil after wetting and saturation. The stability of dry collapse soil is attained primarily due to either the presence of thin film of water under tension (i.e. capillarity) or cementation bridge of fine material or presence of salts between large soil particles.

The capillary tension in unsaturated soils seems to vanish after soil pores saturated (i.e. filled with water). In the case of dry soils, the bond between soil particles of fine

* Correspoding author: hattam@hu.edu.jo

 DOI: 10.1051/matecconf/201712006008

materials and salts is washed away during saturation process, causing soils to collapse. Several criteria are available in the literature for evaluation of the prone of soil to collapse (Jennings and Knight, 1975).

The main objective of this paper is focus on the evaluating of the collapsibility of soil due to the change of pore water fluids.

2 Experimental program

2.1 Material

The material used for the tests was natural soil deposits from Hashemite town, Jordan. The Liquid limit and plastic limit of the soil are 21% and 15%, respectively while the specific gravity of the Soil is 2.63. The grain size distribution of the soil sample indicates that it contains 83% of fine particles. Based on the consistency test of clean, washed, and dry samples along with grain size distribution, the soil is classified as CL-ML according to the Unified Soil Classification System (USCS). Table 1 contains more information regarding the indices properties of the soil.

Table 3.1 Physical properties of hashemite soil.

Index Properties		Value (unit)
Specific gravity		2.63
Natural Water Content (%)		5.3
Liquid limit (%)		21
Plastic limit (%)		15
Plasticity index (%)		6
Grain Size Distribution (%)	Gravel	2
	Sand	15
	Silts	45
	Clay	38

2.1.1 Soil preparation for testing

The soil was pulverized using rubber-tipped pestle until they pass sieve # 40 and larger particles was discarded. The soil samples were mixed with different concentrations of brine (Salty water solution). The brine was diluted with distilled water to give a concentration of total dissolved solids of 25% (undiluted brine), 15%, 10%, 5% and 0%. Thereafter, soil samples were subjected to a different soil characterization tests.

DOI: 10.1051/matecconf/201712006008

3 Test results and analysis

Three sets of double - Oedometer soil specimens (20 mm in height and 76 mm in diameter) were prepared for each testing state. Figure 1 indicates the state of compaction, percent of brine, and number of specimen tested. The chosen soil sample states, as depicted in the Figure 1, were selected to allow soil specimen's to alter its structure from flocculated or bookhouse structures to dispersed or turbostratic structures (Lambe, 1961). The collapse potential tests were conducted according to the ASTM D 5333 – 03. Applied stresses, initial and final heights, thereafter, ratios of initial and final voids were determined for each tested specimen. Voids ratios versus applied stresses for the set of pair samples were plotted. The collapse potential for each pair specimens was determined for all tested cases.

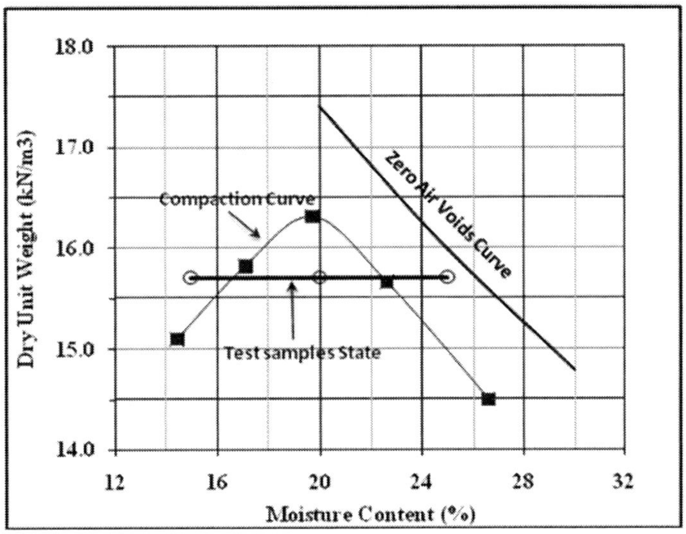

Fig. 1. Compaction curve.

The Oedometer test results for wet and dry, samples of the clay specimens with pore fluid of zero brine at a dry unit weight of 15.7 kN/m3 and a molding moisture content of 15, 20, and 25 were shown in Figures 2a to 2c. It is clearly shown from Figures 2a to 2c that the soil in both dry and water-inundated samples were hardening with the applying stresses, and a substantial volume decrease occurs. This occurs primarily due to the fact that the soil initial structure was flocculated. With the increasing induced pressure on the samples, the soil structure tries to reach into a new equilibrium structure and tries allied into parallel structure named as disperse structure.

 DOI: 10.1051/matecconf/201712006008

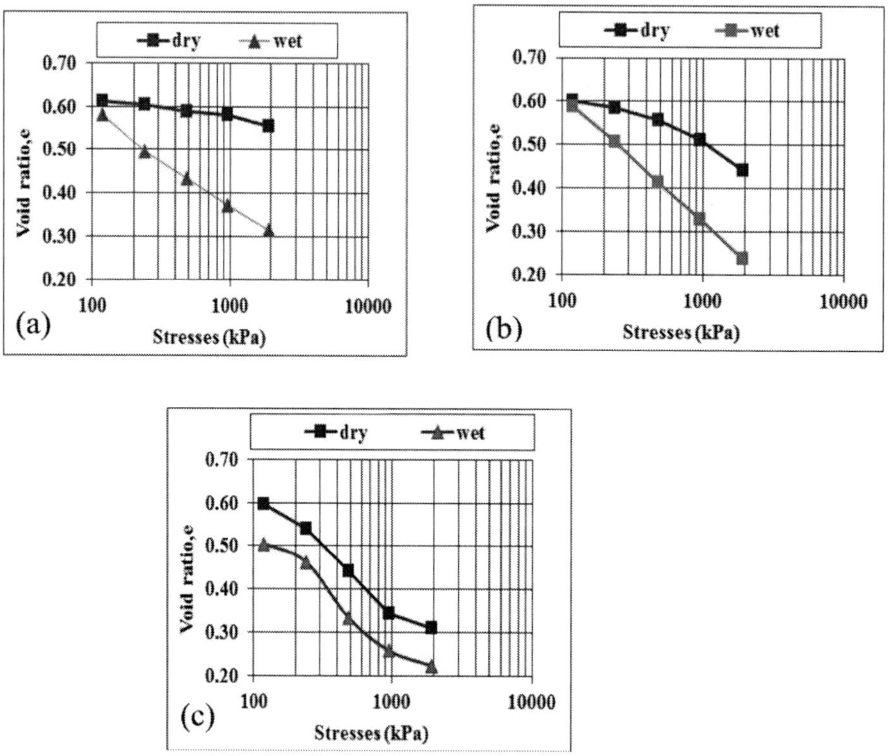

Fig. 2. Compressibility results at dry unit weight of 15.7 kN/m^3 and moisture content of (a) 15% (b) 20% and (c) 25%.

Figures 3 to 4 show the compressibility curves for the clay sample molded with different percentage of brine fluid. It is obvious from the set of curves depicted in Figures 3 to 4 that the added amount of brine changes the characteristics of collapsibility curve despite the fact that the curves tend to a hardening behavior when the stress applied. Thus the curves typify with those curves of soils with sensitive structure and highly disturbance. The increases of the amount of brine fluid in molding the test specimen obviously increase the concentration of ions. Thus, it will increase the thickness of the diffuse double layer. Therefore, clays structure will favor dispersing structure.

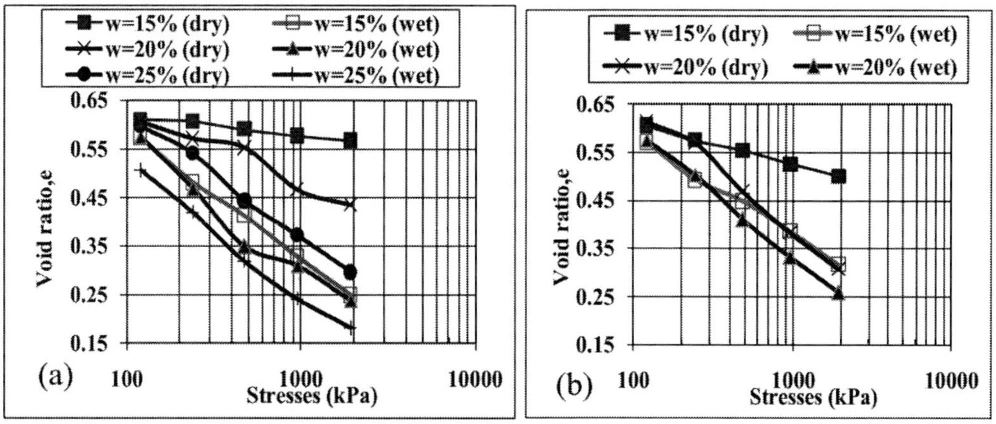

Fig. 3. Compressibility results at dry unit weight of 15.7 kN/m^3 and moisture content had shown in the legend and brine content of (a) 5% (b) 10%.

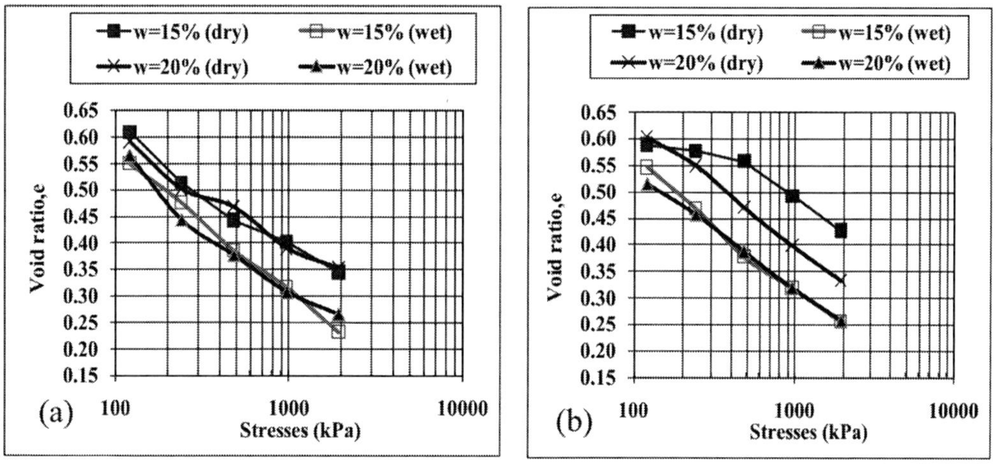

Fig. 4. Compressibility results at dry unit weight of 15.7 kN/m^3 and moisture content had shown in the legend and brine content of (a) 15% (b) 25%.

Collapse potential was evaluated after the dry specimen, while in Oedemeter, reaches the required stress level inundated of brine solution for 24 hours. Thus the collapse potential given as:

$$C_p = \frac{e_f - e_i}{1 + e_o} \tag{1}$$

Where, e_f = final void ratio at the appropriate stress level after wetting, e_i = void ratio at the appropriate stress level before wetting, and e_o = initial void ratio.

Figure 5 shows the variation of collapse potential with applied stresses for different amounts of brine. It is clearly indicated that as the stresses value increases, the amount of collapse increases. Moreover, increasing the amount of salts by weight is not necessary to cause the amount of collapse due to the fact that the salts are not fully washed away when the sample is inundated, which adds an amount of solids to soil solids part, which in this range of applied stress is incrushable. Moreover, the amount of remolding water is deficient to reach an equilibrium amount of water that is fair enough to dissolve all the salt solids.

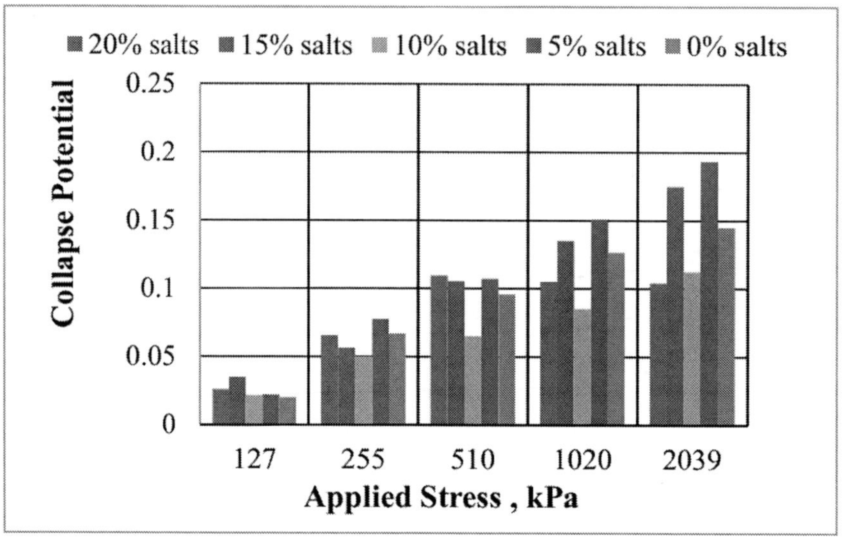

Fig. 5. Variation of collapse potential with stresses for different brine percentage at molding brine content of 20%.

4 Conclusions

In the present study, a set of soil samples of low plasticity were prepared and tested to examine the effect of brine in pore fluid in the collapse potential. The brine was mixed thoroughly with soil specimens to the desired concentration. Thereafter, sets of double oedometer tests were conducted on the prepared samples. Based on the results obtained, a considerable amount of collapse takes place for the sample with increasing the induced pressure on the samples despite of the molding water content or brine percentage added. Moreover, samples prepared in molding water content in the dry side of optimum show larger collapse potential more than that on the wet side of optimum samples and vice versa. Furthermore, at low applied induced pressure, the higher the brine percentage added to the samples, the larger the collapse occurs. However, at large induced pressure, the opposite tendency was observed. Lastly, the amount of collapse potential tends to increase with increasing the induced pressure to reach a certain maximum value for a given brine percentage, regardless the additional induced pressure applied on the sample.

References

1. Benatti J. C. B. and Miguel M. G., A proposal of structural models for colluvial and lateritic soil profile from southwestern Brazil on the basis of their collapsible behaviour. *Engineering Geology* **153**, 1–11 (2013)
2. Mouria G., Shinodab S., Golosovc V., Chalovd S., Shiibae M., Horif T., Okig T., Estimating the collapse of aggregated fine soil structure in a mountainous forested catchment. *Journal of Environmental Management* **138**, 24–31 (2014)
3. Shalaby S.S., The Assessment of the Collapse Potential of Fills during Inundation using Plate Load Tests. *Life Science Journal* **11**, 1001-1006 (2014)
4. Omar H Al Hattamleh, Odeh A. Odeh, Hussein Aldeeky, and M. N. Akhtar, Characterizations of Soil Collapsibility: Effect of Salts Dilution. *Journal of Materials and Engineering Structures* **2**, 33–43 (2015)

5. Jennings, J.E. and Knight, K., A guide to construction on or with materials exhibiting additional settlement due to collapse of grain structure. *Proc. 6th Reg. Conf. Africa on Soil Mechanics and Foundation Engineering* **6,** 99-105 (1975)
6. Lambe, TW. 1961. Residual pore pressures in compacted clay. *Proceeding 5th International Conference on Soil Mechanics and Foundation Engineering*. Paris, France.
7. ASTM . Annual Book of ASTM Standards – Section 4: Construction, Volume 04.08 Soil and Rock (1): D420-D5550. *American Society for Testing and Materials*, USA.

Resilient modulus for unbound granular materials and subgrade soils in Egypt

Rabah Mousa[1], *Alaa* Gabr[1], *Mohamed* G. Arab[2, 3*], *Abdelhalim* Azam[1],and *Sherif* El-Badawy[1]

[1]Public Works Eng. Dept., Faculty of Engineering, Mansoura University, Mansoura, Egypt
[2]Structural Eng. Dept., Faculty of Engineering, Mansoura University, Mansoura, Egypt
[3]Department of Civil and Environmental Engineering, University of Sharjah, Sharjah, UAE

Abstract. Mechanistic Empirical (ME) pavement design methods started to gain attention especially the last couple of years in Egypt and the Middle East. One of the challenges facing the spread of these methods in Egypt is lack of advanced properties of local soil and asphalt, which are needed as input data in ME design. Resilient modulus (M_r) for example is an important engineering property that expresses the elastic behavior of soil/unbound granular materials (UGMs) under cyclic traffic loading for ME design. In order to overcome the scarcity of the resilient modulus data for soil/UGMs in Egypt, a comprehensive laboratory testing program was conducted to measure resilient modulus of typical UGMs and subgrade soils typically used in pavement construction in Egypt. The factors that affect the resilient modulus of soil/UGMs were reviewed, studied and discussed. Finally, the prediction accuracy of the most well-known M_r Prediction models for the locally investigated materials was investigated.

1 Introduction

Mechanistic Empirical (ME) pavement design methods has to be calibrated using specific pavement types, materials, specific local traffic and environmental conditions, which limit the possibility of using them, unless local calibration is conducted. In the Mechanistic-Empirical Pavement Design Guide (MEPDG), a pavement system is analyzed by computing the structural responses (stresses, strains, and deflections) based on the mechanical properties of different pavement materials. Then, the pavement distresses are predicted by empirical models using the computed critical strains and deformations.

The proper characterization of Unbound Granular materials (UGMs) and subgrade soils is essential in the design and rehabilitation of pavement structures. The resilient modulus (M_r) is used as a fundamental engineering property to describe stress–strain relationship of soil/UGM under cyclic loading. M_r is defined as the deviator stress (σ_d) divided by axial recoverable strain (ε_r) as shown in Equation (1).

$$M_r = \frac{\sigma_d}{\varepsilon_r} \tag{1}$$

* Corresponding author: marab@sharjah.ac.ae

The MEPDG recommends the use of resilient modulus as level 1 input data (the highest level of accuracy) instead of California Bearing Ratio (CBR) for pavement design. Thus, the objectives of this paper are to:

- Present the state of the art regarding some factors affecting the resilient modulus of soil/UGMs and the available M_r predictive models for soil/UGMs
- Determine the resilient modulus of four typical UGMs and five subgrade soils that are used for pavement construction in Egypt.
- Apply some of the well-known resilient modulus models exist in the literature on the investigated materials to obtain material constants and recommend the most accurate model for the local materials.

2 Literature review

Several research studies have been conducted to characterize the resilient behavior of UGMs and fine grained subgrade soils [1]. Granular pavement layers demonstrate a nonlinear, time dependent and elastoplastic response under traffic loading [2, 3, 4]. On the other hand, traditional elasticity theories consider the response of granular materials as linear-elastic, which requires resilient modulus and Poisson's ratio [5].

Previous studies have shown that many factors affect the resilient modulus of UGMs and subgrade soils. Lekarp et al [6] and Azam et al [7] reported that the aggregate gradation, fine content, applied stress level, moisture content, matric suction, and density had significant effect on M_r. In the following subsections, the influence of these factors is briefed.

2.1 Soil index properties

Several studies investigated the effect of gradation, fines content, particle shape, maximum nominal aggregate size, liquid limit (LL), plasticity index (PI), coefficient of uniformity (C_u) and coefficient of curvature (C_c) on resilient modulus [8, 9, 10]. Raad et al. [8] studied the behavior of typical granular base materials with different gradations under Repeated Load Triaxial Test (RLTT) conditions. Results indicated that the dense graded aggregate exhibited the highest resilient modulus values compared to the open graded aggregate. In a similar study, Tian et al. [9] investigated the significance of UGMs gradation on M_r. Three different types of aggregate (Richard spurbase (RS) limestone with the value of Los Angeles Abrasion (LAA) of 24% ,Sawyer sandstone subbase aggregate with LAA of 28% and one base material) were investigated in this study. Test results show that the medium gradation of RS produced substantially higher M_r values than finer gradation, and slightly higher than coarser gradation. On the other hand for sandstone, the coarser gradation produced the highest M_r values than finer and medium limits. Similarly, Uthus et al. [11] noted that the soils with higher fine content, showed significant reduction in resilient modulus with slight increase in water content.

2.2 Stress State

The stress state is the most significant factor on the resilient modulus of UGMs and subgrade soils. Several studies have reported that M_r of untreated UGMs is dependent on confining pressure and deviator stresses, which increases considerably with the increase of those stresses [7, 10, 2, 12, 13, 14]. Kolisoja [15] reported that Poisson's ratio of UGMs increases with increasing deviator stress and decreasing confining pressure. Cabrera [16] reported that M_r decreased significantly with the increase in maximum cyclic stress amplitude. Results also showed that M_r generally increases with the increase in confining pressure level.

2.3 Moisture Content and Matric Suction

Numerous studies stated that the M_r of UGMs and subgrade soils fundamentally depends on moisture content or matric suction in both laboratory and in-situ condition [14, 17].

Hicks and Monismith [18] showed that the resilient modulus significantly influenced by moisture content. Moreover, they found that a loss in modulus value was observed as the moisture content increases over the Optimum Moisture Content (OMC). Lekarp [6] concluded that the pore pressure controls deformational behaviour as opposed to the level of saturation. Thom and Brown [19] demonstrated that moisture has some greasing impact on particles, and consequently builds an increase in the deformation of the aggregate structure with a loss in resilient modulus even without a generation of any pore water pressure. While, Dawson et al. [20] found that the stiffness of well-graded UGM underneath the optimum moisture content tends to increase with the decrease of moisture due to development of suction, then diminished with increasing the moisture content in the wet side of the compaction curve. An obvious diminishment of resilient modulus as well as Poisson's ratio has been found by Lekarp [6] with the increase in moisture content, particularly at high level of saturation. Ekblad [21] demonstrated the impact of water content on resilient behavior of different gradations by changing the maximum particle size and the grading shape . The author reasoned that for the loss of resilient modulus has been more articulated by increasing the water content in the higher stress levels. Andrie et al [22] studied the effect of water content on both UGMs and subgrade soils. They observed that water content had little impact on the resilient modulus of base materials compared to the subgrade soils. Studies conducted by [23, 24, 17], found that the matric suction has been appeared to be a vastly improved predictor of engineering behavior than moisture content.

2.4 Resilient Modulus Prediction Models

Different models have been developed to predict the resilient modulus of UGMs and subgrade soils based on stress state, soil index properties and moisture content. The following subsections present most of the developed models found in the literature for predicting resilient modulus of coarse and fine-grained soils.

2.4.1 Models based on stress state

Numerous researchers have developed resilient modulus models to predicted (M_r) values depending on stresses for both coarse and fine-grained soils. For example, Hicks and Monismith [18] developed the very well-known K-θ model for UGMs based on bulk stress as described in Equation 2.

$$M_r = k_1(\theta)^{K_2} \tag{2}$$

Where M_r = resilient modulus, K_1, and K_2 = regression coefficients, θ = bulk stress = ($\sigma_1 + \sigma_2 + \sigma_3$), σ_1 = major principal stress, σ_2 = intermediate stress, σ_3 = minor principal stress.
May and Witczak [25] proposed a model that correlates the resilient modulus with the normalized measures of the mean stress and octahedral shear stress as follows:

$$M_r = K_0 * \left(\frac{\sigma_m}{P_a}\right)^{K_1} * \left(\frac{\tau_{oct}}{\tau_{ref}}\right)^{K_2} \tag{3}$$

Where P_a = atmospheric pressure, σ_m = mean normal stress = $((\sigma_1 + 2\sigma_3)/3)$, τ_{oct} = octahedral shear stress = $(\sqrt{2}/3)(\sigma_1 - \sigma_3)$, τ_{ref} = reference shear stress = $\frac{\sqrt{2}}{3}*q_f$; q_f = peak shear strength= ($d+\sigma_m*\tan\beta$); d and β are Druker–Prager failure parameters and K_o, K_1, K_2 = regression parameters.
Uzan [2] simplified Equation 3 by excluding the reference shear stress as follows:

$$M_r = K_1 pa\left(\frac{\theta}{pa}\right)^{K_2}\left(\frac{\tau_{oct}}{pa}\right)^{K_3} \quad (4)$$

Equation 4 was incorporated in the MEPDG with adding one to the shear stress term and it is well-known as the universal Witczak model [26].

$$M_r = K_1 pa\left(\frac{\theta}{pa}\right)^{K_2}\left(\frac{\tau_{oct}}{pa}+1\right)^{K_3} \quad (5)$$

Rahim & George [13] proposed two sets of models for predicting resilient modulus for coarse and fine-grained soils depending on stress state as described in Equations 6 and 7, respectively:

For coarse-grained soils:

$$M_r = K_1 Pa\left(\frac{\theta}{(\sigma_d+1)}+1\right)^{K_2} \quad (6)$$

For fine-grained soils:

$$M_r = K_1 Pa\left(\frac{\sigma_d}{(\sigma_c+1)}+1\right)^{K_2} \quad (7)$$

Where σ_c = confining pressure and K_1, K_2 are regression model parameters.

2.4.2 Models based on soil index properties

It is desirable to develop simple models for the estimation of M_r based on the simple index properties to overcome the complexity of the resilient modulus test as well as the cost of testing, which requires expensive equipment and well trained technicians. Many correlations were developed to predict resilient modulus depending on materials properties and soil characteristics such as CBR, PI, LL, water content (WC), dry density (γ_{dry}), percentage passing sieve No 200 (p#200), percentage passing sieve No (40) (p#40), C_c and C_U. The MEPDG suggests that resilient modulus of fine-grained soils can be predicted using the following Equation that was modified by Heukelom and Klomp [27] for materials having CBR of < 10%:

$$M_r\,(psi) = 1500\ CBR \quad (8)$$

For materials with CBR > 10%, Equation 9 is used instead by MEPG [26] as follows:

$$M_r\,(psi) = 2555\ (CBR)^{0.65} \quad (9)$$

Rahim [28] developed correlations for coarse and fine grained soils as given in Equations (10) and (11) as follow:

For coarse-grained soils:

$$M_r = 324.14\left(\frac{\gamma dry}{WC+1}\right)^{0.8998} * \left(\frac{P\#200}{Log\ CU}\right)^{0.4652} \quad (10)$$

For fine-grained soils:

$$M_r = 17.29(\left(\frac{\gamma dry}{WC+1}\right)^{2.18} + \left(\frac{P\#200}{100}\right)^{-0.609} \quad (11)$$

Where γ_{dry} is the relative dry density (%) to MDD), WC is moisture content (%)), P#200 is percentage of fines passing from sieve No.200, and (LL) is liquid limit (%).

Yan et al. [29] proposed two predictive models for subgrade soils based on gene expression programming (GEP) to correlate M_r with routine properties of subgrade soils as follows:

$$M_r = \text{atan}\ \{\gamma_{dry}\ (\frac{\gamma_d - U_C}{pi})\} + \{2 + (\frac{\sqrt{PI}}{P\#200})\} + \sigma_d + \{2 * \sin (\frac{\gamma_d * \exp(\text{atan}(\sin P\#200))}{P\#200})\} + \{\sigma_d(\text{atan}(\text{sqrt}(P\#200)) - (\sigma_d * P\#200)/\gamma_{dry})\} + \{\text{atan}(\text{sqrt}(U_C) - \gamma_{dry}) + \text{atan}\ \gamma_{dry} \quad (12)$$

Where Uc = unconfined compressive strength

2.4.3 Models based on moisture content/degree of saturation

Several models have been developed to consider the impact of moisture content or degree of saturation on resilient modulus. For example, Li and Selig [30] predicted resilient modulus in terms of moisture content as follows:

$$M_r/M_{r\,OPT} = 0.98 - 0.28 * (W - W_{OPT}) + 0.29 * (W - W_{OPT})^2 \quad (13)$$

Where (Mr_{OPT}) is the resilient modulus at OMC, (W) is water content after test and (W_{OPT})is the OMC.

Andrie et al [22] modified the universal model (Equation 5) to include the moisture content effect. They developed two models, one for UGMs and one for the fine-grained soils as follow:

For coarse-grained soils:

$$M_r = 10^{a+\frac{b-a}{1+EXP(\beta+K_s(S-S_{opt}))}} K_1 p_a \left(\frac{\theta}{p_a}\right)^{K_2} \left(\frac{\tau_{oct}}{p_a}+1\right)^{K_3} \quad (14)$$

For fine-grained soils:

$$M_r = 10^{a+\frac{b-a}{1+EXP(\beta+K_w(W-W_{opt\_std}))}} K_1 p_a \left(\frac{\theta}{p_a}\right)^{K_2} \left(\frac{\tau_{oct}}{p_a}+1\right)^{K_3} \quad (15)$$

Where K_S: regression parameter for moisture and density effect, a is regression parameter at minimum of Log (M_r/M_{ropt}), b is regression parameter at maximum of Log ($M_r/M_{ropt.}$), β=Ln.((-b)/a), S - S_{opt} is the variation in degree of saturation, W is gravimetric moisture content, W_{opt}. gravimetric OMC corresponding to standard compaction and K_W is a regression parameter.

Recently, Garcia et al. [31] developed a model taking into account the effect of moisture content and stress state on predicting resilient modulus as shown in the following Equation:

$$M_r = e^{1.98-0.0714(W-W_{OPT})} * \left(\frac{\sigma_d}{\sigma_3}\right)^{-0.2} \quad (16)$$

Where (W) moisture content after testing,

3 Investigated materials

In this paper, four typical limestone aggregates typically used in road construction in Egypt were investigated (3 bases, and one subbase). These materials were sourced from Ataqqa quarry in Suez governorate, Egypt. In addition, five subgrades soils (three clayey soils and two silty soils) were collected from subgrade soil around Mansoura city, Egypt.

3.1 Laboratory testing

The testing program was conducted on the investigated UGMs/soils to determine the index soil properties and resilient modulus. Lab testing included Particle Size Distribution (PSD)[32], AASHTO classification[33], specific gravity[34] , Atterberg limits [35], modified Proctor compaction [36], CBR [37], and RLTT [38].

3.2 Index engineering properties

Sieve analysis tests for UGMs were conducted in accordance with the ASTM C 136. Figure 1 presents the PSD for UGMs. The basic engineering properties of the investigated materials are summarized in Tables 1 and 2 for UGMs and subgrade soils, respectively along with the specification limits and the test standards.

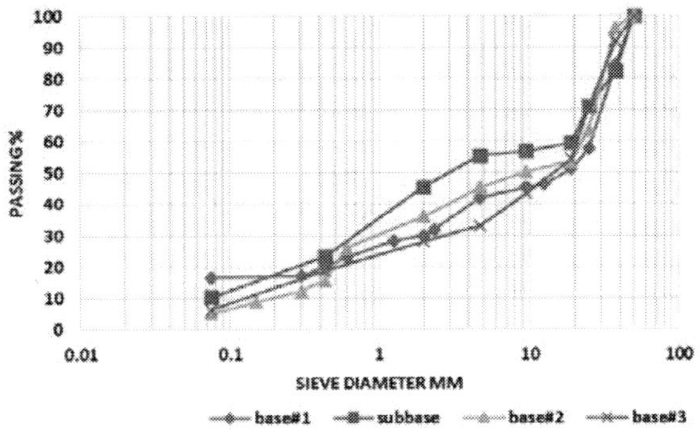

Fig. 1. Particle size distribution for the investigated UGMs.

Table 1. Basic engineering properties of the investigated UGMs.

Material	**Proctor Test**		**Atterberg limits**			**CBR**	**Specific gravity**		**Soil Class.**
	ASTM D 1557		**ASTM D 4318**			**ASTM D 1883**	**ASTM C 127**		**AASHTO M 145-91**
	MDD*gm/ cm^3	OMC, %	LL, %	PL, %	PI, %	%	Gs_b	%Water absorption	
Base #1	2.22	7.2	24	20.4	3.6	73	2.35	3.3	A-1-b
Base #2	2.181	7.5	23	18	5	60.4	2.46	1.6	A-1-b
Base#3	2.245	7.5	24.4	19.1	5.3	84	2.54	1.5	A-1-a
Subbase	2.225	7.2	29	NP		25	2.68	0.1	A-1-a

*MDD = Maximum Dry Density; OMC = Optimum Moisture Content; Gs_b= bulk specific gravity

Table 2. Basic engineering properties of the investigated subgrade soils

Material	**Proctor test**		**Atterberg limits**			**CBR**	**Soil type**	**Soil classification**
	ASTM D 1557		**ASTM D 4318**			**ASTM D 1883**		**AASHTO M 145-91**
	MDD g/cm^3	OMC, %	LL, %	PL, %	PI, %	%		
SG#1	1.762	18	68	42.2	25.8	5.61	Clay	A-7-5
SG#2	1.795	15	39.2	31.5	7.72	16.18	Silt	A-4
SG#3	1.437	18	59.1	32.9	26.2	5.3	Clay	A-7-6
SG#4	1.65	16.5	32	18.1	13.9	16.1	Silt	A-4
SG#5	1.57	20	70	45.4	24.6	5.6	Clay	A-7-6

3.3 Sample preparation for resilient modulus testing

A steel split mold of 300 mm high and internal diameter of 150 mm was used for sample preparation of the UGMs for the RLTT. For the clay and silt subgrade soils, a split mold of 200 mm height and 100 mm diameter was used. Duplicate specimens were compacted according to the modified Proctor compaction effort in six layers at MDD and OMC for each material.

Generally, samples were left one day for curing before de-molding (Figure 2) and testing. The final height, diameter and weight of each specimen were recorded before testing. A rubber membrane was stretched around the specimen by the membrane expander and then the membrane was sealed to the top and bottom caps by means of O-rings as shown in (Figure 2).

DOI: 10.1051/matecconf/201712006009

Resilient modulus testing was conducted according to [34] using the Universal Testing Machine (UTM-25) located at the Highway and Airport Engineering Laboratory (H&AE-LAB), Faculty of Engineering, Mansoura University, Egypt. Two Linear Variable Differential Transducers (LVDTs) were mounted externally to the load cell in the UTM to measure deformations. The AASHTO T307 protocol for UGM and subgrade soils consists of a pre-conditioning sequence and 15 loading sequences. The number of load repetitions is 500 cycles for the conditioning stage and 100 cycles for each loading sequence. The load shape was haversine with 0.1 second loading duration and 0.9 second as rest period for UGMs, while it was 0.2 second loading duration and 0.8 second as rest period for subgrade soils. In the conditioning sequence, a confining pressure of 103.4 kPa and 93.1 kPa deviator stress were applied for UGMs, while the confining pressure and deviator stresses for subgrade soils were 41.4 kPa and 24.8 kPa, respectively. The resilient modulus was calculated as the average of the last five cycles of each sequence. The final moisture content was determined after each test. The properties of the tested samples after testing are presented in Table 3.

Fig. 2. Sample preparation and test set up.

Table 3. Final moisture content and relative compaction for test specimens.

Soil type	**Speciemn**								
	Base#1	**Base#2**	**Base#3**	**subbase**	**SG#1**	**SG#2**	**SG#3**	**SG#4**	**SG#5**
	A-1-b	**A-1-b**	**A-1-a**	**A-1-a**	**A-7-5**	**A-4**	**A-7-6**	**A-4**	**A-7-6**
WC, %	7.15	7.42	7.1	7.18	17.54	14.56	17.54	16.15	19.53
γ_{dry} , gm/cm^3	2.199	2.175	2.215	2.195	1.715	1.735	1.425	1.61	1.521
Relative Compaction, %	99.08	99.7	98.6	98.65	97.33	96.7	96.7	97.5	96.9

3.4 Results and analysis of resilient modulus testing

The relationships between the resilient modulus and the applied deviator stress under different confining pressures for base#1, SG#2 and SG#5 as an example are illustrated in (Figures 3, 4, and 5), respectively. It can be concluded from these figures that both of deviator and confining stresses have significant effect on resilient modulus values for both UGM and subgrade soils. As both of deviator and confining stresses increase, the resilient modulus was increased for UGM, while it was decreased for both subgrade soils. However, the impact of confining pressure on resilient moduli was less pronounced for clayey soils (SG#5). This finding agreed with previous studies [39, 28]. The variation in behavior between UGM and subgrade soils could be attributed to the sample physical properties such as percentage of passing sieve no. #200, fine and moisture contents as given previously in Tables 1 and 2.

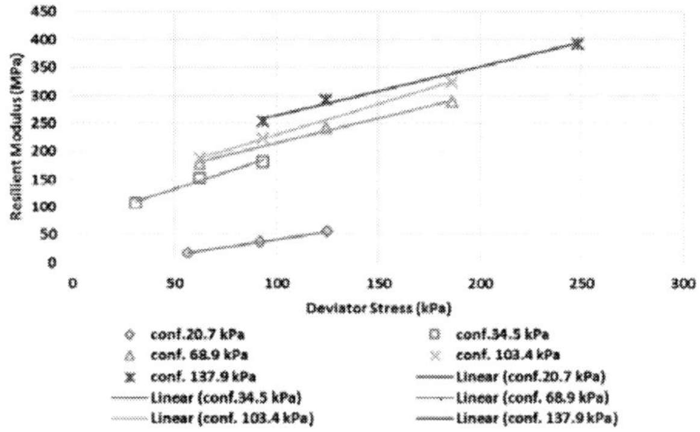

Fig. 3. Effect of deviator stress and confining pressure on resilient modulus for base#1.

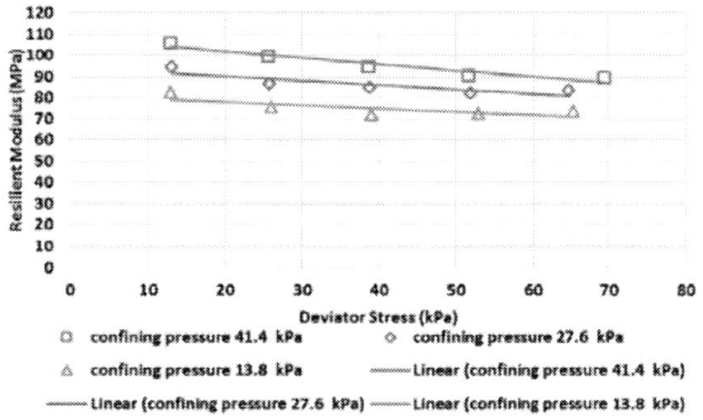

Fig. 4. Effect of deviator stress, and confining pressure on resilient modulus for SG#2.

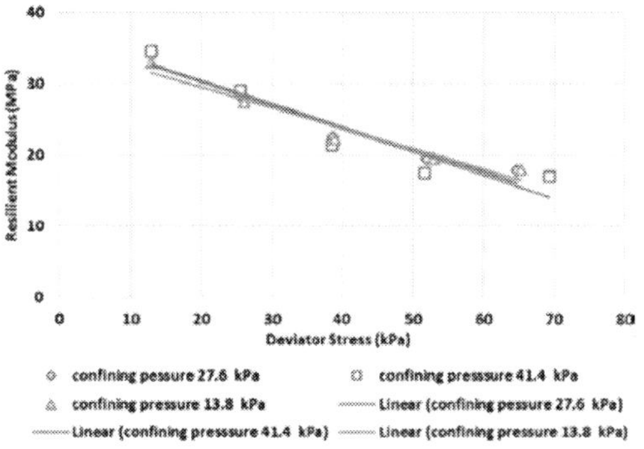

Fig. 5. Effect of deviator stress, and confining pressure on resilient modulus for SG#5.

4 Resilient modulus modelling

Some of the well-known published models in the literature was applied to the RLTT results i.e., K-θ, universal, and Rahim& George models to predict resilient modulus for UGM and subgrade soils. The K-Θ, universal, and Rahim &George stress models, which presented in Equations (2, 5, 6 and 7), respectively were used to predict resilient modulus for each material. Table 4 presents the regression constants of Equations 2, 5, 6, and 7 for each material. Resilient modulus prediction accuracy in terms of coefficient of determination, R^2

for UGMs was excellent for the four equations, however for subgrade soils prediction accuracy was ranged from fair to excellent.

Table 4. Regression coefficients and prediction accuracy of Equations 2, 5, 6, and 7.

Soil type	Resilient modulus predictive models									
	K-θ model			Universal model				Rahim& George stress models		
	K_1	K_2	R^2	K_1	K_2	K_3	R^2	K_1	K_2	R^2
Base#1	3.91	0.69	0.98	0.91	0.467	0.797	0.99	3.242	0.69	0.99
Base#2	17.54	0.489	0.98	1.779	0.404	-0.211	0.86	7.51	0.60	0.98
Base#3	33.66	0.385	0.99	1.878	0.237	0.565	0.95	20.367	0.451	0.99
Subbase	3.64	0.67	0.98	0.965	0.737	-1.85	0.99	3.64	0.67	0.98
SG#1-	7.91	0.34	0.98	0.57	0.152	-0.59	0.81	0.585	-0.13	0.63
SG#2-	32.56	0.20	0.99	1.02	0.321	-1.36	0.99	1.09	-0.26	0.78
SG#3-	110.1	-0.09	0.97	0.511	-0.297	2.195	0.954	0.5142	0..364	0.78
SG#4-	37.02	0.12	0.98	0.05	-0.06	1.365	0.87	0.55	0.22	0.83
SG#5-	8.75	0.21	0.98	0.39	-0.06	-3.16	0.87	0.38	-0.53	0.98

It can be seen from the Table that there is no clear trend was noted for the values of the regression coefficients.

Equations (10 and 11) for coarse and fine grained materials, respectively were also utilized to predict the resilient modulus based on soil properties such as LL, PI, P#200, Cc and Cu.

Figures 6 and 7 present the prediction of the resilient modulus for the four UGMs and five subgrade soils, respectively using Equations 10 and 11 along with the values of the regression coefficients. It can be observed from Figures 6 and 7 that the prediction accuracy in terms of R^2 was fair and poor for UGMs and subgrade soils, respectively.

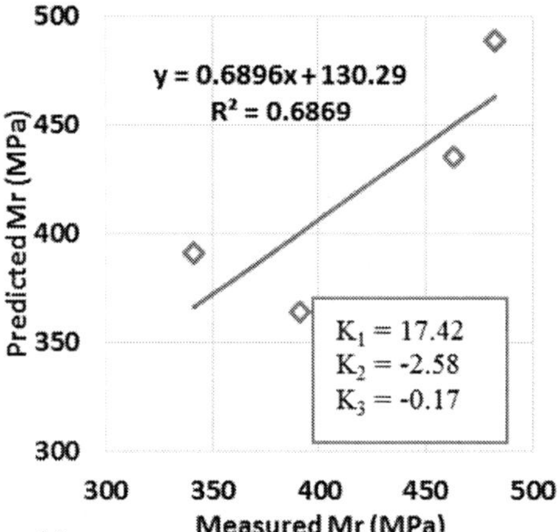

(a)

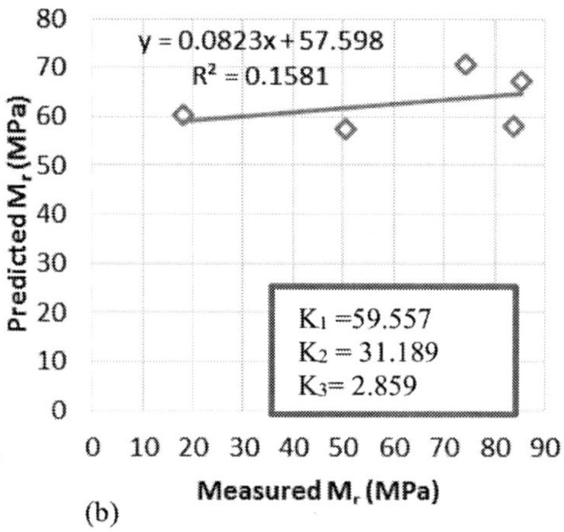

(b)

Fig. 6 (a) M_r Prediction for UGMs using Eq. 10 (b) M_r Prediction for subgrade soils using Eq.11.

Moreover, Equation (16) was used to predict the resilient modulus values for fine-grained soils. The values of regression coefficients as well as the values of R^2 of Equation (16) are summarized in Table 5.

It can be seen from the Table that Equation (16) yielded excellent fit for all investigated subgrade soils.

Table 5. Values of regression coefficients of Equation 16.

Soil types	K_1	K_2	a	b	R^2
SG#1	53.89	0.06	3.49	1.026	0.99
SG#2	89.456	-0.140	4.344	0.99	0.99
SG#3	66.62	0.2156	3.702	1.0354	0.94
SG#4	64.09	0.101	4.64	1.01	0.99
SG#5	25.61	0.27	2.802	0.959	0.97

Finally, Equation (13) was used in order to predict resilient modulus for all materials including UGMs and subgrade soils and obtain one set of materials constants. Figure (7) shows the relationship between measured and predicted resilient moduli for all materials including UGMs and subgrade soils using Equation (13). It is clear from the figure that Equation (13) gave excellent fit with R^2 of 0.99 using the same values of regression coefficients presented previously in Equation 13. This means that Equation 13 is a general model and can be used to predict resilient modulus for UGMs and subgrade materials samples collected in this study.

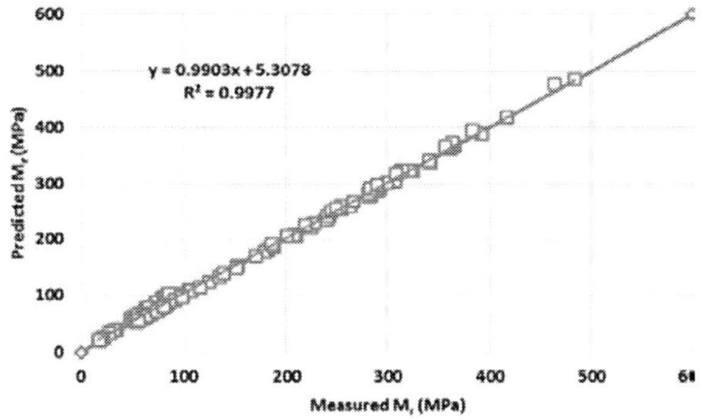

Fig. 7. Resilient modulus prediction for UGMs and subgrade soils using Equation 13

5 Summary and conclusions

Based on the literature review presented in this paper and the engineering properties of the investigated materials, the following conclusions can be drawn:

- There are many parameters that influence the behavior of UGMs and fine-grained soils under repeated loads. These factors are stress level, density, gradation, fines content, grain size, aggregate type, particle shape, and moisture content or matric suction. Researchers seem to agree that the resilient modulus is mostly influenced by the level of applied stresses and the amount of moisture content in the material.
- RLTT was conducted on three base materials, one subbase material, and five types of subgrade soils in order to determine resilient modulus values. Resilient modulus increased significantly with the increase in both deviator and confining stresses for UGMs, while it decreased for subgrade soils. Minimal influence of confining pressure on resilient modulus was observed for clayey soils.
- Number of well-known models was used to obtain the regression coefficients for each investigated material. Equation (13) was used to predict resilient modulus for all materials including UGMs and subgrade soils and get single set of regression coefficients. Models that used for predicting resilient modulus for each material fitted reasonably resilient modulus data with

accuracy in terms of R^2 varied from fair to excellent. Li and Selig model provided very excellent prediction accuracy with R^2 of 0.99 proving that it can be used as a general model for predicting resilient modulus for the nine investigated materials using Li and Selig regression coefficients.

References

1. E. J. Yoder and M. W. Witczak, John Wiley & Sons, (1975).
2. J. Uzan, J., 1985. Transportation Research Record, *Transp. Res. Rec.*, no. 1022, (1985)
3. M. R. Thompson and R. P. Elliott, *Transp. Res. Rec.*, no. 1043, (1985).
4. M. Kim, A. M., E. Tutumluer, M., and J. Kwon, *Resilient modulus of Flexible Pavements,* October, pp. 195–208, (2009)
5. A. Adu-Osei, *Characterization of unbound granular layers in flexible pavements*, (2001)
6. F. Lekarp, U. Isacsson, and A. Dawson, *J. Transp. Eng.*, vol. 126, no. **1**, pp. 66–75, (2000)
7. a M. Azam, D. a Cameron, and M. M. Rahman, *Can. Geotech. J.*, vol. 50, no. November 2012, pp. 1143–1158, (2013)
8. L. Raad, G. H. Minassian, and S. Gartin, no. 1369. (1992).
9. P. Tian, M. Zaman, and J. Laguros, *Transp. Res. Rec. J. Transp. Res. Board*, no. 1619, pp. 75–84, (1998)
10. R. G. Hicks, PhD thesis, University of California, Berkeley, Calif, (1970).
11. L. Uthus, Å. Hermansson, I. Horvli, and I. Hoff, *Curr. Pract. Cold Reg. Eng.*, pp. 1–13, (2006)
12. G. T. H. Sweere, TU Delft, Delft University of Technology, (1990).
13. a. M. Rahim and K. P. George, *Int. J. Pavement Eng.*, **vol. 6**, no. 2, pp. 89–96, (2004)
14. A. R. Gabr, K. G. Mills, and D. A. Cameron, *Geotech. Geol. Eng.*, **vol. 31**, no. 1, pp. 119–132, (2013)
15. P. Kolisoja, Tampere University of Technology, (1998)
16. A. Cabrera, "Evaluation of the laboratory resilient modulus test using a New Mexico subgrade soil," (2012)
17. a M. Azam, D. a Cameron, a G. Gabr, and M. M. Rahman, Matric Suction in Recycled Unbound Granular Materials, pp. 1367–1376, (2014)
18. R. G. Hicks and C. L. Monismith, *Highw. Res. Rec.*, no. 345, (1971)
19. N. H. Thom and S. F. Brown, *Effect of moisture on the structural performance of a crushed-limestone road base*, no. 1121. (1987)
20. A. R. Dawson, N. H. Thom, and J. L. Paute, *Gomes Correia, Balkema, Rotterdam*, pp. 35–44, (1996)
21. J. Ekblad, "Water in coarse granular materials : Resilient and retentive properties," pp. 117–123, (2008)
22. D. Andrei, M. W. Witczak, and W. N. Houston, "Resilient modulus predictive model for unbound pavement materials," in *Contemporary Topics in Ground Modification, Problem Soils, and Geo-Support*, pp. 401–408, (2009)
23. N. Khoury, R. Brooks, and C. Khoury, *Int. J. Geotech. Eng.*, **vol. 3**, no. 2, pp. 303–311, (2009)
24. C. E. Cary and C. E. Zapata, *Road Mater. Pavement Des.*, **vol. 12**, no. 3, pp. 615–638, (2011)
25. R. W. May and M. W. Witczak, no. 810. (1981)
26. I. ARA, *Guide for mechanistic-empirical design of new and rehabilitated pavement*

structures, Part, (2004)
27. W. Heukelom and A. Klomp, *International conference on the structural design of asphalt pavements*, vol. 203, no. 1, (1962)
28. a. M. Rahim, *Int. J. Pavement Eng.*, **vol. 6**, no. 3, pp. 163–169, (2005)
29. K. Yan, H. Xu, and G. Shen, *Int. J. Geomech.*, **vol. 14**, no. 6, p. 4011-4025, (2013)
30. D. Li and E. T. Selig, *J. Geotech. Eng.*, vol. 120, no. 6, pp. 939–957, (1994)
31. N. Perez-Garcia, D. Fredlund, P. Garnica-Anguas, and N. Mestas-Martinez, *Infraestruct. Vial*, vol. 17, no. 30, pp. 23–30, (2015)
32. American Society for Testing and Materials (ASTM), *Annu. B. ASTM Stand. Des. C 136*, (2001)
33. American Association Of State Highway And Transprtation Officials AASHTO, *Stand. Specif. Transp. Mater. Methods Sampl. Testing, M 145-91*, (2009)
34. American Society for Testing and Materials (ASTM), *Annu. B. ASTM Stand. Des. C 127*, (2004)
35. American Society for Testing and Materials (ASTM), *Annu. B. ASTM Stand. Des. D4318-93*, (1995)
36. American Society for Testing and Materials (ASTM), *Annu. B. ASTM Stand. Des. D 1557*, (2012)
37. American Society for Testing and Materials (ASTM), *Annu. B. ASTM Stand. Des. D 1883*, vol. 4, (2004)
38. American Association Of State Highway and Transprtation Officials AASHTO, *Des. T 307-99*, vol. 99, (2012)
39. A. Yau and H. L. Von Quintus, *Study of LTPP laboratory resilient modulus test data and response characteristics*, (2002).

MATEC Web of Conferences 120, 06010 (2017) DOI: 10.1051/matecconf/201712006010
ASCMCES-17

Effect of soil subgrade modulus on raft foundation behavior

Omer Mughieda[1,*] *Mohamed Sherif* Mehana[2], and *Kenan* Hazirbaba[3]

[1]Dept. of Civil Engineering, Abu Dhabi University, Abu Dhabi, UAE
[2]UNii Engineering Consultancy, 47378, Abu Dhabi, UAE
[3]Dept. of Civil Infrastructure and Environ. Eng, Khalifa University, Abu Dhabi, UAE

Abstract. The present study is carried out to investigate the effect of soil subgrade modulus on bending moment, shear, and deformation characteristics of raft foundation. Subgrade models are an approximate representation for the actual load-displacement behavior of the supporting soil. One of the widely-used methods to model soil subgrade modulus is "Winkler" method where the interaction between soil and foundations is represented by a number of springs. The main flaw of this model is the lack of coupling between springs and representation of the non-linearity of load settlement behavior of soil. In this work an attempt is made to analyze "Winkler "method through a commercial software (SAFE V2014) in terms of the effect of soil subgrade modulus on the behavior of raft foundation.

1 Introduction

The type of foundation used for any structure depends on various parameters such as geometric properties of the structure, super-structure loads, foundation material properties, foundation thickness and supporting soil sub-grade modulus. "Raft" or "Mat" foundation is commonly used in Abu Dhabi, UAE, especially to support low-rise building and light structures. A raft foundation is suitable where large settlements or differential settlements are anticipated. In general, use of a raft foundation proves to be more economical compared to spread footings in case spread footings would cover more than one third of structure footprint area [1].

A clear understanding of the behavior of raft foundation in relation to the supporting subgrade is mandatory to ensure the economy and safety in design, hence the objective of this paper is to examine the effect of supporting soil subgrade modulus on raft design. More specifically, this study focuses on the changes in raft straining actions and deflections due to the change in the value of soil subgrade reaction "*Ks*".

* Corresponding author: omer.mughieda@adu.ac.ae

2 Design of raft foundation

The structural design of a raft foundation can be carried out by two common methods: the conventional rigid method and the non-rigid method (e.g. Winkler method). In addition, finite difference and finite element methods can be used. As non-rigid methods consider the effects of raft deformations on the distribution of bearing pressure, relationship between settlement and bearing pressure needs to be defined. This can be done using the coefficient of subgrade modulus ($Ks=q/\Delta$, where: q= bearing pressure and Δ=settlement).

3 Soil subgrade modulus

In Winkler method, bed springs are used to represent the interaction between soil and raft. The stiffness of springs depends on the value of coefficient of subgrade modulus (*Ks*). The magnitude of *Ks* depends on several factors such as width, shape, position and depth of the foundation. There is no single *Ks* value even if the aforementioned factors are well-defined, because the q-Δ relationship is nonlinear [1].

Various methods, such as field testing as well as empirical relations, have been developed to estimate *Ks*. In this study, *Ks* was determined from Standard Penetration Tests (SPT) performed in four different locations in Abu Dhabi, UAE.

Meyerhof (1956) proposed correlation between net bearing pressure and "SPT" N60. According to Meyerhof theory, for 25mm of estimated settlement [2]:

$$q_{net}\,(\mathrm{KN/m2}) = N60/0.08 \text{ (for } B < 1.22 \text{ m)} \quad (1)$$

$$q_{net}\,(\mathrm{KN/m2}) = N60/0.125\,((B+0.3)/B)2 \text{ (for } B > 1.22 \text{ m)} \quad (2)$$

Where:
Fd = depth factor = $1+0.33\,(Df/B)$
B = Foundation depth in meter.
Se = settlement in mm.

Researchers [2-6] have observed that Meyerhof (1956) results are conservative. Later Meyerhof (1965) suggest that the net bearing pressure should be increased by 50%. Bowles (1977) proposed that the modified form of bearing equations to be expressed as:

$$qnet\,(\mathrm{KN/m2}) = (N60/2.5) * Fd * (Se/25) \text{ (for } B < 1.22 \text{ m)} \quad (3)$$

$$qnet\,(\mathrm{KN/m2}) = N60/0.08\,((B+0.3)/B)2 * Fd * (Se/25) \text{ (for } B > 1.22 \text{ m)} \quad (4)$$

Generally shallow foundations are designed for a maximum settlement of 25 mm and differential settlement of 19 mm. However, the width of rafts is larger than isolated spread footings, hence the depth of zone of influence is much larger than that of spread footing. Hence the loose soil pockets under Raft may be more evenly distributed. Accordingly, maximum settlement can be assumed to be 50 mm and differential settlement to be 19 mm [2].

Using the above-mentioned assumptions and conservatively assuming Fd = 1.0, (Eq.4) can be assumed as following:

$$qnet\,(\mathrm{KN/m2}) = 25 * N60 \quad (5)$$

Using actual N60 values obtained from "SPT" performed in five different locations in Abu Dhabi. Figure 1 shows the studied locations. Using Meyerhof (1965) theory modified by Bowles; soil subgrade modulus were calculated and tabulated in Table 1.

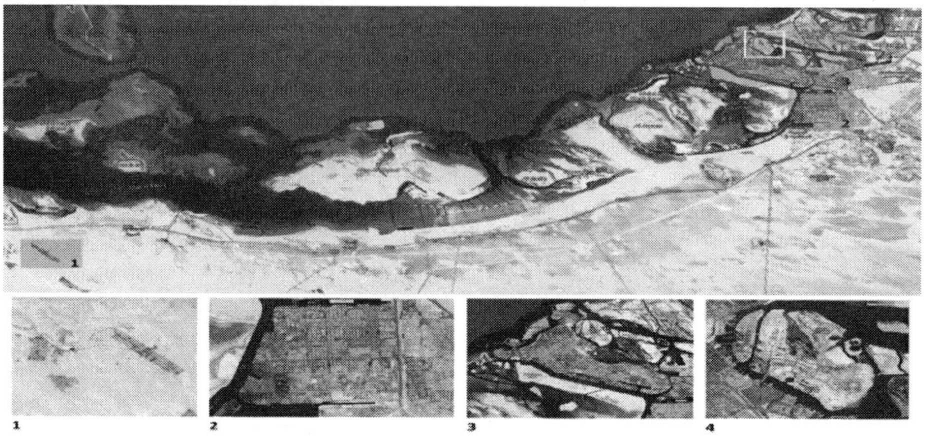

Fig. 1. Areas in Abu Dhabi considered in the study.

Table 1. Subgrade modulus values for different areas in Abu Dhabi.

	Location	*Ks* (kN/m^3)
1	Madinat Zayed-Werstern reigion	25000
2	Mussafah city	21500
3	Abu Dhabi island	10000
4	Reem island	15000

4 Analysis of raft foundation

In this study, the effect of soil subgrade modulus on raft has been investigated using finite elements method. Shell element have been used to define raft foundation shown in Figure 2. Area springs used to represent soil subgrade modulus based on "*Ks*" values illustrate in Figure 1.

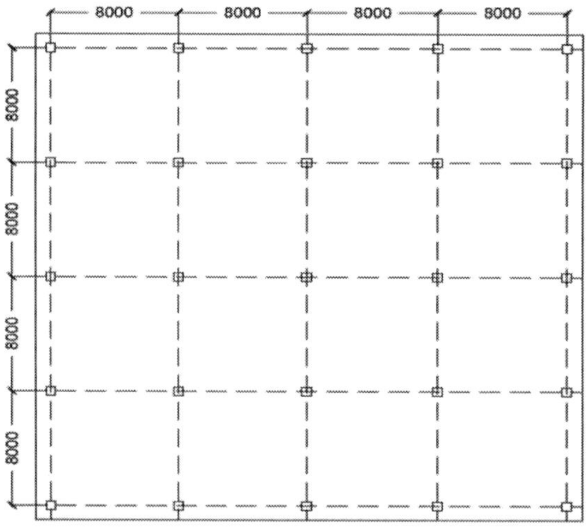

Fig. 2. Plan view of raft foundation for 10 story building

5 Effect of soil subgrade reaction on bending moment and shear forces

Figures 3, 4 & 5 represent bending moment for column strip, bending moment for middle strip and shear force diagram respectively corresponding to "*Ks*" equal to 1.0x104, 1.5x104, 2.15x104 & 2.5x104 kN/m^3. Bending moment and shear force are not very susceptible to the changes in values of soil "*Ks*".

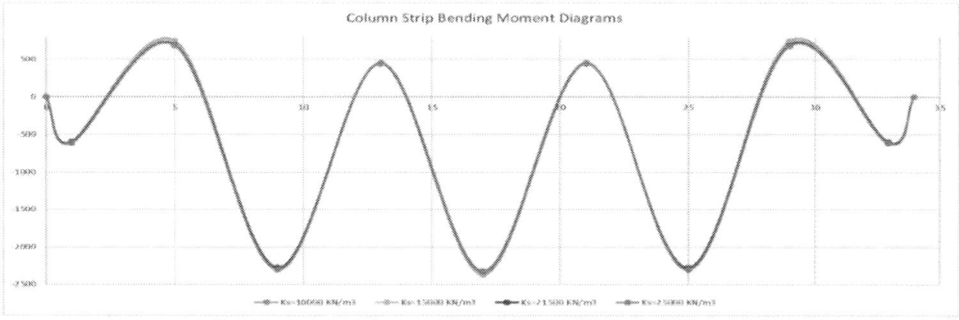

Fig. 3. Column strip bending moment diagram

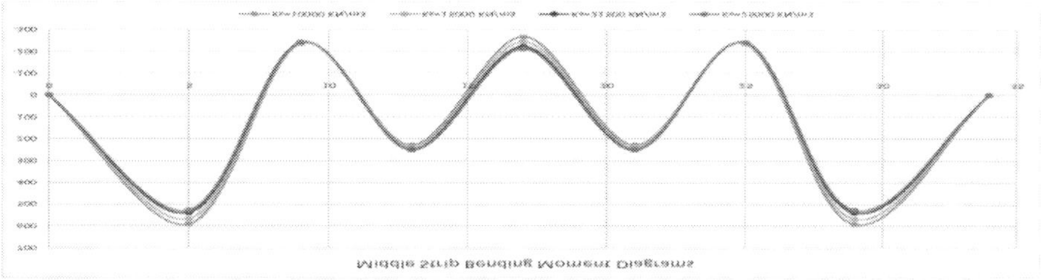

Fig. 4. Middle strip bending moment diagram

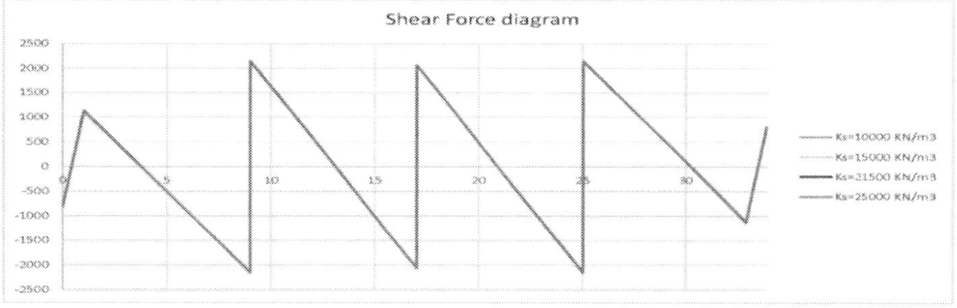

Fig. 5. Column strip Shear force diagram

6 Effect of soil subgrade reaction on raft deformation

Figures 6, 7, 8 & 9 represent Raft deformations corresponding to "*Ks*" equal to 1.0x104, 1.5x104, 2.15x104 & 2.5x104 kN/m^3.

 DOI: 10.1051/matecconf/201712006010

By definition; settlement of Raft is inversely proportional to soil subgrade modulus ($Ks=q/\Delta$). In Figures 6, 7, 8 & 9; settlement contour lines obtained by "SAFE" analysis indicate that settlement values decrease when "Ks" increase conforming the theoretical definition.

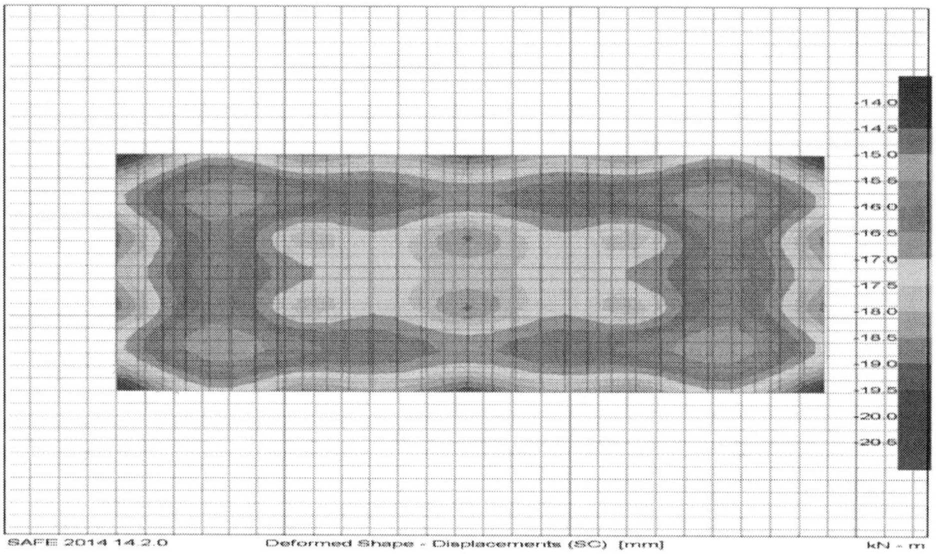

Fig. 6. Raft deformations (Ks=10000 kN/m^3).

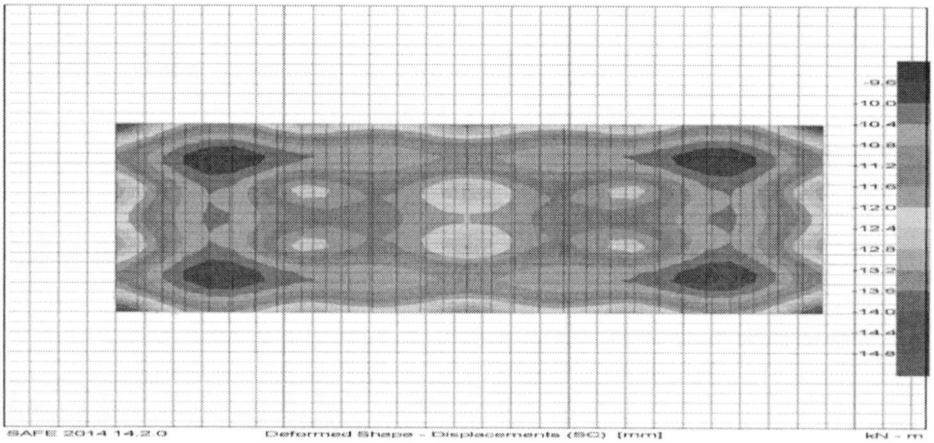

Fig. 7. Raft deformations (Ks=15000 kN/m^3).

DOI: 10.1051/matecconf/201712006010

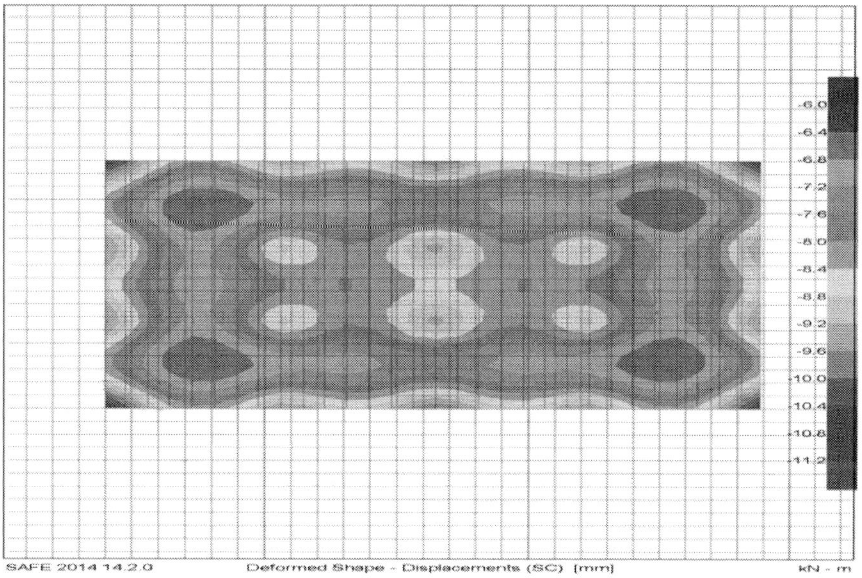

Fig. 8. Raft deformations (*Ks*=21500 kN/m^3).

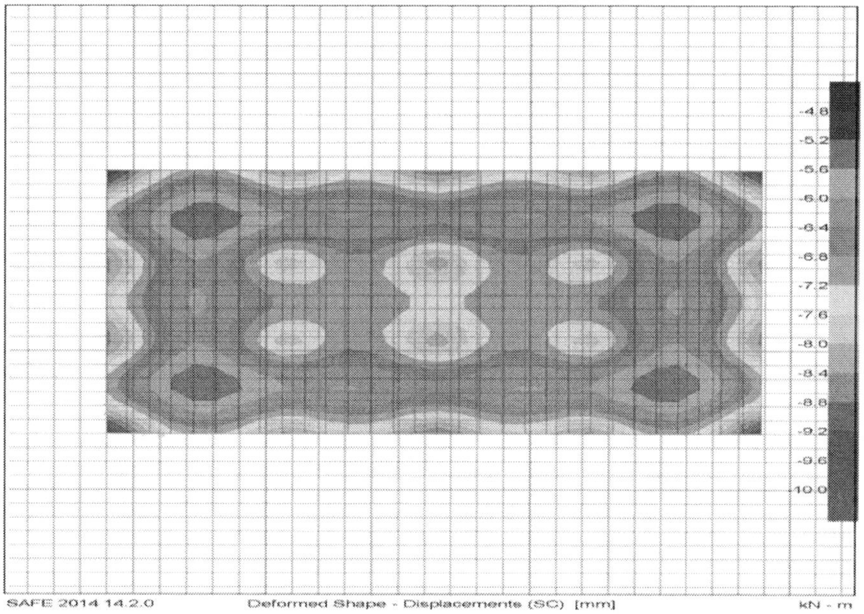

Fig. 9. Raft deformations (*Ks*=25000 kN/m^3).

7 Conclusions

In the present work, parametric study is carried out on the effect of soil subgrade modulus on behavior of Raft foundation. In the light of the preceding discussions the following conclusions can be drawn:

- Changes in the value of soil subgrade reaction have no significant impact on values of Raft straining actions. (Bending moment and shear force)
- Raft deflection decreases with increasing the value of soil subgrade reaction. However, it is recommended that these values should be verified using either empirical methods or field tests.
- Changing the value of soil subgrade modulus has significant effect on pressure distribution on soil below foundation. As soil subgrade modulus increases, bearing pressure is concentrated in springs below columns; while springs in between columns are less susceptible to pressure distribution. Hence, soil behavior tends to "rock" for high values of soil subgrade modulus.

References

1. D. P. Coduto, Foundation design, principles and practices, Prentice Hall, Englewood Cliffs, NJ, (1994).
2. B. Das, Principles of Foundation Engineering, seventh ed., Stamford, (2007).
3. J. E. Bowles, (1974), Analytical and Computer Method in Foundation Engineering, Mc Graw Hill Book Co., (1974)
4. J. E. Bowles, Foundation Analysis and Design, 5th Edition, Mc Graw Hill Book Co., New York, (1977).
5. S. C. Gupta, Raft Foundation Design and Analysis with A Practical Approach, New Age International. (2007).
6. A. S. Morshed, M. Monzurul, A Design Rationale for Mat Foundation Based on Finite Element Analysis. (1996).

Numerical analysis of shallow foundations on geogrid reinforced soil

Mohamed G. Arab[1,2], Maher Omar[1] and Ali Tahmaz[1]

[1]Civil & Environmental Engineering Department, University of Sharjah, 27272 Sharjah, UAE
[2]Structural Engineering Department, Mansoura University, 27272 Sharjah, UAE

Abstract. Geogrids have been increasingly used as soil improvement technique over the past decade. This increasing trend is attributed to its ease of application and low cost compared to other improvement techniques. This paper provides a review of the published results related to the ultimate and allowable bearing capacities of shallow foundations on geo-grid-reinforced homogeneous sand soil. A numerical analysis is also conducted to investigate soil-geogrid interaction considering several critical parameters that affect the overall behavior of geogrid-reinforced soil. The critical parameters such as the width and length of geogrid layers, spacing between the reinforcement layers, and the depth of reinforcement and geosynthetic stiffness have been investigated.

1 Introduction

There are several geogrid-reinforcement techniques that gained wide popularity in the recent years for its proven performance, ease of installation and low cost compared with other improvement techniques. Mechanically stabilized earth technology for example is now well established in the heavy construction industry as a reliable method in the construction of structures such as retaining walls, embankments over soft soil, and steep slopes.

The first systematic study using mechanically stabilized earth to improve the bearing capacity of shallow foundations was reported by Binquet and Lee [1, 2]. In their study, laboratory model tests were used to develop bearing capacity relationships for a strip foundation supported by sand reinforced with metallic strips (aluminum foil).

Since the original work by Binquet and Lee, several laboratory model test studies have been published relating to the improvement of the load-bearing capacity of shallow foundations supported by sand reinforced with various materials such as geogrids [3, 4], geotextiles [5], rope fibers [6], metal strips and metal bars [7-12].

Over the last decade, the use of geogrids for soil reinforcement has increased greatly, primarily because geogrids are dimensionally stable and combine features such as high tensile modulus (low strain at high load), open grid structure, positive shear connection characteristics, light weight, and long service life.

This paper summarizes the important results of the existing studies on the load-bearing capacity of shallow foundations on geogrid-reinforced soil. Most of the results presently available have been obtained primarily from small-scale laboratory model tests. Moreover, a numerical study is conducted and results are presented in this paper. The numerical study shows some fundamental aspects of the behaviour of the geogrid reinforced sand. Different parameters that affect the behaviour of the geogrid reinforced sand is discussed.

2 Failure mechanisms in geogrid reinforced soil at ultimate load

Several researchers have studied the different geometrical parameters effect on the behaviour of the reinforced soils in a non-dimensional form (Figure 1a). These parameters are the following: a) location of the top layer of reinforcement measured from the bottom of the foundation, u; b) the depth of reinforcement, d

$$d = u + (N - 1)h \tag{1}$$

where N is the number of reinforcement layers and h is the vertical distance between two consecutive reinforcement layers;
c) width of each reinforcement layer, b; and d) length of each reinforcement layer, l.

The enhancement in bearing capacity due to soil reinforcement is generally expressed in terms of a non-dimensional parameter called bearing capacity ratio, BCR. The bearing capacity ratio can be expressed either in the form of ultimate bearing capacity or allowable bearing capacity at a given settlement level of the foundation [13]. Figure 1b shows typical load-settlement curves of a foundation over soil with and with-out reinforcement. Based on this concept, the bearing capacity ratio can be defined as,

$$\mathrm{BCR}_u = \frac{q_{u(R)}}{q_u} \quad \text{and,} \quad \mathrm{BCR}_s = \frac{q_R}{q} \tag{2}$$

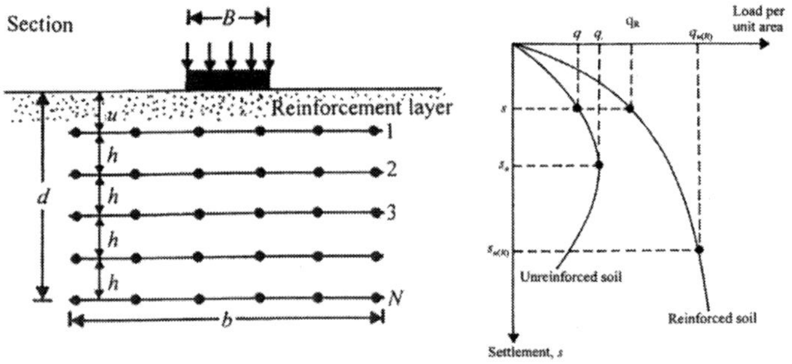

Fig. 1. (a) Geometric parameters of foundation supported by reinforced soil; (b) typical load-settlement curves for unreinforced and reinforced soil supporting a foundation.

Binquet and Lee [1, 2] studied the bearing capacity of shallow foundations resting on reinforced earth. Based on this study, Binquet and Lee proposed failure mechanisms for geogrid-reinforced soils depending on the depth of the reinforcement and number of reinforcement layers as shown in Figs. 2a, 2b, and 2c.

The nature of failure in the soil mass shown in Fig.2a generally occurs when the first layer of reinforcement is located at a depth, u, greater than about 0.67B (B = width of the foundation). The reinforcement may act in this arrangement as a rough rigid base located at a limited depth.

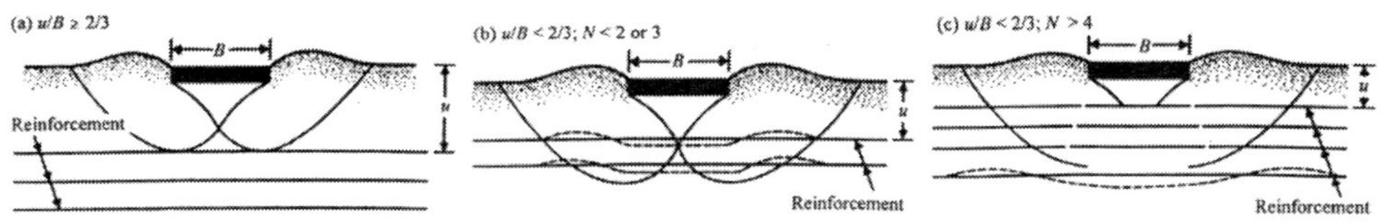

Fig. 2. Three modes of bearing capacity failure of a strip foundation on reinforced earth (after Binquet and Lee, 1975b)

The type of failure shown in Fig. 2b could occur if u/B is less than about 0.67 and the number of layers of reinforcement, N, is less than about 2–3. In this type of failure, reinforcement pull-out occurs. According to this study, the most beneficial effect of reinforced earth is obtained when u/B is less than about 0.67 and the number of reinforcement layers is greater than 4 but not more than 6–7 (Fig. 2c).

Based on the theory advanced by Binquet and Lee [2], Fig. 3 shows an idealized condition for the development of the failure surface in soil for the condition shown in Fig. 2c. It consists of a central zone— Zone I—immediately below the foundation that settles along with the foundation with the application of load. On each side of Zone I, the soil is pushed outward and upward—this is Zone II. The points A′, A″, A‴, . . . and B′, B″, B‴, . . define the limiting lines between Zones I and II, can be obtained by considering the shear stress distribution, τ_{xz}, in the soil caused by the foundation load. The term τ_{xz} refers to the shear stress developed at a depth d below the foundation at a distance x measured from the centre line of the foundation. The nature of variation of τ_{xz} at any given depth z is shown by the broken lines in Fig. 3. Points A′ and B′ refer to the points at which the value of τ_{xz} is maximum at $z = z_1$. Similarly, A″ and B″ refer to the points at which the value of τ_{xz} is maximum at $z = z_2$. The distance at which the maximum value of τ_{xz} occurs can be given as $x = X_o$.

It can be expected that failure in soil reinforced with geo-textile or geogrid layers will be, in general, similar to that described by Binquet and Lee [2] for the case of metal strip reinforcement. However, the magnitude of the ultimate bearing capacity and the allowable bearing capacity for a given level of settlement (and thus BCR_u and BCR_s) will be a function of several factors, such as,

a. The degree of compaction of soil;
b. Relative stiffness of the geogrid or geotextile; and
c. u/B, b/B, l/B, h/B, and d/B.

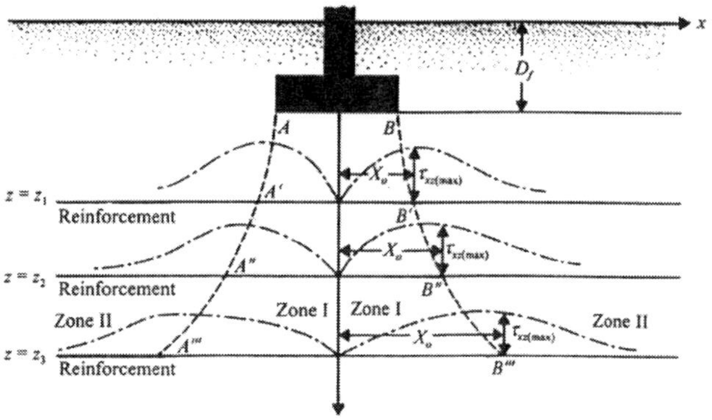

Fig. 3. Failure mechanism under a strip foundation supported by reinforced earth based (after Binquet and Lee (1975b))

3 bearing capacity on geogrid-reinforced soil

3.1 Geogrid general

Geogrids are made with high-modulus polymer materials, such as polypropylene and polyethylene through tensile drawing. Geogrids are high tensile stiffness netlike material with large openings called apertures. These apertures enable interlocking with the surrounding soil to reinforce soils. Geogrids are generally of two types: (i) biaxial and (ii) uniaxial. The commercial geogrids currently available for soil reinforcement have a nominal rib thickness of about 0.5–1.5 mm and junctions of about 2.5–5 mm. The grids used for soil reinforcement usually have apertures that are rectangular or elliptical in shape. The dimensions of the apertures vary from about 25–150 mm. Geogrids are manufactured so that the open areas of the grids are greater than 50% of the total area. Due to their high tensile stiffness, these geogrids develop reinforcing strength at strain levels as low as 2%.

3.1 Laboratory experiments on geogrid reinforced soil

Omar et al. 1993 conducted a series of laboratory scale tests to estimate the bearing capacity for strip footing. Omar et al. 1993 tested an aluminum strip with 76.2 mm width on geogrid reinforced soils. The tests were conducted in box measuring 1100 mm length × 304 mm width × 914 mm depth. Rounded silica sand used in the test has the following properties: dry unit weight of 17.14 kN/m^3, 70% relative density and an average peak friction angle of 41°. Omar conducted parametric study varying the following parameters, number of geogrid layers *N*, the depth of first geogrid layer *u*, the width of the geogrid layer *b*, and the vertical distance between two consecutive reinforcement layers, *h*.

3.1.1 Variation of number of geogrid layers, N

Figure 4 shows results of load per unit area versus settlement of surface model strip foundations (D_f = 0) on geogrid-reinforced sand with the following geometrical features u/B = h/B = 0.33; b/B = 10. The effect of increasing number of geogrid layers on the bearing capacity of the geogrid reinforced soils is clearly shown in Figure 4.

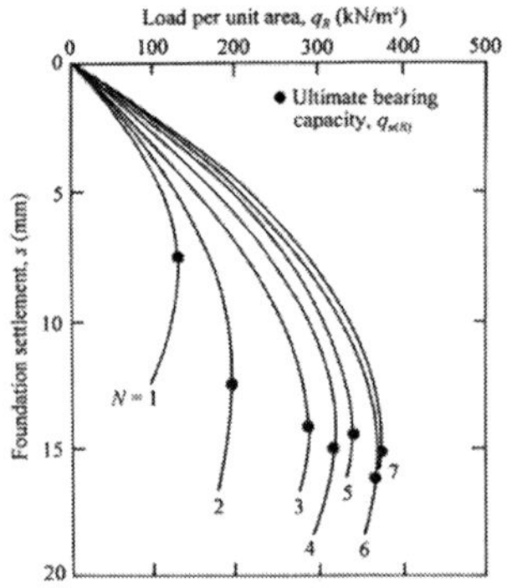

Fig. 4. Load-settlement curves result for a surface strip foundation on geogrid-reinforced sand (after Omar et al., [14]).

3.1.2 Variation of u/B

Figure 5 shows typical relationship of normalized depth of the first geogrid layer (*u/b*) based on laboratory test results of Guido et al. [5], Akinmusuru and Akinbolande [6], Omar et al. [15] and Yetimoglu et al. [16] for bearing capacity tests conducted on surface foundations supported by multi-layered reinforced sand. It appears from these tests results that $(u/B)_{max} \approx 0.9$ to 1. It is also interesting to note that the magnitude of $(u/B)_{max}$ was recommended by Binquet and Lee [1, 2] to be about 0.67. Also, laboratory test results suggest that $(u/B)_{cr}$ as defined by Fig. 5 is about 0.25 to 0.5. An analysis of the test results of Schlosser et al. [17] yields a value of $(u/B)_{cr} \approx 0.4$. Large-scale model tests of Adams and Collin [18] showed that $(u/B)_{cr}$ is approximately 0.25. Miyazaki and Hirokawa [19] also observed a similar value of (u/B)cr for the case of N = 1 (that is, single layer of geogrid reinforcement).

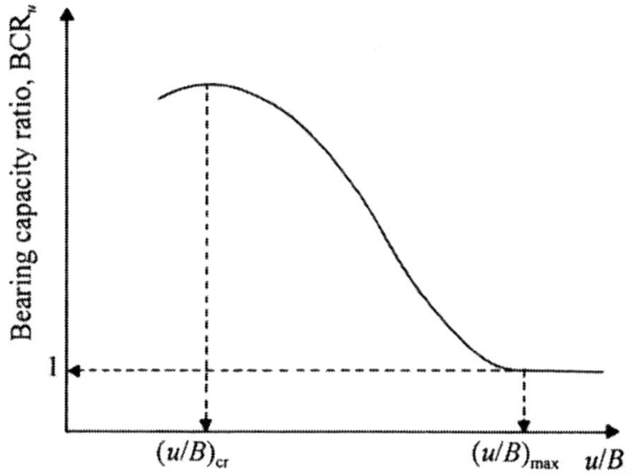

Fig. 5. Typical variation of BCR_u with u/B for given values of b/B, h/B, and d/B.

3.1.3 Variation of b/B and l/B

Figure 6 shows a relationship between the normalized width of the geogrid reinforcement versus the bearing capacity ratio BCR_u. The figure suggests that the bearing capacity increases with geogrid width increase until a critical value Omar et al. [20] suggested the following empirical relationships for nondimensional parameters of the geo-grid-reinforced soil based on results of laboratory model shown earlier,

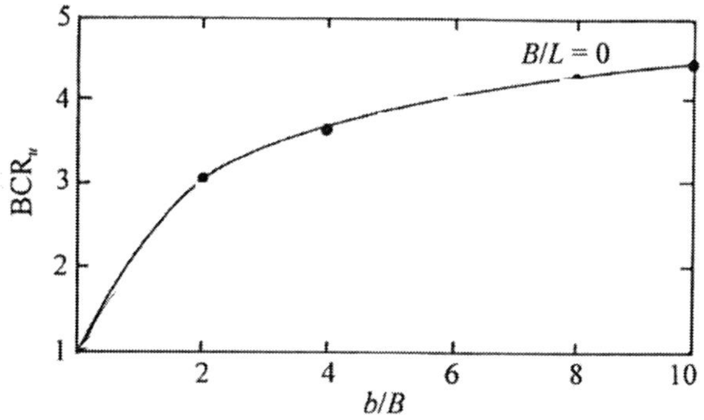

Fig. 6. Variation of BCR_u with b/B for strip footing—u/B = h/B = 0.33 (after Omar et al., 1993c).

3.1.3 Relative Stiffness of Geogrids

It was noted earlier in this paper that the relative stiffness of the geogrid reinforcement may contribute to the magnitude of BCR_u. This will be more noticeable in the case of small-scale laboratory model tests. Guido et al. [5] tested two types of geogrids with different tensile stiffness. The results of bearing capacity ratio BCR_u versus the normalized width of the geogrid is shown in Figure 7. The results shows a limited increase in bearing capacity with increasing the tensile stiffness of the geogrid. The tensile strength of the geogrids used for Guido et al. [5] tests shown in Fig. 7 were as follows:

	Tensile strength (kN/m)	
Geogrid	Machine direction	Cross-machine direction
SS1	16.8	24.8
SS2	23.0	26.8

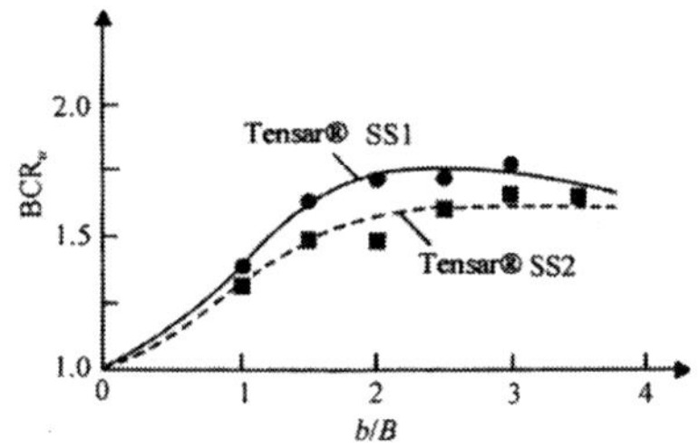

Fig. 7. Variation of BCR_u with b/B for two geogrids—square foundation; u/B = 0.5; h/B = 0.25; B = 0.305 m; N = 3 (after Guido et al., 1987)

3.1.3 Scale effects

Das and Omar [21] conducted several laboratory model tests on surface strip foundations supported by geogrid-reinforced sand in which the foundation width B was varied. For all tests u/B = h/B = 0.33, b/B = 8, and d/B = 2. Figure 8 shows the variation of BCR_u with B and relative density of sand, D_r, obtained from the tests. For a given D_r, the magnitude of BCR_u de-creases with the increase in B and reaches an approximate minimum value at B ≈ 150 mm. Figure 8 shows the plots of BCR_u versus D_r for various values of B. From this figure it appears that B should be at least 150–180 mm to obtain BCR_u from laboratory model tests.

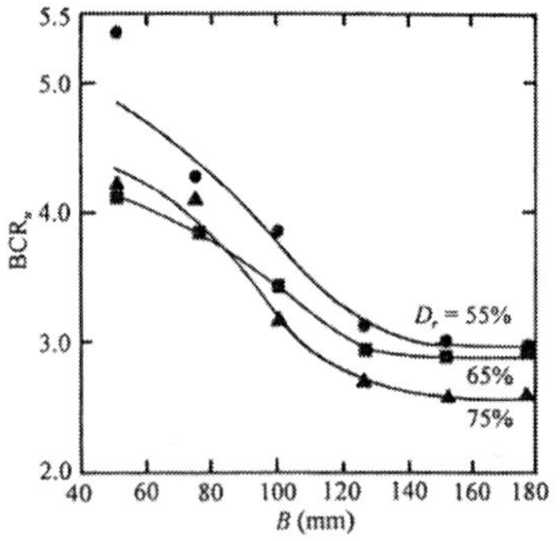

Fig. 8. Variation of BCR_u with B—strip foundation; Tensar® BX1000 geogrid; u/B = h/B = 0.33; b/B = 8; d/B = 2 (after Das and Omar, 1994).

3.1.4 Bearing capacity for limited settlement

Shallow foundations are often designed for limited settlement levels and, hence, the magnitude of BCRu become meaningless for B > about 1 m. Therefore, it is necessary to determine the bearing capacity ratio at various levels of settlement. The non-dimensional quantity BCR_s defined by Eq. 2 will be used to define the bearing capacity at specific settlement (s). Omar et al. [22] and Khing et al. [13] conducted limited model studies to determine the relationship between BCR_u and BCR_s for strip and square model foundations supported by geogrid-reinforced sand. Figure 9 shows the variation of BCR_u and BCR_s with d/B (for s/s_u = 0.25, 0.5, and 0.75).

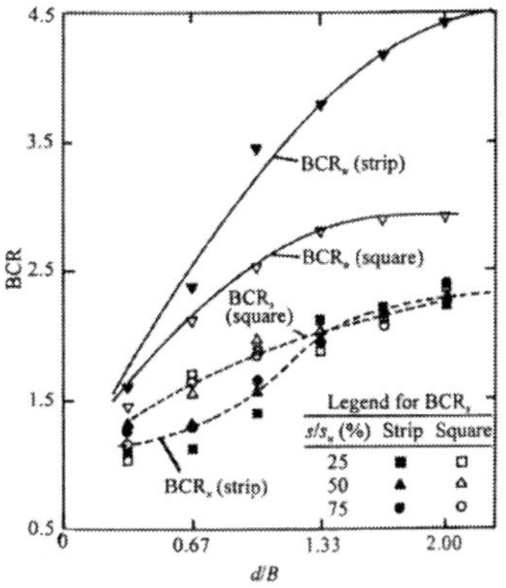

Fig. 9. Variation of BCR_u and BCR_s with d/B for strip and square foundations on sand (after Omar et al., [22])

It is important to note that s_u is the settlement of the foundation at ultimate load when supported by unreinforced sand (also see Fig.1). Based on the plots shown in Fig. 9, the following general observations can be made:

a. For a given foundation on geogrid reinforced soil the magnitude of BCR_u is higher than that of BCR_s at s/s_u less than or equal to 0.75.
b. The ratio between the ultimate bearing capacity ratio (BCR_u) and bearing capacity ratio at specific settlement (BCR_s) is about 0.56 for strip foundation.
c. The BCR_s magnitude decreases with decrease of the value of the settlement (s).

4 Numerical study

A series of two-dimensional finite element analyses (FEA) of footing resting on geogrid reinforced soil is performed in order to understand the behavior trends of foundations on geogrid reinforced soils. The analysis is performed using the finite element program Plaxis software package (version 2016, 2016).

The geometry of the numerical model was assumed to be 10 times the laboratory model in Omar et al., [22] (the footing width B = 0.76m and thickness, t = 0.2 m, soil height (sand layer and soft clay layer) is assumed to be 10B. The software generates the numerical mesh automatically with fifteen node triangle plane strain elements for the soil, and five node beam elements for the footing while five node elastic elements are used for the geogrid elements.

DOI: 10.1051/matecconf/201712006011

4.1 Finite element model

Hardening soil model available in Plaxis was used to simulate the nonlinear behavior of the sand. Hardening soil model is hyperbolic stress–strain model with limiting states similar to that of Mohr-Coulomb by means of the friction angle ϕ, cohesion c and the dilatancy angle ψ. However, soil stiffness is stress dependent by using three different input stiffness: the triaxial stiffness E_{50}, the triaxial unloading stiffness E_{ur} and the oedometer loading tangent stiffness E_{oed}. The foundation was treated as elastic beam elements based with flexural rigidity (EI) and normal stiffness (EA). The interaction between the geogrid and soil is modeled at both sides by means of interface elements, which allow for the specification of a reduced friction compared to assumed friction of the soil. A refined mesh was adopted to minimize the effect of mesh dependency on the finite element modeling of cases involving changes in the number, length, and the location of geogrid layers. In the finite element modeling, the initial stress condition of the slope was established first by applying the gravity force due to soil. A prescribed footing load (beyond the footing ultimate stress) is applied in increments accompanied by iterative analysis up to failure (the iterations stop when failure is reached). The modeled boundary conditions were assumed such that the vertical boundaries are free vertically and constrained horizontally while the bottom horizontal boundary is fully fixed. The general model for the footing on reinforced soil mass, generated mesh, and the boundary conditions are shown in Fig. 10. The model shows the general dimensions of the numerical model with b, d, u, and h is as defined in Fig. 1. The width of the strip footing is assumed to be 0.76 m while the dimensions and number of layers for the geogrid elements are changed to examine the effect of the geogrid configurations on the footing behavior. The hardening soil constitutive parameters are assumed for a dense soil similar to the laboratory testing conducted by Omar et al. [22], As mentioned earlier, Omar et al. used sand with following properties: dry unit weight of 17.14 kN/m^3, 70% relative density and an average peak friction angle of 41°. The dilating angle ψ was assumed to be 10° and the rest of the soil parameters are shown in Table 1.

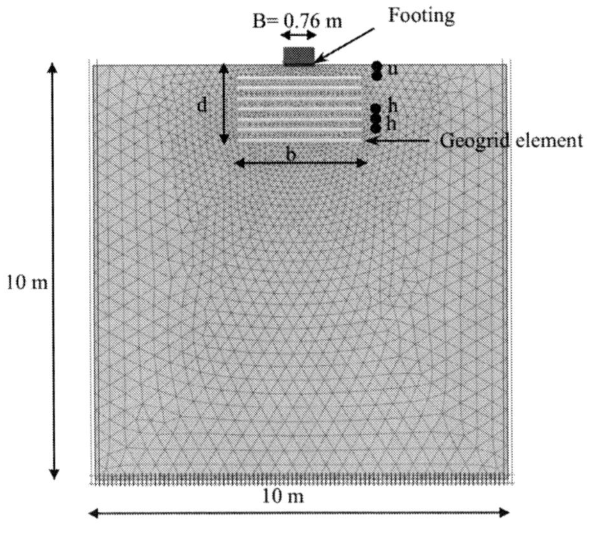

Fig. 10. Prototype of the footing model, generated mesh and boundary conditions.

Table 1. Finite element model parameters for different parts of the numerical model.

Parameter	Sand	Footing	Geogrid
Tri-axial stiffness E_{50} (kN/m^2)	40 000	NA	NA
Cohesion (c) (kN/m^2)	0.00	NA	NA
Friction angle (φ) (degree)			
Dilatancy angle (ψ) (degree)	10.00	NA	NA
Soil Unit Weight (γ) (kN/m^2)	17.14	NA	NA
Poisson's ratio (υ)	0.3	NA	NA
Interface reduction factor (R_{int})	0.67	NA	NA
EA (kN/m)	NA	5 000 000	2000
EI (kN m^2/m)	NA	8500	Deform with soil mesh

4.1 Results and discussion

All the parameters of a control case is tabulated in Table 1. A drained condition is used for all the numerical models. A parametric study was conducted to investigate the effect of different parameters on the bearing capacity of geogrid reinforced soil. For most numerical models, only one parameter is changed and plotted in the x-axis.

4.2.1 Bearing capacity behavior of unreinforced sand

A numerical model similar to the numerical model introduced in Fig 10 with no geogrid elements was developed as a control model. The numerical model mesh was generated automatically, though Plaxis automatic mesh scheme, with the same mesh size used for reinforced sand models. A hardening constitutive model with the same properties introduced in Table 1 was used for the control numerical model. Conventional Mohr-Coulmb linear plastic model with linear stiffness similar to hardening model of 40 000 kN/m^2 was used to compare results with the hardening nonlinear model. The footing was loaded until failure was reached. The footing settlement is expressed in non-dimensional form as ratio of footing width B (S/B, %). Figure 11 shows a comparison of bearing stress q vs S/B curve compared to model with same properties only changing the constitutive model from hardening model to Mohr-Columb model. The friction angle of the model was changed to 30 degrees with zero dilatancy with same hardening model. The results shows a much lower bearing capacity with lower stiffness compared to control model. The Mohr-Columb model shows a much stiffer response with a higher bearing capacity compared to the control model (hardening non-linear model).

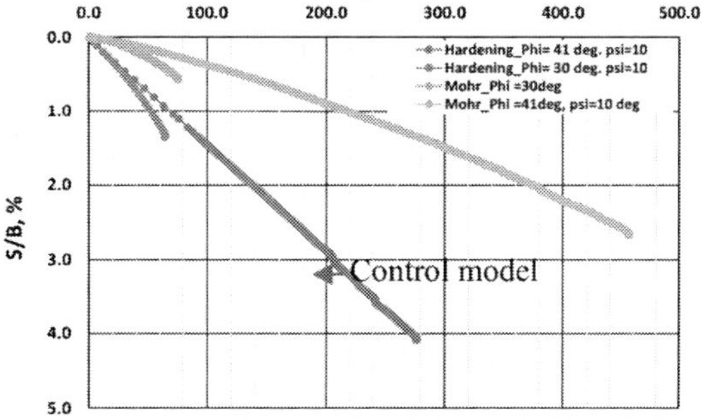

Fig. 11. Variation of footing stress q with S/B percentage for the model footing with different soil constitutive model.

4.2.1 Effect of number of geogrid reinforcement layers, N

Two series of numerical studies were conducted in order to study the effect of varying the number of geogrid layers on the bearing capacity of strip footing on geogrid-reinforced soil. In the first study, the control model properties and geometry was used changing only the number of the geogrid with u/B = h/B = 0.5; b/B = 4. Figure 12 shows results of load per unit area versus settlement expressed as S/B ratio of surface numerical model strip foundations (D_f = 0) on geogrid-reinforced sand with the following geometrical features u/B = h/B = 0.5; b/B = 4. The effect of increasing number of geogrid layers on the bearing capacity of the geogrid reinforced soils is clearly shown in Figure 12. The increase in the number of geogrid layers results in an increase in the bearing capacity and slight increase in the overall stiffness of the reinforced-sand. Figure 13 shows that the ultimate bearing capacity ratio (BCR_u) for the soil increases with the number of grid layers, N. The BCR_u reaches value of 6 when the number of geogrid layers reaches seven layers. Which means that numerical model predicts a significant increase in ultimate bearing capacity of geogrid reinforced soil with the increase of number of geogrid layers. Figure 13 shows also a relationship of the BCR_s at a settlement ratio (S/B%) about 3.3% corresponding to a settlement of about 2.5 cm versus the increase in the number of the geogrid layers. The figure shows that the value of the BCRs reached a value of approximately 1.5 for seven layers of geogrid reinforcement compared to a value of 6 for BCRu for the same number of geogrid layers. This value is much less than values reported in the literature for BCRs. Omar et al., [22] reported values of BCRs up to about 2.5 for S/Su=25%. Moreover, the increase in the number of geogrid layers does not have a significant effect on the BCRs. This is attributed to the fact that the calculated load per unit area versus settlement for reinforced soils did not show a significant increase in stiffness compared to unreinforced soils especially at low settlement. The tangent modulus of the reinforced soil did not significantly improve with the soil reinforcement. This is why the calculated BCR_s did not show significant increase with the increase in the number of the reinforcement layers. This does not agree with laboratory results reported by Omar et al., [22]. This is may be attributed to that numerical model did not capture the effect of pre-tention effect in the geogrid layers during compaction.

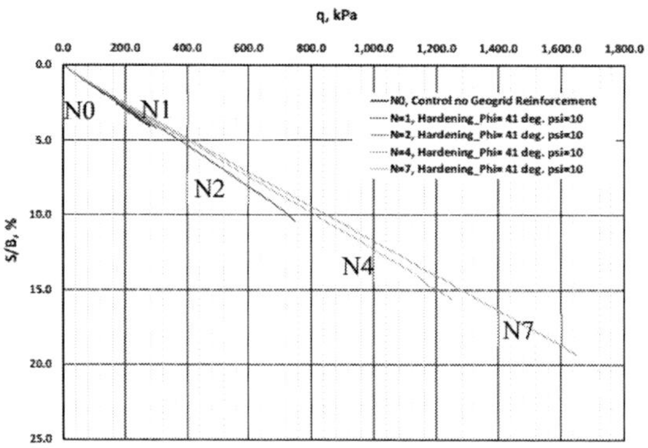

Fig. 12. Load-settlement curves results for a numerical model of strip footing on geogrid-reinforced sand with variation of number of geogrid layers.

4.2.1 Effect of constitutive model on the behavior of geogrid-reinforced soil

The effect of the constitutive model on the calculated behavior of the geogrid-reinforcement soil numerical model was investigated. For this study a parametric study was conducted using control model with the same geometry and properties shown in Fig.10 changing only the constitutive model from hardening model to the conventional Mohr-Coulomb model. The Mohr- Coulomb constitutive model is elastic perfectly plastic model compared while the hardening model is nonlinear model. The Mohr-Columb model was used with the same properties similar to the base model with elastic modulus of 40 000 kN/m2 with a friction angle of 41° and a dilatancy angle of 10°. In the parametric study, the number of geogrid layers was changed while keeping the rest of the parameters constant. Figure 13 shows that for a soil modeled as Mohr-Coulomb model the increase in number geogrid layers the calculated BCR did not show any significant increase with the increase of the number of Geogrid layers.

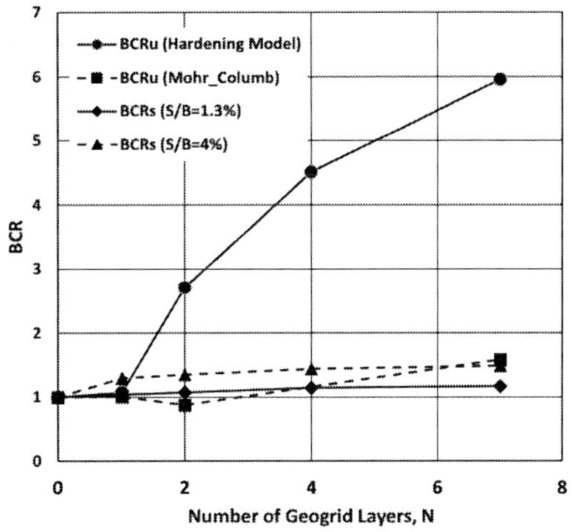

Fig. 13. Variation of BCR_u and BCR_s with number of georid layers.

4.2.1 Effect of the geogrid stiffness on the bearing capacity of the geogrid-reinforced soil

A study was conducted to investigate the effect of the geogrid stiffness on the behavior of the geogrid-reinforced soil. In this study, control model was used with the following properties u/B = 0.5; b/B = 4 with only one layer of reinforcement. The modulus of the geogrid EA was changed keeping all the other factors constant. A relationship was plotted between the geogrid modulus EA and BCR_u shown in Figure 14. Figure 14 shows that increasing the geogrid modulus one order of magnitude increased the BCR_u from 1.07 to 1.62. However, increasing the stiffness one more order of magnitude did not significantly increase the BCR_u.

 DOI: 10.1051/matecconf/201712006011

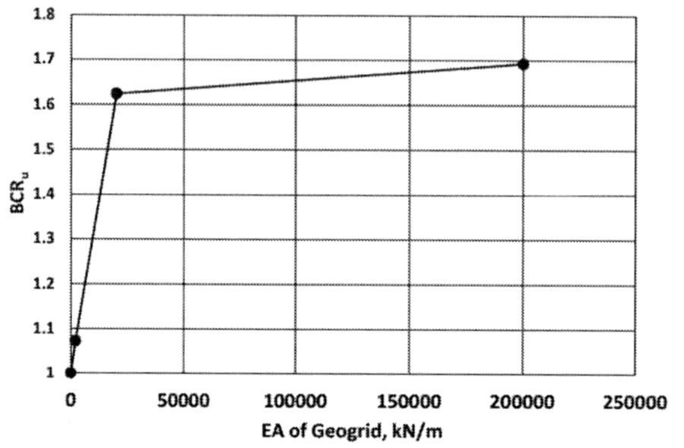

Fig. 14. Variation of BCR_u with geogrid stiffness.

4.2.1 Effect of the geogrid width (b) on the bearing capacity of the geogrid-reinforced soil

A study was conducted to investigate the effect of the geogrid width on the behavior of the geogrid-reinforced soil. In this study, control model was used with the following properties u/B = h/B= 0.5; b/B = 4, N=7 with seven layers of reinforcement. The width of the reinforcement was changed while keeping the rest of the parameters constant. Figure 15 shows a relationship between the BCR_u with the geogrid width. In the figure, the BCR_u increases with the increase of the geogrid width until the width reaches a value of about 3B the increase in width become insignificant in terms of bearing capacity increase.

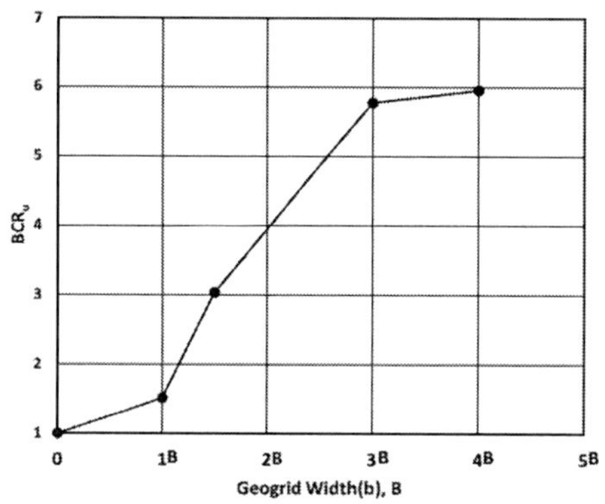

Fig. 15. Variation of BCR_u with width of the geogrid (b).

5 Conclusions and future work

The bearing capacity behavior of strip footing on geogrid reinforced soil was investigated. The effect of geogrid reinforcement on the sand bearing capacity was investigated numerically. Based on the results of the investigation the following conclusions can be drawn:

1- A significant increase in soil bearing capacity with geogrid reinforcement in the soil was shown experimentally and numerically.
2- Hardening non-linear constitutive model is suitable to capture the main characteristics of the geogrid reinforced soils. However, Mohr-Columb is not a suitable constitutive model to capture the main characteristics for geogrid-reinforced soil.
3- The numerical model was not able to capture the behavior of the geogrid-reinforced soil at low settlement.
4- An agreement between the experimental and numerical results on the general trend of critical parameters on the geogrid-reinforced soil bearing capacity. In all cases, BCR_s calculated from the numerical analysis appears to be smaller than obtained from the experimental results.

References

1. Binquet, J., & Lee, K.L. Bearing capacity tests on reinforced earth mass. Journal of Geotechnical Engineering Division, ASCE, 101(12), 1241-55 (1975a).
2. Binquet, J., & Lee, K.L. Bearing capacity analysis of reinforced earth slabs. Journal of Geotechnical Engineering Division, ASCE, 101(12), 1257-76 (1975b).
3. Guido, V.A., Biesiadecki, G.I., & Sullivan, M.J. Bearing capacity of a geotextile reinforced foundation. Proceedings of the XI International Conference on Soil Mechanics and Foundation Engineering, 3, A.A. Balkema, The Netherlands, 1777-80 (1985).
4. Guido, V.A., Chang, D.K., & Sweeny, M.A. Comparison of geogrid and geotextile reinforced slabs. Canadian Geotechnical Journal, 23, 435-40 (1986)..
5. Guido, V.A., Knueppel, J.D., & Sweeny, M.A. Plate load tests on geogrid-reinforced earth slabs. Proceedings of Geosynthetics '87, 216-25 (1987).
6. Akinmusuru, J.O., & Akinbolande, J.A. Stability of loaded footings on reinforced soil. Journal of Geotechnical Engineering Division, ASCE, 107, 819-27. (1981).
7. Fragaszy, R.J., & Lawton, E.C. Bearing capacity of reinforced sand subgrades. Journal of Geotechnical Engineering Division, ASCE, 110, 1500-07. (1984).
8. Fragaszy, R.J., Lawton, E.C., & Asgharzadel-Fozi, Z. Bearing capacity of reinforced sand. Proceedings of the VIII European Conference on Soil Mechanics and Foundation Engineering, Helsinki, 1, 51-84. (1983).
9. Huang, C.C., & Hong, L.K. Ultimate bearing capacity and settlement of footings on reinforced sandy ground. Soils and Foundations, 49(5), 65-73. (2000).
10. Huang, C.C., & Meng, F.Y. Deep footing and wide-slab effects on reinforced sandy ground. Journal of Geotechnical and Geoenvironmental Engineering, ASCE, 123(1), 30-36. (1997).
11. Huang, C.C., & Tatsuoka, F. Prediction of bearing capacity in level sandy ground reinforced with strip reinforcement. Proceedings of the International Geotechnical Symposium on Theory and Practice of Earth Reinforcement, Fukuoka, Japan, A.A. Balkema, The Netherlands, 191-96. (1988).
12. Huang, C.C., & Tatsuoka, F. Bearing capacity of reinforced horizontal sandy ground. Geotextiles and Geomembranes, 9, 51-82. (1990).
13. Khing, K.H., Das, B.M., Puri, V.K., Cook, E.E., & Yen, S.C. The bearing capacity of a strip foundation on geogrid-reinforced sand. Geotextiles and Geomembranes, 12(4), 351-61. (1993).

14. Omar, M.T., Das, B.M., Puri, V.K., & Yen, S.C. Ultimate bearing capacity of shallow foundations on sand with geogrid reinforcement. Canadian Geotechnical Journal, 30(3), 545-59. (1993a).
15. Omar, M.T., Das, B.M., Puri, V.K., Yen, S.C., & Wright, M.A. A comparison of the ultimate bearing capacity of square and strip foundations on geogrid-reinforced sand. Proceedings of Numerical Models in Geomechanics, A.A. Balkema, 967-76. (1992).
16. Yetimoglu, T., Wu, J.T.H., & Saglamer, A. Bearing capacity of rectangular footings on geogrid-reinforced sand. Journal of Geotechnical Engineering, ASCE, 120(12), 2083-99. (1994)
17. Schlosser, F., Jacobsen, H.M., & Juran I. Soil reinforcement. General Report, VIII European Conference on Soil Mechanics and Foundation Engineering, Balkema, Helsinki, 83-103. (1983).
18. Adams, M.T., & Collin, J.C. Large model spread footing load tests on geogrid-reinforced soil foundations. Journal of Geotechnical and Geoenvironmental Engineering, 123(1), 66-72. (1997).
19. Miyazaki, K., & Hirokawa, F. Fundamental study of reinforcement of sand layer in model test. Proceedings of the International Symposium on Earth Reinforcement Practice, Fukuoka, Japan, 1, 647-52. (1992).
20. Omar, M.T., Das, B.M., Puri, V.K., Yen, S.C., & Cook. E.E Ultimate bearing capacity of rectangular foundations on geogrid-reinforced sand. Geotechnical Testing Journal, ASTM, 16(2), 246-52. (1993c).
21. Das, B.M., & Omar, M. T. The effect of foundation width on model tests for the bearing capacity of sand with geogrid reinforcement. Geotechnical and Geological Engineering, 11, 135-41. (1994).
22. Omar, M.T., Das, B.M., Puri, V.K., Yen, S.C., & Cook. E.E. Shallow foundations on geogrid-reinforced sand. National Academy of Sciences, Washington, D.C., Transportation Research Record 1414, 59-64 (1993b).

DOI: 10.1051/matecconf/201712007001

Sustainability assessment and physical characterization of pervious concrete pavement made with GGBS

Hilal El-Hassan[1,*] and *Peiman* Kianmehr[2]

[1]Department of Civil and Environmental Engineering, United Arab Emirates University, Al Ain, United Arab Emirates, P.O. Box 15551
[2]Department of Civil Engineering, American University in Dubai, Dubai, United Arab Emirates, P.O. Box 28282

Abstract. The increasing use of pervious concrete as sustainable and environment-friendly paving materials is primarily owed to its ability to reduce pavement runoff. The mechanical and transport properties of pervious concrete with 50% ground-granulated blast furnace slag (GGBS) replacement are examined in this paper. Open-graded 10 mm and 20 mm aggregates were used to attain porosity of 10%, 15%, and 20%. Polypropylene short cut fibers were added to the mix. The clogging potential of pervious concrete exposed to dust was also investigated. The results indicated that increasing the porosity led to a decrease in compressive and tensile strength. Similar findings were reported when smaller aggregates were used. The fiber addition was only effective in low-porosity concrete. Permeability, characterized by its coefficient k, was proportional to porosity and inversely proportional to aggregate size. After 40-year simulated dust exposure, the concrete permeability could be restored with water flushing maintenance process. In comparison to ordinary Portland cement (OPC) concrete, pervious concrete incorporating GGBS is a more sustainable paving solution, offering a reduction in heat island effect and electricity consumption while also alleviating carbon emissions.

1 Introduction

Porous or pervious pavements have been recommended by the U.S. Environmental Protection Agency (EPA) and the U.S. Green Buildings Council (USGBC) as sustainable alternatives for the current paving systems [1]. A typical cross-sectional schematic of pervious concrete pavement is presented in Figure 1. Past research described five main advantages for employing pervious pavements compared to traditional dense pavements. It permits natural water treatment with an increased retention time before discharging it into the ecosystem, reduces the demand for storm water collection systems, reduces energy consumption by nearly 20%, increases skid resistance and replenishes ground water aquifers below the pavement [2, 3]. Additionally, their use could lessen the glare on road surfaces,

* Corresponding author: helhassan@uaeu.ac.ae

DOI: 10.1051/matecconf/201712007001

decrease tire-pavement noise, preserve natural resources, and reduce lifecycle maintenance costs [4].

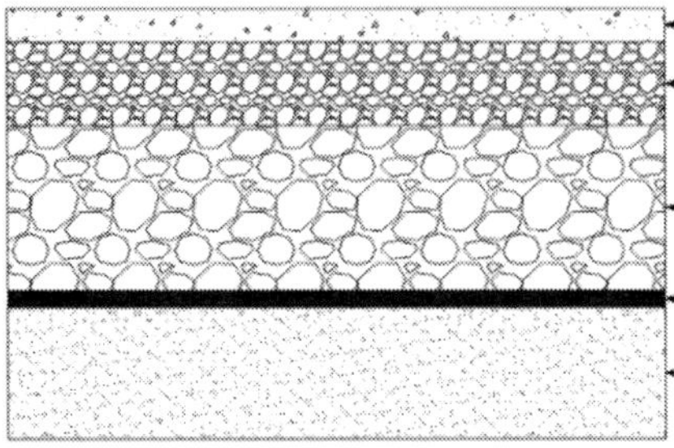

Fig. 1. Typical cross-sectional schematic of pervious concrete pavement [5].

The main benefit of employing pervious pavements in some applications is eliminating storm water collection systems by replenishing pavement runoff. In areas experiencing blockage in roadway drainage, such as the United Arab Emirates (UAE), this would be especially valuable. In the UAE, heavy rain occurs, on average, four times per year as per precipitation charts from 2010–2012 [6], inducing excessive traffic disturbances, road overflowing, car accidents, economic losses, and life causalities [7]. It is estimated that traffic incidents, including those due to rain, caused more than 2.9 billion Dirhams (AED) of losses on Dubai's economy during the year of 2013 and would incur more than 4 billion AED during the year of 2020 [8].

Studies have shown that concrete is the most environment-friendly and efficient road material despite the fact that most roadways are asphaltic [9]. Pervious concrete is different than ordinary concrete in that it includes no fine aggregates, has a void content between 15 and 25%, and experiences compressive strengths of 2.8 to 28 MPa [10]. From an economic standpoint, the Portland Cement Association (PCA) reported that concrete is in fact cheaper than asphalt based on initial and lifecycle maintenance costs given its superior durability [11]. Others noted 30% savings in initial, energy, and maintenance costs, and 8 to 9% reduction in carbon dioxide emissions [12]. With a higher albedo value, concrete is able to reduce electricity and lighting consumption while also decreasing the heat island effect of compared to asphaltic pavement [1, 9]. A study by National Resources Council of Canada showed 0.8% to 6.9% in fuel savings and associated CO_2 emissions per 100 km roadway [13, 14]. Furthermore, 1 km of high volume roadway paved with conventional concrete produced 674 tons of CO_2, while asphalt concrete generated 738 tons [15].

This paper aims to investigate the potential use of self-draining pervious concrete pavements in the UAE as a mechanism counter the societal and economic consequences of rainwater runoff. Ordinary Portland Cement (OPC) and ground-granulated blast furnace slag (GGBS) with 1:1 ratio by mass are used to cast 9 concrete mixes. GGBS, a byproduct of iron and steel making industry, is utilized as a sustainable cement replacement to reduce the carbon footprint while providing a brighter and improved surface finish to so-produced concrete. Based on previous studies, fibers are also added to improve mechanical performance of pervious concrete [16]. The objective is to optimize mechanical and transport properties with various concrete mixture proportions. A simulation over the 40-year lifespan of concrete is carried out to study its clogging potential in the UAE's desert-dominant terrain. Analyses of embodied energy and carbon footprint are also conducted to compare pervious concretes with a 100% OPC concrete.

2 Materials and methods

2.1 Sample preparation

Pervious concrete mixes incorporated ASTM Type I ordinary Portland cement (OPC), ground granulated blast-furnace slag (GGBS), 10 mm and 20 mm aggregates, and Asota MCP/AFC 18 fibers (micro crack preventer/anti fire crack). The chemical composition and physical properties of cement and GGBS are given in Table 1 and 2, respectively. Figure 2 shows the particle size distribution of aggregates following ASTM C136 [17]. The mixing constituents were maintained at room temperature (22±1°C) prior to mixing, according to ASTM C192 [18]. After mixing in a drum mixer, fresh concrete was cast into 150 x 300 mm cylinders. The obtained specimens were demolded after 24 hours and then water-cured in a tank at 23±1°C until testing.

Table 1. Chemical composition of cement and GGBS.

	Chemical composition (%)							
Material	**CaO**	**SiO_2**	**Al_2O_3**	**Fe_2O_3**	**MgO**	**NaO**	**SO_3**	**LOI**
Cement	63.2	19.9	4.9	2.3	2.5	0.8	3.8	2.6
GGBS	42.1	35.4	10.6	0.4	8.1	-	0.3	3.1

Table 2. Physical properties of cement and GGBS.

	Physical properties	
Material	**Specific gravity**	**Surface area (cm^2/g)**
Cement	3.14	3670
GGBS	2.90	4550

ACI 522R-10 was followed in proportioning the concrete mixes [4]. The mix design for 9 concrete batches are summarized in Table 3. Each mix was labeled based on the size of aggregates used and porosity; for instance, C/10/20 was made with 10 mm aggregate and had a porosity of 20%. A control batch (Ref-OPC) was prepared with 10% porosity, 10 mm aggregates, and OPC as binder. The concrete's experimental porosity was compared to its design value as per Equation 1 [19].

$$\textit{Experimental Porosity}\ (\%) = [1 - \left(\frac{\textit{Oven-dried Weight}\ (\text{kg}) - \textit{Submerged Weight}\ (\text{kg})}{\textit{Sample Volume}\ (\text{m}^3)\ .\ \textit{Density of Water}\ (\text{kg m}^{-3})}\right)]100\% \quad (1)$$

The amount of aggregate and water-to-cementitious material (w/cm) ratio was fixed to 1833.6 kg/m^3 and 0.30, respectively, whereas the cementitious material content was reduced to increase the porosity. In all mixes, the cementitious binder incorporated cement and GGBS at a 1:1 ratio. The density of the hardened concretes ranged between 2030 kg/m^3 and 2275 kg/m^3 compared to 2359 kg/m^3 for the controlled reference. Similar values have been reported in other work [20]. For 1 m^3 of concrete, 0.6 kg of 18 mm polypropylene short cut fibers were added. The resulting concretes were labeled, for example, C/10/20/f. The mixes were designed to study the effect of aggregate size, porosity and fibers addition on mechanical and transport properties.

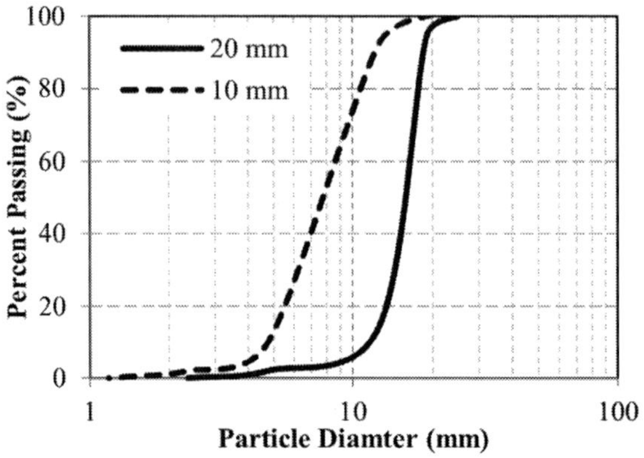

Fig. 2. Particle size distribution of aggregates.

Table 3. Mix design of pervious concretes

Sample	Aggregate (kg/m^3) 10 mm	Aggregate (kg/m^3) 20 mm	Cementitious Material (kg/m^3) OPC	Cementitious Material (kg/m^3) GGBS	Cementitious Material (kg/m^3) Total	Water (kg/m^3)	Fibers (kg/m^3)	Porosity (%)	Density (kg/m^3)
C/10/10	1833.6	0.0	192.6	192.6	385.2	115.5	0.0	10.0	2191.8
C/10/10/f	1833.6	0.0	192.6	192.6	385.2	115.5	0.6	10.0	2200.2
C/10/15	1833.6	0.0	185.2	185.2	370.4	111.1	0.0	15.0	2100.4
C/10/15/f	1833.6	0.0	185.2	185.2	370.4	111.1	0.6	15.0	2105.2
C/10/20	1833.6	0.0	178.0	178.0	355.9	106.8	0.0	20.0	2036.3
C/10/20/f	1833.6	0.0	178.0	178.0	355.9	106.8	0.6	20.0	2040.1
C/20/10	0.0	1833.6	200.1	200.1	400.1	120.0	0.0	10.0	2259.1
C/20/10/f	0.0	1833.6	200.1	200.1	400.1	120.0	0.6	10.0	2261.5
Ref-OPC	1833.6	0.0	420.1	0.0	420.1	126.0	0.0	10.0	2359.3

2.2 Performance evaluation

2.2.1 Mechanical testing

Pervious concrete specimens were removed from the water curing tank and maintained in a moist condition up to mechanical testing. In accordance with ASTM C39, the 7 and 28-day compressive strength was measured [21]. The splitting tensile strength of cylindrical concrete samples was determined after 28 days following ASTM C496 [22]. For each experiment, three replicate cylinders per batch were tested to obtain an average.

2.2.2 Falling head permeability test

The permeability coefficient, k, represents the permeability of concrete. It is the amount of fluid passing through a unit area per unit time under a unit hydraulic gradient. As specified by ACI 522R-10, the falling head apparatus (see Figure 3) was used to estimate the permeability of pervious concrete [4]. While the mechanism remains the same in major investigations, the procedure is often modified depending on the type of experimentation. A graduated plastic pipe of cross-sectional area A_p (m^2) was attached to the top cap of the vessel. The concrete cylinders with cross-sectional area A_c (m^2) were supported halfway through the vessel. The samples employed in this test were sealed around the circumference with epoxy mortar to maintain an axial water flow without lateral leakage [23]. As the water level dropped from a height h_0 (m) to h_1 (m), it passed a distance, L (m), in time, t (s). For each concrete batch, three replicate cylinders were tested and averaged. Darcy's law, given by Equation 2, was used to calculate the permeability coefficient, k (m/s) [24]:

$$k = \frac{A_p L}{A_c t} \ln\left(\frac{h_0}{h_1}\right) \tag{2}$$

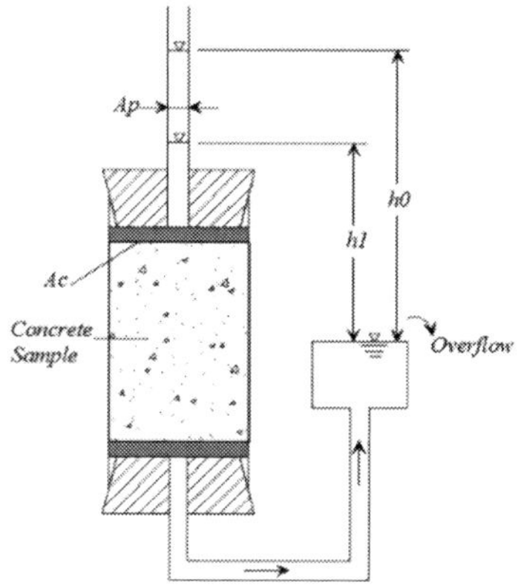

Fig. 3. Falling head permeability apparatus [5].

2.2.3 Dust clogging test

The service life of pervious pavements is majorly affected by the clogging due to dust exposure [25]. Recently, a study showed that, on average, dust deposited and accumulated at 9.34 g/m^2.day and peaked at 21.31 g/m^2.day [26]. The dust used in this study was collected an urban area in Dubai. Sieve analysis was employed to obtain the particle size distribution, shown in Figure 4. The falling head apparatus was modified to examine the clogging potential of the various concrete mixes. To simulate rain activity in dusty weather, a shower head was installed in the vessel cap and a mixture of water and dust, proportioned based on the reported accumulations, surged through. The permeability coefficient was then recorded over a 40-year simulation representing pavement service life.

2.2.4 Albedo test

Albedo values represent the ratio of reflective to incident light discharged onto a surface. According to study by Kaloush et al., weathered and new asphalt pavement had an albedo ranging between 0.05 and 0.15 [27]. On the other hand, concrete made with ordinary Portland cement (OPC) had an albedo value in the range of 0.20 and 0.40 [27]. Apparently, brighter material resulted in higher albedo and a reduction in heat island effect. Additionally, replacing dark asphalt pavements with concrete pavements showed a reduction in electricity demand and lighting costs by nearly 30% [3, 9]. To measure the albedo of pervious concrete specimens, a digital camera was paired with a software package to obtain the luminance value and Red/Green/Blue (RGB) histogram of a selected area. They are then used to determine the albedo value.

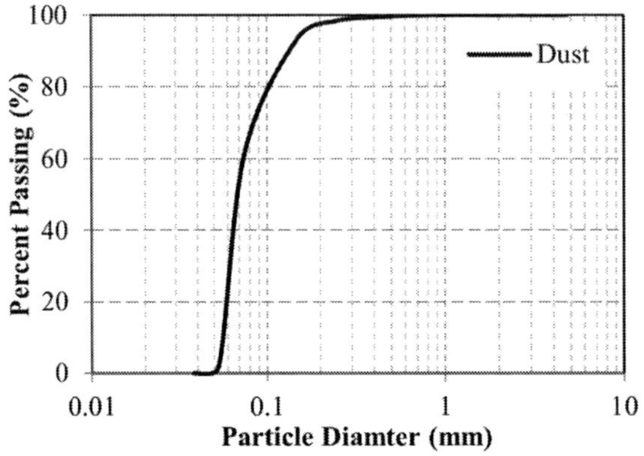

Fig. 4. Particle size distribution of dust.

3 Results and discussion

3.1 Compressive and splitting tensile strength

The 7-day compressive strength results are presented in Figure 5. By comparing batches C/10/10, C/10/15 and C/10/20, the effect of porosity on compressive strength was assessed with constant aggregate size of 10 mm. The aggregate-to-cementitious material (A/CM) ratio was increased to produce a more pervious or porous concrete. As a result, the compressive strength was reduced due to more voids and lesser hydration products. Another comparison of the 7-day compressive strength concrete C/10/10 and C/20/10 investigated the effect of aggregate size. Typically, concretes made with open-graded aggregates produce more voids if a larger particle size is used [9]. In fact, to maintain a porosity of 10%, more cementitious material was used in 20 mm mixes compared to 10 mm counterparts. It was clear that concrete with 20 mm aggregates experienced higher compressive strength owing to the higher cementitious material content and, as a result, more hydration products. In addition, the effect of adding Asota MCP/AFC 18 fibers to concrete is studied after 7 days in Figure 5. While their addition seemed to improve strength in concretes with 10% porosity, it reduced the mechanical performance for 15% and 20% samples. Apparently, the fibers utilized in this work could only increase compressive strength of concrete with low porosity (10%).

After 28-day subsequent hydration, concrete specimens presented higher compressive strength. Mechanical performance of concretes with 10% porosity and 20 mm aggregates was superior to other samples at 54 MPa. It is much higher than the minimum 28-day strength required by the Virginia Department of Transportation (VDOT) of 20.7 MPa [28]. With 15 MPa, concrete C/10/20/f was considered inadequate for dense roadways but could be recommended for areas of low traffic congestion such as local roads, parking lots and road shoulders.

Figure 5 also shows the splitting tensile strength after 28 days. With 10 mm aggregates, an increase in the porosity of concrete from 10% to 20% led to a 24% decrease in tensile strength. Changing the aggregate size from 10 mm to 20 mm, however, resulted in an increase in tensile strength by 25%. The addition of fibers did not significantly enhance the tensile strength. While the length of the fibers possibly rendered them ineffective in the produced pores, it is apparent that porosity was the major governing factor. Similar results have been noted in other studies on pervious concrete [29].

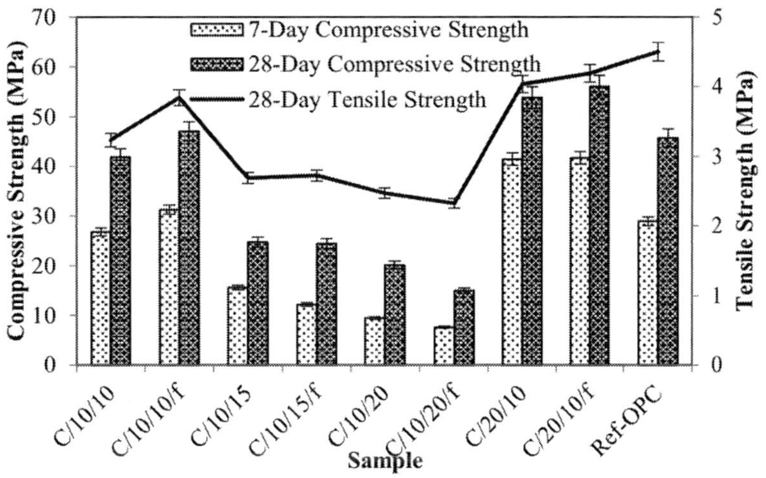

Fig. 5. Compressive and tensile strength of concrete samples.

3.2 Permeability

The permeability of pervious concrete was investigated through its coefficient, k. The falling head apparatus and Equation 2 were collectively used to determine the experimental values of k, presented in Table 4. The results show a proportional relation between permeability and porosity, as has also been reported in other work [19, 29]. The lower cementitious material content in concretes incorporating 50% GGBS resulted in a higher permeability compared to the control sample. Larger aggregates required more CM and, consequently, led to a less dense concrete with a lower k value. Furthermore, the addition of fibers seemed to have decreased the permeability by occupying empty space in the concrete structure. To better explain the experimental values, an area of 1 m^2 is considered across which water is drained. The resulting permeability for C/20/10 of 1.8 mm/s is equivalent to draining 1.8 L of water per second or 108 L per minute.

3.3 Dust clogging

Based on the mechanical and transport properties, concrete C/20/10 was selected to evaluate its potential use as pervious concrete pavement on the roads of the UAE. Samples were exposed to dust for a 40-year simulation of the concrete lifespan. Dust was chosen as it is finer than sand, could lead to further transportation into the concrete, and could cause a more noticeable reduction in the value of k [30]. As such, the effect of dust exposure on the permeability of pervious concrete was investigated following the accumulation reported by Hamza [26]. The results, depicted in Figure 6, present a continuous decrease in the permeability coefficient over 40 years.

A maintenance procedure was proposed by water flushing the concrete specimens after 12 years of simulated exposure. While the initial permeability was 1.8 mm/s, it could be restored to 1.69, 1.45, and 1.2 mm/s due to subsequent maintenance at 12, 24, and 36 years, respectively. Accordingly, a 67% recovery of k was recorded after 36 years, indicating that water flushing or cleansing could recycle the concrete while disposing of clogging dust particles. It is possible that surface clogging, as noted by visual inspection, could be responsible for the reduction in permeability. In fact, it is for this reason that cleaning the surface with water was effective in partially restoring the permeability.

Table 4. Permeability coefficient of concrete mixes.

Sample	Permeability coefficient, k (mm/s)
C/10/10	1.94 ± 0.18
C/10/15	2.50 ± 0.21
C/10/20	7.50 ± 0.74
C/10/20/f	7.22 ± 0.79
C/20/10	1.80 ± 0.19
Ref-OPC	1.58 ± 0.29

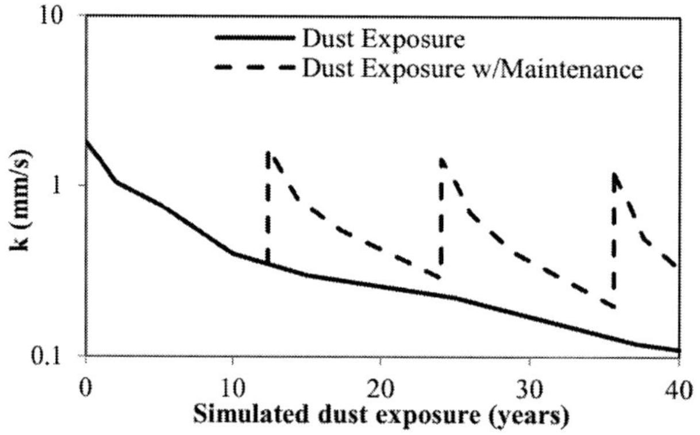

Fig. 6. Dust clogging test of concrete C/20/10 [5].

3.4 Albedo values

The albedo values of different materials are summarized in Table 5. A white paper had an albedo value of 0.60 and served as a reference. The sustainable concrete mixes with 50% GGBS presented an albedo of 0.52, while that of the reference made with 100% OPC was 0.40. Other work on OPC concrete reported a value of 0.45 [27]. It seemed that incorporating GGBS, which had an off-white color, contributed to a brighter surface finish of the pervious concrete. If such lighter-colored concrete is used in UAE cities, it could aid in reducing the reported 6°C heat island effect [31].

Table 5. Albedo values of different paving materials.

Material	Albedo Value
Asphalt [27]	0.10
OPC Concrete [27]	0.45
OPC+GGBS Concrete	0.52
Ref-OPC	0.40
White Paper	0.60

3.5 Embodied energy, carbon footprint and cost analysis

The embodied energy and carbon dioxide emissions of pervious concretes are presented in Figure 7. OPC, GGBS, aggregates and fibers have respective embodied energies of 5.5, 1.6, 0.083, and 99.2 MJ/kg [32]. OPC and GGBS discharge 0.913 tons and 0.067 tons of CO_2 per ton of production, respectively [33]. Analyses findings show that producing 1 m^3 of pervious concrete (C/10/10) emits 188.7 kg of CO_2 and consumes 1519.5 MJ of energy. In comparison, the use of OPC as the sole cementitious material used (Ref-OPC) results in carbon emissions of 383.6 kg/m^3 with an embodied energy of 2310.6 MJ/m^3. The incorporation of GGBS in the mix reduced the embodied CO_2 and energy by 51% and 34%, respectively.

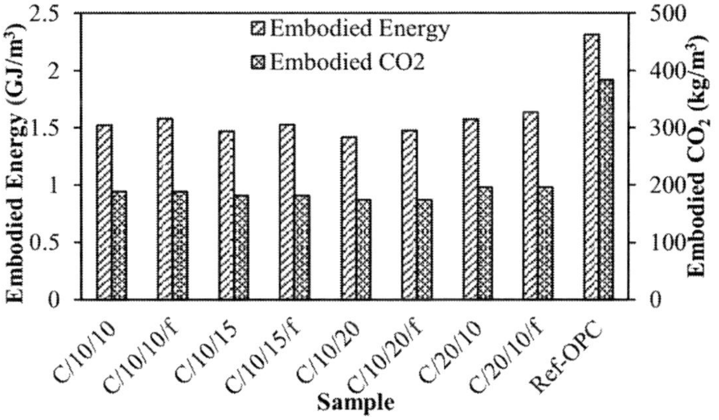

Fig. 7. Embodied Energy and carbon footprint of concrete mixes.

4 Conclusions

The applicability of employing pervious concrete pavement made with cement and GGBS (1:1) was investigated. The mechanical and transport properties and environmental impact were studied. Based on the findings of this study, the following conclusions may be drawn:

(1) A decrease in the porosity of cast concrete resulted in an increase in mechanical performance. Concretes made with larger aggregates (20 mm) exhibited higher compressive strength compared to those made with 10 mm. Both findings owed to higher cementitious material content, thus producing a denser concrete with more hydration products. The compressive strength of low porosity concrete (10%) was improved with the addition of fibers, but the tensile strength was unaffected.

(2) The concrete permeability was characterized by its coefficient, k. Results showed a higher permeability with higher porosity. The optimum concrete mix with 10 mm aggregates and 20% porosity could drain 450 L of water per minute over a surface area of 1 m^2. It was apparent that the use of 20-mm aggregates could reduce the permeability owing to a denser concrete with higher cementitious material content.

(3) A 40-year dust exposure showed a consistent decrease in permeability. A maintenance process was proposed to flush the concrete with water during its service life. It proved the recyclability of such concretes by restoring 67% of the initial permeability.

(4) The use of GGBS as 50% cement replacement increased the albedo value to 0.52 compared to 0.40 and 0.10 for OPC concrete and asphalt, respectively. This would decrease the heat island effect as well as reduce electricity demand and lighting costs for paved roadways. Concrete with 10 mm aggregates and 20% porosity (C/10/20) required the least amounts of cement and GGBS. Compared to OPC concrete, the associated

carbon emissions of the least produced concrete were 55% less. The use of fibers could not be advocated given the increase in embodied energy with minor improvement in mechanical properties.

References

1. EPA. Reducing urban heat islands: compendium of strategies. USA: United States Environment Protection Agency, (2008).
2. Lee M-G, Huang Y-S, Chang T-K, Pao C-H, editors. *Experimental Study of Pervious Concrete Pavement*. Emerging Technologies for Material, Design, Rehabilitation, and Inspection of Roadway Pavements; (2011); Hunan, China: ASCE.
3. Ashley E. Environmental and Cost Benefits of High Albedo Concrete. Concrete in focus. (2008):2.
4. ACI Commitee 522. Pervious Concrete: Report No. 522R-10. American Concrete Institute: American Concrete Institute, (2010).
5. El-Hassan H, Kianmehr P. Pervious concrete pavement incorporating GGBS to alleviate pavement runoff and improve urban sustainability. Road Materials and Pavement Design. (2016):1-15.
6. Spark W. Historical Weather for 2012 in Dubai, United Arab Emirates United States of American: Cedar Lake Ventures, Inc; (2012) [cited 2014].
7. Carroll L. One dead amid flooding and traffic chaos after heavy rains in UAE. The National UAE. (2014).
8. Shahbandari S. Traffic congestion costs more than Dh700,000 per kilometre in Dubai. Gulf News. (2015).
9. Ferguson BK. *Porous Pavements*. United States of America: Taylor & Francis; (2005). 600 p.
10. Chindaprasirt P, Hatanaka S, Chareerat T, Mishima N, Yuasa Y. Cement paste characteristics and porous concrete properties. Constr Build Mater. (2008);**22**(5):894-901.
11. Sullivan EJ, Moss A. Paving Cost Comparisons: Warm-Mix Asphalt Versus Concrete. USA: Portland Cement Association, (2014).
12. Gadja JW, VanGeem MG. A Comparison of Six Environmental Impacts of Portland Cement Concrete and Asphalt Cement Concrete Pavements. USA: Portland Cement Association, (2001) Contract No.: PCA R&D Serial No. 2068.
13. Taylor GW, Farrell P, Woodside A. Additional Analysis of the Effect of Pavement Structures on Truck Fuel Consumption. Canada: Government of Canada, (2002).
14. Taylor GW, Patten JD. Effects of Pavement Structure on Vehicle Fuel Consumption – Phase III. Canada: National Research Council Canada, (2006) Contract No.: CSTT-HVC-TR-068.
15. Athena Institute. A Life Cycle Perspective on Concrete and Asphalt Roadways: Embodied Primary Energy And Global Warming Potential. USA: Cement Association of Canada, (2006).
16. Hesami S, Ahmadi S, Nematzadeh M. Effects of rice husk ash and fiber on mechanical properties of pervious concrete pavement. Constr Build Mater. (2014);**53**:680-91.
17. ASTM. Standard Test Method for Sieve Analysis of Fine and Coarse Aggregates. C136. USA: ASTM International; (2014).
18. ASTM. Standard Practice for Making and Curing Concrete Test Specimens in the Laboratory. C192. USA: ASTM; (2015).
19. Montes F, Valavala S, Haselbach LM. New test method for porosity measurements of Portland cement pervious concrete. Journal of ASTM International. (2005);**2**(1).

20. Ćosića K, Koratb L, Ducmanb V, Netingera I. Influence of aggregate type and size on properties of pervious concrete. Constr Build Mater. (2015);**78**:69-76.
21. ASTM. Standard Test Method for Compressive Strength of Cylindrical Concrete Specimens. C39. USA: ASTM; (2015).
22. ASTM. Standard Test Method for Splitting Tensile Strength of Cylindrical Concrete Specimens. C496. USA: ASTM; (2011).
23. Hall C, Hoff WD. *Water Transport in Brick, Stone and Concrete* USA: Taylor & Francis; (2011).
24. Das B. *Principles of Geotechnical Engineering*. 7 ed: Cengage Learning; (2009).
25. Coughlin JP, Campbell CD, Mays DC. Infiltration and Clogging by Sand and Clay in a Pervious Concrete Pavement System. J Hydrol Eng. (2012);**17**(1):68-73.
26. Hamza W, Enan MR, Al-Hassini H, Stuut J-B, de-Beer D. Dust storms over the Arabian Gulf: a possible indicator of climate changes consequences. Aquatic Ecosystem Health & Management. (2011);**14**(3):260-8.
27. Kaloush KE, Carlson JD, Golden JS, Phelan PE. The Thermal and Radiative Characteristics of Concrete Pavements in Mitigating Urban Heat Island Effects. USA: Portland Cement Association, (2008) Contract No.: PCA R&D SN2969.
28. Virginia Department of Transportation. *Road and Bridge Specifications*. USA: VDOT; (2007).
29. Neithalath N. Development and characterization of acoustically efficient cementitious materials. USA: Purdue University; (2004).
30. Tegen I, Fung I. Modeling of mineral dust in the atmosphere: Sources, transport, and optical thickness. J Geophys Res. (1994);**99**(D11):22897-914.
31. Frey CM, Rigo G, Parlow E, editors. *Investigation of the daily urban cooling island (UCI) in two coastal cities in an arid environment: Dubai and Abu Dhabi (U.A.E)*. 25th EARSeL Symposium Global Developments in Environmental Earth Observation from Space; (2005); Porto, Portugal.
32. Hammond G, Jones C. Inventory of Carbon & Energy (ICE) Version 2.0 United Kingdom: University of Bath, (2011).
33. MPA Cement. Embodied CO2e of UK cement, additions and cementitious material. United Kingdom: MPA Cement, (2015) Contract No.: ST/FS/18.

Investigation on the effect of phase segregation on the mechanical properties of polymer modified bitumen using analytical and morphological tools

Noor Zainab Habib[1,*], *Ibrahim* Kamaruddin[2] , *Isa M* Tan[2] and *Masaharu* Komiyama[3]

[1]Heriot-Watt University Dubai Campus, Dubai International Academic City, Dubai, U.A.E
[2]Universiti Teknologi PETRONAS, Bandar Sri Iskandar, Tronoh, Perak, Malaysia
[3]University of Yamanashi, Japan Chome-3-11 Takeda, Kofu, Yamanashi Prefecture 400-8510, Japan

Abstract. The mechanical strength of polymer modified bituminous mix which is evaluated in terms of stiffness reflects the deformation behaviour of bituminous mixture. The stiffness of the binder strongly depends on its chemical composition besides aggregate structure of the bituminous mixture. Compatibility of polymer with the base bitumen is considered as major strength enhancing factor in the case of polymer modified bituminous mix. To relate the factors affecting the mechanical strength of the binder a new approach in this research study was adopted in order to relate the compatibility of 80/100 pen bitumen with polymer Polypropylene (PP) using analytical and morphological tools. Morphological tools like Atomic Force Microscopy (AFM) and Transmission Electron Microscope (TEM) were used to study the changes due to the difference in compatibility of polymer with bitumen. It was observed that the presence of phase segregated layer as observed by AFM surface morphology was considered as the main strength enhancing factor of the bituminous mix. While the presence of lamella as observed in TEM scanned images of PP PMB revealed that addition of polymer forms localized network in the phase segregated layer which induces the stiffening effect in PMB. From the chromatographic method (SARA) analytical analysis Saturate, Aromatics, Resin and Asphaltene fractionate were determined in the blend. It was observed that chemical composition of the blend also has a profound effect on the rheological properties and on the morphology of polymer modified bituminous (PMB) blend. It was concluded that PP PMB gets benefited by phase segregated layer in the blend when the polymer concentration was kept below 3% as it sufficiently enhances the mechanical strength of the PMB binder as observed by dynamic creep results.

* Corresponding author : n.habib@hw.ac.uk

1 Introduction

Bitumen modification with polymer has now become the most widely used alternate method to improve the pavement performance properties. Polymer modified bituminous binder (PMB) are used because of its better engineering properties which are considered as a major element in extending the service life of the pavement. Enhanced engineering properties due to PMB modification were observed in terms of reduction in pavement deterioration by improvement in high temperature rutting resistance, reduction in low temperature thermal cracking and reduction in fatigue crack propagation [1].

In order to get the best results, compatibility of the blend is considered an important issue as it has a strong influence on its morphology and rheology, as low molecular weight bitumen when mixed with high molecular weight polymer compatibility, is difficult to achieve as it depends on the solubility parameters of bituminous blend [2]. Thermoplastic polymers because of their higher viscosity at high temperature and lower density than bitumen usually tends to cream. The blend thus produced doesn't have any chemical interaction and should be treated only as a physical system consisting of two partially miscible phases [3].

Oxidation of bitumen during blending or in-service life of pavement also considered as one of the major cause of stiffening or strength affecting factor of the binder. Thus the study of the chemical composition in understanding and effect of modification on the chemical fractions present before and after modification is important to know [4]. The evolution of chemical structure because of modification usually enhances the mechanical strength of binder and ultimately the modified bituminous mix [5]. Since polymer and bitumen are two immiscible component of the blend because of their difference in molecular weight [6], this difference has a profound effect on the physical, chemical and mechanical properties of the PMB blend. The characteristics of the mix were due to thermodynamic instability as for thermodynamically stable blend the Gibbs free energy of the mix must be lower than zero [7].

2 Objective of study

This objective of this study was to discuss the behaviour of polymer Polypropylene used as a modifier for 80/100 pen bitumen with its effect on rheology and morphology. The morphological study is considered incomplete and thus supplemented with the chemical analysis using SARA fractionate analysis which provides the information about the change in the content of saturate, aromatics, resins and asphaltene in the blend. The unique behaviour of polymer in the blend as revealed by the morphological and chemical analysis helps in understanding the thermodynamics of partially dispersed polymer and to understand the effect of polymer bitumen incompatibility on mechanical strength. The mechanical strength will be evaluated by Dynamic creep test results.

3 Methodology

3.1 Material

The material used includes, 80/100 penetration grade base bitumen obtained from PETRONAS refinery, Melaka, Malaysia and Polypropylene (PP) in the powder form was used for modification supplied by PETRONAS Polypropylene Sdn. Bhd. Malaysia. The polymer has melt flow index of 8 g/10 min and density of 0.887g/cm^3.

3.2 Sample preparation and test methods used for analysis

PP modified blend was prepared by mixing 80/100 Pen bitumen with polypropylene at 120 rpm. Slow rate of mixing was adopted intentionally as it was known that compatibility of polymer can't be achieved. Thus only for physical dispersion of polymer in bitumen, slow speed was adopted and also to prevent the agglomeration. The concentration of PP was kept between 1-3% by weight of bitumen. Morphological analysis was done to study the evolution of structure brought by chemical changes in the virgin bitumen after modification using atomic force microscopy (AFM) using SII NANO NAVI E- Sweep and transmission electron microscopy (TEM) by using ZEISS LIBRA 200 FE.

Chemical analysis for the determination of asphaltene was done according to ASTM D-3279 while SARA chromatographic method ASTM D 4124 - 01 was used to fractionate bitumen. For mechanical analysis of bituminous mix, Dynamic Creep test was performed according to BS D 226. The creep deformation of Standard Marshall Specimens prepared at optimum bitumen content for virgin and PP modified mix was measured as a function of pulse counts or rather time. The load used during the test on the specimens was uniaxial and dynamic, representing the repeated application of axle loads on the pavement structure. The results obtained were used to develop the relationship between mixture stiffness to the binder stiffness to predict mechanical performance of mix.

4 Results & discussion

4.1 Morphological analysis

AFM and TEM were used to study the internal morphology of the PMB. The analysed samples describe the extent of the continuous phase in the blend and compatibility of the bitumen–polymer within the blend. The compatibility of the bitumen with the polymer depends on molecular weight, as higher molecular weight polymers are incompatible with lower molecular weight bitumen [6].

AFM images were taken in tapping mode, which revealed information in three - dimensional topography and phase shift, as shown in the images of virgin and modified bitumen in Figure 1-5. Different surface compositions were observed in phase shift contrast. Figure1(a) shows a two - dimensional phase image of virgin bitumen in which "bee" structure can be observed, which was believed to be asphaltene micelles [6, 8]. The three - dimensional phase image of the same bee structure is shown in Fig.1(b), in which protruding tubers from the surface of maltene fraction can be observed representing asphaltene [6, 9], most polar fraction of the bitumen while the surrounding phase around bee structure formed by lighter or less polar fraction maltene composed mainly of resin and aromatics. The surrounding flat surface may be concluded as non-polar saturates or alkanes. Within this protruding structure two different phases can be observed with altering dark and bright region [10]. These alternating light and dark regions in bee represents portion of different relative stiffness (different viscoelastic properties) having contrasting mechanical properties. The same protrudes having different viscoelastic properties were observed in Figure 2, in TEM image of virgin bitumen which under prolong exposure of electron beam disappears after some time. Figure 3(a) -5(a) shows the evolution of phase change with the addition of PP in bitumen. Phase evolution was observed with the formation of phase separated polymer layer on the surface of bitumen whose thickness seems to be increasing with concentration of polymer in blend as higher molecular weight polymer seems to be incompatible with the lower molecular weight bitumen thus causing the phase separation [6]. The TEM images of the same samples shows fine streaks of lamella, which was supposed to be due to presence of isotactic polypropylene in the PMB blend. The 3D AFM phase images of PP modified bitumen shows variation in surface roughness which decreases with the increase in polymer concentration in the blend. From Figure 3 (a), 1%PP PMB phase image shows phase separated polymer layer totally covering the underneath bitumen layer, as evidenced by the disappearance of the "bee" structures. It is noted in Figure 3(a) that the phase image of 1% PP PMB is very rough, indicating that the surface is micro-mixture of two materials having different viscoelastic properties. TEM scanned image of same sample (Figure 3(b), 4(b) & 5(b) shows helical entangled chain of PP where dispersion and thickness of entangled chain decreases with the increase in polymer content indicating that higher polymer content in bitumen brings segregation of polymer and bitumen. The fine streaks of polymer totally covers the underneath bitumen layer. This interpretation was also confirmed by AFM scanned images (Figure 3(a), 4(a) & 5(a) where PP segregated layer clearly visible forming a surface that is a micro-mixture of bitumen and PP. Figure 4(a) of 2% PP PMB although shows surface roughness, but to much less degree compared to 1% PP PMB, indicating that the cover- up of the surface by segregated PP is more complete when PP concentration is increased to 2%. 3% PP PMB as shown Figure 5 (a) where phase image expose a complete flat phase separated layer of polymer above the bitumen. These AFM images unfold the surface evolution by the segregated PP.

The morphological phase segregation in the PP PMB blend as observed by AFM images were due to thermodynamic instability in the blend as the weak van der Wall forces in the blend were just attractive forces unable to generate heat increasing enthalpy of the blend, while the high molecular weight polymer at higher concentration level tends to agglomerate due to presence of dispersed polymer particles increasing the molar volume and decreasing the entropy of the blend which ultimately leads to phase segregation due to the instability of blend on cooling [7].

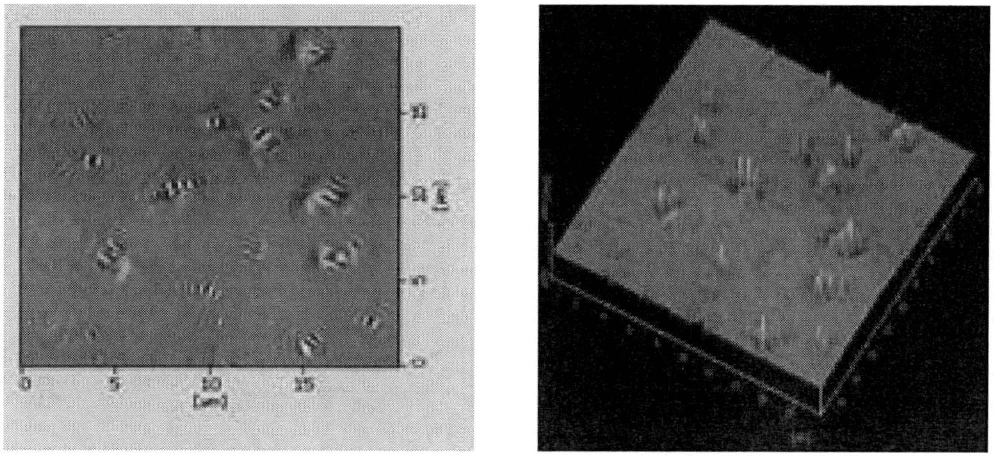

(a) (b)

Fig. 1. Virgin Bitumen AFM images (a) Phase image (b) 3 D image.

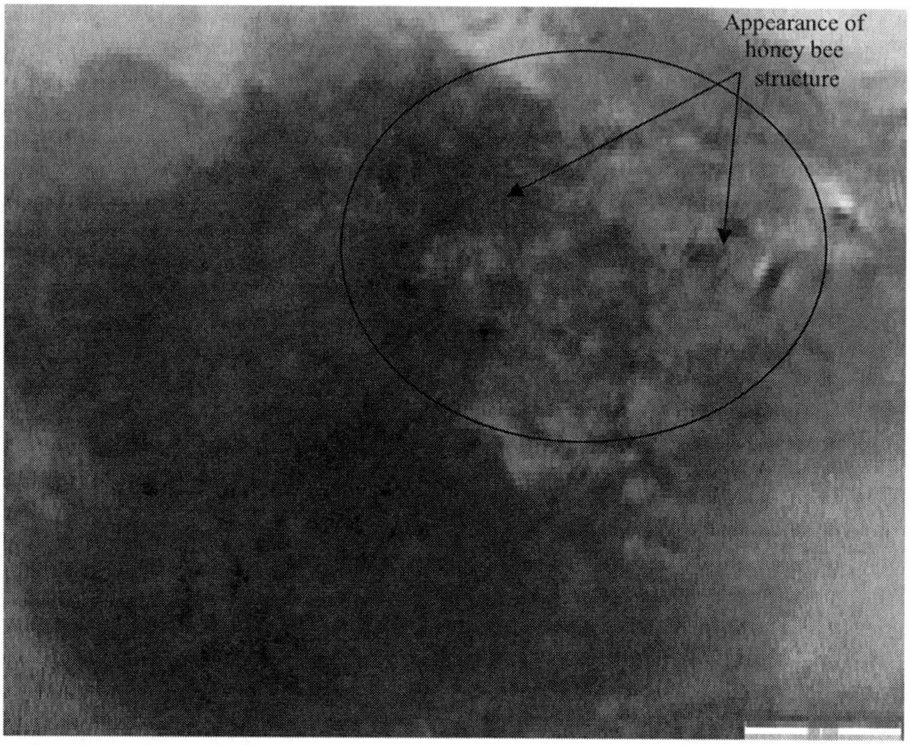

Fig. 2. TEM image of Virgin Bitumen.

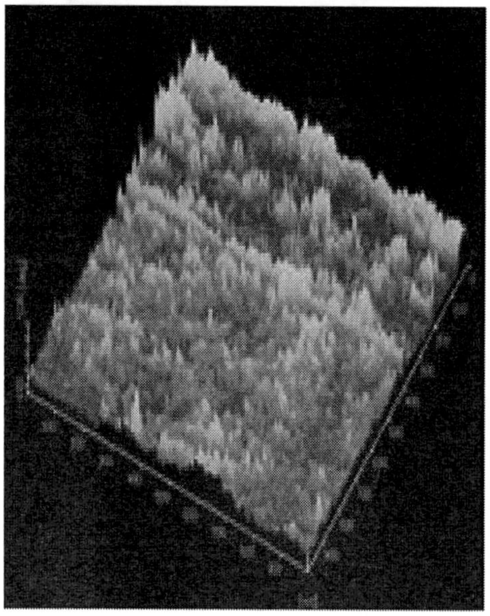

(a)

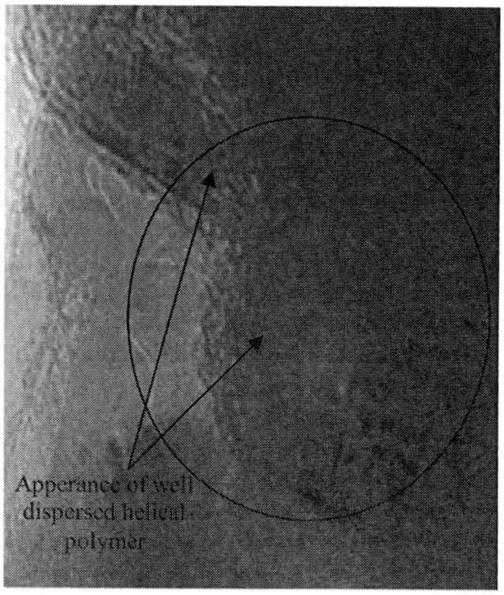

(b)

Fig. 3. 1% PP PMB (a) AFM image (b) TEM image.

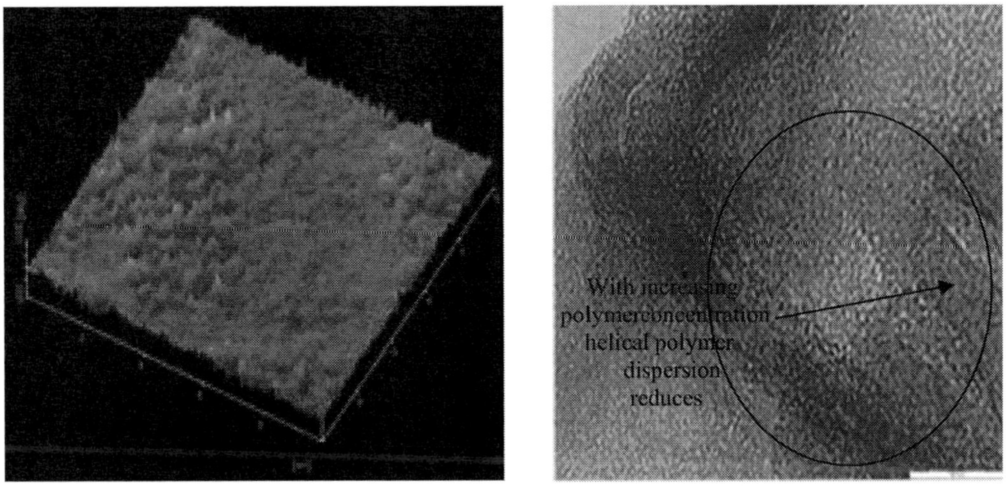

(a) (b)

Fig. 4. 2% PP PMB (a) AFM image (b) TEM image.

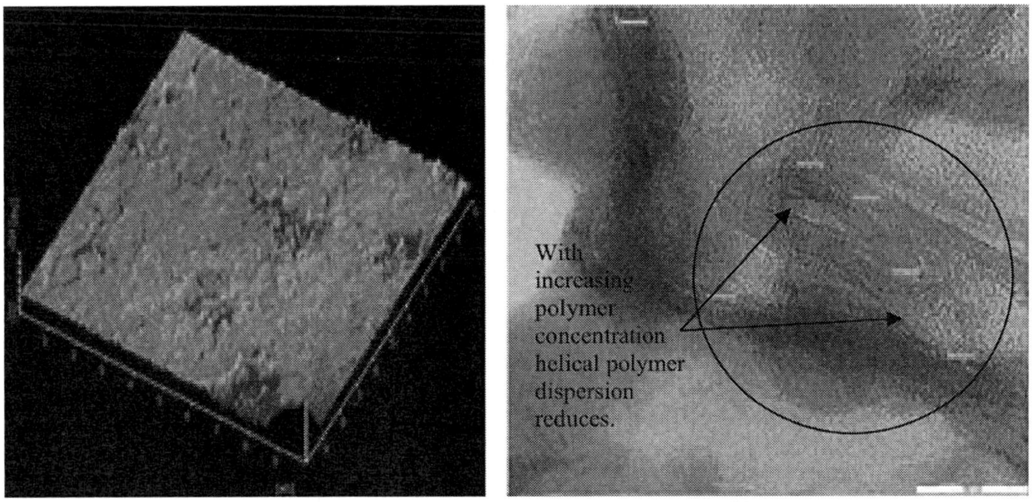

(a) (b)

Fig. 5. 3% PP PMB (a) AFM image (b) TEM image.

The morphological segregation can be better understood on the concept of solubility parameter which describe the interaction between the molecules on the basis of solubility parameter to predict the extent of mix ability when two different materials being mixed [2]. The colloidal stability of the mix depends on dispersive interaction between molecules due to the dispersive component, the polar component, and hydrogen component. Bitumen saturates has the solubility parameter between 17.4 and 20 MPa 0.5 [2] when mixed with polypropylene with solubility parameter between 16 and 19.2 MPa0.5 [11] interact with bitumen saturate oils, initiate penetration of bitumen saturate oils into the polymer with the formation of swollen polymer particles within PMB blend. Thus even a small addition of polymer in the blend increases the viscosity of PP PMB [12] and induces the polymer phase in the blend which soon segregates upon cooling as observed by AFM scanned images (Figure 3-5).

4.2 Asphaltene content analysis (ASTM D 3279)

ASTM n-heptane precipitation method was used to determine asphaltene content in virgin bitumen and PP modified bitumen. Asphaltene is the most difficult to define as till now it was noted by many researchers that it is very difficult to isolate them alone and still their chemical formula can't be established yet. Thus the only way to characterized them based on their insolubility in pentane or heptane [13]. The asphaltene content represented by an insoluble content in n-heptane is presented in Table 1. From previous research, it was found that the concentration of asphaltene usually varies from 5 to 25% in bitumen [14] which in turn depends on the origin of the oil crude source. Usually, asphaltene content in the range of between 5 -25% is considered as moderate range, which exhibit sol-gel characteristics basically responsible for their viscoelastic properties [15].The sol-gel characteristic of bitumen is considered beneficial as characterized by delay in the elastic deformability that will help in accommodating larger stresses in pavement because of the considerable degree of elastic deformation [16].From Table 1 it is inferred that aspahltene content for virgin bitumen was found to reduces gradually from 15% to 13.56% with the addition of polymer in bitumen. This decrease in asphaltene content indicates that the high molecular weight micelle becomes more peptized [16] or stabilized by the addition of polymer in the blend as observed by decrease in colloidal Index (CI), as presented in Table 1 enhancing the pseudo plastic behaviour of the blend which is very commonly observed for polymeric solution [17].

4.3 Asphaltene content analysis (ASTM D 3279)

The fractionate analysis of bitumen using SARA chromatography helps in understanding the chemistry of bitumen, where the analysed fractions namely saturates, aromatics, resins represent the maltenic fraction or oily stabilized base while asphaltene (n- heptane's insoluble) are dispersed in it. SARA is helpful in determining the stability of in term of Colloidal Index "Ic" defined as ratio of the dispersed phase to dispersing phase [15] where lower value of Ic represent that it would offer compatibility when blended with bitumen

$$Colloidal\ Index\ I_c = \frac{Asphaltene + Saturates}{Resins + Aromatics} \tag{1}$$

The results of SARA for virgin bitumen are shown in Table 1. From Table 1 it was observed that there is a decrease in asphaltene with the increase in resin by the addition of PP in the virgin bitumen. Bitumen with sufficient quantities of resins and aromatics shows adequate solvating power, where the asphaltenes are fully peptized, well dispersed and does not form extensive associations. Such bitumen's are known as solution of ''SOL'' type bitumen's [18]. When polymer is induced in such bitumen the higher molecular weight asphaltene with the similar molecular weight PP, would try to compete for solvency and thus absorb oils from maltene fraction (saturate & aromatics) and induces phase separation due to insufficient maltene being left in the blend.

From the SARA results, it can be inferred that composition of each fraction depends on the solubility parameter of the fraction to the particular solvent used. There is a decrease in asphaltene content observed in PP modified bitumen in comparison to virgin bitumen. One of the reasons for this decrease might because of the solvent, n-heptane which was used for asphaltene precipitation seems to be weaker as a solvent or the amount of solvent used for precipitation of asphaltene of PP PMB was not sufficient to completely precipitate the asphaltene. Thus the maltene solution obtained after separating asphaltene does contain a substantial portion of the asphaltene in maltene which was observed by the increase in resin content and decrease in aromatics for all PP PMB. Because of the polar nature of resin, it sticks together with asphaltene or may form chelate due to presence of heteroatoms in bitumen [19]. The increase in resin content was also due to the transition of asphaltene to resin during blending, which causes the enhancement in viscosities due to PP modification as mention in an earlier paper [12] causing agglomeration of suspended miscible polymer particles in the blend. The presence of saturates in bitumen are responsible for fluidity in the bitumen. A decrease in saturates and aromatics indicates that oily fraction of the bitumen was absorbed by the addition of polymer. The decrease in aromatics in PP PMB also indicates the formation of dispersed polymer in the blend as observed by the decrease in colloidal index and segregation of polymer phase as observed in all AFM scanned images (Fig. 3 - 5). Decrease in aromatics also indicates the oxidation at mixing by the formation of the carbonyl group as mentioned in earlier work [20].

Table 1. SARA analysis results.

	Asphaltene %	**Saturates %**	**Aromatics %**	**Resins %**	**CI**
Virgin Bitumen	15	12	41	31	0.38
1% PP PMB	11.05	6.77	39.63	41.53	0.22
2% PP PMB	12	7.52	35.20	43.50	0.33
3% PP PMB	13.56	8.09	31.78	46.10	0.28

4.4 Mechanical analysis of bituminous mix

Dynamic creep test was conducted on triplicate samples prepared at optimum bitumen content obtained by Marshall Test results for control and PP modified samples [21]. The results are presented in the form of permanent accumulated strain and creep stiffness. The average mix stiffness modulus (S_{mix}) obtained from the creep modulus of three prepared mix specimen at optimum binder content for particular bituminous mix is plotted in double logarithmic scale against stiffness modulus of bituminous binder (S_{bit}), evaluated by using Van der Poel's nomograph as shown in Figure 6. The resistances to permanent deformation from the creep tests can be observed from the slope of the line of the log-log relationship of mixture stiffness versus binder stiffness. The slope of the line for each particular mix indicates sensitivity of the mix to loading time and hence bitumen stiffness [22].

A comparison of the mix stiffness with control mix at start and end of 1 hour test is presented in Table 2. When compared to control mix it was observed that there was an increase of 83% and 71% in creep stiffness for 1%PP and 2% PP respectively. 3% PP modified samples failed to show any improved performance as presented in Table 2. The higher percentage of polymer in blend although does enhances the viscosity but the agglomeration of polymer particles failed to coat the aggregate surface properly, thus unable to increase the modified bituminous concrete mix stiffness. The phase separated layer for 1% PP PMB as observed in AFM image (Figure.3) shows a higher surface roughness consisting of two different microstructure provides a better coating to the aggregate particle and enhanced viscoelastic property of binder. The combined effect of it is increased in stiffness value as observed by dynamic creep results for 1% & 2% PP PMB tested samples.

Table 2. Mix stiffness at the start and end of the test.

Mix Type	Stiffness at Start (MPa)	Stiffness at End(MPa)	Increase↑/ Decrease↓ in Stiffness at End w.r.t Control mix (%)
Control Mix	192.05	18.98	
1% PP PMB	788.88	111.06	82.91↑
2% PP PMB	822.7	66.02	71.25↑
3% PP PMB	212.3	29.89	36.50↑

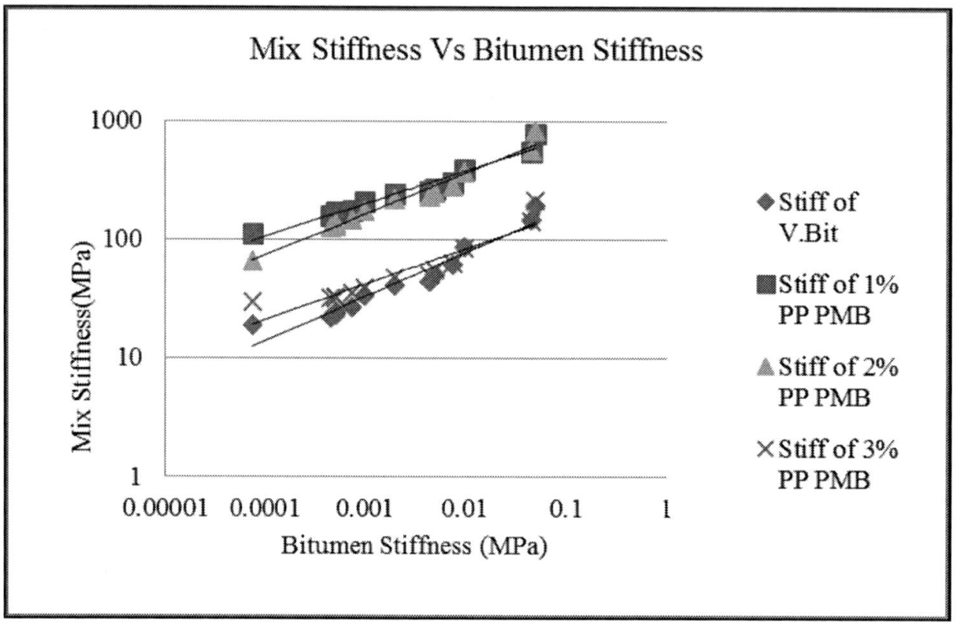

Fig. 6. Mix stiffness vs bitumen stiffness.

Results of the dynamic creep test can also be used to estimate strain accumulation in the sample against repeated loading cycles. The strain rate can effectively represent the viscoelastic behaviour of bituminous materials under repetitive loading cycle [23]. From Figure 7, it is observed that first 100 cycles are considered as compaction zone followed by a stable zone where stress is directly proportional to strain or can be defined as linear zone, where material show resistance to permanent deformation.

The stress-strain relationship in bituminous material can also be studied in view of dissipated energy concept [24] where, work done on the pavement by each application of the wheel load. During application of the wheel load a certain part of the work done is utilized in deforming the surface layer of the pavement. From Figure 7 it is observed that 2% PP & 3% PP modified bituminous mixtures showed higher strain level till the end in comparison to 1% PP and virgin bitumen.

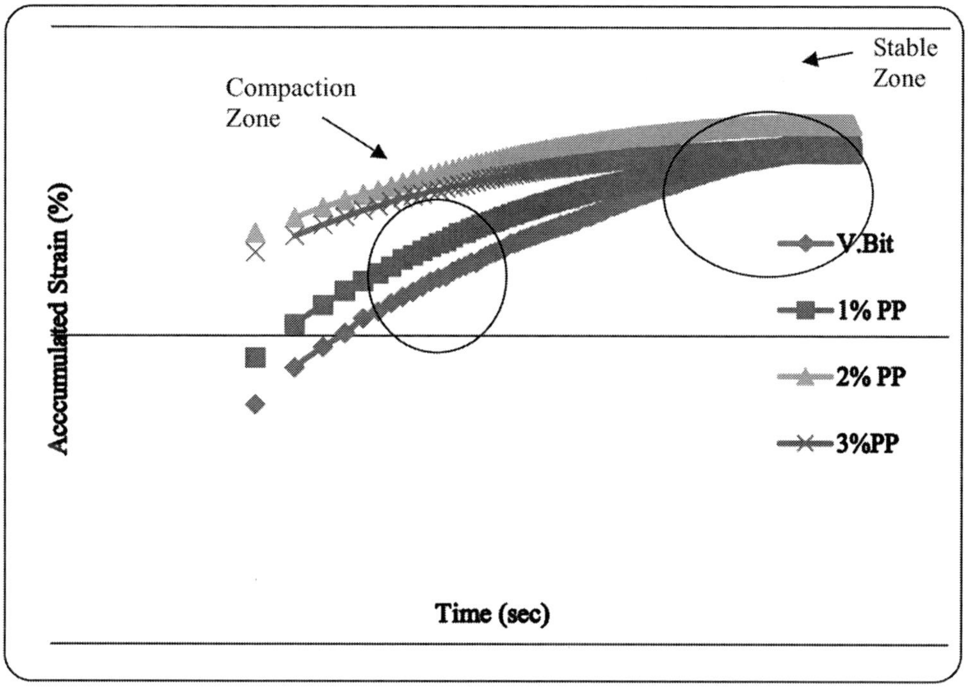

Fig. 7. Log accumulated strain vs log time of loading.

This was due to the viscoelastic behaviour, which is considered to be major component as some part of the dissipated energy was recovered during elastic recovery of the binder where polymer entangled chain realign themselves to resist the deformation of the pavement while the rest of the energy is released as crack initiation followed by plastic deformation of the material [23].It can be concluded that 2% & 3% PP PMB mixtures dissipate higher energy than control mix which may be considered due to stiffening effect of the polymer, because of its dispersion ability in blend forming the three dimensional network which was being proved by the significant effect on material behaviour to deformation and also to the strain level achieved till the end of test

5 Conclusion

The phase separated polymer layer for all percentages of polymer in the blend is considered as beneficial as it enhances material stiffness as observed by the dynamic creep results.

The chemical analysis of blend revealed that asphaltene are not only responsible for inducing the stiffening effect in the blend but the phase separated polymer layer as observed by AFM scanned images with lower colloidal index considered responsible for enhanced mechanical properties of polypropylene modified bituminous mix. The TEM images revealed the well-defined polymer network represented by lamella in all scanned images are mainly responsible for enhanced mechanical properties of modified bituminous mix.

It is concluded that modification of bitumen will help in improving the deformation behaviour of the bituminous mix when the concentration of the polymer in the blend is kept below 3%.

References

1. W. Uddin, *Appl. Rheo*, **13**, 191-199 (2003)
2. P.G. Redelius, *Fuel*, **79**, 27-35 (2000)
3. C. Giavarini, et al., *Fuel*, **75**, 681-686 (1996)
4. L. Michon, D Martin, JP Planche, B Hanquet, *Fuel*, **76**, 9-15 (1997)
5. D. Sybilski, *Mat and Struct*, **30**,182-187 (1997)
6. L. Didier, *Advances in Colloid and Interface Sci*, **145**, 42-82(2009)
7. B. Stuart, *Polymer Analysis*. United Kingdom: John Wiley & Sons, Ltd (2002)
8. Loeber, et al., *J. Micro*, **182**, 32-39 (1996)
9. S.P. Wu, et al., *Const. Build. Mat.*, **23**,1005-1010 (2009)
10. E. R. Dourado, et al., *J. Micro.*, **245**,119-128 (2012)
11. A. S. Michaels, et al., *J. Appl. Poly. Sci.* **12**, 1621-1624 (1968)
12. I. Kamaruddin. N.Z. Habib, M. Napiah and I. M. Tan, Int. *J. Civil and Envi. Eng.*, **3**, 96-100 (2011)
13. R. R. Coelho, et al., *Fuel Proces. Tech.*, **87**, 325-333 (2006)
14. D. W. John Read, *The Shell Bitumen Handbook*, Fifth ed.: Thomas Telford Publishing, (2003)
15. S. Haddadi, et al., *Const. Build. Mat.*, **22**, 1212-1219 (2008)
16. L. Loeber, et al., *Fuel*, **77**, 1443-1450 (1998)
17. S. N. Bhattacharya, *"Rheology Fundamentals and Measurement,"* R. M. I. o. T. M. Australia., Ed., ed. Melbourne: Royal Melbourne Institute of Technology (1997)
18. GD Airey, TM Singleton, AC Collop, *J. Mat in Civil Eng.*, **14**, 344-354 (2002)
19. J. F. Masson, et al., *J. of Micro.* **221**, 17-29 (2006)
20. N. Z. Habib, et al., National Postgraduate Conference (NPC), 1-7 (2011)
21. I. Kamaruddin, N. Z. Habib, M. Napiah and I. M. Tan, *World Acad. Sci. Eng. and Tech.*, **58**, 830 - 835 (2011)
22. I. Kamruddin, *PhD Thesis*, Civil Engineering Department, University of Leeds (1998)
23. I. Widyatmoko, et al., Materials and Structures, vol. 32, pp. 304-310, 1999
24. I. Widyatmoko, C. Ellis, J. M. Read, *Mat.and Struct.*, **32**, 218-223 (1999)

 DOI: 10.1051/matecconf/201712007003

Comparison of Witczak NCHRP 1-40D & Hirsh dynamic modulus models based on different binder characterization methods: a case study

Ahmed M. Khattab[1], *Sherif* M. El-Badawy[2],* *Mahmoud* Elmwafi[3], *Al Abbas* Al Hazmi[4]

[1]Mansoura University, Public Works Engineering Department, Faculty of Engineering, Mansoura University, Mansoura 35516, Egypt
[2]Associate Professor, Mansoura University, Public Works Engineering Department, Faculty of Engineering, Mansoura University, Mansoura 35516, Egypt
[3]Professor, Mansoura University, Public Works Engineering Department, Mansoura 35516, Egypt
[4]Projects Manager, Ministry of Transport, General Directorate for Material and Research, Riyadh 11178, Kingdom of Saudi Arabia

Abstract. The Pavement ME Design method considers the hot mix asphalt (HMA) dynamic modulus (E*) as the main mechanistic property that affects pavement performance. For the HMA, E* can be determined directly by laboratory testing (level 1) or it can be estimated using predictive equations (levels 2 and 3). Pavement-ME Design introduced the NCHRP1-40D model as the latest model for predicting E* when levels 2 or 3 HMA inputs are used. This study focused on utilizing laboratory measured E* data to compare NCHRP1-40D model with Hirsh model. This comparison included the evaluation of the binder characterization level as per Pavement ME Design and its influence on the performance of these models. E*tests were conducted in the laboratory on 25 local mixes representing different road construction projects in the kingdom of Saudi Arabia. The main tests for the mix binders were dynamic Shear Rheometer (DSR) and Brookfield Rotational Viscometer (RV). Results showed that both models with level 3 binder data produced very similar accuracy. The highest accuracy and lowest bias for both models occurred with level 3 binder data. Finally, the accuracy of prediction and level of bias for both models were found to be a function of the binder input level.

1 Introduction

Dynamic modulus (E*) characterizes in part the viscoelastic nature of hot mix asphalt (HMA). E* is a measure of the resistance of the HMA to deformation under compressive sinusoidal loading [1]. Several mix parameters were explored, in the NCHRP 9-19 project, in order to find which parameter strongly affects the mix performance [2].

* Corresponding author: sbadawy@mans.edu.eg

Based on the reported results, a strong correlation was found between E* and pavement field performance such as rutting and cracking over a diverse range of traffic and climatic conditions [2].

The Mechanic-Empirical Pavement Design Guide (MEPDG), now the latest version is called Pavement ME-Design, offers three hierarchical input levels for material and traffic characterization. For the three HMA material input levels, E* is the major material property input to calculate the stress-strain relationships of the HMA layer(s) at the critical locations considering the influence of truck load, truck speed and the environmental conditions. In addition, the stiffness of the HMA layer also affects the state of stress of underlying layers.

E* values for level 1 are measured in the laboratory using the Asphalt Materials Performance Tester (AMPT) or a Universal Testing Machine. However, the E* laboratory test is time-consuming, requires costly and advanced equipment, and skilled technicians. Consequently, over years, researchers have tried to establish E* predictive models using a number of laboratory E* databases. The most well-known models are the ones developed by Professor Witczak and his students over the last few decades [3-5]. Hirsh model is another well-known model for E* prediction [6]. In MEPDG or Pavement ME Design, level 2 and 3 inputs for the HMA layer(s), two different E* predictive models (Witczak's models) were adopted in the Pavement ME software. The first model is called "NCHRP 1-37A" which follows a sigmoid function and predicts E* of the mix as a function of mix aggregate gradation, volumetric properties, binder viscosity and loading frequency [3]. The second model which is the NCHRP1-40D is a modified version of the latest model developed by Bari and Witczak [4]. Following is a comprehensive overview of both Witczak NCHRP1-40D and Hirsh models.

2 NCHRP 1-40D G_b*-based E* model (revised Witczak model)

Bari and Witczak [7] presented a model in 2006 to predict the HMA dynamic modulus. The model is considered as a modified version of the well-known Witczak- Andrei predictive E* model (NCHRP 1-37A model). In this model the asphalt binder is expressed in terms of G_b* and δ not the conventional viscosity (η), as G_b* can better characterize binder stiffness as a function of changes in temperature and loading frequency. G_b* is also a part of the Superpave system for the binder characterization. In 2007, this model was modified under the NCHRP 1-40D project and incorporated since version 0.95 of MEPDG under the name "NCHRP 1-40D G_b*-based E* model" [4]. The model was developed based on a large database containing 7,400 E* data points from 346 mixtures [4]. The E* database for Witczak model contains non-aged, short-term oven aged for 4 hours at 135 °C, plant aged mixes, field aged cores as well as rubber modified mixes. The mathematical structure of the model is a sigmoidal function as presented in Eq. (1) [4].

$$\log_{10} E^* = 0.02 + 0.758\left(|G_b^*|^{-0.0009}\right)\times\begin{pmatrix}6.8232 - 0.03274\rho_{200} + 0.00431\rho_{200}^{2} + 0.0104\rho_4 - 0.00012\rho_4^{2}\\ + 0.00678\rho_{38} - 0.00016\rho_{38}^{2} - 0.0796V_a - 1.1689\left(\dfrac{V_{beff}}{V_a + V_{beff}}\right)\end{pmatrix}$$
$$+\frac{1.437 + 0.03313V_a + 0.6926\left(\dfrac{V_{beff}}{V_a + V_{beff}}\right) + 0.00891\rho_{38} - 0.00007\rho_{38}^{2} - 0.0081\rho_{34}}{1 + e^{(-4.5868 - 0.8176\log|G_b^*| + 3.2738\log\delta)}} \tag{1}$$

where:
E*=HMA dynamic modulus, in 10^5 psi; $|G_b^*| = G_b^*$ =binder complex shear modulus, psi; δ= binder phase angle, degrees; f= loading frequency in compression, in Hz; V_a= % air voids in the mix,; V_{beff} = % effective binder content, by volume; ρ_{34}= % cumulative retained weight on the ¾ in, sieve; ρ_{38}= % cumulative retained weight on the 3/8 in. sieve; ρ_4=% cumulative retained weight on the No.4 sieve; and ρ_{200}= % passing No.200 sieve.

The coefficient of determination (R^2) of this model is 0.91 in logarithmic scale. It should be noted that G_b* and δ values contained in the database used for the model development are mostly estimated from regression models based on viscosity rather than measured in the laboratory. Thus, the error in the model's predictions may be caused by the predicted G_b* and δ values. Furthermore, several studies have shown that this model significantly over predicts the dynamic modulus, particularly at the highest and lowest temperatures [8-10]. It was also reported that this model generally tends to overvalue the influence of temperature and underestimate the influence of other mixture characteristics such as the mix gradation [10].

3 Hirsh model

Christensen et al. [6] modified the original Hirsch model to predict E* of HMA. The alternate formulation of the Hirsch model "from here on will be called Hirsh Model" is shown in Eqs. (2) & (3) [6]:

$$|E*| = Pc\left[4{,}200{,}000\left(1-\frac{VMA}{100}\right)+3|G_b^*|\left(\frac{VFA*VMA}{10{,}000}\right)\right] + (1-Pc)\times\left[\frac{1-\frac{VMA}{100}}{4{,}200{,}000}+\frac{VMA}{3\times VFA\times|G_b^*|}\right] \quad (2)$$

where,
|E*|= dynamic modulus of the mixture, in psi; |G_b*| =|G*| = complex shear modulus of the binder, in psi; VMA= % voids in the mineral aggregates; VFA= % voids filled with asphalt; Pc= contact volume estimated from the following equation:

$$\mathbf{Pc} = \frac{\left(20+\frac{VFA\times 3|G_b^*|}{VMA}\right)^{0.58}}{650+\left(\frac{VFA\times 3|G_b^*|}{VMA}\right)^{0.58}} \quad (3)$$

Hirsh model was established using a much lower database containing only 206 E* measurements representing 18 HMA mixtures with only eight different binders. The R^2 of the model is 0.98. As reported by the researchers, in the above equations G_b* can be determined in the laboratory using Dynamic Shear Rheometer (DSR) or a similar device [6]. It can also be determined from mathematical models. Additionally, similar to Witczak model, it should be determined at the same temperature and loading frequency of the mixture modulus, E*, and in consistent units [6]. This model is simpler in form and number of inputs compared to Witczak model. However, it was reported to lose its prediction accuracy when used to predict E* values for Witczak database [4, 7]. It was also reported by several researchers that Hirsh model, similar to Witczak model, yields significantly biased E* predictions at the extreme low and high temperatures [4, 7, 11-13].

4 Comparison studies of Witczak and Hirsh models

Several studies focused on using laboratory measured E* data to compare and evaluate the accuracy of the two well-known Witczak models which are the NCHRP 1-37A and NCHRP 1-40D [14-16]. The studies conducted by Al-Khateeb et al. [14] and Yousefdoost et al. [15] showed that the NCHRP 1-37A model produced more accurate E* predictions compared to the NCHRP 1-40D model, particularly at the higher temperature range [. Whereas a recent study in China by Wang et al. [16] showed that the NCHRP 1-40D model yielded accurate E*estimates compared to the NCHRP 1-37A model.

Few other studies compared between the prediction accuracy of the NCHRP 1-37A and Hirsh models. For example, Kim et al. [17] at North Carolina tested 42 asphalt mixtures and showed that the NCHRP 1-37A model yielded better E* values compared to Hirsh model at cooler temperatures than at warmer temperatures. However, both models showed similar behavior at intermediate and high temperatures. Obulareddy et al. [18] tested 11 asphalt concrete mixtures from different projects in Louisiana and results showed that Hirsh model had excellent correlation with measured dynamic modulus values at each traffic level in comparison with to the NCHRP 1-37A model. Awed et al. [19] concluded that Hirsh and NCHRP 1-37A models generated more accurate E* values compared to the NCHRP 1-40D model based on 1032 E* measurements from 25 mixes from the state of Idaho.

The prediction accuracy of the NCHRP 1-40D and Hirsh models was also investigated by Sakhaeifar et al at based on 3180 E* measurements. The results showed that the NCHRP 1-40D and Hirsh models produced undesirable bias at low and high E* values [20]. Recently, Georgouli et al. [21] tested 15 asphalt base and wearing course mixtures (1350 E* measurements) from Greece to evaluate the accuracy of the NCHRP 1-37A, NCHRP 1-40D, and Hirsch models. For asphalt base mixtures the results showed that the NCHRP 1-37A model produced accurate results. On the other hand, Hirsh model under-predicted E* values while Bari and Witczak model over predicted E* values. The researchers also reported that the large bias in the estimated E* values of Greece mixes using Bari and Witczak or the NCHRP 1-40D E* models is due to the associated equations used for the estimation of G* and δ which in most cases did not yield acceptable values [21].

Few studies focused on the investigation of the effect of binder characterization input level, as per Pavement ME Design or MEPDG, on the prediction accuracy of the E* models [5, 9, 13]. Other studies evaluated the effect of the MEPDG binder input level on both Witczak E* predictive models based on 27 Idaho asphalt mixes [9, 13]. MEPDG binder characterization input levels 1 and 2 were obtained from laboratory testing while level 3 binder input was based on the default MEPDG values. The researchers reported that the NCHRP 1-37A model with level 3 binder input produced the most accurate and least biased E* predictions compared to the other levels of binder input and the NCHRP 1-40D model at all three binder input levels [10]. Similar results on mixes from the Kingdom of Saudi Arabia (KSA) were also reported by Khattab et al. [5]. Thus, based on the multitude of the literature research studies covering the presented E* models, one may surmise that the prediction accuracy of these models is not consistent and vary depending of the properties of HMA and binders.

5 Objectives and scope of work

The main objective of this study is to compare the accuracy of Witczak NCHRP1-40D model with Hirsh model for laboratory measured E* values of 25 local KSA Superpave HMA mixtures considering the influence of the binder input level as per Pavement ME Design (level 1 versus level 2/3). Dynamic Shear Rheometer (DSR) and Brookfield Rotational Viscosity (RV) tests were conducted on the mixtures' binders using Rolling Thin Film Oven (RTFO) aged specimens. E* testing was conducted on 25 local mixes at six loading frequencies and four different temperatures. Finally, measured E* values were compared with predicted values using both The NCHRP1-40D and Hirsh models at the different binder input levels of the Pavement ME Design.

6 Materials

Based on the climatic regions in KSA, there are three Superpave Performance Grades (PG) that are typically used. These grades are PG 64-10, PG 70-10, and PG 76-10 which were used in this study. Eight types of modifiers (shown in Table 1) were used to achieve the required PG grades of the modified binders (PG70-10 and PG76-10). A total of 25 KSA mixes were investigated. The investigated field mixes cover the different climatic regions of KSA (i.e., central, northern, southern and, eastern regions of KSA) [5, 22]. The mixtures contain14 laboratory mixes and 11 plant-produced mixes that were collected from ongoing road construction projects in KSA. Each laboratory mix had three different gradations (coarse, medium, and fine) and three percentages of air voids (2%, 4%, and 8%).

7 Binder testing

Laboratory tests were performed on samples from the asphalt binders contained in the investigated mixtures to determine the binder properties. Brookfield Rotational Viscosity (RV) testing, according to AASHTO TP48-97, was conducted on short term aged samples from the investigated binders using the Rolling Thin Film Oven Test (RTFO) [23]. The RV tests were conducted at three temperatures of 135, 150, and 165 °C. The Dynamic Shear Rheometer (DSR) test was also conducted on RTFO aged samples in order to determine the complex shear modulus (G*) and phase angle (δ). Controls DSR model 81-PV6002 was used for the DSR testing in accordance with AASHTO T 315-06 [24]. The test was conducted at three different temperatures according to the binder PG grade and one loading frequency of 10 radians per second (1.59 Hz) in accordance with the Superpave specifications.

8 HMA dynamic modulus sample preparation and testing

A Gyratory Compactor was used to prepare E* samples of 150 mm in diameter and 170 mm in height. The samples were sawed from the top and bottom to obtain a final height of 150 mm and then cored in the middle to achieve a final diameter of 100 mm.

The E* tests were conducted according to AASHTO T342-11 at four different temperatures of -10, 4.4, 21.1, and 54.4 °C. At each temperature, the test was conducted at frequencies of 25, 10, 5, 1, 0.5, and 0.1 Hz [25]. Asphalt Mixture Performance Tester (AMPT) was used for all E* tests. The majority of the E* tests were performed on two replicate samples per mix.

9 E* Predictions

For both E* predictive models, G_b* and δ for the binder need to be estimated at the same test temperature and frequency of E*. However, based on the Superpave requirements G_b* and δ are determined at different temperatures and only one loading frequency (1.59 Hz, which is equivalent to 10 rad/s). In order to overcome this problem, the following methodology, which is used by the Pavement ME Design, was followed to estimate G_b* and δ at the loading frequency and temperature of the measured E* values. Eq. (4) was first used to compute binder viscosity at different temperatures based on the laboratory measured G_b* and δ values at the 10 rad/sec frequency [26, 27].

$$\eta = \left(\frac{G_b\,^*}{10}\right)\left(\frac{1}{\sin\delta}\right)^{4.8628} \quad (4)$$

where:

G_b* = dynamic shear modulus of binder, psi; η = binder viscosity, in 10^6 poise; and δ = phase angle of the binder, degrees.

Then, Eq. (5) which represents a linear fit to the viscosity-temperature data (from Equation 4) was used to estimate the ASTM A and VTS parameters for each investigated binder, [28].

$$\log\log \eta = A + VTS \ \log T_R \tag{5}$$

where:
η = binder viscosity, in cp; T_R= testing temperature, in Rankine; A= regression intercept; and VTS= viscosity-temperature susceptibility (regression slope of the viscosity-temperature relationship).

Finally, Eqs (6 through 11) were used to find the binder G_b* and δ for both investigated models at the temperature and frequency at which E* needs to be determined [8]:

$$\log\log\eta_{f_s,T} = A' + VTS' \log T_R \tag{6}$$

$$A' = 0.9699 * f_s^{-0.0527} * A \tag{7}$$

$$VTS' = 0.9668 * f_s^{-0.0575} * VTS \tag{8}$$

$$f_s = f_c/2\pi \tag{9}$$

$$\delta = 90 - 0.1785 * \log(\eta_{f_s,T})^{2.3814} * (f_s)^{(0.3507+0.0782VTS')} \tag{10}$$

$$|G_b^*| = 1.469 * 10^{-9} * \log(\eta_{f_s,T})^{12.0056} * f_s^{0.7418} (\sin\delta)^{0.6806} \tag{11}$$

where:

f_c = loading frequency in dynamic compression loading as used in the E* testing, in Hz; f_s= loading frequency in dynamic shear loading mode as used in the G_b* testing, in Hz; A'= adjusted "A" (adjusted for loading frequency); VTS'= adjusted "VTS" (adjusted for loading frequency); η =binder viscosity as a function of both loading frequency (f_s) and temperature (T_R), in cP; T_R = temperature, in Rankine; |G_b*|= G_b* = complex binder shear modulus, in Pa.

10 Pavement ME-design binder input levels

For input level 1a (Pavement ME Design level 1 conventional binder data), the RV test results on the RTFO aged binders were fit with Eq. (5) to find the A-VTS values. For input level 1b (Pavement ME Design level 1 Superpave performance grade (PG) binder data), the DSR results on the RTFO aged binders were used to calculate the viscosity at each E* tested temperature using Eq. (4). The A and VTS values for each binder were calculated using Eq. (5). Finally, Equations (6 through 11) were used to estimate G_b* and δ at the same frequency and test temperature of E*. Lastly, for input level 3 (Pavement ME Design level 3 default A-VTS parameters), the Pavement ME Design default A-VTS values stored in the software were selected based on the Superpave binder PG grade.

Table 1 summarizes the A and VTS parameters estimated for levels 1a and 1b. The coefficient of determination (R^2) as well as the default A and VTS values (level 3) recommended by the Pavement ME Design for the investigated binders are also shown in the table. The high R^2 values indicate excellent fit.

The bias in prediction was assessed by determining the closeness of the slope and intercept of the regression line of measured and predicted E* values to one and zero, respectively [29]. The effect of the binder input level on the predicted E* values using the

NCHRP1-40D model is shown in **Figure 1 (a through c)**. The figure shows relatively very good prediction accuracy as indicated by the goodness of fit statistics summarized in **Table 2**. The prediction accuracy (lowest scatter) was the highest for the predictions based on level 3 binder characterization (Pavement ME Design default A-VTS values based on the PG grade of the binder) with R^2 of 0.84 and S_e/S_y of 0.40.

Table 1. A-VTS Values for RTFO-Aged Binders based on Brookfield, DSR, and Pavement ME Design Level 3 Values [5].

Mix ID	Binder PG Grade	Level 1a (Brookfield) Pavement ME Level 1 Conventional			Level 1b (DSR@10rad/s) Pavement ME Level 1 Superpave			Level 3 (Pavement ME Default)	
		A	VTS	R^2	A	VTS	R^2	A	VTS
H, T6(F&M&C)	64-10	11.990	-4.025	0.999	8.815	-2.897	0.995	11.432	-3.842
H, T7(F&M&C)	76-10	10.630	-3.536	0.999	11.680	-3.908	0.998	10.059	-3.331
FW	70-10	11.480	-3.838	0.999	8.674	-2.844	0.999	10.690	-3.566
SJ	76-10	12.420	-4.181	0.998	8.542	-2.787	0.999	10.059	-3.331
SP	76-10	11.230	-3.743	0.999	9.699	-3.200	0.999	10.059	-3.331
HG	70-10	11.690	-3.906	0.999	10.200	-3.386	0.999	10.690	-3.566
FQ	76-10	9.465	-3.133	0.996	8.542	-2.787	0.999	10.059	-3.331
AJM (K&D)	76-10	10.590	-3.528	0.999	9.493	-3.125	0.999	10.059	-3.331
AUYN	76-10	10.460	-3.471	0.999	10.060	-3.327	0.999	10.059	-3.331
BIN JAR	70-10	10.130	-3.371	0.997	9.590	-3.160	0.999	10.690	-3.566
HH	70-10	10.128	-3.371	0.994	9.788	-3.245	0.999	10.690	-3.566
A	64-10	12.668	-4.262	0.998	9.501	-3.143	0.997	11.432	-3.842
D	76-10	11.501	-3.835	0.999	12.123	-4.065	0.999	10.059	-3.331
NH	76-10	11.551	-3.855	0.999	10.916	-3.652	0.980	10.059	-3.331

However, all three cases showed variable degrees of bias in the predictions as indicated by the slope and intercept of the regression line shown in the figures. Again, the lowest scatter occurred when level 3 binder values was used (slope of 0.9402 and intercept of 0.365). The unconstrained regression lines shown in **Figure 1 (a to c)** also show that the bias in the predictions based on level 1a binder data was mostly in all E* values while for the case of level 1b and level 3 binder data it was mostly in the low and intermediate E* values.

For Hirsh model the comparison of measured and predicted E* values at the different binder characterization levels is shown in **Figure 1 (d to f)**. The goodness of fit statistics for the E* predictions shown in Figure 1 indicates a very good accuracy (low scatter). Similar to Witczak model, the data suggests that the highest accuracy ($R^2 = 0.85$ and $S_e/S_y = 0.39$) and lowest bias (slope = 0.862 and intercept = 0.4107) occurred when level 3 binder data was used. For Hirsh model the bias in the predictions was mostly at the high and low temperatures (low and high E* values) as reported by many research studies [4, 7].

The figure shows that the accuracy of both models when E* is predicted based on either level 1b or level 3 binder is very close. If level 1a binder data is used, then Hirsh model is more accurate compared to Witczak model. However, the bias of Hirsh model is higher compared to Witczak model at all binder input levels especially at the high and low E* values.

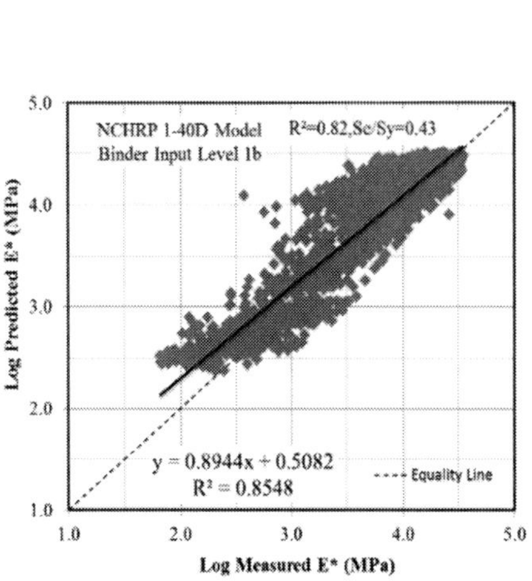

a. NCHRP 1-40D (binder input level 1a)

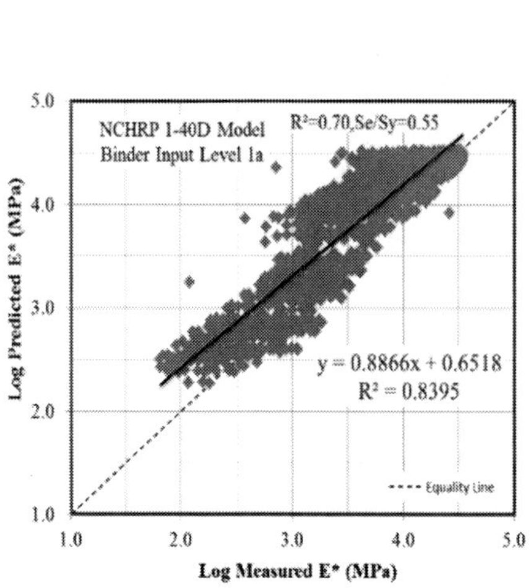

d. Hirsh (binder input level 1a)

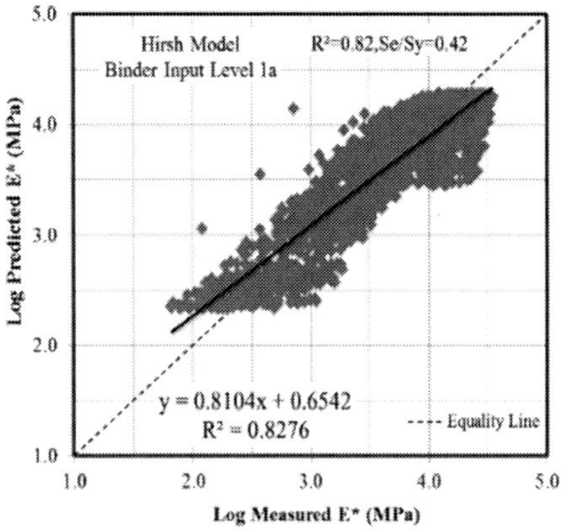

b. NCHRP1-40D (binder input level 1b)

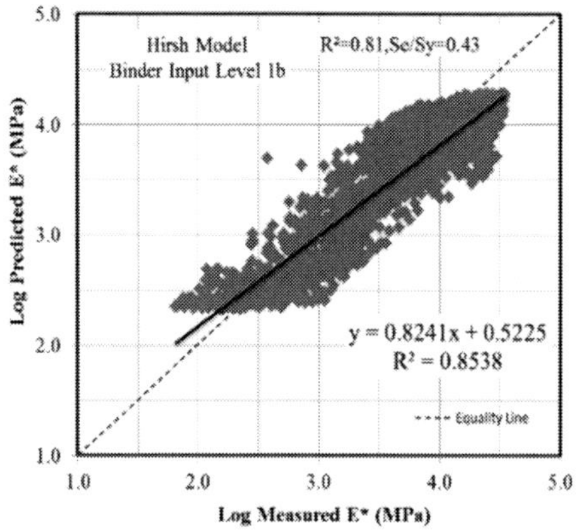

e. Hirsh (binder input level 1b)

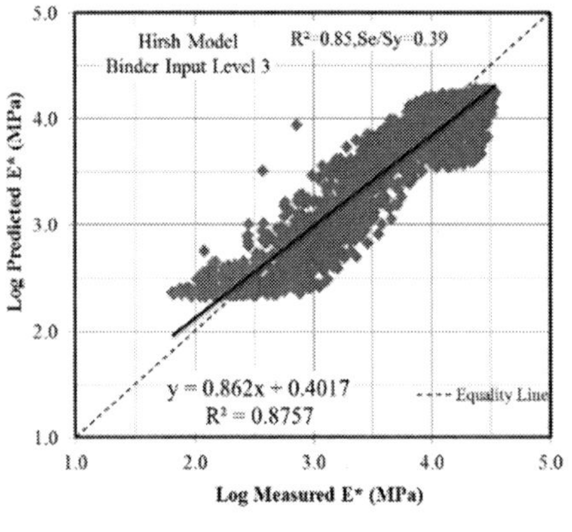

c. NCHRP1-40D (binder input level 3)

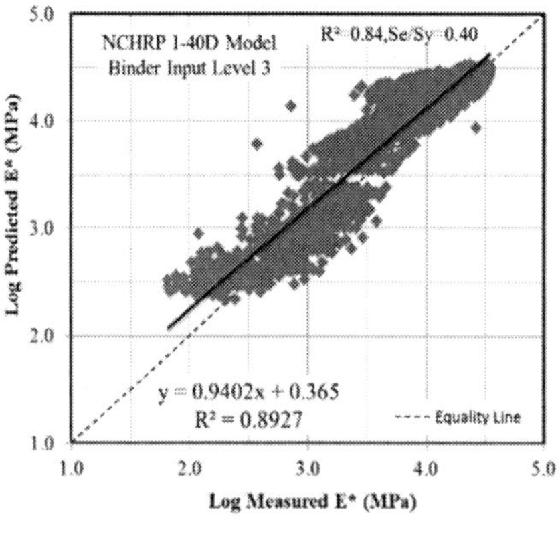

f. Hirsh (binder input level 3)

Fig. 1. Measured versus predicted E* values using Witczak NCHRP1-40D and Hirsh models at the different binder input levels

12. Conclusions

The accuracy and level of bias of both Witczak NCHRP1-40D and Hirsh dynamic modulus prediction models with three Pavement ME Design binder characterization input levels were studied for 25 local KSA Superpave mixtures. The results of this study suggest that the performance (in terms of accuracy and bias) of the investigated models is greatly influenced by the binder characterization method. Both Witczak NCHRP 1-40D and Hirsh models yielded relatively accurate E* predictions for the KSA mixes with very similar goodness of fit statistics for the predictions based on level 1b and level 3 binder data. However, Hirsh model along with binder data based on level 1a (A-VTS values based on RV data) was more accurate compared to the NCHRP 1-40D model. Both models showed some bias in the predictions which varied with the binder input data. For Hirsh, the bias was more towards the lower and higher E* values while for Witczak the bias was more towards the lower and intermediate E* values. Generally, the NCHRP1-40 model yielded lower bias compared to Hirsh model at the three binder characterization input levels. Thus, Hirsh model needs some improvements in terms of bias. Both Witczak and Hirsh models with level 3 binder data produced very similar accuracy based on the goodness of fit statistics. Finally, the highest accuracy and lowest bias for both models occurred with level 3 binder data.

References

1. Kim, Y. R., Editor (2009): "Modeling of Asphalt Concrete", ASCE Press, McGraw Hill.
2. Witczak, M.W., (2005): "Use of the Dynamic Modulus (E*) Test as a Simple Performance Test for Asphalt Pavement Systems (AC Permanent Deformation Distress)", Volume I of IV Preliminary Draft Final Report, NCHRP Project 9-19: Superpave Support and Performance Models Management, Arizona State University.
3. Andrei, D., Witczak, M. W., and Mirza, M. W, (1999): "Development of A Revised Predictive Model for the Dynamic (Complex) Modulus of Asphalt Mixtures", Inter Team Technical Rep. prepared for the NCHRP 1-37A Project, Dept. of Civil Engineering, Univ. of Maryland, College Park, MD.
4. Bari, J., and Witczak, M. W., (2007): "New Predictive Models for the Viscosity and Complex Shear Modulus of Asphalt Binders for Use with the Mechanistic-Empirical Pavement Design Guide", Transportation Research Record 2001, Transportation Research Board, Washington, D.C., 9–19.
5. Khattab A. M., El-Badawy S. M., Al Hazmi A. A., Elmwafi M., (2014): "Evaluation of Witczak E* Predictive Models for the Implementation of AASHTOWare-Pavement ME Design in the Kingdom of Saudi Arabia", Construction and Building Materials, V (64): pp. (360-9).
6. Christensen Jr., D. W., Pellinen, T. K., & Bonaquist, R. F., (2003): "Hirsch Model for Estimating the Modulus of Asphalt Concrete", Asphalt Paving Technology, Lexington, KY: Journal of the Association of Asphalt Paving Technologists, pp. (97- 121).
7. Bari, J., and Witczak, M. W., (2006): "Development of A New Revised Version of the Witczak E* Predictive Models for Hot Mix Asphalt Mixtures", J. Assoc. Asphalt Paving Technol., Vol. 75, pp. (381–417).
8. Witczak M., El-Basyouny M., El-Badawy S., (2007): "Incorporation of the New (2005) E* Predictive Model in the MEPDG", NCHRP 1-40D Final Report Addendum, Arizona State University, Tempe, Arizona.
9. El-Badawy, S., Bayomy, F., and Awed, A., (2012): "Performance of MEPDG Dynamic Modulus Predictive Models for Asphalt Concrete Mixtures - Local Calibration for Idaho", ASCE's Journal of Materials in Civil Engineering, Volume 24, Issue 11:1412-1421.

10. Ceylan, H., Schwartz, C. W., Kim, S., and Gopalakrishnan, K.,(2009): "Accuracy of predictive Models for Dynamic Modulus of Hot-Mix Asphalt," Journal of Materials in Civil Engineering, Vol. 21, No. 6, pp. 286-293.
11. Martinez, F. O., and Angelone S. M., (2009) "Evaluation of Different Predictive Dynamic Modulus Models of Asphalt Mixtures Used in Argentina, Bearing Capacity of Roads, Railways, and Airfields", Tutumluer & Al-Qadi (eds), Taylor & Francis Group, London.
12. Bayomy, F., Abu Abdo, A., Weaver, T., Darveshi, P., Nielsen, R., Jung, S.J., and Baek, S., (2010): "Development and Evaluation of Performance Tests to Enhance Superpave Mix Design and its Implementation in Idaho. Phase A: Evaluation of Mix Resistance to Deformation," Final Report-Part 1 of ITD project No. RP 181, USDOT Project No. DTOS59-06-G-00029, NIATT Report N09-10A, University of Idaho, pp. 160.
13. El-Badawy, S., Awed, A., and Bayomy, F., (2011): "Evaluation of the MEPDG Dynamic Modulus Prediction Models for Asphalt Concrete Mixtures", Proc., 1st Integrated Transportation and Development Institute Congress, I. Al-Qadi and S. Murrell, eds., ASCE, Reston, VA, pp. (576–585).
14. Azari, H., Al-Khateeb, G., Shenoy, A., and Gibson, N. (2007). "Comparison of Simple Performance Test E* of Accelerated Loading Facility Mixtures and Prediction E* Use of NCHRP 1-37A and Witczak's New Equations." Transportation Research Record 1998, Transportation Research Board, Washington, D.C., 1–9.
15. Yousefdoost, S., Vuong, B., Rickards, I., Armstrong, P., and Sullivan, B. (2013): "Evaluation of Dynamic Modulus Predictive Models for Typical Australian Asphalt Mixes", Delivering New Age Solutions: 15th AAPA International Flexible Pavements Conference, 22-25 September, Royal International Conference Centre, Brisbane.
16. Hou, H., Wang, T., Wu, S., Xue, Y., Tan, R., Chen, J., and Zhou, M., (2016): "Investigation on the Pavement Performance of Asphalt Mixture Based on Predicted Dynamic Modulus", Construction and Building Materials; 106:11-17.
17. Kim, R., King, M., and Momen, M. (2005). "Typical Dynamic Moduli Values of Hot Mix Asphalt in North Carolina and their Prediction." Transportation Research Board 84th Annual Meeting Compendium of Papers 05-2568 (CD-ROM), Washington, D.C.
18. Obulareddy, S., (2006): "Fundamental Characterization of Louisiana HMA Mixtures for the 2002 Mechanistic-Empirical Design Guide", Master Thesis, Louisiana State Univ., Baton Rouge, LA.
19. Awed, A, (2010): "Material Characterization of HMA for MEPDG Implementation in Idaho", Master's Thesis, Univ. of Idaho, Moscow, ID.
20. Sakhaeifar, M., Kim, R., and Kabir, P., (2015): "New Predictive Models for the Dynamic Modulus of Hot Mix Asphalt", Construction and Building Materials; 76:221-231.
21. Georgouli, K., Loizos, A., and Plati, C. (2016): "Calibration of Dynamic Modulus Predictive Model", Construction and Building Materials; 102:65-75.
22. Ministry of Transport (MOT) in the Kingdom of Saudi Arabia. Hot Asphalt Mix Design System Using Superpave System Detailed in Asphalt Institute SP-2 and The AASHTO 2005 Standards. SUPERPAVE Coding System in KSA, Version: 1/2006.
23. AASHTO, (2006): "Standard Method of Test for Viscosity Determination of Asphalt Binder Using Rotational Viscometer", AASHTO Designation: T48-97, Washington, D.C., USA.
24. AASHTO, (2006): "Standard Method of Test Determining the Rheological Properties of Asphalt Binder Using a Dynamic Shear Rheometer (DSR)", AASHTO Designation: T315-06, Washington, D.C., USA.
25. AASHTO, (2011). Standard Method of Test for Determining Dynamic Modulus of Hot Mix Asphalt Concrete Mixtures, AASHTO Designation: TP 342-11, American Association of State Highway and Transportation Officials, Washington, D.C., USA.

26. ARA, Inc., ERES Consultants Division (2004). Guide for Mechanistic-Empirical Design of New and Rehabilitated Pavement Structures, NCHRP 1-37A Final Report, Transportation Research Board, National Research Council, Washington, D.C.
27. Garcia G, Thompson M. (2007). HMA Dynamic Modulus Predictive Models –A Review. Research Report FHWA-ICT-07-005, Urbana, IL.
28. ASTM. (2009). "Standard Viscosity-Temperature Chart for Asphalts." ASTM D2493/D2493M-09, West Conshohocken, PA.
29. Ceylan H., Gopalakrishnan K., Kim S. (2008), Advanced Approaches to Hot-Mix Asphalt Dynamic Modulus Prediction. Canadian Journal of Civil Engineering. 35(7):699-707.
30. Pellinen TK. (2001), Investigation of the Use of Dynamic Modulus as an Indicator of Hot-Mix Asphalt Performance: PhD Dissertation, Arizona State University, Tempe, Arizona.

Development of a mobile application for manual traffic counts

Mohammad Ghanim[1,*] and *Khalid* Khawaja[2]

[1]Department of Civil and Architectural Engineering, Qatar University, Doha, Qatar
[2]Office of Academic Affairs, RIT Dubai, Dubai, UAE

Abstract. The use of technology in performing traffic counts has become widely used by many transportation agencies and private sectors. Different technologies have been employed to perform such counts, such as inductive loops, image processing, Bluetooth technologies, and laser and infrared technologies. However, the use of these advanced technologies come with extra cost to develop, install, and maintain. While the use of advanced technologies provide reliable traffic counts and vehicle classifications data, the use of manual traffic is an inevitable task. Manual traffic counts can serve different purposes, such as performing quality control or conducting short-term traffic survey. Manual traffic counts are usually conducted with the use of the traditional paper-and-pencil approach or the use of hand-held devices that are specifically developed to assess performing manual traffic counts. In this paper, the development of a mobile app that can be used to assist those who need to perform manual traffic counts. The app can be used by many public agencies, private sectors, and college students to collect their traffic data. It also provides users with the capabilities of sharing their traffic counts instantly, where traffic counts can be extracted and processed quickly by other parties.

1 Introduction

The advances of technology nowadays allow transportation and engineers to collect traffic counts using many different technologies that vary in their complexity, type of collected traffic data, and price, and mobility. However, there is always a need to perform manual traffic counts for different reasons, such as the associated costs and quality control needs. There are many portable products and hand-held devices that are used to facilitate manual traffic counts. Usually these devices are bundled with a special software to abstract, process, and summarize traffic counts.

The use of mobile apps nowadays has spread widely and rapidly over the last 10 years. There are over three million mobile apps that are developed and operated under the leading app stores, for different smartphone operation system platforms (i.e., iOS, Android, Amazon, Windows Phone, and BlackBerry) [1-4]. Furthermore, the hardware components themselves are developed with integrated devices that will increase their usability, such as

* Corresponding author: mghanim@qu.edu.qa

global positioning systems (GPS), front and rear high resolution cameras, and accelerometers.

The use of smart phone and mobile applications have gained popularity over the last decade, as can be seen in these examples [5-8].

Despite the exponential growth of using mobile apps to do many different tasks, its usability to perform manual traffic counts are very limited. Although this can be partially associated with the fact that there is no comprehensive mobile app that can do several traffic counts functionalities.

This paper illustrates the development of a mobile app that is comprehensive enough to assist performing most of the manual traffic count needs, such as mainstream flow counts, pedestrians' counts, turning movement counts, and vehicle classification.

2 Mobile app development motivation

Although there are several hand-held devices that are used to collect manual traffic data, most of the portable traffic counters are usually expensive, when compared to widely spread tablets and smartphones. Traffic counts data cannot be easily transferred for further processing. Their size is big when compared to the area of hand palms. Moreover, they have limited functionalities that does not exceed recording traffic counts. Therefore, the need for a manual traffic count system that is price-insensitive, highly portable, with on-demand availability is needed.

3 Traffic abacus functionalities

In this project, a mobile app called "Traffic Abacus" that works under Android's OS is being developed [9]. The Android OS is used due to its wide usage. The developed app is designed to allow several manual traffic counts needs, covering most of the scenarios where manual traffic counts are suited. These functionalities are:

- 4-leg Intersections: w/o U-Turns and w/o pedestrians.
- 4-leg Intersections: w/o U-Turns and with pedestrians.
- 4-leg Intersections/Roundabouts: with U-Turns and w/o pedestrians.
- 4-leg Intersections/Roundabouts: with U-Turns and with pedestrians.
- Vehicle Classification: according to the FHWA Scheme "F".
- Directional Mainstream Traffic Counts: can be classified per lane.

Although roundabouts are not specifically listed as an option, a four-leg intersection with U-Turns. As can be seen in Fig. 1, a snapshot is taken for each of the previous functionalities.

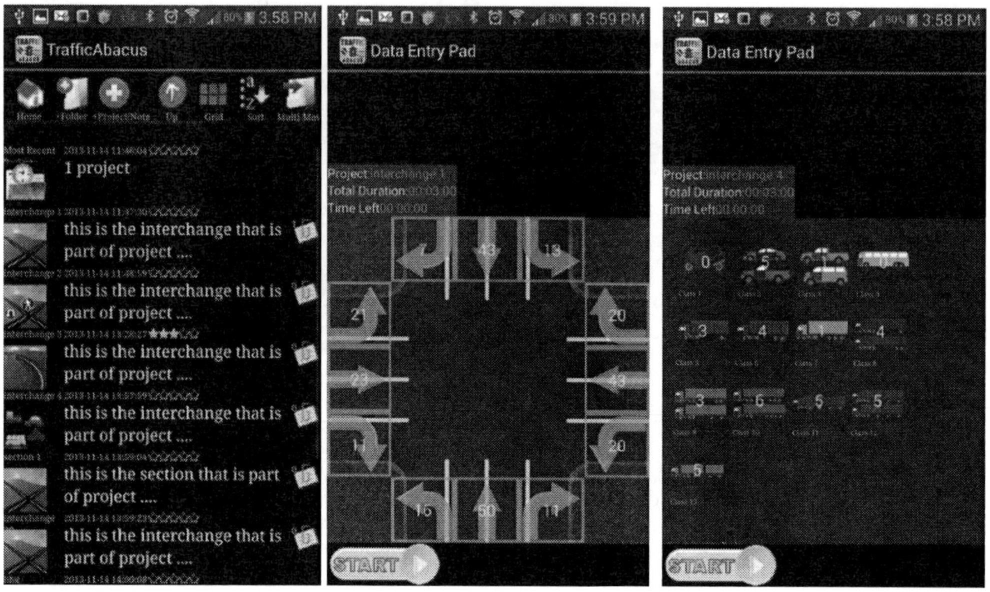

Fig. 1: Traffic Abacus Application Interface – Sample Snapshots.

4 Traffic abacus architecture

This application is developed to work on Android devices. The following broad objectives were stated for the application project:

- The application must support all the basic traffic count scenarios:
- The application must function properly on a single device without internet connection
- The application must support ability to link notes and media files (audio, photo and video) to each counting session
- The application must support ability to group various sessions together
- The application must allow sharing of collected data

As a result, all the scenarios that were presented in the previous section were implemented in a standalone application with the architecture shown in Fig. 2 and Fig. 3. The user can initiate a project for a scenario data collection and group projects in a folder. The user can also attach notes and media files to any project. A local database was used to store collected data and any associated notes. The database allows for the operation of the application independent from any network connection. An export operation leads to the generation of an XML file with all the data. The XML file is able to represent a single project, a folder that contains several projects, or even several folders. This XML file can then be sent to any user of the Traffic Abacus application who can import the file to get the same view. This allows and promote team collaboration and simultaneous team data collection. In addition, this architecture lends itself to future enhancement that allows server based real-time synchronization of team data utilizing the XML file as the data exchange medium. The next section gets into some more details related the add-ons gained from data collection on the mobile.

 DOI: 10.1051/matecconf/201712007004

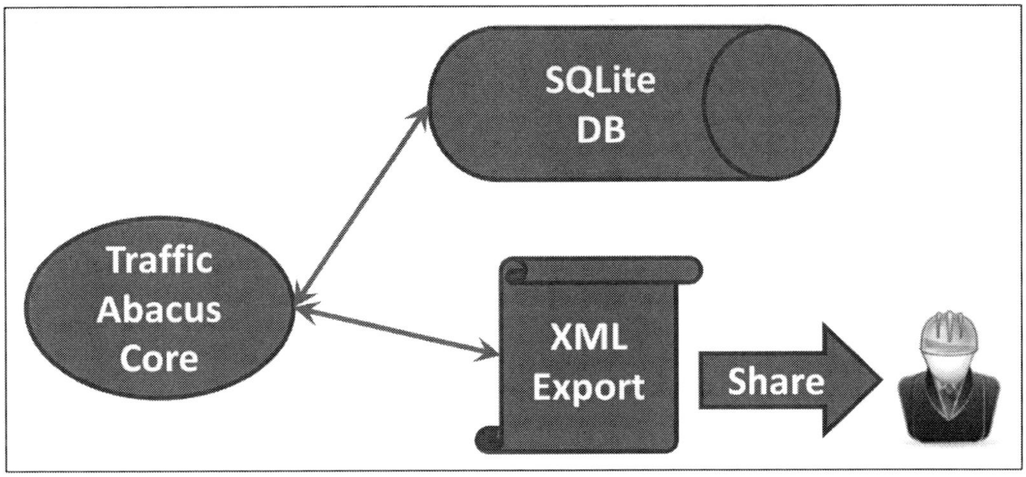

Fig. 2: Application Architecture.

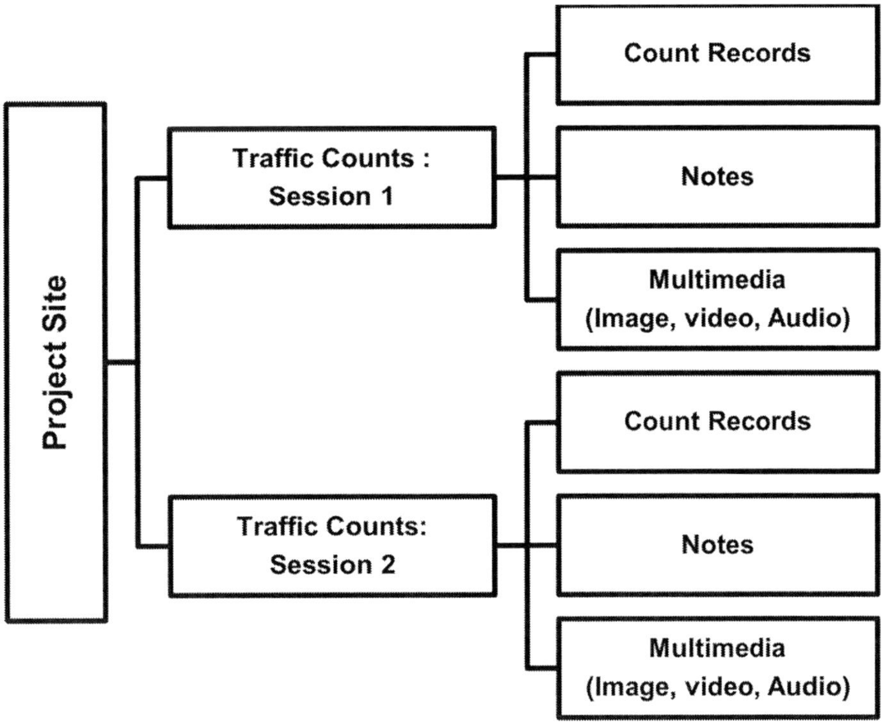

Fig. 3: Hierarchical Presentation of a Project Site Contents.

5 Traffic abacus application add-ons

In order to take full advantage of technologies available in mobile apps, the following features are also implemented:

5.1 Multiple sessions records:

Once a project is initiated, the app allows the user to record different count sessions, such as morning, mid-day, and evening peak periods. Therefore, the count sessions can be deleted or processed. They also can be different traffic counts category, such intersection counts and vehicle classifications.

5.2 Traffic counts timestamp:

A timestamp is created for each click. This timestamp is stored in the local database. Each timestamp contains the time, movement type, and project ID. This data can be viewed in the application or, as explained earlier, exported in an XML file that can be viewed using any web browser or XML viewer. Furthermore, traffic counts can be processed using any available software, such as Excel and Google Sheets. One of the advantages of using the timestamp is that data can be processed and recorded at a user-defined interval. For instance, it can be aggregated every five-minute, 15-minute or one hour intervals, as can be seen in Fig. 4.


```
- <dataPoints class="java.util.ArrayList">
    <dataPoint createDate="2016-02-06 20:12:16.8 GMT+04:00" clickedButton="ebt_t"/>
    <dataPoint createDate="2016-02-06 20:12:17.1 GMT+04:00" clickedButton="wbt_t"/>
    <dataPoint createDate="2016-02-06 20:12:17.3 GMT+04:00" clickedButton="ebl_t"/>
    <dataPoint createDate="2016-02-06 20:12:17.4 GMT+04:00" clickedButton="wbr_t"/>
    <dataPoint createDate="2016-02-06 20:12:17.7 GMT+04:00" clickedButton="sbr_t"/>
    <dataPoint createDate="2016-02-06 20:12:18.0 GMT+04:00" clickedButton="sbl_t"/>
    <dataPoint createDate="2016-02-06 20:12:18.2 GMT+04:00" clickedButton="sbt_t"/>
    <dataPoint createDate="2016-02-06 20:12:18.9 GMT+04:00" clickedButton="ebr_t"/>
    <dataPoint createDate="2016-02-06 20:12:19.2 GMT+04:00" clickedButton="nbt_t"/>
    <dataPoint createDate="2016-02-06 20:12:19.4 GMT+04:00" clickedButton="nbl_t"/>
    <dataPoint createDate="2016-02-06 20:12:19.5 GMT+04:00" clickedButton="nbr_t"/>
    <dataPoint createDate="2016-02-06 20:12:20.4 GMT+04:00" clickedButton="wbl_t"/>
    <dataPoint createDate="2016-02-06 20:12:20.6 GMT+04:00" clickedButton="wbl_t"/>
    <dataPoint createDate="2016-02-06 20:12:20.9 GMT+04:00" clickedButton="wbl_t"/>
    <dataPoint createDate="2016-02-06 20:12:21.1 GMT+04:00" clickedButton="wbl_t"/>
    <dataPoint createDate="2016-02-06 20:12:21.3 GMT+04:00" clickedButton="wbl_t"/>
    <dataPoint createDate="2016-02-06 20:12:21.7 GMT+04:00" clickedButton="ebr_t"/>
    <dataPoint createDate="2016-02-06 20:12:21.9 GMT+04:00" clickedButton="ebt_t"/>
    <dataPoint createDate="2016-02-06 20:12:22.3 GMT+04:00" clickedButton="ebt_t"/>
    <dataPoint createDate="2016-02-06 20:12:22.9 GMT+04:00" clickedButton="ebl_t"/>
    <dataPoint createDate="2016-02-06 20:12:23.8 GMT+04:00" clickedButton="nbt_t"/>
    <dataPoint createDate="2016-02-06 20:12:24.0 GMT+04:00" clickedButton="nbl_t"/>
```


Fig. 4: Data points presentation in the Traffic Abacus XML file.

5.3 Online sharing:

Once the traffic count session is over and the XML file is created and saved on the smartphone or tablet, it can be instantly shared with anyone. Therefore, processing and extracting the data can be done almost immediately.

5.4 Note attachment:

This functionality allows the traffic surveyor to attach notes associated with a particular count session. These notes can be anything related to the project, such as incident occurrences and weather condition. These notes can be a valuable reference if proper notes are taken, and an unexpected traffic trend is observed based on data records.

5.5 Multimedia records:

In addition to typed notes, the app allows the user to take images, audio videos records and attach them to a particular project. These types of records are important to identify several site features, such as geometric layout. Voice and video records are also helpful and they can serve many other purposes.

Fig. 5 shows the panel that allows addition of projects as well as addition of attached data. While notes and media files can be selected to include with any folder including any number of projects (second panel in the figure), they are the only objects that can be selected when we are adding material as attachments to a particular project (first panel in the figure). Figure 5 provides a hierarchical presentation of attached data to any project.

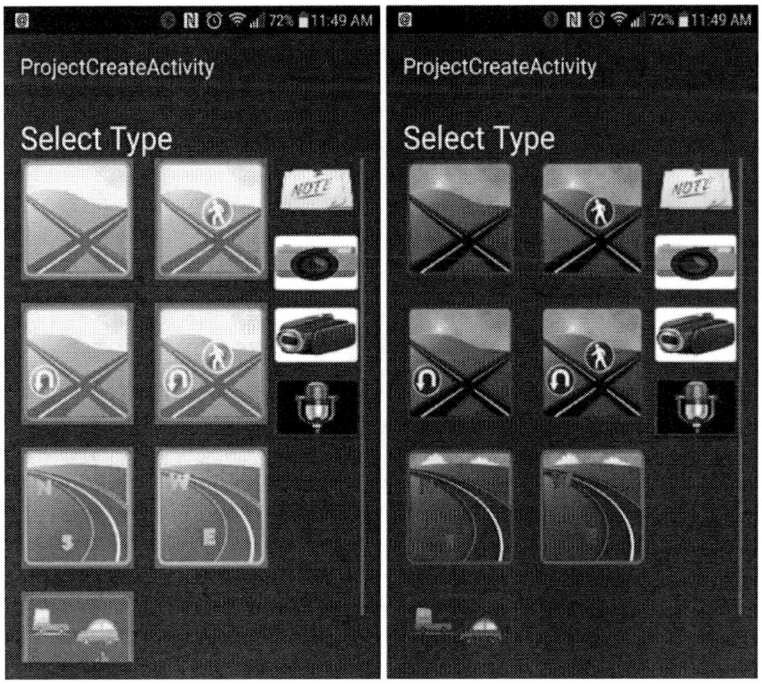

Fig. 5: Panel that allows for the attachment of various data types.

6 Students' feedback

This mobile app has been used by students from two different universities in the Middle East, namely the American University in Dubai and Qatar University. The students chose to use this application due to the unavailability of professional traffic counters developed by the industry, or the lack of proper training on these counters and extract count data. Some of the students chose to use this application because of the ease of use. The students used the app to collect traffic counts for their classroom and research projects.

A special survey is currently being developed to collect student's feedback on their experience of using such a mobile application to collect traffic data manually. This survey will assess the usefulness and the convenience of collecting traffic data. It will also be used overcome any shortage that the students might have experienced. Based on the survey results, more enhancement can be developed to assure full functionality for the mobile application.

7 Conclusion

In this project, the development of a mobile app called "Traffic Abacus" that works under Android's operating system has been described. This app is designed to provide a practical tool for traffic engineering students and professional personnel's to collect manual traffic counts. During the development phase of this app, the different needs of manual traffic counts and data collections were considered. This includes the need for counting turning movements at intersections, counting pedestrians' movements, stream traffic counts, and vehicle classification studies.

This paper demonstrates the fundamental assistance that the mobile application can offer to traffic engineering students and professionals when manual traffic counts are needed. Additionally, it provides additional add-ons that are important to facilitate the process of collecting traffic data. Some of these features are the ability to add multiple count sessions under one project, the ability to attach images, videos, and notes. Furthermore, this app provides the ability share and export project immediately once the count session is over.

8 Recommendations

While this paper demonstrates the functionalities of first phase in this project, further developments will follow. Some of the future plans is to add the Global Positioning System information to the traffic count sessions, through the XML file. Furthermore, the research team is evaluating the potential of expanding the platform of this project to include other operation systems, such as iOS and Windows Phone OS.

References

1. Allan, A., *Learning iOS Programming: From Xcode to App Store*. 2013: " O'Reilly Media, Inc.".
2. Nurmi, D., et al. *The eucalyptus open-source cloud-computing system*. in *Cluster Computing and the Grid, 2009. CCGRID'09. 9th IEEE/ACM International Symposium on*. 2009. IEEE.
3. Allen, S., V. Graupera, and L. Lundrigan, *Pro smartphone cross-platform development: iPhone, blackberry, windows mobile and android development and distribution*. 2010: Apress.
4. Yan, T., et al. *Fast app launching for mobile devices using predictive user context*. in *Proceedings of the 10th international conference on Mobile systems, applications, and services*. 2012. ACM.
5. Shi, C., et al. *Routine driving infotainment app: Gamification of performance driving*. in *Adjunct Proceedings of the 4th International Conference on Automotive User Interfaces and Interactive Vehicular Applications*. 2012.
6. Wang, T., et al. *WalkSafe: a pedestrian safety app for mobile phone users who walk and talk while crossing roads*. in *Proceedings of the Twelfth Workshop on Mobile Computing Systems & Applications*. 2012. ACM.
7. Ozbay, K., et al., *Big data and calibration and validation of traffic simulation models*. Traffic and Transportation Simulation: Looking Back and Looking Ahead: Celebrating, 2014. **50**: p. 92-122.
8. Gavin, M., et al. *A cycle route planner mobile-app for Dublin city*. in *Irish Transportation Research Network Conference (ITRN2011), 31 Aug-1 Sep 2011, University College Cork, Cork, Ireland*. 2011. Irish Transportation Research Network.
9. Rogers, R., et al., *Android application development: Programming with the Google SDK*. 2009: O'Reilly Media, Inc.

Modeling travel behavior based on household characteristics

Sameer Abu-Eisheh[1], *Mohammad* Ghanim[2,*] and *Alaa* Dodeen[1]

[1]Department of Civil Engineering, An-Najah National University, Nablus, Palestine
[2]Department of Civil and Architectural Engineering, Qatar University, Doha, Qatar

Abstract. This paper presents the relationship between travel behaviour based on demographic and socioeconomic characteristics of households in the City of Jericho, within the context of trips generation. The significance of this study is being the first comprehensive study to correlate a city-wide household travels to its characteristics in Palestine. Another significant part for the study is the consideration of different land uses to assess the correlation. This study uses a multivariable regression analysis (MRA) model to analyse the household collected data in order to correlate generated trips to the demographic and socioeconomic characteristics of households. The study results revealed the most significant socioeconomic and demographic characteristics that are associated with quantifying the overall household's travel behaviours in the City of Jericho. The importance of this paper is derived from being the first comprehensive large-scale travel behaviour study that was conducted for an entire city in Palestine. The paper also paves the road to investigate the transferability of its findings to other Palestinian cities with different sizes.

1 Introduction

Transportation planning processes are extensively used to forecast future travel demand [1]. The estimated travel demand is usually used by planners and traffic engineers as benchmarks for future transportation studies, roadway facilities, and traffic services [2]. Following the traditional four-step transportation planning model, the first step to forecast travel demand usually starts with trip generation models [3]. Therefore, trip generation models get special attention in transportation planning studies for both developed and developing countries [4-6].

The State of Palestine has a unique condition because of the economic, social, and political challenges it faces [7-9]. These challenges have made it difficult in the past to perform comprehensive large-scale travel behavior studies. In this study, The City of Jericho is chosen as the study area to develop trip generation models for its local conditions [7, 10]. The City of Jericho is an example of urban areas in Palestine. It is located in the eastern part of the middle of the West Bank as shown in Fig 1.

* Corresponding author: mghanim@qu.edu.qa

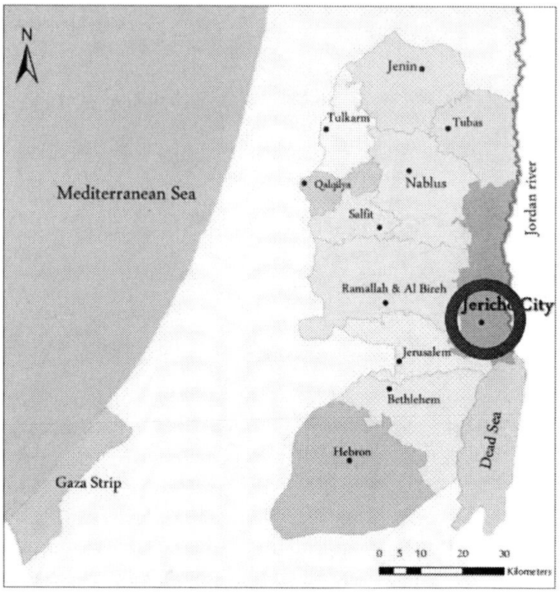

Fig 1. The Location of the City of Jericho in the West Bank, Palestine.

2 Research objectives

The objective of the study is to correlate generated trips for the City of Jericho to the demographic and socioeconomic characteristics of households for different traffic analysis zones. This relationship can be integrated with the transportation planning process for the City of Jericho. Multivariable regression analysis approach is used to develop this relationship between socioeconomic and demographic characteristics of traffic analysis zones and number of daily trips generated by households.

3 Data collection

In order to proceed with this study, the City of Jericho was divided into different traffic analysis zones (TAZs). The selection of these zones considered different factors, such as the existing and proposed land uses, population distribution, residential densities, and the existing natural or man-made boundaries [11]. Accordingly, the City was divided into 14 different zones, as can be found in Table 1.

With respect to the household socioeconomic and demographic characteristics, face-to-face interviews with different households from the different TAZ areas were done. This approach is used to assure the highest respond rates and data accuracy level as well. The collected data included several related factors such as household size, age and gender distribution, and household income. All the considered factors can be found in Table 2.

Table 1. Traffic Analysis Zones and Corresponding Characteristics [11].

Zone	Land Use Type								Area, Km^2	Population	Number of Housing Units
	Residential	Commercial	Public Facilities	Agricultural	Cultural Heritage	Parking	Industrial	Parks			
1	x	x	x						0.547	946	188
2	x	x	x	x					2.479	2458	489
3	x			x	x				2.409	958	190
4	x	x	x	x		x	x		1.254	3,688	733
5	x	x	x	x	x				3.726	1,453	289
6	x	x	x	x	x			x	2.049	3256	647
7	x	x	x	x	x			x	1.668	2816	560
8	x	x	x	x					2.142	3047	606
9	x		x	x					2.113	275	55
10	x	x	x	x					4.185	1512	301
11	x	x	x	x					1.262	2915	580
12	x	x	x	x		x	x		3.864	814	162
13	x	x	x	x		x	x		10.182	3179	632
14	x	x	x	x		x			3.005	8,838	1703
Total									40.885	35,886	7,134

Table 2. Independent Variables and Their Descriptions.

Variable	Description
SIZE	Number of persons in the household
EMP	Number of employed persons in the household
EDU	Number of persons receiving education in the household
AGE_A	Number of persons under 16 years in the household
AGE_B	Number of persons between 17 and 30 years in the household
AGE_C	Number of persons between 31 and 50 years in the household
AGE_D	Number of persons between 51 and 64 years in the household
AGE_E	Number of persons above 65 years in the household
INC	Monthly household income (Thousand New Israeli Shekel)

The independent variable that is discussed in this study is the number of daily trips made by household (ALLTRIP). It should be made clear that are many other trip-related dependent variables (such as trips per activity and trips per time-of-day) that are considered, but they are beyond the scope of this study, and are not described in this manuscript.

Household characteristics were collected from 713 respondents in the form of personal interviews of randomly selected household for each analysis zone. This sample size is chosen to meet the minimum sample size required for this study at 95% confidence level and ±0.02 margin of error. The sample size also meets the U.S. Bureau of Public Roads (BPR) guidelines, which states that 10% of the dwelling units should be examined, if the population is less than 50,000 persons [3].

4 Data analysis

4.1 Multivariable regression analysis

A stepwise multivariable regression analysis is used using SPSS. Under this method, the fitness of the regression analysis is evaluated based on different statistical tests that would quantify the significance of each independent variable that is assumed to be relevant. Accordingly, independent variables with coefficients that are least significant are excluded. The model only returns coefficients that have significant t-test values.

Equation 1 below shows the coefficients of independent variables that are significant. The dependent variable is *ALLTRIP*. Table 3 summarizes the regression results for the estimated general trip generation model.

$$\boldsymbol{ALLTRIP} = 0.569SIZE + 1.069EDU + 1.384EMP - 0.245AGE_B + 0.116INC \quad (1)$$

Equation 1 suggests that the best independent variables (among others) that would explain the number daily trips per household are the variables that are associated with the household size, education, employment, young family members (between 17 and 30), and monthly income. For all of the independent variable (except AGE_B), the directly proportional relationship between the independent and dependent variables are logical, and the number of daily trips tends to increase with the increase of household size, education, employment, and income. However, it was noticed that the number of household's daily trips tend to be less for families with more members who are aged between 17-30 years. Although this finding seems to contradict the common sense, but it is consistent with the general economic, political, and social conditions in Palestine. For instance, the unemployment rate is high for this age group in Palestine, especially for females within this age group, where they even have lower employment opportunities.

Table 3. Regression Results for the General Trip Generation Model (Number of Daily Trips per Household).

Intercept & Variables	Coefficient	Standard Error	t-Value
SIZE	0.569	0.061	9.297
EDU	1.069	0.072	14.795
EMP	1.384	0.109	12.703
AGE_B	-0.245	0.064	-3.844
INC	0.116	0.034	3.371
Summary: *R^2: 0.944*	*F-value: 2397.622*	*Sample Size: 713*	

5 Conclusions and recommendations

This study demonstrates the process of quantifying household's travel behaviour for the City of Jericho in Palestine based on socioeconomic and demographic characteristics. It uses a multivariable regression analysis approach in identifying the most significant variables in determining the number of daily trips per household.

The study has experienced several political, social, and economic challenges, but the study team proposed countermeasures to overcome these challenges, which lead to the success of the study. These measures can be deployed for other city-wide studies targeting easier and efficient deployment. The significance of this study is being the first initiative toward comprehensive trip generation models for Palestinian cities. The models will then be integrated within a large-scale transportation planning strategies at the national level. Furthermore, this study is envisaged to be in its potential for transferability. Since the Palestinian cities share similar demographic, social, and economic characteristics. However, further studies are needed to assure the applicability of transferability process.

References

1. Hutchinson, B.G., *Principles of Urban Transport Systems Planning.* (1974).
2. Manheim, M.L., *Fundamentals of Transportation Systems Analysis; Volume 1: Basic Concepts.* (1979).
3. U.S. Bureau of Public Roads, *Guidelines for Trip Generation Analysis,* , U.S. Government Printing Office, Editor. June 1967: Washington D.C.
4. Sofia, G.G., A.H.A. Ali, and H.A.N. Al-Zubaidy, *Trip Generation Modeling for Selected Zone in Al-Diwaniyah City.* Journal of Engineering and Development, (2012). **16**(4).
5. Sarsam, S. and S. Al-Hassani, *Modeling Household Trip Generation for selected Zones at Al-Karkh Side of Baghdad City.* Journal of Engineering, (2011). **17**(6): p. 1462-1472.
6. Al-Taei, A.K. and M. Taher, *Trip Attraction Development Statistical Model in Dohuk City Residential Area.* Al-Rafidain Engineering, (2006). **14**(2): p. 12-23.
7. The European Investment Bank and the Ministry of Transportation, *Systematica, Politecnica, Idrotec, and Air Support, Road and Transportion Master Plan-West Bank and Gaza Strip: Draft Final Report.* June (2016): Ramallah, Palestine.
8. Hassan Moussa, *Development of A Trip Generation Model for Gaza City*, in *Civil Engineering Department.* 2013, M.Sc. Thesis, Islamic University of Gaza.
9. Unviersal Group for Engineering and Consulting, *Analytical Study on Population Projection, Palestine 2015 and 2050: Final Report.* Feb (2012): National Spatial Plan, Ramallah, Palestine. .
10. Palestinian Central Bureau of Statics. *Localities in Jericho & Al Aghwar Governorate by Type of Locality and Population Estimates, 2007-2016.* (2015) [cited 2013; Available from: http://www.pcbs.gov.ps/Portals/_Rainbow/Documents/jerich.htm.
11. Cit of Jericho, *Master Plan for the City of Jericho.* 2010: Jericho.

 DOI: 10.1051/matecconf/201712007006

Road users' opinion about pedestrian safety in the emirate of Sharjah, UAE- survey results

Mariam Alhmoudi[1],* *Muamer* Abuzwidah[2], and *Khaled* Hamad[3]

Sustainable Civil Infrastructure Systems Research Group, University of Sharjah, Sharjah City, United Arab Emirates.

Abstract. This paper is to investigate the pedestrian safety in Sharjah (UAE), and suggest recommendations to improve safety in the emirate. A survey data was collected from 570 participants in December, 2016, and included the following informations: measure the awareness of drivers and pedestrians, determine the behavior of road users, take general views from the road users, and measure the level of users` satisfaction. The results showed that characteristics of road users including (gender, age group, nationality, social status, education level and the income) can affect their behavior at different levels and that may lead to pedestrian accidents. Also, most participants noticed that there are differences in the behavior of drivers and pedestrians based on their nationalities, and they indicated that pedestrian mistakes are main cause of this type of accidents. Moreover, the data showed that most drivers confirmed that they give more attention to pedestrians in mixed land used than other areas and they mentioned that "traffic signal + marked lines" is the clearest place for them to be attentive to pedestrians crossing the road. Overall, based on the results of this paper, there is an urgent need to re-evaluate the pedestrian facilities in the Emirate such as (Design, locations, and their availability) as well as focusing more on the education and the law enforcement.

1 Introduction

The United Arab Emirates (UAE) has achieved remarkable development at all levels since the 1980s, particularly the Emirate of Sharjah which is characterized by a renaissance in urban, culture and economics. However, this significant development was accompanied by an increase in population and, consequently, the number of vehicles moved around the emirate. Unfortunately, this rapid increase has led to an increase in traffic problems, especially pedestrian accidents. Actually, pedestrian accidents are a global problem, which were a one of leading causes of death, injuries, as well as increased financial costs worldwide. Statistics showed that an average of over 270,000 pedestrians are killed in highway accidents every year worldwide, which is about 22% of all road fatalities (WHO, 2013). Also, global statistics indicated that the fatality rate for road users is significantly higher in Arab Gulf countries comparing to western countries, for example, in 2007 mortality rates because of road traffic accidents were 37.1, 29 and 16.9 per 100,000

* Corresponding author: U00041081@sharjah.ac.ae

population in the United Arab Emirates (UAE), Saudi Arabia and Kuwait respectively (Who.int, 2013), compared to the rates of 6.1 for the UK and 14.8 for the USA (WHO, 2013). Although a significant difference in the levels of fatalities being observed between Arab Gulf countries and western countries, studies on traffic safety that represent our region are very limited.

Therefore, there is an essential need for such studies to improve road safety standards I n our region, which in turn will increase the awareness among road users, thus will eliminate the heavy losses in lives and money. The study area of this research is the Emirate of Sharjah in the United Arab Emirates. The Emirate of Sharjah, stretching along the Arabian Gulf and the Indian Ocean coast, it is the third emirate in the United Arab Emirates in terms of area, density of population, and popularity. It covers 2590 Km2 and has an official estimated population of over 793,573 residents in 2005. (Federal Competitiveness and Statistics Authority). This paper aims to investigate pedestrian safety in the Emirate of Sharjah by conducting and analyzing a survey study and then suggest strategies to improve pedestrian safety in the emirate.

2 Literature review

This section addresses some studies which investigated subjects in traffic safety in the UAE. Bener and Alwash (2005), evaluated the impacts of over speeding on traffic safety in the United Arab Emirates. They found that characteristics of road users strongly attributed to 90% of road accidents. Also, the authors indicated that the speed rates decreased significantly as a result of installing the radar systems and speed cameras in 1994; thus significantly reduced the number of deaths and injuries caused by traffic accidents. However, the number of human losses were still higher than in USA and UK, knowing that the traffic accidents was the second cause of death in the UAE. In addition to that, the authors observed that, although the number of speeding vehicles decreased, the fatalities rate remained the same, the interpretation of this finding is that there was a strong impact of behavior of local drivers on fatalities rate. Bener and Crundall (2005), studied traffic accidents in the UAE during 2000 to identify the major factors attributed to these accidents. The authors found that 38% of the accidents occurred because of careless driving, 13% caused by excessive speed, and 32% because of other human factors. Abdalla (2002), identified causes of increasing traffic fatalities and injuries in the Emirate of Dubai, UAE. The findings of this study indicated that the risk of traffic fatality was higher in Dubai, compared to some other developed countries, also it is found that the probability of a driver causing an accident is considerably higher for those driving goods vehicles, however it is also associated with other factors. Moreover, many countermeasures were applied to improve pedestrian safety but the result did not well evaluated. Abuzwidah and Abdel-Aty (2015) countermeasures must be evaluated to ensure the maximum benefits from the treatments and to avoid the unexpected mistakes. Hamad (2016), described the extent of the road traffic accidents problem in the Emirate of Sharjah, UAE for the period (2001-2014). The findings of this paper revealed several useful facts which could be used to find directions for future research, policies, and programs on highway safety in the Emirate of Sharjah. Hamad (2016) indicated that Sharjah's population almost doubled during the period 2001-2014, however the number of injury/fatality caused by traffic accidents decreased by more than half during the same period. Also, the author indicated that the annual number of road accidents and injuries increased until 2008 then the number sharply decreased thereafter remained stable in recent years, he attributed this finding to the introduction of the new traffic law of the UAE. In addition to that the author found that, although both accidents and injuries per 100,000 population extremely decreased annually, the fatalities per 100,000 population only slightly decreased, he attributed this to the increasing of severity levels of accidents. Generally, Hamad (2016), concluded that Sharjah showed a better rate of fatalities per 100,000 population than that for the whole UAE. El-Sadig et al. (2004), evaluated the effectiveness of seatbelt legislation in the Eastern District of Abu Dhabi (UAE). The authors found that the implementing of this legislation resulted in statistically significant reduction in morbidity in those arriving alive at hospital following a accidents, including reduced severity of injury, admission rate to hospital and the duration of hospital stay. Bendak & Al-Saleh (2013), evaluated seat belt (SB) wearing rates for drivers and Front Seat Passengers (FSPs) in the UAE through an observational field study. They also investigated perceptions and behavior of drivers on this issue as well as human factors that affect wearing rate through a randomly distributed questionnaire. The findings of this paper showed that the overall SB wearing rate in the country was 61% for drivers and 43.4% for FSPs, indicating that there were considerable differences between the seven emirates in the country. The questionnaire results show that age, education level, gender, social status and nationality of drivers affect wearing habits and perceptions.

While some efforts has been completed in studying traffic safety in the UAE, this paper will contribute with them by investigating pedestrian safety in the Emirate of Sharjah, aiming to provide friendly environment for this important category of road users around the emirate. This research should be considered as a first step in research into this important area, and recommendations for further research can be concluded from this paper.

3 Methodology

In the beginning of this research we aimed to collect, investigate and analyze pedestrian accident records from all available sources in Sharjah. These sources include traffic department, insurance companies, government/private hospitals and Roads and Transport Authority (RTA). Whoever, there is a lack of the data sources and therefore, collection of pedestrian accidents data has proven more challenging than anticipated. Instead, a questionnaire was designed as an alternative way to collect data and investigate pedestrian safety in Sharjah. Finally, policies and measures to tackle the identified problems are investigated and discussed. The questionnaire was randomly distributed for unspecified sample of public, private sector and government organizations across the Emirate of Sharjah after clarifying the purpose of the study and getting their attention.

Google Forms was used to create two links (in Arabic and English) for this questionnaire, and then social media including Instagram, WhatsApp, emails, Facebook in addition to hard copies were used in publication of the questionnaire. Questionnaire distribution took place in December, 2016 and was stopped after the number reached 570 participants. It was thought that such a sample size would reflect the behavior and attitudes of the whole road users population and is within the norm of sample sizes used in similar studies.

The main objectives of this questionnaire are to determine factors that affect pedestrian safety in the Emirate of Sharjah, study and analyze road users' perceptions and attitudes, specifying level of satisfaction, and other general opinions in this area. In the beginning, the questionnaire enquired about road users' personal characteristics including (age, gender, education level, nationality, social status, and income) and then, it followed by questions that clarified respondents' views and facts. Descriptive statistics, tables and Fig.s including pie charts and bar charts were used to explore and assess questionnaire results.

4 Results and discussion

4.1 The Identities of respondents

Determining identities of participants is very important to measure the differences among different categories. This questionnaire started with questions on identity, Fig.1 shows the identities of the participants includes the following: gender, age group, nationality, social status, educational level and income.

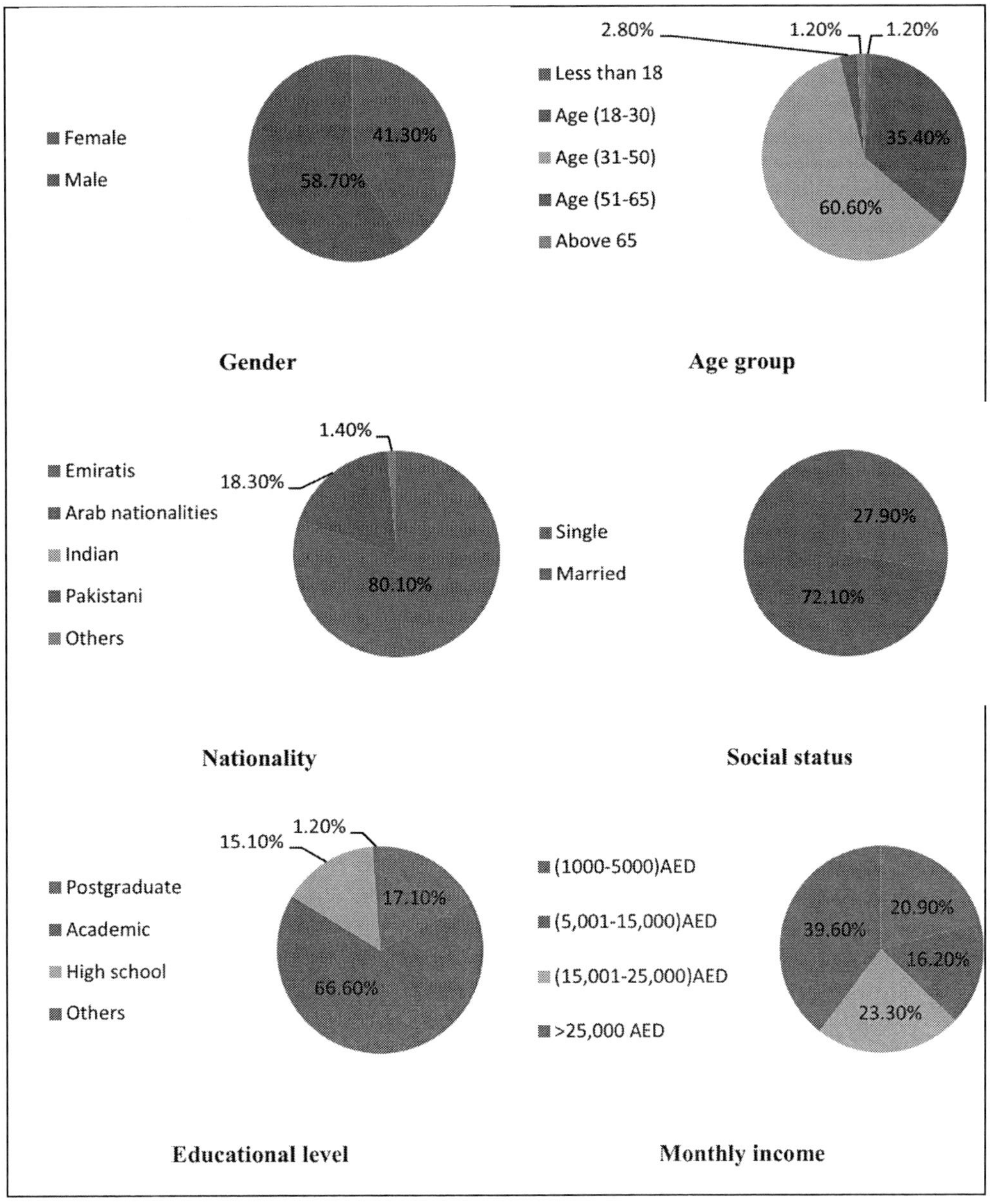

Fig. 1. Identities of respondents.

Based on Fig. 1, it could be concluded that this survey showed the opinion of both genders almost equally, regarding the ages, the largest age groups of respondents were (18-30) and (31-50) where they formed 35.4% and 60.6% respectively, Emiratis were the primary respondents, with other Arabs at 18.3%, married persons constituted 72.1% of respondents and 27.9% were single, the majority of respondents were educated, and finally, the respondents were from high, middle, and low income.

4.2 The summary of respondents' views and facts

Initially, respondents were asked if they have a pedestrian accident before. As shown in Fig. 2, 10.2% of respondents reported that they experienced pedestrian accidents, while 89.8% did not. This result revealed the fact that Emirate of Sharjah faces the problem of pedestrian accidents, where it is despite the fact that this sample represents only 0.07 from the emirate's residents according to the 2005 census, we found that 10.2% exposed to pedestrian accidents.

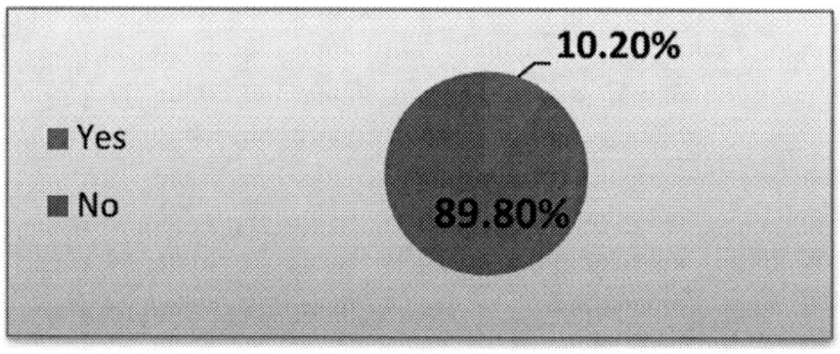

Fig. 2. Did you have a pedestrian accident before?

The characteristics of respondents that exposed to pedestrian accidents were examined to determine related significant factors using χ^2- test and 95% of significance level, Table.1 clarifies the results.

Table 1. Characteristics of respondents that exposed to pedestrian accidents.

Characteristics	Categories	Significance
Gender	Female=21 Male=36 Undefined=1	Not significant
Age group	18-30=17 31-50=39 51-65=1 Less than 18=1	Significant
Nationality	UAE=39 Arab Nationalities=17 Pakistani=2	Significant
Social status	Married=42 Single=16	Significant
Education level	Postgraduate=11 Academic=33 High school=11 Other=3	Significant
Income	1,000-5,000=12 5001-15,000=14 15,001-25,000=13 >25,000=19	Not significant

Respondents were then asked to report if they exposed to hazardous situations while walking or crossing the roads. The answer presented in Fig. 3, indicated that 48% of the sample was exposed to hazardous situations while walking or crossing the roads, which is a high percentage. On the other hand, 52% responded that they were not exposed to such situations. This result indicates that the likelihood of pedestrian accidents occurring is very high in the emirate. Also, this result, particularly we mean (48% who answering "yes") attributed to the behavior of drivers and pedestrians. In Table 2, characteristics of respondents who exposed to hazardous situations while walking or crossing the roads were examined to determine related significant factors using χ^2- test/α=0.05.

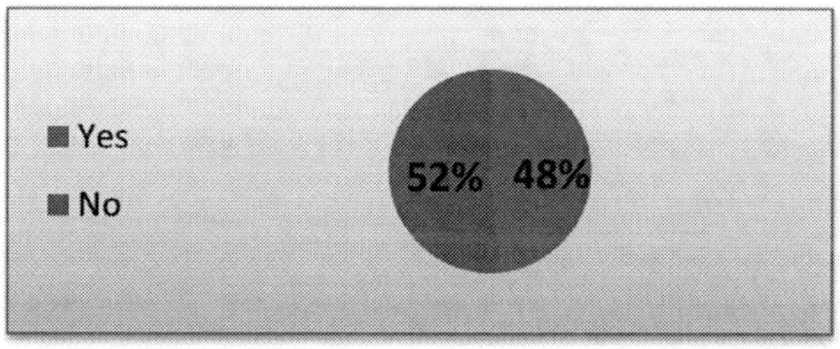

Fig. 3. Are you exposed to hazardous situations while walking or crossing the roads?

Table 2. Characteristics of respondents who exposed to hazardous situations while walking or crossing the roads.

Characteristics	Categories	Significance
Gender	Female=80 Male=190 Undefined=2	Significant
Age group	18-30=109 31-50=148 51-65=9 Less than 18=6	Significant
Nationality	UAE=208 Arab Nationalities=60 Others=4	Significant
Social status	Married=190 Single=80 Undefined=2	Significant
Education level	Postgraduate=46 Academic=178 High school=45 Others=3	Significant
Income	1,000-5,000=50 5001-15,000=43 15,001-25,000=69 >25,000=100 Undefined=10	Significant

In Fig. 4 respondents were asked as drivers, if they pay attention to pedestrians when they drive their vehicles. It is concluded that the respondents have a very good degree of awareness and attention to pedestrian traffic, where the percentage of respondents who answered "yes" reached 94.6%. However, the percentage of answering "yes" could be less, since some people might have been shy or embarrassed to report the truth. In Table 3,

DOI: 10.1051/matecconf/201712007006

characteristics of respondents who do not pay attention to pedestrians when they drive were examined to determine related significant factors using χ^2- test/α=0.05.

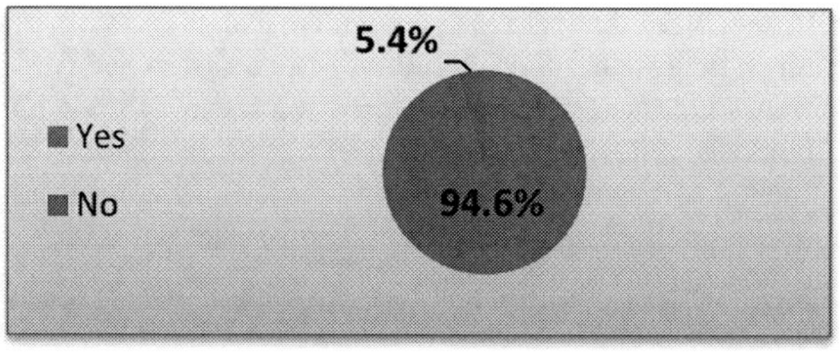

Fig. 4. Do you pay attention to pedestrians when you drive your vehicle?

Table 3. Characteristics of respondents as drivers who do not pay attention to pedestrians.

Characteristics	Categories	Significance
Gender	Female=9 Male=19 Undefined=2	Not significant
Age group	18-30=14 31-50=13 Less than 18=1 Undefined=2	Significant
Nationality	UAE=21 Arab Nationalities=6 Pakistani=1 Undefined=2	Significant
Social status	Married=17 Single=11 Undefined=2	Not significant
Education level	Postgraduate=6 Academic=17 High school=4 Others=3	Significant
Income	1,000-5,000=8 5,001-15,000=2 15,001-25,000=6 >25,000=11 Undefined=3	Not significant

In Fig. 5, respondents were asked if they give way to pedestrians in case of crossing places not designated to them. The result shows that 75.4% of the drivers stopped for pedestrians even though they were crossing wrong. On the other hand, 24% of the drivers retained their priority for the road and did not open the way for pedestrians.

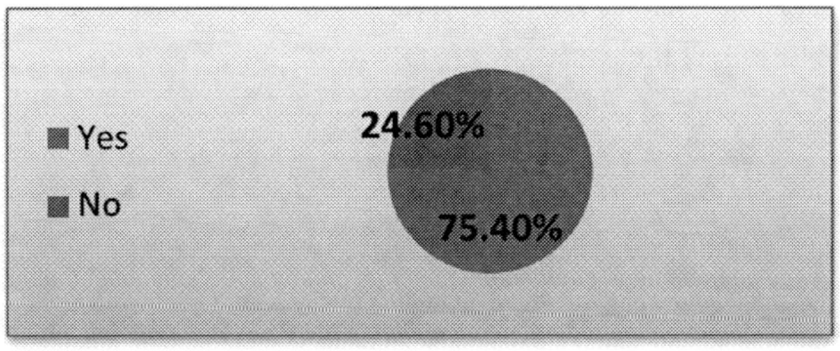

Fig. 5. Do you give way to pedestrians in case of crossing places not designated to them?

In Fig. 6 respondents were asked about their commitment to speed limits in congested pedestrian movement places. The result shows that 85% of the drivers are committed to speed limits, which is good thing; unfortunately, 15% said they are not, where this percentage could be higher, since some people might have been shy or embarrassed to report the truth. In Table 4, characteristics of respondents who are not committed to speed limits in congested pedestrian movement places were examined to determine related significant factors using χ^2- test/α=0.05.

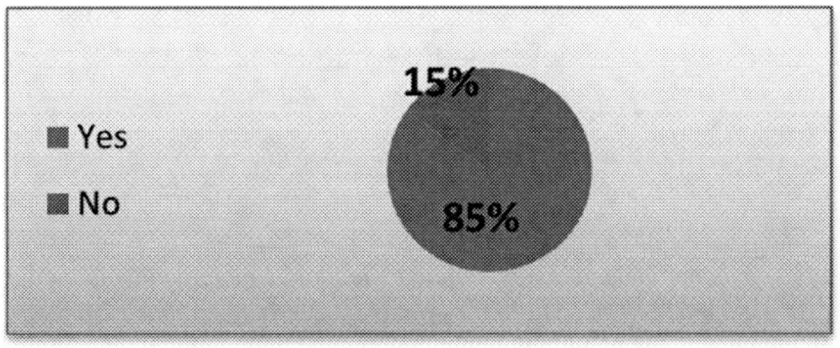

Fig.6. Are you committed to speed limits in congested pedestrian movement places?

Table 4. Characteristics of respondents as drivers who are not committed to speed limits.

Characteristics	Categories	Significance
Gender	Female=28 Male=57	Significant
Age group	18-30=54 31-50=29 Less than 18=2	Significant
Nationality	UAE=69 Arab Nationalities=16	Significant
Social status	Married=45 Single=40	Not significant
Education level	Postgraduate=8 Academic=61 High school=14 Others=2	Significant
Income	1,000-5,000=21 5,001-15,000=9 15,001-25,000=22 >25,000=30 Undefined=3	Significant

In Fig. 7 respondents were asked if they give priority to pedestrians in their designated places to cross. From Fig. 7, it is observed that 97% of the drivers are committed to giving priority to pedestrians in their designated places, which is very good thing; unfortunately,

3% they are not, where this percentage could be higher, since some people might have been shy or embarrassed to report the truth.

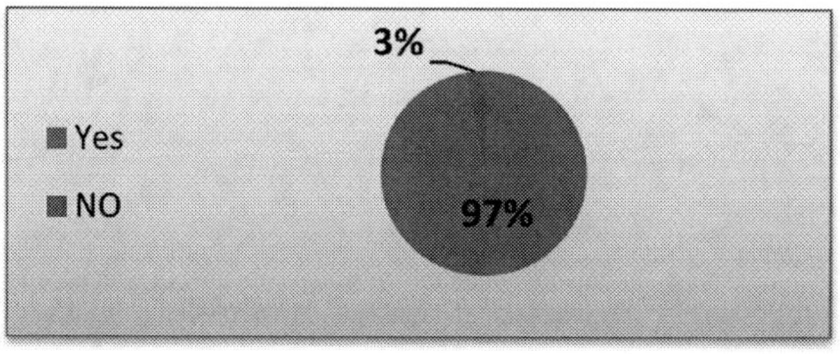

Fig.7. Do you give priority to pedestrians in their designated places to cross?

In Fig. 8 respondents were asked to answer yes or no with the stated reason if possible if they feel resentful of pedestrian traffic on the roads. As shown in Fig. 8, it is observed that 77.1% were not happy with pedestrian movement in the emirate, and their reasons were as follows: causing traffic jams, confusing drivers, increasing the likelihood of accidents through slow-motion, wrong estimation, crossing on high-speed roads, and sudden crossing. 17.8% of the drivers saw that pedestrian movements as normal, and they have the right to share the roads with vehicles with limitations. 5.1% of the drivers answered "sometimes".

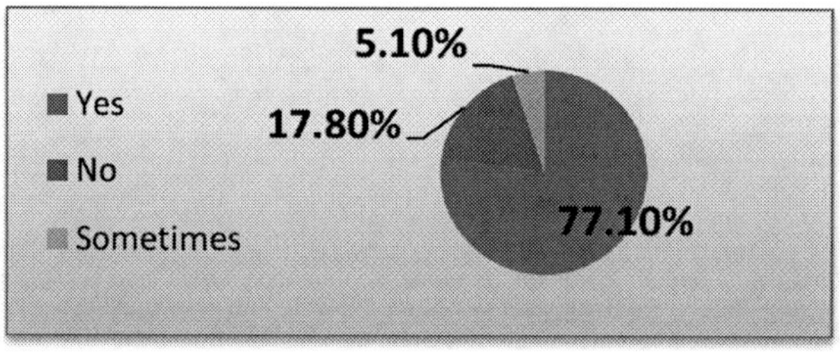

Fig.8. Do you feel resentful of pedestrian traffic on the roads?

In Fig. 9 respondents were asked about the clearest crossing point for them as drivers. The result shows that the largest percentage in the sample agreed that "traffic signal + marked lines" is the clearest place for drivers to be attentive to pedestrians crossing the road. "Raised marked lines" came in second in terms of importance, and finally marked lines.

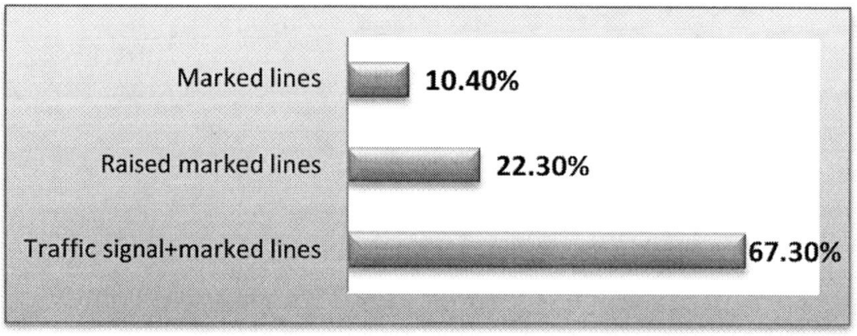

Fig. 9. What is the crossing point that is very clear for you as a driver?

In Fig. 10 respondents were asked if they expect pedestrian errors and braced for them. The result in Fig. 10 showed that 85.7% of the drivers expect pedestrian errors and pay attention to them, which is a good thing. On the other hand, 14.3% of the respondents do not care about pedestrian movement. In Table 5, characteristics of respondents as drivers who are not expect pedestrian errors and braced for them were examined to determine related significant human factors using χ^2- test/α=0.05.

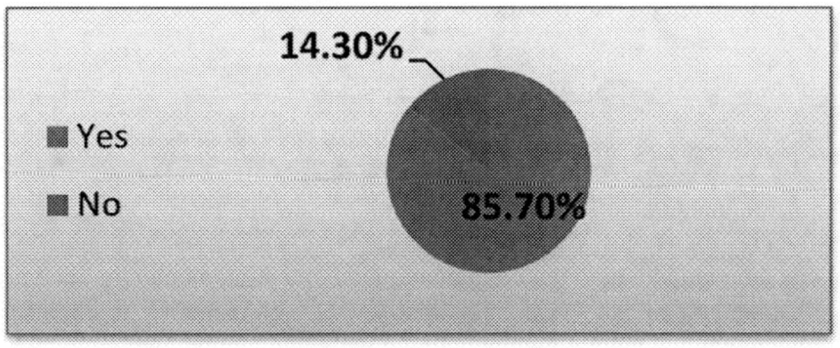

Fig.10. Do you expect pedestrian errors and braced for them?

Table 5. Characteristics of respondents as drivers who are not expect pedestrian errors and braced for them.

Characteristics	Categories	Significance
Gender	Female=38 Male=42	Not significant
Age group	18-30=44 31-50=34 Less than 18=2	Significant
Nationality	UAE=62 Arab Nationalities=15 Others=3	Significant
Social status	Married=45 Single=35	Not significant
Education level	Postgraduate=15 Academic=53 High school=11 Others=1	Significant
Income	1,000-5,000=22 5,001-15,000=13 15,001-25,000=17 >25,000=24 Undefined=4	Not significant

In Fig. 11 respondents were asked about the places that they give more attention to pedestrian traffic. From Fig. 11, we could concluded that drivers give maximum attention to pedestrians in mixed land used with 39%, then in school areas with 33.9%, followed by residential areas with 21.3%. They gave industrial areas 5.3%, and in last place agricultural areas with 0.5%. It is noted here that the drivers ignored pedestrian traffic in the industrial zones, note that these areas in Sharjah are vital and busy of pedestrian movement and workers, which require more attention and focus.

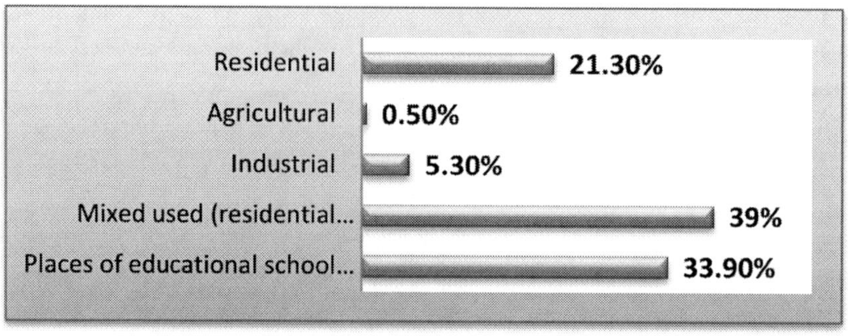

Fig. 11. Where do you give more attention to pedestrian traffic?

In Fig. 12 respondents were asked about the causes of pedestrian accidents in their views. From Fig. 12, it could be concluded that the main reason for pedestrian accidents is "neglect of pedestrian and errors committed by them" with 49%. "Driver negligence and lack of attention" came in second place with 24%, followed by "Lack of safe crossing points" with 16.6%, in last "Defects in road design" with 10.4%. It is noted here that respondents attributed pedestrian accidents mainly to human element including pedestrian and drivers with 73%, and 27% percentage distributed among other reasons. This result seems very logical and must be taken in consideration, where if we assumed that the roads designs have no defects and there are enough safe crossing points, the problem will remain in place and it is attributed to the primary factor which is the human element.

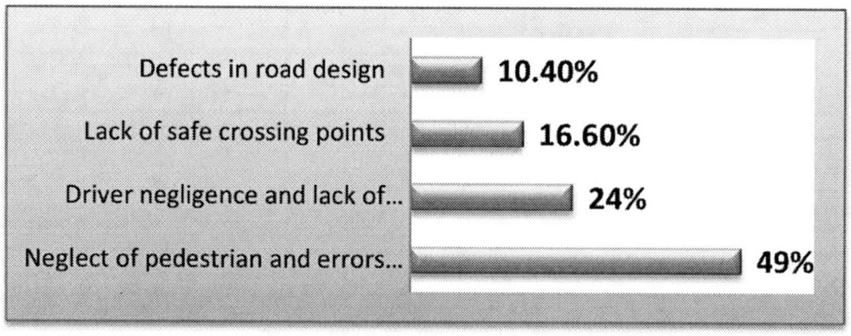

Fig. 12. Causes of pedestrian accidents.

In Fig. 13 respondents were asked as pedestrians if they will cross in places not intended for that. Fig. 13 shows that 69.8% were committed to crossing rules, while 30.2% are not. In Table 6, characteristics of respondents as pedestrians who will cross in places not intended for that were examined to determine human related significant factors using χ^2- test/α=0.05.

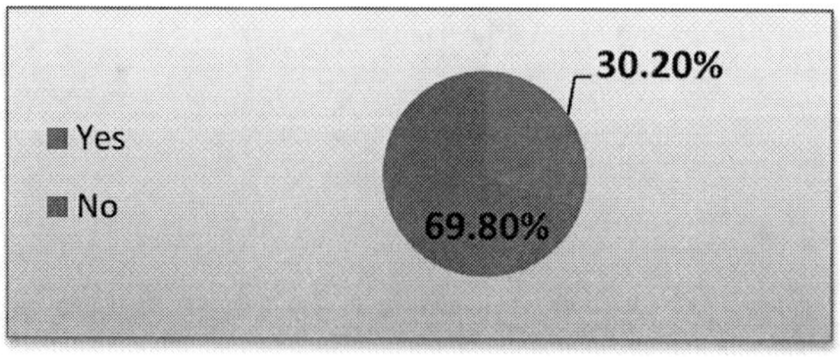

Fig. 13. If you are a pedestrian will you cross in places not intended for that?

Table 6. Characteristics of respondents as pedestrians who will cross in places not intended for that.

Characteristics	Categories	Significance
Gender	Female=48 Male=122	Significant
Age group	18-30=74 31-50=91 51-65=2 Less than 18=3	Significant
Nationality	UAE=119 Arab Nationalities=45 Pakistani=1 Others=5	Significant
Social status	Married=62 Single=108	Significant
Education level	Postgraduate=23 Academic=123 High school=20 Others=4	Significant
Income	1,000-5,000=44 5,001-15,000=22 15,001-25,000=42 >25,000=54 Undefined=8	Significant

In Fig. 14 respondents were asked as pedestrians to answer yes or no with the reasoning if they feel safe and satisfied during transit in designated place. Fig. 14 shows that 69.2% feel safe while crossing the roads in designated places; on the other hand, 28.3% feel unsafe because of feeling fear of reckless drivers and excessive speeds.

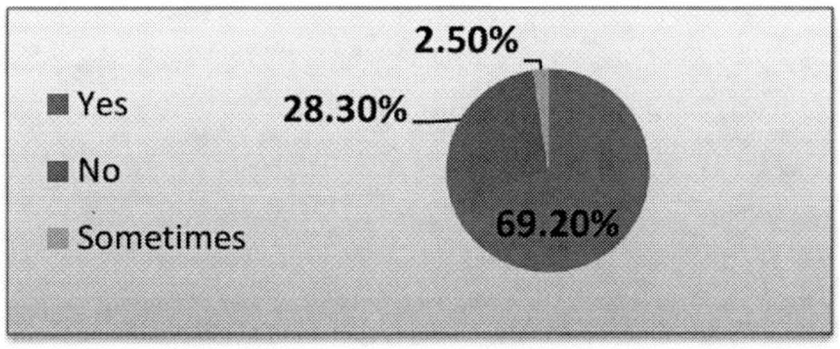

Fig. 14. Do you feel safe and satisfied during transit in designated places?

In Fig. 15 respondents were asked if they feel differences in the behavior/attitudes of drivers and pedestrians according to their nationalities. Fig. 15 shows that almost all participants in this questionnaire feel differences in the behavior of drivers and pedestrians according to their nationalities.

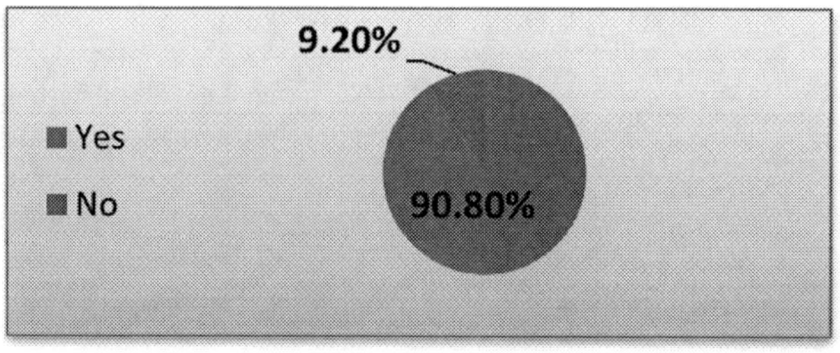

Fig. 15. Do you feel differences in the behavior/attitudes of drivers and pedestrians according to their nationalities?

In Fig. 16 respondents were asked to evaluate the drivers' behaviors in Sharjah towards pedestrians. The result is shown in Fig. 17, indicates to dissatisfaction in large percentage.

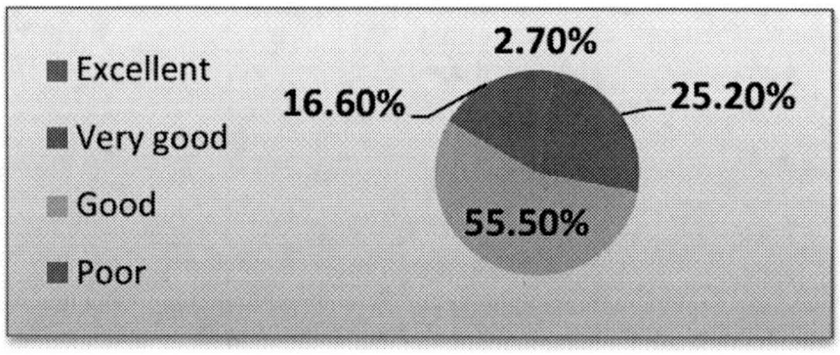

Fig. 16. Assessment of drivers' behaviors in Emirate of Sharjah towards pedestrians.

Finally, in Fig. 17 respondents were asked to evaluate pedestrian's behavior in the Emirate of Sharjah. The result is shown in Fig. 18, shows that this assessment reflects a bad impression, where almost three quarters of sample evaluated pedestrian's behavior in Sharjah between good and poor.

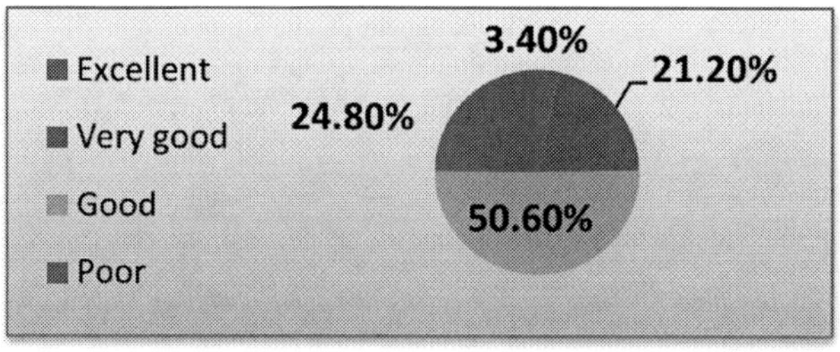

Fig. 17. Assessment of pedestrian's behavior in Emirate of Sharjah.

5 Conclusion and recommendations

5.1 Conclusion

This paper investigated pedestrian safety in the Emirate of Sharjah, UAE by conducting and analyzing a survey including the following targets: measuring the awareness of drivers and pedestrians regarding this area, determine the behavior of road users in some situations, take general views from the public, and measure the level of satisfaction. After analyzing the survey results, several interesting findings were revealed including the following points:

- Characteristics of road users including (gender, age group, nationality, social status, education level and income), affect their behavior at different levels, specifically behavior that may lead to pedestrian accidents.
- The largest percentage of respondents agreed that there are differences in the behavior of drivers and pedestrians according to their nationalities.
- The largest percentage of respondents attributed pedestrian cashes mainly to the negligence and mistakes committed by pedestrian.
- The largest percentage of respondents as drivers confirmed that they give more attention to pedestrians in mixed land used than other areas.
- The largest percentage of respondents agreed that "traffic signal + marked lines" is the clearest place for drivers to be attentive to pedestrians crossing the road.
- Based on questions that measure the level of satisfaction for drivers and pedestrian regarding this issue, it is found that a considerable percentage of drivers in the sample are not happy with pedestrian traffic in the emirate, and their reasons were as follows: causing traffic jams, confusing drivers, increasing the likelihood of accidents through slow-motion, wrong estimation, crossing on high-speed roads, and sudden crossing. On the other hand pedestrian in the sample, pointed to their feeling of uncomfortable while crossing the roads in their intended places because of feeling fear of reckless drivers and excessive speeds.
- The assessments of pedestrian's behavior and drivers' behaviors towards pedestrians in the Emirate of Sharjah reflect a bad impression, where almost three-quarters of both evaluations were between poor and good.

5.1 Recommendations

According to the results that have been reached through this research of pedestrian safety in the Emirate of Sharjah, we recommend that decision-makers consider the following suggestions in order to increase the comfort and safety of pedestrian environment around the emirate:

- Behavior and Education: increasing awareness of traffic culture among road users is the perfect solution to this issue, this can be achieved by educating road users through awareness and guidance campaigns, distribution of awareness brochures in several languages, and educational lectures at universities, schools, and institutions. More attention and focus should be given to some categories that committed errors lead to pedestrian accidents significantly and repeatedly.Improvement of Infrastructure, Road Design in busy Pedestrian Areas: there was a need shown to increase pedestrian access to services places, by installing pedestrian crossing points (whether foot bridges, subways, raised crosswalks, normal marked crosswalks, or others) according to the road needs and population density around it. In addition, speed reduction signs and warning signs should be installed in crowded areas which may also support pedestrian safety. Also, installing a pedestrian barrier in roads will force pedestrians to use the safe crossing points. Construction of speed bumps before pedestrian crossing points is a good measure which will force drivers to drive more slowly and give way for pedestrians to cross safely. The research noticed that there are available sidewalks in the Emirate of Sharjah, but they are not adequate and in some cases misused; therefore, we need to intensify control on the use and installation of a sufficient number of sidewalks in order to reduce the amount of time pedestrians are exposed to vehicles. Also, road marking must be improved, especially in places where pedestrians cross.

- Legislation: More stringent laws should be enacted for pedestrians and drivers concerning the crossing points and adherence to these rules. For example we can impose immediate fines on violators of pedestrians or drivers or install cameras in pedestrian crossing points and areas busy with pedestrians to monitor traffic violations.

Overall, based on the results of this paper, there is an urgent need to re-evaluate the pedestrian facilities in the Emirate such as (Design, locations, and their availability) as well as focusing more on the research to reach a clear conclusions for the decision makers.

References

1. Abdalla, I. Fatality risk assessment and modeling of drivers responsibility for causing traffic accidents in Dubai. *Journal of Safety Research*, 33, pp.483-496. (2002)
2. Abuzwidah, M & Abdel-Aty, M. "Safety assessment of the conversion of toll plazas to all-electronic toll collection system. *Accident Analysis and Prevention* 80, pp. 153-161. (2015)
3. Bendak, S.& Al-Saleh, K. Seat belt utilisation and awareness in UAE, *International Journal of Injury Control and Safety Promotion*, 20(4), pp.342-348. (2013)
4. Bener, A. and Alwash, R. A perspective on motor vehicle accident injuries and speeding in the United Arab Emirates, *Journal of Traffic Injury Prevention*, Vol. 3, pp.61-64. (2002)
5. Bener,A. and Crundall, D. Road traffic accidents in the United Arab Emirates compared to Western countries. *Advances in Transportation Studies an international Journal* Section A 6. (2005)
6. El-Sadig et al. Evaluation of effectiveness of safety seatbelt legislation in the United Arab Emirates. *Accident Analysis & Prevention*, 36(3), pp.399-404. (2004)
7. Federal Competitiveness and Statistics Authority. Population by the emirate 1975-2005 [online] Available at: http://original.fcsa.gov.ae/Home/ReportDetails1/tabid/90/Default.aspx?ItemId=1880&PTID=129&MenuId=1 [Accessed Dec, 2016]. (2010)
8. Hamad, K. Road traffic accidents trends in Sharjah, United Arab Emirates during 2001-2014. *International Journal of Vehicle Safety*, 9(1), pp.24-38. (2016)
9. World Health Organization. [online] Available at: http://www.who.int/mediacentre/news/notes/2013/make_walking_safe_20130502/en/ [Accessed Dec. 2016]. (2013)

Modelling of traffic safety of traditional mainline toll plazas

Muamer Abuzwidah[1,*], *Mohamed* Abdel-Aty[2]

[1] Department of Civil and Environmental Engineering, University of Sharjah, Sharjah, UAE
[2] Department of Civil, Environmental and Construction Engineering, University of Central Florida, Orlando, Florida

Abstract. Freeways and expressways are considered an important part of any successful transportation system. These facilities carry the majority of daily trips on the transportation network. Although expressways (Toll Roads) offer high level of service, and are considered the safest among other types of roads, traditional toll plazas may have both safety and operational challenges. The traditional toll facilities still experience many crashes, many of which are severe. Therefore, the main focus of this paper is to evaluate the effect of tollbooths' number and types (i.e. cashier lanes, coin lanes and Electronic Toll Collections (ETC) lanes) on safety performance at traditional toll plazas. The literature showed that about 30 to 40 percent of the total crashes on expressways occur at these facilities. An extensive data collection was conducted that included 42 traditional mainline toll plazas located on approximately 750 miles of expressways in Florida, and multiple analytical techniques were used. Based on the finding in this paper, choosing locations for the toll plazas that have safe distances from the interchanges, installing flashing lights at night time, and finding ways to improve signage and road marking are potential means of reducing lane changes before tollbooths. More research is needed and researchers may build on the results of this study.

1 Introduction

The use of toll roads has risen dramatically in many countries around the world, and in some countries, toll roads' miles have almost doubled in the past decade. In the United States, there are currently many tolled road facilities; these facilities vary in type, size, ownership, and tolling systems deployed (Abuzwidah et al., 2014). Some of these facilities are private along with those owned and operated by various public agencies around the States (Abuzwidah, 2011) (Yang et al., 2014).

* Corresponding author: mabuzwidah@sharjah.ac.ae

Even though toll roads offer high mobility benefits, traditional toll facilities may pose high traffic safety risk (Abuzwidah et al., 2014). The literature also showed that different toll collection systems have been adopted by different toll agencies around the world (Mohamed et al., 2000) (Schaufler, 1997). The most common toll collection systems can be summarized as follows:

1. Traditional Mainline Toll Plaza (TMTP); this design requires vehicles to rapidly decelerate, navigate through different fare transaction options, and then accelerate and merge with traffic (Miami-Dade Expressway, 2014) (McDonald, and Stammer, 2001). These confusing maneuvers constitute safety challenges and form high risk locations on toll roads.

2. Hybrid Mainline Toll Plaza (HMTP); this system retrofits existing tollbooths with express open Electronic Toll Collection (ETC) lanes. This design is allowing more than 81 percent of the vehicles in Florida to travel at full speeds using electronic transponders or license plate recognition technology in an open road environment with fewer diverge and merge maneuvers before and after the toll plaza. The HMTP design combines either express Open Road Tolling (ORT) lanes on the mainline and separate traditional toll collection to the side, or traditional toll collection on the mainline and separate ORT lanes to the side (FTE, 2014). However, the HMTP is widely deployed by many toll authorities such as in Florida, Illinois, New Jersey, and many other states.

3. All-Electronic Toll Gantry or All-Electronic Toll Collection (AETC); this system is completely barrier-free that replaces all tollbooths with regular express ETC lanes to change the toll plaza to be similar to regular segments. The AETC system allows driving straight through an open road without needing to change lanes, stop the vehicle, or even slow down to pay a toll. The payment will be done automatically, instantly and accurately by using the automatic toll collection transponder known as prepaid transponder (FTE, 2015).

Past studies and the current data have indicated that certain locations at the TMTP and HMTP are more likely to experience traffic crashes than the regular segments on the expressway. In April 2006 in Washington, D.C., investigators for the U.S. National Traffic Safety Board (NTSB) revealed that the most dangerous locations on the highways are toll plazas. In the same year, the NTSB reported that 49 percent of all crashes on expressways in Illinois occurred at toll plazas, and three times as many people died in them as in crashes on the rest of the same roadways. Also, 30 percent of all crashes on the Pennsylvania Turnpike happen at toll plazas and 38 percent of all collisions on New Jersey toll highways are toll plaza incidents (NTSB, 2014). An older study (Mohamed et al., 2000) found that about 32 percent of the total crashes that occurred on the Orlando expressway system were located at the traditional mainline toll plazas.

However, there are some obstacles to the use of the HMTP and AETC systems in many countries because these systems require good arrangements between the tolling agencies and the department of motor vehicles database. These arrangements are needed to identify and bill drivers who do not have the prepaid transponder. This processing is called Toll-By-Plate program, which is an image based electronic toll collection system that uses photographic images of the vehicle's license plate to identify the customers responsible for payments and bill them (FTE, 2015).

The use of the HMTP and AETC systems has demonstrated measured improvements in traffic operations and environmental issues. Also, it was proved that some simple and quick treatments could improve safety at TMTP (Sze et al., 2008). For example, (Wong et al.,

2006) evaluated the effects of a traffic guidance scheme for auto-toll lanes on traffic safety at TMTP and they found that the overall lane-changing rate decreased significantly by 23 percent and the pooled conflict count decreased sharply by 44 percent, also the crash count decreased sharply by 38 percent. However, there is a lack of research that compares and evaluates the safety impacts of using different toll plaza types. So, there is an urgent need to assess the traffic safety effects of these facilities.

To the best of our knowledge, there were no studies that evaluated the safety impacts of tollbooths number on the mainline toll plazas. Therefore, the main focus of this paper is to evaluate the effect of tollbooths number and types (i.e. cashier lanes, coin lanes and Electronic Toll Collections (ETC) lanes) on safety performance at traditional toll plazas; as well as providing safety guidelines for these facilities to promote safety on expressways.

2 Methodology

Negative Binomial Regression Model: Count data is usually modeled using a Poisson distribution; the main characteristic of the Poisson distribution is that its mean is equal to its variance. Several studies found that a negative binomial distribution fits crash frequency data better (Brown et al., 2006) (Abdel-Aty and Radwan, 2000). The Negative Binomial (NB) is similar to a Poisson distribution, though its variance is greater than its mean (Agresti, 2002). The NB model can be used to estimate crashes at locations, such as toll plaza segments (Chang, 2005).

P(yi) ~ Negbin (λi, k):

$$\frac{\Gamma(1/_k + y_i)}{y_i\,!\,\Gamma(1/_k)} \left(\frac{k\lambda_i}{1 + k\lambda_i}\right)^{y_i} \left(\frac{1}{1 + k\lambda_i}\right)^{1/_k} \tag{1}$$

Where, y = number of crashes on a certain location at a certain period,

λ= Expected number of crashes on a certain location at a certain period

k= over-dispersion parameter.

The NB estimation equation is given as follows;

$$\text{Log}\,\lambda_i = \log(EV_i) + (\beta_0 + \beta X_i + \varepsilon_i). \tag{2}$$

Where λ_i is the estimated number of crashes at location *i*; β is a vector of explanatory variables, and EV_i is the exposure variable which is included in models as an offset term to account for unequal exposure to crashes along toll plazas. Count models such as NB need a mechanism to deal with different exposure scenarios. For example the Annual Average Daily Traffic (AADT), or expected number of crashes, can be included as an exposure offset term in NB models (Chen, 1997). By offsetting the exposure, the response variable becomes the log crash frequency at a certain location. ε_i is the term adjusting for over dispersion. It is an error term due to a combination of variables omitted from the model and pure randomness. Maximum Likelihood Estimation (MLE) is used to estimate the coefficients of the explanatory variables and the over dispersion parameter. The goodness of fit of the NB model is measured by analyzing the over dispersion parameter.

3 Data description

Different tolling systems are expanding on the Florida expressways in the past decade. Multiple sets of data maintained by Florida Department of Transportation (FDOT, 2015) were used in this study including:
Crash Analysis Reporting (CAR) system, Roadway Characteristics Inventory (RCI) system, TranStat IView aerial mapping system, and Financial Project Search Database. Moreover, Google Earth and the publication reports of Florida Turnpike and Central Florida Expressway Authority (CFX, 2015) were used to verify the toll plazas' characteristics. A 42 mainline toll plazas located on approximately 750 miles of toll roads in Florida were selected. Crash data from four-year period (2011-2014) were investigated. Crashes that occurred within the influence areas (1 mile before and 0.5 mile after the centerline of the mainline toll plaza) (MUTCD, 2015) were extracted from the crash database.

4 Analysis and results

4.1 Part I : Effect of the total number of tollbooth on the safety at TMTP

Data from 42 Traditional Mainline Toll Plazas (TMTP) were used in this approach. Crash data from a four-year period (2011, 2012, 2013 and 2014) was investigated. Figure 1 shows Traditional Mainline Toll Plazas. This design combines express lanes and traditional cash lanes.

Fig. 1. Traditional Mainline Toll Plazas. (Source: (FTE, 2015))

A series of Negative Binomial (NB) models were fitted to establish the relationships between traffic and roadway characteristics, and crash frequency at the vicinity of TMTPs. Exploratory modeling indicated that the crash frequency is not significantly associated with Reversible Electronic Toll Collections (ETC) lanes, number of entry and exit ramps, and distance to entry and exit ramps.

Table 1 shows the parameter estimates of the final model. It should be noted that in some TMTPs, there are a Reversible ETC lanes based on the peak hours. Only log Annual Average Daily Traffic (AADT) and the number of tollbooths came out to be significant with the expected signs. The coefficient of both variables have a positive sign, which indicates that as they increase, the frequency of TMTP-related crashes increases.

Table 1. Estimates of coefficients for Effect of the number of tollbooths on the safety.

Negative Binomial					
Negative Binomial: Analysis Of Maximum Likelihood Parameter Estimates					
Crash Type	Injury Level	Parameter	Estimates	Pr > ChiSq	AIC
All	All	Intercept	-11.245	<0.0001	186.648
		Log AADT	1.6272	<0.0001	
		# of tollbooths	0.169	0.0225	
		Dispersion	0.143		

This finding could be attributed to the fact that the AADT is considered as the exposure factor; more traffic at the TMTP would lead to higher chances of crash occurrence. This is expected, since higher AADT results in higher exposure of vehicles to weaving manoeuvres to avoid delay at some lanes which in turn would result in higher crash frequencies. Also, the number of tollbooths was found to be significant, as the number of tollbooths increase the crashes are more likely to be increased, and that may be because of the increase of multiple lane changes.

4.2 Part II : Effect of the number of the ETC and cash lanes on the safety at TMTP

Data from 42 Traditional Mainline Toll Plazas (TMTP) also was used in this approach. Crash data from a four-year period (2011, 2012, 2013 and 2014) was investigated. It should be noted that the number of ETC lanes and coin lanes were not significant, therefore it was excluded from the final model. Figure 2 shows Tollbooths Types of Traditional Mainline Toll Plazas. This design combines express lanes and traditional cash lanes.

Fig. 2. Tollbooths Types of Traditional Mainline Toll Plazas. (Source: *(*FTE, 2015*))*

A series of NB models also were fitted to establish the relationships between traffic, roadway characteristics and crash frequency at the vicinity of TMTPs. The results indicated that the crash frequency was not significantly associated with number of ETC and coin lanes, number of entry and exit ramps, and distance to entry and exit ramps. Table 2 shows the parameter estimates of the final model.

Table 2. Estimates of coefficients for Effect of the number of tollbooths on TMTP safety.

Negative Binomial					
Analysis Of Maximum Likelihood Parameter Estimates					
Crash Type	Injury Level	Parameter	Estimates	Pr > ChiSq	AIC
All	All	Intercept	-18.356	<0.0001	168.214
		Log AADT	1.846	<0.0001	
		# of cash lanes	0.113	0.0322	
		Dispersion	0.126		

Only log Annual Average Daily Traffic (AADT) and the number of cash lanes came out to be significant with the expected signs. The coefficient of both variables have a positive sign, which indicates that as they increase, the frequency of TMTP-related crashes increases.

This finding could be attributed to the fact that the AADT is considered as the exposure factor; more traffic at the TMTP would lead to higher chances of crash occurrence. This is expected, since higher AADT results in higher exposure of vehicles to weaving maneuvers (change lanes) between cash lanes and ETC lanes, which in turn would result in higher crash frequencies. Also, the number of cash lanes found to be significant.

Therefore, as the number of cash lanes increase, the crashes are more likely to be increased, and that may be because of the increase of the lane changes between the different types of lanes. The number of ETC lanes and coin lanes were not significant and that maybe because most of the locations in this study have similar number of the ETC and coin lanes. Therefore, it is strongly recommended to be reevaluated with different data to reach a constant conclusion about the effect of the number of the coin lanes at the traditional mainline toll plaza.

5 Conclusions

Expressways (toll roads) and freeways are considered an important part of any successful transportation system. These facilities carry the majority of daily trips on the transportation network. Although toll roads offer a high level of service, traditional plazas still experience many crashes, many of which are severe. Therefore, this study examines for the first time the traffic safety impact of the number and types of tollbooths on the traditional mainline toll plaza. And that by helping understand the relationship between the crash frequency and several important crash-related factors and circumstances such as traffic volume, total number of tollbooth and the number of cash lanes.

Crash data from four-year period was investigated, and 42 traditional mainline toll plazas were evaluated. The current data has indicated that the number of tollbooths and the number of the cash lanes were found to be significant in the final models. Therefore, as the number of toll booths increase the crashes are more likely to be increased, and that may be because of the increase of the potential of making multiple lane changes. Moreover, this study showed that as the number cash lanes increase, the crashes also are more likely to be increased.

The number of ETC lanes and coin lanes were not significant and that maybe because most of the locations in this study have similar number of the ETC and coin lanes. Therefore, it is strongly recommended to be reevaluated with different data to reach a constant conclusion about the effect of the number of the ETC and coin lanes at the traditional mainline toll plaza.

The results indicated significant relationships between the crash frequency and the number of tollbooths, number of the cash lanes and the annual average daily traffic. This means all of these variables significantly affect the frequency of the traditional toll plazas-related crashes. For those agencies that need to adopt this design of the toll collection system, improving traffic safety at these facilities should take a priority.

Overall, choosing locations for the toll plazas that have safe distances from the interchanges, installing flashing lights at night time, and finding ways to improve signage and road marking are potential means of reducing lane changes before tollbooths. These finding are very important as a first step to evaluate these facilities and researchers may build on these results to reach a clear conclusion.

References

1. Abdel-Aty, M. and Radwan, E. "Modeling Traffic Accident Occurrence and Involvement." *Accident Analysis and Prevention* No. 32, 633–642. (2000)
2. Abuzwidah, M. "Evaluation and Modeling of the Safety of Open Road Tolling System" M.S. Thesis. Orlando, University of Central Florida. (2011)
3. Abuzwidah, M., Abdel-Aty, M., and Ahmed, M. "Safety Evaluation of Hybrid Mainline Toll Plazas". *Journal of the Transportation Research Board, in press.* (2014)
4. Agresti, A. Categorical data analysis. New York: Wiley-Interscience. (2002)
5. Brown, L., D.R., McDonald, and E.J. Myers. Developing Traffic Control Strategies at Toll Plazas. *ITE Journal*, November, pp. 22–26. (2006)
6. Central Florida Expressway Authority (CFX). https://www.oocea.com . Last access Jan 2017.
7. Chang, L. Analysis of freeway accident frequencies: Negative binomial regression versus artificial neural network. *Safety Science*, (8), 541. (2005)
8. Chen, C. Statistical analysis of the effect of demographic and roadway factors on traffic crash involve ment. (1997)
9. Federal Highway Administration (FHWA), Manual on Uniform Traffic Control Devices (MUTCD) Chapter 2F. http://mutcd.fhwa.dot.gov/knowledge/faqs/faq_part2.htm Last access Jan 2017.
10. Florida Department of Transportation (FDOT). http://www.dot.state.fl.us/ . Last access Jan 2017.
11. Florida Turnpike Enterprise TOLL-BY-PLATE. https://www.tollbyplate.com/index . Last access Jan 2017. (FTE, 2015)
12. Kim, K., Nitz, L., Richardson, J. and Li, L. Personal and behavioral predictors of automobile crash and injury severity. *Accident Analysis and Prevention* 27, 4, 469–481. (1995)
13. McDonald, D. and R. Stammer. Contribution to the Development of Guidelines for Toll Plaza Design. *Journal of Transportation Engineering*, Vol. 127, No. 3, pp. 215–222.
14. Miami-Dade Expressway. http://www.mdxort.com/. Last access Jan 2017.(2001)
15. Mohamed, A., Abdel-Aty, M. and Klodzinski, J. Safety Considerations in Designing Electronic Toll plazas: Case Study. *ITE Journal.* (2000)

16. National Transportation Safety Board (NTSB). http://www.ntsb.gov/. Last access Jan 2017.
17. Schaufler, A.E. NCHRP Synthesis of Highway Practice 240: Toll Plaza Design. Transportation Research Board of the National Academies, Washington, D.C., (1997)
18. Sze, NN, SC Wong, and WF Chan. n.d. "Traffic crashes at toll plazas in Hong Kong." Proceedings Of The Institution Of Civil Engineers-Transport 161, no. 2: 71-76. (2008)
19. Wong, S., Sze, N., Hung, W., Becky, P., & Lo, H. (n.d). The effects of a traffic guidance scheme for auto-toll lanes on traffic safety at toll plazas. *Safety Science*, 44(9), 753-770.
20. Yang, H., Ozbay, K., Bartin, B., and Ozturk, O. "Assessing the Safety Effects of Removing Highway Mainline Barrier Toll Plazas." J. Transp. Eng., 10.1061/ (ASCE) TE.1943-5436.0000681, 04014038. (2014)

Determining u-turn adjustment factor for signalized intersections in Doha, Qatar

Khaled Hamad[1*] and *Hassan* Abu-Hamda[2]

[1] Sustaimanle Civil Infrastructure Systems Research Group, University of Sharjah, Sharjah City, United Arab Emirates
[2] Al-Jaber Engineering, Qatar

Abstract. This paper summarizes the work conducted to estimating a U-turn adjustment factor at selected signalized intersections in Doha, Qatar. Unfortunately, the Highway Capacity Manual (HCM) 2010 does not consider the effect of U-turn traffic on the capacity of signalized intersection; therefore, a few researchers attempted to develop U-turn adjustment factors to be added to the HCM analysis of signalized intersections when heavy U-turn traffic exists. To achieve this, the average headway was measured for U-tuning and left-turning vehicles at three different intersections in Doha. To estimate the U-turn adjustment factor, 198 queues with 2,327 U-turn vehicles and 1,564 left-turn vehicles were used to develop a regression model with the headway as the dependent variable and the U-turn percentage as the independent variable. The resulting quadratic model shows an increase in the average headway as the percentage of U-Turn traffic increases, with an R^2 of 0.591. This indicates that the presence of U-turning vehicles has a significant effect on increasing the average headway thus reducing the capacity of signalized intersection. The result shows the U-Turn adjustment factor ranges from 0.76 for 100% of U-turn traffic to 0.95 for 10% of U-turn traffic.

1 Introduction

Traffic signals are traffic control devices used to manage the flow of traffic through intersections in a safe and efficient manner. Whether it is for designing new timing plans (to allocate green times) or operational analysis of existing conditions (to determine level of service), traffic engineers require an accurate estimate of the capacity of the signal.

A traffic signal's capacity, defined as the maximum number of vehicles that can reasonably be expected to pass through the intersection under prevailing traffic, roadway, and signalization conditions during a 15-minute period, can be computed as follows [1]:

$$c = s \cdot \frac{g}{C} \tag{1}$$

where:

* Corresponding author: khamad@sharjah.ac.ae

c = Capacity (vehicle/hour)
s = Saturation flow rate (vehicle/hour)
g = Green interval time (second)
C = Cycle length (second)

Therefore, saturation flow rate is a key parameter in estimating a signalized intersection's capacity. The HCM defines saturation flow rate as the equivalent hourly rate at which previously queued vehicles can cross an intersection approach under prevailing conditions, assuming that a green signal is available at all times and no lost time is experienced [1]. Since measuring saturation flow rate in the field is challenging, many engineers opt to follow the HCM procedure to estimate saturation flow rate using the following model [1]:

$$s = s_o \, N \, f_w \, f_{HV} \, f_g \, f_{bb} \, f_a \, f_{LU} \, f_{LT} \, f_{RT} \, f_{Lpb} \, f_{Rpb} \qquad (2)$$

where:

s = Saturation flow rate for subject lane group, expressed as a total for all lanes in lane group (vehicle/hour),
s_o = Base saturation flow rate at ideal condition (passenger car/hour/lane)
N = Number of lanes in a lane group,
f_w = Adjustment factor for lane width,
f_{HV} = Adjustment factor for heavy vehicles in traffic stream,
f_g = Adjustment factor for approach grade,
f_p = Adjustment factor for existence of a parking lane and parking activity adjacent to lane group,
f_{bb} = Adjustment factor for blocking effect of local buses that stop within intersection area;
f_a = Adjustment factor for area type,
f_{LU} = Adjustment factor for lane utilization,
f_{LT} = Adjustment factor for left turns in lane group,
f_{RT} = Adjustment factor for right turns in lane group,
f_{Lpb} = Pedestrian adjustment factor for left-turn movements, and
f_{Rpb} = Pedestrian-bicycle adjustment factor for right-turn movements.

As shown in Eq. 2, there are many adjustment factors to account for the various non-ideal geometric, traffic, and environmental conditions that may influence the headway of vehicles departing the intersection. Two of these factors, the left-turning traffic (LT) and the right-turning traffic (RT), deal with turning vehicles which can significantly impact the saturation flow rate thus the operational performance of signalized intersections. In GCC countries, U-turning (UT) traffic not only it is allowed to be made from the left lane(s), it can also be significant amount. Making a UT usually takes longer time to complete than a simple left turn because of the reduction of speed the driver needs to make due to the smaller turning radius needed); therefore, assuming that they have the same effect on the capacity of a signalized intersection is not appropriate. Unfortunately, the HCM does not take into consideration the effect of UT traffic on the saturation flow rate; instead UT traffic is considered as normal LT traffic. This is clearly a simplification as the speed of U-turning vehicles is less than the speed of left-turning vehicles and this will reduce the capacity of the intersection. Moreover, the mixture of UT traffic in a LT lane will disturb the saturation headway. Thus, instead of having a stabilized discharge headway after the 4th or 5th vehicle, the discharge headway will be unpredictable changing as the order of left and U-turning vehicles changes. This will certainly affect the flow of LT traffic and as a result affect the capacity of the signalized intersection.

There is no widely-accepted method that predicts the effect of the U-turning traffic on the capacity of a signalized intersection. Moreover, few researchers attempted to study this issue and studying the local conditions to take into consideration the local driver behavior is important and has been suggested by the HCM 2010. Therefore, it is important to analyze and estimate the effect of UT under local conditions. This study is going to explore the effect of permitted UT traffic in a protected LT on the capacity of signalized intersection, which will make this study an important topic for Qatar.

2 Literature review

Several researchers have investigated the effect of U-turning traffic on signal operation. The following paragraphs summarize some of the recent research effort in this topic.

Adams and Hummer studied the effect of U-turns on the left turn saturation flow rate in 1993 in North Carolina [2]. In their study, four signalized intersections with protected left-turn were selected. They recorded the saturation flow rate and the percentage of U-turn traffic for 198 queues during midday peak periods on weekdays. After analyzing the data collected and using t-tests and regression model, the study found that there is no correlation between the saturation flow rate and the percentage of U-turns for intersections where U-turn traffic in the left-turn lane is less than 50% of the total traffic. However, for 50 to 65% of U-turn traffic, their study was inconclusive due to small sample size. They also found that a reduction factor for the saturation flow rate is necessary for left-turns when more than 65% of the traffic is U-turn traffic. From the regression model they recommended a reduction factor for U-turn traffic of 65% to 85% to be 0.9, and for 85% or more of U-turn traffic they recommended a reduction factor of 0.8.

Tsao and Chu investigated the effect of U-turning traffics on the saturation flow rate in left-turn lanes in Taiwan [3]. They recorded 600 headways of left-turning vehicles and 160 headways of U-turning vehicles from a few traffic signals. They found, as can be expected, that the headway of U-turning traffic is larger than that of left-turning traffic. Their study showed that when a vehicle is preceded by left-turning vehicle, the average headway of a U-turning passenger car is 1.27 times that of a left-turning passenger car. However, when it is preceded by U-turning vehicle the average headway of U-turning passenger cars is 2.17 times that of left-turning passenger cars.

In a study by Liu et al. in 2005, the effect of U-turning vehicles on the left-turn saturation flow rate was analyzed [4]. The data was collected from three signalized intersections in Tampa Bay, Florida. A total of 260 discharge queues were recorded, that included 1,441 left turning vehicles and 571 U-turning vehicles. From the data collected, a regression model was developed to estimate the relationship between the average queue discharge headway and the various percentages of U-turning vehicles in the left-turn traffic stream. Using a regression model, adjustment factors for various percentages of U-turning vehicles were developed. It was recommended that a capacity reduction should be done when the flow on the left-turn lane has at least 40% U-turning vehicles.

Carter et al studied the operational and safety effects of U-turns at signalized intersections in North Carolina [5]. In this study, vehicle headways at 14 signalized intersections were measured. Using regression analysis, they found that the saturation flow rate in left-turn lane is reduced by 1.8 % for every 10% of increase in U-turn percentage.

Hossain estimated the UT, LT, and RT turning adjustment factors for some signalized intersections in Malaysia [6]. Using regression analysis, the results indicated that the LT and RT adjustment factors decrease as the proportion of turning vehicles increases and the turning radius decreases. He also showed that the U-turn adjustment factor decreases with the increase in the percent of U-turning vehicles, specifically the U-turn adjustment factor decreases 2.25% for every 10% increase of U-turn percentage. It is worth mentioning that the UT traffic adjustment factor was limited to a single LT lane only.

Jobair bin Alam et al. estimated adjustment factors for U-turn, heavy vehicles, number of lanes, and lane width in Makah, Saudi Arabia [7]. Five signalized intersections were selected for this study. The study indicated that while many of these adjustment factors have been developed by the HCM, the operation and performance of signalized intersections rely on the roadway's environmental features and users' behavioral characteristics that significantly differ among locations. Based on 1,589 headways measurements, it was found that the adjustment factor for the UT vehicles with 30% of UT vehicles is 0.95 in Makah. Moreover, the adjustment factor for heavy vehicles was found as 0.78 for 30% of heavy vehicles.

In general, these studies showed that the increase in U-turn traffic decreases the saturation flow rate which will decrease the capacity of the traffic signal. However, the magnitude of the reduction reported by the various studies was not the same. This can be justified by the different driver behavior, operational traffic conditions, and road characteristics between the different locations where these studied were conducted. In fact, the HCM clearly adds a disclaimer emphasizing the need for adapting locally measured values outside of North America [1].

The purpose of this paper is to summarize the procedure undertaken and results achieved to quantifying the impact of U-turn traffic on the operation of signalized intersections in a region known for its aggressive driver behavior, namely Doha, Qatar. This will assist local engineers to better estimate the saturation flow and capacity of signalized intersections in the city and the region, which will ultimately improve functionality of signals and reduce delays for the vehicles.

3 Methodology

As it was revealed in the literature review, and unlike other factors, the HCM does not provide much indo on how to adjust for U-turn traffic at signalized intersections. Few research has developed a UT adjustment factor in their local conditions and all of them used the headway and UT percentage to develop the factor; however, some used the average discharge headway and others considered the headway of each vehicle depending on the car that preceded it, left-turning or U-turning.

In this study, an adjustment factor that takes into consideration the effect of U-turning vehicles on the capacity of signalized intersection under local conditions is developed. This is accomplished by analyzing the relation between the percent of U-turning vehicles and the average headway. Even though, there are other factors that can affect the average headway besides the U-turning vehicles, like the turning radius, they have not been considered in this study. The reason for not considering the turning radius is that all the intersections used in this study have the same turning radius.

To develop the relation between the percent of U-turning vehicles and the average headway, various models have been attempted. To find the best-fit model, the ANOVA test was implemented. After finding the best fit, the model is used to calculate the UT adjustment factor for a certain percent of U-turning vehicle. This adjustment factor can be directly used to estimate the capacity reduction due to the presence of U-turn traffic.

The following information was collected using video recordings for sample intersections:

1. Number of U-Turning vehicles in each lane
2. Number of left-turning vehicles in each lane
3. Number of heavy vehicles (buses and trucks) making a LT or UT in each lane

4. Queue discharge time, which is the time that elapses from the start of the green interval till the last vehicle in the previously queued vehicles passes the stop line) for each lane

Once this information it obtained, the average queue discharge time for each signal phase is calculated using:

$$h = \frac{T}{N_u + N_l} \tag{3}$$

where

h = the average discharge headway for each turning vehicle
T = the queue discharge time
N_u = the number of U-turning vehicles in the queue
N_L = the number of left-turning vehicles in the queue

To account for the presence of heavy vehicles, a passenger-car equivalent factor of 2.0 was used to convert the mix traffic with heavy vehicles into passenger-car only traffic.

Using the percentage of the UT traffic and the average discharge headway, a regression model is developed to relate the two variables. The following four models were attempted: Linear, Logarithmic, Quadratic, and Exponential. The best-fit model is then used to calculate the discharge headway for the various percentages of UT traffic. The model is also used to compute the base discharge headway (h_o) using a UT percentage equal to zero.

Finally, the UT adjustment factor is calculated as follows:

$$f_{UT} = \frac{s}{s_o} = \frac{3600/h}{3600/h_o} = \frac{h_o}{h} \tag{4}$$

where

f_{UT} = Adjustment factor for the U-turning vehicles
h_o = Average discharge headway without U-turn traffic, obtained from the best-fit model when UT percentage is zero
h = Average discharge headway with U-turn traffic, obtained from the best-fit model
s = Saturation flow rate
s_o = Base saturation flow (when there is no UT traffic in the lane)

The developed U-turn adjustment factor should be used, along with other factors, in calculating the saturation flow rate (Eq. 2). Including this factor will lead to better estimation of the saturation flow rate since it takes the effect of UT traffic into consideration.

4 Data collection and processing

To measure the saturation flow rate at signalized intersection, three signalized intersections in Doha, Qatar were selected. To separate the effect of other factors that can affect the capacity of a signalized intersection, the following criteria were used to select the sample intersections to be used in the study:

1. Lane with of 12 feet (3.65 m)
2. No parking adjacent to a travel lane within 250 feet (75 m) of the stop line
3. No disturbance from heavy vehicles
4. No disturbance form right turning vehicles on the other approaching lane
5. No disturbance form left turning vehicles on the other approaching lane
6. No disturbance from pedestrians
7. Approach's grade is level
8. Intersection has exclusive LT lane

9. LT lane has a protected phase
10. UT traffic is allowed
11. LT lane has significant UT traffic

Figure 1 shows the location of the study sites. These intersections have long queues for extended periods of time, especially during peak hours.

All data were recorded during peak hours on weekdays, during the period from February 2013 to May 2013. Traffic data were collected via video with each video record was expected to capture all turning movements: left turning movement, through movement, and right turning movement. Ideally, the camera has to be placed in a way that enables the viewer to see the start and end of each phase, the end of the queue for each lane, and the stop line of the lane. Figure 2 shows, for example, the location of the two video recording cameras used to collect the data needed at Al-Sadd intersection. Figure 3 is a screenshot taken from one of the video recording at the same intersection. In a few cases where it was impossible to see the end of the queue, we assumed that the end of the queue happens when there was large time gap (more than 2 seconds) between two consecutive vehicles.

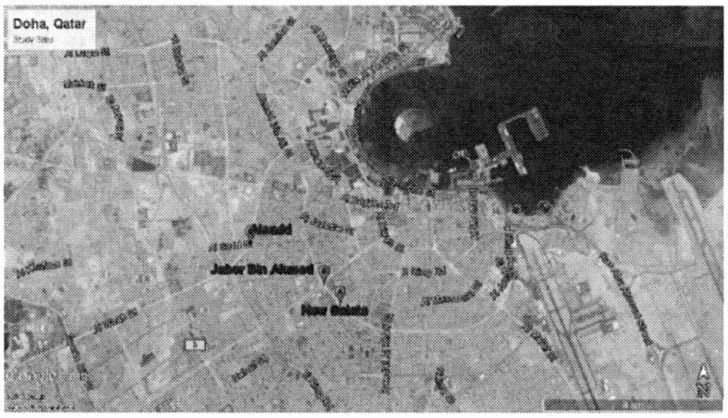

Fig. 1. Location of selected sites.

Fig. 2. Placement of video cameras at Jaber Bin Ahmed Intersection.

DOI: 10.1051/matecconf/201712007008

Fig. 3. Sample screen capture from the video recording.

Table 1 shows the sample size from the different intersections. The total number of queues considered from the three selected intersections combined was 198 and the total number of U-turning vehicles considered was 2,329. This sample size is comparable to other similar studies; for example, Adams and Hummer study utilized 198 queues.

Table 1. Chosen intersections and collected data from each intersection.

Intersection	No. of Queues	No. UT Vehicles	No. LT Vehicles	UT% in the lane
New Salata	46	512	424	55%
Jabr Bin Ahmed	105	1,135	1,056	52%
Al Sadd	47	680	84	89%
All	198	2,327	1,564	60%

Using the video recordings of the selected signalized intersection, the time frame has to be selected so that it (1) has the highest percentage of UT traffic, (2) occurred during peak hours of weekdays, and (3) has no weather disturbance.

5 Results and discussion

As previously mentioned, the first step towards developing the U-turn adjustment factor is to model the relation between the percentage of UT traffic and the average discharge headway. For each site, the following four regression models were attempted: Linear, Logarithmic, Quadratic, and Exponential. Table 2 shows the summary and the parameter estimates of the developed best-fit models for the three sample sites, as well as for all the sites combined. Figure 4 shows these models plotted against the dataset from all intersections combined, as an example.

From Table 2, it is clear that all the four models, regardless of the site, have shown acceptable results Moreover, all of the models have reasonable R^2 values, ranging from around 0.50 to 0.64. The R^2 value indicates the percentage of the variability in the headway explained by the variability of the percent of UT traffic. Except the model for Jaber Bin Ahmed Intersection, the quadratic model outperformed the other models for the other two sites as well as for the all intersections combined. Nevertheless, the difference between the best-performing model at Jaber Bin Ahmed, i.e. exponential, and the quadratic model is only minor (R^2 of 0.590 vs. 0.581). Therefore, we adopted the quadratic model (Eq. 5) to relate the discharge headway and UT percentage at the selected sites in Doha, Qatar.

$$h = -0.340 \times UT^2 + 0.887 \times UT + 1.721 \quad (5)$$

where

h = Average discharge headway for U-turning vehicles
UT = Percentage of U-turning vehicles

Table 2. Developed models for each intersection.

	Parameter Estimates			**Model Summary**		
	Constant	b_1	b_2	R^2	F	P value
New Salata Intersection						
Linear	1.811	0.505		0.640	79.871	0.000
Quadratic	*1.761*	*0.700*	*-0.176*	*0.642*	*39.403*	*0.000*
Logarithmic	2.255	0.260		0.637	79.040	0.000
Exponential	1.826	0.243		0.636	78.502	0.000
Jaber Bin Ahmed Intersection						
Linear	1.750	0.665		0.580	142.23	0.000
Quadratic	1.727	0.788	-0.136	0.581	70.76	0.000
Logarithmic	2.264	0.227		0.542	122.07	0.000
Exponential	*1.763*	*0.327*		*0.590*	*147.92*	*0.000*
Alsadd Intersection						
Linear	1.664	0.624		0.524	49.447	0.000
Quadratic	*2.641*	*-1.762*	*1.427*	*0.556*	*27.53*	*0.000*
Logarithmic	2.283	0.505		0.504	45.762	0.000
Exponential	1.725	0.282		0.531	50.945	0.000
All Intersections Combined						
Linear	1.826	0.481		0.569	258.996	0.000
Quadratic	*1.721*	*0.887*	*-0.340*	*0.591*	*141.061*	*0.000*
Logarithmic	2.250	0.224		0.574	264.193	0.000
Exponential	1.836	0.232		0.572	262.354	0.000

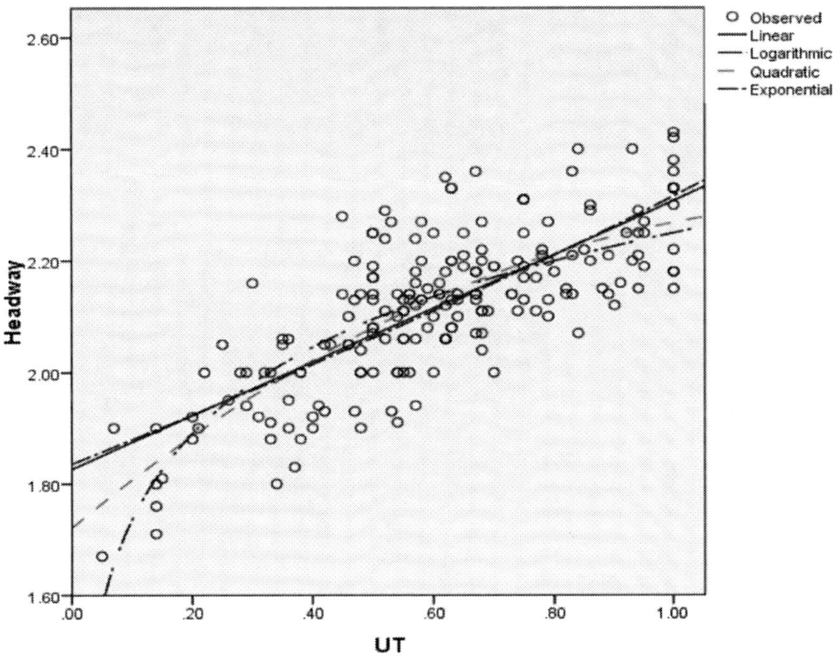

Fig. 4. Regressed models for all intersections combined.

Table 3 summarizes the results of the ANOVA test for the selected quadratic model, which also confirms that the null hypothesis can be rejected because the P-value is equal to zero. In other words, the probability that the headway was not affected by UT percentage and instead the variance was just random event is zero. To verify the basic assumption that the variance is homogeneous, the unstandardized residuals were plotted for the independent variable (Figure 5). It is clear that the variance is spread evenly across the 0-axis and that it does not have any clear pattern, Therefore, the basic assumption about the homogenous variance is valid.

Table 3. ANOVA test results for the selected quadratic model.

	Degree of Freedom	$\sum$ of Squares	Mean^2	F	P value
Regression	2	2.264	1.132	141.061	0.000
Residual	195	1.565	0.008		
Total	197	3.830			

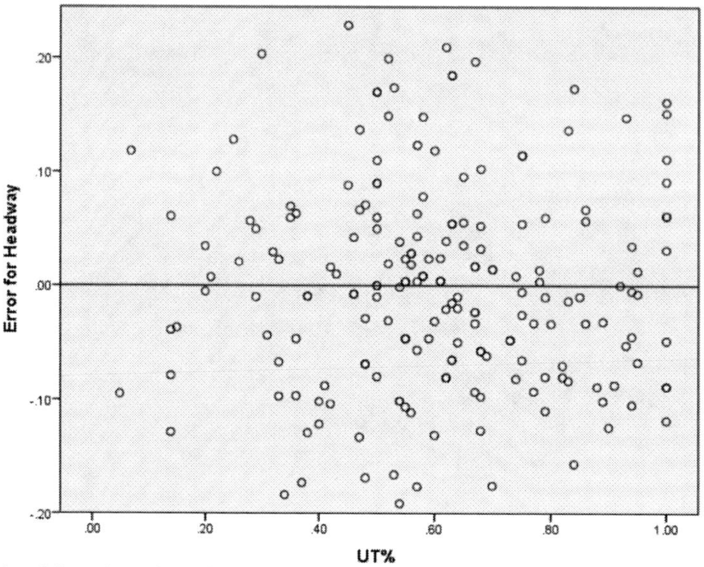

Fig. 5. Residuals of the selected quadratic model.

Using Eq. 4 and Eq. 5, the UT adjustment factor for the selected intersections in Doha, Qatar can be calculated using:

$$f_{UT} = \frac{1.721}{-0.340 \times UT^2 + 0.887 \times UT + 1.721} \qquad (6)$$

where

f_{UT} = Adjustment factor for U-turning vehicles

UT = Percentage of U-turning vehicles

Using Eq. 6, we calculated the UT adjustment factor for different UT percentages ranging from 0% to 100%, with 10% incremental, as shown in Table 4.

Table 4: UT Adjustment factors for different percentages of U-turn traffic.

UT	**0%**	**10%**	**20%**	**30%**	**40%**	**50%**	**60%**	**70%**	**80%**	**90%**	**100%**
h	1.72	1.81	1.88	1.96	2.02	2.08	2.13	2.18	2.21	2.24	2.27
h_o	1.72	1.72	1.72	1.72	1.72	1.72	1.72	1.72	1.72	1.72	1.72
f_{UT}	1.00	0.95	0.91	0.88	0.85	0.83	0.81	0.79	0.78	0.77	0.76

As previously mentioned, a few studies developed adjustment factors for UT traffic in many cities around the world. These studies differ in the location of the study and the methodology used. In addition, driver behavior could vary significantly from one place to the other. This has led to differences among the reported results. Therefore, it is important to compare the results of this study to other studies, namely Tsao and Chu [3], Carter el al [5], and Wang et al [8]. Figure 6 plots the values of f_{UT} vs. UT percent estimated in this study as well as their corresponding values reported in the above mentioned studies.

DOI: 10.1051/matecconf/201712007008

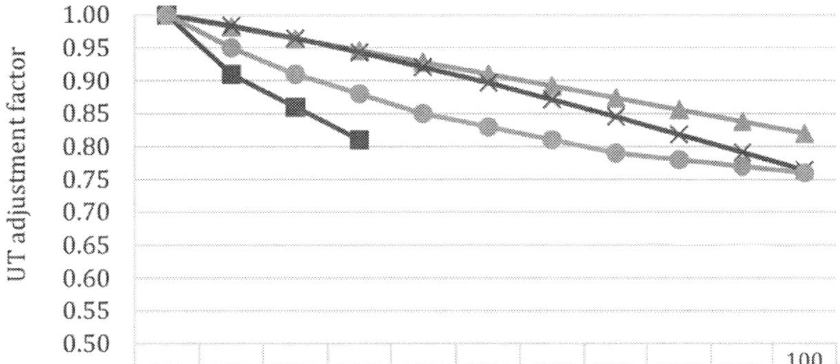

	0%	10%	20%	30%	40%	50%	60%	70%	80%	90%	100 %
Tsao & Chu (1996)	1.00	0.91	0.86	0.81							
Carter et al (2005)	1.00	0.98	0.96	0.95	0.93	0.91	0.89	0.87	0.86	0.84	0.82
Wang et al (2008)	1.00	0.98	0.96	0.94	0.92	0.90	0.87	0.84	0.82	0.79	0.76
This Study	1.00	0.95	0.91	0.88	0.85	0.83	0.81	0.79	0.78	0.77	0.76

Fig. 6. Comparison of U-turn adjustment factor vs. U-turn percentages from different studies.

As can be seen in Figure 6, it is clear that this study shows a higher effect of the UT traffic on the UT adjustment factor, especially at lower percentages of U-turning vehicles, as compared to the other studies. It also shows that previous studies by Wang et al and Carter et al showed that their UT adjustment factors decrease linearly with the increase of UT percentage; whereas this study and Tsao's show a quadratic decrease concaving upward as the UT percentage increases. However, Tsao's study is limited in the range of the UT% [0% to 30%].

When comparing this study and Carter and Wang's study, we can see that the effect of the U-turning traffic in low percentages is higher in this study than the other two studies. However, as the UT% increases from 50% to 100% the difference in the UT adjustment factor between this study and the other two studies decreases. This might seem unusual at first but this could be explained by the following observation: as the U-turning vehicles increases, the movement of the vehicles becomes more synchronized thus decreasing the rate at which the adjustment factor is effected by the UT%. Furthermore, this difference between the results of the various studies is expected because Carter's and Wang's studies were conducted in North Carolina and Florida, respectively, while this study was conducted in Doha, Qatar. Comparing this study to Tsao and Chu's study conducted in Taiwan, we can see that the effect of UT traffic in low UT percentages is lower than those in Doha. There is no widely-accepted method that predicts the effect of the U-turning traffic on the capacity of a signalized intersection. Finally, the results of this study reaffirm the need to take into consideration local traffic conditions when modeling signal operation.

6 Conclusion

The HCM 2010 offers no adjustment factor to take into consideration the effect of U-turning vehicles on saturation flowrate at signalized intersection. A few researchers have studied this issue and developed such an adjustment factor. Furthermore, the HCM itself advices to consider the local traffic conditions and driver behavior when estimating the capacity of a signalized intersection. This study developed a U-turn adjustment factor for signalized intersections in Doha, Qatar. Towards this end, traffic data were collected from three different signalized intersections in the City.

Specifically, the discharge headways of a total of 2,327 U-turning vehicles and 1,564 left-turning vehicles from 198 queues were analyzed. To develop the U-turn adjustment factor for Doha, the first step was to model the relation between the percentage of UT traffic and the average discharge headway. Four regression models were attempted: Linear, Logarithmic, Quadratic, and Exponential. While all the four models have shown acceptable results with R^2 values ranging from around 0.50 to 0.64, the quadratic model outperformed the other models; and therefore, it was adopted. Using this model, the UT adjustment factor was computed and ranged from 0.95 at 10% U-turn traffic to 0.76 when 100% of turning traffic is making U-turns. Compared to the other studies, it was found that a higher effect of the UT traffic on the UT adjustment factor, especially at lower percentages of U-turning vehicles. Unlike two other previous studies, this study showed that the UT adjustment factors show a quadratic decrease concaving upward as the UT percentage increases.
The results of this study will help local engineers to have better estimates of the capacity of signalized intersections. Eventually, this will benefit the road users as there will be less delay at traffic signals; and therefore, they will reach their destination faster. Nevertheless, the finding of this study resulted from analyzing only three selected signalized intersections in Doha, Qatar. Prior to generalizing these results whether in Doha itself or other cities in the GCC region, validating these results using a larger sample size may be necessary. Moreover, this study did not take into consideration other potentially contributing factors such as turning radius, turning speed, size of turning vehicle. All of these factors could be important factors affecting the headway of turning vehicles. Therefore, it is recommended to investigate their effect on signal capacity in a future study.

References

1. Highway Capacity Manual, Transportation Research Board of the National Academies, Washington D.C., 2010.
2. Adams, J. C., and J. E. Hummer. Effects of U-Turns on Left-Turn Saturation Flow Rates. Transportation Research Record 1398. Transportation Research Board, National Research Council, Washington, D.C., 1993.
3. Tsao, S., and S. Chu. A Study on Adjustment Factors for U-Turns in Left- Turn Lanes at Signalized Intersections. Journal of Advanced Transportation, Vol. 29, No. 2, 1996, pp.183–192.
4. Liu, P., J. Lu, J. Fan, J. Pernia, and G. Sokolow. Effect of U-turns on the capacity of signalized Intersections. TRR 1920, Journal of Transportation Research Board of the National Academies, Washington D.C., 2005, pp. 74-80.
5. Carter, D., J. Hummer, R. Foyle, and S. Phillips. Operational and safety effects of U-turns at signalized Intersections. TRR 1920, Journal of Transportation Research Board of the National Academies, Washington D.C., 2005.
6. Quazi Hossain. Estimation of Turning Adjustment Factors at Signalized Intersections According to Malaysian Traffic Conditions. Master Thesis, University Sains, Malaysia, 2007.
7. Md. Jobair Bin Alam, Khalid A. Osra, Hamid O. Al-Bar and Shaher Z. Zahran. Signalized Intersection Capacity Adjustment Factors for Makkah, Saudi Arabia. Canadian Journal of Transportation, Volume 4, August 2011.
8. Wang, Xiaodong. Effects of U-Turns on Capacity at Signalized Intersections and Simulation of U-Turning Movement by Synchro. Master Thesis, University South Florida, Florida, 2008.

MATEC Web of Conferences **120**, 07009 (2017)
ASCMCES-17
DOI: 10.1051/matecconf/201712007009

The impact of transit signal priority: case study in Dubai, UAE

Hadeel Al Khateeb[1]* and *Akmal* Abdulfatah[2]

[1] Director of Planning & Survey Dept. – Ajman Municipality, Ajman - UAE
[2] Associate Professor, Civil Engineering – American University of Shrajah, Sharjah – UAE

Abstract. Abdulfatah & Alkhateeb proposes on this paper one of the Intelligent Transportation Systems- ITS, the Transit Signal Priority- TSP, which involves real time tracking. The United States is considered one of the lead countries in implementing this strategy since decades ago. Europe and Germany have also implemented TSP quite extensively to improve rider-ship and buses efficiency. TSP is a strategy that supports the transit system through easing the movement of the service. The operational technique used for TSP is controlling the signal system at the intersections. A detector system must be integrated in the implementation stage. It will detect the preference of public buses among the other vehicles in the traffic queue. Determining of optimal detector location for the TSP with queued lanes is the main feature. It must be studied and decided on before starting different signalization proposals for the intersection. Controlling traffic signal in favor of the public bus system will improve reliability and efficiency for buses. Meanwhile, the extreme impact of the TSP on general traffic as well as on public buses became the basis for many recent studies and thesis tested by many authors. The scope of the project as addressed in this paper is to report the delivered results of different scenarios tested, and then compare these proposed scenarios to select the best signal consideration, which has a potential impact on bus and traffic performance. After applying TSP, the results show that the impact of actuating traffic signals in Al Mussallah / Al Maktoum corridor after extending green time for bus approach (15 sec) through minimum reduction on average travel time for busses by 7.04%and tremendous increase in travel time for cars by 38.68%. A reduction in cars Avg. Speed by 28.04% in comparison to 35.52% buses Avg. Speed. On the other hand, a reduction in buses delay received by 14.11% and a decrease in Avg. Stopped delay per bus of about 18.56%, while cars received an increase in delay of around 57.00% with increased Avg. Stopped delay per car of 55.13%.

* Corresponding author: Hadeel.akhateeb@gmail.com

1 Introduction

In the last three decades, the demand for using different modes of transportation in many countries became a tremendous need. The main objective of many governmental transport authorities all over the world is to supply and/or improve different transport services to accommodate the increasing demand for daily travels. Accordingly, recommending new strategies and/or replicating the most successful practices based on travelers' preferences became a target. Implementing and improving road network infrastructure is also recommended as it encourages more efficient and reliable transit system operations in the city. All these goals require policies/ strategies and need to be operated through implementation plans.

This paper proposes one of the Intelligent Transportation Systems- ITS, the Transit Signal Priority- TSP, which involves real time tracking. The United States is considered one of the lead countries in implementing this strategy since decades ago. Europe and Germany have also implemented TSP quite extensively to improve rider-ship and buses efficiency.

1.2 TSP considerations

The TSP is designed to benefit the public transport (buses) in the signalized intersections. If it is successful, then the impact will be reflected on the buses performance, and will increase the service reliability and efficiency. Thus, it will improve travel time on overall traffic. Accordingly, surrounding traffic environment must be considered when implementing TSP in any signalized intersection. TSP in brief has been proven through many case studies all over the world to have either positive/negative impact on the traffic stream or both on some cases. Both cases happen when assuming that we have Red truncation (Early green priority) strategy (Garrow, M. and Machemehl 1998, and A. Skabardonis, 2000) [1] which was in favor of bus approach through early green; before the completion of the full signal cycle for the whole system. As a result, the early green for the detected buses will be improved compared to opposite directions which will receive more delays and lead to vehicles queue.

Advanced control systems and the new Auto Vehicle Identification system – AVI, which are available in the market nowadays, have been incorporated in the TSP strategy to achieve the successfulness (Strategies for Transit: A Review of Selected Experiences in the United States, The Seventh World Congress on Intelligent Transportation Systems, 2000) [5]. Criteria demanded when implementing TSP:

- Vehicle volumes other than buses (Vehicular volume)
- Traffic signal coordination
- Bus headways and volumes (frequency and occupancy)
- Bus stops location (i.e. near v/s far-side traffic signal)
- Pedestrian

1.2.1 Bus stops

As per recorded in the TSP planning and Implementation Handbook, and the results of the surveys conducted by different Transportation Agencies, it is most preferable to locate all bus stops on far- side; as it reduces the green extension interval that is required to be granted to the bus. (Manual of Uniform Traffic Control Devices, 2000) [4]. It minimizes the difficulty when signal controller is predicting the bus arrival. Accordingly, it will give more accuracy for the prediction and will facilitate the operation of bus checking out of the system. (Transit Signal Priority System Performance Monitoring and Optimization, 2001) [8]. Figure (1) illustrates the two different bus stop types.

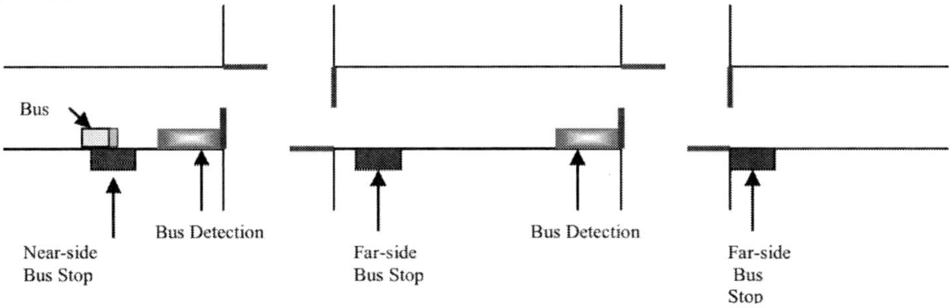

Fig. 1. Location of Bus Stops and Distance from TSP Detector.

1.2.2 Pedestrians

TSP strategy offers special treatment for transit vehicles (buses) crossing a signalized intersection. However, some challenges need to be studied/ considered when implementing TSP. Since most of signalized intersections, which will become TSP signalized intersections, have at the same time special signals for pedestrians. The pedestrian signals are always timed to automatically offer pedestrian pass signal (Green light in their special designed signals). The time of this green is to be consistent with the Red Truncation of the same vehicles signals, to allow them safely crossing in the street. This issue is still not extensively discussed since it is fully dependent on the pedestrian volume at the time the bus reaches the detection zone. (A. Skabardonis, Paper Presented at the 79th Annual Meeting of Transportation Research Board, 2000) [3].

1.3 TSP route characteristic

The best route for implementing TSP application is where the road is operating under multi-modal traffic system with Four to Six Lanes in Arterials. Currently, minimum number for routes deploying TSP in the US are Two and few cities have Six Routes and above and it will increase in the future. In North America, Bus Transit Signal Priority- BRT is deployed for only Buses/ BRT. Both concepts, TSP and BRT, are two faces for the same coin. However, Transit vehicles are those vehicles in service (The Seventh World Congress on Intelligent Transportation Systems, Turin, Italy, 2000) [4]. In some other cities, such as Philadelphia, the streetcar system operates on regular streets with other traffic vehicles. While Light Rail Transit – LRT application is implemented on dedicated Right-of-Way - ROW separate from other traffic vehicles (Collura, J. and others, 2001). [7].

The number of signalized intersections varies between the two-utilizing left-turn phase to more than 1,500 intersections in large metropolitans (i.e. Houston has 1563). Bus stops maybe located either near the intersection or far away from it. Meanwhile, it is preferable to have the bus stop far away from the intersection; this would help to more accurately predict the arrival time for the transit vehicle at the intersection. Thus, it will lead to a reduction in the green extension (i.e. Transit vehicle checking out). When route is operating under TSP, the headway will range from 90- seconds to 30-minutes. But the peak hour headway ranges from 5- minutes to 30- minutes according to the route and service type, according to the Manual of Uniform Traffic Control Devices, 2000) [4].

1.3.1 Bus transit lines

At this paper, a Transit line is defined as the Bus Transit Lines designed to serve buses or Transit Vehicles. It is a sequence of multiple transit stops fixed according to a specified time table (Bill Kloos, 2002) [9]. The determination of the stop times is basically made by dwell time distributions or according to the service times for the passengers. By the time of drafting this paper all the stated lines were operated by RTA, Bus Operation section must verify the location of the new bus stops, code the bus lines on each route, and add accumulated bus frequencies (bus per peak hour per direction). When using VISSIM similarly to the private vehicles generated volumes, the transit line codes must be entered by external station per route not the arrival distribution.

TSP strategy requires 'Signal controller' to change the traffic light and 'Confirmation device' to allow traffic movement in the intersection (Figure 2). An operation traffic signal is dictated by traffic signal system capability. Traffic signals are designed to operate in a way to minimize total traveler delay. In other words, the traveler delay is eventually the transit vehicle delay as the case of TSP (A. Skabardonis, 2000) [3].

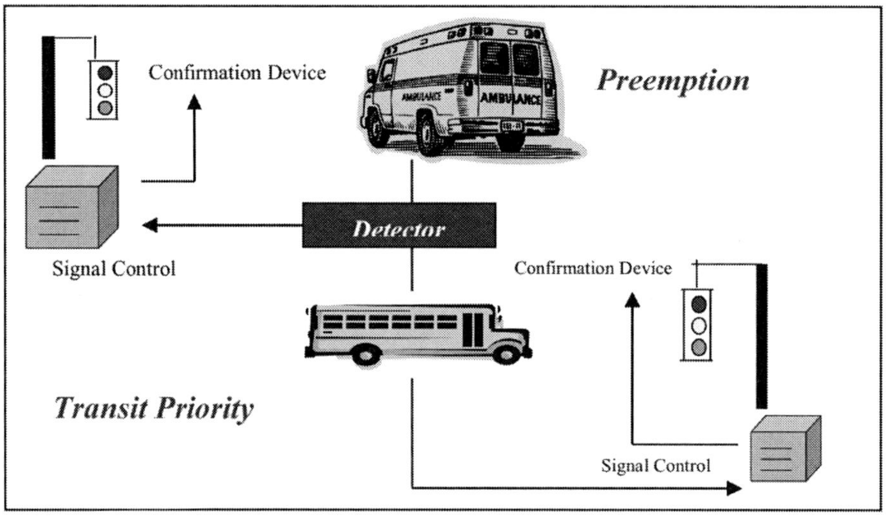

Fig. 2. Example of the transit and preemption operation.

1.3.2 Waiting Time

Waiting time is a component specified for only public transport mode. It is based on the frequency of the public buses available for travel, according to the Final Report, prepared to St. Cloud Metropolitan Transit Commission, Minnesota, 2000[6]. The bus users are either considered to have no idea about the timetable and may arrive at any time at a bus stop and wait for the bus arrival, or they may have a reasonable idea about bus timetables. The majority are considered to be from the second type of bus users/ passengers. The regular passengers are most likely to reach the bus stop within a reasonable time limit before bus services arrival. The following formula is showing how the waiting time is calculated for those passengers:

$$WT = \begin{cases} \dfrac{Headway}{2} & , \text{if } Headway \leq 9 \text{ minutes} \\ 1.5 \times \sqrt{Headway} & , \text{if } Headway \geq 9 \text{ minutes} \end{cases} \quad (1)$$

Where, WT is the waiting time

Bus Headway: is the amount of time (usually in minutes) when buses are passing certain point on lane in a successive manner. This point is measured by a reference point on the buses, according to the Manual of Uniform Traffic Control Devices, 2000[4].

2 Literature review

In the last three decades, the demand for using different modes of transportation in many countries became a tremendous need. The main objective of many governmental transport authorities all over the world is to supply and/or improve different transport services to accommodate the increasing demand for daily travels. Accordingly, recommending new strategies and/or replicating the most successful practices based on travelers' preferences became an essential target. Improving roads infrastructure is another means to encourage more efficient and reliable transit system operations. All these goals require multiple policies on a strategic level, and need to be operated through implementation plans.

This project proposes one of the Intelligent Transportation Systems- ITS, the Transit Signal Priority- TSP, which involves real time tracking. The United States is considered one of the lead countries in implementing this strategy since decades ago. Europe and Germany have also implemented TSP quite extensively to improve rider-ship and buses efficiency. This is as per stated in the report issued by Transportation Research Board of National Academies, Bus and Rail Transit Preferential Treatments in Mixed use Traffic, 2010[16].

2.1 TSP considerations

The TSP is essentially designed to benefit the public transport (buses) in the signalized intersections. If this is achieved successfully, then the impact will be reflected on the buses performance, and will increase the service reliability and efficiency. On the other hand, the system performance will impact the travel time on overall traffic. Accordingly, the surrounding traffic environment must be considered when implementing TSP in any signalized intersection. Hereon, traffic considerations will directly impact the overall traffic conditions and are supposed to be analyzed to evaluate the impact of the TSP. TSP in brief is as proven through many case studies all over the world to have either positive or negative impact on the traffic stream or both in some cases. The question that could be raised is when these two extremes "Negative and Positive" happen at the same time. The answer is clearly justified as follows: by assuming that we have Red truncation (Early green priority) strategy (Garrow, M. and Machemehl 1998, and A. Skabardonis, 2000) [1]:

- When bus passes by the detection zone then the detector will distinguish between it and other vehicles passing the same corridor.
- The detector will send certain algorithms to the signal controllers.
- The signal controller will translate the received algorithms to actuated parameters for each signal in the intersection

- In the corridor, the approach of the detected bus will be receiving the signal priority, which is in this case the early green before the completion of the full signal cycle for the whole system in the corridor.
- Thus, the signal will be turned to green as the bus approaches while the approach that is supposed to get the green signal will receive expansion in the red truncation which will increase the delay on that approach.
- As a result, the early green for the detected bus approach will be the fortunate direction. While the opposite directions (any of the other approaches), will receive extended red signal timing which will cause more delay on these directions and lead to vehicles queue.

High advanced control systems and the new Auto Vehicle Identification system – AVI, which are available in the market nowadays, have been incorporated in the TSP strategy to achieve the successfulness (Strategies for Transit: A Review of Selected Experiences in the United States, The Seventh World Congress on Intelligent Transportation Systems, 2000) [5]. Criteria demanded when implementing TSP:

- Vehicle volumes other than buses (Vehicular volume)
- Traffic signal coordination
- Bus headways and volumes (frequency and occupancy)
- Bus stop locations (i.e. near v/s far-side traffic signal)
- Pedestrians

2.2 TSP benefits and impacts – case studies

Implementing TSP will generate multiple benefits for the city if it decides to adopt the ITS strategy. Many countries tend to implement it because of the following general impacts such as reduction in: travel time for Transit vehicles, Transit delay, stops, and schedule un-reliability. There are minor impacts to cross street traffic and buses. Bearing in mind the impact of the TSP is difficult to be measured when the difference between before and after scenarios are small. Thus, TSP impact is approached when higher passengers and operational benefits for the Transit vehicles tempt to become faster, and when there is less variable travel time and improved on-time performance (Ghada, S. and Newland, 2002) [10].

Although, controlling before/after TSP condition requires good knowledge and understanding of the traffic condition in the targeted corridor, it is still difficult to measure the impact when the following variations appear in the study corridor (i.e. traffic volume changes, traffic accidents, weather weekends, and arrival time for the data collected by personnel. Many studies tested the impact of turning on/off the TSP for the sake of gathering information and studying the impact by comparing the outcome. These experiments impact transit vehicle drivers negatively, since they usually benefit from this strategy. The whole experiment was not ideal. But the simulation could predict the impact through testing different scenarios which is summarized in 5 published case studies, justifying the general impact of the TSP applications on Buses performance. (Bill Kloos, 2002) [9].

3 Methodology

3.1 Project scope

Multiple site visits were carried out to record observations and to understand all measures related to system operations of TSP by 2007. Some interview protocols, and some findings commissioned by RTA pervious relevant studies such as Bus Priority Study. Useful information about the new proposals introduced by PTA as a short- term strategies.

The scope of the project is to analyze the impact of the Transit Signal Priority (TSP) application using VISSIM software. The project is considered the Bus Route in Deira Area which will connect "Al Mussalla Street - 15" with "Al Maktoum Road - 12" to implement the proposed TSP. This corridor will be evaluated according to many reasons such as its mixed-use area and the corridor lies in the CBD area and it includes (4) signalized intersections, where the TSP can be implemented (Figure -3)

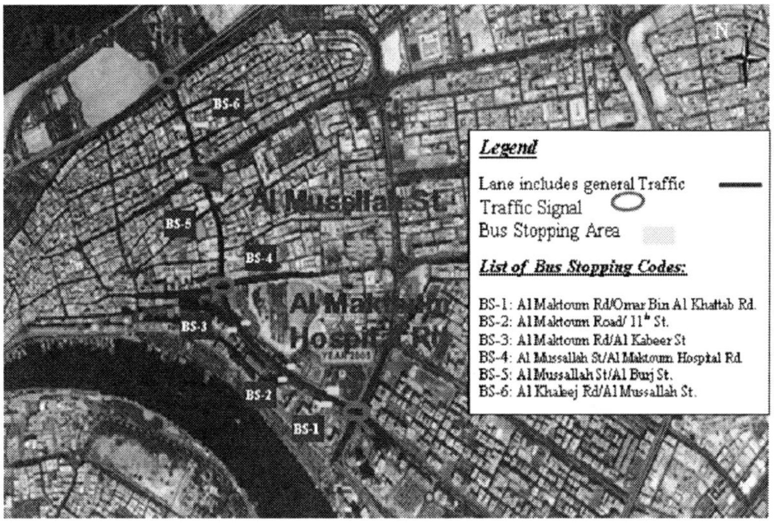

Fig. 3. Al Mussallah /Al Maktoum Corridor.

3.2 Project methodology

The project designed to evaluate the impact of TSP on the study corridor and tent to evaluate the impact of it on the (Buses) versus (Private vehicles) crossing this corridor in 3 main traffic conditions (Morning Peak–AM, Lunch Peak- LT, and Evening Peak – PM). As per Dubai R-1000 model, it is specified as short-term (2010) – all the proposed scenarios will be tested and analyzed for the short term horizon, medium term (2015), and target year (2020).Volumes extracted from R-1000 Dubai transportation model and (Road capacity, Public transport lines, Buses stops, Alignments, Headways) in accordance with the future policies, network improvements, and metro operation, the results expected to be not much differing from the analysis of the short term period – 2010. Adopted methodology to develop TSP scenarios was through evaluating the impact of the TSP by using VISSIM on different traffic conditions for the different horizons; through comparing results among different scenarios for Buses versus private vehicles, in order to recommend choice for the TSP strategy.

RTA previous Mass Transit Plan for Dubai (PS – 700) 1997, (PS- 002) 1999, (PS – 007) 2000, and (R-1000) 2005, and Public Transport Bus Master Plan were considered the bases for data collections. The methodology also relied on the compilation of data from secondary sources such as the updated (VISUM- 9.4) model [12], run (R-1000) model for both public transport and private vehicles and extracting volumes from the model. In addition; collected data for Public transport existing routes and networks, conducted interviews with public transport market stakeholders. Using micro-simulation model "VISSIM- 9.4" [13] a software tool to experience the different scenarios. Synchro [11] is a signalized model and is used to extract the signals coordination for the fixed- time signals to be used in the VISSIM. Finally, VisVAP [14] which is the flow chart program which includes multiple loop statements with specified functions has been used in order to implement the TSP. Which in turn is required to create (*.PUA) [15] file for inter-stages descriptions that are considered the parameters and the definitions which are all referred by the VisVAP in order to actuate the program as illustrated in Figure 4.
All stated tools presented the impact of the different signal timings on different measures:

- Average mean speed
- Travel time
- Delay
- Network performance
- Queue length
- Number of stops, and
- Link evaluation

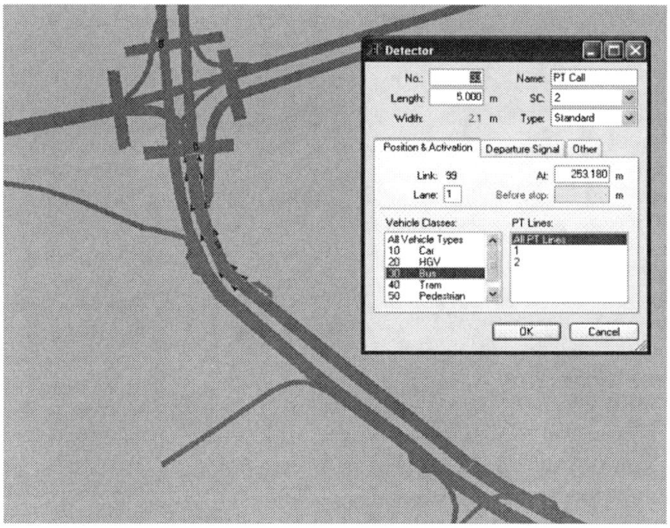

Fig. 4. Model Simulated through VISUM/ VISSIM/ VIS VAP.

3.4 Experimental design

Accordingly, the project final design is summarized in figure (5).

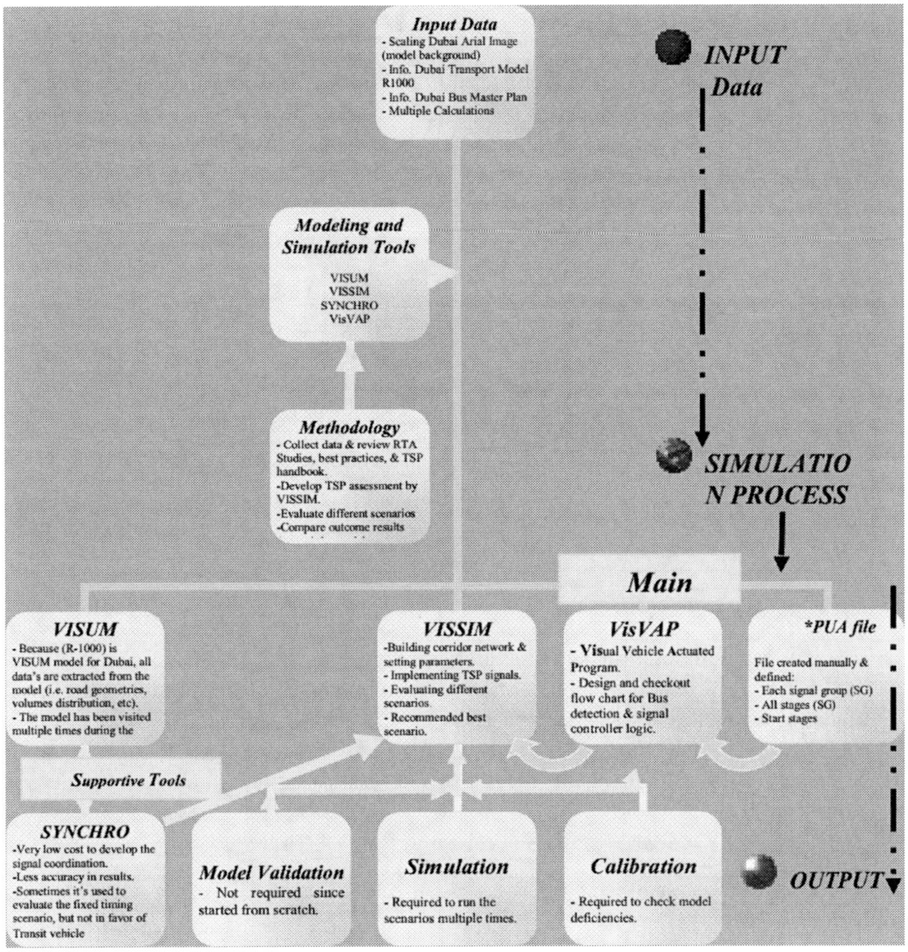

Fig. 5. Experimental Technical Design.

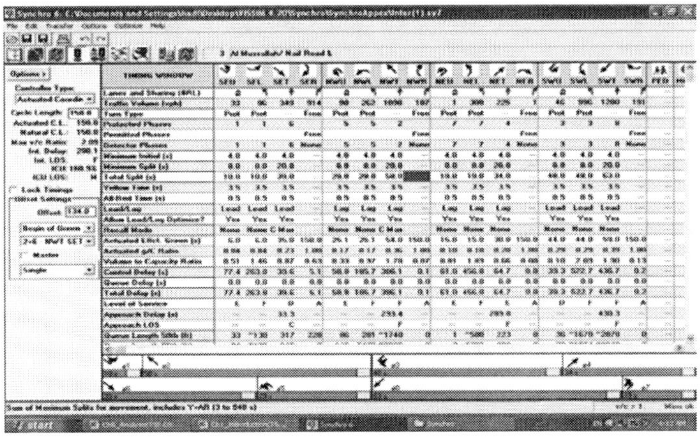

Fig. 6. Green Extension Method is considered in the Evaluation. (Syncro 6).

5 Results and discussion

Developing the U-turn adjustment factor requires to model the relation between the percentage of UT traffic and the average discharge headway. For each site, four regression models were attempted: Linear, Logarithmic, Quadratic, and Exponential.

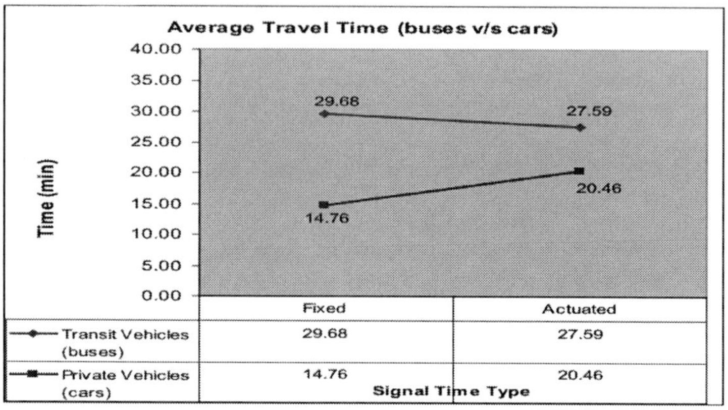

Fig. 7. Measurements Analysis.

After analysis implementation as illustrated in Figure 7, the results of the AM_2010 peak for both Fixed &Actuated time scenarios were analyzed after simulating/ running the model 5 times under different seeds to reach consistent results. LT_2010 and PM_2010 peaks were also simulated but not analyzed since the results were too close to the AM_2010. Tables (1-2-3-4) shows the results of Fixed and Actuated scenario in favor of Buses.

Table 1. Average Travel Time Rate of Change.

Vehicle Type	Fixed (min)	Actuated (min)	Rate of Change
Private Vehicle (Car)	14.76	20.46	38.68
Transit Vehicle (Bus)	29.68	27.59	-7.04

Table 2. Average Speed Rate of Change.

Vehicle Type	Fixed (Kph)	Actuated (kph)	Rate of Change
Private Vehicle (Car)	6.622	4.765	-28.04 %
Transit Vehicle (Bus)	7.735	10.482	35.52 %

Table 3. Average Delay Rate of Change.

Vehicle Type	Fixed (s)	Actuated (s)	Rate of Change
Private Vehicle (Car)	426.889	670.212	57.00 %
Transit Vehicle (Bus)	695.162	597.042	-14.11 %

Table 4. Average Stopped Delay Rate of Change.

Vehicle Type	Fixed	Actuated	Rate of Change
Private Vehicle (Car)	678.592	1052.679	55.13 %
Transit Vehicle (Bus)	478.344	389.560	-18.56 %

Figures (8 and 9) with Table (5) represent the queue lengths in general is getting higher when using actuated signals especially in the last 4 sections in the link. Whereas the queue counters number 1, 2, 3, 6, and 7 implemented in the bus approaches are increased. However, queue counters number 4, 5, and 8 show reductions in the queue length after treating bus approaches with the green extension. The logical reason is because the average queue length is already very high when the transit vehicle (bus) is detected by the check-in detector in order to start its treatment (green extension). Basically, the queue already started since this sub network is a part of the whole Deira road network, which means it is impacted by previous queues occurring due to other interchanges/ intersections connected to the corridor. Moreover, the scenario is testing the peaks period where the vehicle volumes is the highest. It is to be expected that in the off peak period the queue lengths will be less which will lead to a reduction in the number of stops within the queue.

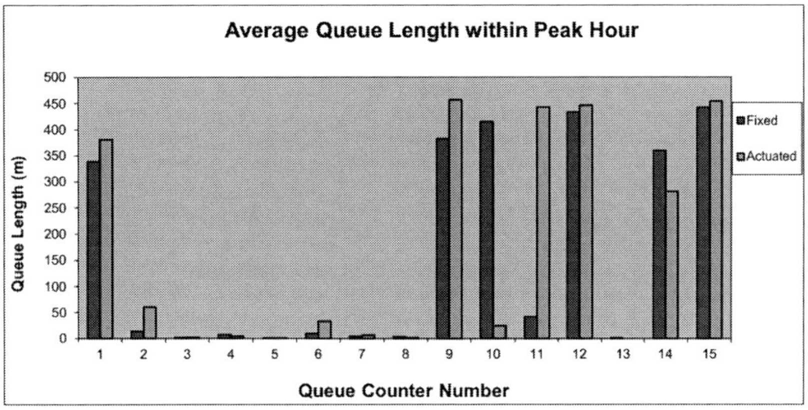

Fig. 8. Average Queue Length within Peak Hour.

Table 5. Average queue length (fixed v/s actuated).

QC	Fixed (m)	Actuated (m)	Rate of Change
1	339	381	12.60
2	14	61	323.26
3	1	2	50.00
4	7	5	-36.36
5	1	0	-66.67
6	10	33	241.38
7	4	6	72.73
8	3	2	-44.44
9	382	457	19.44
10	414	24	-94.13
11	41	443	970.97
12	433	446	3.00
13	1	0	-100.00
14	360	282	-21.69
15	442	454	2.87

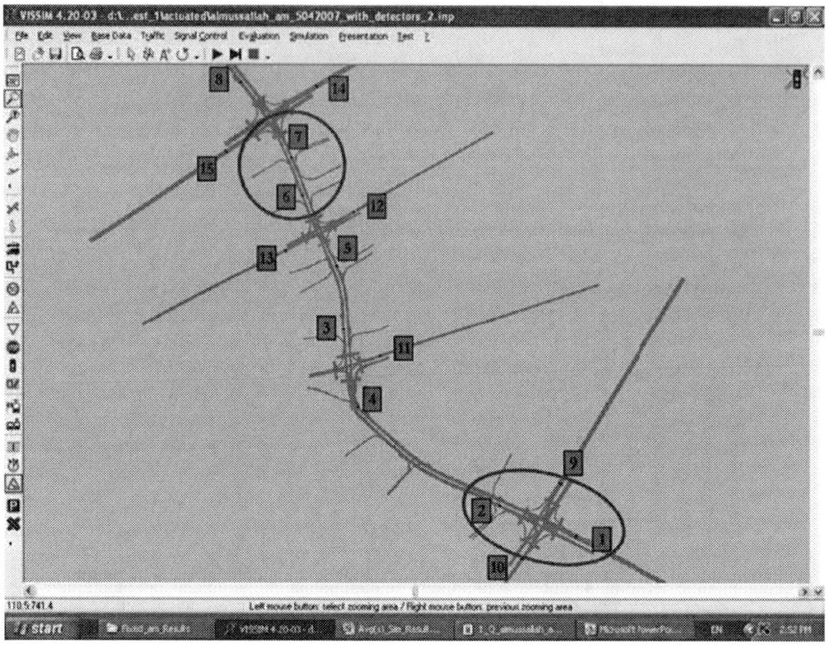

Fig. 9. Locations of the highest queue length (fixed v/s actuated).

In summary, this study investigated the impact of the TSP on Al Mussallah/ Al Maktoum corridor. Hence, Buses received reduction in travel time by 7.04 % when actuating traffic signals. On the other hand, cars receive about 38.68% as an increase on its travel time. Also, buses receive delay reduction of about 14.11%, whereas cars delay is increased by 57.00%. Which is quite good compared to the international practices. For instant, active TSP improved bus operation through reduced average bus delay from 35.9 to 28.0 seconds (by 22.1%) against no scenario as well as improved bus speed by 4.1% which improved bus performance as proved by (Lee & Shalabi, 2011) [17] when they studied the strategy implementation at one of City of Brampton main streets which is crossing the city south-north with 2-lane per each direction at queensland- Australia.

In general; the Green extension and the Early green strategy (TSP actuated strategy), provide slight improvement in the bus route in term of travel time, delay, and stop points. However, there is still an increase in the queue length. (Table 5) shows that the higher the green extension, the better the results that are generated according to the common practices. It is most likely to introduce the TSP strategy at first in the off peak in order to evaluate the impact of it in the bus service since it normally receives more delays in the peak hours. Improving the headways also would generate better results as to improve the bus services. In addition, to dedicate a special lane for buses would also impact its travel time and headway since these lanes will not be competing with other traffic flows running through these intersections.

6 Conclusion

6.1 TSP impact summary

TSP strategy also will essentially influence the public transport service in Dubai by the following:

- Extending the green time to accommodate bus movement though the signalized intersection. Since the study corridor contains 4 signalized intersections, then the bus route will be impacted positively in the whole study corridor because the green extension will be coordinated in the 4 intersections.
- Improving Buses reliability/ efficiency though adherence to transit schedules, leads private vehicle drivers to shift their mode of transport (model split) into PT.
- When private vehicle driver becomes a bus passenger, then he will gain peaceful time to relax, while the signal movements frustrate other private vehicle drivers.
- Because of the yield and drive status caused by the red truncation in the non-transit vehicle priority approaches, vehicles will reach the wear condition very fast. Thus, vehicle owners will suffer from regular maintenance money lose.
- In term of the environmental aspect, reducing number of private vehicles in a corridor will reduce vehicle fuel consumption and emissions.

6.2 Dubai case study results

Impacts of actuating traffic signals in Al Mussallah/ Al Maktoum corridor after extending green time for bus approach (15 sec):

- Minimum reduction occurs on buses average travel time along the study corridor by 7.04 % when applying TSP.
- Total travel time of cars in the network will be tremendously increased by 38.68% when applying TSP.
- Reduction in cars average speed is received when applying TSP by 28.04%. Whereas, buses average speed is increased by 35.52%
- Reduction in buses delay is received when applying TSP by around 14.11%. In addition, there is decreased average stopped delay per bus of about 18.56%. On the other hand, cars received an increase in a delay about 57.00%, with increased average stopped delay per car of about 55.13%.
- Overall network performance for all vehicles is similar to the passenger car performance.
- TSP sometimes increases the queue length in links including bus routes, or it shows a reduction. On other links, the TSP in general increases queue lengths.
- As a result, public transport buses reliability and efficiency will be increased. This will lead to encourage people to shift their mode of transport.
- Overall conclusion is to achieve better rider-ship in Dubai.

6.3 Recommendations

It is worthwhile to consider that the transit vehicle (Buses) carries significantly higher occupancy than Private Vehicles (cars). A reasonable balance must be maintained between transit vehicles and other vehicles running in the corridor in terms of its travel time and delays in order to deliver a successful TSP strategy for all stakeholders involved in the implementation of this strategy. In other words, the private vehicle travelers might have other alternatives when they receive delays in travel time, so their decision might be challenging. As an example, they could choose a dynamical route as an alternative. Accordingly, this leads RTA to think more comprehensively about the TSP associated services. Then RTA is responsible to integrate the existing and the future public buses services with other transit services proposed by their studies such as the metro and ferry system. In addition, to improve the quality of the public bus service RTA is recommended to adopt the most recent intelligent transportation system such as TSP. However, to achieve better results when implementing TSP strategy, other aspects should be considered and integrated with the proposed strategy such as:

- Improving public bus system according to its frequencies and stop locations and integrate the full network in traffic signals integration system.
- Improving other intelligent transportation system-ITS incorporated with the Transit vehicles (i.e. GPS, Smart Phone Bluetooth, real-time systems .. etc).
- Considering the waiting time for transit vehicle passengers to ride in the bus stops by preparing adherence time tables for buses as well as shortening bus headways and frequencies running in the corridor (i.e. from every 20 minutes to 15 or less).

Signals data and intersection approaches are also recommended to be reviewed and evaluated under different traffic conditions (i.e. on/ off peak periods). Reviewing road networks and bus routes based on the data collection methodologies in different horizons such as surveys, interviews, technical standards and specifications is also recommended. This is needed in order to establish future guidance in UAE adaptation for some implemented successful established case studies for Transit and Transportation/ Traffic agencies which worked together to support local agencies when exploring the implementation of TSP. Overall Guideline for a successful implementation is when its considered in the Planning phase which will continuously strengthen and develop it during each project phase. Very advanced system and design of traffic signal cabinets allow for transit agencies to access the equipment. To encourage the involvement of all stakeholders in the TSP strategy, more reliable data collected with real-time platforms, and to carry on maximize financial benefit reports for all is recommended. Florida had the best practice on successful guidelines for the same (Diane Quigley, 2014) [18].

Although the capital costs to implement TSP per intersection will approximately cost a total of ($5,000), the cost of software installation is ($20,000) and controller equipment is ($30,000). It has a high impact and is recorded when it is implemented on projects planning stage as per Planning and Implementation Handbook, FTA [2].

Overall conclusion is if Dubai RTA considers all the above-mentioned recommendations then the rider-ship in Dubai will be improved and roads network congestion will be reduced.

References

1. S. R. Sunkari, Beasley, "Model to Evaluate the Impacts of Bus Priority on Signalized Intersections", P.S. Urbanik II, Transportation Research, (1995).
2. Garrow, M. and Machemehl, "Development and Evaluation of Transit Signal Priority Strategies", Research Report, SWUTC/97/472840-00068-1, Southwest Region University Transportation Center, Center for Transportation Research, University of Texas at Austin, (1998).
3. Peter Koonce, W. C. & K. Turner "Implementing Traffic Signal Priority for A. Skabardonis A. Skabardonis A. Skabardonis A. Skabardonis A. Skabardonis Buses", Compendium of Technical Papers. Institute of Transportation Engineers, (1999).

4. A. Skabardonis, "Control strategies for transit priority". Paper presented at the 79th Annual Meeting of the Transportation Research Board, (2000).
5. Manual of Uniform Traffic Control Devices, Millennium Edition, Federal Highway Administration, US, Dept. of Transportation, Washington DC, US, (2000).
6. Vahidi, H. "Transit Signal Priority: A Comparison of Recent and Future Implementations" Presented at 70th Annual ITE Meeting in Nashville, TN, (2000).
7. Westwood Professional Services, Inc. "St. Cloud Metropolitan Transit Commission Transit Priority Evaluation Project", Final Report, prepared to St. Cloud Metropolitan Transit Commission, Minnesota, November (2000)
8. Kiel Ova, Ayman Smadi, "Evaluation of Transit Signal Priority Strategies for Small- Medium Cities", December (2001).
9. Washington State Transportation Centre (TRAC), "Transit Signal Priority System Performance Monitoring and Optimization", January (2001).
10. Ghada, S. and Newland, "Effectiveness of Bus Signal Priority", National Center for Transit Research, Center for Urban Transportation Research, University of South Florida, Tampa, FL, January (2002).
11. Synchro software program, User Guide 6, David Husch, John Albeck, Traffic-ware Publisher, (1993-2003)
12. VISUM, User Manual version 9.4, PTV Version, 2006
13. VISSIM, User Manual version 4.2, PTV Version, 2006
14. VisVAP, Appendix A: Add-On VAP, VISSIM Manual, PTV, VAP version 2.15, (2006)
15. VisVAP, Appendix B: Add-On VisVAP, VISSIM Manual, PTV, VisVAP version 2.15, (2006)
16. TCRP Synthesis 83-bus and Rail Transit Preferential Treatments in Mixed use Traffic, Asynthesis of Transit Practice, Consultant Alan R.Danher, PB Americas, Inc. , Orlando - Florida, (2010)
17. Rule based Transit Signal Priority Control Method using a Real-Time Transit Travel Time Prediction Model Queensland university of Technology, Australia – Lee, Jinwoo & Shalaby, Amer , -, (2011)
18. Diane Quigley, Florida Dept. of Transportation, Transit Signal Priority Implementation Guidance, Final Report, July 7, (2014)

Assessing the energy performance of modern glass facade systems

Abdelsalam Aldawoud *

Architectural Engineering Department, University of Sharjah
Sharjah, United Arab Emirates

Abstract. The design and construction of the building envelope have a significant effect on the energy performance of buildings because the building envelope physically separates the exterior from the interior environment. The use of glass as a component of the building envelope has been significantly growing since its initial introduction as a building material. The glass is necessary for aesthetics, ventilation, and daylight. Despite the many benefits to the building, the glazing still provides the lowest insulating value. Of all envelope components, windows and skylights represent major sources of thermal losses and gains which impact the thermal comfort in buildings. Therefore, window size, type, and design are the most important consideration for reducing the need for space heating and cooling. This study investigates the thermal performance and the effectiveness of various modern glass facade systems to improve building efficiency levels. The main objectives of this study are to provide a comparison and evaluation of glazing systems and to establish guidelines for energy-efficient envelope design that provide architects and designers with a basic framework for the selection of the most appropriate glazing systems. It has been noted that the spectrally selective system and the electrochromic glazing system show high energy efficiency compared to other tested glazing systems.

1 Introduction

The building construction sector in the UAE has witnessed a fast track growth in last three decades. According to the 2015 Buildings Energy Data Book, energy usage in the UAE has grown at an annual average of 4 percent over the past six years, with projections that it will increase to 5 percent through 2020 [4]. Overall electricity consumption has more than doubled in the last 10 years, at a pace that will be difficult to provide for over the long term.

The building sector in UAE is by far the largest energy consumer and a major contributor to Greenhouse gasses (GHG) emissions. To reduce the growth of energy consumption, a comprehensive energy efficiency measures should be adopted to improve building efficiency. The design of a building façade is considered a key design element that

* Corresponding author: aaldawoud@sharjah.ac.ae

could significantly influence the building energy performance and its environmental impact. In UAE, most of the new buildings adopt the concept of a fully glazed façade because of the advancement of façade construction technology and physical properties of glass. However, the interest among architects and engineers to design light and transparent buildings with fully glazed envelopes to access more daylight [2]. This trend could lead to unintended consequences of increasing the energy consumption of the construction due to higher solar heat gain.

Despite the fact that fenestration systems are used as a method to provide a visual connection to the outside environment, they are the weakest energy link of the building envelope [1]. Careful choice of glazing type could improve the building envelopes energy performance and reduce the consumption by decreasing the lighting and heating and cooling loads in buildings. This could provide the desired natural daylight and the visual connection to the outside in well-designed buildings [3].

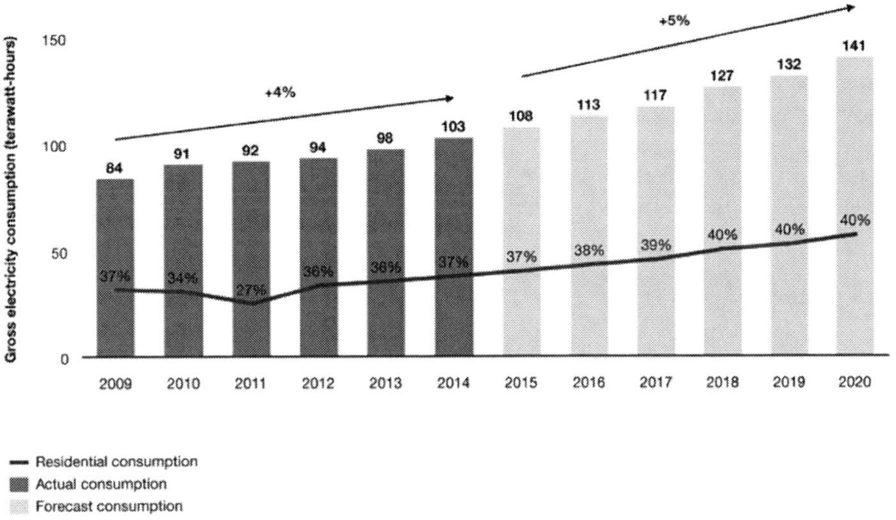

Fig. 1. UEA electricity consumption and the forecast in the next five years.
Source: Economist Intelligence Unit (EIU), 2015; International Energy Agency (IEA), 2015.

Despite the fact that fenestration systems are used as a method to provide a visual connection to the outside environment, they are the weakest energy link of the building envelope [1]. Careful choice of glazing type could improve the building envelopes energy performance and reduce the consumption by decreasing the lighting and heating and cooling loads in buildings. This could provide the desired natural daylight and the visual connection to the outside in well-designed buildings [5].

This study investigates the thermal performance and the effectiveness of various modern glass facade systems to improve building energy efficiency levels. The main objective of this study is to provide a comparison and evaluation of glazing systems. The intention is to establish some strategies and guidelines for energy-efficient envelope design that provide architects and designers with a basic framework for the selection of the most appropriate glazing systems. This study also aims to improve the energy performance of buildings in UAE by examining and evaluating different glazing systems which could have a direct impact on solar radiation and heat gain performance.

2 Methodology

For the purpose to investigate the thermal energy and the effectiveness of various glazing systems, a model which represents an office building is created in DesignBuilder simulation program which is the only software used in the analysis. DesignBuilder is an energy modeling software developed using EnergyPlus (U.S. DOE, 2011) as the analysis engine. An energy performance calculation for a whole building is made to compare the performance of the Façade system to other types of façades used. In all simulation scenarios, assumptions and simplifications are made; however, these assumptions and simplifications are maintained for each scenario, so that the glazing type was the only variable that changed.

2.1 Geometry

The reference building model is eight stories with a total floor area of 16335 m2. The model has the same geometrical proportion, length, width, and height. The construction template used for the building was the same for all glazing alternatives. The glazing type was the only variable that changed throughout the simulation process. Figure 2 shows the model created in DesignBuilder. Since the only variable is the glazing type and the high level of detail will not affect the final results, the template for "Medium Weight, is used for the construction of the walls and moderate Insulation" is used for the walls, floors, ceilings and roofs.

Fig. 2. Building model created using Design Builder.

2.2 Classification of the climate in Sharjah

The energy performance analyses are conducted by assuming buildings located in Sharjah, 25°33′N 55°52′E. The climate in Sharjah as in UAE is a hot sunny and characterized by high levels of temperature, humidity, and sunshine. The summer months, from June to September, are too hot for comfort. The temperatures range from 10°C in winter to a high 48°C at the height of summer. The weather file used in the simulation is obtained by measurements at the Sharjah International Airport station.

2.3 Type of glazing investigated

A wide range of glazing technologies is selected in the study. The average window-wall ratio of the whole building is about 30%. The reference model scenario has the same opening area, and this is kept for all generated alternatives throughout the simulation process. The window wall ratio and the openings distribution on different façade's orientation are demonstrated in Figure 2. Table 1 demonstrates the numeric distribution which shows that north and south facing windows area is about 60% of total windows area of the building.

Table 1. Window-wall ratio.

	Total	North	East	South	West
Gross Wall Area [m^2]	5600	1960	840	1960	840
Window Opening Area [m^2]	1618.86	568.51	240.91	568.51	240.91
Gross Window-Wall Ratio [%]	28.91	29.01	28.68	29.01	28.68

The glazing systems can be divided into several categories. Several glazing technologies are selected and described below. The thermal and optical properties are summarized in Table 2.

- Basic Standard Single Clear Glass: These products are comprised a single pane of standard glass.
- Insulating Glass Units: These products use two panes of glass separated by an air space to control heat transfer. A common configuration is two ¼" panes separated by a ½"airspace. The primary principle behind this technology is that conduction is reduced due to the airspace.
- Glazing filled with Inert Gas: Gas other than air such as Argon or Krypton is used to suppress conduction and convection behavior between the two panes.
- Electrochromic Glazing: These products operate similarly to Liquid Crystal Devices. However, they allow for the user to vary the amount of current through the device to vary the degree of tint in the window.
- Coatings: These products that are applied directly to the glazing control solar heat gain by limiting those wavelengths of light that are allowed to pass through and those that are reflected. These products are commonly referred to as Low-E coatings.
- Reflective Glazing Systems: reflective glass has a mirror-like appearance and reflects and absorbs a major proportion of the sun's direct short wave solar radiation. The degree of reflectivity is dependent on the type of coating and the orientation of the glass.
- Spectrally Selective Systems: These products typically involve films, tints, or coatings; which reflect selected wavelengths of light, while allowing others to pass through. There is a broad range of performance qualities that can be achieved using these products.

Table 2. Thermal and optical properties for investigated glazing systems.

Variables	SHGC	Direct Solar transmission	Light transmission	U-value (W/m^2-K)
Single Clear glass 6mm	0.819	0.775	0.881	5.778
Double Clear 3mm-6mm Air	0.762	0.8	0.812	3.159
Triple clear 3mm-6mm Air	0.684	0.595	0.738	1.757
Double metallic reflective coating Clear 6mm-13mm Arg	0.179	0.8824	0.127	2.109
Double Electrochromic absorptive colored 6mm-6mm Air	0.197	0.08718	0.114	2.429
Double Low-emissivity LoE (e2=.1) Clear 3mm-13mm Air	0.598	0.538	0.769	1.786
Triple low-e filled with argon	0.474	0.351	0.661	0.786
Double Electrochromic reflective colored 6mm-6mm Air	0.182	0.078	0.137	2.429
Double Low-e Spectrally selective Clear 3mm-13mm-6mm Air	0.43	0.359	0.691	1.646
Triple Clear 3mm-13mm Air	0.474	0.358	0.661	0.982
Double Reflective-low-transmittance Tint 6mm-6mm Air	0.168	0.056	0.073	2.788

3 Results and discussion

The results of different glazing systems alternatives energy performance have been collected and compared with benchmark condition. The results showed that the energy performance varies with the glazing type. In general, results show that the "Double Low-E Spectrally selective Clear 3mm-13mm-6mm Air" and the "Double Electrochromic absorptive colored 6mm-6mm Air" Energy performances are the best among all investigated glazing systems (Figure 3). These two systems have the potential to save up to almost 60% compared to the single glazing energy performance.

Accordingly, single clear glass energy performance is the worst among all investigated glazing systems and therefore served as the baseline by which the improvements of the other systems were measured. Heat losses through the glass unit are reduced by optimizing the glazing gap between the two glass panes when filled with air. The glazing with the gap width of 13 mm, performs better than glazing with the gap width of 3 mm or 6 mm. While double clear glazing reduction is about 24% less than single glazing energy performance, Double; reflective and Low-E glazing reduces the energy transmission by 37% compared to the single glazing energy performance.

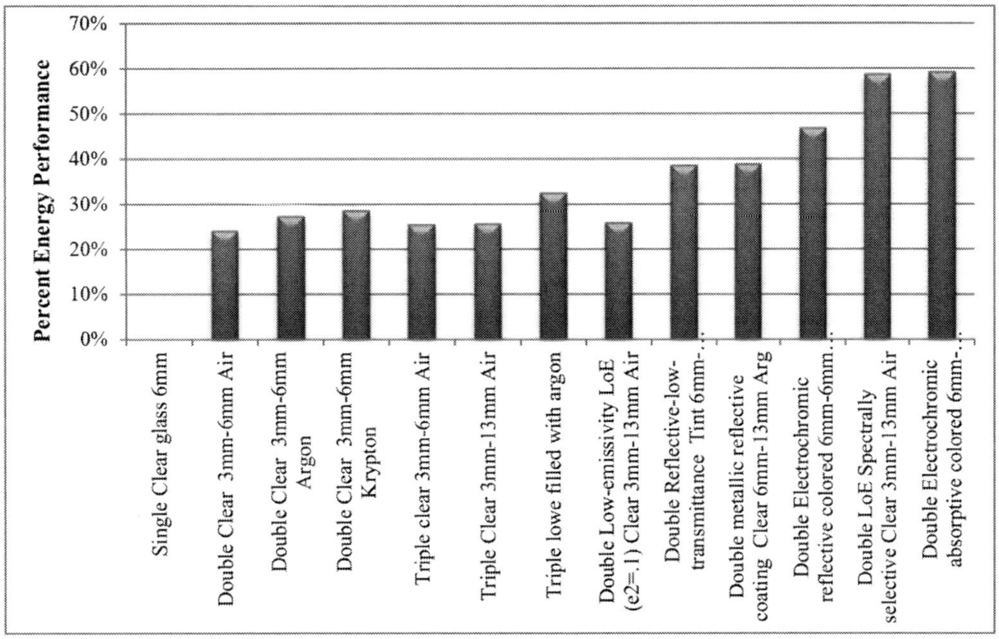

Fig. 3. Energy performance of different glazing systems alternatives.

The results also show that the type of gas fills significantly improve the energy efficiency of the glazing system. Compared with glazing that has a gap filled with air (density:1.29 kg/m3 at 0°C), nearly 15% performance increase is achieved by adding argon gas (density:1.78 kg/m3) to glazing unit. Adding krypton gas (3.71 kg/m3) to the glazing gap improves the performance of the glazing system by almost 5% over the argon filled glazing gap and about 20% over air filled glazing gap. The energy performance of different glazing systems with different gas fills is demonstrated in Figure 4.

In addition, it has been found that using clear triple glazing enhances the energy performance by almost 26% compared to single glazing base case. On the other hand, triple glazing with Low-E filled with argon gas has even better performance with a potential of a reduction of almost 33%.

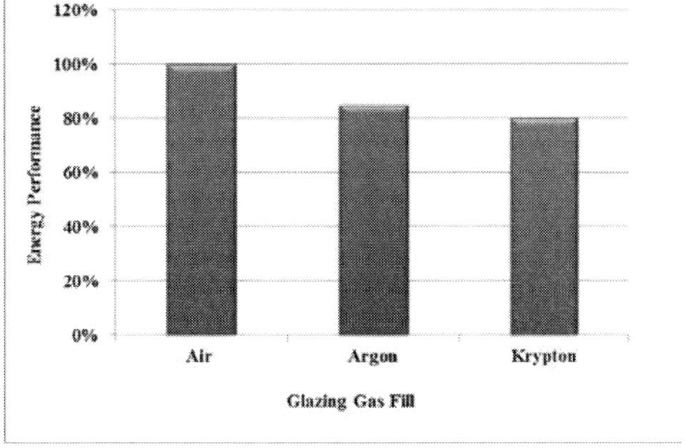

Fig. 4. Energy performance of different glazing systems with different gas fill.

4 Conclusions

This analysis involved the comparison of the impact of different glazing systems on the annual energy consumption of buildings. This is mainly related in hot humid climate to solar heat gain which impacts the cooling load and the total energy consumption of buildings. Results show that the performances of the investigated glazing systems vary and the best glazing performers are those with the greatest insulating properties. Significant reductions in energy consumption due to the heat gain of solar transmission could be reached by using spectral selective systems that focus on reducing heat gains and blocking direct sun, while permitting daylight and views to the outdoors. Based on the results of this analysis, the following concluding remarks can be made:

- The amount of energy of heating and cooling consumed in buildings depends on greatly on the glazing type which impacts solar heat transmission. Based on energy performance, single glazing is a very poor. Double and triple glazing has better energy performance.
- In hot humid climate, the glazing solar heat gain coefficient (SHGC) and the U-value of have a significant impact on the energy performance of each glazing type. Low solar heat gain coefficient and low U-value glazing are important factors to consider when selecting the appropriate glazing type. They are greatly effective in reducing the amount of heat entering the building.
- Smart glazing technologies including electrochromic and spectral selective systems can control the direct solar gain and help reduce the heat gain due to solar radiation.
- The size of the gap between glazing panes, and the type of the gas fill impact the performance of glazing. Argon or Krypton gas fills show better performance than air.

- Tint layers and Low-e coatings improve the performance of the glazing. Tinted and Low-E double glazed exhibits better performance in blocks some of the solar radiation.

References

1. Aldawoud, A. (2012). Conventional fixed shading devices in comparison to an electrochromic glazing system in hot, dry climate. Energy and Buildings 59 (2013), pp. 104–110.
2. Arasteh, D., Goudey, H., Huang, J., Kohler, C. & Mitchell, R. (2006). Performance Criteria for Residential Zero Energy Windows. Lawrence Berkeley National Laboratory, Berkeley CA 94720.
3. Chi-Ming, L. & Yao-Hong, W. (2011). Energy-Saving Potential of Building Envelope Designs in Residential Houses in Taiwan. Energies 4, (2011), pp. 2061-2076.
4. International Energy Agency (IEA), 2015. World Energy Outlook 2015 - United Arab Emirates Energy Efficiency.
5. Ochoa, C.E., Aries, M.B.C., Loenen, E.J.V. & Hensen, J.L.M. (2012). Considerations on design optimization criteria for windows providing low energy consumption and high visual comfort. Applied Energy 95 (2012), pp. 238–245.

Towards evaluation and prediction of building sustainability using life cycle behaviour simulation

Mohamed Marzouk[1,*] and *Mostafa* El-hawary[2]

[1]Professor of Construction Engineering and Management, Structural Engineering Department, Faculty of Engineering, Cairo University, Egypt.
[2]PhD Candidate, Structural Engineering Department, Faculty of Engineering, Cairo University, Egypt.

Abstract. Nowadays researchers and practitioners are oriented towards questioning how effective are the different building life cycle activities contribution to preserving the environment and fulfilling the need for equilibrium. Terminologies such as Building sustainability and Green Buildings have long been adopted yet the evaluation of such has been driven through the use of rating systems. LEED of the United States, BREEAM of the United Kingdom, and Pearl of the United Arab Emirates are namely good examples of these rating systems. This paper introduces a new approach for evaluation of building life cycle sustainability through simulation of activities interaction and studying its behaviour. The effort focuses on comprehending impact and effect of suitability related activities over the whole building life cycle or period of time. The methodology includes gathering a pool of parameters through benchmarking of five selected rating systems, analytical factorization for the gathered parameters is used to elect the most influencing parameters. Followed by simulation modelling using System dynamics to capture the interaction of the considered parameters. The resulting behaviour obtained from simulation is studied and used in designing a tool for prediction of sustainability.

1 Introduction

The Buildings Sustainability has long been defined as the process of implementing an environmentally efficient resources management. Sometimes referred to as Green building with a structured approach where practice expands and complements the classical building design concerns of economy, utility, durability, and comfort [1]. Green building design involves finding the balance between homebuilding and the sustainable environment. This requires close cooperation of the design team, the architects, the engineers, and the client at all project stages [2]. Having a tool that can efficiently evaluate practicability and sustainability of buildings has been a vital concern by practitioners, and efforts for interventions resulted in using holistic concepts in introducing rating systems such as

* Correspoding author: mm_marzouk@yahoo.com

Leadership in Energy and Environmental Design (LEED) of the United States, BREEAM of the United Kingdom, and Pearl of the United Arab Emirates.

Developed by the United States Green Building Council, LEED was one of the early attempts to tackle this evaluation issues, through setting rating systems for the design, construction, operation, and maintenance of green buildings. LEED, similar to most successive rating systems, has a main objective which is escalating reduction in the overall impact of the building processes on human health and the natural environment. However, beside this new technologies were constantly developed to enhance production of greener buildings. Rating systems nowadays forms a cornerstone in most industry projects; even owners are adopting its concepts as additional requirements. All rating systems efficiently evaluated environmental impact and resources use for an instance of the building in its life cycle as it is difficult to assess the relative improvements in sustainability of one decision versus another [3]. Studying the different activities interaction rather than individual influence could provide a different prospective fortunately systems thinking process facilitates such approach.

In concept, the goal of all rating systems is to produce sustainable buildings throughout the whole processes of construction, operation, and maintenance. From project idea generation and until destruction, while taking account for surrounding parameters like air, light, energy, water, construction materials and methods. While the applicability of each rating system differs, all rating systems draws upon occupants health, energy, and resources use, reducing waste, pollution and environmental degradation, and use of natural materials that are available locally. A wide range of mathematical techniques were developed to examine economic, environmental and ecological impacts of various pollution control actions, and thus aid decision makers in formulating effective sustainable development policies [4-10].

Systems thinking (System Dynamics) is defined as the process of understanding how things, regarded as systems, influence one another within a whole system, and fortunately has proved to be an efficient process for understanding the complex problems behaviour. The typical steps towards system dynamics involves primarily identification of the problem statement or question, followed by development of a dynamic hypothesis according to explanation of the problem causes. Then a system model is assembled with focus on reflecting behaviour in real world. With such model, insights are developed and conclusions are generated to enhance or propose solutions.

This paper investigates the use of system dynamics to evaluate building life cycle sustainability through behaviour of activities interaction. In other words, a way for comprehending impact and effect of sustainability related activates over whole building life cycle or period of time. In addition, the paper presents a structured approach towards interpreting building sustainability behaviour over its life cycle span, and paves the road to a proper debate over rating systems imposed activities.

2 Proposed framework

The proposed framework consists of four main steps as proposed by Forrester [11] with six supporting activities. The first step is "Problem Articulation (boundary selection)" which involves gathering information, on four worldwide rating systems. Within the same scope of problem articulation step, the parameters of the rating systems are then scaled in a table form through the first activity of “Parameters Selection”. The resulting scale is then recorded as the base for benchmarking activity of the four rating systems to select a final list of potential parameters. The Following activity “Key Parameters Identification” involved parameters comparison with results from one of the most recent studies within the Middle East region that focused on selection of potential parameters for a new rating system introduced in the Kingdom of Saudi Arabia [12]. The output from this activity was based on identification of parameters with scores equal to 3 or 4.The second step was "Dynamic Hypothesis" through which correlation ideas for the key parameters were investigated. Using historical certification data reasonable correlation of the different parameters was derived. The following supporting activity involved mapping of the correlation ideas in a causal loop Diagram (CLD) form to properly grasp the understanding of the system. The Third step in the effort concerned "Formulation of Simulation Model" with all decision rules specification, and estimation of behavioural relationships, and consistency tests. The final step in this effort was interpreting the behaviour through "Policies Evaluation" which involved three activities "Behaviour Interpretation, Behaviour Analysis, and Adoption Rationale". A clear benefit for using system thinking approach was having feedback through closed loops as shown in Figure 1.

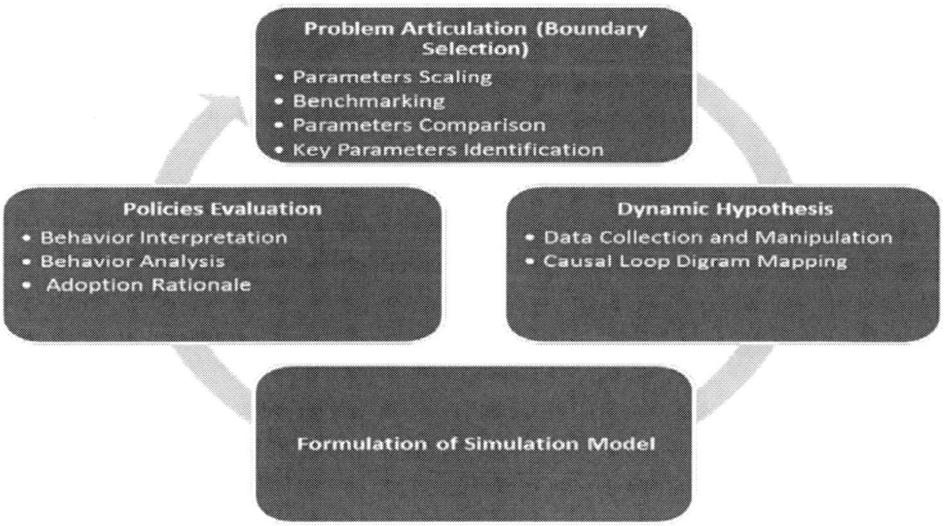

Fig. 1. Proposed building life cycle sustainability investigation framework.

3 Problem articulation

There are more than 600 rating system worldwide for evaluating the sustainability in construction industry (BRE 2008). The rating systems evaluate the sustainability of buildings through awarding points for satisfying green building criteria or parameters. The sustainability parameters and their importance weights usually differ from system to another according to differences in the environmental and regional conditions, and also due to cultural differences. Accordingly, each region needs to identify sustainability parameters that fit its conditions.

The sustainability rating systems that were considered in this study included are LEED, and BREEAM as an example from international rating systems. While for the regional examples GPRS and Pearl were selected. The aim of selecting such regional and international diverse rating systems helps in reflecting a wide range of environmental conditions.

Using simple approach for identification of parameters importance, each factor within context of named rating system is given a score only if exists in an alternative rating system. Accordingly the study boundaries were adopted from the selected rating systems factors available, and assumptions were made to consider the scaled parameters only as

 DOI: 10.1051/matecconf/201712008002

influencing factors. In order to equally evaluate parameters importance from each rating system, the sum of scores assigned in the previous sub-step was calculated. And a final table for parameters of interest was produced. Additionally, parameters were compared with results from one of the most recent studies within the Middle East region that focused on selection of potential parameters for a new rating system introduced in the Kingdom of Saudi Arabia [12].

4 System dynamic modelling

4.1 Causal loop diagram

One efficient method for interpretation of philosophies are the Causal loop Diagram (CLD) method, while most system thinking approaches deal with CLD as method of illustration it is also recorded as a mode of dilemma definition. The effort here involves mainly four loops as shown in Figure 2.

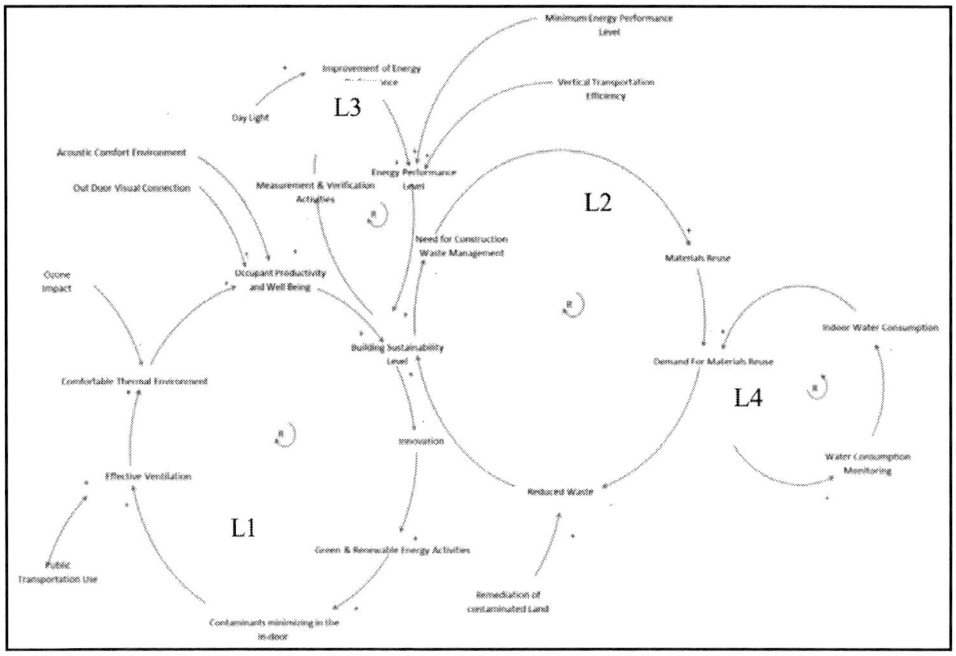

Fig. 2. Formulation of dynamic hypothesis using causal loop diagram.

4.2 Formulation of simulation model

Assumptions related to correlation ideas were made based on generated relations from the statistical analysis and assumed CLD. Due to complexity of studding all interaction between all the different key factors, the model has presented interactions assumed in the CLD. These interactions were: Correlation between occupants of the building and the different environmental impact factors, construction and maintenance activities dynamics, energy effect, and effect of water consumption.

Using VenSIM software the proposed simulation model was developed based on the resulting CLD and generated relationships from USGC data. Mainly the model was composed of five stocks with core stock of "Building sustainability level. Figure 3 illustrates the model with all stocks in the model. The simulation runs were performed for 100 years as a proof of concept towards using the technique in addition to illustration of expected behaviour.

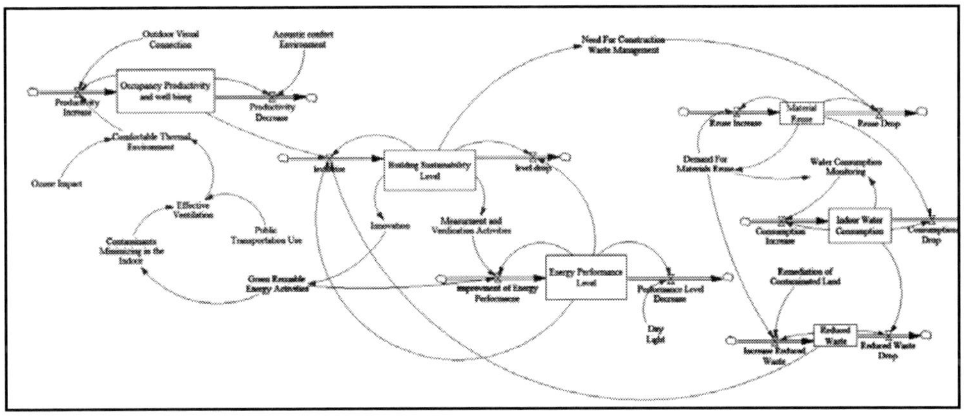

Fig. 3. Graphical representation of simulation model.

Conventionally through the construction period most sustainability activities and precautions are not yet implemented fully and accordingly behaviour hesitates, while through the early period of operation, sustainability plans will be well implemented resulting in a dramatic increase in behaviour, followed by less interest in enhancement of sustainability activities. The same conceptual understanding can be mapped over the rest of the illustrated behaviours. For instance, Figure 4-b illustrates the behaviour of Occupancy Productivity and Well Being (OPWB), from which the smooth increase can indicates how adoption of sustainability activities may results in better working environment. The behaviour also is recorded as a desired on for such factor. Additionally, Figure 4-c shows the behaviour of Indoor Water Consumption (IWC) which is, initially, has no water consumption record through construction stage, while a sudden rise of consumption occurs as soon as the building start its operation, followed by a continuous steady decrease in consumption as occupants continue applying sustainability techniques. On the other hand, Figure 4-d depicts Energy Performance Level (EPL) behaviour which starts with specific value and drops through period of construction due to sustainability measures and techniques implementation. As soon as the operation stage, starts a sudden increase in energy performance level occur. This may add to the increase in energy usage and accordingly this increase will drops gradually due to inadequate implementation of sustainability measures and techniques.

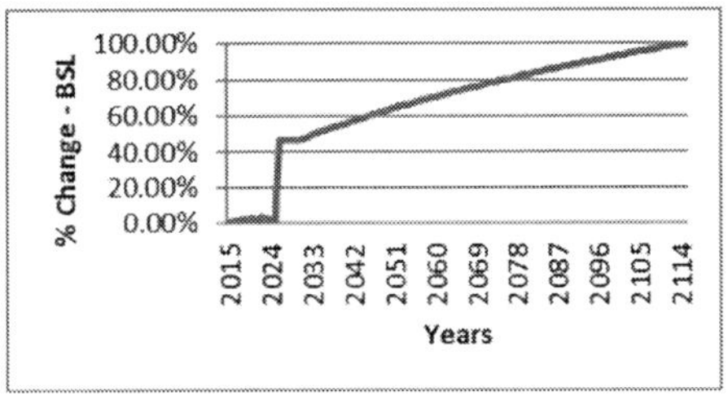

a. Building Sustainability Level Behaviour

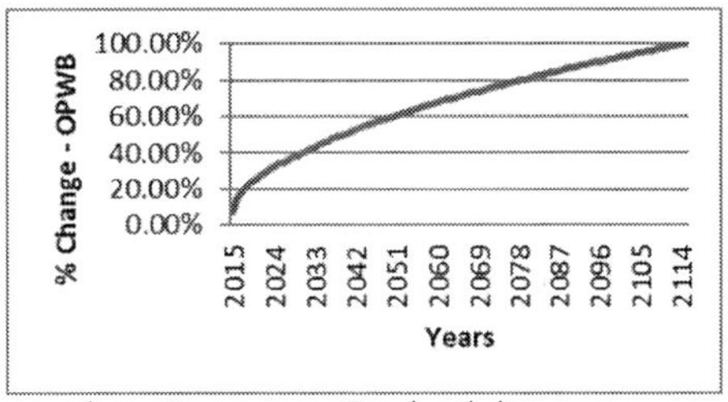

b. Occupancy Productivity and Well Being Behaviour

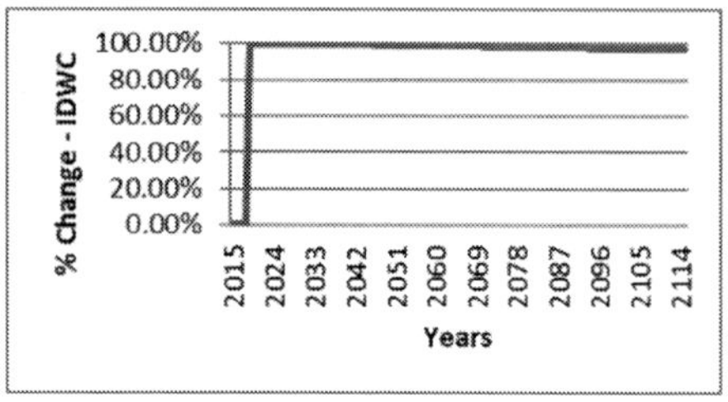

c. Indoor Water Consumption Behaviour

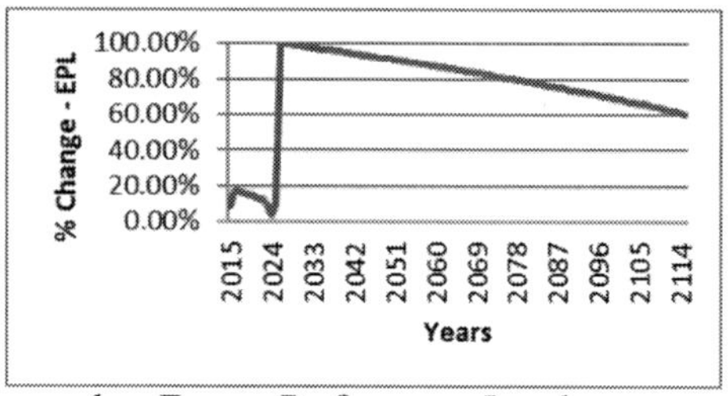

d. Energy Performance Level Behaviour

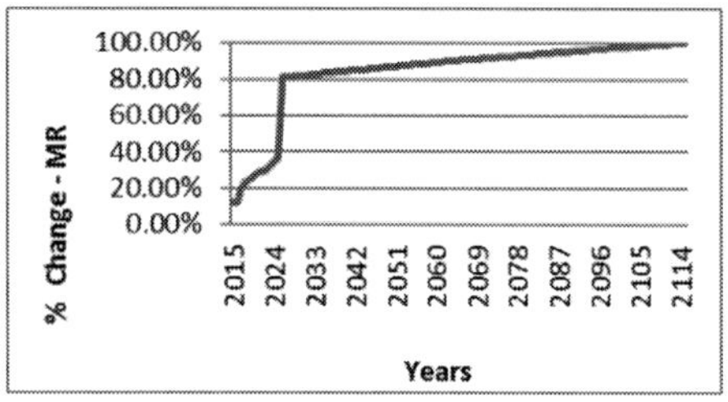

e. Materials Reuse

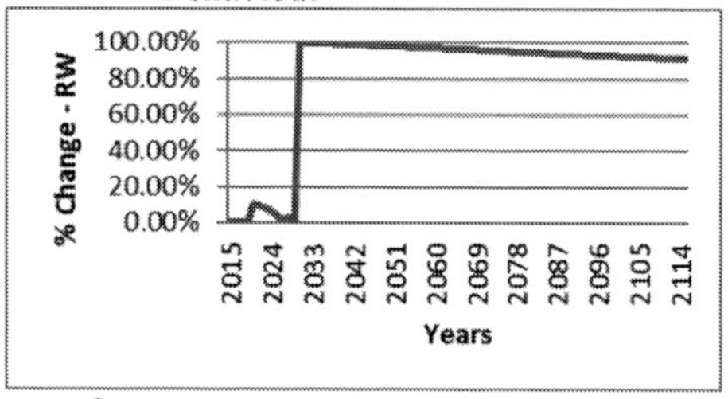

f. Reduced Waste

Fig 4. Building sustainability level behaviour.

5 Summary and conclusion

This research presented a generic framework towards implementing system thinking based approach for building life cycle sustainability enhancement. The effort methodology has included four main steps with six supporting activities, starting by problem articulation and moving to dynamic hypothesis, then formulation and modelling, and finally policies evaluation. One major contribution from implementing this effort over traditional holistic approaches was identifying the benefit of considering interaction between system factors. The benefit was clear in the spotted negative effect recorded by modelling building lifecycle sustainability behaviour. Moreover the use of system thinking in building life cycle comprehension added number of dimensions such as sustainability stretched analysis, factors interaction prospective, and possibility for performance prediction. Future enhancement for the framework shall involve lowering the threshold for parameters selection to score of 2 or less to increase the number of factors studied, also inclusion of additional causes correlations shall enhance the produced behaviour and will provide more robust model for policies recommendation. . An additional important fragment shall be the inclusion of “delay” for some factors that may cause burden shift after completion of the construction period.

References

1. EPA, *Green Building Basic Information U.S.* Environmental Protection Agency, (2009) available at <http://www.epa.gov/greenbuilding/pubs/about.htm>.
2. Ji, Y., and Plainiotis, S. *Design for Sustainability*. China Architecture & Building Press, Beijing, China (2006).
3. Anastas, P.T., and Zimmerman, J.B. Design Through the 12 Principles of Green Engineering. *Environmental Science & Technology*, 37(5), 94A – 101A (2003).

4. Zhang, N., Li, Y.P., Huang W.W., and Liu, J. An Inexact Two-Stage Water Quality Management Model for Supporting Sustainable Development in a Rural System. *Journal of Environmental Informatics*, 24(1) 52-64 (2014).
5. Chang, N.B., Chen, H.W., and Ning, S.K. Identification of river water quality using the fuzzy synthetic evaluation approach. *Journal of Environmental Management*, 63(3), 293-305 (2001).
6. Karmakar, S., and Mujumdar, P.P. A two-phase grey fuzzy optimization approach for water quality management of a river system. *Advances in Water Resources,* 30(5), 1218-1235 (2007).
7. Deviney Jr, F.A., Brown, D.E., and Rice, K.C. Evaluation of bayesian estimation of a hidden continuous-time Markov chain model with application to threshold violation in water-quality indicators. *Journal of Environmental Informatics*, 19(2), 70-78 (2012).
8. Graveline, N., Loubier, S., Gleyses, G., and Rinaudo J.D. Impact of farming on water resources: Assessing uncertainty with Monte Carlo simulations in a global change context. *Agricultural Systems*, 108, 29-41 (2012).
9. Kisi, O., Akbari, N., Sanatipour, M., Hashemi, A., Teimourzadeh, K., and Shiri, J. Modeling of dissolved oxygen in river water using artificial intelligence techniques. *Journal of Environmental Informatics*, 22(2), 92-101 (2013).
11. Xu, T.Y., and Qin, X.S. Solving water quality management problem through combined genetic algorithm and fuzzy simulation. *Journal of Environmental Informatics*, 22(1), 39-48 (2013).
12. Forrester, J. W. *Industrial Dynamics*, Wiley, New York. U.S (1961).
13. Al-Gahtani, K., AlSulahi, I., El-Hawary, M., and Marzouk, M. Investigating Sustainability Parameters of Administrative Buildings in Saudi Arabia. *Technological Forecasting & Social Change Journal*, 105, pp. 41-48 (2016).

Thermoelectric generator experimental performance testing for wireless sensor network application in smart buildings

Mohamed Al Musleh[1,*], *Evangelia* Topriska[1,*], *Lynne* Jack[2,*], *David* Jenkins[2,*]

[1] Heriot Watt University – Dubai Campus, Dubai International Academic City, Dubai, United Arab Emirates
[2] Heriot Watt University, EH14 4AS, Edinburgh, United Kingdom

Abstract. In order to make a conventional building more efficient or smarter, systems feedbacks are essential. Such feedbacks can include real-time or logged data from various systems, such as temperature, humidity, lighting and CO2 levels. This is only possible by the use of a network of sensors which report to the building management system. Conventional sensors are limited due to wiring and infrastructure requirements. Wireless Sensor Networks (WSN) however, eliminates the wiring limitations but still in certain cases require periodical battery changes and maintenance. A suitable solution for WSN limitations is to use different types of ambient energy harvesters to power battery-less sensors or alternatively to charge existing batteries so as to reduce their changing requirements. Such systems are already in place using various energy harvesting techniques. Thermoelectric Generators (TEG) are one of them where the temperature gradient is used to generate electricity which is conditioned and used for WSN powering applications. Researchers in this field often face difficulty in estimating the TEG output at the low-temperature difference as manufacturers' datasheets and performance data are not following the same standards and in most cases cover the high-temperature difference (more than 200C°). This is sufficient for industrial applications but not for WSN systems in the built environment where the temperature difference is much smaller (1-20C° is covered in this study). This paper presents a TEG experimental test setup using a temperature controlled hotplate in order to provide accurate TEG performance data at the low-temperature difference range.

1 Introduction

Energy used within buildings and building services accounts for about 80% of the total energy use in the United Arab Emirates with 73% residential consumers [1, 2], with the introduction of green buildings regulation and policies, the local authorities' plan for

* Corresponding author: m.al-musleh@hw.ac.uk

energy rating in buildings. To reduce energy use in general, effective building energy management systems are essential. For the precise control feedback from the different energy systems and buildings areas, WSN is used. Wireless communications and the internet is the main topic for research in a variety of applications, in relation building energy systems and the use of WSN, wireless sensor nodes are distributed around the building and continuously measures, log, and transfer data to a central location. Powering these sensor nodes is still a challenge and ambient energy harvesting is a promising solution for such applications. Thermoelectric generators (TEG) is one of the energy harvesting technologies that can be used. Their operation is based on the concept of thermoelectricity, which is the direct conversion of temperature difference to electricity via a temperature gradient (Seebeck effect). Thermoelectric (TE) devices were originally made of two different metals and later with the introduction of semiconductors variety of doped materials were used to make TE modules. A module normally consists of multiple TE elements connected electrically in series and thermally in parallel in such a way to maximize the generated voltage. One side of the module is to be the hot side where heat is applied, while the other side is called the cold side where heat gets dissipated or rejected, Figure 1 below shows a schematic of a TEG module. [3,4]

The generated voltage from a TEG is directly proportional to the temperature difference between both sides of the module and the Seebeck coefficient of the material used is shown in Equation 1.

$$\boldsymbol{E}_{emf} = S\,\Delta T \tag{1}$$

Where Eemf represents the generated electromotive force, S is the Seebeck coefficient which is material dependent, and ΔT is the temperature gradient difference. [6]

Commercial sensor nodes which use TEG as a power source are available commercially but mostly for high-temperature difference applications such as boilers and industrial processes. Examples of commercial sensor nodes are shown in Figure 2 below. When it comes to buildings applications the temperature difference available is much lower. If boilers, hot water, and space heating systems which are not the main energy consumers in the UAE are excluded, the available temperature difference can be as low as 20C° or even lower in certain cases. The selection of suitable TEG for such task requires performance data within this range of temperature difference which is not currently available. [7, 8]

DOI: 10.1051/matecconf/201712008003

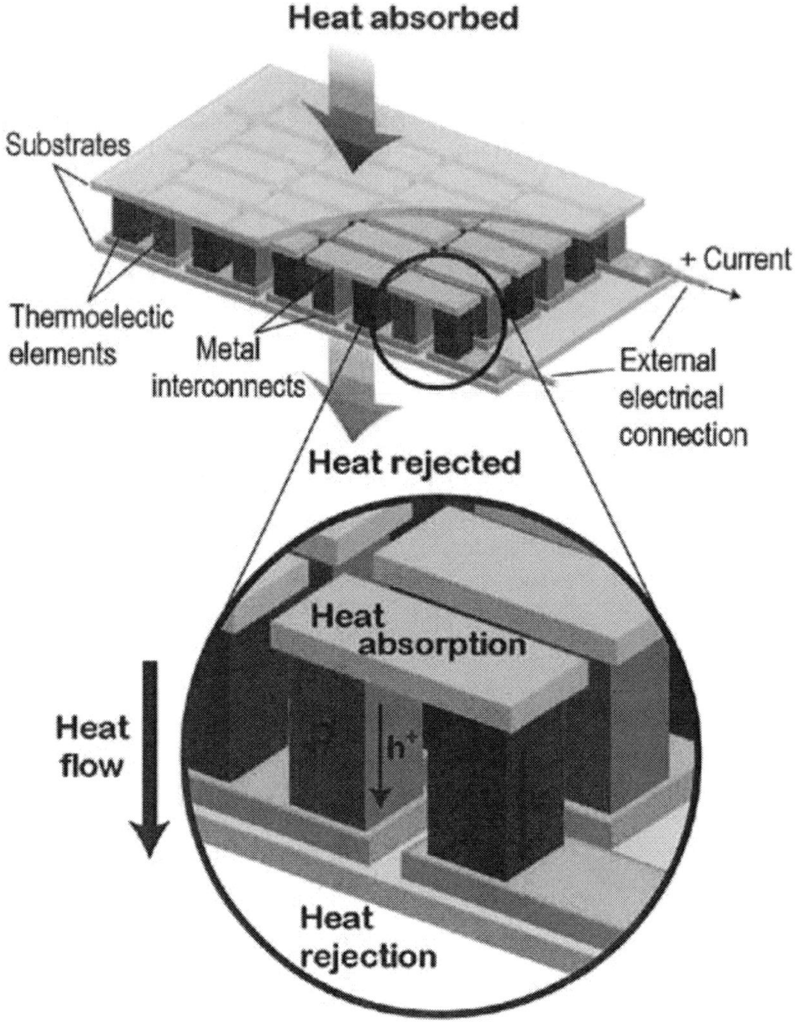

Fig. 1. Schematic of a TEG, TE elements of different types connected together to form a module. [5]

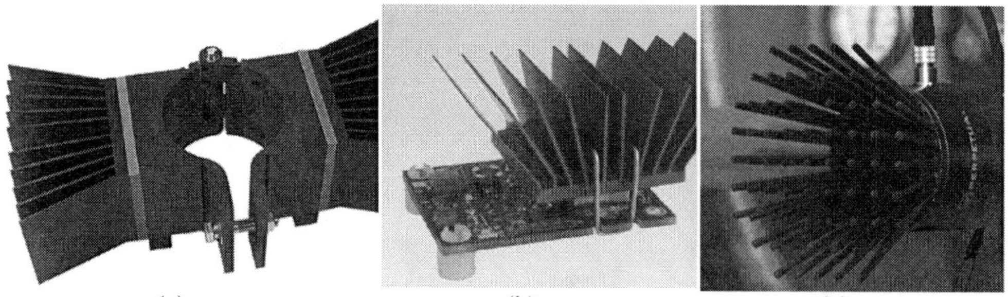

(a) (b) (c)

2 Experimental setup

The experimental setup consisted of a high precision feedback controlled hot plate where the TEG being tested is placed on top with the hot side of the TEG facing the hot plate. K-Type thermocouples were used on both sides of the TEG to measure the actual temperature from each side. Aluminum heat sink with copper core was used to cool down the cold side of the TEG by forced convection using an electric axial fan. Thermal paste with a thermal conductivity of 0.925 W/m-k was used between all the different layers to ensure proper heat transfer, finally, the setup was fastened to the hot plate to maximize the accuracy of the results, Figure 3 below shows a schematic of the setup.

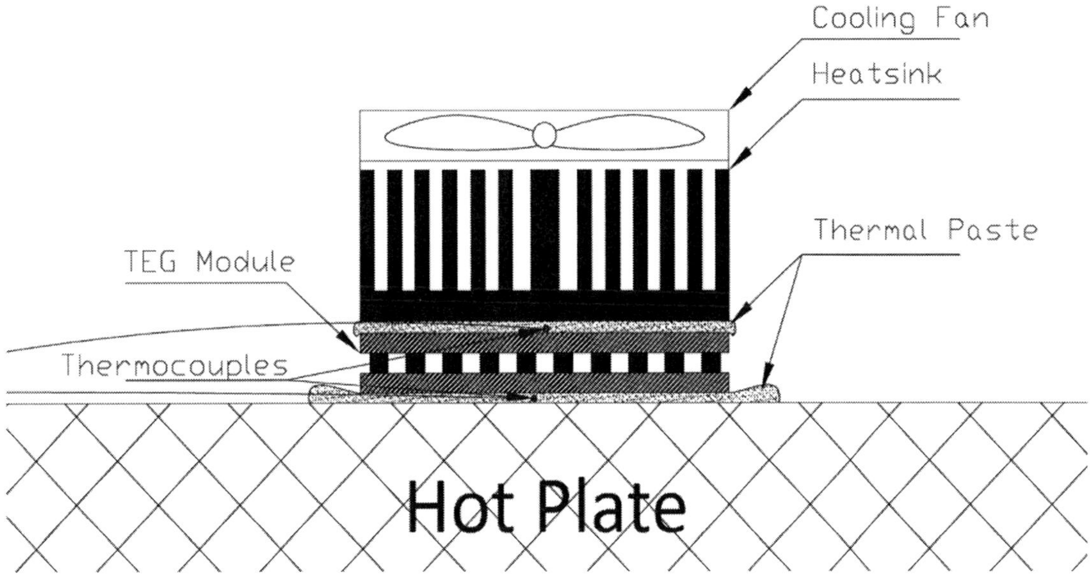

Fig. 3. Schematic illustration of the testing setup.

A TG12-4 TEG module was acquired for this test, which is a Marlow Industries Inc. product, designed as a universal TEG, rated at 4 watts at a temperature difference of 180C°. [12] Manufacturer's data sheet provides performance data with 50C° as cold side temperature and a minimum of 70C° as hot side temperature. It is common practice for manufacturers and suppliers of TEG to provide performance data at high-temperature difference which normally covers the optimum operation condition for the TEG to generate the rated power. In practice, however, different applications have a variety of working temperature ranges, and often users and researchers face the issue of not having performance data at the low-temperature difference. This is the main motivation for the construction of this setup and conduction of these experimental measurements. Figure 4 shows the module used and its positioning on the hot plate. [12]

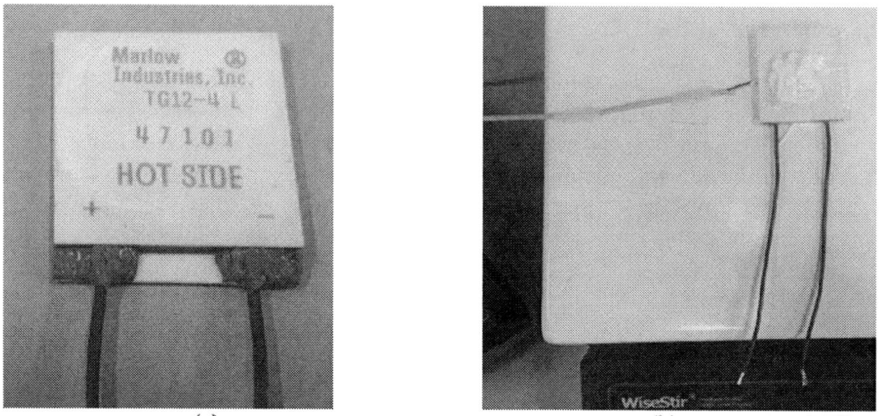

(a) (b)

Fig. 4. (a) TEG module obtained for testing (b) TEG module installed on the hotplate with the hot side thermocouple.

The K-Type thermocouple was placed between the TEG and the hot plate to measure the hot side temperature, thermal paste was used to ensure thermal conductivity from the plate to the TEG and the thermocouple. Figure 5 shows an example performance data for the device from the manufacturer's datasheet.

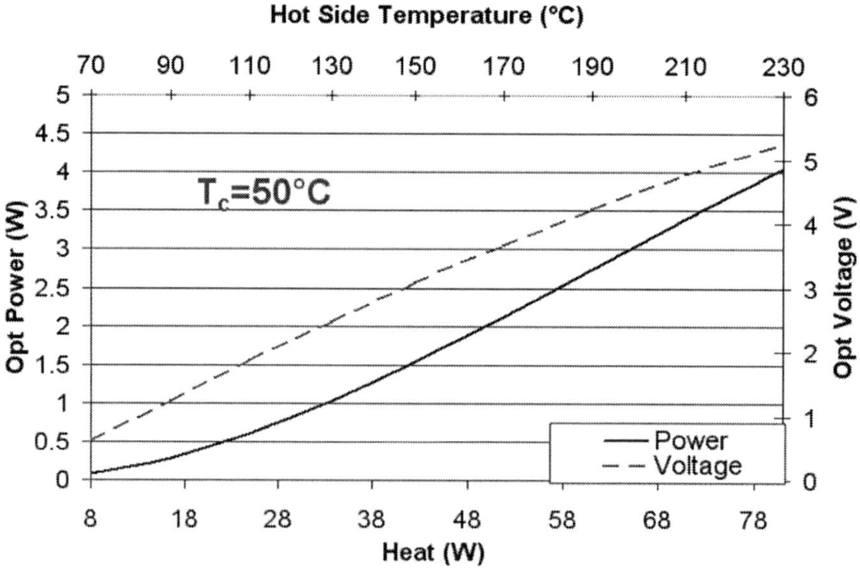

Fig. 5. Performance data from Manufacturer's data sheet [12].

The second thermocouple was placed on top of the TEG and attached to the cold side of it, thermal paste was used to ensure thermal conductivity. The heat sink was then placed on top with the copper core centered over the TEG. To maximize the cooling effect of the heat sink, an axial electric cooling fan was used which is powered by an independent power source of 12V DC. The fan speed was maintained at the same level

 DOI: 10.1051/matecconf/201712008003

during the experiment, Figure 6 next shows the heat sink used and then the heat sink with the fan attached to it.

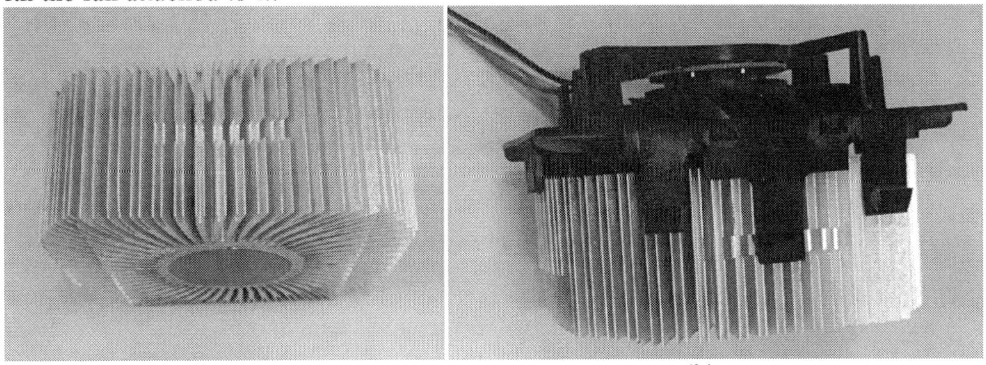

(a) (b)

Fig. 6. (a) Heat sink used (b) Heat sink with the electric axial fan attached.

The complete setup was then secured to the hot plate as any movement will affect the accuracy of the measurement. Both thermocouples were attached to dual channel temperature data logger, and the TEG output to a digital multimeter set to measure open circuit voltage, or short circuit current, Figure 7 below shows the complete setup of the experiment where the hot plate with the TEG and heat sink in the middle, temperature data logger on the left, and digital multimeter on the right. The hot plate temperature was then varied from 40C° down to 0C° with 2.5C° steps, allowing enough time between changes for the TEG temperature to reach steady state and the generated electricity to stay at a fixed level. The values were recorded, and a next setting was applied. It was noted that the TEG hot side temperature was a couple degrees Celsius lower than the hotplate temperature but this has no effect on the experiment results as thermocouples were installed directly to the hot and cold sides of the TEG.

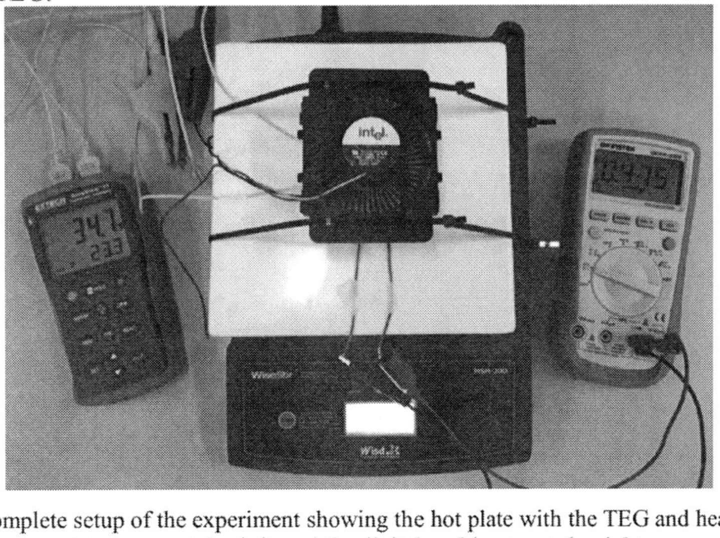

Fig. 7. The complete setup of the experiment showing the hot plate with the TEG and heat sink in the middle, temperature data logger at the left, and the digital multimeter at the right.

3 Experiment results

The temperature of the hot plate was varied from 40C° down to 0C° on 2.5C° steps. On each step, the open circuit voltage and the short circuit current were measured using a digital multimeter. Enough time was allowed between measurements for the TEG output values to reach steady state. The test was carried in a room with the following conditions; room temperature 20.3 C°, relative humidity of 56%, and an atmospheric pressure of 1010 mbar. The temperature of the cold side of the TEG was slightly rising due to the increase in hot side temperature, but values were nearby the room temperature due to the heat sink and the cooling fan. The hot side temperature was increasing based on the changes on the hot plate set temperature, with a slight difference. Both sides temperatures are presented in the graph shown in Figure 8. The temperature difference ranged from 1C° to 20C° which covers a good range of temperature difference available in buildings for WSN applications and energy harvesting.

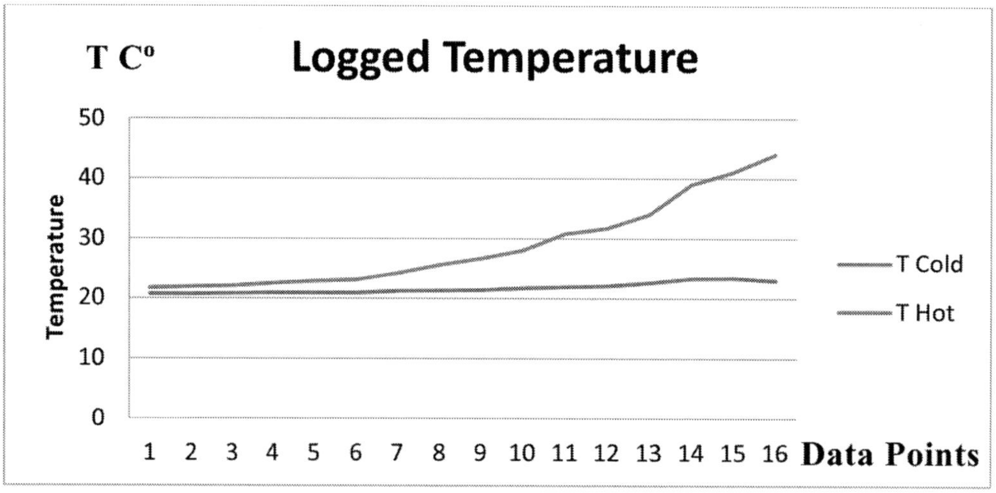

Fig. 8. Graph showing the logged temperature data from both sides of the TEG.

The multimeter was permanently attached to the TEG for continuous measurement of the open circuit voltage. The voltage values were recorded every time both sides of TEG temperature values reached a steady state. After every voltage reading, the short circuit current was recorded at the same temperature difference. Using Equation 1 of the Seebeck effect the Seebeck coefficient was calculated to be 41mV/C° for this specific TEG. Open circuit voltage was observed to be directly proportional to the temperature difference, it was closely following the increase in the TEG hot side temperature as shown in Figure 9.

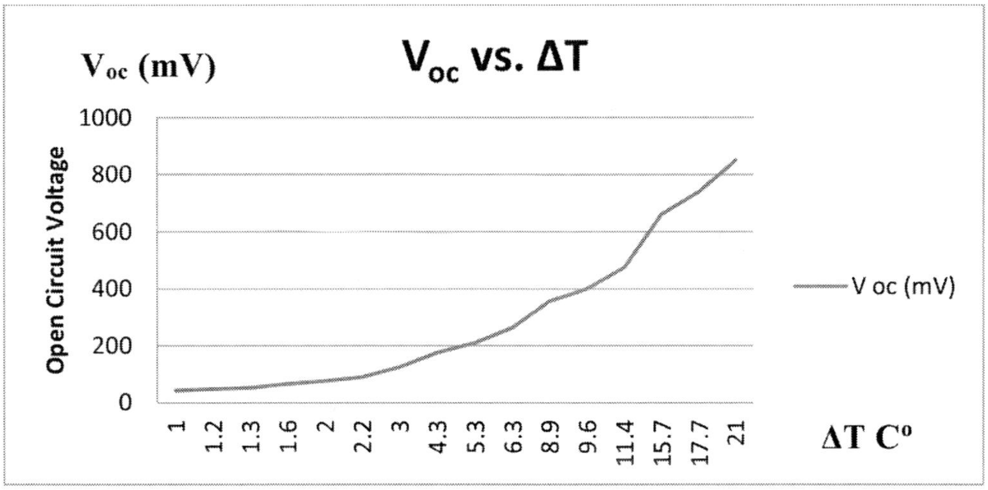

Fig. 9. Graph showing open circuit voltage vs. temperature difference.

Short circuit current followed the temperature increase with similar behavior to the open circuit voltage as shown in Figure 10, this indicates that the generated power of the TEG is increasing in a comparable scheme.

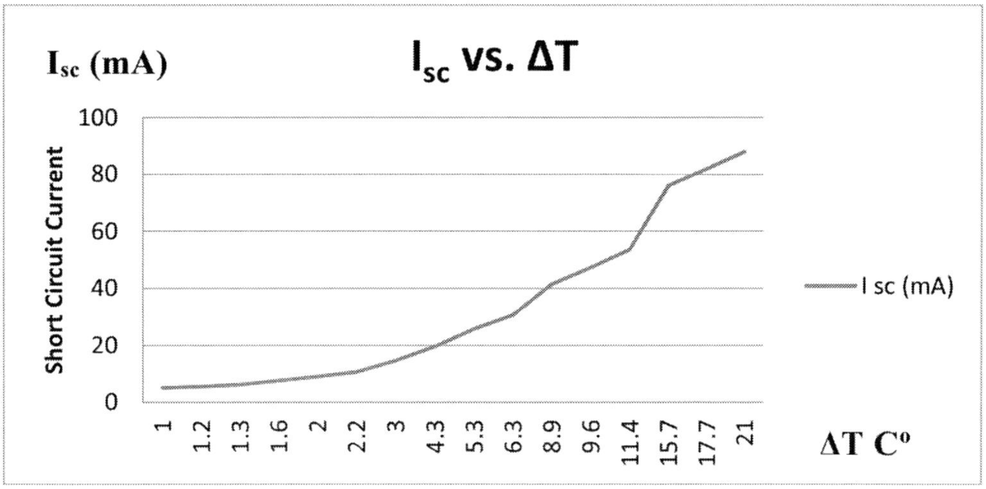

Fig. 10. Graph showing short circuit current vs. temperature difference.

4 Results discussion

Generated electricity from the TEG due to the temperature difference showed a directly proportional to the increase in the hot side temperature, which is matching the behavior described in the performance data provided by the manufacturer. The experimental test results showed the electrical output capabilities of the subject TEG at the very-low-temperature difference as low as 1 C°, such performance data is not covered in the product datasheet, this is exactly what designers and researchers in the field of ultra-low-temperature difference are looking for.

The Seebeck coefficient was calculated using Equation 1 by substituting the measured values of the open circuit voltage at each data point, results were varying within a narrow range from 39.6 mV/C° to 41.8 mV/C° with the average at 41 mV/C° including the extremely low-temperature point.

Furthermore, taking in consideration that the subject TEG module is a general purpose TEG, which means considering the operation condition at the design stage of the TEG module could maximize the module performance at such low-temperature difference.

5 Conclusion

The experiment showed that testing of TEG using this method is possible, while further testing and validation is required. The use of an air-cooled heat sink makes it rather complicated to accurately calculate the heat flux through the TEG in order to estimate efficiency values. A suitable solution for this is to use heat exchanger instead in which a refrigerant flow at a cretin flow rate and temperature. Measuring the inlet and outlet temperature of the refrigerant together with the flow rate will provide more accurate figures on heat dissipation from the cold side of the TEG. The presented setup is suitable for performance measurement of TEG. In order to measure efficiency, a heat flux meter needs to be installed on either side of the TEG, or use the heat exchanger method mentioned earlier.

References

[1] Kelly Clarke,'80% energy consumed by buildings', Khaleej Times, (2016).
[2] "2015 Annual Statistics" Dubai Electricity and Water Authority, (2016).
[3] Qian Huang, Chao Lu, Mark Shaurette, and Robert F. Cox, "Environmental thermal energy scavenging powered wireless sensor network for building monitoring", Proceedings of the 28th international association for automation and robotics in construction (ISARC), PP 1376-1380, (2011).
[4] Wei He, Gan Zhang, Xingxing Zhang, Jie Ji, Guiqiang Li, Xudong Zhao, "Recent development and application of thermoelectric generator and cooler", Journal of Applied Energy 143, PP1–25, (2015).
[5] G. Jeffrey Snyder," Small Thermoelectric Generators", The Electrochemical Society Interface, (2008).
[6] Raşit Ahıska, Hayati Mamur, "A review: thermoelectric generators in renewable", International journal of renewable energy research, **Vol.4**, No.1, (2014).
[7] Wensi Wang, Victor Cionca, Ningning Wang, Mike Hayes, Brendan O'Flynn, and Cian O'Mathuna, "Thermoelectric energy harvesting for building energy management wireless sensor networks", International journal of distributed sensor networks, Volume (2013).

[8] David Salerno, "ultralow voltage energy harvester uses thermoelectric generator for battery-free wireless sensors", LT Journal of analog innovation, **Vol. 20**, No. 3, (2010).
[9] Joshua Moczygemba, 'Energy harvesting TEG power strap for industrial, chemical, oil and gas applications product overview and performance', Marlow Industries, Inc.,(2015).
[10] Product datasheet,"TE-Core ThermoHarvesting power module", Micropelt GmbH, (2015).
[11] Product technical brochure, 'Power Puck; Plug-and-play renewable thermoelectric energy for wireless sensor networks', Perpetua Power Source Technologies Inc., (2014).
[12] Product technical datasheet, 'TG12-4 Thermoelectric Generator', Marlow Industries, Inc.

As a clean, sustainable and renewable energy—hydropower in Turkey

Ibrahim Yuksel [1,*], *Hasan* Arman [2], and *Ibrahim Halil* Demirel [3]

[1]Yildiz Technical University, Faculty of Civil Engineering, Department of Civil Engineering, Hydraulics Division, 34220 Davutpasa, Istanbul, Turkey
[2]United Arab Emirates University, College of Science, Geology Department, P.O. 15551, Al-Ain, UAE
[3]Yildiz Technical University, Civil Engineering Ph.D. Student, 34220, Istanbul, Turkey

Abstract. At the beginning of 2000's, Turkish government began privatizing some of its electricity generating and distribution networks, and allowing for more private construction and ownership in the sector. Owing to Turkey's regions, most of which are hilly, it can be possible to develop relatively higher heads without expensive civil engineering works, so that relatively smaller flows are required to develop for the desired power. In these cases, it may be possible to construct a relatively simple diversion structure and to obtain the highest drop by diverting flows at the top of a waterfall. There are intensive investigations to improve and development hydropower in Turkey. In recently, electricity demand has increased significantly and it is the fastest growing end-use of energy. Therefore, in the developing countries such as Turkey, the role of hydropower energy is very important. Because benefits of hydroelectric power make it an important contributor to the future world energy, particularly in the less developed countries. In this paper, as a clean, sustainable and renewable energy, hydropower and its benefits have been investigated in Turkey and some proposals which are to develop hydropower in this country have been presented.

1 Introduction

Energy is essential to economic and social development and improved quality of life in Turkey as in other countries. Much of the world's energy, however, is currently produced and consumed in ways that could not be sustained if technology were to remain constant and if overall quantities were to increase substantially.

Hydropower contributes one-fifth of the world's power generation. In fact, it provides the majority of supply in 55 countries. For several countries, hydropower is the only domestic energy resource. Its present role in electricity generation is therefore substantially greater than any other renewable energy technology, and the remaining potential, especially in the less developed countries such as Turkey, is vast.

* Corresponding author: yukseli2000@yahoo.com

While it is not a panacea, in that it is restricted to sites with available water and appropriate geomorphology, hydropower's flexibility and proven technology sets it apart from other renewable energy sources [1, 2].

2 Hydropower potential in Turkey

Turkey has a total gross hydropower potential of 433 GWh/yr, but only 125 GWh/yr of the total hydroelectric potential of Turkey can be economically used. So it can be said that, hydropower (especially small hydropower) potential should be used very widely and economically in Turkey.

Turkey's annual total gross, technically feasible and economically feasible hydropower potentials calculated by General Directorate of State Hydraulics Works (DSI) are 435, 215 and 128 tWh, respectively. 35 % of the economically feasible hydropower, total 45 155 gWh/year is in operation, 8 % (10 129 gWh/year) is under construction and 57 % (72 339 gWh/year) is being designed. Turkey has important hydropower potential as shown in Table 1 [3, 4].

Table 1. Hydroelectric potential in Turkey by river basins.

Name of river basin	Installed capacity (MW)	Average generation (GWh)	Energy Firm (GWh)
Susurluk	373.26	1336.00	955.00
Gediz	123.00	359	144.00
B. Menderes	278.80	1053.00	254.00
B. Akdeniz	596.79	2321.00	769.00
Antalya	1432.12	5262.00	1762.00
Sakarya	1177.54	2523.00	1515.00
B.Karadeniz	507.35	1747.00	995.00
Yeşilırmak	1270	5321.00	4171.00
Kızılırmak	2060.51	6181.00	4004.00
D.Akdeniz	1632.12	6039.00	3188.00
Seyhan	2048.29	8012.00	3847.00
Ceyhan	1663.20	5558.00	2824.00
Fırat	9672.93	38072.00	30096.00
D.Karadeniz	3462.55	11346.00	5288.00
Coruh	3178.90	10706.00	6217.00
Aras	834.92	2539.00	1775.00
Van Kapalı	62.06	260.00	157.00
Dicle	4969.88	16912.00	10641.00
Total Turkey	**35539.31**	**126109.00**	**78770.00**

In Turkey, the distribution of the hydro power plants, that are under design level, is presented in Table 2 according to their hydro capacity [5]. As can be seen, 30.34 % of all of the annual energy will be generated by SHP. There is 80 installed SHP in Turkey 5 % of which with medium head and 95 % with high head [6]. Being generally a mountainous country, Turkey's SHP potential is high. There is installed 80 SHP with 177 mW capacity. However, the remaining economically feasible potential is nearly 22 000 gWh/year [5].

Table 2. Distribution of under design hydro power plants according to their hydro capacity.

Classification	Number of HEPP	Total Capacity (mW)	Total Annual Energy (gWh)	Percentage of Total Annual Energy
<5 MW	139	312	1 568	2.17
5 to 10 MW	79	548	2 135	2.95
10 to 50 MW	186	4 595	18 244	25.22
50 to 100 MW	54	3 824	13 524	18.70
100 to 250 MW	36	5 527	18 179	25.13
250 to 500 MW	11	3 500	11 657	16.11
500 to 1000 MW	3	1 791	3 199	4.42
>1000 MW	1	1 200	3 833	5.30
TOTAL	**509**	**21 297**	**72 339**	**100**

In Turkey, development of small hydropower began in the year 1902. Since then, many small hydropower plants have been installed by government organizations, private sector, and local municipalities in many parts of the country [7, 8].

In Turkey, during last three decades the average annual increase of SHP capacity was 5-10%. As of 2002, the total development of SHP capacity that is accepted as small hydropower according to the Electrical Power Resources Survey and Development Administration which is less than 10 MW, in Turkey was 849,1 MW and total annual energy production was 3,623 (Table 3) [3, 5].

Table 3. Small hydropower development in Turkey.

Status of SHP	Number of SHP	Installed Capacity (MW)	Energy Generation (GWh)
In operation	70	175.4	654
Under construction	6	21.7	130
In final design	7	38.8	168
Infeasibility and pre-feasibility	120	613.2	2,671
Total	**203**	**849.1**	**3,623**

At the beginning of 2004, total number of SHP plants in operation throughout the country was 62, with a total installed capacity of 190 MW, about 1.5% of the total hydropower potential in Turkey [3]. Table 3 shows the summary of the total situation of SHP development in Turkey.

Total installed projects capacity of SHP is 2.45% and the total energy potential is about 2.96%, which have installed capacity less than 10 MW. At the end of 2004, about 96% of the all ready-exploited potential is from dams and HEPPs, and the remainder is from run-off river and canal SHP. Neglecting the geothermal, wind, and solar generation, about 65% of the electricity is produced by thermal power plants, and hydropower plants produce about 35% of the remaining electricity [9].

3 The advantages of small hydropower

Small hydropower plants combine the advantages of hydropower with those of decentralized power generation, without the disadvantages of large-scale installations. Small hydropower is [10]:

- A sustainable resource,
- An efficient resource,
- A secure resource,
- A clean resource and
- A renewable resource.

Small hydropower can be economically attractive, sometimes even offering the least-cost method of generating electricity. Small hydropower has the ability to generate electricity instantly, to supply both base and peak load generation, has a long life is easy to maintain and is highly reliable.

In addition, small hydropower development brings prosperity and raises the standard of living due to educational facilities, public health benefits, roads, electricity power and other infrastructure developments.

4 Cost of small hydropower

Small hydropower, especially the very small and the low-head plants, can normally only compete where allowance are made for the external costs associated with fossil fuels and nuclear power. Small hydropower projects are generally considered to be more environmentally favorable and sustainable than both large hydro and fossil fuel powered plants. So, small-scale hydropower is more economically competitive than small-scale fossil fuel/steam-electrical power.

The capital required for small hydropower plants depends on the effective head, flow rate, geological and geographical features, continuity of water flow, equipment such as turbines and civil engineering works. Making use of existing weirs, dams, storage reservoirs and ponds can significantly reduce both environmental impact and costs. Sites with low heads and high flows require a greater capital outlay, as larger civil engineering works and turbine machinery will be needed to handle the larger flow of water.

Apart from the investment and production costs, the other principal cost element is operation and maintenance, including repairs and insurance, which can account from 1.5-5% of investment costs. Both the production and investment costs differ considerably depending on the plant's head height.

5 Conclusion

Hydropower is a proven, well understood technology based on more than a century of experience. Its schemes have the lowest operating costs and longest plant lives. Upgrades and refurbishment of generating plant can readily extend scheme life. Hydropower plants also provide the most efficient energy conversion process. Modern plants can convert more than 95% of moving water's energy into electricity, while the best fossil-fuel power plants are only about 60% efficient. Hydropower also has the highest energy payback ratio. During the lifetime of a scheme, it can produce more than 200 times the energy needed to build it.

Turkey is an energy importing nation with more than 65 percent of our energy requirements met by imported fuels. Air pollution is becoming a significant environmental concern in the country. In this regard, hydropower, especially small hydropower (SHP) is becoming attractive solution for clean and sustainable energy future of Turkey. Because Small hydropower can be successfully developed as long as it produces electricity at competitive prices and under conditions that respect the environment. On the other hand, small hydropower's main challenges relate to both economics and ecology.

Turkey as developing country has abundant SHP resources, which are economically feasible for development. In this country the majority of people live in rural areas.

In most rural areas of this country, because of lack of potable water, electricity and convenient transportation, the local economy remained underdeveloped and people were very poor. So, if it is developed of SHP projects in rural areas of this country, a lot of problems discussed above could be solved.

References

1. Yuksel I., Kaygusuz K., *The İnternational Conference of Maximizing The Benefits of Hydropower*, 25-27 September, Porto Carras, Greece, Abstract of Conference. (2006)
2. Kaygusuz, K., Jour. of En. Sou., Issue **26** (2004)
3. DSI, State Hydraulic Works, *Statistics on hydropower*, Ankara, Turkey (2004)
4. Yuksel I., Yuksek O. and Onsoy H., *International Symposium Water for Development Worldwide*, September, Istanbul/Turkey, General Directorate of State Hydraulics Works, Vol.: **1**, pp. 115-123, (2005)
5. EIE, Electrical Power Resources Survey and Development Administration, *Hydroelectric Power Activities of the EIE*, Available from http://www.eie.gov.tr, (2004)
6. Punys, P., *Hidroenergia 04, Sweden. RTLED (Report of Turkey Long Term Electric Energy Demand)*, Ministry of Energy and Natural Resources, Ankara, Turkey, (in Turkish) (2004)
7. Kaygusuz, K. Ener.Sour., Vol: **21**, pp. 581-588, (1999)
8. Yuksel, I., Jour. of En. Explor. & Exploit., Vol. **26**, Number 1, pp. 35 – 52, (2008)
9. Yuksek, O., Komurcu, MI., Yuksel, I and Kaygusuz, K., Jour. of En. Pol., Vol. **34**, Number 17, pp. 3093- 3103, (2006)
10. Yuksel, I., Jour. of En. Sour., Part B: Econ., Plan., and Pol.", Vol. **2**, pp. 113–121, (2007)

Small hydropower for renewable energy and water efficiency in Turkey

Ibrahim Yuksel [1],* *Hasan* Arman [2], and *Ibrahim Halil* Demirel [3]

[1]Yildiz Technical University, Faculty of Civil Engineering, Department of Civil Engineering, Hydraulics Division, 34220 Davutpasa, Istanbul, Turkey
[2]United Arab Emirates University, College of Science, Geology Department, P.O. 15551, Al-Ain, UAE
[3]Yildiz Technical University, Civil Engineering Ph.D. Student, 34220, Istanbul, Turkey

Abstract. Because of social and economic development, the demand for energy and particularly for electricity is growing rapidly in Turkey. The main indigenous energy resources are hydro and lignite. Turkey has no big oil and gas reserves. Almost all oil natural gas (NG) and high quality coal are imported. It has a large potential for renewable energies. Hydroelectricity is well established as one of the principal energy-producing technologies around world, providing some 20 % of the world's electricity. In the developing countries, the proportion rises to around 40-50%. The capacity of large hydroelectric schemes can be several times that of a conventional power station. They are highly efficient, reliable, and long lasting. Turkey has an abundant hydropower potential to be used for generation of electricity and must increase hydropower production in the near future. For these reasons, hydropower and especially small hydropower are emphasized as Turkey's renewable energy sources. In this paper, to increase production of hydroelectric energy, some streams which are in the different basins in Turkey, have been analysed and some conclusions obtained from this study have been presented.

1 Water resources in Turkey

The annual average precipitation in Turkey is estimated at 643 mm, corresponding to a volume of 500 km^3 (Table 1). The average runoff coefficient is 0.37, and the annual runoff is 186 km^3 (2400 m^3/ha). Subtracting from this figure the estimated water rights of neighbouring countries, minimum stream flow requirements for pollution control, aquatic life and navigation, and topographic and geologic constraints, the annual consumable water potential of 12 km^3 should be added to this, bringing the total annual consumable potential to 107 km^3 [1].

Precipitation differs considerably both from year to year and among the river basins. The annual depth of precipitation is as high as 250 cm in the Eastern Black Sea Region, and as low as 30 cm in some parts of Central Anatolia. Most of the country's water potential

* Corresponding author: yukseli2000@yahoo.com

lies in the south-east (28 %) and the Black Sea region (8 %). Turkey's water resources can be considered in 26 drainage basins.

Table 1. Water resources in Turkey.

Water Resources	**Value**
Mean annual precipitation	642.6 mm
Volume of the mean annual precipitation	501 km^3
Surface Water	
Annual flow	186.05 km^3
Annual run-off coefficient	0.37
Present annual consumption	27.5 km^3
Groundwater	
Annual safe yield	12.2 km^3
Allocated amount	7.6 km^3
Present annual consumption	6.0 km^3

2 General review of hydroelectric power potential

Turkey's geographic location has several advantages for extensive use of most of the renewable energy sources. It is on the humid and warm climatic belt, which includes most of Europe, the near East and western Asia [2]. After 2001, the Turkish government began privatizing some of its electricity generating and distribution networks, and allowing for more private construction and ownership in the sector.

Owing to Turkey's regions, most of which are hilly, it can be possible to develop relatively higher heads without expensive civil engineering works, so that relatively smaller flows are required to develop for the desired power. In these cases, it may be possible to construct a relatively simple diversion structure and to obtain the highest drop by diverting flows at the top of a waterfall.

There are intensive investigations to improve the small and large hydropower development in Turkey and some of small hydropower plants are still under construction in order to put this aim into practice.

Turkey's annual total gross, technically feasible and economically feasible hydropower potentials calculated by General Directorate of State Hydraulics Works (DSI) are 435, 215 and 128 tWh, respectively. 35 % of the economically feasible hydropower, total 45 155 gWh/year is in operation, 8 % (10 129 gWh/year) is under construction and 57 % (72 339 gWh/year) is being designed.

According to findings of a study carried out by Bakir [3], in which a new criterion was developed related to key concept of "the economical feasibility", by taking into consideration some undervalued benefits of hydro plants and some overvalued benefits of thermal power plants; economically feasible hydropower potential goes up 188 tWh/year, with an increase ratio of 47 % compared to DSI value. Turkey's hydropower potential according to DSI and the new developed criteria (by Bakir) together with installed power values are given in Table 2.

Table 2. Turkey's annual hydropower potential according to DSI and Bakir.

Basin	Gross Pot. (gWh)	Potential According to DSI		Potential Acc. to Bakir	
		Econ. Feasib. Pot. (gWh)	Installed Power (mW)	Econ. Feasib. Pot. (gWh)	Installed Power (mW)
Firat (Euphrates)	84 122	39 375	10 345	46 267	12 176
Dicle (Tigris)	48 706	17 375	5 416	24 353	7 610
Eastern Black Sea	48 478	11 474	3 257	24 239	6 925
Eastern Meditt.	27 445	5 216	1 490	10 978	3 137
Antalya	23 079	5 355	1 537	9 232	2 638
Coruh	22 601	10 933	3 361	12 431	3 825
Ceyhan	22 163	4 825	1 515	8 865	2 860
Seyhan	20 875	7 853	2 146	9 394	2 609
Kizilirmak	19 552	6 555	2 245	7 821	2 697
Yesilirmak	18 685	5 494	1 350	8 408	2 213
Western Black Sea	17 914	2 257	669	7 166	2 108
Western Meditt.	13 595	2 628	723	5 438	1 511
Aras	13 114	2 372	631	5 246	1 418
Sakarya	11 335	2 461	1 175	3 967	1 984
Susurluk	10 573	1 662	544	2 643	881
Others (Total)	30 744	1 788	546	1 721	507
Total	432 981	127 623	36 950	188 169	55 099

3 Prediction of Turkey's long term electric energy demand

A demand prediction model, called "Model for Assessment of Energy Demand" (MAED) was used to predict Turkey's long term electric energy demand. This model, prepared by International Atom Energy Agency (IAEA), is a simulation model which evaluate mid and long term energy demands. MAED evaluates future energy needs based on mediumto long term scenarios of socioeconomic, technological and demographic development in a country or region.

The electric energy demand varies according to various parameters. The main parameters to affect the energy demand are: Gross National Product (GNP); population and demographic variations; development in housing, industry, agriculture and transportation sectors; income per capita; climatic conditions; employment, technological development, etc.

Three kinds of planning scenario were applied in the model and MAED Model was executed for three scenarios and Turkey's annual electric energy demand values were predicted from 2004 to 2020. According to the results of the study, Turkey's annual electric energy demand in 2020 is predicted between 407 to 571 tWh [4].

4 Small hydropower in Turkey

The development of hydro-electricity in the 20th century was usually associated with the building of large dams. Hundreds of massive barriers of concrete, rock and earth were placed across river valleys world-wide to create huge artificial lakes. While they created a major, reliable power supply, plus irrigation and flood control benefits, the dams necessarily flooded large areas of fertile land and displaced many thousands of local inhabitants. In many cases, rapid silting up of the dam has since reduced its productivity and lifetime. There are also numerous environmental problems that can result from such major interference with river flows [5].

Small, mini and micro hydro plants (usually defined as plants less than 10 mW, 2 mW and 100kW, respectively) play a key role in many countries for rural electrification. Small-scale hydro is mainly 'run of river,' so does not involve the construction of large dams and reservoirs. Small hydro currently contributes over 40 gW of world capacity. The global small hydro potential is believed to be in excess of 100 gW. The industry believes that small hydro will have a strong resurgence in Europe in the next 10 years, after 20 years of decline [5].

The distribution of the hydro power plants that are under design level is presented in Table 3 according to their hydro capacity [7]. As can be seen, 30.34 % of all of the annual energy will be generated by SHP. There is 80 installed SHP in Turkey 5 % of which with medium head and 95 % with high head [6]. Being generally a mountainous country, Turkey's SHP potential is high. There is installed 80 SHP with 177 mW capacity. However, the remaining economically feasible potential is nearly 22 000 gWh/year [7].

Table 3. Distribution of under design hydropower plants according to their hydro capacity.

Classification	Number of HEPP	Total Capacity (mW)	Total Annual Energy (gWh)	Percentage of Total Annual Energy
<5 MW	139	312	1 568	2.17
5 to 10 MW	79	548	2 135	2.95
10 to 50 MW	186	4 595	18 244	25.22
50 to 100 MW	54	3 824	13 524	18.70
100 to 250 MW	36	5 527	18 179	25.13
250 to 500 MW	11	3 500	11 657	16.11
500 to 1000 MW	3	1 791	3 199	4.42
>1000 MW	1	1 200	3 833	5.30
TOTAL	**509**	**21 297**	**72 339**	**100**

5 Focus on the eastern Black Sea region

Among 26 hydrological basins in Turkey, the Eastern Black Sea Basin (EBSB) has great advantages from the view point of SHP potential. Because, the annual average precipitation is the highest in the country going up to 2329 mm in Rize Province. Also, the basin covers sharp valleys and there are a lot of steep streams with considerable discharges and heads.

A study is being carried out by EIE to evaluate energy potential of small streams in Turkey. The preliminary results of this study are given in Table 4. The preliminary results of only 8 basins out of 26 basins in Turkey have been obtained yet and the other basins are being studied. As can be seen, 59 projects out of 132 (44.7 %) projects are in EBSB and annual energy of EBSB projects comprises of 886.56 gWh (52.18 %) out of 1698.86 gWh for all of the projects. The potential and potential percentage of SHP projects with various intervals of coefficient of profitability (CP) are also given in Table 4. It is also obvious that, most of (81 %) the projects in EBSB are profitable with CP>1. The profitable projects in EBSB cover 56.4 % of all of the profitable projects discussed in this study [7].

Table 4. Preliminary results of studied SHP projects by EIE.

Basin	Num. of Project	Capacity (mW)	Annual Energy Potential (gWh)	Potential (Percentage) of Projects According to Coefficient of Profitability (CP)		
				CP>1	0.7<CP< 1	CP<0.7
East. Black Sea	59	157.75	886.56	718 (81 %)	130 (15 %)	38 (4 %)
Mid Mediter.	20	69.09	278.52	260 (94 %)	10 (3 %)	8 (3 %)
Gediz	7	41.76	166.20	116 (70 %)	27 (16 %)	23 (14 %)
West. Mediter.	9	23.51	111.78	99 (89 %)	8 (7 %)	4 (4 %)
Susurluk	15	23.74	110.55	23 (21 %)	77 (69 %)	11 (10 %)
Wes.Black Sea	15	21.90	108.86	40 (37 %)	47 (43 %)	22 (20 %)
Aegean	5	4.76	20.33	0 (0 %)	14 (67 %)	7 (33 %)
B. Menderes	2	3.38	16.06	16 (100 %)	0 (0 %)	0 (0 %)

According to tables presented by EIE [7], there are 166 projects being studied with capacity of 9 742 MW and annual energy potential of 33 793 gWh. 73 projects with 8 644 MW and 29 375 gWh/year, are planned with various kinds of dams and remaining 93 projects with 1 098 MW and 4 418 GWh/year are planned as "run of river", without storage and dams. Within the run of water projects, 47 projects (50.5 %) with 149.71 MW (13.6 %) capacity and 768.41 gWh/year (17.3 %) will be made in EBSB. All of these data emphasize the importance of EBSB on the total SHP potential in Turkey.

In addition to the above considerations, it should be emphasized that, there are more SHP potential to be studied in Turkey, and especially in EBSB. In the following, brief results of two case studies, carried out in the EBSB, are summarized. None of these projects are taken into consideration in the calculation Turkey's hydro power potential and both of them are very economical projects.

One of the projects was carried out in Rize – Ikizdere - Ruzgarli Basin. A feasibility report was prepared [8] for a hydro power plant in which water is being diverted by a weir and no storage dam is necessary. Design flow discharge of the stream is 2.74 m^3/s and net head is 184.5 m, resulting in 4 MW hydropower and 8.8 gWh/year electric energy. This plant will be operated by private sector by build-own-operate (BOO) system.

The other project was carried out in Rize Municipality Domestic Water Facilities. A preliminary report was prepared [9] for a hydro power plant. This is a complicated project, which comprises connecting two streams, generating hydro power energy, water flowing in a pipe, at the downstream of the pipeline again generating hydro power and letting the domestic water to flow where it will be used. It was predicted that totally 20 gWh/year electric energy would be obtained, without creating any negative effects on the quality and quantity of the drinking water. Since most of the facilities are already built, the cost will be less than new-built plants.

Similar to the above two projects, various local projects may be easily and economically developed to improve Turkey's electric energy potential and therefore the economy. Many of the SHP projects are under investigation and the preliminary reports of some of them are either under preparation or ready. By enlarging SHP potential, the economic status of the rural people, most of whom are unemployed and poor, will be improved by constructing various kinds of structures (weirs, canals, etc) and thus by diminishing unemployment and by providing cheaper electricity for domestic usage (lighting, heating, using electronic apparatus etc.). SHP structures will also diminish deforestation, because nearly all of the rural people have used wood in heating.

6 Cost of small hydropower

Small hydropower, especially the very small and the low-head plants, can normally only compete where allowance are made for the external costs associated with fossil fuels and nuclear power. Small hydropower projects are generally considered to be more environmentally favourable and sustainable than both large hydro and fossil fuel powered plants. So, small-scale hydropower is more economically competitive than small-scale fossil fuel/steam-electrical power.

The capital required for small hydropower plants depends on the effective head, flow rate, geological and geographical features, continuity of water flow, equipment such as turbines and civil engineering works. Making use of existing weirs, dams, storage reservoirs and ponds can significantly reduce both environmental impact and costs. Sites with low heads and high flows require a greater capital outlay, as larger civil engineering works and turbine machinery will be needed to handle the larger flow of water.

Apart from the investment and production costs, the other principal cost element is operation and maintenance, including repairs and insurance, which can account from 1.5-5% of investment costs. Both the production and investment costs differ considerably depending on the plant's head height.

7 Recommendations for sustainable development of SHP

Looking to the future, there are good reasons to support small hydropower in Europe and worldwide. First and foremost, it is a source of renewable energy, which, if used on a small scale and handled sensitively, has few environmental risks. Increased use will help to reduce CO_2 emissions and help countries to achieve their Kyoto obligations as well as to stave off global warming. Secondly, the depletion of oil and natural gas deposits will lead to higher generation costs for thermal plants, helping to improve the economics of SHP. It will also serve to enhance economic development and living standards, especially in remote areas with limited or no electricity. Rural communities have been able to attract new industries (mostly related to agriculture) owing to their ability to draw power from SHP stations. In countries such as China, India, Turkey and South Africa, rapid SHP development has also boosted the development of local manufacturers to support these hydropower plants. In addition, the ability of SHP to be combined with water infrastructure projects will allow it to become a regular feature in developing countries as they overhaul their irrigation, water supply and sewerage systems [10].

Some recommendations for a sustainable development of hydropower include:

- Governments should establish an equitable, credible and effective environmental assessment process that takes into account both environmental and social concerns, with a predictable and reasonable schedule.
- Developing countries should develop energy policies that clearly set out objectives regarding the development of power generation options, including small hydropower.

- Project designers should apply environmental and social criteria when comparing project alternatives, in order to eliminate unacceptable schemes early in the planning process.
- Project design and operation should be optimised by ensuring the proper management of environmental and social issues through the project cycle.
- Local communities should benefit from a small hydropower project, both in the short and the long-term.

8 Concluding remarks

In this paper, Turkey's water resources, hydroelectric and especially small hydroelectric potential are researched, focusing on the Eastern Black Sea Region. The main conclusions obtained from this study are summarized as follows:

From the viewpoint of energy sources such as petroleum and natural gas reserves, Turkey is not a rich country, but has an abundant hydropower potential to be used for generation of electricity. Turkey must base its energy strategy on developing the whole hydroelectric potential as soon as possible. Turkey's main indigenous energy resources are hydro and almost all oil natural gas and high quality coal are imported.

Therefore, in order to avoid foreign dependency both in sources and funds, Turkey must discover new and renewable energy resources. Turkey's annual total gross, technically feasible and economically feasible hydropower potentials calculated by DSI are 435, 215 and 128 tWh, respectively. However, according to findings of another study, in which a new criterion is developed related to key concept of the economic feasibility, economically feasible hydropower potential goes up 188 tWh/year, with an increase ratio of 47 % compared to DSI value.

Small hydropower represents an alternative to fossil fuel generation, and doesn't contribute to either greenhouse gas emissions or other atmospheric pollutants. However, developing the remaining hydropower potential offers many challenges and pressures from some environmental action groups over its impact has tended to increase over time. Moreover, in the context of the restructuring of the electricity sector, markets may favour more polluting and less costly options. On the other hand, small hydropower's main challenges relate to both economics and ecology. Small hydropower can be successfully developed as long as it produces electricity at competitive prices and under conditions that respect the environment.

Small hydropower can provide a positive social and economic contribution to the community through employment creation and good quality of life by contributing to an assured supply of electricity.

Small hydro utilizes a source of energy (water) which is both renewable and abundant throughout the year in many parts of Turkey.

References

1. DSI, State Hydraulic Works, *Statistics on hydropower*, Ankara, Turkey (2004).
2. Yuksek, O., Komurcu, MI., Yuksel, I and Kaygusuz, K., Jour. of En. Pol., Vol. **34**, Number 17, pp. 3093- 3103, (2006)
3. Bakir, N.N., *Turkey's hydropower potential and review of electricity generation policies from EU perspective*, http://www.ere.com.tr, (2005)
4. RTLED., *Report of Turkey Long Term Electric Energy Demand*, Technical Report, Min. of En. and Natu. Res., Ankara, Turkey, 1-81 (in Turkish), (2004)
5. Paish, O., Renew. and Susta. En. Rev., Vol: **6(6)**, pp. 537-556, (2002)

6. Punys, P., *Small hydro power in the New EU Member States, Hidroenergia 04, Sweden. RTLED (Report of Turkey Long Term Electric Energy Demand)*, Minis. of En. and Nat. Res., Ankara, Turkey, (in Turkish) (2004)
7. EIE, Electrical Power Resources Survey and Development Administration, *Hydroelectric Power Activities of the EIE,* Available from http://www.eie.gov.tr (2004)
8. Onsoy, H., Yuksek, O., Yuksel, I., *Feasibility Report of Rize – Ikizdere – Ruzgarlı HEPP, Feasibility Report,* KTU Civil Eng. Dept., Trabzon, Turkey, 1-131 (in Turkish), (2004)
9. Onsoy, H., Yuksek, O., Yuksel, I., *Preliminary Report of Rize Municipality Domestic Water Facilities, Preliminary Report,* KTU Civil Eng. Dept., Trabzon, Turkey, 1-4 (in Turkish), (2004)
10. Yuksel, I., Jour. of En. Expl. & Explo., Vol.**24**, Number 4-5, pp. 361-370, (2006)

Cost benefit analysis for failure of sewer pipelines

Mohamed Elmasry[1,*], *Alaa* Hawari[1], and *Tarek* Zayed[2]

[1] Dept. of Civil and Architectural Engineering, Qatar University, P.O. Box 2713, Doha, Qatar
[2] Dept. of Building, Civil and Environmental Engineering, Concordia University, Quebec, Canada

Abstract. Sewer pipelines failure in sewage networks can have adverse potential impacts on socio-economic aspects in any community. Due to the fact that it's difficult to capture the relationship between the physical and economical aspects as a result of critical sewer pipelines failure, economic concepts are used to evaluate the economic loss as a result of these failures. In this paper an analysis for the costs resulting from sewer pipelines failure and the benefits achieved from avoiding failures are presented. The costs included in the cost benefit analysis are the direct costs used to reinstate failed pipelines and the indirect costs, borne by the society and economy. In the benefits analysis, only the tangible and measurable benefits limited to the health sector and preventing diseases are addressed in this paper. It is expected that the proposed approach could help in estimating the economic losses due to sewer pipelines failure especially for the intangible factors that are difficult to measure. In addition it could help decision makers in taking necessary measures to preserve critical assets that could have adverse potential impacts on valuable natural resources such as surface and groundwater and soil surrounding failed pipelines.

1 Introduction

The research addressing buried infrastructure in general and sewer pipelines in particular raises the concerns about lack of condition assessment tools, cheap non-destructive inspection technologies, and data for use in risk assessment models[1]. These challenges are considered as impediments to better asset management for which interest has been increasing lately. Consequences of failure of sewer pipelines which is part of risk assessment models could have adverse potential impacts on business and environment. Due to the uncertainty accompanied to estimating the costs of these failures, little research have been conducted to valuate them. Since the end of the 1990s, numerous studies have been carried out with the aim of defining the costs of failure of infrastructures that are not considered direct costs. The studies first agreed to describe costs incurred by owners of underground infrastructure to reinstate the failed assets [2]. Then, started to identify costs that are invoiced as social or indirect costs; borne by the whole society [3-6].

*Corresponding author: m_elmasry@aucegypt.edu

To determine the consequences of failure of sewer pipelines, authorities in Seattle in USA developed a risk assessment tool that determines the consequences of failure for sewer pipelines in monetary values [7]. In another study and as a part of Computer Aided Rehabilitation of Sewer and Storm Water Networks (CARE-S) project in Europe, a study was carried out to investigate the social and economic costs of sewer failures [8]. As part of the grand central model developed to determine the societal costs for pipeline failure, traffic valuation as a result of pipeline failures has been addressed by Cromwell and Pearson [9]. Although the grand central model is considered as one of the best efforts to date to estimate the societal costs for pipeline failure, it requires good knowledge of traffic flow and is dependent on many parameters that are hard to identify such as the remaining distance, the type of routes taken, the weather and most of all; the behaviour of commuters. To overcome the difficulties that might accompany estimating the consequence of failure in monetary amounts, another approach was used; in which determining the consequence of failure values was considered as a multi-criteria decision making problem, and different pipes were considered as the alternatives with different factors' weights having different importance levels as the criteria [10,11].

Cost Benefit Analysis (CBA) is a quantitative approach used by economists where monetary and non-monetary aspects can be incorporated in the decision making process. CBA estimates the changes in utility resulting from changes in conditions that customers would face as a result of changes in environmental quality, and/or availability of products and labour. The analysis conducted in CBA is clear in its requirement where different parameters are either categorized in costs or benefits with the goal of maximum utility - based on the nature of the problem- is considered the goal. In addition, CBA should show the costs and benefits accruing to different social groups of beneficiaries and non-beneficiaries. Also, CBA has the ability to express costs and benefits either in monetary units or weighted by an index related to the importance of beneficiary or non-beneficiary groups. Several cost benefit analysis approaches have been addressed in literature to valuate decision in terms of benefits gained in the field of infrastructures and funds allocation [12-14]. Approaches such as: benefit cost ratio [15, 16] and cost utility [17, 18] differing based on the objectives for evaluation of decision alternatives. The objective of this paper is to analyze costs - including direct and indirect ones - resulting from failures of sewer pipelines and the benefits returning on consumers and community as a result of avoiding such failures in an attempt to determine the economic loss as a result of these failures.

2 Economic loss of infrastructures failure

Critical infrastructures also known as critical assets are usually identified using a risk based assessment methodology as stated by the American Electric Reliability Corporation [19]. Risk analysis methodologies help in assessing critical infrastructures from which proper assessment management plan can be carried out for these infrastructures. This is usually carried out by performing a what-if scenario for the outcome of the event of failure of infrastructure and assessing the outcome in a qualitative or quantitative manner [20]. Economic loss is defined as the difference between the economic output in a baseline scenario representing the business as usual and the output in a disaster scenario which represents post infrastructure failure. The economic output in the disaster scenario models the consequence of failure of an asset. Estimating the economic loss requires integrating lost output overtime. Therefore the process of estimating the impacts resulting from a disaster starts with analyzing the various hazards, the exposure and vulnerability of physical, social and environmental systems.

3 Cost Benefit Analysis (CBA) of sewer pipelines failure

3.1 Cost of failure of sewer pipelines

To determine the consequences of sewer pipeline failures, direct and indirect costs as a result of these failures are estimated. The consequences of failures are then determined based on the total costs, where a pipe has a higher consequence if the total costs of failure are higher than when compared to another pipe with less cost. Total costs of failure can be determined by adding direct costs and societal costs (indirect costs).

3.1.1 Direct Costs

The direct costs can be categorized as all costs related to the reconstruction of underground infrastructure. These costs are easily spotted and identified because they correspond mostly costs that are tangible which include costs of materials, resources (equipment and labour) involved in restoration works in addition to management and/or administrative costs [21].

3.1.2 Indirect Costs

It is difficult to give an exact definition regarding the indirect costs associated with the failure of an asset in general and sewer pipelines in particular. However, the indirect costs can be arbitrarily defined as costs associated with loss of productivity of other functioning parties and lost wages as a result of failure. The following section describes the cost estimation for different categories of sewer pipelines failure considered in the proposed model.

Costs related to service interruption

Service outage: Infrastructure failure could have a cascading effect, which means that if a stretch in a network fails, this failure might affect an adjacent stretch in the same network or other utilities. The disruption in these services conveyed by the affected utilities represents a cost to society. The cost can be assigned to the service interruption as per Eq. (1).

$$\mathbf{C_s} = D * \sum_{i=1}^{N} C_h * n \tag{1}$$

Where: C_s: Cost of an interruption of service, D: Duration of the interruption (h), C_h: Average hourly Cost of the service ($/h), n: the number of affected customers of Type (i) and N: the number and total number of affected customers.

Emergency interventions: In some cases lives of citizens might be jeopardized as a result of sewer pipelines failure. Thus, emergency interventions such as public services including paramedics or hospital services, police forces and civil defence department might be needed to help threatened lives to ensure safety of publics. The costs related to the use of these resources shall be taken into account as indirect costs even though it's not necessarily present in all cases as per Eq. (2).

$$\mathbf{C_{IS}} = \sum_{i=1}^{N} c_v^i * t_R^i \tag{2}$$

Where: C_{IS}: Total cost of intervention of public security services , c_v^i: Hourly rate of vehicle of type (i) ($/h), t_R^i: Repsonse time of vehicle of type (i) (h) and N is the number of emergency vehicles deployed in the accident.

Costs related to traffic disruption and road closure

Road closure and traffic disruption: Sewer pipelines failure may cause partial or total closure of roadways in the affected zones and might further extend outside the premises of the failure, resulting in numerous costs [22].

$$\mathbf{C_{Con}} = d * f_c * \sum_{i=1}^{N}(Con_{dis}^{i} - Con_{nor}^{i}) * \sum_{j=1}^{M} n_{V(i)}^{j} \quad (3)$$

Where: C_{Con}: Cost of overconsumption due to congestion,d: Disruption distance (km), f_c: Fuel price ($ / L),Con_{dis}^{i}: Average consumption of vehicles of type (i) during disruption (L/km),Con_{nor}^{i}: Average consumption of vehicles of type (i) during normal cases (L/km),$n_{V(i)}^{j}$: Number of vehicles of type (i) impacted per day (j) (Vehicles/day) and N and M: Total number of vehicles and total number of days, respectively.

Loss of parking spaces: During the restoration of failed sewer pipelines located in the premises of roads in urban areas, it may be important to take into account the aspect of reducing the space on ground which would also affect areas dedicated to car parking [23]. This cost can be calculated as per the model of Pucker et al. [24] as per Eq.(4).

$$\mathbf{C_{lp}} = N_{na} * c_p * R_o * d * t \quad (4)$$

Where: C_{lp}: Cost of loss of parking spaces, N_{na}: Number of non-accessible parking spaces, c_p: Hourly cost of parking ($ / h), R_o: Rate of occupancy (%), d: Length of the reconstruction period (day) and t : Number of operating hours per day (h / day)

Loss of time as a result of increased traveling distances for vehicles: The most crucial point related to the disruption of traffic in studying failure of sewer pipelines, is the traffic congestion. The loss of time can be calculated using extra time needed to travel the same distance which is the most close to the real time situation allowing quantification of costs of congestion correctly and easily [6].

$$\mathbf{C_{TD}} = \sum_{i=1}^{N}(n_v^i * o.r_v^i * r_p^i) * t * d \quad (5)$$

Where: C_{TD}: Cost of time loss due to traffic diversion, n_v^i: Number of vehicles of type (i), $o.r_v^i$: Occupancy ratio of vehicle of type (i), r_p^i: Hourly rate of passenger in vehicle of type (i), t: Detour time (h), d: Number of days (days) and N is the total number of vehicles of type (i)

Loss of time as a result of increased traveling distances for pedestrian: Zhang et al, [25] developed a model to estimate the loss of time in case of pedestrians using the same previous approach as per Eq.(6).

$$\mathbf{C_{PD}} = r_p * t * n * d \quad (6)$$

Where: C_{PD}: Cost of lost time due to the detour for pedestrians, r_p: Average hourly rate of a pedestrian ($ / h), t: Detour time (h), n: Number of pedestrians per day and d: Number of days (Days)

Increased running costs for vehicles: Traffic congestion causes an increase in vehicle operating costs due to utilizing detours which would consequently cause an increase in vehicles' running costs [6]. Thus, it is possible depending on the speed and type of vehicles to establish an average maintenance cost per kilometre that is considered as a cost of the road network disruption [22].

$$\mathbf{C_{dis}} = \sum_{i=1}^{N} c_v^i * \sum_{j=1}^{M} n_v^j * d \quad (7)$$

Where: C_{dis}: Additional cost due to additional distance, d: Additional distance, c_v^i: Running cost per Kilometer for vehicle of type (i) ($/km), n_v^j: Number of vehicles of type (i) impacted per day (j) (Vehicles/day) and N and M, are the total number of vehicles of different types and the total number of days spent in restoration works respectively.

Costs related to environmental impacts

The environmental impact of a failed sewer pipelines could be during the event of failure and during reinstatement of the failed asset. Although the disturbances might not have visible impacts, many of these disturbances might be costly and irreversible. The flooding of failed sewer pipes affects the surface and ground water quality and soil in which the waste effluent is discharged. Wastewater discharged due to flooding or leakage in sewage pipelines to water bodies and soil surrounding these pipes could have adverse health issues and complications. Medical costs could be generated despite all precautionary measures that could be made, because of the diseases or infections that increase the health care costs [4]. To calculate the volume of contaminated groundwater and/or soil, the advancement of plume is calculated using dispersion equations. The rate of advancement of contaminant plume can be retarded if there is a reaction between its components and groundwater constituents as per Eq. (8) [26].

$$\mathbf{R} = 1 + \frac{\gamma_b}{\varepsilon} * k_d \quad (8)$$

Where: $k_d = \beta f k_{ow}$, f = 0.01 in sandy soil to 0.10 in muck, β= 0.41 and k_{ow}: is a constant.

The effluent that leaks from broken sewer pipelines are usually untreated raw sewage that seeps to groundwater and soil surrounding the location of failure. This can result in introduction of chemicals such as nitrates phosphates, and microorganisms [27]. To determine the concentration of chemicals intrusion from which volume of contaminated soil and/or groundwater can be estimated. Eq.(9) can be used to calculate the maximum concentration just downstream the contaminant source using parameters related to the medium in which it is desposed [26].

$$\mathbf{C_{max}} = \frac{M}{8 * porosity * \left(\frac{\pi t}{R}\right)^{3/2} * \sqrt{D_x D_y D_z}} \quad (9)$$

Where, D_x = Bulk velocity * diameter of soil particles and $D_y = D_z = 0.1 * D_x$ and the spread of contaminant can be determined using Eq. (10) and (11) [26].

$$\mathbf{4\sigma_x} = \frac{4 * \sqrt{2 * D_x t / R}}{3} \quad (10)$$

$$\mathbf{4\sigma_y} = \frac{4 * \sqrt{2 * D_y t / R}}{3} \quad (11)$$

Costs related to economic impacts: General activity of businesses and/or industries is often impacted by asset failures and disturbances which can then be reflected on the economy in terms of reduced productivity, and delays to work.

Reduced productivity: When shops or businesses are affected by failure of an asset, employees' productivity is affected consequently. Gilchrist and Allouche, [4] developed a model to estimate the cost as a result of decreased productivity as a result of noise in a work place as shown in Eq. (12).

$$\mathbf{C_{nw}} = T * \sum_{i=1}^{N} (F * r_h * n) \quad (12)$$

Where: C_{nw}: Cost of noise pollution in the workplace, T: duration of the project (hours), F: Reduction factor for worker (i), r_h: Average hourly rate of worker (i) ($ / h) , n: number of workers affected and N: number of types of workers affected.

Delays and absences from work: Lost productivity can be related to health disorders, and psychological disorders that can be linked to the failure and restoration works of an asset. Eq.(13), estimates the cost of absences from work as a result of restoration works.

$$\mathbf{C_a} = \sum_{i=1}^{N} (r_h^i * n_i * t) \quad (13)$$

Where: C_a: Cost of absence from work, r_h^i:Hourly rate of employee of type (i) ($/hour), n_i: Number of employees of type (i), t: Time away from work by employee of type (i) (hours) and N: is the number of total employees of different types

3.2 Benefits from avoiding failure of sewer pipelines

The impacts as a result of sewer pipelines failure can be described as increased disease burden due to contaminated surface and ground water bodies that could lead to an increase in illness rates in areas that suffer from waste water contamination. The increased disease burden can result in medical expenditures spent on treatment, indirect costs resulting from illness, which includes the cost of time lost from work, decreased human productivity, and premature deaths and disabilities. As such, the costs paid by patients seeking health care and increased running costs could increase the costs paid by the society for that purpose. As per a study done by World Human Organization [28], the benefits from proper sanitation would be avoiding diarrheal diseases that would result in direct economic benefits. Table 1 shows the different variables included when studying how would the health sector and patients benefit from avoiding the failure of assets and consequent contamination of wastewater.

4 Cost benefit analysis

An economic cost benefit analysis is crucially important for effective resource allocation to evaluate the economic costs of interventions and the resulting benefits especially when there are many criteria to determine where these resources need to be allocated such as social and environmental considerations. Although it may appear that estimating the benefits resulting on consumers from the improved health or wastewater network is impossible, there are factors that can be measured which would result in the improvement of these services. Cost benefit analysis aims to better understand the social and economic wellbeing in communities after restoration or upgrading failed pipelines. The procedure for a CBA usually comprise determining costs and benefits, quantifying the non-market impact, including indirect costs and calculation of economic performance indicators (i.e. Economic Net Present Value (ENPV), Internal Rate of Return (IRR),...etc.) which is usually done by analyzing macro-economic and social conditions in a community. Eq. (14) shows the $({}^{B}/_{C})$ formula used in the cost benefit analysis adopted in this research.

$$ {}^{B}/_{C} = \frac{\sum_{t=1}^{T} \frac{B_t}{(1+r)^t}}{\sum_{t=1}^{T} \frac{C_t}{(1+r)^t}} \tag{14} $$

Where T: is the total number of years of the study period, B_t: is the benefits per year (Values of harm avoided each year) and C_t: is the total costs paid

Table 1. Variables included in estimating benefits from avoiding sewer pipelines failure.

Beneficiary		Variable	Value
Health Sector	Expenditures saved, due to less illness	Unit cost per treatment	Cost per visit (US$7) [28]
			Cost per day (US$28) [28]
		Number of cases	Variable by region
		Visits or days per case	1 Outpatient visit per case [28]
			5 Days for hospitalized cases [28]
		Hospitalization rate	92% of cases ambulatory [28]
			8% of cases hospitalized [28]
Patients	Expenditures saved due to less illness	Transport cost per visit	Variable by region
		Number of cases	Variable by region
		Expenditures saved, due to less illness	
	Money saved by avoiding days lost from work	Days off work	2 days (1-4) [28]
		Number of people of working age	Variable by region
		Cost of time	Minimum wage rate [28]
	Value of loss of life avoided (life expectancy)		

5 Case study

To determine the applicability of the previously proposed equations to determine direct and indirect costs, an actual case study with real data is used for a sewer pipeline located in Monteal city in Quebec, Canada. Table 2 shows the relevant data that were used in calculating the costs. The case study comprises a 200 mm - vitrified clay pipe with a total length of 60 meters.

Figure 1 shows the different indirect costs, where the groundwater and soil remediation costs represented almost 60% of the total indirect costs. Figure 2 shows the ratio between direct and indirect costs as a result of the sewer pipeline failure, it can be seen that the values of indirect costs are almost equal the direct costs indicating that they can't be neglected and that they are as important as direct costs.

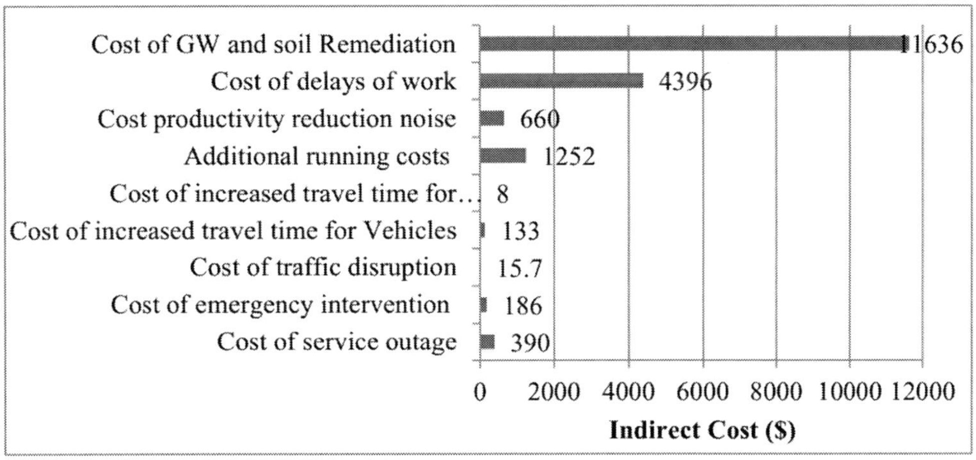

Fig. 1. Indirect Costs for Applied Case Study.

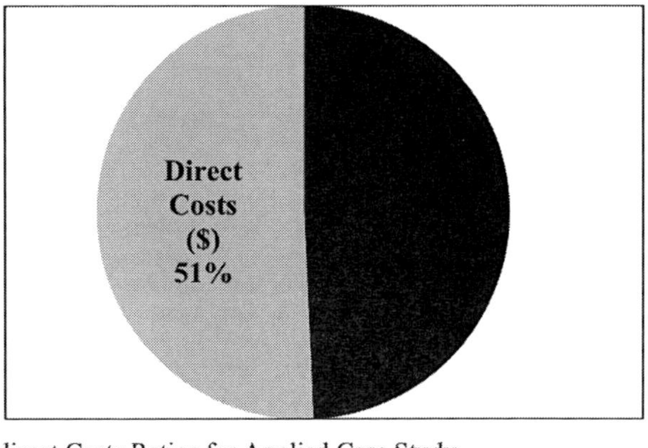

Fig. 2. Direct and Indirect Costs Ratios for Applied Case Study.

To determine the benefits as a result of proper sanitary service, Table 3 shows the different benefits and cost saved as a result of better sanitary service as a result of avoiding failure of sewer pipelines. It should be noted that these costs could be considered as Quality Adjusted Life Years (QALYs) which represents the marginal utility for customers in case of better quality sanitation.

By applying Eq. (15) with study period of one year, the benefit to cost ratio resulted in a value of $\frac{26340}{37882} = 0.7$, which indicates that the costs borne by the society as a result of failure would be more than the benefits. In other words the 0.7 ratio indicates that the failure of sewer pipeline is catastrophic and more significant than other pipes that would have a ratio greater than 1.

Table 2. Parameter values used in application of case study.

Parameter	Value		
Duration of outage (hours)	72		
Cost of service ($/hr)	0.028		
Number of customers affected	195		
Cost of emergency vehicle	62		
Number of emergency vehicles	3		
Disrupted distance (km)	1.77		
Fuel price ($/Litre)*	1.07	1.01	1.01
Average consumption of vehicle in disruption (L/Km)*[29]	0.15	0.42	0.50
Average consumption of vehicle in normal cases (L/Km)*[29]	0.1	0.28	0.33
Number of vehicles impacted (Vehicles/day)*[29]	567	13	9
Number of inaccessible parking spaces	0		
Hourly cost of parking ($/hr)	4		
Occupancy ratio (%)[29]	1.63		
Number of operation hours for parking lot (hr/day)	24		
Hourly rate of passengers ($/hr)[29]	11		
Detour time (h)	0.0042		
Number of pedestrians affected [29]	57		
Running cost per kilometer for vehicle ($/km)*	0.7	0.8	1.1
Number of workers affected in work place	17		
hourly rate for workers in work place ($/hr)	11		
F reduction factor	0.05		

*Values represent different types of vehicles namely light, medium and heavy

Table 3. Parameter values used in calculating benefits of avoiding sewer pipeline failures.

Parameter	Value
Number of customers affected	195
Number of probable infections (14%)[28]	27
92% for 1 day (7$/day)	175
8% for 5 days (28$/day)	280
Costs saved by individuals ($)	860
Costs saved by avoiding work absence ($)	385
Cost of lost opportunity (1 Year)	24640
Quality Adjusted Life Years (QALYs) (Benefits)	26340

6 Conclusion

A CBA for the costs as a result of sewer pipelines failure and health benefits that return on the community as a result of avoiding that failure was presented in this paper. CBA was used with the intent of determining the economic loss of sewer pipelines failure. Direct costs spent to reinstate the failed pipelines were defined as monetary amounts that can be distinctively shown in the bills of municipalities. While, indirect costs were defined as any costs borne by the society and/or environment as a result of that failure. These costs included: costs as a result of road closures and traffic diversion, additional consumption of fuel, additional running costs, costs for remediation of contaminated groundwater and soil, delays to work, loss of parking spaces and others. Avoiding the failure of sewer pipelines was considered as an enhancement for the quality of living as a result of better sanitation and avoiding costs spent in case of illnesses. A case study with real data was used to examine the applicability of the proposed method for a sewer pipeline in Montreal city. By calculating the different costs, it was found that the indirect costs represented a significant portion of the cost of failure which can't be neglected. It was concluded that as the benefit to cost ratio decreases the significance of failure increases indicating catastrophic levels of failure. There are several limitations and areas of enhancement for this research, such as adding other factors to be included in calculating the costs of failure. Also, other concepts could be included in the CBA such as the willingness to pay and acceptance. In addition, several enhancements in calculating the disruption distance and contaminated groundwater and soil volumes are also recommended.

This publication was made possible by NPRP grant # (NPRP6-357-2-150) from the Qatar National Research Fund (a member of The Qatar Foundation). The statements made herein are solely the responsibility of the authors. Also the authors would like to thank the public works authority of Qatar (ASHGAL) for their support in the data collection.

References

1. Gaewski, P., Tata and Howard, Inc., and Blaha,F., AwwaRF, *Analysis of Total Cost of Large Diameter Pipe Failures (*2007).
2. Jones, C.. Paper presented at the Life Extension of Technologies Workshop, New York, *Incentives and Barriers to the use of Trenchless Technology (*1999)
3. Allouche, E., Ariaratnam, S., and AbouRizk, S. *Construction Congress VI: ASCE, VA.* 543-553, Multi-Dimensional Utility Model for Selection of a Trenchless Construction Method (2000).
4. Gilchrist, A., and Allouche, E. N.. Tunnelling and underground space technology, **201,** 89-104, *Quantification of social costs associated with construction projects: state-of-the-art review (*2005).
5. Manuilova, A., Dormuth, D., and Vanier, D. *National Research Center, Canada,* MIIP Report: A Case Study of Use and External Components of Social Costs that are related to Municipal Infrastructure Rehabilitation, (2009).
6. Rahman, S., Vanier, D.J., and Newton, L. A. *National Research Center, Canada,* MIIP Report: Social Cost Considerations for Municipal Infrastructure Management. (2005)
7. Martin, T., Johnson, D., and Anschell, S. Proceeding, Leading Edge Conference on Strategic Asset Management, IWA, Lisbon, Portugal, *Using historical repair data to create customized predictive failure curves for sewer pipe risk modeling.*(2007).

8. Torterotot, J. P., Sousa e Silva, D., Barbier, R., Werey, C., Pereira A., Konig, A., Montginoul, M., and Waechter V. *IWA Publishing,* CARE-S Computer Aided Rehabilitation of Sewer and Storm Water Networks.73-88, (2006).
9. Cromwell, J. E., and Pearson, N. *AWWARF, Denver,* Cost of infrastructure failure. (2002).
10. Salman, B. and Salem, O. *Journal of pipeline systems engineering and practice, Risk* Assessment of Wastewater Collection Lines Using Failure Models and Criticality Ratings., **3**, 68-76. (2012)
11. Ana, E., V., Ph.D. dissertation, University of Brussels, Belgium. Sewer asset management - sewer structural deterioration modelling and multi-criteria decision making in sewer rehabilitation projects prioritization (2009).
12. Higgins, A.M. and Harris, A.H. *Critical Care Clinics,* Health economic methods: cost-minimization, cost-effectiveness, cost-utility, and cost-benefit evaluations, **281**, 11-24.(2012)
13. Polinder, S., Toet, H., Panneman, M. and Beeck, E. *World Health Organization Regional Office for Europe, Denmark,* Methodological Approaches for Cost-effectiveness and Cost-utility Analysis of Injury Prevention Measures, (2011).
14. Moayyedi, P. and Mason, J. ,*European Journal of Gastroenterology and Hepatology,* Cost-utility and cost-benefit analyses: how did we get here and where are we going?, 16-**6**, 527-534 (2004).
15. Adey, B.T. and Hajdin, R. *Journal of Structure and Infrastructure Engineering.* Methodology for determination of financial needs of gradually deteriorating bridges with only structure level data, 7 -**8**, 645-660 (2011).
16. Vacheyroux, G. and Corotis, R., *Journal of Structure and Infrastructure Engineering* Strategies of investment in the management of urban bridges: a life-cycle approach illustrated for Paris, , 9-**11**, 1080-1093 (2013).
17. Marinoni, O., Adkins, P. and Hajkowicz, S. *Journal of Environmental Modelling and Software*, Water planning in a changing climate: join application of cost utility analysis and modern portfolio theory, 26 -**1**, 18-29 (2011).
18. Hajkowicz, S., Spencer, R., Higgins, A. and Marinoni, O., *Journal of Environmental Management, Evaluating* water quality investments using cost utility analysis, , 884, 1601-1610, (2008)
19. North American Electric Reliability Corporation, Standard CIP-002-1, Cyber Security, Critical Asset Identification. Washington, DC (2006).
20. European Commission, Proposal for a Directive of the Council on the Identification and Designation of European Critical Infrastructure and the Assessment of the Need to Improve Their Protection, COM2006787 Final, Brussels, Belgium (2006).
21. Khogali, W. E., and Mohamed, E. H. Managing utility cuts: Issues and considerations. Paper presented at the NCRR/CPWA Seminar Series: Innovations in Urban Infrastructure. APWA International Public Works Congress NRCC/CPWA Seminar Series Innovations in Urban Infrastructure, 1-11, (1999).
22. Gourvil, L. and Joubert, F. Évaluation de la congestion routièredans la région de Montréal: Québec: Transports Québec, (2004).
23. Boyce, G., and Bried, E., *Proceedings of North American No-Dig,* Benefit–cost analysis of micro-tunneling in an urban area. **94**,36-01 (1994)
24. Pucker, J., Allouche, E., and Sterling, R. *Proceeding of NASTT No-Dig Conference,* Social Costs Associated with Trenchless Projects: Case Histories in North American and Europe. C4-04 (2006).
25. Zhang, A., Boardman, A. E., Gillen, D., and Waters, I., *Report for Transport Canada* Towards estimating the social and environmental costs of transportation in Canada. (2004)

26. Gulliver, J., *Springer-Verlag New York, 10.1007/978-1-4614-5731-2*, Transport and Fate of Chemicals in the Environment: Selected Entries from the Encyclopedia of Sustainability Science and Technology (2012).
27. Rail, D., *Technomic Publishing Company, Inc., Lancaster, PA.* Groundwater Contamination: Sources, Control, and Preventative Measures. 37-40, (1989)
28. World Bank, *Urban Development Sector Unit, Vietnam Country Department of the World Bank* , Project appraisal document on a proposed loan to the Socialist Republic of Vietnam for the Ho Chi Minh City environmental sanitation project, (2001).
29. http://www.statcan.gc.ca/eng/start, accessed 15 November 2016

Retrofitting residential buildings in the University of Sharjah

Emad Mushtaha[1,*], *Sondos* Ayman Jara[1], *Nour* Ali Tahmaz[1]

[1, 2] Architectural Engineering Department, University of Sharjah, UAE

Abstract. The aim of the research is to investigate best retrofitting strategies that minimizing energy consumption within the context of thermal comfort. The research has targeted two residential compounds in the University of Sharjah: Al Khwarizmi and Al Baironi. Climate Consultant and ECOTECT were used in the simulation process in order to select best passive strategies and tools for the retrofitting. Several parameters such as building orientation, shading devices, and glazing types, were tested during the simulation process. As a result, the study determined the most suitable tools for the existing compounds. The study showed that the use of double glazing with shading devices reduced cooling loads in Al Baironi with 5.42 MWh and 8.13 MWh in Al Khwarizmi. Also, addressed that U-shaped residential layout is more efficient than liner-shaped layout.

1 Introduction

1.1 Brief description

Al Khwarizmi and Al Al Baironi are faculty compounds located in the University City of Sharjah (Fig.1.1-2.2). The residence has several types of two, three, and four-bedroom villas where they are designed similar in both compounds, but oriented and formed differently regarding the surrounding opened spaces. The villas in Al Baironi compound are formed linearly unlike Al Khwarizmi compound formed in U blocks. Four- bedroom villas have an approximate area of 216.330 sq.m designed for families with a minimum number of 5 members. The villa consists of a living room connected directly with a dining room with its service area, i.e. (toilet, kitchen, maid room with its own toilet and a small store on the ground floor). The first floor consists of 3 bedrooms with one shared bathroom and one master bedroom besides an office (Fig. 3.1-3.3). As the number of students is in an increase in the University of Sharjah, therefore more faculty members are expected to join the university. Providing more villas and residences to new faculties is the trend in the university. As faculties work for almost 9 hours a day during weekdays, so when they return home they need to rest and feel comfortable. On the other hand, the university pays around AED 70 Million yearly to Sharjah Electricity and Water Authority (SEWA) for its water and electricity consumption. Therefore, providing healthy and efficient buildings can contribute in solving such issue. However, existing buildings, including residences are a

* Corresponding author: emushtaha@sharjah.ac.ae

part of the solution. Retrofitting existing buildings in the context of thermal comfort is urgently needed. Villas need to be carefully analyzed and improved using passive design tools such as shading devices, glazing, and natural ventilation. Building orientation and form should also be considered because it plays a major role in affecting the energy gained and lost by buildings and this has a large influence on the thermal performance of buildings.

Comfort is the result of the interaction of physical exchanges, physiological, psychological, social and cultural rights; it depends on the architecture, the clothing, the eating habits and the climate as defined in the article (K. Fabbri, 2015). Thermal comfort studies started in the twentieth century when Gagge AP started to solve particular problems due to stressful situations in the workplace. In the twentieth century the architectural theories (Movement modernity, , Functionalism, Bauhaus, Le Corbusier with Le Modulor, De Stijl, CIAM International style, etc.) and technical manuals have put man at the center, as an individual with a physical dimension, founding an interest in the design and construction of residential buildings.

1.2 Photos, top views and floors plans

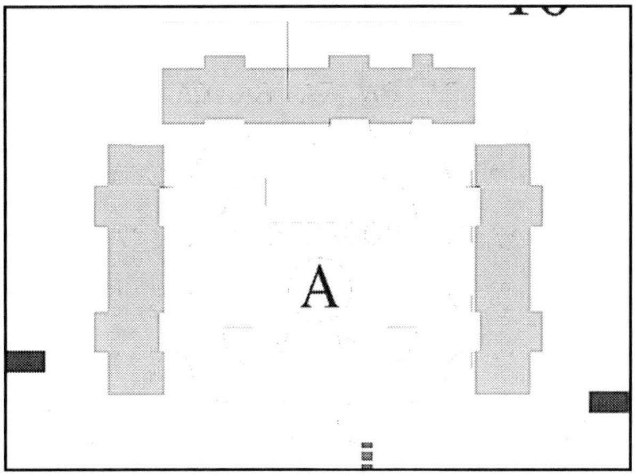

Fig. 1.2.1. Al Khawarizmi Compound – Top View.

Fig. 1.2.2. Al Khwarizmi Compound – Street View.

 DOI: 10.1051/matecconf/201712008007

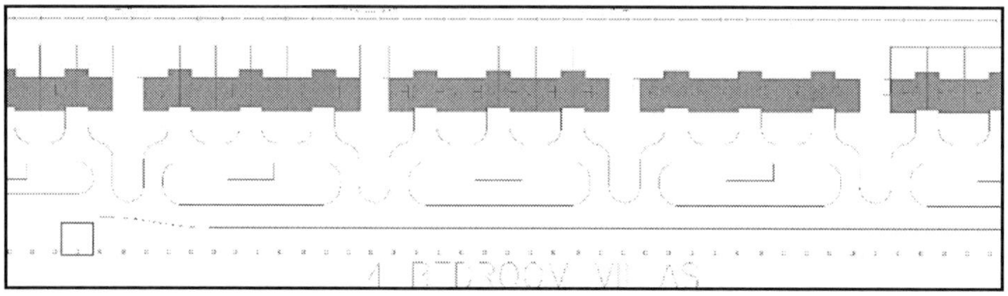

Fig. 1.2.3. Al Bayrouni Compound – Top View.

Fig. 1.2.4. Al Baironi Compound – Street View.

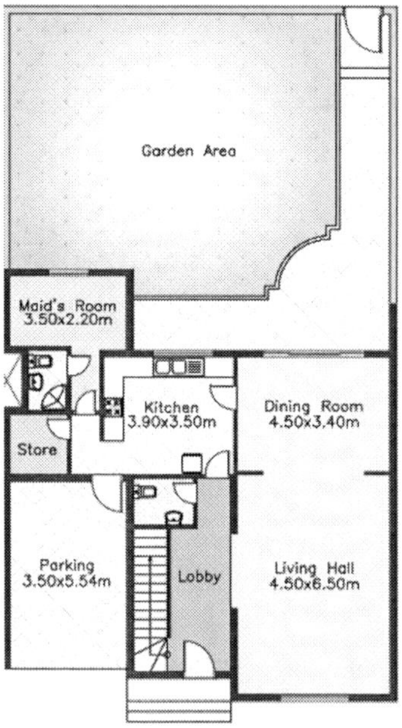

Fig. 1.2.5. Typical Ground Floor Plan.

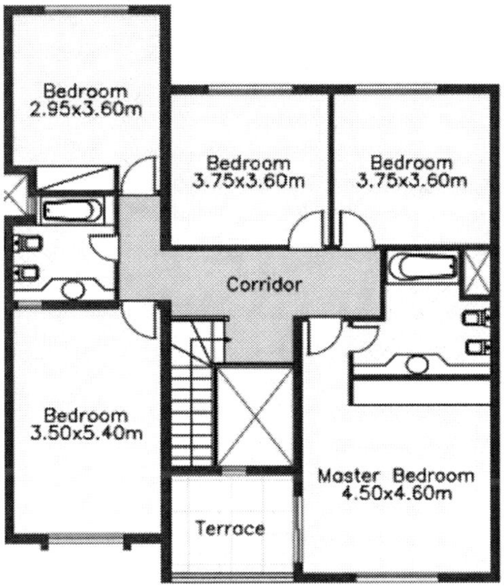

Fig. 1.2.6. Typical First Floor Plan.

Fig. 1.2.7. Typical Front Elevation.

1.3 Photos and top views

UAE is one of the GCC countries in the Middle East that has high-recorded temperature and high humidity ratio, which makes the environment not comfortable. Thus, most of the activities have been done indoor as residents prefer to escape from the annoying weather. Studying the thermal comfort during the year, the authors have gone through certain analysis and researches for each month taking into consideration which months are cold and warm. December, January, February and March are cold months but not discomfort. From April to May there will be a slight stress in temperature if people prolonged their exposure to sun. During the summer starting from late May to September temperature is extremely high with an average of temperature above 41 °C (Fig. 1.3.1).

 DOI: 10.1051/matecconf/201712008007

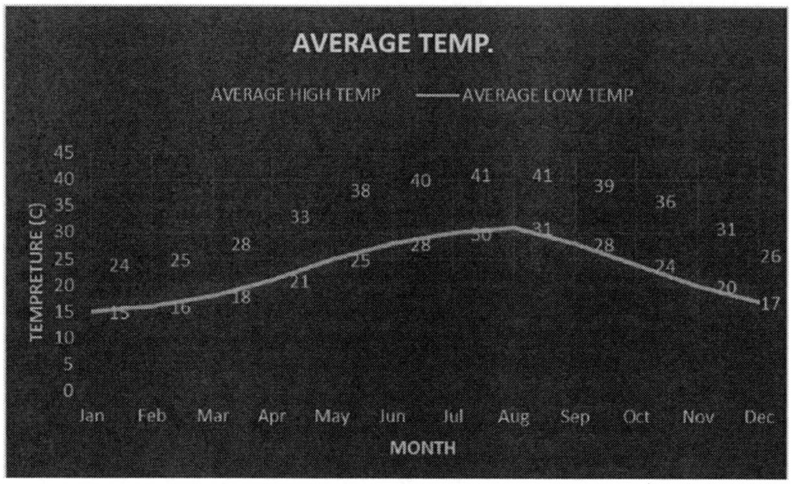

Fig. 1.3.1. Sharjah Average Temperature for Sharjah, UAE.

Figure 1.3.2 shows that 1633 hours are warm and need shading in order to achieve thermal comfort, while 901 hours are located within and slightly below thermal comfort. From the chart, it is obvious that warm hours are almost double than comfort hours in the year.

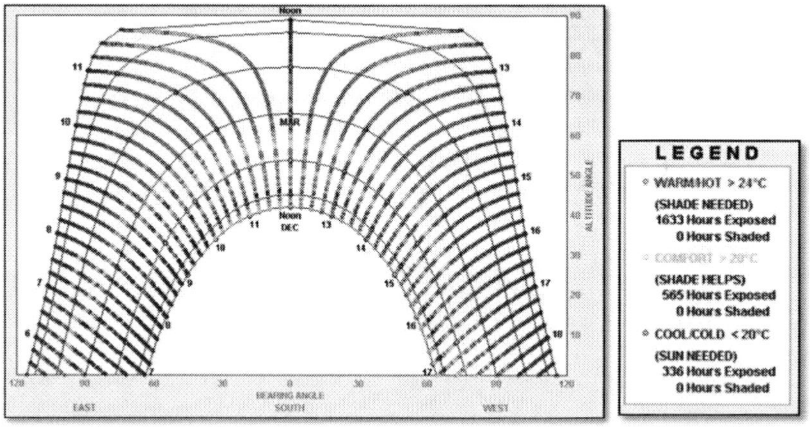

Fig. 1.3.2. Thermal Comfort on South elevation and Sun Shading Chart.

2 Research objectives

This paper aims to evaluate the planning and designing of housing environment in the compound by studying the impact of residences' forms on its thermal performance. Also, it aims at deriving strategies and suggestions that help in reducing energy consumption for four- bedrooms villas.

3 Methodology

The analysis of the two compounds in context of the thermal comfort went through the following:

1. Understanding UAE climate and the strategies required to help achieve thermal comfort by integrating Climate Consultant software into the study.
2. Analyzing buildings orientation, heat gain breakdown, and heating and cooling loads by using Ecotect software.

4 Research framework

The research addresses the baseline case first followed by several cases to improve the original case. Deciding the required cooling strategies for an area requires an integration of Climate Consultant into the study. From the study, several strategies were listed to improve thermal environment shown in Fig.4.1. Out of the strategies, building orientation, shading devices, and glazing were investigated in both compounds as shown in Table 1.

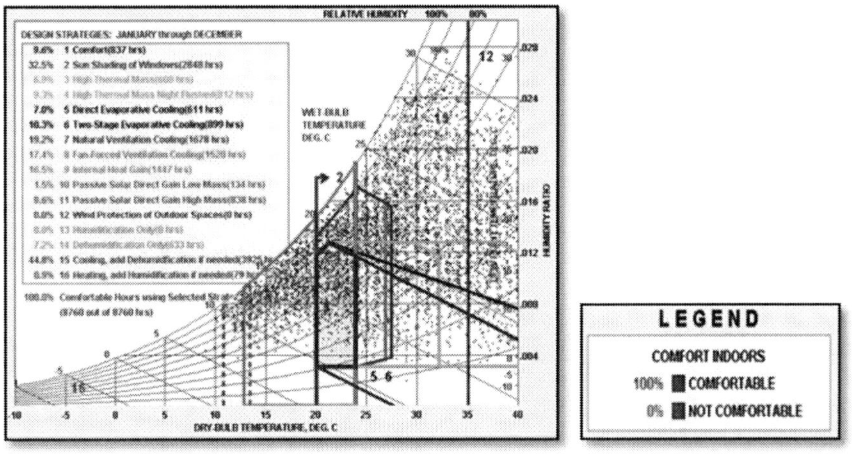

Fig. 4.1. Design Strategies by Climate Consultant.

Table 1. Simulation Cases.

Case 1 :Al Khawarizmi Compound	
Case 1	Baseline Case with Single Glazing
Case 1A	Double Glazing
Case 1B	Shading Device
Case 1C	Double Glazed + Shading Device

Case 2 : Al Al Baironi Compound	
Case 2	Baseline Case with Single Glazing
Case 2 A	Double Glazing
Case 2 B	Shading Device
Case 2 C	Double Glazed + Shading Device

4.1 Building orientation

▪ ***Al Khawarizmi Compound***

The main façade of the compound that shows the windows of the living rooms and two bedrooms is facing northeast. Therefore, this façade does not need shading devices unlike the backside façade of two bedrooms, office and kitchen that needs shade as it is facing southwest where heat gain is maximum due to sun low angle in the sky and to accumulated heat during the day. To improve the orientation of this building, an analysis using ECOTECT suggested the best orientation to be 180° degrees from north for Al Khwarizmi compound.

▪ ***Al Al Baironi Compound***

The main façade of the compound that show windows of the living room and two bedrooms is facing northwest. Therefore, this façade does not need shading devices unlike the backside façade of two bedrooms, the office and the kitchen needs to be shaded since it is facing southeast. To improve the orientation of this case, the best suggested orientation for Al Baironi compound was also set to 180.0° degrees from North.

4.2 Shading devices

▪ ***Al Khwarizmi and Al Baironi Compounds***

The front façade does not have any shading devices, while the backside of the villa has horizontal shading devices for the kitchen and the living room at ground floor unlike the first floor has no shading devices. To improve the case, shading devices were added to the simulation for the backside façade in the first floor since it is facing southwest. Same simulation process was performed for the case of Al Baironi.

4.3 Glazing Type

▪ ***Al Khwarizmi and Al Baironi Compounds***

The existing glazing in both villas is Single Glazed Aluminum Frame, which has high ratio of heat gain and loss. To improve this case, Double Glazed Aluminum Frame has been added to the simulation.

5 Analysis and result

5.1 Building orientation

The orientation of the building plays a major role in affecting the thermal comfort and the consumption of energy in buildings. The weather tool in the Ecotect software analyzes the solar radiation and calculates the best building orientation according to the solar heat absorbed in annual super-heated and under- heated periods Fig.5.1. The green line represents the orientation of the building that gets the highest solar radiation during winter; the thin red line represents the orientation of the building that gets the highest solar radiation in summer, while the yellow line represents the best orientation after comprehensive consideration of local meteorological condition. The best orientation for the Al Khwarizmi and Al Baironi compound villas was 180.0° degrees from north (Fig.5.1).

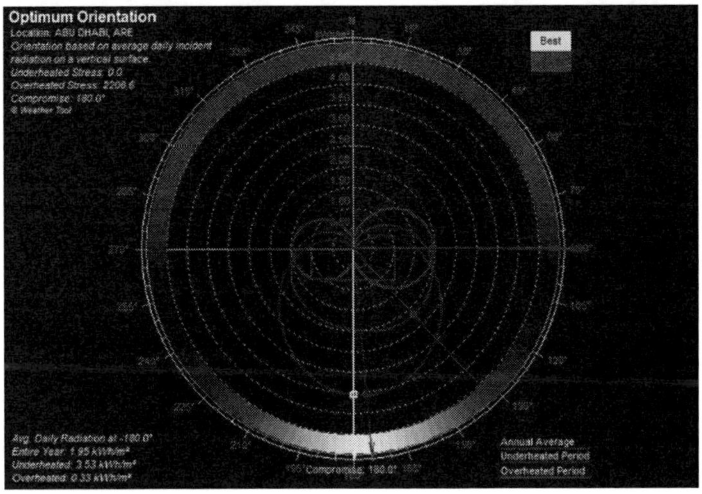

Fig. 5.1. Analysis of the Best Building Orientation of Al Khwarizmi and Al Baironi Compound.

5.2 HVAC

HVAC is the amount of energy that is required to maintain the zones at a comfortable indoor temperature. Heat gains breakdown consist of different categories starting with building conduction, which is heat gains/losses that occur through the fabric of the building itself. The Sol-Air gains are due to indirect solar exposure and known as Sol-Air temperature. These heat gains are caused by the molecular excitation within the building materials when exposed to solar radiation. Direct solar heat gains occur through transparent surfaces such as skylights and windows. On the other hand, natural ventilation represents gain and losses that happen due to infiltration. Other factors affecting cooling loads beside earlier parameters like internal gains due to artificial lighting, occupancy by people and equipment. For the Inter-zonal gains and losses that occur between adjacent zones, heat flows from one zone to another.

5.3 Simulation Analysis

This section highlights the analysis of earlier simulation cases for both compounds as follows:

▪ ***Case1: Baseline Case with Single Glazing Panes***

Using ECOTECT, several simulations were run for the baseline case from 1 June to 30 Sep., i.e. summer months. Different percentages of heat gains were received through fabric, Internal, Sol-Air, etc. for both compounds (Figs.5.3.1 & 5.3.2). As most of these heat gains come from fixed elements that not easy to retrofit in reality. Therefore, the study has focused on possible and feasible strategies that are doable and quick in installation such as replacing single pane windows with double pane windows and adding external shading devices.

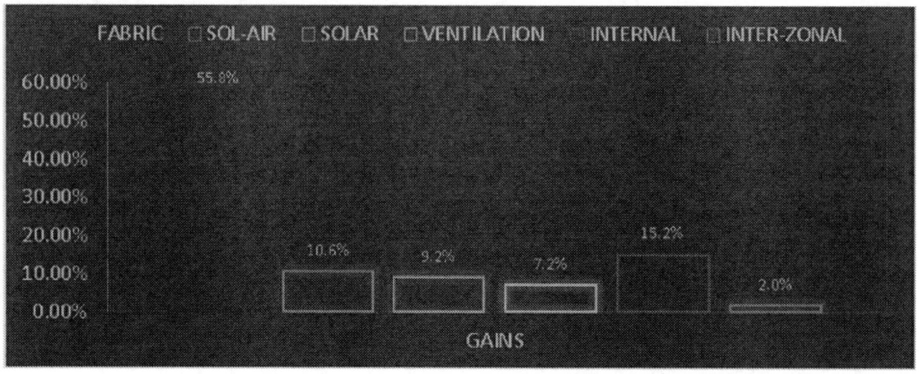

Fig. 5.3.1. Gains Breakdown for Baseline Case with Single Glazing Type in summer (Al Khwarizmi Compound).

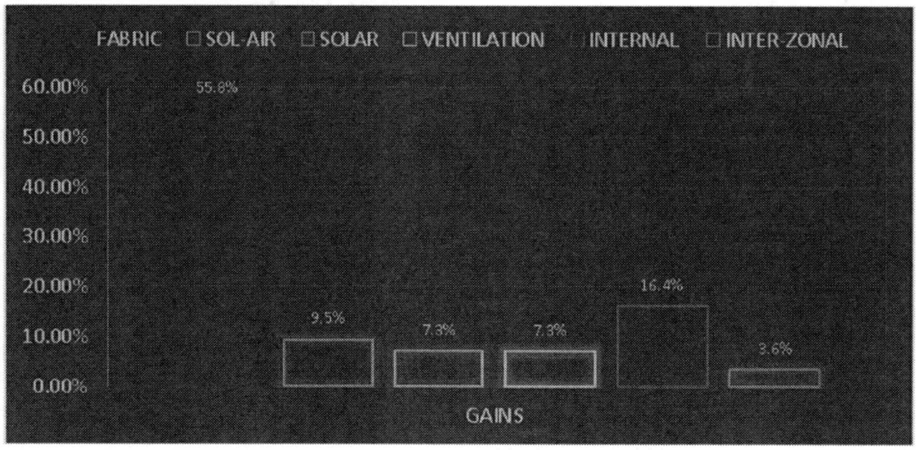

Fig. 5.3.2. Gains Breakdown for Baseline Case with Single Glazing Type in summer (Al Baironi Compound).

- ***Cases from 1A-1C***

Double glazing panes have replaced single glazing panes in the simulation of Case1A in order to reduce heat transfer through convention, while in the Case of 1B shading devices have been added to the windows to block sun radiation penetrating indoor spaces. One more case combing Case1A with Case1B was introduced in Case1C in order to reduce indoor temperature and accordingly decreasing cooling loads. According to the simulation shown in figure (5.3.3), the reduction is dropped from 9.20 % to 4% with an improvement of 5.20%. Similar investigation has been done for the case of Al Baironi and it is found in that Case2 has a smaller heat reduction than Case1 Figure (5.3.2 & 5.3.4). Going through the simulation for all cases, the reduction was dropped from 7.5% to 5 % with an improvement of 2.30% (Fig.5.3.4).

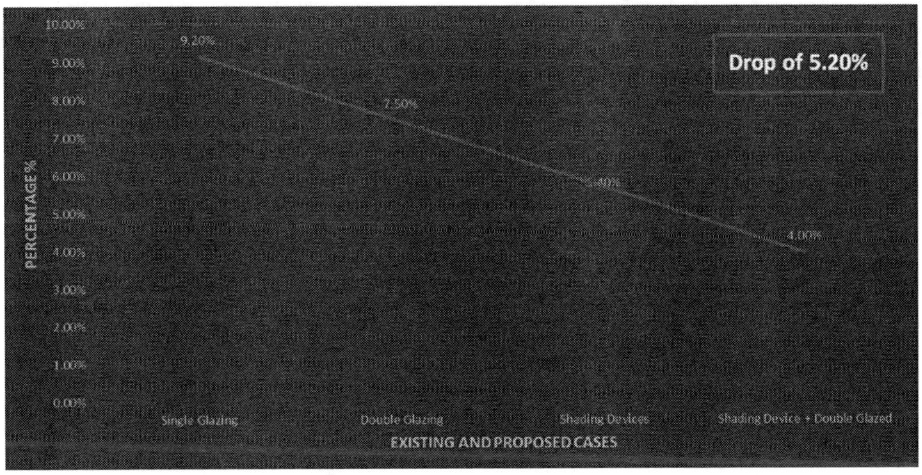

Fig. 5.3.3. Solar Heat Reduction in summer for the case of Khwarizmi Compound.

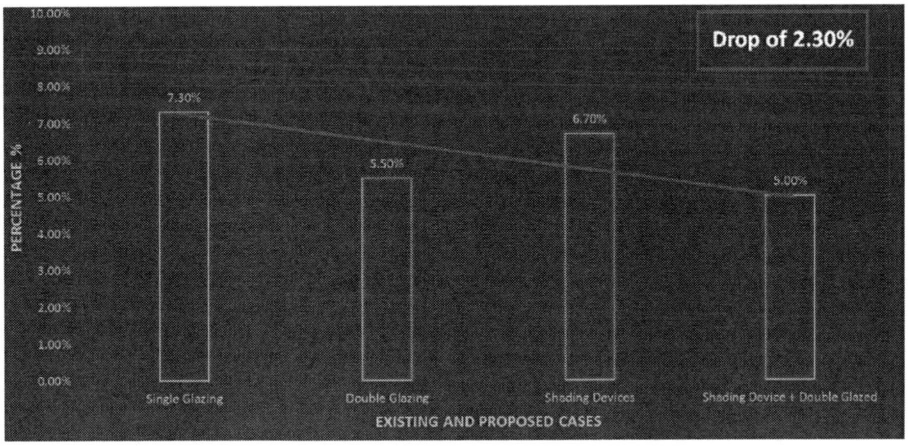

Fig. 5.3.4. Solar Heat Reduction in summer for the case of AL Baironi Compound.

5.4 Solar gain reduction comparison

Comparing the compounds of Al Khwarizmi and Al Baironi, The analysis in Figure.5.4.1 shows that the double glazing proposal is more effective for Al Baironi Compound than the shading devices proposal, while both proposals are effective in Al Khwarizmi Compound, because of its form (U-Formed) and compound's orientation.

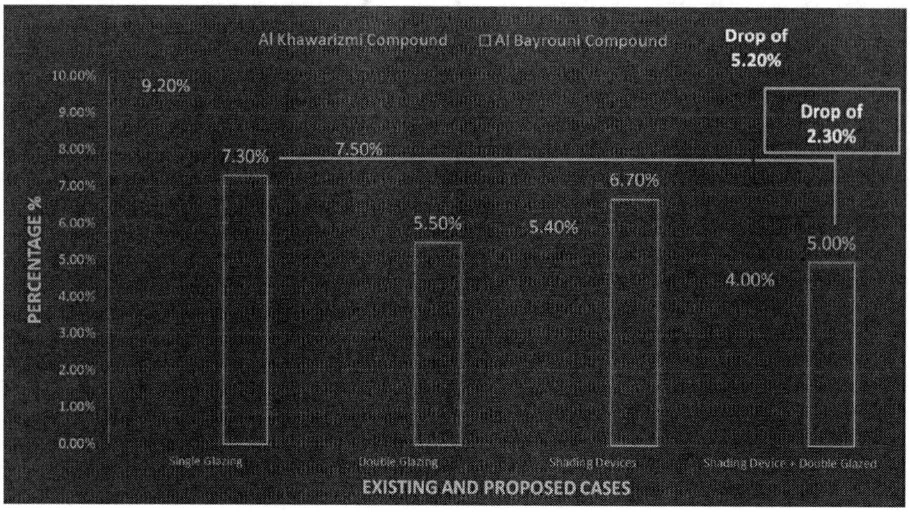

Fig. 5.4.1 Solar Heat Reduction in summer for both compounds.

5.5 Monthly cooling loads for both compounds

Analysis of the cooling loads of all cases, including the existing case for each compound were simulated separately, and then combined in one chart as in Fig.5.1.1 in order to compare the impact of each parameter on cooling loads for both compounds.

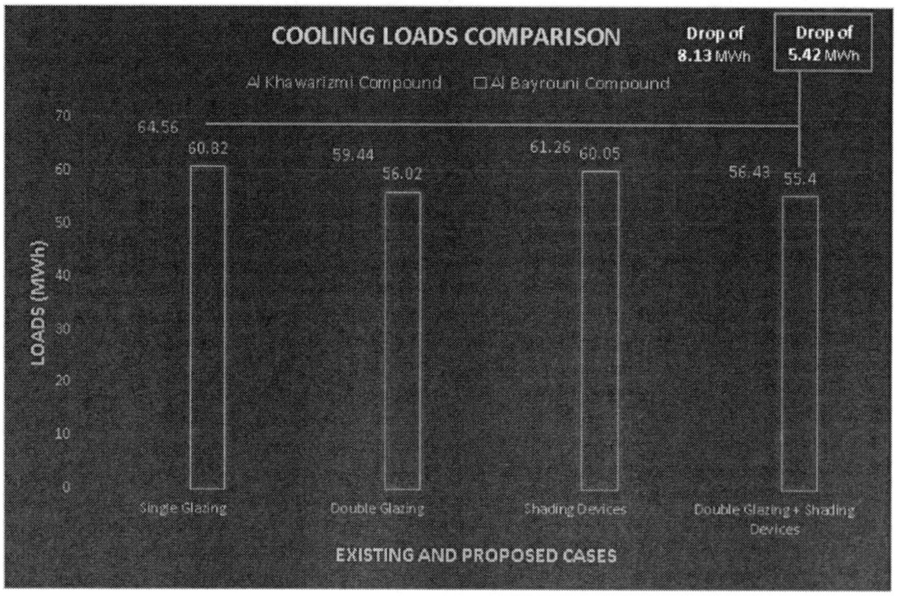

Fig. 5.1.1. Cooling Loads Reduction Comparison Chart.

From the analysis, it shows that the case of Al Khwarizmi consumed more cooling load than the case of Al Baironi with a difference of 3.73 MWh to reach to 64.56MWh. The total reduction in last case 1C reached to 55.64 MWh from 60.84 MWh with an improvement of 8.13 MWh. The other compound the Al Baironi has dropped from 60.82 MWh to 55.4 MWh with an improvement of 5.42 MWh.

6 Conclusion

To build a comfortable indoor environment, many decisions should have been taken into consideration when designing residential compounds. Building forms and shapes besides the orientation of buildings and the materials used in construction are very important aspects to consider. In the university city of Sharjah single glazed windows are used where windows play an important role in reducing solar radiation. It is necessary to introduce more effective windows materials with better insulation values. From the study, after changing the material of windows from single to double glazing panes, the results were showed an improvement in the thermal performance of the building and helped in energy saving. Another strategy applied was adding shading device to the building facades helped in improving the indoor environment. Therefore, combining several passive design tools together to get the maximum benefit would definitely positively impact buildings. Thus, a number of other points that have not been covered in this should be explored in future studies. Further techniques and applications have to be investigated and integrated to the study in order to help the university reduce energy consumption in all residential and educational buildings.

References

1. Anthony, C. M. Indoor Climate of Multi-storey Hospital Buildings in Hot Humid Regions. An attemt to achieve comfortable indoor climate through passive climatic design in Ampara, Sri Lanka, p. 21.
2. Climate, Sharjah Climate History. (n.d.). Retrieved from My Weather: http://www.myweather2.com/City-Town/United-Arab-Emirates/Sharjah/climate-profile.aspx?month=1
3. ECOTECT TUTORIAL: Calculate Passive Gains Breakdown. (n.d.). Retrieved from wiki.natural frequency: http://wiki.naturalfrequency.com/wiki/Passive_Gains_Breakdown
4. Lin, J. (2007, Nov. 18). Introduction to ECOTECT V5.6. Retrieved from Tools for Sustainablity: http://www.toolsforsustainability.com/files/downlo (lin, introduction to ecotect v5.6, nov.18, 2007)ad/tutorials/ecotect/ecotecttutorial-thermalanalysis.pdf
5. Chapter 2 A Brief History of Thermal Comfort: From Effective Temperature to Adaptive Thermal Comfort

State-of-the-art in development of diffuser augmented wind turbines (DAWT) for sustainable buildings

Arouge Agha[1,*], *Hassam* Nasarullah Chaudhry[1]

[1]EGIS, Heriot-Watt University, Edinburgh, UK

Abstract. In this paper, a review of the development of Diffuser Augmented Wind Turbines (DAWT's) into the Built Environment has been presented. DAWT's offer a lot of potential as electricity providers in areas where it is needed, such as in the built environment. Research into DAWT's has revealed that flow along a building significantly affects inlet conditions to the rotor due to flow separation at the leading edge. Adjusting the area ratio, length/diameter ratio as well as diffuser design can improve the performance of the free-standing DAWT significantly. Placing the turbine at the centre of a roof is found to allow the best wind conditions to the wind turbine. It was found that a turbine can be placed at a height 1.3 times the height of a small building for the best results. It has been found that vaulted roof's encourage acceleration of air flow better than other topologies. Furthermore, a recent approach to building-integrated wind turbines involves a flow-enhancing architectural design for buildings to improve favourable inlet conditions to a DAWT.

1 Introduction

With depleting fossil fuel supplies, there is a growing need to address the replacement of our current energy production methods as much as possible. Recent efforts in research have shown realistic power generation techniques from renewable energy are emerging successfully such as power from wind energy, a competitive and efficient alternative to non-renewable power generation. There is an abundance of available wind energy on-shore and off-shore for power conversion. Horizontal Axis Wind Turbines (HAWT) and Vertical Axis Wind Turbines (VAWT) are the most fundamental wind turbine configurations that dominate the wind energy market. The three-bladed HAWT is currently leading the Wind Energy Conversion Systems (WECS) sector, in terms of efficiency, large-scale installation and manufacturing robustness.

The exergetic performance of a HAWT or a VAWT can be improved with the use of passive technology such as a Diffuser (sometimes part of a shroud) to form the Diffuser Augmented Wind Turbine (DAWT). This aerodynamic device can 'funnel' incident air onto a rotor accelerating flow to achieve a greater power extraction for the turbine over a more controllable, continuous period of operation than is currently possible. The Diffuser's

* Corresponding author: aa182@hw.ac.uk

capability for effectively capturing incoming wind means it can work in turbulent wind conditions and low speeds which originally made it a uniquely valuable tool in the aerospace industry from aircraft to gas turbine technology. However, the costly additional weight of the Diffuser can be an issue in terms of load bearings. DAWT's can be installed on small to large scales, in the built environment (building integrated) where wind can be turbulent and hard to capture and potentially in open field environments and off-shore applications.

Recently, a multitude of investigations have been performed to assess wind energy potential in built environments. The conclusion from these studies have recommended the use of wind turbines (Ayhan and Sağlam, 2012 [1]; Ishugah et *al.* 2014 [2]; Toja-Silva et *al.* 2013 [3]). As Nishimura et *al.* (2013) [4] explains, understanding the building topologies and velocity profiles available for a given location is crucial. This will identify site-specific locations for the DAWT's. From Computational analysis, it was found that wind flow accelerates through spaces between buildings up to distances of 60m after which speeds drop and slowly recover. It was also found that wind flow accelerates up along the free side of buildings reaching the roof. Any wind turbines should therefore be installed behind buildings to capture accelerated flow or along the centreline between buildings. This approach agrees with one of the types of integrated wind turbines discussed by Abohela et *al.* (2011) [5], where the wind turbine is on free standing tower away from the building. The second type includes building-mounted wind turbines and the third type is where the building is architecturally designed for wind capture.

In line with this, Abohela et *al.* (2013) [6] conducted a study to identify a roof shape that encourages the best wind profiles. Comparing vaulted, domed, flat, gabled, pyramidal and wedged roofs for a range of wind directions, it was found that the vaulted roof showed the best results especially when flow was parallel to the roof from the side of the building along its curvature. Furthermore, a study on the placement of a HAWT was tested. It was recommended that at a height of 1.3H (H being the height of the building) the turbine should be placed at the midpoint. Similarly, Tabriz et *al.* (2014) [7] found that small scale wind turbines for a sloped roof with a small inclination angle, are subject to optimum wind conditions at the middle of a roof but given a tall tower the rotor can be placed near the edge of the roof to overcome the turbulent flow and capture skewed flow as it deflects off the edge and along the roof.

In theory, the operational capacity of DAWT's and the wind conditions in most built environments (whether densely urban or not) seems to match well. This paper aims to provide a state-of-the-art review of research and developed approaches to design integration and analysis of DAWT's into the built environment. Section 2 covers computational studies that have been conducted to test and analyse DAWT's based on Horizontal- and Vertical- Axis Turbines for various building applications, in line with the second approach of integrated wind turbine types and Section 3 describes how diffuser aerodynamic design can be used in building design, in line with the third approach of integrated wind turbine types Abohela et *al.* (2011) [5]. Section 4 introduces some further work with the conclusion in Section 5.

2 Design-based computational studies for integration of DAWT's into the built environment

Installing wind turbines in the built environment has an added advantage. Electricity can be supplied where needed and fed into the grid directly (Abohela et *al.* 2011 [5]). However, since these turbines are often small scale, a collection of turbines may be needed to achieve enough electricity for a given building especially for high-rise buildings and in densely populated areas. With the use of a DAWT, the number of required turbines can be decreased, and at higher altitudes it may be possible to better protect the rotor itself from adverse weather conditions. This is subject to the additional weight and loading of the diffuser. Currently, there exists a good selection of commercially available DAWT's. They are nearly all DAWT's based on Horizontal Axis Turbines. Nonetheless, research continues to assess ways of optimising and improving performance through better fluid flow properties. Computational Fluid Dynamic (CFD) analysis provides the best accurate simulation and analysis of air flow through wind turbines. Therefore, most research on DAWT's is CFD based with experimental validation. The examples in this section are for studies that designed and tested DAWT's specifically for the application in the built environment. Therefore the approach differs slightly compared to free standing DAWT's because the presence of building(s) can significantly affect the inflow conditions to the diffuser and so this has to factored into the design.

2.1 Diffuser action using HAWT's

There have been many studies on diffuser design and analysis with some of the most comprehensive work starting in the 1970's. Nearly all studies have focussed on the Horizontal Axis Turbine as the rotor. The analysis of literature has bought forward a range of design techniques and computational approaches for enhancing air mass flow through the diffuser, decreasing exit pressures as much as possible (sub-atmospheric), achieving optimum tip-speed ratios, turbine disk loadings and power coefficients. Some studies have now applied theory to the application of DAWT's on buildings.

Wang et *al.* (2007) [8] conducted a parametric analysis on the optimisation of blade design and rotor design to better suit the conditions of the built environment. A CFD analysis was then performed in a virtual wind tunnel. It was explained that in wind turbines with small rotor diameters, small torques can be achieved posing a problem therefor for starting rotation in low wind speeds common urban areas. A larger torque can be accomplished with an increased number of blades. A 6-bladed rotor with SD2030 airfoil cross section was designed. It was found that the flow and pressure were much higher at the tip than at the root as seen in fig.1, indicating greater torque generation. Different nose cone and nacelle designs were also tested. It was found that the nose and nacelle should be designed for a smallest possible diameter and volume. But this can be hard to achieve as the generator will require adequate space.

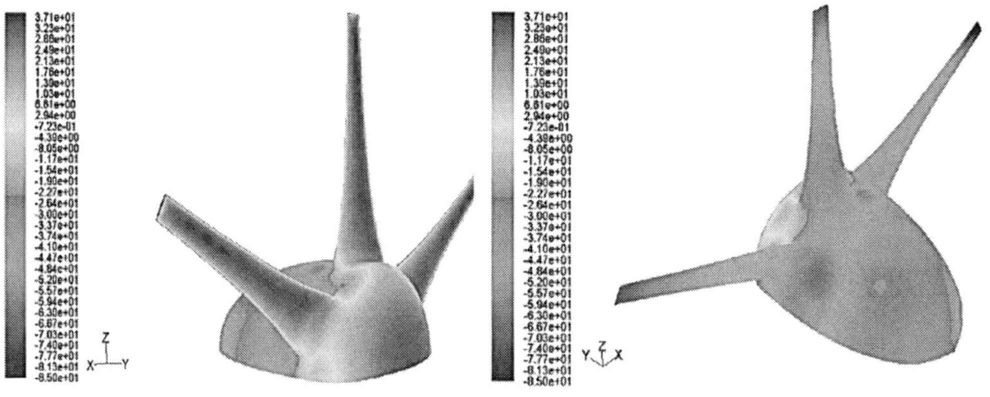

Fig. 1. Contours of Static Pressure (Pa) a) In front of the rotor and b) Behind the rotor (Wang et *al.* 2007) [8].

Wang et *al.* (2008) [9] then developed a methodology to study the aerodynamics of a small domestic wind turbine with a specially designed shroud for use on building rooftops. A Rutland 913 Windcharger turbine was chosen from tests carried out in a wind tunnel. The pressure distribution and power outputs were measured and validated against a CFD analysis. BEM theory was used to optimise the blades and the DAWT was then re-evaluated for its performance in the CFD analysis. The shroud consisted of a converging inlet (concentrator) with large area ratio, a constant diameter test section and a diffuser with smaller area ratio than the inlet. The RNG $k-\varepsilon$ turbulent model and Single Reference Frames (SRF) in GAMBIT and ANSYS Fluent was used for simulation of the virtual wind tunnel. The results showed that the wind speed was increased by a factor of 1.5 through the shroud alone, maximum power augmentation was about 2.2. In all cases there existed an optimum tip speed ratio for every incident wind speed tested. With the addition of the shroud, a 50% increase in power output was found in low speeds and 25% in higher wind speed above 10m/s, fig. 2. The results from computational work agreed well with the experimental testing, any errors encountered fell within 5%. Fig. 3. shows that optimum tip speed ratios were in the range of 4-5 for the DAWT in all computed wind speeds. At lower wind speeds the power coefficient was higher within the optimum tip speed range. And, for a large range of wind speeds, the range for the power coefficients remained quite narrow.

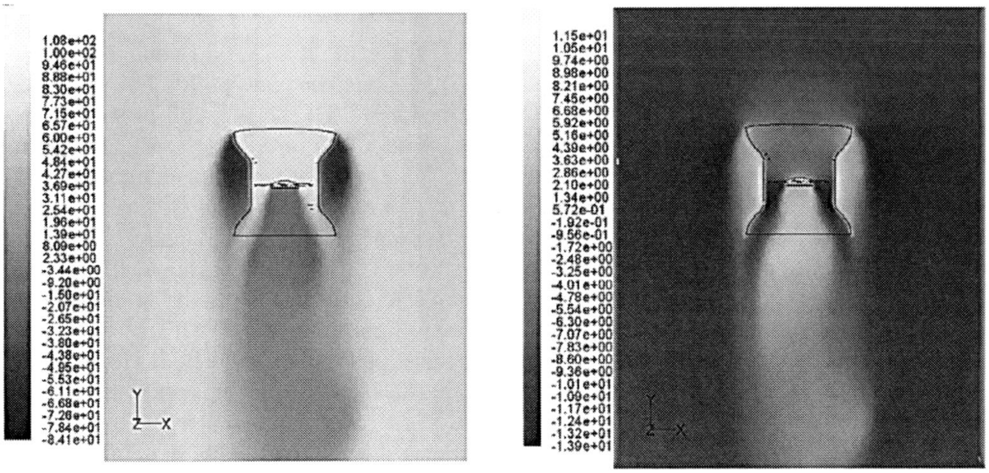

Fig. 2. Result Contours a) Static pressure (Pa) and b) Axial Velocity (m/s) (Wang et *al.* 2008) [9].

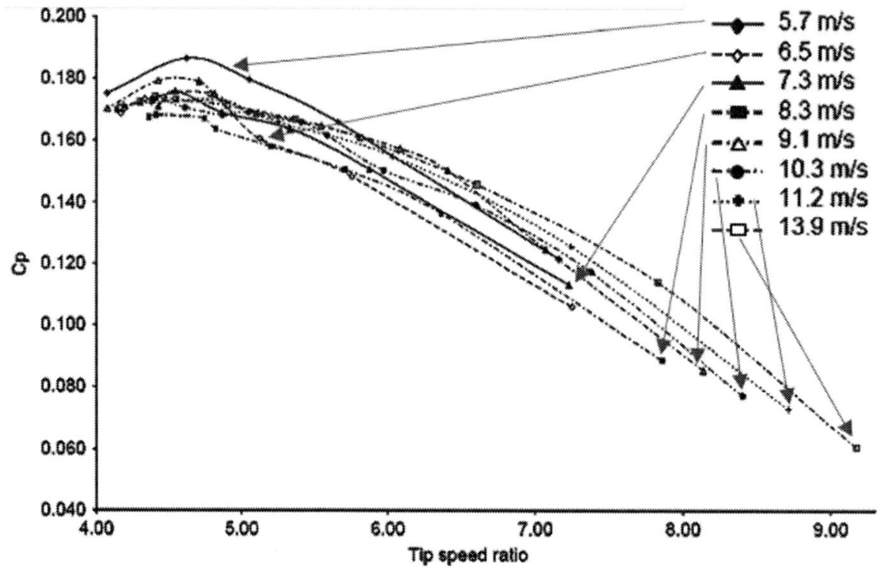

Fig. 3. Computed tip-speed ratio's and power coefficients. (Wang et *al.* 2008) [9].

Based on work by Abe et *al.* (2005) [10]; Ohya et *al.* (2008) [11] and Ohya and Karasudani (2010) [12], Kosasih and Jafari (2014) [13] designed and built a DAWT using the 3-bladed Ampair 300W Turbine. The blades had NACA 63-210 airfoil profiles. The rotor diameter was 190 mm with length of 120mm with a L/D of 0.63, area ratio (outlet to inlet) of 1.61 and diffuser expansion angle of 12°. Tip clearance was kept minimal to 2-3-mm. The wind tunnel dimensions were 450 mm x 450 mm x 1500 mm with wind speed achieving up to 25 m/s. The DAWT was tested and compared with manufacturer's data and was then simulated in CFD software. Seen in fig.4, the CFD model for the AMPAIR was modelled in a virtual wind tunnel with dimensions 4x10m and the rotor diameter was 1300mm.The maximum power achieved rom the bare turbine was 210W. It was found that as H/D and L/D were increased independently, more power was extracted. However, at increasing L/D ratio, smaller H/D ratios were found to increase power outputs because if both ratios increased there as more flow separation and viscous losses from the diffuser surface. The maximum C_p for the bare turbine was 0.257 and with the diffuser the maximum C_p achieved was 0.31 for small H/D ratios and large L/D ratios.

Fig. 4. The AMPAIR 300. Left: Actual rotor and Right: CFD model (Kosasih and Jafari (2014) [13].

2.2 Diffuser action using VAWT'S

Zanflorin and Letizia (2014) [14] conducted a 2D study on 1.2kW 3-bladed H-Darrieus DAWT in ANSYS Fluent v.15. The URANS (Unsteady Reynolds Averaged Navier-Stokes) implicit model was used along with the SST (Shear Stress Transport) $k-\omega$ model due to its ability to solve cases with high adverse pressure gradients and simulate detailed vortices exhibited during dynamic tall at low tip-speed ratios. The diffuser dimensions were defined and it was found that the converging section prevented the flow from diverging towards the exit and realigned flow well before the approach to the rotor. The model was then placed on a dual-pitched roof of a building (fig. 5) with height 16.6m and an inlet velocity coefficient of 0.35 (typical to urban areas). A 44% increase in power out was achieved at a speed of 6m/s and a 50% increase at a wind speed of 10.6m/s. To achieve the same results, the turbine installed without the diffuser, would need to be increase through a height of 1m.

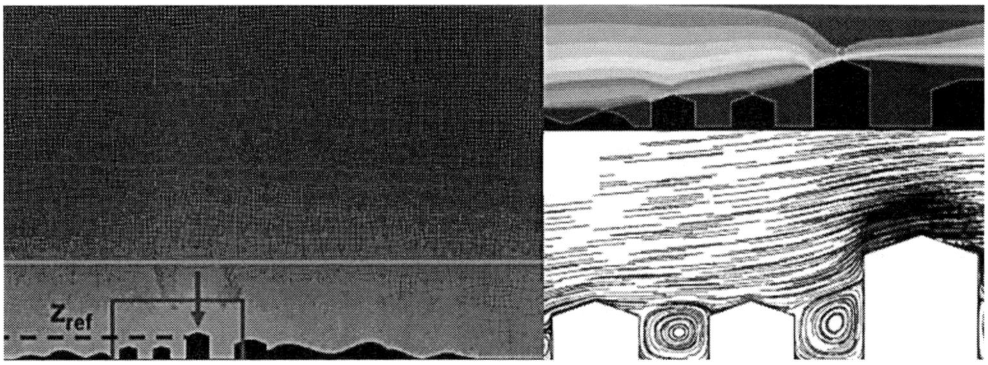

Fig. 5. **Left:** The red arrow indicates the location of the DAWT, **Right**: Velocity contours showing accelerated flow towards the top of the dual-pitched roof (Zanflorin and Letizia 2014) [14].

There have also been studies on design DAWT's based on the wind conditions in the built environment for use on existing buildings. One or two studies have been conducted recently that questioned the use of a HAWT inside a diffuser. Krishnan and Paraschivoiu (2015) [15] conducted a 3D CFD analysis of a building integrated DAWT using a vertical axis rotor for the turbine. The most appropriate inlet geometry for the DAWT was assessed leading to a final design. The coefficient of power increased from 0.135 to 0.34 with the use of the shroud and the optimum diffuser inlet angle with reference to the flat roof tested was 5° as seen in fig.6a). Although the study proved that the addition of the diffuser improved performance of the bare turbine, the power coefficient was below the Betz limit and the augmentation factor was very close to one. The air inlet condition was 6m/s and it can be seen from the final design in fig.6b) (with the turbine's optimum tip-speed ratio at 0.38) that the maximum speeds in the turbine were not much higher than the air speed. It can be seen from literature that the use of the vertical axis turbine may not be able to succeed over the use of a horizontal axis turbine in DAWT design.

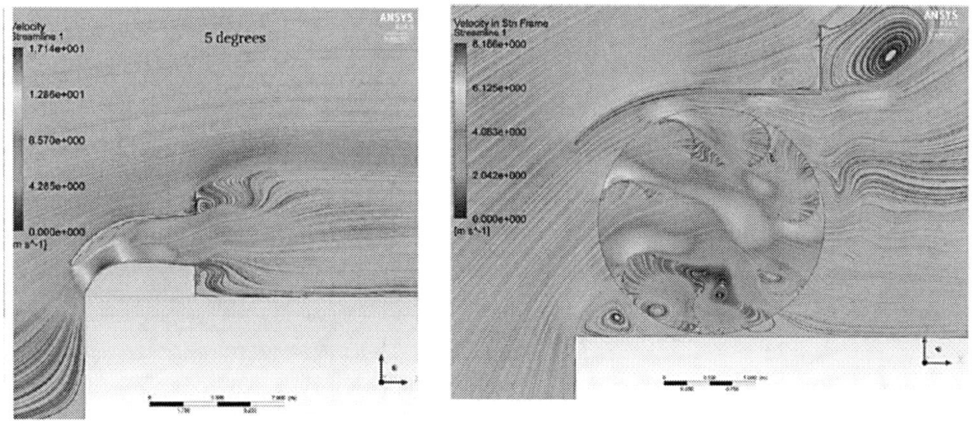

Fig. 6. **a)** Streamline flow for the 5° inlet diffuser angle showed the largest maximum velocity compared to 0°, 10° and 15° and **b)** streamline flow for the final design of the turbine (Krishnan and Paraschivoiu, 2015) [15].

Chong et *al.* 2013 [16] designed a novel Power-Augmentation-Guide-Vane (PAGV) shroud that covered a Sistan wind turbine. It was integrated into a wind-solar-rain harvester, seen in fig.7). It was designed to aesthetically blend into the building and a mesh would cover the exposed rotors to prevent bird strike. In wind tunnel testing, under free-running conditions) rotor inertia and bearing friction were applied and it was found that the rpm of the rotor was increased by 75.16%. In a CFD study, torque was increased by 2.88 times and the power output increased 5.8 times at a wind speed of 3m/s.

Fig. 7. Computer Aided Design of the PAGV showing its intended roof installation (Chong et al. 2013) [16].

Wong et *al.* 2014 [17] conducted a 2D study in ANSYS Fluent 14.0 on a DAWT with NACA 0015 blades. Tip-speed ratio was set at 5.1 and velocity 0.091m/s from experimental data. The SST $k-\omega$ model was used with a sliding mesh. The original design for the shroud used four pairs of ODGV (Omni-Directional Guide Vanes) tilted at angles of 20° and 55° whereby the outer diameter of the ODGV was twice the rotor diameter of the bare VAWT. The new design divided the guide vanes into two segments ad were bent at an angle of 10°. Torque qualities were significantly improved with the new design and the power coefficient increased by 31.65% compared to the original design and the cut-in speed was reduced. Compared to the bare VAWT, the power coefficient was increased by 147.1%.

3 Diffuser action in building design

Fig.8a) shows a conceptual design for the use of a large scale diffuser effect for the built environment. The building is itself, a diffuser. The company, Norwin from Denmark set-up three 225kW, 29m diameter rotors between the Bahrain World Centre towers. Each turbine was installed on a dedicated steel frame fixed between the towers as seen in fig.8b) (Smith et *al.* 2012) [18]. Another example is The Strata in London which is topped with three 9m diameter wind turbines, fig.8c). Using aerodynamic building designs almost as if they are diffusers themselves is a very good way of funnelling a larger mass flow of air in the built environment. This potentially means greater power extraction to cover more of a buildings electricity load.

Fig. 8.a) A conceptual aerodynamic design for a building (Mertens, 2006) [19], b) the Bahrain Tower and c) the Strata in London (Smith et *al.* 2012) [18].

Akesson (2012) [20] produced a feasibility report on the potential to incorporate DAWT's into bridges. E39 is a road that runs along the west coast of Norway with a number of fjords with ferry crossings along its entire length. The fjord crossings range from 1.5km to 25km in length and have depths reaching down to 1.3km. Typically, these fjords have suspension bridges, floating bridges and submerged floating tunnels. Based on the design of a 10kW DAWT with a maximum diameter of 3.1m, 215 turbines were installed per kilometre at 1.5D spacing's, as seen in the drawing in fig. 9). This translates to 80% of the length of the bridges installed with these turbines. Turbines were installed mainly on two types of bridges, suspension bridges and floating bridges. Turbines were installed in the former at heights of 50m and the latter at 10m. It is stipulated that the DAWT's used cope with turbulent conditions much more successfully than HAWT's typically chosen for this application. They are easier to integrate into the bridge infrastructure, are well protected. To give an example, at the Boknafjord Bridge the installed power is 14.5MW with local wind speeds between 6 and 8m/s. This innovative concept is in its early stages and while it successfully exploits the use of available wind energy, there are a number of structural challenges that need to be addressed.

Fig. 9. Computer Aided Design depicting the intended installation of DAWT's below the road of the bridge covering its width and length completely (Akesson 2012) [20].

Watson et *al.* 2007 [21] conducted a CFD study on two diffusers with similar geometries but with different lengths on a building, seen in fig.10). The model was simulated in ANSYS CFX using the SST $k-\omega$ model because flow separation could be accurately predicted than with the $k-\varepsilon$ model. The building had a half-width of 5m and height 3m. The diffuser inlets in both cases were designed to allow for the smoothest flow possible. The velocity was 5m/s. It was found that the building accelerated air along the leading face but separated at the leading edge. The separation point however, was displaced into the inlet of the duct due to its placement at the leading edge. The design of the diffuser inlet depended heavily on being able to transfer the vertical component of the incoming wind to horizontal flow because it would adversely affect efficiencies. Compared to the free-standing diffuser, it was found that for low contraction angles there was a performance drop for the building mounted diffuser due to flow separation along the bottom of the diffuser. It was also found however, that with increasing rotor resistance, performance improved.

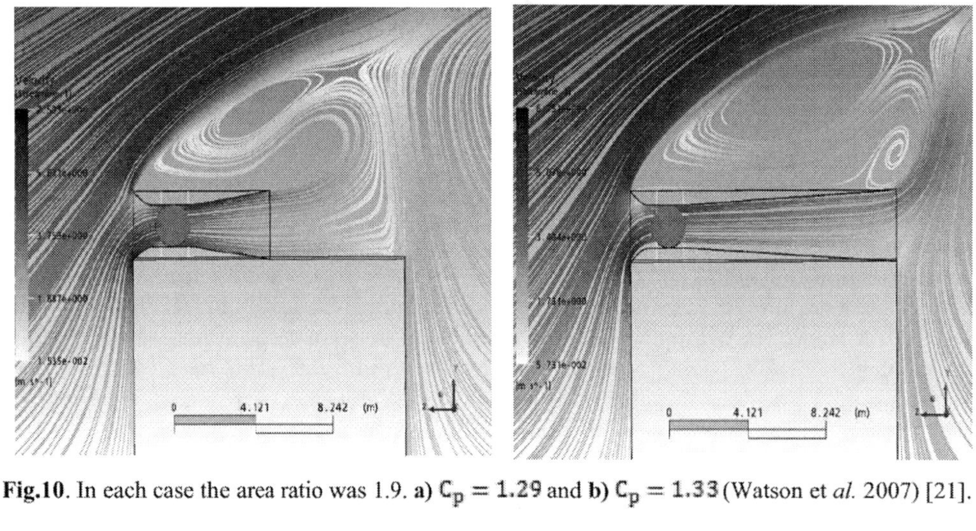

Fig.10. In each case the area ratio was 1.9. **a)** $C_p = 1.29$ and **b)** $C_p = 1.33$ (Watson et *al.* 2007) [21].

4 Further work

Further work is required in assessing the real difference between the use of Horizontal- and Vertical- Axis Turbines in Diffusers. There haven't been many experimental studies on the actual operation of a roof-mounted DAWT. This would address a variety of other factors such as load bearing, fatigue and operational factors of DAWT's in real-time electricity production. An important factor to consider is safety. It has been stipulated many times in literature that compared to conventional WECS, DAWT's are considered safer due to their smaller size, protecting shroud, lower rpm and larger number of blades Kosasih and Tondelli (2012) [22]. So bird and bat strikes are anticipated as highly unlikely which is especially favourable considering the implementation of DAWT's into the built environment. There have not as yet been any studies in this area.

Below are the results from two case studies conducted by the author, modelled under steady state, free stream conditions. The wind speed used 9.1m/s. The RNG $k - \varepsilon$ model was chosen with standard wall function. The streamlines show strong vortices that formed in the wake of the shroud for the DAWT case in fig.11a). This implies a drop of pressure as energy is extracted. This should therefore have the effect of pulling in more air mass flow through the shroud. This is similar to the B-DAWT (Building–DAWT) in fig.11b) where the flow is accelerated. However, the flow seen is not symmetrical for the B-DAWT as it is for the DAWT. Due to the presence of the building, the air flow is forced to change direction bending as it reaches the roof edge. This means the air doesn't flow directly into the shroud and the flow pattern for the B-DAWT shows larger acceleration along the bottom half of the inner shroud volume. In both cases, the increase in wind speed was around 47% based on the maximum velocity computed at the rotor. The augmentation factors were computed as 2.46 for the DAWT and 1.38 for the B-DAWT compared to the benchmark bare rotor.

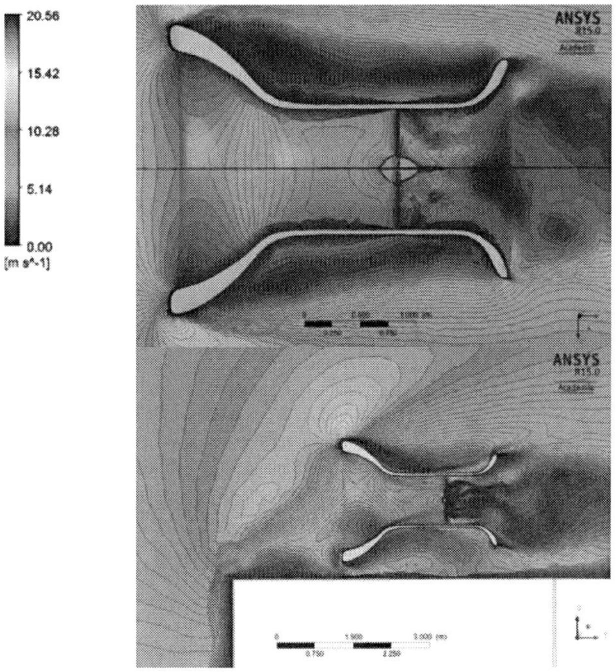

Fig. 11. Velocity Contours. Air flow is from the left. a) DAWT and b) B-DAWT.

5 Conclusion

Due to its aerodynamic shape the diffuser can accelerate air mass flow through a rotor thereby enabling greater power extraction. Many parametric and experimental studies have been performed for a range of diffuser parameters and different methodologies for making performance assessments are available. The uses of CFD simulations are crucial to accurately simulating flow at low cost. It is a flexible tool that is useful for optimisation exercises especially when many different parameters are involved. It can be concluded from literature so far that in the application of a DAWT to a building, the design of the diffuser is specific to the chosen rotor and the rotor chosen can be specific to the location of the building due to the wind conditions it sits in. Where possible, a building can be aerodynamically designed to encourage better flow conditions into the DAWT inlet and the diffuser can then be designed accordingly to the specifications of the rotor. In this way, a DAWT can be installed to its optimum operating conditions with material, loads on the turbine system and costing permitted.

The study presented has been supported by the funding from Heriot-Watt University, at the School of Energy, Geoscience, Infrastructure and Society.

Nomenclature

C_p	Power Coefficient
ε	Turbulent Dissipation Rate
D	Diffuser Diameter
H	Flange Height
k	Turbulent Kinetic Energy
L	Length of Diffuser
$\boldsymbol{P}$	Power
P	Pressure
V	Velocity
ω	Specific Dissipation Rate

References

1. D. Ayhan, and Ş. Sağlam, "A technical review of building-mounted wind power systems and a sample simulation model." Ren. Sust. Energy Rev. **16**(1): 1040-1049, (2012).
2. T. F. Ishugah, et *al.* "Advances in wind energy resource exploitation in urban environment: A review." Ren. Sust. Energy Rev. **37**: 613-626, (2014).
3. F. Toja-Silva, et *a*l. "Urban wind energy exploitation systems: Behaviour under multidirectional flow conditions—Opportunities and challenges." Ren Sust. Energy Rev. **24**: 364-378, (2013).
4. A. Nishimura, et *al.* "Wind Turbine Power Output Assessment in Built Environment." Smart Grid Ren. Energy **04**(01): 1-10, (2013).
5. I. Abohela, et *al.* "Urban wind turbines integration in the built form and environment." FORUM Ejournal, School of Architecture, Planning and Landscape, Newcastle University, UK **10**: 23-39, (2011).
6. I. Abohela, et *al.* "Effect of roof shape, wind direction, building height and urban configuration on the energy yield and positioning of roof mounted wind turbines." Ren. Energy **50**: 1106-1118, (2013).
7. A. B. Tabrizi, et *al.* "Performance and safety of rooftop wind turbines: Use of CFD to gain insight into inflow conditions." Ren. Energy **67**: 242-251, (2014).
8. F. Wang, et *al.* "Development of small domestic wind turbine with scoop and prediction of its annual power output." Ren. Energy **33**(7): 1637-1651, (2007).

9. F. Wang, et *al.* "The methodology for aerodynamic study on a small domestic wind turbine with scoop." J. Wind Eng.Ind.Aero. **96**(1): 1-24, (2008).
10. K. Abe, et *al.* "Experimental and numerical investigations of flow fields behind a small wind turbine with a flanged diffuser." J.Wind Eng. Ind. Aero. **93**(12): 951-970, (2005).
11. Y. Ohya, et *al.* "Development of a shrouded wind turbine with a flanged diffuser." J. Wind. Eng. Ind. Aero. **96**(5): 524-539, (2008).
12. Y. Ohya, and T. Karasudani, "A Shrouded Wind Turbine Generating High Output Power with Wind-lens Technology." Energies **3**(4): 634-649, (2010).
13. B. Kosasih, and S. A. Jafari, "High-Efficiency Shrouded Micro Wind Turbine for Urban-Built Environment." Appl. Mech. and Mat. **493**: 294-299, (2014).
14. S. Zanflorin and S. Letizia, "Improving the performance of wind turbines in urban environment by integrating the action of a diffuser with aerodynamics of the rooftops" Energy and Procedia **82**: 774-781, (2015).
15. A. Krishnan and M. Paraschivoiu "3D analysis of building mounted VAWT with diffuser shaped shroud." Sus. Cities. Society **27**: 160-166, (2015).
16. W. T. Chong et *al.* "Performance investigation of a power augmented vertical axis wind turbine for urban high-rise application." Ren. Energy **51**: 388-397, (2013).
17. K. H. Wong, et *al.* "The Design and Flow Simulation of a Power-augmented Shroud for Urban Wind Turbine System." Energy Procedia **61**: 1275-1278, (2014).
18. J. Smith, et *al.* "Built environment wind turbine roadmap". Colorado, USA, *National Renewable Energy Laboratory (NREL)*: 1-58, (2012).
19. S. Mertens, "Wind Energy in the Built Environment - Concentrator Effects of Buildings", *Delft University of Technology*: 1-180, (2006).
20. L. Åkesson, "Technology survey for renewable energy- Integrated to bridge constructions. Statens vegvesens", *SP Technical Research Institute of Sweden*: 1-43, (2012).
21. S. J. Watson, D. G. Infield, J. P. Barton, S. J. Wylie, "Modelling of the Performance of a Building-Mounted Ducted Wind Turbine" J. Phys. Conf. Seris **75**:1-10, (2007).
22. B. Kosasih and A. Tondelli "Experimental Study of Shrouded Micro-Wind Turbine." Procedia Eng. **49**: 92-98, (2012).

The effect of span length and girder type on bridge costs

Mustafa Batikha[1,*] , *Osamah* Al Ani[1], and *Taha* Elhag[1]

[1] School of Energy, Geoscience, Infrastructure and Society; Heriot-Watt University, Dubai, U.A.E

Abstract. Bridges have an important role in impacting the civilization, growth and economy of cities from ancient time until these days due to their function in reducing transportation cost and time. Therefore, development of bridges has been a knowledge domain in civil engineering studies in terms of their types and construction materials to confirm a reliable, safe, economic design and construction. Girder-bridge of concrete deck and I-beam girder has been used widely for short and medium span bridges because of ease and low-cost of fabrication. However, many theoretical and practical investigations are still undertaken regarding the type of beam girder; i.e steel composite or prestressed concrete. This paper evaluates the effect of bridge span and the type of girder on the capital cost and life cycle costs of bridges. Three types of girders were investigated in this research: steel composite, pre-tensioned pre-stressed concrete and post-tensioned pre-stressed concrete. The structural design was analyzed for 5 span lengths: 20, 25, 30, 35 and 40m. Then, the capital construction cost was accounted for 15 bridges according to each span and construction materials. Moreover, the maintenance required for 50 years of bridge life was evaluated and built up as whole life costs for each bridge. As a result of this study, the influence of both span length and type of girder on initial construction cost and maintenance whole life costs were assessed to support the decision makers and designers in the selection process for the optimum solution of girder bridges.

1 Introduction

Bridges have an important role in impacting the civilization life due to their function in reducing transportation cost and time. Therefore, the demand of bridges has been expanded dramatically these days in most countries. For example, it is registered that almost 1500 bridges per year were built in the USA between years 1996 to 2006 [1]. In Dubai, it is recorded that bridges reaches a number of 347 at the end of 2012[2]. Subsequent to this considerable increase in bridges number, low cost bridge construction is a big challenge facing the designer.

In this study, the cost of simply supported I-beam girder-bridge is investigated since it is the most popular because of ease and low-cost of fabrication together with its suitability for short and medium span. Furthermore, Boatman (2010)[3] explores that the service life for

* Corresponding author: m.batikha@hw.ac.uk

steel composite I-Beam or pre-stressed I-beam is about 45 years which is 17 years more than using box-girder beam. This is another reason behind choosing I-beam for this study. On the other hand, three kinds of construction materials are selected for the girders of bridge: steel composite girder, pre-tensioned pre-stressed concrete girder and post-tensioned pre-stressed concrete girder. Then, the effect of bridge span and the type of girder on the capital cost and life cycle costs of bridges is evaluated.

Many researchers have tried to answer the question of girder-bridge cost in respect of both span length and life cycle. For example, Jagtab and Shahezad (2016) [4] studied various spans of I-girder bridge to compare between the initial cost of using pre-stressed concrete girder or steel girder. In their work, it was found that steel beam is preferable for short span length up to 15m. Whereas, pre-stressed concrete girder is more economical in 37% for 24m span length. The saving may reach 46% for 36m span length. In 2016, Singh etal.[5] discussed two different span lengths for I-beam girder bridge: 20 m and 25m. Life service of 100 years was taken into consideration in the cost estimation. As a conclusion of their work, the pre-stressed I-girder bridge is the lowest cost compared to the composite steel I-girder. The concluded saving was 10% and 15% for 20m-span and 25m-span respectively. It is worth mentioning that Florida Department Of Transportation (FDOT) estimates the average cost of construction Pre-stress concrete I-girder simple span bridge as 12% less than using steel girder [6]. However, more studies still need to be undertaken to answer the bridge cost question according to span and materials in use where this paper analysed these costs for different spans.

2 Structural analysis of different simple-span bridges

2.1 Geometry, material properties and loading

Five simply supported Bridges of different span-lengths were studied in this paper: 20, 25, 30, 35 and 40m. The section of girder-bridge is shown in Figure 1. The distance among girders is 2.25m with slab thickness of 200mm and carriageway of 6.95m. The concrete barrier has a self-weight of 6.22kN/m. While, 5kN/m2 was assumed for both asphalt self-weight and service loading. Using BD 37/01 (2002) [7], live loads was applied to bridge as HA-UDL of 12.4 kN/m2 plus HA-KEL of 32.9kN/m and HB-45 unit as described in Figure 2. In this study, The total length of the HB vehicle was taken as 9.6 m which leads to most sever affect to the bridge.

The concrete material was considered to have a cylindrical characteristic compressive strength equals to 40MPa and modulus of elasticity of 35GPa. While, the steel reinforcement has 460MPa yield strength and 200GPa elastic modulus. For pre-stressed section design, super low relaxation 7-wire strands were used. Each strand has 12.9mm diameter and 186kN ultimate load.

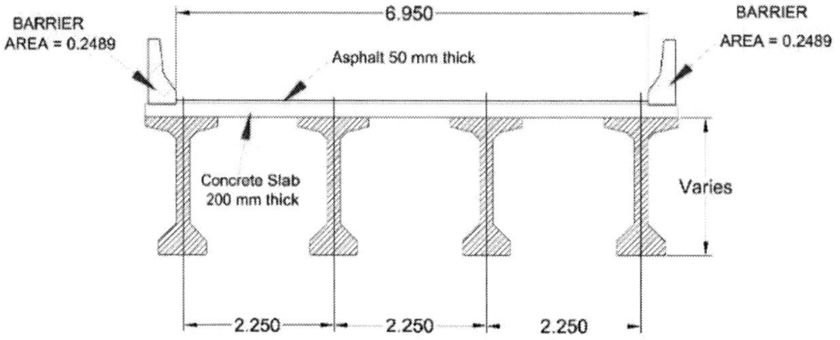

Fig. 1. Bridge cross section of 4 girder-system

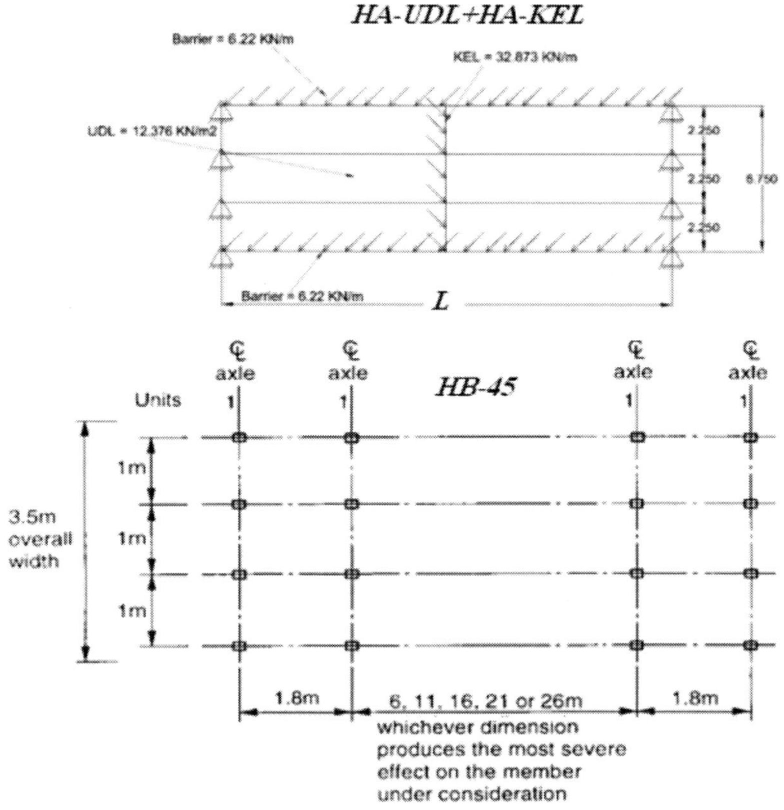

Fig. 2. Traffic live load applied to the bridge[7].

2.2 FE modeling and results

Using SAP2000-Ver 15.0.1[8], elastic analysis was performed for all bridges using estimated sections (Appendix A). Table 1 presents the dimensions properties, the ultimate internal forces produced by the current analysis and the capacity (resistance) of each section. Both shear resistance and bending resistance for each section were obtained using BS 5950-3.1:1990 [9] for composite section and BS 8110-1:1997 [10] for pre-stressed concrete section. From Table 1, it can be seen clearly that the section estimation was proper since the ultimate bending moment is almost same as bending capacity.

Figure 3 describes the relationship between the span length and depth of the girder for the three cases used in this research: composite section, pre-tensioned pre-stressed concrete and post-tensioned pre-stressed concrete. It can be explored that steel composite girder offers always lower depth up to 41% less for 40m bridge span.

Table 1. Dimensions properties, ultimate internal forces and capacity for each section.

Section Details and Properties of Girders														
Sr No.	Girder Length (m)	Girder Type	Girder Depth (m)	Girder & Salb Depth (m)	Section Area with Slab (m)²	Section Area without Slab (m)²	Moment of Inertia (mm)⁴.(10)¹²	No. of Tendons			Force in Tendons (KN)	Moment Resis. (KN.m)	ULS Moment (KN.M)	Shear Resis. (KN)	ULS Shear (KN)
								Bototm	Top	Total					
1	40 m	Pre-Tension	2	2.2	1.095	0.645	0.7163	69 B	3 T	72	9375	16770	16186	922	1951
2	40 m	Post-Tensiom	2	2.2	1.06	0.6099	0.5151	74 B		74	9635	16702	15798	1727	1951
3	40 m	Composite	1.1	1.3		0.0701						13607	12228	7514	1361
4	35 m	Pre-Tension	1.8	2	1.055	0.605	0.5635	58 B	4 T	62	8073	12956	12602	812	1793
5	35 m	Post-Tensiom	1.8	2	1.035	0.585	0.3981	74 B		74	9635	12977	12336	1585	1793
6	35 m	Composite	0.92	1.12		0.0494						9552	9340	4080	1246
7	30 m	Pre-Tension	1.6	1.8	1.015	0.565	0.4313	49 B	3 T	52	6767	9885	9647	689	1628
8	30 m	Post-Tensiom	1.6	1.8	0.975	0.5249	0.2953	54 B		54	7031	9857	9367	1289	1628
9	30 m	Composite	0.98	1.18		0.0316						6234	6010	3347	1158
10	25 m	Pre-Tension	1.4	1.6	0.925	0.475	0.2967	43 B	3 T	46	5999	7663	7055	594	1466
11	25 m	Post-Tensiom	1.4	1.6	0.915	0.4648	0.2108	46 B		46	7163	7465	6879	1183	1466
12	25 m	Composite	0.91	1.11		0.0286						5390	5230	3083	1100
13	20 m	Pre-Tension	1.2	1.4	0.91	0.46	0.2187	30 B	3 T	33	4279	4795	4676	481	1323
14	20 m	Post-Tensiom	1.2	1.4	0.855	0.4047	0.1435	31 B		31	4036	4686	4567	859	1323
15	20 m	Composite	0.835	1.035		0.0224						4108	3584	2490	1036

3 Cost analysis of different simple-span bridges

This study carried out and compared two types of cost analysis for each bridge span: (a) comparison analysis for only the initial capital costs; and (b) cost analysis including the initial costs and 50-year life cycle costs encompassing the maintenance costs. The life cycle costs are derived using Eqs. (1)-(2) [11]:

$$F_V = P_V(1+i)^n \tag{1}$$

$$P_V = F_V[1/(1+r)]^n \tag{2}$$

where,

F_V is Future Value of n number of years.

P_V is Present Value.

i is the inflation rate.

r is the interest rate.

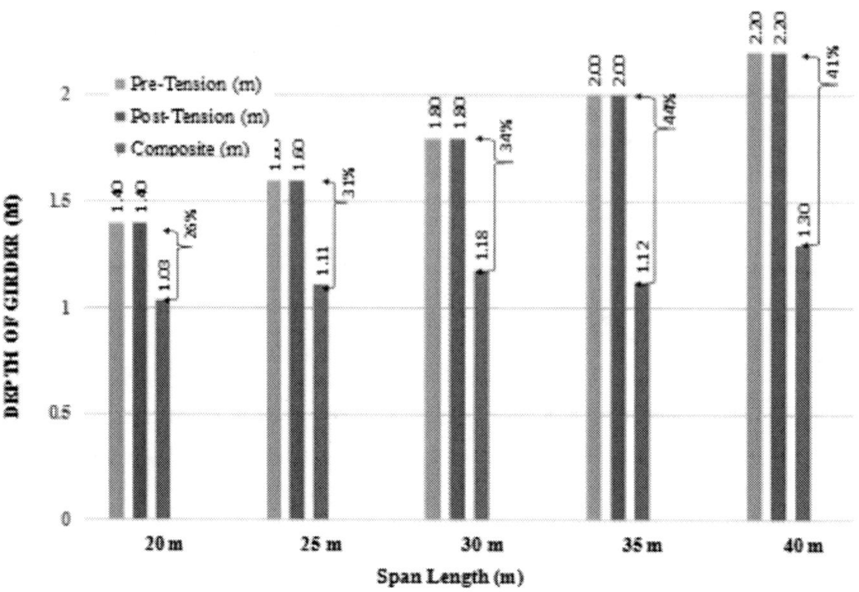

Fig. 3. The relationship between span and depth of the girder.

Following the preparation of the strutrual design for the three types of bridges with five lengths for each type, the material quantities have been calculated for each girder and listed in bill of quantities. These materials include: concrete, steel, strands, grounting, etc. In addition, the rate for each item in the BoQ has been based on average rates of previous projects in United Arab Emirates. The results of the analysis have been illustrated by the following bar-chart figures.

Figure 4 shows the total capital costs in Emirati Dirhams (DHS) of 4-girder bridges of different span and different section type. It is evident that for spans between 20m and 30m, the composite girder has the least capital cost compared with the pre-tensined and post-tensioned girders. However, for spans between 35m and 40m the composite girder is much expensive. The reason for that might be because of the structural loads and depths of girders required for short, medium and long spans.

Whereas, Figure 5 exhibits the 50-year life cycle costs for same bridges. It is worth mentioning that the daily cleaning, which is required for all types of bridges, was not included in the maintenance analysis, because it will not make any difference for the comparison analysis of this study. The daily cleaning will only add the same constant cost for all types of girders. From Figure 5, it is clear that, except for the 20m span, all other spans are more expensive with a composite girder compared to pre-tensined and post-tensioned girders. For both cases in Figures 4 and 5, the growth percentage in cost is increased as the span length is increased.

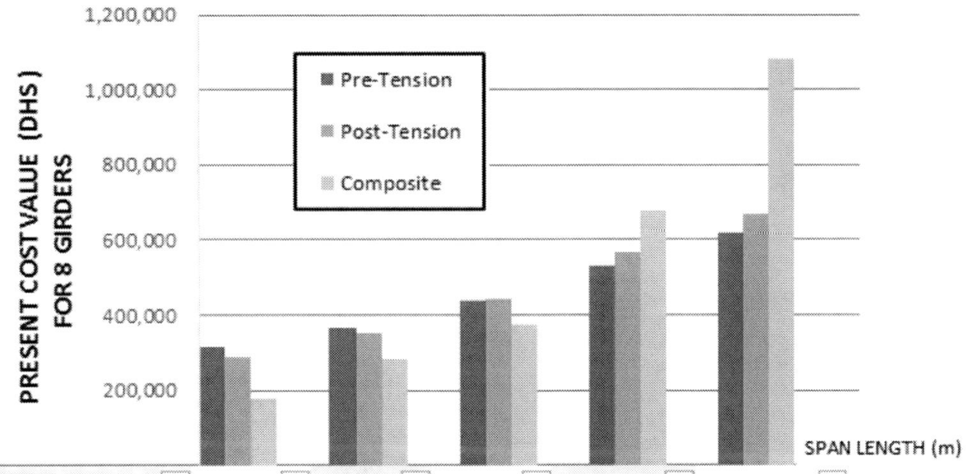

Type of Girder	20 m	25 m	30 m	35 m	40 m
Pre-Tension	315,625.76	367,261.12	440,518.96	529,010.56	616,093.44
Post-Tension	288,480.00	351,379.76	444,634.08	566,702.72	669,506.40
Composite	176,960.00	280,800.00	374,560.00	675,840.00	1,084,000.00

Fig. 4. The total capital costs (present value) for 8 girders of different span and section type.

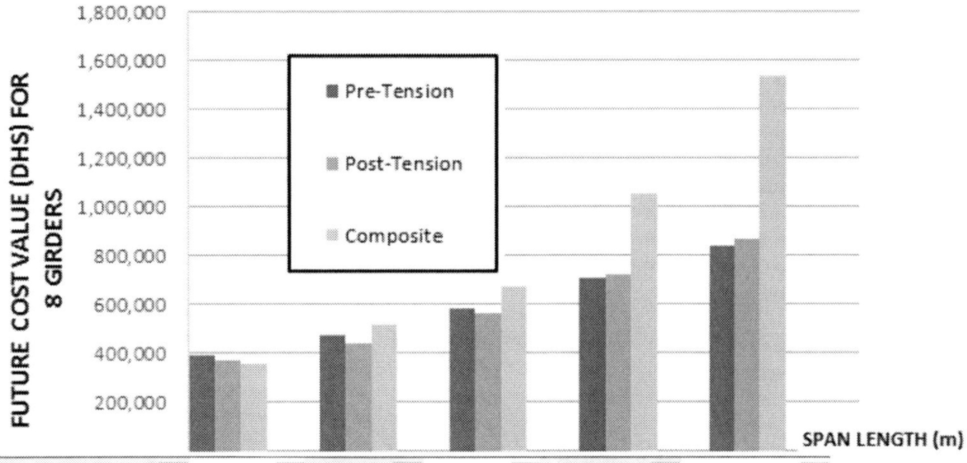

Type of Girder	20 m	25 m	30 m	35 m	40 m
Pre-Tension	390,618.40	472,052.88	579,431.28	706,485.76	836,013.20
Post-Tension	366,991.36	438,544.96	562,423.12	719,546.96	862,266.08
Composite	353,759.12	516,647.92	672,582.80	1,053,057.68	1,531,381.60

Fig. 5. The total 50-year total life cycle cost (future value) for 8-girders of different span and section type.

The conclusion can be seen in Figure 6 which demonstrates the relative differences in cost for both steel composite section and pre-stressed concrete section for all spans. Figure 6 clarifies the considerable impact of maintenance and life cycle costs. Without maintenance consideration, the cost of composite-girder bridge is lower than the pre-stressed one up to span 30m. While, the 50-year life cycle cost with maintenance is always lower for the pre-stressed bridge. It reaches 45% less than the steel composite bridge for 40m span. On the other hand, the FDOT[6] cost can be applicable only for 35m bridge span for capital cost and 25m bridge span in case of maintenance costs. Otherwise, having pre-stressed bridge provides much lower cost.

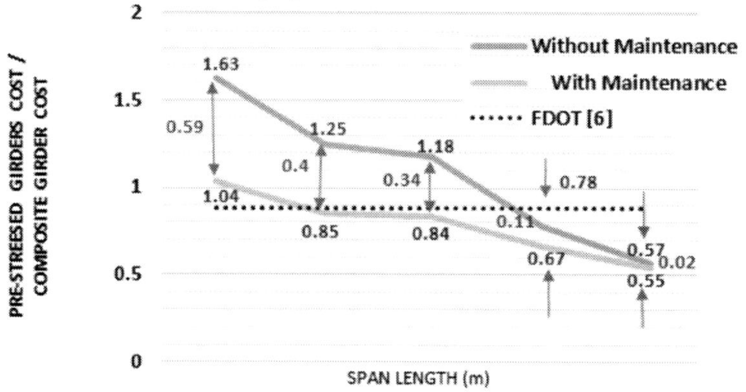

Type of Girder	20 m	25 m	30 m	35 m	40 m
Without Maintenance	1.63	1.25	1.18	0.78	0.57
With Maintenance	1.04	0.85	0.84	0.67	0.55
FDOT [6]	0.88	0.88	0.88	0.88	0.88

Fig. 6. The relative differences in cost for both steel composite section and pre-stressed section.

4 Conclusions

In this paper, three types of girders were investigated: steel composite, pre-tensioned pre-stressed concrete and post-tensioned pre-stressed concrete. The structural design was analyzed for 5 span lengths: 20, 25, 30, 35 and 40m. Then, the capital construction cost and 50-year life cycle cost were accounted for 15 bridges according to each span and construction materials.

It was found that although steel composite section proposes lower section depth, but the life cycle cost of pre-stressed section is always lower and can reach 45% of the composite section cost in some cases. It was also evident that the longer the span of the girder the less expensive the pre-stressed section will be.

References

1. J. Memmott, *Highway Bridges in the United States—an Overview*, U.S. Department of Transportation, Bureau of transportation statics, special report (2006).
2. Road and Transport Authority (RTA), *RTA statistics 2012* (2012).
3. B. Boatman, *Prestressed vs. Steel Beams: Expected Service Life*, Michigan Department of Transportation, State of Michigan (2010).
4. B. C. Bhagyashree C Jagtap and M. Shahezad , *Comparative Study of Prestressed Concrete Girder and Steel Plate Girder for Roadway Over Bridge*, International Journal of Scientific Research in Science, Engineering and Technology (IJSRSET), **2(1)**, pp.113-117 (2016).

5. N. Singh, N.P. Devgan, A.M. Kalra and S. Surinder Pal, *Effect on Economy on Successive Increase in the Span of Bridges*, Proceedings of the 2nd International Conference on Civil, Structural and Transportation Engineering (ICCSTE'16), Ottawa, Canada, May 5 – 6, **Paper No. 114**, pp. 1-8 (2016).
6. Florida Department Of Transportation (FDOT), *Bridge Costs, Transportation costs reports*, April 29 (2014).
7. Highway Agency, *BD37/01: Design manual for road and bridges: Loads for Highway Bridges,* London (2002).
8. Computer and structures, Inc., *SAP2000: Structural Analysis Program*, V15.0.1, Berkeley (2011).
9. British Standards Institution (BSI), *BS 5950-3.1: 1990, Structural use of steel work in building, Part 3: Design in composite construction, Section 3.1 Code of practice for design of simple and continuous composite beams*, London (2002).
10. British Standards Institution, *BS 8110-1: 1997 Structural use of concrete, Part 1: Code of practice for design and construction*, London (2002).
11. H.K.Baker and G.E. Powell, *Understanding Financial Management*, Blackwell Publishing Ltd., USA (2005).

DOI: 10.1051/matecconf/201712008009

APPENDIX A: Estimated sections for five different spans girder-bridge

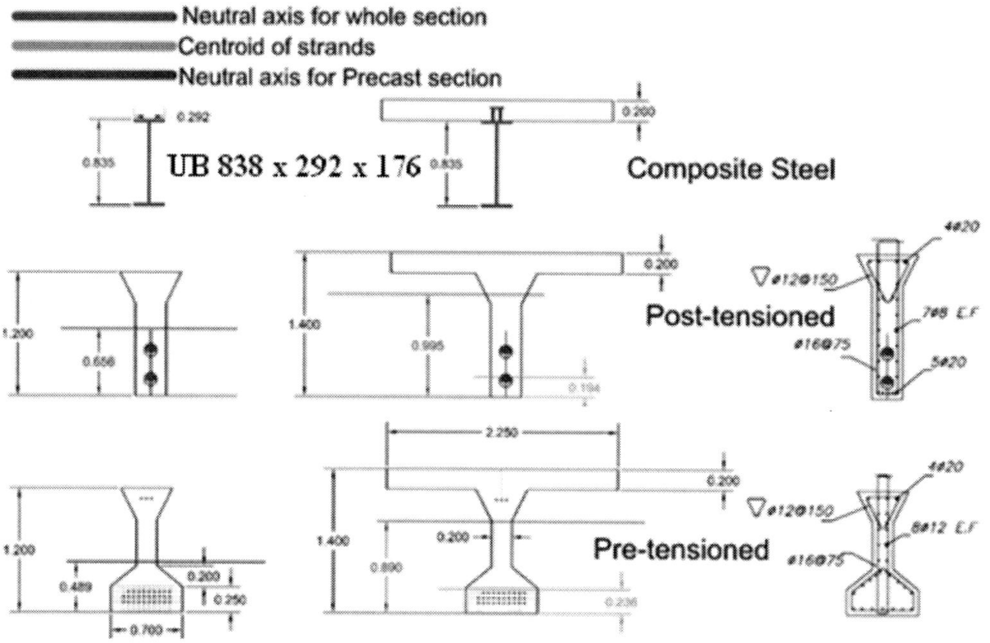

Fig. A-1. Estimated sections for 20m-span bridge.

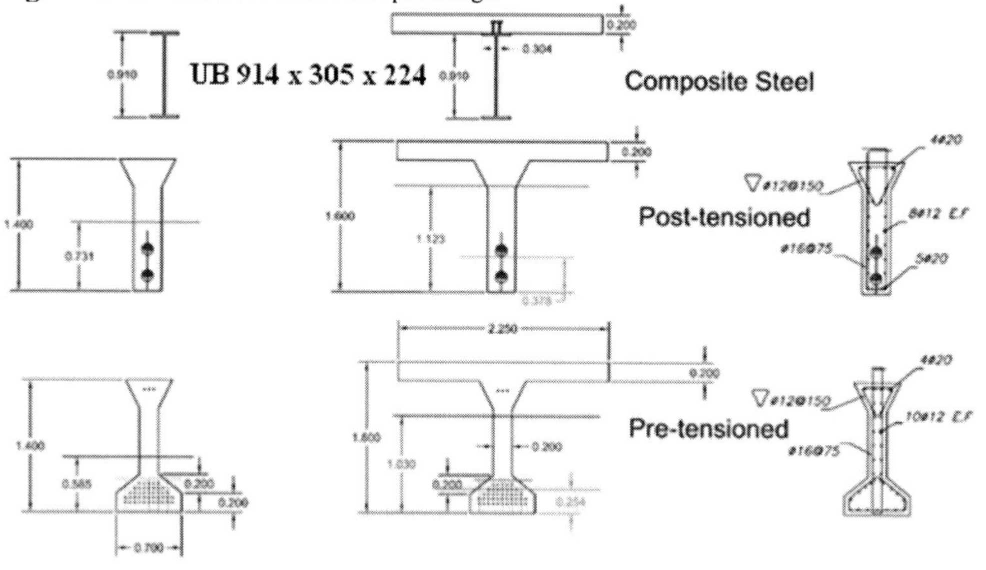

Fig. A-2. Estimated sections for 25m-span bridge.

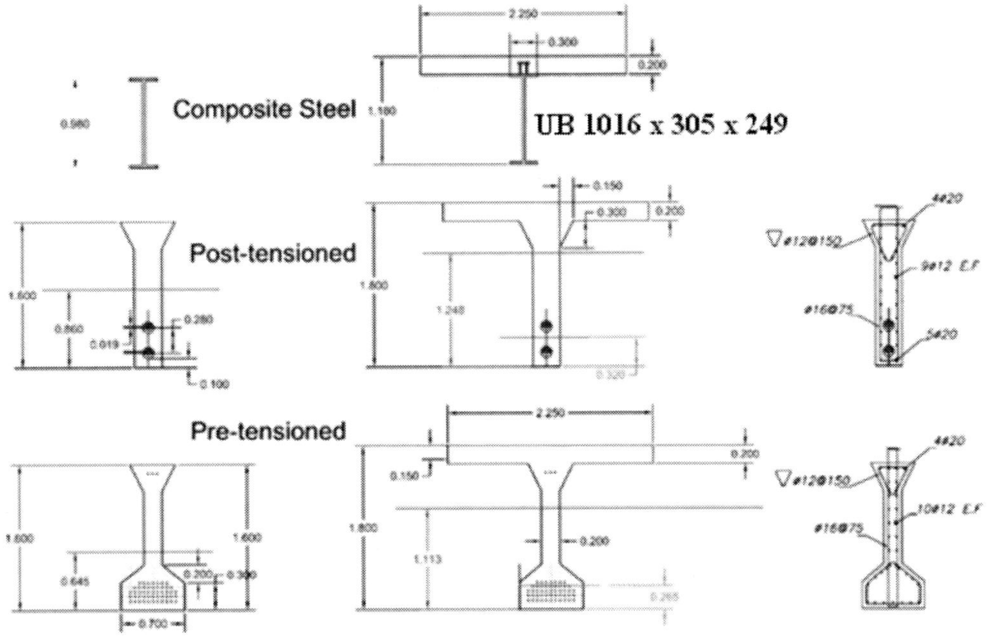

Fig. A-3. Estimated sections for 30m-span bridge.

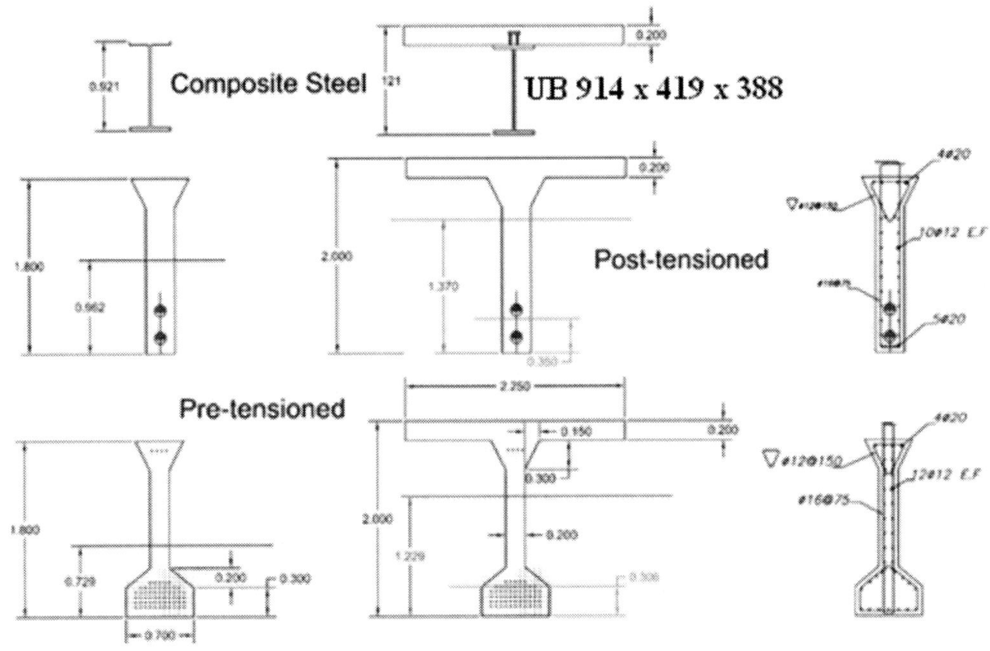

Fig. A-4. Estimated sections for 35m-span bridge.

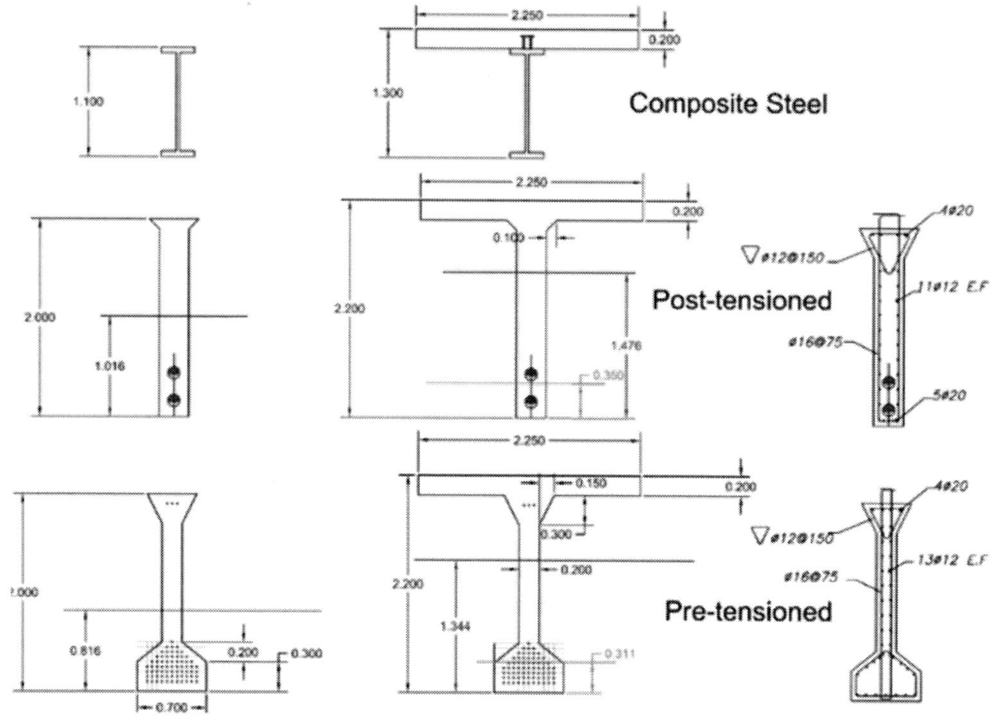

Fig. A-5. Estimated sections for 40m-span bridge.

Studying the impact of construction dewatering discharges to the urban storm drainage network(s) of Doha city using infoworks integrated catchment modeling (ICM)

Mohammed Sameer[1,*], and *Rabee* Rustum[2]

[1]WorleyParsons Qatar, Al Wosail Tower, West Bay, Doha, Qatar
[2]Heriot Watt University, Dubai International Academic City, Dubai, UAE

Abstract - The discharge of construction dewatering flows to the storm drainage network for disposal is a common activity in Qatar. The Dupuit empirical approach was utilized to establish various hypothetical dewatering scenarios on the basis of site classifications, which were modeled on 4 Case Study Areas of Doha's Existing Surface Drainage Network in order to study the impact of dewatering discharge against an established baseline. The simulations were undertaken using InfoWorks Integrated Catchment Modeling (ICM) software for critical and non-critical rainfall events. The results indicated significant localized flooding in excess of the baseline conditions for scenarios exceeding 0.5 m^3/sec flows, while individual catchments demonstrated variations and sensitivities on the basis of catchment properties and rainfall events. It is evident that dewatering discharge under unpredictable rainfall events poses various levels of risk to the city's infrastructure, which is further exacerbated due to the massive scale of construction activity in the country and the rising ground water table in Greater Doha Area basin.

1 Introduction

Although the groundwater tables in Qatar have been depleting in its two major inland basins, namely the northern and southern basins, the Doha basin area has seen a significant increase in the groundwater table [1]. This is primarily due to two major reasons, the first is due to rapid urbanization, causing land use changes & over-irrigation using treated wastewater (TSE) [2, 3], the second is due to increased saline intrusion from the sea [1, 4]. Land use changes for Doha were further studied by Hashem and Balakrishnan between 1997 to 2010 [5]. The resulting rise of Groundwater table in the Doha basin has yilded increased dewatering flows at construction sites in Doha, as well as flood incidences for larger projects, such as the Doha Metro [6].

The overseeing Authority for Drainage Infrastructure in Qatar, the Public Works Authority, ASHGHAL presently recommends, the following four disposal options for

* Corresponding author: ms@msameer.engineer

construction dewatering flows, as prescribed in The Construction Dewatering Guidelines, Qatar [7]:

1. Discharge to sea via Surface and Groundwater Network
2. Direct Discharge to Sea
3. Discharge to lagoons (covered by geotextile is preferred or else lagoon not covered by geotextile).
4. Discharge by re-injection (Deep Aquifer Recharge).

According to the guidelines, Option 1 – Discharge to sea via Surface and Groundwater Network is described as the most common practice in Qatar. However, on small scale projects, direct discharge to the sea or re-injection may not be feasible or cost effective. These flows are usually discharged to the sea or inland lagoons via the storm water drainage network [8].

The practice of disposal of dewatering discharge via the storm-sewer lines is not new. Cox III [9] in his study of dewatering of a construction site in a semi-urban area suggested tapping into the Storm Sewer lines for disposal of construction dewatering discharges. More recently, this practice of disposal to sewer/storm drains has become quite common in many urbanized areas where other disposal options are limited or cost prohibitive. Dewatering practices in Qatar, including the Construction Dewatering Guidelines of Qatar have been formed by adapting parts of various regulations, on the basis of their applicability in Qatar.

Hydraulic modelling tools, such as InfoWorks CS have been used regularly to assess the overflows and flooding of combined sewers [10]. However there hasn't been any significant research undertaken in regards to studying the impacts of construction dewatering flows on the surface network under an active rainfall event in order to assess potential back flooding, overflows or exacerbation of existing sensitive flooded areas.

2 Methodolgy

2.1 Dewatering Discharge Flows for Doha: Field Data vs Empirical Estimation

Dewatering discharge data was obtained from KOOP Water Services Qatar through site visits, for two sites of the Doha Metro Project, the Al Azizyah and Sports City stations of the Gold line. The steady state pumping readings for two subsequent days to maintain current drawdown depths were obtained for each site, which were used to dereive the daily dewatering discharges. For comparison, the dewatering discharge for both the sites is estimated using the Dupuit-Forchhiemer [11-13] approach using the given properties. The properties of these sites, the daily discharge readings and the empirical estimations are summarized in Table 1 below.

Using the Dupuit-Forchheimer approach, the theoretical discharge estimation is very close to the actual discharge flowmeter readings from the sites, as shown in Table 1. It should also be noted that for the Al Azizyah station the estimation is slightly higher than the actual. The degree in variation arises due to various factors from the hydraulic conductivity to the fluctuations in the water table as a result of ongoing dewatering. However overall the Dupuit-Forchheimer approach can be utilized to accurately estimate the flows in gross quantities, with a high degree of accuracy.

Table 1: Site Properties for Al Azizyah and Sports City Metro Stations.

Site Property	Sports City Station	Al Azizyah Station
Excavation Area (m^2)	12,768	7,200
Ground Elevation (m) QNHD	+21.00	+18.5
Existing Ground water level (m) QNHD / m BGL	+15.95 / 5.05	+11 / 7.5
Maximum Excavation level (m) QNHD / m BGL	-9.65 / 30.65	-4.7 / 23.2
Ground water level required (m) QNHD / m BGL	-10.65 / 31.65	-5.7 / 24.2
Total Drawdown required (m) BGL	26.6	16.7
Average Hydraulic Conductivity, K (m/s)	3.11×10^{-05}	10^{-05}
Flowmeter Reading on Day 1 (m^3/day)	2,250,699 (18/03/2016)	702,895 (25/06/2015)
Flowmeter Reading on Day 2 (m^3/day)	2,257,877 (19/03/2016)	704,073 (26/06/2015)
Daily Discharge (Day 1 – Day 2) (m^3/day)	7,178	1,178
Estimation using Dupuit-Forchheimer Approch (m^3/day)	7,486	2,066

2.2 Design Storm Events: Rainfall Data for Qatar

The average annual rainfall in Doha, measured at Doha International Airport from 1962 to 2011 is 76.6mm [14]. The analysis of surface drainage networks are commonly done by utilizing 'Design Rainfall' data. These are derived from the IDF curve for the particular region. For Qatar, the current recommended IDF curve is based on the research by Bazaraa and Ahmed for the rainfall dataset available from DIA from 1962-1989 [15]. Since then the IDF has been updated due to availability of further data [14].

2.3 Baseline Model Overview: Hydrology and Hydraulics

The Storm Water Model for the Existing Surface Drainage Network in Doha city was obtained from the Public Works Authority – Ashghal, Qatar. The model was built in InfoWorks CS as a 1D model, consisting of piped networks, pumps, reservoirs, outfalls and catchments. The entire model has multiple catchments and outfalls, which have been modelled individually as part of various projects and combined into a large central model.A total of 4 selected catchment areas were simulated for this study.

To run an extensive study comprising of various scenarios and simulations, the model was converted to an InfoWorks ICM model in order to utilize ICM's improved functionalities such as the Scenario Manager and the Group Run Scheduler. The Scenario Manager and the Run Scheduling tools were effectively employed to create over 20 different scenarios for each Case Study Area (CSA) model, and run a total of 800+ unique simulations.The model for each CSA was simulated using their existing parameters and no changes were made with the exception of the addition of dewatering flows as per the defined scenarios.

DOI: 10.1051/matecconf/201712008010

2.4 Overall model parameters:

Combined, the 4 CSA's consists of approximately 738 ha of public realm area comprising of roads, plots and network elements. These elements consists of a total of 1670 nodes, 1660 pipes, 5 outfall, 2 storage tank, 6 penstocks (modeled as sluice), 3 screens, 3 weir, 7 pumps and 670 sub-catchments.

All the pipes have the Colebrook-White attribute associated with them, for estimating the frictional losses. The majority of the pipes have a roughness value of 1.5mm and 3mm assigned, which is acceptable given the networks in the model are not new [3]. Some pipes also have roughness of 0.6mm, indicating newer pipes.

2.5 Case Study Areas: Approch & Considerations

The 4 studied Case Study Area's are presented in Figure 1 and further described briefly in table 2.

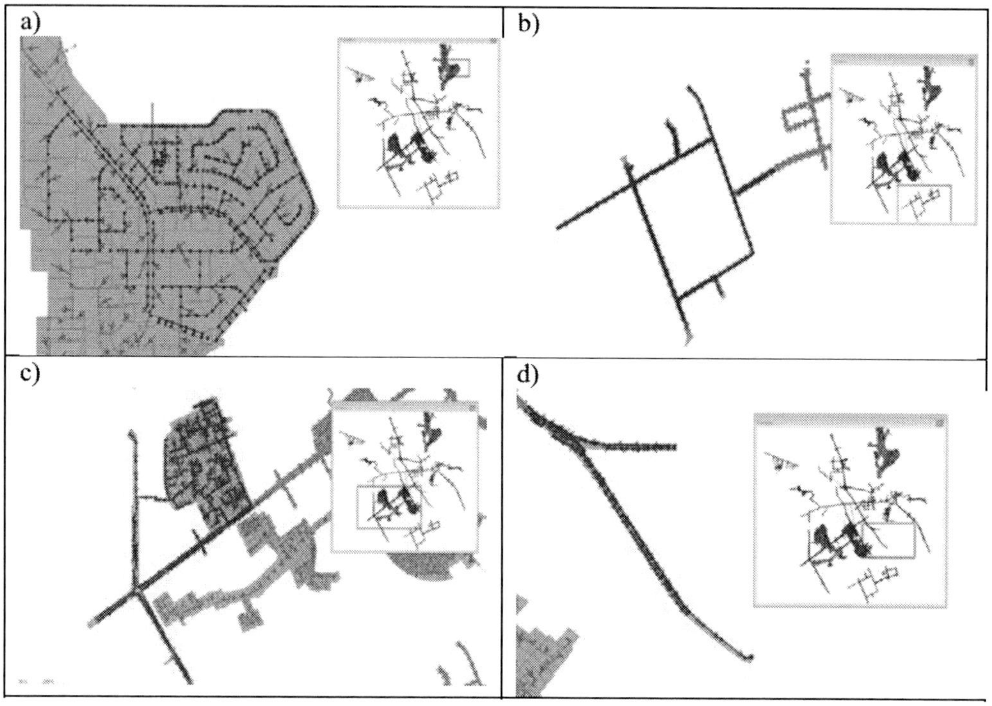

Fig. 1: Model View for a) CSA 1 – West Bay Catchment, b) CSA 2 - Barwa City Catchment, c) CSA 3 - Al Azizya Subcatchment and d) CSA 4 – Section of Doha Expressway.

Table 2: Description of Case Study Areas.

	CSA 1	**CSA 2**	**CSA 3**	**CSA 4**
Water Table	High (1 – 5m)	Average (5-7m)	Average (5-7m)	High (3-7m)
Land use	Highrises, Commercial, with Basements	Residential Complex & Commercial	Mixed Residential & Commercial	Residential & Commercial
Catchment Type	Roads and Plots	Roads Only	Roads and Plots	Expressway Only
Discharge	To Sea via Outfall	To Lagoon via Outfall	To Sea via Outfall	To Lagoon via Outfall

2.6 Location of dewatering sites:

A total of 3 new redundant Subcatchments are added to each CSA to simulate the Dewatering scenarios. The Subcatchments are labelled as Site1, Site2 and Site3. The hypothetical scenario site Subcatchments are located on actual empty plots within CSA catchment area. The flows are added as 'tradeflow' to each site, corresponding to the simulated scenarios and discharge is allocated to the nearest manhole in the catchment.

Subcatchment 'Site1' is simulated for all the single site scenarios. In these simulations, all the other sites have no flows assigned to them and are rendered inactive. The location of 'Site1' in all the CSA are on the empty plot farthest from the outfall. This enables the dewatering flows to remain longer in the system and travel through longer lengths of pipes within the network. The scenarios were built and assigned to the relevant Sites using the 'Scenario Manager' function of InfoWorks ICM. Overall, there are a total of 20 scenarios simulated for each CSA per storm event, 10 for low K ($3x10^{-5}$) and 10 for average K ($5x10^{-4}$). This brings out a total of 20x4x9 = 720 scenarios across the 4 CSA areas or 180 dewatering scenarios per Case Study Area. These scenarios do not include the initial baseline runs.

3 Results and analysis

The model was run for a total of 16 rainfall events to analyze and define the baseline conditions, prior to the addition of any dewatering flows. These events and the flood analysis of the baseline model runs are shown in Table 3 for flooding above the kerb level. The results in the baseline model are only considered for the Case Study Areas.

Table 3: Number of flooded nodes above kerb level for Baseline Model (Combined)

Duration/Frequency	15 Minutes	30 Minutes	60 Minutes	120 Minutes
1 in 2 years	1	4	7	5
1 in 5 years	3	23	52	49
1 in 10 years	15	59	131	85
1 in 25 years	56	142	193	182

The results of the baseline model run shows flooding in all the case study areas for all durations of 1 in 25 year storm event. Majority of the flooding is in CSA 1 and 3, while CSA 2 and 4 indicated negligible baseline flooding. It was also observed that the flooding peaks for the 60 minutes duration storms. The number of flooded nodes can be seen receding for the 120 minutes duration rainfall. This is observed for all rainfall frequency events. The least flooding was observed in 15 and 30 minutes duration storms for 1 in 2, 1 in 5 and 1 in 10 year events.

It is clear that the 60 minutes duration storm is critical storm for the baseline model, having the most number of flooded nodes. Similarly 15 and 30 minutes duration storms are non-critical but will cause flash floods. The 1 in 25 year frequency storms have been observed to be severe for all durations. The 120 minutes duration storms are non-critical and are excluded for dewatering scenarios. Furthermore the PWA Ashghal lists the basis of these storm events and their associated flood protection requirements, as illustrated in Table 4 below [16].

It is understood that if a rain event causes flooding in the catchment, any additional flows will only further exacerbate the flooding, thus 1 in 25 year storm events are also excluded from the study. Based on the results of the baseline runs, the model is re-run with the subsequent dewatering scenarios for only the critical and non-critical events, i.e., for 15, 30 and 60 minutes durations of the 1 in 2 years, 1 in 5 years and 1 in 10 years frequency storm events. The baseline is used as a benchmark to study the impact of dewatering flows under the same rainfall events.

Table 4: Flood Protection Criteria for Various Areas in Qatar [16].

Event	Area
1 in 2 Years Storm	Parks, Playgrounds, natural areas and minor roads
1 in 5 Years Storm	Low cost housing, Major roads
1 in 10 Years Storm	Government, institutional and other office developments, technically sensitive property, basements, power equipment, High cost housing
1 in 25 Years Storm	High prestige or Ceremonial development

For analysis, the results were further constructed as column charts on the basis of total number of flooded nodes for both baseline and dewatering scenarios. This sorting of data resulted in 3 charts for each CSA, with each chart illustrating the impact of dewatering discharge in comparison to the baseline for the respective storm event frequencies. These charts and their subsequent analysis for CSA 1 is presented below.

3.1 Analysis of CSA-1 Dewatering Scenarios

The run result graphs for CSA 1 – West Bay Catchment is presented in Figure 2 to Figure 4 below.

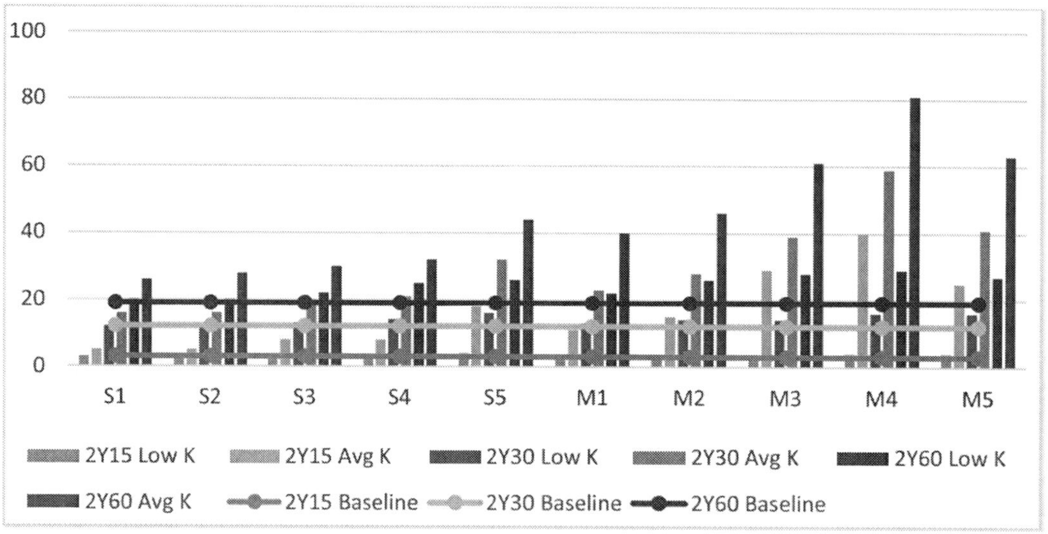

Fig. 2: Number of Flooded Nodes for CSA 1 – 2 Year Storm Events.

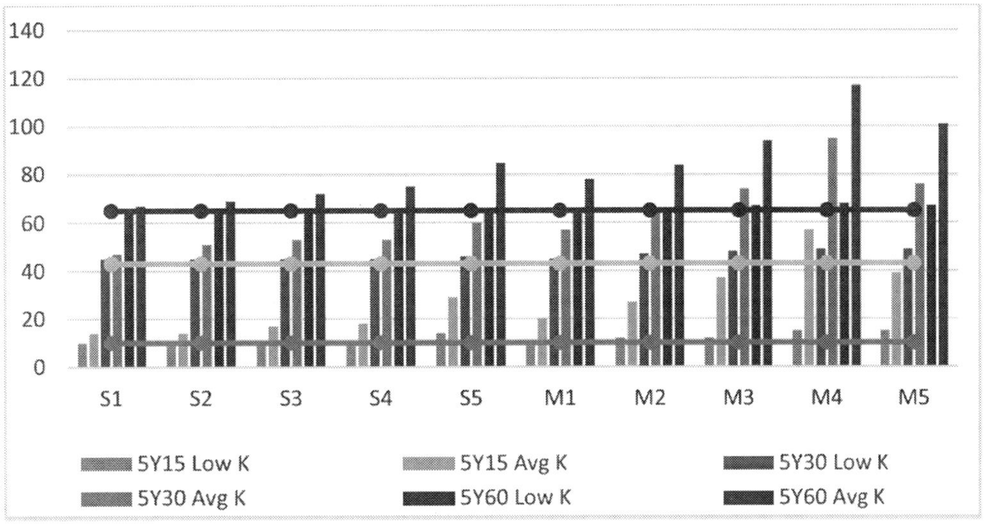

Fig. 3: Number of Flooded Nodes for CSA 1 – 5 Year Storm Events

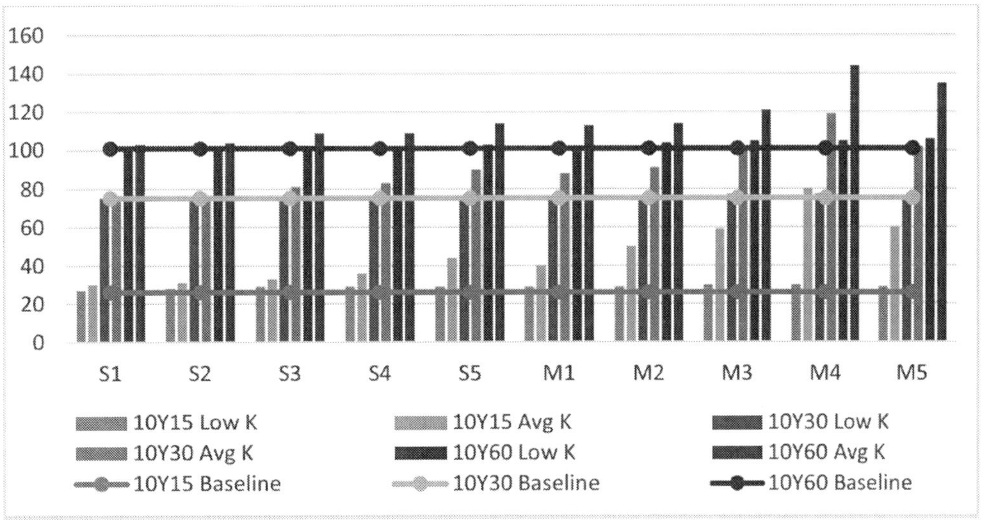

Fig. 4: Number of Flooded Nodes for CSA 1 – 10 Year Storm Events

The dewatering scenario for 2Y15 Low K has no significant impact on the network across the various site types and stays at par with the baseline. In contrast, the dewatering flows yielded by 2Y15 Average K indicates increased flooding, which increases with the site type for single site scenarios (S1 to S5) and varies with multi-site scenarios (M1 to M5). Similar observations are made for the 5Y15 and 10Y15 Low K scenarios, with only slight increase in flooding noted. The impact is more significant in the 2Y30 minute and the 2Y60 minute Low K duration storm scenarios, as increased flooding is noted in comparison to the corresponding baseline. This is further exacerbated for the 2Y30 & 2Y60 Average K scenario types. On the contrary, this is not the case for the 5 year and 10 year storm event frequency of 30 and 60 minutes durations, each of which has a higher baseline flooding to begin with.

3.2. General assessment for dewatering discharges to CSAs:

The results indicates that impact of Dewatering Discharge within CSA 1 area is critical for shorter duration rainfall and is likely to cause more flash flooding for storm events under 30 minute duration storms of shorter frequency of 1 in 2 year to 1 in 5 year. Overall CSA-1 catchment area is more sensitive to flooding and the impact of dewatering discharge within this catchment even under non-critical storm events can result in an increase in flooding incidences. However, as more residential areas contribute to the catchment, the resultant baseline flood incidences will rise and any further dewatering activities will only excarberate the baseline flooding, as seen in CSA 3.

In regards to flooded areas within the catchment, the baseline flooded locations are the most sensitive areas. All the nodes flooded for the baseline, are also flooded in subsequent dewatering scenarios. However, there are additional flooded nodes in dewatering scenarios which are not flooded in the baseline. These nodes are located within the vicinity of the dewatering site (Subcatchment) where additional flows are added, indicating immediate localized flooding during a rainfall event. The flooding pattern indicates that the type of site and the number of active sites, i.e. the dewatering flow quantities play a vital role for changes in flood patterns under simulated rainfall events.

Since CSA-2 is a road only catchment, the catchment is very sensitive to any additional flows beyond its design limits. Similar observations are made in CSA-4,which has a higher design service. The pattern for road only catchments indicate that they are not very sensitive to flooding for non-critical events or events of shorter frequencies and durations, but of larger stroms beyond the design capacity. Although CSA 2 and CSA 4 may present an ideal opportunity to discharge smaller dewatering flows to the catchment, it should be noted that the outfalls are not to the sea but to lagoons and thus controlled discharge has to be undertaken with respect to the capacity of the lagoon itself with enough contingency for the runoff from any potential rainfall event, if the dewatering activity is expected to coincide.

4 Discussion

4.1 Variations in impact

1. Variations due to Site Type Category: Throughout the simulations, variation in impact of dewatering discharge is observed due to site category. For single site scenarios, it is noted that S1 (Small Residential) and S2 (Compounds) type sites, have minimal impact at low flow rates (Low K) and moderate impact at high flow variations (High K). Similarly, flooding incidences rise in case of S3 (High Rises) and S4 (Commerical) categories, which have moderate (Low K) to high impact (Avg K) in comparison to the baseline for the given storm event. In all the scenarios the largest impact noted was from the S5 (Mega Projects) category. When Multi-site scenarios are evaluated the results are dependent on the contributing sites and the CSA. Overall the impact in multi-site scenario is higher, however the dewatering flows from multiple sites (2+) can actively be discharged and managed, given the contributing flows from two or more sites are low.
2. Variations in Impact due to Location of Construction Site: In all the CSA's analyzed any dewatering induced flooding beyond the baseline was highly localized to the contributing Subcatchment, which is the location of the hypothetical construction site. Since flooding is highly localized the impact will vary depending on the sensitivity of the catchment within the vicinity of the dewatering site.
3. Variations in Impact due to Hydraulic Conductivities & other Ground Water Properties: The resulting flows yielded by changes in ground properties varies significantly and play

a role in the contributing flows from dewatering activities. The Dupuit approach provides a range of flows from 0.01 m^3/s to 0.9 m^3/s for the simulated scenarios. The variation is largely in part due to the variation in hydraulic conductivity, K of the soil and the water table level (m) BGL. The range is further divided and varies for the dewatering site properties on the basis of the total excavation area and the excavation depth, both of which have been used to define and categorize sites for the scenarios studied. It should also be noted that if the hydraulic conductivities were greater than the values utilized (Low K at $3x10^{-5}$ m/s and Avg K at $5x10^{-4}$ m/s), the impact or its variation may be more severe. For smaller catchments, variation in geotechnical properties are expected to remain within a minor range.

4. Variations in Impact due to Storm Events: Variations occur due to both the frequencies of the storm events and the duration of the event. Since the runoff from the catchments will vary for different storm events, the resulting impact from dewatering is also highly dependent on them. It is evident that flooding will occur when runoff generated from contributing catchments exceeds the network's design service levels, which is highly dependent on the Time of Concentration, i.e., the longest run length from upstream to downstream and the properties of the components of the network such as pipe sizes, roughness, leakages, age of network, etc. This is implicated in the form of overall sensitivity of the catchment due to these contributing factors.

5 Overall impact of dewatering discharge

The overall impact of dewatering discharge to the surface drainage network for Doha under rainfall conditions is summarized in Table 5 below on the basis of the dewatering flow quantities. The table lists the number of nodes flooded beyond the baseline flooding, for the given storm events. The summary excludes scenarios where there is no baseline flooding in order to deduce the impact of dewatering discharge due to the existing sensitivity of the catchment.

Table 5: Impact Risk Assessment for Dewatering Discharge

Total Dewatering Discharge	Impact with 2Y Storm Events (Durations, minutes)			Impact with 5Y Storm Events (Durations, minutes)			Impact with 10Y Storm Events (Durations, minutes)		
Q (m^3/s)	2Y15	2Y30	2Y60	5Y15	5Y30	5Y60	10Y15	10Y30	10Y60
0.01 – 0.05	0	7	11	0	6	2	6	0	0
0.06 – 0.1	2	21	42	10	19	12	18	3	14
0.11 – 0.25	12	25	46	20	30	20	20	6	19
0.26 – 0.5	22	29	36	23	30	29	29	22	23
0.51 – 0.75	32	39	60	35	46	44	54	45	36
0.76 – 1.00	59	66	86	61	63	66	79	63	54
>1.00	83	102	132	98	105	115	122	110	116

Impact Risk Categorization based on Scoring:

	0-10:	Negligible Impact
	11-25:	Low Impact Risk
	25-50:	Moderate Impact Risk
	50-75:	High Impact Risk
	75+:	Severe Impact Risk, discharge not recommended.

It is evident that as the flow quantity discharged to the network increases, so does the corresponding flooding regardless of the storm event. The only significant difference is the variation in the level of flooding on the basis of the rainfall event. It is also noted that the flooding is negligible for smaller flow quantities below 0.1 m^3/s. However Low Impact Risk exists for frequent storm events of extended durations such as the 2Y60. The exacerbation is minimal for 2Y15 and 10Y60, the two ends of the spectrum. Both these events show similar behavior, though the latter will have increased flood duration for the incidences than the 2Y15. In case of larger flow quantities in excess of 0.5 m^3/s, the Impact Risk increases significantly from moderate to high and severe flooding. Flows larger than that quantity will expose the network to high Impact risk, with potential localized flooding incidences even without an active rainfall event. It should also be noted that the chart indicates flooding in excess of the baseline, and is essentially a chart depicting the exacerbation of flooding if there is existing potential of flood incidences within the catchment.

5.1 Recommendations for discharge management

1. Recycle & Reuse Dewatering Effluent: This practice falls under 'green site' concept and has been recommended under various frameworks and standards. However recycling of water is not fully utilized on most small scale projects in Qatar [7].
2. Flow management by Relevant Authorities: Potential Flooding due to dewatering discharge on storm drainage can be controlled by effective flow management and setting discharge limits as part of O&M. Limits will vary depending on the capacity of the network the flows are being discharged to, the number of contributing discharge flows and rainfall conditions.
3. Point of discharge: In addition to controlled discharge, the allocated manhole/point of discharge in relation to its location from outfall can impact the localized flooding and downstream conditions. Allocation of discharge point can help prevent minor flood incidents during rainfall events. Locations prone to flooding during rainfall events and those of high importance should be avoided for dewatering discharge, even for small quantities. If the construction site is large, multiple point of discharges can be allocated to minimize the impact and reduce localized flood incidences.
4. Ground Water Re-injection as a Disposal Option: The Inland Groundwater Basins have reduced significantly over the years [9]. This has forced the country to rely on Desalination to meet potable water demands. Artificial Groundwater Recharge is one of the grand challenges listed out by the Qatar National Vision 2030. There have already been research to identify areas of natural groundwater recharge in Qatar with watershed modelling and recharge using desalinated seawater to maintain current groundwater reserves [17, 18]. This principle has already been adopted in some sites of Doha Metro Project by Qatar Rail. Outskirts of Doha also provide the opportunity to use lagoons for re-injection instead of pumped recharge, in a very cost effective and natural manner.
5. Groundwater Control via Horizontal Drainage: Horizontal drainage via symbiotic networks of perforated pipes within the existing surface drainage network should be explored. It is understood that some parts of Doha may already have such systems in place [3]. Horizontal drainage can be utilized to maintain a service level of the groundwater table and thus indirectly reduce the discharge flows from construction dewatering activities. These flows are expected to increase as the water table increases in the near future.

6 Conclusion

The Dupuit approach was successfully utilized to estimate the construction dewatering flow quantities. The validity of these estimates were also tested by obtaining field data from current ongoing sites. Finally, the impact of Construction Dewatering Discharge to the Urban Storm Drainage Networks of Doha city was successfully evaluated using InfoWorks ICM for a total of four different Case Study Areas. The modelling simulated various storm events of varying frequencies and durations. The results arising from these simulations were analyzed and the impact correlation with the total quantity of flows discharged to the network was developed. It is noted that each catchment behaved uniquely on the basis of its runoff surfaces and the total contributing dewatering discharge. There are noticeable differences in the impact due to variations in properties such as hydraulic conductivities, water table levels, catchment types, runoff coefficients, location of site, etc. These properties were also examined and the difference in impact highlighted for the same. On the basis of the results obtained and the data presented, it can be concluded that Dewatering Discharge to the Surface Drainage Network is an overlooked activity that has various risks associated with it. The impact of dewatering discharge can yield significant localized flooding incidences with flows above 0.5 m^3/sec and minor incidences with flows below 0.5 m^3/sec. The scenarios studied are plausible and to a very large degree pre-existing within the Gulf Region.

References

1. K.M. Amer, *Groundwater for sustainable development: problems, perspectives and challenges*. Chapter 3, CRC Press (2008)
2. Ministry of Development Planning and Statistics, *Water Statistics In the State of Qatar 2013* (2016)
3. MWH, *Qatar Integrated Drainage Master Plan. Final Master Plan Report*, **Issue 1**. Doha (2013)
4. K. Al-Akhras, and A. Al-Miraikhi, *From Fundamentals to Applications in Geotechnics Proceedings of the 15th Pan-american Conference on Soil Mechanics and Geotechnical Engineering*, IOS Press, 15-18 (2015)
5. N. Hashem, and P. Balakrishnan, *Annals of GIS*, **21(3)**, pp.233-247 (2015)
6. J. Bambridge, Rising water table 'challenging' Doha construction. *Construction Week Online* (2015)
7. Public Works Authority, *Construction Dewatering Guidelines Qatar* (2014)
8. G. Perez Russo, Definition and implementation of the dewatering work method statement for the Red Line North underground project in Doha-Qatar (2014)
9. G.C. Cox III, *International Conference on Case Histories in Geotechnical Engineering.* **Paper 12** (1984)
10. P. Koudelak, and S. West, Water and Environment Journal, **22(2)**, pp.81-87 (2008)
11. J.P. Powers, and C.J. Herridge, *Construction dewatering and groundwater control: new methods and applications*. John Wiley & Sons (2007)
12. M. Preene, *Groundwater control. Design issues*, **1185**, pp.80-7 (2000)
13. M. Preene, *Groundwater lowering in construction: a practical guide to dewatering* **Vol. 6**. CRC Press (2012)
14. A. Al Mamoon, N.E. Joergensen, A. Rahman, and H. Qasem, International Journal of Sustainable Built Environment, **3(1)**, pp.111-118 (2014)
15. A.S. Bazaraa, and S. Ahmed, Rainfall Characterization in an arid area (1991)
16. Public Works Authority, *Qatar Sewerage and Drainage Design Manual*, **Vol 3** (2005)
17. H. Baalousha, MODSIM2015, *21st International Congress on Modelling and Simulation*. pp. 2026-2032 (2015)
18. J.O. Kimrey, Proposed artificial recharge studies in northern Qatar No. **85-343**. (1985)

User-inspired design methodology using Affordance Structure Matrix (ASM) for construction projects

J. Uma Maheswari[1,*], *V. Paul C.* Charlesraj[2], and *Soma* Battacharya[3]

[1]Indian Institute of Technology Delhi, New Delhi 110016 India
[2]RICS School of Built Environment, Delhi NCR 201313 India
[3]Samsung Heavy Industries, Noida, Uttar Pradesh 201304 India

Abstract. Traditionally, design phase of construction projects is often performed with incomplete and inaccurate user preferences. This is due to inefficiencies in the methodologies used for capturing the user requirements that can subsequently lead to inconsistencies and result in non-optimised end-result. Iterations and subsequent reworks due to such design inefficiencies is one of the major reasons for unsuccessful project delivery as they impact project performance measures such as time and cost among others. The existing design theories and practice are primarily based on functional requirements. Function-based design deals with design of artifact alone, which may yield favourable or unfavourable consequences with the design artifact. However, incorporating other interactions such as interactions between user & designer is necessary for optimised end-result. Hence, the objective of this research work is to devise a systematic design methodology considering all the three interactions among users, designers and artefacts for improved design efficiency. In this study, it has been attempted to apply the theory of affordances in a case project that involves the design of an offshore facility. A step-by-step methodology for developing Affordance Structure Matrix (ASM), which integrates House of Quality (HOQ) and Design Structure Matrix (DSM), is proposed that can effectively capture the user requirements. HOQ is a popular quality management tool for capturing client requirements and DSM is a matrix-based tool that can capture the interdependency among the design entities. The proposed methodology utilises the strengths of both the tools, as DSM compliments HOQ in the process. In this methodology, different affordances such as AUA (Artifact-User-Affordance), AAA (Artifact-Artifact-Affordance) and DDA (Designer-Designer-Affordance) are captured systematically. Affordance is considered to be user-driven in this context that is in contrast to prevailing design methodologies. The initial feedback from industry on the proposed ASM-based design methodology is encouraging.

* Corresponding author: umapaul@civil.iitd.ac.in

1 Introduction

A genuine question which arises in modern days to design science is: "Is design only a predictive input-output process?" In other words, "Is it sufficient to have the archetypal design where an end result is certain at the end of execution of a defined algorithm?" The answers to these questions are no longer "Yes" and a broader arena of design science is an area of modern day research. A designer designs the product by considering all functional requirements, which describe closed system only without considering user actions. But same product can cause unexpected behaviour when it enters into the environment. Hence, before entering into the construction phase, the design has to be checked for positive and negative user actions. These user actions are nothing but "affordance". Affordance is what one system interacts with another system (user, artifact). Therefore, design earlier, which was treated as a closed system design, should now be considered as an open system design (involving environment). Complex interactions arise between users and artifacts when open system design is considered. Wegner [1] established a proof showing that "interaction is an inherently more powerful paradigm for computing than algorithms". Hence, in addition to algorithmic based design approaches, user action based design, which is also known as "affordance based design" should also be considered at design phase [2]. Many theories and methodologies have evolved in order to cater to the modern day design requirement. When it comes to construction design, the requirement is even complex and amorphous.

Therefore, affordance based design, which supports such complex and open system design has been considered for investigation for the design of construction projects. Unlike functional design, affordance based design captures two way relationships. Thereby, its suitability was tested in an ongoing project involving the design of Solid Control Room/Shale Shaker Room as part of an offshore project. The complex and interactive feature of construction design has been represented in the form of an Affordance Structure Matrix (ASM). With the help of a real life case study, an endeavour has been made to represent various interactions and a path forward for the downstream design has been proposed.

2 Background

2.1 Design changes

The biggest challenge which the construction industry is facing as of today is change management. The changes that are surfaced during the execution stage has a great bearing on quality, schedule and the cost of any project. The root cause analysis of the major changes can be greatly attributed to the poor design. One can say that the main reason for this could be, there is no systematic and scientific process for design and design management, which is being followed for construction design. Till today, construction design is more of experience oriented rather than any process orientation. It is mainly human driven and so it is error prone. The construction design can be considered as a conglomeration of many subsystems. So, in simple terms, in the definition of construction design, the two most important aspects are:

- System and subsystem requirements
- The interactions among subsystems and users

If we consider the timeline of entire construction project design phase as shown in Figure 1, though the requirements are normally considered in the initial part of the design stage of the project, as the requirements understood completely, various teams starts interacting each other in order to generate a solution to the given problem. However, there is no methodology to capture these interactions which are an integral part of a complex system design like that of construction design. A greater need is to find out a single platform or tool which can simultaneously cater to both requirements and system interactions and thereby generate design guidelines, which will ultimately result in a better design and thereby minimizing the cost and schedule impact of construction projects. Affordance based design is a methodology which is built around considering functions (requirements) as well as every type of interactions present in the system. The suitability of this methodology is tested with a real life example.

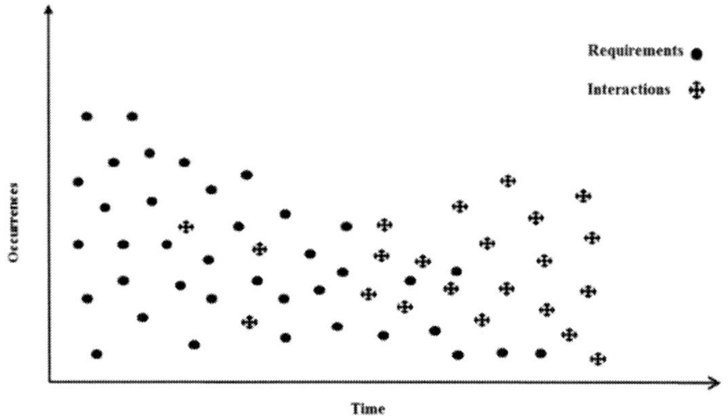

Fig. 1. Timeline of design phase of a construction project.

2.1.1 Affordance

The word "Affordance" does not exist in the Oxford dictionary. However the genesis of Affordance is the word "afford" and in functional sense it means "To afford". Different researchers and scientists have defined affordance in their own terms and understanding as follows:
Affordance is "Perception of the environment leads to a course of action" [3];
Affordance can be defined as "the actionable relation between the environment and an actor" (a person or animal or an object/artifact) [4];
"Affordances are properties of the environment taken relative to an observer."[5].

In the perspective of design, Affordance can be defined how a function is interacting with its surroundings. Maier [6], has put forward few adjectives which connot be affordance in design. According to him a design can't be separated from certain attribute like semantic (by literal), semi-rational, fuzzy-algorithmic, predicative, subjective, and unpredictable nature of humanity, because everything is always designed for human use, usually designed by humans themselves), and situated within an extended context of a complex world.
Artifact - An object which is not naturally present but occurs as a result of the preparative or investigative procedure is termed as artifact in general in design. In simpler term artifact what we are interested to design or what is the desire of the end user.

User - The ecological surroundings where and artefact will be used or operated consists of its users. The users could be the human who will be personally using the product, the environment where it is situated, the surrounding objects which might interact with it in the course of its operation are all considered as the user for that particular artifact.

Artifact-Artifact Affordance (AAA) - In a complex multi artifact system the different artifacts need to remain in self-organized, ordered and provide all the affordance to each other. They shall be in harmony and in an equilibrium in order to achieve a sustainable design. All requirements which one artifact has to provide to each others are defined as the artifact-artifact affordance. In construction design this definition finds an important significance as AAA is the core of construction design. AAA has a great significance in construction design as there are extensive numbers of interactions in construction design. Hence, it is imperative that this interaction is considered at an early stage and due importance is given at each and every stage of design. The design improvements and identifying the area of improvements can be identified from the AAA relationship. The essence of affordance is reciprocities. Hence, when we consider AAA, it actually invokes the reciprocity or the trade-off of design, which brings about an improved design. The number of such interactions is captured in early design stage would result into a refined and improved design and execution. AAA expresses a relationship that is indirectly useful to users. Floors must support users walking on them, however, walls must support roofs, but this is ultimately to protect users within the building. AAA is significant when there is a multi-designer interface. It is of great importance that at least the AAA are identified so that designers are in a position to identify the area where this interaction can be reviewed. In case, if there are any negative affordances and the artifacts are not in harmony and equilibrium, then this can be investigated and negative affordances can be mitigated by making design improvements. For ascertaining this requirement, it is important to identify how the artifacts are related to one another and how best these relations can be established. In construction design, these play an inescapable feature and can be the leading factor for increasing design efficiency.

Artifact-User Affordance (AUA) - The essence of affordance primarily focuses on the user and endeavors to identify and maximize the artifact user affordance. AUA is the expectations or the integrated functionality the artifact provides to its users in the entire life cycle of the artifacts. A better and improved design is the one where there is an overall increase in the affordance. This could be achieved by three ways. First, ideate new and additional affordances. Second, is to improve the value of the existing positive affordances without minimizing other positive affordances. Third, is to decrease the effect of negative affordance. Figure 2 indeed shows in simple terms how four artifacts are offering the affordance to each other as required by extending the puzzle semicircle and by a cut-out on the puzzle body. Similarly, this artifact A, B, C and D can be held by the user P, Q, R and S conveniently to finally make the Jigsaw complete.

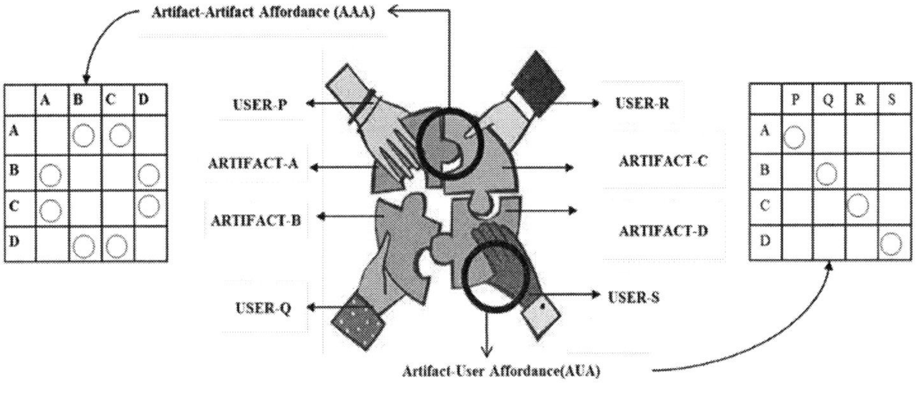

Fig. 2. AAA and AUA.

2.1.2 Affordance structure matrix (ASM)

All underlying theories and concepts ultimately need to evolve as an implementable form. Affordance Structure Matrix (ASM) developed by Maier et al [7] is that very form, which embodies the concepts of affordance and inspired by matrix tools of House of Quality (HOQ) and Design Structure Matrix (DSM) as shown in Figure 3. The ASM is a tool to represent interactions and compare requirements information with physical structure at the conceptual stage. Requirements information on design is termed as the affordance of the systems.

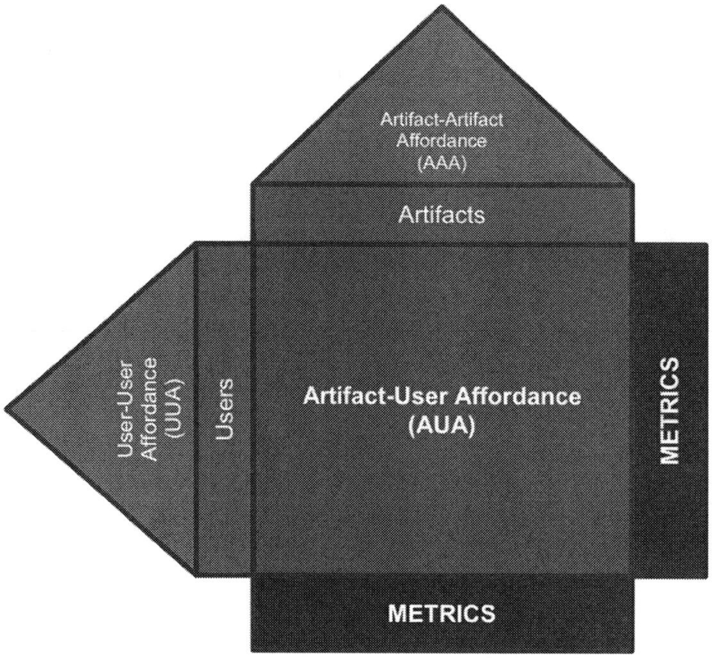

Fig. 3. Generic Representation of ASM.

The top part or the "Roof of HOQ" of the ASM is a Design Structure Matrix (DSM) that depicts the relationships among components. This is the most important part as this depicts AAA. Since these interactions play a significant role in the design of a construction project, building this metric is one of the prime steps in developing the ASM. The left side of the ASM similarly captures the relationships between affordances. However the interaction matrices and their organizations may vary depending upon the design problem and its complexity. The utility of ASM is that it is a good platform to represent various interactions and also it is possible to numerically represent those interactions.

3 Solution methodology

The traditional flow chart of affordance based design developed by Maier and Fadel [8] has been modified to suit the requirement of construction design and presented in Figure 4. The flow of execution this methodology is matched with the construction project design requirements. The step by step process to be followed, starting from the desire and subsequent steps have been shown in the Figure 4. The generic affordances can further be refined on specific design cases. However, these are some requirements, which are generic and is applicable in almost all types of construction design. This can be further be expanded and all micro level affordances can be identified and depicted for considering them into the design.

It is imperative that some steps may not be exactly relevant in some cases, however, the options can be evaluated in case to case basis. A systematic method to identify affordances using the interaction between functions and tasks that shall be performed by the users have been introduced in ASM for evaluating what affordances are embedded in each component of a product and thus grading. This matrix can illustrate correlations of affordances and also of components. This modified methodology is suitable for the construction design where multi artifact interactions are of prime importance. It is taken care that all different kinds of interactions up to the designer level has been captured into the methodology so that the design cycle is complete.

4 Case study

An ongoing offshore project was considered for demonstration. This project is involving the construction of a jackup rig, which is a type of mobile platform primarily used for drilling operations in the sea for an oil major of Norway that had intended to develop a jackup rig facility in the Norwegian Continental Shelf (NCS). The main study is on a "Solid Control Room/Shale Shaker Room" which is a component of jackup rig. The shale shaker room (SS room) was in its design stage and will be soon resuming its fabrication phase. SS room used as solid control system to separate solids from drilling mud. Drilling mud is used for drilling process on sea bed which is to lubricate and cool the drill bit. However, while drilling, solids generating from seabed cutting obstructs the drilling process which has to separate in shale shaker machine and reuse the drilling mud. Hence, many interactions are developed between users and machine (Artifact) inside SS room and artifact to artifact in this operation. Bad design results in inefficient operations. Thus, during design phase using affordance based design approach, all interactions have to be captured for better operations at execution phase.

DOI: 10.1051/matecconf/201712008011

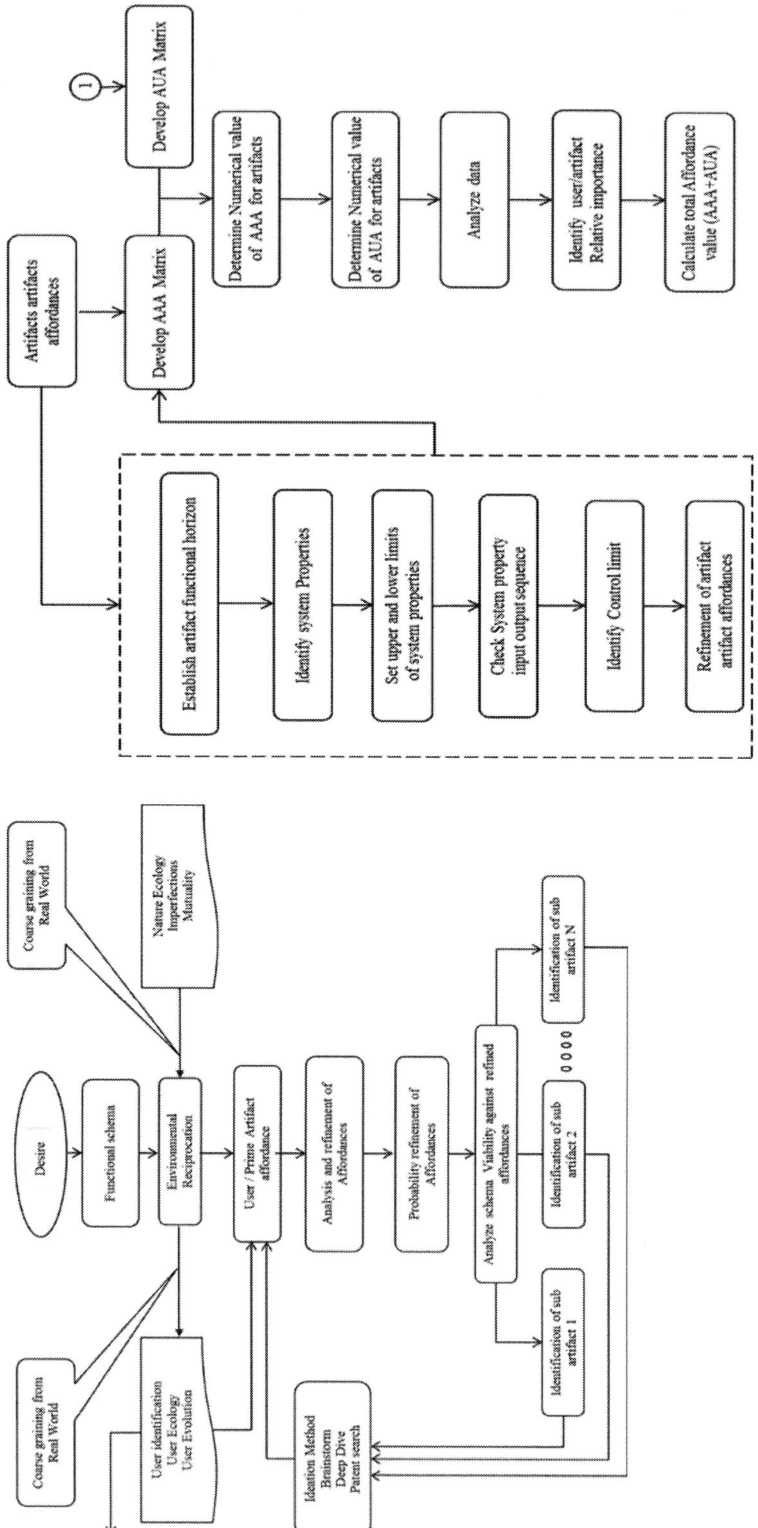

Fig. 4. Design process methodology *(Adapted and modified from [8])*.

The major function of the main artifact, which is a solid control room, where solid required/generated in drilling operation will be handled. This facility comprises of many sub-artifacts, which are listed as a part of data collection. Data was collected by meeting the users (operators) in the course of design review and discussed on the affordances and what is crucial to them. The list of artifacts was found as part of the earlier design and form the affordances. The second set of data was collected by visiting the owner's consultant office. The project details were reviewed. Based on the discussions with engineers involved, the final users required for case study were finalized. List of affordances was prepared with the help of generic list of affordances and also discussion with the owner's consultant team and finalised.

4.1 ASM testing and results

The step by step process for ASM development is presented in this section, which is based on the solution methodology proposed earlier (Figure 4).

4.1.1 Steps for development of ASM

Step 1: Desire

Design phase begins with clients or users "Desire ". In this case study, desire is design of a SS room of a jackup rig drilling platform.

Step 2: Functional schema

A SS room to house six solid controls namely shale shaker machine (SS machine) with all its accessories with entry, exit and to enable the machine to operate for a life of 25 years. While deciding the functional schema of SS room, more information has to be coarse grain from the real world to check whether any artifacts available in the market performing the same function. If so, then how it is functioning and take reference of the same for deciding the functional schema and define the functional schema as shown in Fig. 5.

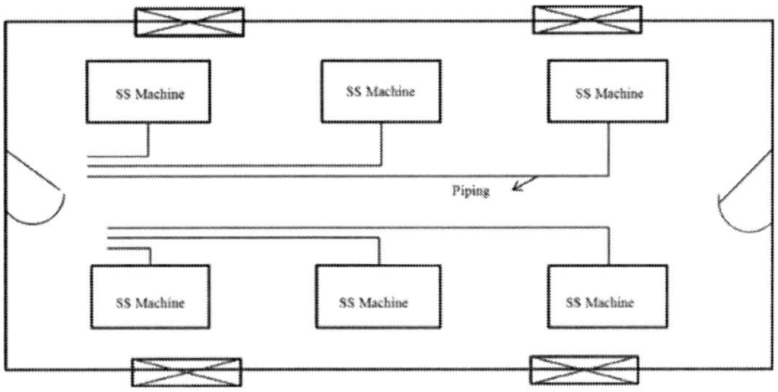

Fig. 5. Functional schema of SS room.

Step 3: Environmental reciprocation

While designing the artifact, designers have to check in what environment this artifact is finally going to build and use thereafter. Therefore, the surroundings mainly consist of the user of the room, all other physical entities around the room and the environment as it is. Here, it is a marine environment and it is possible to design a room in this environment with its users, hence there is a positive reciprocation of environment. While considering the environmental reciprocation, designer has to check the following :

- Mutuality (Can such a room coexist in this environment mutually without harming each other ?);

- Ecology (Will the ecological change allow this artifact's sustainability ?); and
- Imperfection (Consider that the functional schema decided in step 2 is not perfect and needs improvement).

Step 4: User identification

The ultimate aim of any design is how well the human can use artifact functionally. Therefore, list of users as shown in Table 1 has to be identified at design phase itself to understand their needs. Many methods are available such as by ideation, deep drive, and brainstorm methods to identify users.

Table1. List of users

S. No	Users
1	Drilling Operator
2	Maintenance User
3	HVAC Maintenance User
4	Electrical Maintenance User
5	International Maritime Authority
6	NORSOK
7	Owner
8	Fabricator

Step 5: User Artifact Affordance

At design phase, all users were asked for the specific requirement and what needs to be afforded by the artifacts are collated as shown in Table 2.

Step 6: Analysis and refinement of affordances

Are all the affordances viable, affordable, and justifiable? If yes, then proceed. If not, then refine the affordances and prepare the affordances accordingly. For example, is it sufficient if the SS machines were accessible from three sides, if yes then change the affordance no 1 accordingly.

Step 7: Probabilistic refinement of affordances

Are all the requirements expected in the lifetime of the artifacts? If not refine the affordances accordingly. For example, what is the probability of fire hazard? If the probability is very low eliminate requirement of passive fire protection (S. No 11 in Table 2) and keep only active fire protection arrangement.

Step 8: Analyse schema viability against refined affordances

Once all affordances are defined and refined, designer has to check again whether the functional schema is still valid for all the affordances. For instance, as discussed in step 6, slight modification in affordance no.1 is done (accessibility made only from three sides only) and the affordance no.11 as mentioned earlier in step 7 has been removed.

Table 2. List of User-Artifact Affordances.

S.No	User-Artifact Affordances	S.No	User-Artifact Affordances
1	All SS machines are to be accessible from every side	17	One side of the room will have a ballast tank. Pressure for the same to be afforded
2	Diverter house box to be accommodated	18	Pipe supports can be installed
3	Access to be provided to the diverter house	19	Weight to be within prescribed limit
4	Noise level outside - less than 60 Db	20	Escape shall be minimum 2.3 meter height as per norsok s002
5	Life - 25 years	21	Solas (safety of life at Sea) requirement met
6	2 Door size 2.5x3 m	22	Satisfy PSA (Petroleum safety Authority) requirements
7	Provide adequate ventilation	23	NMA (national Maritime Authority) regulation to be met
8	Provide adequate illumination	24	Local law requirement to be met
9	Trolley movement within room to be afforded	25	Economy/cost to be optimized
10	Dynamic loading due to trolley movement to be afforded	26	Corrosion in marine environment to be afforded
11	Passive fire protection to be afforded	27	Corrosion under insulation to be avoided
12	Window to the shaker control room to be afforded	28	Clean design to be afforded
13	A monorail with two ton capacity to be accommodated	29	Easy cleaning to be afforded
14	Two safe escape to be provided	30	Vibration generated in shale shaker room to be within limits
15	A stretcher movement to be afforded	31	Welding to be afforded
16	Firewater piping to be accommodated	32	Erection to be afforded

Step 9: Define the sub-artifacts based on the affordances and functional schema

Finally, sub-artifacts that have direct interactions with users and other sub artifacts have to be generated based on affordances generated previously and functional schema (shown in the Table 3).

Step 10: Establish artifacts functional horizon

From the functional schema and the affordances, matrix of artifacts can be established. For each artifacts its functions and extent of functionality is defined. For example, in this case study, the Shale Shaker machine (SS machine) has to perform the solid control activity for a certain quantity and quality of solid per day or per hour. Apart from its main function, it also needs to be accommodated and all its other affordances listed are to be defined.

Step 11: Identify system property

For each of the artifacts, the required affordances are converted to system property. For example the affordance is that all sub-artifacts are to be accommodated. Then dimension is the system property.

Table 3. List of Sub artifacts.

S. No	Sub artifacts
1	Six Shale Shaker Machines
2	Two doors
3	Control Room Window
4	Diverter Box
5	Access platform to Diverter Box
6	Trolley Movement Space
7	Firewater Piping
8	Mud Piping
9	Fire Extinguishers
10	HVAC Ducts
11	Electrical Trays
12	Instrument Cable Trays
13	2 Ton Capacity Monorail
14	Pipe/Tray/Duct Penetrations
15	Noise Insulation Panel
16	Lights
17	Bulkheads/Decks

Step 12: Define system property upper and lower limits

For each system property, define what could be the upper and lower limit of the property. For example, it is established what could be the minimum and maximum dimension for the SS machine that can be accommodated in SS room.

Step 13: Check system property input/output sequence

The system property of different artifacts depends on the system output of other artifacts. For example, the size of the SS room would depend on the size of the SS machine. The output (dimension) of SS machine is the input to the size of the SS room.

Step 14: Refinement of artifact-artifact affordance

Once the artifact-artifacts affordances are defined, it may be reviewed for further refinement.

Step 15: Develop roof of ASM (AAA interaction matrix)

First of all, the sub artifacts have been considered and they have been placed in the Top Main row area as shown in Fig. 6. The roof of the ASM is considered for developing AAA interaction matrix, which is DSM of artifacts. Here it is considered how one artifact is related to other artifact.

Step 16: Develop main body of ASM (AUA interaction matrix)

All users have been listed in the left main column of ASM and related to artifacts to develop AUA interaction matrix as shown in Fig. 6. It is decided to find out how each and every user is related to different sub artifacts.

Step 17: Numerically simulate the AAA index

A numerical simulation is done by assigning relationship index (S- Strongly Related, M – Moderately Related, L- Weakly Related). Numerical value of 9, 3, and 1 has been assigned to this relationship. Total index per artifact has been calculated to establish an AAA index for that artifact.

Step 18: Numerically simulate the AUA index

Similar relationship index has been found from data and numerical values assigned as explained in step 17. Total index per user-artifact relationship has been calculated to establish an AUA index for that artifact.

Step 19: Analyse data and identify user/artifact relative importance

From this DSM, the AAA index such as weak relation for each artifact can be found out and further improvement in the design can be made. Similar way, the weak relation between users and artifact in AUA interaction matrix can be worked out for design improvement. This step considerably increases the positive affordance as all AAA and AUA have been captured and hence, the total AAA and AUA value can increase.

Step 20: Total Affordance Index:

Total Affordance Index (Sum of AAA and AUA) calculated to find out which artifact or a group of artifacts will be the key governor and will lead the design.

4.2 Discussion

4.2.1 Polarity

Affordances have already been classified based on the type of reciprocities as AAA and AUA. However, affordances have also been classified based on their polarity. If the affordance is beneficial to the user of the artifact, it is considered to a positive affordance. In case the affordance is harmful to the user, it is considered as the negative affordance. This concept has a prime importance in the design. As a designer, we are expected to increase the positive affordance and reduce the negative affordance. This has been explained by an example in the Figures 7 & 8.

In Figure 7, the open window had two positive affordances, day light and ventilation. At the same time this also had a negative affordance of entry of burglar or animal. The total affordance of the design was only 1. In the Figure 8, iron grills are provided on the window and the negative affordance has been converted to a positive affordance and the total affordance of the window design has been increased to 3. In this way, design can be improved by eliminating the negative design or providing solutions to negative design.

4.2.2 Designer- designer affordance

In construction design, it is imperative that when AAA exists, the relevant designer also have an affordance to other designers so that an improved design can be achieved. So from AAA matrix, affordance list and from the list of designers, the Designer-Designer Affordance (DDA) matrix can be generated. The relative importance of the designer can be established from the numerically simulated index value. From the DDA matrix, input and output requirement of parameters, relevant design documents can be scheduled. DSM techniques may apply for efficient scheduling of the documents.

Relationship indicator	Value
"S" Strongly Related	9
"M" Moderately related	3
"L" Weekly realated	1

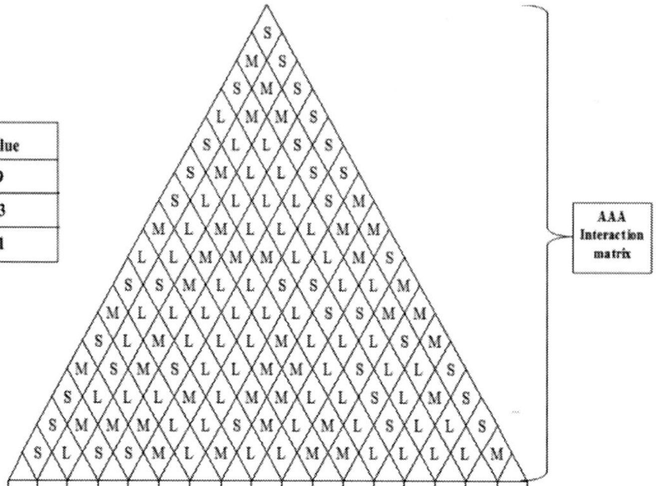

	Users	Relative importance of users	Six shale shaker Machine	2 doors	Control room Window	Diverter box	Access platform to Diverter box	Trolly movement Speae	Firewater Piping	Mud Piping	Fire extinguishers	HVAC ducts	Electrical Trays	Instrument cable trays	2 ton capacity Monorial	Pipe/Tray/Duct Penetrations	Noise inculation panel	Lights	Bulkhaeds/Decks	
	Drilling operator	103	S	S	S	S	S	S	S	S	S	M	L	L	L	L	S	S	L	
	Shaker maintenance user	54	S	S	L	L	S	S	L	L	L	L	L	L	S	L	L	S	L	
	IMO as an user	72	L	S	L	L	L	S	S	L	S	L	L	L	L	S	S	S	S	
	Norsok as an user	93	S	S	S	L	S	S	S	M	M	L	L	L	S	L	S	S	M	
	Electrical item maintenance user	75	L	S	L	S	S	S	L	L	L	L	S	M	M	S	L	S	M	
	HVAC item maintenance user	74	L	S	L	S	S	S	L	L	L	S	M	M	M	S	L	M	M	
	Fabricators	96	L	S	S	L	S	M	M	M	M	S	M	M	S	S	S	L	S	
	Owner	87	S	M	M	S	L	L	S	M	S	L	M	M	M	L	S	S	S	
AAA	Numerical value of AAA		104	56	60	70	50	64	48	56	48	40	52	46	46	66	34	70	108	
AUA	Numerical value of AUA		40	66	34	40	56	58	42	22	36	26	22	16	38	40	48	58	38	
AUA	Scaled value of AUA		71	117	60	71	100	103	75	39	64	46	39	28	68	71	85	103	68	
Total	**Numerical index of AAA & AUA**		**175**	**173**	**120**	**141**	**150**	**167**	**123**	**95**	**112**	**86**	**91**	**74**	**114**	**137**	**119**	**173**	**176**	**= 2227**

AUA Interaction matrix

Fig. 6. ASM for a Shale Shaker Room.

Note:
A lower index of AAA matrix can have a higher value of AUA; A lower value of the AUA index can have a higher index of AAA.

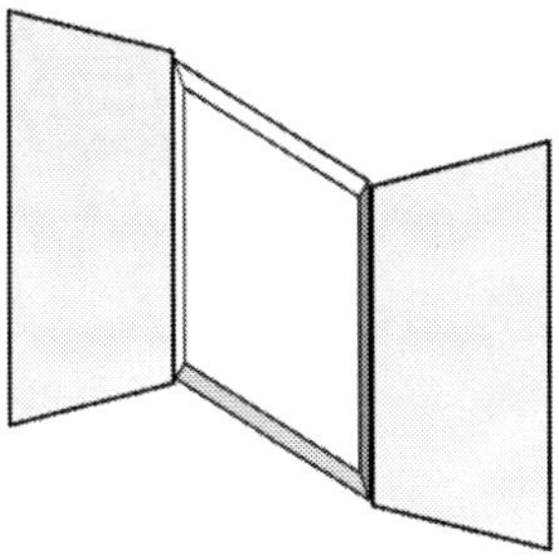

Affordances	Polarity	Score
Air ventilation	(+)	1
Daylight	(+)	1
Entry of Burglar/Animal	(-)	-1
	Total affordance =	1

Fig 7. Design with negative affordance.

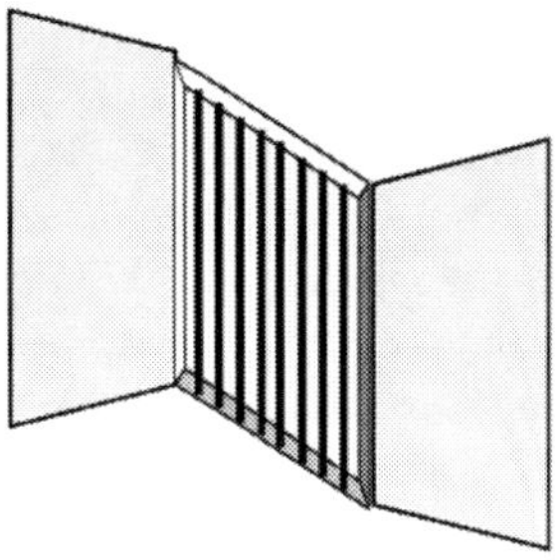

Affordances	Polarity	Score
Air ventilation	(+)	1
Daylight	(+)	1
Stop entry to burglar/animal	(+)	1
	Total affordance =	3

Fig. 8. Change of negative to positive affordance.

5 Conclusion

The concept of affordance definitely paves a way for a systematic and successful methodology, which can be implemented in construction design. The early capture of all affordances and the concept of AAA, AUA and DDA in DAU (Designer-Artifact-User) System adds a process approach to the conventional design hierarchy. The simple understanding that increases in positive affordance will yield a better design can be acceptable. However, whether this methodology is the best approach compared to other concepts that are on-course is questionable. Expert's comments align with this concept. Time required for this approach would be a hindrance to the acceptance of this process. But implementing this process can substantially lead to efficient design for construction projects. The salient conclusions are:

- The concept of affordance initiates the process of identifying expectations /requirements from a design
- The ASM framework enables capturing all interaction and relations by matrix approach and providing a common language among designers
- Systematic trading off of affordance (increasing positive affordance and decreasing negative affordance) yields an alternate design
- Micro level design improvement, where multi artifacts and multidisciplinary design is involved can be achieved by AAA and DDA
- This methodology also facilitates information cataloguing, leading to archived knowledge for use in future projects.

References

1. P. Wegner, Communications of ACM, **40**, 80 (1997)
2. J. R. A. Maier, and G. M. Fadel, *Research in Engineering Design*, **20**, 13 (2009)
3. J. J. Gibson, in: Shaw, R. and Bransford, J. (Eds.), *Perceiving, acting, and knowing. toward an ecological psychology*, 67 (Lawrence Erlbaum Associates, Hillsdale, NJ, 1977)
4. J. J. Gibson, *The ecological approach to visual perception*, (Pyschology Press, New York, 1986)
5. A. J. Wells, *Ecological Psychology*, **14**, 140 (2002)

6. J. R. A. Maier, *Affordance based design: theoretical foundations and practical applications* (VDM Verlag Dr. Muller GmbH & Co, Saarbrucken, Deutschland, 2011)
7. J. R. A. Maier, T. Ezhilan, and G. M. Fadel, *Proceedings of the ASME 2007 International Design Engineering Technical Conferences and Computers and Information in Engineering Conference*, 277 (2007)
8. J. R. A. Maier, and G. M. Fadel, *Proceedings of ICED 2007, the 16th International Conference on Engineering Design* (2007)

Performance-based contracting for maintaining transportation assets with emphasis on bridges

Mohammed Alsharqawi[1,*], *Saleh* Abu Dabous[2], and *Tarek* Zayed[1]

[1]Concordia University, Department of Building, Civil and Environmental Engineering, Montréal, Québec, Canada
[2]University of Sharjah, Department of Civil and Environmental Engineering, Sharjah, United Arab Emirates

Abstract. With a large number of aging transportation infrastructure assets in North America and the growing problem of deterioration across the globe, managing these assets have been the subject of ongoing research. There is an overwhelming amount of maintenance and rehabilitation works to be done and selecting a suitable maintenance, repair or replacement (MRR) strategy is one of the most challenging tasks for decision makers. Limited budget and resources are even making the decision making process more challenging. Maintaining infrastructure to the highest possible condition while investing the minimal amount of money has promoted innovative contracting approaches. Transportation agencies have increased private sector involvement through long term performance-based maintenance contracts or what is called Performance-Based Contracting. PBC is a type of contract that pays a contractor based on the results achieved, not on the methods for performing the maintenance work. By looking into the literature, it is observed that agencies are expanding the amount of contracting they do in order to maintain and achieve a better standard of infrastructure facilities. Therefore, the objective of this paper is to study and review performance-based contracting for transportation infrastructure with emphasis on bridge assets.

1 Introduction

Civil infrastructure systems contribute to social and economic welfare through serving a large number of population and businesses. Bridges are vital components of this civil infrastructure system in terms of mobility, economy, and development of communities. Any failure in these structures may result in both human life and economic loss. Therefore, maintaining these bridges at functional and safe levels is necessary. In North America, large parts of bridges infrastructure were constructed during the 1960's and 1970's. Accordingly, they are facing an increasingly deteriorating problem. There is an overwhelming amount of maintenance, repair, and replacement (MRR) activities to be done

*Corresponding author: mohammed.alsharqawi@concordia.ca

but the resources (funding, staffing, equipment) available are too limited and selecting a suitable MRR strategy to achieve a better standard of infrastructure facilities is one of the most challenging tasks for decision makers. Limited budget is even making the decision-making process more challenging. The growing needs and limited resources to maintain the transportation networks have resulted in motivating transportation agencies to expand the amount of contracting they do.

With the conventional traditional method-based contracts, the owner agency specifies techniques, methods, materials, quantities, along with the time that the maintenance activities should be executed [1]. This causes a limitation on the use of new materials and innovative techniques and technologies while maintaining the infrastructure asset [2]. Contractors have better competencies of fulfilling required performance level through innovating in such new and unproven technologies [3]. This can be achieved by extending the short maintenance period to a longer period so that the performance levels can be satisfactorily measured over a longer duration. This type of agreement is called Performance-Based Contracting (PBC) which is a type of contract that pays a contractor based on the results achieved, not on the methods for performing the work. Such contracts allow the maintenance contractors to innovate and improve the efficiency of the services provided to public[4]. Incentives are introduced to the contractors, in this type of contact, as an increase of payments owing to exceeding on a pre-defined performance measure. However, the maintenance contractors will be penalized for failing to comply with performance measures or to promptly rectify revealed deficiencies. The objective of this paper is to study and review performance-based contracting for transportation infrastructure with emphasis on bridge assets as an alternative method for maintenance contracting.

2 Performance-based contracting for maintenance overview

There has been some movement over the past decade toward a performance-based contract model for maintaining and managing transportation infrastructure. Performance-Based Contracting (PBC) is defined as "a contracting method that provides incentives and/or disincentives to the maintenance contractor to achieve desired outcomes or results; in its purest form, PBC does not detail how, when, or where to do the work." [5]. The hallmark of PBC is that maintenance contractors are paid based on the results achieved and not on the methods utilized for performing the work. By paying contractors incentives and/or disincentives for measured performance instead of the quality of inputs and giving them the flexibility in choosing materials and methods in delivering maintenance services, contractors are encouraged to use new methods and mix quality which fosters innovation and improves the performance of the structure throughout its life. As in the case of PBC, the long-term planning provides the contractor with enough time horizon to deploy new technologies. Further, the contractors are encouraged to eliminate the future need to perform costly corrective maintenance actions by taking timely preventive maintenance ones [6].

Types of performance-based contracts differ according to scope and coverage. The scope refers to the service activities and the addressed assets while the coverage relates to the amount of highway covered and the geographic area [5]. For instance, a PBC can cover either single asset such as only traffic signs, only pavement, only bridges, ... etc. or all assets along a highway corridor. In terms of service activities, PBC may deal with only a single activity (e.g. sign replacement) or a set of related activities (e.g. rest area maintenance). Stankevich et al. [7] stated that the complexity level of a PBC can range from "simple" to "comprehensive" depending on the number of assets and range of services included where a simple PBC may cover a single service and a comprehensive one typically covers all assets and comprise the full range of services needed to manage and maintain the contracted corridor. Usually, transportation agencies that are newly practicing PBC find the best approach to first gain experience is by contracting a single activity, a single asset, or one set of related activities in a single maintenance area. Once the agency has acquired experience, it is likely to expand the number of assets under contract, coverage area, and the period of performance. Performance contracts are not limited to maintenance including preventive, routine, periodic, and demand-responsive maintenance as rehabilitation of roads and bridges has been part of many performance-based contracts [5].

2.1 Historical review of performance-based contracting

PBC for maintenance was first implemented in the province of British Columbia in Canada. Then, provinces of Ontario and Alberta followed by developing their own performance-based contracts. Later, it has become the backbone of maintenance contracting in Australia, New Zealand, some European countries as Finland and England, and to an increasing degree in other countries including the United States where states of Virginia, Texas, and Florida are leading in this area. Gradually, PBC trends spread to other developed and developing countries worldwide. Anastasopoulos et al. [8] illustrate a map of countries (dark highlighted areas) which implemented PBCs as can be seen in Figure 1. Moreover, transportation agencies in North America and around the globe have developed a variety of methods for undertaking PBC, known by other names such as Performance-Specified Maintenance Contracts in Australia and New Zealand; Asset Management Contract in the US; Area Maintenance Contract in Finland and Ontario, Canada; Managing Agent Contract in the UK; and *Contrato de REcuperacion y MAntenimiento* (CREMA), which means Contract for Rehabilitation and Maintenance in Argentina [5].

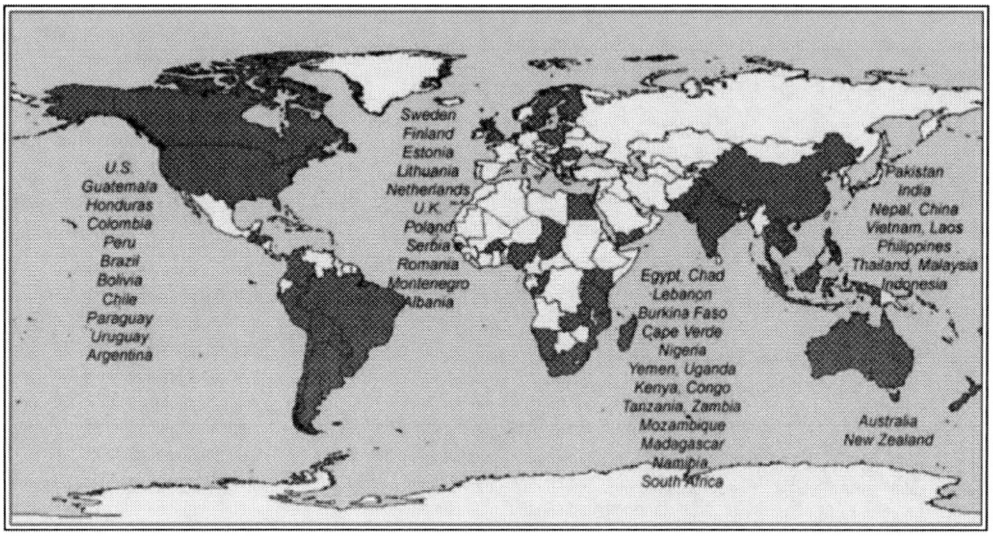

Fig. 1. PBCs implementation worldwide [8].

2.2 Performance-based contracting framework

As discussed above, there is a high tendency towards PBC in the area of transportation. In road maintenance, Sultana, M. et al. [9] introduced seven main issues that should be considered by the transportation agencies before applying PBC as shown in Figure 2. The first issue is the transportation agency obligation to define the performance specifications and set-up a standard for these performance measures. After that, the agency should check the private sector capability of handling the maintenance to reach the desired LOS quality. Then, the implementation stage takes place where an initial project has to be decided for the performance-based contracts to be applied on. Afterward, a detailed risk analysis has to be done in order to define the events that are out of the maintenance contractors' control and share those risks with the maintenance contractor. Hence, the performance monitoring process takes place where the maintenance contractors are evaluated according to their performance within the contract period. The sixth issue is related to the agency employees as the agency requires fewer people for administrative work and supervision in PBC. The fear of losing job is considered an important issue for the employees and the transportation agency. Finally, the seventh issue is the proper definition of the payment and termination clauses in the contract to avoid any conflicts or disputes that may arise during the contract period.

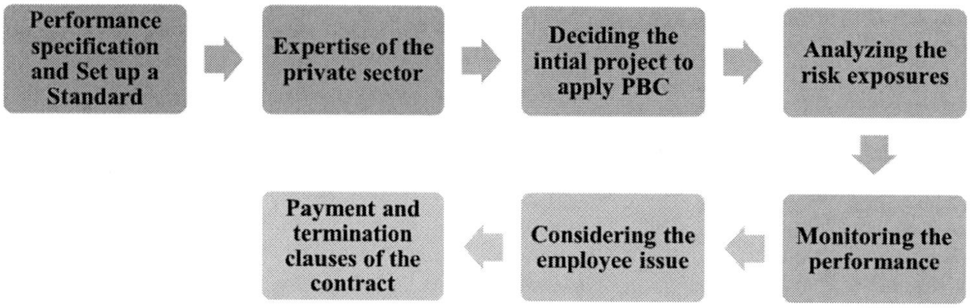

Fig. 2. The seven main issues before applying PBC [9].

2.3 Benefits of performance-based contracting

Agencies that have converted to PBC claim cost saving between 10-50%. Table 1 shows the cost savings of different countries under PBC over the traditional contracts [7]. Moreover, many government agencies reported an 18% improvement in satisfaction with the maintenance contractors' performance. The agencies added that this price reduction and customer satisfaction took place in several areas from the non-technical, technical, and professional services [10]. Some other advantages found in the literature [5] include but not limited to:

- Improved level of service (could cost more)
- The transfer of risk to the contractor
- More innovation
- More integrated services
- Enhanced asset management
- Ability to reap the benefits of partnering
- Building a new industry
- Achieving economies of scale

2.4 Risks in Performance-Based Contracting

In the conventional traditional-based contracts, transportation agencies, called as a client normally, prescribe the techniques, methods, materials specifications, and the time to execute maintenance activities which reduce risks of the maintenance contractors.

However, this is not the case in performance-based contracts where a contractor is generally free to make decisions of "what to do," "how," and "when" as long as the specified performance measures are achieved [11, 12]. Hence, the entire risk of any failure or shortcomings of made decisions is transferred to the contractor. Figure 3 depicts the distribution of risk moving from traditional contracting to different forms of contracting. As shown in the figure, agency risks tend to decrease while contractor risks increase when long-term contracts are adopted [13]. The best way of managing the risk associated with such long-term maintenance contracts is to properly i) predict deterioration rates of contracted assets; ii) determine appropriate design, specifications, and materials; iii) plan needed maintenance interventions; iv) estimate quantities [1, 7, 14]; additionally v) inspect and measure the current condition state of contracted assets; and vi) select suitable MRR actions.

Table 1. Cost savings of different countries under PBC over the conventional traditional contracts [7]

Country	Cost Savings (%)
Norway	About 20% - 40%
Sweden	About 30%
Finland	About 30% - 35%; about 50% less cost/km
Holland	About 30% - 40%
Estonia	20% - 40%
England	10% minimum
Australia	10% - 40%
New Zealand	About 20% - 30%
USA	10% - 15%
Ontario, Canada	About 10%
Alberta, Canada	About 20%
British Columbia, Canada	Some, but might be in the order of 10%
*Source:*P. Pakkala.	

Maintenance contracting has various risks. Types of risks, among others, are identified as follows [5]:

-Poor quality of construction
-Unexpectedly severe weather
-Unanticipated environmental problems
-Emergencies
-Unanticipated legislative change
-Unexpected traffic growth
-A short-term focus that fails to minimize long-term life-cycle costs
-Difficulty in acquiring the resources needed to perform the work (e.g., subcontractors)
-The possibility of having to correct problems covered under a warranty.

However, it is worth noting that the allocation of risks in PBC varies from one country to another [15]. It is essential to identify all the risk variables in order to avoid or mitigate the undesirable consequences such as the first identified risk, poor quality of construction. Thus, the concept of warranty is implemented. The next section aims to provide an overview of the key aspect that is needed before applying PBC which is the performance measures (PMs) that are used to define the levels of service (LOS).

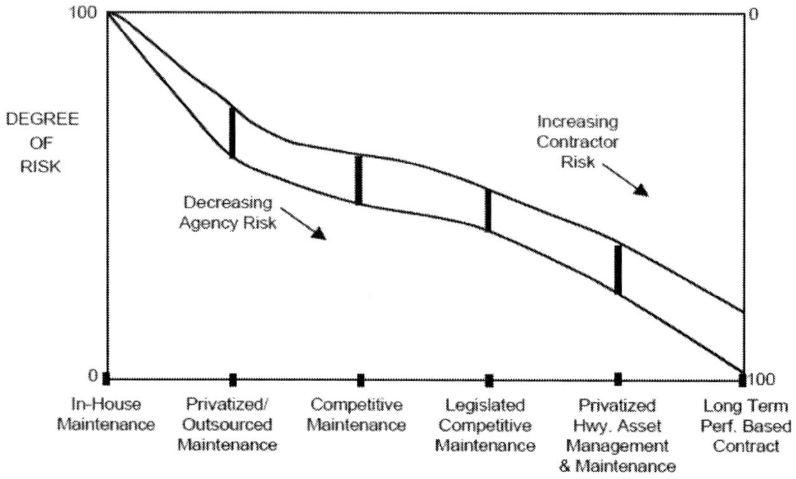

Fig. 3. Distribution of risk with different contract approaches [16].

3 Performance measures in the PBC

PBCs can address different types of assets and operations including: pavements, bridges, roadsides, traffic operations, traffic services, incident response, hazardous materials clean up, and emergency repairs [5]. For every type of maintenance and operations, performance measures should be clearly defined and objectively measurable in order to avoid ambiguity and risk disputes [17]. In some countries, measures of performance are often expressed in terms of levels of service represented by specific rating scales corresponding to the condition of different assets. Performance measures (PMs) and performance level of service (LOS) are the basis of a PBC which are expected to be achieved by the maintenance contractor [11]. By looking into the literature, it is observed that many transportation agencies established and defined many performance measures and targets extensively for roads maintenance. However, in the area for bridges management, only a few distinct has addressed these measures. Haas et al. [18, 19] suggested performance indicators for bridges to be the remaining life in years and safety. Many experts on performance measurement argue the benefit of using a few key performance measures instead of many because of the associated simplicity and manageability of those few performance measures [5]. According, performance measures can be defined as a set of outcome-based levels that an agency uses to evaluate the success of the contractor and performance goals are the minimum acceptable levels to be achieved for each performance measure [11].

Performance measures and targets are prescribed in the contract specifications as performance indicators or measures. Performance indicators are measures of quality; quite often the distress indicators which are monitored annually, or more frequently. It is assumed that these indicators give an accurate picture of the asset condition [3]. In general, for a performance measure to be effective, the following questions should be addressed [20]:

- Is the performance measure specific?
- Is the performance measure measurable?
- Is the performance measure achievable?
- Is the performance measure results- oriented?
- Is the performance measure timely?
- Does the measurement meet with the agency's objectives and desires?
- Has the performance been measured before?

- Does the measurement conflict with the agency's standard specifications?
- Does the measurement aim to improve performance?

A bridge performance measures in a PBC could be categorized into two groups: structural performance and non-structural performance measures. Non-structural performance measures include these non-structural components, such as: bridge guard rails, sound barriers, lighting, signs, drainage system, and bridge cleaning. Structural performance measures are these attributes that indicate the condition or performance of the bridge components, such as cracking, concrete disintegration, steel reinforcement corrosion and expansion joint problems, these assembled can represent the structural condition reliability which is a component's physical condition fitness, an indicator of a component performance state. Therefore, in this research, there liability of the bridge condition is suggested to be one of the key performance indicators or measures to express a bridge's level of service (LOS). Reliability is the base measure for the bridge remaining service life, an indicator that is suggested by Haas et al. [18, 19]. Other PMs can include but not limited to, bridge railings, lighting, and cleaning. After defining the performance measures, it is equally important to define performance thresholds such that the LOS never exceeds the threshold limit. The client agency specifies these performance thresholds to be met or exceeded during the contract period. This research suggests a threshold of 40% representing a poor performance state. Further, the client agency can define performance targets or standards where a LOS rating of 80% for primary systems and 75% for the secondary systems is desired targets [21].

4 Performance monitoring of the PBC

In order to assess the maintenance contractors' performance, de la Garza et al. [4], presented five key components for monitoring PBC for maintenance and their direct relationship with the overall performance as shown in Figure 4. The identification of these five key components is based on de la Garza et al. [4] study to existing approaches commonly used in the public and private sector for measuring and monitoring performance including the following six approaches: i) ISO 9001:2000 Criteria for Performance Excellence, ii) Malcolm Baldrige National Quality Program, iii) Kaplan and Norton's Balance Scorecard Approach, iv) Mark Graham Brown's Scorecard Approach, v) Department of Energy Performance Measurement Program, and vi) NCHRP 14-12: Highway Maintenance Quality Assurance Program.

Fig. 4. The five components for monitoring the PBC for road maintenance [4].

As shown from the figure, the five main components for monitoring the PBC are the quality of service, timeless of response, safety procedures, cost-efficiency, and LOS effectiveness [4]. A brief description of each component is as follows:
1- Quality of service evaluates the customers' perception with respect to the condition of the asset and the performance of the maintenance contractor. Customers are the ultimate evaluators of the quality of service provided; therefore, it is extremely important to assess their satisfaction.
2- Timeliness of response assesses the response time of the maintenance contractor to service events or deficient elements as requested by users and need to be attended to in a timely manner.
3- Safety procedures ensure that the maintenance contractor is properly implementing a safety program. It also checks that the maintenance crews performing the work are exposed to minimum risk of accidents.
4- Cost efficiency assesses the cost savings, if any, accrued by the government as a result of engaging a contractor to perform performance-based maintenance services.
5- Level of service effectiveness indicates how far the maintenance contractor is meeting the defined performance criteria and performance targets throughout the contract period.

For instance, if the suggested key performance indicator i.e. bridge reliability fails to meet performance threshold (e.g., LOS < 40%) within the warranty period, the maintenance contractor has to repair or rehabilitate the bridge and may also have to pay a penalty for failing to maintain the pre-defined bridge LOS.

5 Payments system for the PBC

Payments in performance-based contracts differ significantly from traditional-based contracts. In traditional contracts, payments to contractors are based on the amount of inputs such as number of working hours or quantities; for example, cubic meters of concrete [11]. In contrast, performance-based contracting is a type of contract that pays a contractor based on the results achieved, not on the methods for performing the work. It provides incentives, disincentives, or both to the contractor to achieve desired LOS. In other words, payments for the management and maintenance of assets in this type of contract are explicitly linked to the contractor successfully meeting or exceeding these minimum performance indicators. Failing to comply with the pre-defined performance measures or to promptly rectify revealed deficiencies affects the contractor's payment adversely through a series of clearly defined penalties. Accordingly, a proper payment system that provides incentives, disincentives, or both is needed in PBC.

6 Summary and conclusions

Performance-Based Contracting (PBC) is a type of contract that specifies certain clearly defined minimum performance measures that the contractor is required to meet or exceed when delivering maintenance services during the contract period. It has been successfully developed and applied in different countries, showing great cost savings and better level of service. However, in this type of contract, the entire risk of any failure or shortcomings to achieve the desired targets is transferred to the maintenance contractor. Thus, it is important to develop proper performance modeling to avoid or mitigate such risk. The main challenge in PBC is to define proper performance measures and their thresholds to guarantee an acceptable LOS while performance monitoring. This study suggests bridge condition reliability as a key performance indicator or measure with a minimum LOS rating of 40 as a performance limit. Other six performance measures are identified as well. Since maintenance contractors are assessed in terms of performance, not in terms of level of exerted efforts, it is necessary to establish payment criteria including incentives and/or disincentives. The disincentives or incentives can consist of reductions or increases in payments for respectively falling short or exceeding the desired LOS.

References

1. Z. Alyami, S. Tighe, TRR Journal. **2**, 1-10 (2013)
2. K. Panthi, R. U. Farooqui, S. M. Ahmed, *ICCIDC-I*, 456-463 (2008)
3. K. Panthi, *A Methodological Framework for Modeling Pavement Maintenance Costs for Projects with Performance-Based Contracts* (2009)
4. J. de la Garza, J. Piñero, M. Ozbek, *A Framework for Monitoring Performance-Based Road Maintenance Contracts*, 433-441 (2009)
5. W. A. Hyman, *NCHRP Synthesis 389: Performance-Based Contracting for Maintenance* (2009)
6. D. Gupta, A. Vedantam, J. Azadivar, *Optimal contract mechanism design for performance-based contracts* (2011)
7. N. Stankevich, N. Qureshi, C. Queiroz, *Performance-based Contracting for Preservation and Improvement of Road Assets* (2009)
8. P. C. Anastasopoulos, B. G. McCullouch, K. Gkritza, F. L. Mannering, K. C. Sinha, J Infrastruct Sys.**16**, 251-263 (2010)
9. M. Sultana, A. Rahman, S. Chowdhury, World Acad Sci Eng Technol.**62**, 350-355 (2012)
10. Office of Federal Procurement Policy (OFPP), *Report on the performance-based service contracting pilot project* (1998)
11. Z. Alyami, *A Two-Phase Maintenance and Rehabilitation Framework for Pavement Assets under Performance Based Contracts* (2012)
12. A. S. Al-Kathairi, *Performance Based Road Asset Management System, with a case study: Abu Dhabi* (2014)
13. C. Queiroz, The World Bank (1999)
14. A. S. Al-Kathairi, A. A. El Halim, Int JRes Sci Manag.11-16 (2014)
15. G. F. Segal, A. T. Moore, S. McCarthy, The Institute. **21** (2003)
16. R. Haas, J. Yeaman, C. Raymond, L. C. Falls, *Fifth International Conference on Managing Pavements* (2001)
17. G. Zietlow, *cutting costs and improving quality through performance-based road management and maintenance contracts* (2005)
18. R. Haas, S. Tighe, J. Yeaman, L. C. Falls, *Annual Conference of theTransportation Association of Canada* (2008)
19. R. Haas, G. Felio, Z. Lounis, L. C. Falls, *Annual Conference of the Transportation Association of Canada* (2009)
20. SAIC, *Performance contracting framework fostered by highways for LIFE* (2006)
21. K. L. Smith, M. Stivers, T. E. Horner, A. R. Romaine, *NCHRP Web Document 8: Highway Maintenance Quality Assurance* (1997)

MATEC Web of Conferences **120**, 08013 (2017) DOI: 10.1051/matecconf/201712008013
ASCMCES-17

Modelling process integration and its management – case of a public housing delivery organization in United Arab Emirates

Senthilkumar Venkatachalam[1*], *Fikri* Diweiri[2], and *Noura*Al Suwaidi[3]

[1]Sustainable Construction Materials and Structural Systems Research Group, Department of Civil and Environmental Engineering, University of Sharjah, Sharjah, United Arab Emirates
[2]Department of Industrial Engineering and Engineering Management, University of Sharjah, Sharjah, United Arab Emirates.
[3]Innovation and Futurism Office, Directorate of Housing, Sharjah, United Arab Emirates

Abstract. Huge volume of project information are generated during the life cycle of an AEC projects. These project information are categorized in to technical and administrative information and managed through appropriate processes. There are many tools such as Document Management Systems, Building Information Modeling (BIM) available to manage and integrate the technical information. However, the administrative information and its related processes such as the payment, status, authorization, approval etc. are not effectively managed. The current study aims to explore the administrative information management process of a local housing delivery public agency. This agency manages more than 2000 housing projects at any time of a year. The administrative processesare characterized withdelivery inconsistencies among various project participants. Though there are many commercially available process management systems, there exist limitations on the customization of the modules/ systems. Hence there is a need to develop an information management system which can integrates and manage these housing projects processes effectively. This requires the modeling of administrative processes and its interfaces among the various stakeholder processes. Hence this study aims to model the administrative processes and its related information during the life cycle of the project using IDEF0 and IDEF1X modeling. The captured processes and information interfaces are analyzed and appropriate process integration is suggested to avoid the delay in their project delivery processes. Further, the resultant model can be used for effectively managing the housing delivery projects.

*Corresponding author: svenkat@sharjah.ac.ae

1 Introduction

Construction projects are complex endeavors that require multiple stakeholder's involvement to accomplish the expected project outcomes [1]. It is further characterized by its components, people and processes variations and its complex relationships and interfaces [2]. This characterization brings a socio technical complexities to the project delivery. These complexities are characterized in all aspects of projects, including organization structure, procedures, technology etc., [1]. Further these complexities influence the management capabilities of project stakeholders in terms of project team selection, planning, coordination and selection of procurement arrangements, characterizing them as one of the most complex undertakings for any project delivery. Therefore, understanding and managing complexities is essential to achieve satisfactory performance of construction project delivery [3]. Millions and millions of project information are produced during the construction projects. Management of these information play a vital role on addressing the complexities and its knock-on challenges appeared technically and administratively.

The use of information technology (IT) has been demonstrated to be capable of managing the project information, however prior to recent times has been largely discrete. However, the recent CAD working platforms such as Building Information Modeling (BIM) to exchange the information has enhanced the communication and collaboration in the technological side mostly [4][5][6]. The administrative information exchange and collaboration is been managed through companywide ERP systems and other document management systems but pose challenges on the data security and customization to the local needs. Hence there exists a need to the development of new information management system to improve productivity on the delivery of information through effectively processing and providing necessary information. Further, it also supports organizations to achieve efficient work performance. An information system is a structured collection of information resources, information technology, people and all activities that works in coordination to achieve definite objectives in the business organization. As information technology is essential and vital to business, the development of an information system in the organization becomes critical for the success of the organization. However extracting the various relationships such as functional, technical and administrative is vital to the success of this Information System. This paper explains the step by step procedure adopted in a governmental Housing Delivery Organization called Department of Housing (DH) with the aim of identifying the information and process flow among the various stakeholders with DH projects and develop an appropriate method to effectively manage the information to avoid inefficiencies in their housing project delivery. The following section outlines the current status of DH's project delivery process and its need towards an Information Management System. This is followed by the development of IDEFO models to capture the various interfaces among the project processes. The theoretical validation is reported to get the initial feedback against the developed systems followed by the summary and conclusions.

2 Housing development organization background

DH is one of the various establishments under the state government. DH's core business is to provide people with financial aid. DH consists of three main departments, which are Housing Applications Department (HAD), Engineering Services Department (ESD) and Project Management Department (PMD). The housing application department receive, review, study and approve the application from the beneficiaries. Then the engineering services department aids the approved applicant to choose a suitable design and to establish a contractor among the three parties. Then the project management department supervisesthe work on the approved applicant's site to assure quality and performance of the contractor. One of the sub objectives of the DH includes the assistance of obtaining low cost contract for its beneficiaries, hence theDHprojects are considered as low cost and small scale projects. Currently the DH manages more than 2000 running projects with limited number of employees, there exists challenges on managing the project progress information and its related transactions during the execution of their government sponsored (Grant) projects. The timely delivery of these transactions are critical and delays will result in contractors/consultants dissatisfaction which leads to the non-participation of contractors/consultants to Grant based DH projects. The project management department has been advocated to address this problem. In order to understand the root cause of the problem, the PMD conducted interviews and questionnaire survey followed by the statistical analysis on the responses obtained to identify the root cause of the problem. However due to the length restrictions of the paper, the detailed statistical analysis is not been included within the scope of this paper. However, a brief about the results obtained is been discussed to justify the improvement processes.

3 Analysis on the current status and processes of project delivery in DH

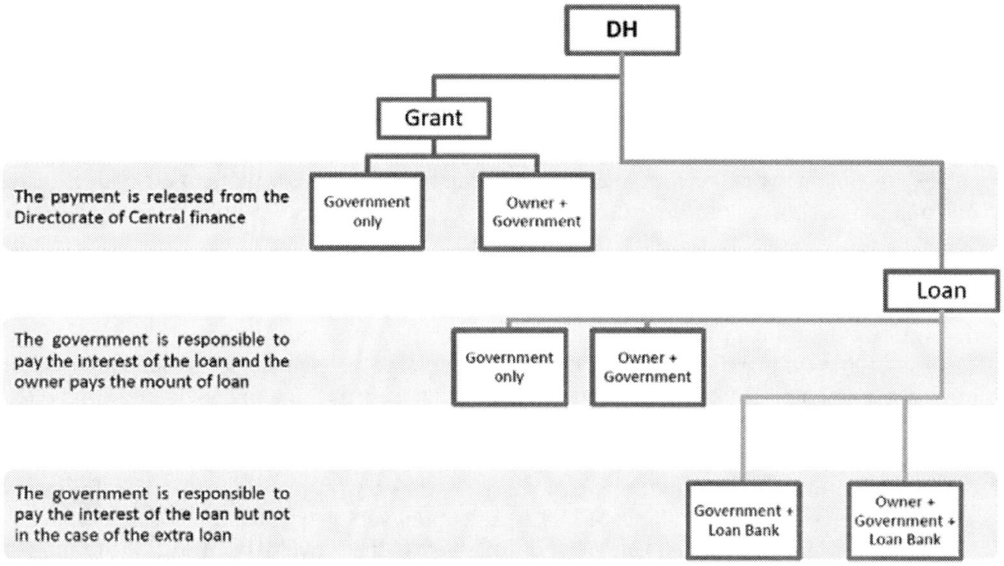

Figure 1. Types of project delivery in DH based on various financing options

Figure 1 illustrates the types of project delivery that the DH is practicing, mainly based on the type of financing options such as loan and grants at the higher abstraction level. Many projects were executed through the Grants route and the same were experiencing the problem mentioned in the previous section. The current project processes and its delivery status on the DH's projects were asked to the contractors and consultants who works for the DH projects. The information is obtained through semi structured focus group interviews about the various processes and project delivery status. Based on the interview, along with the documental evidences the various processes and its delivery status were summarized. The summarized data were presented in the form of a process diagram to the contractor and the consultants and revised according to the real world practice. Further based on the focus group interview it was revealed that the payment approval processes were delayed a most to impede the project delivery. Various reasons/ causes were listed by the respondents, the same were captured through a cause and effect diagram as shown in figure 2.

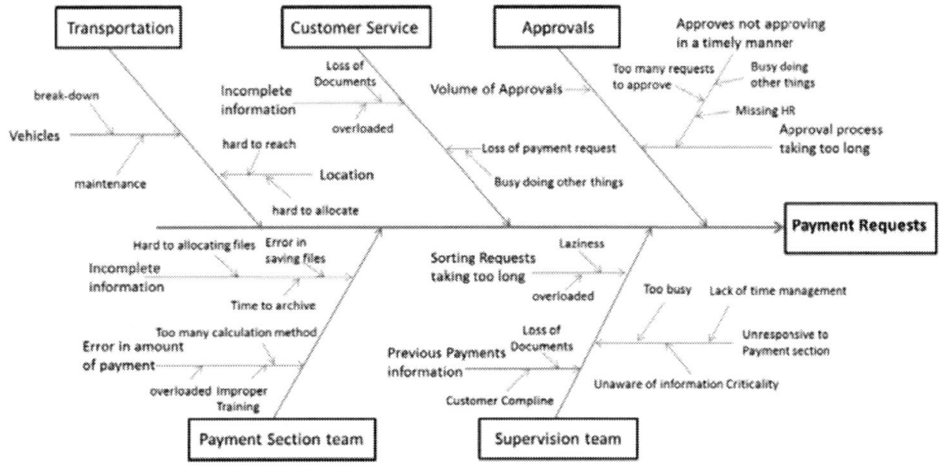

Fig. 2. Cause and effect diagram on factors influencing the payment processes.

Further to identify the important influential causes, a sample of 50 completed projects consists of 24 loan projects and the 26 grants projects were analyzed statistically based on its time taken on various project life cycle processes on each of the above depicted causes. The descriptive statistics results showed that the approval process taken more than the prescribed 10 days by the DH. Further, one-way ANNOVA test was conducted to identify the correlation between the different causes and the delay. For example, customer service employees are accountable for delivering the completed form on time, including full documents requested by the payment section. A one-way ANOVA Test was conducted between "the time taken to send the completed document from the customer service section to the payment section" versus total time spent on the payment process to measure the customer service team's impact of delay in their processes. The results of the test, has a p-value of 0.000 and R-sq.of 65.07 percent. Which confirms the moderate correlations. Hence the customer service process need improvement. Further with the same One-way ANOVA tests with respect to the other causes, revealed that the Approval and the Supervision processes had a significant correlation to the delay in project payment delivery time and hence need improvement.

4 Development of the framework for Information management system

To improve the existing processes the DH processes are captured through the functional modeling method called IDEFO [7]. In order to develop a new system with the improvedprocesses flow, the system need to be defined by its requirements and specify the various functions in order to design and develop the system that performs the intended purpose. With this intention the Different IDEFO models of the existing DH system is captured at various hierarchical levels and depicted in figures 3 to10. After this, the information modeling at various processes are captured through IDEF1X, it is a semantic data modeling technique which captures information and its interactions [8] to modify the process flow effectively.

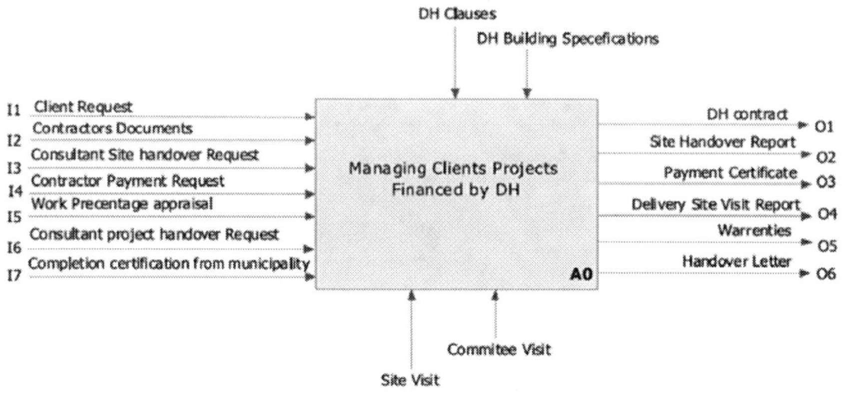

Fig. 3. Top level context diagram Node-AO managing client project financed by DH.

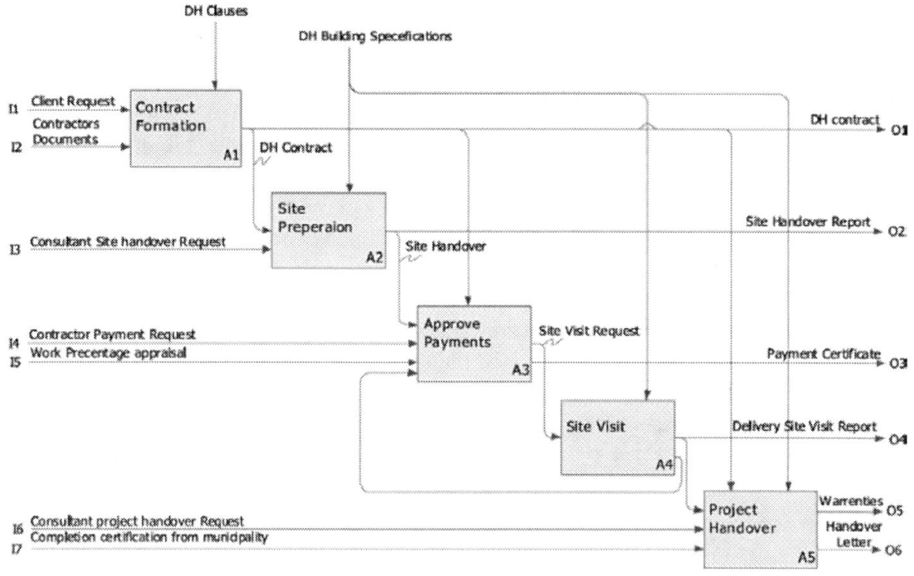

Fig.4. Decomposition of Node-AO- managing client's projects financed by DH.

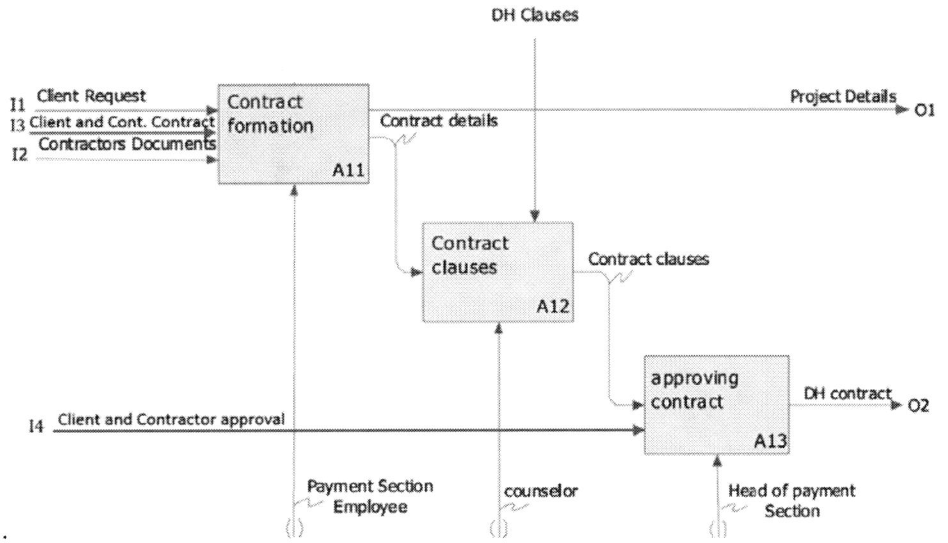

Fig.5. Decomposition of Node-A1.

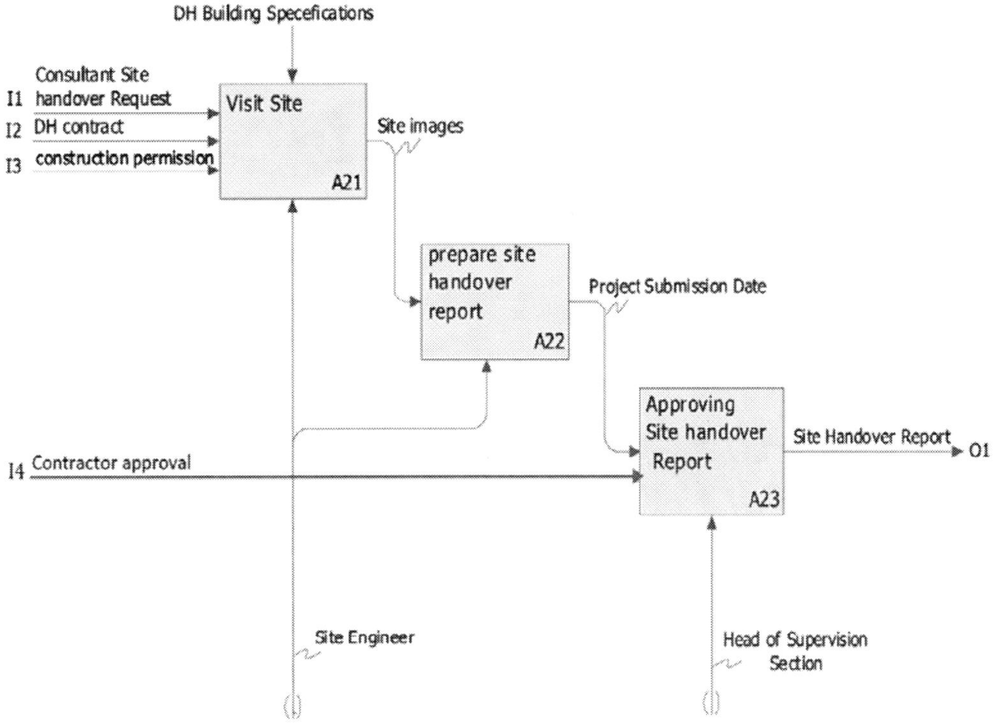

Fig.6. Decomposition of Node-A2.

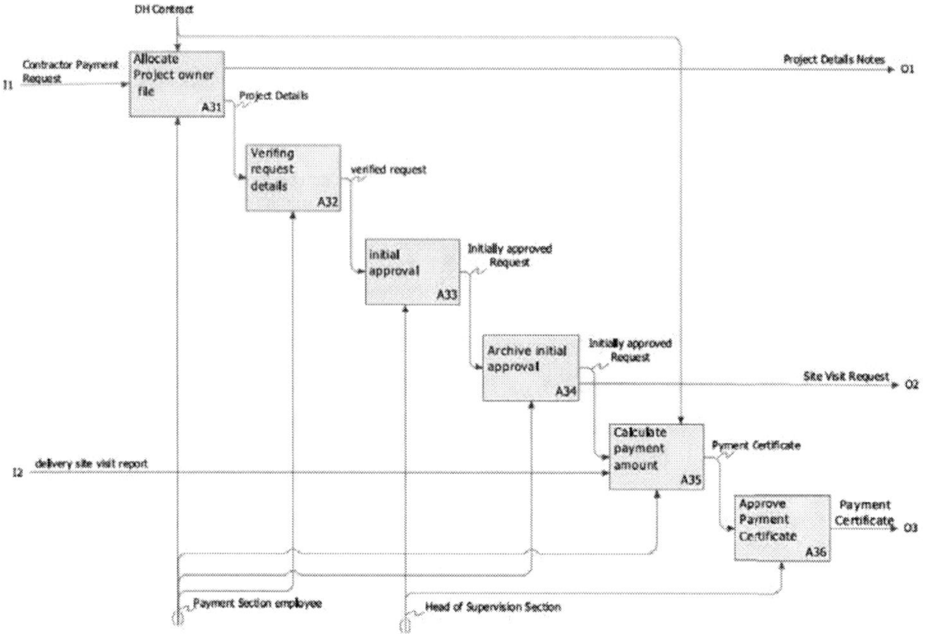

Fig.7. Decomposition of Node –A3.

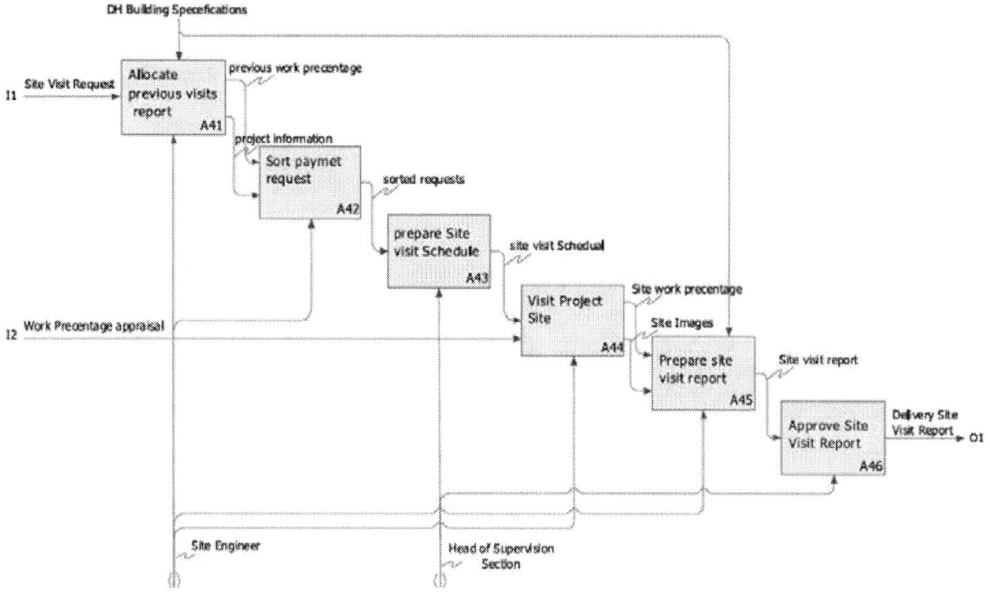

Fig.8. Decomposition of Node –A4.

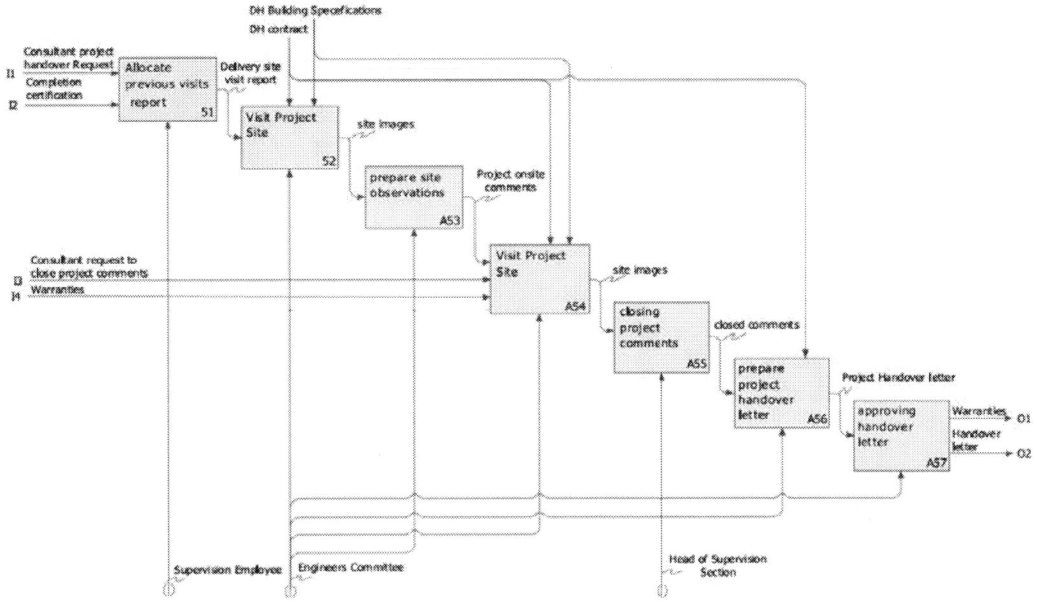

Fig.9. Decomposition of Node –A5.

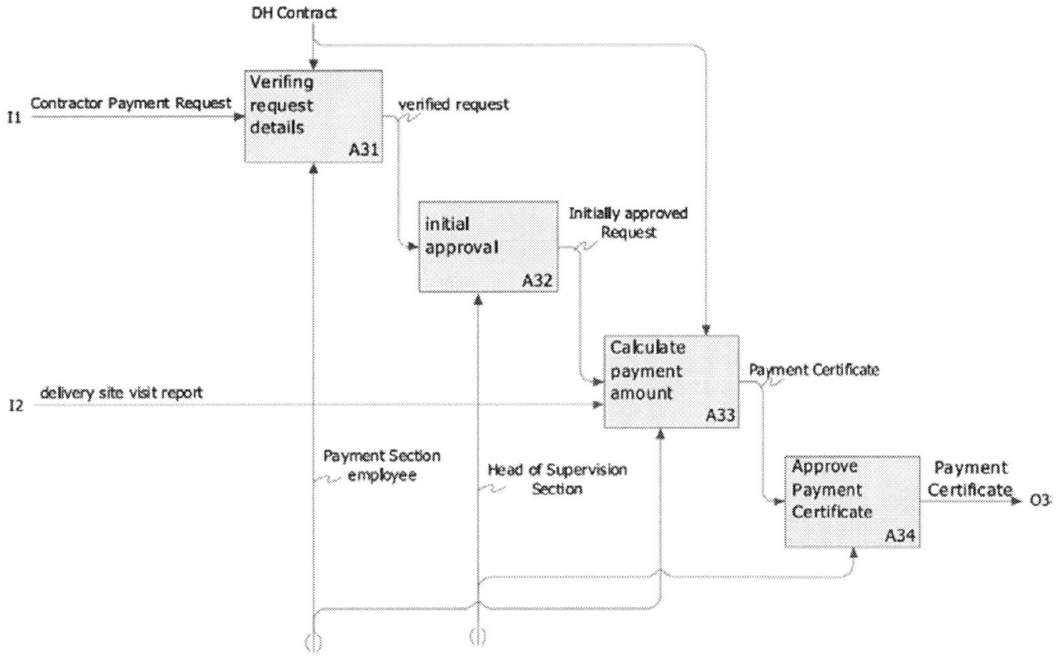

Fig.10. Modified Node –A3.

The overall activity is described in Figure 3 as managing clients' projects. The client request to manage the project is input to this activity along with contractor payment request during the construction, which is input for the payment certificates in order to finally handover the project (handover letter). In addition, the activity provides an information output, Site HandoverReport and Delivery Site Visit Reports. DH Clauses and Building Specifications control this activity the decomposition of node A0 is shown in Figure 4. Activity Contract Formation (Node A1), is the first project preparation step. The DH Clauses controls this activity, which yields to the final contract between three parties DH, Client and Contractor (DH contract) as shown in Figure 5. Activity Site preparation (Node A2), is the second preparation step for a project financed by DH as shown in Figure 6. The DH building specifications controls this part, which leads to Site handover to the contractor (Site Handover Report). The subsequent activities of Approve Payments, Site Visit and Project Handover are represented in a similar manner during the following stages of the project. Each has an information input and output requirements. Also, each activity is guided by the clauses given in the DH contract as shown in Figures 6 to 9 respectively.

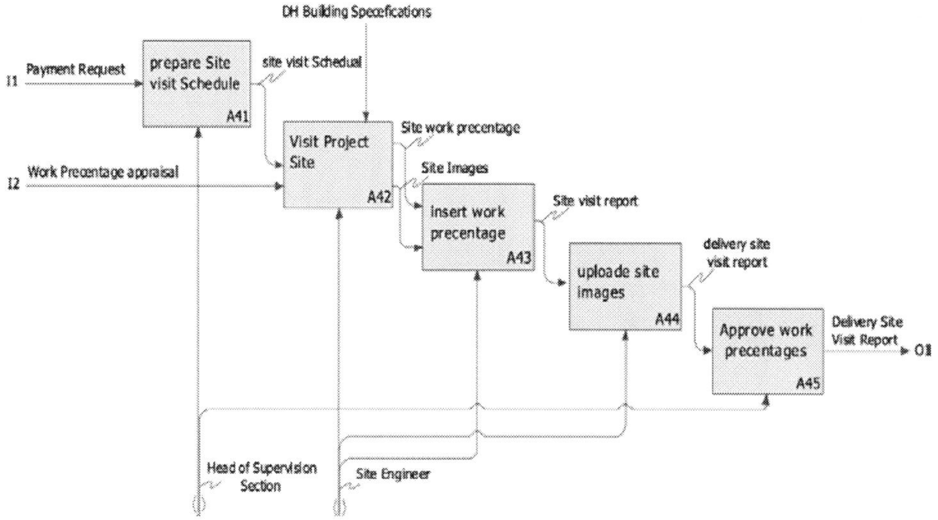

Fig.11. Modified Node –A4.

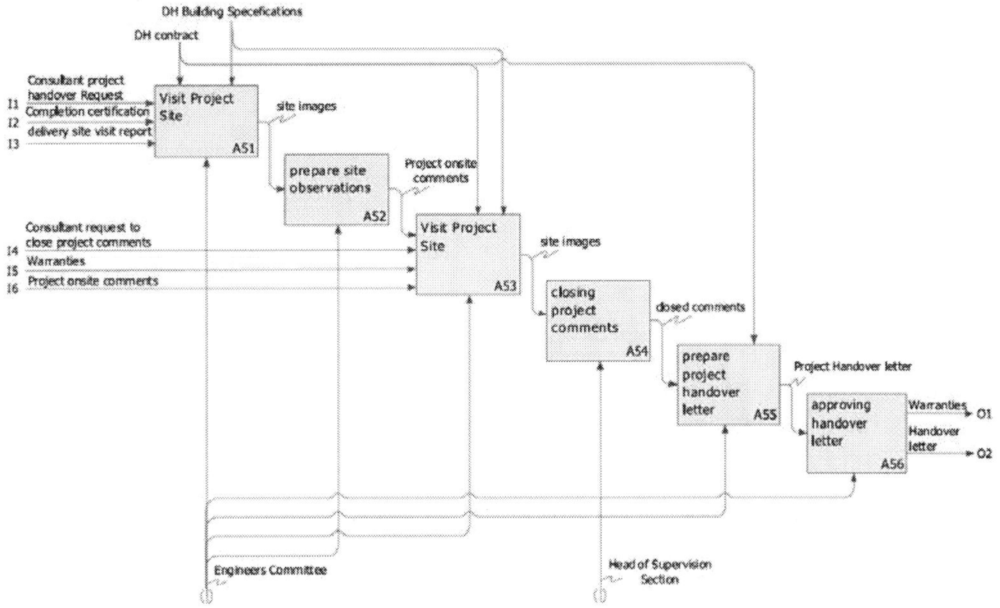

Fig.12. Modified Node –A5.

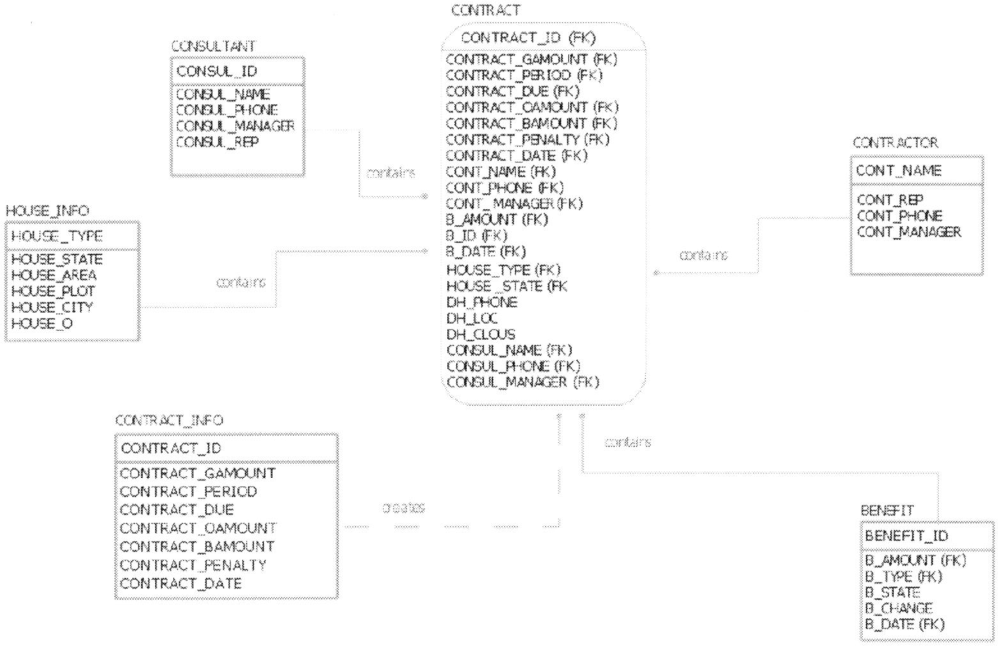

Figure 13. IDEF1X for contract information.

In activity A11, Forming Contract, the payment section employee forms the DH contract after he/she receives a request from the client with the respective contractor documents. Then the contract details is written according to the Client and Contractor contract. In activity A12, Contract Clauses, the Counselor forms the contract according to the contract details, where the DH Clauses controls this activity. Finally, contract is approved in A13, upon contract approval, the payment section employee requests client and contractor to sign the contract in presences of Head of Payment Section to finally establish a contract between the mentioned three parties (DH Contract), and a copy is sent to Supervision Section.

The decomposition of node A2 is shown in figure 6, called Site Preparation. The process begins with A21, Visit Site, where site engineer receives a request from consultant to handover the site. Site engineer visit the site and take site images to make sure that there is no existing structure/ construction at the site. In A22, Prepare Site handover Report, the site engineer prepares the handover report, which is an information output; the date to handover the project is indicated along with other project information. In A23, head of supervision section approve the report along with the contractor approval. Then the Site Handover Report is submitted to the consultant and the contractor of the project.

Figure 7 shows the decomposition of node A3. In A31, the contractor request for payment of his project is the input to this activity. Payment section employee allocatesa project file through the existing contract and named after the owner. In activity A32, verifying request details, the details written in request is verified based on the information recorded in project folder. The initial approval for payment request is obtained in A33 from head of payment section. In A34, payment section employee archives initial payment approval. In activity A35, calculate payment amount, delivery site visit report is an input to this activity where the payment employee have to calculate the payment amount and project DH contract controls this activity. Head of payment section approve payment amount in A36 to finally have an approved payment certificate for the contractor request.
Figure 8 shows the decomposed activity steps of node A4, Site Visit. In A41, site visit request is the input to this activity. The supervision section have to allocate the project file and the output of this activity is previous work percentage and project information. In A42, the site engineer sort the site visit requests according to his allocated area. In A43, prepare site visit schedule, in which, head of supervision section establish a site visit schedule according to the sorted site visit requests. Site engineer visit the project site in A44 to determine initial work percentage and verify it by capturing site image. In activity A45, site engineer prepares site visit report to specify each work percentage category and attach site images for each category, in which DH building specification controls the activity. Head of supervision section modifies and approve site visit report in A46 and the output of this activity is delivery site visit report which is sent to payment section.

Figure 9 shows the decomposition of node A5, Project Handover. In A51, consultant project handover request along with completion certificate from the municipality, is an input to this activity. Supervision employee allocates previous delivery site visit report to verify that work percentage appropriate for handover. In activity A52, engineers committee visits the project site, in which again DH contract and building specification controls to this activity. In A53, committee prepares site observation and the output is project onsite comments. Comments are drafted by head of committee and if all the necessary progress is been made then the head prepares project handover letter.

The IDEF0 model of the basic elements of the functional process "as is "in DH is presented in the previous section. To verify the same, the authors showed the same functional model in the form of a flow chart with appropriate details to the DH employees involved in the housing delivery processes and necessary modifications and additions were administered accordingly. Further, they also agreed that the boundaries of the system defined by the IDEF0 model are acceptable. Although, it is possible to carry out the decomposition of the IDEFO to further smaller abstraction level to for further more detailed modeling, the authors decided to stop this level as the model has given an appropriate level of understanding of the DH's housing delivery process. With these input, the information management system is developed through an IDEF1X model [8]for its relational database design for the proposed system as shown in figure 13. The figure 13outlines the contract formation and the relationship between different information and also gives an insight into the data that has to be captured with each contract and its relationship/ interfaces with other data generated during the delivery process.

Based on the data information modeling through IDF1X, the "to be IDEFO" functional modeling is developed in order to have an integrated process to manage the DH delivery process efficiently. After the development of the complete information model (IDEF1X), the final functional model (IDEF0) is developed. In the DH case, the nodes A3, A4 and A5 are modified as the A1 and A2 doesn't require any modifications. Some of the redundant functions are removed in the "to be" model and some of the necessary functions are added accordingly to encourage the process and information integration. Figures 10, 11 and 12 depicts the modified IDEFO functional model. The revised model removed the redundant functions/ processes as a result of the proposed information management systems and its information integration. For example, the redundant function such as "Allocate Project owner File" and "Archive Initial Approval" in the as is model is been removed in the A3 node as the information system database has the details in the central database. Similarly in the A4 node the processes had been modified as the information system has the inbuilt automated reporting function through uploading the percentage complete and the related site images.

5 Discussions and conclusions

This project demonstrates the identification of the process inefficiencies in the housing delivery organization through a cause and effect diagram. The analysis revealed that the customer services approval and supervisory process in the DH need improvement in their delivery process. The DH project delivery is then explored further by capturing the processes using IDEFO, which helped the authors to understand the functions, participant processes and their interrelationships. The developed functional model is then used to develop an IDEFX1 information / data model to assist the development of relational data base to the proposed information management system. With the IDEFX1 models the functional models are modified to avoid the redundant processes which exists in the DH delivery process and the appropriate "to be" functional models are generated for the efficient project process integration and the delivery. Though there are many information system available in the market to use and tailor made the need, the development of an in-house system is advantageous due to many reasons such as the opportunity for its customization/ modifications, data security etc. Further study on decomposing the functions in to lower level may improve the system, as well as the interoperability among the different technical tool with the proposed system need to be explored further. Due to the length constraints of the paper, the authors are limited to only present the process integration and management steps. The development of Information Management System and its validation can be demonstrated in future work.

References

1. C.Fang, and F. Marle, Dealing with project complexity by matrix-based propagation modelling for project risk analysis. *Journal of Engineering Design*, **24**(4), 239–256.(2013)
2. D. Antoniadis, F. Edum-Fotwe, A. Thorpe, and R. McCaffer, R,*Exploring Complexity in Construction Projects*.1–6.(2008).
3. D. D. Gransberg, J.S. Shane, K. Strong,and C.L. del Puerto, Project Complexity Mapping in Five Dimensions for Complex Transportation Projects. *Journal of Management in Engineering*, **29**(4), 316–326 (2013).

4. C. Dossick, and G. Neff, Organizational Divisions in BIM-Enabled Commercial Construction.*Journal of Construction Engineering and Management*, **136**(4), 459–467, (2010).
5. A. Grilo, and R. Jardim-Goncalves, Value proposition on interoperability of BIM and collaborative working environments,*Automation in Construction*, **19**(5), 522–530, (2010).
6. A. Porwal, and K.N. Hewage, Building Information Modeling (BIM) partnering framework for public construction projects. Automation in Construction, **31**, 204–214, (2013).
7. G. Varun, William J. Kettinger,*Process Think: Winning Perspectives for Business Change in the Information Age*. **168**, (2000).
8. T. Bruce, *Designing Quality Databases with IDEF1X Information Models*. New York: Dorset House, (1992).

An integrated approach for soil classification: The Kingdom of Bahrain study case

Khalil Al-Joburi [1*]

[1]University of Bahrain, Department of Civil Engineering, Isa Town, Bahrain

Abstract. Soil classification is a complicated and difficult process; the level of complexity depends on the extent of the information needed. The objective of this paper is to produce a digital soil map for the Kingdom of Bahrain. The Kingdom does not have such a map at the present time. Borehole data of more than 300 sites is collected form the Ministry of Work. Old maps are obtained from the Surveying Land Registration Bureau (SLRB). Multispectral LandSat images from the years 1973 to 1976 are downloaded from online sources. All paper format maps are scanned, geo-referenced and converted to GIS format. LandSat images are classified using supervised image classification; pixel value refers to a specific soil type. Borehole data is treated as a point feature and soil related information is entered as attribute data. A hybrid model that performs classification based on weighted average mean for the raster and vector data is developed. The model consists of calibration and execution module. The model is tested and a digital soil map for the Kingdom of Bahrain is produced. The map encompasses the historical Awal land and the newly reclaimed areas. Analysis revealed that there are solid waste dumping activities in the sea and lowland areas.

1 Introduction

Soil information is critical for Civil Engineering and construction activities, agricultural, land resource management, policy making, environmental monitoring, and all aspects of infrastructure development and planning. More than ninety percent of the world's food production is dependent on soil [1]. Soil and terrain information is needed for policy-making, land resource management, and for monitoring the environmental impact of development [2]. In fact, soil information constitutes a precious resource inventory associated with life on earth. Knowledge of soil types and spatial distribution is a prerequisite for any rational land use planning for agriculture, irrigation, drainage, etc. Soil information records provide an insight into the potentials, and limitations, of soil for effective and sustainable utilization. Soil information provides a truthful and scientific account of a soil's physical properties and the extent of distribution so that predictions can be made about their characters and possible effective utilization. It also provides adequate information about characteristics of soils (e.g., texture, depth, particle-size characteristics,

* Corresponding author: kaljuboori@uob.edu.bh

liquid limit, plasticity index, structure, stoniness, drainage, acidity, and salinity) which can be utilized for planning and development. Lack of comprehensive information about global, national or local land resources increases the risk of uninformed policy decisions, continued degradation of land and water resources, excessive carbon emissions to the atmosphere, and that the Millennium Development Goals will not be achieved [3]. Global and regional models that address climate change, land degradation, and hydrological processes need soil input parameters with complete area coverage, but currently there are only few spatially exclusive datasets available [4,5]. Recently professionals and researchers are using soil information in much more sophisticated ways.

The objective of this paper is to produce a digital soil map for the Kingdom of Bahrain based on a hybrid practical procedural model that integrates traditional and advanced methods. The new DSM will include the unclassified soil and the newly reclaimed areas.

2 Soil mapping methods

The goal of soil mapping is to classify soils into groups with similar properties and behaviour. The process depends for the most part on the intended applications. For Civil Engineering applications, bearing capacity is the most critical. For agriculture scientists, water content and organic properties are vital. Literature search shows that there are many classification methods available globally and they are used daily for engineering and scientific applications. However, they can be divided into two categories:

- Traditional methods;
- Advance methods.

A brief description of each follows.

2.1 Traditional methods

These methods attempt to classify soils based primarily on soil grain size, texture, degree of plasticity (for engineering purposes), soil weathering, chemistry, profile thickness, etc. for Pedological classifications. The process involves laboratories and field tests, the most common and widely used methods are Unified Soil Classification System (USCS), The American Association of State Highway and Transportation Officials (AASHTO) and Agricultural Soil Classification System (ASCS).

2.2 Advance methods

More information about soil properties in a digital format is increasingly needed for agricultural, engineering, and environmental applications. The concept of digital soil mapping is adopted. The term is defined as 'the creation and population of spatial soil information by the use of field and laboratory observational methods coupled with spatial and non-spatial soil inference systems' [6-8]. The new advanced methods utilize remote sensing, Geographic Information Systems (GIS) and Digital Elevation Model (DEM) and others to provide soil information in a Digital Soil Map (DSM) format that can be easily integrated and spatially analyzed. These methods attempt to classify soils based primarily on spectral data reflected or emitted energy in visible, thermal, and microwave zones. A number of factors generally govern the spectral response patterns of soil. The properties of soils that govern spectral reflectance are color, texture, structure, mineralogy, organic matter, free carbonates, salinity, moisture, and the oxides-hydroxides of iron and manganese. Chemical compositions of the soil influence spectral signature of soils through the absorption processes in near infrared (NIR) and middle infrared (MIR) domains. The process can be done using satellite imaging, aerial photography, and field spectroscopy and it is mostly performed based on local scale.

Soil properties are derived based on physical and empirical methods that have been measured using remote sensing. These include mineralogy, texture, soil iron, soil moisture, soil organic carbon, soil salinity, and carbonate content [2]. In complex landscapes, combinations of spectral and Digital Elevation Models (DEM) can significantly improve soil mapping. Typically, the surface is parameterized by attributes such as elevation, slope, aspect, plan and profile curvature, and flow accumulation [9]. Remote sensing presents potential for extending existing soil information in the two major ways:

- Remote sensing method reduces the need for costly filed work to map inaccessible areas.
- Remotely sensed data can be spatially analyzed to derive soil characteristics.

Moreover, remotely sensed imagery can be used as a data source supporting digital soil mapping [10, 11]. Although remote sensing and soil spectroscopy have been recognized as a potentially effective and cost-efficient technology, they are not yet routinely used in soil surveys. Our knowledge of how to apply advances in remote sensing to soil and terrain mapping is still incomplete [10]. There are many methods of deriving soil properties from remotely sensed images; the most common methods will be briefly reviewed. A list of other methods can be found in [2, 12].

2.3 Automated classification

In these methods, multispectral passive imagery and active Radar images are often used. Pixels of similar spectral signatures are grouped together. Their spectral properties indicate that they are very likely to present one practical land cover. The process can be done fully automated (unsupervised classification), or signature data for particular features spatially distributed is added as an input to enhance the classification process (supervised classification). Research shows that several important soil characteristics can be derive from satellite imagery, such as mineralogy, iron content, soil salinity, soil moisture, and vegetation indices [2]. Many commercial satellite platforms in visible, thermal and microwave ranges are available and are used for soil classification as shown in Table 1. Many researchers, to meet certain classification criteria based on field collection and production of the substance signature or spectral properties, continuously enhance the methods, example, fuzzy classification [13], object-based classification [14], expert driven multi-level approach, LIDR DEM and a combination of DEM classification and spectral data [15].

Table 1. Current most common used satellites for soil classification.

Sensors	Category	Bands	Resolution (m)	Spectral range (μm/cm)	country
LandSat	Passive	8	15-60	0.45-12.5	USA
MODIS	Passive	36	250-1000	0.4-14.40	USA
MERIS	Passive	15	300	0.39-1.040	Europe
ASTER	Passive	15	15-90	0.52-11.65	Europe
Hyperion	Passive	242	30	0.40-2.500	USA
LISS II	Passive	4	36	0.45-0.86	India
SRTM	Active	C -X	30	2.5- 8 (cm)	USA
RADARSAT2	Active	C	30-100	4-8 (cm)	Canada

2.4 Vegetation indices

Vegetation index is one of the most effective methods to estimate the status of vegetation in a certain area. The method is based on using the red and infrared channels of the Advanced Very High Resolution Radiometer (AVHRR) 4-km global resolution. The Normalized Difference Vegetation Index (NDVI) is calculated as NDVI = (Near Infrared Red - Red) / (Near Infrared Red +Red) [16]. It has become the most used product derived from NOAA AVHRR data. The techniques are primarily used for predicting the status of vegetation. Research shows it can be successfully used for soil organic matter.

2.5 Predictive soil mapping methods

Predictive Soil Mapping (PAM) can be defined as the development of a numerical or statistical model of the relationship among environmental variables and soil properties, which is then applied to a geographic database to create a predictive map [12]. It is a computer-assisted production of digital maps of soil types and soil properties. The objectives are to: (1) produce an efficient collection of soil data; (2) produce and present data that better represents soil; (3) explicitly incorporate expert knowledge in the model; (4) provide insight and understanding into the soil forming processes.

The methods use raster data modeling to better represent the actual soil reality. A raster grid cell presented in form of X, Y, Z, where Z represents any physical soil attributes, while X and Y represent spatial location. The cell size presents the spatial resolution, which can change to acuminate for any scale depending on data availability. The raster grid cell is continuous in nature and a more realistic presentation of soil than traditional methods. It allows for a scientific way of accuracy assessment. These methods utilize the new advances in remote sensing, GIS, DEM, and fuzzy logic to greatly improve the way that soil maps are produced.

2.6 Global and national soils and terrain digital database (SOTER)

SOTER is a land resource information system based on the concept that features the land - in which terrain and soil occur - incorporate processes and systems of inter relationships between physical, biological and social processes overtime [17]. The central goal of the SOTER program is to set up a worldwide geospatial database for soils and terrain, including digital geo-referenced soil map units and their attribute information. GIS is used to hold the necessary data manipulation, analysis, mapping, management, and potentially monitoring changes of soil characteristics and terrain resources. The SOTER method enforces a strict set of rules for delineation and definition of soil and terrain units.

The methodology can be used at any scale ranging from global to local. The SOTER method was initially developed in Russia and Germany and was then steadily adapted worldwide. The method is similar in many aspects to the traditional physiographic soil mapping approach, but it focuses on both the soil and terrain properties and dynamic relationships. Moreover, it focuses on creating a universal geo-referenced geospatial database. Underlying the SOTER methodology is the identification of areas of land with a distinctive pattern of soil properties, parent materials, lithology, surface form, slope, parent material, and soil. Tracts of land distinguished in this manner are named SOTER units. Each unit has its own unique combination of soil and terrain characteristics. The Food and Agriculture Organization of the United Nations (FAO) adopted the method and joined the World Soil Information (ISRIC) to develop a soil and terrain global database at a scale of 1:1,000,000. Choosing the proper most effective method(s) depends on several factors, such as the prime objective of soil classification, the scale or spatial at which the soil information is needed, the soil characteristics that are under investigation, cost and data accessibility.

3 The study area

The Kingdom of Bahrain is an archipelago of around 33 islands; the largest is the main island of Bahrain, all are low-lying islands between latitudes 26.100° and 25.22° North and longitudes 51.070° and 50.160° East [18]. They lay in an arm of the Arabian Gulf, known as the Gulf of Salwa, which separates the Qatar peninsula and Saudi Arabia (Figure 1).

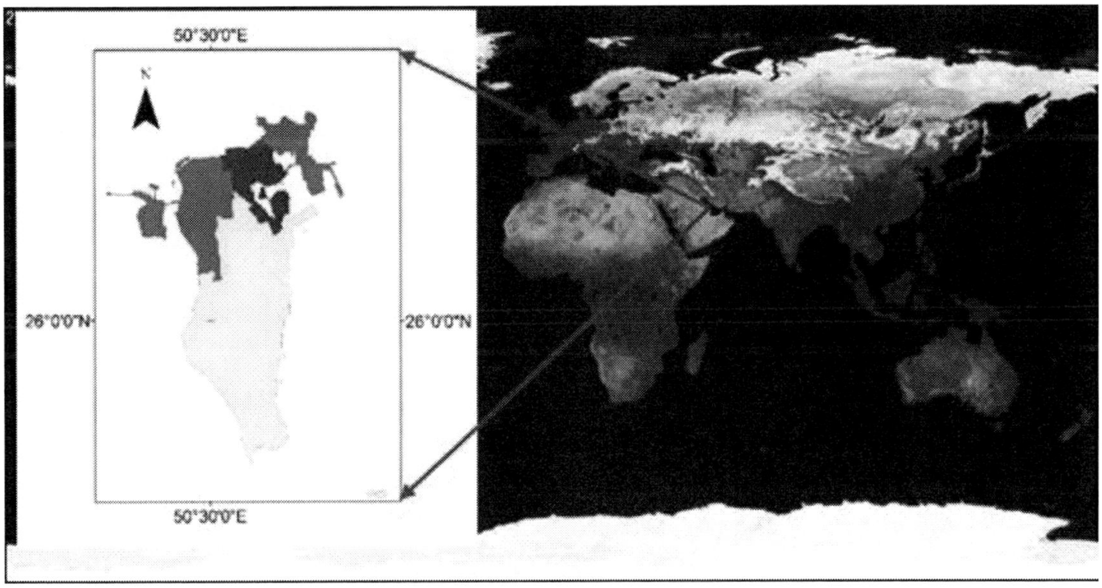

Fig. 1. Kingdom of Bahrain geographic location.

The Kingdom's total land area was about 620 km^2 in 1965 [19]. Due to significant land reclamation activities since 1965, the estimated total area is 756 km^2 [18]. This represents an increase of 22% in total square area, making the Kingdom the most dynamic country in the world in terms of land reclamation.

The surface soils of Bahrain were thoroughly investigated by the Bahrain Surface Material Resources Survey between December 1974 and April 1975. The team was made up of more than 20 scientists, geologists, surveyors, and engineers. They adapted the USCS approach to categorize soil map. Their work was presented in a book titled "GEOLOGY, GEOMORPHOLOGY AND PEDOLOGY OF BAHRAIN" published by Geo Abstracts Ltd. in 1980 [20]. Later it was published by E. M. BRIDGES and C. P. BURNHAM in 1980 in the Journal of Soil Science in an article titled "SOILS OF THE STATE OF BAHRAIN." The soil of Bahrain was placed in four groups: Cultivated Solonchaks, Natural Solonchaks, Regosols, and Rock dominated .The team produced a soil map for Bahrain (Figure 2).

4 Methodology

To achieve the research objective a hybrid practical procedural model that integrated traditional and advanced methods and utilize paper format maps, classified satellite images, and borehole information. Since the Kingdom soil is partially classified, several challenges need to be addressed and examined thoroughly to decide the best and most effective approach. Among these issues are:

- Should the classification process follow the same schema that was used in 1975 or is new one needed? To ensure consistency and to be able to add the newly classified soil to that of 1975, the USCS classification schema is adapted.

- Should the classification process follow the same schema that was used in 1975 or is new one needed? To ensure consistency and to be able to add the newly classified soil to that of 1975, the USCS classification schema is adapted.

- To what extent can the boreholes data that is available for construction purposes can be used for soil classification? The borehole reports are mostly in PDF and some in excel format. Careful examination of a typical borehole report sample shows that it contains soil description, depth of water table and Standard Penetration Test (SPT-N). Accordingly, the borehole information is used for soil classification, as it provides an accurate field data that are similar to the ones used in 1975. Borehole data availability and accessibility are major problems that are encountered in this research. Large parts of the unclassified zones do not contain boreholes. Incorporating data within geospatial data base is a real challenge since researchers do not agree as yet on the best possible way of utilizing such data within GIS platform.

- How to classify the areas where there is no borehole data available? One option is to look into agriculture, geomorphology, geology, and drainage maps produced by the 1975 team. Vigilant inspection of this paper format shows that they are useful, but not enough to complete the classification process. The 1969 topographic map provides reasonable information about land use at that time, especially the agricultural areas. Another option is to utilize LandSat images. LandSat 1 and 2 of 1973 provide valuable information since they are within the same time frame. Supervised classification of LandSat 2 of 1973 is performed. Incorporating all these data in meaning and effective manner required special algorithm that is not currently available within ArcGIS environment.

- How to classify the newly reclaimed lands that is added to the the Kingdom and it estimated to be 136 km^2. To classify the entire Kingdom soil a new approach has to be developed. The approach must incorporate the pervious work of 1975 and classify the remaining soil using the same schema. The outcome of the proposed approach must be a complete digital soil map that is to be used as a stand alone or along other spatial data.

5 The proposed technique

As such, a hybrid method that incorporates more than one technique and makes the most use of available sources would be a better option. Design such model and in a manner that can encompass all available forms of available input data is a challenge. All input data have to be converted into digital raster format and geo-referenced. Raster cells values represent soil type. Each input data is assigned different weight based on ground truth. Example borehole information is assigned the highest weight and drainage information is the least. Further more each output raster cell is determined individually based on its own spatial location. The model consists of calibration and execution phases. The calibration phase is designed to compute the appropriate weight. Execution phase determines the actual soil type based on weight average mean. Figure 3 illustrates the conceptual model.

DOI: 10.1051/matecconf/201712009001

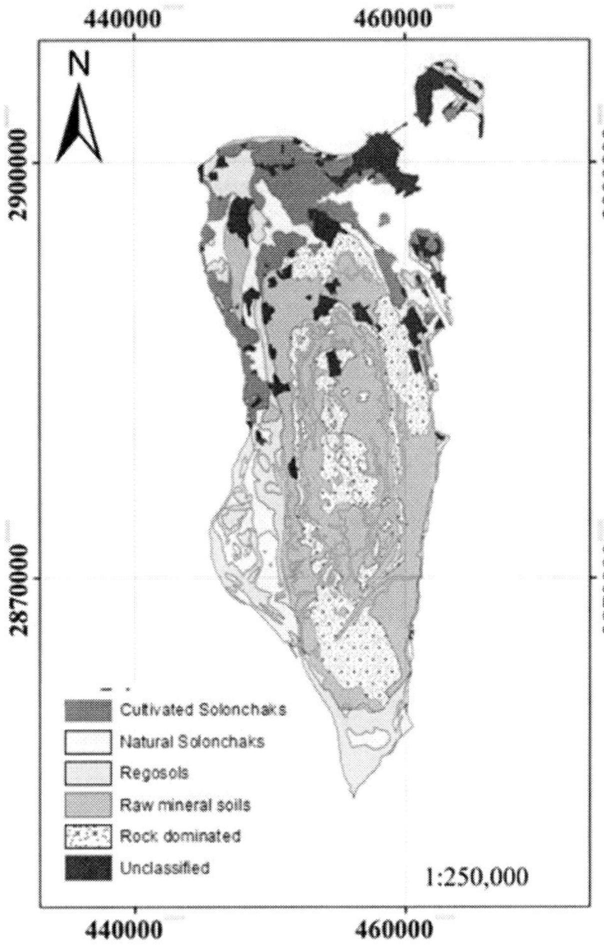

Fig. 2. Bahrain Soil Classification of 1975, note urban areas are not classified.

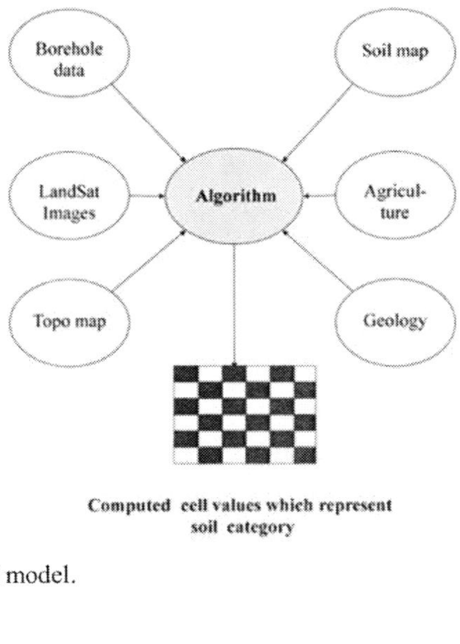

Fig. 3. Algorithm conceptual model.

DOI: 10.1051/matecconf/201712009001

6 Results

Following the classification process outlined above, a complete digital soil map for the Kingdom of Bahrain is produced (Figure 4). Accuracy of the obtained results is checked against ground truth for selected areas. Matching rate between the determined soil category and field data is ranged between 85% to 95%. The zones where there was low match rate are the areas that have no borehole data. The soil boundaries were not clearly identified due to the lack of borehole data in some areas. Digital soil map for the Kingdom of Bahrain is produced. The map encompasses the historical Awal land and the newly reclaimed areas. The produced digital map has many advantages over the current paper format map. The most important is the ability to be utilized for spatial analysis along with other theme layers and non-spatial statistical data. It can be used in conjunction with other spatial and attribute information for many construction and infrastructure activities. Analysis revealed that there are solid waste dumping activities in the sea and lowland areas in Manama and Muharraq that contains domestic and building-construction municipal solid waste.

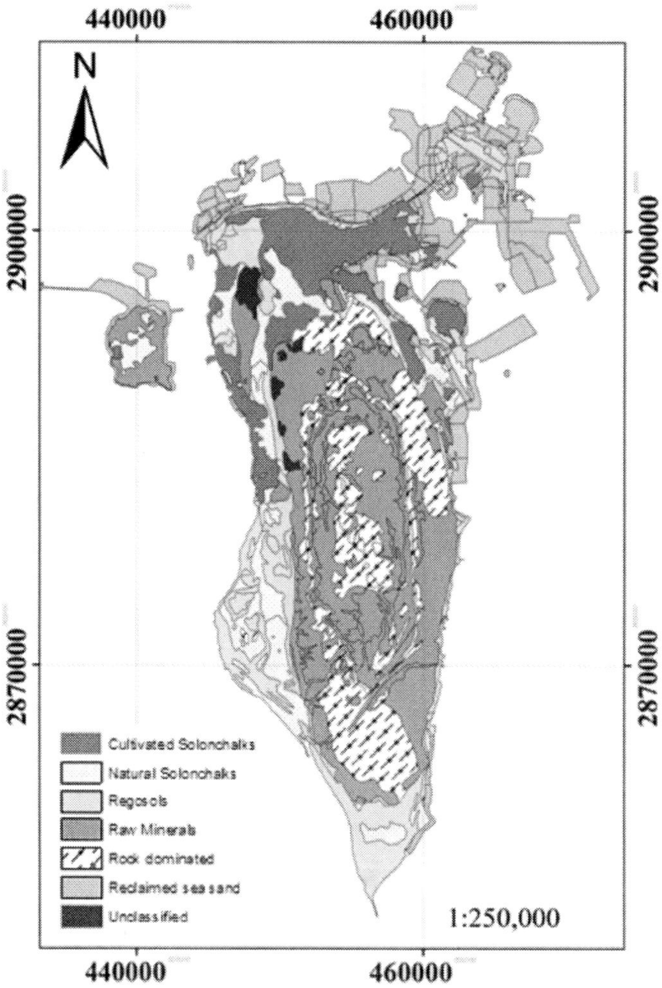

Fig. 4. Completed Bahrain digital soil map.

7 Conclusion

Soil information records provide an insight into the potentials, and limitations, of soil for effective and sustainable utilization. Soil information provides a truthful and scientific account of a soil's physical properties and the extent of distribution so that predictions can be made about their characters and possible effective utilization. It also provides adequate information about characteristics of soils (e.g., texture, depth, particle-size characteristics, liquid limit, and plasticity index, structure, stoniness, drainage, acidity, salinity) which can be utilized for planning and development. Producing digital soil map that can be easily integrated with other data is difficult and challenges. The Kingdom of Bahrain does not have such map. This research produced the first digital GIS ready soil map. Soil map of the reclaimed lands is mostly sea sand and shell deposit with no trace of clay or silt.

References

1. Venkataratnam L. & M.L. Manchanda. Remote sensing in soil resource management. ISPRS Workshop on Application of Remote Sensing and GIS for Sustainable Development. National Remote Sensing Agency, Hyderabad, Nov. 24-25 (1997).
2. Mulder V.L. et al The use of remote sensing in soil and terrain mapping — A review, Geoderma, (2011) . doi:10.1016/j.geoderma.2010.12.018.
3. Van Engelen V. e-SOTER, Annex 1 — Description of Work. EU 211578, ISRIC. Wageningen, The Netherlands report, (2008).
4. Anderson M.C. A thermal-based remote sensing technique for routine mapping of land-surface carbon, water and energy fluxes from field to regional scales. Remote Sens. Environ, 4227–4241, **112**,12 (2008).
5. Bastiaanssen W.G.M. et al. SEBAL model with remotely sensed data to improve water-resources management under actual field conditions. J. Irrig. Drain. Eng. , 85–93, **131**,1 (2005).
6. Lagacherie P., McBratney A.B., Voltz M. Digital Soil Mapping: An Introductory Perspective. Elsevier, Amsterdam, (2006).
7. McBratney A.B., Santos M.L., Minasny B. On digital soil mapping. Geoderma, 3–52, **117**,1-2 (2003).
8. Carré F., McBratney A.B., Mayr T., Montanarella L. Digital soil assessments: Beyond DSM. Geoderma, 69–79, **142**,1–2 (2007).
9. Moore A.B., Morris K.P., Blackwell G.K., Jones A.R., Sims P.C. Using geomorphological rules to classify photogrammetrically-derived digital elevation models. Int. J. Remote Sens. , 2613– 2626, **24**,13 (2003).
10. Ben-Dor E. et al Imaging spectrometry for soil applications. Adv. Agron. 97, 321–392 (2008).
11. Slaymaker O. The role of remote sensing in geomorphology and terrain analysis in the Canadian Cordillera. Int. J. Appl. Earth Obs. Geoinf. , 7, **3**,1 (2001).
12. Sculla P., Franklina J., Chadwickb O., McArthura D. Predictive soil mapping: a review. Progress in Physical Geography, 171–197, **27**,2 (2003).
13. MacMillan R.A., Jones R.K., McNabb D.H. Defining a hierarchy of spatial entities for environmental analysis and modeling using digital elevation models (DEMs). Comput. Environ. Urban Syst. 175–200, **28**,3 (2004).
14. Irvin B.J., Ventura S.J., Slater B.K. Fuzzy and isodata classification of landform elements fromdigital terrain data in Pleasant Valley, Wisconsin. Geoderma, 137-154, **77**,2–4 (1997).
15. Klingseisen B., Metternicht G., Paulus G. Geomorphometric landscape analysis using a semi-automated GIS-approach. Environ. Modell. Softw. , 109–121, **23**,1(2008).

16. Tucker J., Pinzon J., Brown M. An extended AVHRR 8-km NDVI dataset compatible with MODIS and SPOT vegetation NDVI data. International Journal of Remote Sensing 4485–4498, **26**,20 (2005).
17. [17] Van Engelen W.P.V. Global and national soils and terrain digital databases (SOTER) procedures manual. International Soil Reference and Information Centre, Wageningen, The Notherland, report (2013).
18. Zainal K., Al-Madany I., Al-Sayed H. The cumulative impacts of reclamation and dredging on the marine ecology and land-use in the Kingdom of Bahrain, Marine Pollution Bulletin, 1452-1458, **64** (2012).
19. Noor Al-Nabi M. The History of Land use and Development in Bahrain. Manama, Kingdom of Bahrain, 16 (2012).
20. Doornkamp J. C, et al. GEOLOGY, GEOMORPHOLOGY AND PEDOLOGY OF BAHRAIN, 314-320 (Norwich, England 1980).

A qualitative assessment of desertification change in the Tarfaya basin (Morocco) using panchromatic data of Landsat ETM^+ and oli: sand encroachment approach

Ali Aydda[1,*], *Ahmed* Algouti[1], and *Abdellah* Algouti[1]

[1]University of Cadi Ayyad, Faculty of Sciences Semlalia, Department of Geology, Prince My Abdellah Boulevard, BP 2390, 40000 Marrakesh, Morocco.

Abstract. The purpose of the present work is to assess desertification change in the Tarfaya basin (Morocco) based on quantifying sand dunes mass change at the corridor scale using two Panchromatic bands of Landsat ETM^+ and OLI with 15 m of resolution covering the study area for ten years (2005-2016). In this work, the sand dunes quantification is qualitative and is based on automatic extraction and classification of sand dunes shape using co-occurence texture filters and Support Vector Machine (SVM) classifier. The statistical results show that the area covered by sand was increased during the last ten years, which reveal that desertification becomes more intense.

1 Introduction

Desertification is one of the most common and severe environmental problems in the world. Approximately 42 % of the world's land surface is threatened by this phenomenon [1]. 20 % of these areas are covered by sand materials [2], which indicate an advanced stage of this phenomenon.

Quantifying desertification change with time is necessary and essential step to assess this phenomenon in order to improve the adopted fighting strategies. Mainly, several studies have shown the importance of remotely sensed data to quantify desertification change at a large scale for long and short time period. The majority of these studies have been focused on green mass (vegetation) [3-5]. Furthermore, where the sand presents a high spectral reflectance than vegetation in the desert areas (Figure 1), other studies have been concentrated on this material to quantify and track sand dunes encroachment variability with time in order to assess and track desertification [6-8].

* Corresponding author: ayddaali@gmail.com

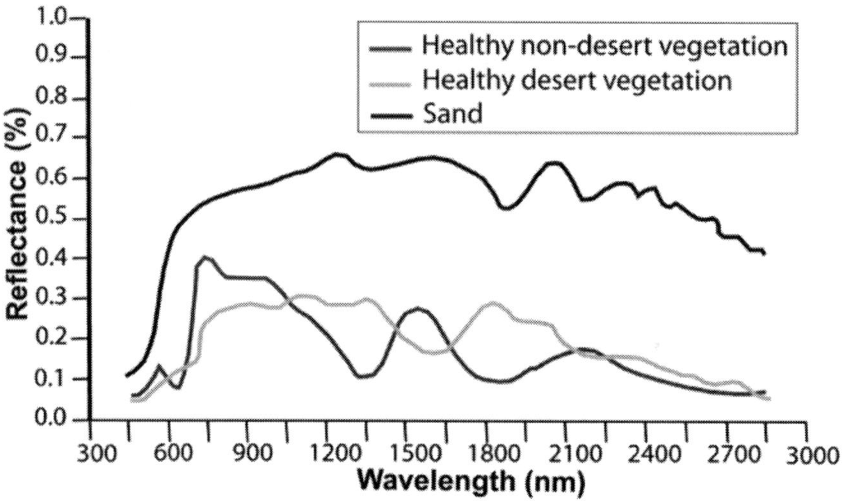

Fig. 1. Spectral response curves for healthy non-desert and desert vegetation, and sand (After [9]; in [8]).

According to [7], we can assess desertification intensity, easily from remotely sensed data, using one of the following indicators:
(i) blown sand area percentage in a region's total land area;
(ii) annual expanded area percentage of moving sand in a region's total land area;
(iii) vegetation cover percentage referring to grassland and forest land;
(iv) biomass annual reduction percentage in a region's total land area.

In this research, we used expanded sandy area indicator to quantify sand mass change in order to assess desertification change in the Tarfaya basin (Morocco). Practically, the sand mass quantification is qualitative and is based on automatic extraction and classification of sand dunes shape from two Panchromatic bands of Landsat ETM+ and OLI using co-occurence texture filters and Support Vector Machine (SVM) classifier.

2 Study area

The study area is the central dune corridor of Tarfaya basin, Morocco (Figure 2). This latter is a vast coastal platform covered by a thick Maghrebian Sandstone-Limestone flat bed (hard rock). This flagstone is overlain by movable sand dunes and broken in another place by sabkhas (Figure 2).

In this basin, the sand take origin in the beaches between Akhfinir and Tarfaya from coastal erosion [10], and is mostly transported towards the SSW in the form of barchans dunes (Figure 3), by maritime trade wind which mostly blows from NNE (Figure 3), follow three corridors of sand dunes. The eastern corridor extended from Foum Agoutir at wadi of Sakia El Hamra in south that is around 150 km in length. The shortest central corridor disappears in the Tah sabkha that is around ten kilometers in length. The western corridor is one of the longest rivers of sand in the world with 310 km in length and around 12 km in width.

 DOI: 10.1051/matecconf/201712009002

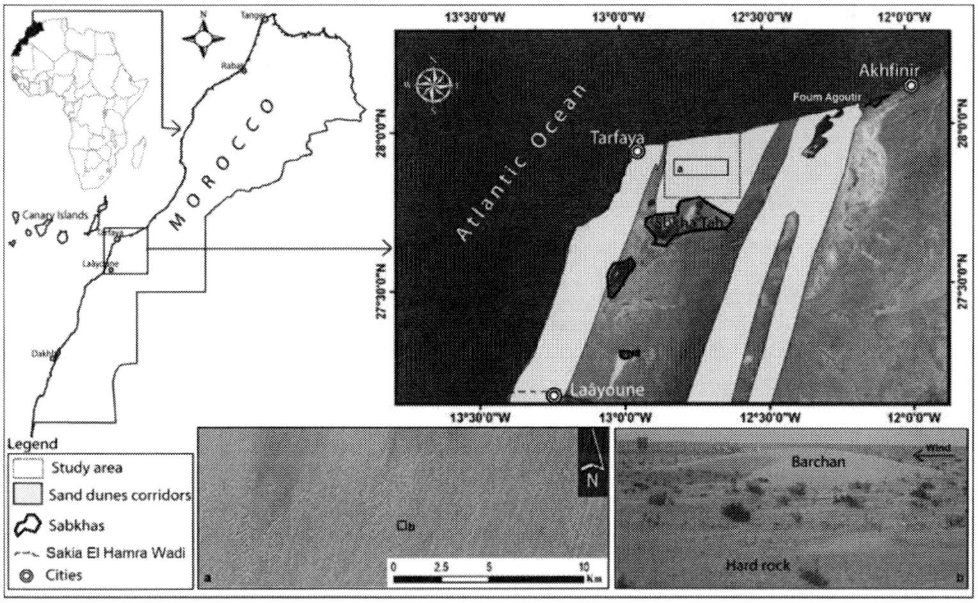

Fig. 2. Location of the study area.

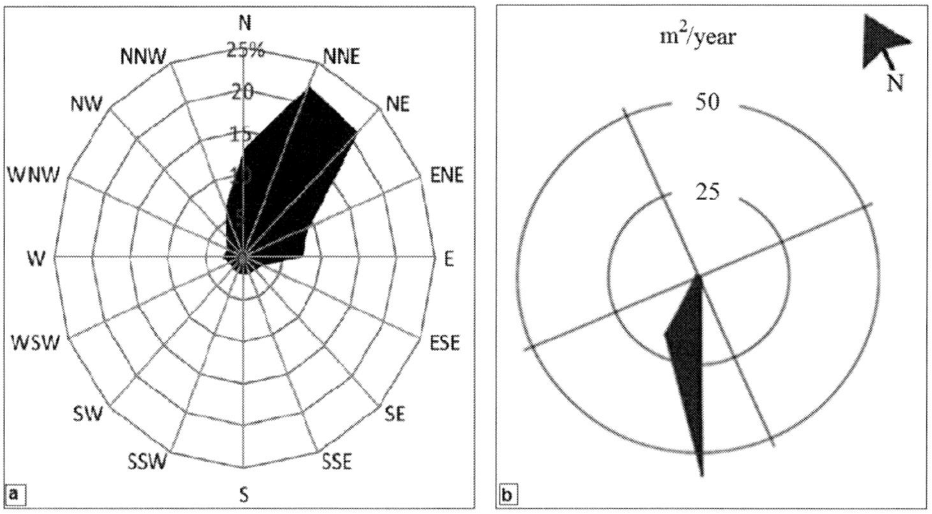

Fig. 3. Wind and sand flux roses of Tarfaya basin, a) wind rose according to [11], b) sand flux rose after [12].

3 Data and methodology

For its availability without cost and with primary process in various websites we decide to use Landsat data.

In this research, two scenes of Landsat ETM+ and OLI covering the study area for ten years (acquired in 2005 and 2016 respectively) were obtained from two different data sources at no charge, which are processed by the Level 1 Product Generation System (LPGS). These two images are geometrically and radiometrically corrected through incorporating ground control points and Digital Elevation Model (DEM) [13]. This level of correction, called Standard Terrain Correction (Level-1T), provides a highest quality and suitable data for pixel-time series analysis [13]. For its spatial resolution (15 m), we used Panchromatic bands of these scenes to create this study. Table 1 lists the characteristics of used data sets including capture, sensor, acquisition date, data type, and data source.

Table 1. Data set characteristics.

Satellite	Sensor	Acquisition date	Data type	Data source
Landsat 7	Enhanced Thematic Mapper Plus (ETM^+)	31.12.2005	L1T	Global Land Cover Facility (GLCF)
Landsat 8	Operational Land Imager (OLI)	27.01.2016	L1T	United States Geological Survey EarthExplorer (USGS EE)

The aim of the present work is to assess desertification change in the Tarfaya basin based on quantifying expanded sandy area from 2005 to 2016 at central corridor scale. The sand quantification is based on automatic extraction and classification of sand dunes shape. The detail of methodology applied in this study is presented in Figure 4. The all steps cited in this methodology were achieved using ENVI® software.

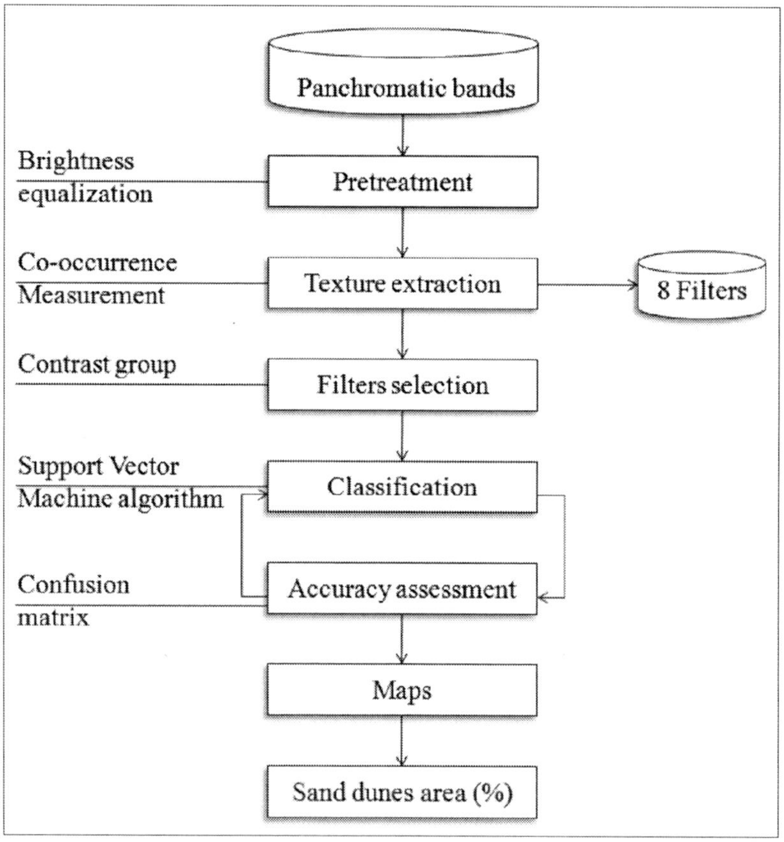

Fig. 4. Flow chart showing the steps used in this work.

Firstly, in term of improving bands visualization the interactive stretching and histogram matching were used. The sand dunes shapes extraction is the second step; it was achieved by applying co-occurrence measurement algorithm. This algorithm provides eight filters according three groups: contrast group (contrast, homogeneity and dissimilarity), orderliness group (angular second moment and entropy) and statistics group (mean, variance and correlation). The third step involves combining contrast group filters (Figure 5). The classification is the next step; it was adopted for same data sets of combined filters by delimiting polygons around barchans (sand dunes) and space between them (hard rock) (Figure 6), and was executed using Support Vector Machine (SVM) classifier. The classification accuracy assessment is the next step; it was realized by confusion matrix by which six parameters were computed: User's Accuracy (UA), Producer's Accuracy (PA), Commission Errors (CO), Omission Errors (OE), Overall Accuracy (OA) and Kappa Coefficient (K). In order to obtain the good results the two last steps were repeated. The sand dunes area was calculated from final classified maps of 2005 and 2016. The desertification change is difference of sandy area computed from both classified maps.

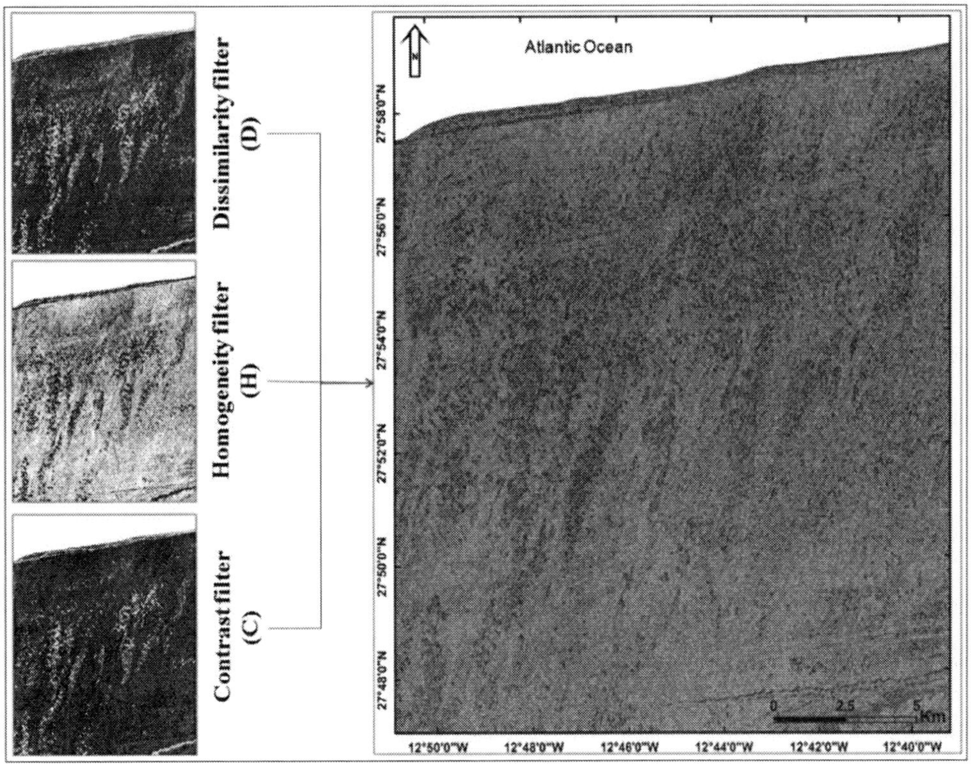

Fig. 5. Utilized filters, RGB: DHC.

 DOI: 10.1051/matecconf/201712009002

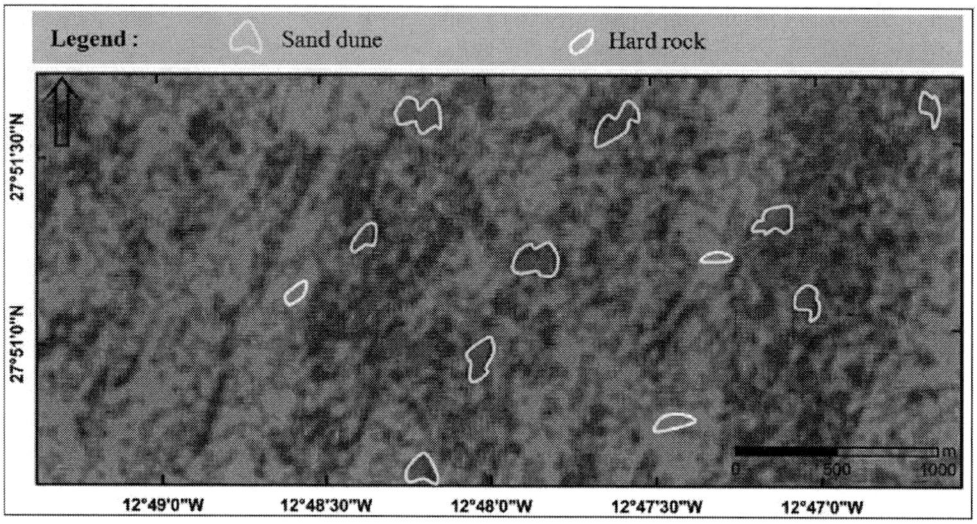

Fig. 6. Example of test areas used for classification.

4 Results and discussion

The classification maps from Panchromatic bands of Landsat ETM$^+$ and OLI are shown in Figure 7. The accuracy statistics of these classifications are shown in Table 2.

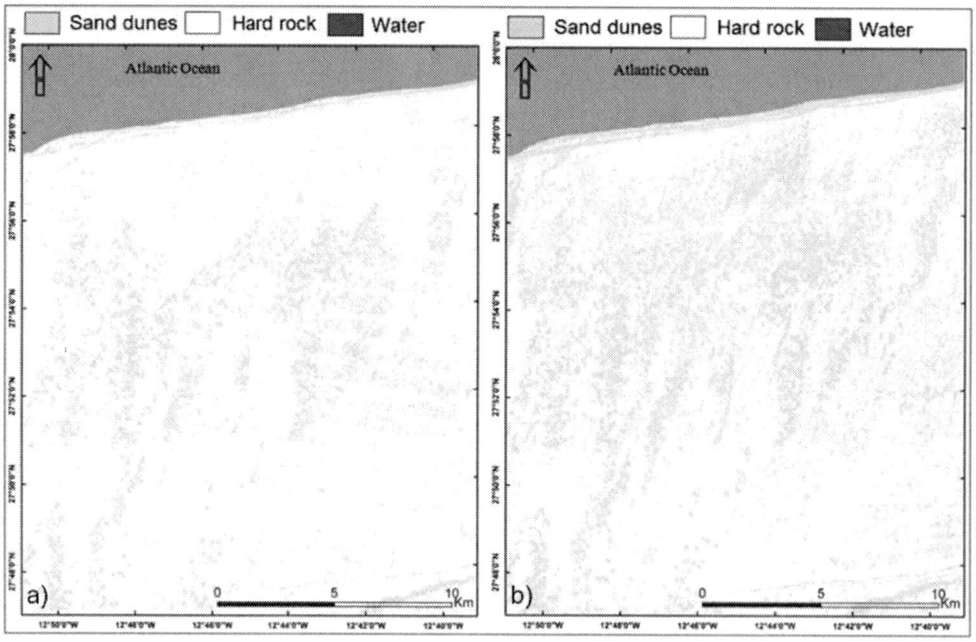

Fig. 7. Final version of classified maps for the years: a) 2005 and b) 2016.

Table 2. Accuracy statistics for the final version classified maps. UA: User's Accuracy, PA: Producer's Accuracy, CO: Commission Errors, OE: Omission Errors, OA: Overall Accuracy, and K: Kappa Coefficient.

		Sand(%)	Hard rock(%)	**UA(%)**	**PA(%)**	**CE(%)**	**OE(%)**	**OA(%)**	**K**
Map 2005	Sand(%)	97.20	0.25	99.49	97.20	0.51	2.80	98.8	0.97
	Hard rock(%)	2.80	99.75	98.62	99.75	1.38	0.25		
Map 2016	Sand(%)	99.96	1.59	98.11	99.96	1.89	0.04	99.1	0.98
	Hard rock(%)	0.04	98.41	99.96	98.41	0.04	1.59		

To ensure that the classification results are dependable and acceptable for evaluating change their overall accuracies must be higher than 85% [14]. In this work, the overall accuracy exceeds 98% for both classification maps, which meet the limit for classification results acceptance. This accuracy explained by clarity of lithological data in the study area (desert zone). However, even if the classifications indicate a very high Kappa coefficient for the both maps (equal 0.97 and 0.98 for the years 2005 and 2016 respectively), the commission and omission errors point out low rates of misclassification (Table 2). This later may be explained by the sand incorporation in the small fissures observed on the hard rock surface.

A closer observation on the classified maps reveals an increase of surface covered by sand dunes and a decrease of surface occupied by hard rock. The detail of sand dunes and hard rock surfaces change for the years 2005 and 2016 is shown in Figure 8.

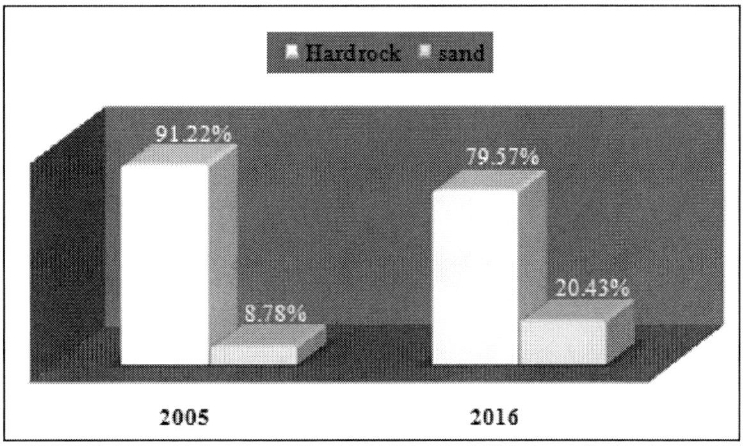

Fig. 8. Percentages of surface covered by sand dunes and hard rock for the years 2005 and 2016.

The sand dunes were covered 8.78% and 20.43% of the total zone respectively in 2005 and 2016, which indicate that the surface occupied by these dunes was increased by 11.65% during the last ten years. The sand dunes increasing indicate that desertification becomes more intense in the study area.

The hard rock was decreased by same value of sand dunes increasing, which was occupied 91.22% in 2005 and 79.57% in 2016. The hard rock change is in relation with sand dunes contribution.

According to [10], the sand dunes increasing is partially due by marine contributions, which is indicated by an active coastal erosion and by a decrease of suspended sediments in the sea surface, and mainly attributable to the continental contributions which is manifested by erosion of composed sand dunes and by barchans migration towards the interior of the continent.

5 Conclusions

The aim of this study is to assess desertification change during the last ten years in the Tarfaya basin (Morocco) through assessment of sand dunes expansion using Panchromatic bands of Landsat ETM$^+$ and OLI collected in 2005 and 2016, respectively. In this work, the classification technique was used after sand dunes automatic extraction using co-occurence texture filters.

The results show that the area covered by sand dunes has increased between 2005 and 2016 by 11.65%, which indicate that desertification becomes more intense in the last ten years. Those findings emphasize the need for improving the old techniques or proposing a new strategy to fight against this problem.

References

1. UNEP, John Wiley & Sons, Inc. and Arnold (Second edition), 182 p (1997)
2. K. Pye, H.Tsoar, Unwin Hyman, London, p. 396 (1990)
3. U. Hellden, Ambio, **20**, 372-383 (1991)
4. M.C. Runnstrom, Ambio, **29**, 468-476 (2000)
5. J. Li, X. Yang, Y. Jin, Z.Yang, W. Huang, L. Zhao, T. Gao, H. Yu, H. Ma, Z. Qin, B. Xu, Remote Sensing of Environment, **138**, 19-26 (2013)
6. M. Kumar, E. Goossens, R. Goossens, International Journal of Remote Sensing, **14**, 1689-1703 (1993)
7. T. Wang, W. Wu, X. Xue, Q. Sun, G. Chen, Science in China Series D: Earth Sciences, **47**, 78-88 (2004)
8. D.K. Lam, T.K. Remmel, T.D. Drezner, Remote Sensing, **3**, 1-13 (2010)
9. J.R. Ehleringer, Oecologia, **49**, 366-370 (1981)
10. A. Aydda, Thèse de doctorat, Université de Cadi Ayyad de Marrakech, Maroc, 234 p (2016)
11. K. Selouane, Thèse de doctorat, Ecole des Mines de Paris, France, 437 p (2008)
12. H. Elbelrhiti, Thèse de Doctorat, Université Paris VII, France, 160 p (2005)
13. USGS, http://landsat.usgs.gov/Landsat_Processing_Details.php
14. G.G. Wright, J.G. Morrice, International Journal of Remote Sensing, **18**, 3811-3834 (1997)

Mapping urban impervious surfaces from an airborne hyperspectral imagery using the object-oriented classification approach

Rahim Aguejdad[1,*], *Aziz* Serradj[1], and *Christiane* Weber[2]

[1] Laboratoire Image Ville Environnement UMR7362-CNRS, 3 Rue de l'Argonne, 67000 Strasbourg, France
[2] Maison de la Télédétection UMR TETIS, 500 Rue Jean-François Breton, 34000 Montpellier, France

Abstract. The objective of this research is to explore the capabilities of the hyperspectral imagery in mapping the urban impervious objects and identifying the surface materials using an object-oriented approach. The application is conducted to Toulouse city (France) within the HYEP research project in charge of using hyperspectral imagery for the environmental urban planning. The method uses the multi-resolution segmentation and classification algorithms. The first results highlight a high potential of the hyperspectral imagery in land cover mapping of the urban environment, especially the extraction of impervious surfaces. They, also, illustrate, that the object-oriented approach by means of the fuzzy logic classifier yields promising results in distinguishing the mean roofing materials based only on the spectral information. Conversely to the red clay tiles and metal roofs, which are easily identified, the concrete, gravel and asphalt roofs are still confused with roads.

1 Introduction

Remote sensing technology has a great potential to afford consistent and useful spatial information at different spatial and temporal scales. It can help understanding the links between land use and infrastructure change and a variety of social, economic, demographic and environmental processes [1, 2]. This information derived from remotely sensed datasets leads to an improved observation and monitoring that can benefit applied urban planning and management [2]. However, land cover mapping of the urban environment by means of the remote sensing remains a distinct challenge due to the strong diversity of materials and objects, leading to spectral heterogeneity and geometric complexity of the urban areas as well as the presence of the shadow scenes [3]. In fact, several studies have showed that current satellite data provide limited information to deal with the richness, heterogeneity and complexity of the materials and forms in the urban areas. For instance, [4] has showed the limits of the multispectral sensors for the urban impervious surfaces extraction compared to

* Corresponding author: rahim.aguejdad@live-cnrs.unistra.fr

the hyperspectral imagery capabilities. In addition to that, intra-urban discrimination of land cover and land use types is needed for many local scales remote sensing applications requirements [2]. Accordingly, several studies have shown that a spatial resolution of at least 5 m is necessary to accurately extract the land cover objects [5, 6]. In addition to the spatial accuracy of the sensor, the definition of the thematic mapping classes and the image classification accuracy require specific spectral characteristics ranging from the Visible Near-Infrared (VNIR) to Short Wave Infrared (SWIR) wavelength domains [7, 8].

The recent introduction of advanced high spatial resolution hyperspectral sensors provides new perspectives [9]. Thanks to its high spectral capacities in identifying different urban objects [10], the hyperspectral imagery is used to address key science and application questions in a wide array of fields, ranging from ecosystem function and diversity to human health, precision agriculture and urbanization. For example, [11] have used hyperspectral aerial images for mapping asbestos cement roofs and developed a spectral index related to the roof weathering status. At the same time, the recent developments in hyperspectral domain have stressed on the need for advanced classification methods [12]. Recently, [13] proposed a generic method for fusion of hyperspectral and VHR multispectral image classifications in urban areas. The proposed fusion model relies on posterior class probabilities which are integrated through an energy minimization process [13]. Furthermore, [12] have proposed a new methodology for spectral-spatial classification of hyperspectral images. This method, which is as an alternative to current pixel-wise methods, simultaneously considers the spectral and spatial information [12].

The hyperspectral users' community in France has addressed very detailed science needs for a high-resolution hyperspectral mission in the following themes: the study of vegetation, the coastal and inland water ecosystems, geosciences, the urban environment, the atmospheric studies, the security and defence [14]. Accordingly, the HYPXIM task specification work has been conducted to define a hyperspectral camera with high spatial resolution (8 m) coupled to a panchromatic camera with 2 m spatial resolution. To simulate the images acquired by such space instrument, airborne acquisition missions were carried out. The objective is to assess the gain provided by the HYPXIM mission compared to existing missions such as the multispectral or hyperspectral Pleiades coming from EnMAP (30 m) and Prisma (20 m).

The aim of this paper is to analyze the potential of the simulated datasets provided by the airborne hyperspectral imagery in mapping urban impervious surfaces using an object-oriented classification approach compared to an analysis on per-pixel basis. The study is applied to Toulouse city (France) within the HYEP (Hyperspectral imagerY for Environmental urban Planning) research project in charge of using the hyperspectral imagery for the environmental urban planning.

2 Materials and Methods

2.1 Study area and data

The study area is the city of Toulouse (43°36'19'' N, 1°26'34''E) which is located in the South-Western of France (Fig. 1). The used hyperspectral imagery is derived from the simulation of the future HYPXIM sensor using data acquired by the HySpex hyperspectral camera (www.hyspex.no) over the city centre in 2015. It records data with a high number of narrowly defined spectral bands (192). It provides a continuous spectrum ranging from 400 nm to 2500 nm, covering both VNIR as well as SWIR domains (Table 1) with 4 m ground geometric resolution.

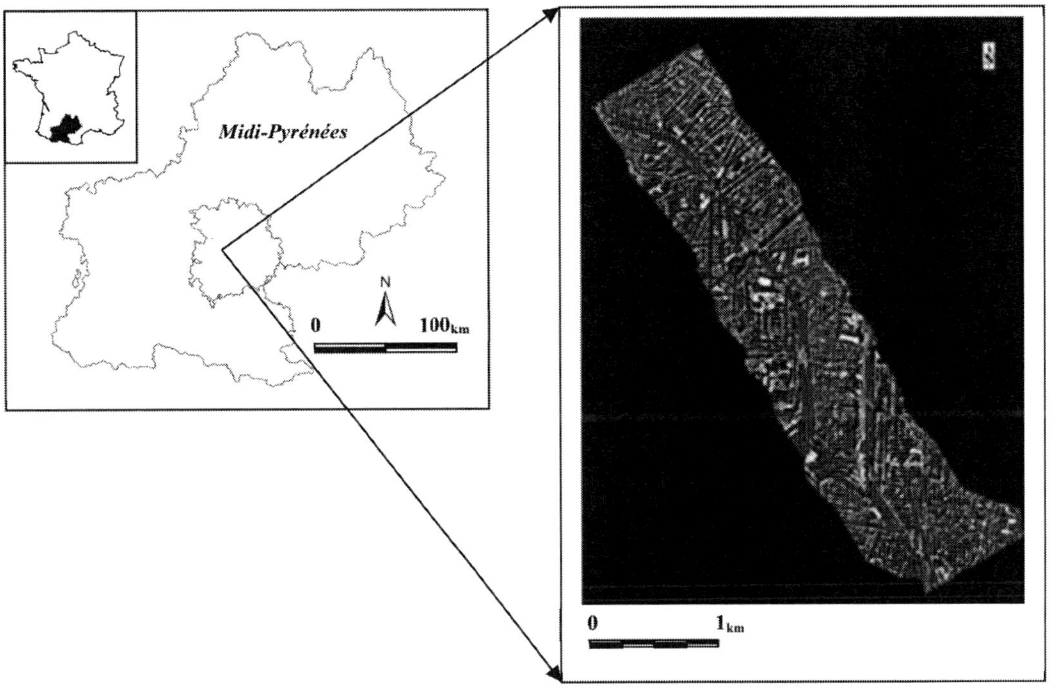

Fig. 1. RGB hyperspectral image centred over the dense city centre of Toulouse (Toulouse ONERA, 2015).

The HySpex hyperspectral VNIR-SWIR georeferenced radiances image (0.4μm-2.5μm; GSD=2m) is used to simulate different spatial and spectral resolutions, corresponding to the HYPXIM sensor which is under study and the operative PLEIADES and SENTINEL-2 instruments. The HYPXIM sensor is the future hyperspectral space borne instrument leaded by the CNES (*Centre National d'Etudes Spatiales*). It is expected to provide data with a spatial resolution of 8 m in the VNIR-SWIR range.

Table 1. Hyperspectral data spectral characteristics.

Domain	Spectrum (nm)	Band width Δλ (nm)	Bands list
Visible (Vis)	418 - 691	10.9	1 - 26
Near-InfraRed (NIR)	702 - 1095	10.9	27 - 63
Short Wave InfraRed (SWIR)	1105 - 2502	10.9	64 - 192

The BDTOPO® topographic database will be used as a reference map to validate the segmentation and classification results. The BDTOPO® is a 3D topographic database that provides a vector and structured description of the spatial objects with a metric mapping accuracy designed to scales ranging from 1/5000 to 1/50000. The segmentation accuracy will be assessed by comparing the created image objects with the building blocks from the BDTOPO®. The validation feedbacks will allow setting the adequate input segmentation parameters and then refining the delineation of the borders of objects.

2.2 Method

An image based supervised classification is performed using an object-oriented approach by means of the eCognition software. First, a routine image pre-processing including radiometric, atmospheric, and geometric corrections is performed. The atmospheric correction is carried out using the COCHISE (atmospheric correction code for hyperspectral images of remote-sensing sensors) algorithm [15]. Second, the image segmentation is performed using all the spectral bands except those presenting an excessively noise. A multi-resolution segmentation algorithm is performed to create meaningful image objects using only the spectral information. The scale factor is set to 0.4, whereas the shape criterion is not taken into consideration. These segmentation parameters are derived based on several tests, which are performed to deal with the characteristics of the target objects present in the study area. Then, spectral difference segmentation is applied to merge spectrally similar neighboring image objects if the difference between their layer mean intensities is below the user-defined threshold value given by the maximum spectral difference. The last step consists in classifying the image objects with respect to a predefined hierarchical network. The supervised classification procedure uses the fuzzy logic classifier which is based on the use of membership functions.

3 Results and discussion

Fig. 2 shows the image objects as the output of the segmentation process. The highly heterogeneous patterns of development (residential vs commercial, individual vs collective housing, low and high density, geometry, roof materials…) limits the application of the same segmentation parameters. In addition to that, the presence of "mixed" pixels that compose the borders of objects leads to their inaccurate delineation. One of the solutions is to apply the process of unmixing. Another alternative consists in using different input segmentation criteria after splitting the study area into homogeneous zones.

Image from Google Maps©, 2017

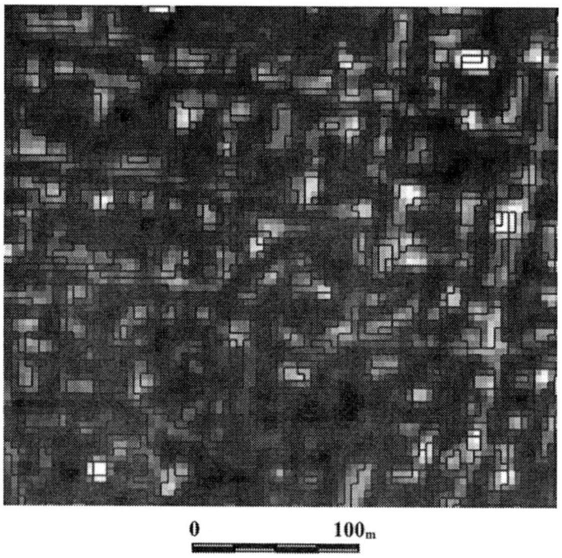

Fig. 2. Image objects in a residential area of low density and highly dominated by individual houses.

Fig. 3 shows the map of the buildings that are extracted by means of the classification of the image objects. It illustrates that the classification algorithm performs in mapping red clay tiles and metal building' roofs. These well-classified segments consist in bright construction materials (Zinc, Wood, PVC).

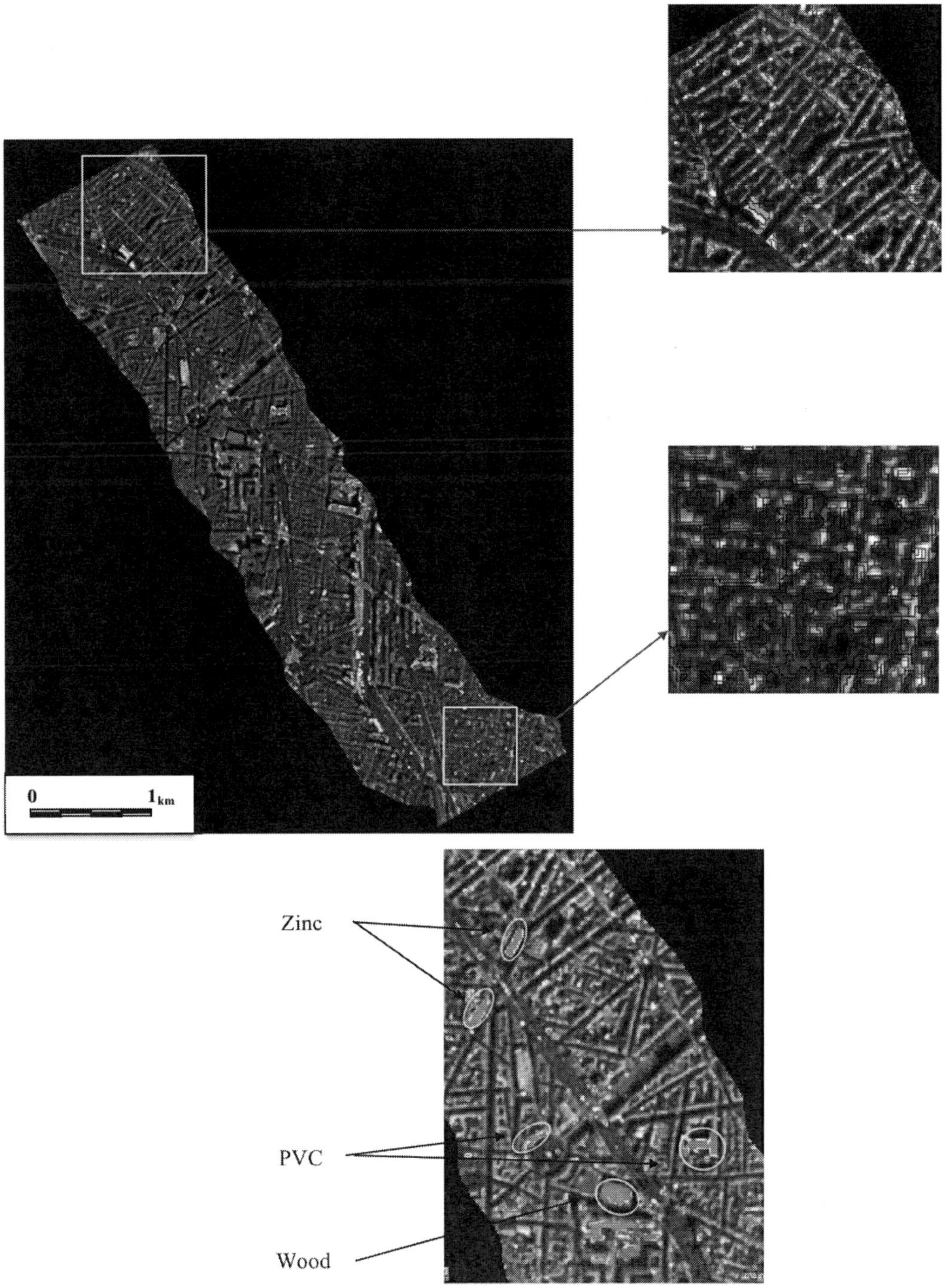

Fig. 3. Extraction of red clay tiles and metal roofs buildings.

Despite the obvious merits that the used hyperspectral imagery may offer based on the 192 provided spectral bands, great challenges need to be addressed such as the observed confusion between the spectral signatures similarity between some urban land use and cover objects. For instance, buildings with concrete, gravel or bitumen roofs present similar reflectance behavior compared to the road network and associated structures such as car parks (Fig. 4).

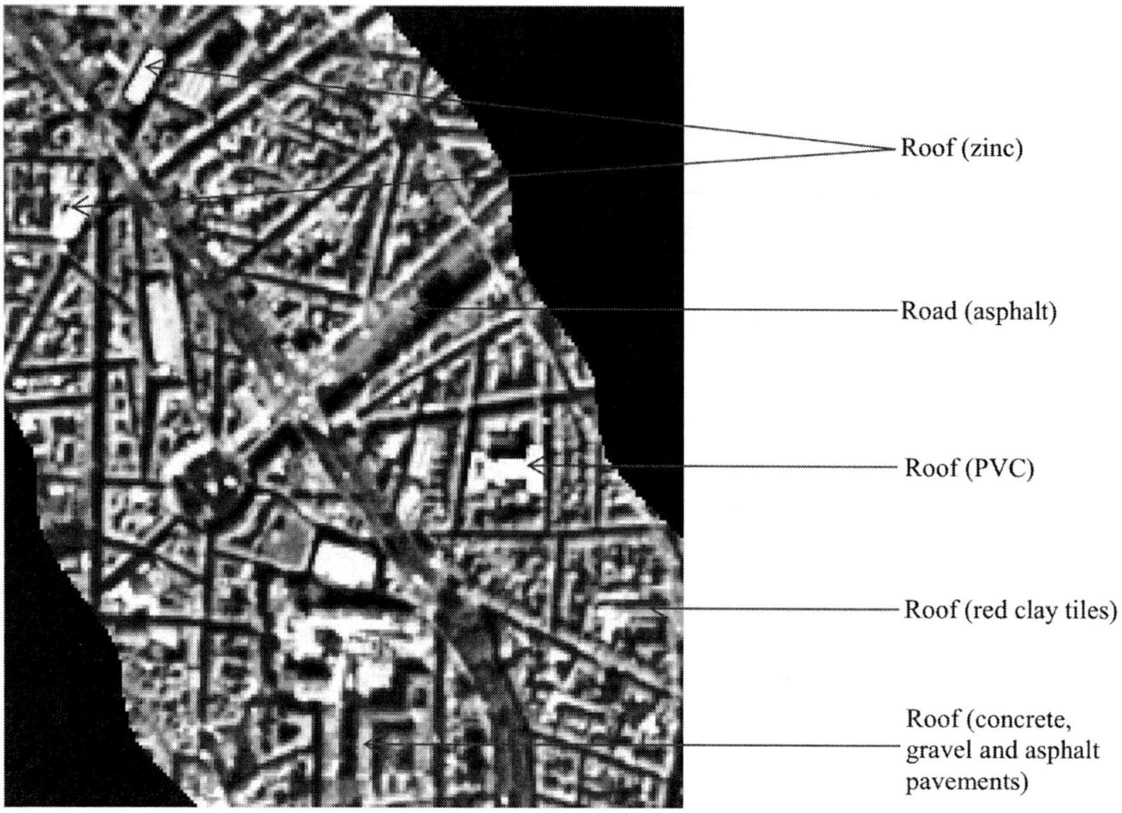

Fig. 4. Spectral similarity problem between roads and asphalt or concrete grey pavements roofs.

Fig. 5 illustrates the extent to which the reflectance curves of roads and grey roofs over the VNIR and SWIR wavelength spectral domains are close to each other.

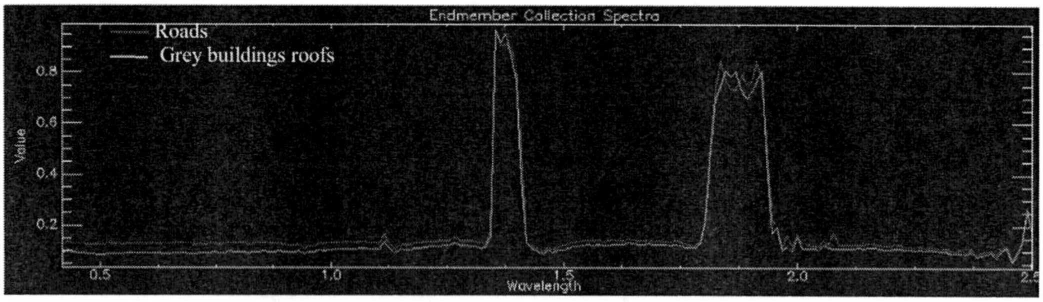

Fig. 5. Spectral profiles similarity between roads and dark roofs.

In addition to the spectral information, one solution consists in the use of complementary information (3D information, texture, morphology and contextual measurements) that will probably help to improve the fuzzy classification outputs. The use of the elevation information such as LiDAR data as a thematic layer in combination with a hyperspectral imagery in the classification process will allow to deal with the omnipresence of confusion between roads and concrete, gravel or bitumen buildings roofs.

4 Conclusion

This study highlights the high potential of the simulated HYPXIM hyperspectral imagery for the urban land cover mapping, especially the extraction of impervious surfaces. It, also, illustrates that the object-oriented classification approach by means of the fuzzy logic method yields promising results in identifying the mean construction roof' materials based on the spectral information, only. Indeed, the red clay tiles and metal roofs are easily distinguished compared to the grey roofs that are frequently confused with the roads. One of the future perspectives is to compare the object-oriented method to the Super Angular Mapper (SAM) classifier that will be trained through a set of samples (roads, vegetation, water, and buildings) selected from an external spectral library. The comparison and accuracy assessment of the two methods will be based on quantitative measurements such as the Kappa statistic using the BDTOPO® as the reference map. Besides, this work, which is in progress, will be applied to different spatial resolutions datasets (2, 4, 8 and 10 m).

This research is supported by the French ANR HYEP (*Hyperspectral imagerY for Environmental urban Planning)* project. The authors would like to thank the anonymous reviewer for his comments and suggestions that helped us to significantly improve this paper.

References

1. R.R. Rindfuss, P.C. Stern, *Linking remote sensing and social science.* In S. Liverman et al. (Eds.), People and pixel-linking remote sensing and social science (pp. 1–28). Washington, DC: National Academic Press (1998)
2. M. Herold, H. Couclelis, K.C. Clarke, The Role of Spatial Metrics in the Analysis and Modeling of Urban Land Use Change. Computers, Environment and Urban Systems **29**, 369–399 (2005)
3. F. Priem, F. Canter, Synergistic Use of LiDAR and APEX Hyperspectral Data for High-Resolution Urban Land Cover Mapping. Remote sensing **8**, 787 (2016)
4. D.S. Lu, Q.H. Weng, Use of impervious surface in urban land-use classification. Remote Sensing of Environment **102**, 146–160 (2006)
5. R. Welch, Spatial resolution requirements for urban studies. International Journal of Remote Sensing **3**(2), 139–146 (1982)
6. R.R. Jensen, P.J. Hardin, A.J. Hardin, Classification of urban tree species using hyperspectral data. Geocarto International **27**, 443–458 (2012)
7. E. Ben-Dor, *Imaging spectrometry for urban applications. In Imaging Spectrometry: Basic Principles and Prospective Applications*; Van Der Meer, F.D., De Jong, S.M., Eds.; Kluwer Academic Publishers: Dordrecht, The Netherland; pp. 243–281 (2001)
8. D.A. Roberts, M. Herold, *Imaging spectroscopy of urban materials.* In Infrared Spectroscopy in Geochemistry, Exploration and Remote Sensing; King, P., Ramsey, M.S., Swayze, G., Eds.; Mineral Association of Canada: Ottawa, ON, Canada; pp. 155–181 (2004)

9. S. Gadal, W. Ouerghemmi, *Morpho-spectral objects classification by hyperspectral airborne imagery*, 8th Workshop in Hyperspectral Image and Signal Processing: Evolution in Remote Sensing, Aug. 2016, LosAngeles, USA (2016)
10. S. Pascucci, C. Bassani, R.M. Cavalli, L. Fusilli, A. Palombo, S. Pignatti, F. Santini, *Hyperspectral remote sensing capability for mapping near-surface asbestos deposits and pollutants dispersion in soils*. Proc. Hyperspectral 2010 Workshop, Frascati, Italy, 17–19 March (2010)
11. C. Cilia, C. Panigada, M. Rossini, G. Candiani, M. Pepe, R.Colombo, Mapping of asbestos cement roofs and their weathering status using hyperspectral aerial images. ISPRS International Journal of Geo-Information **4**, 928–941(2015)
12. Z. Miao, W. Shi, A new methodology for spectral-spatial classification of hyperspectral images. Journal of Sensors, Vol. 2016, 12 pages (2016)
13. A. Hervieu, A. Le Bris, C. Mallet, *Fusion of hyperspectral and VHR multispectral image classifications in urban areas*. ISPRS Annals of Photogrammetry, Remote Sensing and the Spatial Information Sciences, Vol. III-3, Prague, Czech Republic, 12-19 July (2016)
14. S. Michel, M.J. Lefevre-Fonollosa, S. Hosford, *HYPXIM – A hyperspectral satellite defined for science, security and defence users*. Proc. 'Hyperspectral 2010 Workshop', Frascati, Italy, 17–19 March (2010)
15. Y. Boucher, L. Poutier, V. Achard, X. Lenot, C. Miesch, *Validation and robustness of an atmospheric correction algorithm for hyperspectral images*. Algorithms and Technologies for Multispectral, Hyperspectral, and Ultraspectral Imagery VIII, N° 4725-54, SPIE AeroSense 2002, Orlando, FL., 1-5 April (2002)

Satellite based seasonal land use classification and change detection analysis of landsat-8 operational land imager

Wasim Pervez [1,*], *Shoab Ahmad Khan* [1], *Ejaz Hussain* [1], *and Faisal Amir* [1]

[1]National University of Sciences and Technology H-12, 44000, Islamabad, Pakistan

Abstract. This study investigated the Landsat-8 Operational Land Imager (OLI) to determine its suitability for change detection analysis and mapping applications due to its enhanced signal-to-noise ratio (SNR), spectral band configuration, technical superiority, improved system design and high radiometric resolution. Earlier Landsat series had a lower SNR, comparatively smaller radiometric resolution and limitations in spectral band configurations. Pre-classification methods e.g., image differencing; image rationing, vegetation indices and principal components used for change detection analysis do not indicate type of the change due to its limitations. Therefore, better technique of post classification change detection was used on OLI data in the study area which provided information of the type of the change i.e., from-to change detection. This paper evaluated the OLI support vector machines (SVM) classification suitability using data covering four different seasons (i.e., spring, autumn, winter, and summer) after pre-processing and atmospheric correction. After classification, the major change detection results of OLI SVM-classified data for the four seasons were compared to change detection results of six cases: (1) winter to spring; (2) winter to summer; (3) winter to autumn; (4) spring to summer; (5) spring to autumn; and (6) summer to autumn. Seasonal change in the shoreline resulted with the corresponding change in categories. The OLI data classifications were made by applying SVM classifier and due to its improved features, OLI data is found suitable for seasonal land cover classification and post classification change detection analysis.

1 Introduction

The Landsat-8 Operational Land Imager (OLI) was launched in 2013. Intended to provide timely and high quality visible and infrared data, Landsat-8 includes several new features compared to ALI and Landsat 7, specifically: (1) two sensors (30 m resolution with eight spectral bands) and a Thermal Infrared Scanner (TIRS) (100 m resolution with two thermal bands) [1]; (2) a new Coastal/Aerosol (CA) band and a Cirrus band; (3) improved technical superiority with more bits (i.e., 12 bits as compared to 8 bits) and an improved signal-to-noise (SNR) performance over ALI and Landsat-7 [2]; and (4) improved instrument design

* Corresponding author: wasim_pervaiz@pnec.nust.edu.pk, wasim_pervaiz@yahoo.com

compared to ALI and Landsat 7 [3–5]. The width of OLI band 5 is refined to exclude atmospheric absorption features at 825 nanometers (nm) and its panchromatic bandwidth is reduced to improve detection of vegetation vs. non-vegetation compared with previous Landsat series. In OLI, the new Cirrus band detects thin clouds more accurately compared to previous satellite-based sensors (i.e., ALI and Landsat-7) [6]. Landsat 8 provides regional monitoring through remote sensing with substantial improvements in data quality due to advances in its noise reduction and spectral coverage designs [7, 8]. Studying different satellite sensors and their parameters (i.e., scanning systems, sensor characteristics, data systems, resolution, and so on) is important. However, studies examining the characteristics of Landsat-8 and its pre- and post-flight calibrations are rare in the literature [9–15].

Change detection remains a challenging problem and is used different for classification and change detection analysis [16–18]. Different satellite based change detection methods have different advantages and disadvantages [19, 20]. In some change detection methods, no training data is required [21] e.g., in image differences [22] and clustering based methods [23] provide only change vs no change. If training data is available, then post classification change detection analysis [24] supervised change vector analysis [25, 26] and direct classification [27] provide "from-to" information and are preferred methods. However, selecting a change detection method still depends heavily upon its intended application [28–32]. Land use/cover change information is used for various applications [33]. Different change detection types include: urban land cover and change detection [34, 35] impervious surface change detection [36], land cover change detection [37, 38]. Post classification change detection analysis is one of the most applied techniques due to its advantages of minimizing environmental and atmospheric differences between images and matrix of change detection. Hence this technique is applied in this paper.

ALI (30 m resolution with nine spectral bands) has a swath width of 37 km and lower technical superiority than OLI but better technical superiority than Landsat-7 [6, 39–42]. Hyperion (30 m resolution with 12 bits) has a swath width of 7.7 km and a swath length of 185 km and contains hundreds of bands, but it requires atmospheric correction during preprocessing [43, 44]. IKONOS, Quick Bird and WorldView-2 are useful in applications that have detailed image spatial requirements; however, their sensors have limitations for long-term or regular monitoring over large areas [45–47]. Hyperspectral remote sensing, due to its technical characteristics, has been considered suitable for use in mapping applications [48–65]. Classifiers for mapping application can be categorized as iterative and non-iterative. SVM is an iterative classifier, and is superior to SAM [66, 67] and SID classifiers which are non-iterative [68]. The SVM classifier provides better results even when only small amounts of training data are available or with high-dimensional data, because it focuses on training data near class boundaries [69–72].

This study conducted satellite-based change detection analysis using Landsat-8 OLI images. The main goals of this paper were: (i) to evaluate the results of SVM classifications on OLI data over four different seasons (i.e., spring, autumn, winter, and summer) after preprocessing and atmospheric correction; (ii) to use the post-classification change detection results from OLI SVM-classified data of four seasons to perform a change detection analysis of six cases: (1) from winter to spring, (2) winter to summer, (3) winter to autumn, (4) spring to summer, (5) spring to autumn, and (6) summer to autumn; (iii) to compare the classification accuracies achieved on the OLI data by applying SVM classifier.

2 Study area, sensors, and its characteristics

This study's main objective was to perform change detection analysis and mapping of the study area (Figure 1). The study area was classified into six distinct classes (i.e., bare soil, built-up area, mixed forest, shrub, deep water and shallow water). Table 1 shows the imaging geometry conditions, sensor altitude, sun azimuth, sun elevation, and scene center latitude and longitude for the Landsat-8 OLI, ALI, and Hyperion satellites. Table 2 shows the band names, wavelengths, center wave length, ground resolution, and swath width of the Landsat-8, ALI and Hyperion bands [73–75]. Excluded from this study are the OLI panchromatic and cirrus bands, the ALI, and the Hyperion bands. Landsat-8 data covering four seasons (i.e., autumn, summer, spring and winter) were acquired and used for change detection analysis (Table 3).

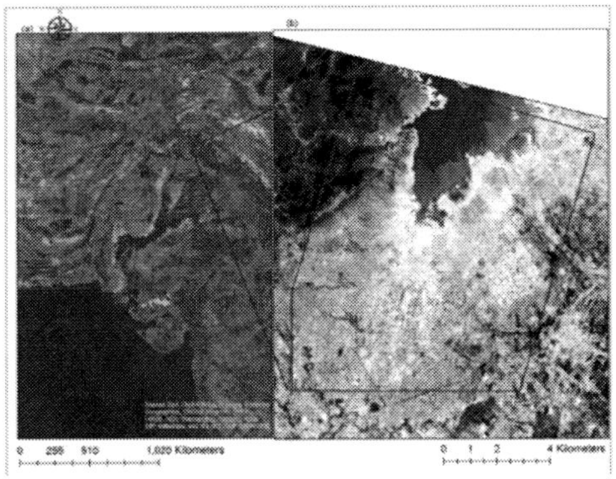

Fig. 1. (a) Study area location (regional) and (b) study area (outlined in red).

Table 1 Imaging geometry conditions and scene center latitudes and longitudes for Landsat-8

	L8 OLI[a] 22 October 2015	L8 OLI[a] 16 June, 2015	L8 OLI [a] 13 April, 2015	L8 OLI [a] 6 December, 2014
Sensor Altitude	705 km	705 km	705 km	705 km
Off-nadir/Nadir	Nadir	Nadir	Nadir	Nadir
Sun Azimuth	156.08°	111.55°	135.07 °	159.10 °
Sun Elevation	42.78 °	68.45°	58.58 °	31.57 °
Scene center latitude	33.17 °	33.27°	33.17 °	33.17 °
Scene center longitude	72.88 °	72.87°	72.85 °	72.88 °

a. L8 OLI = Landsat-8 Operational Land Imager

Table 2. Characteristics and center wavelength Landsat 8, ALI, and Hyperion sensors [73–75]

Band Name[a]	Band Number	Wavelength (nm)	CW[a] (nm)	Band width (nm)	Ground Resolution (m)	Swath Width (km)
Landsat-8 sensor						
Coastal/ Aerosol	1	433–453	443	20	30	185
Blue	2	450–515	482	65		
Green	3	525–600	562	75		

 DOI: 10.1051/matecconf/201712009004

Red	4	630–680	655	50		
NIR	5	845–885	865	40		
SWIR-1	6	1560–1660	1610	100		
SWIR-2	7	2100–2300	2200	200		
Pan	8	500–680	590	180	15	
Cirrus	9	1360–1390	1375	30	30	
TIR-1	10	10300–11300	10800	1000	100	
TIR-2	11	11500–12500	12000	1000		
ALI sensor						
Pan	10	480–690	585	210	10	37
Blue	1	433–453	443	20	30	
Blue1	2	450–515	482	65		
Green	3	525–605	565	80		
Red	4	633–690	662	67		
NIR-1	5	775–805	790	30		
NIR-2	6	845–890	867	45		
SWIR-1	7	1200–1300	1250	100		
SWIR	8	1550–1750	1650	200		
SWIR	9	2080–2350	2215	270		
Hyperion sensor						
VNIR	1–70	355–1057	706	702	30	7.7
SWIR	71–242	851–2577	1714	1726		

a. VNIR = visible-infrared, SWIR = short-wave infrared, and TIR = thermal infra-red, CW = center wavelength

Table 3. Landsat-8 data for four seasons acquired for change detection analysis

Season	Date of Acquisition	Landsat-8 OLI[a] (Path/Row 150/37)
Autumn	22 October, 2015	LC81500372015295LGN00
Summer	16 June, 2015	LC81500372015167LGN00
Spring	13 April, 2015	LC815003720015103LG00
Winter	6 December, 2014	LC81500372014340LG00

a. OLI = Operational Land Imager

3 Data preprocessing

3.1 Atmospheric correction of OLI data

OLI data was pre processed by atmospherically correction and conversion to reflectance values [76–78]. Previous studies used different techniques for pre processing of satellite imagery [79–87]. This study used FLAASH to perform atmospheric correction of OLI images.

4 Results and discussion

4.1 Support vector machines design

The SVM classifier is used to determine a decision function [42]. Linear SVM is used to separate linearly separable classes. When training data is represented by x_i (with n numbers of samples), then y_i, i = 1, 2… n represents the corresponding classes, such that $y \in \{-1, +1\}$
Case 1: For two-dimensional data, the spatial function is shown in Equation 1:

$$f(X) = m \cdot x + c , \quad (1)$$

where m is a vector used to orient the differentiating/dividing plane and c is a scalar that determines the offset of the discriminating plane.
When $m \cdot x + c > 0$, x_i is assigned to the 1st class, while when $m \cdot x + c < 0$, x_i belongs to the 2nd class. When $|f(x)| \geq 1$, the classification interval ($2/\|m\|$, $\|m\|$ or $\|m\|^2$) will be the smallest. Equation (2) shows the optimal function:

$$Min_m = 1/2\|m\|^2 = 1/2(m.m), \quad (2)$$

this should satisfy Equation (3) to maximize the classification margin and thus form the optimal hyper plane, L:

$$y_i(m \cdot x + c) - 1 \geq 0 , \qquad i = 1,2,....n. \quad (3)$$

Equation (4) defines the space denoted by the set of functions for hyperplanes L1 and L2:

$$f_{m,c} = sign(m \cdot x + c) . \quad (4)$$

For a nonlinear case of the dual optimization problem, Equation 5 is used:]

$$\max L_{(w,c,\gamma)} = \sum_{i=1}^{k} \gamma_i - 1/2 \sum_{i=1}^{k} \sum_{j=1}^{k} \gamma_i \gamma_j y_i y_j K(x_i, x_j) , \quad (5)$$

subject to the following constraints:

$$\sum_{i=1}^{k} \gamma_i y_i = 0 \text{ and } C \geq \gamma_i \geq 0. \quad (6)$$

The penalty parameter (C) and the kernel parameter (γ) values were set to optimum values for the radial basis function (RBF).

4.2 Operational land imager support vector machine classifications and change detection analysis

4.2.1 Experimental Setup

The pixel-based supervised classifier (i.e., SVM) was implemented on OLI data e.g., Figure 2) acquired on 22 Oct, 2015 (Autumn), 16 June, 2015 (Summer), 13 April, 2015 (Spring) and 6 December, 2014 (Winter) using the following steps:
Step 1: Identifying and formulating six classes in the study area.
Step 2: Selection of the regions of interest (ROIs) from the image data, which included 202 samples (6 partitions) of bare soil, 243 samples (2 partitions) of built-up area, 62 samples (2 partitions) of mixed forest, 47 samples (2 partitions) of shrub, 278 samples (5 partitions) of

deep water and 69 samples (4 partitions) of shallow water. These ROIs and reference data were selected using several different methods: (1) surveys of Pakistan maps, (2) high resolution (1 m resolution) panchromatic OrbView-3 imagery, (3) high resolution ALI panchromatic data (10 m spatial resolution), (4) high resolution Google Earth software, (5) acquaintance with the study area, (6) selected field visits, and (7) prominent and consistent land use/cover locations (8) high resolution Sentinel 2 imagery (9) analysis of original Landsat imageries.

Step 3: For the OLI data, using the SVM classifier, the following values were assigned to parameters: kernel parameter gamma (γ) = 0.143 was calculated as inverse of the number of bands, penalty parameter C = 100 maximum value so that all training pixels to be assigned certain class, pyramid parameter = zero value to process imagery at full resolution, and the classification probability threshold = zero to assign entire pixel to certain class.

The OLI data were classified (Figure 2) using the SVM classifier for all four seasons (i.e., winter, spring, summer and autumn). Figure 2 (a) to (d) shows the land use/land cover classifications from the Landsat-8 Operational Land Imager (OLI) data for winter (6 December, 2014), spring (13 April, 2015), summer (16 June, 2015), and Autumn (22 October, 2015) made by the SVM classifier. The OLI classified data for the four seasons were used to perform change detection analysis for six cases: (1) winter to spring; (2) winter to summer; (3) winter to autumn; (4) spring to summer; (5) spring to autumn; (6) summer to autumn.

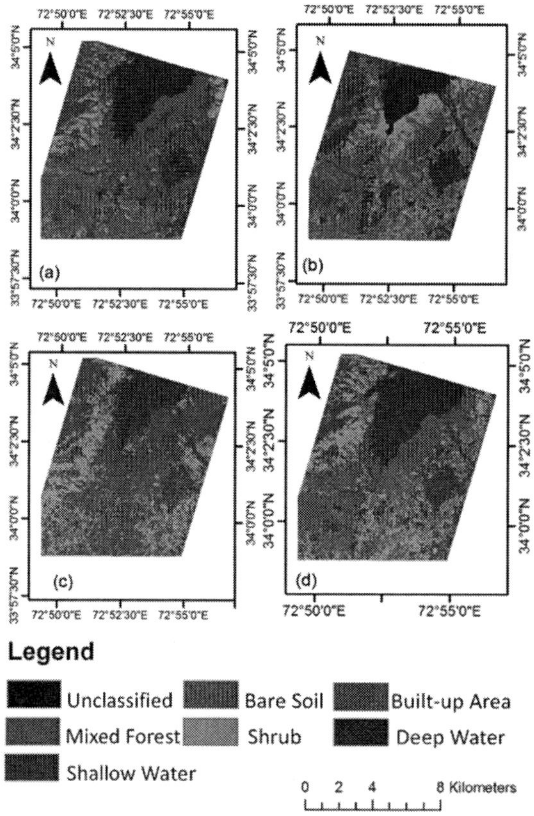

Fig. 2. Land use/land cover classifications of Landsat-8 Operational Land Imager (OLI) data made with the support vector machine (SVM) classifier (a) winter (6 December, 2014), (b) spring (13 April, 2015) (c) summer (16 June, 2015), and (d) autumn (22 October, 2015)

4.2.2 Change detection from winter to spring (case 1)

The change detection technique was implemented in ENVI and executed on SVM-classified data of Landsat-8 OLI from winter (6 December, 2014) to spring (13 April, 2015). In the change detection analysis from winter to spring, it is found from the change detection matrix (Table 4), decrease of spatial extensions of deep water 34.45%, bare soil 15.92%, mixed forest 44.97% and increase of shallow water 141.63 %, shrub 109.75% and built-up area 31.21%. The major shoreline change detection analysis from winter to spring (Figure 3) shows a change of category with a reduction in dam water mapping from deep water to shallow water, deep water to shrub, and deep water to bare soil. These changes are due to shrub increases in spring compared to winter. Similarly, a change of category from bare soil to shrub resulted from shrub increases in spring as compared to winter. A change of category from bare soil to shallow water and shrub to shallow water near the shoreline resulted from an increase in shallow water in the channels in spring as compared to winter.

Table 4. Landsat-8 Operational Land Imager (OLI) data change detection percentage from winter to spring (case1)

Class	Bare Soil	Built-up Area	Mixed Forest	Shrub	Deep Water	Shallow Water
Bare Soil	64.90	17.44	44.39	25.48	8.85	4.58
Built-up Area	6.74	64.39	0.88	31.11	0.52	7.16
Mixed Forest	1.81	0.14	40.78	0.23	0.69	0.00
Shrub	24.82	11.08	13.45	39.29	20.73	7.54
Deep Water	0.00	0.05	0.10	0.00	65.42	0.05
Shallow Water	1.71	6.89	0.40	3.88	3.78	80.67
Class Total	100.00	100.00	100.00	100.00	100.00	100.00
Class Changes	35.09	35.60	59.22	60.70	34.58	19.33
Image Difference	-15.92	13.21	-44.97	109.75	-34.45	141.63

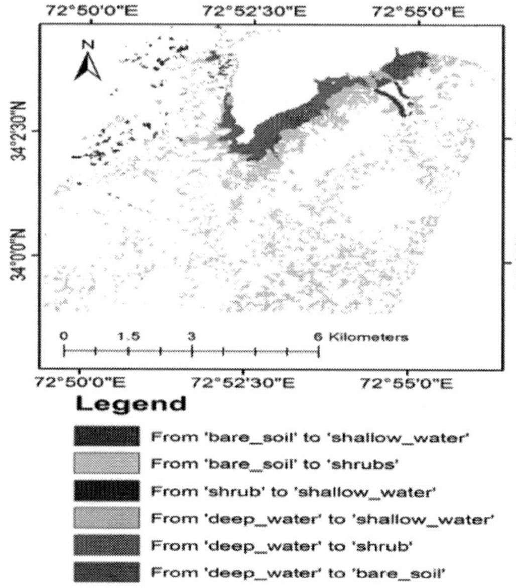

Fig.. 3 Change detection results of Landsat-8 Operational Land Imager (OLI) data using the support vector machine (SVM) classifier from (a) winter (6 December, 2014) to (b) spring (13 April, 2015)

4.2.3 Change detection from winter to summer (case 2)

The change detection technique was implemented in ENVI and executed on SVM-classified data of Landsat-8 OLI from winter (6 December, 2014) to summer (16 June, 2015). In the change detection analysis from winter to summer, it is found from the change detection matrix (Table 5), decrease of spatial extensions of deep water 30.49%, bare soil 15.04%, built-up area 39.90% and increase of shallow water 187.34 %, shrub 119.12% and mixed forest 20.14%. The major shoreline change detection analysis from winter to summer (Figure 4) shows a change of category with a reduction in dam water mapping from deep water to shallow water, deep water to shrub, deep water to bare soil and deep water to mixed forest. These changes are due to shrub and mixed forest increases in summer compared to winter. Similarly, a change of category from bare soil to shrub and bare soil to mixed forest resulted from shrub and mixed forest increases in summer compared to winter. A change of category from bare soil to shallow water near the shoreline was due to an increase of shallow water in the channels during the summer compared to winter.

Table 5. Landsat-8 Operational Land Imager (OLI) data change detection percentage from winter to summer (case2)

Class	Bare Soil	Built-up Area	Mixed Forest	Shrub	Deep Water	Shallow Water
Bare Soil	60.02	42.52	10.26	47.78	5.27	3.773
Built-up Area	3.64	37.93	0.25	2.20	6.31	13.35
Mixed Forest	4.00	0.56	69.70	6.94	6.64	1.4
Shrub	29.99	12.78	18.59	39.17	4.42	2.26
Deep Water	0.00	0.00	0.05	0.00	69.46	0.11

Shallow Water	2.35	6.20	1.13	3.89	7.78	79.11
Class Total	100.00	100.00	100.00	100.00	100.00	100.00
Class Changes	39.98	62.07	30.29	60.83	30.54	20.89
Image Difference	-15.04	-39.90	20.14	119.12	-30.49	187.34

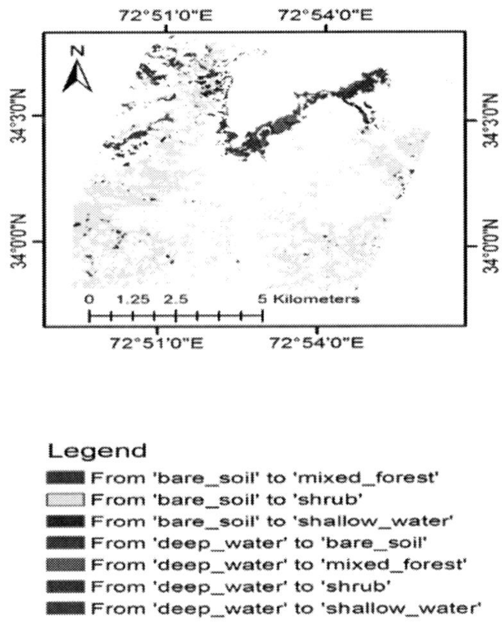

Fig. 4. Change detection of Landsat-8 Operational Land Imager (OLI) data using the support vector machine (SVM) classifier from (a) winter (6 December, 2014) to (b) summer (16 June, 2015)

4.2.4 Change detection from winter to autumn (case 3)

The change detection technique was implemented in ENVI and executed on SVM-classified data of Landsat-8 OLI from winter (6 December, 2014) to autumn (22 October, 2015). In the change detection analysis from winter to autumn, it is found from the change detection matrix (Table 6), decrease of spatial extensions of mixed forest 6.47%, bare soil 31.32% and increase of shallow water 144.37 %, deep water 12.86%, shrub 113.32% and built-up area 1.23%. The major shoreline change detection analysis (Figure 5) shows category changes with little increase in dam water mapping from bare soil to deep water and from bare soil to shallow water in area near the shoreline. Similarly, category changes from mixed forest to shrub resulted from winter to autumn season.

Table 6. Landsat-8 Operational Land Imager (OLI) data change detection percentage from winter to autumn (case3)

Class	Bare Soil	Built-up Area	Mixed Forest	Shrub	Deep Water	Shallow Water
Bare Soil	58.99	15.08	12.19	16.08	0.17	7.59
Built-up Area	5.95	69.26	0.34	12.88	0.00	9.04
Mixed Forest	2.38	0.01	76.34	0.65	0.00	0.00
Shrub	25.92	9.16	10.25	69.29	0.00	1.40
Deep Water	3.23	0.01	0.39	0.00	99.80	2.96
Shallow Water	3.52	6.47	0.48	1.10	0.03	78.99
Class Total	100.00	100.00	100.00	100.00	100.00	100.00
Class Changes	41.01	30.74	23.65	30.71	0.19	21.00
Image Difference	-31.32	1.23	-6.467	113.32	12.86	144.37

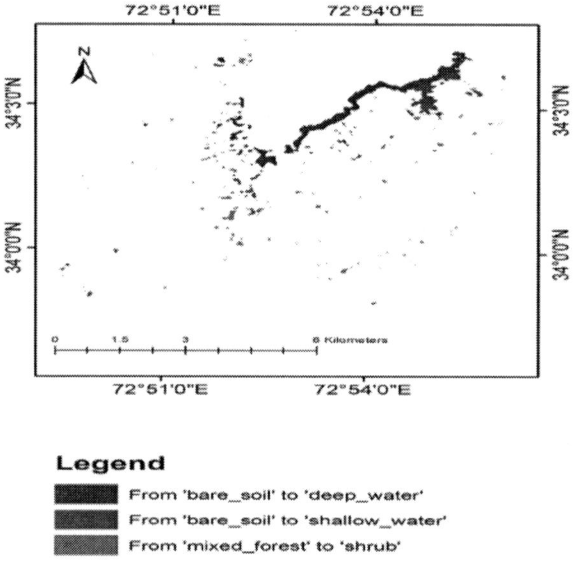

Fig. 5. Change detection of Landsat-8 Operational Land Imager (OLI) data using the support vector machine (SVM) classifier from (a) winter (6 December, 2014) to (b) autumn (22 October, 2015)

4.2.5 Change detection from spring to summer (case 4)

The change detection technique was implemented in ENVI and executed on SVM-classified data of Landsat-8 OLI from spring (13 April, 2015) to summer (16 June, 2015). In the change detection analysis from spring to summer, it is found from the change detection matrix (Table 7), decrease of spatial extensions of built-up area 46.92% and increase of shallow water 18.92 %, deep water 6.04%, shrub 4.46%, mixed forest 118.315% and bare soil 1.03%. The major shoreline change detection analysis (Figure 6) shows category changes with very little increase in dam water mapping from bare soil to deep water and from bare soil to shallow water in areas near the shoreline from spring to summer. Similarly, category changes from shrub to bare soil and shrub to shallow water resulted from conversion of shrub to bare soil and shallow water respectively from spring to summer.

Table 7. Landsat-8 Operational Land Imager (OLI) data change detection percentage from spring to summer (case4)

Class	Bare Soil	Built-up Area	Mixed Forest	Shrub	Deep Water	Shallow Water
Bare Soil	59.09	37.50	11.64	50.80	0.00	1.87
Built-up Area	2.81	34.70	1.28	6.41	0.00	7.02
Mixed Forest	10.56	0.29	75.90	4.21	0.07	7.76
Shrub	25.18	25.01	7.66	34.53	0.00	10.14
Deep Water	0.59	0.00	1.76	0.53	99.85	2.25
Shallow Water	1.76	2.48	1.76	3.50	0.07	70.96
Class Total	100.00	100.00	100.00	100.00	100.00	100.00
Class Changes	40.90	65.29	24.09	65.46	0.15	29.03
Image Difference	1.03	-46.92	118.31	4.46	6.04	18.92

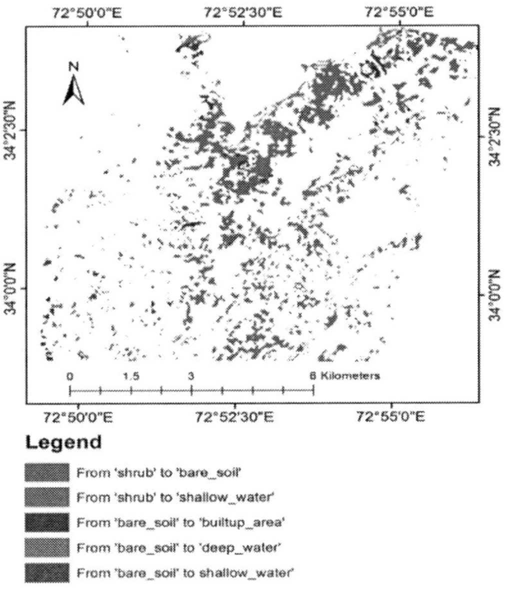

Fig. 6. Change detection of Landsat-8 Operational Land Imager (OLI) data using the support vector machine (SVM) classifier from (a) spring (13 April, 2015) to (b) summer (16 June, 2015)

4.2.6 Change detection from spring to autumn (case 5)

The change detection technique was implemented in ENVI and executed on SVM-classified data of Landsat-8 OLI from spring (13 April, 2015) to autumn (22 Oct 2015). In the change detection analysis from spring to autumn, it is found from the change detection matrix (Table 8), decrease of spatial extensions of bare soil 18.32%, built-up area 10.59% and increase of shallow water 1.14%, deep water 72.20%, shrub 1.7% and mixed forest 69.96%. The major shoreline change detection analysis (Figure 7) shows category changes with increases in dam water mapping from shrub to deep water, bare soil to deep water and from shallow water to deep water in areas near the shoreline. These changes were due to a decrease in shrub from spring to autumn. Similarly, a small category change from shrub to bare soil was due to the decrease of shrub from spring to autumn.

Table 8. Landsat-8 Operational Land Imager (OLI) data change detection percentage from spring to autumn (case5)

Class	Bare Soil	Built-up Area	Mixed Forest	Shrub	Deep Water	Shallow Water
Bare Soil	61.03	5.11	21.08	29.34	0.02	14.49
Built-up Area	3.09	65.99	0.18	8.46	0.00	13.86
Mixed Forest	7.75	0.179	73.50	2.70	0.02	0.11
Shrub	20.65	27.19	1.80	39.18	0.00	14.29
Deep Water	3.88	0.50	2.30	17.29	99.85	13.94
Shallow Water	3.60	1.00	1.12	3.02	0.11	43.28
Class Total	100.00	100.00	100.00	100.00	100.00	100.00
Class Changes	38.97	34.00	26.49	60.82	0.15	56.72
Image Difference	-18.32	-10.59	69.96	1.7	72.20	1.14

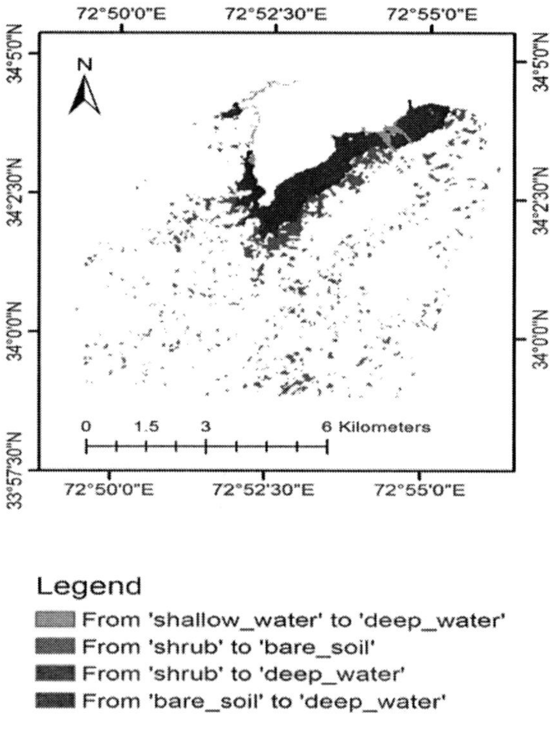

Fig. 7. Change detection of Landsat-8 Operational Land Imager (OLI) data classified using the support vector machine (SVM) classifier from (a) spring (13 April, 2015) to (b) autumn (22 October, 2015)

4.2.7 Change detection from summer to autumn (case 6)

The change detection technique was implemented in ENVI and executed on SVM-classified data of Landsat-8 OLI from summer (16 June, 2015) to autumn (22 Oct 2015). In the change detection analysis from spring to autumn, it is found from the change detection matrix (Table 9), decrease of spatial extensions of bare soil 19.15%, mixed forest 22.15%, shrub 2,65%, deep water 62.39%, shallow water 14.96% and increase of built-up area 68.44%. The major shoreline change detection analysis (Figure 8) shows a category changes with an increase in dam water mapping from shrub to deep water, bare soil to deep water and from shallow water to deep water in areas near the shoreline. The changes are due to decreases of shrubs and bare soil from summer to autumn. Similarly, category changes from shrub to bare soil, mixed forest to shrub and mixed forest to deep water were the result of conversions of shrub and mixed forest areas to bare soil and shrub/mixed forest areas from summer to autumn. A small change of category from bare soil to shallow water resulted from changes of some bare soil to shallow water near the shoreline from summer to autumn. A change of category from bare soil to deep water resulted from the conversion of bare soil to deep water from summer to autumn near the shoreline.

Table 9. Landsat-8 Operational Land Imager (OLI) data change detection percentage from summer to autumn (case6)

Class	Bare Soil	Built-up Area	Mixed Forest	Shrub	Deep Water	Shallow Water
Bare Soil	56.93	11.54	4.62	34.18	0.00	19.66
Built-up Area	13.98	64.57	0.65	9.67	0.00	11.79
Mixed Forest	0.14	0.00	61.28	6.06	0.00	0.19
Shrub	21.3	2.46	22.56	46.01	0.00	10.06
Deep Water	3.96	14.60	10.15	2.99	99.97	23.37
Shallow Water	3.69	6.82	0.74	1.09	0.03	34.93
Class Total	100.00	100.00	100.00	100.00	100.00	100.00
Class Changes	43.07	35.43	38.72	53.99	0.03	65.07
Image Difference	-19.15	68.44	-22.15	-2.65	62.39	-14.96

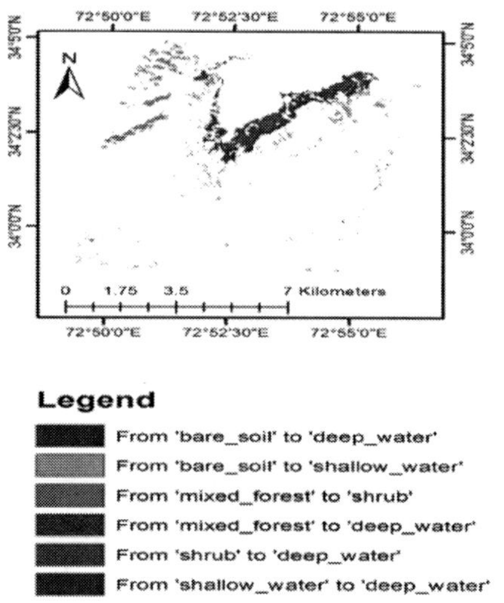

Fig. 8. Change detection of Landsat-8 Operational Land Imager (OLI) data classified using the support vector machine (SVM) classifier from (a) summer (16 June, 2015) to (b) autumn (22 October, 2015).

4.3 Classification accuracy

Classification accuracy was assessed based on the confusion matrix on the OLI data (for the SVM classifier) half the samples were used as test points, resulting in an overall accuracy and kappa coefficient (Tables 10–13).

Table 10. Landsat-8 Operational Land Imager (OLI) performance winter data using support vector machine (SVM) vs. the ground truth (%)[a]

Class	Bare Soil	Built-up Area	Mixed Forest	Shrub	Deep Water	Shallow Water
Bare Soil	99.54	4.53	0.00	19.61	0.00	0.00
Built-up Area	0.46	94.65	0.00	1.96	0.00	17.65
Mixed Forest	0.00	0.00	98.39	7.84	0.00	0.00
Shrub	0.00	0.00	1.61	70.59	0.00	0.00
Deep Water	0.00	0.00	0.00	0.00	100.00	0.00
Shallow Water	0.00	0.82	0.00	0.00	0.00	82.35
Total	100.00	100.00	100.00	100.00	100.00	100.00

a. Overall Accuracy = 95%, Kappa Coefficient = 0.94

Table 11. Landsat-8 Operational Land Imager (OLI) performance spring data using support vector machine (SVM) vs. the ground truth (%)[a]

Class	Bare Soil	Built-up Area	Mixed Forest	Shrub	Deep Water	Shallow Water
Bare Soil	99.11	0.00	22.58	0.00	0.00	2.94
Built-up Area	0.00	99.18	0.00	0.00	0.00	1.47
Mixed Forest	0.89	0.00	77.42	0.00	0.00	0.00
Shrub	0.00	0.82	0.00	98.04	0.00	0.00
Deep Water	0.00	0.00	0.00	0.00	100.00	0.00
Shallow Water	0.00	0.00	0.00	1.96	0.00	95.59
Total	100.00	100.00	100.00	100.00	100.00	100.00

a. Overall Accuracy = 97%, Kappa Coefficient = 0.97

Table 12. Landsat-8 Operational Land Imager (OLI) performance summer data using support vector machine (SVM) vs. the ground truth (%)[a]

Class	Bare Soil	Built-up Area	Mixed Forest	Shrub	Deep Water	Shallow Water
Bare Soil	97.52	3.29	0.00	6.38	0.00	0.00
Built-up Area	0.00	89.71	0.00	0.00	0.00	10.14
Mixed Forest	0.00	0.00	100.00	0.00	0.00	2.90
Shrub	2.48	5.76	0.00	87.23	0.00	10.14
Deep Water	0.00	0.00	0.00	0.00	100.00	0.00
Shallow Water	0.00	1.23	0.00	6.38	0.00	76.81
Total	100.00	100.00	100.00	100.00	100.00	100.00

a. Overall Accuracy = 94%, Kappa Coefficient = 0.92

Table 13. Landsat-8 Operational Land Imager (OLI) performance autumn data using support vector machine (SVM) vs. the ground truth (%)[a]

Class	Bare Soil	Built-up Area	Mixed Forest	Shrub	Deep Water	Shallow Water
Bare Soil	99.53	0.00	0.00	0.00	0.00	0.00
Built-up Area	0.00	99.59	0.00	1.96	0.00	4.41
Mixed Forest	0.00	0.00	100.00	0.00	0.00	0.00
Shrub	0.47	0.41	0.00	98.04	0.00	0.00
Deep Water	0.00	0.00	0.00	0.00	100.00	0.00
Shallow Water	0.00	0.00	0.00	0.00	0.00	91.18
Total	100.00	100.00	100.00	100.00	100.00	100.00

a. Overall Accuracy = 99%, Kappa Coefficient = 0.98

The overall accuracy and kappa coefficient for OLI data classifications with SVM were, respectively 94% and 0.94 for winter (Table 10), 97% and 0.97 for spring (Table 11), and 94% and 0.92 for summer (Table 12) and 94% and 0.98 for autumn (Table 13). It is found

(Table 10–13) that OLI data classifications with SVM resulted higher class discrimination accuracy, overall accuracy (more than 94%) and kappa coefficient. Hence it is suitable for change detection analysis and mapping applications.

5 Summary and conclusions

The results of this study confirmed the potential utility of seasonal satellite based land cover classification of OLI data for all four seasons (i.e., autumn, summer, spring and winter) and post-classification change detection analysis. The OLI data (using SVM classifier) was successfully classified with regard to all six test classes (i.e., bare soil, built-up area, mixed forest, shrub, deep water and shallow water) after pre-processing and atmospheric correction. It is found (Table 10–13) that OLI data classifications with SVM resulted higher class discrimination accuracy, overall accuracy (more than 94%) and kappa coefficient and thus suitable for mapping application.

The OLI SVM-classified data for four seasons i.e., winter (6 December, 2014), spring (13 April, 2015), summer (16 June, 2015), and autumn (22 Oct 2015) were used for change detection analysis of six cases: (1) winter to spring, (2) winter to summer, (3) winter to autumn, (4) spring to summer, (5) spring to autumn, and (6) summer to autumn. The major shoreline change detection analysis from winter to spring (case 1) resulted in an increase of shrub and the amount of shallow water in channels in spring as compared to winter and shows category changes with a reduction in dam water mapping. The major shoreline change detection analysis from winter to summer (case 2) resulted in increases of shrub, mixed forest categories and shallow water in the channels in summer as compared to winter and show category changes with the reduction in dam water. The major shoreline change detection analysis (case 3) resulted in a decrease of bare soil and mixed forest from winter to autumn and showed category changes with little increase in dam water mapping in areas near the shoreline. The major shoreline change detection analysis (case 4) resulted small change of bare soil due to conversions of shrub to bare soil and shallow water from spring to summer and showed category changes with very little increase in dam water mapping in areas near the shoreline. The major shoreline change detection analysis (case 5) resulted in a decrease of shrub due to conversion of shrub to deep water, bare soil to deep water, shallow water to deep water from spring to autumn and showed category changes with an increase in dam water mapping in areas near the shoreline. The major shoreline change detection analysis (case 6) resulted in a decrease of shrub, mixed forest and bare soil due to conversion of shrub and mixed forest to bare soil from summer to autumn and showed category changes with increases in dam water mapping in areas near the shoreline.

These results confirmed that the new OLI technology, with its superior radiometric performance and improved design, resulted in accurate inland water mapping and suitable for post classification change detection analysis. Remote sensing provides an important method of Earth observation and change detection analysis. The change in dam water reservoir capacity determined from the OLI data due to seasonal change was clearly shown during the water classification. In conclusion, the improved OLI technology is both suitable for post-classification change detection analysis and resulted in accurate mapping of the study area.

The first author is thankful to Dr Tariq Mairaj, Dr Hammad Raza, Dr Salman Atif, Dr Mohammad Salman, Junaid Aziz Khan of NUST and Dr Mohammd Ali Maud of UET Pakistan universities for their help and support.

References

1. B. L. Markham, et al., The Landsat data continuity mission operational land imager (OLI) radiometric calibration. *In IEEE Int. Geoscience and Remote Sensing Symp. (IGARSS)*, pp. 2283–2286 (2010).
2. U.S. Geological Survey, *"Landsat Data Continuity Mission"*, U.S. Geological Survey, Washington, DC (2012).
3. D. Gerace, Demonstrating Landsat's new potential to monitor coastal and inland Water. Ph.D dissertation, Dept. Imaging Science, Rochester Inst. Technol., Rochester, NY (2010).
4. A. Gerace, and J. Schott, The increased potential of the Landsat data continuity mission to contribute to case 2 water quality studies. In *Proc. SPIE, Earth Observing Systems XIV*, 7452, 74520U, San Diego, CA (Aug 2009).
5. N. Pehlevan, and J. R. Schott, Investigating the potential of the operational land imager (OLI) for monitoring case II waters using a look-up-table approach. In *Pecora 18: Forty Years of Earth Observation-Understanding a changing World*, Herndon, VA (2011).
6. S. N. Kloiber, P. L. Berezonik, L. G. Olmanson, and M. E. Bauer, A procedure for regional lake water clarity assessment using Landsat multispectral data. Remote Sens. Environ. **82**, 38–47, (2002).
7. W. Pervez, Valiuddin, S. A. Khan, J. A. Khan, 2016. Satellite-based land use mapping: comparative analysis of Landsat-8, Advanced Land Imager, and big data Hyperion imagery. J. Appl. Remote Sens. **10** (2), 026004 (2016); doi: 10.1117/1.JRS.10.026004
8. D. Roy, M. Wulder, T. Loveland, C. Woodcock, R. Allen, , M. Anderson, D. Helder, J. Irons, D. Johnson, R. Kennedy, et al. Landsat-8: Science and product vision for terrestrial global change research. Remote Sens. Environ. **145**, 154–172 (2014).
9. N. Flood, Continuity of Reflectance Data between Landsat-7 ETM+ and Landsat-8 OLI, for both top-of-atmospheric and surface reflectance. A Study in the Australian landscape, Remote Sens. **6**, 7952–7970 (2014).
10. B. Markham, J. Barsi, G. Kvaran, L. Ong, E. Kaita, S. Biggar, J. Czapla-Myers, N. Mishra, D. Helder, Landsat-8 operational land imager radiometric calibrationand stability. Remote Sens. **6**, 12275–12308 (2014).
11. J. Czapla-Myers, N. McCorkel, Anderson, K. Thome, S. Biggar, D. Helder, D. Aaron, L. Leigh, N. Mishra, The ground-based absolute radiometric calibration of Landsat 8 OLI. Remote Sens. **7**, 600–626 (2015).
12. R. Morfitt, J. Barsi, R. Levy, B. Markham, E. Micijevic, L. Ong, P. Scaramuzza, K. Vanderwerff, Landsat-8 Operational land imager (OLI) radiometric performance on orbit. Remote Sens.**7**, 2208–2237 (2015).
13. E. Knight, G. Kvaran, Landsat-8 operational land imager design, characterization and performance. Remote Sens.**6**, 10286–10305, (2014).
14. P. Li, Jiag, Z. Feng, Cross-Comparison of Vegetation Indices Derived from Landsat-7 Enhanced Thematic Mapper (ETM_) and Landsat-8 Operational Land Imager (OLI) sensors. Remote Sens.**6**, 310–329 (2014).
15. Y. Ke, J. Im, J. Lee, H. Gong, Y. Ryu, Characteristics of Landsat 8 OLI-derived NDVI by comparison with multiple satellite sensors and in-situ observations. Remote Sens. Environ.**164**, 298–313 (2015).
16. J. Rogan, D. Chen, Remote sensing technology for mapping and monitoring land cover and land use change. Progr. Plan. **61**(4), 301–325 (2004).

17. P. Sinha, L. Kumar, Indedpendent t wo step thresholding of binary images in inter-annual land cover change/no cange identification. ISPRS J. Photogram. Remote Sensing **81**, 31–43 (2013).
18. C. Benedek, M. Shadaydeh, Z. Kato, J. Sziranyi, Multilayer Markov Random Field models for change detection in optical remote sensing images. ISPRS J. Photogram. Remote Sensing (2015).
19. T. Blasche, Object based image analysis for remote sensing. ISPRS J. Photogram. Remote Sensing **65**(1), 2–16 (2010).
20. J. Chen, M. Lu, X. Chen, J. Chen, L. Chen, A spectral gradient difference based approach for land cover change detection ISPRS J. Photogram. Remote Sensing **85**, 1–12 (2013).
21. X. Chen, J. Chen, Y. Shi, Y. Yamaguchi, An automated approach for updating land cover maps based on integrated change detection and classification methods ISPRS J. Photogram. Remote Sensing **71**, 86–95 (2012).
22. G. Metternicht, Change detection assessment using fuzzy sets and remotely sensed data: an application of topographic map revision. ISPRS J. Photogram. Remote Sensing **54**(4), 221–233 (1999).
23. I. Bruzzone, D. F. Prieto, Automatic analysis of the difference image for unsupervised change detection . IEEE Trans. Geosci. Remote Sensing **38**(3), 1171–1182 (2000).
24. J. L. Silvan-Cardenas, I. Wang, On quantifying post- classification subpixel land cover changes. ISPRS J. Photogram. Remote Sensing **98**, 94–105 (2014).
25. F. Bovolo, L. Bruzzone, M. Marconcini, A novel approach to unsupervised change detection based on a semisupervised SVM and a similarity measure. IEEE Trans. Geosci. Remote Sensing **45**(7), 2070–2080 (2008).
26. C. He, A. Wei, P. Shi, Q. Zhang, Y. Zhao, Detecing land use/land cover change in rural-urban fringe areas using extended chabge-vector analysis. Int. J. Appl. Easth Observ. Geoinform. **13**(4), (2011).
27. L. Bruzzzone, R. Cossu, G. Vernazza, Detection of land cover transitions by combining multidate classifiers, Pattern Recogn. Lett. **25**(13) 1491–1500 (2004).
28. M.J. Canty, Image Analysis, and Change Detection in Remote Sensing: With Algorithms for ENVI/IDL, 2nd ed.; CRC Press : Boca Raton, FL, USA, (2010).
29. A. Almutairi, T.A. Warner, Change Detection Accuracy and Image Properties: A Study Using Simuated Data. Remote Sens.**2**, 1508–1529, 2010.
30. A. Hecheltjen, F. Thonfeld, G. Menz, Recent advances in remote sensing change detection- A review. In Land Use and Land Cover Mapping in Europe, Manakos, I., Braun, M., Eds.; Springer Netherlands: Dordecht, The Neterlands, pp. 145–178, (2014).
31. D. Lu, P. Mausel, E. Brondizio, E.Moran, Change detection techniques. Int. J. Remote Sens. **25**, 2365–2401 (2004).
32. P. Coppin, I. Jonckheere, K. Nackaerts, B. Muys, E. lambin, Review Article Digital Image change detection methods in ecosystem monitoring : A review. Int. J. Remote Sens. **25**,1565–1596, (2004).
33. M. Hussain, D. Chen, A. Cheng, H. Wie, and Stanley, D., Change detection from Remotely Sensed Images: From Pixel-Based to Object –Based Approches. ISPRS J. Photogramm and remote Sensing **80**: 91–106. Doi:10.1016/j.isprs.2013.03.006. (2013).
34. J. Unger, et al., Modelling of the Urban Heat Island Pattern Based on the Relationship between Surface and Air temperature. Quarterly Journal of the Hungarian Meterlogical Service **114**, 287–302 (2010).

35. L. Musci, L. Muladi, A. Henits, Farsang, and V. Albrecht, Large scale UHI Mapping Based on Spatial Information Provided by Young Volunteers. Carpathian Journal of Earth and Environmental Sciences 9 (2): 31–43 (2014).
36. D. Lu, E. Moran, and S. Hetrick, Detection of Impervious Surface Change with Multitemporal Landsat Images in an Urban-Rural Frontier. ISPRS J. Photogramm and remote Sensing, **66**: 298–306. doi: 10.1016/j.isprsjprs.2010.10.010. (2011).
37. A. Schneider, Monitoring Land Cover Change in Urban and Peri-Urbanareas Using Dense Time Stacks pf Landsat Satellite Data a data Mining Approach. Remote Sens. Environ, **124**,689–704. Doi:10.1016/j.rse.2012.06.006. (2012).
38. F. Yuan, K.E. Swaya, B. C. Loefffelholz, and M.E. Bauer, Land Cover Classification and Change Analysis of the Twin Cities (Minnesota) Metropolitan Area by Multitemporal Landsat Remote Sensing. Remote Sens. Environ, **98**, (2–3) : 317–328. Doi:10.1016/j.rse.2005.08.006. (2005).
39. B.E. Hubbard, J.K. Crowley, Mineral mapping on the Chilean-Bolivian Altiplano using co-orbital ALI, ASTER and Hyperion imagery: data dimensionality issues and solutions, Remote Sens, **99**, 173–186, 2005.
40. National Aeronautics and space administration, Earth Observing-1 Advanced Land Imager, 2002 *http://eo1.gsfc.nasa.gov/Technology/ALIhome1.htm*
41. National Aeronautics and Space Administaration, Earth Observating-1 Eo1 General Mission, 2004 *http://eo1.gsfc.nasa.gov/new/general/*
42. M.A. Wulder, J.C. White, S.N. Goward, G.M. Jeffrey, J.R. Irons, M.Herold, W.B. Cohen, T.R. Loveland, C.E. Woodcock, Landsat continuity: issues and opportunities for land cover monitoring. Remote Sens. Environ. **112**, 955–969, (2008).
43. M. Folkman, J. Pearlman, L. Liao, and P. Jarecke, EO-1/ Hyperion hyperspectral imager design, development, characterization, and calibration in W. L. Smith and Y. Yasuoka, (Eds.). Hyperspectral Remote Sensing of the Land and Atmosphere **4151**, 40–51, (2001).
44. L. C. Sanders, J. R. Schott, and R. Raqueno, A VNIR/SWIR atmospheric correction algorithm for hyperspectral imagery with adjacency effect. Remote Sens. Environ. **78**, 252–263, (2001).
45. E. J. Hochberg, S. Andrefouet, and M. R. Tyler, Sea surface correction of high surface resolution Ikonos images to improve bottom mapping in near shore environment. IEEE Trans. Geosci. Remote Sens. **41**, 1724–1729, (2003).
46. D. Mishra, S. Narumalani, D. Rundquist, and M. Lawson. Benthic habit mapping in tropical marine environments using Quick bird multispectral data. Photogramm. Eng. Remote Sens. **72**, 1037–1048, (2006).
47. S. Andrefouet et al., Multi-site evaluation of IKONOS data for classification of tropical coral reef environments. Remote Sens. Environ. **88**, 128–143, (2003).
48. M. Pal., Ensemble of support vector machines from land cover classification. Int. J. Remote Sens. **26**(5), 1007–1011, (2008).
49. Y. Kavzoglu, and I. Colkesen., A kernel function analysis for support vector machines for land cover classification. Int. J. Appl. Earth Obs. Geoinf. **11**, 352–359, (2009).
50. A. Mathur, and G.M. Foody, Crop classification by support vector machine with intelligently selected training data for nonoperational application. Int. J. Remote Sens. **29**(8), 2227–2240, (2008).
51. H. Hakvoort, et al., Towards airborne remote sensing of water quality in the Netherlands-Validation and error analysis. ISPRS J. Photogramm. Remote Sens. **57**, 171–183, (2002).

52. S. Thiemann, and H. Kaufmann. Lake water quality monitoring using hyperspectral airborne data-A semiempirical multispectral and multitemporal approach for the Mecklenburg Lake District, Germany. Remote Sens. Environ. **81**, 228–237, (2002).
53. M. J. Canty, Boosting a fast neural network for supervised land cover classification. Comput. Geosci. **35**(6), 1280–1295, (2009).
54. B. Dixon, and N. Candade, Multispectral land use classification using neural networks and support vector machines, one or the other, or both? Int. J. Remote Sens. **29**(4), 1185–1206, (2008).
55. P. Barry, , EO-1/Hyperion Science Data User's Guide, Level 1_B. May 2001, http://www.eoc.csiro.au//hswww/oz_pi/doc/EO1HSDataUsersGuide.pdf (18 March 2016).
56. R.Beck, , *EO-1 User Guide-Version 2.3, Satellite Systems Branch.* USGS Earth Resources Observation Systems Data Center, Cincinnati, (2003).
57. B. W. Pengra, C. A. Johnston, and T. R. Loveland, Mapping an invasive plant, Phragmites australis, in coastal wetlands using the EO-1 Hyperion hyperspectral sensor. Remote Sens. Environ. **108**(1), 74–81 (2007).
58. D. D. Xu, G.-Q. Ni, L.-L. Jiang, Y.-T.Shen, T.-L. S. Li, Ge, and X.-B. Shu, Exploring for natural gas using reflectance spectra of surface soils. Adv. Space Res. **41**, 1800–1817 (2008).
59. R. Zhang, and J. Ma, Feature selection for hyperspectral data based on recursive support vector machines. Int. J. Remote Sens. **30**(14), 3669–3677 (2009).
60. P. Du, K. Tan, and X. Xing, Wavelet SVM in reproducing kernel Hilbert space for hyperspectral remote sensing image classification. Opt. Commun. **283**, 4978–4984 (2010).
61. S. Pignatti, R. M. Cavalli, V. Cuomo, V. Fusilli, S. Pascucci, M. Poscolieri, et al., Evaluating Hyperion capability for land cover mapping in a fragmented ecosystem: Pollino National Park, Italy. Remote Sens. Environ. **113**, 622–634, (2009).
62. J. Wang, Y. Chen, T. He, C. Lv, A. Liu, Application of geographic image cognition approach in land type classification using Hyperion image: A case study in China, Int. J. Appl. Earth Obs. Geoinf. **12S**, S212–S222, (2010).
63. S. J Walsh, A. L. McCleary, C. F. Mena, Y. Shao, J. P. Tuttle, A. González, et., Quick Bird and Hyperion data analysis of an invasive plant species in the Galapagos Islands of Ecuador: Implications for control and land use management. Remote Sens. Environ. **112**, 1927–1941, (2008).
64. C. Huang, L. S. Davis, and G. Jr. Townshend, An assessment of support vector machines for land cover classification. Int. J. Remote Sens. **23**, 725–749, (2002).
65. H. Nemmour, and Y. Chibani, Multiple support vector machines for land cover change detection: An application for mapping urban extension. ISPRS J. Photogramm. **61**, 125–133, (2006).
66. M. L. Clark, D. A.Roberts, and D. B. Clark, Hyperspectral discrimination of tropical rain forest tree species at leaf to crown scales. Remote Sens. Environ. **96**, 375–398, (2005).
67. D. Vyas, N. S. R. Krishnayya, K. R. Manjunat, S. S. Ray, and S. Painigraphy, Evaluation of classifiers for processing Hyperion (E-O1) data of tropical vegetation. Int. J. Appl. Earth Obs. Geoinf. **13** 228–235, (2011).
68. D.G. Googenough et al., Processing Hyperion and ALI for forest classification. IEEE Trans. Geosci. Remote Sens **41**, 1321–1331, (2003).
69. M. Dalponte, L. Bruzzone, L. Vescovo, and D. Gianelle, The role of spectral resolution and classifier complexity in the hyperspectral images of forest areas. Remote Sens. Environ. **113**, 2345–2355, (2009).

70. A. Plaza, et al., Recent advances in techniques for hyperspectral image processing. Remote Sens. Environ. **113**, S110–S122, (2009).
71. F. Melgani, and L. Bruzzone, Classification of hyperspectral remote sensing image with support vector machines. IEEE T. Geosci. Remote **42**, 1778–1790, (2004).
72. M. Pal, and P. M. Mather, Some issues in the classification of DAIS hyperspectral data. Int. J. Remote Sens. **27**, 2895–2916, (2006).
73. S. G. Ungar, J. S. Pearlman, J. A. Mendenhall, Reuter, D., Overview of the earth observing one (E-O1) mission. IEEE Trans. Geosci. Remote. **41**, 1149–1159, (2003).
74. D. Roy, M. Wulder, T. Loveland, R. Allen, M. Anderson, D. Helder, J. Irons, D. Johnson, R. Kennedy, T. Scambos, Landsat-8: Science and product vision for terrestrial global change research. Remote Sens. Environ. **145**, 154–172, (2014).
75. D. R. Hearn, C. J. Digenis, D. E. Lencioni, Mendenhall, J. A., Evans, J. B., Walesh, R. D., E-O1 advanced land imager overview and spatial performance. IEEE Trans. Geosci. Remote Sens. **2**, 897–899, (2001).
76. *Landsat 8 (L8) Data Users Handbook (2015)*, LSDA-1574 Version 1.0, Department of the Interior U.S. Geological Survey, EROS Sioux Falls, South Dokota.
77. P. S. Thenkabail, E. A. Enclona, M. S. Ashton, and B. Van Der Meer, Accuracy assessments of hyperspectral waveband performance for vegetation analysis applications. Remote Sens. Environ. **91**(3–4), 354–376, (2004a).
78. P. S. Thenkabail, E. A. Enclona, M. S. Ashton, C. Legg, and M. J. D. Dieu, Hyperion, IKONOS, ALI and ETM+ sensors in the study of African rainforests. Remote Sens. Environ. **90**, 23–43, (2004b).
79. B. Datt, et al., Preprocessing EO-1 Hyperion hyperspectral data support the application of agriculture indexes. IEEE Trans. Geosci. Remote Sens. **41**, 1246-1259, (2003).
80. M. G. Allan, D. Hamiton, B. J. PHicks, and L. Brabyn, Landsat remote sensing of chlorophyll a concentration in central North Island lakes of New Zealand. Int. J. Remote. Sens., **32**, 2037–2055, (2011).
81. Y. H. Ahn, P. Shanmugam, and J. E. Moon, Retrieval of ocean colour from high resolution multi-spectral imagery for monitoring highly dynamic ocean features. Int. J. Remote. Sens., **27**, 367–392, (2006).
82. B. L. Markham, M. O. Haque, J. A. Barsi, E. Micijevic, D. L. Helder, K. J. Thome, D. Aaron, and J. S. Czapla-Myers, Landsat-7 ETM+: 12 years on-orbit reflective-band radiometric performance. IEEE Trans. Geosci. Remote Sens. **50**(5), 2056–2062, (2012).
83. D. A. Palandro, et al., Qualification of two decades of shallow-water coral reef habitat decline in the Florida Keys National Marine Sanctuary using Landsat data (1984–2002). Remote Sens. Environ. **112**, 3388–3399, (2008).
84. V. E. Brando, and A. G. Dekker, Satellite hyperspectral remote sensing for estimating estuarine and coastal water quality. IEEE Trans. Geosci. Remote Sens. **41**, 1378–1387, (2003).
85. C. D. Mobley, and L. K. Sundman, *Hydrolight 5, Ecolight User Guide*, Bellevue, WA: Sequoia Scientific, Inc., (2008).
86. Z. Lee, Y. H. Ahn, C. Mobley, and R. Arnone, Removal of surface reflected light for the measurement of remote-sensing reflectance from an above-surface platform. Opt. Express, **18**, 26313–26324, (2010).
87. K. G. Ruddick, F. Ovidio, and M. Rijkeboer, Atmospheric correction of SeaWiFS imagery for turbid coastal and inland waters. Appl. Opt. **39**, 897–912, (2000).

UAV-Borne photogrammetry: a low cost 3D surveying methodology for cartographic update

Gabriella Caroti[1,*], *Andrea* Piemonte[1], and *Riccado* Nespoli[2]

[1]Civil and Industrial EngineeringDepartment (DICI), University of Pisa, Largo Lucio Lazzarino 1, Pisa, Italy
[2]3D A.P.S. (Aero PhotogrammetricSurveys), Via F.Redi 78/a3, Arezzo, Italy

Abstract. Territorial management requires the most possible up-to-date mapping support of the status quo. Regional scale cartography update cycle is in the medium term (10 to 15 years): therefore, in the intervening time between updates relevant Authorities must provide timely updates for new works or territorial changes. Required surveys can exploit several technologies: ground-based GPS, Terrestrial Laser Scanning (TLS), traditional topography, or, in the case of wider areas, airborne photogrammetry or laser scanning. In recent years UAV-based photogrammetry is becoming increasingly widespread as a versatile, low-cost surveying system for small to medium areas. This surveying methodology was used to generate, in order, a dense point cloud, a high resolution Digital Surface Model (DSM) and an orthophotograph of a newly built marina by the mouth of the Arno river in Pisa, Italy, which is not yet included in cartography. Surveying activities took place while the construction site was in operation. Case study issues surfaced in the course of the survey are presented and discussed, suggesting 'good practice' rules which, if followed in the survey planning step, can lessen unwanted effects due to criticalities. Besides, results of quality analysis of orthophotographs generated by UAV-borne images are also presented. Such results are discussed in view of a possible use of orthophotographs in updating medium- to large-scale cartography and checked against existing blueprints.

1 Introduction

Technological evolution referred to methods and methodologies for furthering knowledge of land and its components has gained an ever increasing importance in the evolution of many applied sciences. New technologies and methodologies in surveying have led, in the last decade, to major innovations in Geomatics. Such innovations have disclosed new boundaries and solutions for the multidisciplinary integrated approach to collect 2- and/or 3-D metrical data and related information for informative theming [1]. Digital surveying represents one of the main innovations brought along by evolution in surveying

*Correspondingauthor: gabriella.caroti@unipi.it

methodology. Terrestrial and Airborne Laser Scanning and the image-based modeling techniques yield dense 3-D point clouds of both natural and built areas by means of simplified procedures, so that potential users are numerous and with vast cultural differences. Point clouds derived from each of these methodologies differ deeply as to definition and properties, which on turn define the best application field: point clouds derived from laser scanning are best suited to generation of high-definition geometrical models, whereas digital photogrammetry is best suited for 3D models with high-definition textures [2]. This approach has been tested on different occasions on landscapes [3-6] as well on architecture and cultural heritage [7-9] and infrastructures [10, 11].

Lately, Unmanned Aerial Vehicles (UAVs) have been introduced in Geomatics as data collection systems for low altitude photogrammetric surveying [12]. High-resolution imagery, acquired using UAVs and processed by structure from Motion (SfM) and Multi-View Stereo (MVS) photogrammetry produce 3D point-clouds whose accuracy ranges from sub-meter level to a few centimeters [13-15]. New researches in this field introduce this methodology as a promising technology, providing new opportunities also for map updating [16, 17].

Goals for this work include:

1. Check of the usability of the UAV-based survey for updating of medium- to large-scale cartography and of the match with dockyard drafts;

2. Emphasize critical issues of UAV-based surveys in dockyard areas close to bodies of water.

2 Materials and methods

2.1 Test area

The case study under investigation refers to the area where the touristic port of the Marina di Pisa resort currently stands.

Fig. 1. Area intended for marina construction, before work – October 2007 (image from Google Earth)

The marina is located on the Pisa seaside at the mouth of the Arno River, and is included in both the urban fabric and the Regional Park of Migliarino, San Rossore and Massaciuccoli. The area where the marina currently rises was home to a FIAT industrial plant for car manufacturing. Following the cessation of activities in 1988, the decommissioned plant was left in a state of ever worsening decay (Figure 1). At the end of the 2000s, a recovery and rehabilitation plan was approved for the area, providing its reclamation and securing, as well as its new intended use as marina with related commercial and residential spaces.

This work has dramatically altered the appearance of the mouth of the Arno River and the urban layout of Marina di Pisa (Figure 2), and medium- to large-scale cartography of the area is no longer representative of its actual conditions (Figure 3).

Fig. 2. Marina area in 2016 (image from Google Earth)

Fig. 3. 1:2000 regional cartogaphy abstract highlighting the industrial plant area where the marina has been built

The present investigation, framed in the foresaid context, aims at assessing the potential, in terms of detail and accuracy, of UAV-based photogrammetry for timely updates of new works and landscape modifications during the period intervening between updates of medium- to large-scale cartography.

In order to assess product quality, Ground Control Points (GPs) have been compared against project documents provided by the dockyard site management; besides, the UAV-based survey has been compared against Google Earth.

2.2 Survey methodologies

Actual survey operations have been carried out when the construction site was still operating, following a site inspection in order to check the status quo of the area and dockyard organization. Flight area coverage and times, as well as take-off and landing sites, have been defined in collaboration with works direction to ensure maximum safety. The flight has taken place during the lunch break, covering the north pier area, as some vessels were already moored in other zones of the marina (Figure 4).

The flight plan was designed for a restitution scale greater than 1:1000, providing for a coverage of approximately 9 hectares with a total flight length of about 1200m

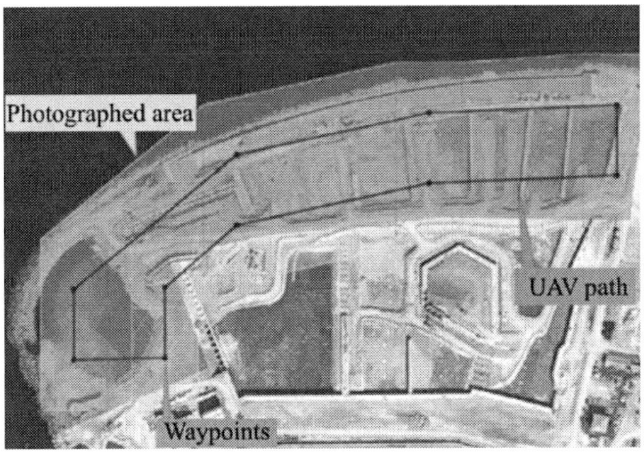

Fig. 4. Covered area.

The airborne photogrammetry survey has been carried out by means of an OnyxStarhexacopter (Figure 5) fitted with a Sony NEX 7 digital camera, whose technical features are listed below:

1. Weight: 300g
2. Sensor dimensions: 23.5mm x 15.6mm
3. Sensor definition: 6000pixel x 4000 pixel
4. Optics: 18-55mm zoom, set in Wide mode (f≅19mm)

Planning of image collection provided for pseudo-nadiral shots, with horizontal strips; flight height was ≅ 80m, with Ground Sampling Distance (GSD)≅2cm/px, 85% overlap along both axes, and theoretical coverage of about 100m x 65m.

Fig. 5. Flight preparation.

Photogrammetric processing of the collected images was carried out by means of Agisoft'sPhotoScan commercial software. PhotoScan implements SfM and MVS photogrammetric algorithms, and its potential in generating 3D models from point clouds are well documented in literature [18].

In this case, the software has been used to generate a Dense Digital Surface Model (DDSM) and an orthophotograph of the object area. Control Points (CPs) have been uniformly spread across the area and signalized; their coordinates have been surveyed by RTK GPS.These points have been used for the definition of external and internal orientation parameteres of the images, by means of a self-calibration procedure built into the bundle block adjustment, in order to scale and georeference the photogrammetric survey.

In the processing step, some of these points were used solely for accuracy checks as CPs, while the remaining were used as GCPs to compute the photogrammetric model [19-21].

3 Results and discussion

The present case study has highlighted some application peculiarities of potential interest for implementation of the survey methodology in construction and coastal sites.

Airborne image collection has pointed out some critical issues in the survey area which were not detected in previous inspections.Firstly, large construction sites are complex contexts, where works of different kind pursue time and cost optimization, which on turn leads to adapt local situations and logistics on an almost daily basis. In the present case, this required on site modification of the spot chosen for takeoff and landing, as well as of the flight plan. The spot chosen for takeoff and landing was a yard on the edge of the site, with dusty, sandy ground, which entailed a harmful event in the first flight. In fact, the turbulence due to the UAV rotors lifted the dust, impeding correct operation of two components of the surveying system, i.e. the proximity sensor (range meter) and the camera infrared sensor. As for the former, a ground proximity signal continuously fed to the autopilot system caused the UAV to rise uncontrollably, until manual flight mode was switched on, allowing for landing. On the other hand, the infrared sensor malfunction caused a stop in automatic image shooting, since the procedure was controlled by the sensor itself. The problem was solved by using a raised stand as takeoff platform.

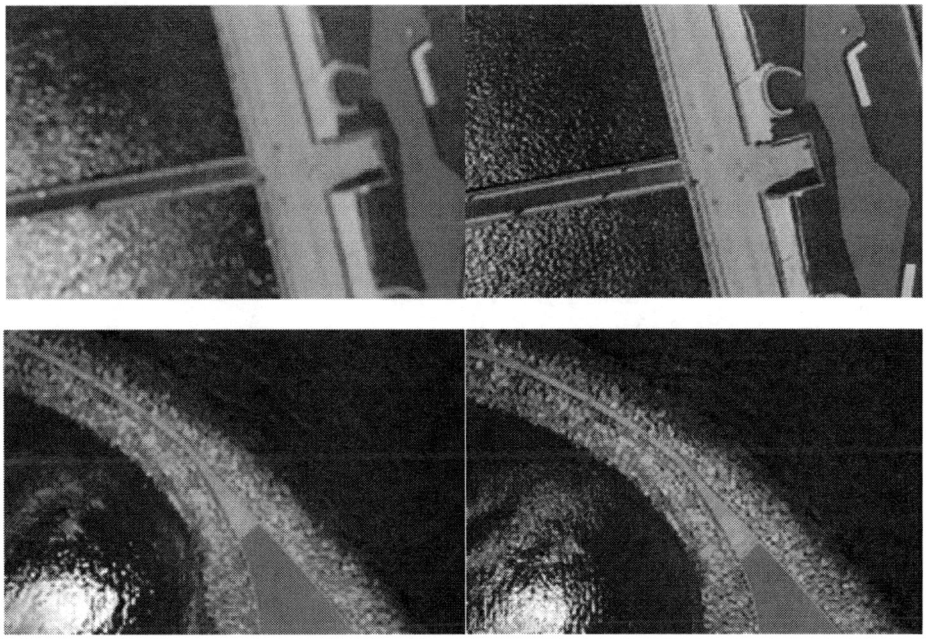

Fig. 6. Focusing issues on water bodies

Another early problem referred to use automatic focus mechanisms on mostly submersed areas (Figure 6). Water surface is in fact too homogeneous for automatic focusing to work correctly, which caused blurred images in some cases and image skipping in others. This problem has been solved by setting manual focus at infinity, as the flight level was greater than the hyperfocal distance.

A further issue, providing useful directions for flight planning, is wind assessment, even more so in this case. In fact, the marina area often features breezes, which can on occasion grow stronger; these, in turn, have even greater influence on lighter aircrafts, such as UAVs. An early test showed that strong crosswinds caused higher battery consumption, due to corrections applied by autopilot to keep the aircraft in route, reducing flight range and coverable area. Planning strips so to avoid crosswind optimizes range and flight plan compliance. The final flight focused on the north piers and large vessels mooring dock.

Image orientation errors amounted to about .5 pixels. Subsequent input of GCPs and CPs in the processing allowed for quality assessment of the DDSM obtained by photogrammetric processing. Maximum error values in the 10cm range are compliant with the research goals and in accordance with literature on the subject.

Given good quality as regards flight planning, image collection and GCPs survey, expected planimetric errors are one to three times pixel size on the object, with elevation error roughly doubling the former. The orthophotograph of the test area has been generated by the PhotoScan software (Figure 7) and georeferenced in the Technical Regional Cartography system.

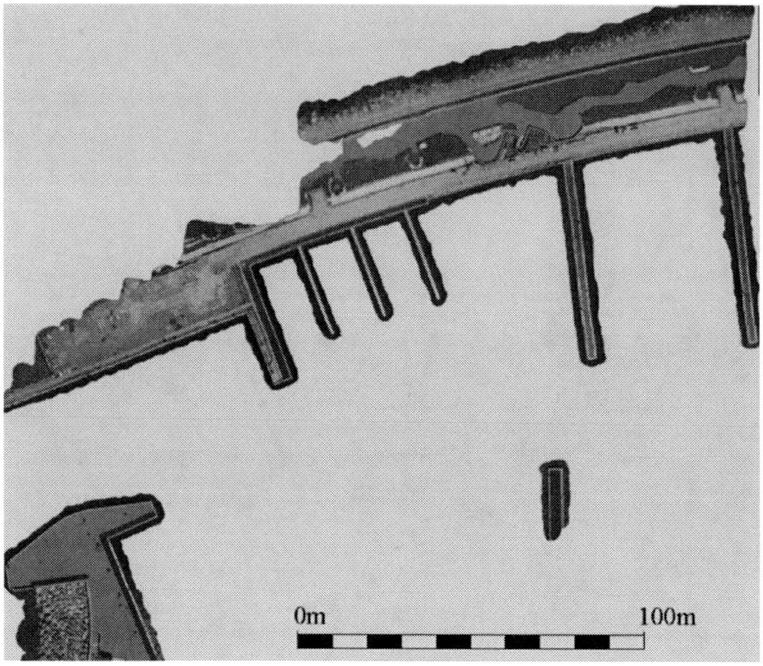

Fig. 7. Orthophotograph of the North-West part of the surveyed area (Figure 2)

A further comparison has been made against the marina blueprint, also georefenced in the Technical Regional Cartography system. This displayed an excellent compliance within the foresaid limitations (Figure 8). The area of the shorter piers showed a planimetric deviation amounting to about 80cm, which could suggest a systematic error (Figure 9). On the other hand, planimetric positioning of other structures posed no similar issues, as shown for the longer piers located in close proximity (Figure 8 right).

In order to detect the origin of this anomaly, a RTK-GPS survey has been carried out on 10 points, homogeneously spread across the test area and detectable in both documents. Its results showed that the actual position of the short piers differs from their planned position, so that any deviation detected by the comparison is not due to survey methodology

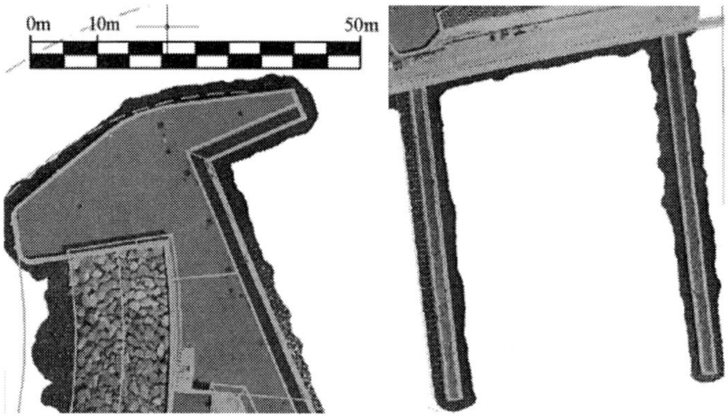

Fig. 8. Comparison areas with 10cm range compliance. On the left the pier entrance in the West part of the surveyed area (Figuera 2). On the right the long (~60m) piers in the North part of the surveyed area (Figure 2).

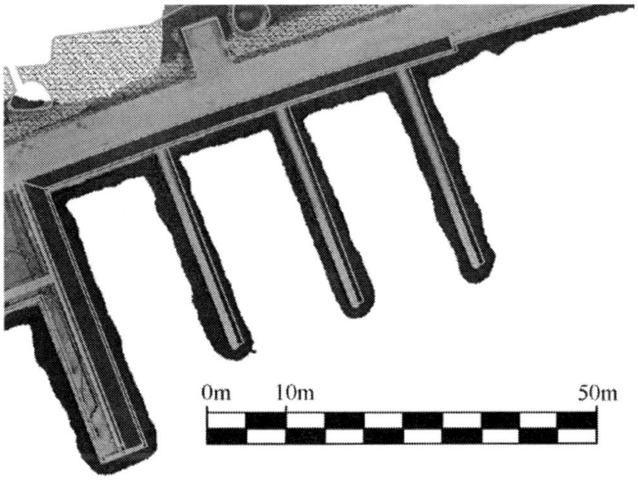

Fig. 9. Short (~30m) piers: deviations ≅80cm. North-West of the surveyed area (Figure 2).

The discordance between blueprint and actual position is corroborated by overlaying the orthophotograph with the corresponding Google Earth image (Figures 10 and 11).

In order to register the model orthographic projection in the same reference system of Google Earth service, it was exported by SfM software in "kmz" format.

It is of course advisable to keep in mind that accuracy Google Earth renderings is in the 1:10.000 range, anyway a first qualitative analysis denotes good compliance. Availability of up-to-date images in Google Earth can provide a tool for updating medium-scale cartography. When instead higher detail levels are required, tools as UAV-based photogrammetry may allow updates of small cartography portions with comparatively low costs. The UAV-based survey has been checked for accuracy by means of the CPs, while images shown below highlight the greater resolution of images obtained by low-altitude surveys.

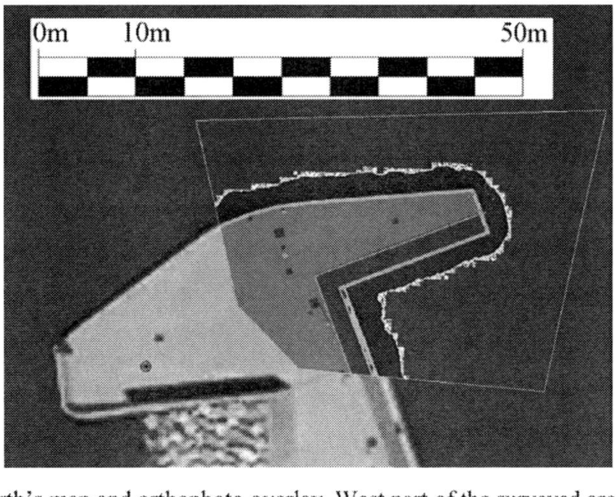

Fig. 10. Google Earth's map and orthophoto overlay. West part of the surveyed area.

Fig. 11. Google Earth's map and orthophoto overlay. North part of the surveyed area.

4 Conclusion

UAV-based surveys in seaside construction sites face peculiar operating issues. The present investigation has shown that application of the methodology with proper precautions allows for medium- to large-scale map updates in the intervening time required for official updates. The different checks, carried out on the GCPs, on dockyard blueprints and on Google Earth maps, have shown that orthophotographs produced from airborne photogrammetry meet accuracy and detail requirements for cartography, aside from providing high radiometric resolution which facilitate reading of landscape changes due to new works.

Thanks are due to ASTRO (ApplicazioniScientifiche e Topografiche per ilRilievoOperativo – Scientific topogaphic applications for field surveying) laboratory technicians Dr. Andrea Bedini and Dr. Jessica Micheloni, to Ph.D. candidate Dr. Isabel Martínez-Espejo Zaragoza and to Pisa Marina dockyard management.

References

1. E.M. Cepolina, G. Caroti, A. Piemonte, P.G. Rojas,*17th Int Conference on Harbor, Maritime and Multimodal Logistics Modelling and Simulation*, 51 (2015).
2. F. Chiabrando, V. Di Pietra, A. Lingua, P.Maschio, F.Noardo, G.Sammartano, and A. Spanò,*Int Arch Photogramm, Remote Sensing and Spatial Information Sciences*,**XLI-B5**, 413 (2016)
3. J. Roelensa, S. Dondeynea, J. Van Orshovenb, J. Dielsa,*Int JApplEarth Obs*, **53**, 64 (2016)
4. G. Caroti, A. Piemonte, M. Redini, *ApplGeomatics*, **7 (4)**, 243 (2015)
5. G. Caroti, A. Piemonte, M.Redini, *Int Arch Photogramm, Remote Sensing and Spatial Information Sciences*,**XL-5/W3**, 99 (2013)
6. G. Caroti, F.Camiciottoli, A. Piemonte, M.Redini,*Int Arch Photogramm, Remote Sensing and Spatial Information Sciences*,**XL-5/W3**, 51 (2013)
7. P.G. Robleda, G. Caroti, I.Martínez-Espejo Zaragoza, A.Piemonte,*Int Arch Photogramm, Remote Sensing and Spatial Information Sciences*,**XLI-B5**, 391 (2016)

8. M.G. Bevilacqua, G. Caroti, I. Martínez-Espejo Zaragoza, A. Piemonte,*Remote Sensing*, **8 (3)**, 1 (2016)
9. G. Caroti, I. Martínez-Espejo Zaragoza, A. Piemonte, *Int Arch Photogramm, Remote Sensing and Spatial Information Sciences*,**XL-5/W4**, 285 (2015)
10. M. Castro, S. Lopez-Cuervo, M.Paréns-González, C. de Santos-Berbel,*Surv Rev*, **48 (350)**, 309 (2016)
11. G. Caroti, A. Piemonte, *SurvRev*, **42 (315)**, 92 (2010)
12. I. Colomina, P. Molina, *ISPRS J Photogramm Remote Sensing*, **92**, 79 (2014)
13. M.Gerke,H. J. Przybilla,*PhotogrammFernerkun*, **1**, 17 (2016)
14. S.Harwin, A.Lucieer, *Remote Sensing*, **4**, 1573 (2012)
15. F. Nex,F. Remondino, *Applied Geomatics*, **6 (1)**, 1 (2014)
16. S. Crommelinck, R. Bennett, M.Gerke, F.Nex, M.Y. Yang, G.Vosselman,*Remote Sensing*, **8 (8)**, 1 (2016)
17. M. Cramer, S. Bovet, M.Gültlinger, E.Honkavaara, A. McGill, M.Rijsdijk, M. Tabor, V.Tournadre, *Int Arch Photogramm, Remote Sensing and Spatial Information Sciences*,**XL-1/W2**, 93 (2013)
18. F. Remondino, M.G. Spera, E. Nocerino, F. Menna, F.Nex,*PhotogrammeRec*, **29(146)**, 144 (2014)
19. K. Thoeni, D.E.Guccione, M.Santise, A. Giacomini, R.Roncella, G. Forlani, *IntArchPhotogramm, Remote Sensing and Spatial Information Sciences*,**XLI-B5**, 909 (2016)
20. M. Jarzabek- Rychard, M.Karpina, *Int Arch Photogramm, Remote Sensing and Spatial Information Sciences*, **XLI-B1**, 1121 (2016)
21. G. Caroti, I.Martínez-Espejo Zaragoza, A.Piemonte, *Int Arch Photogramm, Remote Sensing and Spatial Information Sciences*, **XL-1/W4**, 103 (2015)

Multi-Temporal satellite imagery for infrastructure growth assessment of Dubai City, UAE

Mohamad Ali Khalil[1]*, *Rami* Al-Ruzouq[1], *Khaled* Hamad[1],and *Abdallah* Shanableh[1]

[1]Sustaimanle Civil Infrastructure Systems Research Group, University of Sharjah, Sharjah City, United Arab Emirates

Abstract. Throughout the past few decades, Dubai City has witnessed massive growth in its urban area and infrastructure facilities. The discovery of oil and gas in the Emirate significantly played a role for such a rapid growth. Given this fact and the short time-period for such an expansion, it is crucial to develop an understanding of the patterns of the development in the City, where policy-makers, researchers, and concerned authorities would gain better vision and strategy for the future. Recent advances in satellite imagery in terms of improved spatial and temporal resolutions are allowing for efficient identification of change patterns and the prediction of areas of growth. This study aims to quantify and analyse the spatial–temporal urbanization that took place in Dubai City throughout the past decades (specifically from early 1970's until 2015). Multi temporal satellite images with various geometric and radiometric resolutions will be utilized for this purpose. The suggested methodology consists of a sequence of image processing techniques that include supervised and unsupervised classification.Subsequently, the classified images were utilized to quantify the urbanization of the City. The results show that since 1970, the urbanization and population have been dramatically increasedby 5 and 12 times respectively.The resulting trend can be potentially usedto evaluate the consequences of massive urban development, such as City infrastructure, water, environmental and the social impact.

1 Introduction

Over the last few decades, the United Arab Emirates and Dubai City in particular have witnessed significant economic growth since the oil discovery in the region. Such a growth resulted in a rapid urbanization across the City, where a lot of residential, commercial, sports and tourism projects have been implemented. This expansion can be observed by many of the mega scale world's landmarks such as Burj Khalifa (the tallest building in the world), Dubai Mall (the largest mall in the world by the total area), and Miracle Garden (the world's largest natural flower garden). The previous fact can be clearly noticed by observing that approximately 25% of the world's construction cranes are located in Dubai

*Correspondingauthor:mkhalil@sharjah.ac.ae

[1]. The development extends to include land and coastal areas where artificial islands were constructedsuch as the Palm Jumeirah, the Palm Deira and the World's Islands.

The present research aims to quantify and analyze the spatial-temporal aspects of the infrastructure growth in Dubai City. In this study, Geographic Information System (GIS) and Remote Sensing (RS) techniqueswill be utilized to detect and quantify urban development in Dubai City. The last 40 years has witnessed remarkable advances in remote sensing and GIS technologies (i.e., satellites' sensor, and image processing and analysis algorithm) [2]–[5]. Refereeing to literature and previous studies,ithas been found that GIS and RS techniques were not used efficiently for change detection and urban development for the study area under consideration. Therefore, the main objectives of this research are as follow:

- Inspect, quantify, and assess the spatial-temporal urbanization growth in Dubai City, and
- Correlate the results from the previous step to the population growth of the City.

This paper is formatted in six sections, as follows:It starts with an introduction (which is this section) to explain the motivation and objectivesof this research. Section two (i.e., theatrical background and literature review) describe few basic concepts of change detection, and illustrate few techniques that can be used to perform it. Further, it reviews list of previous research that were conducted on change detection. Section three discuss the selected study area and few attributes about it. Section four (i.e., data collection and pre-processing) presents the data collection effort in details and the pre-processing steps need to be done before applying change detection to the study area (i.e., Dubai City). Section five (i.e., change detection) is split into twofold. First fold would explain the algorithm of the utilized image classification technique. On the other hand, the second fold will present and discuss the results of the quantification of urban development.Finally, section six will be concluding the research by illustrating on the learned lessons, and state few limitation of the presented research.

2 Theoretical background and literature review

Change detection can be defined as the process of identifying differences in a state of an object (or phenomenon) by observing it at different times [6]. It involves the ability to quantify changes using multi-spectral, multi-temporal and/or multi-source imagery having different resolution captured at different times. Traditional change detection techniques are based on visual (in other words manual) comparison of temporal imagery. However, the increasing availability and complexity of accessible imagery makes the visual change detection very hard, and even impractical. Thus, the need of the adoption (and sometimes even development) of an automotive, reliable, and fast change detection techniques is obvious. Several advanced change detection techniques have been developed previously by other scholars [6]–[9]. Example of those techniques are Canny, Prewitt, Roberts, and fuzzy logic [10].

Many research utilized satellite images to detect spatial and temporal changes of objects (e.g. cites or lands)—keeping in mind the different criteria and tools that were used. For instance, Nassar et el. quantified the land cover change in neighboring Dubai City using time series of remotely sensed data for the period between 1972 to 2011 [1]. Their results indicated that a dramatic increase in the expansion of Dubai's urban area has occurred. Specifically, the compound annual growth rate over the whole study period was 10.03%, where the peak growth happened at a rate of 13.03% during 2003-2005. Furthermore, they noted that the urban growth of Dubai included substantial increase in the vegetation and water bodies, as well as extraordinary rate of construction of offshore islands. Their study provided novel insights into the pace and process of urban growth in Dubai. Just as important, the authors noted the importance to evaluate the environmental consequences of this form of rapid urban development in future research. Nasser et al. research and our present study are considered one of the few studies in this topic for Dubai.

Ji et al. studied the spatial and temporal distribution of expressway and its relations to land cover and population in Beijing City, China [3]. They collected data of roads, land cover and population from 2000 to 2010. Their results showed that the transfer of land cover near expressways was more intensive and frequent compared to parts of the City. Further, these regions also witnessed the most rapid rates of population growth. An interesting point the authors highlighted is the environmental pollution and health risks to residents living nearby expressways.

In a similar context, Al-sharif et al. studied the spatio-temporal urban and population growth in Tripoli City, Libya using remotely sensed data and Geographic Information System (GIS) [5]. They collected four-satellite-imageries and population censuses of Tripoli for the period from 1987 to 2010. The study findings demonstrated that there is an obvious indication of urban dispersion and urban sprawl for the metropolitan of Tripoli City, i.e. sprawl is increasing with time. The authors came up with models to give guidance to identify and quantify the change likely to happen if the historical trends of urban growth continue.

Yu and Ng combined remote sensing images, landscape metrics and gradient analysis to analyze and compare spatial and temporal dynamics of the urban sprawl in Guangzhou, China [11]. Their results showed that the spatial landscape change in Guangzhou exhibits different patterns at its urban center compared to its rural areas. Moreover, the recent urbanized areas showed a higher fragmentation than the other areas. In terms of characterizing the urban expansion, they found that the urban area expanded towards the north and south of their city. They attributed this to the population increase and rapid economic development. Moreover, they emphasized in their study that in order to reveal the complexity of landscape pattern, temporal data and spatio-temporal dynamics of landscape changes are needed along with the gradient analysis to capture the baseline.

Aljoufie et al. analyzed the spatial urban growth and transportation in Jeddah City, Saudi Arabia using temporal satellite images ranging from 1964 to 2007 [12]. To do so, they developed eight urban growth and transportation indices to quantify the urban growth and transportation patterns in the City of Jeddah. Based on their results, they concluded that urban growth in Jeddah is having an outward expansion and sprawl development. Moreover, they claim that their results improve the understanding of the relationship between the urban growth and transportation infrastructure in the City.

In general, and as illustrated from the previously reviewed paper;the use of GIS and RS is well practiced to detect changes for large-scale study areas. In fact, different techniques, approaches, and indiceswere developed and utilized by others to accomplish the task in the best manner. However, our local area still lacks the saturation of such kind of studies, especially Dubai City—which is as indicated earlier, one of the fastest growing cities around the world.

In this study, we utilized a supervised and unsupervised classification image technique to detect temporal changes for Dubai City over the last 40 years.Given the characteristics and nature of the urbanization that took place in Dubai, it is understandably that the outcome of this research shall presents valuable information not only for Dubai's officials (policy makers, researchers, and concerned authorities), but extend to benefit other governmental and research authorities around the world.

DOI: 10.1051/matecconf/201712009006

3 Study area

The proposed algorithm in this paper will be applied to the entire City of Dubai (including the offshore development). Figure 1presents the selected study area at the year of 2015 (the image was retrieved from Landsat Satellite, and then clipped to highlight the boundary of the City only).

Dubai is one of the seven emirates that forms the United Arab Emirates, UAE. Dubai (as an Emirate) is considered the second largest Emirate after the Emirate of Abu Dhabi (in terms of both population and area). However, Dubai as a City is actually the largest City in the UAE with the highest population. Dubai has a creek that runs south from the Arabian Gulf for 13 km, dividing the city into Deira to the east and Bur Dubai to the west. As for the climatological aspect, Dubai, and UAE in general, are characterized as a hyperarid climate. In other words, its land is characterized as dry with low rainfall rates (average yearly rainfall is 120mm). Above that, evaporation rates are high (the weather temperature in the City can rise more than 45^{0}C in summer the period). The humidity is also high; the average humidity throughout the year is around 60%.

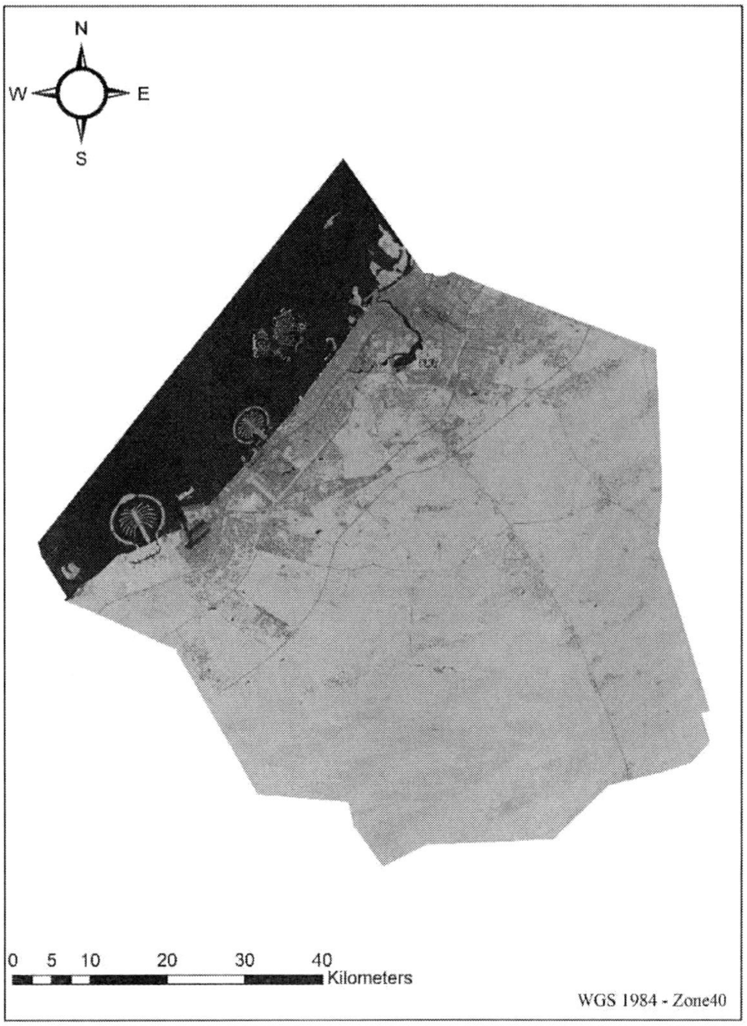

Fig. 1. Dubai City at the year of 2015—acquired from Landsat8.

4 Data collection and processing

Figure 2 shows the designed flow chart for this research. At a glance, the very first step we needed to do is to collect multi-temporal images for Dubai City form RS satellite. The main source for the images for this research was Landsat from the United States Geological Survey (USGS). After acquiring the images, a pre-prepared mask was used to crop the images for the selected study areas (i.e., Dubai City). It is worth mentioning here that all the processing was done using ArcMap©. Subsequently, the prepared images were usedto quantify the changes that occurred throughout the years. Moreover, and whenever possible, the population growth will be presented in conjunction to the urbanization growth. Finally, and at the conclusion section, the results will be summarized.

Table 1 shows a summary of the collected satellite images that have different radiometric and geometric resolutions. The table shows the satellite name and the images' acquired years. Figure 3 presents the prepared images that were used later for the supervised and unsupervised classification.

Table 1 : Collected images.

Satellite Name	Year Acquired	Source
L1-5 MSS	1973	USGS
LT 4-5 TM	1985	USGS
LT 4-5 TM	1995	USGS
Landsat 7	2003	USGS
Landsat 8	2015	USGS

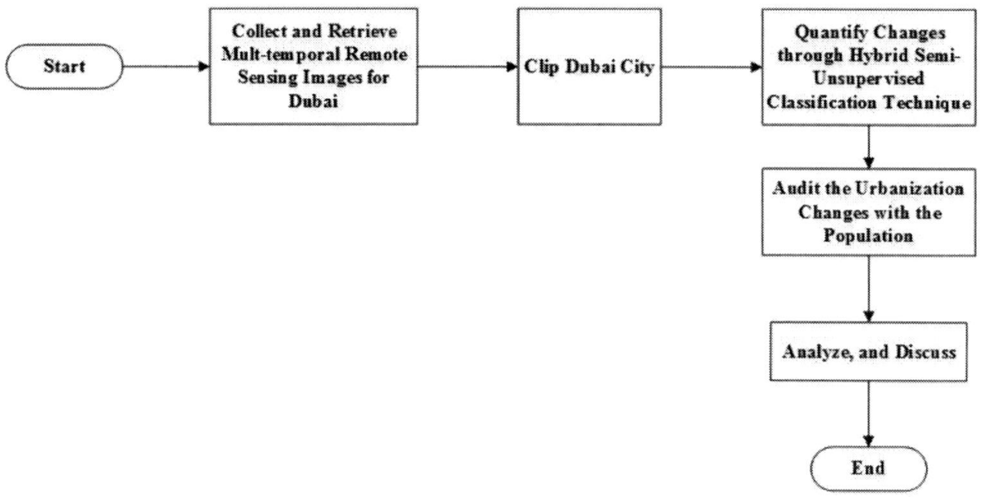

Fig. 2. Research flow chart.

Fig. 3. Multi-temporal images for the study area.

5 Change detection

This section is divided into two parts, first explains the adopted algorithm to classify the image (i.e., the *semi-supervised classification—SUC*) and second, presents and discuss the outcome results of the urbanization quantification of the classified images.

5.1 Semi-Unsupervised Classification, SUC

Digital image classification simply means grouping pixels to represent land cover features. Land cover could be sand, vegetation, urban, or any other types of features. Mainly, there are three techniques to do that, which are: 1) supervised image classification, 2) unsupervised image classification and finally 3) object-based image analysis.

In a nutshell, the difference between the supervised and unsupervised image classification is that the supervised classification asks the user (which is you) to build a *training* database that contains representative sample of all features the user wants to classify. On the other hand, the unsupervised classification groups features based on the reference properties of pixels. In other words, the user is not supposed to interfere at all in the classification—in contrast to the supervised classification. Finally, the object-based image analysis is an advanced method used to segment a pixel-based image into map objects that can then be classified as a whole. This type of analysis is ideal for mapping with high-resolution imagery. For more information about the three types of classifications, readers are encouraged to explore the following references: [13]–[18]

In this research, we applied a hybrid classification technique that fuses supervised and unsupervised classification. We called that method as "*Semi-Unsupervised Classification, SUC*". In specific, the natural breaks (Jenks) method was utilized to classify the images (unsupervised) into three classes, which are, sand and vegetation, water, and urbanization. The images were audited later visually by the user to correct any pitfall that might occur because of radiometric error and/or distortion (supervised interference).Other than this, the remaining steps were done without the intervention of the user. It worth noting here that the result of the unsupervised classification images were converted to vector shapefile to make them amendable—for the supervised intervention, if needed. For that, the method was called "semi".Figure 4 illustrates the flowchart of the proposed technique.

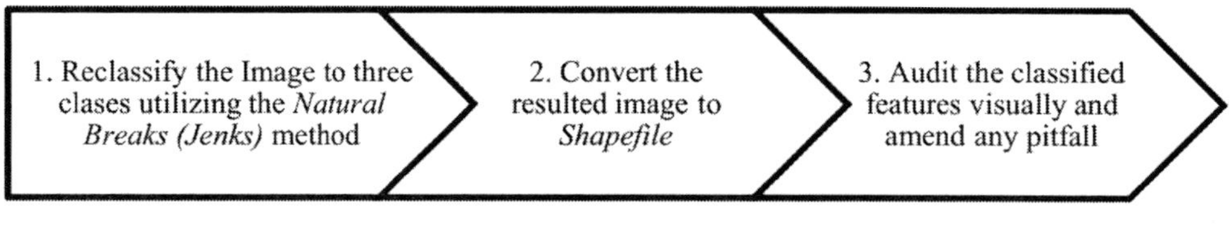

Fig. 4. Semi-Unsupervised Classification flow chart.

5.2 Results and discussion

Figure 5 shows a sample result of the 2015 and 1995 semi-unsupervised classified images. As can be seen in the Figure, the SUC was more or less capable of capturing the urbanization trend of the City. Moreover, it is worth emphasizing here that the corrected features by the user were minimal.

DOI: 10.1051/matecconf/201712009006

Fig. 5. 2015 and 1995 SUC for Dubai City

To further our analysis quantitatively; we obtained the urbanization area from the classified images and present it in conjunction to the population growth of the City. Moreover, we opted the Annual Urban Spatial Expansion Index (AUSEI) that was proposed by [12]. The AUSEI would describe the temporal changes of Dubai in terms of its annual urbanization growth rate. The AUSEI is computed as follows:

$$AUSEI_t = \frac{(U_t - U_{t-1})/U_t}{(N_t - N_{t-1})} \times 100 \quad (1)$$

Where, U_t and U_{t-1} are the total urbanized areas at the current year under study and former year; and N is the total number of years from time t (current year) to time t-1 (former year).

Table 2 and Figure 6 show how the City evolved in terms of area, AUSE, and population.

Table 2 : Dubai city development during the last 45 years—SUC.

Year	Urbanization (Km^2)	AUSEI		Population [19] and [20]
		Period	Index Value (%)	
2015	661.0932	2003-2015	4.0	2,446,675
2003	343.4742	1995-2003	6.0	1,141,610
1995	177.9633	1985-1995	1.4	689,424
1985	153.5247	1973-1985	2.9	370,791
1973	Less than 100	-	-	183,000 @1975

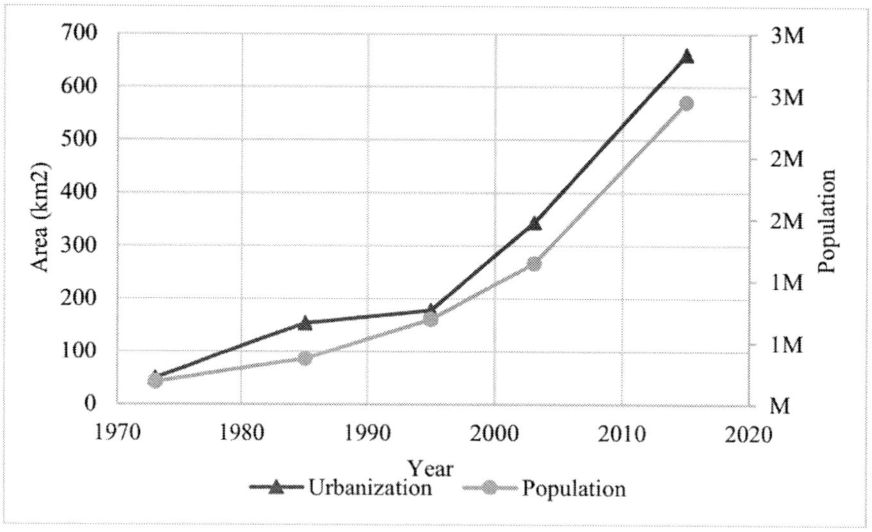

Fig. 6. Urbanization and population growth of dubai city

Out of those results, the following points were drawn out:

- Overall, Dubai City has witnessed a substantial expansion of its urban area during the period 1973 to 2015. The urban area increased from less than 100 Km^2 in 1973 to more than 600 Km^2 in 2015, which is an increase of 500% in less than 50 years.
- In terms of the AUSEI, the index shows that the highest urbanization pace occurred at the period between 1995 and 2003 (i.e., AUSEI equals to 6%). In

contrast, the lowest pace happened during the period of 1985 and 1995, AUSEI of 1.4%. As for the period of 2003 to 2015, though the world witnessed a harsh economic crisis, the City maintained a moderate urbanization rate, i.e., AUSEI of 4%.

- The population of the City jumped from 183,000 at 1975 to more than 2M in 2015, which is an increase of more than 1,200% in less than 50 years.
- As can be seen in Figure 6, the population and urbanization trends almost coincide perfectly to each other. In some other Cities, the trend between both might differ at a certain point of time because of horizontal expansion of those cities—that results from high-rises buildings. Indeed, Dubai City has many high-rise buildings; yet, such a differentiation does not appear to occur in our results. This actually entice us to further investigate this matter in future studies.
- In summary, the results of this analysis highlight the massive development of Dubai City in both urbanization and population aspects during the past four decades. It is worth noting here that one of the major drawback of this approach is that it only demonstrated the overall urbanization of the City. In other others, it could not differentiate between growth in buildings and roads. It might be important to have this detailed differentiation in more advanced studies.

6 Conclusions

This study is one of the rare studies that attempted to quantify the rapid urbanization of the City for the last four decades. This was achieved by using different types of Landsat satellite images, such as Landsat 8, and Landsat 7. This effort would be much difficult without the utilization of such remote sensing images. In this study, one approaches was attempted to quantify such changes (though there are multiple in the literature). The method basically is a hybrid unsupervised classification approach (we called it semi-unsupervised classification), where the user has the flexibility to audit the resulted classification, and amend any pitfalls that could occur regarding the any color distortion of the image.

Generally, there was a rapid pace of urbanization development over the entire study period, especially in the late 90's and beginning of the 2000's—regarding Dubai City. And despite the economic crises that occurred in 2007/8, Dubai City continued to evolve but in a slower rate. By comparing the urbanization quantitative analysis with the population growth of the City, we found there was a high correlation trend between both (see Figure 6). This is a little bit strange results to us, since we were expecting a small differentiation at a certain point time because of the horizontal expansion of the City (e.g. high-rise buildings). Indeed, this outcome shall open the door for further research to study this matter.

Despite the appealing results presented in this study, there are certain limitations the authors did not consider. The following points illustrate the main limitation in this study:

- There was no ground truth the researchers relied on to validate the attempted approaches,
- The utilized approach (the semi-unsupervised classification) did not separate between roads and buildings development. Such a differentiation might be needed to special micro-case-studies, and
- The study lacked to correlate between the urbanization rate and the environmental status and dashboards of the City.

References

1. A. K. Nassar, G. Alan Blackburn, and J. Duncan Whyatt, "Developing the desert: The pace and process of urban growth in Dubai," *Computers, Environment and Urban Systems*, vol. **45**, pp. 50–62, (2014).
2. H. S. Sudhira, T. V. Ramachandra, and K. S. Jagadish, "Urban sprawl: Metrics, dynamics and modelling using GIS," *International Journal of Applied Earth Observation and Geoinformation*, vol. **5**, no. 1, pp. 29–39, (2004).
3. W. Ji, Y. Wang, D. Zhuang, D. Song, X. Shen, W. Wang, and G. Li, "Spatial and temporal distribution of expressway and its relationships to land cover and population: A case study of Beijing, China," *Transportation Research Part D: Transport and Environment*, vol. **32**, pp. 86–96, (2014).
4. [M. K. Jat, P. K. Garg, and D. Khare, "Monitoring and modelling of urban sprawl using remote sensing and GIS techniques," *International Journal of Applied Earth Observation and Geoinformation*, vol. **10**, no. 1, pp. 26–43, (2008).
5. A. a a Al-sharif, B. Pradhan, H. Zulhaidi, M. Shafri, and S. Mansor, "Spatio-temporal Analysis of Urban and Population Growths in Tripoli using Remotely Sensed Data and GIS," *Indian Journal of Science and technology*, vol. **6**, no. 8, pp. 5134–5142, (2013).
6. A. Singh, "Review article digital change detection techniques using remotely-sensed data," *International journal of remote sensing*, vol. **10**, no. 6, pp. 989–1003,(1989).
7. I. Dowman, "Automated procedures for integration of satellite images and map data for change detection: the archangel project," *International Archives of Photogrammetry and Remote Sensing*, vol. **32**, pp. 162–169, (1998).
8. R. I. Al-Ruzouq and A. F. Habib, "Linear features for automatic registration and reliable change detection of multi-source imagery," *Journal of Spatial Science*, vol. **57**, no. 1, pp. 51–64,(2012).
9. L. Bruzzone and D. F. Prieto, "Automatic analysis of the difference image for unsupervised change detection," *IEEE Transactions on Geoscience and Remote Sensing*, vol. **38**, no. 3, pp. 1171–1182, (2000).
10. MathWorks, "Edge Detection." [Online]. Available: http://www.mathworks.com/discovery/edge-detection.html. [Accessed: 03-Aug-2016].
11. X. J. Yu and C. N. Ng, "Spatial and temporal dynamics of urban sprawl along two urban-rural transects: A case study of Guangzhou, China," *Landscape and Urban Planning*, vol. **79**, no. 1, pp. 96–109, (2007).
12. M. Aljoufie, M. Zuidgeest, M. Brussel, and M. van Maarseveen, "Spatial-temporal analysis of urban growth and transportation in Jeddah City, Saudi Arabia," *Cities*, vol. **31**, pp. 57–68, (2013).
13. T. Blaschke, "Object based image analysis for remote sensing," *ISPRS Journal of Photogrammetry and Remote Sensing*, vol. **65**, no. 1, pp. 2–16, (2010).
14. R. C. Weih and N. D. Riggan, "Object-based classification vs. pixel-based classification: Comparitive importance of multi-resolution imagery," *The International Archives of the Photogrammetry, Remote Sensing and Spatial Information Sciences*, vol. **XXXVIII**, pp. 1–6, (2010).
15. J. Norton, D. Gallant, and N. Scotia, "Supervised Image Classification," no. March, (2014).
16. G. M. Foody and A. Mathur, "Toward intelligent training of supervised image classifications: Directing training data acquisition for SVM classification," *Remote Sensing of Environment*, vol. **93**, no. 1–2, pp. 107–117, (2004).
17. T. W. Lee and M. S. Lewicki, "Unsupervised image classification, segmentation, and enhancement using ICA mixture models," *IEEE Transactions on Image Processing*, vol. **11**, no. 3, pp. 270–279, (2002).

18. A. E. M. Omran A. Salman, "Differential evolution methods for unsupervised image classification," *Proc. of the IEEE Congress on Evolutionary Computation*, vol. **2**, pp. 966–973, (2005).
19. Dubai Statistical Center, "Population-Clock @ www.dsc.gov.ae." [Online]. Available: https://www.dsc.gov.ae/en-us/EServices/Pages/Population-Clock.aspx.
20. Archived from Tedad.ae, "background @ web.archive.org." [Online]. Available: https://web.archive.org/web/20100516221856/http://www.tedad.ae/english/about_census/background.html.

MATEC Web of Conferences **120**, 09007 (2017) DOI: 10.1051/matecconf/201712009007
ASCMCES-17

Impact of urban expansion on potential flooding, storage and water harvesting in the city of Sharjah, United Arab Emirates

Abdallah Shanableh[1,*], *Rami* Al-Ruzouq[1], *Mohsin* Siddique[1], *Tarek* Merabtene[1], *Abdullah* Yilmaz[1], and *Monzur* Imteaz[2]

[1]Department of Civil and Environmental Engineering, University of Sharjah, United Arab Emirates
[2]Swinburne University of Technology, Australia

Abstract. The work presented in this article is an extension of an earlier study [1] in which runoff in the city of Sharjah, United Arab Emirates (UAE), was assessed for the period 1967-2010. In this study, the impacts of urban expansion on potential flooding, storage and water harvesting were assessed for the period 1976-2016 using rainfall data recorded during the same period. Sharjah is a generally a flat coastal city that is located in a dry hot region average rainfall of approximately 100 mm/y. Remote sensing and GIS techniques were used to assess the expansion of built areas in the city using multi-temporal Landsat satellite images. The built areas in the city increased during the study period by approximately four folds while the population increased by more than 10 folds. The study results demonstrate the historical increase in runoff potential, flooding potential and runoff coefficient and decline in water storage potential. The results are consistent with the observed incidence of significant local flooding episodes in urban areas during rainfall events. Furthermore, an assessment of potential urban water harvesting based on expansion of built areas in the city is provided and appropriate urban storm water management practices for the city are recommended.

1 Introduction

Sharjah city is the capital and largest city in the Emirates of Sharjah, one of seven emirates that make the United Arab Emirates UAE). Sharjah is a coastal city located in a dry hot region, with mean daily high temperatures of 24-42°C and daily mean temperatures of 18-34°C. The rainy months are typically November to April, with average rainfall of approximately 100 mm/y. Figure 1 shows the UAE map and the Landsat image of Sharjah city, with the study area identified.

The UAE has experienced rapid growth and development during the study period, 1976-2016. Urbanization leads to expansion of built areas and land cover changes, which have significant impacts on runoff quantity and quality. Urbanization can also leads to changes in

* Corresponding author: shanableh@sharjah.ac.ae

evapotranspiration, interception and sediments suspension and transport. It also impacts infiltration and stream flow and changes the time of concentration and runoff speed.

Multi-temporal remote sensing has applications in many fields and is especially useful for detecting and analysing historical land use changes. Remote sensing data integrated into an appropriate geographic information system (GIS) allow analysis and visual display of the data. Integration of remote sensing and GIS allows assessment of historical land use changes, identification of urban features, building a database suitable for analysing runoff change, and estimating hydrological model data in a practical and cost-effective manner [2-8].

Traditional change detection studies are based on visual/manual comparison of temporal images. Recently, however, automatic, reliable, and fast land feature change detection techniques have been developed. Such techniques that rely on identifying feature differences at different times [9] are essential to reduce the high cost associated with spatial data updating activities. In this study, both traditional and automatic feature detection were used to assess land use changes during the study period, 1976-2016.

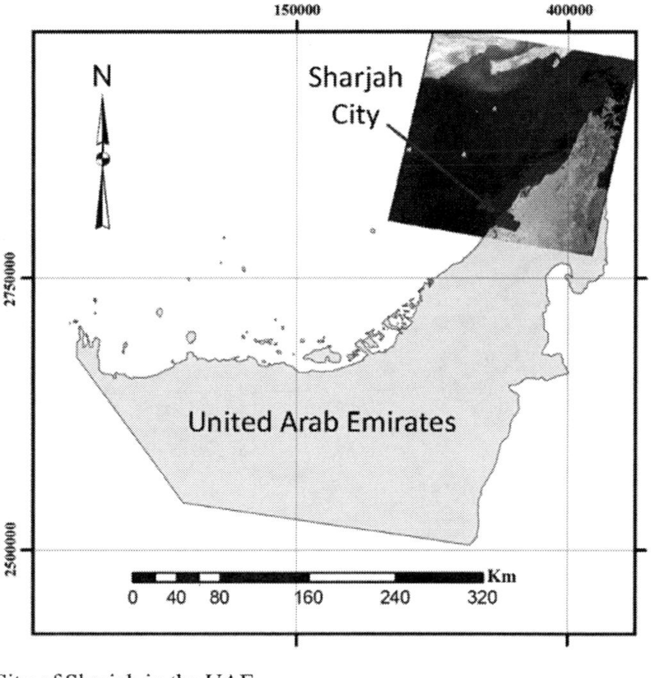

Fig. 1. Location of City of Sharjah in the UAE.

The main objective of this study was to assess the historical impact of land use change on potential flooding, storage and water harvesting in Sharjah during the period 1976-2016 using rainfall data recorded during the same period. The study was based on detecting land use changes using multi-temporal Landsat satellite images and GIS techniques. Furthermore, an assessment of potential urban water harvesting based on expansion of built areas in the city is provided and appropriate urban storm water management practices for the city are recommended. The study documents the historical expansion of the city and presents valuable information to help guide urban storm water management and future expansion of the city.

2 Methodology

2.1 Landcover change detection

Two methods were used to assess land use changes in Sharjah during the study period 1976-2016: (1) multi-temporal image analysis, and (2) manual digitization of land features. In the first method, land cover changes were assessed through detecting and registering linear features in multi-temporal Landsat satellite images obtained for the study period. Image analysis involved linear feature extraction, image registration and pixel–pixel subtraction using Landsat images with different spatial and temporal resolutions. Straight-line segments have high semantics and can be reliably extracted from the images. These linear features represent the streets and buildings in urban areas and are used as the base for change detection. Select interest and well distributed points were manually digitized for registration to drive the transformation parameters needed for the image registration.

The second method was based on classifying the different land use features and digitizing them manually based on visual inspection and knowledge of the area. Older images had lower resolution but fortunately many of the old features remained standing and appeared in recent images. As the digitization process is tedious, saturated urban blocks were digitized as units and the digitized areas were distributed on the different land uses by ratio. Examples of digitized areas using the two methods are shown in Figure 2.

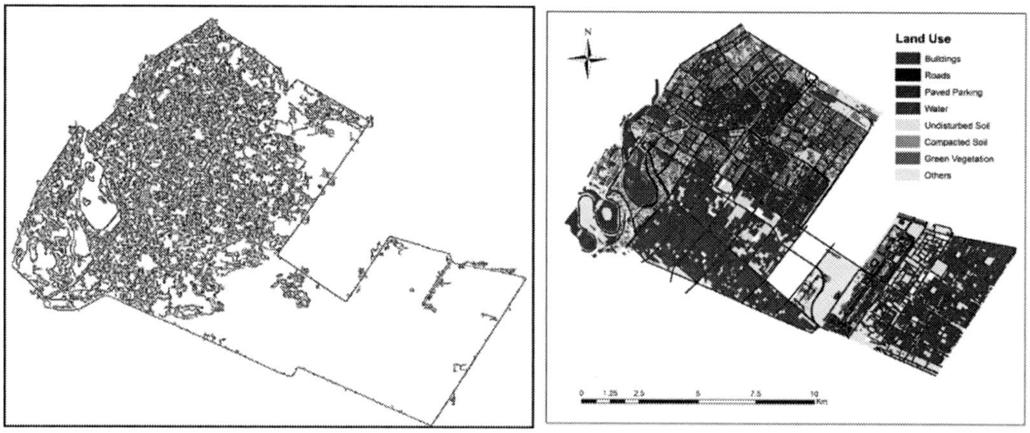

Fig. 2. Examples showing estimation of land use using: edge detection of built features; and direct digitization of features.

2.2 Runofff analysis

The methodology used to estimate the distributed runoff in the city was based on the procedure outlined by [10] and [11]. The US Soil Conservation Service (SCS) model was used to estimate runoff changes in Sharjah. The model is based on assigning curve numbers (CN) representing the runoff characteristics of the different land uses. The SCS model equations are presented below:

$$Q = \frac{[P - 0.2S]^2}{P + 0.8S} \qquad (1)$$

$$S = 25.4\left(\frac{1000}{CN} - 10\right) \tag{2}$$

where, S=potential maximum storage (mm), Q=runoff depth (mm), P=cumulative precipitation (mm); CN=curve number.

Solving Equation 1 requires two parameters; P and CN. The SCS Equation 1 suggests that the minimum rain to generate runoff is P=0.2S. For a given antecedent moisture condition, the curve number is determined by the type of urban land cover and extend of urbanization. For purposes of estimating appropriate CN values, the manually digitized features were divided into five land use areas: high population density; residential/commercial mixed-use; residential; industrial; and public. Each of the five areas was further subdivided into five sub-areas; buildings, roads, paved parking/space, undisturbed soil, compacted soil, vegetation, and others. The undeveloped open area included the undisturbed soil, the compacted soil and vegetation, and the built areas included the remaining land features. Curve numbers (Table 1) were assigned to the different land uses based on standard SCS tables [12].

Table 1. Curve numbers assigned to the different land uses.

Built Area	Compact Soil	Undisturbed Soil	Vegetation	Roads	Paved Parking	Others
90	80	75	90	90	87	87

Composite CN values ($\overline{CN}$) were weighed according to their corresponding areas (A_i), as in Equation 3:

$$\overline{CN} = \frac{\sum CN_i * A_i}{\sum A_i} \tag{3}$$

The runoff coefficient, C, can be estimated from runoff, Q, per precipitation event, P, as in Equation 4:

$$C = \frac{Q}{P} \tag{4}$$

Potential water harvesting was quantified based on recovery of 80% of precipitation from 80% of built areas (i.e., 80% of built areas was assumed to be rooftops of buildings). The potential water harvest per capita was estimated using estimated population of the city, as in Equation 5.

$$Harv. = \frac{0.64 * P * A_{Buildings}}{Population * No.\ of\ Days} \tag{5}$$

DOI: 10.1051/matecconf/201712009007

2.3 Rainfall

Based on data obtained from Sharjah Airport Authority's weather station, Figures 3 and 4 show the annual and average monthly rainfall in Sharjah during 1977-2014. The statistical average, maximum, and minimum values of rainfall in Sharjah during the same period are 108.8 mm, 318.9 mm and 8.6 mm, respectively.

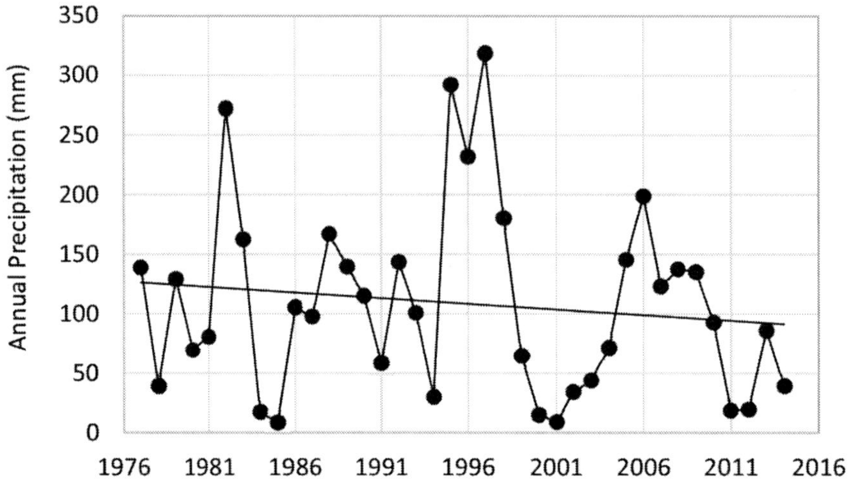

Fig. 3. Annual rainfall recorded at Sharjah International Airport during 1977-2014 and average rainfall during the same period.

The rainy season in Sharjah, as shown in Figure 4, may be considered to span from November to April (6 Months). An analysis of month rainfall shows that 93% of total annual rainfall is received during the Rainy season.

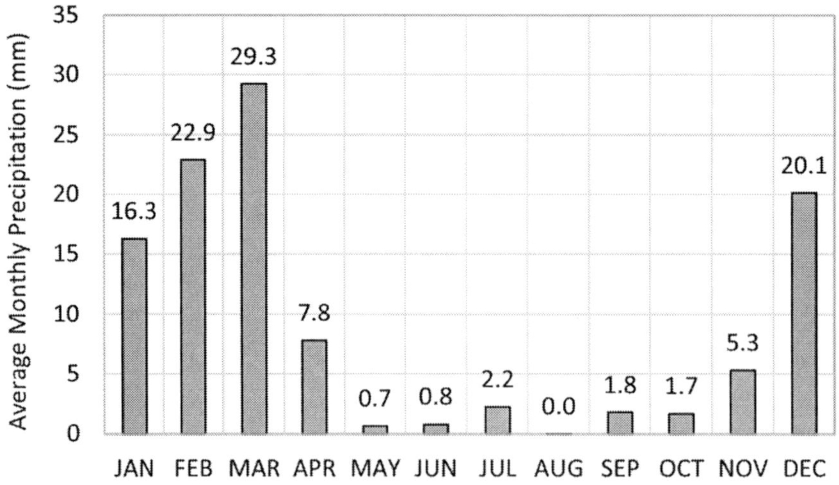

Fig. 4. Average monthly rainfall recorded at Sharjah International Airport during 1977-2014.

Sherif et al. (2013) produced intensity-duration-frequency (IDF) curves for the United Arab Emirates. A range of IDF and rainfall values based on Sherif's work are presented in

Table 2. For all practical reasons, the data in Table 2 show that the range of precipitation per event is less than 100 mm. Therefore, the runoff characteristics in the city were evaluated for precipitation between 17 mm to 120 mm, with 17 mm being the minimum precipitation required to generate runoff based on the choice of CNs for the open areas.

Table 2. Range of rainfall intensity-duration-frequency and rainfall for the UAE desert foreland [13].

	1000 Years Return Period		2 Years Return Period	
Duration (min)	**Intensity (mm/hr)**	**Rainfall (mm)**	**Intensity (mm/hr)**	**Rainfall (mm)**
5	250	21	50	4
25	130	54	25	10
50	75	63	18	15
100	50	83	10	17
150	35	88	7	18
200	30	100	6	20

3 Results and discussion

3.1 Growth and expansion of Sharjah city

The data in Figure 5 show that the built areas within the various landuses expanded during the period 1976 to 2016. Overall, the built areas expanded by approximately four times; from approximately 25 km^2 to about 120 km^2. At the same time, the undeveloped open areas declined from approximately 100 km^2 in 1976 to about 10 km^2 in 2016.

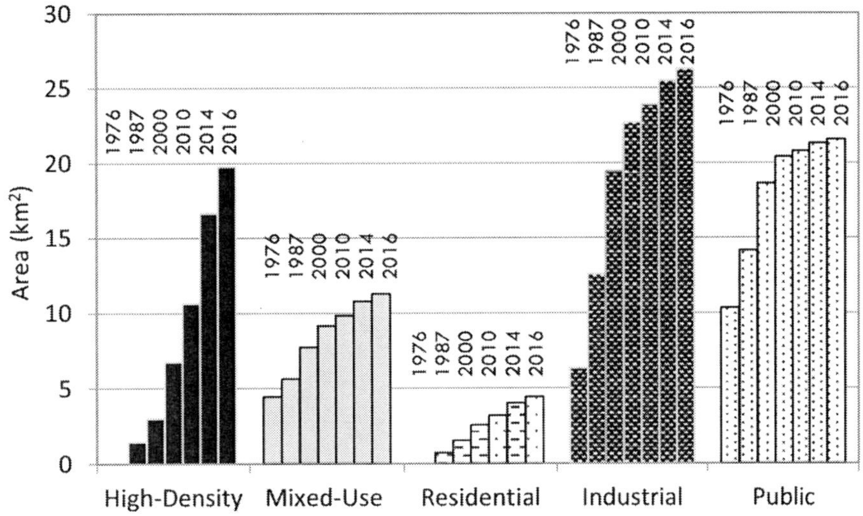

Fig. 5. Expansion of the various landuses in the city of Sharjah during 1976-2016.

Comparison of built and open areas estimated from Landsat images using the direct digitization and edge-detection techniques are shown in Figure 6. Compared to the digitization technique, the edge-detection technique significantly over-estimated built areas and underestimated open areas. It should be noted that the open and built areas estimated using the digitization technique were used in this study. The estimated population of Sharjah city grew from approximately 80,000 people in 1976 to approximately 1 million in 2016, which represents more than 10-fold increase. The expansion of built areas and dramatic population growth reflect the rapid economic growth of Sharjah and the UAE.

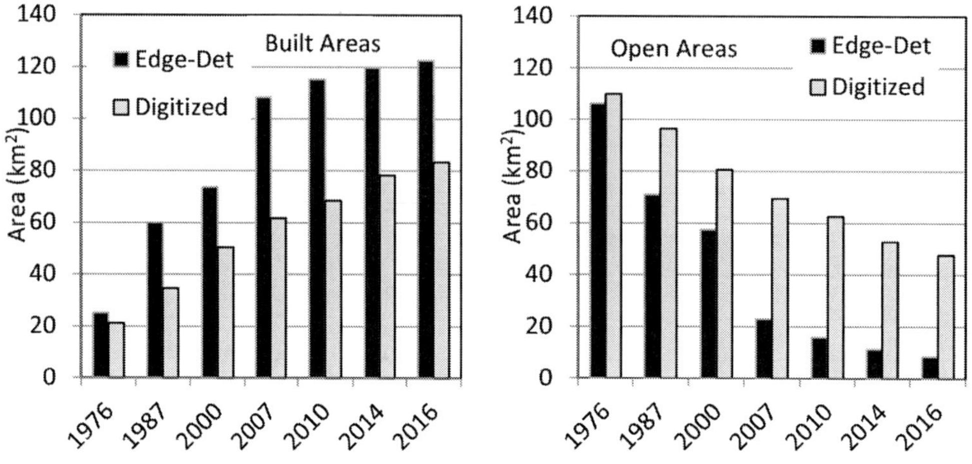

Fig. 6. Comparison of built and open areas estimated from satellite images using either of two methods: (1) direct digitization of features; and (2) edge-detection.

3.2 Flooding, storage and water harvesting potentials

Due to rain scarcity, Sharjah city authorities did not invest in a comprehensive drainage system. However, as the city grew, localized flooding events, although infrequent, affected many neighbourhoods and disrupted transportation and business activity. Therefore, the community increasingly demanded and concerned government authorities realized the need for a proper drainage system.

The historical expansion of built areas (Figures 5 and 6) resulted in at least three impacts that increase potential flooding, which are: (1) increase in imperviousness of land cover; (2) reduction in quantity of precipitation required to generate runoff; and (3) reduction in the maximum potential storage. As a result, local flooding severity and extent increased significantly.

The changes in flooding characteristics during the period 1976 to 2016 are shown graphically in Figure 7. The minimum precipitation required to generate runoff (P_{Min}) in the various zones of the city is also shown in Figure 7. The data show that P_{Min} declined in the various zones with time as the built areas, and consequently impervious areas, expanded. The minimum precipitation declined from about 13-17 mm in 1976 to about 7-11 mm in 2016. For open areas alone, P_{Min} was about 17 mm, and theoretically no excess rain is expected from such areas below P_{Min} = 17 mm, as shown in Figure 7. However, at P =17 mm, excess rain from built areas increased with time. At P = 34.4 mm, which is the 24-hours maximum daily rainfall [13], excess rain from open areas declined over the years as the built areas expanded, Figure 7.

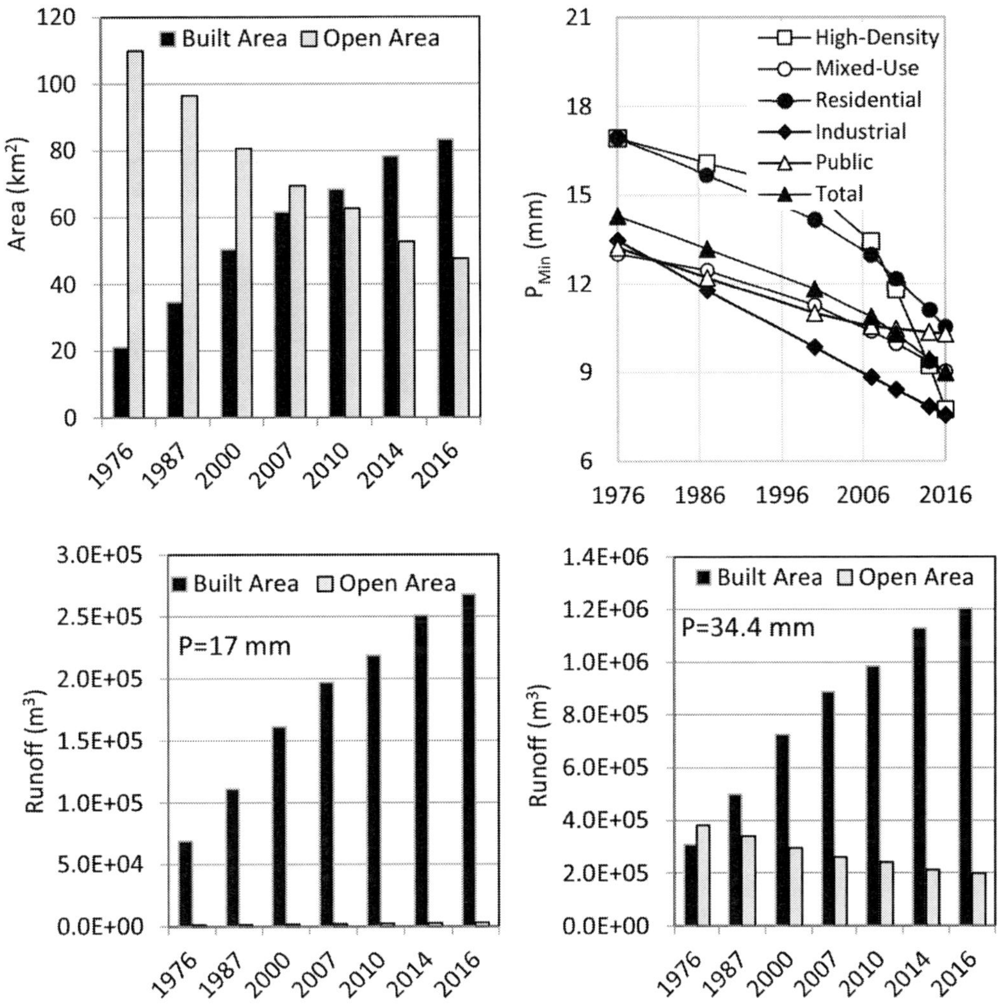

Fig. 7. Expansion in built areas in Sharjah city and impacts on runoff: (a) historical change in open/built areas; (b) historical change in minimum rain required to generate runoff (PMin); (c) excess rain at P = 17 mm; and (d) excess rain at 24-hours maximum.

The historical increase in runoff (Q) in the Sharjah city is shown in Figure 8. With the fraction of built areas increasing from 16% in 1976 to approximately 65% in 2016, the runoff coefficient increased from approximately 0.1 in 1976 to 0.6 in 2016. The results in Figure 8 show that change in runoff coefficient with urbanization declines as rainfall increases. According to the SCS model, the increase in runoff with urbanization results from the decrease in the maximum potential storage, S, as shown in Figure 9. The potential storage is determined by the curve number, CN, as in Equation 2, and changing the land cover and expanding urban areas increases the CN and therefore decreases S. The data in Figure 9 show the historical maximum storage declining in the various zones of the city between 1976 and 2016, with the overall city storage declining from about 71 mm in 2016 to about 45 mm in 2016.

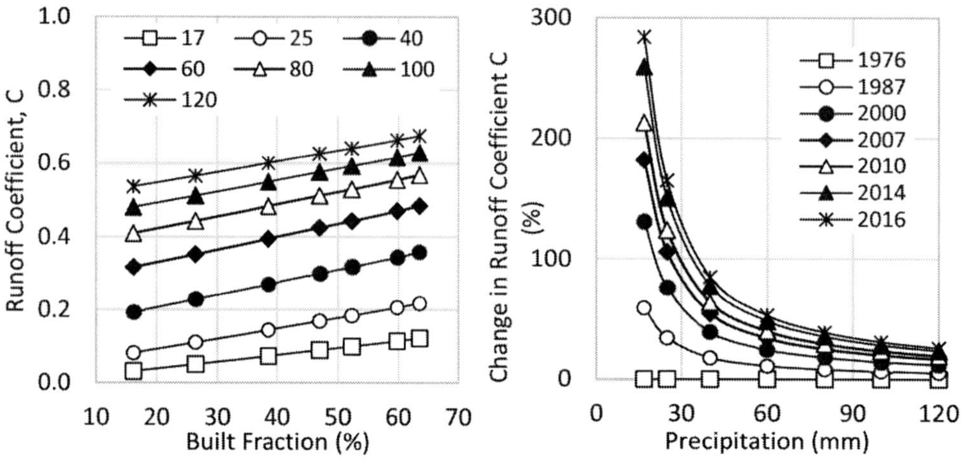

Fig. 8. Increase in runoff and runoff coefficient as a result of expansion in built areas in Sharjah city.

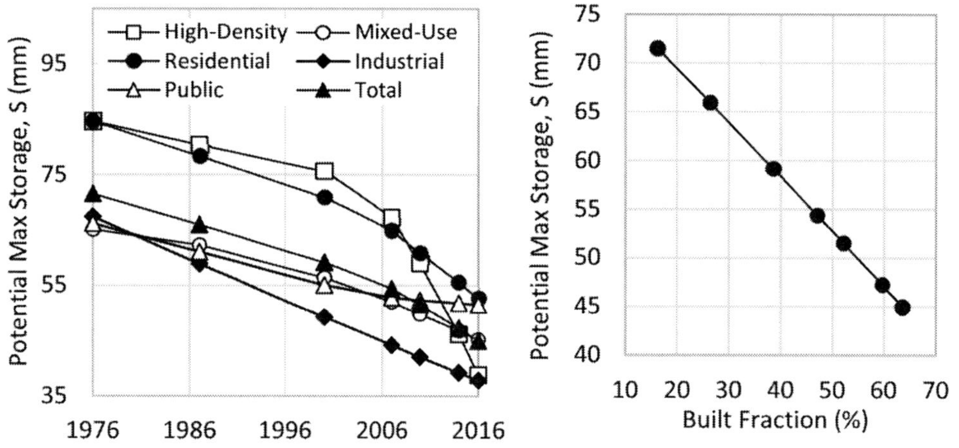

Fig. 9. Decline in potential maximum storage as a result of expansion in built areas in Sharjah city.

The results in Figure 10 show the historical per capita water harvesting potential from rooftops in the city averaged over the whole year and over the rainy season. In the calculations, the percentage of rooftops from the built area was estimated to be 80% and the fraction of rain that was assumed to be collectable was 80%. The data in Figure 10 show that the potential water harvest initially declined between 1976 and 1986 as the built area to population ratio was relatively high in 1976, then the ratio declined dramatically in 1986, then increased gradually thereafter. The data show that the annual water harvest reached about 15 L/capita per day averaged over 365 days and about 30 L/capita per day averaged over 180 days of the rainy season. If only the winter season is considered, the water harvest potential can reach about 45 L/capita per day for 90 days. These numbers are significant and constitute approximately 10% to 30% of the water consumption rate of residents in Sharjah city of approximately 200 L/capita/day.

DOI: 10.1051/matecconf/201712009007

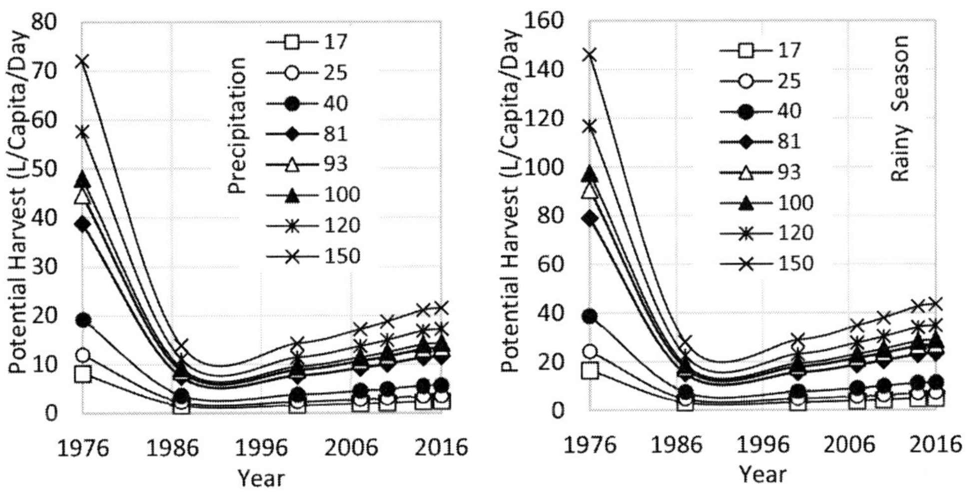

Fig. 10. Potential water harvest from rooftops in Sharjah during a typical year or during the rainy season (180 days) based on 100 mm average annual rain of which 93% occurs during the rainy season.

Water harvesting, combined with strategically located local excess rain water storage ponds and promotion of pervious land cover practices are potentially suitable management practices for Sharjah. Such practices support the sustainability goals of the city and may prove feasible in the long-term.

4 Summary and conclusions

During the period 1976 to 2016, the built area in Sharjah expanded by four folds while the population increased by about 10 folds. The expansion of built areas significantly increased the following parameters: fraction of impervious land; runoff coefficient; and excess rain, or potential runoff. On the other hand, expansion of built areas significantly decreased the following parameters: fraction of impervious land; maximum potential storage; and minimum rain required to generate runoff.

This study documents the historical expansion of Sharjah city and presents valuable information to help guide urban storm water management and future expansion of the city. Historically, city planners saw no practical value in investing in drainage infrastructure to deal with the occasional and localized road and neighbourhood flooding problems. However, the seriousness of impacts continued to grow as the city turned into a modern metropolitan center. The results from this study suggest that water harvesting can satisfy a significant portion of the water demand in Sharjah city. Water harvesting, combined with strategically located local excess rain water storage ponds and promotion of pervious land cover practices are potentially suitable management practices for Sharjah. Such practices support the sustainability goals of the city and may prove feasible in the long-term.

References

1. A. Shanableh, and R. Al-Ruzouq, Integration of Multi-Temporal Satellite Imagery in Modelling Urban Expansion Effects on Runoff in the City of Sharjah, United Arab

Emirates, In: Proceedings of 35th *Hydrology & Water Resources Symposium*, Perth, Australia, (2014)

2. E.T. Engman, and R.J. Gurney, *Remote Sensing in Hydrology,* Chapman and Hall, London, UK, (1991)
3. G.A. Schultz, Remote Sensing Application to Hydrology: Runoff. *Hydological Sciences*, 41 (4), 453-475, (1996)
4. M. Jain, U. Kothyari, and K. Raju, A GIS-Based Distributed Rainfall-Runoff Model. *J. Hydrology*, 299 (1 & 2), 107-135, (2004)
5. D. Kumar, and R. Bhattacharjya, Distributed Rainfall Runoff Modelling, *Int. J. Earth Sci. Eng.*, 4 (6), 270-275, (2011)
6. D.S. Lu, Q.H. Weng, Spectral Mixture Analysis of the Urban Landscape in Indianapolis with Landsat ETM Plus Imagery, Photogramm. *Eng. Remote Sensing*, 70, 1053–1062, (2004)
7. Q.H. Weng, X.F. Hu, and D.S. Lu, Extracting Impervious Surfaces From Medium Spatial Resolution Multispectral and Hyperspectral Imagery: A Comparison. *Int. J. Remote Sens.* 29, 3209–3232 (2008)
8. C.S. Wu, and A.T. Murray, Estimating Impervious Surface Distribution by Spectral Mixture Analysis. *Remote Sens. Environ.*, 84, 493–505 (2003)
9. A. Singh, Digital Change Detection Techniques Using Remotely Sensed Data. *Int. J. Remote Sensing*, 10 (6), 989-1003 (1989)
10. Q.H. Weng, Modeling Urban Growth Effects on Surface Runoff with the Integration of Remote Sensing and GIS. *Environ. Manage.*, 28, 737–748 (2001)
11. R. Al-Ruzouq, A. Habib, Linear Features for Automatic Registration and Reliable Change Detection of Multi-Source Imagery, *Journal of Spatial Science*, 57 (1), 51-64 (2011)
12. USDA, United States Department of Agriculture, Urban Hydrology for Small Watersheds, Soil Conservation Service, Engineering Division. *Technical Release* 55 (TR-55) (1986)
13. M. M. Sherif, A. Mohamed, A. Shetty, and R. K. Chowdhury, Analysis of Rainfall, PMP and Drought in United Arab Emirates, *International Journal of Climatology*, DOI:10.1002/joc.3768 (2013)

AUTHOR INDEX

AUTHOR INDEX

AUTHOR INDEX